NMR of Quadrupolar Nuclei in Solid Materials

EMR Handbooks

Based on the *Encyclopedia of Magnetic Resonance* (EMR), this monograph series focuses on hot topics and major developments in modern magnetic resonance and its many applications. Each volume in the series will have a specific focus in either general NMR or MRI, with coverage of applications in the key scientific disciplines of physics, chemistry, biology or medicine. All the material published in this series, plus additional content, will be available in the online version of EMR, although in a slightly different format.

Previous EMR Handbooks

NMR Crystallography
Edited by Robin K. Harris, Roderick E. Wasylishen, Melinda J. Duer
ISBN 978-0-470-69961-4

Multidimensional NMR Methods for the Solution State
Edited by Gareth A. Morris, James W. Emsley
ISBN 978-0-470-77075-7

Solid-State NMR Studies of Biopolymers
Edited by Ann E. McDermott, Tatyana Polenova
ISBN 978-0-470-72122-3

Forthcoming EMR Handbooks

RF Coils for MRI
Edited by J. Thomas Vaughan and John R. Griffiths
ISBN 978-0-470-77076-4

UTE Imaging
Edited by Graeme M. Bydder, Gary Fullerton and Ian R. Young
ISBN 978-0-470-68835-9

Encyclopedia of Magnetic Resonance

Edited by Robin K. Harris, Roderick E. Wasylishen, Edwin D. Becker, John R. Griffiths, Vivian S. Lee, Ian R. Young, Ann E. McDermott, Tatyana Polenova, James W. Emsley, George A. Gray, Gareth A. Morris, Melinda J. Duer and Bernard C. Gerstein.

The *Encyclopedia of Magnetic Resonance* (EMR) is based on the original printed *Encyclopedia of Nuclear Magnetic Resonance*, which was first published in 1996 with an update volume added in 2000. EMR was launched online in 2007 with all the material that had previously appeared in print. New updates have since been and will be added on a regular basis throughout the year to keep the content up to date with current developments. Nuclear was dropped from the title to reflect the increasing prominence of MRI and other medical applications. This allows the editors to expand beyond the traditional borders of NMR to MRI and MRS, as well as to EPR and other modalities. EMR covers all aspects of magnetic resonance, with articles on the fundamental principles, the techniques and their applications in all areas of physics, chemistry, biology and medicine for both general NMR and MRI. Additionally, articles on the history of the subject are included.

For more information see: www.wileyonlinelibrary.com/ref/emr

NMR of Quadrupolar Nuclei in Solid Materials

Editors

Roderick E. Wasylishen
University of Alberta, Edmonton, Canada

Sharon E. Ashbrook
University of St Andrews, St Andrews, UK

Stephen Wimperis
University of Glasgow, UK

WILEY
A John Wiley and Sons, Ltd., Publication

This edition first published 2012

Registered office
John Wiley & Sons Ltd, The Atrium, Southern Gate, Chichester, West Sussex, PO19 8SQ, United Kingdom

For details of our global editorial offices, for customer services and for information about how to apply for permission to reuse the copyright material in this book please see our website at www.wiley.com.

Library of Congress Cataloging-in-Publication Data

NMR of quadrupolar nuclei in solid materials / editors, Roderick E. Wasylishen, Sharon E. Ashbrook, Stephen Wimperis.
p. cm.
Includes Index.
ISBN 978-0-470-97398-l (cloth)
1. Nulclear quadrupole resonance spectroscopy. 2. Nuclear spin. 3. Solids–Analysis. I. Wasylishen, Roderick E. II. Ashbrook, Sharon E. III. Wimperis, Stephen.
QD96.N84N57 2012
538'.362–dc23

2012002021

A catalogue record for this book is available from the British Library.

ISBN-13: 978-0-470-97398-1

Set in 9.5/11.5 pt Times by Laserwords (Private) Limited, Chennai, India
Printed and bound in Singapore by Markono Print Media Pte Ltd

Encyclopedia of Magnetic Resonance

International Advisory Board

Contents

Part C: Applications 211

Contributors

Jean-Paul Amoureux
Université de Lille, Villeneuve d'Ascq 59650, France
Chapter 10: MQMAS NMR: Experimental Strategies

Sharon E. Ashbrook
School of Chemistry and EaStCHEM, University of St Andrews, St Andrews KY16 9ST, UK
Chapter 3: Quadrupolar Coupling: An Introduction and Crystallographic Aspects
Chapter 11: STMAS NMR: Experimental Advances
Chapter 18: Oxygen-17 NMR of Inorganic Materials

Robert J. Attrell
Department of Chemistry and Centre for Catalysis Research and Innovation, University of Ottawa, 10 Marie Curie Private, Ottawa, ON K1N 6N5, Canada
Chapter 19: Chlorine, Bromine, and Iodine Solid-State NMR

Frédéric Blanc
Department of Chemistry, University of Cambridge, Lensfield Road, Cambridge CB2 1EW, UK
Chapter 20: Quadrupolar NMR of Ionic Conductors, Batteries, and other Energy-Related Materials

David L. Bryce
Department of Chemistry and Centre for Catalysis Research and Innovation, University of Ottawa, 10 Marie Curie Private, Ottawa, ON K1N 6N5, Canada
Chapter 4: Quadrupolar Nuclei in Solids: Influence of Different Interactions on Spectra
Chapter 19: Chlorine, Bromine, and Iodine Solid-State NMR

Rebecca P. Chapman
Department of Chemistry and Centre for Catalysis Research and Innovation, University of Ottawa, 10 Marie Curie Private, Ottawa, ON K1N 6N5, Canada
Chapter 19: Chlorine, Bromine, and Iodine Solid-State NMR

Nicholas J. Curro
Department of Physics, University of California, Davis, CA 95616, USA
Chapter 23: Quadrupolar NMR of Superconductors

Michael Deschamps
Département de Chimie, Université d'Orléans, BP 6759, 1 Rue de Chartres, 45067 Orléans cedex 2, France
Chapter 12: Correlation Experiments Involving Half-Integer Quadrupolar Nuclei

Luminita Duma

Ecole Normale Supérieure, Départment de Chimie, Laboratoire des BioMolécules, UMR 7203 CNRS-ENS-UPMC, 24 rue Lhomond, 75005 Paris, France
Université Pierre et Marie Curie Paris 6, 4 Place Jussieu, 75005 Paris, France
Chapter 16: Nitrogen-14 NMR Studies of Biological Systems

Ray Dupree

Department of Physics, University of Warwick, Coventry CV4 7AL, UK
Chapter 9: Double Rotation (DOR) NMR

Paul D. Ellis

Biological Sciences Division, K8-98, Pacific Northwest National Laboratory, Richland, WA 99352, USA
Chapter 25: Quadrupolar NMR of Metal Nuclides in Biological Materials

Gillian R. Goward

Department of Chemistry and Brockhouse Institute for Materials Research, McMaster University, Hamilton, ON L8S 4M1, Canada
Chapter 20: Quadrupolar NMR of Ionic Conductors, Batteries, and other Energy-Related Materials

Philip J. Grandinetti

Department of Chemistry, The Ohio State University, Columbus, OH 43210-1185, USA
Chapter 8: Dynamic Angle Spinning

Frank Haarmann

Institute of Inorganic Chemistry, RWTH Aachen University, Aachen D-52074, Germany
Chapter 28: Quadrupolar NMR of Intermetallic Compounds

Mohamed Haouas

Tectospin, Institut Lavoisier de Versailles, Université de Versailles-St. Quentin en Yvelines, 78035 Versailles, France
Chapter 21: Quadrupolar NMR of Nanoporous Materials

Arno P.M. Kentgens

Radboud University Nijmegen, Institute for Molecules and Materials, Heyendaalseweg 135, 6525 AJ Nijmegen, The Netherlands
Chapter 7: Quadrupolar Nutation Spectroscopy

Scott Kroeker

Department of Chemistry, University of Manitoba, Winnipeg, MB R3T 2N2, Canada
Chapter 26: Nuclear Waste Glasses: Insights from Solid-State NMR

Olga B. Lapina

Boreskov Institute of Catalysis, Russian Academy of Sciences, Prospect Lavrent'eva 5, Novosibirsk 630090, Russia
Chapter 27: Quadrupolar Metal NMR of Oxide Materials Including Catalysts

Andrew S. Lipton

Biological Sciences Division, K8-98, Pacific Northwest National Laboratory, Richland, WA 99352, USA
Chapter 25: Quadrupolar NMR of Metal Nuclides in Biological Materials

Pascal P. Man
Université Pierre et Marie Curie, Paris 94200, France
Chapter 1: Quadrupolar Interactions

Charlotte Martineau
Tectospin, Institut Lavoisier de Versailles, Université de Versailles-St. Quentin en Yvelines, 78035 Versailles, France
Chapter 21: Quadrupolar NMR of Nanoporous Materials

Dominique Massiot
CNRS-CEMHTI, Site Hautes Températures, 1D Avenue de la Recherche Scientifique, 45071 Orléans cedex 2, France
Chapter 12: Correlation Experiments Involving Half-Integer Quadrupolar Nuclei

Thomas T. Nakashima
Department of Chemistry, University of Alberta, Edmonton, AB T6G 2G2, Canada
Chapter 6: Sensitivity and Resolution Enhancement of Half-Integer Quadrupolar Nuclei in Solid-State NMR

Luke A. O'Dell
Steacie Institute for Molecular Sciences, National Research Council of Canada, 100 Sussex Drive, Ottawa, ON K1A 0R6, Canada
Chapter 14: Quadrupolar NMR to Investigate Dynamics in Solid Materials

Tatyana Polenova
Department of Chemistry and Biochemistry, 036 Brown Laboratories, University of Delaware, Newark, DE 19716, USA
Chapter 25: Quadrupolar NMR of Metal Nuclides in Biological Materials

Marek Pruski
Department of Chemistry, Ames Laboratory, Iowa State University, Ames, IA 50011, USA
Chapter 10: MQMAS NMR: Experimental Strategies

Christopher I. Ratcliffe
Steacie Institute for Molecular Sciences, National Research Council of Canada, 100 Sussex Drive, Ottawa, ON K1A 0R6, Canada
Chapter 14: Quadrupolar NMR to Investigate Dynamics in Solid Materials

Robert W. Schurko
University of Windsor, Department of Chemistry and Biochemistry, Windsor, ON N9B 3P4, Canada
Chapter 5: Acquisition of Wideline Solid-State NMR Spectra of Quadrupolar Nuclei

Mark E. Smith
Department of Physics, University of Warwick, Coventry CV4 7AL, UK
Chapter 18: Oxygen-17 NMR of Inorganic Materials

Leigh Spencer
Department of Chemistry and Brockhouse Institute for Materials Research, McMaster University, Hamilton, ON L8S 4M1, Canada
Chapter 20: Quadrupolar NMR of Ionic Conductors, Batteries, and other Energy-Related Materials

Jonathan F. Stebbins
Department of Geological and Environmental Sciences, Stanford University, Stanford, CA 94305, USA
Chapter 22: Quadrupolar NMR in the Earth Sciences

Francis Taulelle
Tectospin, Institut Lavoisier de Versailles, Université de Versailles-St. Quentin en Yvelines, 78035 Versailles, France
Chapter 21: Quadrupolar NMR of Nanoporous Materials

Victor V. Terskikh
Steacie Institute for Molecular Sciences, National Research Council Canada, Ottawa, ON K1A 0R6, Canada
Chapter 27: Quadrupolar Metal NMR of Oxide Materials Including Catalysts

Alexander J. Vega
Department of Chemistry and Biochemistry, University of Delaware, Newark, DE 19716, USA
Chapter 2: Quadrupolar Nuclei in Solids

Roderick E. Wasylishen
Department of Chemistry, University of Alberta, Edmonton, AB T6G 2G2, Canada
Chapter 4: Quadrupolar Nuclei in Solids: Influence of Different Interactions on Spectra
Chapter 6: Sensitivity and Resolution Enhancement of Half-Integer Quadrupolar Nuclei in Solid-State NMR

Cory M. Widdifield
Department of Chemistry and Centre for Catalysis Research and Innovation, University of Ottawa, 10 Marie Curie Private, Ottawa, ON K1N 6N5, Canada
Chapter 19: Chlorine, Bromine, and Iodine Solid-State NMR

Stephen Wimperis
School of Chemistry and WestCHEM, University of Glasgow, Glasgow G12 8QQ, UK
Chapter 3: Quadrupolar Coupling: An Introduction and Crystallographic Aspects
Chapter 11: STMAS NMR: Experimental Advances

Gang Wu
Department of Chemistry, Queen's University, Kingston, ON K7L 3N6, Canada
Chapter 15: Alkali Metal NMR of Biological Molecules
Chapter 17: Oxygen-17 NMR Studies of Organic and Biological Molecules

James P. Yesinowski
Chemistry Division, Naval Research Laboratory, Washington, DC 20375-5342, USA
Chapter 24: Quadrupolar NMR of Semiconductors

Josef W. Zwanziger
Department of Chemistry, Dalhousie University, Halifax, NS B3H 4J3, Canada
Chapter 13: Computing Electric Field Gradient Tensors

Series Preface

The *Encyclopedia of Nuclear Magnetic Resonance* was published in eight volumes in 1996, in part to celebrate the fiftieth anniversary of the first publications in NMR in January 1946. Volume 1 contained an historical overview and ca. 200 short personal articles by prominent NMR practitioners, while the remaining seven volumes comprise ca. 500 articles on a wide variety of topics in NMR (including MRI). Two "spin-off" volumes incorporating the articles on MRI and MRS (together with some new ones) were published in 2000 and a ninth volume was brought out in 2002. In 2006, the decision was taken to publish all the articles electronically (i.e. on the World Wide Web) and this was carried out in 2007. Since then, new articles have been placed on the web every three months and a number of the original articles have been updated. This process is continuing. The overall title has been changed to the *Encyclopedia of Magnetic Resonance* to allow for future articles on EPR and to accommodate the sensitivities of medical applications.

The existence of this large number of articles, written by experts in various fields, is enabling a new concept to be implemented, namely the publication of a series of printed handbooks on specific areas of NMR and MRI. The chapters of each of these handbooks will comprise a carefully chosen selection of Encyclopedia articles relevant to the area in question. In consultation with the Editorial Board, the handbooks are coherently planned in advance by specially selected editors. New articles are written and existing articles are updated to give appropriate complete coverage of the total area. The handbooks are intended to be of value and interest to research students, postdoctoral fellows, and other researchers learning about the topic in question and undertaking relevant experiments, whether in academia or industry.

Robin K. Harris
University of Durham, Durham, UK

Roderick E. Wasylishen
University of Alberta, Edmonton, Alberta, Canada

November 2009

Volume Preface

In August 1950, the classic paper by R. V. Pound, "Nuclear Electric Quadrupolar Interactions in Crystals", appeared in *Physical Review* and opened the door for NMR studies of quadrupolar nuclei in solids. Looking back at this 18-page masterpiece (*Phys. Rev.*, **79**, 685–702) one is struck by the numerous theoretical and experimental insights provided by Professor Pound. Apart from discussing the ^{7}Li, ^{23}Na, and ^{27}Al NMR spectra of single crystals of $Li_2SO_4{\bullet}H_2O$, $NaNO_3$, and Al_2O_3, respectively, powder lineshapes and relaxation effects (including the results of saturating satellite transitions) are provided. Many outstanding papers quickly followed and a 1957 review by M. H. Cohen and F. Reif, published in *Solid State Physics—Advances and Applications*, summarized the early quadrupolar NMR literature.

Over the next 15 years, many papers dealt with the analysis of quadrupolar NMR powder lineshapes complicated by anisotropic magnetic shielding, dipolar interactions, and so on. A comprehensive review of magnetic resonance lineshapes in polycrystalline solids appeared in 1975 (P. C. Taylor, J. F. Baugher and H. M. Kritz, *Chem. Rev.*, **1975**, *75*, 203–240). It is interesting to mention that in the section of this review discussing spinning techniques, the following statement appears: "Little has been done with nuclei possessing quadrupole moments. Because the quadrupolar interaction tensor is traceless, spinning the sample will eliminate first-order quadrupolar effects. Second-order effects will remain in a modified form, however, and these could be studied in the absence of dipolar broadening" (two references to E. R. Andrew and coworkers follow). We think it is fair to say that up to this time almost all contributions in the field of quadrupolar NMR of solids had been made by physicists working largely with "home-built equipment". In the early 1980s, there was an explosion of activity involving NMR studies of quadrupolar nuclei with nonintegral spins in solids using magic angle spinning techniques. Many chemists and chemical physicists (E. Oldfield, A. Samoson, E. Lippmaa, R. K. Harris, C. A. Fyfe and others) were responsible for demonstrating the advantages of magic angle spinning (MAS) for investigating noninteger quadrupolar nuclei in solids. At the same time, the potential of quadrupolar nuclei, in particular ^{2}H, for studying dynamics in solids was demonstrated by several research groups.

It is not the purpose of this preface to present a review of NMR activity involving quadrupolar nuclei in solids; however, it is important to recognize that the pace at which techniques and applications in this area of research has developed in recent years has not subsided. In fact, the number of relevant publications continues to expand. There are several reasons for this trend, including the following: First, the availability of high magnetic field strengths has had an enormous impact on the nature of the problems that one can tackle. In particular, because the second-order quadrupolar interaction scales as the inverse Larmor frequency, the quadrupolar perturbation of the central NMR transition, $m_I = \frac{1}{2} \leftrightarrow m_I = -\frac{1}{2}$, decreases at higher fields. Second, there have been considerable technological improvements in producing rotors for sample spinning experiments that spin rapidly, stably, and reliably. As outlined in this handbook, this has led to the development and use of several techniques (e.g., double-rotation (DOR), dynamic-angle spinning (DAS), multiple-quantum magic-angle spinning (MQMAS), satellite transition magic angle spinning (STMAS)). At the same time, commercial vendors of NMR equipment have made available double- and triple-resonance probes capable of MAS in standard-bore magnets. Moreover, improvements in spectrometer hardware and software have provided experimentalists with unprecedented flexibility in designing pulse sequences, and so on. Finally, computers and quantum mechanical techniques for computing electric field gradient tensors, magnetic shielding tensors, and so on for nuclei embedded in crystal lattices as well as in "isolated" molecules are making important contributions to science in this area.

The purpose of the present handbook is to provide under a single cover the fundamental principles, techniques, and applications of quadrupolar NMR as it pertains to solid materials. The chapters herein have been taken from or will appear as individual articles in the *Encyclopedia of Magnetic Resonance* (both the online and printed versions). Each chapter has been prepared by an expert who has made significant contributions to our understanding and appreciation of the importance of NMR studies of quadrupolar nuclei in solids. The text is divided into three sections: (i) Basic Principles, (ii) Advanced Techniques, and (iii) Applications. The first section provides the reader with the background necessary to appreciate the challenges in acquiring and interpreting NMR spectra of quadrupolar nuclei in solids. The second section presents cutting-edge techniques and methodology for employing these techniques to investigate quadrupolar nuclei in solids. The final section explores applications of solid-state NMR studies of solids ranging from investigations of dynamics, characterizations of biological samples, organic and inorganic materials, porous materials, glasses, catalysts, semiconductors, and high-temperature superconductors.

As mentioned above, the articles can also be found with minimal differences in format in the online *Encyclopedia of Magnetic Resonance*, at www.wileyonlinelibrary.com/ref/emr. The online versions also include brief autobiographies of the authors, a list of related encyclopedia articles, and in some cases, acknowledgements by the authors. They also have cross-references to encyclopedia articles that are not part of this handbook. Additionally, article abstracts and keywords can be found online.

We hope that readers find the contributions here instructive and that the knowledge acquired advances their research. Also, the related handbooks, *NMR Crystallography* and *Solid-State NMR of Biopolymers* should provide complementary information about NMR of solids. Finally, we wish to thank all authors for their contributions to this handbook, and Professor Robin K. Harris for many helpful suggestions. We also thank people at Wiley, particularly Stacey Woods, Elizabeth Grainge and Rosanna Curran, for their efforts, patience in assembling author contributions as well as Martin Rothlisberger for his support and leadership.

Roderick E. Wasylishen
University of Alberta, Edmonton, Canada

Sharon E. Ashbrook
University of St Andrews, St Andrews, UK

Stephen Wimperis
University of Glasgow, UK

June 2012

PART A
Basic Principles

Chapter 1
Quadrupolar Interactions

Pascal P. Man

Université Pierre et Marie Curie, Paris 94200, France

1.1 INTRODUCTION

Nuclei are characterized by an atomic number Z, a mass number A, and a nuclear spin I. The value of I depends on those of A and Z (Table 1.1). Nuclei with spin $I > 1/2$ are multiple energy level systems and are called quadrupolar nuclei. They represent more than 70% of those in the Periodic Table. However, they are not as frequently investigated in NMR as other elements, because of their quadrupole moments Q, which interact with the electric field gradient (EFG) generated by their surroundings. This coupling, called the quadrupolar interaction and denoted by $\mathcal{H}_{\mathrm{Q}}$, may be much stronger than the amplitude of the rf excitation pulse. As a result, it affects the line intensity and alters the lineshape. These effects make the interpretation of spectra more difficult. Usually, only the first two expansion terms of $\mathcal{H}_{\mathrm{Q}}$ are considered: the first-order, ($H_{\mathrm{Q}}^{[1]}$), and second-order, ($H_{\mathrm{Q}}^{[2]}$), quadrupolar interactions, in the vocabulary of standard perturbation theory. $H_{\mathrm{Q}}^{[1]}$ splits the spectrum of a half-integer quadrupole spin system in a single crystal into $2I - 1$ satellite lines, but the central line remains at the Larmor frequency ω_0. The additional effect of $H_{\mathrm{Q}}^{[2]}$ is to shift further all the lines, including the central line.

When the sample is in powder form, as it usually is, it is mainly the central line that is observed. Moreover, its lineshape becomes nonsymmetrical when $H_{\mathrm{Q}}^{[2]}$ is large. In favorable cases, the powder pattern of the satellite spinning sidebands is detected using the popular MAS technique. The powder pattern of the central line is characterized by three parameters: the quadrupolar coupling constant $\chi = e^2qQ/\hbar$, which is the product of a nuclear property (eQ) and a crystal property (eq), the asymmetry parameter η and the center of gravity of the experimental line, $\delta_{\mathrm{CG}}^{\mathrm{exp}}$ (in ppm). χ is a measure of the strength of the quadrupolar interaction and η a measure of the deviation of the EFG from axial symmetry. The true chemical shift δ_{CS} of the central line is related to

NMR of Quadrupolar Nuclei in Solid Materials
Edited by Roderick E. Wasylishen, Sharon E. Ashbrook and Stephen Wimperis
 ISBN: 978-0-470-97398-1

Table 1.1. Value of nuclear spin I as a function of atomic number Z and mass number A

	Z	
A	Odd	Even
Odd	Half-integer I	Half-integer I
Even	Integer I	$I = 0$

these three parameters:[1,2]

$$\delta_{CS} = \delta_{CG}^{exp} + \frac{1}{30}\left(1 + \frac{1}{3}\eta^2\right) \times \left[I(I+1) - \frac{3}{4}\right]\left[\frac{3\chi}{2I(2I-1)\omega_0}\right]^2 \quad (1.1)$$

A precise determination of δ_{CS} is required if its value has to be correlated with bond lengths and bond angles. Several methods are available for determining χ and η. They can be grouped into two categories:

1. there is a series of techniques, especially the mechanical spinning of the sample,[1–5] based on the frequency domain response of the spin system (see Chapter 9);
2. the second series deals with the time domain response of the spin system to rf excitation[6–8] (see Chapters 7, 10 and 11).

The experimental center of gravity δ_{CG}^{exp} is determined by spectral simulation. However, spectra acquired with DAS or DOR probes provide this value directly[5] (see Chapter 8). Books dealing with these modern techniques are available.[9–11]

In the present chapter, we focus on the frequency domain response of half-integer quadrupolar spin larger than 1. (Jellison and co-workers[12] calculated perturbation terms up to third order for integer spins $I = 1$ and 3.) The first part is devoted to a derivation of the Hamiltonians corresponding to first- and second-order perturbations, with the emphasis on the different conventions used in the literature, namely, the asymmetry parameter, the components of spherical tensors in their principal axis system, the Larmor frequency, transitions, and the transition frequency. With this in mind, the Magnus expansion is applied instead of standard perturbation theory. For simplicity, Hamiltonians are expressed in angular velocity units and relaxation phenomena are not taken into account. In the second part, NMR parameters related to single crystal spectra and powder patterns in static and MAS measurements are presented (see Chapters 19 and 22), in particular, the second-order quadrupolar shift, the critical points and the lineshapes of the powder patterns for various values of η, and the second-order quadrupolar shift for the center of gravity of a powder pattern. In the appendix, the commonly used Euler angles as well as those used by Baugher and co-workers[13–15] are given in graphical form. The Wigner rotation matrix, expressing the components of the same spherical tensor in two different coordinate frames, is also given.

1.2 QUADRUPOLAR HAMILTONIAN IN A UNIFORM SPACE

Slichter[16] and others[17,18] introduce the quadrupolar interaction from the classical concept of the charge density for a nucleus in a space where the three coordinate axes x, y, and z are equivalent. Then, the quantum mechanical form of this interaction is obtained using operators. Thanks to the Wigner–Eckart theorem, the Hamiltonian representing the quadrupolar interaction independently of the Cartesian coordinate frame is defined:

$$\hbar\hat{\mathcal{H}}_Q = \frac{eQ}{6I(2I-1)} \sum_{\alpha,\beta=x,y,z} V_{\alpha\beta}[\tfrac{3}{2}(\hat{I}_\alpha\hat{I}_\beta + \hat{I}_\beta\hat{I}_\alpha) - \delta_{\alpha\beta}I(I+1)] \quad (1.2a)$$

with

$$V_{\alpha\beta} = \left.\frac{\partial^2 U}{\partial\alpha\,\partial\beta}\right|_{r=0} \quad (1.2b)$$

$\delta_{\alpha\beta}$ is the Kronecker delta symbol, U is the electrostatic potential at the origin (inside the nucleus) generated by external charges, and $V_{\alpha\beta}$ are the Cartesian components of the EFG at the origin, $\boldsymbol{V}$, which is a second-rank symmetrical tensor. In the principal axis system Σ^{PAS} of the EFG, $\boldsymbol{V}$ is diagonal:

$$\boldsymbol{V} = \begin{bmatrix} V_{XX} & 0 & 0 \\ 0 & V_{YY} & 0 \\ 0 & 0 & V_{ZZ} \end{bmatrix} \quad (1.3)$$

with the convention $|V_{ZZ}| \geq |V_{YY}| \geq |V_{XX}|$. Furthermore, the Laplace equation $V_{XX} + V_{YY} + V_{ZZ} = 0$ holds for $\boldsymbol{V}$. Thus, only two independent parameters are required:

$$eq = V_{ZZ} \quad (1.4a)$$

$$\eta = \frac{V_{XX} - V_{YY}}{V_{ZZ}} \quad (1.4b)$$

the largest component and the asymmetry parameter, respectively, with $1 \geq \eta \geq 0$.

In the coordinate frame Σ^{PAS}, the Cartesian tensor representation of the quadrupolar interaction [equation (1.2a)] takes the form

$$\hbar\hat{\mathcal{H}}_{\mathrm{Q}} = \frac{e^2qQ}{4I(2I-1)}[3\hat{I}_Z^2 - I(I+1) + \eta(\hat{I}_X^2 - \hat{I}_Y^2)] \quad (1.5a)$$

In terms of the operators

$$\hat{I}_+ = \hat{I}_X + \mathrm{i}\hat{I}_Y, \quad \hat{I}_- = \hat{I}_X - \mathrm{i}\hat{I}_Y \quad (1.5b)$$

equation (1.5a) becomes

$$\hbar\hat{\mathcal{H}}_{\mathrm{Q}} = \frac{e^2qQ}{4I(2I-1)} \times[3\hat{I}_Z^2 - I(I+1) + \tfrac{1}{2}\eta(\hat{I}_+^2 + \hat{I}_-^2)] \quad (1.5c)$$

Sometimes, the opposite convention is adopted for η:

$$\eta = \frac{V_{YY} - V_{XX}}{V_{ZZ}} \quad (1.6)$$

which is associated with the condition $|V_{ZZ}| \geq |V_{XX}| \geq |V_{YY}|$.[19,20] As a result, a negative sign appears in front of η in equations (1.5a) and (1.5c) and in subsequent expressions containing η.

1.3 SPHERICAL TENSOR REPRESENTATION FOR THE QUADRUPOLAR HAMILTONIAN

The passage from one coordinate frame to another is more conveniently realized if the quadrupolar interaction is expressed as a second-rank irreducible spherical tensor, according to Mehring:[21]

$$\hbar\hat{\mathcal{H}}_{\mathrm{Q}} = \frac{eQ}{2I(2I-1)}\sum_{q=-2}^{2}(-1)^q V^{(2,q)}T^{(2,-q)} = \frac{eQ}{2I(2I-1)}\sum_{q=-2}^{2}(-1)^q V^{(2,-q)}T^{(2,q)} \quad (1.7)$$

In any Cartesian coordinate frame Σ, the spherical tensor and Cartesian tensor components of $\boldsymbol{V}$ are related by

$$\left.\begin{aligned} V^{(2,0)} &\equiv V_0 = 3\sqrt{\tfrac{1}{6}}V_{zz} \\ V^{(2,1)} &\equiv V_1 = -V_{xz} - \mathrm{i}V_{yz} \\ V^{(2,-1)} &\equiv V_{-1} = V_{xz} - \mathrm{i}V_{yz} \\ V^{(2,2)} &\equiv V_2 = \tfrac{1}{2}(V_{xx} - V_{yy}) + \mathrm{i}V_{xy} \\ V^{(2,-2)} &\equiv V_{-2} = \tfrac{1}{2}(V_{xx} - V_{yy}) - \mathrm{i}V_{xy} \end{aligned}\right\} \quad (1.8)$$

and those of $\boldsymbol{T}$ as

$$\left.\begin{aligned} T^{(2,0)} &= \tfrac{1}{6}\sqrt{6}[3\hat{I}_z^2 - I(I+1)] \\ T^{(2,1)} &= -\tfrac{1}{2}(\hat{I}_z\hat{I}_+ + \hat{I}_+\hat{I}_z) \\ T^{(2,-1)} &= \tfrac{1}{2}(\hat{I}_z\hat{I}_- + \hat{I}_-\hat{I}_z) \\ T^{(2,2)} &= \tfrac{1}{2}\hat{I}_+\hat{I}_+ \\ T^{(2,-2)} &= \tfrac{1}{2}\hat{I}_-\hat{I}_- \end{aligned}\right\} \quad (1.9)$$

with $\hat{I}_+ = \hat{I}_x + \mathrm{i}\hat{I}_y$ and $\hat{I}_- = \hat{I}_x - \mathrm{i}\hat{I}_y$. These two operators are different from those of equation (1.5b) despite the same notation. It is worth noting that the numerical factors in the components of $\boldsymbol{V}$ and $\boldsymbol{T}$ [equations (1.8) and (1.9)] differ from author to author.

Using equations (1.7)–(1.9), the spherical tensor representation of the quadrupolar interaction in the coordinate frame Σ becomes

$$\hbar\hat{\mathcal{H}}_{\mathrm{Q}} = \frac{eQ}{4I(2I-1)}\{\tfrac{1}{3}\sqrt{6}[3\hat{I}_z^2 - I(I+1)]V_0 + (\hat{I}_z\hat{I}_+ + \hat{I}_+\hat{I}_z)V_{-1} - (\hat{I}_z\hat{I}_- + \hat{I}_-\hat{I}_z)V_1 + \hat{I}_+^2V_{-2} + \hat{I}_-^2V_2\} \quad (1.10)$$

Slichter[16] uses nearly the same relationship, apart from a negative sign due to another choice of V_1. From equations (1.4a), (1.4b), and (1.8), the spherical tensor components of $\boldsymbol{V}$ in Σ^{PAS} are obtained:

$$V_0^{\mathrm{PAS}} = \sqrt{\tfrac{3}{2}}eq, \quad V_1^{\mathrm{PAS}} = V_{-1}^{\mathrm{PAS}} = 0, \qquad V_2^{\mathrm{PAS}} = V_{-2}^{\mathrm{PAS}} = \tfrac{1}{2}eq\eta \quad (1.11a)$$

If the other convention for η, namely, equation (1.6) is used then the spherical tensor components of $\boldsymbol{V}$ in Σ^{PAS} are[20]

$$V_0^{\mathrm{PAS}} = \sqrt{\tfrac{3}{2}}eq, \quad V_1^{\mathrm{PAS}} = V_{-1}^{\mathrm{PAS}} = 0, \qquad V_2^{\mathrm{PAS}} = V_{-2}^{\mathrm{PAS}} = -\tfrac{1}{2}eq\eta \quad (1.11b)$$

1.4 QUADRUPOLAR INTERACTION AS A PERTURBATION OF ZEEMAN INTERACTION

A nuclear spin possesses a magnetic moment $\boldsymbol{\mu}$ and an angular momentum $\hbar \boldsymbol{I}$, which are related by the gyromagnetic ratio γ:

$$\boldsymbol{\mu} = \gamma \hbar \boldsymbol{I} \tag{1.12}$$

In the laboratory frame Σ^{lab}, the direction of the strong static magnetic field $\boldsymbol{B}_0$ is taken as the z axis. The coupling of the magnetic moment with $\boldsymbol{B}_0$ is the Zeeman interaction $\hat{\mathcal{H}}_Z$:

$$\hbar \hat{\mathcal{H}}_Z = -\boldsymbol{\mu} \cdot \boldsymbol{B}_0 = -\hbar \omega_0 \hat{I}_z \tag{1.13a}$$

$$\omega_0 = \gamma B_0 \tag{1.13b}$$

where $\omega_0/2\pi$ is the Larmor frequency. Sometimes this frequency is defined as $\omega_0/2\pi = -\gamma B_0/2\pi$. As a result, the Zeeman interaction takes the form $\hbar\hat{\mathcal{H}}_Z = \hbar\omega_0 \hat{I}_z$. As with η, the choice of ω_0 changes the sign of some expressions below.

We deal with the case where $\mathcal{H}_Q$ can be treated as a weak perturbation of the Zeeman interaction. It is then more convenient to express interactions in the frame Σ^{obs} rotating relative to Σ^{lab} with an angular velocity ω_0 so that the spherical tensor representation of the quadrupolar interaction expressed by equation (1.10) becomes time-dependent:[22]

$$\begin{aligned}\hbar\hat{\mathcal{H}}_Q(t) &= \exp(\mathrm{i}\hat{\mathcal{H}}_Z t)\hbar\hat{\mathcal{H}}_Q \exp(-\mathrm{i}\hat{\mathcal{H}}_Z t)\\ &= \frac{eQ}{4I(2I-1)}\{\tfrac{1}{3}\sqrt{6}[3\hat{I}_z^2 - I(I+1)]V_0\\ &\quad + \hat{I}_+(2\hat{I}_z+1)V_{-1}\exp(-\mathrm{i}\omega_0 t)\\ &\quad - \hat{I}_-(2\hat{I}_z-1)V_1\exp(\mathrm{i}\omega_0 t)\\ &\quad + \hat{I}_+^2 V_{-2}\exp(-\mathrm{i}2\omega_0 t) + \hat{I}_-^2 V_2 \exp(\mathrm{i}2\omega_0 t)\}\end{aligned} \tag{1.14}$$

However, the first term in the curly brackets (i.e., the secular term) remains time-independent. In order to make the quadrupolar interaction completely time-independent, $\mathcal{H}_Q(t)$ is averaged over one Larmor period $2\pi/\omega_0$ up to first-order, using the Magnus expansion:[23]

$$\begin{aligned}\langle\hat{\mathcal{H}}_Q(t)\rangle &= \frac{\omega_0}{2\pi}\int_0^{2\pi/\omega_0} \mathrm{d}t\, \hat{\mathcal{H}}_Q(t)\\ &\quad - \frac{\mathrm{i}\omega_0}{4\pi}\int_0^{2\pi/\omega_0}\mathrm{d}t \int_0^t\\ &\quad \times \mathrm{d}t'\, [\hat{\mathcal{H}}_Q(t), \hat{\mathcal{H}}_Q(t')] = H_Q^{(0)} + H_Q^{(1)}\end{aligned} \tag{1.15}$$

with

$$H_Q^{(0)} = \frac{eQ}{4I(2I-1)\hbar}\frac{\sqrt{6}}{3}[3\hat{I}_z^2 - I(I+1)]V_0 \tag{1.16}$$

$$\begin{aligned}H_Q^{(1)} &= -\frac{1}{\omega_0}\left[\frac{eQ}{4I(2I-1)\hbar}\right]^2\\ &\quad \times\{\sqrt{6}V_0V_{-1}\hat{I}_+(2\hat{I}_z+1)^2\\ &\quad - \sqrt{6}V_0V_1\hat{I}_-(2\hat{I}_z-1)^2\\ &\quad + 2\sqrt{6}V_0V_{-2}\hat{I}_+^2(\hat{I}_z+1)\\ &\quad + 2\sqrt{6}V_0V_2\hat{I}_-^2(\hat{I}_z-1)\\ &\quad + 2V_{-1}V_1\hat{I}_z[4I(I+1) - 8\hat{I}_z^2 - 1]\\ &\quad + 2V_{-2}V_2\hat{I}_z[2I(I+1) - 2\hat{I}_z^2 - 1]\}\end{aligned} \tag{1.17}$$

Usually, only the secular terms that commute with $\hat{I}_z$ (i.e., the last two terms in the curly brackets of $H_Q^{(1)}$ are considered. With this simplification, $H_Q^{(0)}$ and $H_Q^{(1)}$ are equivalent to the first-order, $H_Q^{[1]}$, and second-order, $H_Q^{[2]}$, terms in standard perturbation theory,[24] i.e.,

$$H_Q^{[1]} = H_Q^{(0)} = \frac{eQ}{4I(2I-1)\hbar}\frac{\sqrt{6}}{3}[3\hat{I}_z^2 - I(I+1)]V_0 \tag{1.18}$$

$$\begin{aligned}H_Q^{[2]} = H_Q^{(1)} &= -\frac{1}{\omega_0}\left[\frac{eQ}{4I(2I-1)\hbar}\right]^2\\ &\quad \times\{2V_{-1}V_1\hat{I}_z[4I(I+1) - 8\hat{I}_z^2 - 1]\\ &\quad + 2V_{-2}V_2\hat{I}_z[2I(I+1) - 2\hat{I}_z^2 - 1]\}\end{aligned} \tag{1.19}$$

respectively. Equations (1.18) and (1.19), derived in the rotating frame Σ^{obs}, are unchanged in the laboratory frame Σ^{lab}. This is because they commute with the Zeeman interaction. In other words, they commute with the operator $\hat{I}_z$. From now on, we shall use the language of standard perturbation theory. The first-order quadrupolar interaction $H_Q^{[1]}$ is independent of ω_0, whereas the second-order quadrupolar interaction $H_Q^{[2]}$ is inversely proportional to ω_0. Therefore, a strong static magnetic field is required to reduce the effects of $H_Q^{[2]}$.

1.5 ENERGY LEVELS AND THE SPECTRUM OF A SINGLE CRYSTAL

When a free spin I is introduced into a strong static magnetic field, the Zeeman interaction splits its $2I + 1$ energy levels $|m\rangle$, whose energy is defined by

$$\langle m|\hat{\mathcal{H}}_Z|m\rangle = -m\omega_0 \tag{1.20}$$

and the difference between two consecutive energy levels $(m - 1, m)$, expressed in angular velocity units, is

$$\omega^{(Z)}_{m-1,m} = \langle m-1|\hat{\mathcal{H}}_Z|m-1\rangle - \langle m|\hat{\mathcal{H}}_Z|m\rangle = \omega_0 \tag{1.21a}$$

We choose the same convention as Abragam[18] for the pair $(m - 1, m)$ and equation (1.21a) to represent the transition and the transition frequency, respectively, but other authors choose $(m, m - 1)$, $(m, m + 1)$, $(m + 1, m)$, equation (1.21a) or its negative

$$\omega^{(Z)}_{m-1,m} = \langle m|\hat{\mathcal{H}}_Z|m\rangle - \langle m-1|\hat{\mathcal{H}}_Z|m-1\rangle \tag{1.21b}$$

Of course, these choices affect some later relationships dealing with transitions and transition frequencies. Equation (1.21a) implies that the energy levels $|m\rangle$ of a free spin in a strong static magnetic field $\boldsymbol{B}_0$ are equally spaced. The separation between two adjacent levels is ω_0. In the spectrum, a single line is located at ω_0. However, these energy levels may be shifted by other interactions, including the quadrupolar interaction discussed in this chapter.

The first-order quadrupolar interaction $H^{[1]}_Q$ shifts the energy levels $|m\rangle$ by an amount

$$\langle m|H^{[1]}_Q|m\rangle = \frac{eQ}{4I(2I-1)\hbar}\frac{\sqrt{6}}{3}[3m^2 - I(I+1)]V_0 \tag{1.22}$$

and in the spectrum, its contribution to the line position, i.e., the first-order quadrupolar shift $\omega^{(1)}_{m-1,m}$ of the line position associated with the transition $(m - 1, m)$, is

$$\begin{aligned}\omega^{(1)}_{m-1,m} &= \langle m-1|H^{[1]}_Q|m-1\rangle - \langle m|H^{[1]}_Q|m\rangle \\ &= \frac{3eQ}{4I(2I-1)\hbar}\frac{\sqrt{6}}{3}(1-2m)V_0\end{aligned} \tag{1.23}$$

The spectrum consists of $2I$ lines, the central one of which, associated with the transition $(-1/2, 1/2)$, is still located at ω_0. The other $2I - 1$ lines are called satellite lines.

When the second-order quadrupolar interaction $H^{[2]}_Q$ is taken into account, the energy levels $|m\rangle$ are shifted further:[25]

$$\begin{aligned}\langle m|H^{[2]}_Q|m\rangle = &-\frac{1}{\omega_0}\left[\frac{eQ}{4I(2I-1)\hbar}\right]^2 \\ &\times \{2V_{-1}V_1 m[4I(I+1) - 8m^2 - 1] \\ &+ 2V_{-2}V_2 m[2I(I+1) - 2m^2 - 1]\}\end{aligned} \tag{1.24}$$

and its contribution $\omega^{(2)}_{m-1,m}$ to the line position, i.e., the second-order quadrupolar shift of the line, is[26]

$$\begin{aligned}\omega^{(2)}_{m-1,m} &= \langle m-1|H^{[2]}_Q|m-1\rangle - \langle m|H^{[2]}_Q|m\rangle \\ &= -\frac{2}{\omega_0}\left[\frac{eQ}{4I(2I-1)\hbar}\right]^2 \\ &\quad\times \{V_{-1}V_1[24m(m-1) - 4I(I+1) + 9] \\ &\quad+ \tfrac{1}{2}V_{-2}V_2[12m(m-1) - 4I(I+1) + 6]\}\end{aligned} \tag{1.25}$$

Therefore, the line associated with the transition $(m - 1, m)$ is located in the spectrum at

$$\omega_{m-1,m} = \omega_0 + \omega^{(1)}_{m-1,m} + \omega^{(2)}_{m-1,m} \tag{1.26}$$

In the following two subsections, we apply equation (1.26) to two experiments, in which the single crystal is either static or is spinning at the magic angle.

1.5.1 Spectrum of a Static Single Crystal

We have to express V_0 in equation (1.23), and V_1, V_{-1}, V_2, and V_{-2} in equation (1.25) in terms of the components of $\boldsymbol{V}$ in Σ^{PAS}, equation (1.11a). For this purpose, the following relationship is used:

$$V_i = \sum_{j=-2}^{2} \mathcal{D}^{(2)}_{j,i}(\alpha,\beta,\gamma)V^{\mathrm{PAS}}_j \tag{1.27}$$

where the Euler angles α, β, and γ describe the direction of the strong static magnetic field in Σ^{PAS} (Figure 1.1) and $\mathcal{D}^{(2)}_{j,i}$ (α, β, γ) is the Wigner rotation matrix defined in the appendix. For example,

$$V_0 = \sqrt{\tfrac{3}{2}}\, eq\left[\tfrac{1}{2}(3\cos^2\beta - 1) + \tfrac{1}{2}\eta\sin^2\beta\cos 2\alpha\right] \tag{1.28}$$

Its substitution into equation (1.18) yields

$$H^{[1]}_Q = \tfrac{1}{3}\omega_Q[3\hat{I}_z^2 - I(I+1)] \tag{1.29}$$

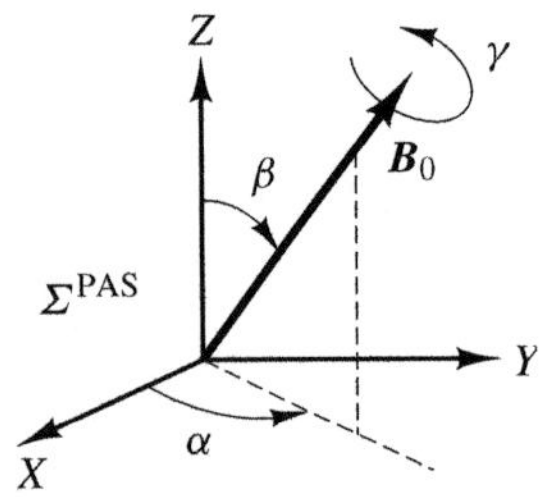

Figure 1.1. Euler angles defining the direction of $\boldsymbol{B}_0$ in the principal axis system Σ^{PAS} of the EFG during a static experiment.

with

$$\omega_Q = \frac{3\chi}{4I(2I-1)} \times \left[\tfrac{1}{2}(3\cos^2\beta - 1) + \tfrac{1}{2}\eta \sin^2\beta \cos 2\alpha\right] \quad (1.30)$$

A negative sign will appear in front of η if the other convention for η, equation (1.6), is chosen or the Euler angles used by Baugher and co-workers[13–15] are used. The definitions of $H_Q^{[1]}$ by equation (1.29) or of ω_Q by equation (1.30) are not unique. Other definitions can be found in the literature. The first-order quadrupolar shift of the lines $(m-1, m)$, equation (1.23), becomes

$$\omega_{m-1,m}^{(1)\mathrm{static}} = (1-2m)\omega_Q \quad (1.31)$$

The lines in the spectrum are separated by the same quantity $2\omega_Q$, but the central line is not shifted.

The other two factors V_1V_{-1} and V_2V_{-2} in equation (1.25) are

$$2V_1V_{-1} = -\tfrac{3}{2}e^2q^2[(-\tfrac{1}{3}\eta^2\cos^2 2\alpha + 2\eta\cos 2\alpha - 3)\cos^4\beta + (\tfrac{2}{3}\eta^2\cos^2 2\alpha - 2\eta\cos 2\alpha - \tfrac{1}{3}\eta^2 + 3)\cos^2\beta + \tfrac{1}{3}\eta^2(1-\cos^2 2\alpha)] \quad (1.32a)$$

$$V_2V_{-2} = \tfrac{3}{2}e^2q^2[(\tfrac{1}{24}\eta^2\cos^2 2\alpha - \tfrac{1}{4}\eta\cos 2\alpha + \tfrac{3}{8})\cos^4\beta + (-\tfrac{1}{12}\eta^2\cos^2 2\alpha + \tfrac{1}{6}\eta^2 - \tfrac{3}{4})\cos^2\beta + \tfrac{1}{24}\eta^2\cos^2 2\alpha + \tfrac{1}{4}\eta\cos 2\alpha + \tfrac{3}{8}] \quad (1.32b)$$

The second-order quadrupolar shift of the central line, using equation (1.25), is given by

$$\omega_{-1/2,1/2}^{(2)\mathrm{static}} = -\frac{1}{6\omega_0}\left[\frac{3\chi}{2I(2I-1)}\right]^2 [I(I+1) - \tfrac{3}{4}] \times [A(\alpha,\eta)\cos^4\beta + B(\alpha,\eta)\cos^2\beta + C(\alpha,\eta)] \quad (1.33)$$

with

$$\left.\begin{aligned} A(\alpha,\eta) &= -\tfrac{27}{8} + \tfrac{9}{4}\eta\cos 2\alpha - \tfrac{3}{8}(\eta\cos 2\alpha)^2 \\ B(\alpha,\eta) &= \tfrac{30}{8} - \tfrac{1}{2}\eta^2 - 2\eta\cos 2\alpha + \tfrac{3}{4}(\eta\cos 2\alpha)^2 \\ C(\alpha,\eta) &= -\tfrac{3}{8} + \tfrac{1}{3}\eta^2 - \tfrac{1}{4}\eta\cos 2\alpha - \tfrac{3}{8}(\eta\cos 2\alpha)^2 \end{aligned}\right\} \quad (1.34)$$

When the EFG has axial symmetry ($\eta = 0$), equation (1.33) becomes simply

$$\omega_{-1/2,1/2}^{(2)\mathrm{static}} = -\frac{1}{16\omega_0}\left[\frac{3\chi}{2I(2I-1)}\right]^2 [I(I+1) - \tfrac{3}{4}] \times (1-\cos^2\beta)(9\cos^2\beta - 1) \quad (1.35)$$

It is worth noting that the third Euler angle γ does not appear in equations (1.30), (1.33), and (1.34); this is because $\boldsymbol{B}_0$ is a symmetry axis for the spins. Our results are identical to those of Narita and co-workers[27] [note that their paper contains a typographical error concerning the expression of cos 2α in $C(\alpha, \eta)$]. Subsequently, Baugher and co-workers[15] obtained expressions similar to equation (1.34), except that their terms containing η have the opposite sign. Their comment 23, concerning the sign in front of all the terms in cos 2α, is explained in our appendix using the Euler angles (Figure 1.10) defined by Goldstein.[14] Another way to obtain the same results as those of Baugher and co-workers[15] is to employ the usual Euler angles (Figure 1.9) and to replace η by $-\eta$ (the other convention for η). This point is confirmed by Hirshinger and co-workers[28] and by Chu and Gerstein.[29] Wolf and co-workers[30] have determined the third-order perturbation term, and shown that it is proportional to $(2m-1)/\omega_0^2$. Therefore, the position of the central line is not shifted further by this new term.

1.5.2 Spectrum of a Rotating Single Crystal

First of all, the expressions for V_0, V_1, V_{-1}, V_2, and V_{-2} must be expressed in terms of the components of $\boldsymbol{V}$ in the coordinate frame Σ^{MAS} of the rotor. To do this, the Wigner rotation matrix is applied

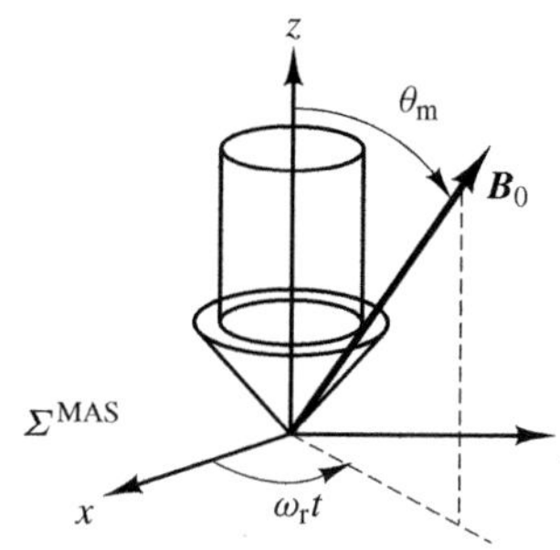

Figure 1.2. Euler angles defining the direction of $\boldsymbol{B}_0$ in the rotor coordinate frame Σ^{MAS} during a MAS experiment. In Σ^{MAS}, $\boldsymbol{B}_0$ rotates around the rotor with the angular velocity ω_r; θ_m is the magic angle; the third angle is $\gamma = 0$. $\boldsymbol{B}_0$ performs a right-hand, positive rotation in $\sum^{\mathrm{MAS}}$. Therefore, the rotor performs a right-hand, negative rotation in $\sum^{\mathrm{lab}}$.

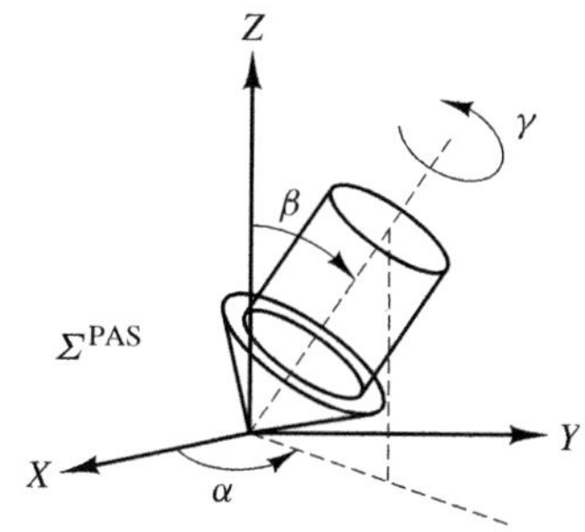

Figure 1.3. Euler angles defining the direction of the rotor in the principal axis system Σ^{PAS} of the EFG during a MAS experiment. In Σ^{PAS}, the rotor containing the sample appears static.

once more:

$$V_i = \sum_{j=-2}^{2} \mathcal{D}^{(2)}_{j,i}(\omega_r t, \theta_m, 0) V_j^{\mathrm{MAS}} \quad (1.36)$$

where ω_r is the angular velocity of the rotor and $\theta_m = 54.73°$ is the magic angle (Figure 1.2). Then, V_j^{MAS} must be expressed in terms of the components of $\boldsymbol{V}$ in Σ^{PAS}:

$$V_j^{\mathrm{MAS}} = \sum_{k=-2}^{2} \mathcal{D}^{(2)}_{k,j}(\alpha, \beta, \gamma) V_k^{\mathrm{PAS}} \quad (1.37)$$

where the Euler angles α, β, and γ describe the direction of the rotor in Σ^{PAS} (Figure 1.3).

The first step, equation (1.36), yields

$$\omega^{(1)\mathrm{MAS}}_{m-1,m} = \sqrt{\tfrac{2}{3}}(1-2m)\frac{3eQ}{8I(2I-1)\hbar} \times V_0^{\mathrm{MAS}}(3\cos^2\theta_m - 1) \quad (1.38)$$

The second step, equation (1.37), yields the first-order quadrupolar shift:

$$\omega^{(1)\mathrm{MAS}}_{m-1,m} = \tfrac{1}{2}(1-2m)\omega_Q(3\cos^2\theta_m - 1) \quad (1.39)$$

This shift is zero when the crystal rotates at the magic angle. In other words, all the energy levels become equally spaced. Therefore, a single line instead of $2I$ lines appears in the spectrum at ω_0.

For the second-order quadrupolar shift, the first step, equation (1.36), yields

$$\begin{aligned}\omega^{(2)\mathrm{MAS}}_{m-1,m} = &-\frac{2}{\omega_0}\left[\frac{eQ}{4I(2I-1)\hbar}\right]^2 \\ &\times \{-\tfrac{1}{6}V_2^{\mathrm{MAS}}V_{-2}^{\mathrm{MAS}}[50m(m-1) \\ &- 6I(I+1)+17] \\ &+ \tfrac{1}{3}V_1^{\mathrm{MAS}}V_{-1}^{\mathrm{MAS}}[8m(m-1)+2] \\ &- \tfrac{1}{2}V_0^{\mathrm{MAS}}V_0^{\mathrm{MAS}} \\ &\times [14m(m-1) - 2I(I+1)+5]\}\end{aligned} \quad (1.40)$$

The second step, equation (1.37), yields, in the fast rotation regime,[1]

$$\begin{aligned}\omega^{(2)\mathrm{MAS}}_{m-1,m} = &-\frac{3}{32\omega_0}\left[\frac{\chi}{I(2I-1)}\right]^2 (1+\tfrac{1}{3}\eta^2) \\ &\times [2I(I+1) - 14m(m-1) - 5] \\ &+ \frac{3}{128\omega_0}\left[\frac{\chi}{I(2I-1)}\right]^2 \\ &\times [6I(I+1) - 34m(m-1) - 13] \\ &\times g(\alpha, \beta, \eta)\end{aligned} \quad (1.41)$$

with

$$\begin{aligned}g(\alpha,\beta,\eta) = &\tfrac{1}{2}(1+6\cos^2\beta - 7\cos^4\beta) \\ &+ \tfrac{1}{3}\eta(1-8\cos^2\beta+7\cos^4\beta)\cos 2\alpha \\ &+ \tfrac{1}{18}\eta^2[-7(1-\cos^2\beta)^2 \\ &\times \cos^2 2\alpha + 8 - 4\cos^2\beta]\end{aligned} \quad (1.42)$$

For the central line, the second-order quadrupolar shift is[31]

$$\omega^{(2)\mathrm{MAS}}_{-1/2,1/2} = -\frac{1}{6\omega_0}\left[\frac{3\chi}{2I(2I-1)}\right]^2 [I(I+1) - \tfrac{3}{4}] \times \left[D(\alpha,\eta)\cos^4\beta + E(\alpha,\eta)\cos^2\beta + F(\alpha,\eta)\right] \quad (1.43)$$

with

$$\left.\begin{aligned} D(\alpha,\eta) &= \tfrac{21}{16} - \tfrac{7}{8}\eta\cos 2\alpha + \tfrac{7}{48}(\eta\cos 2\alpha)^2 \\ E(\alpha,\eta) &= -\tfrac{9}{8} + \tfrac{1}{12}\eta^2 + \eta\cos 2\alpha - \tfrac{7}{24}(\eta\cos 2\alpha)^2 \\ F(\alpha,\eta) &= \tfrac{5}{16} - \tfrac{1}{8}\eta\cos 2\alpha + \tfrac{7}{48}(\eta\cos 2\alpha)^2 \end{aligned}\right\} \quad (1.44)$$

As in the case of a static sample, the Euler angle γ does not appear in equations (1.43) and (1.44). This is because the experimental conditions correspond to the fast rotation regime. However, this angle does appear in the intermediate regime where the angular velocity of the rotor is of the same order of magnitude as the linewidth. As a result, spinning sidebands appear in the spectrum. Samoson and co-workers[23] established a general expression for $\Omega^{(2)\mathrm{MAS}}_{m-1,m}$ that clearly shows the presence of modulations due to the rotation of the rotor:

$$\Omega^{(2)\mathrm{MAS}}_{m-1,m} = \omega^{(2)\mathrm{MAS}}_{m-1,m} + \sum_{n=1}^{4}[A_n\cos(n\omega_r t) + B_n\sin(n\omega_r t)] \quad (1.45)$$

Two typographical errors appear in the *annexe* of their paper:[23] the expressions for A_1^λ and A_2^λ should be

$$\left.\begin{aligned} A_1^\lambda &= -\tfrac{1}{2}\sqrt{2}\sin 2\beta^\lambda\rho_{20}^\lambda + \tfrac{1}{3}\sqrt{3}\cos 2\alpha^\lambda\sin 2\beta^\lambda\rho_{22}^\lambda \\ A_2^\lambda &= \tfrac{1}{2}\sin^2\beta^\lambda\rho_{20}^\lambda + \tfrac{1}{6}\sqrt{6}\cos 2\alpha^\lambda(1+\cos^2\beta^\lambda)\rho_{22}^\lambda \end{aligned}\right\} \quad (1.46)$$

Equation (1.45) allows us to investigate the spinning sidebands.

1.6 POWDER SPECTRUM

In most cases, the sample is in powder form, because the growth of single crystals of significant size is not always possible. As a result, only the central line is detected in NMR. However, satellite lines can be detected without spinning the sample if $\chi/2\pi < 300$ kHz; for example, for ^{23}Na ($I = 3/2$) in $NaNO_3$ or ^{7}Li ($I = 3/2$) in $LiNbO_3$. When the MAS technique is applied, the spinning sidebands of the satellite lines are detected; for example, for iodine ^{127}I ($I = 5/2$) in KI or the two isotopes of bromine ($I = 3/2$) in KBr. These two compounds are used for setting the magic angle of the MAS probe in the vicinity of ^{29}Si and ^{27}Al frequencies, respectively.

In a powder sample, the principal axes of the EFG associated with each crystallite are randomly oriented with respect to $\boldsymbol{B}_0$. The transition frequencies are not unique, but depend on the distribution of the Euler angles α and β describing the direction of the rotor in the coordinate frame Σ^{PAS} in a MAS experiment at high spinning rate, or the direction of $\boldsymbol{B}_0$ in the coordinate frame Σ^{PAS} in an experiment without spinning.

The resonance condition $\omega(\alpha, \cos\beta)$ represents a surface in a 3D space described by the parameters ω, α, and $\cos\beta$. The critical points of this surface define divergences and shoulders in the spectrum. They are roots of the two coupled equations[12,13]

$$\left.\begin{aligned} \frac{\partial}{\partial\alpha}\omega(\alpha,\cos\beta)\Big|_{\alpha=r,\cos\beta=s} &= 0 \\ \frac{\partial}{\partial\cos\beta}\omega(\alpha,\cos\beta)\Big|_{\alpha=r,\cos\beta=s} &= 0 \end{aligned}\right\} \quad (1.47)$$

The nature of the critical point (r, s) is related to the sign of the Wronskian determinant D_W:

$$D_W = \left[\left(\frac{\partial^2\omega}{\partial\alpha\,\partial\cos\beta}\right)^2 - \left(\frac{\partial^2\omega}{\partial\alpha^2}\right)\left(\frac{\partial^2\omega}{\partial^2\cos\beta}\right)\right]_{\alpha=r,\cos\beta=s} \quad (1.48)$$

If D_W is positive then the critical point (r, s) represents a divergence; if D_W is negative then the critical point represents a shoulder. Unfortunately, this method does not allow us to determine the lineshape. Therefore, numerical calculations are required.

1.6.1 Powder Pattern for a Pair of Satellite Lines

For static samples,[27] numerical calculations are based on the summation of signals for the direction of each crystallite. The two Euler angles are redefined by $\alpha = 2\pi p/300$ and $\cos\beta = p/300$, where $p = 0, \ldots, 300$.

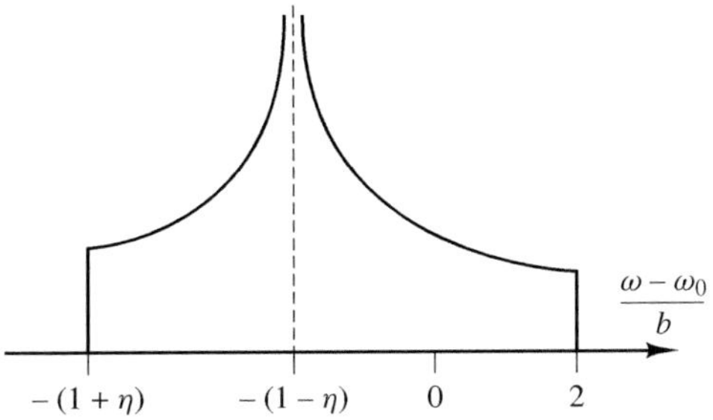

Figure 1.4. Critical points[13] of a satellite powder pattern associated with the transition $(m - 1, m)$; $b = 3\chi(1 - 2m)/8I(2I - 1)$.

For each value of η, the summation is performed on

$$\omega_{m-1,m}^{(1)\text{static}} \Big/ \frac{3\chi(1-2m)}{8I(2I-1)} = 3\cos^2\beta - 1 + \eta\sin^2\beta\cos 2\alpha \tag{1.49}$$

A satellite lineshape is shown in Figure 1.4. The positions of the shoulder and divergence[13] are derived from equations (1.47) and (1.48). The powder pattern of a pair of satellite lines for several values of η is plotted in Figure 1.5.

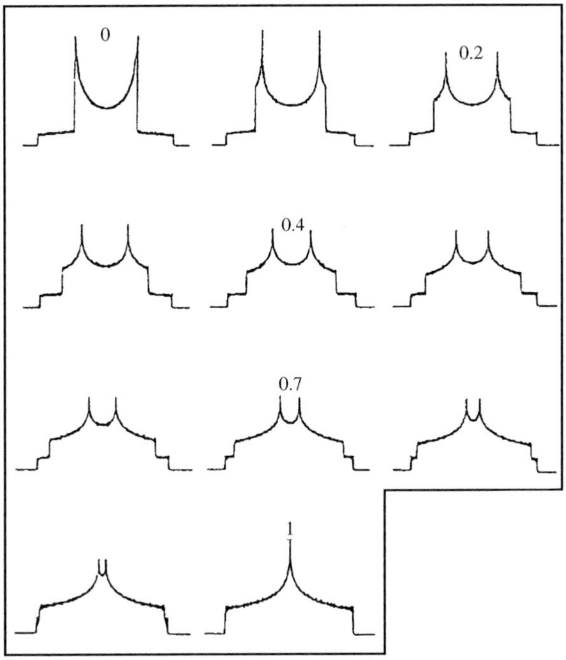

Figure 1.5. Simulated powder pattern of a pair of satellite lines for increasing values of the asymmetry parameter η from 0 to 1 in steps of 0.1.

1.6.2 Powder Pattern for the Central Line

The critical points in the powder pattern of the central line have been determined for the two experiments. For a static sample, they have been determined by Stauss.[32] Two patterns (Figure 1.6) appear according to the value of η compared with 1/3; for example, for ^{39}K in inorganic potassium salts[33] or ^{139}La in lanthanum salts.[34] It is worth noting that these critical points depend neither on the convention for η nor on the choice of Euler angles: changing η to $-\eta$ yields the same set of critical points. For a rotating sample, these points have been determined by Müller.[31] As previously, two powder patterns (Figure 1.7) appear, depending on the value of η compared with 3/7. These patterns are clearly observed in ^{27}Al compounds.[35]

The powder pattern of the central line is obtained as in the previous case using equation (1.33) for a static sample or equation (1.43) for a rapidly rotating sample. To facilitate the comparison, the two sets of spectra are superposed as shown in Figure 1.8. We can see that the linewidth is reduced by a factor ranging from 2 to 4, depending on the value of η when a MAS experiment is performed. From a practical point of view, the asymmetry parameter of an experimental spectrum can be estimated by comparing the lineshape with those in Figure 1.8. Then the quadrupolar coupling constant can be calculated using the positions of the experimental critical points and those represented in Figures 1.6 and 1.7. Other simple procedures are also defined for extracting quadrupolar parameters from the spectra of static[36] or rotating[37] samples.

The second-order quadrupolar shift of the center of gravity $\omega_{\text{iso}}(m - 1, m)$ of the powder pattern is determined as follows:[1]

$$\begin{aligned}\omega_{\text{iso}}(m-1,m) &= \frac{1}{4\pi}\int_0^{\pi} d\beta \sin\beta \int_0^{2\pi} d\alpha\, \omega_{m-1,m}^{(2)\text{MAS}} \\ &= -\frac{3}{40\omega_0}\left[\frac{\chi}{I(2I-1)}\right]^2 (1 + \tfrac{1}{3}\eta^2) \\ &\quad \times [I(I+1) - 9m(m-1) - 3]\end{aligned} \tag{1.50}$$

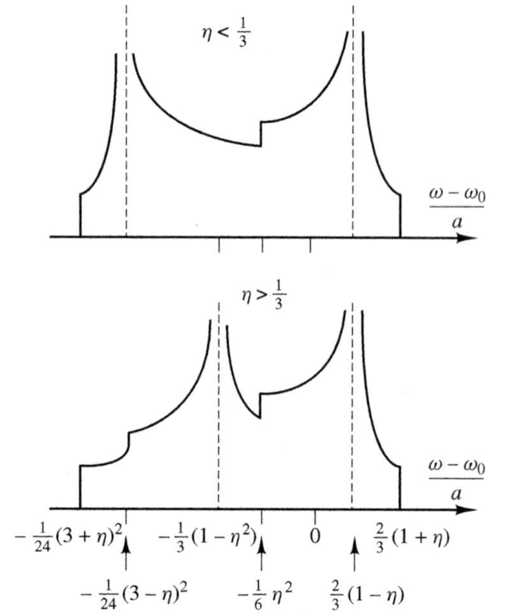

Figure 1.6. Critical points determined by Stauss.[32] They are associated with the powder pattern of the central line in a static experiment;

$$a = -\frac{1}{6\omega_0}\left[\frac{3\chi}{2I(2I-1)}\right]^2\left[I(I+1)-\frac{3}{4}\right]$$

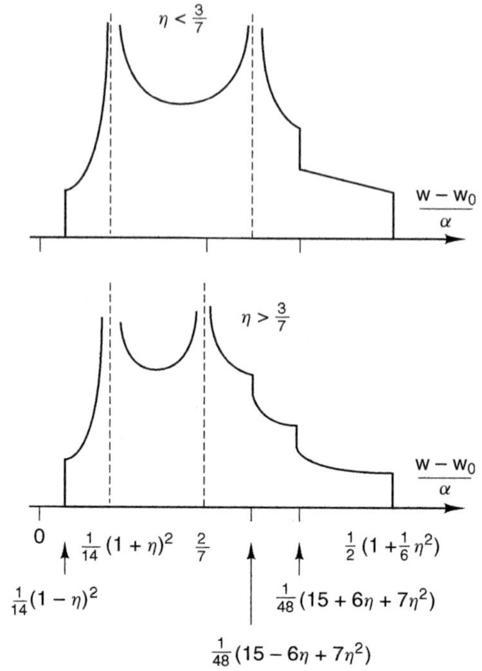

Figure 1.7. Corrected critical points[9,11] determined by Müller.[31] They are associated with the powder pattern of the central line in a MAS experiment;

$$a = -\frac{1}{6\omega_0}\left[\frac{3\chi}{2I(2I-1)}\right]^2\left[I(I+1)-\frac{3}{4}\right]$$

From a practical point of view, the experimental chemical shift of the center of gravity associated with the transition $(m-1, m)$, $\delta_{CG}^{exp}(m-1, m)$, consists of two terms: the true chemical shift $\delta_{CS}(m-1, m)$ and the contribution from the second-order quadrupolar shift $\omega_{iso}(m-1, m)/\omega_0$. Thus,

$$\delta_{CS}(m-1, m) = \delta_{CG}^{exp}(m-1, m) - \omega_{iso}(m-1, m)/\omega_0 \quad (1.51)$$

For the central line, equation (1.51) becomes equation (1.1).

1.7 APPENDIX

The Euler angles are extensively used in the study of the quadrupolar interaction, especially in MAS, VAS, DAS, and DOR. They are defined as three successive positive angles of rotation for the coordinate frame (c.f.) as described in Figure 1.9. First (Figure 1.9a), the starting c.f. (x, y, z), called the old c.f., in which we know the components V_k^{OLD} of the spherical tensor $\boldsymbol{V}$, is rotated counterclockwise around the z axis by an angle α. This rotation generates a new c.f. (X, Y, z). Then (Figure 1.9b) the counterclockwise rotation of this intermediate c.f. around the Y axis by an angle β generates a second intermediate c.f. (X', Y, z'). Finally (Figure 1.9c), this second intermediate c.f. is rotated counterclockwise by an angle γ around the z' axis, resulting in a c.f. (x', y', z'), called the new c.f., in which we wish to know the components V_j^{NEW} of the spherical tensor $\boldsymbol{V}$. It is worth noting that, in this definition of the Euler angles, β and α also represent the polar angles of the z' axis in the old coordinate frame.

As explained by Spiess,[20] the mathematical tool for expressing the components of the same spherical tensor $\boldsymbol{V}$ in two different coordinate frames, where the new c.f. is obtained by three positive angles of rotation (α, β, γ) of the old one, is the Wigner rotation matrix $\mathcal{D}_{p,q}^{(2)}(\alpha, \beta, \gamma)$ reported in Table 1.2. This

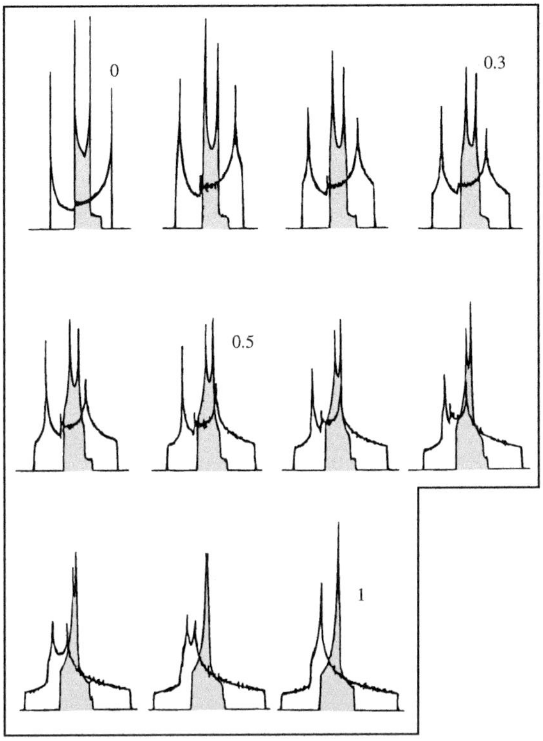

Figure 1.8. Superposition of simulated powder patterns of the central line during static and MAS (shaded spectra) experiments for increasing values of the asymmetry parameter η from 0 to 1 in steps of 0.1.

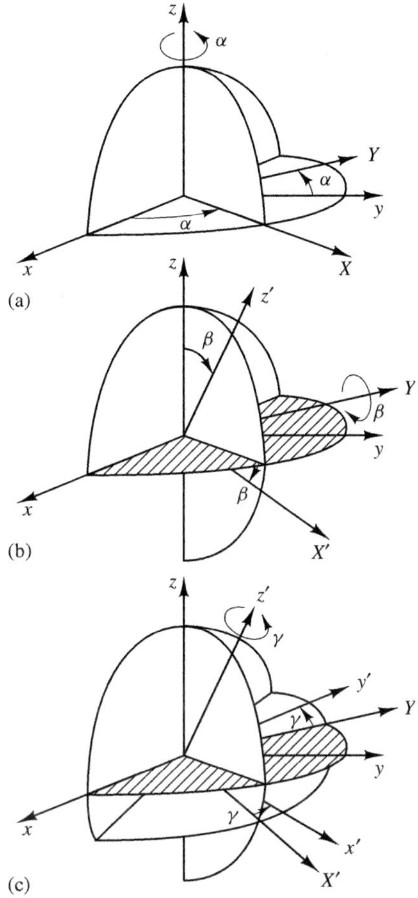

Figure 1.9. Euler angles defining positive rotation of the coordinate frame (x, y, z). They are used by Narita and co-workers,[27] Spiess,[20] Doane,[38] and Freude and Haase.[9] The angles β and α are the polar angles of the z' axis in (x, y, z).

matrix relates the components V_j^{NEW} of the spherical tensor in the new c.f. to the known components V_k^{OLD} of the same tensor in the old one by the relationships[20,21,38,39]

$$V_j^{\text{NEW}} = \sum_{k=-2}^{2} \mathcal{D}_{k,j}^{(2)}(\alpha, \beta, \gamma) V_k^{\text{OLD}} \qquad (1.52)$$

with the summation over the first subscript.

Figure 1.10 shows the Euler angles used by Baugher and co-workers.[13,15] First (Figure 1.10a), as in the previous definition, the old c.f. (x, y, z) is rotated counterclockwise by an angle ϕ around the z axis, generating an intermediate c.f. (X, Y, z). Then (Figure 1.10b) this intermediate c.f. is rotated counterclockwise, this time around the X axis, by an angle θ, generating a second intermediate c.f. (X, Y', z'). Finally (Figure 1.10c), a counterclockwise rotation of this second intermediate c.f. around the z' axis by an angle ψ produces the new c.f. (x', y', z'). In contrast to the previous definition, θ and ϕ are not the polar angles of the z' axis in the old coordinate frame. In Figure 1.9, α is the angle relating the y axis and the line of nodes Y, whereas in Figure 1.10, ϕ is the angle relating the x axis and the line of nodes X. These two angles are related by $\alpha + 1/2\pi = \phi$. In other words, $\alpha + 1/2\pi$ and ϕ correspond to the angle connecting the x axis with the line of nodes. As a result, $\cos 2\phi = -\cos 2\alpha$. This explains the change of sign for the terms containing η in equations (1.30) and (1.34), as in the paper of Baugher and co-workers.[15]

Table 1.2. Wigner rotation matrix $\mathcal{D}^{(2)}_{p,q}$ (α, β, γ), identical to that used by Spiess,[20] Mehring,[21] and Doane,[38] but the transposed and complex conjugate of that used by Haeberlen,[19] equal to $\mathcal{D}^{(2)}_{p,q}$ $(-\alpha, -\beta, -\gamma)$ used by Freude and Haase[9]

p	q = 2	1	0	−1	−2
2	$\frac{1}{4}(1 + \cos\beta)^2\, e^{-2i(\alpha+\gamma)}$	$-\frac{1}{2}(1 + \cos\beta)\sin\beta\, e^{-i(2\alpha+\gamma)}$	$\sqrt{\frac{3}{8}}\sin^2\beta\, e^{-2i\alpha}$	$-\frac{1}{2}(1 - \cos\beta)\sin\beta\, e^{i(-2\alpha+\gamma)}$	$\frac{1}{4}(1 - \cos\beta)^2\, e^{-2i(\alpha-\gamma)}$
1	$\frac{1}{2}(1 + \cos\beta)\sin\beta\, e^{-i(\alpha+2\gamma)}$	$[\cos^2\beta - \frac{1}{2}(1 - \cos\beta)]\, e^{-i(\alpha+\gamma)}$	$-\sqrt{\frac{3}{8}}\sin 2\beta\, e^{-i\alpha}$	$[\frac{1}{2}(1 + \cos\beta) - \cos^2\beta]\, e^{i(-\alpha+\gamma)}$	$-\frac{1}{2}(1 - \cos\beta)\sin\beta\, e^{-i(\alpha-2\gamma)}$
0	$\sqrt{\frac{3}{8}}\sin^2\beta\, e^{-2i\gamma}$	$\sqrt{\frac{3}{8}}\sin 2\beta\, e^{-i\gamma}$	$\frac{1}{2}(3\cos^2\beta - 1)$	$-\sqrt{\frac{3}{8}}\sin 2\beta\, e^{i\gamma}$	$\sqrt{\frac{3}{8}}\sin^2\beta\, e^{2i\gamma}$
−1	$\frac{1}{2}(1 - \cos\beta)\sin\beta\, e^{i(\alpha-2\gamma)}$	$[\frac{1}{2}(1 + \cos\beta) - \cos^2\beta]\, e^{i(\alpha-\gamma)}$	$\sqrt{\frac{3}{8}}\sin 2\beta\, e^{i\alpha}$	$[\cos^2\beta - \frac{1}{2}(1 - \cos\beta)]\, e^{i(\alpha+\gamma)}$	$-\frac{1}{2}(1 + \cos\beta)\sin\beta\, e^{i(\alpha+2\gamma)}$
−2	$\frac{1}{4}(1 - \cos\beta)^2\, e^{2i(\alpha-\gamma)}$	$\frac{1}{2}(1 - \cos\beta)\sin\beta\, e^{i(2\alpha-\gamma)}$	$\sqrt{\frac{3}{8}}\sin^2\beta\, e^{2i\alpha}$	$\frac{1}{2}(1 + \cos\beta)\sin\beta\, e^{i(2\alpha+\gamma)}$	$\frac{1}{4}(1 + \cos\beta)^2\, e^{2i(\alpha+\gamma)}$

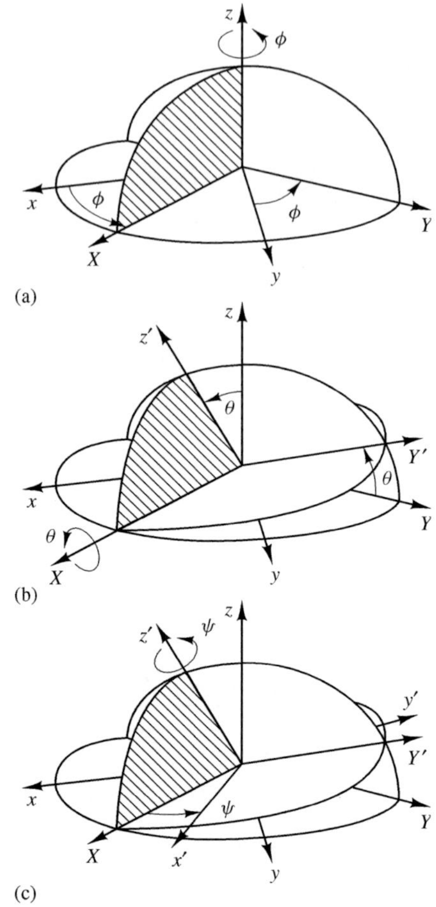

Figure 1.10. Euler angles used by Baugher and co-workers[13,15] and Stauss,[32] and defined by Goldstein.[14] The angles θ and ϕ are not the polar angles of the z' axis in (x, y, z).

RELATED ARTICLES IN THE ENCYCLOPEDIA OF MAGNETIC RESONANCE

Average Hamiltonian Theory

Double Rotation

High Speed MAS of Half-Integer Quadrupolar Nuclei in Solids

Internal Spin Interactions and Rotations in Solids

Line Narrowing Methods in Solids

Magic Angle Spinning

Magic Angle Spinning: Effects of Quadrupolar Nuclei on Spin-1/2 Spectra

Multiple-quantum Magic-angle Spinning Experiments on Half-integer Nuclei: Fundamentals

Overtone Spectroscopy of Quadrupolar Nuclei

Relaxation Theory for Quadrupolar Nuclei

Rotating Solids

Satellite Transition NMR Spectroscopy of Half-Integer Quadrupolar Nuclei under Magic-angle Spinning

Tensors in NMR

Variable Angle Sample Spinning

REFERENCES

1. A. Samoson, *Chem. Phys. Lett.*, 1985, **119**, 29.
2. E. Lippmaa, A. Samoson, and M. Mägi, *J. Am. Chem. Soc.*, 1986, **108**, 1730.
3. F. Lefebvre, J. P. Amoureux, C. Fernandez, and E. G. Derouane, *J. Chem. Phys.*, 1987, **86**, 6070.
4. D. Massiot, A. Kahn-Harari, D. Michel, D. Müller, and F. Taulelle, *Magn. Reson. Chem.*, 1990, **28**, S82.
5. B. Q. Sun, J. H. Baltisberger, Y. Wu, A. Samoson, and A. Pines, *Solid State Nucl. Magn. Reson.*, 1992, **1**, 267.
6. P. P. Man, *Chem. Phys. Lett.*, 1990, **168**, 227.
7. B. C. Sanctuary and T. K. Halstead, in *Advances in Magnetic and Optical Resonance*, ed. W. S. Warren, Academic Press, San Diego, 1990, Vol. 15, p. 79.
8. A. P. M. Kentgens, *J. Magn. Reson. A*, 1993, **104**, 302.
9. D. Freude and J. Haase, in *NMR Basic Principles and Progress*, eds P. Diehl, E. Fluck, H. Günter, R. Kosfeld, and J. Seelig, Springer-Verlag, Berlin, 1993, Vol. 29.
10. J. L. Dye, A. S. Ellaboudy, and J. Kim, in *Modern NMR Techniques and their Application in Chemistry*, eds A. I. Popov and K. Hallenge, Marcel Dekker, New York, 1991.
11. G. Engelhardt and D. Michel, *High-Resolution Solid-State NMR of Silicates and Zeolites*, Wiley, Chichester, 1987.
12. G. E. Jellison, Jr., S. A. Feller, and P. J. Bray, *J. Magn. Reson.*, 1977, **27**, 121.
13. P. C. Taylor, J. F. Baugher, and H. M. Kriz, *Chem. Rev.*, 1975, **75**, 203.
14. H. Goldstein, *Classical Mechanics*, 2nd edn., Addison-Wesley, Reading, MA, 1980.
15. J. F. Baugher, P. C. Taylor, T. Oja, and P. J. Bray, *J. Chem. Phys.*, 1969, **50**, 4914.
16. C. P. Slichter, *Principles of Magnetic Resonance*, 3rd edn., Springer-Verlag, New York, 1990.
17. M. H. Cohen and F. Reif, in *Solid State Physics*, eds F. Seitz and D. Turnbull, Academic Press, New York, 1957, Vol. 5.
18. A. Abragam, *The Principles of Nuclear Magnetism*, Oxford University Press, Oxford, 1961.
19. U. Haeberlen, *High Resolution NMR in Solids. Selective Averaging*, Academic Press, New York, 1976.
20. H. W. Spiess, in *NMR Basic Principles and Progress*, eds P. Diehl, E. Fluck, and R. Kosfeld, Springer-Verlag, Berlin, 1978, Vol. 15.
21. M. Mehring, *Principles of High Resolution NMR in Solids*, 2nd edn., Springer-Verlag, Berlin, 1983.
22. A. Samoson and E. Lippmaa, *Phys. Rev. B*, 1983, **28**, 6567.
23. A. Samoson, E. Kundla, and E. Lippmaa, *J. Magn. Reson.*, 1982, **49**, 350.
24. H. J. Behrens and B. Schnabel, *Physica B*, 1982, **114**, 185.
25. P. P. Man, H. Théveneau, and P. Papon, *J. Magn. Reson.*, 1985, **64**, 271.
26. B. A. Huggins and P. D. Ellis, *J. Am. Chem. Soc.*, 1992, **114**, 2098.
27. K. Narita, J. Umeda, and H. Kusumoto, *J. Chem. Phys.*, 1966, **44**, 2719.
28. J. Hirschinger, P. Granger, and J. Rosé, *J. Phys. Chem.*, 1992, **96**, 4815.
29. P. J. Chu and B. C. Gerstein, *J. Chem. Phys.*, 1989, **91**, 2081.
30. F. Wolf, D. Kline, and H. S. Story, *J. Chem. Phys.*, 1970, **53**, 3538.
31. D. Müller, *Ann. Phys. (Leipzig)*, 1982, **39**, 451.
32. G. H. Stauss, *J. Chem. Phys.*, 1964, **40**, 1988.
33. T. J. Bastow, *J. Chem. Soc., Faraday Trans.*, 1991, **87**, 2453.

34. B. Herreros, P. P. Man, J. M. Manoli, and J. Fraissard, *J. Chem. Soc., Chem. Commun.*, 1992, 464.

35. J. Skibsted, E. Henderson, and H. J. Jakobsen, *Inorg. Chem.*, 1993, **32**, 1013.

36. P. Granger, *Magn. Reson. Chem.*, 1990, **28**, 156.

37. G. Engelhardt and H. Koller, *Magn. Reson. Chem.*, 1991, **29**, 941.

38. J. W. Doane, in *Magnetic Resonance of Phase Transitions*, eds. F. J. Owens, C. P. Poole, Jr., and H. A. Farach, Academic Press, New York, 1979.

39. D. M. Brink and G. R. Satchler, *Angular Momentum*, 2nd edn., Clarendon Press, Oxford, 1968, p. 51.

Chapter 2
Quadrupolar Nuclei in Solids

Alexander J. Vega

Department of Chemistry and Biochemistry, University of Delaware, Newark, DE 19716, USA

2.1 INTRODUCTION

Nuclei with spin quantum number $I \geqslant 1$ have an electric quadrupole moment that couples with the inhomogeneous internal electric fields existing in molecules and solids. Since this quadrupolar interaction is usually stronger than other interactions such as chemical shift and dipole–dipole couplings, it dominates the NMR spectra of quadrupolar nuclei in solid materials. Liquid-state NMR spectra of quadrupolar nuclei are not affected to a comparable extent, because the fast isotropic tumbling of the molecules greatly diminishes the impact of the quadrupolar interaction. While the NMR signals of quadrupolar nuclei in solids are split into multiplets that at times broaden the lineshapes of powder samples over many megahertz, the quadrupolar interaction does not split the NMR lines of liquid samples. Instead, it has a pronounced effect on the T_1 and T_2 relaxation times of nuclei in liquids. The NMR spectra of quadrupolar nuclei in liquid crystals, where the quadrupolar splitting is partially reduced, provide a wealth of information concerning molecular orientation and dynamics.

This chapter is devoted to the solid-state aspects of the quadrupolar interaction. Since many subtopics from this area of NMR are discussed elsewhere in the *Encyclopedia of Magnetic Resonance*. The aim of this chapter is to provide a general overview rather than comprehensive discussions of each specific method. The first two sections following this introduction summarize the quantum mechanical properties of quadrupolar spins and their response to radiofrequency (RF) pulses. This survey is given without theoretical derivations and with only a few mathematical formulas. The important experimental approaches are cataloged in Section 2.4. Some of the fundamental theoretical concepts underlying quadrupolar spin properties are presented in Section 2.5.

Quadrupolar effects in NMR of solids were first reported and analyzed in 1950 by Pound.[1] Introductions to NMR of quadrupolar nuclei are provided in the classic 1957 review article by Cohen and Reif[2] and in the textbooks by Abragam[3] and Slichter.[4] Kanert and Mehring also covered the basics of quadrupole NMR in their 1971 review of quadrupole NMR of disordered cubic solids.[5] Nuclei with half-integer spins have distinctive quantum mechanical properties that have led to the development of special methods for their detection. Progress in this area was summarized in a 1993 review article by Freude and Haase.[6] NMR

NMR of Quadrupolar Nuclei in Solid Materials
Edited by Roderick E. Wasylishen, Sharon E. Ashbrook and Stephen Wimperis

ISBN: 978-0-470-97398-1

of integer-spin nuclei is almost exclusively limited to ^2H and ^{14}N. Although both are $I = 1$ spins, the disparate orders of magnitude of their quadrupolar interaction energies necessitate different methodologies for their study in the solid state. Deuterium NMR spectroscopy was reviewed by Spiess[7] in 1985.

2.2 BASIC SPIN PROPERTIES

2.2.1 Nuclear Electric Quadrupolar Interaction

The nuclear quadrupolar interaction is the coupling of the electric quadrupole moment of the nucleus with the gradient of the electric field generated by the other charges in the system. The quadrupole moment is usually denoted by eQ and the magnitude of the electric field gradient (EFG) by eq, where e is the elementary charge. Since it is impractical to produce NMR-detectable EFGs by means of external charged conductors, the gradients are exclusively generated by the electrons and nuclei of molecules and crystals. Thus, the size of the quadrupolar interaction experienced by a particular nucleus is a constant that is characteristic of the molecular or crystalline environment. Its value in frequency units, e^2qQ/h, is called the *nuclear quadrupolar coupling constant* (*NQCC*). In the Encyclopedia of Magnetic Resonance, we use the letter χ as the shorthand notation for e^2qQ/h. A variety of different symbols (C_q, C_Q, C_{qcc}) are found in other publications.

Closely related to the NQCC is the so-called quadrupolar frequency ν_Q (in hertz) or ω_Q (in radians per second^{-1}), in this chapter defined as

$$\nu_Q = \frac{\omega_Q}{2\pi} = \frac{3\chi}{2I(2I-1)} \qquad (2.1)$$

The use of ν_Q is often preferred to that of χ, because it simplifies equations. Moreover, the quadrupolar frequency describes the actual strength of the quadrupolar interaction more closely than does the NQCC. It should be noted that equation (2.1) is not a universally accepted definition of ν_Q, although in literature dealing with half-integer spins it is often (but not always) found in this form. However, in the case of $I = 1$ authors usually prefer $\nu_Q = 3/4\chi$ to the $\nu_Q = 3/2\chi$ that follows from equation (2.1). For the sake of uniformity, we adhere to equation (2.1) for all values of I throughout this chapter.

The EFG is a 3D entity with the properties of a tensor. To describe it fully, we need to specify its size, shape, and orientation. The quantity eq was introduced above as a parameter of the size. The shape is characterized by the asymmetry parameter η, which is a measure of the deviation of the EFG from axial symmetry. η can have any value between 0 and 1, with $\eta = 0$ corresponding to axial symmetry. The orientation of the EFG with respect to the molecular or crystalline structure is defined by three Euler angles. The tensor properties are more fully defined in Section 2.5.1. In the literature, quadrupolar interactions are commonly reported by specification of their NQCC and η. The angular parameters are not usually provided unless orientational information is of special interest.

The quadrupolar interaction vanishes in three general cases.

1. No quadrupolar interaction is ever associated with an $I = 1/2$ nucleus, because eQ vanishes for all subatomic particles with spin quantum number $I = 0$ or 1/2.
2. The NQCC is zero when a quadrupolar nucleus is positioned at a cubic (octahedral or tetrahedral) site, because then $eq = 0$ by symmetry.
3. The quadrupolar interaction of a nucleus belonging to a molecule in an isotropic liquid or a gas is averaged to zero by the rapid tumbling motion of the molecule.

2.2.2 NQR and NMR Spectra of Quadrupolar Nuclei

The nature of magnetic resonance spectra of quadrupolar nuclei in solids depends to a large extent on the size of the electric quadrupolar interaction relative to that of the Zeeman interaction with the externally applied magnetic field $\boldsymbol{B}_0$. The strength of this magnetic interaction is given by the Larmor frequency $\omega_0 = 2\pi\nu_0 = \gamma \boldsymbol{B}_0$, where γ is the gyromagnetic ratio. In this section, we review the general features of magnetic resonance spectra for the various magnitude ranges of the ratio ν_0/ν_Q.

We begin with the case where no magnetic field is applied at all ($\nu_0 = 0$). The quadrupolar interaction then splits the magnetic energy levels of the nuclear spins into patterns like those shown in Figure 2.1. The energy levels are associated with particular orientations of the nuclear spin axis with respect to the EFG

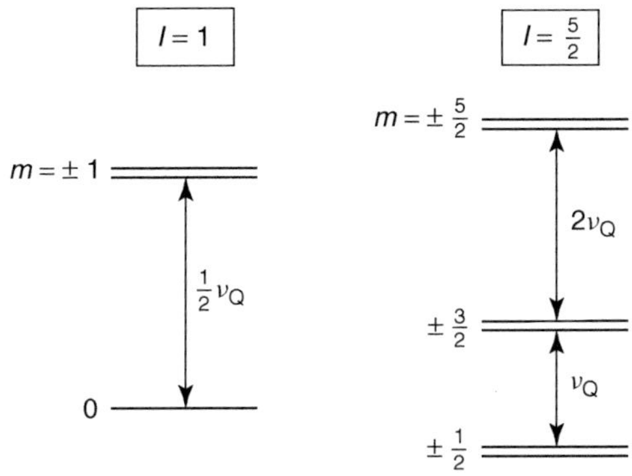

Figure 2.1. Quadrupolar energy levels in zero magnetic field. These two examples are for axial symmetry of the EFG. If the EFG is not axially symmetric, the transition frequencies are more complicated functions of the quadrupolar frequency ν_Q and the asymmetry parameter η, while the eigenstates are linear combinations of m states. Furthermore, the $m = \pm 1$ degeneracy of the $I = 1$ states is lifted when $\eta = 0$. However, the eigenstates of half-integer spins remain degenerate in pairs for any value of η.

axes, and can thus be identified by magnetic quantum numbers m. The spectroscopic transitions among these magnetic states can be detected with the regular RF methods of magnetic resonance. This branch of spectroscopy is known as *nuclear quadrupole resonance* (*NQR*). Its first spectrum was observed by Dehmelt and Krüger.[8] Unlike the extremely broad high-field NMR spectra of quadrupolar nuclei in randomly oriented powder samples, the NQR resonances are sharp. In fact, they provide the most precise measurements of the quadrupolar frequency and the NQCC. Although NQR employs the RF methods of magnetic resonance, it is not an NMR technique in the proper sense, since NMR is defined as a spectroscopy associated with a Zeeman field. NQR is therefore outside the scope of the *Encyclopedia of Magnetic Resonance*, although some aspects of it are discussed in the articles on. For further literature on the subject, the reader is referred to the classic monograph by Das and Hahn,[9] several other books and tabulations,[10–12] and the review articles in the series 'Advances in Nuclear Quadrupole Resonance'.[13]

A weak magnetic field, corresponding to a Larmor frequency ν_0 smaller than ν_Q, shifts and splits the NQR lines. The frequency shifts of this so-called Zeeman effect of NQR are functions of the orientation of the magnetic field with respect to the crystal axes. Consequently, in randomly oriented powder samples, the effect is observed as a broadening of the peaks.

When ν_0 is much larger than ν_Q, we are in the regime of NMR. In the extreme limit of vanishing ν_Q, the energy levels are the $2I + 1$ Zeeman levels, $E_m = m\omega_0$, giving rise to $2I$ coinciding transitions of frequency ν_0. A relatively small quadrupolar interaction, $\nu_Q \ll \nu_0$, shifts the eigenvalues of the Zeeman levels and splits the NMR spectrum into $2I$ peaks. Examples for spins $I = 1$ and 5/2 are shown schematically in Figure 2.2. For the theoretical description of these effects, we follow the methods of perturbation theory and express the quadrupolar corrections to the energy levels as the sums of first-order terms of order ω_Q and second-order terms of order ${\omega_Q}^2/\omega_0$ (see Section 2.5.3). Third- and higher-order terms do not need to be considered.

The main features of quadrupolar NMR spectra are governed by the equation describing the orientation dependence of the first-order energy corrections ${E_m}^{(1)}$ of the Zeeman levels m:

$$\Delta E_m^{(1)} = \frac{1}{2}\Delta\omega_Q(\theta, \phi)\left[m^2 - \frac{1}{3}I(I+1)\right] \quad (2.2)$$

where $\Delta\omega_Q$ is a fraction of ω_Q and is a function of the polar angles θ and φ which relate the Zeeman field direction to the EFG principal axes system (see Section 2.5.1 and Figure 2.11):

$$\begin{aligned}\Delta\omega_Q(\theta, \phi) &= 2\pi\Delta\nu_Q(\theta, \phi) \\ &= \frac{1}{2}\omega_Q\left[3\cos^2\theta - 1 - \eta\sin^2\theta\cos 2\phi\right]\end{aligned} \quad (2.3)$$

$\Delta\nu_Q$ (or $\Delta\omega_Q$) is called the *quadrupolar splitting* because the $2I$ lines in the spectrum are, to first order, equally spaced by $\Delta\nu_Q$. This follows from equation (2.2), which predicts that the first-order frequency shifts $\Delta\nu^{(1)}{}_{m\leftrightarrow m+1}$ of the allowed transitions $m \leftrightarrow m+1$ are given by

$$\Delta\nu_{m\leftrightarrow m+1}^{(1)} = \Delta\nu_Q(\theta, \phi)\left(m + \frac{1}{2}\right) \quad (2.4)$$

In powder samples, the orientation dependence of $\Delta\nu_Q$ broadens the individual transitions and causes them to overlap. However, the $-1/2 \leftrightarrow 1/2$ transition of noninteger spins ($I = 3/2, 5/2, 7/2, 9/2$) does not experience a first-order broadening ($m = -1/2$ in equation (2.4)). Consequently, this "central transition" stands out as a relatively sharp peak at the center of the "satellite transitions". It is clear from equation (2.4) that there is no central transition associated with integer-spin nuclei. The dependence of

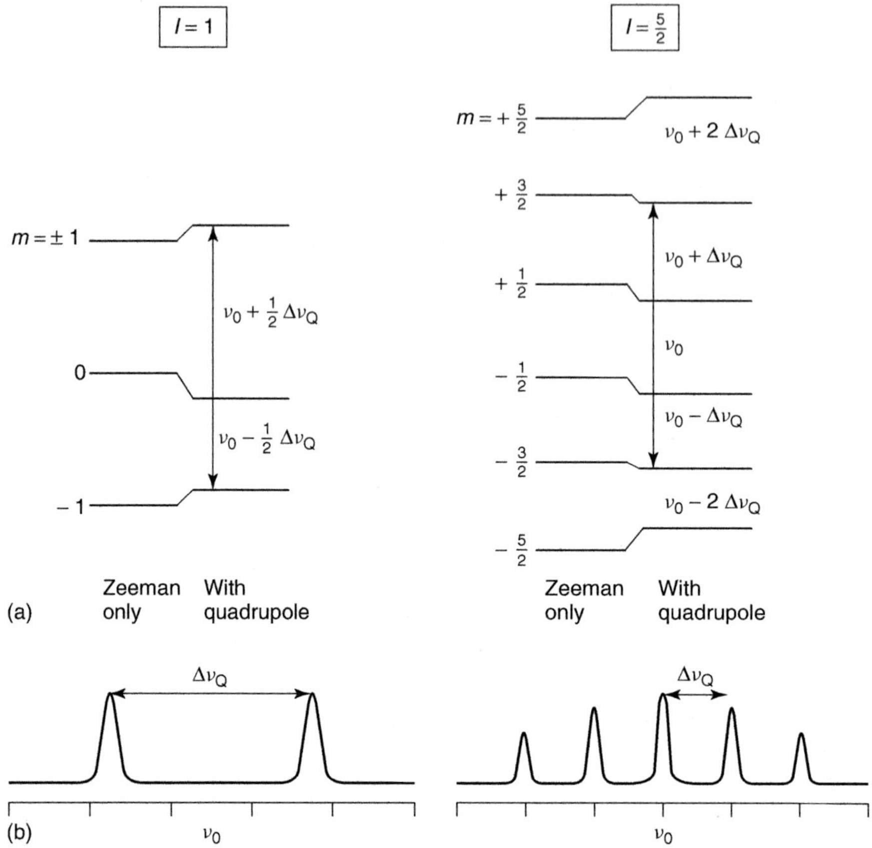

Figure 2.2. (a) Energy levels of spins $I = 1$ and 5/2 in a Zeeman field, in the absence and in the presence of a quadrupolar interaction. The frequencies of the allowed ($\Delta m = 1$) transitions are indicated with first-order quadrupolar corrections. The arrows mark forbidden ($\Delta m = 2$, 3) transitions. (b) The corresponding quadrupole-splits the NMR spectra. $\omega_0 = 2\pi\nu_0$ is the Larmor frequency and $-\omega_Q = 2\pi\Delta\nu_Q$ is the quadrupolar splitting.

$\Delta\nu_Q(\theta, \phi)$ on the orientation is similar to that of the anisotropic chemical shift. However, unlike the chemical shift, the first-order quadrupolar effect has no isotropic contribution, implying that the centers of gravity of the spectral distributions are not shifted. Thus, each satellite transition has a powder lineshape characteristic of the value of η, and is centered around the Larmor frequency. Figure 2.3 shows a few simulated examples for $I = 1$ and 5/2.

The broadening of the satellites is often too large to be captured within the bandwidth of the NMR spectrometer. In such a case, we only observe the central transition, the lineshape of which is dominated by second-order quadrupolar shifts. The orientation dependence of this shift is of a different nature to that of the first-order satellite shift. A consequence of this is that it contributes to an isotropic shift of the order of ${\nu_Q}^2/\nu_0$ (see equation (2.15) in Section 2.4.3). Equation (2.2) shows further that levels m and $-m$ have identical first-order energy shifts. Therefore, in addition to the central transition, there are forbidden transitions $\Delta m = 2, 3, \ldots$ (indicated in Figure 2.2 by arrows) with transition frequencies that are not affected to first order by the quadrupolar interaction. The experimental methods of double quantum excitation and overtone spectroscopy of $I = 1$ nuclei take advantage of this special property (see Section 2.4).

The energy-level diagrams of Figure 2.4 illustrate the transition from the NQR limit to the NMR limit. The figure shows the variation of the energy levels of a spin when the ratio ν_0/ν_Q changes gradually from 0 to ∞. This is shown for two cases, both with

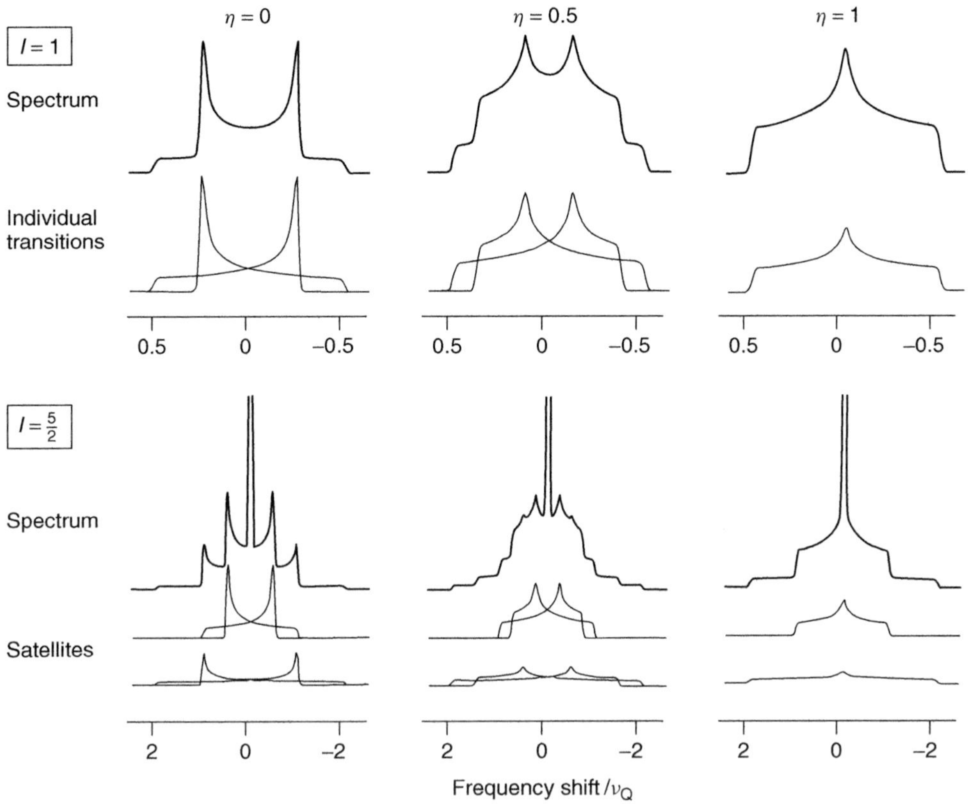

Figure 2.3. First-order quadrupolar spectra of powder samples simulated for $I = 1$ and 5/2 and for three values of the asymmetry parameter η. The lineshapes of individual transitions are drawn below the spectra. The central transition peaks of $I = 5/2$ are off scale. The frequency scale is in units of the quadrupolar frequency ν_Q as defined in equation (1). The Larmor frequency is at the center of the spectra.

an axially symmetric EFG—one with the EFG symmetry axis $\boldsymbol{q}$ parallel to the field $\boldsymbol{B}_0$ 4(b), and one with $\boldsymbol{q}$ perpendicular to $\boldsymbol{B}_0$ 4(c). The difference between the two sets of energy curves demonstrates the strong dependence of the transition frequencies on the relative orientations of the EFG and $\boldsymbol{B}_0$. When ν_0 is of the order of ν_Q, the powder spectra are sufficiently broad that detection of magnetic resonance becomes impractical within the limitations of bandwidth and sensitivity of the equipment currently in use. Figure 2.4 also indicates the spin states in the limiting cases. Note that these eigenstates are quantized along the symmetry direction of the prevailing field, as is indicated by the Q and Z subscripts on the quantum numbers m. In general, these states are not identical. In fact, they are linear combinations of each other. A certain amount of mixing of the pure Zeeman states occurs even when the quadrupolar interaction is small in comparison with the Zeeman interaction. We say that the spins are then "no longer quantized along the Zeeman direction" (see Section 2.5.3). This mixing controls a number of the NMR phenomena reviewed below, including overtone NMR, heteronuclear dipolar splittings, and zero-field NMR.

2.2.3 Level Populations

Since $I > 1/2$ spins have more than two Zeeman levels, it is necessary to use two or more independent parameters for the description of their relative populations. In thermal equilibrium, the high-field population pattern is as illustrated in Figure 2.5(a) for $I = 1$ or Figure 2.5(c) for $I = 3/2$.

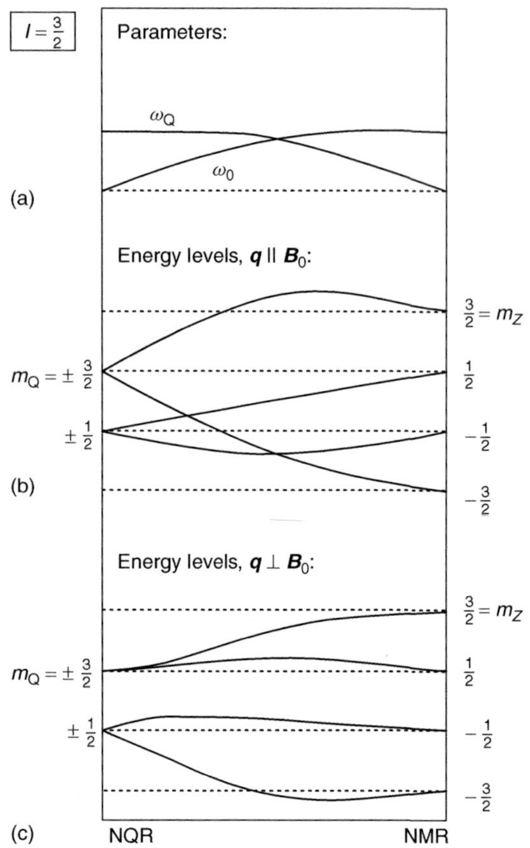

Figure 2.4. Energy levels of a spin $I = 3/2$ in a coexisting electric field gradient and a magnetic Zeeman field, calculated for varying Larmor frequency ω_0 and quadrupolar frequency ω_Q. The gradual change of these parameters, from the NQR limit ($\omega_0 = 0$) on the left to the NMR limit ($\omega_Q = 0$) on the right, is shown in the curves in (a). The EFG is assumed to be axially symmetric. The corresponding energies of the spin states are shown for two relative orientations of the symmetry axis $\boldsymbol{q}$ of the EFG and the Zeeman field $\boldsymbol{B}_0$ (b and c). The magnetic quantum numbers are indicated for the NQR and NMR limits (m_Q is quantized along $\boldsymbol{q}$; mZ is quantized along $\boldsymbol{B}_0$).

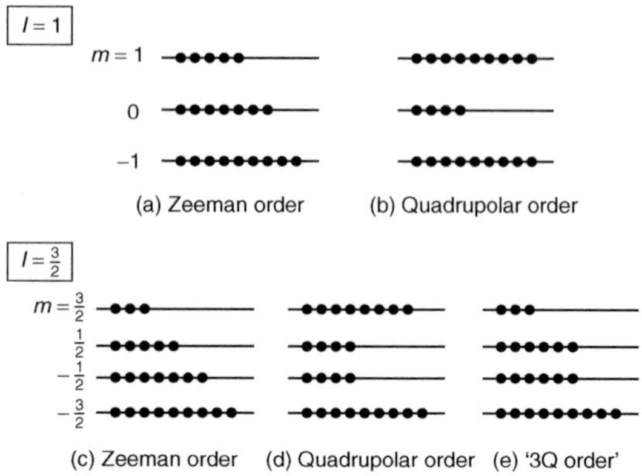

Figure 2.5. Level-population patterns of spins 1 and 3/2 according to Zeeman order, quadrupolar order, and 'triple quantum order'.

The populations of the levels m are then given by $(1 - m\hbar\omega_0/kT)/(2I + 1)$, which is the Boltzmann distribution in the high-temperature approximation. This is called *Zeeman order*, because the populations are determined by the Zeeman interaction. It is characterized by uniform population increments between adjacent levels. The application of RF pulses causes deviations from the Boltzmann distribution. When these deviations are such that the Zeeman pattern is retained, we can introduce a meaningful "spin temperature" replacing T in the distribution expression above. The distribution pattern can also be of the type depicted in Figure 2.5(b) and (d). This is called *quadrupolar order*, because the deviations from the average population follow the $m^2 - I(I + 1)/3$ dependence associated with the quadrupolar interaction (equation (2.2)). In general, the population distribution of spins $I = 1$ is a combination of the two. Unlike Zeeman order, quadrupolar order does not contribute to the z magnetization. In thermal equilibrium, quadrupolar order is negligible, because the population differences associated with it are ν_0/ν_Q times smaller than those of equilibrium Zeeman order.

For spins with $I \geqslant 3/2$, more types of population distribution need to be considered. An important example is the $I = 3/2$ configuration illustrated in Figure 2.5(e). This arrangement can be viewed as Zeeman order in which the $m = \pm 1/2$ populations are equilibrated and do not contribute to the z magnetization (e.g., as a result of selective saturation of the central transition), whereas the equilibrium population difference of the $m = \pm 3/2$ levels and their contribution to the z magnetization are retained. For later reference, we call this *triple quantum order*.

2.2.4 Relaxation

Spin relaxation is caused by randomly fluctuating interactions. In the case of quadrupolar nuclei, the fluctuations can be electric (due to a time-dependent size or orientation of the EFG) or magnetic (due to fluctuating chemical shifts, dipole–dipole interactions, or interactions with unpaired electrons). When caused by random molecular motions or lattice vibrations, the electric fluctuations usually dominate, since in most cases the quadrupolar interaction is stronger than the magnetic dipolar interactions and chemical shifts.[6] On the other hand, a magnetic relaxation mechanism mostly prevails in the presence of conduction electrons[14] or in structures containing paramagnetic centers such as superconducting oxide materials.

Longitudinal spin–lattice relaxation is the process by which nonequilibrium level populations (see Figure 2.5) revert to thermal equilibrium. Unlike relaxation of spin-1/2 nuclei, which involves a single transition probability between two energy levels, relaxation of quadrupolar nuclei is generally characterized by several transition rates. Hence, the relaxation decay is multiexponential, and cannot be quantified by a single T_1. For instance, the quadrupole-induced spin–lattice relaxation of a spin $I = 1$ has two principal spin–lattice relaxation times: the conventional T_1 for the return of Zeeman order to the equilibrium populations, and a different time T_{1Q} for the decay of quadrupolar order (see Section 2.2.3). Spin–lattice relaxation patterns of higher spins are more complex. Provided the initial population distribution has the form of Zeeman order, spin–lattice relaxation has one decay constant for $I = 1$, two for $I = 3/2$ or 2, three for $I = 5/2$ or 3, etc.[15] In powders, the complexity is compounded with a dependence of the relaxation time on the orientation of the EFG with respect to the Zeeman field. However, the transition rates are mostly of comparable magnitude, such that an approximate assessment of a global T_1 is not unwarranted. The dependence of T_1 on the correlation time τ_c of the motion follows the general Bloembergen–Purcell–Pound (BPP) or Redfield behavior, given by

$$\frac{1}{T_1} = \omega_Q^2 \left(\frac{a\tau_c}{1+\omega_0^2\tau_c^2} + \frac{b\tau_c}{1+4\omega_0^2\tau_c^2} \right) \quad (2.5)$$

where the numerical coefficients a and b reflect the amplitudes of the EFG fluctuations, the geometric details of the motion, a dependence on I, and the multiexponential character of the decay. A log–log plot of T_1 versus τ_c has the familiar shape shown in Figure 2.6, with its minimum around ω_0^{-1}. For relaxation due to lattice vibrations, the equation is cast in

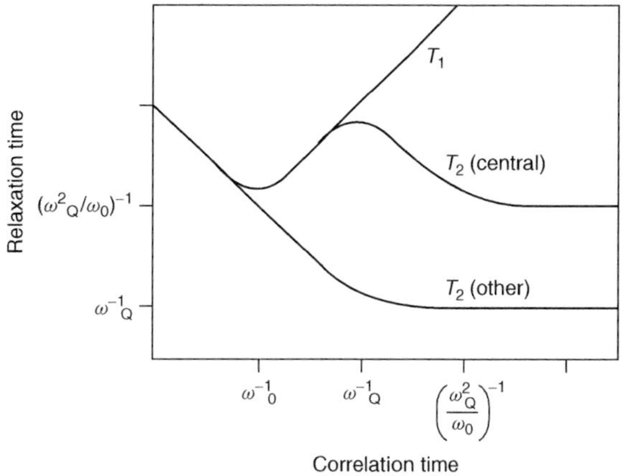

Figure 2.6. Idealized log–log plot of T_1 and T_2 relaxation times versus the correlation time of the motion for a model where the spin relaxation is caused by large-scale fluctuations of the EFG. T_2 (central) is the inverse halfwidth of the central transition spectra of half-integer spins; T_2 (other) is that of satellite transitions or transitions of integer spins.

different forms reflecting the thermal behavior of the phonons.[6] A similar expression, with a hyperfine coupling constant instead of ω_Q, holds for magnetically induced relaxation.

As is usually done in BPP-type discussions of the transverse relaxation time, we define T_2 as a measure of the inverse linewidth of the spectrum. (The signal decay in echo experiments, albeit important in deuterium NMR, is outside the scope of this chapter.) In the slow motion limit, the powder spectra have widths and shapes like those of the examples shown in Figure 2.3. Hence, $1/T_2$ is approximately equal to ω_Q, except for the central transition of half-integer spins, which has $1/T_2$ of the order of $\omega_Q{}^2/\omega_0$. In the fast motion limit, the spectra are reduced to motionally narrowed Lorentzian lines having full linewidth at half-maximum equal to $1/\pi T_2$. The majority of the spectral components follow the familiar BPP pattern schematically drawn as the T_2(other) curve in Figure 2.6. If the relaxation mechanism is quadrupolar, the condition for motional narrowing is $\tau_c \ll \omega_Q^{-1}$, and the T_2 of the narrowed spectrum varies with the correlation time according to

$$\frac{1}{T_2} = c\omega_Q^2\tau_c \quad (2.6)$$

where the numerical coefficient c depends on the motional model.

The effect of quadrupolar relaxation on T_2 of the central transition is markedly different,[16,17] as is indicated by T_2 (central) in Figure 2.6. A gradual increase in the motional rate (decreasing correlation time) begins to narrow the rigid limit spectrum when τ_c becomes smaller than the reciprocal linewidth ($\tau_c < (\omega_Q{}^2/\omega_0)^{-1}$), but for shorter τ_c, the line broadens again, and T_2 passes through a minimum when $\tau_c \approx \omega_0{}^{-1}$. This T_2 behavior was observed for ^{23}Na in amorphous polymer electrolytes above the glass transition temperature.[18]

The T_2 minimum is associated with a dynamic frequency shift.[16,17] For half-integer spins in viscous liquids where the correlation time is larger than $\omega_0{}^{-1}$, the divergence of a "slow" T_2(central) and a 'fast' T_2(other) is observed as a biexponential relaxation decay and is the basis for discrimination of sodium ions in the various compartments of tissues. Transverse relaxation of the central transition due to unpaired electrons does not differ substantially from the relaxation of the other transitions.

2.2.5 Typical Values of the Nuclear Quadrupolar Coupling Constant

Table 2.1 lists typical values of the NQCC (χ), the quadrupolar frequency (ν_Q as defined in equation

Table 2.1. Typical values of nuclear quadrupole coupling constants (NQCC), quadrupole frequencies (ν_Q), and Larmor frequencies (ν_0)

Nucleus	I	Natural abundance (%)	Typical NQCC (MHz)	Typical ν_Q (MHz)	ν_0 at 7.05 T (MHz)
Integer spin:					
^2H	1	0.015	0.16	0.24	46.1
^6Li	1	7.4	0.001	0.0015	44.2
^{10}B	3	19.6	1	0.1	32.2
^{14}N	1	99.6	4	6	21.7
Half-integer spin:					
^7Li	3/2	92.6	0.05	0.025	116.6
^9Be	3/2	100	0.4	0.2	42.2
^{11}B	3/2	80.4	2	1.5	96.3
^{17}O	5/2	0.037	7	1	40.7
^{23}Na	3/2	100	2	1	79.4
^{27}Al	5/2	100	1–20	0.15–3	78.2
^{35}Cl	3/2	75.5	60	30	29.4
^{37}Cl	3/2	24.5	(×0.788)	(×0.788)	24.5
^{39}K	3/2	93.1	4	2	14.0
^{45}Sc	7/2	100	3	0.2	73.0
^{51}V	7/2	99.8	3	0.2	78.9
^{55}Mn	5/2	100	60	9	74.0
^{63}Cu	3/2	69.1	60	30	79.5
^{65}Cu	3/2	30.9	(×0.925)	(×0.925)	85.2
^{71}Ga	3/2	39.6	150	75	91.5
^{75}As	3/2	100	160	80	51.4
^{87}Rb	3/2	27.8	20	10	98.2
^{79}Br	3/2	50.5	400	200	75.2
^{81}Br	3/2	49.5	(×0.835)	(×0.835)	81.0
^{91}Zr	5/2	11.2	20	3	28.0
^{93}Nb	9/2	100	50	2	73.6
^{96}Mo	5/2	15.7	2	0.3	19.6
^{133}Cs	7/2	100	0.7	0.02	39.4
^{121}Sb	5/2	57.2	500	70	71.8
^{127}I	5/2	100	2000	300	60.0
^{209}Bi	9/2	100	500	20	48.2

(2.1)), and the Larmor frequency (ν_0 in the magnetic field of a 300-MHz spectrometer) for selected nuclei. The data were gathered from various sources in the literature. Unless the quadrupole interaction is drastically reduced by cubic symmetry, the actual NQCCs for a particular nucleus do not usually differ by more than a factor of two or three from the quoted value. The multiplication factors in parentheses indicate the ratios between the NQCC of isotopes of the same element. These ratios are fixed by the ratios of their *eQ* values. The relative magnitudes of ν_Q and ν_0 serve as a guide for determining whether a nucleus is better studied by NQR or by NMR.

2.3 INTERACTION WITH RADIOFREQUENCY FIELDS

The manner in which RF pulses affect quadrupolar spins in a high magnetic field depends strongly on the relative magnitudes (in frequency units) of three parameters. One is the RF amplitude $\omega_1 = 2\pi\nu_1 = \gamma \boldsymbol{B}_1$, where $\boldsymbol{B}_1$ is the strength of the rotating magnetic component of the RF field. The other two are the first-order quadrupolar splitting $\Delta\omega_Q = 2\pi\Delta\nu_Q$ and the difference $\Delta\omega_0 = 2\pi\Delta\nu_0$ between the resonance frequency of the spins and the carrier frequency of the RF pulse. Here, the resonance frequency is the combination of Larmor frequency, chemical shift, and second-order quadrupolar shift. While there are no other ways in which second-order quadrupolar effects impact the performance of RF pulses to a noticeable extent, first-order splittings make a large difference. Consequently, the orientation dependence of $\Delta\nu_Q$ can cause a wide range of responses to a uniform RF pulse when it is applied to a powder sample. To avoid this complication, we limit the discussion in this section to samples with a uniform $\Delta\nu_Q$ as in a single crystal. The distinct experimental conditions and their respective effects on quadrupolar spins are categorized below. The summary is descriptive in nature and is presented without theoretical explanations. For the latter, see Section 2.5.5.

2.3.1 Nonquadrupolar Nuclei

To introduce the RF-related concepts of nutation, spin locking, and population transfer, we first consider the simple example of a vanishing NQCC ($\nu_Q = 0$). In the rotating frame (the axes system that rotates with the frequency of the RF carrier), the RF field is represented by a constant vector ν_1 of length $\boldsymbol{\nu}_1$ along x, while the Zeeman field is effectively reduced to an offset vector $\Delta\nu_0$ of length $\mathbf{1}\boldsymbol{\nu}_0$ along z (see the axis diagram in Figure 2.7a). The magnetization precesses about the "effective field", which is represented by the vector $\boldsymbol{\nu}_{\text{eff}}$. We distinguish the two limiting cases of (i) irradiation close to resonance, $\Delta\nu_0 \ll \nu_1$, and (ii) irradiation far-off-resonance, $\Delta\nu_0 \gg \nu_1$. These cases are also represented in the more schematic illustration in Figure 2.7(b), where the effective range of the RF field is represented in the frequency domain by a shaded rectangle of width of order ν_1 centered at the carrier frequency. In NMR terminology, the precession induced by a pulse of type (i) is referred to as *nutation*. In general, this term relates to precessions caused by RF irradiation and those which begin with the spins in thermal equilibrium. In case (i), the nutation is in the (y, z) plane, and takes place with a nutation frequency equal to the RF amplitude, $\nu_{\text{nut}} = \nu_1$. If before the application of the pulse the spins have been prepared so that they point in the x direction of the rotating frame, they will be spin locked by a pulse of type (i). No nutation or excitation is induced by a far-off-resonance pulse, but we can speak of spin locking in its presence. Namely, the direction of the effective field ν_{eff}, which is nearly parallel to z in case (ii), can be viewed formally as a spin-locking field for z magnetization. This notion is a useful starting point for the description of an adiabatic passage that occurs when $\Delta\nu_0$ is slowly changed from one side of resonance to the other. During an adiabatic passage, the spins remain spin locked along ν_{eff} and rotate together with it from z through x to $-z$. Eventually, this results in population inversion of the Zeeman levels. In the case of $I = 1/2$ the passage transfers the population of the 1/2 level to the $-1/2$ level and vice versa. The criterion for adiabaticity is that the parameter

$$\alpha = \frac{\omega_{\text{nut}}^2}{\mathrm{d}\Delta\omega_0/\mathrm{d}t} \tag{2.7}$$

must be larger than 1. If the passage is sudden ($\alpha \ll 1$), the magnetization remains in the original direction and no populations are transferred. If it is intermediate ($\alpha \approx 1$), the magnetization ends up in a direction that is not spin locked.

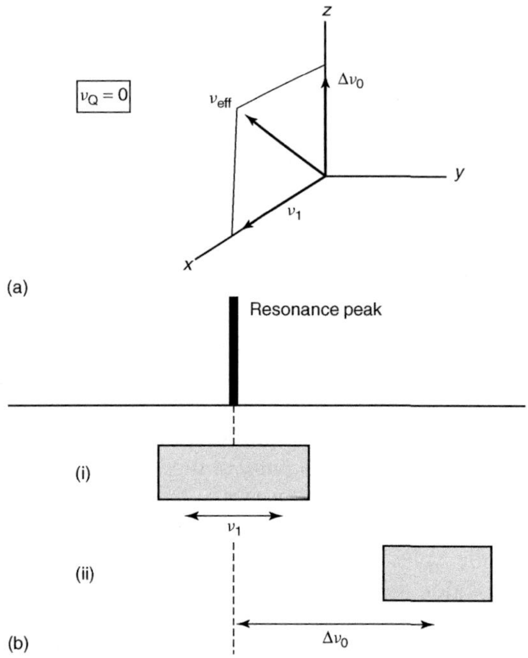

Figure 2.7. (a): Rotating frame representation of an RF field of amplitude ν_1 (in frequency units) and a resonance offset $\Delta\nu_0$. This representation is adequate for a nonquadrupolar spin system, either $I = 1/2$ or $I > 1/2$. (b): Schematic frequency domain representation of RF irradiation of nonquadrupolar spins. The RF excitation profile is indicated by a shaded rectangle centered at the RF carrier frequency and extending over the approximate excitation range of the RF field. The limiting cases are (i) on-resonance RF ($\Delta\nu_0 \ll \nu_1$) and (ii) off-resonance RF ($\Delta\nu_0 \gg \nu_1$).

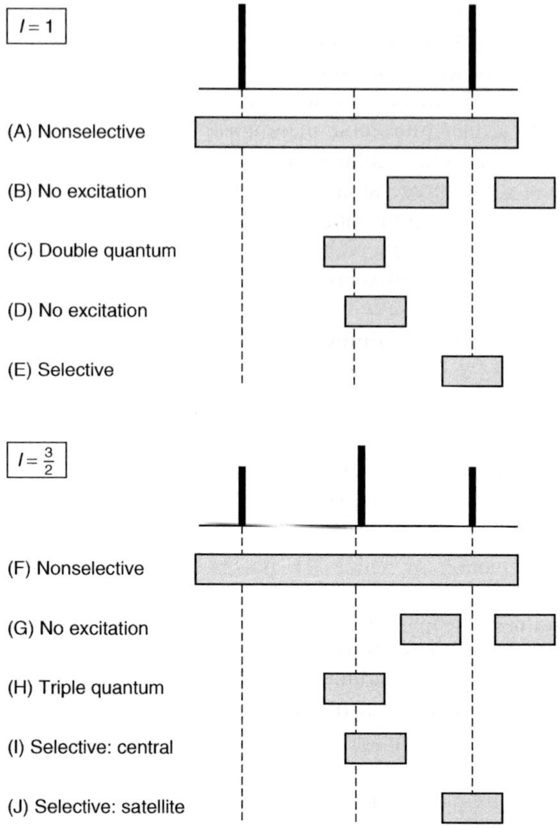

Figure 2.8. Schematic frequency domain representation of RF irradiation of quadrupolar spins $I = 1$ and 3/2. The RF ranges are indicated as in Figure 2.7. The response of the spins to the irradiation depends on the position of the carrier frequency and on the width of the excitation range with respect to the quadrupole-split NMR peaks shown at the top.

2.3.2 Spin-1 Nuclei

For the visualization of RF fields in the presence of quadrupolar interactions, we can no longer resort to a simple 3D vector picture. Instead, we shall review the various aspects of RF irradiation with the help of the illustrations in the frequency domain shown in Figure 2.8.

2.3.2.1 Nutation

When $I = 1$, we distinguish five special cases denoted (a)–(e):

Case (a)
The excitation is nonselective when ν_1 is larger than both $\Delta\nu_Q$ and $\Delta\nu_0$. The two allowed transitions (see the $I = 1$ portion of Figure 2.2) are then simultaneously excited. As in the $\nu_Q = 0$ case (Figure 2.7), the nonselective pulse induces nutation with frequency $\nu_{nut} = \nu_1$. However, following the pulse, the spin system will not continue to behave like a nonquadrupolar nucleus. Excitation by two or more pulses, with quadrupolar interactions acting in the intervals, can create as many as eight distinct spin-state configurations, of which polarizations along x, y, and z are only three examples.

Two other kinds of spin states are of particular interest. These are "quadrupolar order", which was discussed in the previous section (see Figure 2.5), and "double quantum coherence", which is a quantum mechanical state that can be thought of as a linear combination of $m = +1$ and -1 states. Although double quantum coherence is not observable in the form of nuclear magnetization, it has some similarity to regular transverse magnetization ("single quantum coherence") in that it has x and y components that undergo precession under frequency offset. Unlike single quantum coherence, the double quantum precession frequency is twice the offset frequency. For density matrix representations of these coherences, see Section 2.5.4.

The most prominent nonselective pulse sequence is the quadrupolar echo sequence, also called *solid echo sequence*. It consists of two out-of-phase 90° pulses separated by an interval τ: $90°_x - \tau - 90°_y$. The spin dephasing due to first-order quadrupolar effects is refocused by the second pulse, and an echo is formed at time τ following the second pulse.

Case (b)
When ν_1 is small ($\nu_1 \ll \Delta\nu_Q$) and the RF profile does not overlap with any of the transitions (Figure 2.8b shows two examples), the spins are not excited, except in case (c) below.

Case (c)
A weak RF field ($\nu_1 \ll \Delta\nu_Q$) applied at the exact midpoint between the two spectral lines ($\Delta\nu_0 = 0$) induces $\Delta m = 2$ transitions directly between the $m = +1$ and -1 states, while leaving the $m = 0$ state unaffected. This effect was indicated in Figure 2.2 as the forbidden double quantum transition. It can be understood in terms of a second-order perturbation of the first-order quadrupolar interaction by the RF interaction (see Section 2.5.5). It results in a double quantum nutation frequency given by[19]

$$\nu_{2Q} = 2\nu_1^2/\Delta\nu_Q \tag{2.8}$$

A pulse of duration $\tau_p = 1/4\nu_{2Q}$ is a double quantum 90° pulse. It transforms Zeeman order into a state of double quantum coherence. A double quantum 180° pulse inverts the populations.

Case (d)
When the RF carrier is slightly off-resonance, no double quantum excitation occurs, despite the fact that the offset may be less than the RF amplitude ($\nu_{2Q} \ll \Delta\nu_0 < \nu_1 \ll \Delta\nu_Q$).

Case (e)
Selective excitation of one of the allowed transitions ($-1 \leftrightarrow 0$ or $0 \leftrightarrow +1$) is caused by a weak RF field applied at its resonance frequency. The corresponding nutation frequency is given by $\nu_{nut} = \nu_1\sqrt{2}$.

2.3.2.2 Spin Locking

Each of these five forms of RF irradiation is associated with one or more spin-locked spin configurations. If the RF field is applied along x in the rotating frame, the following spin states are spin locked under the various conditions: nonselective irradiation (A) spin locks transverse spin polarization along x; off-resonance irradiation (B) and (D) spin lock Zeeman and quadrupolar order; double quantum irradiation (C) spin locks double quantum coherence of type x; and selective irradiation of an allowed transition (E) spin locks the corresponding single quantum coherence.

2.3.2.3 Population Transfer

Adiabatic passages, similar to the population inversion described in section 2.3.1, can occur for quadrupolar spins $I = 1$ when the RF profile crosses over from one side of a transition to the other. For instance, a slow passage between the two situations drawn for case (b) in Figure 2.8 causes population exchange between the levels $m = 0$ and -1, and thus transforms pure Zeeman order to a combination of Zeeman and quadrupolar order. Such a passage is materialized by a slow sweep of $\Delta\nu_0$, which can in turn be done by sweeping of the carrier frequency or of the magnetic field. Another possibility is the sweeping of $\Delta\nu_Q$. Because of its orientation dependence, the quadrupolar splitting can easily be varied by simply turning the sample. The passage is adiabatic if the change in $\Delta\nu_Q$ is sufficiently slow for the condition $\alpha > 1$ to be satisfied, with the adiabaticity parameter defined as $\alpha = \omega_{nut}{}^2/(d\omega_Q/dt)$. Magic-angle spinning (MAS) is a particularly effective method for $\Delta\nu_Q$ sweeping, since $\Delta\nu_Q$ of any crystallite experiences two or four zero-crossings per rotation cycle. (This is consistent with the $\Delta\nu_Q$ averaging to zero by MAS.) If the spinning rate is denoted by ν_R, the adiabaticity parameter is roughly equal to

$$\alpha \approx \frac{\nu_1^2}{\nu_R \nu_Q} \tag{2.9}$$

2.3.3 Nuclei with Half-Integer Spin

2.3.3.1 *Nutation (I = 3/2)*

The response of a spin $I = 3/2$ to the application of RF pulses is in many respects similar to that of $I = 1$, but there are a few differences to be pointed out. We again distinguish among five special cases illustrated in the bottom half of Figure 2.8.

Case (f)
When ν_1 covers the entire spectrum, the excitation is nonselective and the nutation frequency is $\nu_{\mathrm{nut}} = \nu_1$. The spin dynamics of sequences with more than one pulse is even richer than for $I = 1$ nuclei, since there are 15 independent spin states for $I = 3/2$ (see Section 2.5.4). The nonselective solid echo sequence $90^\circ{}_x - \tau - 90^\circ{}_y$ refocuses first-order quadrupolar dephasing as in the case of $I = 1$.

Case (g)
There is no excitation when the RF profile does not overlap with any transition (but see Section 2.5.5).

Case (h)
A weak RF pulse applied at the exact midpoint between the satellite transitions induces a triple quantum excitation[20,21] between $m = -3/2$ and 3/2 with a nutation frequency given by $\nu_{3Q} = 3{\nu_1}^3/2\Delta{\nu_Q}^2$. The resonance conditions for triple quantum transition and for the allowed central transition do not coincide exactly, because the two transitions have different second-order shifts. The triple quantum excitation is quenched when the carrier frequency is slightly off-resonance ($\Delta\nu_0 > \nu_{3Q}$), but when $\Delta\nu_Q$ is not much larger than ν_1, the excitation is effective for $\Delta\nu_0 \lesssim \nu 1$.

Case (i)
A weak pulse ($\nu_1 < \Delta\nu_Q$) with an RF profile that overlaps with the central transition induces selective excitation of the latter. The nutation frequency for this transition is $\nu_{\mathrm{nut}} = 2\nu_1$.

Case (j)
Selective excitation of the satellites occurs with a nutation frequency $\nu_{\mathrm{nut}} = \nu_1\sqrt{3}$. Note that, unlike selective excitation of the narrow central transition, selective excitation of satellite transitions cannot be achieved simultaneously for all crystallites in a powder sample.

2.3.3.2 *Spin Locking (I = 3/2)*

The spin states of $I = 3/2$ nuclei that are spin locked under the various modes of RF irradiation follow a pattern similar to that of $I = 1$, and do not need to be discussed in detail. However, the case of selective irradiation of the central transition (case (i)) deserves some special attention. During an RF pulse of this type, two spin-state configurations can be spin locked. One is the population difference of the $m = \pm 3/2$ levels illustrated as *triple quantum order* in Figure 2.5(e). For convenience, we give it the shorthand notation T_z to indicate that it is associated with the triple quantum transition and that it contributes to z magnetization. The other is single quantum coherence of the central transition, denoted by C_x, which is a linear combination of states $m = 1/2$ and $-1/2$ contributing to x magnetization. Density matrix representations of T_z and C_x are given in Section 2.5.5. When a type (I) pulse is applied to spins $I = 3/2$ that are initially in thermal equilibrium, two things happen simultaneously: the T_z portion of the spin state is spin locked while the central transition portion undergoes nutation. However, if the central transition portion is prepared in the C_x state by appropriately chosen preparatory pulses, simultaneous spin locking of T_z and C_x can be achieved.

2.3.3.3 *Population Transfer (I = 3/2)*

An adiabatic passage caused by zero-crossing of $\Delta\nu_Q$ in a slowly rotating sample transfers a spin-locked T_z state to C_x, and vice versa. As mentioned above for $I = 1$, MAS is an efficient method for inducing these passages. The criterion for adiabaticity is again $\alpha > 1$, with α as defined as in equation (2.9).[22] The populations of the $\pm 3/2$ states can also be transferred to the $\pm 1/2$ states by adiabatic sweeping of $\Delta\omega_0$.[23]

2.3.3.4 *Central Transition (I ≥ 3/2)*

Similar concepts can be applied to half-integer spins with $I > 3/2$, but, other than the general case of central transition excitation, they will not be discussed further. The frequency of nutation of the central transition induced by nonselective irradiation, as exemplified by case (f) in Figure 2.8, is given by $\nu_{\mathrm{nut}} = \nu_1$. On the other hand, when the excitation is selective as in case (i), the general formula for the nutation frequency is

$$\nu_{\text{nut}} = \left(I + \frac{1}{2}\right)\nu_1 \qquad (2.10)$$

This effective enhancement of the RF amplitude has obvious consequences for the choice of pulse length for obtaining optimum signal intensity in a selective single-pulse excitation experiment. For $I = 3/2$ the apparent 90° pulse is a nominal 45° pulse. Likewise, for $I = 5/2$, it is 30°, and so on. An additional result of selective excitation is that it reduces the intensity of the resulting central transition signal. If the signal (not including the satellites) following a nonselective pulse of length τ_p can be described by

$$S(\tau_p) = S_0 \sin \omega_1 \tau_p \qquad (2.11)$$

then the signal following a selective pulse is

$$S(\tau_p) = \frac{S_0}{I + \frac{1}{2}} \sin\left[\left(I + \frac{1}{2}\right)\omega_1 \tau_p\right] \qquad (2.12)$$

The flip angle dependence according to equations (2.11) and (2.12) is plotted in Figure 2.9 for $I = 3/2$ and 5/2. So far, we have avoided discussion of RF amplitudes that are intermediate between selective and nonselective excitations ($\nu_1 \approx \nu_Q$). However, this situation is often encountered in practice. The response of the spin system to intermediate RF amplitudes is more complex than in the limiting cases. For details see Chapter 7. A significant feature is that for relatively small flip angles ($\omega_{\text{nut}}\tau_p < \pi$), the intermediate $S(\tau_p)$ functions fall between the limits of selective and nonselective pulses. Hence, since the initial slopes of the functions of equations (2.10) and (2.11) are identical, the signal intensity following very short pulses is independent of the size of the quadrupolar interaction. This result has important implications for the quantitative interpretation of NMR signal intensities.

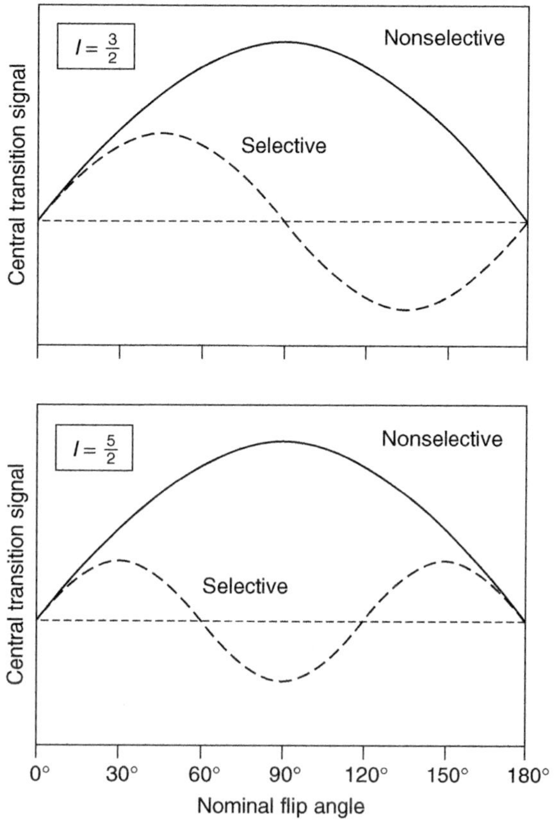

Figure 2.9. Central transition signal intensity following excitation by a single RF pulse of nominal flip angle $\omega_1\tau_p$, where τ_p is the duration of the pulse, plotted for the limits of nonselective ($\omega_1 \ll \Delta\omega_Q$) and selective ($\omega_1 \gg \Delta\omega_Q$) irradiation.

2.4 EXPERIMENTAL METHODS

Methods for NMR detection of quadrupolar nuclei in solids are surveyed in this section. The emphasis is not on the achievements of the applications of these techniques, but rather on the underlying spectroscopic principles and their interrelations. Consequently, this section frequently refers to the basic spin properties introduced in the previous sections. The survey is divided in subsections on deuterium, ^{14}N, and half-integer spins, reflecting the fact that in practical applications the choice of a workable NMR method is usually dictated by the size of the NQCC and by the presence or absence of a central transition. The section concludes with a brief introduction to zero-field NMR.

2.4.1 Deuterium

Detailed theoretical and experimental aspects of solid-state ^2H NMR are covered in articles listed in the Related Articles section of this chapter as well as Chapter 14 of this book. Applications to polymers were reviewed by Spiess.[7] Key references to the literature can also be found in these articles.

2.4.1.1 First-Order Spectra

The most intensively studied first-order quadrupolar spectra are those of deuterium. This is partially due to the chemical importance of the hydrogen atom and the advantages of selective deuteration, but there is also a fortuitous combination of spectroscopic conditions. Deuterium is among the few quadrupolar nuclei for which the total width of the spectra is never much larger than 200 kHz. This ensures that the deuterium spectrum of any chemical structure can be detected within a practical bandwidth and the spins can be excited with essentially nonselective pulses. In addition, the spectra are not appreciably affected by second-order quadrupolar shifts, dipolar broadenings, or chemical shifts, because in the case of deuterium all these contributions are at least two orders of magnitude smaller than ν_Q. Another advantage is that the powder lineshapes of the first-order spectra of spin-1 nuclei are composed of only two spectral transitions, and are thus less complex than those of higher spin nuclei (see Figures 2.2 and 2.3). These circumstances allow the observation of the sometimes subtle lineshape changes caused by nonrandom distributions of molecular orientation (as in liquid crystalline materials) or by rapid random reorientations of the EFG tensor. In fact, the bulk of solid-state deuterium NMR work is focused on the elucidation of molecular motions, particularly in polymers, liquid crystals, and adsorbed molecules. In combination with a judicious use of relaxation times and the application of 2D methods, motions with correlation times from nanoseconds to seconds can be characterized.

2.4.1.2 Zeeman Order and Quadrupolar Order

In the so-called spin-alignment experiment,[7] the spins are prepared in a state of quadrupolar order (see Section 2.2.3), where they can be held for a time as long as T_{1Q} permits. During that time, chemical exchange can modify the level populations. 2D methods use this and a similar Zeeman-order scheme for characterization of slow molecular dynamics.

2.4.1.3 Pulse Sequences

Although deuterium spins can be excited with a nonselective pulse, the FID following a single 90° excitation pulse is short and largely undetectable because of the receiver dead time. Therefore, the signals need to be created with the solid echo sequence, $90°_x - \tau - 90°_y$, where the second half of the echo serves as the FID for further data processing. The dependence of the signal on τ provides an extra experimental parameter for the study of motions. Quadrupolar order is created with the Jeener–Broekaert sequence, $90°_x - \tau - 45°_y$,[24] and detected with a 45° read-out pulse.

2.4.1.4 Double Quantum Transitions

Since the double quantum transition frequency is not shifted by the quadrupolar interaction, it is useful for measurements of the chemical shift tensor[25] or, in conjunction with MAS, the isotropic chemical shift,[26] and also for achieving high resolution in imaging.[27] In fact, the double quantum coherence precesses in the rotating frame with twice the offset frequency, and thus has an enhanced sensitivity to small changes in the resonance frequency. Because double quantum coherences are not directly observable, their time dependence has to be measured in a 2D-type experiment where the spins are allowed to evolve for a stepwise incremented time t_1, at the end of which they are detected with a nonselective read-out pulse that transfers double quantum coherence to observable signals. The double quantum state can be prepared by a variety of methods[28]: a double quantum excitation pulse (Figure 2.8c), double quantum cross-polarization (a method where Hartmann–Hahn contact between, say, protons and deuterons is established by adjusting ν_1 of the protons to be equal to ν_{2Q} given by equation (2.8)),[29] two nonselective pulses,[26] or three nonselective pulses.[27]

2.4.1.5 Double Quantum Decoupling

Effective dipolar decoupling of deuterium from other nuclei by nonselective irradiation of the allowed transitions requires a very high RF intensity, which is difficult to produce in practice. However, decoupling can also be achieved through stirring of the $m = +1$ and -1 states by double quantum excitation.[19] It does not matter that this method does not stir the $m = 0$ state, because the latter is magnetically "neutral" and does not contribute to dipolar broadening. The provision that the irradiation must be close to the double quantum resonance frequency (see Section 2.3.2.1) presents no practical problem, since the resonance condition is only affected by chemical shifts and second-order quadrupolar shifts.

2.4.1.6 *Magic-Angle Spinning*

The similarity between the orientational dependence of the first-order quadrupolar splitting and that of CSA implies that MAS removes quadrupolar broadening in deuterium spectra of powders in the same way in which it narrows CSA broadening. However, since rotation speeds are much smaller than the width of the spectrum, the centerband of the MAS spectrum is always accompanied by a large number of sidebands, the envelope of which resembles the static lineshape.[30,31] Owing to the absence of an isotropic first-order shift, the position of the centerband is entirely determined by isotropic chemical shifts and second-order quadrupolar shifts.

2.4.2 Nitrogen-14

With the exception of chemical structures where near-cubic site symmetry reduces the NQCC to small values,[32] first-order ^{14}N spectra of powders are too broad for detection by NMR. Consequently, direct measurement of ν_Q and η of most compounds is feasible only by NQR spectroscopy. The spin-1/2 isotope ^{15}N is usually preferred for solid-state NMR studies, despite its low natural abundance (0.365%). Nevertheless, ^{14}N nuclei can be detected by several indirect NMR methods that circumvent the bandwidth problems related to the first-order broadening.

2.4.2.1 *Double Quantum Transistors*

The absence of first-order quadrupolar effects on double quantum resonance frequencies reduces the spectral width to within practical detection limits. However, double quantum spectroscopy of ^{14}N is much more difficult to perform than the corresponding deuterium experiments. For instance, the method of detection of double quantum coherence by way of coherence transfer to the allowed transitions is not applicable to ^{14}N, because the allowed transitions are inaccessible. Instead, double quantum coherence can be observed indirectly via cross-polarization to neighboring protons.[33]

2.4.2.2 *Overtone NMR*

As was mentioned in connection with Figure 2.4 in Section 2.2.2, a strong quadrupolar interaction causes the admixture of, e.g., some $m = 0$ character in the $m = 1$ and -1 states. As a result, the nominally forbidden $\Delta m = 2$ transition between the $m = -1$ and 1 levels acquires some degree of $\Delta m = 1$ character and becomes weakly allowed. Overtone NMR is the direct excitation and observation of this transition. It is performed at twice the Larmor frequency, $2\nu_0$, which is the frequency corresponding to the energy difference between the $m = \pm 1$ levels (see Figure 2.2). Although both overtone excitation and double quantum excitation (see Section 2.2.3) induce forbidden transitions between the $m = \pm 1$ levels, they are based on entirely different principles: overtone NMR makes use of the second-order quadrupolar perturbation of Zeeman levels, but is otherwise a direct detection method. On the other hand, double quantum NMR does not rely on mixing of the Zeeman levels, but rather on a second-order RF perturbation of the quadrupole energy levels (see Section 2.5.5). The practical incentive for overtone NMR is obviously that the resonance frequencies are not shifted by first-order quadrupolar effects, an advantage shared with double quantum NMR.

2.4.2.3 *Heteronuclear Dipolar Splitting*

The single-crystal NMR spectrum of, say, a ^{13}C nucleus coupled by dipolar interaction to a nearby ^{14}N nucleus is split into three lines corresponding to the magnetic states $m = -1$, 0, 1 of the $I = 1$ spin. In first approximation, the magnetic moments associated with these ^{14}N states are aligned with the Zeeman field and have values proportional to m. This has two consequences: the ^{13}C triplet is symmetric, and the dependence of the spectral positions on the crystal orientation is such that MAS removes the dipolar broadening in powders. However, in the presence of a large NQCC, the ^{14}N spins are no longer quantized along the Zeeman field, as was pointed out in our discussion of Figure 2.4. This changes the directions and magnitudes of the magnetic moments of the three eigenstates, and, hence, the symmetry of the triplet and the orientation dependence of its peak positions. These effects were first observed in a single crystal.[34] The modified orientation dependence also prevents complete narrowing of the dipolar broadening by MAS.[35] Provided ν_Q is sufficiently small in comparison to the ^{14}N Larmor frequency ν_0, the residual splitting of the ^{13}C MAS peak can be evaluated by perturbation theory. It was shown to be of the order $\nu_D\nu_Q/\nu_0$, where ν_D is the magnitude of the dipolar interaction. The literature dealing with these phenomena is summarized in a review article by Harris and Olivieri.[36]

2.4.2.4 Population Transfer

Slow MAS rotation under continuous RF irradiation of ^{14}N nuclei induces spin flips through adiabatic transformations among the m states (see Section 2.3.2). This effect has found application in a REDOR-type method for $^{13}C-^{14}N$ distance determination.[37] The principle of regular REDOR experiments as applied to $I = S = 1/2$ spin pairs is as follows. The FID signal of the S spins decays as a result of the dipolar interaction between I and S, but MAS refocuses the dipolar dephasing and creates a rotational echo at the end of a full rotor period. The refocusing is undone if a 180° pulse is applied to the I spins, flipping them from one m state to another at some time during the rotor cycle. This causes a rotational echo signal reduction of S, which can be analyzed to determine the $I-S$ atomic distance. The REDOR experiment cannot be applied in this form when the I spins are ^{14}N, because the 180° pulse must be nonselective to be effective. Instead, one can bring about the desired m flips by population transfer under adiabatic sample spinning conditions (transfer of populations in double resonance or TRAPDOR).[37]

2.4.3 Nuclei of Half-Integer Spin

Detailed theoretical and experimental aspects of solid-state NMR of half-integer spins are covered elsewhere in this book. The literature on this subject before 1993 is summarized in an extensive review article by Freude and Haase.[6] Key references to the literature can also be found in these articles.

2.4.3.1 Static First-Order Spectra

Examples of simulated first-order quadrupolar lineshapes of half-integer nuclei in static samples are shown in Figure 2.3. Experimental spectra are, however, rarely reported. Their detection necessitates excitation by more than one pulse, since the receiver deadtime renders the FID following a single pulse largely undetectable. If nonselective pulses are feasible, a quadrupolar echo sequence identical to that described above for deuterium can be applied to refocus the signal and to allow "zero-time resolution". This has been demonstrated for $I = 3/2$[38] and 5/2.[39] Another approach is the so-called two-pulse free induction decay.[6] It is a 2D experiment consisting of two nonselective pulses. The first creates coherences of $\Delta m = 1$ transitions that evolve with frequencies equal to multiples of $\Delta\nu_Q$. After a time t_1, the second pulse is applied to transfer the coherences to the central transition, which can easily be detected. Central transition coherences during t_1 are suppressed by phase cycling, and the 2D spectra of spins 5/2 are simplified in that they are dominated by the first satellite as a result of less effective coherence transfer from the second satellites.

2.4.3.2 MAS of First-Order Spectra

The orientational dependence of $\Delta\nu_Q$ is such that it allows narrowing of first-order powder spectra by MAS. Compared with deuterium, this experiment is more demanding in terms of spinning stability, because the spectra are generally broader. Jakobsen and co-workers have succeeded in obtaining highly resolved sideband patterns with ν_Q as large as 1 MHz for $I = 3/2$ and 0.5 MHz for $I = 5/2$. Fine structure of the sidebands due to second-order shifts and sideband envelopes can be analyzed to obtain detailed information on the spin system.

2.4.3.3 Multiple Quantum Coherence

Triple quantum coherence can be excited with an on-resonance selective pulse,[20] as indicated in Figure 2.8(h). The method of pulsed multiple quantum NMR (much-practised in one- and multidimensional NMR[28]) was first demonstrated in 1975 by Hatanaka *et al.*,[40] who created ^{27}Al ($I = 5/2$) double quantum coherence in a single crystal by the application of two consecutive selective pulses at different allowed transition frequencies.

2.4.3.4 Echoes and Multiple Pulse Experiments

Numerous combinations of nonselective and selective, on- and off-resonance pulses have been reported for the excitation, refocusing, coherence transfer, or selective detection of single and multiple quantum coherences. This diverse subfield of quadrupolar NMR has been reviewed by Sanctuary and Halstead[41] and by Freude and Haase.[6]

2.4.3.5 *Spin Counting*

In NMR, the signal intensity is proportional to the number of spins that give rise to it. It can thus be used for quantitative analysis, provided the signals of the unknown sample and a reference sample are excited and detected under comparable conditions. In the case of NMR of half-integer spins, three complicating factors need to be considered.

1. When only the central transition is observed, the signal is reduced by a factor reflecting the relative intensities of the central and satellite transitions. For a given spin I these are given by

$$S_{m\leftrightarrow m+1} \propto I(I+1) - m(m+1) \quad (2.13)$$

   The relative intensities and the percentages of the total intensity represented in the central transition are listed in Table 2. The appropriate reduction factor needs to be accounted for when signal intensities of solids are compared with a liquid reference sample, because in liquids all the transitions are observed. A corresponding signal reduction is also observed when defects are introduced in cubic crystals: at perfectly cubic sites the full signal is detected because ν_Q vanishes, but defects lower the symmetry, increase ν_Q, and wipe out the satellites.[5]
2. The NQCC can be so large that even the second-order effects broaden the central transition beyond the detection limit. Aluminum-27 NMR is particularly susceptible to this effect, since minor chemical modifications can drastically enhance the NQCC (Table 2.1). A classic example is the disappearance of ^{27}Al signal due to atoms near the surface of high-surface-area alumina, resulting in an inverse correlation between the signal intensity and the specific surface area.[42] In studies of disordered systems, it is always good practice to supplement ^{27}Al NMR spectra with a quantitative assessment of the percentage of nuclei that are represented in the spectrum. Sometimes, more useful information on the nature of a sample is revealed by a determination of the amount of "NMR-invisible" Al than by the interpretation of an observed but nonrepresentative lineshape.
3. For quantitative comparison between different signals, it is imperative to work with excitation pulses of sufficiently small flip angle to ensure that the signal intensity does not depend on $\Delta\nu_Q$ (compare Figure 2.9). The largest deviation is between purely selective and nonselective excitations. To keep it under 5%, the nominal flip angle (i.e., the nutation angle if the pulse were applied to a liquid sample) must be smaller than 18°, 11°, and 8° for $I = 3/2$, 5/2, and 7/2, respectively. For deviations less than 10%, the flip angles must be limited to 25°, 15°, and 11°, respectively.

2.4.3.6 *Nutation Spectroscopy*

Nutation spectroscopy in its simplest form is the study of the signal intensity following a single pulse, measured as a function of the length of the pulse. Fourier transformation in 2D fashion yields a nutation spectrum that reflects the distribution of nutation frequencies ν_{nut} of the spins during the pulse. Nutation spectroscopy applied to the central transition provides indirect information on the first-order quadrupolar parameters: when $\nu_1 \gg \nu_Q$, the nutation spectrum has a peak at ν_1; when $\nu_1 \ll \Delta\nu_Q$ the peak is at $(I + 1/2)\,\nu_1$ (see equation (2.12) and Figure 2.9). Nutation spectra of powders obtained with intermediate RF amplitudes ($\nu_1 \approx \nu_Q$) feature characteristic

Table 2.2. Relative intensities of the transitions of half-integer quadrupolar nuclei

I	Relative intensities	Central transition (%)
3/2	3 : 4 : 3	40
5/2	5 : 8 : 9 : 8 : 5	25.7
7/2	7 : 12 : 15 : 16 : 15 : 12 : 7	19.0
9/2	9 : 16 : 21 : 24 : 25 : 24 : 21 : 16 : 9	15.2

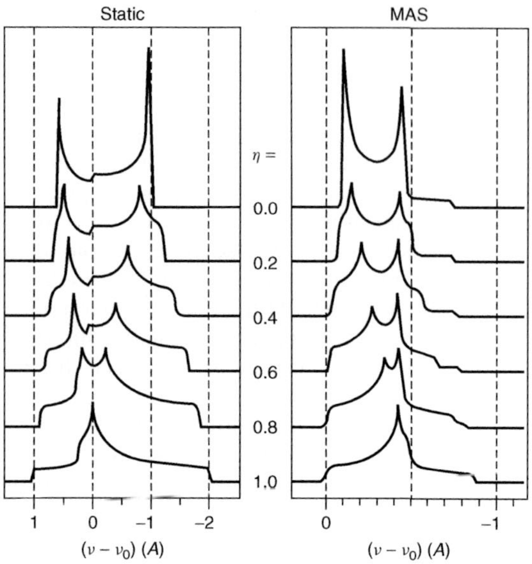

Figure 2.10. Central transition lineshapes of half-integer spins broadened by the second-order quadrupolar effect, calculated for static and fast MAS conditions. The frequency scale is in units of A, which is defined in equation (14).

powder lineshapes that are sensitive to the ratio ν_Q/ν_1 and to η. For further details, see Chapter 7.

2.4.3.7 *Second-Order Spectra: STATIC, MAS, and VAS*

The central transition spectra of powder samples have received a great deal of attention. Figure 2.10 shows the spectral lineshapes for various values of η, without and with MAS. The shapes of the spectra do not depend on I, but their widths do. For that reason, the lineshapes in Figure 2.10 are plotted on a universal frequency scale, where one unit represents a frequency increment of

$$A = \frac{1}{9}\left[I(I+1) - \frac{3}{4}\right]\frac{\nu_Q^2}{\nu_0} \tag{2.14}$$

The symmetry of the orientation dependence of the second-order shift differs from that of first-order effects such as CSA, dipolar interaction, and first-order quadrupolar splitting. Consequently, complete line narrowing is not achieved with MAS, as may be seen in Figure 2.10. Nevertheless, MAS spectra are still preferable to static spectra, not only because the quadrupolar pattern is three to four times narrower, but also because CSA and dipolar broadenings are removed. Chemical shift resolution of MAS spectra improves dramatically when the magnetic field is increased. Since second-order broadening and chemical shift are proportional to $1/\nu_0$ and ν_0, respectively, the resolution scales as ν_0^2. High spinning speeds also improve resolution, because they prevent overlap of sidebands. Another approach is the technique of variable angle spinning (VAS), in which the samples are rotated about an axis that does not necessarily make the magic angle with the Zeeman field. Certain angles give narrower second-order spectra, but complete narrowing is not obtained.

2.4.3.8 *Narrowing of Second-Order Broadening*

Two general approaches were developed to remove the second-order quadrupolar broadening of the central transition. The first aproach is based on the particular properties of orientational dependence of the second-order shift. It led to the inventions of two composite sample rotation techniques: DAS[43] and DOR.[44] For details, see Chapter 8. It should be noted that sample rotation does not change the center of gravity of the spectra. Consequently, the spectral position of the narrowed peak is the combination of a chemical shift and an isotropic second-order quadrupolar shift, where the latter is given by

$$\Delta\nu_{\text{iso}}^{(2)} = -\frac{1}{30}\left[I(I+1) - \frac{3}{4}\right]\frac{\nu_Q^2}{\nu_0}\left(1 + \frac{1}{3}\eta^2\right) \tag{2.15}$$

The second approach uses the fact that the orientational dependence of the second-order shift of the central transition is, except for a numerical factor, equal to that of multiple quantum coherences. It led to the invention of MQMAS, which is a 2D multiple quantum MAS experiment where the spins are first excited to a triple (or quintuple) quantum coherence state (see Section 2.3.3.1) and are subsequently transferred to central transition coherence where second-order dephasing is refocused.[45]. This method has the advantage that it avoids the instrumentation demands and bandwidth limitations involved in composite sample rotation methods. On the other hand, the quality of MQMAS spectra depends on the efficiency of the multiple quantum excitation and that of the transfer to central transition coherence. The many advances made in this respect are described in Chapter 10. Related to MQMAS is the method of STNMR, which utilizes the proportionality of the second-order shifts of central and satellite transitions (Chapter 11). Like in DAS and DOR, an isotropic second-order quadrupolar shift contributes to the frequency shift of the narrowed signal obtained by MQMAS or STMAS.

2.4.3.9 Heteronuclear Dipolar Splitting

Modifications of dipolar splitting patterns similar to those of nuclei coupled to ^{14}N (see Section 2.4.2) are also observed when the neighboring nucleus is of half-integer spin and has a large NQCC. Examples of coupling to $^{63}Cu/^{65}Cu$, $^{35}Cl/^{37}Cl$, and other nuclei have been documented.[36]

2.4.3.10 Relaxation

Spin–lattice relaxation does not usually follow a single exponential behavior for the reasons outlined in Section 2.4.3.10. Additional complications arise when the T_1 of a central transition is measured by monitoring the signal following saturation. The results depend on whether a single selective saturation pulse or a long saturation comb is applied, and on whether the measurements are done under MAS or static conditions. The apparent T_1 can vary by more than an order of magnitude, depending on the measurement method.[46] The differences are caused by variations in initial population distributions of the energy levels (compare Figure 2.5c and e) and by variability of the effectiveness of spin diffusion between the central transition and satellite transitions of a neighboring nucleus.[46]

2.4.3.11 Spin Locking and Population Transfer

The x magnetization formed by the application of a selective y pulse to the central transition corresponds to the spin state C_x (see Section 2.3.3). When spin locked by RF irradiation in the x direction, it decays with the relaxation time $T_{1\rho}$. However, under MAS at slow rates ($\alpha > 1$ with α defined as in equation (2.9)), the relaxation decay is interrupted by adiabatic population transfer from C_x to T_z (see Section 2.3.3). As a result, the x magnetization of every nucleus in the sample disappears at the first zero-crossing of its oscillating quadrupolar splitting $\Delta\nu_Q$ and then reappears again at the next zero-crossing. Since different nuclei have their zero-crossings at different times, the total signal decays gradually, but because every nucleus has an even number of zero-crossings per rotor cycle (two or four), the signal grows back toward the end of the first rotor cycle. This pattern is repeated for successive rotor cycles until $T_{1\rho}$ relaxation causes the signal to decay. However, when the rotation rate is so fast that $\alpha \ll 1$, there is no population transfer, and the signal does not decay other than by relaxation. Under rotation at intermediate rates ($\alpha \approx 1$), the passages transform C_x to spin states that are not spin locked, resulting in an irreversible signal decay.[22] Another method of population transfer is slow sweeping of the RF carrier frequency. For instance, it can be applied to static $I = 5/2$ nuclei for transfer of the populations of the $m = \pm 5/2$ states to the $m = \pm 1/2$ states in order to obtain a fivefold increase of central transition signal intensity.[23]

2.4.3.12 Cross-Polarization and CPMG

Cross-polarization (CP) is the transfer of spin-locked polarization (magnetization) from nuclei S to neighboring nuclei I by simultaneous RF irradiation of the two spin systems under matched conditions of the two RF amplitudes ν_{1S} and ν_{1I}. When both I and S are spins 1/2, the matching requirement is the familiar Hartmann–Hahn condition, $\nu_{1S} = \nu_{1I}$. However, if one or both are quadrupolar nuclei, the appropriate matching condition is that the two nutation frequencies be equal, $\nu_{\mathrm{nut},S} = \nu_{\mathrm{nut},I}$. Thus, for CP of the central transition of half-integer spins I from protons with spin $S = 1/2$, the condition is $\nu_{1S} = (I + 1/2)\nu_{1I}$ (see Section 2.3.3). The resulting I polarization has the spin-locked spin configuration C_x. If the cross-polarization is done under MAS conditions (CP MAS), complications arise as a result of the zero-crossings of $\Delta\nu_Q$. The same RF irradiation that establishes the CP matching also serves as a spin lock field for the newly formed central transition polarization. Hence, if the adiabaticity parameter α is in the intermediate range, the signal enhancement is frustrated by the irreversible decay of C_x caused by the nonadiabatic passages (see the preceding paragraph). This can be a reason for poor performance of CP MAS of the central transition.[47] Another approach to sensitivity enhancement of the central transition signal is the Quadrupole Carr–Purcell–Meiboon–Gill echo sequence (QCPMG).

2.4.4 Zero-Field NMR

The technique of zero-field NMR makes use of the connectivity, illustrated in Figure 2.4, between the spin states at zero Zeeman field (the NQR limit) and at high field (the NMR limit). A sample is mechanically shuttled back and forth between a position where the field is strong and another position where the field vanishes. In this way, the NQR spectrum can be indirectly detected as an NMR signal, with the combined advantages of the higher sensitivity of NMR and the higher spectroscopic resolution of NQR (see Section 2.2.2).

2.5 THEORY

2.5.1 Electric Field Gradient

The nuclear quadrupole interacts with an EFG. This is the gradient of the electric field created by the charges other than the nucleus under consideration. The isotropic portion of the EFG, i.e., the part that originates from the s electrons, which have a non-vanishing charge density at the site of the nucleus, has no relevance to NMR, because the energy of its interaction with the nuclear charge distribution does not change when the nuclear spin axis changes orientation.[4] To be sure, the isotropic EFG contributes to isotope shifts in atomic spectra in the form of a "volume effect" or "field effect".[48,49] However, in the context of NMR, it is customary to ignore its existence. The remaining EFG tensor **V** is thus purely anisotropic. It has three principal tensor components, V_{XX}, V_{YY}, and V_{ZZ}, which are associated with a principal axis system (PAS) X, Y, Z (in the present notation, we reserve capital indices for the PAS and let the lower case x, y, z stand for a general axis system). The off-diagonal elements, V_{XY}, V_{XZ}, etc., are zero in the PAS. The diagonal elements satisfy the Laplace equation

$$V_{XX} + V_{YY} + V_{ZZ} = 0 \tag{2.16}$$

reflecting the fact that we ignore the isotropic component. Following the convention

$$|V_{ZZ}| \geq |V_{XX}| \geq |V_{YY}| \tag{2.17}$$

for the assignment of the three PAS directions, we define the quantity

$$eq = V_{ZZ} \tag{2.18}$$

and the asymmetry parameter

$$\eta = \frac{V_{YY} - V_{XX}}{V_{ZZ}} \tag{2.19}$$

The convention ensures that $0 \leq \eta \leq 1$. When $\eta = 0$, the EFG tensor is axially symmetric about the Z axis: $V_{XX} = V_{YY} = -V_{ZZ}/2$. For arbitrary η, equations (2.18) and (2.19) are solved to give

$$V_{XX} = -\frac{1}{2}(1 + \eta)eq \tag{2.20}$$

$$V_{YY} = -\frac{1}{2}(1 - \eta)eq \tag{2.21}$$

One should be aware that many authors prefer the convention $|V_{ZZ}| \geq |V_{YY}| \geq |V_{XX}|$ instead of equation (2.17) for the assignment of the X and Y axes. The appropriate definition of η is then $(V_{XX} - V_{YY})/V_{ZZ}$, and the signs of η-containing terms in formulas such as equations (2.3), (2.20), (2.21), and (2.22)–(2.27) are reversed.

Two kinds of local symmetry at the site of the nucleus dictate the symmetry of the EFG tensor.

1. Cubic point symmetry (including eightfold cubic, sixfold octahedral, and fourfold tetrahedral coordinations) results in $V_{XX} = V_{YY} = V_{ZZ}$, and, hence, by the Laplace equation, $eq = 0$.
2. Axial point symmetry (including structures where the nucleus lies on a threefold, fourfold, fivefold, or sixfold symmetry axis) results in $\eta = 0$.

In ionic crystals, the EFG tensor can be calculated from the known positions and charges of the surrounding ions. However, the actual EFGs experienced by the nucleus are many times larger than the calculated values as a result of distortions of the local electron cloud. The Sternheimer antishielding factor accounts for this correction (see Chapter 1).

Below are formulas for the components of the EFG tensor in a more general axis system x, y, z. To be completely general, we ought to specify three Euler angles for the orientation of x, y, z with respect to X, Y, Z. However, in most NMR experiments, we do not need to know more about the laboratory frame than the direction of the z axis specifying the $\boldsymbol{B}_0$ field orientation. In such bases, it is sufficient to define the two polar angles θ and ϕ of z with respect to the PAS. In fact, the polar angles represent two of the Euler angles, as is demonstrated in Figure 2.11, where the directions of x, y, and z are seen to be obtained from

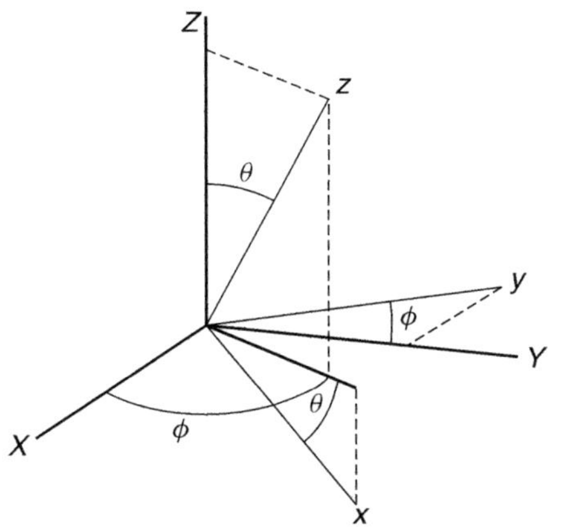

Figure 2.11. Transformation from a principal axis system X, Y, Z to a more general axis system x, y, z. The transformation is accomplished by a rotation through ϕ about Z followed by a rotation through θ about y. θ and ϕ are the polar and azimuthal angles of z with respect to X, Y, Z.

X, Y, and Z by rotating the system first about Z over angle ϕ and then about the new y over angle θ. The third Eulerian rotation (about z) is not executed. The transformation depicted in Figure 2.11 leads to

$$V_{xx} = \frac{1}{2}eq(3\sin^2\theta - 1 - \eta\cos^2\theta\cos 2\phi) \quad (2.22)$$

$$V_{yy} = \frac{1}{2}eq(-1 + \eta\cos 2\phi) \quad (2.23)$$

$$V_{zz} = \frac{1}{2}eq(3\cos^2\theta - 1 - \eta\sin^2\theta\cos 2\phi) \quad (2.24)$$

$$V_{xy} = V_{yx} = \frac{1}{2}eq\eta\cos\theta\sin 2\phi \quad (2.25)$$

$$V_{xz} = V_{zx} = -\frac{1}{2}eq\sin\theta\cos\theta(3 + \eta\cos 2\phi) \quad (2.26)$$

$$V_{yz} = V_{zy} = \frac{1}{2}eq\eta\sin\theta\sin 2\phi \quad (2.27)$$

There are, however, situations where three parameters must be specified to define the relative orientations of the PAS and the laboratory frame. For instance, when the EFG changes direction during the NMR experiment because of molecular motions or sample spinning, the full information concerning relative orientations at different times is generally needed for analysis of spin dynamics and relaxation rates.

2.5.2 Hamiltonian

$$\begin{aligned}\hat{H}_Q = \frac{eQ}{4I(2I-1)\hbar}\{&V_{zz}[3\hat{I}_z^2 - I(I+1)] \\ &+ (V_{xx} - V_{yy})(\hat{I}_x^2 - \hat{I}_y^2) + 2V_{xy}(\hat{I}_x\hat{I}_y + \hat{I}_y\hat{I}_x) \\ &+ 2V_{xz}(\hat{I}_x\hat{I}_z + \hat{I}_z\hat{I}_x) + 2V_{yz}(\hat{I}_y\hat{I}_z + \hat{I}_z\hat{I}_y)\end{aligned} \quad (2.28)$$

This formula contains the tensor elements of **V**, which depend on the orientation of the axes with respect to the PAS, as was pointed out above. Another representation of the Hamiltonian is in terms of irreducible tensor operators:

$$\hat{H}_Q = \frac{1}{2\hbar}\sum_{k=-2}^{2}(-1)^k\hat{Q}_k V_{-k} \quad (2.29)$$

where

$$\begin{aligned}\hat{Q}_0 &= \frac{eQ}{2I(2I-1)}\frac{1}{\sqrt{6}}[\hat{I}_z^2 - I(I+1)] \\ \hat{Q}_{\pm 1} &= \mp\frac{eQ}{2I(2I-1)} + (\hat{I}_z\hat{I}_\pm + \hat{I}_\pm\hat{I}_z) \\ \hat{Q}_{\pm 2} &= \frac{eQ}{2I(2I-1)}\hat{I}_\pm^2 \\ \hat{I}_\pm &= \hat{I}_x \pm i\hat{I}_y\end{aligned} \quad (2.30)$$

and

$$\begin{aligned}V_0 &= \sqrt{\frac{3}{2}}V_{zz} \\ V_{\pm 1} &= \mp V_{xz} - iV_{yz} \\ V_{\pm 2} &= \frac{1}{2}(V_{xx} - V_{yy}) \pm iV_{xy}\end{aligned} \quad (2.31)$$

The transformation properties of the irreducible tensors make this form of the Hamiltonian particularly useful for description of rotations of the coordinate system. The notation of equation (2.29) is also helpful for writing out the Hamiltonian in matrix form.

2.5.3 Hamiltonian in Matrix Form

Since many aspects of the spin dynamics of quadrupolar nuclei can readily be explained by inspection of the structures of the matrices of the various contributions to the total Hamiltonian, we give here a few representative examples. The matrices are given in the standard representation of spin I, i.e., the matrix elements are $\langle m'|\hat{H}_Q|m\rangle$, where $|m\rangle$ are the $2I+1$ eigenstates of $\hat{I}_z$: $|I\rangle$, $|I-1\rangle$, ..., $|-I+1\rangle$, $|-I\rangle$. The spin operator properties of the $\hat{Q}_k$ components in equation (2.29) place the coefficient V_0 in the diagonal elements, $V_{\pm 1}$ in elements one position removed from the diagonal, and $V_{\pm 2}$ two positions removed. As two examples, we give the matrices for $I = 1$,

$$\hat{H}_Q = \frac{3eQ}{2I(2I-1)\hbar} \times \begin{pmatrix} \frac{1}{6}V_{zz} & \sqrt{\frac{1}{2}}V_{-1} & V_{-2} \\ -\sqrt{\frac{1}{2}}V+1 & -\frac{1}{3}V_{zz} & -\sqrt{\frac{1}{2}}V_{-1} \\ V_{+2} & \sqrt{\frac{1}{2}}V_{+1} & \frac{1}{6}V_{zz} \end{pmatrix} \tag{2.32}$$

and for $I = 3/2$,

$$\hat{H}_Q = \frac{3eQ}{2I(2I-1)\hbar} \times \begin{pmatrix} \frac{1}{2}V_{zz} & \sqrt{\frac{1}{3}}V_{-1} & \sqrt{\frac{1}{3}}V_{-2} & 0 \\ -\sqrt{\frac{1}{3}}V_{+1} & -\frac{1}{2}V_{zz} & 0 & \sqrt{\frac{1}{3}}V_{-2} \\ \sqrt{\frac{1}{3}}V_{+2} & 0 & -\frac{1}{2}V_{zz} & -\sqrt{\frac{1}{3}}V_{-1} \\ 0 & \sqrt{\frac{1}{3}}V_{+2} & \sqrt{\frac{1}{3}}V_{+1} & \frac{1}{2}V_{zz} \end{pmatrix} \tag{2.33}$$

In high-field NMR, the quadrupolar Hamiltonian is considered to be a perturbation of the Zeeman Hamiltonian $\hat{H}_Z = \omega_0 \hat{I}_z$. The first-order perturbation is determined by the secular part of $\hat{H}_Q$, i.e., the portion $\hat{H}_Q^{(1)}$ of $\hat{H}_Q$ that commutes with $\hat{H}_Z$. Since the latter is represented by a diagonal matrix with nondegenerate eigenvalues, $\hat{H}_Q^{(1)}$ retains only the diagonal elements of equations (2.32) and (2.33), 'truncating' $\hat{H}_Q$ to

$$\hat{H}_Q^{(1)} = \frac{1}{2}\Delta\omega_Q\left[\hat{I}_z^2 - \frac{1}{3}I(I+1)\right] \tag{2.34}$$

${H_Q}^{(1)}$ determines the first-order energy-level corrections ${E_m}^{(1)}$ given in equation (2.2). The second-order corrections ${E_m}^{(2)}$ of the eigenvalues involve double products of the off-diagonal elements of $\hat{H}_Q$ divided by differences between Zeeman energies:

$$E_m^{(2)} = -\frac{\omega_Q^2}{9\omega_0} m \left\{2[I(I+1) - 2m^2 - \tfrac{1}{4}]\left(\tfrac{|V_1|}{eq}\right)^2 - [I(I+1) - m^2 - \tfrac{1}{2}]\left(\tfrac{|V_2|}{eq}\right)^2\right\} \tag{2.35}$$

These eigenvalue equations are used to calculate first- and second-order quadrupolar shifts of transition frequencies. The expressions $(|V_1|/eq)^2$ and $(|V_2|/eq)^2$ in equation (2.35) are functions of the orientation of the Zeeman field with respect to the EFG principal axes and of η (see equations (2.31) and (2.22)–(2.27)). For instance, the second-order shift of the central transition is

$$\Delta\omega^{(2)}_{-1/2\leftrightarrow 1/2} = -\frac{\omega_Q^2}{9\omega_0}[I(I+1) - \tfrac{3}{4}] \times \left[2\left(\frac{|V_1|}{eq}\right)^2 - \left(\frac{|V_2|}{eq}\right)^2\right] \tag{2.36}$$

which for $\eta = 0$ reduces to

$$\Delta\omega^{(2)}_{-1/2\leftrightarrow 1/2} = -\frac{\omega_Q^2}{16\omega_0}[I(I+1) - \tfrac{3}{4}] \times (1 - \cos^2\theta)(9\cos^2\theta - 1) \tag{2.37}$$

The eigenfunctions of the Hamiltonian are also perturbed. In other words, the vectors representing them in wavefunction space are slightly tilted. The general formula for these tilted states is

$$|m\rangle \to |m\rangle + \sum_{m'\neq m} \frac{\langle m'|\hat{H}_Q|m\rangle}{(m-m')\omega_0}|m'\rangle \tag{2.38}$$

which can be used to evaluate the Zeeman state mixing that determines the detectability of overtone spectra and the dipolar splittings in spectra of neighboring nuclei. The correction terms have the forms $(\omega_Q/\omega_0)(V_{\pm 1}/eq)|m \pm 1$ and $(\omega_Q/\omega_0)(V_{\pm 2}/eq)|m \pm 2$ multiplied by numerical coefficients. Because the corrections involve off-diagonal elements of $\hat{H}_Q$ divided by ω_0, they are considered to be second-order perturbations.

2.5.4 Density Matrix

The state of an ensemble of mutually noninteracting nuclei of spin I is described by a density matrix $\hat{\rho}$ having $2I+1$ rows and columns. Since this is a Hermitian matrix, it has in addition to a unit-matrix term,

$(2I+1)^2-1$ independent traceless components (3 when $I=1/2$; 8 when $I=1$; 15 when $I=3/2$; etc.). It is convenient to choose a basis set of mutually independent matrices or operators for the description of the density matrices. In the case of spin-1/2, the three operators $\hat{I}_x$, $\hat{I}_y$, and $\hat{I}_z$ are the natural selection for that purpose, but for higher spin numbers the choices are not so obvious. A number of formalisms exist. Many authors present their theory in terms of irreducible tensor operators. Others employ fictitious spin-1/2 operators[21,50] or specially adapted operators.[51] Since the spin dynamics are determined by the Liouville–von Neumann equation

$$\frac{d\hat{\rho}}{dt} = i[\hat{\rho}, \hat{H}] \tag{2.39}$$

the preferred formalism depends on the nature of the Hamiltonian to be analyzed, and is often the one that offers the simplest set of commutation relations. Not infrequently, however, other considerations such as ease of visualization, relaxation properties, or personal taste of the theoretician determine the choice of formalism.

Below, we give the traceless parts of the density matrices corresponding to the special spin states that were mentioned in Sections 2.2.3, 2.3.2, and 2.3.3. Rather than referring to a particular basis set, we reproduce here the full matrices in the standard representation. For $I=1$, we have mentioned Zeeman order $\hat{I}_z$ and quadrupolar order $\hat{Q}_z$,

$$\hat{I}_z = \begin{pmatrix} 1 & 0 & 0 \\ 0 & 0 & 0 \\ 0 & 0 & -1 \end{pmatrix}, \quad \hat{Q}_z = \frac{1}{\sqrt{3}}\begin{pmatrix} 1 & 0 & 0 \\ 0 & -2 & 0 \\ 0 & 0 & 1 \end{pmatrix} \tag{2.40}$$

single quantum coherences $\hat{I}_x$ and $\hat{I}_y$ corresponding to x and y magnetization,

$$\hat{I}_x = \frac{1}{\sqrt{2}}\begin{pmatrix} 0 & 1 & 0 \\ 1 & 0 & 1 \\ 0 & 1 & 0 \end{pmatrix}, \quad \hat{I}_y = \frac{1}{\sqrt{2}}\begin{pmatrix} 0 & -i & 0 \\ i & 0 & -i \\ 0 & i & 0 \end{pmatrix} \tag{2.41}$$

single quantum coherences $\hat{S}_x$ and $\hat{S}_y$ associated with selective excitation of the (0, −1) transition,

$$\hat{S}_x = \begin{pmatrix} 0 & 0 & 0 \\ 0 & 0 & 1 \\ 0 & 1 & 0 \end{pmatrix}, \quad \hat{S}_y = \begin{pmatrix} 0 & 0 & 0 \\ 0 & 0 & -i \\ 0 & i & 0 \end{pmatrix} \tag{2.42}$$

and double quantum coherences with x and y phases,

$$\hat{D}_x = \begin{pmatrix} 0 & 0 & 1 \\ 0 & 0 & 0 \\ 1 & 0 & 0 \end{pmatrix}, \quad \hat{D}_y = \begin{pmatrix} 0 & 0 & -i \\ 0 & 0 & 0 \\ i & 0 & 0 \end{pmatrix} \tag{2.43}$$

The $I=3/2$ spin states relevant to the discussions in this chapter are Zeeman order $\hat{I}_z$ and quadrupolar order $\hat{Q}_z$,

$$\hat{I}_z = \frac{1}{2}\begin{pmatrix} 3 & 0 & 0 & 0 \\ 0 & 1 & 0 & 0 \\ 0 & 0 & -1 & 0 \\ 0 & 0 & 0 & -3 \end{pmatrix}, \quad \hat{Q}_z = \frac{1}{2}\begin{pmatrix} 1 & 0 & 0 & 0 \\ 0 & -1 & 0 & 0 \\ 0 & 0 & -1 & 0 \\ 0 & 0 & 0 & 1 \end{pmatrix} \tag{2.44}$$

three coherences associated with the central transition,

$$\hat{C}_x = \frac{1}{2}\begin{pmatrix} 0 & 0 & 0 & 0 \\ 0 & 0 & 1 & 0 \\ 0 & 1 & 0 & 0 \\ 0 & 0 & 0 & 0 \end{pmatrix}, \quad \hat{C}_y = \frac{1}{2}\begin{pmatrix} 0 & 0 & 0 & 0 \\ 0 & 0 & -i & 0 \\ 0 & i & 0 & 0 \\ 0 & 0 & 0 & 0 \end{pmatrix}, \quad \hat{C}_z = \frac{1}{2}\begin{pmatrix} 0 & 0 & 0 & 0 \\ 0 & 1 & 0 & 0 \\ 0 & 0 & -1 & 0 \\ 0 & 0 & 0 & 0 \end{pmatrix} \tag{2.45}$$

and three coherences associated with the triple quantum transition,

$$\hat{T}_x = \frac{1}{2}\begin{pmatrix} 0 & 0 & 0 & 1 \\ 0 & 0 & 0 & 0 \\ 0 & 0 & 0 & 0 \\ 1 & 0 & 0 & 0 \end{pmatrix},$$

$$\hat{T}_y = \frac{1}{2}\begin{pmatrix} 0 & 0 & 0 & -i \\ 0 & 0 & 0 & 0 \\ 0 & 0 & 0 & 0 \\ i & 0 & 0 & 0 \end{pmatrix}$$

$$\hat{T}_z = \frac{1}{2}\begin{pmatrix} 1 & 0 & 0 & 0 \\ 0 & 0 & 0 & 0 \\ 0 & 0 & 0 & 0 \\ 0 & 0 & 0 & -1 \end{pmatrix} \quad (2.46)$$

Note that $\hat{T}_z$ is equivalent to the 'triple quantum order' represented in Figure 2.5. The nonvanishing matrix elements in equations (2.42)–(2.46) form 2×2 submatrices. They are examples of fictitious spin-1/2 operators.

2.5.5 Hamiltonian in Rotating Frame

An RF field of amplitude ω_1 and carrier frequency ω is best described in the axis frame that rotates with frequency ω with respect to the laboratory frame. The corresponding rotating-frame Hamiltonian is

$$\hat{H}_{\mathrm{rf}} = \omega_1 \hat{I}_x \quad (2.47)$$

whose matrix elements are proportional to those of $\hat{I}_x$:

$$\langle m|\hat{I}_x|m \pm 1\rangle = \frac{1}{2}\sqrt{I(I+1) - m(m \pm 1)} \quad (2.48)$$

In the same rotating frame, the Zeeman Hamiltonian is reduced to the resonance-offset Hamiltonian

$$\hat{H}_{\mathrm{os}} = (\omega_0 - \omega)\hat{I}_z = \Delta\omega_0 \hat{I}_z \quad (2.49)$$

The matrix elements of the quadrupolar Hamiltonian in the rotating frame are transformed to oscillating functions obtained by replacing V_k in equations (2.29), (2.32), and (2.33) with $V_k\exp(-ik\omega t)$. In the presence of RF fields, it is usually permissible to neglect the rapidly oscillating terms. This reduces $\hat{H}_{\mathrm{Q}}$ in the rotating frame to $\hat{H}_{\mathrm{Q}}^{(1)}$ of equation (2.34). In summary, we consider a rotating-frame Hamiltonian consisting of three terms:

$$\hat{H}_{\mathrm{R}} = \hat{H}_{\mathrm{Q}}^{(1)} + \hat{H}_{\mathrm{os}} + \hat{H}_{\mathrm{rf}} \quad (2.50)$$

In matrix form, it is for $I = 1$,

$$\hat{H}_{\mathrm{R}} = \frac{1}{2}\begin{pmatrix} \frac{1}{3}\Delta\omega_{\mathrm{Q}} + 2\Delta\omega_0 & \omega_1\sqrt{2} & 0 \\ \omega_1\sqrt{2} & -\frac{2}{3}\Delta\omega_{\mathrm{Q}} & \omega_1\sqrt{2} \\ 0 & \omega_1\sqrt{2} & \frac{1}{3}\Delta\omega_{\mathrm{Q}} - 2\Delta\omega_0 \end{pmatrix} \quad (2.51)$$

and for $I = 3/2$,

$$\hat{H}_{\mathrm{R}} = \frac{1}{2}\begin{pmatrix} \Delta\omega_{\mathrm{Q}} + 3\Delta\omega_0 & \omega_1\sqrt{3} & 0 & 0 \\ \omega_1\sqrt{3} & -\Delta\omega_{\mathrm{Q}} + \Delta\omega_{\mathrm{Q}} & 2\omega_1 & 0 \\ 0 & 2\omega_1 & \Delta\omega_{\mathrm{Q}} - 3\Delta\omega_0 & \omega_1\sqrt{3} \\ 0 & 0 & \omega_1\sqrt{3} & \Delta\omega_{\mathrm{Q}} - 3\Delta\omega_0 \end{pmatrix} \quad (2.52)$$

The corresponding matrices for higher values of I can be evaluated similarly. Here, we only mention that when m and $m \pm 1$ equal $-1/2$ and $1/2$, the square root in equation (2.48) reduces to $I + 1/2$, resulting in

$$\langle -\tfrac{1}{2}|\hat{H}_{\mathrm{rf}}|\tfrac{1}{2}\rangle = \langle \tfrac{1}{2}|\hat{H}_{\mathrm{rf}}| -\tfrac{1}{2}\rangle = \tfrac{1}{2}\omega_1(I + \tfrac{1}{2}) \quad (2.53)$$

The nutation frequencies of selective central transition nutation, equation (2.10), are determined by these matrix elements.

The various ways in which RF pulses affect a quadrupolar spin system were reviewed in Section 2.3. In an effort to keep that presentation succinct, a quantum mechanical explanation of the effects was not given. The following discussion is intended to show how the effects are related to the Hamiltonian in the rotating frame, equation (2.50). As in Section 2.3, we begin with an example of a nonquadrupolar spin to introduce the concepts of nutation, spin locking, and population transfer. Taking $I = 1/2$ for simplicity, we have the Hamiltonian matrix

$$\hat{H}_{\mathrm{R}} = \frac{1}{2}\begin{pmatrix} \Delta\omega_0 & \omega_1 \\ \omega_1 & -\Delta\omega_0 \end{pmatrix} \quad (2.54)$$

If the RF field vanishes, the dependence of its eigenvalues on the offset is represented by two straight lines crossing at zero offset, as indicated by the broken lines in Figure 2.12(a). The associated eigenstates are $|1/2\rangle$ and $|-1/2\rangle$. Introduction of the RF term changes the eigenvalues and eigenstates provided ω_1 (occupying the off-diagonal elements in equation (2.54)) is not much smaller than $\Delta\omega_0$ (the difference between the diagonal elements). This leads to an avoided level crossing in the eigenvalue diagram, with a residual level splitting of ω_1 at the center. This result follows immediately from the solution of the eigenvalue equation of the matrix of equation (2.54) with vanishing diagonal elements. When $\Delta\omega_0 = 0$, the eigenstates of $\hat{H}_R$ are the linear combinations

$$|c+\rangle = (|+\tfrac{1}{2}\rangle + |-\tfrac{1}{2}\rangle)/\sqrt{2} \qquad (2.55)$$

$$|c-\rangle = (|+\tfrac{1}{2}\rangle - |-\tfrac{1}{2}\rangle)/\sqrt{2} \qquad (2.56)$$

Zeeman order corresponds to populations of the untilted $|\pm 1/2\rangle$ wavefunctions. Since these are eigenstates of the Hamiltonian when $\Delta\omega_0 \gg \omega_1$, Zeeman order is spin locked far-off-resonance. Another way of expressing this is by saying that the density matrix $\hat{I}_z$, which represents Zeeman order, commutes with the Hamiltonian far-off-resonance. On resonance, the spin-locked density matrix is $\hat{I}_x$, which corresponds to populations of the eigenstates $|c\pm\rangle$. Since $\hat{I}_z$ does not commute with the Hamiltonian on resonance, a spin system in thermal equilibrium is not spin locked on resonance, but rather undergoes nutation in the RF field with a frequency ω_{nut} equal to the level splitting, which in this case is ω_1. Population transfer occurs when $\Delta\omega_0$ is swept slowly from above to below resonance. If the passage is sufficiently slow such that the adiabatic condition of equation (2.7) is satisfied, an eigenstate of the Hamiltonian is changed into the state that is connected to it by continuity in the level diagram. The diagram (a) in Figure 2.12 shows that, when going from above to below resonance, the $|+1/2\rangle$ state connects via $|c+\rangle$ to $|-1/2\rangle$. (Compare the $z \to x \to -z$ trajectory of the magnetization vector in the axis system of Figure 2.7.) In this way, the density matrix is converted from $\hat{I}_z$ to $-\hat{I}_z$, and the populations of the Zeeman levels are exchanged.

These concepts are readily extended to quadrupolar spins. Examples of offset-dependent eigenvalues

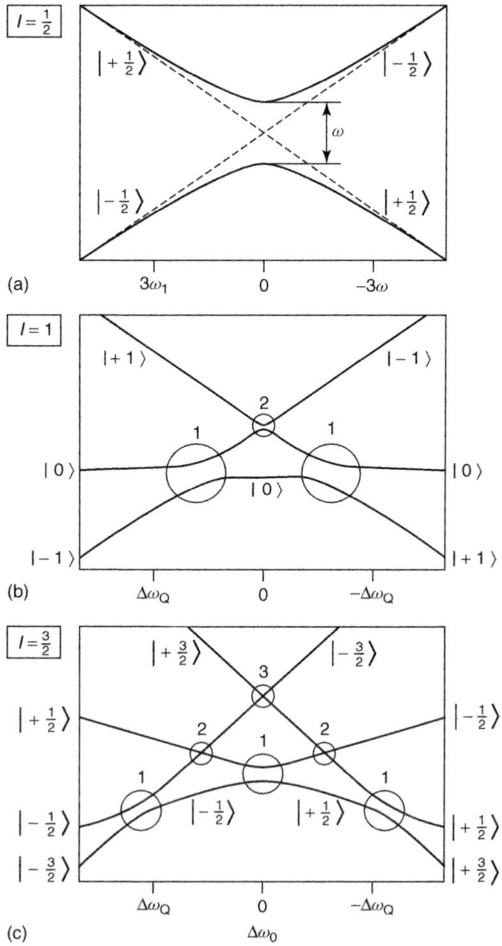

Figure 2.12. Eigenstate diagrams of a rotating frame Hamiltonian consisting of a first-order quadrupolar term, a frequency offset term, and an RF term. The eigenvalues are plotted versus the offset $\Delta\omega_0$. (a) The $I = 1/2$ diagram is representative of the case of a nonquadrupolar nucleus ($\Delta\omega_Q = 0$). (b) and (c) The $I = 1$ and $I = 3/2$ diagrams represent cases of weak irradiation ($\omega_1 < \Delta\omega_Q$). The actual $\Delta\omega_Q\omega_1$ ratios used in the simulations were 5 and 10 for $I = 1$ and 3/2, respectively. Eigenfuctions $|m\rangle$ outside the circled regions are indicated. The numbered circles mark regions of avoided level crossings and the respective values of Δm.

are plotted in the $I = 1$ and $I = 3/2$ diagrams in Figure 2.12. They were calculated by numerical diagonalization of the matrices of equations (2.51) and (2.52) for selective RF excitation conditions ($\omega_1 < \Delta\omega_Q$). Avoided level crossings are indicated by the

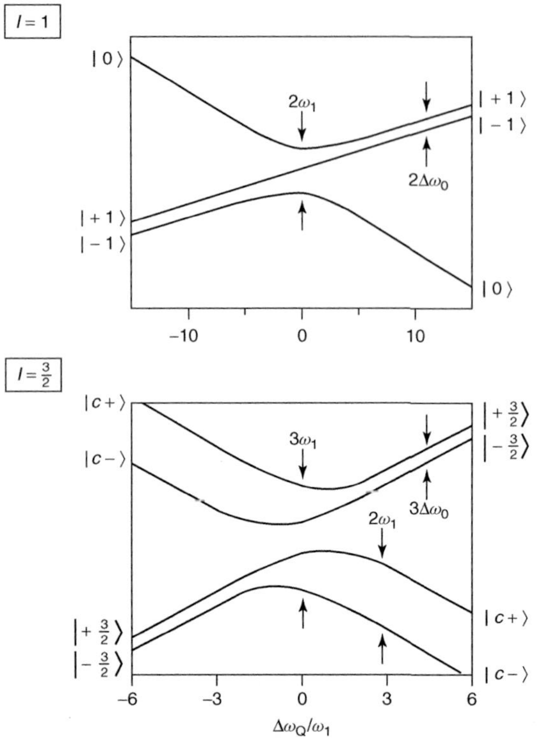

Figure 2.13. Eigenstate diagrams of a rotating frame Hamiltonian consisting of a first-order quadrupole term, a frequency offset term, and an rf term. The eigenvalues are plotted versus the quadrupole splitting $\Delta\omega_Q$. The diagrams represent cases of small frequency offsets ($\Delta\omega_Q < \omega_1$). The actual $\omega 0/\omega 1$ ratios used in the simulations were 0.25 and 0.15 for $I = 1$ and 32, respectively. The eigenvalue levels for vanishing $\Delta\omega_Q$ are equally spaced by $\omega 1$. The eigenfuctions are indicated for large $\Delta\omega_Q$. $|c\pm\rangle$ are linear combinations of $|\pm 1/2\rangle$ (see text).

circles in the figure. They occur at values of $\Delta\omega_0$ for which two diagonal elements of the Hamiltonian are equal. The numbers above the circles mark the differences Δm of the m values of the crossing states. The residual level splittings are the nutation frequencies ω_{nut} of the corresponding excitations. For single quantum crossings ($\Delta m = 1$), they are determined by the respective off-diagonal elements in the Hamiltonian. For multiple quantum transitions, the residual splittings are to be calculated by higher order perturbation theory, resulting in the general expression

$$\omega_{\text{nut}}(m' \leftrightarrow m) = M_{m' \leftrightarrow m}\,\omega_1 \left(\frac{\omega_1}{\Delta\omega_Q}\right)^{\Delta m - 1} \qquad (2.57)$$

Examples of the coefficients are, for $I = 1$, $M_{0\leftrightarrow+1} = \sqrt{2}M_{-1\leftrightarrow+1} = 2$; for $I = 3/2$, $M_{-1/2\leftrightarrow+1/2} = 2$, $M_{+1/2\leftrightarrow+3/2} = \sqrt{3}$, $M_{-1/2\leftrightarrow+3/2} = 2\sqrt{3}$, $M_{-3/2\leftrightarrow+3/2} = 3/2$. Furthermore, $M_{m'\leftrightarrow m} = M_{m\leftrightarrow m'} = M_{-m'\leftrightarrow -m}$. The coefficients for $I = 5/2$ are tabulated elsewhere.[23] The various excitation conditions listed in Figure 2.8 are easily located in Figure 2.12. Note that the double quantum transition of $I = 3/2$ at $\Delta\omega_0 = 1/2\Delta\omega_Q$ was omitted in Figure 2.8.

In the areas outside the circles, i.e., when the irradiation is off-resonance, the eigenstates are essentially pure Zeeman states $|m\rangle$. The corresponding spin-locked spin configurations are populated states $|m\rangle$ (see Figure 2.5), which are represented by diagonal density matrices (e.g., $\hat{I}_z$, $\hat{Q}_z$, $\hat{C}_z$, and $\hat{T}_z$). An adiabatic sweep of $\Delta\omega_0$ interchanges the diagonal matrix elements by population transfer at the avoided level crossings. Finally, examples of the dependence of the eigenvalues on $\Delta\omega_Q$ are shown in Figure 2.13. They were calculated for a relatively small resonance offset ($\omega_1 > \Delta\omega_0$). The case of nonselective excitation (Figure 2.8a and f) is represented at the centers of the diagrams ($\Delta\omega_Q \ll \omega 1$), where adjacent levels are split by ω_1. A nonselective pulse affects the entire spin system, with a nutation frequency $\omega_{\text{nut}} = \omega_1$. At large quadrupolar splittings ($\Delta\omega_Q \geq \omega_1$) all the ω_1-containing off-diagonal elements of equations (2.51) and (2.52) are nonsecular, except the central transition elements of $I = 3/2$, which connect essentially degenerate states. Thus, the nutation frequency for the central transition is $(I + 1/2)\omega_1$, even when the quadrupolar splitting is large. The spin-locked states of a spin 1 are seen to be made up of populations of Zeeman states, which can be adiabatically transferred into each other by slow sweeping of $\Delta\omega_Q$ from positive to negative values, or vice versa.

In the case of half-integer spins, the situation is slightly different. The eigenstates associated with the central transition are now the sum and difference of the Zeeman states (equations (2.55) and (2.56)); population of these eigenstates corresponds to the density matrix $\hat{C}_x$. Figure 2.13 further shows that an adiabatic zero-crossing of $\Delta\omega_Q$ transfers the $|c\pm\rangle$ states to the $|\pm 3/2\rangle$ states, i.e., it transfers $\hat{C}_x$ to $\hat{T}_z$.

RELATED ARTICLES IN THE ENCYCLOPEDIA OF MAGNETIC RESONANCE

Cross Polarization in Solids

Deuterium NMR in Solids

Deuteron Relaxation Rates in Liquid Crystalline Samples: Experimental Methods

Double Rotation

Echoes in Solids

High Speed MAS of Half-Integer Quadrupolar Nuclei in Solids

High Temperature Superconductors

Internal Spin Interactions and Rotations in Solids

Line Narrowing Methods in Solids

Liquid Crystalline Samples: Deuterium NMR

Liquid Crystalline Samples: Relaxation Mechanisms

Magic Angle Spinning Carbon-13 Lineshapes: Effect of Nitrogen-14

Magic Angle Spinning: Effects of Quadrupolar Nuclei on Spin-1/2 Spectra

Membranes: Deuterium NMR

Multiple-quantum Magic-angle Spinning Experiments on Half-integer Nuclei: Fundamentals

Nitrogen NMR

Overtone Spectroscopy of Quadrupolar Nuclei

Polymer Dynamics and Order from Multidimensional Solid State NMR

Quadrupolar Metal Nuclides in Bioinorganic Chemistry: Solid-State NMR Studies

Quadrupolar Nuclei in Glasses

Quadrupolar Nuclei in Liquid Samples

Quadrupolar Transition Metal and Lanthanide Nuclei

Quantitative Measurements

Radiofrequency Pulses: Response of Nuclear Spins

REDOR and TEDOR

Relaxation: An Introduction

Relaxation of Quadrupolar Nuclei Measured via Multiple-Quantum Filtration

Relaxation Theory for Quadrupolar Nuclei

Rotating Solids

Satellite Transition NMR Spectroscopy of Half-Integer Quadrupolar Nuclei under Magic-angle Spinning

SQUIDs

Variable Angle Sample Spinning

Zero Field NMR

REFERENCES

1. R. V. Pound, *Phys. Rev.*, 1950, **79**, 685.
2. M. H. Cohen and F. Reif, in *Solid State Physics*, eds F. Seitz and D. Turnbull, Academic Press: New York, 1958, Vol. 5, p. 321.
3. A. Abragam, *Principles of Nuclear Magnetism*, Clarendon Press: Oxford, 1961, Chaps. 6 and 7.
4. C. P. Slichter, *Principles of Magnetic Resonance*, 3nd edn, Springer-Verlag: Berlin, 1990, Chap. 10.
5. O. Kanert and M. Mehring, in *NMR Basic Principles and Progress*, eds P. Diehl, E. Fluck, and R. Kosfeld, Springer-Verlag: Berlin, 1971, Vol. 3, p. 1.
6. D. Freude and J. Haase, in *NMR Basic Principles and Progress*, eds P. Diehl, E. Fluck, and R. Kosfeld, Springer-Verlag: Berlin, 1993, Vol. 29, p. 1.
7. H. W. Spiess, *Adv. Polym. Sci.*, 1985, **66**, 23.
8. H.-G. Dehmelt and H. Krü ger, *Naturwissenschaften*, 1950, **37**, 111.
9. T. P. Das and E. L. Hahn, in *Solid State Physics*, eds F. Seitz and D. Turnbull, Academic Press: New York, 1958, p. 1.
10. E. A. C. Lucken, *Nuclear Quadrupole Coupling Constants*, Academic Press: London, 1969.
11. I. P. Biriukov, M. G. Voronkov, and I. A. Safin, *Tables of Nuclear Quadrupole Resonance Frequencies*, Israel Program for Scientific Translations: Jerusalem, 1971.
12. G. K. Semin, T. A. Babushkina, and G. G. Iakobson, *Nuclear Quadrupole Resonance in Chemistry*, Wiley: New York, 1975.
13. J. A. S. Smith, ed, *Advances in Nuclear Quadrupole Resonance*, Heyden: London, 1974–1983, Vols. 1–5.

14. C. P. Slichter, *Principles of Magnetic Resonance*, 3nd edn., Springer-Verlag: Berlin, 1990, p. 157.
15. P. S. Hubbard, *J. Chem. Phys.*, 1970, **53**, 985.
16. A. Baram, Z. Luz, and S. Alexander, *J. Chem. Phys.*, 1973, **58**, 4558.
17. L. G. G. Werbelow, *J. Chem. Phys*, 1979, **70**, 5381.
18. S. G. Greenbaum, Y. S. Pak, M. C. Wintersgill, and J. J. Fontanella, *Solid State Ionics*, 1988, **31**, 241.
19. A. Pines, D. J. Ruben, S. Vega, and M. Mehring, *Phys. Rev. Lett.*, 1976, **36**, 110.
20. S. Vega, *Phys. Rev. A*, 1981, **23**, 3152.
21. A. Wokaun and R. R. Ernst, *J. Chem. Phys.*, 1977, **67**, 1752.
22. A. J. Vega, *J. Magn. Reson.*, 1992, **96**, 50.
23. (a) J. Haase and M. Conradi, *Chem. Phys. Lett.*, 1993, **209**, 287; (b) J. Haase, M. Conradi, C. P. Grey, and A. J. Vega, *J. Magn. Reson.*, 1994, **109**, 90.
24. J. Jeener and P. Broekaert, *Phys. Rev.*, 1967, **157**, 232.
25. S. Vega, T. W. Shattuck, and A. Pines, *Phys. Rev. Lett.*, 1976, **37**, 43.
26. R. Eckman, L. Mü ller, and A. Pines, *Chem. Phys. Lett.*, 1980, **74**, 376.
27. E. Gü nther, B. Blü mich, and H. W. Spiess, *Mol. Phys.*, 1990, **71**, 477.
28. R. R. Ernst, G. Bodenhausen, and A. Wokaun, *Principles of Nuclear Magnetic Resonance in One and Two Dimensions*, Clarendon Press: Oxford, 1987 Chap. 5.
29. S. Vega, T. W. Shattuck, and A. Pines, *Phys. Rev. A*, 1980, **22**, 638.
30. J. L. Ackerman, R. Eckman, and A. Pines, *Chem. Phys.*, 1979, **42**, 423.
31. J. H. Kristensen, H. Bildsøe, H. J. Jakobsen, and N. C. Nielsen, *J. Magn. Reson.*, 1991, **92**, 443.
32. J. Mason, ed, *Multinuclear NMR*, Plenum Press: New York, 1987, p. 339.
33. M. Reinhold, P. Brunner, and R. R. Ernst, *J. Chem. Phys.*, 1981, **74**, 184.
34. (a) D. L. VanderHart, H. S. Gutowsky, and T. C. Farrar, *J. Am. Chem. Soc.*, 1967, **89**, 5056; (b) M. E. Stoll, R. W. Vaughan, R. B. Saillant, and T. Cole, *J. Chem. Phys.*, 1974, **61**, 2896.
35. (a) E. Lipmaa, M. Alla, H. Roude, R. Teeaar, I. Heinmaa, and E. Kundla, Magnetic resonance and related phenomena, *Proceedings of the 20th Congr'es Amp'ere, Tallinn, August 21–26, 1978*, eds E. Kundla, E. Lipmaa, and T. Saluvere, Berlin: Springer-Verlag, 1979; (b) S. J. Opella, M. H. Frey, and T. H. Cross, *J. Am. Chem. Soc.*, 1979, **101**, 5856.
36. R. K. Harris and A. C. Olivieri, in *Progress in Nuclear Magnetic Resonance Spectroscopy*, eds J. W. Emsley, J. Feeney, and L. H. Sutcliffe, Pergamon Press: Oxford, 1992, Vol. 24, p. 435.
37. (a) C. P. Grey and W. S. Veeman, *Chem. Phys. Lett.*, 1992, **192**, 379; (b) C. P. Grey, A. J. Vega, and W. S. Veeman, *J. Chem. Phys.*, 1993, **98**, 7711.
38. K. R. Jeffrey, *J. Chem. Phys.*, 1977, **66**, 4677.
39. E. Oldfield, H. K. C. Timken, B. Montez, and R. Ramachandran, *Nature (London)*, 1985, **318**, 163.
40. H. Hatanaka, T. Terao, and T. Hashi, *J. Phys. Soc. Jpn.*, 1975, **39**, 835.
41. B. C. Sanctuary and T. K. Halstead, *Adv. Magn. Opt. Reson.*, 1990, **15**, 79.
42. D. E. O'Reilly, *Adv. Catal.*, 1960, **12**, 31.
43. (a) A. Llor and J. Virlet, *Chem. Phys. Lett.*, 1988, **152**, 248; (b) B. F. Chmelka, K. T. Mueller, A. Pines, J. Stebbins, Y. Wu, and J. W. Zwanziger, *Nature (London)*, 1989, **339**, 42.
44. A. Samoson, E. Lippmaa, and A. Pines, *Mol. Phys.*, 1988, **65**, 1013.
45. L. Frydman and J. S. Harwood, *J. Am. Chem. Soc.*, 1995, **117**, 5367.
46. D. E. Woessner and H. K. C. Timken, *J. Magn. Reson.*, 1990, **90**, 411.
47. A. J. Vega, *Solid State NMR*, 1993, **1**, 17.
48. H. G. Kuhns, *Atomic Spectra*, Academic Press: New York, 1962, pp. 368.
49. W. H. King, *Isotope Shifts in Atomic Spectra*, Plenum Press: New York, 1984.
50. S. Vega and A. Pines, *J. Chem. Phys.*, 1977, **66**, 5624.
51. A. J. Vega and Z. Luz, *J. Chem. Phys.*, 1987, **86**, 180.

Chapter 3

Quadrupolar Coupling: An Introduction and Crystallographic Aspects

Sharon E. Ashbrook[1] and Stephen Wimperis[2]

[1]*School of Chemistry and EaStCHEM, University of St Andrews, St Andrews KY16 9ST, UK*
[2]*School of Chemistry and WestCHEM, University of Glasgow, Glasgow G12 8QQ, UK*

3.1 INTRODUCTION

As schoolchildren, we were all told that the ocean tides are raised by the pull of the Moon's gravity. Unfortunately, this explanation fails on at least three counts: (i) the Earth rotates on its axis every 24 h, yet there are not one but *two* high tides a day, (ii) the pull of the Sun's gravity on the Earth is much greater than that of the Moon's (and hence the Earth orbits the Sun, not the Moon), and (iii) lakes and inland seas experience no tides whatsoever. The correct explanation is that tides are an induced quadrupolar distortion of the Earth's oceans (hence two high and two low tides a day) caused by the gravitational *field gradient* of the Moon across the diameter of the Earth (which, owing to the much shorter distance between the Earth and the Moon, is far greater than that of the Sun).

Although on a very different length scale, and involving charges rather than masses and electric rather than gravitational fields, this is essentially the same phenomenon as the quadrupolar coupling that dominates NMR spectra of spin $I > 1/2$ nuclei. Atomic nuclei possess a nonspherical distribution of electric charge and hence an electric quadrupole moment (intrinsic in this case, not induced) that can interact with any electric field gradient across the nucleus arising from the surrounding electronic and nuclear environment. This interaction or coupling perturbs the energies of the nuclear spin states, giving rise to spectroscopically observable frequency shifts that can range up to many MHz or even a few GHz in magnitude (see Chapter 1).[1]

Over the last 30 or 40 years, solid-state NMR spectroscopy of quadrupolar nuclei such as ^{2}H, ^{14}N ($I = 1$), ^{7}Li, ^{11}B, ^{23}Na ($I = 3/2$), ^{17}O, ^{27}Al ($I = 5/2$) and the measurement of quadrupolar couplings by solid-state NMR have become well established (see Chapter 2),[2] although continuing interest in the subject has been stimulated by an impressive stream of new experimental techniques for achieving high-resolution spectra[3–8] (see Chapters 8, 10, and 11). In this chapter, we will discuss the quadrupolar interaction in the context of the recent interest in determining crystallographic information by solid-state NMR. First, we briefly review the nature of the quadrupolar interaction and discuss how quadrupolar couplings can be

NMR of Quadrupolar Nuclei in Solid Materials
Edited by Roderick E. Wasylishen, Sharon E. Ashbrook and Stephen Wimperis
 ISBN: 978-0-470-97398-1

calculated and measured experimentally. Then, with these foundations in place, we describe, principally by means of examples, how quadrupolar couplings can be used to provide information about the arrangement of atoms in a crystalline solid.

3.2 THEORY OF QUADRUPOLAR COUPLING

Although small, the atomic nucleus has finite size and, hence, rather than a point electrical charge, we must consider a charge distribution. If this distribution is nonspherical and the centers of mass and charge coincide, the nucleus will possess an electric monopole or point charge (described by a tensor operator of rank $\ell = 0$), an electric quadrupole moment ($\ell = 2$), an electric hexadecapole moment ($\ell = 4$), and so on. There is some experimental evidence for nuclear electric moments with $\ell > 2$ and, to quote Abragam,[9] "there is no reason ... to doubt their existence", but it is the quadrupole moment that is principally responsible for perturbing the energies of the nuclear spin states.

The nuclear quadrupole moment interacts with the electric field gradient (EFG) across the nucleus.[1] Starting with the classical electrostatic interaction of the nuclear and surrounding charge distributions, Slichter replaces the nuclear charge density with its operator equivalent and, after much mathematical manipulation, arrives at the nuclear spin Hamiltonian for the quadrupolar coupling in an arbitrary x, y, z reference frame[10]

$$H_Q = \frac{eQ}{6I(2I-1)\hbar} \sum_{\alpha,\beta=x,y,z} \times V_{\alpha\beta} \left\{ \frac{3}{2}(I_\alpha I_\beta + I_\beta I_\alpha) - \delta_{\alpha\beta} I(I+1) \right\} \quad (3.1)$$

where $\delta_{\alpha\beta}$ is the Kronecker delta and eQ is the nuclear quadrupole moment (units of C m^2). (As noted by Slichter,[10] it is interesting that the rank $\ell = 2$ nuclear quadrupole moment is described by a single parameter, implying that only its cylindrically symmetric component is of relevance here.) The quantity $V_{\alpha\beta}$ (units of V m^{-2}) is a component of the rank $\ell = 2$ Cartesian EFG tensor **V**,

$$V_{\alpha\beta} = \left(\frac{\partial^2 \phi}{\partial\alpha \partial\beta} \right)_{r=0} \quad (3.2)$$

where ϕ is the electrostatic potential at the nucleus generated by the surrounding charge density. In its principal axis system (PAS), **V** is diagonal, with three nonzero elements V_{XX}, V_{YY}, and V_{ZZ}. By convention, we choose $|V_{ZZ}| \geq |V_{YY}| \geq |V_{XX}|$, and define the principal field gradient

$$eq = V_{ZZ} \quad (3.3)$$

and the axial asymmetry parameter

$$\eta_Q = (V_{XX} - V_{YY})/V_{ZZ} \quad (3.4)$$

Using these, the quadrupolar Hamiltonian in equation (3.1) can be written in its PAS as

$$H_Q^{PAS} = \frac{e^2 qQ}{4I(2I-1)\hbar} \left\{ 3I_Z{}^2 - I(I+1) + \eta_Q (I_X{}^2 - I_Y{}^2) \right\} \quad (3.5)$$

It can now easily be verified that $H_Q = 0$ for $I = 1/2$. As a result of the Laplace equation, $V_{XX} + V_{YY} + V_{ZZ} = 0$, another case where H_Q vanishes is that of cubic symmetry in the nuclear surroundings, when $V_{XX} = V_{YY} = V_{ZZ} = 0$. The Laplace equation also restricts the asymmetry parameter to the range $0 \leq \eta_Q \leq 1$.

It is customary to tidy up the constants at the front of equation (3.5) by defining a quadrupolar coupling constant C_Q (in units of Hz)

$$C_Q = e^2 qQ/h \quad (3.6)$$

In this case, the term *coupling* refers to the interaction of the nuclear quadrupole moment eQ and the principal field gradient eq and not, of course, to an interaction between two nuclear spins.

3.3 COMPUTATION OF QUADRUPOLAR PARAMETERS

The nuclear quadrupole moment eQ is an experimentally determined quantity. Computation of parameters such as C_Q and η_Q therefore becomes a question of calculating the $V_{\alpha\beta}$ components of the EFG tensor.

If SI units are used, Slichter[10] states that a point charge e at a position x, y, z generates a $V_{\alpha\beta}$ component at $r = 0$ (the nucleus) of

$$V_{\alpha\beta} = \frac{e}{4\pi\varepsilon_0 r^3} \left(\frac{3\alpha\beta}{r^2} - \delta_{\alpha\beta} \right) \quad (3.7)$$

where α or $\beta = x$, y, z as before and $r^2 = x^2 + y^2 + z^2$. The $V_{\alpha\beta}$ components arising from several point charges may simply be summed to give a total. Therefore, it is quite straightforward (and instructive) to read the *x*, *y*, *z* coordinates of the atoms constituting one or more unit cells of an ionic crystal into a spreadsheet program, assign charges to these atoms (e.g., $+e$ for Na^+, $-e$ for Cl^-, $+2e$ for Ca^{2+}, $-2e$ for O^{2-}, etc.), and compute the full EFG tensor **V** at the origin, which may be placed on the nucleus of interest. The tensor **V** may then be diagonalized, the principal values used to obtain C_Q and η_Q, and the eigenvectors used to obtain the orientation of the EFG tensor in the crystal frame.

The value of C_Q obtained by finding $V_{ZZ} = eq$ using this "point-charge model" and then inserting it into equation (3.6) may be an order of magnitude or more smaller than that measured experimentally, particularly for a nucleus in a heavy atom. This is a consequence of the fact that, so far, the effect of the electrons in the orbitals around the nucleus of interest has been ignored. In a fully ionic solid, these closed-shell electrons will have spherical symmetry (and hence contribute zero-field gradient) until polarized by the EFG arising from the external point charges. If we call this latter field gradient $\mathbf{V}^{\text{ext}}$ and the additional, induced field gradient arising from the distorted closed shell $\mathbf{V}^{\text{ind}}$, then we have $\mathbf{V}^{\text{ind}} = -\gamma_\infty \mathbf{V}^{\text{ext}}$ and

$$\begin{aligned} \mathbf{V}^{\text{tot}} &= \mathbf{V}^{\text{ext}} + \mathbf{V}^{\text{ind}} \\ &= (1 - \gamma_\infty)\mathbf{V}^{\text{ext}} \end{aligned} \quad (3.8)$$

where $\mathbf{V}^{\text{tot}}$ is the total EFG tensor experienced by the nucleus and γ_∞ is a proportionality constant known as the Sternheimer antishielding factor,[10] which has been calculated to be small and positive for Li^+ ($\gamma_\infty \approx +0.25$ and hence shielding, not antishielding) but large and negative for heavier ions, such as Na^+ ($\gamma_\infty \approx -4$), O^{2-} ($\gamma_\infty \approx -14$), and Rb^+ ($\gamma_\infty \approx -50$).

Even including the Sternheimer antishielding factor as a correction, the point-charge model rarely provides results that agree well with experiment as it does not take into account any degree of covalency, i.e., interactions between atoms arising from the outer or valency electrons. As a result, there has been much interest in the use of electronic structure calculations for the numerical computation of EFG tensors. A detailed discussion of these quantum-mechanical methods is outside the scope of this chapter but it can be noted that, from an NMR spectroscopy perspective, they are popularly divided into two basic types:

1. Methods where a cluster of atoms is defined around the atom of interest and either Hartree-Fock theory or density functional theory (DFT) used to obtain the ground-state electronic wavefunction. The EFG tensor is then computed from the charge density (proportional to the square of the wavefunction in the Born interpretation). In order to obtain a good approximation to a crystalline solid, the defined cluster must be as large as possible and/or embedded in a lattice of point charges. This cluster approach can be implemented, for example, using the well-known quantum chemistry program GAUSSIAN.[11]
2. Methods that use a planewave basis set to exploit the periodicity of an infinite crystalline solid made up of repeating unit cells.[12] Normally DFT is used for the electronic structure calculation, with examples including the WIEN2k[13] and CASTEP[14] programs. The former calculates the all-electron wavefunction directly, while the latter replaces the core electrons with a pseudopotential for greater computational efficiency and then reconstructs the all-electron wavefunction using the projector augmented wave (PAW) method. These codes do not require definition of a cluster and have high accuracy as a result of the wavefunction extending throughout the infinite solid.

An important point to note with all these "first-principles" quantum-mechanical methods is that the EFG is a property of the ground-state electronic wavefunction and that, unlike the chemical shift or the indirect spin–spin coupling, its computation does not require knowledge of any excited states. Thus, despite recent improvements in the computation of chemical shifts, it should always be the case that the quadrupolar NMR parameters will be computed with greater time efficiency and (probably) accuracy.

Finally, although very definitely not one of the first-principles methods discussed above, it is worth noting that attempts have been made to calculate EFG tensors, and hence quadrupolar NMR parameters, from electron density maps yielded by advanced X-ray diffraction (XRD) techniques.[15] From the perspective of crystallography, however, it is the reverse of this process that is of greater interest: experimental measurements of quadrupolar NMR parameters yield EFG tensors and knowledge of these could be used to improve the quality of XRD-derived electron density maps.

3.4 EFFECT ON NMR SPECTRA

The effect of the quadrupolar interaction, as expressed by the Hamiltonian H_Q, on nuclear spin energy levels subject to a Zeeman interaction and, hence, on NMR spectra has been extensively discussed in many places (see, for example, Chapter 1)[1] and is only briefly reviewed here.

The general approach is to form the total nuclear spin Hamiltonian H by adding together the Zeeman, H_Z, and the quadrupolar, H_Q, Hamiltonians and then treat the latter as a small perturbation on the former:

$$H = H_Z + H_Q \tag{3.9}$$

where

$$H_Z = \omega_0 I_z \tag{3.10}$$

To proceed further, we now need to express H_Q in the laboratory axis system defined by H_Z and the best way of doing this is to take H_Q^{PAS} in equation (3.5) (where **V** is parameterized by just two terms, eq and η_Q) and rotate it into the laboratory frame. This rotation can be performed in a number of ways but one popular method is to rewrite H_Q^{PAS} in terms of spherical tensor operators, $T_{\ell,m}$,

$$H_Q^{PAS} = \frac{\sqrt{3}\pi C_Q}{2\sqrt{2}I(2I-1)}\left(T_{2,0} + \frac{\eta_Q}{\sqrt{6}}(T_{2,2} + T_{2,-2})\right) \tag{3.11}$$

where we have used the $\{T_{\ell,m}\}$ defined in Chapter 1,[1] and exploit the convenient transformation properties of spherical tensors under rotation,

$$R(\alpha,\beta,\gamma)T_{\ell,m}R^{-1}(\alpha,\beta,\gamma) = \sum_{m'=-\ell}^{\ell} D^{\ell}_{m',m}(\alpha,\beta,\gamma)\,T_{\ell,m'} \tag{3.12}$$

where α, β, γ are the Euler angles and $D^{\ell}_{m',m}(\alpha,\beta,\gamma)$ is a Wigner rotation matrix element.[16] Using equations (3.11) and (3.12) the laboratory-frame quadrupolar Hamiltonian can then be written as

$$H_Q^{Lab} = \frac{\sqrt{3}\pi C_Q}{2\sqrt{2}I(2I-1)}\sum_{m'=-2}^{2}\left(D^2_{m',0}(\alpha,\beta,\gamma) + \frac{\eta_Q}{\sqrt{6}}\left(D^2_{m',2}(\alpha,\beta,\gamma) + D^2_{m',-2}(\alpha,\beta,\gamma)\right)\right)T_{2,m'} \tag{3.13}$$

If the eigenstates, $|i\rangle$, and eigenvalues, $\varepsilon_i^{(0)}$, of H_Z are known, i.e.,

$$H_Z|i\rangle = \varepsilon_i^{(0)}|i\rangle \tag{3.14}$$

and if H_Z dominates H_Q then time-independent perturbation theory[17] states that the energy, ε_i (here in units of s^{-1}), of an eigenstate of H in equation (3.9) is given by

$$\varepsilon_i = \varepsilon_i^{(0)} + \varepsilon_i^{(1)} + \varepsilon_i^{(2)} + \ldots \tag{3.15}$$

where

$$\varepsilon_i^{(1)} = \langle i|H_Q^{Lab}|i\rangle \tag{3.16}$$

and

$$\varepsilon_i^{(2)} = \sum_{i\neq j}\frac{\langle i|H_Q^{Lab}|j\rangle\langle j|H_Q^{Lab}|i\rangle}{\varepsilon_i^{(0)} - \varepsilon_j^{(0)}} \tag{3.17}$$

A first-order correction to an energy, $\varepsilon_i^{(1)}$, in equation (3.16) simply consists of a diagonal element of H_Q^{Lab} in equation (3.13). Since $T_{2,0}$ is the only diagonal $\ell = 2$ tensor operator, we can write a first-order quadrupolar Hamiltonian:

$$H_Q^{(1)} = \frac{\sqrt{3}\pi C_Q}{2\sqrt{2}I(2I-1)}\left(D^2_{0,0}(\alpha,\beta,\gamma) + \frac{\eta_Q}{\sqrt{6}}\times\left(D^2_{0,2}(\alpha,\beta,\gamma) + D^2_{0,-2}(\alpha,\beta,\gamma)\right)\right)T_{2,0} = \omega_Q\left(I_z^2 - \frac{1}{3}I(I+1)\right) \tag{3.18}$$

where

$$\omega_Q = \frac{\omega_Q^{PAS}}{2}(3\cos^2\beta - 1 + \eta_Q\sin^2\beta\cos 2\gamma) \tag{3.19}$$

with

$$\omega_Q^{PAS} = \frac{3\pi C_Q}{2I(2I-1)} \tag{3.20}$$

Note that, unlike H_Q^{PAS} or H_Q^{Lab}, $H_Q^{(1)}$ is orientationally dependent.

Observable NMR transitions have $\Delta m = m_i - m_j = \pm 1$ where m is the magnetic quantum number:

$$I_z|i\rangle = m_i\hbar|i\rangle \tag{3.21}$$

Up to first order, therefore, the frequencies of the observable transitions for a spin I nucleus are given by

$$\begin{aligned}\varepsilon_i - \varepsilon_{i+1} &= \left(\varepsilon_i^{(0)} - \varepsilon_{i+1}^{(0)}\right) + \left(\varepsilon_i^{(1)} - \varepsilon_{i+1}^{(1)}\right)\\ &= \omega_0 + \omega_Q(m_i^2 - m_{i+1}^2) \qquad (3.22)\end{aligned}$$

Thus, in the NMR spectrum of a single crystal, two transitions would be found for a spin $I=1$ with frequencies of $\omega_0 + \omega_Q$ and $\omega_0 - \omega_Q$, three transitions for a spin $I = 3/2$ with frequencies of $\omega_0 + 2\omega_Q$, ω_0, and $\omega_0 - 2\omega_Q$, and so on. In each case, in this "Ernstian"[1,18] convention, the splitting between adjacent resonances in the quadrupolar multiplet is $2\omega_Q$. (In another widely used convention, the splitting between adjacent transitions is defined as ω_Q.) In a powdered sample, all orientations of **V** with respect to B_0 are present simultaneously and the full NMR spectrum is obtained from the average in equations (3.18) and (3.19) of β from 0 to π and (if $\eta_Q \neq 0$) γ from 0 to 2π. Figure 3.1 shows schematic energy levels and simulated single crystal and powder NMR spectra for $I=1$ and 3/2, calculated up to first order. Note that for a half-integer spin quadrupolar nucleus ($I = 3/2, 5/2, 7/2$, etc.), the $m = +1/2 \leftrightarrow -1/2$ or "central" transition is not broadened in the powder spectrum.

Typical values of $\omega_Q^{PAS}/2\pi$ observed in solid-state NMR spectroscopy range up to ~10 MHz while $\omega_0/2\pi$ is typically 100 MHz. Therefore, second-order corrections to the energy levels are often important and equation (3.17) has to be evaluated. After simplification of the products of Wigner rotation matrix elements,[16] the following is obtained:

$$\varepsilon_i^{(2)} = \frac{\left(\omega_Q^{PAS}\right)^2}{2\omega_0}\left(A_{I,m_i}^{(0)} B^{(0)} + A_{I,m_i}^{(2)} B^{(2)}(\beta,\gamma) + A_{I,m_i}^{(4)} B^{(4)}(\beta,\gamma)\right) \qquad (3.23)$$

where the spin and magnetic quantum number-dependent coefficients, $A_{I,m_i}^{(\ell)}$, are given in Table 3.1 for spins $I = 3/2$ and 5/2, and the orientation-dependent coefficients are

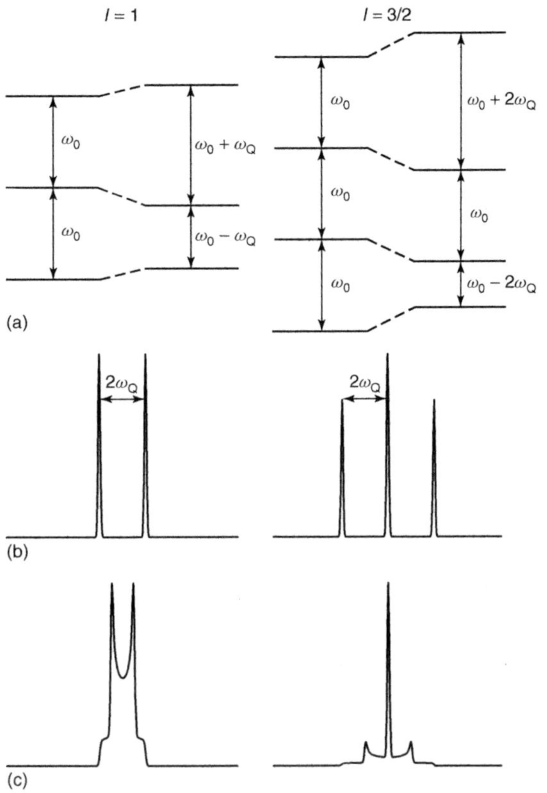

Figure 3.1. (a) First-order perturbation of the Zeeman energy levels by the quadrupolar coupling for spin quantum number $I=1$ and 3/2. (b) Simulations of the resulting (single-crystal) NMR spectra with $2I$ nondegenerate transitions separated by $2\omega_Q$. (c) Corresponding powder NMR spectra.

$$B^{(0)} = \left(1 + \frac{\eta_Q^2}{3}\right) \qquad (3.24a)$$

$$B^{(2)}(\beta,\gamma) = \left(1 - \frac{\eta_Q^2}{3}\right) d_{0,0}^2(\beta) - \frac{2\sqrt{2}}{\sqrt{3}}\eta_Q\, d_{2,0}^2(\beta)\cos 2\gamma \qquad (3.24b)$$

$$B^{(4)}(\beta,\gamma) = \left(1 + \frac{\eta_Q^2}{18}\right) d_{0,0}^4(\beta) + \frac{\sqrt{10}}{3}\eta_Q\, d_{2,0}^4(\beta)\cos 2\gamma + \frac{35}{9\sqrt{70}}\eta_Q^2\, d_{4,0}^4(\beta)\cos 4\gamma \qquad (3.24c)$$

Table 3.1. Spin and magnetic quantum number dependent coefficients $A^{(\ell)}_{I,m_i}$ governing second-order quadrupolar broadening for spin $I = \frac{3}{2}$ and $I = \frac{5}{2}$ (see equation (3.23))

	$A^{(0)}_{I,m_i}$	$A^{(2)}_{I,m_i}$	$A^{(4)}_{I,m_i}$
Spin $I = \frac{3}{2}$			
$m_1 = +\frac{3}{2}$	$\frac{6}{5}$	0	$-\frac{6}{5}$
$m_2 = +\frac{1}{2}$	$-\frac{2}{5}$	$-\frac{8}{7}$	$\frac{54}{35}$
$m_3 = -\frac{1}{2}$	$\frac{2}{5}$	$\frac{8}{7}$	$-\frac{54}{35}$
$m_4 = -\frac{3}{2}$	$-\frac{6}{5}$	0	$\frac{6}{5}$
Spin $I = \frac{5}{2}$			
$m_1 = +\frac{5}{2}$	$\frac{20}{3}$	$\frac{40}{21}$	$-\frac{60}{7}$
$m_2 = +\frac{3}{2}$	$-\frac{4}{5}$	$-\frac{40}{7}$	$\frac{228}{35}$
$m_3 = +\frac{1}{2}$	$-\frac{16}{15}$	$-\frac{64}{21}$	$\frac{144}{35}$
$m_4 = -\frac{1}{2}$	$\frac{16}{15}$	$\frac{64}{21}$	$-\frac{144}{35}$
$m_5 = -\frac{3}{2}$	$\frac{4}{5}$	$\frac{40}{7}$	$-\frac{228}{35}$
$m_6 = -\frac{5}{2}$	$-\frac{20}{3}$	$-\frac{40}{21}$	$\frac{60}{7}$

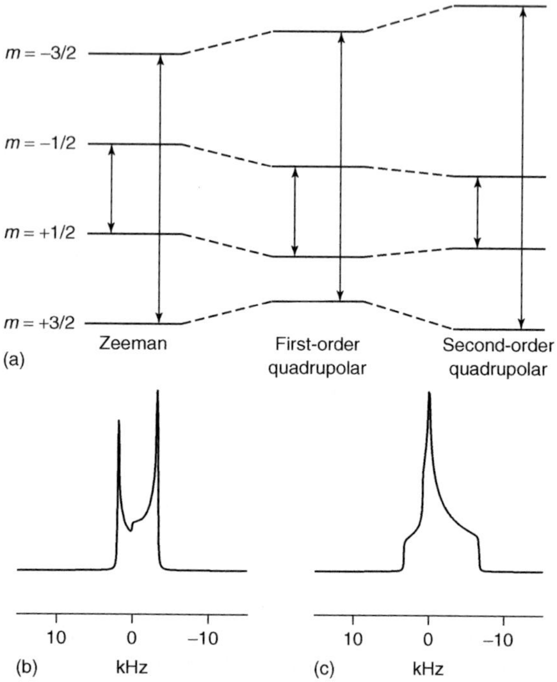

Figure 3.2. (a) Perturbation of the spin $I = 3/2$ Zeeman energy levels by the quadrupolar coupling to first- and second-order approximations, showing that the frequencies of the $m = +1/2 \leftrightarrow -1/2$ central- and $m = +3/2 \leftrightarrow -3/2$ triple-quantum transitions are affected in the second-order (but not first-order) approximation. (b) and (c) Simulations of spin $I = 3/2$ second-order quadrupolar broadened central-transition NMR lineshapes with (b) $\eta_Q = 0$ and (c) $\eta_Q = 1$. Other simulation parameters: $C_Q = 2$ MHz and $\omega_0/2\pi = 100$ MHz.

where $d^{\ell}_{m',m}(\beta)$ is a reduced Wigner rotation matrix element.[16] The two most noticeable effects of the inclusion of this second-order quadrupolar interaction are (i) that the $m = +1/2 \leftrightarrow -1/2$ central transition of a half-integer quadrupolar spin is now broadened in the powder spectrum, although this broadening is typically much smaller (~kHz) than the first-order broadening of the "satellite" transitions (~MHz), and (ii) that its center of mass is shifted away from the Larmor frequency ω_0 by an amount $A^{(0)}_{I,1/2} B^{(0)} (\omega_Q^{PAS})^2/\omega_0$. Figure 3.2 shows schematically the effects of the second-order correction on the spin $I = 3/2$ energy levels and simulations of the resulting powder central-transition lineshape for $\eta_Q = 0$ and $\eta_Q = 1$.

Nowadays, most modern solid-state NMR experiments are performed on powdered materials under conditions of magic angle spinning (MAS). The first-order quadrupolar Hamiltonian in equation (3.18) has a purely rank $\ell = 2$ orientational dependence and so, in the limit of very fast MAS, a single sharp line at the Larmor frequency is predicted. However, in reality, the MAS rate (~kHz) will almost certainly be much smaller than the satellite broadening and in this slow MAS limit the narrow centerband at the Larmor frequency will be replicated in a series of spinning sidebands separated by the MAS rate ω_{MAS}. If second-order quadrupolar effects are considered, then the effects of MAS on the orientation-dependent terms in equation (3.24) must be taken into account: $B^{(0)}$ is unaffected, $B^{(2)}$ vanishes completely if we assume that the MAS rate is faster than $(\omega_Q^{PAS})^2/\omega_0$, and $B^{(4)}$ is scaled by $-7/18$, leaving the MAS-averaged second-order correction to the energy as

$$\langle \varepsilon_i^{(2)} \rangle_{MAS} = \frac{\left(\omega_Q^{PAS}\right)^2}{2\omega_0} \times \left(A_{I,m_i}^{(0)} B^{(0)} - \frac{7}{18} A_{I,m_i}^{(4)} B^{(4)}(\beta', \gamma') \right) \tag{3.25}$$

where β' and γ' are Euler angles relating the PAS of the EFG tensor to the "rotor frame" with its z axis coincident with the MAS axis. Hence, the rank $\ell = 4$ part of the second-order powder broadening survives in the MAS spectrum, as shown in Figure 3.3 for the spin $I = 3/2$ $m = +1/2 \leftrightarrow -1/2$ central transition.

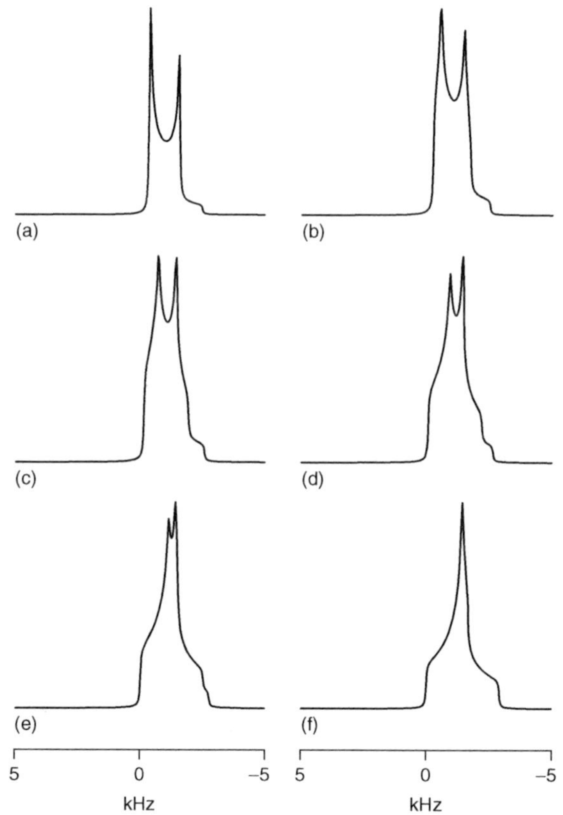

Figure 3.3. Simulations of spin $I = 3/2$ second-order broadened central-transition MAS NMR lineshapes with (a) $\eta_Q = 0$, (b) $\eta_Q = 0.2$, (c) $\eta_Q = 0.4$, (d) $\eta_Q = 0.6$, (e) $\eta_Q = 0.8$, and (f) $\eta_Q = 1.0$. Note that these lineshapes are offset from zero frequency by the isotropic second-order quadrupolar shift. Other simulation parameters are $C_Q = 2$ MHz and $\omega_0/2\pi = 100$ MHz.

3.5 MEASUREMENT BY NMR

Quadrupolar couplings can be measured by a variety of techniques, including microwave rotational spectroscopy, electron spin resonance (ESR), perturbed angular correlation (PAC) of nuclear radiation, Mössbauer spectroscopy, solution-state NMR relaxation measurements, and nuclear quadrupole resonance (NQR), also known as zero-field NMR. These are all outside the scope of this chapter and we concentrate, as in the previous section, on the measurement of quadrupolar couplings in solid-state NMR spectroscopy performed in the high-field limit.

A series of NMR spectra of a single crystal recorded as a function of orientation in a goniometer probehead can be analyzed to yield quadrupolar coupling parameters.[19] Either the satellite transitions or, for a half-integer spin exhibiting second-order effects, the central transition may be followed. Although a powerful method (as it yields the absolute orientation of the EFG tensor in the crystal frame, which is difficult to obtain by other techniques), it is limited by the availability of large crystals of the appropriate material.

The determination of quadrupolar coupling parameters from powder NMR lineshapes is a much more general and widely used approach. Ideally, all transitions are recorded in a single spectrum, most likely using a spin-echo experiment to obtain an undistorted lineshape. This is the approach used for nuclei with small quadrupole moments, such as ^2H, ^6Li ($I = 1$), ^7Li ($I = 3/2$), and ^{133}Cs ($I = 7/2$). The resulting powder spectra can be fitted using a variety of widely available software and, if present, any additional effects of other spin interactions or dynamics can be taken into account.[20] A ^2H NMR example of such a fitting, with dynamic effects included, is shown in Figure 3.4. For most other nuclides, however, the quadrupolar coupling is much larger and the full powder lineshape is too wide to record in a single spectrum. In this case, it can be reconstructed from a series of experiments in which either the radiofrequency or static magnetic field is stepped across the region of interest.[21] Alternatively, for half-integer spin quadrupolar nuclei, the second-order powder lineshape of the central transition can be

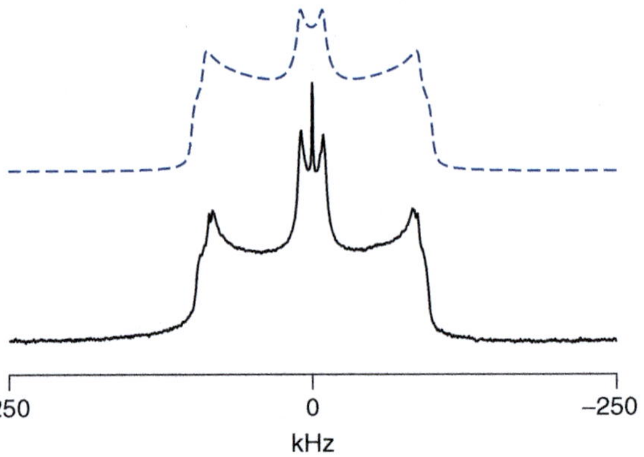

Figure 3.4. ^{2}H ($I = 1$) quadrupolar-echo NMR spectrum of static powdered sodium tetrathionate dihydrate-d_4 ($Na_2S_4O_6.2D_2O$) and computer fitting (dashed line). Experimental parameters: $\omega_0/2\pi = 61.4$ MHz. The fitting includes a dynamic model allowing for 180° flips of the D_2O molecules about their C_2 axes and yields the parameters: $C_Q = 266$ kHz, $\eta_Q = 0$, a D–O–D angle of 105.5°, and rate constant $k = 6.8 \times 10^6$ s^{-1}.

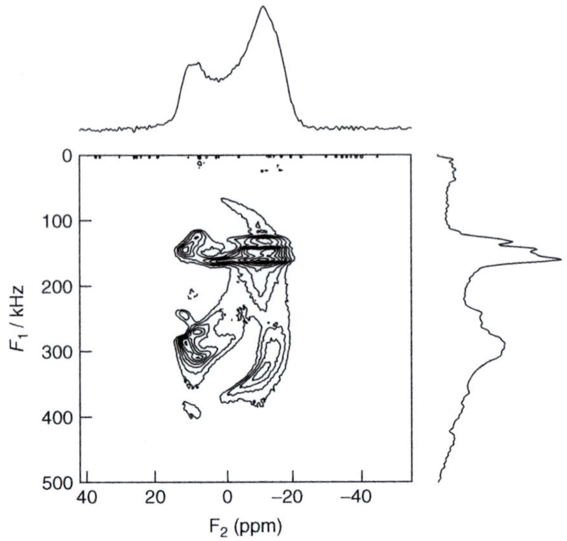

Figure 3.5. 2D nutation ^{27}Al ($I = 5/2$) spectrum of static powdered aluminum acetylacetonate, $Al(C_5H_7O_2)_3$. This correlation spectrum can be analyzed to yield the ^{27}Al quadrupolar coupling parameters $C_Q = 3.0$ MHz and $\eta_Q = 0.15$. Experimental parameters are $\omega_0/2\pi = 104.3$ MHz and radiofrequency field strength $\omega_1/2\pi = 100$ kHz.

recorded (i.e., instead of attempting to record the much broader satellite transitions) and fitted to obtain the quadrupolar coupling parameters. Because the central transition is typically several orders of magnitude narrower than the satellite-transition powder lineshape, it is more important to allow for the presence of other spin interactions, such as the chemical shift anisotropy (CSA), in any fitting. Although popular, the analysis of powder lineshapes of static samples has at least two major drawbacks: (i) the sensitivity is low as the signal is spread across a wide frequency range and, more significantly, (ii) if more than one inequivalent species of the observed nuclide is present their lineshapes will overlap and be unresolved, making analysis difficult.

For half-integer spin quadrupolar nuclei, the nutation behavior of the central transition during a radiofrequency pulse depends largely on the first-order quadrupolar interaction, even though the observed lineshape is unaffected by it.[22] This phenomenon provides a means of measuring quadrupolar parameters from a central-transition powder lineshape even if that lineshape shows no features characteristic of the quadrupolar interaction. 2D nutation NMR spectra are obtained by incrementing an excitation pulse duration as a function of a time t_1 and recording the resulting free induction decays (FIDs) as a function of t_2 and then performing a 2D Fourier transform with respect to $\{t_1, t_2\}$. Computer fitting of the complex lineshape correlating second-order broadening in F_2 with the nutation pattern in F_1 yields the quadrupolar coupling parameters. An ^{27}Al ($I = 5/2$) NMR example of a 2D nutation spectrum is shown in Figure 3.5. As with other methods designed for static samples, nutation spectroscopy is difficult to apply if there are several inequivalent sites for the observed nuclide owing to the lack of site resolution.

In recent years, it has become increasingly popular to study powdered solids by MAS NMR. For nuclei that generally exhibit small quadrupolar couplings, the full static powder lineshape breaks up under MAS into spinning sidebands whose intensities mirror the original static lineshape, thus providing a higher signal-to-noise ratio and often resolving crystallographically inequivalent species, but still allowing extraction of quadrupolar coupling parameters. For nuclei exhibiting larger quadrupolar couplings, full excitation and detection of the satellite transitions is unlikely but, for half-integer spin nuclei like ^{23}Na ($I = 3/2$), ^{17}O and ^{27}Al ($I = 5/2$), quadrupolar parameters can still be extracted by lineshape analysis of the MAS-averaged second-order broadened central transition. Unfortunately, this second-order broadening means that site resolution is not obtained unless the broadening is unusually small or the chemical shift range is large. For spin $I = 5/2$ nuclei, second-order broadened $m = \pm 3/2 \leftrightarrow \pm 1/2$ satellite-transition spinning sidebands are sometimes analyzed as well as, or instead of, the central transition as the broadening is smaller and, hence, there is more chance of resolving multiple sites.[23] Even if MAS lineshapes show no features characteristic of the second-order quadrupolar interaction, the composite quadrupolar parameter $C_Q(1 + \eta_Q{}^2/3)^{1/2}$ can still be extracted by recording spectra at different B_0 field strengths and separating the second-order quadrupolar shift from the chemical shift. Finally, it is worth noting that in very favorable cases it is even possible to extract quadrupolar coupling parameters from MAS NMR spectra of spin $I = 1/2$ nuclei, where powder lineshapes due to second-order "cross terms" between the $I - S$ dipolar interaction and the strong quadrupolar interaction of a "remote" $S \geq 1$ nucleus are sometimes observed.[24]

The problem of second-order MAS lineshapes from inequivalent sites overlapping and hence being unresolved has been tackled with some success in recent years with the development of new experimental methods. The double rotation (DOR)[25] and dynamic angle spinning (DAS)[26] techniques were proposed in 1988 and involve rotating the powdered solid at two angles with respect to B_0 (simultaneously in the case of DOR and sequentially in the case of DAS) to remove the rank $\ell = 2$ and 4 components of the second-order quadrupolar broadening. However, neither DOR nor DAS have achieved wide popularity, as specialist, rather complicated probeheads are needed and spinning rates are limited, severely so in the case of DOR. In contrast, the multiple-quantum (MQ)MAS[27] and satellite-transition (ST)MAS techniques,[28,29] proposed in 1995 and 2000, respectively, are purely MAS-based and so can be performed on standard MAS hardware and at high spinning rates. These methods yield 2D NMR spectra with an "isotropic" dimension (F_1) that is free from first- and second-order quadrupolar broadening, yet retains isotropic chemical and second-order quadrupolar shifts, and an "anisotropic" dimension (F_2) that displays MAS-averaged second-order broadened central-transition lineshapes. A ^{87}Rb ($I = 3/2$) NMR example of a 2D MQMAS spectrum is shown in Figure 3.6(a). Extraction of quadrupolar coupling

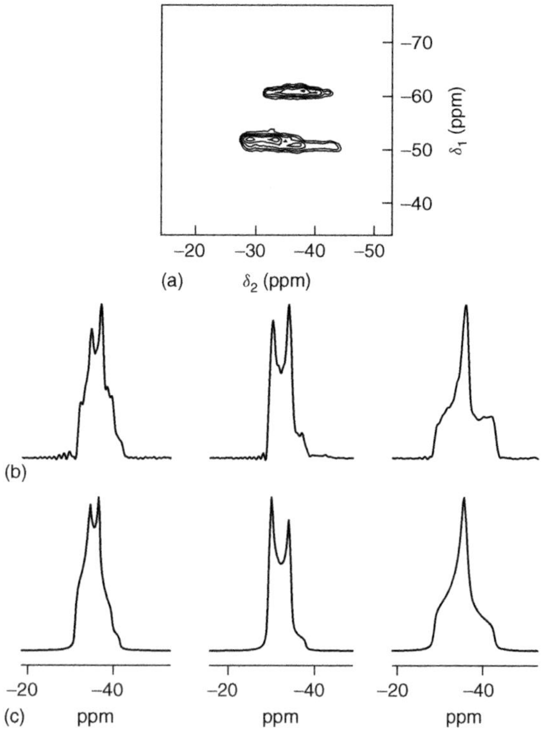

Figure 3.6. (a) 2D ^{87}Rb ($I = 3/2$) MQMAS NMR spectrum of rubidium nitrate, $RbNO_3$. Experimental parameters: $\omega_0/2\pi = 130.9$ MHz. (b) Cross sections parallel to δ_2 through the three second-order broadened ridge lineshapes in (a). (c) Computer fittings of the ^{87}Rb MAS lineshapes in (b), yielding quadrupolar parameters (from left to right): (i) $C_Q = 1.7$ MHz, $\eta_Q = 0.6$; (ii) $C_Q = 1.7$ MHz, $\eta_Q = 0.1$; and (iii) $C_Q = 1.9$ MHz, $\eta_Q = 1.0$.

parameters can be done in a variety of ways, including (i) fitting the central-transition lineshapes in F_2 to obtain C_Q and η_Q (see the examples of ^{87}Rb central-transition lineshapes and their fittings in Figure 3.6b and 3.6c), (ii) measuring the $\{F_1, F_2\}$ coordinates of the center of mass of a 2D peak and deriving $C_Q(1 + \eta_Q{}^2/3)^{1/2}$, and (iii) measuring the peak positions in the F_1 dimension at two different B_0 field strengths and deriving $C_Q(1 + \eta_Q{}^2/3)^{1/2}$. The principal drawbacks of MQMAS and STMAS are that (i) their application is limited to sites with small and moderate quadrupolar couplings, (ii) peak intensities are not quantitative and F_2 central-transition lineshapes can be very distorted, (iii) the overall sensitivity is poor, especially for MQMAS, and (iv) extensive phase cycling and the need for high resolution in F_1 mean that minimum experiment times for these 2D methods can be rather long. This last problem was the inspiration behind the development of the satellite transitions acquired in real time (START)MAS experiment,[30] which provides isotropic spectra in a real-time, 1D experiment; however, it is limited to spin $I = 3/2$ nuclei.

3.6 USE IN NMR CRYSTALLOGRAPHY

3.6.1 Quadrupolar Coupling and Symmetry

The dependence of the quadrupolar coupling parameters upon the EFG tensor, **V**, results in a strong relationship with crystal symmetry.[31] Although this relationship is rarely simple, with a range of factors affecting the EFG, for two types of local symmetry at the site of the nucleus strong restrictions are imposed upon the nature and symmetry of the EFG tensor, providing almost instant insight into the structure.

3.6.1.1 Cubic Point Symmetry

For the case of cubic point symmetry $V_{XX} = V_{YY} = V_{ZZ}$ and, hence, by the Laplace equation, $V_{ZZ} = eq$ is zero. Therefore, C_Q is zero for regular fourfold tetrahedral, sixfold octahedral and eightfold cubic coordination environments. For example, in MgO, which has the rock salt structure (space group *Fm3m*) C_Q is zero for both ^{17}O and ^{25}Mg (both $I = 5/2$).[32] More commonly, however, a quadrupolar coupling of exactly zero is rarely observed as the "perfect" symmetry is often broken either on the short length scale (by distortions and deviations of bond angles and/or bond lengths away from their ideal value) or, on a longer length scale, i.e., when second and third coordination spheres are taken into account. Indeed, even in powdered NaCl, the ^{23}Na NMR resonance, although very narrow, displays a small quadrupolar broadening resulting from the presence of defects and the finite extent of the crystallites and hence a deviation from perfect cubic point symmetry.

3.6.1.2 Axial Point Symmetry

For the case of perfect axial point symmetry $V_{XX} = V_{YY}$ and η_Q is therefore zero. Note this places no additional restriction upon the value of V_{ZZ}. Such a situation occurs when the nucleus in question lies on an *n*-fold rotation axis, where $n = 3$, 4, 5, or 6. For example, in ZnO, which has a dihexagonal pyramidal structure (space group *P63mc*), there is a C_3 axis passing through the Zn and hence η_Q is zero for ^{67}Zn ($I = 5/2$).[33] Similarly, although for TiO_2 the two polymorphs, anatase (space group *I4/amd*) and rutile (space group *P4/mnm*) have tetragonal structures, in anatase η_Q is zero for ^{47}Ti ($I = 5/2$) and ^{49}Ti ($I = 7/2$) as the Ti is situated on a C_4 rotation axis, whereas in rutile the Ti has *m.mm* point symmetry and η_Q is 0.19.[34] A well-known example of local axial symmetry occurs with the ^2H ($I = 1$) nucleus, which, unless part of a strong hydrogen bond, is usually found to have $\eta_Q \approx 0$ in molecular crystals, with V_{ZZ} oriented along the axially symmetric X–H σ-bond.[35]

Apart from the two special cases described above, the crystal symmetry alone may place constraints upon the orientation of the quadrupolar PAS in the crystal frame. For example, in $RbClO_4$ (space group *Pnma*) the Rb sits on a mirror plane, implying that the EFG PAS must have one principal direction perpendicular to this plane.[36] Thus, experimental determination of the absolute orientation of the EFG tensor, usually by single-crystal NMR, can be used to confirm the crystal space group or, if this is unknown, to suggest a possible space group.

3.6.2 Correlation with Local Structure

In addition to symmetry considerations, the determination of quadrupolar parameters can provide a wider insight into local structure. As with the chemical shift, where the correlations are perhaps better known, the quadrupolar parameters C_Q and η_Q display dependences upon local structural features such as coordination number, bond distances, and bond angles.

At a simple level, it might be expected from the point-charge model that the magnitude of the quadrupolar interaction would increase as the deviation from perfect cubic symmetry increases, i.e., that C_Q values associated with coordination environments such as trigonal-bipyramidal or square-pyramidal would be significantly higher than those associated with tetrahedral or octahedral coordination environments.[37] Indeed, there are many examples in the literature where this is found to be the case. For example, the C_Q of a trigonal ^{11}B ($I = 3/2$) species is typically found to be between 2.3 and 2.5 MHz while, in contrast, that of a tetrahedral boron is much smaller, between 0 and 0.5 MHz.[32] This distinction (together with an associated change in chemical shift) has been of great use in the study of borosilicate glasses where it has enabled the relative amounts of each species to be easily determined.

Some care must be taken, however, when attempting to infer coordination number solely from quadrupolar parameters, as many different factors can influence the values observed. For example, the ^{27}Al ($I = 5/2$) MAS NMR spectrum of andalusite (a polymorph of Al_2SiO_5) shows two second-order broadened lineshapes corresponding to C_Q values of 15.6 and 5.9 MHz.[38] However, the larger of these values has been assigned to the six-, not five-, coordinated Al species in this material, as a result of a significant distortion of the octahedral coordination environment. In this context, it is worth noting that attempts have been made to parameterize the distortion of tetrahedral and octahedral Al sites and correlate these with ^{27}Al C_Q and η_Q values, albeit with limited success.[39]

For a range of nuclides, a great many empirical correlations have been proposed between local structure and quadrupolar parameters, although a majority of these are applicable only to series of similar materials. For example, the ^{7}Li ($I = 3/2$) quadrupolar coupling constant has been found to have a linear dependence upon the C–Li–C bond angle in organolithium compounds.[40] However, there are examples of more generally applicable empirical relationships. One important example is the dependence of the ^{17}O C_Q upon the covalency of the M–O bond, which was discovered during the earliest ^{17}O ($I = 5/2$) MAS NMR studies.[41] This dependence has been widely exploited in the study of silicates where "bridging" (Si–O–Si) and "nonbridging" (Si–O–Mg, for example) oxygen species display C_Q values in the ranges 4–6 and 2–3 MHz, respectively.[42] It can be seen in the ^{17}O MQMAS spectrum of orthoenstatite ($MgSiO_3$) in Figure 3.7, for example, that the oxygen species can be grouped into two categories: four with $C_Q \approx 2$–3 MHz (nonbridging oxygens) and two with $C_Q \approx 4$–5 MHz (bridging oxygens).[43] Another example of an apparently general relationship is the strong dependence of the ^{2}H ($I = 1$) quadrupolar parameters (both C_Q and η_Q) and the length, r_{XY}, of a hydrogen bond X–H···Y.[44]

In recent years, first-principles computation of quadrupolar parameters has been increasingly used to provide support for these correlations with local structure and, indeed, as an aid to discovering new ones (see below).

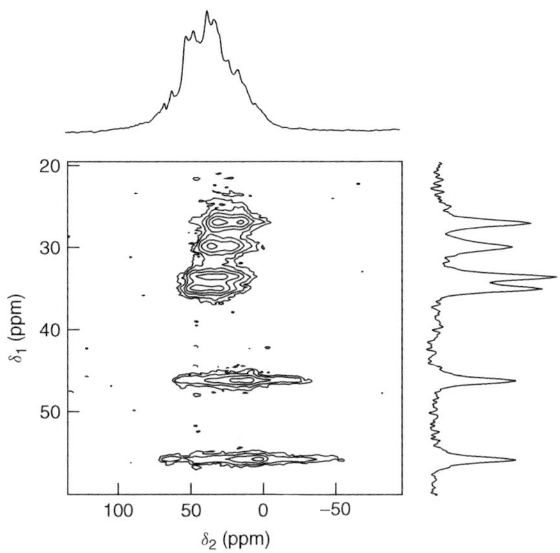

Figure 3.7. 2D ^{17}O ($I = 5/2$) MQMAS NMR spectrum of orthoenstatite, $MgSiO_3$. The ^{17}O MAS spectrum and δ_1 isotropic projection are also shown. Experimental parameters are $\omega_0/2\pi = 54.3$ MHz. Six crystallographically distinct oxygen sites are resolved: four (in the range $\delta_1 =$ 27–37 ppm) with smaller second-order broadenings in δ_2 ($C_Q \approx 2$–3 MHz) and two (with $\delta_1 = 47$ ppm and $\delta_1 =$ 57 ppm) with larger second-order broadenings in δ_2 ($C_Q \approx$ 4–5 MHz). On the basis of these C_Q values, the former are identified as nonbridging oxygens and the latter as Si–O–Si bridging oxygens. (Reproduced from Ref. 43. © American Chemical Society, 2007.)

3.6.3 Comparison with Calculated Parameters

Although very useful, empirical correlations between local structure and quadrupolar parameters have many limitations. These include (i) the additional dependence of C_Q and η_Q upon long-range structure and symmetry, commonly leading to a "fuzziness" or lack of precision in such correlations, (ii) the occurrence of quadrupolar parameters that appear to defy established correlations (e.g., ^{17}O C_Q values of 3.8 and 4.4 MHz in the mineral wadsleyite, β-Mg_2SiO_4,[45] that, based on the well-known correlation discussed above, would indicate that they corresponded to bridging oxygen atoms, whereas, in fact, they arise from nonbridging ones), and (iii) the possible absence of a suitable series of structurally related materials upon which to base the correlation.

These limitations can be overcome by using first-principles computation of quadrupolar parameters, as discussed in Section 3.3. Close agreement between calculated C_Q and η_Q values and experimentally derived ones both confirms the accuracy of the NMR measurements (which is useful if the signal-to-noise ratio is poor or if the lineshapes are distorted, e.g., as occurs with MQMAS and STMAS) and the accuracy of the crystal structure used as the basis for the calculations. This approach has been used, for example, in the ^{17}O MQMAS and STMAS NMR studies of orthoenstatite and wadsleyite referred to above.[43,46] Figure 3.8 shows a comparison of cross sections from the ^{17}O MQMAS NMR spectrum of orthoenstatite in Figure 3.7 and lineshapes simulated using quadrupolar and isotropic chemical shift parameters obtained using the first-principles CASTEP code.

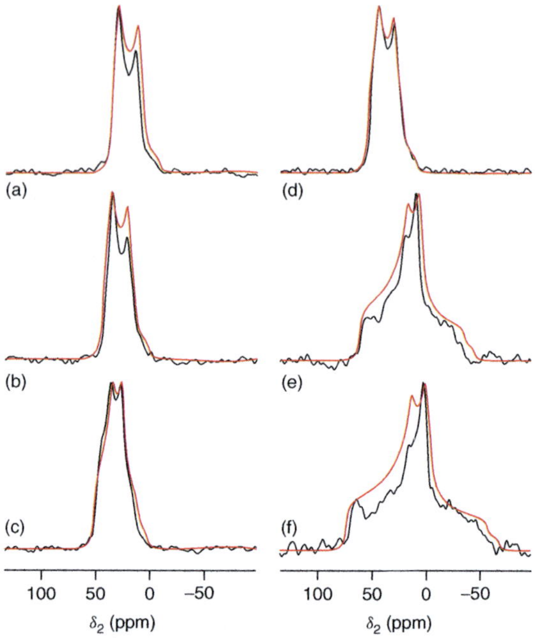

Figure 3.8. Cross sections parallel to δ_2 through the six second-order broadened ridge lineshapes in the ^{17}O MQMAS NMR spectrum of orthoenstatite in Figure 3.7 (black) and lineshapes simulated using quadrupolar and isotropic chemical shift parameters obtained using the first-principles CASTEP code (red). The CASTEP calculated parameters are (a) $C_Q = 3.06$ MHz, $\eta_Q = 0.21$, $\delta_{iso} = 41.0$ ppm; (b) $C_Q = 2.96$ MHz, $\eta_Q = 0.30$, $\delta_{iso} = 47.8$ ppm; (c) $C_Q = 3.03$ MHz, $\eta_Q = 0.61$, $\delta_{iso} = 53.4$ ppm; (d) $C_Q = 3.03$ MHz, $\eta_Q = 0.35$, $\delta_{iso} = 57.4$ ppm; (e) $C_Q = 4.35$ MHz, $\eta_Q = 0.80$, $\delta_{iso} = 62.6$ ppm; and (f) $C_Q = 5.00$ MHz, $\eta_Q = 0.82$, $\delta_{iso} = 74.4$ ppm. (Reproduced from Ref. 43. © American Chemical Society, 2007.)

More interesting from the perspective of NMR crystallography is the possibility that crystal structures could be refined or optimized by minimizing the differences between the calculated and experimental quadrupolar parameters. Electronic structure calculations often reveal the presence of large forces on the atoms and it is usual to minimize these by optimizing the structure, i.e., displacing the atoms from their positions in the diffraction-based structure, but the status of these purely theoretical structures is controversial. To our knowledge no one has yet successfully optimized structures using the experimental quadrupolar parameters as constraints but evidence is emerging that this will soon be possible. For example, in recent work, it was demonstrated that the agreement between experimental and calculated ^{27}Al quadrupolar parameters for the open-framework aluminophosphate AlPO-14 was extremely poor unless optimization of the structure was undertaken prior to the calculation of the NMR parameters.[47] This is probably a consequence of the use of an initial structure derived from powder (as opposed to single-crystal) X-ray diffraction.

In addition to the comparative approach described above, first-principles calculations also provide a flexible way to investigate the systematic dependence of NMR parameters on a particular structural feature. This area was pioneered in a study that used a simple cluster model to determine the dependence of the ^{17}O quadrupolar parameters upon the Si–O–Si bond angle,[48] a trend that was later confirmed experimentally in an ^{17}O DAS NMR study.[49] A range of similar work has appeared more recently, with a number of computational studies discussing the dependence of the ^{17}O C_Q upon the Si–O distance and of η_Q upon the Si–O–Si bond angle.[50] The trends predicted by computation can then be utilized to interpret spectra and ultimately determine the structural moieties present in glasses, where the lack of long-range order restricts the ability of diffraction to provide structural models.

3.6.4 Orientation of Quadrupole/EFG Tensors

First-principles calculations provide the absolute orientation of the quadrupole or EFG tensor in the crystal frame. Experimentally, however, this information is only directly available from single-crystal NMR spectroscopy, with NMR of powdered solids usually only providing the magnitudes of V_{XX}, V_{YY}, and V_{ZZ} (through determination of C_Q and η_Q) and not their direction.

In some circumstances, it is possible to obtain information about the orientation of EFG tensors from powder NMR spectra, hence allowing comparison with calculated tensors. This can be achieved using 2D NMR spectroscopy under either static or spinning conditions, correlating the quadrupole tensor with either a dipole–dipole interaction tensor,[51] a shielding tensor,[52] a paramagnetic shift tensor,[53] or another quadrupole tensor.[54] Only the first of these will provide the absolute orientation of the EFG tensor in the crystal frame, but the last two methods can be used to determine the relative orientation of the tensors, allowing comparison with the results of first-principles calculations and, possibly, absolute orientations to be inferred. For example, Figure 3.9 compares (i) a standard ^{23}Na ($I = 3/2$) MQMAS NMR spectrum of sodium metasilicate pentahydrate ($Na_2SiO_3.5H_2O$) (Figure 3.9a), (ii) an alternative ^{23}Na MQMAS spectrum of the same material

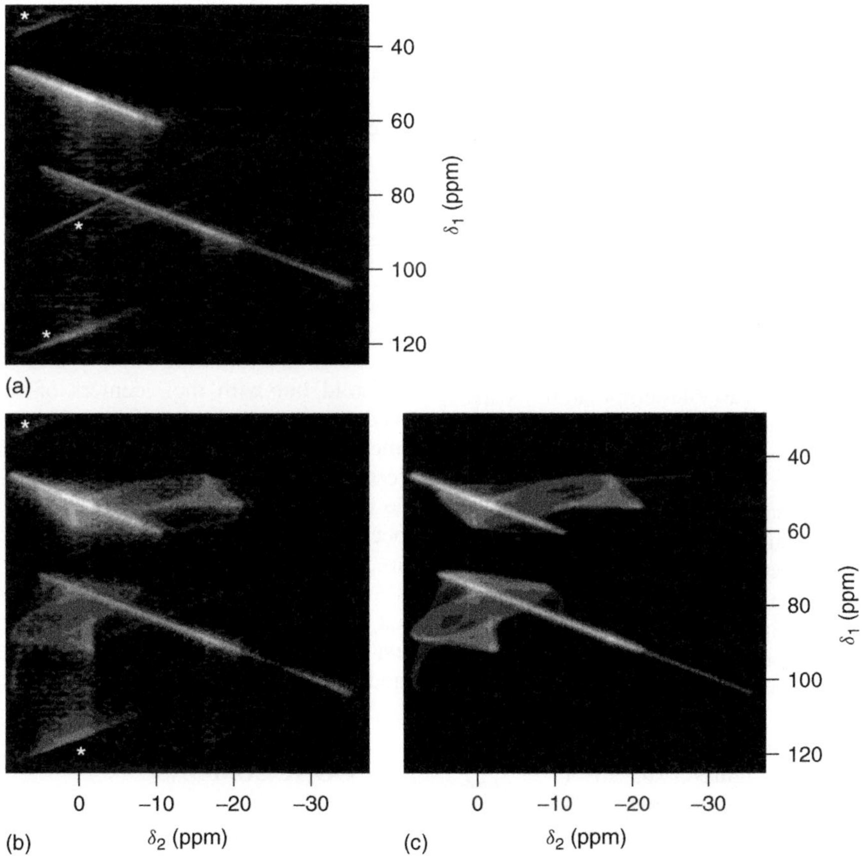

Figure 3.9. (a) 2D ^{23}Na ($I = 3/2$) MQMAS NMR spectrum of sodium metasilicate pentahydrate, $Na_2SiO_3.5H_2O$. (b) As (a), but recorded with a 400-ms mixing time inserted into the pulse sequence. (c) Simulated spectrum with a relative orientation of $\{\alpha, \beta, \gamma\} = \{90°, 35°, 10°\}$ between the two ^{23}Na EFG tensors. Experimental parameters are $\omega_0/2\pi = 105.8$ MHz. (Reproduced from Ref. 54. © American Chemical Society, 2002.)

recorded with a 400-ms mixing time inserted into the pulse sequence, allowing spin diffusion between the two inequivalent Na sites to occur (Figure 3.9b), and (iii) a computer-simulated spectrum with a relative orientation of $\{\alpha, \beta, \gamma\} = \{90°, 35°, 10°\}$ between the two ^{23}Na EFG tensors (Figure 3.9c).[54]

3.6.5 Effect of Dynamics

Solid-state NMR spectra of quadrupolar nuclei are highly sensitive to the presence of dynamics.[55] Because the quadrupolar interaction is large and usually dominates other interactions such as the chemical shift and dipole–dipole coupling, the effects of dynamics on quadrupolar lineshapes are typically much more evident than in NMR of spin $I = 1/2$ nuclei.

The dynamical timescale for which strongly perturbed quadrupolar lineshapes are observed is usually governed by the inverse of the first-order quadrupolar coupling (100 ns–10 μs). For the central-transition lineshapes of half-integer spin quadrupolar nuclei, however, the relevant timescale is the inverse of the second-order quadrupolar interaction (100 μs–1 ms), while under fast MAS conditions (where the MAS rate exceeds the NMR frequency "jump" during the dynamic process), it is the inverse of the MAS rate (10–100 μs). Solid-state dynamics on these "intermediate" timescales typically involve reorientation of complex ions, molecules, or parts of molecules and can be studied both by variable-temperature NMR of static samples (e.g., quadrupolar-echo ^{2}H NMR)[56] or MAS-based techniques (e.g., ^{2}H double-quantum MAS NMR).[57] In favorable cases, the dynamical information obtained can be of a very high order, achieving both elucidation of the mechanism of the motion and measurement of the rate constants involved. As is the case in other areas, it is to be expected that first-principles calculations of NMR parameters will play an increasing role here.

For dynamics on the "fast" timescale, motional effects on the lineshape will not be observed but averaged quadrupolar parameters will be obtained. For example, as shown in Figure 3.10, ^{11}B ($I = 3/2$) MAS NMR of solid orthocarborane, $C_2B_{10}H_{12}$, at room temperature reveals four inequivalent boron sites, each with $C_Q = 0$.[58] The obvious crystallographic conclusion from this is that the four borons are located at sites with cubic point symmetry (see the Section 3.6.1) but this is clearly incompatible with the known icosahedral cage structure of the $C_2B_{10}H_{12}$

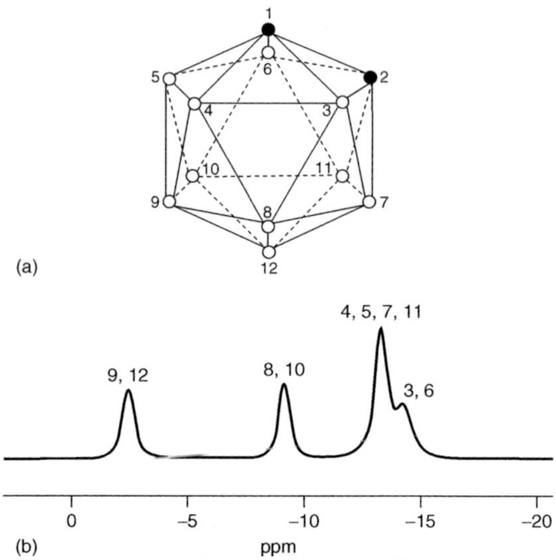

Figure 3.10. (a) Structure of molecule of orthocarborane, $C_2B_{10}H_{12}$, and (b) the ^{11}B ($I = 3/2$) MAS NMR spectrum. Experimental parameters are $\omega_0/2\pi = 128.4$ MHz. As a result of fast dynamics in the solid, none of the four ^{11}B peaks shows evidence of second-order quadrupolar broadening, thus indicating that $C_Q = 0$ for all four distinct boron sites.

molecule. The answer to this apparent conundrum is that orthocarborane is a "plastic crystal" and the molecules are tumbling rapidly and isotropically in the solid but with their centers of mass remaining fixed in a cubic lattice. Thus, if we focus on the time-averaged structure, then the quadrupolar parameters obtained from lineshape analysis are indicating the true long-range crystal symmetry. From the perspective of an instantaneous "snapshot" of the structure, however, then the C_Q values are misleading us as to the local symmetry of the boron sites (but note that ^{11}B spin-lattice relaxation time measurements or first-principles calculations would reveal the unaveraged quadrupolar parameters).

3.7 CONCLUSIONS

If crystallography is defined as the "experimental science of determining the arrangement of atoms in solids" (Wikipedia) then solid-state NMR measurements of quadrupolar couplings clearly have a role to play. There are strong similarities between the chemical shift and the quadrupolar coupling in this respect as both are single-spin centered tensorial interactions that depend on the electronic environment of the nucleus. The key limitation of the quadrupolar interaction compared with the chemical shift is that it can only be measured for nuclei with $I \geq 1$, whereas chemical shift information is available from nuclei with all spin quantum numbers. However, the following points should also be noted: (i) quadrupolar nuclei are much more common than spin $I = 1/2$ nuclei (famously, 75% of all NMR-active nuclides have $I \geq 1$), (ii) as the dominant internal spin interaction, the quadrupolar coupling is usually easier to measure than the chemical shift interaction for spin $I \geq 1$ nuclei, (iii) the quadrupolar coupling, although influenced by local structure to a similar extent to the chemical shift, appears to be a more sensitive reporter of long-range symmetry and order within a crystal, and (iv) as noted in Section 3.3, first-principles calculations of EFG tensors are generally more efficient and (probably) more accurate than those of chemical shift tensors.

Arguably, studies of dynamics represent the most important application of solid-state NMR and, as discussed in Section 3.6.5, the quadrupolar coupling is a uniquely sensitive probe here. The status of dynamic NMR studies in crystallography is harder to define; however, as diffraction-based crystallography can rarely match the rich detail of dynamic information provided by solid-state NMR and the information that it does provide is on a very different timescale. Perhaps, in the future, as NMR crystallography becomes better established, the definition of crystallography may be widened to "experimental science of determining the arrangement and dynamic behavior of atoms in solids".

RELATED ARTICLES IN THE ENCYCLOPEDIA OF MAGNETIC RESONANCE

Computations of Magnetic Resonance Parameters for Crystalline Systems: Principles

Deuterium NMR in Solids

Double Rotation

High Speed MAS of Half-Integer Quadrupolar Nuclei in Solids

Magic Angle Spinning: Effects of Quadrupolar Nuclei on Spin-1/2 Spectra

Multiple-quantum Magic-angle Spinning Experiments on Half-integer Nuclei: Fundamentals

Quadrupolar Nuclei in Glasses

Satellite Transition NMR Spectroscopy of Half-Integer Quadrupolar Nuclei under Magic-angle Spinning

REFERENCES

1. P. P. Man, Quadrupolar interactions, in *Encyclopedia of Magnetic Resonance*, eds R. K. Harris and R. E. Wasylishen, John Wiley & Sons: Chichester, 2007. DOI: 10.1002/9780470034590.emrstm0429.
2. A. J. Vega, Quadrupolar nuclei in solids, in *Encyclopedia of Magnetic Resonance*, eds R. K. Harris and R. E. Wasylishen, John Wiley & Sons: Chichester, 2007. DOI: 10.1002/9780470034590.emrstm0431.
3. Y. Wu, Double rotation, in *Encyclopedia of Magnetic Resonance*, eds R. K. Harris and R. E. Wasylishen, John Wiley & Sons: Chichester, 2007. DOI: 10.1002/9780470034590.emrstm0135.
4. P. J. Grandinetti, Dynamic angle spinning, in *Encyclopedia of Magnetic Resonance*, eds R. K. Harris and R. E. Wasylishen, John Wiley & Sons: Chichester, 2007. DOI: 10.1002/9780470034590.emrstm0137.
5. L. Frydman, Multiple-quantum magic-angle spinning experiments on half-integer quadrupolar nuclei: fundamentals, in *Encyclopedia of Magnetic Resonance*, eds R. K. Harris and R. E. Wasylishen, John Wiley & Sons: Chichester, 2007. DOI: 10.1002/9780470034590.emrstm0333.
6. J. P. Amoureux and M. Pruski, MQMAS NMR: Experimental strategies and applications, in *Encyclopedia of Magnetic Resonance*, eds R. K. Harris and R. E. Wasylishen, John Wiley & Sons: Chichester, 2007. DOI: 10.1002/9780470034590.emrstm0319.
7. Z. Gan, Satellite transition NMR spectroscopy of half-integer quadrupolar nuclei under magic angle spinning, in *Encyclopedia of Magnetic Resonance*, eds R. K. Harris and R. E. Wasylishen, John Wiley & Sons: Chichester, 2007. DOI: 10.1002/9780470034590.emrstm0481.
8. S. E. Ashbrook and S. Wimperis, Advances in STMAS, in *Encyclopedia of Magnetic Resonance*, eds R. K. Harris and R. E. Wasylishen,

John Wiley & Sons: Chichester, 2008. DOI: 10.1002/9780470034590.emrstm1032.

9. A. Abragam, *Principles of Nuclear Magnetism*, Clarendon Press: Oxford, 1961, Chapter 6.
10. C. P. Slichter, *Principles of Magnetic Resonance*, 3rd edn., Springer-Verlag: Berlin, 1990, Chapter 10.
11. M. J. Frisch, *et al. Gaussian 03*, Gaussian Inc., Wallingford, 2004.
12. J. R. Yates and C. J. Pickard, Computations of magnetic resonance parameters for crystalline systems: Principles, in *Encyclopedia of Magnetic Resonance*, eds R. K. Harris and R. E. Wasylishen, John Wiley & Sons: Chichester, 2008. DOI: 10.1002/9780470034590.emrstm1009.
13. P. Blaha, K. Schwarz, G. K. H. Madsen, D. Kvasnicka, and J. Luitz, *Computer code WIEN2k*, Vienna University of Technology, 2001.
14. M. D. Segall, P. J. D. Lindan, M. J. Probert, C. J. Pickard, P. J. Hasnip, S. J. Clark, and M. C. Payne, *J. Phys. Cond. Matt.*, 2002, **14**, 2717.
15. S. Dahaoui, N. E. Ghermani, S. Ghose, and J. A. K. Howard, *Am. Mineral.*, 2001, **86**, 159.
16. R. N. Zare, *Angular Momentum*, Wiley: Chichester, 1996.
17. P. W. Atkins, *Molecular Quantum Mechanics*, Oxford University Press: Oxford, 1983, Chapter 8.
18. R. R. Ernst, G. Bodenhausen, and A. Wokaun, *Principles of Nuclear Magnetic Resonance in One and Two Dimensions*, Clarendon Press: Oxford, 1987.
19. R. V. Pound, *Phys. Rev.*, 1950, **79**, 685.
20. L. S. Batchelder, Deuterium NMR in solids, in *Encyclopedia of Magnetic Resonance*, eds R. K. Harris and R. E. Wasylishen, John Wiley & Sons: Chichester, 2007. DOI: 10.1002/9780470034590.emrstm0111.
21. J. A. Tang, J. D. Masuda, T. J. Boyle, and R. W. Schurko, *Chem. Phys. Chem.*, 2006, **7**, 117.
22. A. P. M. Kentgens, *Prog. Nucl. Magn. Reson. Spectrosc.*, 1998, **32**, 141.
23. C. Jäger, W. Muller-Warmuth, C. Mundus, and L. van Wullen, *J. Non-Cryst. Solids*, 1992, **149**, 209.
24. R. K. Harris, Magic angle spinning: effects of quadrupolar nuclei on spin-1/2 spectra, in *Encyclopedia of Magnetic Resonance*, eds R. K. Harris and R. E. Wasylishen, John Wiley & Sons: Chichester, 2007. DOI: 10.1002/9780470034590.emrstm0285
25. A. Samoson, E. Lippmaa, and A. Pines, *Mol. Phys.*, 1988, **65**, 1013.
26. A. Llor and J. Virlet, *Chem. Phys. Lett.*, 1988, **152**, 248.
27. L. Frydman and J. S. Harwood, *J. Am. Chem. Soc.*, 1995, **117**, 5367.
28. Z. Gan, *J. Am. Chem. Soc.*, 2000, **122**, 3242.
29. S. E. Ashbrook and S. Wimperis, *Prog. NMR. Spectrosc*, 2004, **45**, 53.
30. M. J. Thrippleton, T. J. Ball, and S. Wimperis, *J. Chem. Phys.*, 2008, **128**, 034507.
31. J. W. Akitt and W. S. McDonald, *J. Magn. Reson.*, 1984, **58**, 401.
32. K. J. D. Mackenzie and M. E. Smith, *Multinuclear Solid-state NMR of Inorganic Materials*, Pergamon: Oxford, 2002.
33. G. Wu, *Chem. Phys. Lett.*, 1998, **298**, 375.
34. T. J. Bastow, M. A. Gibson, and C. T. Forwood, *Solid State Nucl. Magn. Reson.*, 1998, **12**, 201.
35. P. Pulay, E. M. Scherer, P. C. A. van der Wel, and R. E. Koeppe, *J. Am. Chem. Soc.*, 2005, **127**, 17488.
36. T. Vosegaard, J. Skibsted, H. Bildsøe, and H. J. Jakobsen, *J. Magn. Reson.*, 1996, **122**, 111.
37. H. Koller, G. Engelhardt, A. P. M. Kentgens, and J. Sauer, *J. Phys. Chem.*, 1994, **98**, 1544.
38. S. F. Dec, J. J. Fitzgerald, J. S. Frye, M. P. Shatlock, and G. E. Maciel, *J. Magn. Reson.*, 1991, **93**, 403.
39. S. Ghose and T. Tsang, *Am. Mineral.*, 1973, **58**, 748.
40. A. Pepels, H. Günther, J. P. Amoureux, and C. Fernandez, *J. Am. Chem. Soc.*, 2000, **122**, 9858.
41. S. Schramm and E. Oldfield, *J. Am. Chem. Soc.*, 1984, **106**, 2502.
42. S. E. Ashbrook and M. E. Smith, *Chem. Soc. Rev.*, 2006, **35**, 718.
43. S. E. Ashbrook, A. J. Berry, D. J. Frost, A. Gregorovic, C. J. Pickard, J. E. Readman, and S. Wimperis, *J. Am. Chem. Soc.*, 2007, **129**, 13213.
44. L. G. Butler and T. L. Brown, *J. Am. Chem. Soc.*, 1981, **103**, 6541.
45. S. E. Ashbrook, A. J. Berry, W. O. Hibberson, S. Steuernagel, and S. Wimperis, *J. Am. Chem. Soc.*, 2003, **125**, 11824.
46. S. E. Ashbrook, L. Pollès, C. J. Pickard, A. J. Berry, S. Wimperis, and I. Farnan, *Phys. Chem. Chem. Phys.*, 2007, **9**, 1587.
47. S. E. Ashbrook, M. Cutajar, C. J. Pickard, R. I. Walton, and S. Wimperis, *Phys. Chem. Chem. Phys.*, 2008, **10**, 5754.

48. J. A. Tossell and P. Lazzeretti, *Chem. Phys. Lett.*, 1987, **112**, 205.

49. P. J. Grandinetti, J. H. Baltisberger, I. Farnan, J. F. Stebbins, U. Werner, and A. Pines, *J. Phys. Chem.*, 1995, **99**, 12341.

50. T. M. Clark and P. J. Grandinetti, *J. Phys. Cond. Matter*, 2003, **15**, S2387.

51. J. D. van Beek, R. Dupree, and M. H. Levitt, *J. Magn. Reson.*, 2006, **179**, 38.

52. J. S. Shore, S. H. Wang, R. E. Taylor, A. T. Bell, and A. Pines, *J. Chem. Phys.*, 1996, **105**, 9412.

53. S. Antonijevic and S. Wimperis, *J. Chem. Phys.*, 2005, **122**, 044312.

54. N. G. Dowell, S. E. Ashbrook, and S. Wimperis, *J. Phys. Chem. A*, 2002, **106**, 9470.

55. M. J. Duer, *Introduction to Solid-State NMR Spectroscopy*, Blackwell Science: Oxford, 2004, Chapter 6.

56. J. H. Davis, K. R. Jeffrey, M. Bloom, M. I. Valic, and T. P. Higgs, *Chem. Phys. Lett.*, **42**, 390.

57. M. Cutajar, S. E. Ashbrook, and S. Wimperis, *Chem. Phys. Lett.*, 2006, **423**, 276.

58. R. K. Harris, J. Bowles, I. R. Stephenson, and E. H. Wong, *Spectrochimica Acta A*, 1988, **44**, 273.

Chapter 4

Quadrupolar Nuclei in Solids: Influence of Different Interactions on Spectra

David L. Bryce[1] and Roderick E. Wasylishen[2]

[1]*Department of Chemistry and Centre for Catalysis Research and Innovation, University of Ottawa, 10 Marie Curie Private, Ottawa, ON K1N 6N5, Canada*
[2]*Department of Chemistry, University of Alberta, Edmonton, AB T6G 2G2, Canada*

4.1 INTRODUCTION

NMR studies of quadrupolar nuclei (i.e., those with nuclear spin quantum numbers $I \geq 1$) in solid materials is a rapidly expanding field. As new experimental techniques and higher magnetic field strengths become available,[1–5] the entire quadrupolar NMR periodic table is becoming available to scientists. As a result, applications of solid-state nuclear magnetic resonance (SSNMR) to quadrupolar nuclei are no longer confined to highly receptive nuclei occupying sites of high symmetry, e.g., ^{27}Al nuclei that occupy tetrahedral sites in zeolites,[6] but now also extend to isotopes such as ^{17}O,[7] ^{33}S,[8] ^{39}K,[9] ^{43}Ca,[10] $^{47/49}$Ti,[11] ^{73}Ge,[12] ^{91}Zr,[13] ^{115}In,[14] $^{135/137}$Ba,[15] etc. that may not occupy symmetric environments.

The unique NMR parameters provided by SSNMR studies of quadrupolar nuclei are the nuclear quadrupolar coupling constant C_Q and the asymmetry parameter η_Q. The nuclear quadrupolar coupling constant is directly proportional to the largest principal component of the electric field gradient (EFG) tensor at the nucleus, $V_{ZZ} = eq_{ZZ}$, and the nuclear quadrupole moment eQ (equation (4.1)).

$$C_Q = \frac{eQV_{ZZ}}{h} = \frac{e^2 q_{ZZ} Q}{h} \tag{4.1}$$

In equation (4.1), h is the Planck constant and C_Q is in hertz. Note, the symbol χ is also sometimes used to denote the nuclear quadrupolar coupling constant (e.g., Ref. 16). "Best values" for nuclear quadrupolar moments were tabulated by Pyykkö in 2008.[17] The quadrupolar asymmetry parameter η_Q is defined in

NMR of Quadrupolar Nuclei in Solid Materials
Edited by Roderick E. Wasylishen, Sharon E. Ashbrook and Stephen Wimperis

equation (4.2):

$$\eta_Q = \frac{V_{XX} - V_{YY}}{V_{ZZ}} \tag{4.2}$$

where $|V_{ZZ}| \geq |V_{YY}| \geq |V_{XX}|$. Note that the trace of the EFG tensor is 0, thus $0 \leq \eta_Q \leq 1$.[16] As outlined in this book, one of the primary objectives of experimental NMR spectroscopists is to characterize the EFG tensor at the nuclear sites of interest in a solid (i.e., obtain values of C_Q and η_Q for each crystallographically distinct site in the structure under investigation). For isotopes with noninteger spin quadrupolar nuclei—approximately 90% of all quadrupolar nuclei—this is generally accomplished by performing NMR measurements on the central transition (CT), $m_I = 1/2 \rightarrow m_I = -1/2$, because it is not broadened by the first-order quadrupolar interaction.[16,18,19] In addition to providing information about the EFG tensors at quadrupolar nuclei, SSNMR measurements are also capable of providing NMR parameters that have been traditionally measured in NMR studies of spin-1/2 nuclei. For example, for noninteger spin quadrupolar nuclei in solids, it is often possible to characterize the chemical shift (CS) tensor at a particular site in addition to the EFG tensor. Sometimes, one can also measure spin–spin coupling tensors where one or both nuclei of a spin pair are quadrupolar. The purpose of this chapter is to provide examples illustrating how the magnetic shielding and spin–spin coupling interactions perturb the NMR spectra of quadrupolar nuclei.

4.2 RULES OF THUMB FOR INTERPRETING THE SOLID-STATE NMR SPECTRA OF QUADRUPOLAR NUCLEI

Before discussing how NMR spectra of quadrupolar nuclei are affected by magnetic shielding and spin–spin coupling interactions, it is useful to briefly review how the nuclear quadrupolar interaction alone affects NMR spectra of solids. Because detailed accounts are available in several textbooks[20,21] and review articles,[16,18,22–27] we only highlight some of the important results.

The total width of the entire powder pattern, $\Delta\nu_{\text{total}}$, for a stationary sample of a noninteger spin quadrupolar nucleus is $(2I - 1)\nu_Q$ where $\nu_Q = 3C_Q/2I(2I - 1)$ is generally known as the *nuclear quadrupolar frequency*. Note, as I increases, the total overall width of the powder pattern decreases for any given value of C_Q (Table 4.1).

As already mentioned and discussed in Ref. 16, the breadth of the CT depends only on the second-order quadrupolar interaction;[16] thus, the width of the CT depends on $C_Q{}^2/\nu_L$, where ν_L is the Larmor frequency (this is a result of second-order

Table 4.1. Relative breadths of powder patterns for noninteger spin quadrupolar nuclei

I	η_Q	$[\Delta\nu_{\text{total}}/C_Q]^a$	$[\Delta\nu_{\text{CT}}/(C_Q^2/\nu_L)]^b$	$[\Delta\nu'_{\text{CT}}/(C_Q^2/\nu_L)]^c$
3/2	0	1	25/192 = 0.13021	3/56 = 0.05357
	0.5	1	145/768 = 0.18880	169/2688 = 0.06287
	1.0	1	1/4 = 0.25000	7/96 = 0.07292
5/2	0	3/5	1/32 = 0.03125	9/700 = 0.01286
	0.5	3/5	29/640 = 0.04531	169/11200 = 0.01509
	1.0	3/5	3/50 = 0.06000	7/400 = 0.01750
7/2	0	3/7	125/9408 = 0.01329	15/2744 = 0.00547
	0.5	3/7	725/37632 = 0.01927	845/131712 = 0.00642
	1.0	3/7	5/196 = 0.02551	5/672 = 0.00744
9/2	0	1/3	25/3456 = 0.00723	1/336 = 0.00298
	0.5	1/3	145/13824 = 0.01049	169/48384 = 0.00349
	1.0	1/3	1/72 = 0.01389	7/1728 = 0.00405

[a] $\Delta\nu_{\text{total}}$ is the total width of the powder pattern from a stationary sample.
[b] $\Delta\nu_{\text{CT}}$ is the total width of the CT powder pattern from a stationary sample.
[c] $\Delta\nu'_{\text{CT}}$ is the total width of the CT powder pattern from a sample spinning rapidly at the magic angle.

perturbation theory). Following Amoureux *et al.*,[22] it is convenient to define the following parameter:

$$a = \left(\frac{\nu_Q^2}{\nu_L}\right)\left[I(I+1) - \frac{3}{4}\right] \quad (4.3)$$

The total breadth of the CT for a stationary powder sample, denoted $\Delta\nu_{CT}$, is given by equation (4.4):

$$\Delta\nu_{CT} = \frac{(25 + 22\eta_Q + \eta_Q^2)a}{144} \quad (4.4)$$

For samples spinning rapidly about the magic angle, the CT line width, denoted $\Delta\nu'_{CT}$, is given by equation (4.5):

$$\Delta\nu'_{CT} = \frac{a(6 + \eta_Q)^2}{504} \quad (4.5)$$

Values for the CT width for $I = 3/2$, 5/2, 7/2, and 9/2 and $\eta_Q = 0.0$, 0.5, and 1.0 are summarized in Table 4.1. Expressions for calculating the line shapes with and without magic-angle spinning (MAS) are given by Man[16] and Amoureux *et al.*[22] Two important points emerge from Table 4.1. First, peak widths are narrower for larger values of I. Second, the breadth of the CT decreases with increasing magnetic field. An experimental example of this inverse scaling of breadth with magnetic field strength may be seen for the ^{139}La NMR spectrum of a stationary sample of lanthanum chloride in Figure 4.1 ($I = 7/2$ for ^{139}La).[28] Note that, in this particular case, C_Q is 15.55 ± 0.05 MHz and $\eta_Q = 0.0$. Since the anisotropy in the axially symmetric lanthanum magnetic shielding tensor is only 50 ppm, the line shape is very nearly completely dominated by the second-order quadrupolar interaction.

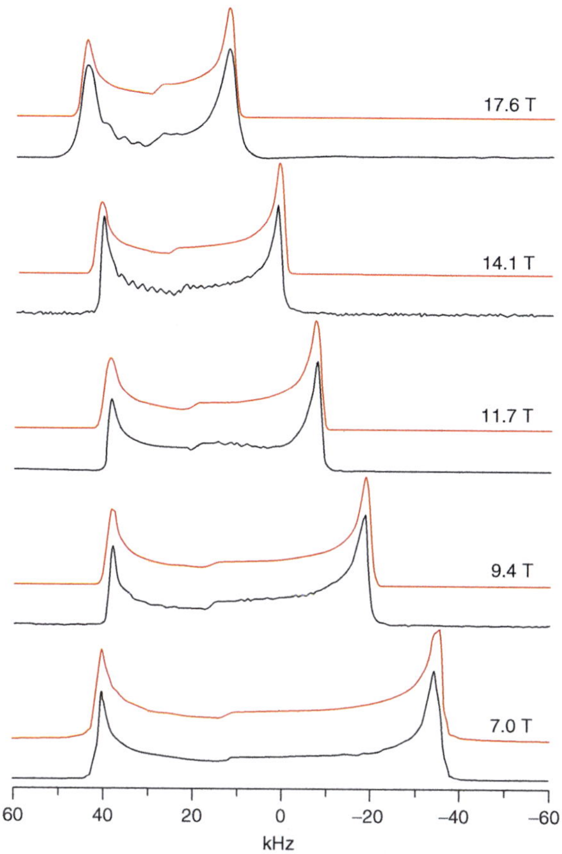

Figure 4.1. La-139 NMR spectra of stationary samples of powdered $LaCl_3$ acquired at five magnetic field strengths. Experimental spectra are plotted below the simulated spectra. (Reprinted from *Solid State Nucl. Magn. Reson.* **28**, K. J. Ooms, K. W. Feindel, M. J. Willans, R. E. Wasylishen, J. V. Hanna, K. J. Pike, M. E. Smith, 125. Copyright (2005), with permission from Elsevier.)

The CT is not perturbed to third order by the quadrupolar interaction;[16,29,30] however, the satellite transitions (STs) are. Third-order effects have been observed on the spectra of the STs.[29,31,32] A recent example illustrating the breakdown of the high-field approximation is shown in Figure 4.2.[31] The $^{185/187}$Re SSNMR spectrum of powdered $NaReO_4$, acquired at 11.75 T, could not be fit using second-order perturbation theory, and in fact, this level of theory incorrectly predicts even the number of discontinuities found in the CT region of the spectrum. An exact treatment of the quadrupolar interaction was required. In this same article,[31] useful guidelines are presented to help assess when second-order perturbation theory is valid, and what magnitudes of errors in C_Q and δ_{iso} may result when the high-field approximation breaks down. For example, for a spin-5/2 nucleus, errors in δ_{iso} become more problematic more rapidly than do errors in C_Q as the ν_Q/ν_L ratio increases. Furthermore, using second-order perturbation theory to fit a spectrum with a large ν_Q/ν_L ratio will always result in values of C_Q and δ_{iso} that are less than the true values. The exact ν_Q/ν_L ratio that "guarantees" a precise fit of the CT powder pattern using second-order perturbation theory depends on what level of error is tolerable in the parameters. The standard rule of thumb that ν_Q/ν_L must be less than 0.1 is a good one. This ratio may be pushed to an upper limit of 0.25 if larger errors are tolerable.

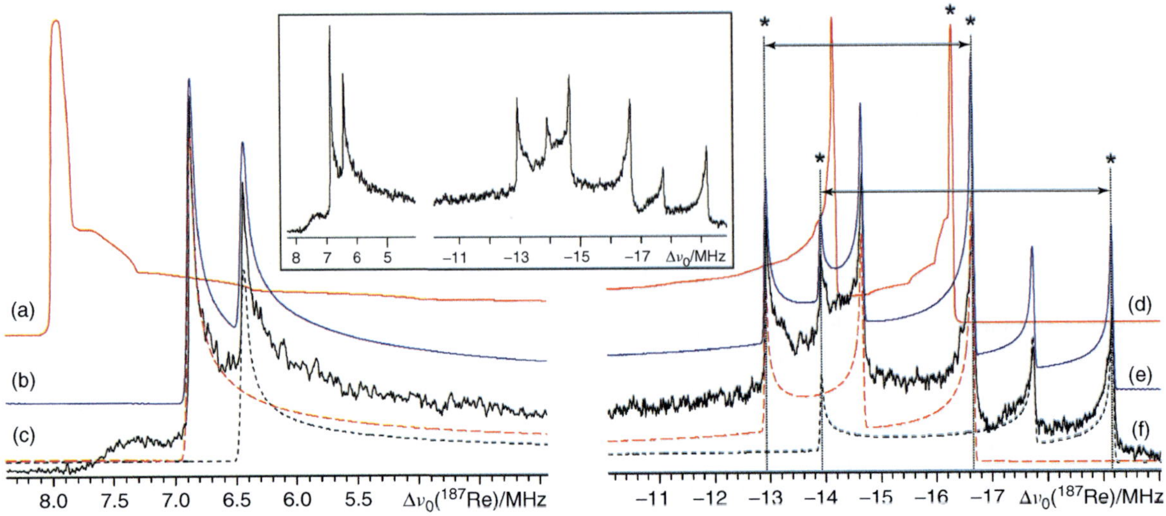

Figure 4.2. Second-order perturbation theory simulation (a, d), exact simulation (b, e), and experimental (c, f) $^{185/187}Re$ SSNMR spectrum of powdered $NaReO_4$ ($B_0 = 11.75\,T$). The second-order perturbation theory spectrum shows only the ^{187}Re signal for clarity. Below, the exact simulation line shapes associated with each of ^{185}Re and ^{187}Re are deconvoluted. The long dashed red trace is ^{187}Re, and the short dashed black trace is ^{185}Re. Low-frequency splittings (denoted by double-headed arrows) are not predicted by the second-order perturbation theory. The discontinuities due to the $m_I = 1/2 \leftrightarrow 3/2$ STs are marked by asterisks. In the inset, the experimental spectrum is shown using equivalent horizontal scaling for both edges of the spectrum. (C. M. Widdifield, A. D. Bain, D. L. Bryce, *Phys. Chem. Chem. Phys.* 2011, **13**, 12413. Reproduced by permission of the PCCP Owner Societies.)

4.3 INFLUENCE OF THE MAGNETIC SHIELDING INTERACTION ON SOLID-STATE NMR SPECTRA OF QUADRUPOLAR NUCLEI

Whenever a quadrupolar nucleus resides at a site where the EFG is nonzero, one can also expect the NMR spectrum of a stationary sample to depend on the magnetic shielding tensor. Jones *et al.*[33] were the first to describe NMR line shapes of quadrupolar nuclei under conditions where both quadrupolar and anisotropic shielding contribute to the line shape. Their treatment considered both tensors to be axially symmetric, with the principal components of each tensor being coincident. A few years later, Baugher *et al.*[34] provided a more general treatment that permitted both interactions to be nonaxially symmetric but again confined the principal components of the EFG and shielding tensor to be coincident. While it was recognized that the EFG and shielding tensors at a nucleus may not be coincident unless required by symmetry, experimental examples were not apparent for many years because of the relatively low applied magnetic fields available to early NMR experimentalists. In the early 1990s, ^{133}Cs NMR studies of powder and single-crystal samples of Cs_2CrO_4 provided examples where the cesium EFG and shielding tensors are noncoincident.[35,36] Since 1990, hundreds of examples have been documented in the literature; the interested reader should consult the references given in the remainder of this section.

It is instructive to first consider the expected CT NMR line shape for a case where crystal symmetry restricts the EFG and shielding tensors to be axially symmetric; hence coincident. First, the largest component of the EFG tensor, V_{ZZ}, could be coincident with δ_{33} (the low-frequency component of the CS tensor, i.e., $\kappa = +1.0$ following the Maryland notation). Second, the unique component of the CS tensor could be the high-frequency component (i.e., V_{ZZ} coincident with δ_{11}; $\kappa = -1.0$). Theoretical CT line shapes for these two situations are depicted in Figure 4.3.[37] Note that, for Case I, the inclusion of magnetic shielding anisotropy results in a second-order powder pattern that increases in breadth. On the other hand, for Case II, the breadth of

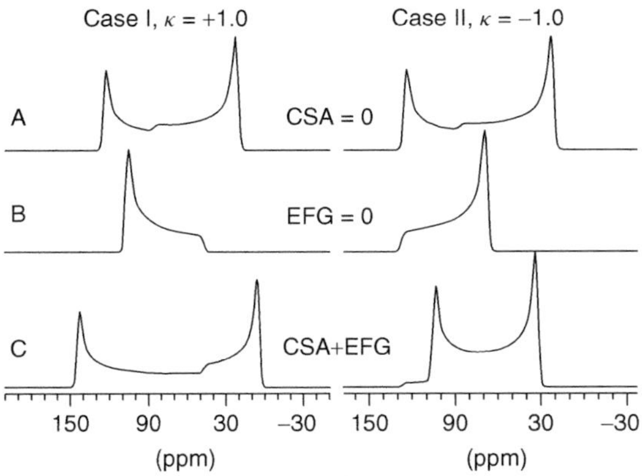

Figure 4.3. Simulations showing the effects of (a) the quadrupolar interaction, (b) chemical shift anisotropy, and (c) the combination of both effects for the central-transition solid-state NMR spectrum of a half-integer quadrupolar nucleus in a powdered sample. Parameters used include $C_Q = 6.0$ MHz; $\eta_Q = 0.0$; $\delta_{iso} = 88$ ppm; $\Omega = 60$ ppm; $\beta = 0°$. (Reprinted from *Magn. Reson.* **133**, R. W. Schurko, R. E. Wasylishen, A. D. Phillips, J. 388. Copyright (1998) with permission from Elsevier.)

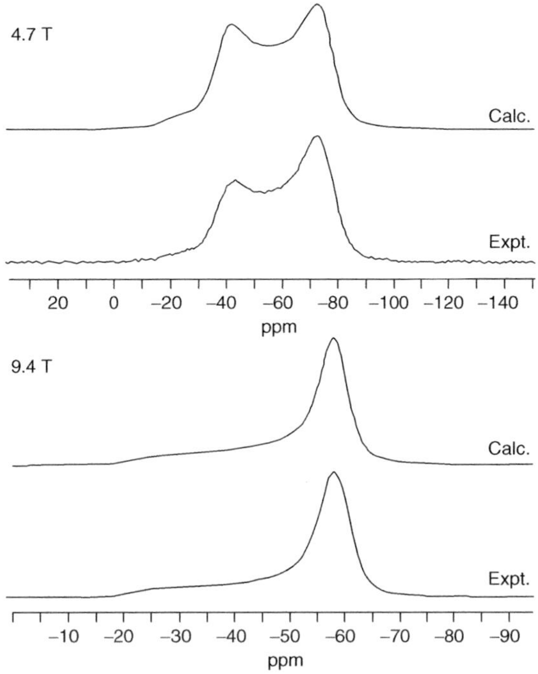

Figure 4.4. Experimental and simulated ^{63}Cu central transition NMR spectra of a stationary powdered sample of $K_3Cu(CN)_4$. Anisotropic quadrupolar and chemical shift tensors are required to reproduce the experimental data. (From S. Kroeker, R. E. Wasylishen, *Can. J. Chem.* 1999, **77**, 1962. © 199 Canadian Science Publishing or its licensors. Reproduced with permission.)

the powder pattern decreases. This behavior is readily understood if one examines the detailed angular dependence of the second-order quadrupolar powder pattern. The discontinuity on the high-frequency side of the spectrum (Figure 4.3a) corresponds to $\theta = 90°$ (i.e., V_{ZZ} perpendicular to the applied magnetic field), while the discontinuity on the right corresponds to $\theta = 41.8°$. The shoulder between the discontinuities, but closer to the high-frequency discontinuity, corresponds to $\theta = 0°$. For Case II, it is obvious that inclusion of anisotropic shielding will shift the discontinuity corresponding to $\theta = 90°$ to lower frequencies and shrink the breadth of the powder pattern. Similarly for Case I, it is clear that the shoulder corresponding to $\theta = 90°$ will be shifted to higher frequencies and increase the breadth of the powder pattern.

The ^{63}Cu NMR spectra of a stationary sample of $K_3Cu(CN)_4$ at 4.7 and 9.4 T (Figure 4.4)[38] provide striking examples of Case II above. Analysis of ^{63}Cu NMR spectra acquired with MAS indicated that $C_Q = 1.125$ MHz and $\delta_{iso} = -49$ ppm; a threefold axis of rotational symmetry at the copper nuclei ensures that $\eta_Q = 0.0$. At 9.4 T, anisotropic magnetic shielding ($\Omega = \delta_{11} - \delta_{33} \approx 42$ ppm and $\kappa = -1$) makes a much larger contribution to the CT breadth than does the second-order quadrupolar interaction. The powder pattern is also broadened by spin–spin coupling to ^{14}N.

Analysis of ^{11}B NMR spectra of a powder sample of trimesitylborane provided the first characterization of a boron magnetic shielding tensor.[39] Boron-11 NMR spectra of MAS samples provided $C_Q = 4.75$ MHz, $\eta_Q \approx 0.0$, and $\delta_{iso} = 77.4$ ppm (note, anisotropic shielding is averaged by MAS). Using these parameters to calculate the ^{11}B NMR spectrum of a stationary sample at 17.63 T provided a spectrum that is about one-third the breadth of the one observed (Figure 4.5). Spectral simulations indicated that the span of the boron shielding tensor is 121 ± 1 ppm.

A subtle example of Case II is provided by the ^{35}Cl NMR spectra of $SrCl_2 \cdot 6H_2O$ (Figure 4.6).[40] For this compound, analysis of ^{35}Cl NMR spectra with MAS yielded $C_Q = 3.91 \pm 0.05$ MHz, $\eta_Q = 0.0$, and

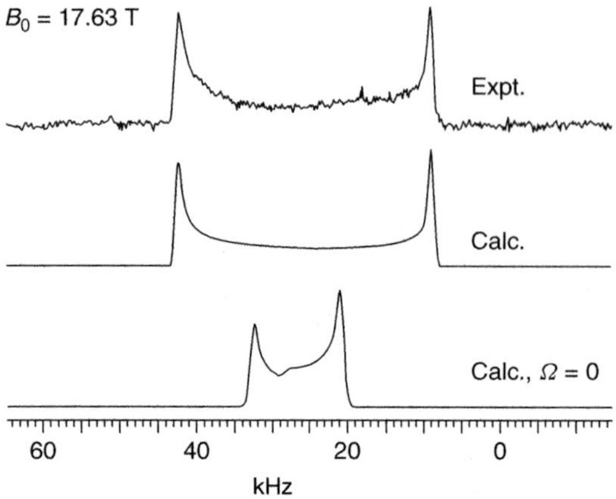

Figure 4.5. Experimental and simulated ^{11}B central transition NMR spectra of stationary powdered trimesitylborane. A span, Ω, of 121 ppm is used for the best-fit simulation. For comparison, a simulated spectrum obtained when the span is set to zero is also shown. (Reprinted with permission from D. L. Bryce, R. E. Wasylishen, M. Gee, *J. Phys. Chem. A* 2001, **105**, 3633. Copyright 2001 American Chemical Society.)

$\delta_{iso} = 90.4 \pm 1.0$ ppm. The experimental ^{35}Cl NMR spectrum of a stationary sample is narrower than that calculated with $\Omega = 0.0$, indicating that $\kappa = -1.0$.

Finally, we consider three examples of ^{115}In NMR spectra of stationary samples of indium triiodide–triarylphosphine 1 : 1 adducts acquired at 21.14 T. The spectra in Figure 4.7(a, b) provide clear examples of where there is no local symmetry requiring the EFG and shielding tensors to be coincident. To properly interpret the spectra, it is essential to acquire spectra at multiple magnetic field strengths, and, in particular, a high field is required to achieve good precision in the EFG and CS tensor parameters. On the other hand, for $I_3In(PPh_3)$ (Figure 4.7c), there is a C_3 symmetry axis along the In–P bond requiring that the indium EFG and shielding tensors be axially symmetric and coincident.

4.4 INFLUENCE OF SPIN–SPIN COUPLING INTERACTIONS ON SOLID-STATE NMR SPECTRA OF QUADRUPOLAR NUCLEI

Quadrupolar nuclei, as for spin-1/2 nuclei, may also be subject to spin–spin coupling interactions. These coupling interactions may be subdivided into direct dipolar coupling (D) and indirect nuclear spin–spin coupling (J) interactions. Perhaps the most frequently encountered effect of spin–spin coupling in the SSNMR spectra of quadrupolar nuclei is the direct dipolar coupling to nearby protons, which typically results in line broadening. As with spin-1/2 nuclei, high-power ^1H decoupling may be applied during signal acquisition to reduce or eliminate the typically unwanted broadening. In many cases, ^1H

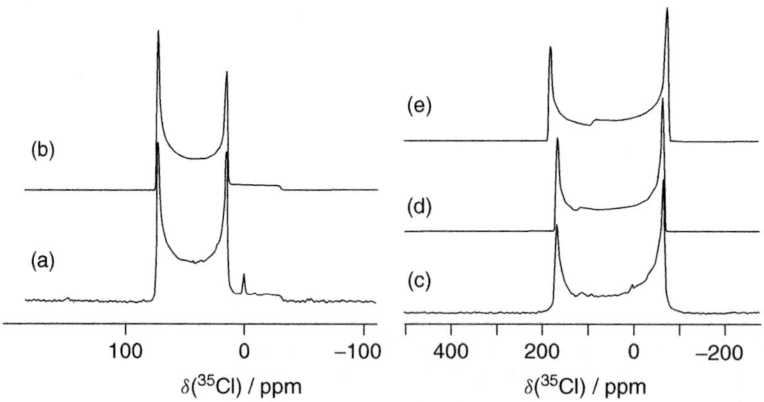

Figure 4.6. Chlorine-35 NMR spectra of powdered $SrCl_2 \cdot 6H_2O$. (a) Experimental spectrum of a MAS sample at 21.1 T. (b) Simulation of the spectrum shown in (a). (c) Experimental spectrum of a stationary sample at 21.1 T. (d) Best-fit simulation of the spectrum shown in (c). (e) Simulated spectrum where the span of the chlorine CS tensor has been set to 0 ppm. (Reproduced with permission from D. L. Bryce, E. B. Bultz, *Chem. Eur. J.* 2007, **13**, 4786. © Wiley-VCH, 2007.)

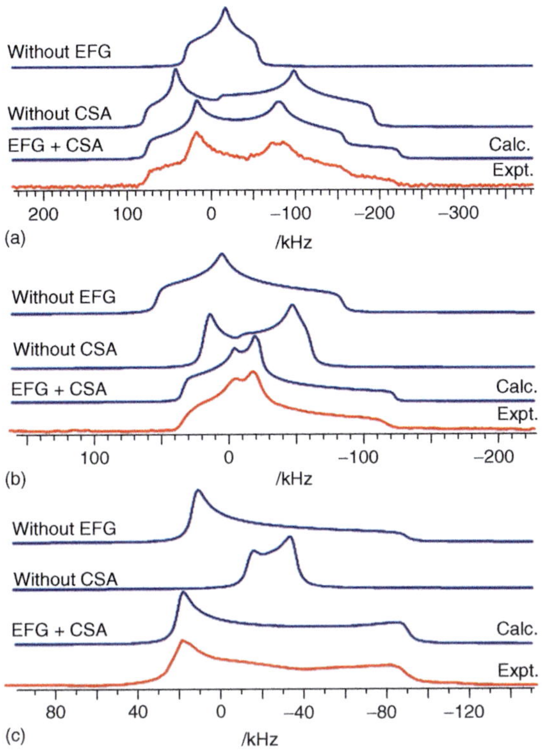

Figure 4.7. Experimental and simulated ^{115}In NMR spectra of stationary powdered samples of (a) $I_3In[P(o\text{-Anis})_3]$, (b) $I_3In[P(p\text{-Anis})_3]$, and (c) $I_3In(PPh_3)$ acquired at 21.14 T. Also shown separately are simulated spectra that demonstrate the separate effects of the indium CSA and EFG. (Reprinted with permission from F. Chen, G. Ma, G. M. Bernard, R. G. Cavell, R. McDonald, M. J. Ferguson, R. E. Wasylishen, *J. Am. Chem. Soc.* 2010, **132**, 5479. Copyright 2010 American Chemical Society.)

decoupling is not critical to obtain usable SSNMR spectra of quadrupolar nuclei since the value of the direct dipolar coupling constant can be orders of magnitude less than the breadth of the CT powder pattern. However, there are certainly cases where ^{1}H decoupling is critical to obtaining accurate line shapes. Below, we present briefly the fundamentals of the dipolar and J-coupling interactions and demonstrate with examples some of the possible effects one may encounter when observing SSNMR spectra of quadrupolar nuclei under stationary or MAS conditions.

Indirect nuclear spin–spin (J) coupling is a bilinear nuclear spin interaction that may be described with a

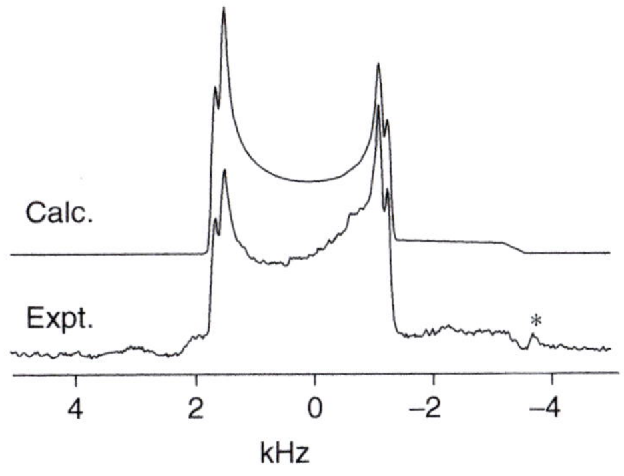

Figure 4.8. Experimental and simulated ^{17}O MAS NMR spectra of the orthorhombic polymorph of triphenylphosphine oxide-^{17}O (37.5%) acquired at 9.4 T. The asterisk denotes a spinning sideband. J-coupling between ^{17}O and ^{31}P is observed; $^1J(^{31}\text{P}, ^{17}\text{O}) = 150 \pm 20$ Hz. (Reprinted with permission from D. L. Bryce, K. Eichele, R. E. Wasylishen, *Inorg. Chem.* 2003, **42**, 5085. Copyright 2003 American Chemical Society.)

second-rank tensor containing up to nine independent components.[41] Most frequently, only the effects of the isotropic value of the tensor (J_{iso}) are observed experimentally. The isotropic coupling gives rise to the familiar splittings of resonances of spin-1/2 nuclei in solution NMR and SSNMR spectra. When the relaxation of a quadrupolar nucleus is sufficiently long and the value of a coupling constant with another nucleus is sufficiently large, analogous "splittings" of the familiar second-order quadrupolar line shape may be observed in the SSNMR spectrum of the quadrupolar nucleus. For example, presented in Figure 4.8 is the ^{17}O ($I = 5/2$) SSNMR spectrum of 37.5% ^{17}O-enriched PPh_3O (orthorhombic polymorph) under MAS conditions.[42] The oxygen is directly bonded to ^{31}P (100% natural abundance) and so the normal second-order quadrupolar line shape is split by the value of $^1J(^{31}\text{P},^{17}\text{O})$, which is 150 Hz in this case. The same value, within experimental error, was observed for a second polymorph of the same compound.

The case of the ^{121}Sb ($I = 5/2$) MAS NMR spectrum of $KSbF_6$ at 21.1 T provides a similar example (Figure 4.9) (Wasylishen and Terskikh, unpublished results), but in this case, the value of the relevant J-coupling constant, $^1J(^{121}\text{Sb}, ^{19}\text{F}) = 1882$ Hz, is comparable to the breadth of the second-order quadrupolar line shape observed under conditions of ^{19}F decoupling (see lower spectrum in Figure 4.9;

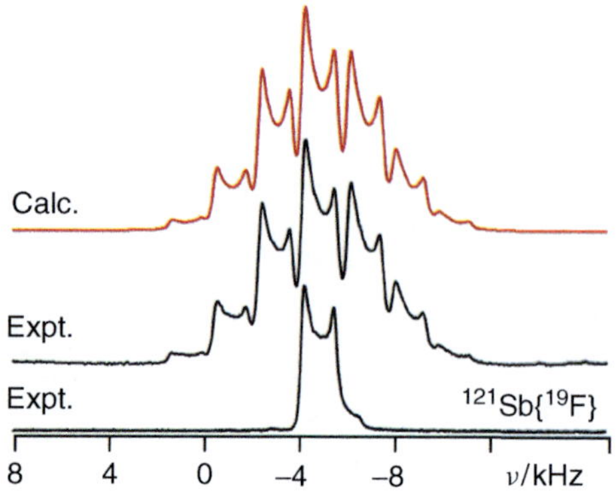

Figure 4.9. Solid-state ^{121}Sb NMR spectra of a MAS sample of $KSbF_6$ at 21.1 T. The lower experimental spectrum was recorded with ^{19}F decoupling, whereas the middle trace was not. (R. E. Wasylishen, V. V. Terskikh, unpublished results.)

about 1300 Hz separate the two high-intensity discontinuities; the CT breadth is approximately 2500 Hz edge to edge). Since there are six equivalent ^{19}F nuclei coupled to ^{121}Sb, one may obtain an excellent simulation of the upper experimental spectrum using the well-known "$n + 1$" approximation, giving in this case seven second-order powder patterns separated from each other by $^{1}J(^{121}\text{Sb}, ^{19}\text{F})$ with relative intensities of 1 : 6 : 15 : 20 : 15 : 6 : 1 (see upper experimental spectrum in Figure 4.9). Simulation of the ^{19}F-decoupled ^{121}Sb MAS NMR spectrum using standard second-order perturbation theory provides the following quadrupolar parameters: $C_Q(^{121}\text{Sb}) = 6.42$ MHz; $\eta_Q = 0.07$.

The ^{71}Ga SSNMR spectrum of $GaCl_3PPh_3$ under MAS conditions (Figure 4.10) provides an interesting example where the second-order quadrupolar broadening of the CT of ^{71}Ga is negligible at 21.1 T; rather, gallium chemical shift anisotropy (CSA) and $^{1}J(^{71}\text{Ga}, ^{31}\text{P})$ coupling are largely responsible for the observed spectral features (Chen *et al.*, unpublished results). The high applied magnetic field and the relatively high symmetry at gallium result in a negligible second-order quadrupolar broadening of the CT. The ^{71}Ga nucleus is characterized by a span of 150 ppm and a J-coupling to ^{31}P of 977 Hz. The spectrum is composed of a series of spinning sidebands due to CSA, each of which is split by the J-coupling to ^{31}P.

As described elsewhere,[43] the direct dipolar coupling is a magnetic interaction between pairs of nuclear magnetic moments. The direct dipolar coupling

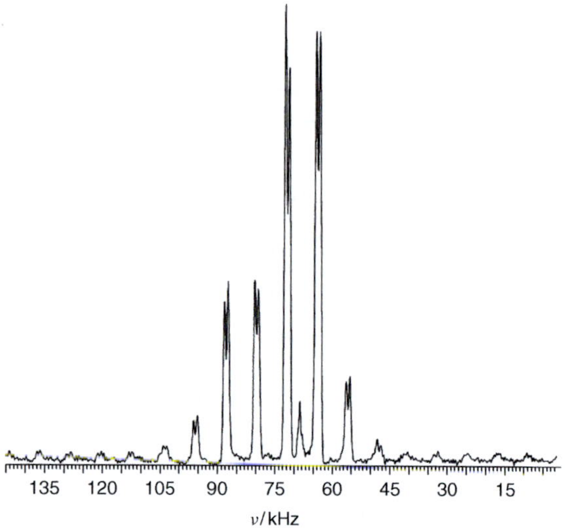

Figure 4.10. Gallium-71 NMR spectrum of powdered $GaCl_3PPh_3$ with 8 kHz MAS at 21.1 T. An analysis of the spectrum provides the values of the isotropic J-coupling constant and of the gallium chemical shift tensor span: $^{1}J(^{71}\text{Ga}, ^{31}\text{P}) = 977$ Hz and $\Omega(\text{Ga}) = 150$ ppm. (F. Chen, R. E. Wasylishen, V. V. Terskikh, unpublished results.)

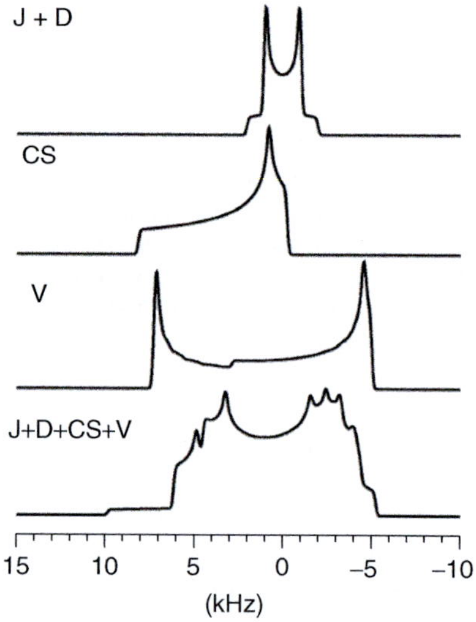

Figure 4.11. Simulated ^{17}O NMR spectra of a powdered sample of the monoclinic polymorph of triphenylphosphine oxide that show the individual contributions of indirect and dipolar coupling (J + D), chemical shift anisotropy (CS), and quadrupolar coupling (V) to the total NMR line shape (bottom trace). (Reprinted with permission from D. L. Bryce, K. Eichele, R. E. Wasylishen, *Inorg. Chem.* 2003, **42**, 5085. Copyright 2003 American Chemical Society.)

constant R_{DD} is defined as follows:

$$R_{DD} = \left(\frac{\mu_0}{4\pi}\right)\left(\frac{\gamma_1\gamma_2\hbar}{2\pi}\right)\langle r_{1,2}^{-3}\rangle \qquad (4.6)$$

where γ_1 and γ_2 are the magnetogyric ratios of the two nuclei, $\hbar$ is the reduced Planck constant, $r_{1,2}$ is the internuclear distance, and μ_0 is the permeability of free space. The observed effective dipolar coupling R_{eff} also always contains a contribution from the anisotropy of the J-coupling tensor ΔJ.[43]

The ^{17}O NMR spectrum of a stationary sample of 35% ^{17}O-enriched PPh_3O provides a useful example where dipolar coupling, J-coupling, CSA, and the quadrupolar interaction all contribute significantly to the observed line shape (Figure 4.11).[42] Despite the number of variables involved in fitting the spectrum, all parameters may be reliably obtained. For example, the values of C_Q, η_Q, δ_{iso}, and $^1J(^{31}P, ^{17}O)_{iso}$ are obtained unambiguously from the ^{17}O MAS NMR spectrum. The value of R_{DD} is inferred from that calculated from the known internuclear distance obtained from X-ray crystallography. Relative tensor orientations are restricted by the symmetry along the P–O bond, and the values of the span (Ω) and skew (κ) of the CS tensor are the final adjustable parameters when simulating the spectrum of the stationary sample. The sensitivity of the parameters to polymorphism in monoclinic and orthorhombic polymorphs of PPh_3O has been discussed.[42] Parameters obtained through the analysis of MAS and stationary powder samples have also been confirmed through ^{17}O DOR NMR experiments.[44]

In contrast to J-coupling, the isotropic value of the direct dipolar coupling tensor is zero. For this reason, direct dipolar coupling effects are not observed directly in solution NMR spectroscopy and are often subtle in fast MAS SSNMR spectra of spin-1/2 nuclei coupled to other spin-1/2 nuclei. However, when a quadrupolar nucleus with a nonzero C_Q is dipolar-coupled to another nucleus, MAS cannot

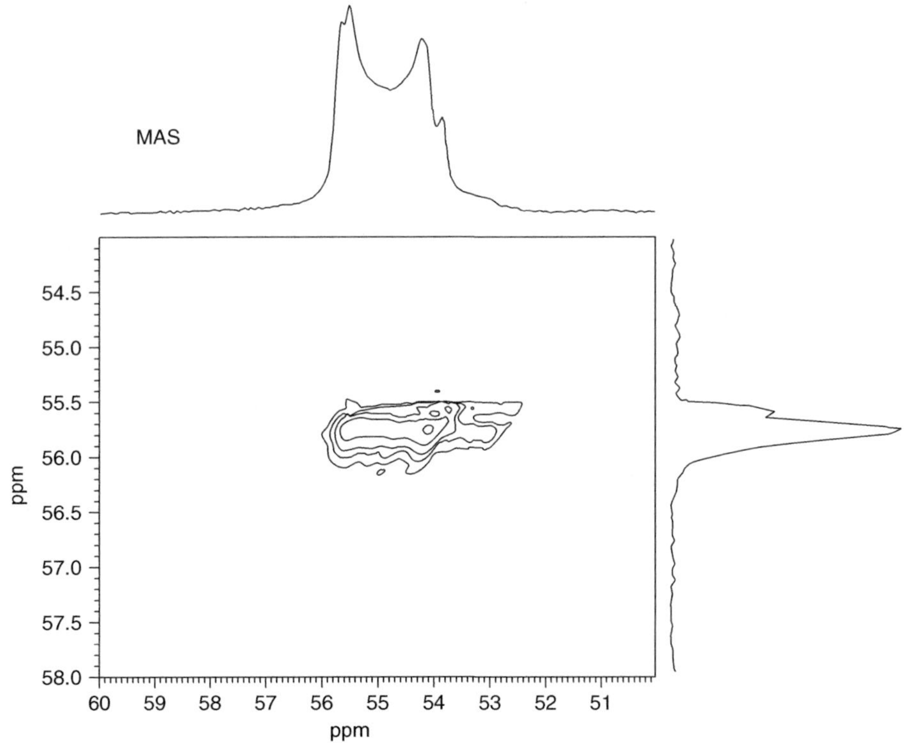

Figure 4.12. 2D ^{11}B MQMAS spectrum of triethanolamineborate obtained at 11.75 T. The ^{11}B MAS NMR spectrum is shown above, and the isotropic projection is shown on the right. (Reprinted with permission from *Chem. Phys. Lett.* **313**, G. Wu, K. Yamada, 519. Copyright (1999) Elsevier.)

completely eliminate the effects of this coupling in the NMR spectra of the coupled nuclei. Perhaps the most well-known example of this effect is the observation of residual dipolar coupling (RDC) in the ^{13}C ($I = 1/2$) MAS SSNMR spectra of $^{13}C-^{14}N$ spin pairs. However, when the high-field approximation applies, it has been shown that RDCs do not generally manifest themselves in the CT MQMAS NMR spectrum of a quadrupolar nucleus, which is dipolar-coupled to a spin-1/2 nucleus.[45] Effects are, however, predicted for the STs.

When two quadrupolar nuclei are dipolar coupled to each other, RDCs will be observed in their MAS and MQMAS SSNMR spectra. Often the effect may be an unresolved broadening, depending on the relative magnitudes of the various interactions. An illustrative example where the RDC is clearly resolved is shown in Figure 4.12. The ^{11}B MAS and MQMAS NMR spectra of ethanolamine borate show residual $^{14}N-^{11}B$ heteronuclear dipolar coupling.[46] Analogous effects may be observed in the case of a homonuclear quadrupolar spin pair. Presented in Figure 4.13 are ^{11}B MQMAS SSNMR spectra of bispinacolatodiborane, where numerical simulations have been used to simulate the line shapes in both dimensions.[47]

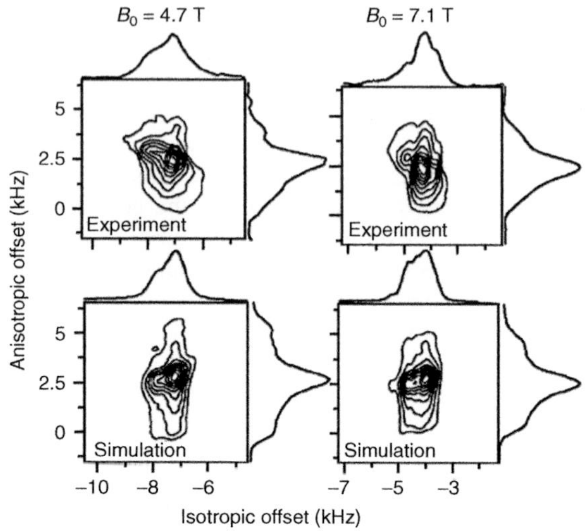

Figure 4.13. Experimental ^{11}B MQMAS NMR line shapes for bispinacolatodiborane at two different magnetic fields and corresponding best-fit simulations. (Reprinted with permission from S. Wi, L. Frydman, *J. Chem. Phys.* 2000, **112**, 3248. Copyright 2000, American Institute of Physics.)

Shown in Figure 4.14 is an example of the effects of homonuclear dipolar coupling on the deuterium MAS NMR spectrum of $Os(D_2)(Cl)_2(CO)(P^iPr_3)_2$.[48] RDC between the two deuterons manifests itself in the line shape of each of the spinning sidebands. Average Hamiltonian theory was used to understand these effects and to simulate the spectrum. Such an analysis allowed for the measurement of the homonuclear deuterium–deuterium dipolar coupling constant and therefore the deuterium–deuterium internuclear distance.

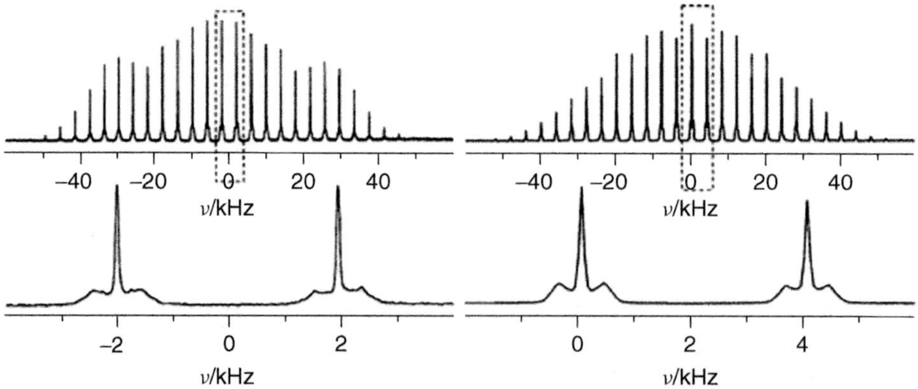

Figure 4.14. Experimental deuterium MAS NMR spectrum of $Os(D_2)(Cl)_2(CO)(P^iPr_3)_2$(top left) and enlargement of the central two sidebands (bottom left). The sidebands exhibit a Pake-like shape. Numerical simulations of the experimental spectra are shown on the right-hand side. (G. Facey, D. Gusev, R. H. Morris, S. Macholl, G. Buntkowsky, *Phys. Chem. Chem. Phys.* 2000, **2**, 935. Reproduced by permission of the PCCP Owner Societies.)

4.5 CONCLUSIONS

In principle, quadrupolar nuclei are almost always affected by interactions in addition to the quadrupolar interaction. Whether such other interactions need to be considered when analyzing the spectrum of a quadrupolar nucleus depends, of course, on the particular nuclear environment, the nuclear spin characteristics of the observed and nearby nuclei, and the type of experiment performed. In this chapter, we have illustrated using examples various possible effects from nuclear magnetic shielding, direct dipolar, and indirect nuclear spin–spin coupling interactions on the spectra of quadrupolar nuclei in powdered samples under simple MAS and stationary conditions. It is important to be aware of the possible effects of these interactions and of "high-order" quadrupolar interactions when analyzing SSNMR spectra of quadrupolar nuclei. The most common effect to be observed is likely to be that of anisotropic magnetic shielding, resulting in an interplay between the quadrupolar interaction and CSA in the NMR spectrum. It is very often advantageous in sorting out the competing effects of various interactions to acquire the NMR spectra at more than one applied magnetic field.

RELATED ARTICLES IN THE ENCYCLOPEDIA OF MAGNETIC RESONANCE

Chemical Shift Tensors

Dipolar and Indirect Coupling: Basics

Dipolar and Indirect Coupling Tensors in Solids

Indirect Nuclear Spin-Spin Coupling Tensors

Quadrupolar Nuclei in Liquid Samples

Tensor Interplay

REFERENCES

1. L. A. O'Dell and R. W. Schurko, *Chem. Phys. Lett.*, 2008, **464**, 97.
2. I. Hung and Z. Gan, *J. Magn. Reson.*, 2010, **204**, 256.
3. A. J. Rossini, H. Hamaed, and R. W. Schurko, *J. Magn. Reson.*, 2010, **206**, 32.
4. J. V. Hanna and M. E. Smith, *Solid State Nucl. Magn. Reson.*, 2010, **38**, 1.
5. S. Bai, W. Wang, and C. Dybowski, *Anal. Chem.*, 2010, **82**, 4917.
6. C. A. Fyfe, *Solid State NMR for Chemists*, C.F.C. Press: Guelph, Canada, 1983, Chapter 9.
7. J. Zhu, E. Ye, V. Terskikh, and G. Wu, *Angew. Chem. Int. Ed.*, 2010, **49**, 8399.
8. L. A. O'Dell and I. L. Moudrakovski, *J. Magn. Reson.*, 2010, **207**, 345.
9. I. L. Moudrakovski and J. A. Ripmeester, *J. Phys. Chem. B*, 2007, **111**, 491.
10. D. L. Bryce, *Dalton Trans.*, 2010, **39**, 8593.
11. J. Zhu, N. Trefiak, T. K. Woo, and Y. Huang, *J. Phys. Chem. C*, 2009, **113**, 10029.
12. V. K. Michaelis and S. Kroeker, *J. Phys. Chem. C*, 2010, **114**, 21736.
13. A. J. Rossini, I. Hung, S. A. Johnson, C. Slebodnick, M. Mensch, P. A. Deck, and R. W. Schurko, *J. Am. Chem. Soc.*, 2010, **132**, 18301.
14. F. Chen, G. Ma, G. M. Bernard, R. G. Cavell, R. McDonald, M. J. Ferguson, and R. E. Wasylishen, *J. Am. Chem. Soc.*, 2010, **132**, 5479.
15. A. Sutrisno, C. Lu, R. H. Lipson, and Y. Huang, *J. Phys. Chem. C*, 2009, **113**, 21196.
16. P. P. Man, Quadrupolar interactions, in *Encyclopedia of Magnetic Resonance*, eds R. K. Harris and R. E. Wasylishen, John Wiley & Sons: Chichester, 2007. DOI: 10.1002/9780470034590.emrstm0429.
17. P. Pyykkö, *Mol. Phys.*, 2008, **106**, 1965.
18. A. J. Vega, Quadrupolar nuclei in solids, in *Encyclopedia of Magnetic Resonance*, eds R. K. Harris and R. E. Wasylishen, John Wiley & Sons: Chichester, 2007. DOI: 10.1002/9780470034590.emrstm0431.
19. S. E. Ashbrook and S. Wimperis, in *NMR Crystallography*, eds R. K. Harris, R. E. Wasylishen, and M. J. Duer, Wiley: Chichester, 2009, Chapter 16.
20. A. Abragam, *The Principles of Nuclear Magnetism*, Oxford Science Publications: Oxford, 1961.
21. M. Duer, *Introduction to Solid-State NMR Spectroscopy*, Blackwell Publishing: Oxford, 2005.
22. J. P. Amoureux, C. Fernandez, and P. Granger, in *Multinuclear Magnetic Resonance in Liquids and Solids*, eds P. Granger and R. K. Harris, Kluwer Academic Publishers: London, 1990, p. 409, Chapter 22.

23. K. J. D. MacKenzie and M. E. Smith, *Multinuclear Solid-State NMR of Inorganic Materials*, Elsevier Science: Kidlington, Oxford, 2002.

24. D. Freude and J. Haase, in *NMR, Basic Principles and Progress*, eds P. Diehl, E. Fluck, H. Günther, R. Kosfeld, and J. Seelig, Springer-Verlag: New York, 1993, Vol. 29, pp. 1.

25. D. Freude, in *Encyclopedia of Analytical Chemistry*, ed R. A. Meyers, Wiley: Chichester, 2000, pp. 12188.

26. P. P. Man, in *Encyclopedia of Analytical Chemistry*, ed R. A. Meyers, Wiley: Chichester, 2000, pp. 12224.

27. A. P. M. Kentgens, *Geoderma*, 1997, **80**, 271.

28. K. J. Ooms, K. W. Feindel, M. J. Willans, R. E. Wasylishen, J. V. Hanna, K. J. Pike, and M. E. Smith, *Solid State Nucl. Magn. Reson.*, 2005, **28**, 125.

29. Z. Gan, P. Srinivasan, J. R. Quine, S. Steuernagel, and B. Knott, *Chem. Phys. Lett.*, 2003, **367**, 163.

30. A. D. Bain, *J. Magn. Reson.*, 2006, **179**, 308.

31. C. M. Widdifield, A. D. Bain, and D. L. Bryce, *Phys. Chem. Chem. Phys.*, 2011, **13**, 12413.

32. C. M. Widdifield and D. L. Bryce, *J. Phys. Chem. A*, 2010, **114**, 10810.

33. W. H. Jones, T. P. Graham, and R. G. Barnes, *Phys. Rev.*, 1963, **132**, 1898.

34. J. F. Baugher, P. C. Taylor, T. Oja, and P. J. Bray, *J. Chem. Phys.*, 1969, **50**, 4914.

35. W. P. Power, R. E. Wasylishen, S. Mooibroek, B. A. Pettitt, and W. Danchura, *J. Phys. Chem.*, 1990, **94**, 591.

36. W. P. Power, S. Mooibroek, R. E. Wasylishen, and T. S. Cameron, *J. Phys. Chem.*, 1994, **98**, 1552.

37. R. W. Schurko, R. E. Wasylishen, and A. D. Phillips, *J. Magn. Reson.*, 1998, **133**, 388.

38. S. Kroeker and R. E. Wasylishen, *Can. J. Chem.*, 1999, **77**, 1962.

39. D. L. Bryce, R. E. Wasylishen, and M. Gee, *J. Phys. Chem. A.*, 2001, **105**, 3633.

40. D. L. Bryce and E. B. Bultz, *Chem. Eur. J.*, 2007, **13**, 4786.

41. R. E. Wasylishen, Indirect nuclear spin-spin coupling tensors, in *Encyclopedia of Magnetic Resonance*, eds R. K. Harris and R. E. Wasylishen, John Wiley & Sons: Chichester, 2007. DOI: 10.1002/9780470034590.emrstm0237.

42. D. L. Bryce, K. Eichele, and R. E. Wasylishen, *Inorg. Chem.*, 2003, **42**, 5085.

43. R. E. Wasylishen, in *NMR Crystallography*, eds R. K. Harris, R. E. Wasylishen, and M. J. Duer, Wiley: Chichester, 2009, pp. 127, Chapter 10.

44. I. Hung, A. Wong, A. P. Howes, T. Anupõld, J. Past, A. Samoson, X. Mo, G. Wu, M. E. Smith, S. P. Brown, and R. Dupree, *J. Magn. Reson.*, 2007, **188**, 246.

45. G. Wu and R. E. Wasylishen, *Mol. Phys.*, 1998, **95**, 1177.

46. G. Wu and K. Yamada, *Chem. Phys. Lett.*, 1999, **313**, 519.

47. S. Wi and L. Frydman, *J. Chem. Phys.*, 2000, **112**, 3248.

48. G. Facey, D. Gusev, R. H. Morris, S. Macholl, and G. Buntkowsky, *Phys. Chem. Chem. Phys.*, 2000, **2**, 935.

PART B
Advanced Techniques

Chapter 5

Acquisition of Wideline Solid-State NMR Spectra of Quadrupolar Nuclei

Robert W. Schurko

University of Windsor, Department of Chemistry and Biochemistry, Windsor, ON N9B 3P4, Canada

5.1 INTRODUCTION

Wideline NMR spectroscopy is a term that has been long applied to NMR experiments involving the acquisition of broad anisotropic powder patterns arising from homonuclear dipole–dipole coupling (e.g., ^{1}H, ^{19}F), large chemical shift anisotropies (e.g., ^{207}Pb, ^{195}Pt, etc.), and even quadrupolar interactions (e.g., ^{2}H). Typically, wideline patterns range from tens of kHz to about 300–400 kHz in breadth and were named as such owing to their large pattern breadths ("line widths") compared to those encountered in most solution NMR experiments. However, there are numerous instances for both spin $I = 1/2$ and quadrupolar nuclei ($I > 1/2$) where their NMR powder patterns may range from hundreds of kHz to many MHz in breadth, putting them well beyond the standard uniform excitation and detection bandwidths of most solid-state NMR (SSNMR) probes.[1] In many cases, the use of composite or other specially shaped pulses is insufficient for uniform excitation. The signal to noise (S/N) ratios of such spectra are extremely low since the total integrated signal intensity is spread over very broad regions; hence, the solution to this problem is not to simply reduce the Q of the probe. In addition, when dealing with quadrupolar nuclei, the nutation behavior of isochromats corresponding to different tensor orientations can be very complex, and uniform excitation of their patterns is further complicated. The combination of these factors makes the acquisition of such spectra very challenging; however, there are techniques, pulse sequences, and hardware that have been developed to address this problem. We have previously suggested the term *ultra-wideline* (UW) *NMR spectroscopy* to describe this set of methodologies,[2] which must be used in cases where standard high-power rectangular pulses are insufficient for the acquisition of uniform, distortionless powder patterns.

NMR of Quadrupolar Nuclei in Solid Materials
Edited by Roderick E. Wasylishen, Sharon E. Ashbrook and Stephen Wimperis
 ISBN: 978-0-470-97398-1

5.2 SYSTEMS FOR STUDY BY ULTRA-WIDELINE NMR SPECTROSCOPY

UW NMR methodologies, in combination with hardware advances and the ever-increasing availability of ultra-high field (UHF) NMR spectrometers, have opened up SSNMR experimentation to many different nuclei in a multitude of materials. UW NMR spectroscopy finds use in applications to four broad categories of NMR nuclides, including (i) half-integer spin quadrupolar nuclei, (ii) integer-spin quadrupolar nuclei, (iii) nuclei in systems with unpaired or conduction electrons, and (iv) spin $I = 1/2$ nuclei of heavy metal elements with large chemical shift anisotropies. Since the focus of this chapter is on quadrupolar nuclei, spin $I = 1/2$ nuclei will not be discussed. Before proceeding to some specific examples of UW NMR spectroscopy, some general considerations for each category are outlined.

Half-integer spin quadrupolar nuclei, which constitute approximately 73% of the NMR-active isotopes in the periodic table, have NMR patterns that may be influenced by both the first-order quadrupolar interaction (FOQI) and second-order quadrupolar interaction (SOQI). There is much interest in acquiring SSNMR spectra of quadrupolar nuclei, since information on the principal components of the electric field gradient (EFG) tensor can be obtained, via comparison of experimental data to spectral simulations.[3] The EFG tensor is a symmetric, traceless, second-rank tensor, which has three principal components defined such that $|V_{33}| \geq |V_{22}| \geq |V_{11}|$, which are more commonly expressed as the quadrupolar coupling constant, $C_Q = eQV_{33}/h$, and the asymmetry parameter, $\eta_Q = (V_{11} - V_{22})/V_{33}$ (here, eQ is the nuclear electric quadrupole moment and h is the Planck constant). The values of C_Q and η_Q are dependent on the size of the quadrupolar interaction and axial symmetry of the EFG tensor, respectively, which are in turn largely determined by the spherical and axial symmetry of the ground-state electron distribution about the nucleus in question. In cases where the quadrupolar interaction is small (i.e., smaller values of C_Q), patterns are typically less than 150 kHz in breadth (for most nuclei), and standard NMR pulse sequences may be applied. For such cases, it is often possible to detect not only the central-transition ($+1/2 \leftrightarrow -1/2$, CT) pattern, which is influenced only by the SOQI, but also the satellite-transition (ST) patterns, which are much broader since they are also influenced by the much larger FOQI. For larger values of C_Q, the ST powder patterns often broadened beyond detection, and only the CT powder pattern is observed. However, the CT pattern for many half-integer quadrupoles can be several MHz in breadth. We note that UW NMR spectra of quadrupolar nuclei may also be influenced by chemical shift anisotropy (CSA, see below for definitions); notably, the CT patterns of half-integer spin nuclei can be particularly complex, especially when the magnitudes of the broadening from the SOQI and CSA are similar in magnitude. These spectra depend not only on the CS and EFG tensor parameters but also on the relative orientation of these tensors, which is normally defined by three Euler angles, α, β, and γ.[4] Fortunately, spectra of most of the half-integer spin quadrupolar nuclei can be acquired using the UW methods discussed herein, which address issues of nonuniform excitation, low S/N and long experiment times.

The anisotropic CT and ST powder patterns can be calculated using average Hamiltonian theory, via the CT and ST Hamiltonians

$$\hat{H}_{\mathrm{CT}} = \hat{H}_{\mathrm{Q}}^{(2)}(\Theta^{\mathrm{Q}}) + \hat{H}_{\mathrm{CS}}(\Theta^{\mathrm{CS}}) \tag{5.1}$$

and

$$\hat{H}_{\mathrm{ST}} = \hat{H}_{\mathrm{Q}}^{(1)}(\Theta^{\mathrm{Q}}) + \hat{H}_{\mathrm{Q}}^{(2)}(\Theta^{\mathrm{Q}}) + \hat{H}_{\mathrm{CS}}(\Theta^{\mathrm{CS}}) \tag{5.2}$$

where the Hamiltonians for the FOQI and SOQI are

$$\hat{H}_{\mathrm{Q}}^{(1)}(\Theta^{\mathrm{Q}}) = C_{\mathrm{Q}}\hat{T}_{20}^{\mathrm{Q}}\hat{R}_{20}^{\mathrm{Q}}(\Theta^{\mathrm{Q}}) \tag{5.3}$$

and

$$\hat{H}_{\mathrm{Q}}^{(2)}(\Theta^{\mathrm{Q}}) = -\frac{C_{\mathrm{Q}}^2}{\nu_0}\sum_{m=-2}^{2}\frac{1}{m}[\hat{T}_{2m}^{\mathrm{Q}}, \hat{T}_{2-m}^{\mathrm{Q}}]\hat{R}_{2m}^{\mathrm{Q}}(\Theta^{\mathrm{Q}}) \times \hat{R}_{2-m}^{\mathrm{Q}}(\Theta^{\mathrm{Q}}), \tag{5.4}$$

and the chemical shift (CS) Hamiltonian is

$$\hat{H}_{\mathrm{CS}}(\Theta^{\mathrm{CS}}) = \gamma\hat{H}_{\mathrm{iso}} + \gamma\hat{T}_{20}^{\mathrm{CS}}\hat{R}_{20}^{\mathrm{CS}}(\Theta^{\mathrm{CS}}). \tag{5.5}$$

The Θ^{Q} and Θ^{CS} are sets of Euler angles that define the orientations of the EFG and CS tensors (the latter of which is discussed below) with respect to the external magnetic field, B_0, the $\hat{R}_{2m}(\Theta)$ are orientation-dependent irreducible spherical tensor elements, and the $\hat{T}_{2m}$ are the irreducible spherical tensor spin operators.[5] Consideration of these Hamiltonians are important in our discussion of acquisition of distortion-free powder patterns that follows.

NMR spectra of *integer-spin quadrupolar nuclei* are made up of only ST powder patterns (see equation (5.2)), since there is no CT. With the exceptions of ^{2}H and ^{6}Li, which have $I = 1$ and small nuclear quadrupole moments, and accordingly small quadrupolar interactions, the NMR spectra of integer-spin quadrupolar nuclei are very challenging to acquire, since the ST powder patterns are dominated by the FOQI term in equation (5.2), and are typically spread across many MHz. Of the remaining NMR-active integer-spin nuclei, the two most important are undoubtedly ^{14}N and ^{10}B, the former for its ubiquity in all areas of chemistry (notably organic and biochemistry) and the latter for its presence in materials such as glasses and minerals. In particular, ^{14}N UW NMR presents a considerable number of challenges to spectroscopists, not only because of its broad patterns but also because of its low gyromagnetic ratio (γ) and correspondingly low Larmor frequencies.[6] Again, UW NMR techniques are well suited for the investigation of such nuclei; to date, the majority of NMR experiments on integer-spin nuclei have been conducted on systems in which these nuclei are in sites of high spherical/Platonic symmetry and therefore possess small quadrupolar interactions.

In many instances, SSNMR experiments are conducted on nuclei in *systems with unpaired or conduction electrons*. In the former case, which includes paramagnetic and ferromagnetic systems, spectra may be influenced by contact and/or pseudocontact shifts. In the latter case, which includes most metals and alloys, it is often possible to observe Knight shifts and associated anisotropies (KSAs). In both cases, it is often very difficult to acquire SSNMR spectra because of large frequency shifts, extreme pattern broadening, and/or extremely fast nuclear spin relaxation processes. The theoretical basis for each of these classes of interactions, and the effects of the unpaired electrons and magnetic susceptibility on relaxation, are well described elsewhere,[7,8] and will not be dwelt upon here. Rather, we note that spectroscopists who acquire the broad NMR patterns associated with such systems are often faced with the exact same problems described above for diamagnetic systems and may apply many of the techniques outlined below.

5.3 METHODOLOGIES FOR ACQUISITION OF ULTRA-WIDELINE NMR SPECTRA

5.3.1 Piecewise Spectral Acquisition

Piecewise (or *stepwise* or *point-by-point*) *spectral acquisition* is utilized in cases where either the excitation bandwidth of a short, high-power, rectangular pulse is insufficient to obtain a uniform powder pattern and/or the detection bandwidth of the probe is limited by its Q value. The term "piecewise" is used to refer generically to techniques in which the total powder pattern is obtained by conducting a series of experiments in which individual points or subspectra are acquired. This can be accomplished in a number of different ways, but all piecewise acquisitions either involve sweeping or stepping the magnetic field in uniform increments while holding the transmitter frequency constant (*field stepping*), or stepping the transmitter frequency in uniform increments at a constant magnetic field (*frequency stepping*). Frequency-stepped acquisitions are more common, since field stepping/sweeping requires specialized hardware that is not available on most modern NMR spectrometers. The nature (i.e., intensities, bandwidths, etc.) of the individual points or subspectra, as well as their frequency of separation, depend predominantly on the excitation and refocusing powers and associated bandwidths of the rf pulses.

Many of the early instances of field-stepped UW NMR spectroscopy were demonstrated by Slichter and co-workers, who utilized this technique to acquire ^{195}Pt UW NMR spectra ($I = 1/2$) of nanoparticles of various sizes.[9] An early illustrative example of frequency-stepped spectroscopy applied to quadrupolar nuclei was reported by Smith and Bastow, where ^{91}Zr ($I = 5/2$) UW NMR spectroscopy was applied to a series of zirconias that were fully and partially stabilized with MgO (Figure 5.1).[10] To acquire these spectra, spin-echo experiments were conducted over a series of equally spaced transmitter frequencies at a fixed magnetic field. For each experiment, the free induction decay (FID) was Fourier transformed and the intensity of the echo spectrum measured; the complete spectrum was generated by plotting the intensity as a function of transmitter frequency. Each of the distinct Zr sites in the unique forms of zirconia are easily identified and differentiated.

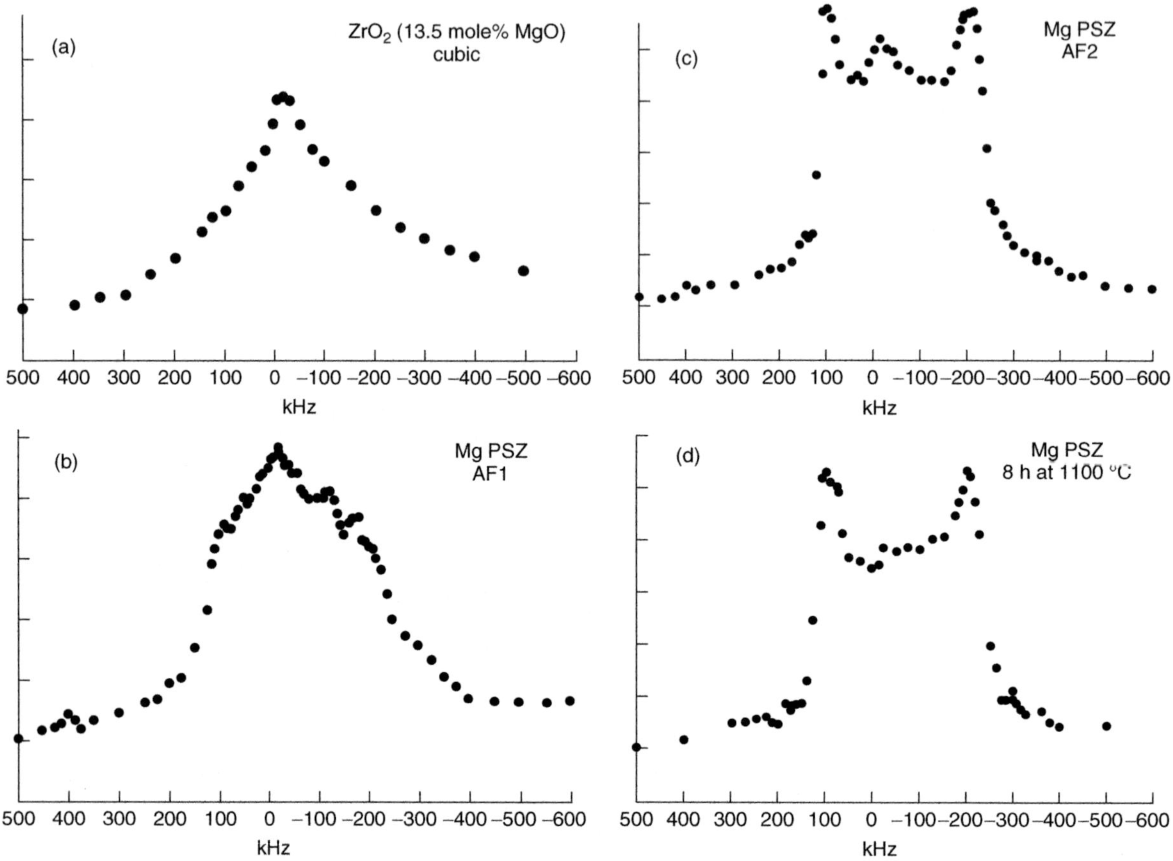

Figure 5.1. ^{91}Zr frequency-stepped UW SSNMR spectra acquired at 9.4 T for (a) ZrO_2 with 13.5 mol% MgO, (b) magnesia partially stabilized zirconia (Mg PSZ) AF1 (as fired #1, fast cooling), (c) Mg PSZ AF2 (slower cooling), and (d) Mg PSZ AF2 annealed for 8 h at 1100 °C. (Figure adapted from Ref. 10. © Elsevier, 1992.)

However, field- and frequency-stepped echo techniques have major limitations. First, the spectral resolution is limited because of the spacings of the transmitter offsets. Second, to increase spectral resolution (which may be necessary for obtaining important spectral discontinuities or for differentiation of multiple sites), one must acquire additional experiments, which adds to the total experimental time. Third, in the absence of automated tuning or other specialized hardware, the probe must be tuned between experiments, which adds additional time and user requirements. Nonetheless, these techniques are used to this day to acquire UW SSNMR spectra for a broad range of quadrupolar nuclei.

In 1995, Massiot, Farnan, and co-workers suggested an alternative means for faster acquisition of UW NMR spectra.[11] In their work on ^{69}Ga and ^{71}Ga (both $I = 3/2$) SSNMR of β-Ga_2O_3, from ^{71}Ga SSNMR data, the ^{69}Ga CT powder patterns were predicted to span 1 MHz in breadth, which is beyond the excitation bandwidth of standard rectangular pulses. Instead of acquiring frequency-stepped echoes and plotting the echo intensity as a function of transmitter frequency, they acquired a series of full-echo FIDs (to increase the total signal intensity), Fourier transformed each to yield a series of subspectra, and added the subspectra to produce the overall powder pattern, which they called a *variable-offset cumulative spectrum* (VOCS). They pointed out that if the pulse power (length) and probe characteristics are known, it is possible to predict the optimal frequency offset to achieve uniform excitation of the entire pattern while minimizing the number of subspectra and total experiment time. At about the same time, Clark *et al.* published a similar work in which field stepping was utilized in place of changing the transmitter frequency, along with a rigorous analytical analysis of the technique.[12] Frydman *et al.* expanded on the work of Massiot, utilizing the VOCS methodology to acquire ^{59}Co ($I = 7/2$) UW SSNMR spectra of CT powder patterns of a series of cobaltophthalocyanines.[13] Through a combination of spectral acquisitions and simulations utilizing finite pulse widths, they illustrated that it is crucial to use low rf powers and short CT selective pulses to avoid orientation-dependent nutation effects and to obtain relatively distortionless powder patterns (Figure 5.2). The VOCS methodology is superior to the simple piecewise plots of intensity as a function of frequency, because of acquisition of fewer subspectra and since the resolution is no longer limited by the frequency/field step size, but rather, the dwell time and number of points for each echo.

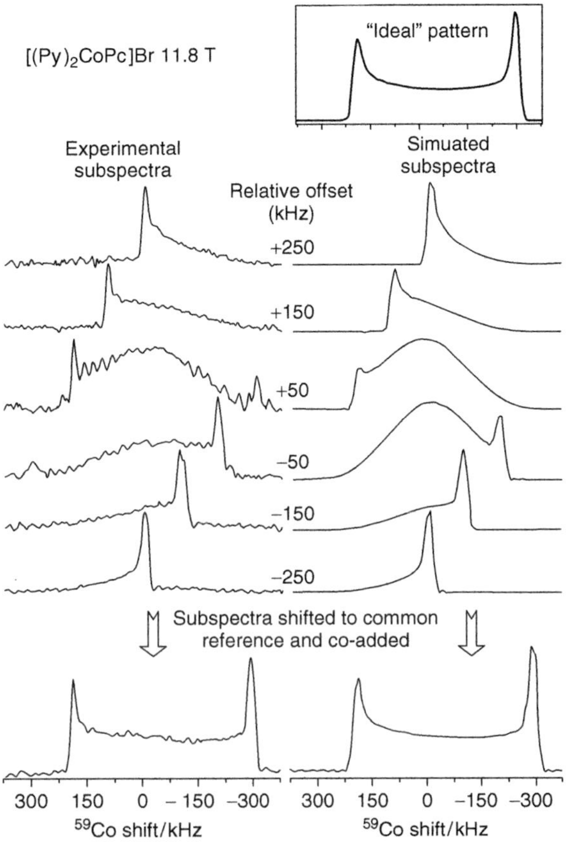

Figure 5.2. ^{59}Co VOCS UW SSNMR spectra of $[(Py)_2CoPc]Br$ acquired at 11.8 T. Subspectra acquired at various transmitter offsets (upper spectra) are co-added to form the total powder patterns (pictured at the bottom). (Adapted with permission from Ref. 13. Copyright 1999 American Chemical Society.)

5.3.2 Spin Echoes

Wideline and UW NMR powder patterns are associated with very fast signal decay in the time domain, e.g., for patterns that are hundreds of kHz in breadth, the signal often decays within tens of microseconds (or less) from the start of the FID. The fast signal decay is a result of the spin precessing at different rates in the transverse plane of the rotating frame due to both inhomogeneous broadening (e.g., distribution of frequencies from the anisotropic chemical shift or quadrupolar interactions) and transverse (spin–spin) relaxation. Conventional Bloch decay experiments (i.e., pulse acquire) are usually insufficient for acquisition of such spectra, since much of the signal may decay during the receiver delay and probe dead time (which are on the order of tens of microseconds or less).

To get around this problem, pulse sequences that refocus the dephasing magnetization via the creation of a *spin echo* are often used. Originally proposed by Hahn,[14] so-called *refocusing pulses* are applied after some delay τ after the initial $\pi/2$ pulse, which effectively refocus the spins that are dephasing as the result of inhomogeneous interactions (the spins that undergo dephasing as the result of transverse relaxation are not refocused—this is known as *decoherence*). Hahn proposed the use of successive $\pi/2$ pulses for this purpose. Carr and Purcell extended Hahn's work by using π pulses for refocusing and utilized a sequence of the form $\pi/2-\tau-\pi-\tau-$acq for measurements of T_2 values (the refocusing portion of the sequence is repeated to observe the echo decay resulting from transverse relaxation).[15] Meiboom and Gill realized that pulse errors could result in a decay of the echo train, which would compromise the accuracy of measured T_2 values. They modified the Carr–Purcell sequence by using π pulses that are phase shifted with respect to the initial $\pi/2$ pulse by 90°, enabling improved refocusing for the accurate measurement of T_2 values.[16]

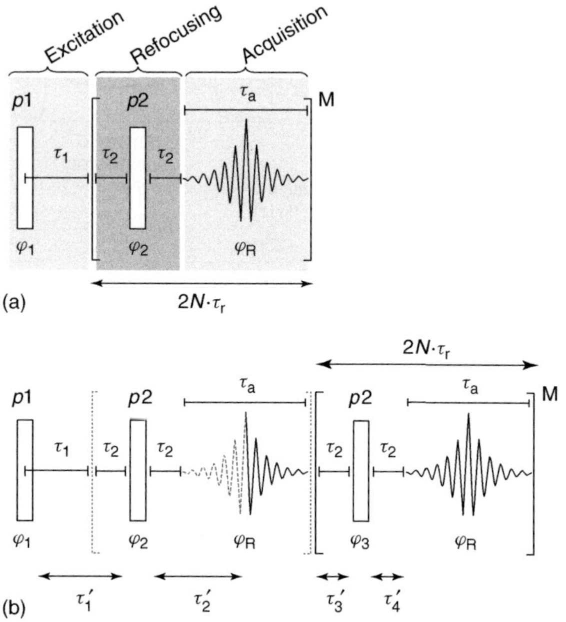

Figure 5.3. (a) CPMG and (b) QCPMG pulse sequences. (Adapted from Ref. 23. Copyright (2010) with permission from Elsevier.)

Spin-echo pulse sequences are commonly applied for the acquisition of the inhomogeneously broadened powder patterns of quadrupolar nuclei. The two most commonly used sequences are the *quadrupolar* or *solid echo sequence* ($\pi/2_x - \tau - \pi/2_y - \tau -$acq, which resembles Hahn's original experiments) and the *Hahn-echo sequence* ($\pi/2_x - \tau - \pi_y - \tau -$acq, which resembles the modified sequence of Carr and Purcell). For half-integer spin quadrupoles, either sequence can be used, but for integer-spin quadrupoles, the former is used almost exclusively.

The *Carr–Purcell Meiboom–Gill (CPMG) sequence* (Figure 5.3a) is also used for the enhancement of S/N in SSNMR spectra; early instances of this were presented by Slichter and co-workers, who utilized this sequence for the acquisition of high S/N spectra of spin $I = 1/2$ nuclides.[17] It is similar to the Hahn-echo sequence, consisting of an initial $\pi/2$ pulse followed by a delay (typically denoted by τ), a refocusing π pulse (phase shifted by 90°), a second delay τ, and an acquisition period. However, the refocusing portion of the sequence is repeated M times, usually until the echoes in the time domain are observed to disappear. The echoes, all contained in a single FID, can be Fourier transformed to produce a spectrum consisting of *spikelets*, which normally has much higher noise than the corresponding Hahn-echo (i.e., $\pi/2 - \tau - \pi - \tau -$acq) spectrum. Alternatively, one may left shift and co-add the echoes to produce a single echo of higher intensity with respect to the individual echoes, which is then Fourier transformed to produce a standard static pattern. The resolutions of spectra derived from CPMG experiments are limited by the delay times between the acquisition periods, as well as the number of points per echo. Hence, if high S/N is desired, the echoes are spaced closely in the time domain, resulting in spikelets that are spaced further apart in the frequency domain (i.e., lower resolution). If higher resolution is desired, the echoes are spaced further in the time domain, resulting in closer spacings in the frequency domain spectrum (i.e., lower S/N). The maximum gain in S/N is related (in part) to the effective T_2 of the observed nucleus, where larger T_2 values lead to more echoes and higher S/N.

In 1997, the CPMG sequence was reintroduced by Larsen *et al.* as quadrupolar CPMG (QCPMG) for application to NMR spectra of CT powder patterns of half-integer spin quadrupolar nuclei (Figure 5.3b).[18] They included examples of ^{87}Rb QCPMG NMR of a series systems with relatively narrow CT powder patterns, as well as ^{59}Co NMR of $(Co(NH_3)_5Cl)Cl$, which has a pattern breadth of ~270 kHz at 9.4 T. The ^{59}Co powder patterns are near the limit of what can be achieved in terms of uniform excitation using rectangular pulses; nonetheless, the S/N gains are significant, enough to convince many other SSNMR research groups to utilize this pulse sequence.

A number of frequency-stepped UW QCPMG NMR experiments have been reported. Some of the earlier examples include a series of ^{67}Zn ($I = 5/2$) SSNMR experiments on systems being utilized as models for biological Zn-containing systems (Figure 5.4a),[19,20] and ^{91}Zr SSNMR experiments on zirconium-containing metallocenes (Figure 5.4b).[21] In both cases, special care was taken to ensure that the transmitter offsets were equal to integer multiples of the spacings between the spikelets. This, and the importance of determining the optimal transmitter spacings to achieve reduced experimental times and uniform excitation, are discussed in detail by Tang *et al.* in their work on ^{27}Al ($I = 5/2$) UW SSNMR of three- and five-coordinate aluminum species,[22] as well as in a comprehensive review by Hung

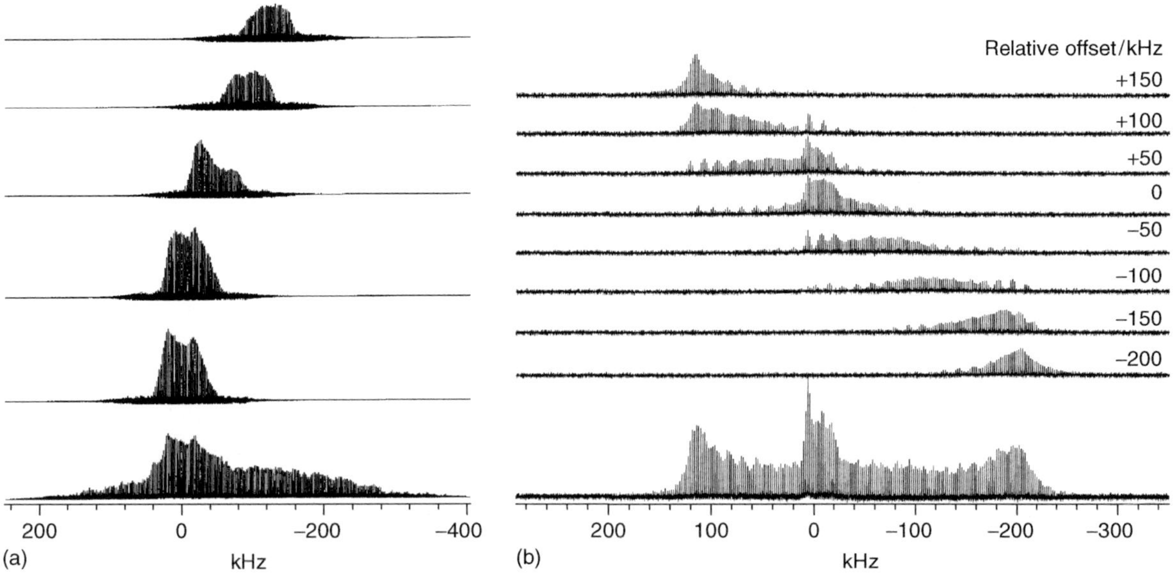

Figure 5.4. (a) ^{67}Zn VOCS QCPMG NMR spectra of $[H_2B(pz)_2]_2Zn$ obtained at 10 K and 18.8 T, with a selection of the subspectra utilized in its construction. (Reproduced from Ref. 19. © American Chemical Society, 2002) (b) ^{91}Zr VOCS QCPMG NMR of Cp2ZrCl2 in a zirconia rotor at 9.4 T, with contributing subspectra. The narrow and broad CT patterns correspond to the metallocene and the rotor, respectively. (Reproduced from Ref. 21. © American Chemical Society, 2004.)

and Gan, which describes many useful hints for acquiring and processing SSNMR spectra acquired with the CPMG sequence.[23]

5.3.3 Uniform Broadband Excitation with Chirped and Shaped Pulses

Obtaining an undistorted CT powder pattern for a half-integer spin quadrupolar nucleus is fraught with challenges. Nuclei in crystallites with different orientations with respect to B_0 have different frequencies arising from the anisotropic NMR interactions and experience distinct nutation frequencies (i.e., the relationship between pulse power and tilt angle is orientation dependent). The full excitation Hamiltonian is given by

$$\hat{H}_{\text{exc}} = \hat{H}_{\text{rf}} + \hat{H}_{\text{Q}}^{(1)}(\Theta^{\text{Q}}) + \hat{H}_{\text{Q}}^{(2)}(\Theta^{\text{Q}}) + \hat{H}_{\text{CS}}(\Theta^{\text{CS}}), \quad (5.6)$$

where $\hat{H}_{\text{rf}} = -\nu_1 \hat{I}_x$. The first and second terms on the right hand side do not commute, and as a result, both the CT and STs experience a complex excitation behavior that is dependent on the ratio $\nu_1/C_Q\hat{R}_{20}^{Q}$.[24–26] In fact, there are only two regimes where the complex nutation behavior is not observed, where $\nu_1 \gg C_Q$ and $\nu_1 \ll C_Q$. In both cases, all of the spins will experience uniform excitation (i.e., effectively, the same nutation angle, θ). The former applies only in cases of very small quadrupolar coupling constants. The latter is by far more common (as is the intermediate case), and CT selective pulses are often utilized (i.e., $\tau_{\text{selective}} = \tau_{\text{nonselective}}(I + 1/2)^{-1}$) to ensure uniform excitation of the CT pattern.[25] Additionally, the use of short pulses and low rf fields have been shown, both experimentally and theoretically, to provide powder patterns free of distortions caused by variable nutation rates.[27] Unfortunately, spectra acquired with low rf fields often have low S/N ratios, which can be problematic for studies of unreceptive nuclei. Furthermore, low rf fields correspond to much longer pulses, which correspond to much narrower excitation bandwidths, and can also result in distortions.

An alternative to the use of conventional rectangular pulses of fixed frequency and phase is the use of *chirped pulses*. Chirped pulses are applied at constant transmitter frequencies and feature controlled modulations of their amplitudes and/or phases for the purpose of exciting broad frequency ranges (among other things). They have found widespread use for broadband decoupling in solution NMR experiments[28–30] and for manipulation of single- and multiple-quantum coherences in SSNMR experiments on quadrupolar nuclei.[31,32] In particular, the wide-band, uniform-rate, and smooth truncation (WURST) pulses of Kupče and Freeman,[30] which are utilized for uniform excitation of broad frequency regions, are of particular use for UW NMR.

In 2007, Bhattacharyya and Frydman published an in-depth investigation of low-power phase-modulated rf pulses and their application for the acquisition of distortionless powder patterns of half-integer spin quadrupolar nuclei.[26] They discuss two distinct methodologies, both involving a Hahn-echo-like $\pi/2-\tau-\pi-\tau-$acq sequence, where the $\pi/2$ and π correspond to WURST-80 type pulses with specifically assigned sweep rates and amplitudes. Using this sequence, it is possible to create an undistorted spin echo, which can be Fourier transformed to yield a distortion-free broadband powder pattern, or an undistorted time-domain echo that is frequency encoded, from which information on the distribution of intensity and frequency can be obtained via a magnitude calculation (i.e., no Fourier transform). A comparison of ^{35}Cl SSNMR spectra acquired with conventional rectangular pulse Hahn-echo, chirped-pulse full-echo, and chirped-pulse frequency-encoded echo experiments is shown in Figure 5.5. The Fourier transformed chirped-pulse full-echo spectra yield the highest S/N of the three experiments; however, there may be interesting applications of the frequency-encoded sequence, since the resulting spectra have resolution characteristics limited by the encoding time.

This article was followed by work from O'Dell and Schurko, who investigated the possibility of utilizing WURST pulses to acquire both individual spin echoes and CPMG echo trains for both integer-spin and half-integer spin quadrupolar nuclei.[33] They found that broad spectra with high S/N could be obtained utilizing a different set of experimental conditions

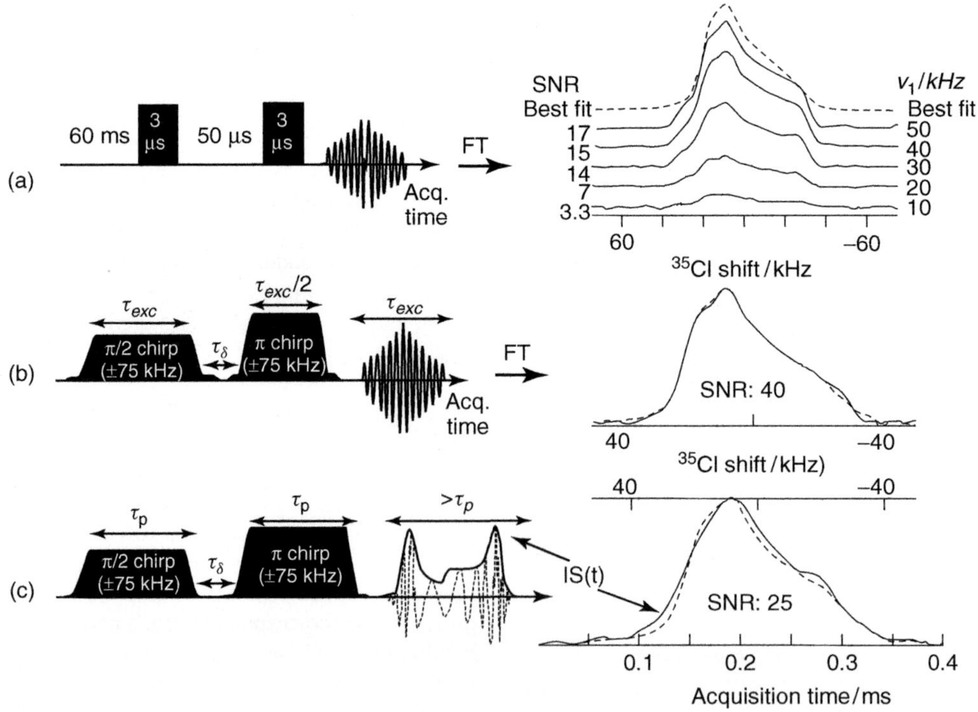

Figure 5.5. ^{35}Cl SSNMR spectra acquired with (a) standard spin-echo sequences with variable rf powers and subsequent Fourier transformation, (b) chirped pulses (frequency swept via phase modulation) to produce a full echo, which is subsequently Fourier transformed, and (c) frequency-progressive echo sequence with chirped pulses followed by magnitude calculation of the time-domain line shape. (Figure adapted from Ref. 26 © American Institute of Physics, 2007.)

from Bhattacharyya and Frydman. Furthermore, they were able to acquire CPMG trains of echoes using a $\pi/2-\tau-\pi/2-(\tau-\text{acq}-\pi/2)_N$ sequence (where N = the number of Meiboom–Gill loops), which when Fourier transformed, yield typical "spikelet" spectra with very high S/N. The pulse sequence, dubbed WURST-QCPMG, was applied to acquire ^{71}Ga and ^{91}Zr UW SSNMR spectra that would normally require frequency-stepped (VOCS) techniques (Figure 5.6), as well as the complete CT and partial ST powder patterns for ^{59}Co in Co(acac)$_3$ (spanning a total breadth of ~1.4 MHz).

This work was followed up by the same authors with an article describing the acquisition of UW SSNMR spectra using frequency-stepped VOCS-style acquisitions using the WURST-QCPMG sequence.[34] Examples of ^{35}Cl ($I = 3/2$) and ^{65}Cu ($I = 3/2$) UW SSNMR spectra of CT powder patterns, a ^{59}Co spectrum of the entire CT and ST manifold for Co(acac)$_3$, and a ^{10}B ($I = 3$) spectrum of the ST transition patterns of ν-B_2O_3 were reported. In this same work, the need for ^{1}H decoupling for samples with protons is discussed, since dipolar dephasing resulting from ^{1}H-X dipolar couplings can drastically reduce the number of echoes in the CPMG echo train (i.e., the effective T_2 is reduced), which results in a corresponding drop in S/N. Further work was published describing a study of broadband excitation using pulses designed using optimal control theory (OCT), where phase- and amplitude-modulated excitation pulses were designed using OC routines included in SIMPSON 2.0;[35] however, to this

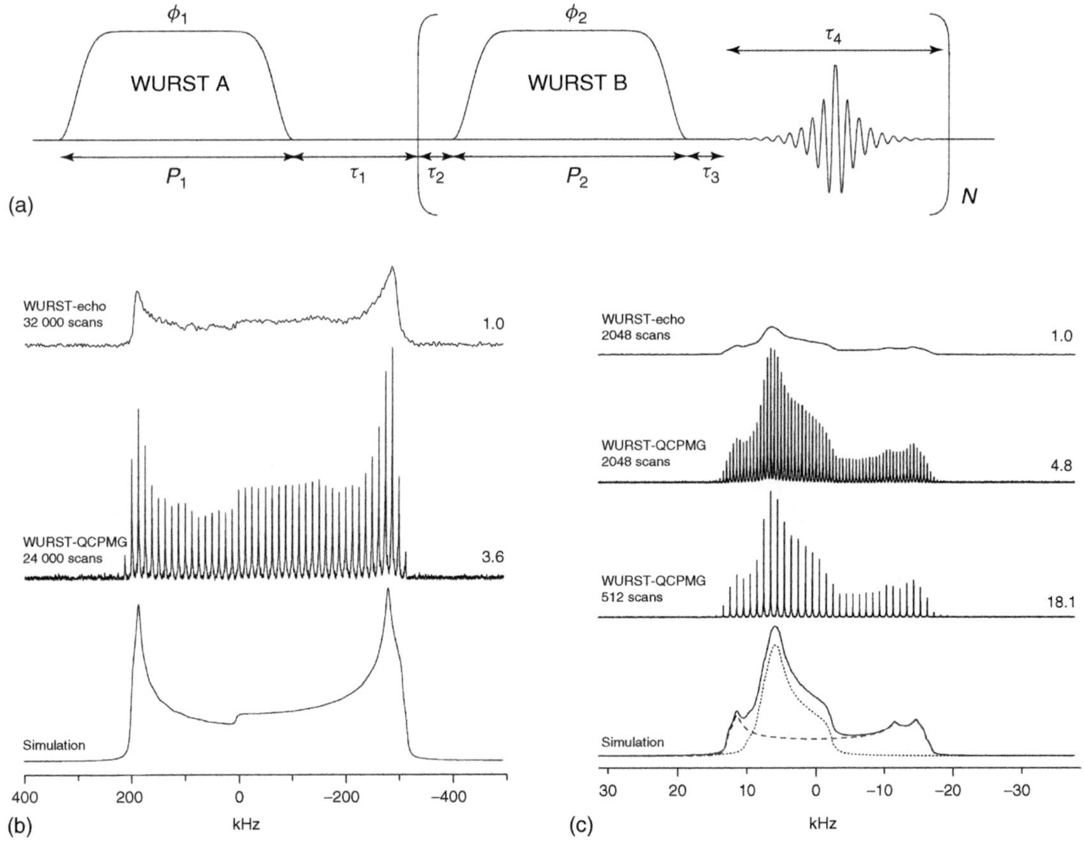

Figure 5.6. (a) The WURST QCPMG pulse sequence. If $N = 1$, this corresponds to the collection of a single echo. (b) ^{71}Ga NMR spectra of gallium phthalocyanine chloride obtained using the WURST echo (top) and WURST QCPMG (middle) sequences. (c) ^{91}Zr WURST echo (top) and WURST QCPMG (second and third). (Figure adapted from Ref. 33. © Elsevier, 2008.)

point, experiments have been limited to $I = 3/2$ nuclei with relatively narrow CT patterns.[36] Additional applications of all of these methods for acquisition of UW SSNMR spectra are discussed in Section 5.4.

5.3.4 Hardware: Microcoils, Probes, and Magnets

Research on the application of microcoils in NMR has largely focused on solution-state NMR samples, with far fewer SSNMR applications.[37] For the purpose of this chapter, microcoils are defined as coils with inner diameters (i.d.) of 1.6 mm or less, covering the range of coil sizes explored in the recent literature. For UW SSNMR experiments, the best case scenario would seem to be the use of a large coil with as much sample as possible, since the detected signal is directly proportional to the number of spins. However, there are two key situations in which the use of microcoils for UW SSNMR may be useful: (i) when the overall amount (or volume) of sample is very limited or (ii) when the nuclide(s) of interest is diluted.

Fortunately, there are some advantages to using microcoils for such experiments. Hoult and Richards have shown that when bulk magnetization arising from the nuclear spins precesses in a plane transverse to the longitudinal axis of a solenoidal coil, a voltage is induced in the coil and is proportional to the magnetic field per unit of current, from which they demonstrate that for coils of fixed lengths but varying diameters, there is an inverse relationship between sensitivity and coil diameter.[38] By the reciprocity theorem, this means that small-diameter coils are capable of producing very large magnetic fields (i.e., fluctuating rf fields) per unit of current they pass. Hence, on a conventional NMR spectrometer, very large rf fields can be generated within the coil using relatively low powers (which has numerous advantages in terms of amplifier powers, sample heating, etc.). Microcoils (and associated probes) typically have low Q values (10–150), which means that they have short recovery times after rf pulses, and very short receiver dead times can be used. Finally, owing to their inherently small size, microcoils typically produce very homogeneous magnetic fields and have much higher filling factors (i.e., fraction of the detection coil volume filled with sample) than conventional coils.

There are very few reports of applications of microcoils to wideline or UW SSNMR experiments. Kentgens and co-workers have designed several styles of microcoils,[39–41] including "piggy-back" microcoils, which are mounted on standard 4 mm rotors and can be used for the acquisition of both MAS and static NMR spectra. They have acquired the ^{27}Al NMR powder pattern of sapphire, in which the CTs and STs are observable within a single experiment.[39] They have also used piecewise experiments to acquire the ^{75}As CT powder patterns of $Al_xGa_{1-x}As$ samples, some of which are greater than 1.0 MHz in breadth at 18.8 T.[37] Sakellariou, Schurko, and co-workers have successfully used microcoils to acquire UW SSNMR spectra for a variety of half-integer quadrupolar nuclei, including ^{87}Rb, ^{71}Ga, ^{91}Zr, and ^{59}Co and have also implemented the QCPMG pulse sequence to further aid in enhancement of S/N.[2] The combination of clever hardware/coil design and implementation of QCPMG and/or chirped pulses may be extremely advantageous for future work in this area.

Aside from microcoil design, some work has been published on the modification of probes for UW NMR experiments.[37] Since chirped-pulse/WURST-type experiments are capable of providing uniform excitation of large bandwidths, in many instances, it is the excitation and detection bandwidths of the probe that are the limiting factors in UW NMR experiments. These bandwidths generally depend on the Q of the probe (and accompanying circuitry); however, there is a trade-off when Q is reduced (i.e., increased bandwidths for decreased S/N), which is undesirable. Another minor problem that arises in the acquisition of UW NMR spectra is the presence of external rf signals (e.g., from FM radio stations) and other spurious background signals,[42] which are observed in cases of low S/N, even when the coil is shielded by a Faraday cage of reasonably high quality. These signals are not easy to remove but are generally much sharper and more incoherent than the typical UW powder patterns and therefore easy to identify.

Finally, the use of UHFs for the acquisition of UW NMR spectra cannot be understated, given that the powder patterns are often spread over several MHz. This is especially true for the case of half-integer quadrupoles, since the breadths of the CT patterns scale as the inverse of the magnetic field. One nice example of this is a ^{55}Mn ($I = 5/2$) UW SSNMR study published by Wasylishen and co-workers,

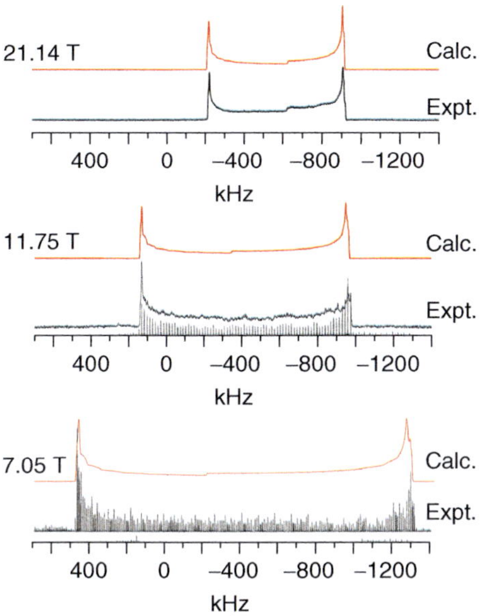

Figure 5.7. ^{55}Mn NMR spectra of the CTs of stationary samples of η^5-CpMn(CO)$_3$ acquired at 7.05, 11.75, and 21.14 T. The top two spectra were acquired using the VOCS QCPMG method, while the bottom spectrum was acquired in a single experiment. (Reproduced from Ref. 43. © American Chemical Society, 2006.)

in which they show QCPMG NMR spectra of η^5-CpMn(CO)$_3$ acquired at 7.05 and 11.75 T, which require the use of VOCS acquisitions.[43] Remarkably, the full CT pattern for this sample can be acquired with a single Hahn-echo experiment at 21.1 T (Figure 5.7). Of course, almost all of the work discussed herein has been conducted on standard superconducting NMR magnets with homogeneous magnetic fields; however, we must give special mention to work by Hung, Gan, and co-workers, who investigated the use of high-field resistive magnet (25 T) for UW SSNMR, and successfully acquired UW QCPMG NMR ^{35}Cl spectra of some metallocene dichlorides.[44]

5.4 APPLICATIONS OF UW SSNMR

Owing to space limitations, this section is not meant to be a comprehensive review of recent UW SSNMR spectroscopy articles, which are estimated to number well over 200 from 2001 to 2010; rather, this section will outline some of the interesting applications and developments in UW SSNMR according to the groupings of nuclides and systems outlined earlier. A series of articles have been selected for this brief review, which discuss NMR of "exotic" nuclei and/or elicit interesting general ideas and facts about UW SSNMR experiments, potential applications, and limitations.

5.4.1 Systems with Half-integer Quadrupolar Nuclei

The majority of UW NMR studies has been reported for systems with half-integer quadrupolar nuclei, which spans areas of inorganic materials, organometallics and coordination compounds, and biological chemistry.

Inorganic materials. Applications to basic inorganic systems are by far the most numerous. Quadrupolar nuclei with large quadrupole moments in four-coordinate metal oxides have been studied with UW NMR techniques, including ^{33}S ($I = 3/2$) NMR of sulfates,[45] ^{75}As ($I = 3/2$) NMR of arsenic oxysalts,[46] ^{53}Cr ($I = 3/2$) NMR of chromate and dichromate salts,[47] and $^{185/187}$Re ($I = 5/2$) NMR of perrhenates.[48] The work on sulfates clearly indicates the necessity for UHF NMR for spin-3/2 nuclei (which generally have the broadest central transitions of all of the half-integer spins), and suggests limitations for current experiments on ^{33}S nuclei in less spherically symmetric environments. The ^{75}As SSNMR is similar in this respect; however, the authors also suggest the utilization of a "histogram" type experiment, in which individual subspectra are collected at widely spaced increments, not to reconstruct a cosmetically appealing spectrum, but rather, to obtain a rough estimate of the quadrupolar parameters. ^{53}Cr, which is considered an unreceptive "low-γ" nucleus, seems to also be currently limited to four-coordinate environments, and frequency-stepped QCPMG experiments will be required for spectral acquisition. $^{185/187}$Re SSNMR experiments on perrhenates conducted at temperatures ranging from −116 to 296 °C yield spectra that reveal dynamical behavior attributed to intramolecular motion and rapid reorientation of the rhenium EFG tensors; these experiments suggest that VT UW NMR could be very useful in examining much higher rates of chemical exchange than conventional SSNMR experiments.

A number of comprehensive studies of half-integer quadrupoles in simple ionic systems, metal oxides, and metals in oxygen coordination environments have also been reported. Bastow and co-workers have published a large amount of work in this vein utilizing piecewise plots of echo intensity as a function of carrier frequency, including ^{69}Ga and ^{139}La ($I = 7/2$) SSNMR experiments on pure and doped $LaGaO_3$ perovskites,[49] and ^{137}Ba ($I = 3/2$) SSNMR of $BaCO_3$,[50] to name only a few. $^{79/81}$Br ($I = 3/2$) VOCS QCPMG NMR experiments were conducted by Bryce *et al.* on a series of ionic bromides,[51] and with the use of UHFs (21.1 T), both bromine EFG and CS tensor parameters were determined (the quadrupolar parameters extracted from simulations of these spectra can be used to correct deficiencies in nuclear quadrupole resonance (NQR) data, from which η_Q values can generally only be approximated for spin-3/2 nuclides). Similar ^{127}I ($I = 5/2$) SSNMR experiments conducted on group 2 metal iodides revealed enormous quadrupolar coupling constants;[52] in some cases, the C_Q values are so large that the CT powder patterns cannot be simulated using first-order perturbation theory (i.e., the high-field approximation fails). However, quadrupolar and CS parameters can be accurately determined via "exact" theoretical modeling and quadrupole parameters refined with ^{127}I NQR measurements. ^{209}Bi ($I = 9/2$) SSNMR data was obtained for a series of Bi-containing complexes by Hamaed *et al.*,[53] and in most cases, WURST and/or QCPMG pulse sequences were necessary for acquisition. Again, the failure of the high-field approximation was observed in this data set, and methodologies combining NMR and NQR are suggested. ^{137}Ba UW SSNMR experiments at 9.4 and 21.1 T, mainly conducted with the QCPMG or WURST-QCPMG pulse sequences, have been shown to be useful for the characterization of a wide array of Ba complexes by Schurko *et al.*[54] and Huang *et al.*[55] Fayon, Massiot, and co-workers have utilized UHFs and VOCS Hahn-echo and QCPMG methods to acquire ^{91}Zr UW SSNMR spectra for a series of zirconium halide systems, from which they have extracted both EFG and CS tensor parameters.[56]

Organometallics and coordination compounds. Schurko and co-workers have used UW SSNMR experiments to examine metal nuclei (^{93}Nb, ^{139}La, ^{91}Zr) at the hearts of metallocenes.[57–59] Conventional Hahn-echo and QCPMG pulse sequences were utilized to conduct a ^{93}Nb ($I = 9/2$) SSNMR study on piano-stool metallocenes. The ^{93}Nb quadrupolar parameters are shown to be very sensitive to oxidation state (i.e., Nb(I) vs Nb(V) species), as well as to the presence of coordinating solvent (THF) molecules. ^{139}La UW QCPMG NMR spectra of bis- and tris-coordinated lanthanocenes reveal simple-to-understand relationships between C_Q and η_Q and the symmetries of the ground-state electron distributions about the ^{139}La nuclei. Zirconocenes and related group 4 metallocene complexes have been extensively studied using ^{91}Zr and ^{35}Cl (Figure 5.8) UW SSNMR methods;[59,60] for each metallocene, the SSNMR spectra act as exquisitely sensitive probes of structure. The utilization of QCPMG or WURST-QCPMG and UHFs is again stressed, and the rapidity with which such spectra can be acquired suggest much potential for investigation of metallocene-based catalysts for polyolefin polymerization (both heterogeneous and homogeneous). Again, the importance of ^{1}H decoupling during the acquisition of the CPMG echo trains is stressed in these works. Wasylishen *et al.* have also investigated metal centers in a number of inorganic and organometallic coordination compounds, including ^{59}Co UW SSNMR of $Co(C_2B_9H_{11})^{2-}$ and $Co(C_5H_5)^{2+}$ salts[61] and ^{115}In ($I = 9/2$) UW SSNMR of triarylphosphine indium(III) trihalide adducts.[62]

Biological systems and model systems. In cases where crystallographic techniques or standard NMR experiments (^{13}C, ^{15}N, etc.) may be insufficient for the characterization of bioinorganic molecules, UW SSNMR experiments may provide complimentary structural information. In this light, Ellis, Lipton, and co-workers have worked extensively on developing and applying ^{67}Zn VOCS QCPMG NMR experiments at UHFs and very low temperatures (down to 5–10 K) with the use of specially constructed probes and ^{67}Zn-enriched samples for the characterization of biological systems containing Zn sites and related model complexes.[63] Among this body of work, they investigated the use of paramagnetic Fe(II) dopants to reduce ^{67}Zn T_1 values in $[H_2B(pz)_2]_2Zn$ (pz = pyrazolylborate) to reduce overall experimental times.[19] They also utilized ^{1}H–^{67}Zn CP/CPMG experiments to acquire UW SSNMR spectra of human carbonic anhydrase[64] and to investigate the influence of pH on ^{67}Zn UW SSNMR spectra of the catalytic Zn sites in the wild type and the H265A mutant of *Aquifex aeolicus* LpxC.[65] Utilizing similar techniques, they also acquired a histogram-type UW SSNMR spectrum for the Cu(I) site in azurin.[66]

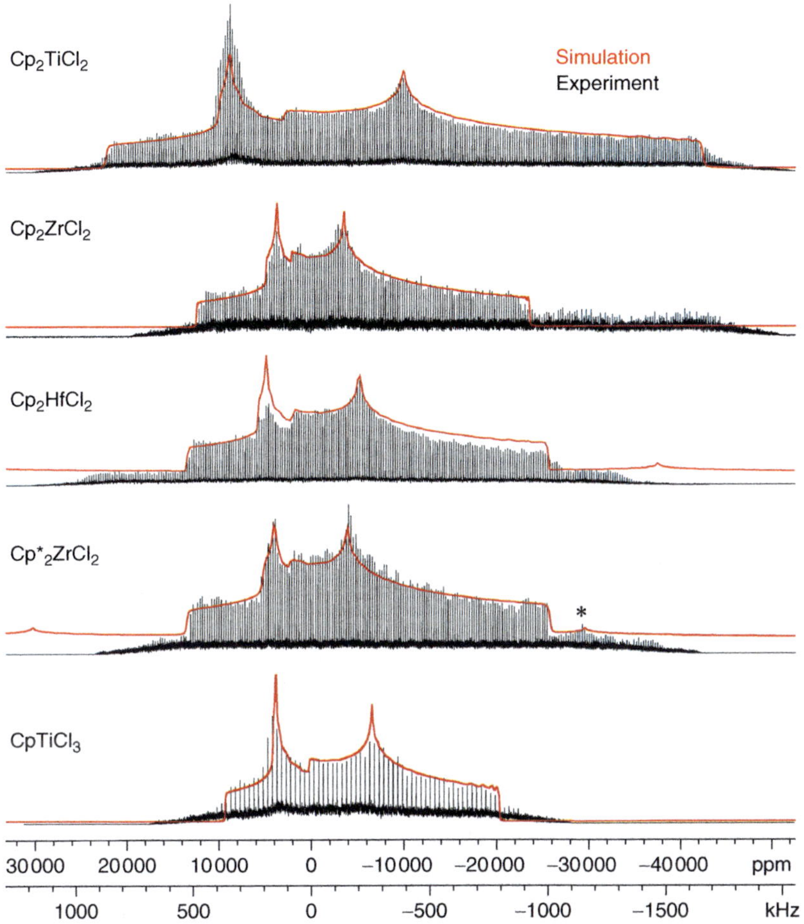

Figure 5.8. ^{35}Cl VOCS QCPMG NMR spectra of metallocene chlorides acquired at 9.4 T, with spectral simulations. (Reproduced from Ref. 60. © American Chemical Society, 2009.)

5.4.2 Systems with Integer-Spin Nuclei

Owing to the relatively small number of NMR-active integer-spin nuclei, there are considerably fewer UW SSNMR studies in the literature. Since ^{2}H and ^{6}Li SSNMR spectra can be acquired with routine quadrupolar echo sequences, they are not discussed here; rather, we focus our attention on ^{14}N ($I = 1$) and ^{10}B ($I = 3$).

Of the two isotopes of nitrogen, ^{14}N and ^{15}N ($I = 1/2$), the latter is much more intensely studied, because of its spin and comparatively higher gyromagnetic ratio. Notwithstanding, the low natural abundance of ^{15}N (0.37%) requires that costly isotopic labeling be done to facilitate reduction of NMR experimental times. Since ^{14}N has $I = 1$, SSNMR spectra comprise two ST patterns corresponding to the ($+1 \leftrightarrow 0$) and ($0 \leftrightarrow -1$) transitions. These patterns are broadened by both the FOQI and SOQI; however, it is the former that results in the extreme broadening that makes ^{14}N SSNMR spectra challenging to acquire. There are two recent reviews providing excellent overviews of the direct[6] and indirect[67] detection of ^{14}N SSNMR spectra. Herein, we shall focus only on a handful of more recent reports of ^{14}N UW SSNMR spectra acquired via methods outlined in Section 5.3.

Until the mid-1990s, the majority of published work on ^{14}N SSNMR was on oriented samples (e.g., single crystals, biological specimens, etc.), with very few examples of UW SSNMR spectra.[6] Sensitivity issues aside, the main problem with routinely investigating samples with ^{14}N SSNMR was the limited excitation bandwidths produced by standard rectangular pulses; as a result, several research groups came up with interesting methods for acquiring such spectra. Yesinowski and co-workers developed two slow-turning techniques dubbed RotIsseRIe (reorientation-induced redistribution of isochromats)[68] and slow-turning echo amplitude modulation and echo reduction (STEAMER),[69] with which they acquired ^{14}N SSNMR spectra of KNO_3 (along with the first ^{14}N frequency-stepped acquisitions). Jakobsen and co-workers utilized ^{14}N MAS NMR experiments to investigate several ionic solids, ammonium salts, and amino acids.[70,71] These experiments are capable of eliciting information on both the anisotropic CS and EFG tensors, but are lengthy and found to be the most successful when the magic angle was precisely tuned (which required specialized probe modifications). Aside from these works, the majority of directly detected UW SSNMR spectra involved frequency-stepped or VOCS spectral reconstructions.[6]

O'Dell, Schurko, and co-workers suggested the use of the WURST-CPMG sequence for the acquisition of wideline ^{14}N SSNMR spectra. In initial work on KNO_3, they noticed an interesting signal enhancement in the Pake doublet patterns, which they named direct enhancement of integer-spin magnetization (DEISM).[72] Interestingly, the population enhancements are not gained from preparatory pulses, but rather from the excitation and refocusing pulses. In addition, since the patterns are so broad, distinct degrees of population transfer can be observed for different crystallite orientations (Figure 5.9). Hence, as excitation of crystallites from one transition commence at one end of the pattern (e.g., for the $+1 \leftrightarrow 0$ transition), the spin polarization of the $0 \leftrightarrow -1$ transition is altered for other crystallite orientations. Using this pulse sequence, they have demonstrated successfully acquired ^{14}N UW SSNMR spectra of amino acids,[73] as well as of samples with a variety of nitrogen structural motifs.[74]

There are only two instances of ^{10}B UW NMR reported in the literature, both with application to glasses, which are routinely studied using ^{11}B ($I = 3/2$) SSNMR. Holland, Smith, and co-workers utilized the VOCS technique to acquire ^{10}B UW NMR spectra of borate glasses.[75] An experiment on 99.62% ^{10}B enriched ν-B_2O_3 glass required 26 frequency steps and over 100 h of experimental time. By comparison, application of the WURST-QCPMG sequence enabled the acquisition of ^{10}B NMR spectrum for a sample of nonenriched glass in only 45 min,[34] suggesting that this methodology may enable the investigation of a wide variety of glasses and ceramics via ^{10}B SSNMR.

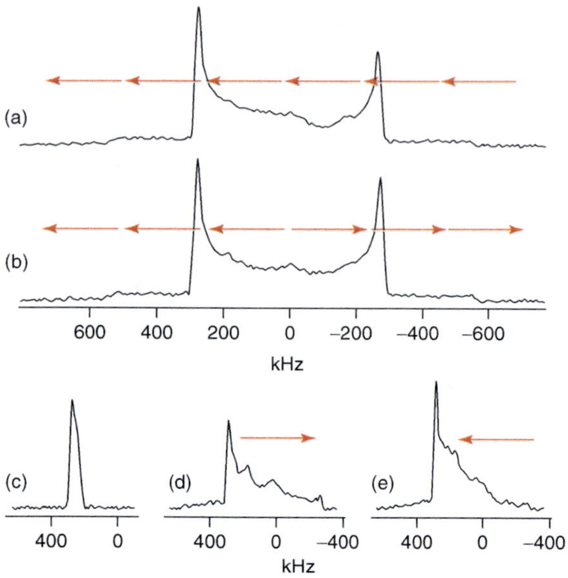

Figure 5.9. ^{14}N VOCS WURST QCPMG NMR spectra of KNO_3. Red arrows indicate the direction of the WURST sweeps. Spectrum acquired with (a) sweeps all in the same direction and that acquired with (b) sweeps in opposite directions from the center. (c) A single subspectrum acquired with QCPMG, and single subspectra acquired with sweeps in the (d) wrong direction and the (e) right direction. (See Ref. 72 for more details.)

5.4.3 Quadrupolar Nuclei in Systems with Unpaired Electrons

There are far too many NMR studies of both paramagnetic systems and metals and alloys with conduction electrons to review in detail here, with many applications directed toward spin $I = 1/2$ nuclei.[8,76] In this section, a few brief highlights of UW NMR techniques applied to quadrupolar nuclei in such systems are given.

Metals, alloys and metal nanoparticles are ideal systems to study with the UW NMR techniques outlined above. Most of the recent publications of UW

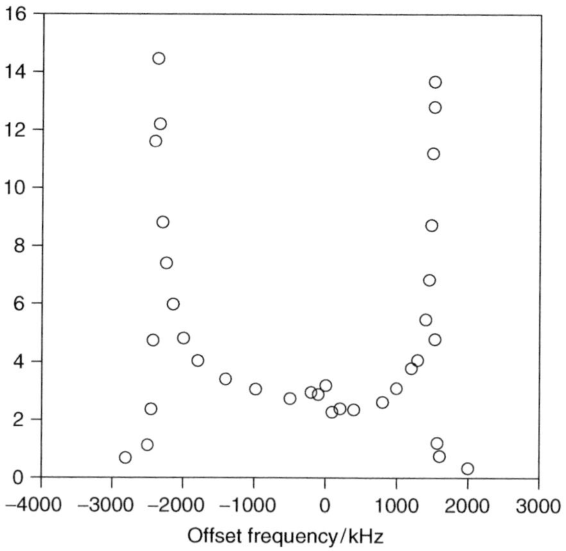

Figure 5.10. The ^{75}As CT powder pattern for arsenic metal, obtained using the frequency-stepped technique. (This spectrum is from Ref. 79.)

NMR spectra for such systems report the use of simple frequency stepping or spin-echo mapping. Bastow has been particularly prolific in this area, publishing articles on a variety of metals and alloys, including an examination of the effects of contact (hyperfine) interactions and Knight shifts on the $^{47/49}$Ti ($I = 5/2$, 7/2) UW NMR spectra of a series of titanium oxides, metals, and alloys,[77] piecewise acquisitions of ^{95}Mo ($I = 5/2$) UW NMR spectra of a $MoSe_2/Mo_3Se_4$ mixture,[78] and ^{75}As ($I = 3/2$) UW NMR of arsenic metal.[79] In many of these works, both the CT and "horns" of the STs are acquired with frequency-stepped techniques (an example for ^{75}As is shown in Figure 5.10).

5.5 CONCLUSIONS

UW NMR experiments can be applied to many nuclei across the periodic table to study a cornucopia of systems, including inorganic materials, inorganic and organometallic coordination compounds, biological systems and related model compounds, metals, alloys, nanoparticles, and so forth. The traditional frequency-stepped "point-by-point" method of acquisition is still widely used; however, recent developments in hardware (UHF magnets, low-Q probes, microcoils, low-γ probes, and accessories, etc.) as well as the application-specialized pulse sequences (chirped WURST pulses, CPMG echo trains, etc.) are emerging as the key techniques for the rapid acquisition of high-quality, high S/N UW SSNMR spectra. Nuclei that were previously referred to as "NMR invisible" are becoming accessible, opening up many new classes of systems that can be investigated with SSNMR.

RELATED ARTICLES IN THE ENCYCLOPEDIA OF MAGNETIC RESONANCE

Echoes in Solids

Inorganic Nuclei: Low Sensitivity Transition Metals

Knight Shift

Quadrupolar Metal Nuclides in Bioinorganic Chemistry: Solid-State NMR Studies

Quadrupolar Nuclei in Liquid Samples

Quadrupolar Transition Metal and Lanthanide Nuclei

Solid State NMR Probe Design

Wide Lines for Nonquadrupolar Nuclei

REFERENCES

1. G. Neue, C. Dybowski, M. L. Smith, and D. H. Barich, *Solid State Nucl. Magn. Reson.*, 1994, **3**, 115.
2. J. A. Tang, L. A. O'Dell, P. M. Aguiar, B. E. G. Lucier, D. Sakellariou, and R. W. Schurko, *Chem. Phys. Lett.*, 2008, **466**, 227.
3. S. E. Ashbrook and M. J. Duer, *Concept. Magn. Res. Part A*, 2006, **28A**, 183.
4. W. P. Power, R. E. Wasylishen, S. Mooibroek, B. A. Pettitt, and W. Danchura, *J. Phys. Chem.*, 1990, **94**, 591.
5. M. J. Duer, *Introduction to Solid-State NMR Spectroscopy*, Blackwell: Oxford, 2004.
6. L. A. O'Dell, *Prog. Nucl. Magn. Reson. Spectrosc.*, 2011. **59**(4), 295.
7. I. Bertini, C. Luchinat, and G. Parigi, *Prog. Nucl. Magn. Reson. Spectrosc.*, 2002, **40**, 249.
8. J. J. van der Klink and H. B. Brom, *Prog. Nucl. Magn. Reson. Spectrosc.*, 2000, **36**, 89.

9. H. E. Rhodes, P. K. Wang, H. T. Stokes, C. P. Slichter, and J. H. Sinfelt, *Phys. Rev. B*, 1982, **26**, 3559.
10. T. J. Bastow and M. E. Smith, *Solid State Nucl. Magn. Reson.*, 1992, **1**, 165.
11. D. Massiot, I. Farnan, N. Gautier, D. Trumeau, A. Trokiner, and J. P. Coutures, *Solid State Nucl. Magn. Reson.*, 1995, **4**, 241.
12. W. G. Clark, M. E. Hanson, F. Lefloch, and P. Segransan, *Rev. Sci. Instrum.*, 1995, **66**, 2453.
13. A. Medek, V. Frydman, and L. Frydman, *J. Phys. Chem. A*, 1999, **103**, 4830.
14. E. L. Hahn, *Phys. Rev.*, 1950, **80**, 580.
15. H. Y. Carr and E. M. Purcell, *Phys. Rev.*, 1954, **94**, 630.
16. S. Meiboom and D. Gill, *Rev. Sci. Instrum.*, 1958, **29**, 688.
17. S. E. Shore, J. P. Ansermet, C. P. Slichter, and J. H. Sinfelt, *Phys. Rev. Lett.*, 1987, **58**, 953.
18. F. H. Larsen, H. J. Jakobsen, P. D. Ellis, and N. C. Nielsen, *J. Phys. Chem. A*, 1997, **101**, 8597.
19. A. S. Lipton, T. A. Wright, M. K. Bowman, D. L. Reger, and P. D. Ellis, *J. Am. Chem. Soc.*, 2002, **124**, 5850.
20. A. S. Lipton, C. Bergquist, G. Parkin, and P. D. Ellis, *J. Am. Chem. Soc.*, 2003, **125**, 3768.
21. I. Hung and R. W. Schurko, *J. Phys. Chem. B*, 2004, **108**, 9060.
22. J. A. Tang, J. D. Masuda, T. J. Boyle, and R. W. Schurko, *Chemphyschem*, 2006, **7**, 117.
23. I. Hung and Z. H. Gan, *J. Magn. Reson.*, 2010, **204**, 256.
24. A. Samoson and E. Lippmaa, *Chem. Phys. Lett.*, 1983, **100**, 205.
25. A. J. Vega, Quadrupolar nuclei in solids, in *Encyclopedia of Magnetic Resonance*, eds R. K. Harris and R. E. Wasylishen, John Wiley & Sons: Chichester, 2007. DOI: 10.1002/9780470034590.emrstm0431.
26. R. Bhattacharyya and L. Frydman., *J. Chem. Phys.*, 2007, **127**, 194503.
27. P. P. Man, J. Klinowski, A. Trokiner, H. Zanni, and P. Papon, *Chem. Phys. Lett.*, 1988, **151**, 143.
28. J. M. Bohlen, M. Rey, and G. Bodenhausen, *J. Magn. Reson.*, 1989, **84**, 191.
29. J. M. Bohlen and G. Bodenhausen, *J. Magn. Reson., Ser. A*, 1993, **102**, 293.
30. Ē. Kupče and R. Freeman, *J. Magn. Reson., Ser. A*, 1995, **115**, 273.
31. A. P. M. Kentgens, *J. Magn. Reson.*, 1991, **95**, 619.
32. D. Iuga, H. Schafer, R. Verhagen, and A. P. M. Kentgens, *J. Magn. Reson.*, 2000, **147**, 192.
33. L. A. O'Dell and R. W. Schurko, *Chem. Phys. Lett.*, 2008, **464**, 97.
34. L. A. O'Dell, A. J. Rossini, and R. W. Schurko, *Chem. Phys. Lett.*, 2009, **468**, 330.
35. Z. Tosner, T. Vosegaard, C. Kehlet, N. Khaneja, S. J. Glaser, and N. C. Nielsen, *J. Magn. Reson.*, 2009, **197**, 120.
36. L. A. O'Dell, K. J. Harris, and R. W. Schurko, *J. Magn. Reson.*, 2010, **203**, 156.
37. A. P. M. Kentgens, J. Bart, P. J. M. van Bentum, A. Brinkmann, E. R. H. Van Eck, J. G. E. Gardeniers, J. W. G. Janssen, P. Knijn, S. Vasa, and M. H. W. Verkuijlen. *J. Chem. Phys.*, 2008, **128**, 052202-1.
38. D. I. Hoult and R. E. Richards, *J. Magn. Reson.*, 1976, **24**, 71.
39. K. Yamauchi, J. W. G. Janssen, and A. P. M. Kentgens, *J. Magn. Reson.*, 2004, **167**, 87.
40. H. Janssen, A. Brinkmann, E. R. H. van Eck, P. J. M. van Bentum, and A. P. M. Kentgens, *J. Am. Chem. Soc.*, 2006, **128**, 8722.
41. P. J. M. van Bentum, J. W. G. Janssen, A. P. M. Kentgens, J. Bart, and J. G. E. Gardeniers, *J. Magn. Reson.*, 2007, **189**, 104.
42. J. A. Tang, B. D. Ellis, T. H. Warren, J. V. Hanna, C. L. B. Macdonald, and R. W. Schurko, *J. Am. Chem. Soc.*, 2007, **129**, 13049.
43. K. J. Ooms, K. W. Feindel, V. V. Terskikh, and R. E. Wasylishen, *Inorg. Chem.*, 2006, **45**, 8492.
44. I. Hung, K. Shetty, P. D. Ellis, W. W. Brey, and Z. H. Gan, *Solid State Nucl. Magn. Reson.*, 2009, **36**, 159.
45. I. Moudrakovski, S. Lang, S. Patchkovskii, and J. Ripmeester, *J. Phys. Chem. A*, 2010, **114**, 309.
46. G. M. Bowers and R. J. Kirkpatrick, *J. Magn. Reson.*, 2007, **188**, 311.
47. M. A. M. Forgeron and R. E. Wasylishen, *Magn. Reson. Chem.*, 2008, **46**, 206.
48. R. W. Schurko, S. Wi, and L. Frydman, *J. Phys. Chem. A*, 2002, **106**, 51.
49. T. J. Bastow, T. Mathews, and J. R. Sellar, *Solid State Ionics*, 2004, **175**, 129.
50. T. J. Bastow, *Chem. Phys. Lett.*, 2002, **354**, 156.

51. C. M. Widdifield and D. L. Bryce, *J. Phys. Chem. A*, 2010, **114**, 2102.
52. C. M. Widdifield and D. L. Bryce, *J. Phys. Chem. A*, 2010, **114**, 10810.
53. H. Hamaed, M. W. Laschuk, V. V. Terskikh, and R. W. Schurko, *J. Am. Chem. Soc.*, 2009, **131**, 8271.
54. H. Hamaed, E. Ye, K. Udachin, and R. W. Schurko, *J. Phys. Chem. B*, 2010, **114**, 6014.
55. A. Sutrisno, C. Lu, R. H. Lipson, and Y. N. Huang, *J. Phys. Chem. C*, 2009, **113**, 21196.
56. O. Pauvert, F. Fayon, A. Rakhmatullin, S. Kramer, M. Horvatic, D. Avignant, C. Berthier, M. Deschamps, D. Massiot, and C. Bessada, *Inorg. Chem.*, 2009, **48**, 8709.
57. A. Y. H. Lo, T. E. Bitterwolf, C. L. B. Macdonald, and R. W. Schurko, *J. Phys. Chem. A*, 2005, **109**, 7073.
58. H. Hamaed, A. Y. H. Lo, D. S. Lee, W. J. Evans, and R. W. Schurko, *J. Am. Chem. Soc.*, 2006, **128**, 12638.
59. A. J. Rossini, I. Hung, S. A. Johnson, C. Slebodnick, M. Mensch, P. A. Deck, and R. W. Schurko, *J. Am. Chem. Soc.*, 2010, **132**, 18301.
60. A. J. Rossini, R. W. Mills, G. A. Briscoe, E. L. Norton, S. J. Geier, I. Hung, S. Zheng, J. Autschbach, and R. W. Schurko, *J. Am. Chem. Soc.*, 2009, **131**, 3317.
61. K. J. Ooms, V. V. Terskikh, and R. E. Wasylishen, *J. Am. Chem. Soc.*, 2007, **129**, 6704.
62. F. Chen, G. B. Ma, G. M. Bernard, R. G. Cavell, R. McDonald, M. J. Ferguson, and R. E. Wasylishen, *J. Am. Chem. Soc.*, 2010, **132**, 5479.
63. A. S. Lipton, R. W. Heck, J. A. Sears, and P. D. Ellis, *J. Magn. Reson.*, 2004, **168**, 66.
64. A. S. Lipton, R. W. Heck, and P. D. Ellis, *J. Am. Chem. Soc.*, 2004, **126**, 4735.
65. A. S. Lipton, R. W. Heck, M. Hernick, C. A. Fierke, and P. D. Ellis, *J. Am. Chem. Soc.*, 2008, **130**, 12671.
66. A. S. Lipton, R. W. Heck, W. A. de Jong, A. R. Gao, X. J. Wu, A. Roehrich, G. S. Harbison, and P. D. Ellis, *J. Am. Chem. Soc.*, 2009, **131**, 13992.
67. S. Cavadini, S. Antonijevic, A. Lupulescu, and G. Bodenhausen, *ChemPhysChem*, 2007, **8**, 1363.
68. E. A. Hill and J. P. Yesinowski, *J. Am. Chem. Soc.*, 1996, **118**, 6798.
69. E. A. Hill and J. P. Yesinowski, *J. Chem. Phys.*, 1997, **106**, 8650.
70. T. Giavani, H. Bildsoe, J. Skibsted, and H. J. Jakobsen, *J. Magn. Reson.*, 2004, **166**, 262.
71. H. J. Jakobsen, A. R. Hove, R. G. Hazell, H. Bildsoe, and J. Skibsted, *Magn. Reson. Chem.*, 2006, **44**, 348.
72. L. A. O'Dell and R. W. Schurko, *J. Am. Chem. Soc.*, 2009, **131**, 6658.
73. L. A. O'Dell and R. W. Schurko, *Phys. Chem. Chem. Phys.*, 2009, **11**, 7069.
74. L. A. O'Dell, R. W. Schurko, K. J. Harris, J. Autschbach, and C. I. Ratcliffe, *J. Am. Chem. Soc.*, 2011, **133**, 527.
75. D. Holland, S. A. Feller, T. F. Kemp, M. E. Smith, A. P. Howes, D. Winslow, and M. Kodama, *Phys. Chem. Glasses-Eur. J. Glass Sci. Technol. Part B*, 2007, **48**, 1.
76. V. I. Bakhmutov, *Chem. Rev.*, 2011, **111**, 530.
77. T. J. Bastow, M. A. Gibson, and C. T. Forwood, *Solid State Nucl. Magn. Reson.*, 1998, **12**, 201.
78. T. J. Bastow, *Solid State Nucl. Magn. Reson.*, 1998, **12**, 191.
79. T. J. Bastow, *J. Phys.: Condens. Matter*, 1999, **11**, 569.

Chapter 6

Sensitivity and Resolution Enhancement of Half-Integer Quadrupolar Nuclei in Solid-State NMR

Thomas T. Nakashima and Roderick E. Wasylishen

Department of Chemistry, University of Alberta, Edmonton, AB T6G 2G2, Canada

6.1 INTRODUCTION

Nuclear magnetic resonance (NMR) is one of the most powerful spectroscopic tools available for characterizing structure in solid materials; however, the sensitivity of these experiments is relatively low. Of the NMR-active nuclei within the NMR periodic table, over 70% possess spin numbers greater than 1/2 and consequently have nuclear quadrupole moments.[1–4] Thus, their NMR response is generally strongly influenced by the nuclear quadrupolar interaction, and hence, their resonant frequencies depend strongly on the orientation of crystallites in the applied magnetic field. A few of these isotopes possess integer nuclear spin, the most popular used in NMR studies being ^{2}H and ^{14}N, but the overwhelming majority has half-integer nuclear spin quantum numbers. The latter nuclei play an important role in studies of advanced materials, biological molecules, catalysts, glasses, metals, semiconductors, superconductors, and of course, in the entire field of chemistry. Many of the materials containing these nuclides are insoluble in typical solvents and/or lose their integrity (i.e., structure) on dissolution; thus, their examination must be done in the solid state. Furthermore, when these materials are soluble in liquids, rapid molecular tumbling averages the nuclear quadrupolar interaction generally resulting in broad, uninformative NMR peaks. Obtaining narrow NMR peaks is important, especially when examining molecules containing the same NMR-active isotope in many different magnetic environments, e.g., ^{13}C NMR studies of proteins.

NMR of Quadrupolar Nuclei in Solid Materials
Edited by Roderick E. Wasylishen, Sharon E. Ashbrook and Stephen Wimperis
 ISBN: 978-0-470-97398-1

When investigating half-integer spin quadrupolar nuclei in solids, it is desirable to focus on "isolated" systems containing one or only a few unique sites or to selectively label materials with the NMR-active nuclei of interest. This is necessary because these nuclides generally yield broad orientation-dependent peaks that may be difficult or impossible to unravel because of many factors, e.g., their breadths and signal-to-noise issues. Experimental techniques to improve the sensitivity and resolution (in certain cases) of NMR spectra of quadrupolar nuclei have been the focus of several research groups.[5–11] If magic angle spinning (MAS) is feasible, it results in averaging most NMR interactions (i.e., anisotropic magnetic shielding, dipolar interactions), but the second-order quadrupolar interaction cannot be completely averaged by MAS. Note that in practice, for MAS to be practical, one generally has to spin the sample at a frequency of at least one-third the total width of the central transition for a stationary sample.

The single-quantum NMR transitions, $\Delta m_I = \pm 1$, of half-integer spin nuclei may be classified into two distinct categories: the central transition (CT), $m_I = 1/2 \rightarrow m_I = -1/2$, and the satellite transitions (STs). A spin-3/2 nucleus has two STs, $m_I = +3/2 \rightarrow m_I = +1/2$ and $m_I = -1/2 \rightarrow m_I = -3/2$, which are generally labeled ST1. A spin-5/2 nucleus possesses four STs, two ST1s plus two ST2s, $m_I = +5/2 \rightarrow m_I = +3/2$ and $m_I = -3/2 \rightarrow m_I = -5/2$ (Figure 6.1). Spin-7/2 nuclei also possess ST1s and ST2s but with additional STs labeled ST3 ($m_I = +7/2 \rightarrow m_I = +5/2$ and $m_I = -5/2 \rightarrow m_I = -7/2$). Of course, spin-9/2 will contain one more set of STs labeled ST4 ($m_I = +9/2 \rightarrow m_I = +7/2$ and $m_I = -7/2 \rightarrow m_I = -9/2$).[12–14] To first order, the CT resonance is unaffected by the nuclear quadrupolar interaction; however, the STs are affected by both first- and second-order quadrupolar effects. Thus, it is the CT resonance that is typically observed for powdered samples of half-integer spin quadrupolar nuclei, even though these resonances may span many kilohertz; the ST resonances, on the other hand, may span tens of megahertz or more. While the broadening of the CT may make NMR observation difficult, the shape of the NMR resonance contains a wealth of molecular structural information. The goal of this chapter is to summarize some of the recent techniques that have been used to enhance the sensitivity of the CT by altering the populations of the nuclear energy levels associated with the STs. Any experiment that starts out with CT magnetization will benefit, e.g., the multiple quantum magic angle spinning (MQMAS) experiment.[15,16] As it turns out, some of these population transfer techniques can also be used for resolution enhancement in cases for which more than one polymorph or a mixture of different compounds exists in the powdered sample.[11,17,18]

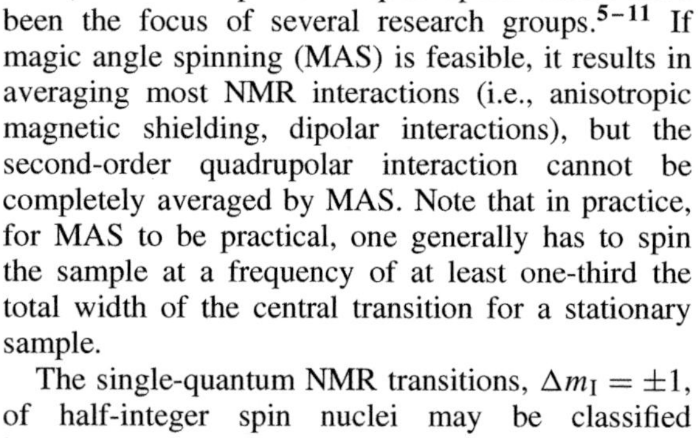

Figure 6.1. Schematic representation of the nuclear energy levels and relative population distributions for a spin-5/2 nucleus: (a) at thermal equilibrium, (b) with inversion of the populations of both ST1s, (c) with inversion of the populations of ST2s followed immediately with inversion of the populations of both ST1s, and (d) with inversion of the populations of both ST1s followed immediately with inversion of both ST2s.

Historically, cross polarization (CP)[19] has been used to increase the sensitivity of solid-state NMR experiments for insensitive, low-abundant, spin-1/2 nuclei that are in close proximity to abundant nuclei with large magnetic moments such as protons (^{1}H). This is because there is a gain in sensitivity in the NMR spectrum of the low-abundant spin nuclei, and generally speaking, the relaxation times for these nuclei are much longer than those of the abundant spins; thus, recycle delays can be much shorter using CP than the normal single-pulse NMR experiment. Analogous CP/MAS experiments for non-integer spin quadrupolar nuclei,[20] however, generally do not provide significant gains in sensitivity in part because the quadrupolar nuclei often have relatively short relaxation times compared to abundant spin-1/2 nuclei. Furthermore, CP is found to cause significant distortions in the CT line shapes of half-integer quadrupolar nuclei in samples spinning at the magic angle.[21,22]

6.2 SINGLE-CRYSTAL ENERGY LEVELS, POPULATIONS, AND DETECTION

The intensities of single-quantum NMR transitions are proportional to the population difference between the two energy levels involved in the transition.[13,14] Consider the spin-5/2 system,[5,23] shown in Figure 6.1(a), where the relative populations of the various m_I states are illustrated with balls. The maximum population difference obtainable between the $+1/2$ and $-1/2$ energy levels (i.e., associated with the CT transition intensity) is shown in Figure 6.1(c). The relative populations shown in Figure 6.1(c) can be obtained by first inverting both ST2s followed by the inversion of both ST1s. Similarly, for a spin-3/2 system, inversion of both ST1s yields the maximum population difference between the $m_I = +1/2$ and $-1/2$ energy levels. For spin-7/2 nuclei, the maximum population difference between the $+1/2$ and $-1/2$ energy levels occurs with inversion of both ST3s, then ST2s, and finally the ST1s. For half-integer quadrupolar nuclei, it is easy to show that the maximum population difference for the $+1/2$ and $-1/2$ levels is $2I$. Thus, the increase in the population difference for the energy levels associated with the CT is 3, 5, 7, and 9 for spin 3/2, 5/2, 7/2, and 9/2, respectively; hence, one might expect analogous enhancements in NMR intensities obtained from the CT, provided it is detected with a selective pulse. These maximum enhancements have been demonstrated for spin-3/2 and -5/2 nuclei obtained from single-crystal samples.[5,7,24] If, on the other hand, the upper and lower levels were saturated (i.e., levels with $m_I \geq +1/2$ have equal populations and levels with $m_I \leq -1/2$ have equal populations), then the maximum enhancements of the CT would be $I + 1/2$, i.e., a factor of 2, 3, 4, and 5 for spin-3/2, -5/2, -7/2, and -9/2 nuclei, respectively.

The importance of using selective pulses in population transfer experiments is illustrated for a spin-3/2 system in Figure 6.2.[25] Shown in Figure 6.2(a) is the ^{23}Na NMR spectrum of a single crystal of $NaNO_3$ obtained using a nonselective 90° pulse for the spin system initially at thermal equilibrium. The CT, labeled $M_x{}^{2,3}$, and the STs, labeled $M_x{}^{1,2}$ and $M_x{}^{3,4}$, have relative intensities 3:4:3 as expected. The superscript numbers represent the nuclear energy levels between each transition. For this particular orientation of the single crystal, the carrier frequency was adjusted so that the CT was positioned at 0 Hz, and the

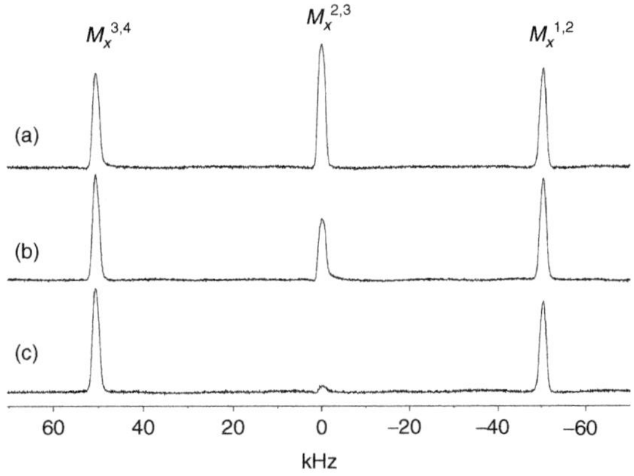

Figure 6.2. Sodium-23 NMR spectra of a single crystal of $NaNO_3$ obtained with a nonselective 90° pulse starting from (a) thermal equilibrium, (b) with inversion of the populations of one of the satellite transitions, and (c) with both satellite transition populations inverted. Populations were inverted with a hyperbolic secant pulse. $M_x{}^{2,3}$ and $M_x{}^{1,2}$ and $M_x{}^{3,4}$ represent the x-magnetization from the central transition and satellite transitions, respectively. (Reprinted with permission, from: T.T. Nakashima, K.J. Harris and R.E. Wasylishen, *J. Magn. Reson.*, 2010, **202**, 162. © 2009 Elsevier Inc.)

STs appear at approximately ±50 kHz. If the crystal was repositioned in the probe, the CT would resonate again at approximately 0 Hz since the ^{23}Na shielding anisotropy is small, while the STs could vary in frequency anywhere from zero to a value equal to ± the quadrupolar coupling constant, C_Q, divided by 2. If the intensity of one of the STs is inverted before applying the nonselective 90° pulse, the central transition, $M_x{}^{2,3}$, appears to be half the intensity of the outer transitions (Figure 6.2b). If both STs are selectively initially inverted, the CT intensity is essentially zero (Figure 6.2c). From simple population considerations (e.g., see Figure 6.1), one would have predicted that the CT intensity is to be enhanced by a factor of 2 and 3 by inverting one and two satellite peaks, respectively. These "unexpected" NMR intensity variations have been observed previously in liquid-state NMR and are loosely referred to as "flip angle effects".[26] As illustrated above, such effects in the solid state are best demonstrated using single crystals as opposed to powdered samples.[27] Density matrix calculations have been performed to reproduce the experimentally determined spectra in the liquid state[28] and recently have also shown success for the case of single-crystal

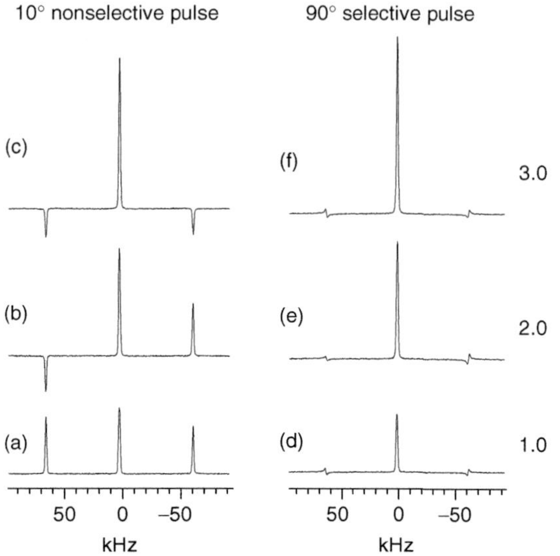

Figure 6.3. Sodium-23 NMR spectra of a single crystal of $NaNO_3$ obtained with a 10° nonselective pulse (a, b, and c) and a 90° selective pulse (d, e, f) at thermal equilibrium (a and d), with the high-frequency transition inverted (b and e), and with both the outer transitions inverted (c and f). (Reprinted with permission, from: T.T. Nakashima, K.J. Harris and R.E. Wasylishen, *J. Magn. Reson.*, 2010, **202**, 162. © 2009 Elsevier Inc.)

solid-state NMR experiments for spin 3/2, 5/2, 7/2, and 9/2.[25] For $I = 3/2$, it was found that the intensities of the STs are only negative with nonselective pulse angles less than approximately 45°. To ensure that inversion of the outer transitions has taken place, flip angles of approximately 10–40° are typically used in the literature.[24] Figure 6.3, on the left, shows the ^{23}Na NMR spectrum from a single crystal of $NaNO_3$ obtained with a nonselective 10° pulse starting at (a) thermal equilibrium, (b) after the high-frequency ST is inverted, and (c) when both STs are inverted prior to applying the nonselective observe pulse. The spectra in Figure 6.3 on the right were obtained with a 90° selective pulse showing the relative enhancements of 2.0 and 3.0 as expected.[29]

6.3 METHODS OF ALTERING POPULATIONS

There are several ways of perturbing the populations of the various energy levels within a nuclear spin system. One simply requires a radiofrequency (RF) at a specific position with the appropriate power level. For NMR single-crystal experiments involving half-integer quadrupolar nuclei, a perturbation that inverts the outermost spin energy levels sequentially leads to the greatest enhancement of the CT. One could accomplish this with a series of well-placed π pulses,[5] but this requires knowledge of the appropriate frequencies. Alternatively, one could apply a linearly varying frequency on one side of the CT frequency and sweep from a frequency far off resonance toward the CT frequency and then change the frequency to the other side of the CT frequency and repeat the process.[5] Obviously, the highest frequency starting point should be greater than that expected for the outermost transition frequency. Haase and Conradi were the first to report this technique for both a single crystal and a powdered sample of Al_2O_3.[5]

Instead of applying the frequencies sequentially on either side of the CT, it is advantageous to apply the frequencies simultaneously to help minimize the effects of spin lattice relaxation in reestablishing the equilibrium Boltzmann populations associated with the CT (i.e., the $m_I = 1/2$ and $-1/2$ levels).[7] The generation of these two frequencies is accomplished by linearly varying the frequency of the amplitude modulation of the centerband. Two modulation methods may be used; first, a converging sweep for which the frequency begins far away from the centerband frequency and sweeps linearly closer to the centerband; or second, a divergent sweep in which the frequency initially is close to the centerband frequency and moves away. Perturbation of the CT frequency should be avoided if one wishes to maximize the population difference between $m_I = 1/2$ and $m_I = -1/2$ levels. This technique is called the *double-frequency sweep* (*DFS*) technique and was first reported by the Kentgens group.[7] Of course, the convergent sweep is desired for optimum sensitivity enhancement used on single crystals as is illustrated in Figure 6.1 by comparing the population distribution shown in Figure 6.1(c) (achieved using ideal convergent DFS) versus Figure 6.1(d) (achieved using ideal divergent DFS).

Another means of modulating the centerband frequency is through the use of a train of very short, hard pulses separated by a short time interval, τ. This technique was developed for selective excitation in liquid-state NMR.[30] The effect of this pulse train is to create RF sidebands about the RF carrier that are positioned at a value $\Delta\nu = \pm n/\tau$ Hz, where n is an integer, and was coined the DANTE pulse train for "delays alternating with nutations for tailored excitation" by Freeman and his group.[30] Similar pulse trains are used to increase the sensitivity of the MQMAS experiment by making conversion of single- to triple-quantum coherence and the reverse more efficient.[8] The pulse train was coined with the acronym FAM for "fast amplitude modulation".[8] Grandinetti *et al.*[9] used the acronym RAPT standing for "rotor-assisted population transfer" when the pulse train is used in conjunction with MAS. The problem with the FAM or RAPT techniques is that only two frequencies on either side of the carrier are generated; thus, lower than optimum sensitivities are obtained when examining stationary single crystals or powdered samples. To improve the performance of the FAM experiment for stationary samples, SW(τ)-FAM and SW($1/\tau$)-FAM experiments have been developed.[31] In the SW(τ)-FAM pulse train, constant time intervals are added to both the pulse durations and interpulse delays, which have the effect of imparting a varying effective frequency of the pulse as the total train time becomes longer. The change in the frequency with time is not linear, and thus the SW($1/\tau$)-FAM pulse train was developed, for which the frequency change is linear with the pulse train length. To accomplish this linearity, both the pulse duration and interpulse delay must be changed via the precalculated algebraic expressions. In most cases of stationary samples, the SW($1/\tau$)-FAM pulse train provides slightly better performance than the SW(τ)-FAM pulse train.[31]

Shaped adiabatic pulses have also been used to invert the populations of the energy levels associated with the STs in the solid state.[11,29,32,33] Such adiabatic pulses are generally created with both amplitude and phase modulation as shown in Figure 6.4 for the hyperbolic secant (HS) pulse.[34–36] The amplitude, B_1, varies as an HS function while the phase, p, varies as a hyperbolic cosine function during the total pulse duration, T_p; the functions are shown on the right side of Figure 6.4. The phase modulation has the effect of changing the effective frequency, FM, of the pulse in the form of a hyperbolic tangential function as shown in Figure 6.4(c) and traverses a frequency range of the bandwidth, Bwd, about the frequency position of the HS pulse. The inversion profile, solid line, generated by the HS pulse is illustrated in Figure 6.5 and is compared to that expected for a similar rectangular "hard" 180° pulse

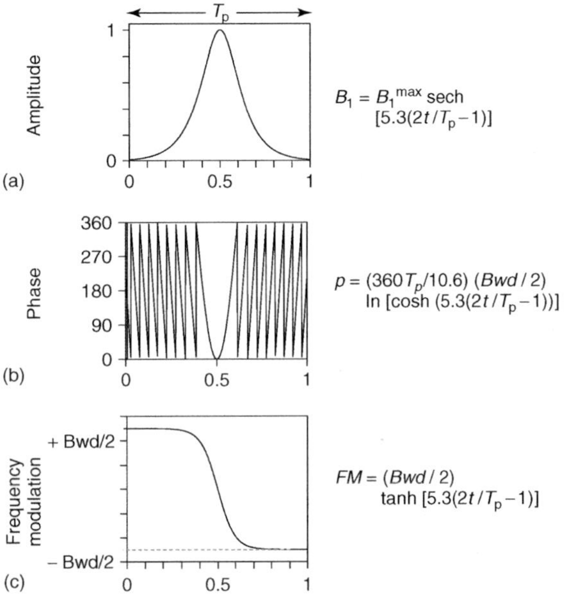

Figure 6.4. Hyperbolic secant pulses are created by varying both (a) the amplitude and (b) the phase of the pulse over the pulse duration, T_p. The phase modulation results in an effective frequency variation in (c) as a hyperbolic tangent function over a frequency range of Bwd.

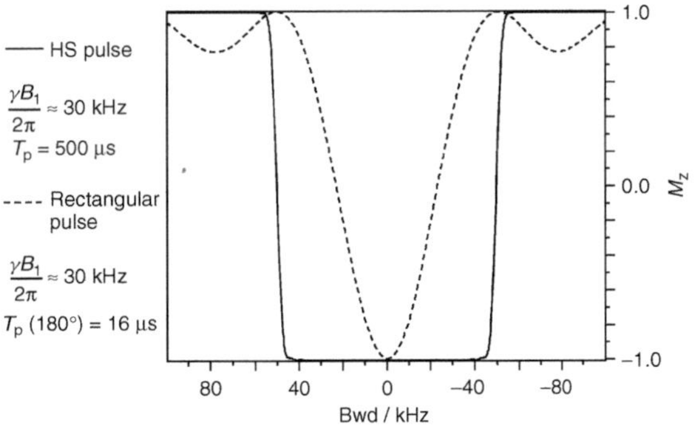

Figure 6.5. z-Magnetization, M_z, inversion profile of a hyperbolic secant pulse of 500 μs duration versus a rectangular pulse of 16 μs.

shown as a dotted line. Clearly, the inversion profile is remarkably uniform over the Bwd frequency for the HS pulse. Any z-magnetization whose resonant frequency is outside of Bwd remains along the z-axis, and any z-magnetization within Bwd is inverted. On the other hand, the effective frequency range for inversion is extremely limited for the rectangular pulse. Also, the HS performance is independent of the RF field once above a certain power level and thus is more or less independent of inhomogeneities in susceptibility and of miscalibration.[36] A further advantage is that it can be made selective by reducing the value of Bwd, as we shall see in experiments to be demonstrated later. Cosine modulation of the amplitude will create two HS pulses about its carrier frequency. Other adiabatic pulse shapes have been used, e.g., WURST[37,38] (wideband uniform-rate smooth truncation) and chirp pulses, both of which employ phase and amplitude modulation.

6.4 FROM SINGLE CRYSTALS TO POWDERS

While single crystals serve as excellent models for demonstrating the use of these population transfer experiments for CT enhancement, the important question is what happens when one is dealing with powder samples. Pinpointing the exact ST frequency location is straightforward for a single crystal but not for a powder sample since the STs are distributed over a wide range of frequencies. The optimum frequency of the inversion pulse for maximum enhancement is not obvious. For noninteger spin quadrupolar nuclei in powder samples, the STs are characterized by an intensity distribution – a powder pattern. Because the CT is independent of the first-order quadrupolar interaction, it is relatively sharp and hence much more intense than the STs. For example, for the ^{87}Rb NMR spectrum of a stationary powder sample of $RbClO_4$, the CT spans a frequency range of about 16.5 kHz while the STs span a frequency range of ±1.6 MHz. For this nucleus, C_Q is 3.221 MHz with an asymmetry value of 0.195.[39]

For a sample spinning at the magic angle, the STs break up into many sharp spinning sidebands, ssbs, each separated by the spinning frequency, but the ST spectra retain the same intensity profile as that of the stationary powder sample. The CT sharpens by a factor of 3–4[40] when spinning and typically gives rise to a few ssbs due to the anisotropy of the magnetic shielding tensor. The ssbs generated by the STs and the CT, however, are generally not coincident but are shifted by what is known as the quadrupolar induced shift or the second-order quadrupolar shift. If one could attain infinite spinning frequencies or perform the rotor-synchronized experiment,[41] the NMR spectrum for a spin-3/2 system would result in a spectrum as in Figure 6.6(a) and for a spin-5/2 system a spectrum as in Figure 6.6(b), which shows that the resonance frequency of the CT is different from those of ST1 and in the case of spin-5/2, also ST2.[4] Samoson[42] has shown that the second-order quadrupolar shift for the m to $m-1$ transition is given by

$$\Delta\nu_{(m,m-1)} = -3/40\left\{C_Q^2/\nu_L\right\} \times \left[\frac{(I(I+1) - 9\text{ m }(\text{m}-1) - 3)}{(I^2(2I-1)^2)}\right][1+\eta^2/3] \quad (6.1)$$

where C_Q is the quadrupolar coupling constant, ν_L is the Larmor precession frequency, and η is the asymmetry parameter, which determines the shape of the powder pattern and is generally determined from simulations. The difference in frequencies for a spin-3/2 nucleus between the CT and the ST can then be calculated from the equation above and is given by

$$\Delta\nu_{(\pm3/2,\pm1/2)} - \Delta\nu_{(+1/2,-1/2)} = (3/40)(C_Q{}^2/\nu_L)(1+\eta^2/3) \quad (6.2)$$

Spinning at a frequency of ν_R at the magic angle then will generate ssbs at frequencies of $\Delta\nu_{(\pm3/2,\pm1/2)} \pm n\nu_R$ and at $\Delta\nu_{(+1/2,-1/2)} \pm n\nu_R$, where n is an integer known as the *order* of the ssb. Clearly, the few ssbs from the CT will not overlap with those from the STs unless the C_Q is very small. Samoson[42] has also shown that the ratio of the width at half-height, $\Delta\nu_{1/2}$, of the STs versus the CT is given by

$$\frac{\Delta\nu_{1/2}(m)}{\Delta\nu_{1/2}(\text{CT})} = \left\{\frac{6I(I+1) - 34m(m-1) - 13}{6I(I+1) - 9/2}\right\} \quad (6.3)$$

For a spin-3/2 nucleus, this ratio is equal to −0.89, which indicates that the ST1 is slightly sharper than the CT and has a shape (i.e., a reversed line shape) opposite that of the CT (see Figure 6.6a).

Shown in Figure 6.6(b) is a calculated NMR spectrum for a spin-5/2 nucleus, in this case, the ^{27}Al NMR spectrum expected for an extremely rapidly spinning powder sample of $Al(acac)_3$.[4,43] The general expressions for the single-quantum transition frequencies relative to ν_L of a spin-5/2 nucleus spinning infinitely fast are[42]

$$\nu_{ST2} = \Delta\nu_{(\pm5/2,\pm3/2)} = (84/4000)(C_Q{}^2/\nu_L)(1+\eta^2/3) \quad (6.4)$$

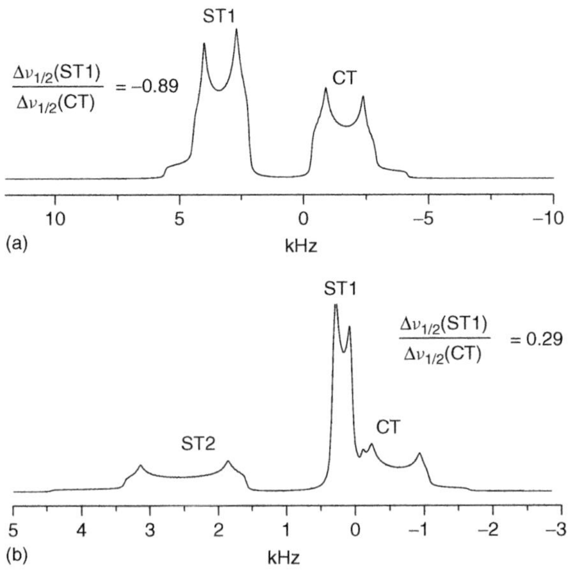

Figure 6.6. (a) Calculated ^{87}Rb NMR spectrum expected for a powdered sample of $RbClO_4$ at infinite spinning speed or with a rotor-synchronized experiment. (b) Calculated ^{27}Al NMR spectrum expected for a powdered sample of $Al(acac)_3$ at infinite spinning speeds or with the rotor-synchronized experiment. Spectra calculated with SIMPSON.

$$\nu_{\mathrm{ST1}} = \Delta\nu_{(\pm 3/2, \pm 1/2)} = (3/4000)(C_Q{}^2/\nu_L)(1 + \eta^2/3) \quad (6.5)$$

$$\nu_{\mathrm{CT}} = \Delta\nu_{(+1/2, -1/2)} = (-24/4000)(C_Q{}^2/\nu_L)(1 + \eta^2/3) \quad (6.6)$$

for the ST2, ST1, and CT, respectively. If the sample is spun at experimentally achievable rotational frequencies, ν_R, typically in the range of 5–60 kHz, then ssbs will be generated about each of the frequencies shown above, i.e., at frequency positions of $\nu_{\mathrm{CT}} \pm n\nu_R$, $\nu_{\mathrm{ST1}} \pm n\nu_R$, and $\nu_{\mathrm{ST2}} \pm n\nu_R$.

6.5 SENSITIVITY ENHANCEMENT FOR POWDERED SAMPLES

Naively one would expect that in order to observe the maximum enhancement for the CT, a frequency swept experiment, such as DFS, starting at a position far off resonance then sweeping toward the CT

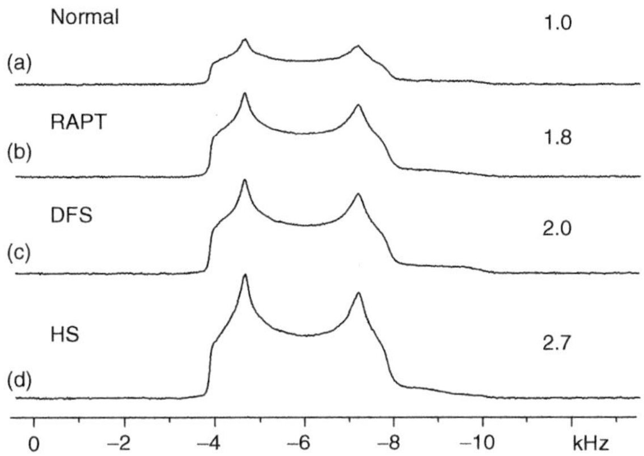

Figure 6.7. Rubidium-87 NMR Hahn-echo spectrum of $RbClO_4$ spinning at 10 kHz (a) normal and with the CT enhanced with (b) RAPT, (c) DFS, and (d) HS. Enhancement factors are shown on the right. (Adapted with permission, from: R. Siegel, T.T. Nakashima and R.E. Wasylishen, *Chem. Phys. Lett.*, 2004, **388**, 441. © 2004 Elsevier B.V.)

to excite all the crystal orientations of the STs would provide the greatest enhancement for the CT. The sweep is stopped before the CT is perturbed. Such experiments have been successfully performed, and the results are shown in Figures 6.7 and 6.8 for the ^{87}Rb NMR spectrum of a powdered MAS sample of $RbClO_4$ as a representative spin-3/2 system[24] and the ^{27}Al NMR spectrum of a powdered MAS sample of $Al(acac)_3$, a representative spin-5/2 system.[23] However, for both $RbClO_4$ and $Al(acac)_3$, it was observed that application of HS pulses to the satellite ssbs provided the largest enhancements. In the HS experiments, the bandwidth of the HS pulse is set to the MAS frequency (vide infra). FAM (or RAPT) and DFS provide enhancements of the order that one expects from saturation of the outer transitions while the application of HS provided enhancements nearer to the theoretical maximum enhancements.

Distortions in the CT enhancement for a spin-3/2 nucleus are observed[10] if the bandwidth of the HS pulse is narrower than that of the total width of the ST1 ssb. Of course, when the HS offset is positioned so that the ST1 ssb is not perturbed, there is no enhancement observed for the CT. When the left portion of one ST1 ssb is irradiated, then the right half of the CT is enhanced and vice versa.[10] This is consistent with Figure 6.6(a), and the experimental

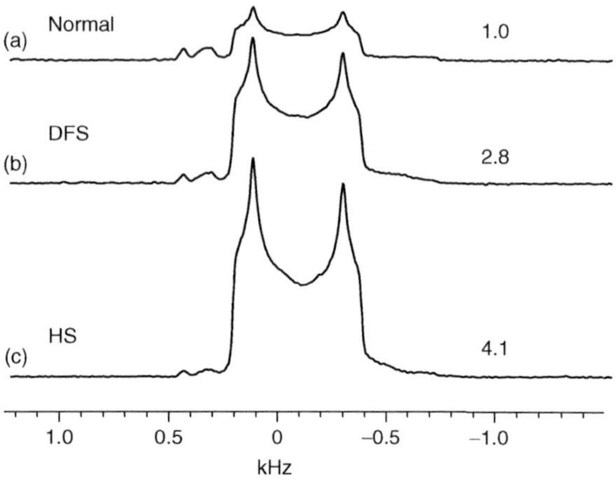

Figure 6.8. Aluminum-27 NMR spectrum of a sample of $Al(acac)_3$ spinning at 20 kHz (a) normal with the CT enhanced by (b) DFS and (c) HS. Enhancement factors are shown on the right. (Reprinted with permission, from: R. Siegel, T.T. Nakashima and R.E. Wasylishen, *Chem. Phys. Lett.*, **421**, 529. © 2006 Elsevier B.V.)

results are confirmed with simulations performed using the program SIMPSON.[44] Maximum undistorted CT enhancement occurs when the HS pulse is effective over one ST1 ssb. So, the conclusion is that only one entire ST1 ssb need be inverted to effect the maximum CT enhancement, a result which may not be entirely expected. However, other researchers have also reported this behavior.[11] It was also found that maximum CT enhancement occurs when the bandwidth of the HS pulse is equal to the spinning frequency, with minimum enhancements occurring at twice the spinning frequency, with another secondary maximum at 3 times the spinning frequency.[10] These results were reproduced theoretically. For a powder sample of $RbClO_4$ spinning at the magic angle, the CT ^{87}Rb NMR enhancements achieved using HS pulses as a function of the HS offset was also examined (Figure 6.9). The experimental and calculated enhancement profile closely mimics that of the relative distributions of the ST1 frequencies due to the different crystal orientations with maximum enhancement occurring at approximately 400–900 kHz from the carrier frequency with an inflection point at about 1000 kHz. There is no enhancement when the offset of the HS pulse is greater than approximately 1600 kHz. Grandinetti *et al.*[45] used a similar approach using the RAPT technique by plotting the enhancement profile versus frequency of the RAPT pulse train to determine the quadrupolar coupling constants of half-integer spin quadrupolar nuclei. Note that the enhancement drops below 1 as the offset frequency approaches the CT resonance because of saturation of the CT.

For spin-5/2 nuclei, the CT enhancements[10] depend on the same parameters noted above for a spin-3/2 system. However, for the ^{27}Al NMR spectrum of $Al(acac)_3$, no oscillation in the enhancement profile was noticed when the bandwidth of the HS pulse was varied, but simply a broad maximum was observed when the bandwidth was equal to the spinning frequency. Also, the enhancement factors increased as the sample spinning frequency increased. Enhancement factors of 2.7, 3.3, 3.5, and 4.1 were obtained at spinning frequencies of 5, 10, 13, and 20 kHz, and it is believed that complete separation of the ssbs is responsible for this. It was concluded again that only one ST ssb should be in the window of the HS pulse and also that within a given ssb, the largest enhancements occur if the ST2 ssb is traversed before the ST1 ssb.[10] In all cases, simulations reproduce the experimental results. For ^{27}Al in $Al(acac)_3$, there is almost a one-to-one correspondence between the HS enhancement and powder pattern profile versus HS frequency. The maximum experimental enhancement of 4.1 occurs at an offset of approximately 150–250 kHz from the carrier, which corresponds to the maximum distribution of the ST1 frequencies.

For both spin-3/2 and -5/2, HS provided maximum enhancements in MAS samples using bandwidths affecting only one ssb, whereas the protocol for studying these samples using DFS is to sweep far off the CT affecting many ssbs. However, both DFS and HS were found to provide similar enhancements involving studies of single crystal or stationary powders.[10] Finally, it has been demonstrated that for MAS samples, CT enhancements obtained by applying DFS can be increased if only one ssb is perturbed[46] just as was found in the HS studies.

6.6 APPLICATION OF HYPERBOLIC SECANT PULSES IN RESOLUTION ENHANCEMENT

As shown in Figure 6.5, the HS pulses are highly selective. This is clearly illustrated in Figure 6.10, which displays the ^{23}Na NMR spectrum of a sample

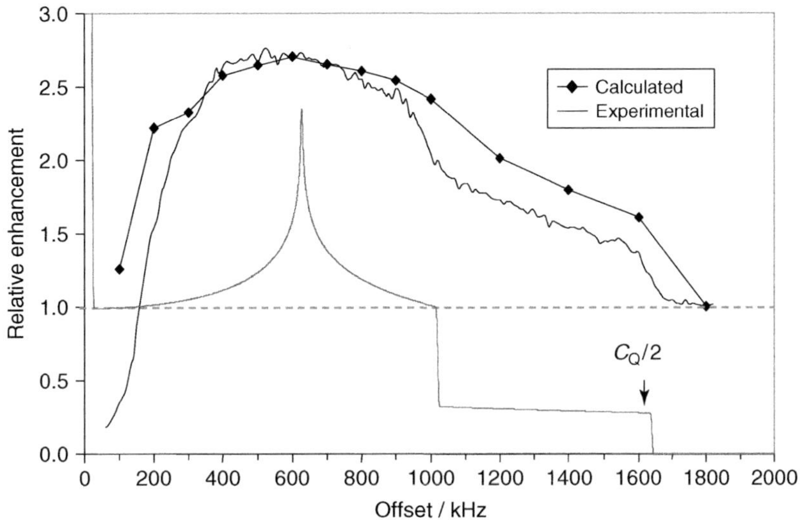

Figure 6.9. Theoretical and experimental plot of the relative CT enhancement versus HS offset of an ^{87}Rb NMR spectrum of powdered $RbClO_4$. The lower trace is one half of the calculated powder pattern for a stationary sample. (Reprinted with permission, from: R.R. Siegel, T.T. Nakashima and R.E. Wasylishen, *J. Magn. Reson.*, 2007, **184**, 85. © 2006 Elsevier Inc.)

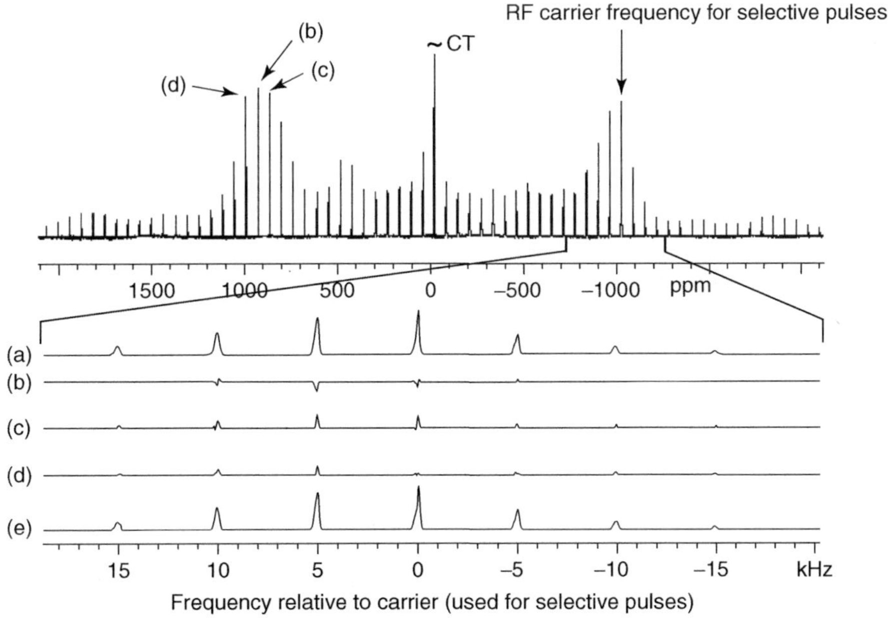

Figure 6.10. Sodium-23 NMR spectra of a powder sample of $NaNO_3$ obtained with MAS at 5.0 kHz. The upper trace is the normal spectrum with the CT off scale, (a through e) with selective excitation, 49 μs, (b through d) with weak HS pulse applied near positions labeled (c), (b), and (d) on top spectrum; (e) is the spectrum obtained when the HS pulse is applied at a frequency between the positions labeled (c) and (d). See text for details. (Reprinted with permission, from: T.T. Nakashima, R. Teymoori and R.E. Wasylishen, *Magn. Reson. Chem.*, 2009, **47**, 465. © 2009 John Wiley & Sons, Ltd.)

of $NaNO_3$ spinning at 5.0 kHz (vide infra).[18] The CT of the normal spectrum in the upper trace is truncated to emphasize the ST1 ssbs. In the series of spectra shown below, labeled (a) through (e), the transmitter was set to a frequency position indicated by the arrow, and the spectra were acquired applying a selective 90° pulse (48 μs). The spectrum displayed in (a) is the normal spectrum, and in (b), (c), and (d), an HS pulse with bandwidth of 1.0 kHz was applied at the frequency locations indicated on the top spectrum immediately before collection of the data with a selective pulse. These spectra indicate that whenever one ssb of the ST1 manifold is perturbed, all of them are affected. The spectrum shown in (e) was obtained when the HS frequency was positioned between the frequencies used in (b) and (c) and yields the normal spectrum obtained in (a). This technique has also been used to eliminate the NMR spectra obtained from spin-1/2 nuclei from samples spinning at the magic angle and has allowed separation of signals arising from different components in a sample.

Use of this information has also allowed the separation of the NMR spectrum into subspectra from each component contained within a mixture[11] but using a WURST pulse,[37,38] which is effective at lower power levels than HS and has a similar excitation profile. Grandinetti *et al.*[11] were able to separate into individual ^{87}Rb NMR subspectra each component from a sample containing a mixture of Rb_2SO_4 (two NMR sites) and $RbClO_4$, by enhancing each individual compound's CT with a selective WURST pulse applied to a particular ST1 ssb of one particular component. Subtraction of this enhanced spectrum from a blank yields the individual subspectrum. By controlling the excitation frequency using a RAPT train composed of Gaussian pulses, Grandinetti and coworkers[17] were also able to separate the same two components of Rb_2SO_4. Wang and Smith[47] used truncated DFS sweeps to excite different regions of the STs to enhance individual CT components of the ^{93}Nb NMR spectrum from a mixture containing compounds having different C_Q values, thus allowing the use of difference spectroscopy to yield subspectra of a mixture of $MgNb_2O_6$ and $LiNbO_3$.

When the C_Q of the nucleus is small but the CSA is large, the NMR spectra obtained from an MAS powder sample is rather complicated and consists of many ssbs as shown in Figure 6.11, which is the ^{51}V NMR spectrum of a powdered sample of a dimer of bis(N,N-dimethylhydroxamido)-

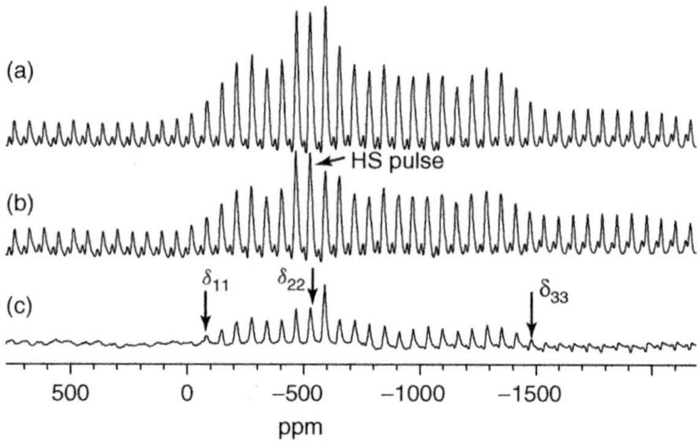

Figure 6.11. Vanadium-51 NMR spectrum of $[V(O)(ONMe_2)_2]_2O$ MAS (5.0 kHz): (a) normal one pulse, (b) with a very weak HS pulse applied as indicated, and (c) subtraction of spectrum (b) from (a). (Reprinted with permission, from: T.T. Nakashima, R. Teymoori and R.E. Wasylishen, *Magn. Reson. Chem.*, 2009, **47**, 465. © 2009 John Wiley & Sons, Ltd.)

hydroxooxovanadate, $\{V(O)(ONMe_2)_2\}_2O$, spinning at 5.0 kHz.[18] It is not obvious from the inspection of the normal spectrum shown in Figure 6.11(a) where the CT begins or ends. Application of a very weak and selective HS pulse to the position of the isotropic resonance (shown in Figure 6.11b) should partially suppress only the CT and by subtraction from the normal spectrum, allows extraction of the principal components of the chemical shift tensor from an analysis of the difference spectrum, shown in Figure 6.11(c).

6.7 CONCLUDING REMARKS

Several methods of enhancing the central transition in NMR spectra of non-integer spin quadrupolar nuclei have been discussed. Application of these techniques will almost always result in significant enhancements of the CT, particularly for spin-3/2 and spin-5/2 systems contained in MAS samples. In case of the HS experiment, one simply adjusts the bandwidth of the HS pulse to the MAS frequency and for spin-3/2 systems, centers the HS pulse near the estimated value of $C_Q/4$. Obviously, the population transfer techniques discussed here will fail if the spin–lattice relaxation times of the quadrupolar nuclei are extremely short (i.e., on the order of the pulse widths used for population inversion of STs). For spin systems with $I = 7/2$ or 9/2, the enhancement techniques described here will result in CT enhancements[48] (Nakashima and Wasylishen, unpublished results), but they are generally much less than the maximum ones because of the complexity of the ssb manifold, which contain differing intensity contributions from the various STs. Research in this area is of course ongoing.

REFERENCES

1. S. E. Ashbrook and M. J. Duer, *Concepts Magn. Reson.*, 2006, **28A**, 183–248.
2. S. E. Ashbrook, *Phys. Chem. Chem. Phys.*, 2009, **11**, 6892–6905.
3. M. E. Smith and E. R. H. van Eck, *Prog. Nucl. Magn. Reson. Spectrosc.*, 1999, **34**, 159–201.
4. R. E. Wasylishen and G. M. Bernard, in *Comprehensive Organometallic Chemistry III*, eds R. H. Carbtree, D. M. P. Mingos, and G. Parkin, Elsevier: Oxford, 2007, Vol. 1, pp. 451–482, Chapter 17.
5. J. Haase and M. S. Conradi, *Chem. Phys. Lett.*, 1993, **209**, 287–291.
6. J. Haase, M. S. Conradi, C. P. Grey, and A. J. Vega, *J. Magn. Reson., Ser. A*, 1994, **109**, 90–97.
7. A. P. M. Kentgens and R. Verhagen, *Chem. Phys. Lett.*, 1999, **300**, 435–443.
8. P. K. Madhu, A. Goldbourt, L. Frydman, and S. Vega, *Chem. Phys. Lett.*, 1999, **307**, 41–47.
9. Z. Yao, H.-T. Kwak, D. Sakellariou, L. Emsley, and P. J. Grandinetti, *Chem. Phys. Lett.*, 2000, **327**, 85–90.
10. R. Siegel, T. T. Nakashima, and R. E. Wasylishen, *J. Magn. Reson.*, 2007, **184**, 85–100.
11. K. K. Dey, S. Prasad, J. T. Ash, M. Deschamps, and P. J. Grandinetti, *J. Magn. Reson.*, 2007, **185**, 326–330.
12. S. E. Ashbrook and S. Wimperis, *J. Magn. Reson.*, 2002, **156**, 269–281.
13. D. Freude and J. Haase, in *NMR Basic Principles and Progress*, eds P. Diehl, E. Fluck, H. Gunter, R. Kosfeld, and J. Seelig, Springer-Verlag: Berlin, 1993, Vol. 29, pp. 1–90.
14. P. P. Man, in *Encyclopedia of Analytical Chemistry*, ed R. A. Meyers, John Wiley & Sons: Tarzana, CA, 2000, Vol. 14, pp. 12224–12265.
15. L. Frydman and J. S. Harwood, *J. Am. Chem. Soc.*, 1995, **117**, 5367–5368.
16. R. Siegel, T. T. Nakashima, and R. E. Wasylishen, *Chem. Phys. Lett.*, 2005, **403**, 353–358.
17. H. T. Kwak, S. Prasad, T. Clark, and P. J. Grandinetti, *J. Magn. Reson.*, 2003, **160**, 107–113.
18. T. T. Nakashima, R. Teymoori, and R. E. Wasylishen, *Magn. Reson. Chem.*, 2009, **47**, 465–471.
19. A. Pines, M. G. Gibby, and J. S. Waugh, *J. Chem. Phys.*, 1973, **59**, 569–590.
20. A. J. Vega, *Solid State Nucl. Magn. Reson.*, 1992, **1**, 17–32.
21. P. J. Barrie, *Chem. Phys. Lett.*, 1993, **208**, 486–490.
22. S. Hayashi and K. Hayamizu, *Chem. Phys. Lett.*, 1993, **203**, 319–324.
23. R. Siegel, T. T. Nakashima, and R. E. Wasylishen, *Chem. Phys. Lett.*, 2006, **421**, 529–533.
24. R. Siegel, T. T. Nakashima, and R. E. Wasylishen, *Chem. Phys. Lett.*, 2004, **388**, 441–445.
25. T. T. Nakashima, K. J. Harris, and R. E. Wasylishen, *J. Magn. Reson.*, 2010, **202**, 162–172.
26. S. Schaublin, A. Hohener, and R. R. Ernst, *J. Magn. Reson.*, 1974, **13**, 196–216.
27. N. M. Trease, K. K. Dey, and P. J. Grandinetti, *J. Magn. Reson.*, 2009, **200**, 334–339.
28. P. Meakin and J. P. Jesson, *J. Magn. Reson.*, 1975, **18**, 411–426.
29. R. Siegel, T. T. Nakashima, and R. E. Wasylishen, *Concepts Magn. Reson.*, 2005, **26A**, 47–61.
30. G. A. Morris and R. Freeman, *J. Magn. Reson.*, 1978, **29**, 433–462.
31. T. Brauniger and P. K. Madhu, *J. Magn. Reson.*, 2008, **193**, 102–109.
32. L. A. O'Dell, K. Klimm, J. C. C. Freitas, S. C. Kohn, and M. E. Smith, *Appl. Magn. Reson.*, 2008, **35**, 247–259.
33. M. R. Hansen, M. Brorson, H. Bildsoe, J. Skibsted, and H. J. Jakobsen, *J. Magn. Reson.*, 2008, **190**, 316–326.
34. M. S. Silver, R. I. Joseph, C. N. Chen, V. J. Sank, and D. I. Hoult, *Nature*, 1984, **310**, 681–683.
35. M. S. Silver, R. I. Joseph, and D. I. Hoult, *Magn. Reson. Med.*, 1984, **1**, 294.
36. M. S. Silver, R. I. Joseph, and D. I. Hoult, *Phys. Rev. A*, 1985, **31**, 2753–2755.
37. Ē. Kupče and R. Freeman, *J. Magn. Reson.*, 1995, **A 117**, 246–256.

38. Ē. Kupče and R. Freeman, *J. Magn. Reson.*, 1996, **A 118**, 299–303.

39. F. H. Larsen, H. J. Jakobsen, P. D. Ellis, and N. C. Nielsen, *J. Phys. Chem. A*, 1997, **101**, 8597–8606.

40. E. Kundla, A. Samoson, and E. Lippmaa, *Chem. Phys. Lett.*, 1981, **83**, 229–232.

41. S. E. Ashbrook and S. Wimperis, *J. Magn. Reson.*, 2005, **177**, 44–55.

42. A. Samoson, *Chem. Phys. Lett.*, 1985, **119**, 29–32.

43. R. W. Schurko, R. E. Wasylishen, and H. Foerster, *J. Phys. Chem. A*, 1998, **102**, 9750–9760.

44. M. Bak, J. T. Rasmussen, and N. C. Nielsen, *J. Magn. Reson.*, 2000, **147**, 296–330.

45. S. Prasad, H. T. Kwak, T. Clark, and P. J. Grandinetti, *J. Am. Chem. Soc.*, 2002, **124**, 4964–4965.

46. T. T. Nakashima, R. E. Wasylishen, R. Siegel, and K. J. Ooms, *Chem. Phys. Lett.*, 2008, **450**, 417–421.

47. X. Wang and L. J. Smith, *Solid State Nucl. Magn. Reson.*, 2009, **36**, 99–102.

48. P. K. Madhu, O. G. Johannessen, K. J. Pike, R. Dupree, M. E. Smith, and M. H. Levitt, *J. Magn. Reson.*, 2003, **163**, 310–317.

Chapter 7
Quadrupolar Nutation Spectroscopy

Arno P.M. Kentgens

Radboud University Nijmegen, Institute for Molecules and Materials, Heyendaalseweg 135, 6525 AJ Nijmegen, The Netherlands

7.1 INTRODUCTION

Quadrupolar nuclei, which possess spin-quantum numbers $I > ^1\!/_2$, have a nonspherically symmetric charge distribution. As a consequence, they interact with the electric field gradient (EFG) resulting from the local charge distribution in a molecule or material as is described in Chapters 1 and 2. The spectral features resulting from this quadrupolar interaction are explained in these chapters. It should be realized, however, that the quadrupolar interaction also influences the way the spin system evolves during a radiofrequency (rf) pulse. In a classical picture, the motion of a magnetic moment M in the rotating frame is a precession about the applied rf field vector $\boldsymbol{B}_1$ with nutation frequency $\Omega_1 = \gamma B_1$. As the spins are also precessing around the external magnetic field B_0, the combined motion is called *nutation*. Radiofrequency pulses are generally denoted as "so many degree" pulses (e.g., 90° or 180° pulse) because for spin $^1\!/_2$ nuclei, depending on the nutation frequency Ω_1, the pulse length determines the angle over which the magnetic moment rotates. In this consideration, it is implicitly assumed that the rf Hamiltonian $\mathcal{H}_{\mathrm{rf}} = \gamma B_1 I_x$ overwhelms all other internal spin interactions, i.e., $\mathcal{H}_{\mathrm{rf}} > \mathcal{H}_{\mathrm{int}}$. Owing to the size of the quadrupolar interaction, this condition is generally not met for spins $I > ^1\!/_2$. Therefore, we must evaluate the time evolution of the spin system taking into account that the quadrupolar Hamiltonian is of the same order of magnitude as the rf Hamiltonian ($\mathcal{H}_{\mathrm{Q}} \approx \mathcal{H}_{\mathrm{rf}}$) or bigger ($\mathcal{H}_{\mathrm{Q}} > \mathcal{H}_{\mathrm{rf}}$). In these situations, the effective pulse angle over which the magnetization component associated with a certain spin transition rotates varies with the strength of the quadrupolar interaction and the rf field strength Ω_1. It is vital to understand these effects in order to get quantitatively interpretable spectra. Furthermore, this is of importance when one tries to implement pulse sequences developed for spin $^1\!/_2$ nuclei for studying quadrupolar nuclei. Moreover, these effects have to be taken into account when setting up cross-polarization experiments. Finally, we can exploit this behavior as a filter for spins experiencing a specific EFG by an appropriate choice of the excitation pulse length.

NMR of Quadrupolar Nuclei in Solid Materials
Edited by Roderick E. Wasylishen, Sharon E. Ashbrook and Stephen Wimperis

ISBN: 978-0-470-97398-1

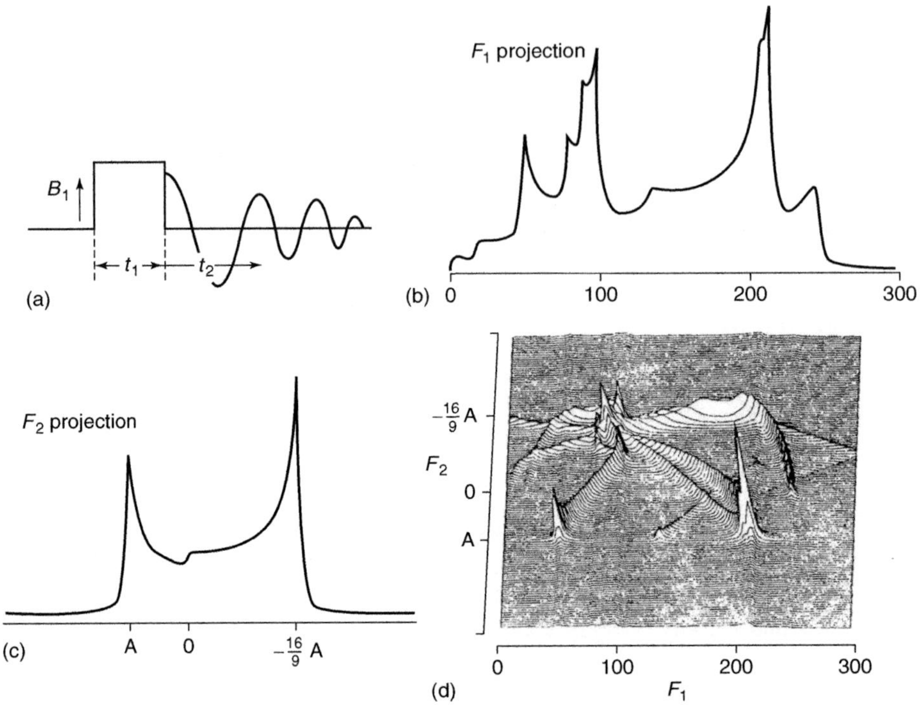

Figure 7.1. (a) Pulse sequence for the 2D nutation NMR experiment; a free induction decay is acquired during t_2 as a function of the pulse length t_1 with an rf field strength $\Omega_1 = \gamma B_1$. Subsequent Fourier transformation of the signals yields a well-featured 2D powder pattern shown in (d). The spectrum was calculated for a spin $I = 5/2$ with $\omega_Q/\Omega_1 = 0.45$ and $\eta = 0$. The F_2 projection of this pattern (c) gives the second-order quadrupolar powder pattern for the central 1/2 and −1/2 transition. The F_1 projection is the 1D nutation spectrum with a characteristic shape for the ratio ω_Q/Ω_1.

As was discussed by Samoson and Lippmaa in the early 1980s,[1–2] the behavior of quadrupolar spin systems under rf irradiation can be put to good use as a spectroscopic tool. The experimental implementation of this 2D nutation experiment is very simple; a series of free induction decays (FIDs) is acquired as a function of the pulse length t_1. So the spin system evolves under the Hamiltonian $\mathcal{H}_{rf} + \mathcal{H}_Q$ during t_1 and subsequently evolves under $\mathcal{H}_Q$, and possible additional interactions (see Chapter 4), during the time period t_2 where the signal is acquired. A double Fourier transform with respect to t_2 and t_1 then gives us a 2D nutation spectrum as depicted in Figure 7.1. From this spectrum, one may ideally infer the number of different species present in the material under study and obtain the size of their quadrupolar interaction relative to the rf field strength applied to the sample. In the remainder of this article, the theoretical and experimental aspects of this experiment are discussed with some examples of its practical use.

7.2 SPIN HAMILTONIANS AND DENSITY MATRIX

During the evolution period t_1 of the nutation experiment (Figure 7.1a), the spin Hamiltonian (all Hamiltonians are expressed in frequency units, i.e., should be read as $\mathcal{H}/\hbar$) for a quadrupolar spin I, in a frame rotating at the rf irradiation frequency ν_1, is given by

$$\mathcal{H}_1 = \mathcal{H}_{off} + \mathcal{H}_{rf} + \mathcal{H}_Q^1 = \Delta\omega I_z - \Omega_1 I_x + \frac{\Omega_Q(\theta, \varphi)}{6}[3I_z^2 - I(I+1)] \qquad (7.1)$$

where $\Delta\omega = 2\pi(\nu_1 - \nu_0)$ is the resonance offset of the irradiation pulse with respect to the resonance frequency of the nuclei and $\Omega_1 = \gamma B_1$ is the nutation frequency determined by the rf field strength. The angular-dependent quadrupolar frequency is defined as $\Omega_Q(\theta, \varphi) = \omega_Q/2[3\cos^2\theta - 1 + \eta\sin^2\theta\cos 2\varphi]$ where $\omega_Q = 2\pi(3C_Q/2I(2I-1))$ is the quadrupolar frequency and θ and φ are the polar angles orienting the magnetic field vector $\boldsymbol{B}_0$ in the principal axis frame of the EFG tensor, with associated principal values V_{XX}, V_{YY}, and V_{ZZ}. The asymmetry of the EFG tensor is represented by $\eta = (V_{XX} - V_{YY}/V_{ZZ})$. The largest principal component of the EFG tensor $V_{ZZ} = eq_{ZZ}$ and the nuclear quadrupolar moment Q determine the quadrupolar coupling constant $C_Q = (e^2 q_{ZZ} Q/h)$. Dipolar interactions and the second-order quadrupolar interaction have been neglected in this analysis. Furthermore, we assumed the presence of a large external magnetic field. Nutation spectroscopy has also reached some popularity in nuclear quadrupolar resonance (NQR); in this case, the angular dependence of the excitation has to be taken into account as the B_1 field has a specific orientation with respect to the quantization axis which is now determined by the EFG tensor. Here we concentrate on the high-field case and refer the reader to Refs 3–7 for a detailed treatment of NQR nutation spectroscopy.

During the detection period t_2 of the nutation experiment, the "regular" FID is acquired for the spin system. The Hamiltonian is now governed by the chemical shift interaction, the first- and second-order terms of the quadrupolar interaction and possible additional spin interactions such as dipolar interactions (see Chapter 4).

$$\mathcal{H}_2 = \mathcal{H}_{\text{off}} + \mathcal{H}_{\text{cs}} + \mathcal{H}^1{}_{\text{Q}} + \mathcal{H}^2{}_{\text{Q}}[+\mathcal{H}_D + \cdots] \quad (7.2)$$

With the knowledge of the Hamiltonians governing the spin system during the different periods of the experiment, we can use the density matrix formalism[8] to calculate the evolution of the spin system. Starting from the density matrix $\rho(0)$ at the beginning of the nutation experiment, generally the density matrix for the spin system in thermal equilibrium, the evolution of the density matrix

$$\rho(t_1, t_2) = e^{-i\mathcal{H}_2 t_2} e^{-i\mathcal{H}_1 t_1} \rho(0) e^{i\mathcal{H}_1 t_1} e^{i\mathcal{H}_2 t_2} \quad (7.3)$$

makes it possible to calculate the signal as $S(t_1, t_2) = \text{Tr}\{\rho(t_1, t_2), I_+\}$ and thus obtain the full 2D nutation spectrum. Considering the signal acquired during t_2 is the regular FID, the F_2-projection corresponds to the ideal 1D spectrum governed by chemical shift and quadrupolar interaction (for the time being, we ignore the possible additional interactions). The intensities and phases of the various frequency components in F_2 are modulated as a function of the length t_1 of the applied rf pulse. The (complex) amplitudes of the various frequency components in t_2 can be obtained from the evaluation of the density matrix at the end of the pulse.

$$\rho(t_1, 0) = e^{-i\mathcal{H}_1 t_1} \rho(0) e^{i\mathcal{H}_1 t_1} \quad (7.4)$$

As the Zeeman states $|I, m>$ are not eigenfunctions of the quadrupolar Hamiltonians, $\mathcal{H}_1$ will not be in diagonal form in the Zeeman basis set and needs to be diagonalized using an orthogonal transformation T

$$\rho(t_1, 0) = T e^{-iEt_1} T^{-1} \rho(0) T e^{iEt_1} T^{-1} \quad (7.5)$$

where E is the diagonalized matrix of $\mathcal{H}_1$ with the eigenvalues

$$E_i = \Sigma_p \Sigma_q T_{p,i} \mathcal{H}_{1p,q} T_{q,i} \quad (7.6)$$

on the diagonal, corresponding to the eigenfunctions

$$|i> = \Sigma_j T_{j,i} |I, m_j> \quad (7.7)$$

which are a linear combination of the Zeeman states. The signal can now be written as

$$\begin{aligned} S(t_1, t_2) &= \text{Tr}\{\rho(t_1, t_2), I_+\} \propto \Sigma_{k,l=k+1} \rho(t_1)_{k,l} \\ &\quad \times \exp\{-i\omega_{2_{kl}} t_2\} \\ &= \Sigma_{i,j,k,l=k+1} (R_{i,j})_{k,l} \exp\{-i\omega_{1_{ij}} t_1\} \\ &\quad \times \exp\{-i\omega_{2_{kl}} t_2\} \end{aligned} \quad (7.8)$$

where $\omega_{1_{i,j}} = E_i - E_j$ are the transition frequencies between the eigenstates $|i>$ and $|j>$ in the rotating frame as given in equation (7.7), and $\omega_{2_{k,l}}$ corresponds to the (single-quantum) transition frequencies detected during t_2. Evaluation of equation (7.5) gives us the amplitude factors for the various transitions:

$$(R_{i,j})_{k,l} = T_{k,i} T_{l,j} \Sigma_{s,t} T_{s,i} T_{t,j} \rho(0)_{s,t} \quad (7.9)$$

where the appearance of $\rho(0)$ depends on the preparation period used in the experiment. Here we only treat the case where the experiment is started with the spin system in thermal equilibrium $\rho(0) \propto I_z$. It should be noted, however, that if the spin system is prepared in a specific state, e.g., by inverting or saturating the satellite transitions, as is done for sensitivity enhancement (see Chapter 6), this has significant effects on the overall nutation behavior as it alters the amplitudes of the various frequency terms $\omega_{1_{i,j}}$.[9–11]

The important step in the calculation of nutation spectra is the diagonalization of the Hamiltonian $\mathcal{H}_1$. This diagonalization has been addressed by numerous authors and can be performed numerically e.g., by using popular software packages that propagate the spin density matrix for a user-defined spin system and pulse sequence such as SIMPSON,[12] or analytically as has been reviewed in Ref. 9. As the most dominant area where quadrupolar nutation spectroscopy is applied is the study of half-integer quadrupolar spins in moderate to large EFGs, we restrict the discussion to this case. Under these conditions, only the central $<1/2; -1/2>$ transition is clearly observed. In combination with the assumption that we start from thermal equilibrium, equation (7.9) reduces to

$$(R_{i,j})_{1/2,-1/2} = T_{1/2,i}T_{-1/2,j}\Sigma_s T_{s,i}T_{s,j}m_s \quad (7.10)$$

Despite the fact that only the central transition is observed, during the evolution time t_1, the spin system develops involving *all* eigenfunctions and eigenvalues depending on the ratio of Ω_Q and Ω_1. The coherences between these levels, with frequencies $\omega_{1_{i,j}}$, are projected onto the $<1/2; -1/2>$ single-quantum coherence when the rf field is switched off. Therefore, the central transition signal that is detected during t_2 is modulated by the frequency terms $\omega_{1_{i,j}}$ evolving during t_1. A 2D Fourier transform of the acquired signal will show a characteristic powder pattern as is shown for an isolated spin $I = 5/2$, subject to quadrupolar interaction only, in Figure 7.1(d). In this case, the F_2 projection of the spectrum equals the second-order quadrupolar powder pattern as the central $<1/2; -1/2>$ transition of a half-integer spin system is not affected by the quadrupolar interaction in the first-order perturbation theory (see Chapters 1 and 2). The F_1 projection of the spectrum is a characteristic powder pattern that depends on the spin-quantum number I, the quadrupolar interaction parameters C_Q and η, and the rf field strength Ω_1. The spectrum is independent of chemical shift, and considering its frequency span that is typically hundreds of kilohertz, dipolar broadenings have only a minor effect. Therefore, this 1D projection, generally referred to as the nutation spectrum will allow the determination of the quadrupolar interaction parameters even if the second-order powder pattern in F_2 is broadened into a featureless line due to, e.g., distributions in the NMR parameters and dipolar interactions as is often encountered in (partially) disordered materials.

To gain more insight into this process, it is worthwhile to study Figure 7.2 describing the effects for on-resonance nutation of a spin $I = 3/2$. Figure 7.2(a) shows the energy level diagram for the spin system during rf irradiation in the rotating frame as a function of the ratio Ω_Q/Ω_1. For $\Omega_Q = 0$, the eigenstates $|i>$ correspond with the eigenfunctions of I_x and we find the four equally spaced energy levels with a transition frequency of Ω_1 as expected. For finite values of Ω_Q/Ω_1, the transition frequencies $\omega_{1_{i,j}}$ become a function of Ω_Q/Ω_1 as can be seen in Figure 7.2(b). With $\Omega_Q/\Omega_1 \gtrsim 6$, one reaches a situation comparable to NQR with a large quadrupolar interaction and a weak Zeeman field present; two energy levels are degenerate and the other two are split by $(I + 1/2)\,\Omega_1$. The amplitudes $R_{i,j}$, as calculated by equation (7.10), associated with the various transitions in the rotating frame contributing to the observed central transition intensity are depicted in Figure 7.2(c) and (d) as a function of the Ω_Q/Ω_1 ratio. With $R_{i,j} = -R_{j,i}$, equation (7.8) shows that the signal detected during t_2 is amplitude-modulated by sine functions of the transition frequencies of the odd-order spin transitions during t_1.

$$S(t_1, t_2) \propto \Sigma_{i,j}(R_{i,j})_{1/2,-1/2} \sin\{\omega_{1_{ij}}t_1\} \times \exp\left\{-i\omega_{1_{1/2,-1/2}}t_2\right\} \quad (7.11)$$

As can be seen from Figure 7.2(c) and (d), the strongest contributions in the amplitude modulation come from the single-quantum transitions 1–2, 2–3, and 3–4, with a small contribution from the triple-quantum 1–4 transition. These observations are reflected in Figure 7.2(e) showing a series of nutation spectra for a single crystal as a function of $\Omega_Q(\theta, \varphi)/\Omega_1$. The nutation spectrum is dominated by a nutation frequency Ω_1 when the quadrupolar frequency is very small, whereas the strongest line in the nutation spectrum corresponds to frequency $(I + 1/2)\,\Omega_1$ for large quadrupolar frequencies. In the region where Ω_1 and Ω_Q are of similar magnitude, the nutation spectrum shows three prominent lines located at the specific single-quantum transition frequencies $\Omega_{i,i+1}$. Note that the spectra are mirrored about the origin as is expected for a pure amplitude-modulated signal. Often, the negative side of the frequency scale is depicted with negative intensities. One can choose to leave out that part of the spectrum, however, as the information is redundant. The nutation spectra for a powdered sample are depicted in Figure 7.3 as a function of the ratio of the quadrupolar frequency ω_Q and the rf field strength Ω_1. Detailed nutation spectra are obtained in the regime $0.1 \leqq \omega_Q/\Omega_1 \leqq 10$ allowing the determination of the quadrupolar parameters when Ω_1 is known. It should be noted that the analysis of the nutation rates as a function of ω_Q/Ω_1 is of equal importance in cross-polarization experiments,

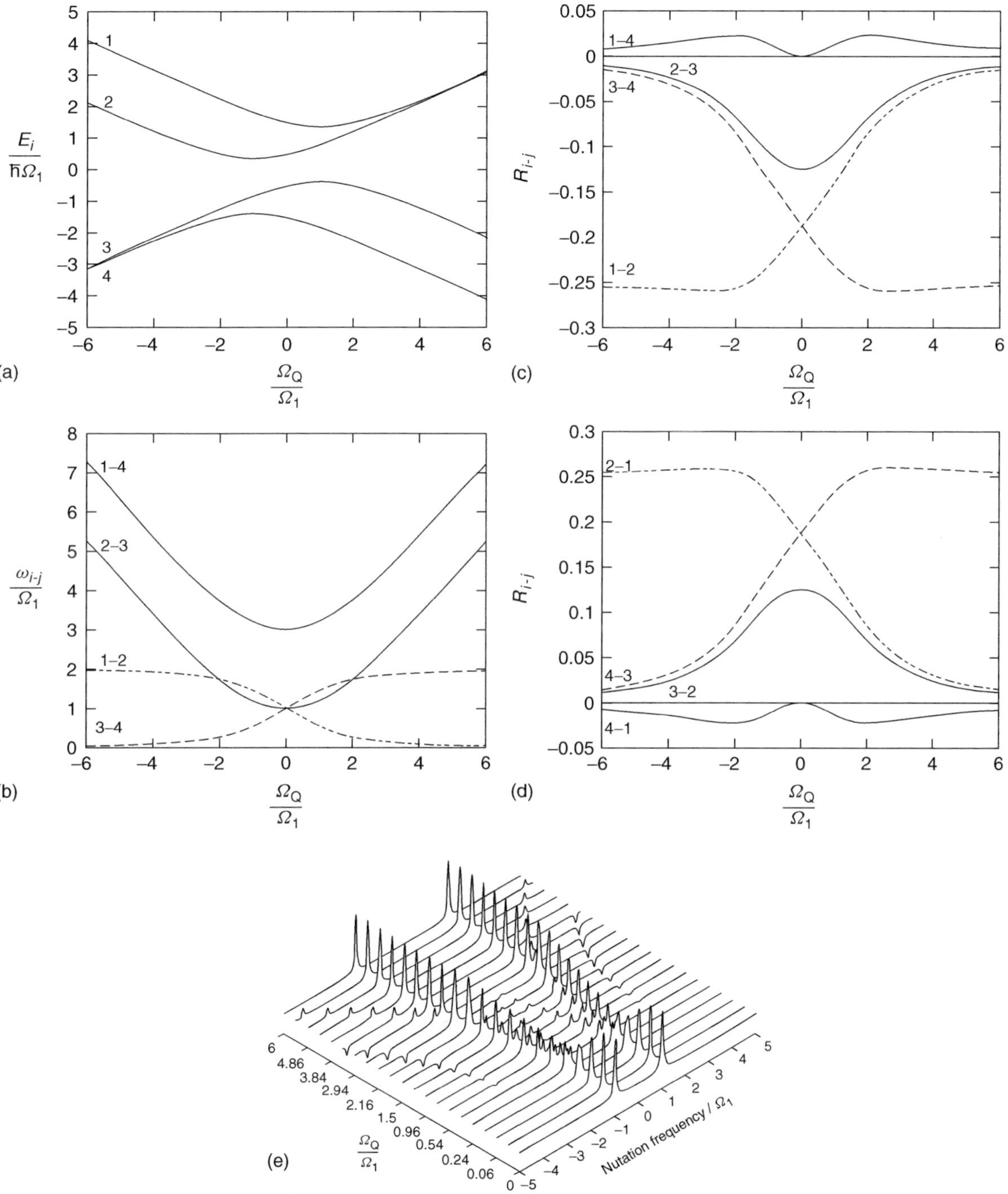

Figure 7.2. (a) Energy level diagram in the rotating frame assuming on-resonance rf irradiation of a spin $I = 3/2$ as a function of Ω_Q/Ω_1. (b) Corresponding transition frequencies in the rotating frame (only positive frequencies are shown). (c, d) Amplitude coefficients $R_{i,j}$ associated with the various spin transitions *i-j*. (e) Simulated nutation spectra for a single crystal as a function of Ω_Q/Ω_1. (Adapted from Ref. 9. © Elsevier B.V., 1998.)

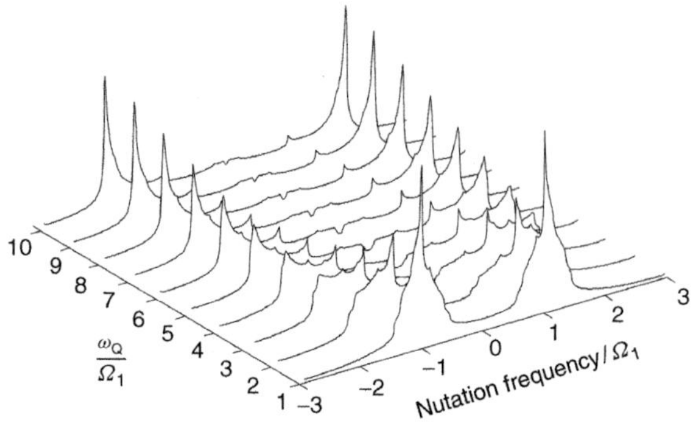

Figure 7.3. Simulated spin $I = 3/2$ nutation spectra for a powder showing well-featured spectra over the range $1 \leq \omega_Q/\Omega_1 \leq 10$. (Adapted from Ref. 9. © Elsevier B.V., 1998.)

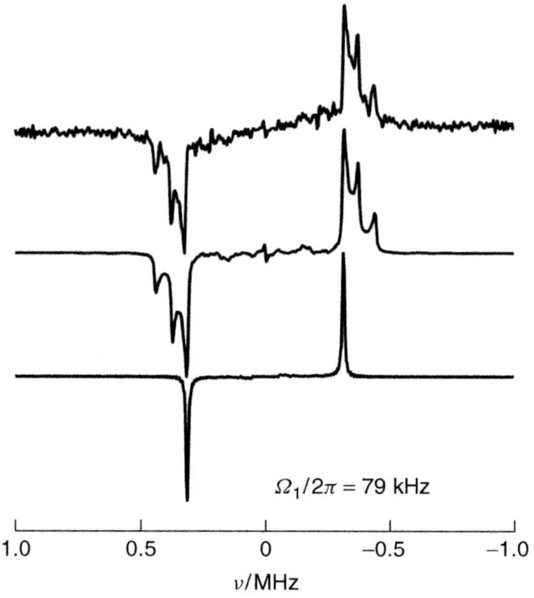

Figure 7.4. ^{139}La nutation spectrum of La_2O_3 using an rf field strength of 79 kHz. The experimental spectrum (top trace) was recorded using CPMG detection employing 0.4 μs refocusing pulses ($\Omega_1/2\pi = 270$ kHz) acquiring only a single point at the top of each echo. The lower trace shows the simulated spectrum including only first-order quadrupolar interactions. In the middle trace, second-order quadrupolar contributions are taken into account.

as it is possible to choose the proton rf field strength to establish a Hartmann–Hahn match to any of the available transitions in the rotating frame as has been analyzed in detail by Ashbrook and Wimperis.[13]

In the rotating frame Hamiltonian $\mathcal{H}_1$, only the first-order term of the quadrupolar interaction was taken into account. However, for very large quadrupolar interactions, higher order terms need to be included. This is especially true for the higher spin-quantum numbers where the second-order term is sizeable with respect to the first-order term. This is demonstrated for ^{139}La ($I = 7/2$) in La_2O_3 in Figure 7.4. ^{139}La in La_2O_3 was first studied by Bastow[14] at 9.4 T using spin-echo mapping because of the large quadrupolar interaction. La occupies a single crystallographic site exhibiting a quadrupolar coupling constant of $C_Q = 58.5$ MHz and $\eta = 0$. Considering the very large quadrupolar frequency $\omega_Q/2\pi = 4.1$ MHz using an rf field strength of only 79 kHz, i.e., $\omega_Q/\Omega_1 \gg 10$, we expect a single nutation frequency at $4 \cdot 79 = 316$ kHz. However, instead of a single resonance, a powder pattern is observed. This can be explained by taking the second-order quadrupolar interaction into account as is shown in the simulations. In this case, the first-order interaction is 4.1 MHz and the second-order term is a sizeable 500 kHz and therefore gives a distinct contribution. Here, the second-order nutation lineshape strongly depends on the rf field strength and the position of the excitation frequency with respect to the resonance line as the rf field strength is much smaller than the second-order powder linewidth in the F_2 dimension.

To obtain easily interpretable nutation spectra of spins with large quadrupolar interaction, it is better to go to much higher rf field strengths as described in section "Applications of Nutation NMR Spectroscopy".

7.3 PRACTICAL CONSIDERATIONS

In order to successfully acquire nutation NMR spectra, it is important to take a number of practical aspects of the experiment into account, as is outlined below.

7.3.1 rf Homogeneity

Considering the fact that the spectral information in a nutation spectrum depends on the ratio of the quadrupolar frequency ω_Q and the rf field strength

Ω_1, it is of utmost importance to avoid a variation of the rf field strength over the sample. rf inhomogeneity does not only lead to line broadening but also changes the appearance of the spectra. The variation of the rf field in a solenoid is the highest in the center of the coil and drops toward the edges leading to an asymmetric spectral profile with lines broadened toward lower frequencies. Additionally, depending on the variation of ω_Q/Ω_1 over the sample, the line broadening may be different in different regions of the spectrum. Therefore, one should use a well-defined coil, preferably with a substantial number of turns and restrict the sample so that it is confined well within the coil.

7.3.2 Resonance Offset

As was demonstrated in equation (7.11), the nutation signal is a pure sine modulation when the spin system is irradiated on-resonance. On-resonance, the symmetry of the Hamiltonian $\mathcal{H}_1$ results in $(R_{i,j})_{1/2,-1/2} = -(R_{j,i})_{1/2,-1/2}$. The introduction of a resonance offset in $\mathcal{H}_1$ lowers its symmetry and therefore the amplitude coefficients are no longer antisymmetric upon interchanging the coefficients i, j i.e., $(R_{i,j})_{1/2,-1/2} \neq -(R_{j,i})_{1/2,-1/2}$. As a result, the modulation of the central transition frequency detected during t_2 becomes a phase modulation that cannot be changed back into an amplitude modulation by phase cycling. The problem of a phase modulation during t_1 is that it will introduce dispersive contributions to the lineshapes, which can severely distort the resulting powder patterns. If resonance offsets are unavoidable, e.g., because the spectrum consists of different resonances, pure phase 2D spectra can be obtained by refocusing the central transition signal in the directly detected (t_2) dimension and record the whole echo. This can be accomplished using a low-power central transition selective 180° pulse at a time τ after the nutation pulse. The advantage of this approach was demonstrated by Shore and coworkers[15] and is shown in Figure 7.5 for their ^{93}Nb nutation spectrum of $PbNb_2O_6$.

It should be noted that the introduction of large resonance offsets can be used to obtain meaningful nutation spectra if the rf field strength is insufficient to bring its ratio to the quadrupolar frequency in the intermediate range $0.1 \leqq \omega_Q/\Omega_1 \leqq 10$. If $\omega_Q/\Omega_1 > 10$, the on-resonance nutation spectrum will consist of a single line at $(I + 1/2)\,\Omega_1$.

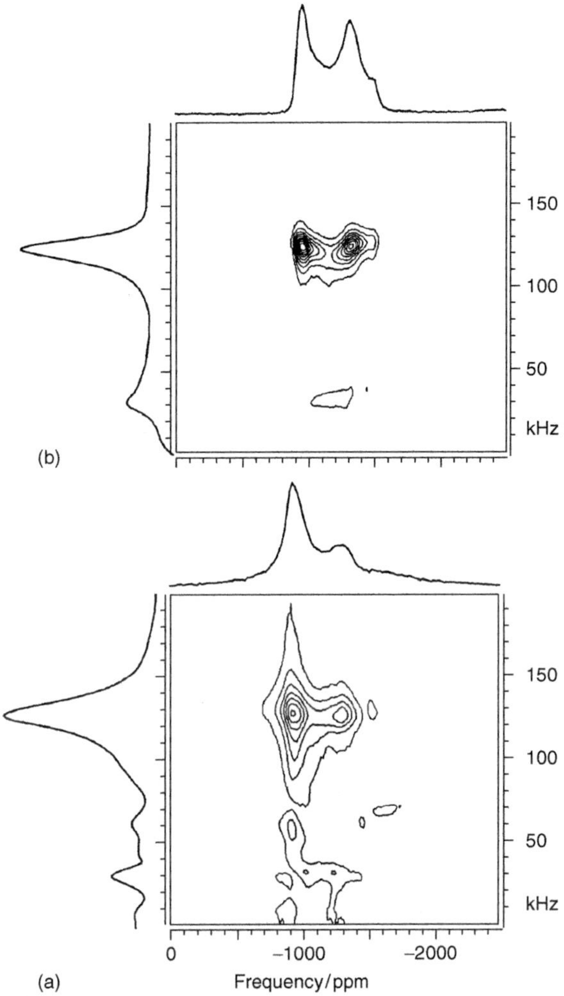

Figure 7.5. ^{93}Nb nutation spectra of $PbNb_2O_6$ acquired using (a) the conventional one-pulse nutation experiment, and (b) whole echo acquisition using a 200 µs delay between the nutation and the refocusing pulse. The latter spectrum shows improved pure absorption lineshapes devoid of dispersive components. An rf field strength of 30 kHz was employed. (Reproduced from Ref. 15. © Elsevier, 1999.)

As has been described in Ref. 9, this can be alleviated by introducing a resonance offset so that the spin system is irradiated near the double quantum transitions ($\Omega_Q/2$) or the satellite transitions Ω_Q. A more versatile solution is, however, to use modern probe hardware allowing the generation of much higher rf field strengths (vide infra).

7.3.3 Recycle Delay

In nutation NMR, extra care has to be taken to ensure that the recycle delay is long enough to allow the spin system to return to thermal equilibrium before repeating the experiment. If the recycle delay is short with respect to T_1, then the build up of the magnetization along the z-axis is incomplete. This will distort the pure sine modulation of the central transition signal (equation (7.11)). In general, Fourier transformation of a distorted $\sin 2\pi\nu_1 t_1$ will not only give a line at frequency ν_1 but also harmonics at $2\nu_1, 3\nu_1, 4\nu_1 \cdots$. The number and amplitude of these harmonics depend on the distortion of the sine wave and can easily be mistaken for components with different quadrupolar frequencies.

7.3.4 QCPMG Detection

A convenient method to enhance the sensitivity of the nutation experiment is the quadrupolar Carr Purcell Meiboom Gill (QCPMG) sequence that was introduced by Nielsen and coworkers.[16] The sensitivity enhancement is accomplished by acquiring the signal as a train of echoes using refocusing pulses. After Fourier transformation, the effect of this operation is to split the central transition into a manifold of sidebands concentrating the signal intensity into a limited number of resonances. If the F_2 dimension of the nutation spectrum is uninformative, one may choose to collapse all intensity into a single line by acquiring only the echo maxima. The rf field strength for the refocusing pulses should be chosen so that optimal refocusing of the central transition takes place. Generally, this field is chosen small enough so that the central transition is selectively inverted, but high enough to efficiently excite the entire central transition powder pattern (of width Δ), yielding the condition $\Omega_{1,\pi} = \Delta/(I + 1/2)$.[17] The sensitivity of the QCPMG experiment is mostly determined by the number of echoes that one can acquire, which in turn depends on the homogeneous T_2 of the spins.

7.3.5 Magic Angle Spinning

Quadrupolar nutation experiments have mostly been confined to static powder studies. Nielsen *et al.*[18] presented the first theoretical treatment and computer simulations of nutation NMR performed under magic angle spinning (MAS). The great advantage of MAS is that it improves the resolution of the nutation spectrum in the F_2 dimension by a factor of $\sim$3 as the second-order quadrupolar powder lineshape is reduced by that factor depending on η. Additionally, dipolar interactions and chemical shift anisotropies can be averaged by MAS if present. Depending on the spinning speed and the ω_Q/Ω_1 ratio, the nutation spectra in the F_1 dimension cannot be interpreted intuitively as compared to data obtained from static samples. Moreover, it appeared that 2D MAS nutation NMR experiments are more susceptible to instrumental errors. A special problem for the MAS experiments is the relatively poor rf homogeneity of most commercial MAS NMR probeheads leading to a variation of ω_Q/Ω_1 over the sample as is described in section "rf Homogeneity". Although the authors addressed this issue in a later publication[19] suggesting alternative sample geometries to minimize rf inhomogeneity effects, MAS nutation NMR experiments did not find many practical applications.

7.3.6 Spectral Processing

Finally, it is worthwhile to look at a few practical aspects of processing 2D NMR experiments. As usual, the first step of processing the acquired data is a Fourier transform in the directly detected dimension. The phase of the entire set of spectra is then set according to the phasing of the first (short pulse) spectrum. Considering the fact that on-resonance nutation spectra consist of a pure sine modulation, a real transformation has to be applied in the t_1 dimension. This can be achieved by blanking the imaginary buffer before the second Fourier transform. In principle, the first point in the t_1-dimension, corresponding to a zero-pulse length, is missing from the data, calling for a first-order phase correction in the F_1 dimension. As nutation spectra consist of broad powder spectra, finding the appropriate phase correction can be demanding. To avoid this problem, it is convenient to put a row of zeros before the first data row to represent the spectrum with zero intensity. For this, it is necessary, however, to choose the length of the first pulse equal to the dwell time in the t_1 dimension.

7.4 APPLICATIONS OF NUTATION NMR SPECTROSCOPY

The introduction of nutation NMR opened the way to obtain structural information from quadrupolar nuclei even if the normal NMR spectrum was not resolved, e.g., because of dipolar interactions or the presence of a distribution in interaction parameters owing to some inherent structural disorder in the sample. As a result, nutation NMR found many practical applications in the study of catalytically relevant materials such as zeolites, aluminophosphates and (amorphous) aluminosilicates, geological glasses and crystals, semiconductors and relaxor ferroelectrics. With the advent of the multiple-quantum MAS[20] (see Chapter 10) and satellite transition MAS[21] (see Chapter 11) experiments, the use of nutation NMR as a spectroscopic tool began to decline. This is mostly due to the fact that the nutation NMR experiment generally lacks resolution if multiple resonances are present, meaning that it can be difficult to uniquely assign overlapping lines. Both multiple quantum magic angle spinning (MQMAS) and satellite transition magic angle spinning (STMAS) have limitations, however. Most importantly, they rely on MAS to narrow the central and satellite transition lineshapes that are broadened by the second-order quadrupolar interaction. If the available spinning speed is less than the residual MAS linewidth, i.e., in cases where very large quadrupolar frequencies are encountered, the central transition and spinning sidebands start to overlap and the resolution is lost. In this regime of large quadrupolar interaction parameters, nutation NMR can be very valuable especially when using probeheads with small diameter coils allowing the generation of very large rf field strength[22] so that it will still be possible to reach the intermediate $0.1 \leqq \omega_Q/\Omega_1 \leqq 10$ regime and obtain well-structured nutation spectra.

This is demonstrated for the ^{23}Na ($I = 3/2$) resonance in Na_2SO_4, with a quadrupolar coupling constant $C_Q = 2.6$ MHz and $\eta = 0.58$. With a quadrupolar frequency $\omega_Q/2\pi = 1.3$ MHz, we need an rf field strength of at least 130 kHz to start to see some detail in the nutation spectra. Figure 7.6 shows a series of ^{23}Na nutation spectra for varying rf field strengths from 165 to 1185 kHz. The spectrum taken with $\Omega_1/2\pi = 165$ kHz is dominated by the resonance at $2\Omega_1$, meaning that at this rf field strength, the excitation is mostly selective to the central transition. The most featured spectra are obtained for rf field strengths between 400 and 600 kHz, but even at $\Omega_1/2\pi = 1185$ kHz we see a powder pattern, meaning we have not yet fully reached the nonselective excitation regime. These spectra clearly show that we can shift the range of nutation spectroscopy by implementing strong rf fields. To elucidate this point, Table 7.1 summarizes the range of quadrupolar coupling constants that will give well-featured nutation spectra at an rf field strength of 100 kHz which is a typical value for a commercial probehead. Looking at the corresponding widths of the central transition, we observe that this width never exceeds 60 kHz. As fast MAS probeheads can spin faster than that, MQMAS or STMAS is a better choice to study quadrupolar spins in this regime. If we look at the possibilities of nutation NMR when employing an rf field strength of 1000 kHz which can be generated using rf coils of ~1 mm diameter, we can study spin systems with quadrupolar frequencies up to 10 MHz. MAS is not useful in this regime, as such high rf field nutation spectroscopy can bridge the gap between MAS NMR and NQR.

A useful application of high rf field is presented for ^{87}Rb ($I = 3/2$) nutation NMR of $RbClO_4$ using QCPMG detection. QCPMG spectra of $RbClO_4$ have been recorded by Larsen *et al.*[16] showing that the ^{87}Rb spectrum is affected by both chemical shift anisotropy (CSA) and second-order quadrupolar broadening. Therefore, analysis of the spectrum is not straightforward where taking spectra at different external field strengths is helpful to find the relative contribution of both interactions and their relative orientation. Alternatively, a nutation spectrum can help in fully analyzing the spectrum as the F_1 projection contains the nutation powder pattern which solely depends on ω_Q/Ω_1, whereas the F_2 projection will depend on CSA and second-order quadrupolar interaction and the relative orientation of the interaction tensors. This is demonstrated in Figure 7.7 showing the experimental and simulated nutation-QCPMG spectrum of $RbClO_4$ employing an rf field strength of 1275 kHz for the nutation pulse.

7.5 QUANTITATIVE SPECTRA AND NUTATION-BASED FILTERING

As the response of a quadrupolar spin system to a pulse of duration τ is sensitive to the ratio of the quadrupolar frequency and the rf field strength, this affects the relative line intensities in a spectrum when

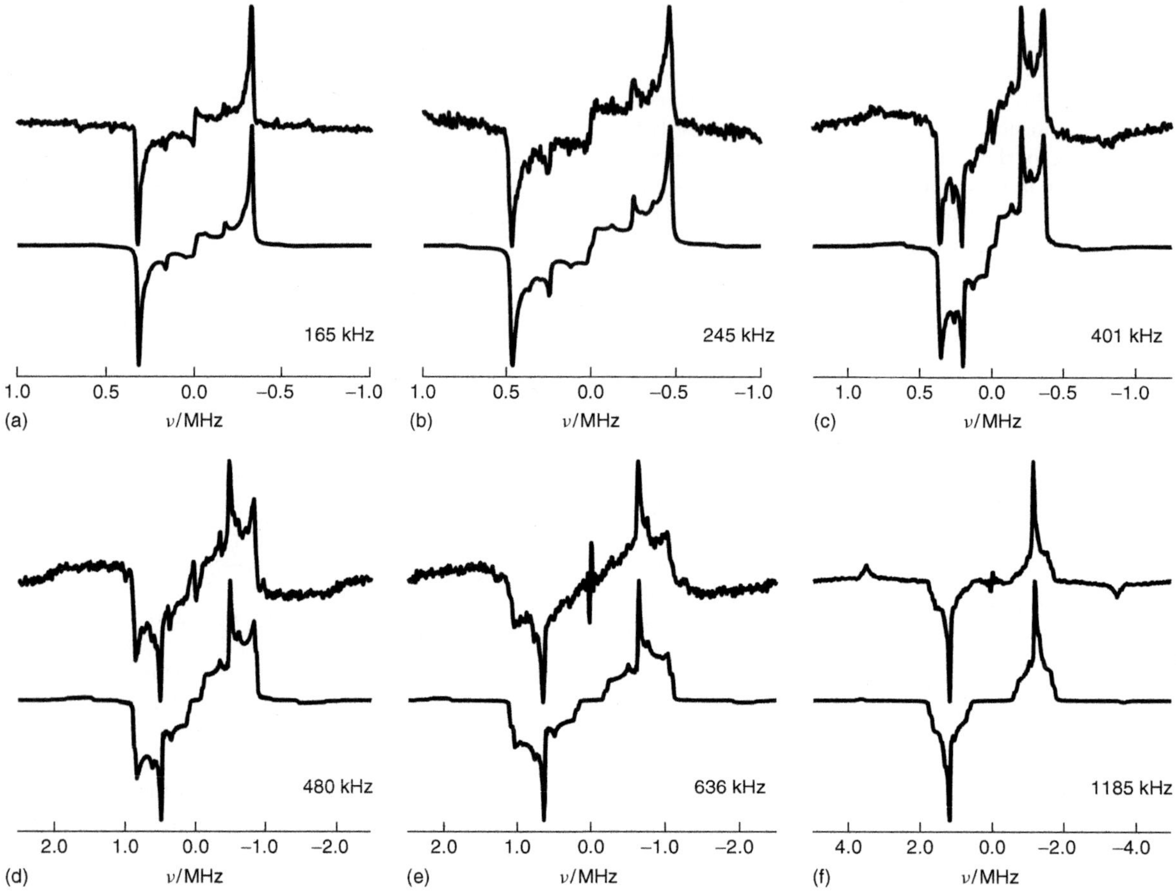

Figure 7.6. Experimental (top traces) and simulated (lower traces) ^{23}Na nutation spectra of Na_2SO_4 at rf field strengths of (a) 165 kHz, (b) 245 kHz, (c) 401 kHz, (d) 480 kHz, (e) 636 kHz, and (f) 1185 kHz as obtained in a homebuilt probehead with a coil with 1.2 mm inner diameter. $C_Q = 2.6$ MHz and $\eta = 0.58$ was used for the spectral simulation.

Table 7.1. Range of quadrupolar coupling constants, along with the central transition second-order quadrupolar line broadening, that will result in well-featured nutation spectra assuming an rf field strength of 100 kHz or 1 MHz

	$\Omega_1/2\pi = 100$ kHz ($\omega_Q/2\pi$ range: 0.05–1 MHz)		$\Omega_1/2\pi = 1000$ kHz ($\omega_Q/2\pi$ range: 0.5–10 MHz)	
Spin (I)	C_Q (MHz)	Central transition linewidth (kHz)	C_Q (MHz)	Central transition linewidth (kHz)
3/2	0.10–2	0.02–7.5	1.0–20	1.9–750
5/2	0.35–7	0.06–22	3.5–70	5.5–2200
7/2	0.7–14	0.09–38	7.0–140	9.4–3750
9/2	1.20–24	0.15–60	12–240	15–6000

[a] At 100 kHz, spin systems with quadrupolar frequencies between $0.05 < \omega_Q/2\pi < 1$ MHz can be studied, at 1 MHz rf field strength this range is $0.5 < \omega_Q/2\pi < 10$ MHz.

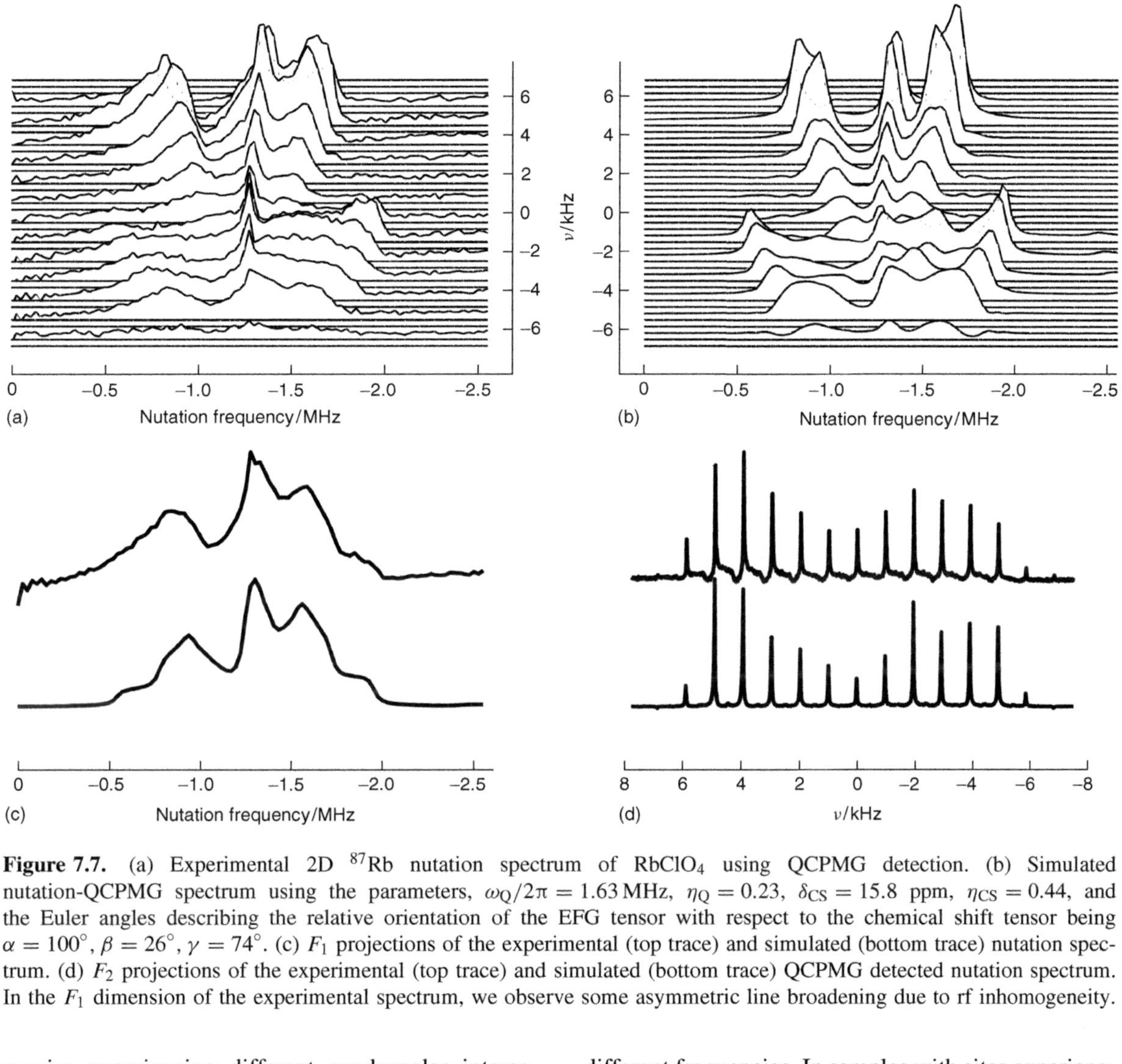

Figure 7.7. (a) Experimental 2D ^{87}Rb nutation spectrum of $RbClO_4$ using QCPMG detection. (b) Simulated nutation-QCPMG spectrum using the parameters, $\omega_Q/2\pi = 1.63$ MHz, $\eta_Q = 0.23$, $\delta_{CS} = 15.8$ ppm, $\eta_{CS} = 0.44$, and the Euler angles describing the relative orientation of the EFG tensor with respect to the chemical shift tensor being $\alpha = 100°$, $\beta = 26°$, $\gamma = 74°$. (c) F_1 projections of the experimental (top trace) and simulated (bottom trace) nutation spectrum. (d) F_2 projections of the experimental (top trace) and simulated (bottom trace) QCPMG detected nutation spectrum. In the F_1 dimension of the experimental spectrum, we observe some asymmetric line broadening due to rf inhomogeneity.

species experiencing different quadrupolar interactions are present. In the case $\omega_Q \ll \Omega_1$, the excitation of the spin system is nonselective where the response equals that of a spin $^1\!/_2$ and the signal amplitude is proportional to $3(I + (1/2))/4I(I + 1)\sin[\Omega_1\tau]$. For the case where $\omega_Q \gg \Omega_1$, the central transition is excited selectively and signal intensity is modulated as $3/4I(I + 1)\sin\left[(I + 1/2)\,\Omega_1\tau\right]$. In the case where ω_Q and Ω_1 are of the same order of magnitude, the intensity modulation is described by the theoretical treatment in section "Spin Hamiltonians and Density Matrix" with the intensity curve as a function of pulse length consisting of a sum of sinusoidals with different frequencies. In samples with sites experiencing different quadrupolar interactions, an appropriate pulse length has to be chosen to get an equal response for each site independent of C_Q. In practice, τ has to be chosen short enough so that $\sin\left[(I + 1/2)\,\Omega_1\tau\right] \approx (I + 1/2)\,\Omega_1\tau\ (\leq \pi/6)$. A detailed account of all the aspects involved in getting quantitative spectra including correcting for possible spectral contributions from the satellite transitions can be found in Ref. 23.

Alternatively, this pulse length effect can be exploited to filter signals on the basis of their quadrupolar interaction parameter by an appropriate choice of pulse length. This is particularly easy to see for a spin

$I = 3/2$, where a 90° pulse for nuclei with $\omega_Q = 0$ corresponds to a 180° pulse for spins experiencing a very large quadrupolar interaction. So using such an excitation pulse will give maximum intensity for spins in a symmetric surrounding, whereas the signal of spins in a very asymmetric surrounding will be strongly attenuated. This was used by Verhagen *et al.*[24] to impose a structural contrast on a magnetic resonance force microscopy (MRFM) image of two different sodium salts and can also be used in regular magnetic resonance images.

A variation of this idea was used by Larsen *et al.*[25] to separate the lineshapes of ^{47}Ti and ^{49}Ti in solid-state NMR spectra. ^{47}Ti ($I = 5/2$) and ^{49}Ti ($I = 7/2$) have similar Larmor frequencies and are therefore simultaneously present in a Ti spectrum. Because of the large quadrupolar interaction that is generally encountered for Ti in most materials, it is possible to selectively excite the ^{47}Ti signal by choosing the condition $(I_{^{47}\text{Ti}} + 1/2)\ \Omega_1\tau = 3\pi/2$, in which case for ^{49}Ti, the effective pulse angle will be $(I_{^{49}\text{Ti}} + 1/2)\ \Omega_1\tau = 2\pi$ and its signal will therefore

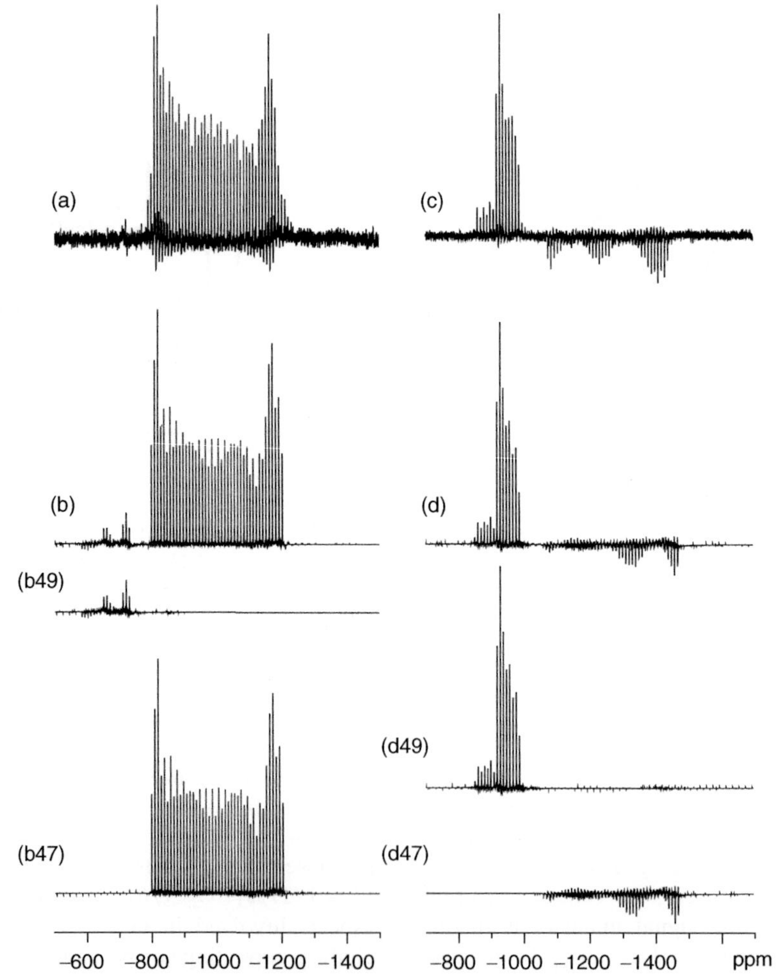

Figure 7.8. Experimental (a and c) and calculated (b and d) spectra of anatase at 21.1 T using a ^{47}Ti-selective pulse sequence (a and b) or ^{49}Ti-selective pulse sequence (c and d), respectively. The calculated spectra are obtained by adding the separately calculated contributions from ^{47}Ti (b47; d47) and ^{49}Ti (b49; d49). (Reproduced from Ref. 25. © Elsevier, 2006.)

not show up in the spectrum. To selectively detect the ^{49}Ti signal, two echo experiments are added with different pulse lengths so that the ^{49}Ti signals add and the ^{47}Ti signals subtract. An example of selective ^{47}Ti and ^{49}Ti spectra of anatase is shown in Figure 7.8.

REFERENCES

1. A. Samoson and E. Lippmaa, *Phys. Rev. B*, 1983, **28**, 6567.
2. A. Samoson and E. Lippmaa, *Chem. Phys. Lett.*, 1983, **100**, 205.
3. J. Dolinsek, F. Milia, G. Papavassiliou, G. Papantopoulos, and R. Rumm, *J. Magn. Reson., Ser. A*, 1995, **114**, 147.
4. N. Velikite, N. Sinyavsky, and M. Mackowiak, *Z. Naturforsch., A*, 1999, **54**, 351.
5. B. Cordier, D. Grandclaude, A. Retournard, L. Merlat, and D. Canet, *Mol. Phys.*, 2005, **103**, 2593.
6. D. Canet, L. Merlat, B. Cordier, D. Grandclaude, A. Retournard, and M. Ferrari, *Mol. Phys.*, 2006, **104**, 1391.
7. M. Ferrari, N. Hiblot, A. Retournard, and D. Canet, *Mol. Phys.*, 2007, **105**, 3005.
8. M. Levitt, *Spin Dynamics: Basic Principles of NMR Spectroscopy*, 2nd ed., John Wiley & Sons Ltd: Chichester, 2008.
9. A. P. M. Kentgens, *Prog. Nucl. Magn. Reson. Spectrosc.*, 1998, **32**, 141.
10. N. M. Trease, K. K. Dey, and P. J. Grandinetti, *J. Magn. Reson.*, 2009, **200**, 334.
11. T. T. Nakashima, K. J. Harris, and R. E. Wasylishen, *J. Magn. Reson.*, 2010, **202**, 162.
12. M. Bak, J. Rasmussen, and N. Nielsen, *J. Magn. Reson.*, 2000, **147**, 296.
13. S. E. Ashbrook and S. Wimperis, *Mol. Phys.*, 2000, **98**, 1.
14. T. J. Bastow, *Solid. State. Nucl. Magn.*, 1994, **3**, 17.
15. S. Prasad, P. Zhao, J. Huang, J. J. Fitzgerald, and J. S. Shore, *Solid. State. Nucl. Magn.*, 1999, **14**, 231.
16. F. H. Larsen, H. J. Jakobsen, P. D. Ellis, and N. C. Nielsen, *J. Phys. Chem. A*, 1997, **101**, 8597.
17. P. R. Bodart, J. P. Amoureux, Y. Dumazy, and R. Lefort, *Mol. Phys.*, 2000, **98**, 1545.
18. N. C. Nielsen, H. Bildsoe, and H. J. Jakobsen, *J. Magn. Reson.*, 1992, **97**, 149.
19. N. C. Nielsen, H. Bildsoe, and H. J. Jakobsen, *J. Magn. Reson.*, 1992, **98**, 665.
20. A. Medek, J. S. Harwood, and L. Frydman, *J. Am. Chem. Soc.*, 1995, **117**, 12779.
21. Z. H. Gan, *J. Am. Chem. Soc.*, 2000, **122**, 3242.
22. K. Yamauchi, J. W. G. Janssen, and A. P. M. Kentgens, *J. Magn. Reson.*, 2004, **167**, 87.
23. D. Massiot, C. Bessada, J. P. Coutures, and F. Taulelle, *J. Magn. Reson.*, 1990, **90**, 231.
24. R. Verhagen, A. Wittlin, C. W. Hilbers, H. van Kempen, and A. P. M. Kentgens, *J. Am. Chem. Soc.*, 2002, **124**, 1588.
25. F. H. Larsen, I. Farnan, and A. S. Lipton, *J. Magn. Reson.*, 2006, **178**, 228.

Chapter 8
Dynamic Angle Spinning

Philip J. Grandinetti

Department of Chemistry, The Ohio State University, Columbus, OH 43210-1185, USA

8.1 INTRODUCTION

Dynamic angle spinning (DAS) is a 2D NMR experiment designed for removing multiple-rank anisotropic broadenings in solid-state NMR. It is a technique that has proven useful for obtaining the high-resolution isotropic solid-state NMR spectra of the central transition of half-integer quadrupolar nuclei broadened to second order. This narrowing is accomplished by using the angle of the sample rotation axis as a dynamic variable in a 2D experiment.

It has long been understood that the second-order anisotropic broadenings of the central transition of half-integer nuclei cannot be removed with the spatial averaging of magic angle or variable angle spinning alone,[1] and it was even thought that no spatial averaging solution existed. In 1988, new averaging approaches that solve this problem were disclosed by both the Pines group and the Virlet group at the *9th European Experimental NMR conference* in Bad-Aussee. In subsequent publications by both the groups,[2,3] the underlying theory for removing second- and higher order broadenings was presented, and two new experimental techniques emerged for removing second-order broadenings: double rotation (DOR) and DAS, both first achieved experimentally at Berkeley (see Chapter 9).[2,4,5] DAS combines the ideas of switched rotation axis[6,7] and discrete averaging experiments[8] to remove the second-order broadenings of the central transition of half-integer nuclei.

In this chapter, the basic principles behind DAS are discussed along with a description of how DAS is implemented and some illustrative examples of its use.

8.2 BASIC PRINCIPLES

In NMR, the Zeeman interaction is normally the dominant nuclear spin interaction, and the NMR transition frequency can be expanded in a perturbation series,[9]

$$\Omega(\theta,\phi) = \Omega^{(0)} + \Omega^{(1)}(\theta,\phi) + \Omega^{(2)}(\theta,\phi) + \dots \tag{8.1}$$

with the Zeeman interaction as the zeroth-order term

$$\Omega^{(0)} = -\gamma B_0 \tag{8.2}$$

where γ is the nuclear gyromagnetic ratio and B_0 is the external magnetic field strength. The higher order terms in equation (8.1) are due to internal

NMR of Quadrupolar Nuclei in Solid Materials
Edited by Roderick E. Wasylishen, Sharon E. Ashbrook and Stephen Wimperis

ISBN: 978-0-470-97398-1

spin interactions (e.g., nuclear shielding, quadrupolar coupling, and so on). These interactions are second rank by nature and thus the higher order terms can be expanded in limited spherical harmonic series,

$$\Omega^{(1)}(\theta,\phi) = \Omega^{(1)}_{\rm iso} + \sum_{k=-2}^{2} c^{(1)}_{2,k} Y_{2,k}(\theta,\phi) \quad (8.3)$$

$$\Omega^{(2)}(\theta,\phi) = \Omega^{(2)}_{\rm iso} + \sum_{k=-2}^{2} c^{(2)}_{2,k} Y_{2,k}(\theta,\phi) + \sum_{k=-4}^{4} c^{(2)}_{4,k} Y_{4,k}(\theta,\phi) \quad (8.4)$$

where (θ, ϕ) are the angles between the external magnetic field direction and the principal axis system of the internal spin interactions. The terms ${\Omega_{\rm iso}}^{(n)}$ and the coefficients $c_{L,k}(n)$ are listed in Tables 8.1 and 8.2 for the nuclear shielding and quadrupolar interactions.

In a liquid sample, the random and rapid molecular reorientations average the NMR transition frequency over all values of θ and ϕ. This incoherent averaging over the full sphere removes all anisotropic broadenings and only the isotropic NMR frequencies are observed. This can be easily seen by averaging equations (8.3) and (8.4) over the sphere.

Table 8.1. Isotropic frequency shifts and spherical harmonic coefficients for the $m \to m-1$ transition of spin I from first-order perturbation theory for nuclear shielding and quadrupolar interactions[a]

	Nuclear shielding	Quadrupolar
$\Omega^{(1)}_{\rm iso}$	$\sigma\gamma B_0$	0
$c^{(1)}_{2,0}$	$\sqrt{\frac{4\pi}{5}}\cdot\gamma B_0(\sigma_{ZZ}-\sigma)$	$\sqrt{\frac{9}{5}}\cdot\frac{\pi^{3/2}\chi}{I(2I-1)}(2m-1)$
$c^{(1)}_{2,\pm2}$	$\sqrt{\frac{2\pi}{15}}\cdot\gamma B_0(\sigma_{YY}-\sigma_{XX})$	$\sqrt{\frac{3}{10}}\cdot\frac{\pi^{3/2}\chi\eta}{I(2I-1)}(2m-1)$

[a] $\sigma_{XX}, \sigma_{YY}, \sigma_{ZZ}$ are the principal components of the nuclear shielding tensor, χ is e^2qQ/h, and η is the quadrupolar asymmetry parameter.

To obtain an isotropic spectrum from a solid sample, it is not necessary to average over all points on the entire sphere.[10] When the anisotropy in the NMR transition frequency is completely described by second-rank spherical harmonics, as in equation (8.3), the anisotropy can be coherently averaged away by reorientating the sample so that the external magnetic field is directed along only the three vertices on the face of an octahedron as shown in Figure 8.1(a). This

Table 8.2. Isotropic frequency shifts and spherical harmonic coefficients for the $m \to m-1$ transition of spin I from second-order perturbation theory for the quadrupolar interaction[a]

$\Omega^{(2)}_{\rm iso}$	$-\frac{3}{10}\frac{\pi^2\chi^2}{I^2(2I-1)^2\gamma B_0}\left(\frac{\eta^2}{3}+1\right)[I(I+1)-9m(m-1)-3]$
$c^{(2)}_{2,0}$	$\sqrt{\frac{36}{245}}\frac{\pi^{5/2}\chi^2}{I^2(2I-1)^2\gamma B_0}\left(\frac{\eta^2}{3}-1\right)[2I(I+1)-9m(m-1)-15/4]$
$c^{(2)}_{2,\pm2}$	$\sqrt{\frac{24}{245}}\frac{\pi^{5/2}\chi^2\eta}{I^2(2I-1)^2\gamma B_0}[2I(I+1)-9m(m-1)-15/4]$
$c^{(2)}_{4,0}$	$\frac{3}{35}\frac{\pi^{5/2}\chi^2}{I^2(2I-1)^2\gamma B_0}\left(\frac{\eta^2}{18}+1\right)[9I(I+1)-51m(m-1)-39/2]$
$c^{(2)}_{4,\pm2}$	$\frac{1}{\sqrt{490}}\frac{\pi^{5/2}\chi^2\eta}{I^2(2I-1)^2\gamma B_0}[9I(I+1)-51m(m-1)-39/2]$
$c^{(2)}_{4,\pm4}$	$\frac{1}{\sqrt{1260}}\frac{\pi^{5/2}\chi^2\eta^2}{I^2(2I-1)^2\gamma B_0}[9I(I+1)-51m(m-1)-39/2]$

[a] χ is e^2qQ/h and π is the quadrupolar asymmetry parameter.

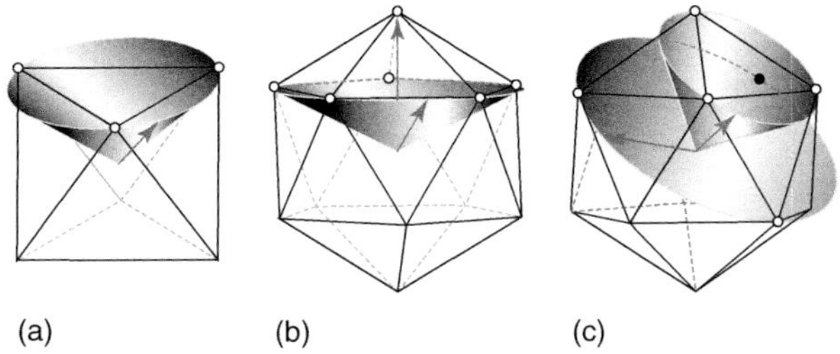

Figure 8.1. (a) Octahedral symmetry can be implemented with a single continuous trajectory in cases where tensors of rank 2 are to be eliminated. (b, c) Icosahedral symmetry can be implemented with just two continuous trajectories in cases where tensors of rank 2 and 4 are to be eliminated. Time spent along one particular trajectory is proportional to the number of vertices. In (b), the two spinning axes are $\beta_1 = 0°$ and $\beta_2 = 63.43°$, and the ratio of times spinning at the two angles is 1:5. In (c), the angles are $\beta_1 = 37.38°$ and $\beta_2 = 79.19°$, and the ratio of times spinning at the two angles is 1:1. (Adapted by permission of Elsevier Science Publishers from B. Q. Sun, J. H. Baltisberger, Y. Wu, A. Samoson, and A. Pines, '*Solid State NMR*', 1992, **1**, 267, and by permission of Clarendon Press from Samoson *et al.*)

is often the case with spin-1/2 nuclei. The discrete reorientation of the sample along these vertices is realized in a magic angle hopping (MAH)[8] experiment, and the continuous rotation on a cone[10] with apex angle of 54.74°, which passes through the vertices of the octahedron, is realized in a magic angle spinning (MAS)[11,12] experiment.

For NMR transition frequencies with higher than first-order terms, octahedral symmetry may not be sufficient for obtaining an isotropic spectrum. For example, if the quadrupolar coupling is sufficiently large that the second-order term is required to describe the spectrum, then the size of the first-order broadening can be on the order of many megahertz; thus it can be virtually impossible to obtain the complete NMR spectrum of a quadrupolar nucleus in a polycrystalline sample, since the bandwidth of conventional NMR spectrometers is typically only a few hundred kilohertz. Fortunately, for half-integer spin quadrupolar nuclei, the central $m = 1/2 \rightarrow -1/2$ transition is unaffected by the first-order quadrupolar term and in many cases is broadened only by the second-order term of equation (8.4), which can be on the order of a few kilohertz. Unfortunately, while the second-rank spherical harmonics in equation (8.4) are removed using octahedral symmetry, the fourth-rank spherical harmonics are not. Therefore, to average away completely the anisotropy in this case, a better approximation to the sphere is required. This can be done by reorientating the sample so that the external magnetic field is directed along the six vertices on the icosahedron as shown in Figure 8.1(b) or in Figure 8.1(c). The discrete reorientation of the sample along these vertices is realized in a dynamic angle hopping (DAH)[13] experiment and the continuous rotation on two cones[10] with apex angles of 0° and 63.43° or 37.38° and 79.19°, which pass through the icosahedral vertices, is realized in a DAS[3–5,10,14–21] experiment.

A DAS reorientation trajectory is implemented as a 2D NMR experiment which correlates a spin's resonance frequency while spinning at one angle with its frequency while spinning at a second angle. In DAS, the sample is spun about a single axis for the first evolution period and then hopped to a second angle for the second evolution period as shown in Figure 8.2(a). Two RF pulses are used to quench the evolution during the time required to hop.

When spinning about a single axis, equations (8.3) and (8.4) are averaged to

$$\overline{\Omega^{(1)}}(\beta, \theta', \phi') = \Omega^{(1)}_{\text{iso}} + P_2(\cos\beta) \times \sum_{m=-2}^{2} c^{(1)}_{2,m} Y_{2,m}(\theta', \phi') \tag{8.5}$$

$$\overline{\Omega^{(2)}}(\beta, \theta', \phi') = \Omega^{(2)}_{\text{iso}} + P_2(\cos\beta) \times \sum_{m=-2}^{2} c^{(2)}_{2,m} Y_{2,m}(\theta', \phi') + P_4(\cos\beta) \sum_{m=-4}^{4} c^{(2)}_{4,m} Y_{4,m}(\theta', \phi') \tag{8.6}$$

where β is the angle between the spinner axis and the magnetic field, $P_n(\cos\theta)$ are Legendre polynomials, and (θ', ϕ') are the angles between the spinner axis and the principal axis system of the internal spin

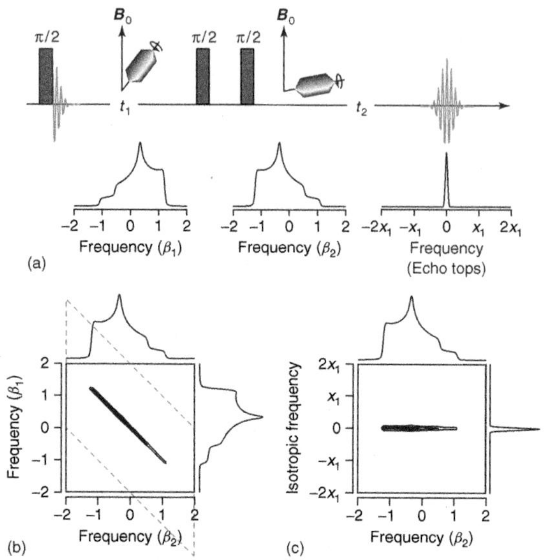

Figure 8.2. (a) Simplest implementation of 2D DAS pulse sequence. (b) 2D variable angle spinning correlation spectrum obtained from the sequence in (a). (c) 2D DAS spectrum obtained from (b) after applying the shearing transformation as described in the text.

interactions. If the two rotor axis angles in a DAS experiment are chosen so that for all the spins the anisotropic frequency at the first angle is mirrored about the isotropic frequency when spinning at the second angle, then all decay of the signal in t_1 due to the anisotropic broadening is refocused into an echo during t_2. For systems broadened up to second order, this mirror image condition is fulfilled when the angle pair (β_1, β_2) satisfies the conditions

$$x_1 P_2(\cos\beta_1) + x_2 P_2(\cos\beta_2) = 0 \qquad (8.7)$$

$$x_1 P_4(\cos\beta_1) + x_2 P_4(\cos\beta_2) = 0 \qquad (8.8)$$

and

$$x_1 + x_2 = 1 \qquad (8.9)$$

The DAS echo forms at a time $(x_2/x_1)t_1$ during t_2 evolution. By collecting the intensity of the echo tops as a function of t_1/x_1 and Fourier transforming this interferogram, the isotropic DAS spectrum is obtained. By collecting all the data in t_2 as a function of t_1, and performing a 2D Fourier transform, a spectrum as shown in Figure 8.2(b) is obtained. This 2D spectrum correlates the NMR spectrum obtained while spinning at β_1 with the NMR spectrum obtained while spinning at β_2. If the angle pair (β_1, β_2) satisfies equations (8.7)–(8.9), then the 2D correlation for each site will be isotropic and linear and given by:

$$\omega_2 = \frac{1}{x_2}(\Omega_{\text{iso}} - x_1\omega_1) \qquad (8.10)$$

There is a continuous set of angle pairs that exist which provide this isotropic 2D correlation, including the angle pairs (0°, 63.43°) where $x_1/x_2 = 5$ and (37.38°, 79.19°) where $x_1/x_2 = 1$. Equations (8.7)–(8.9) may be solved analytically for the angle pairs (β_1, β_2) in terms of x_1 and x_2:

$$\beta_1 = \cos^{-1}\frac{\sqrt{1+\sqrt{\frac{4x_2}{5x_1}}}}{3} \qquad (8.11)$$

$$\beta_2 = \cos^{-1}\frac{\sqrt{1-\sqrt{\frac{4x_1}{5x_2}}}}{3} \qquad (8.12)$$

where $0.8 \le x_2/x_1 \le 5$. Of course, the order of the angle pair may be reversed with a corresponding reversal of the fractional time spent at each angle.

The 2D spectrum of Figure 8.2(b) can be transformed with a shearing transformation to obtain the 2D DAS spectrum shown in Figure 8.2(c), which correlates the isotropic resonance of a site with its anisotropic lineshape. Shearing transformations are well known in NMR.[25] In DAS, the 2D spectrum is sheared by an angle θ_S, given by

$$\theta_S = \tan^{-1}\left(\frac{x_2}{x_1}\right) \qquad (8.13)$$

followed by a scaling of the ω_1 axis by x_1. The shearing transformation may be implemented by applying a t_1-dependent first-order phase correction:

$$S'(t_1', \omega_2') = e^{i\phi(t_1,\omega_2)} S(t_1, \omega_2) \qquad (8.14)$$

where

$$\phi(t_1, \omega_2) = \left(\frac{x_2}{x_1}\right)\omega_2 t_1 \qquad (8.15)$$

before the final Fourier transform with respect to t_1'. This correction removes the tilting in $S(\omega_1, \omega_2)$, transforming it into $S'(\omega_1', \omega_2')$, so that an isotropic spectrum may be obtained from a projection onto the ω_1' axis.

Figure 8.3 depicts the static, MAS, and DAS spectra of polycrystalline $RbNO_3$. $RbNO_3$ contains three inequivalent Rb sites in its unit cell. While the MAS provides considerable narrowing of the static spectrum by removing the second-rank spherical harmonic orientational broadenings, it still contains scaled fourth-rank orientational broadenings.

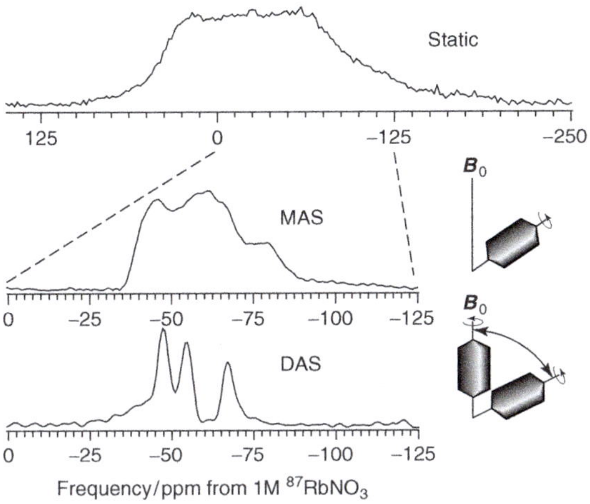

Figure 8.3. Static, MAS, and DAS ^{87}Rb NMR spectra of polycrystalline $RbNO_3$.

With DAS, both second- and fourth-rank orientational broadenings are removed and the three inequivalent Rb sites can be resolved.

Because DAS is a 2D experiment, it is more time consuming than MAS; however, DAS does provide a means of separating the anisotropic lineshapes correlated to the isotropic frequency of each resolved site. For example, Figure 8.4 depicts the 2D DAS spectrum of $RbNO_3$. Since the isotropic resonances of all three inequivalent Rb sites can be resolved, cross sections correlated to each isotropic frequency provide the separated anisotropic lineshape for each site.

These separated lineshapes can be analyzed to obtain the quadrupolar coupling parameters associated with each resonance. This information can be useful in providing structural information. For example, in the case of $RbNO_3$, a simple point charge calculation of the electric field gradients at each Rb site in the unit cell gives quadrupolar asymmetry parameters of approximately 0.9, 0.6, and 0.3. On this basis, the three resonances in the 2D DAS spectrum can be assigned to the three sites in the unit cell as shown in Figure 8.4. This approach can also be extended to amorphous solids where a continuous set of overlapping anisotropic lineshapes can also be separated according to correlated isotropic frequencies and used to quantify continuous structural distributions.

An interesting consequence of the shearing transformation in DAS is that the spinning sidebands can appear at nonintegral multiples of the spinning speed in the isotropic dimension. A schematic example of the behavior of spinning sidebands after a shearing transformation is shown in Figure 8.5 for the two cases $\theta_s = 45^\circ$ and $\theta_s = 38.7^\circ$. The frequency axes in Figure 8.5(a) are in units of Ω_R, the actual spinning speed. The example in Figure 8.5(b) is the familiar situation in 2D echo spectroscopy, where the dephasing and refocusing times are equal (e.g., 2D J spectroscopy). In this example, even though the spectrum is sheared, the sidebands remain aligned such that a projection onto the ω_1' axis contains only sidebands at integer multiples of 0.5 Ω_R. In contrast, the example in Figure 8.5(c), where $x_1 = 0.44$ and $x_2 = 0.56$, describes a 2D echo experiment with unequal dephasing

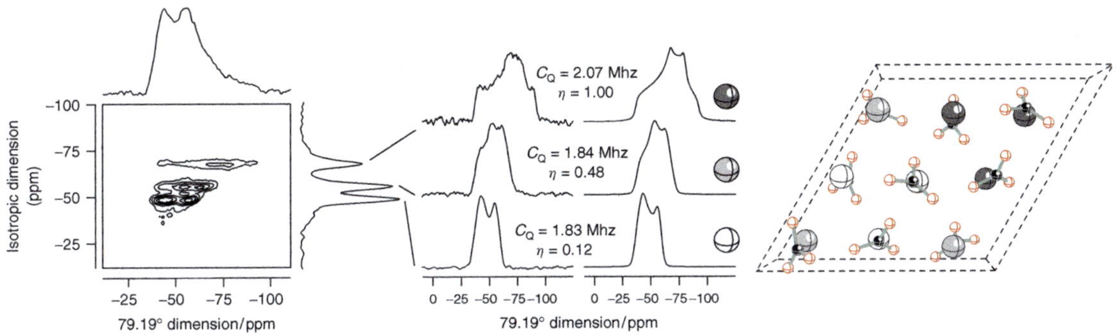

Figure 8.4. On the left is the 2D ^{87}Rb DAS NMR spectrum of $RbNO_3$ at 4.2 T. Cross sections from this spectrum provide the separated 1D anisotropic lineshapes for each ^{87}Rb site. These lineshapes can be fitted to obtain the quadrupolar coupling parameters for each site. In this case, each resonance could be assigned to a Rb site in the unit cell (shown on the right) by comparing the electric field gradients obtained from the measured quadrupolar coupling constants with the electric field gradient at each Rb site obtained from a simple point charge model calculation (see text). (Adapted by permission of Academic Press from Grandinetti *et al.*[20])

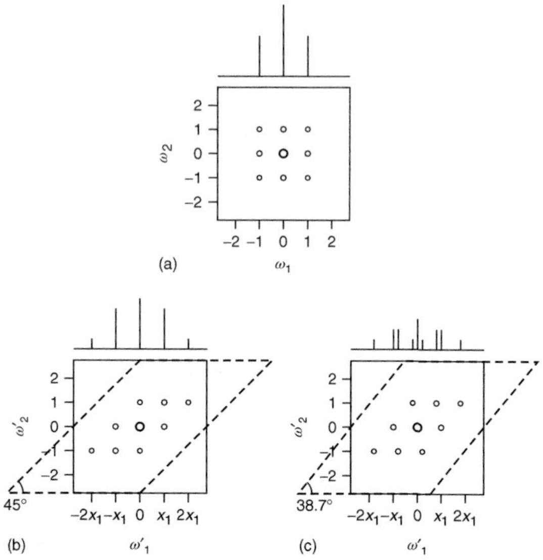

Figure 8.5. Schematic examples of the effects of the shearing transformation on sideband positions in DAS. (a) Unsheared 2D spectrum. (b) Sheared 2D spectrum obtained from the 2D spectrum in (a) by a shearing transformation employing a shearing angle of 45° and a scaling of the ω_1 by $x_1 = 0.5$. In this case, the spinning sidebands in the 2D spectrum are aligned with respect to ω_1' so that a projection onto the ω_1' axis only contains spinning sidebands separated by integer multiples of 0.5 Ω_R. (c) Sheared 2D spectrum obtained from the 2D spectrum in (a) by a shearing transformation employing a shearing angle of 38.7° and a scaling of the ω_1 by $x_1 = 0.56$. In this case, the spinning sidebands in the 2D spectrum are not aligned with respect to ω_1' so that a projection onto the ω_1' axis contains spinning sidebands that are separated by multiples of and also sum and difference frequencies of $x_1\Omega_R = 0.56\ \Omega_R$ and $x_2\Omega_R = 0.44\ \Omega_R$. (Reproduced by permission of Academic Press from P. J. Grandinetti, Y. K. Lee, J. H. Baltisberger, B. Q. Sun, and A. Pines, *J. Magn. Reson.*, 1993, **102**, 71.)

and refocusing times. In this situation, the spinning sidebands are not aligned in ω_1', and consequently a projection onto ω_1' yields a fairly complicated sideband pattern.

Simulated and experimental isotropic projections from the ^{87}Rb DAS of $RbClO_4$, demonstrating the sideband behavior for various pairs of DAS angles, are shown in Figure 8.6. In each isotropic 1D DAS spectrum, the spinning sidebands appear at integral multiples of $x_1\Omega_R$ and $x_2\Omega_R$ and also at the sum and difference frequencies of the integral multiples of $x_1\Omega_R$ and $x_2\Omega_R$. The angle

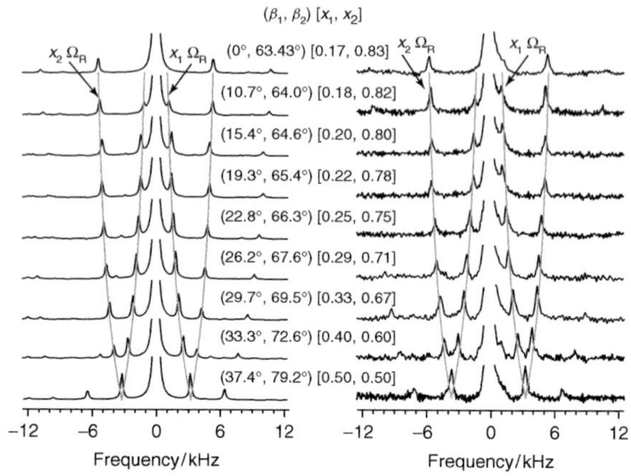

Figure 8.6. Simulated and experimental 1D DAS spectra of $^{87}RbClO_4$ for various rotor angle pairs (β_1, β_2) and fractions [x_1, x_2]. All spectra were obtained with a rotor frequency of $\Omega_R = 6.4$ kHz. The quadrupolar coupling parameters of $RbClO_4$ used in the simulations are $C_q = 3.2$ MHz and $\eta = 0.1$. (Adapted by permission of Academic Press from P. J. Grandinetti, Y. K. Lee, J. H. Baltisberger, B. Q. Sun, and A. Pines, *J. Magn. Reson.*, 1993, **102**, 71.)

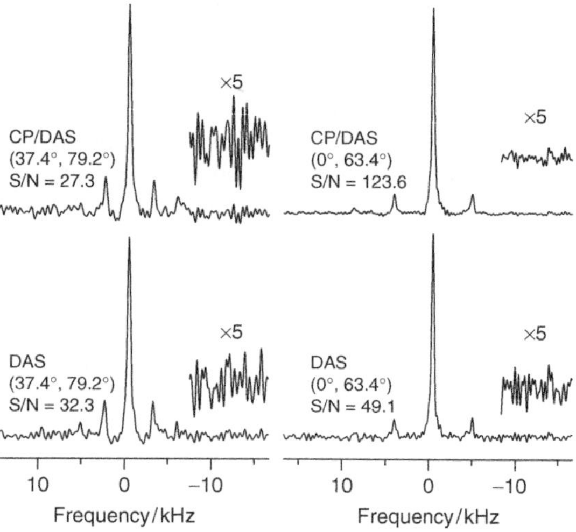

Figure 8.7. DAS and CP DAS spectra of sodium pyruvate acquired using the DAS angle pairs (37.38°, 79.19°) and (0°, 63.43°). With (0°, 63.43°), an improvement of 2.5 in S/N ratio is observed with CP compared with the spectrum taken without CP. In addition, the CP DAS experiment using (0°, 63.43°) has an S/N ratio over 4.5 times that of the CP DAS experiment using (37.38°, 79.19°). (Reproduced by permission of Taylor & Francis Ltd. from Baltisberger *et al.*[24])

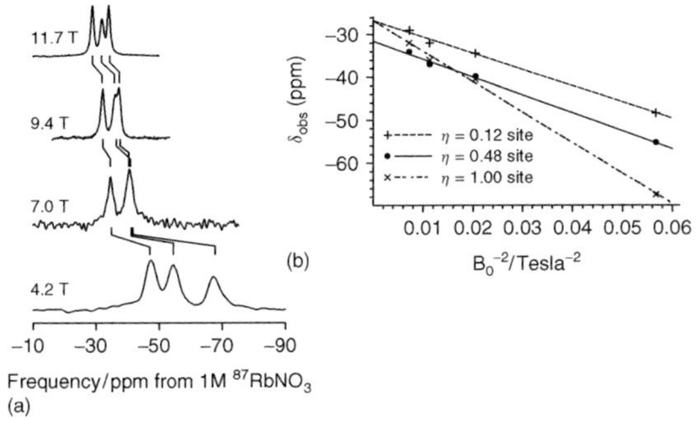

Figure 8.8. (a) ^{87}Rb DAS spectra of $RbNO_3$ collected at (a) 11.7, 9.4, 7.0, and 4.2 T. (b) Isotropic shifts of $^{87}RbNO_3$ plotted versus $1/B_0{}^2$ and fitted using a linear least squares routine. The intercepts and slopes are used to calculate the values of $\delta_{iso}{}^{(CS)}$ (ppm) and $C_q(1+\eta^2/3)^{1/2}$ (MHz), respectively. (Reproduced by permission of Academic Press from Baltisberger *et al.*[16])

pair (37.78°, 79.19°) has an advantage that $x_1\Omega_R = x_2\Omega_R = 0.5\ \Omega_R$ and a simple side band spacing of $0.5\ \Omega_R$ is obtained. Of particular interest is the fact that the 1D DAS spectrum for the (0°, 63.43°) angle pair contains only spinning sidebands at multiples of $x_2\Omega_R = 0.83\ \Omega_R$. The spinning sidebands at $x_1\Omega_R = 0.17\ \Omega_R$ do not exist and thus the (0°, 63.43°) angle pair provides the highest effective spinning speed when removing first- and second-order broadenings with DAS.

The (0°, 63.43°) angle pair offers other advantages, in particular for cross polarization (CP) experiments. Vega[22,23] has shown that it may be difficult or even impossible to obtain efficient CP transfer of all sites in a multisite system when spinning at the magic angle. This problem can be eliminated in a DAS experiment by exploiting the time independence of the spin eigenvalues when spinning at 0° (parallel) to the external magnetic field. By performing the CP step while spinning at 0°, the full static CP intensity can be recovered.[24] In Figure 8.7, the decoupled DAS and CP DAS spectra of sodium pyruvate for the (37.38°, 79.19°) and (0°, 63.43°) angle pairs illustrate the significant increases in signal-to-noise ratio (S/N) that can be obtained with the (0°, 63.43°) angle pair.

From Tables 1 and 2, it is clear that the isotropic frequency shifts of both the first-order nuclear shielding and the second-order quadrupolar shift have different dependences on the external magnetic field strength. These different dependences can be exploited as a means of separating the isotropic chemical and quadrupolar shifts. In Figure 8.8(a), the isotropic ^{87}Rb DAS spectra of $RbNO_3$ measured at four different magnetic fields are depicted. As shown in Figure 8.8(b), the isotropic resonance frequency in parts per million for each site is a linear function of $1/B_0{}^2$ with the isotropic chemical shift given by the intercept and the second-order quadrupolar shift given by the slope.[16]

8.3 IMPLEMENTATION

Four variants of the DAS pulse sequence are depicted in Figure 8.9. The simplest implementation is shown in Figure 8.9(a). In this sequence, DAS can be viewed as a 2D exchange experiment with a rotor reorientation during the mixing time and, as in the 2D exchange experiment, an amplitude-modulated response in t_1 makes it possible to obtain pure absorption-mode 2D spectra using the hypercomplex or TPPI (time proportional phase increment) approach to 2D data acquisition. The hypercomplex data acquired with this sequence are Fourier transformed and phase corrected in a manner similar to that described elsewhere,[25] with the only difference being the shearing transformation described by equations (8.14) and (8.15).

In Figure 8.9(b), a different approach is taken that provides pure absorption-mode lineshapes in addition to a $\sqrt{2}$ increase in the S/N. In this approach, the coherence-transfer echoes, which are formed because of the inhomogeneous broadenings intrinsic to the DAS experiment, are time-shifted by an amount τ with a π pulse to obtain whole echo acquisition in t_2, and thus a pure absorption-mode 2D spectrum. The length of τ is a multiple of the rotor period, and typically large enough so that the echo in t_2 begins at zero. Only the $p = 0 \rightarrow -1 \rightarrow 0 \rightarrow 1 \rightarrow -1$ pathway is selected in this case. The complex 2D data acquired with this sequence are processed in the same manner as conventional phase-modulated 2D data, with the only difference being a τ-dependent first-order phase correction of

$$\phi(\tau, \omega) = \omega_2\tau \qquad (8.16)$$

to remove the phase modulation because of the time-shifted echo, in addition to the shearing transformation described by equations (8.14) and (8.15). When there is an inhomogeneous broadening associated with the isotropic DAS dimension, both pathways shown in Figure 8.9(b) can be acquired using the hypercomplex approach for an additional $\sqrt{2}$ increase in S/N. This often occurs when DAS is applied to materials in which there is a continuous distribution of atomic environments resulting in a continuous distribution of second-order quadrupolar and isotropic chemical shifts.

Mueller *et al.*[18] pointed out that extracting the quadrupolar coupling parameters from the anisotropic DAS dimension can be difficult when additional chemical and dipolar anisotropies are present. Chemical shift and dipolar anisotropic broadenings can be removed when spinning at 54.74°, leaving only quadrupolar anisotropic broadening; 54.74°, unfortunately, is not a solution of equations (8.11) and (8.12). The solution of Mueller *et al.*[18] was to incorporate a second rotor reorientation into the DAS sequence so that the final angle for detection is the magic angle. This approach and its shifted echo variant are shown in Figure 8.9(c) and (d). With the evolution times defined in Figure 8.9(c) and (d), a shearing transformation is not necessary to obtain the isotropic/anisotropic 2D DAS spectrum; however, for the shifted echo experiment, the phase correction of equation (8.16) is still required.

Using rotor-synchronized π pulses, it is possible to eliminate all sidebands in a DAS experiment using the DAH-180 sequence of Gann *et al.*,[13] shown in Figure 8.10. Under this pulse sequence, all evolution at 63.43° that does not contribute to t_1 evolution is refocused by spending equal time as +1 and −1 coherences, while all t_1 evolution occurs only as −1 coherences. This sequence has the advantage

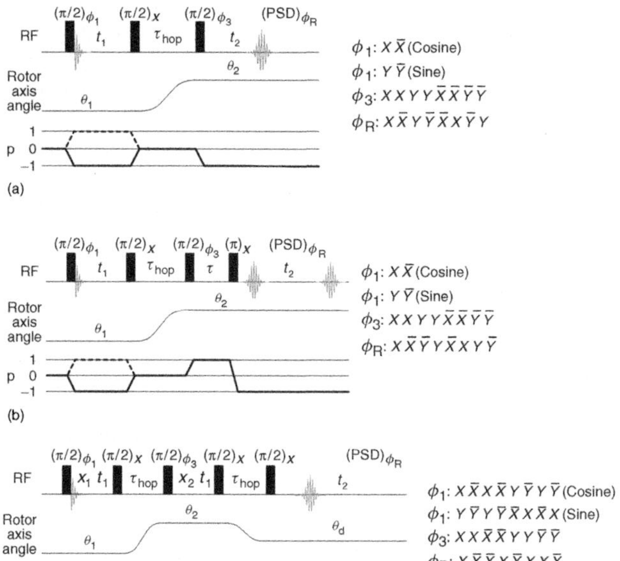

Figure 8.9. Pulse sequences, coherence-transfer pathways, and minimal phase cycles for (a) dynamic angle spinning, (b) shifted echo dynamic angle spinning, (c) MAS-detected dynamic angle spinning, and (d) MAS-detected shifted echo dynamic angle spinning.

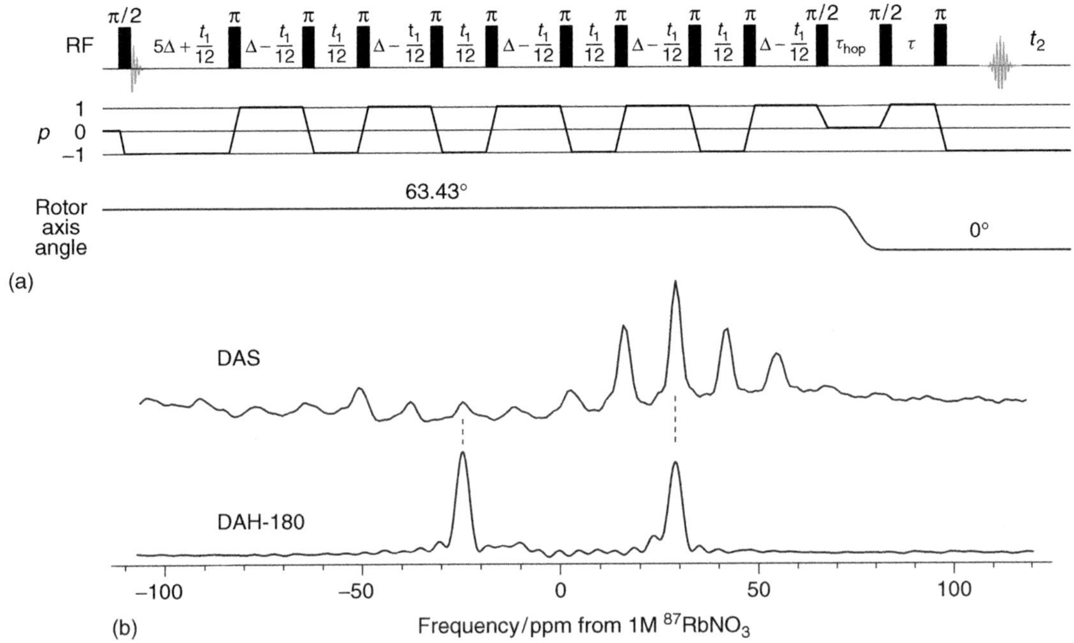

Figure 8.10. (a) DAH-180 pulse sequence that produces sideband-free dynamic angle spinning spectra. The experiment is performed over N rotor cycles with a rotor period of τ_r. $\Delta = N\tau_r/10$. (b) Comparison of the ^{87}Rb spectra of Rb_2SO_4 taken at 9.4 T at a spinning frequency of 1.8 kHz using DAS (with side bands) and using DAH-180 (sideband-free). (Reproduced by permission of Elsevier Science Publishers from Gann *et al.*[13])

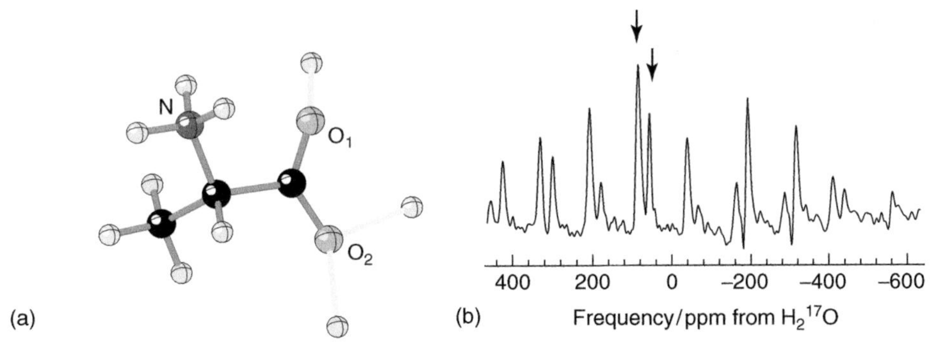

Figure 8.11. (a) The structure of L-alanine showing the differences in hydrogen bonding on the two oxygen sites. (b) Isotropic ^{17}O (0°, 63.43°) CP DAS spectrum of L-alanine at 7.04 T with the two resolved centerbands for the two sites marked with arrows. (Adapted by permission of the International Society of Magnetic Resonance from S. L. Gann *et al.*[28])

that the intensities of the suppressed sidebands are completely transferred to the centerband, unlike other sideband suppression techniques such as total suppression of sidebands (TOSS).[26] In contrast to the DAS sequences in Figure 8.9, the DAH-180 sequence has a trade-off between sensitivity and resolution. DAH-180 is a constant time experiment, so in order to increase the resolution, a larger constant time is required which then leads to increased intensity losses via T_2 relaxation processes.

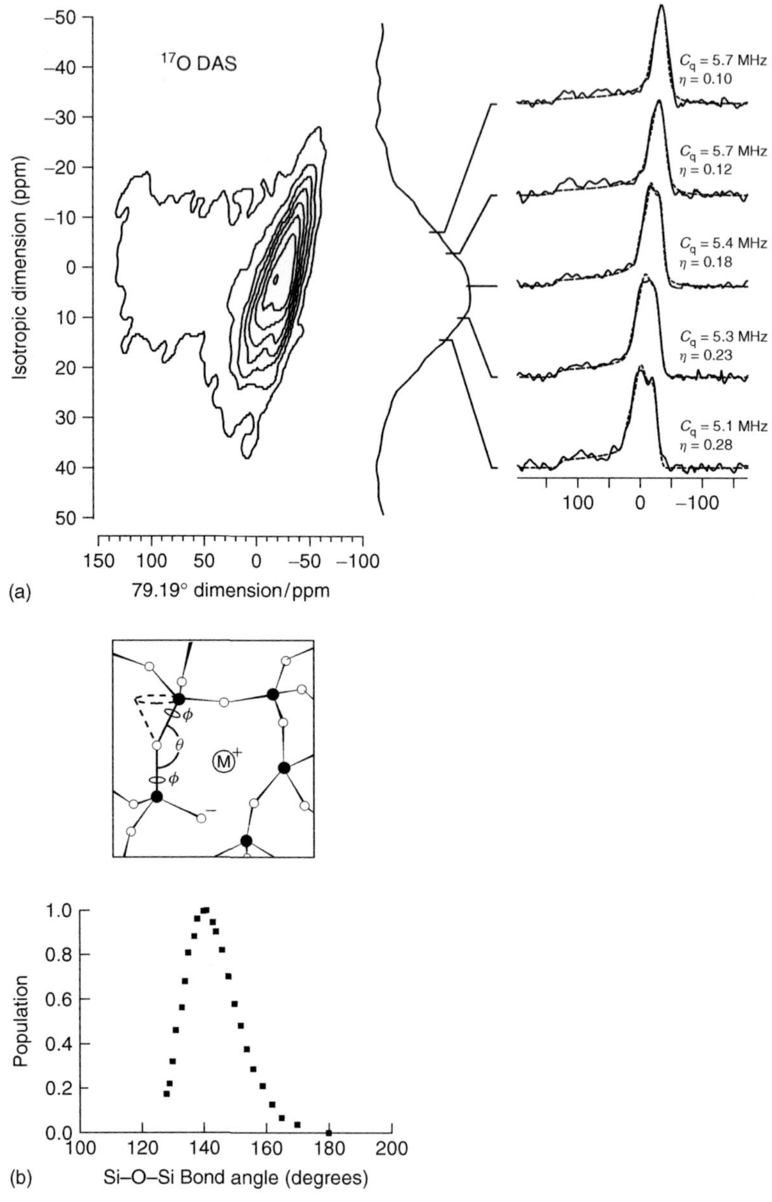

Figure 8.12. (a) ^{17}O 2D DAS spectrum of the bridging oxygen (Si–O–Si) resonances in $K_2Si_4O_9$ glass, shown together with the isotropic lineshape, and selected anisotropic cross sections correlated to specific frequencies in the isotropic lineshape. Because the glass contains a continuous distribution of Si–O–Si bond angles, and the ^{17}O isotropic frequency varies continuously with Si–O–Si bond angle, the bridging oxygen isotropic linewidth in the glass is ~25 times wider than a bridging oxygen resonance in a crystalline silicate. Since the ^{17}O anisotropic lineshapes for each Si–O–Si angle are separated in a 2D DAS spectrum, the separated anisotropic lineshapes can be fitted to obtain the isotropic lineshape as a function of the quadrupolar coupling parameters. With the help of experimental as well as ab initio-based correlations between quadrupolar coupling parameters and the Si–O–Si bond angles, the isotropic lineshape can then be mapped into the Si–O–Si bond angle distribution for the glass. (b) The Si–O–Si bond angle distribution in $K_2Si_4O_9$ glass derived from its ^{17}O DAS spectra. (Adapted by permission of MacMillan Magazines Ltd. from Farnan *et al.*[29])

There are a few limitations of the DAS technique, most notably being its current inability to refocus homonuclear dipolar broadenings. This is a consequence of the storage pulses that fail to store higher than first-rank coherences. These broadenings are largest with abundant nuclei with high gyromagnetic ratios (e.g., ^{27}Al, ^{11}B, ...). Baltisberger *et al.*[24] have shown that homonuclear dipolar broadenings depend on the DAS angle pair. The (0°, 63.43°) angle pair produces the spectra with the minimum dipolar broadenings. For certain nuclei, it is possible to eliminate the homonuclear dipolar broadenings using the isotopic depletion approach of Youngman and Zwanziger.[27] By enriching glassy B_2O_3 to 97% in ^{10}B, they obtained increased resolution by eliminating the ^{11}B dipolar broadenings from the ^{11}B DAS spectrum.

Another limitation arises from the finite time required to hop the rotor axis between the DAS angle pairs. When the spin–lattice relaxation time T_1 is on the order or less than the hop time ($T_1 \leq \tau_{hop}$), there will be significant losses in signal intensity. For this reason, integrated intensities in DAS may not accurately reflect the site populations.

8.4 APPLICATIONS

Since its development, DAS has been successfully applied to a number of multisite crystalline compounds. Mueller and coworkers[14,15] have applied ^{17}O DAS to a number of crystalline silicates, resolving inequivalent bridging (Si–O–Si) and nonbridging (Si–O–M) oxygens. Baltisberger *et al.*[16] have applied ^{87}Rb DAS to a number of crystalline inorganic rubidium salts and obtained isotropic spectra along with correlated anisotropic spectra from which quadrupolar coupling parameters could be obtained. Gann *et al.*[28] obtained the spectrum shown in Figure 8.11 where they could resolve the two crystallographically distinct oxygen sites of L-alanine in a (0°, 63.43°) CP DAS experiment.

Farnan *et al.*[29] showed that DAS is particularly advantageous when applied to amorphous samples where a continuous structural distribution results in a continuous distribution of anisotropic lineshapes. In the 2D DAS spectrum, these anisotropic lineshapes can be separated according to their isotropic frequencies, analyzed, and then used to help map the isotropic lineshape into structural distributions. Figure 8.12(a) depicts a 2D ^{17}O DAS spectrum of the bridging oxygen (Si–O–Si) resonances in $K_2Si_4O_9$ glass. In this case, the Si–O–Si bond angle distribution in Figure 8.12(b) was derived from the 2D DAS spectrum.

RELATED ARTICLES IN THE ENCYCLOPEDIA OF MAGNETIC RESONANCE

Amorphous Materials

Cross Polarization in Rotating Solids: Spin-1/2 Nuclei

Double Rotation

Line Narrowing Methods in Solids

Magic Angle Spinning

Magic Angle Turning and Hopping

Multidimensional Spectroscopy: Concepts

Quadrupolar Nuclei in Glasses

Quadrupolar Nuclei in Liquid Samples

Sideband Analysis in Magic Angle Spinning NMR of Solids

Variable Angle Sample Spinning

REFERENCES

1. S. Ganapathy, S. Schramm, and E. Oldfield, *J. Chem. Phys.*, 1982, **77**, 4360.
2. A. Samoson, E. Lippmaa, and A. Pines, *Mol. Phys.*, 1988, **65**, 1013.
3. A. Llor and J. Virlet, *Chem. Phys. Lett.*, 1988, **152**, 248.
4. B. F. Chmelka, K. T. Mueller, A. Pines, J. Stebbins, Y. Wu, and J. W. Zwanziger, *Nature (London)*, 1989, **339**, 42.
5. K. T. Mueller, B. Q. Sun, G. C. Chingas, J. W. Zwanziger, T. Terao, and A. Pines, *J. Magn. Reson.*, 1990, **86**, 470.
6. T. Terao, T. Fujii, T. Onodera, and A. Saika, *Chem. Phys. Lett.*, 1984, **107**, 145.
7. A. Bax, N. M. Szeverenyi, and G. E. Maciel, *J. Magn. Reson.*, 1983, **55**, 494.
8. A. Bax, N. M. Szeverenyi, and G. E. Maciel, *J. Magn. Reson.*, 1983, **52**, 147.

9. M. Goldman, P. J. Grandinetti, A. Llor, Z. Olejniczak, J. R. Sachleben, and J. W. Zwanziger, *J. Chem. Phys.*, 1992, **97**, 8947.
10. A. Samoson, B. Q. Sun, and A. Pines, in *Pulsed Magnetic Resonance: NMR, ESR, and Optics—A Recognition of E. L. Hahn*, ed. D. M. S. Bagguley, Clarendon Press: Oxford, 1992.
11. E. R. Andrew, A. Bradbury, and R. G. Eades, *Nature (London)*, 1958, **182**, 1659.
12. I. J. Lowe, *Phys. Rev. Lett.*, 1959, **2**, 285.
13. S. L. Gann, J. H. Baltisberger, and A. Pines, *Chem. Phys. Lett.*, 1993, **210**, 405.
14. K. T. Mueller, Y. Wu, B. F. Chmelka, J. Stebbins, and A. Pines, *J. Am. Chem. Soc.*, 1991, **113**, 32.
15. K. T. Mueller, J. H. Baltisberger, E. W. Wooten, and A. Pines, *J. Phys. Chem.*, 1992, **96**, 7001.
16. J. H. Baltisberger, S. L. Gann, E. W. Wooten, T. H. Chang, K. T. Mueller, and A. Pines, *J. Am. Chem. Soc.*, 1992, **114**, 7489.
17. M. A. Eastman, P. J. Grandinetti, Y. K. Lee, and A. Pines, *J. Magn. Reson.*, 1992, **98**, 333.
18. K. T. Mueller, E. W. Wooten, and A. Pines, *J. Magn. Reson.*, 1990, **92**, 620.
19. K. T. Mueller, G. C. Chingas, and A. Pines, *Rev. Sci. Instrum.*, 1991, **62**, 1445.
20. P. J. Grandinetti, J. H. Baltisberger, A. Llor, Y. K. Lee, U. Werner, M. A. Eastman, and A. Pines, *J. Magn. Reson., Ser. A*, 1993, **103**, 72.
21. P. J. Grandinetti, Y. K. Lee, J. H. Baltisberger, B. Q. Sun, and A. Pines, *J. Magn. Reson., Ser. A*, 1993, **102**, 195.
22. A. J. Vega, *J. Magn. Reson.*, 1992, **96**, 50.
23. A. J. Vega, *Solid State NMR*, 1992, **1**, 17.
24. J. H. Baltisberger, S. L. Gann, P. J. Grandinetti, and A. Pines, *Mol. Phys.*, 1994, **81**, 1109.
25. R. R. Ernst, G. Bodenhausen, and A. Wokaun, *Principles of Nuclear Magnetic Resonance in One and Two Dimensions*, Oxford University Press: Oxford, 1987.
26. W. T. Dixon, *J. Chem. Phys.*, 1982, **77**, 1800.
27. R. E. Youngman and J. W. Zwanziger, *J. Non-Cryst. Solids*, 1994, **168**, 293.
28. S. L. Gann, J. H. Baltisberger, E. W. Wooten, H. Zimmerman, and A. Pines, *Bull. Magn. Reson.*, 1994, **16**, 68.
29. I. Farnan, P. J. Grandinetti, J. H. Baltisberger, J. F. Stebbins, U. Werner, M. A. Eastman, and A. Pines, *Nature (London)*, 1992, **358**, 31.

Chapter 9
Double Rotation (DOR) NMR

Ray Dupree

Department of Physics, University of Warwick, Coventry CV4 7AL, UK

9.1 INTRODUCTION

The anisotropic interactions present in a solid mean that the position of the resonance depends on the orientation of the crystal with respect to the magnetic field. Thus, for a polycrystalline solid the NMR linewidth of a static sample is usually much larger than that for liquids. The three main interactions that broaden the line are the dipole coupling, the chemical shift and, for nuclei with spin $I > 1/2$, which is about 70% of the NMR active isotopes, the quadrupole interaction. All of these interactions have a $(3\cos^2\theta - 1)$ angular dependence to the first order in the interaction and therefore magic angle spinning (MAS),[1] where the sample is spun about an angle where $\cos^2\theta = 1/3(\sim 54.74^\circ)$ to the magnetic field is widely used. However, the quadrupole interaction is often sufficiently large that second-order perturbations to the Zeeman interaction must be taken into account. The quadrupole interaction is described in terms of the electric field gradient (EFG), a tensor that, in its principal axis system, is described by the three components V_{zz}, V_{yy}, and V_{xx}, where $|V_{zz}| \geq |V_{yy}| \geq |V_{xx}|$, and the quadrupole moment eQ. It is conventional to recast these into the largest component V_{zz}, called the EFG, and to write $V_{zz} = eq$, and the asymmetry parameter η_Q with $\eta_Q = (V_{xx} - V_{yy})/V_{zz}$ $(0 \leq \eta_Q \leq 1)$. The quadrupole coupling constant $C_Q = e^2qQ/h$ expresses the size of the interaction in frequency and Figure 9.1 shows a quadrupolar perturbed energy level diagram for spin $I = 5/2$. Since the first-order perturbation is proportional to m^2, the $\pm 1/2$ levels are equally affected and their separation is the same as with no quadrupole interaction. However, the second-order quadrupole interaction (proportional to C_Q^2/ν_0) changes their separation in an angularly dependent way and MAS only partially removes the broadening. Figure 9.2 shows the angular dependence of the first- and second-order terms, which are proportional to the Legendre polynomials $P_2(\cos\theta)$ and $P_4(\cos\theta)$, respectively, where

$$P_2(\cos\theta) = \frac{1}{2}(1 - 3\cos^2\theta)$$

$$P_4(\cos\theta) = \frac{1}{8}(35\cos^4\theta - 30\cos^2\theta + 3) \quad (9.1)$$

It can be seen that they have no common root, i.e., there is no angle that will enable both terms to be reduced to zero. At the "magic" angle where

NMR of Quadrupolar Nuclei in Solid Materials
Edited by Roderick E. Wasylishen, Sharon E. Ashbrook and Stephen Wimperis
 ISBN: 978-0-470-97398-1

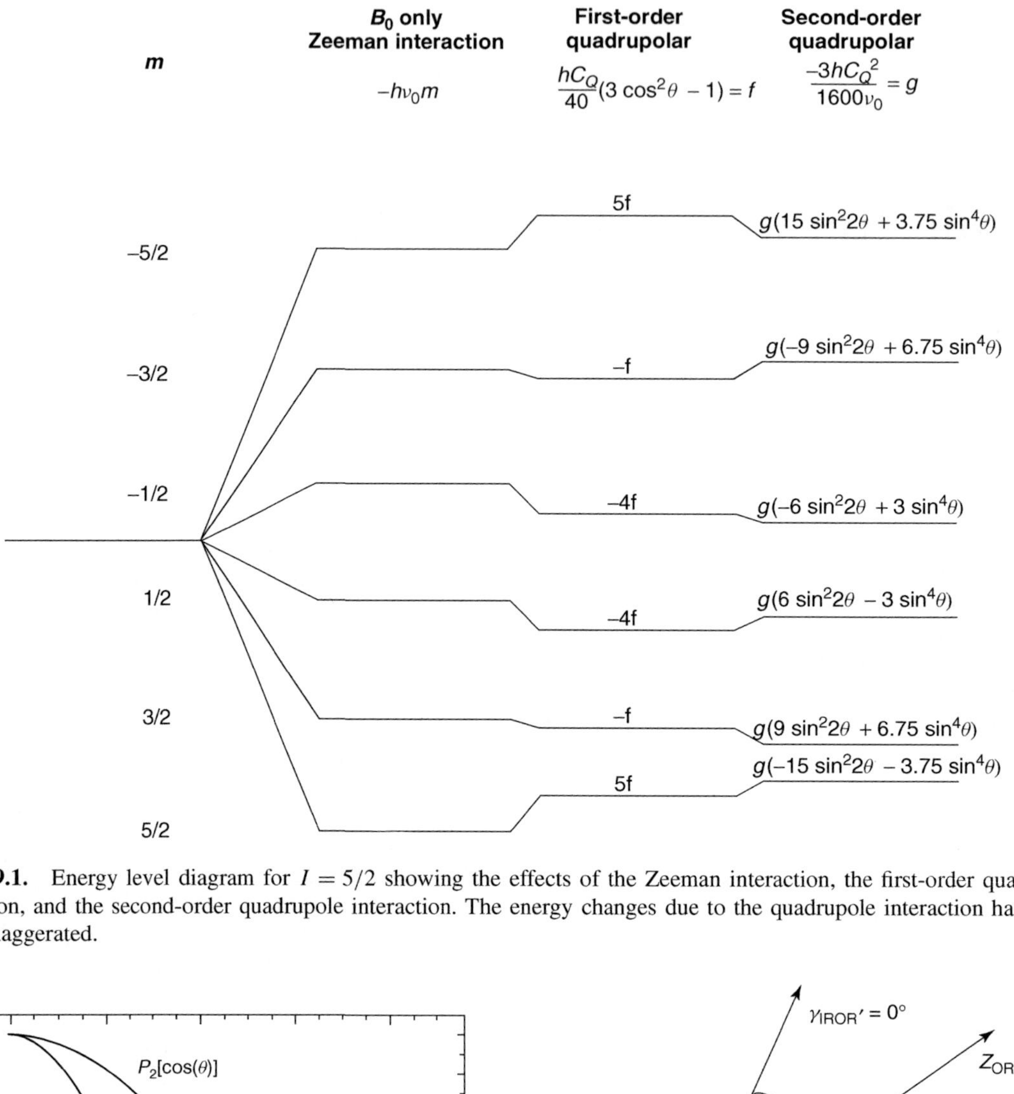

Figure 9.1. Energy level diagram for $I = 5/2$ showing the effects of the Zeeman interaction, the first-order quadrupole interaction, and the second-order quadrupole interaction. The energy changes due to the quadrupole interaction have been much exaggerated.

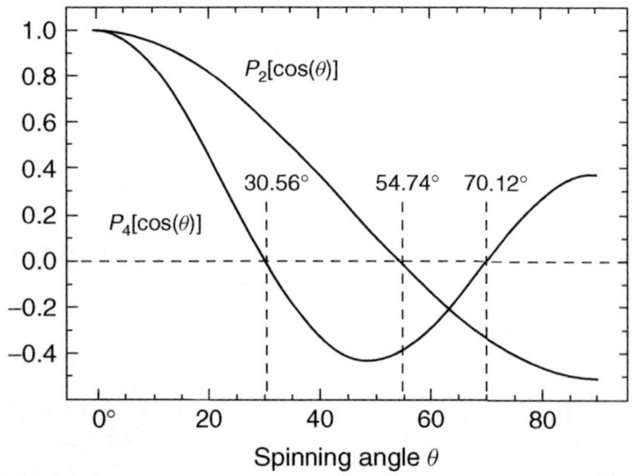

Figure 9.2. Angular dependence of the Legendre polynomials P_2 and P_4.

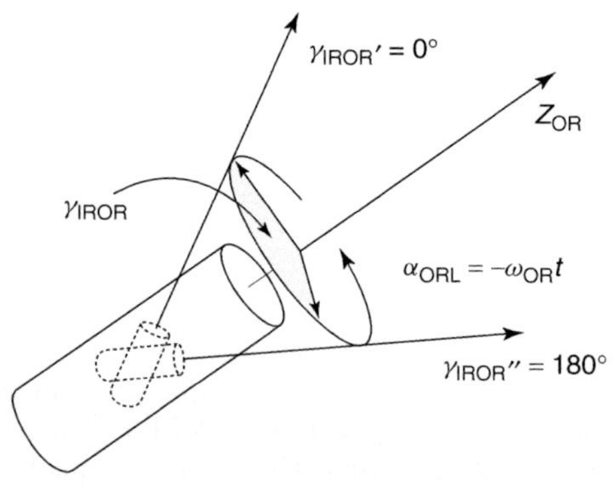

Figure 9.3. Schematic diagram of the DOR rotor showing the outer (OR) and inner (IR) rotors. γ_{IROR} is the phase of the inner rotor, which plays an important role in odd-order sideband suppression. The inner rotor is at 30.56° to the outer rotor. (Adapted from Ref. 13. © Elsevier, 2009.)

$P_2(\cos\theta_{magic}) = 0$ $P_4(\cos\theta_{magic}) = -7/18$ and, for a line where broadening by the quadrupolar interaction is dominant, MAS only reduces the width by ~3. In double rotation (DOR), there are two rotors, one inside the other, such that the outer rotor is at the magic angle ($P_2(\cos\theta_1) = 0$) and the inner rotor at an angle such that $P_4(\cos\theta_2) = 0$, where $\theta_2 = 30.56°$. This is shown schematically in Figure 9.3. DOR was first proposed independently by two groups[2,3] in 1988 and first implemented by Samoson *et al.*[3] Because of the mechanical difficulties of spinning one rotor inside another, the first rotors were very large, 20 mm outer diameter, with a maximum outer rotor frequency of ~400 Hz. However, this was rapidly followed by a design with an outer rotor diameter of ~14 mm able to spin up to nearly 1 kHz,[4] which was sufficient, on a 400 MHz spectrometer, to narrow the ^{23}Na line of sodium oxalate from ~10 kHz (static) to 85 Hz and that of ^{17}O in wollastonite ($CaSiO_3$) to 70 Hz showing the dramatic increase in resolution obtainable with DOR.

The ability of one rotor to spin inside another depends upon the torque exerted by the inner rotor upon the outer, which has to be small for successful spinning. The torque is zero if

$$\frac{\omega_{IR}}{\omega_{OR}} = \cos\beta_{IR}\left(\frac{I_{tr} - I_{ax}}{I_{ax}}\right) \quad (9.2)$$

where ω_{IR} and ω_{OR} are the angular frequencies of the inner and outer rotor, respectively, β_{IR} is the angle between the inner and outer rotors (30.56° for DOR), and I_{tr} and I_{ax} are the transverse and axial moments of inertia of the inner rotor. Typically, this frequency ratio, determined by the dimensions of the inner rotor through I_{tr}/I_{ax}, is chosen to be ~5 (which means the length is ~3 times the diameter). For stable spinning, the ratio ω_{IR}/ω_{OR} should be slightly above the torque free value since a random slowing of the inner rotor ω_{IR} would then lower the torque, tending to increase the speed, and a random rise in frequency would increase the torque, slowing the rotation.

As with MAS, sidebands are formed, separated by the outer rotor spinning frequency from the isotropic line and, for multisite samples, the relatively slow spinning speed can complicate the DOR spectrum. However, the sample is effectively axially oriented relative to the outer rotor (Figure 9.3) and thus the modulation of the free induction signal, which leads to sideband formation, depends on the position of the internal rotor. As shown by Samoson and Lippmaa,[5] if the start of the data acquisition is alternately synchronized at outer rotor positions corresponding to $\gamma_{IROR'} = 0°$ and $\gamma_{IROR''} = 180°$ (Figure 9.3), while each phase cycle step is kept fixed for two acquisitions rather than one, then all odd-numbered sidebands are cancelled, effectively doubling the outer rotor speed. An example is shown in Figure 9.4(a), where a ^{71}Ga spectrum

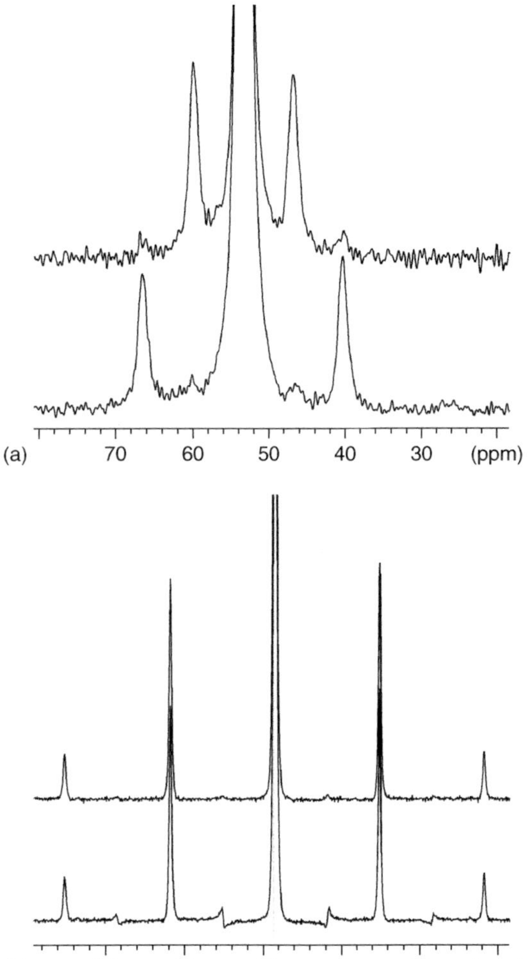

Figure 9.4. (a) ^{71}Ga DOR spectrum with (lower) and without (upper) odd-order sideband suppression. (b) ^{23}Na DOR spectrum of NaCl with the outer rotor spinning at 1.5 kHz, showing the effect of a small (3 μs) timing error on the effectiveness of sideband suppression. For the upper spectrum, the time delay between the rotor synchronisation trigger and when the inner rotor is in the "up" position was approximately correct at 308 μs, whereas for the lower spectrum it was set as 311 μs.

acquired normally (i.e., with no synchronization) is superposed over one with sideband suppression. It can be seen that the first-order sidebands are almost invisible when the spectrum is acquired in this way. The efficiency of the suppression can depend very sensitively upon correct timing, which is adjusted by having a variable delay after the tachometer signal so that the signal is acquired at the correct rotor positions. Figure 9.4(b) shows that incomplete cancellation occurs even for a timing error of only 3 μs in the setup sample, ^{23}Na in NaCl. Further sideband suppression can be achieved by, e.g., applying an additional 180° inversion pulse and adding spectra when $\gamma_{IROR'} = 90°$ and 270°,[6,7] which, in principle, eliminates the second-order sidebands and effectively quadruples the apparent rotation frequency; however, inversion of quadrupolar nuclei is not straightforward and this has been little used.

Soon after the implementation of DOR, commercial probes were produced with a maximum outer rotor frequency of ~1.2 kHz. However, they were difficult to use with realistic spin rates of ~800 Hz. Furthermore, the stability was such that the operation of these first-generation probes was limited to ~1 h and interest in DOR waned when it was demonstrated that high-resolution spectra can be obtained for half-integer quadrupolar nuclei using standard MAS probes by means of the 2D multiple-quantum (MQ) MAS (1995) experiment. Nonetheless, the Samoson group have gradually improved the performance such that the present generation DOR probes, with ~10 mm outer rotor diameter and computer-controlled stabilization, have outer rotor frequencies of ~2 kHz and are able to spin stably for long periods (>2 days), allowing 2D DOR experiments to be undertaken.

9.2 INFORMATION FROM ONE DIMENSIONAL DOR EXPERIMENTS

The DOR line occurs at δ_{DOR}, the center of gravity of the MAS spectrum, i.e., it is displaced from the chemical shift δ_{iso} by what is termed the quadrupole shift, δ_{QIS}, i.e.,

$$\delta_{DOR} = \delta_{iso} - \delta_{QIS} = \delta_{iso} - \frac{3}{40} f(I) \frac{P_Q^2}{\nu_0^2} \quad (9.3)$$

where $P_Q = C_Q(1 + \eta_Q^2/3)^{1/2}$ is known as the quadrupolar product and ν_0 is the resonant frequency. $f(I)$ is a spin-dependent factor $[I(I+1) - 3/4]/I^2(2I-1)^2$ which equals 1/3 for $I = 3/2$, 2/25 for $I = 5/2$, etc. Therefore, to obtain the chemical shift δ_{iso} DOR measurements at multiple fields are required, where a plot of

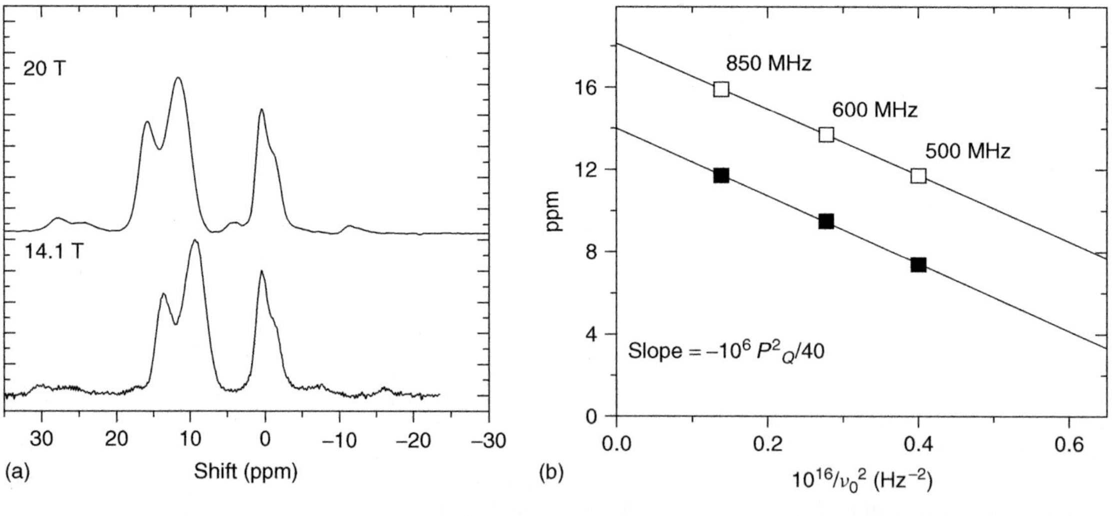

Figure 9.5. (a) ^{11}B DOR spectrum of Pyrex at two fields. The positions of the three coordinated boron, B^{III}, resonances, 10–16 ppm, are strongly dependent on field, whereas the four coordinated boron, B^{IV}, resonances near 0 ppm have a small EFG (and thus quadrupolar shift) and their position is independent of magnetic field. (b) A plot of the position against $1/\nu_0^2$ for the B^{III} peaks. The intercept gives the isotropic chemical shift δ_{iso} and the slope P_Q (see equation 9.3).

the observed shift against $1/\nu_0^2$ will give both δ_{iso} and P_Q. An example of such a plot is shown in Figure 9.5 for ^{11}B in Pyrex glass. The resonances from four coordinated boron, B^{IV}, near 0 ppm have a small EFG and their position is hardly affected by the magnetic field, whereas the position of the three coordinated boron, B^{III}, between 16 and 10 ppm is field dependent. The accuracy of such a plot can sometimes be improved in cases where it is possible to obtain a 3QMAS spectrum since the quadrupole shift here is of opposite sign and (10/17) of that in equation 9.3.[8,9]

Although the presence of spinning sidebands (SSBs) can be a nuisance, particularly for multisite samples where DOR is especially advantageous, they contain information on all the anisotropic interactions present. Thus, a further advantage of DOR over MAS for quadrupolar nuclei is that the chemical shift anisotropy (CSA) can be obtained from the sideband intensity in a somewhat analogous, but more complicated, manner to that for spin-1/2 nuclei under MAS. The difference is that the sideband intensity depends upon *all* the anisotropic interactions present and the angles between them. For a detailed discussion, see Hung *et al.*[10] They point out that if the CSA is small, then C_Q and η_Q can be determined from the sideband intensities; however, once there is a significant CSA, this must be included to account for the intensities. Figure 9.6 (top) shows the ^{17}O DOR spectrum of L-alanine at 14.1 T; two resonances are observed at $\delta_{DOR} = 227$ ppm and 217 ppm corresponding to the

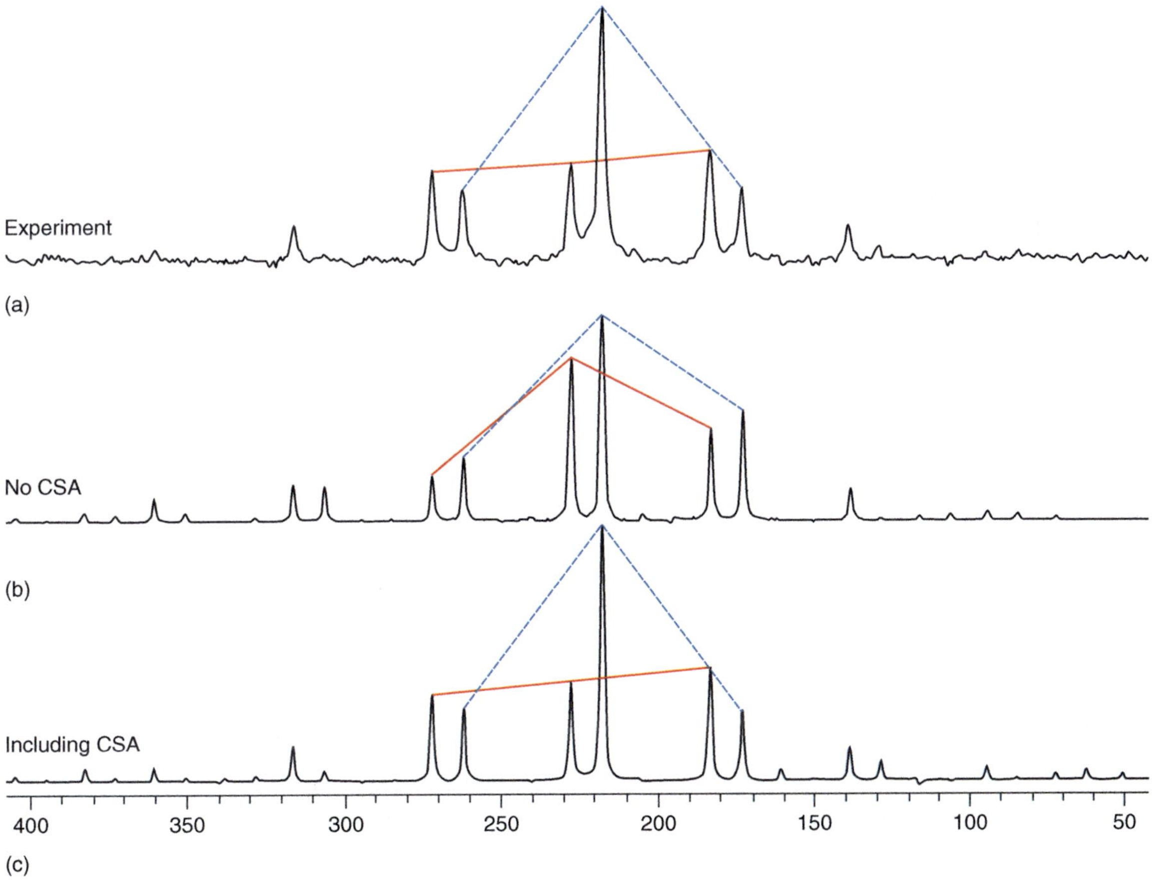

Figure 9.6. ^{17}O DOR NMR spectrum of L-alanine (in zwitterionic form) at 14.1 T (a), together with simulations without (b), and with CSA (c). The outer rotor $\nu_{OR} = 1805$ Hz and sideband suppression was used. (Reproduced from Ref. 10. © Elsevier, 2007.)

two oxygen sites (O1 and O2) of the CO_2^- moiety together with their SSBs. It can be seen that the two isotropic lines are quite different in intensity, even though their site ratio is 1:1, and furthermore, the relative height of their sidebands is also very different despite a fairly similar C_Q. A simulation excluding CSA, Figure 9.6 (middle), has very different main peak and sideband intensities from those observed. It is necessary to include both the CSA and the relative orientation of the EFG and CS tensors to obtain the correct spectrum. The sideband intensities are quite sensitive to these parameters and the simulation shown in Figure 9.6 (bottom) allowed the CS tensor (δ_{11}, δ_{22}, δ_{33}) and the relative orientation to be obtained to a similar accuracy (~20 ppm in CS span, Ω, ($\Omega = \delta_{11} - \delta_{33}$, where $\delta_{11} \geq \delta_{22} \geq \delta_{33}$), 5° in angle) to that from multifield static and MAS measurements.[10]

Since the linewidths under DOR can be very narrow, J-coupling can sometimes be measured directly from the splitting of the line. An example is the $^{17}O-^{31}P$ J-coupling in $OPPh_3$, whose slow spinning ^{17}O DOR spectrum is shown together with simulations in Figure 9.7.[10] The splitting of the isotropic resonance is clearly visible and directly gives $^1J_{OP}$ (= 161 ± 2 Hz) to high accuracy. However, there is a difference in the relative intensities of the two components of the line (for both the centerband and the SSB), which can only be accounted for by inclusion of the $^{17}O-^{31}P$ dipolar interaction in any simulation, since, as shown in the top two simulations, if only the quadrupole interaction, Q, or Q and the CSA are included in the simulation, the two components have the same relative intensities in both the centerband and the sidebands. The effect of including the dipolar coupling, D (calculated from the O–P distance), and neglecting CSA (shown in the middle simulation) changes the relative intensities but in the wrong direction. It is only when all interactions are included that it is possible to obtain agreement with the experimental spectrum shown at the bottom. In this case, the CS span can be determined to ~10 ppm and the Euler angles to 5°. Furthermore, since the relative intensities of the centerband peaks are also sensitive to the angle between the $^{17}O-^{31}P$ internuclear vector and the principal axis (V_{zz}) of the EFG tensor at the ^{17}O nucleus, its orientation in the molecular frame can be determined without the necessity of single crystal measurements (see Hung *et al.*[10] for further details).

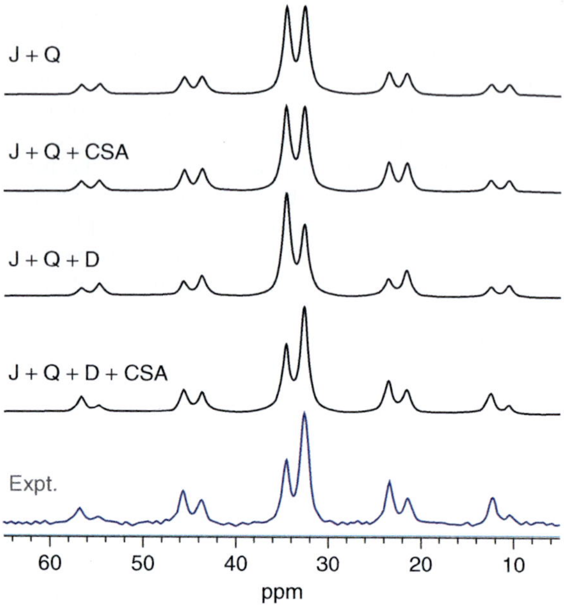

Figure 9.7. Bottom slow spinning ^{17}O DOR NMR spectrum of $OPPh_3$ at 14.1 T showing the effect of $^{17}O-P$ J-coupling. Above simulations with various interactions including quadrupole interaction ($J + Q$) top, quadrupole + CSA, quadrupole + dipolar (D), and all interactions ($J + Q + D +$ CSA). (Reproduced from Ref. 10. © Elsevier, 2007.)

Although the DOR linewidths in glasses are much broader than those in crystalline materials, it is still often possible to resolve the different sites present in the glass and to use the sideband intensities to aid their assignment. An example is ^{17}O in vitreous B_2O_3 where the boroxol ring oxygens have a much larger CSA than the non-ring oxygens.[9]

Provided the dominant cause of linewidth is the quadrupole interaction, DOR experiments have a sensitivity similar (or higher) to MAS since the poor filling factor necessitated by the large outer rotor is more than compensated for by the reduction in linewidth.

9.3 TWO DIMENSIONAL DOR EXPERIMENTS

As DOR gives isotropic spectra in a direct 1D experiment, and since the linewidths under DOR are comparable with those for spin-1/2 nuclei under MAS, it is possible to envisage similar homo- and heteronuclear correlation experiments between quadrupolar nuclei to those for spin-1/2 nuclei. In addition, new 2D experiments, such as MQ DOR, can be implemented that give in direct fashion information difficult to obtain otherwise.

9.3.1 Homonuclear Correlation Experiments

The dipolar-coupled proximities between quadrupolar nuclei have been probed by high-resolution, 2D homonuclear, NOESY-like, spin-diffusion, single-quantum (SQ) DOR experiments,[11] with this experiment being used to assign the two 5 coordinated aluminum sites in $9Al_2O_3{\cdot}2B_2O_3$ on the basis of their relative spatial proximity to the 4 and 6 coordinated sites[12] and to assign the boron sites in B_2O_3 glass.[13] The pulse sequence and coherence pathway is shown in Figure 9.8(a). The magnetization is stored along I_z for a mixing time τ_{mix} and is transferred between dipolar-coupled nuclei when they are strongly coupled (close in distance), giving rise to cross-peaks in the 2D contour plot. Figure 9.8(b) shows an example of ^{23}Na in Na_2SO_3, which has three Na sites with distances from a sodium to its neighbor varying from 3.09 Å for Na1–Na2 to 3.76 Å for

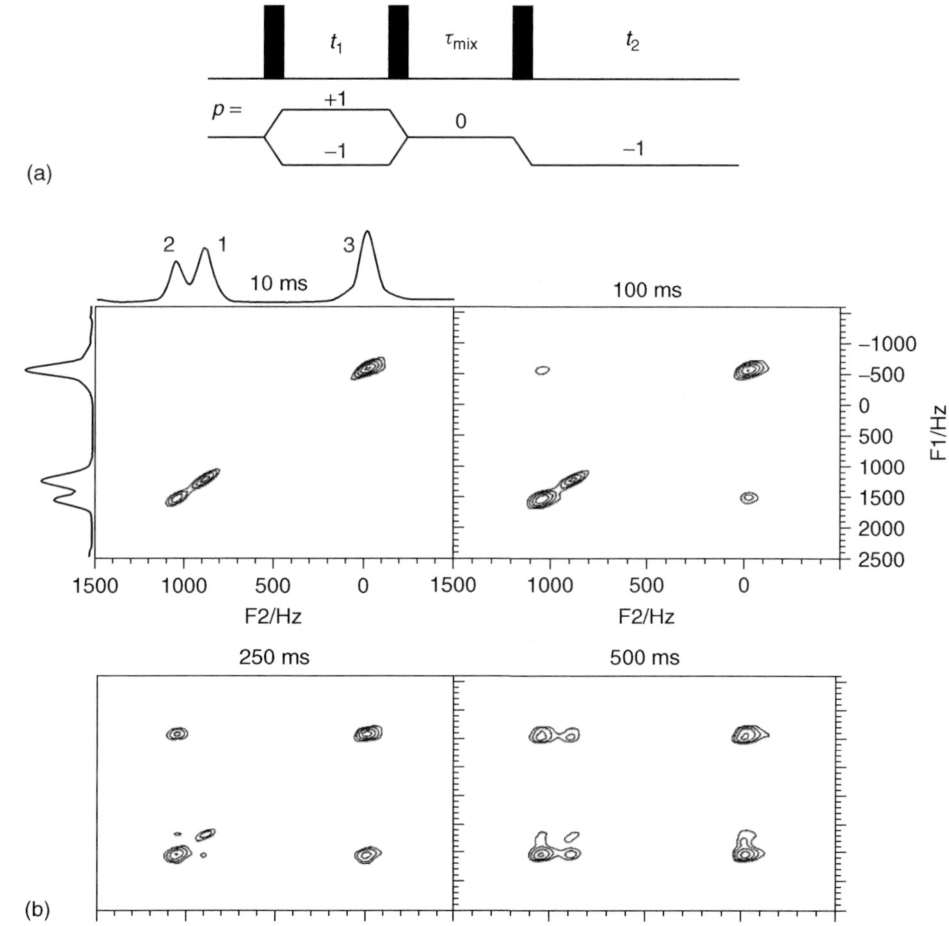

Figure 9.8. (a) Pulse sequence and coherence pathway for a 2D DOR spin-diffusion (NOESY) experiment, (b) 2D DOR SD spectrum of Na_2SO_3 at 14.1 T for different mixing times showing the buildup of the cross-peaks, which allows assignment of the ^{23}Na spectrum.

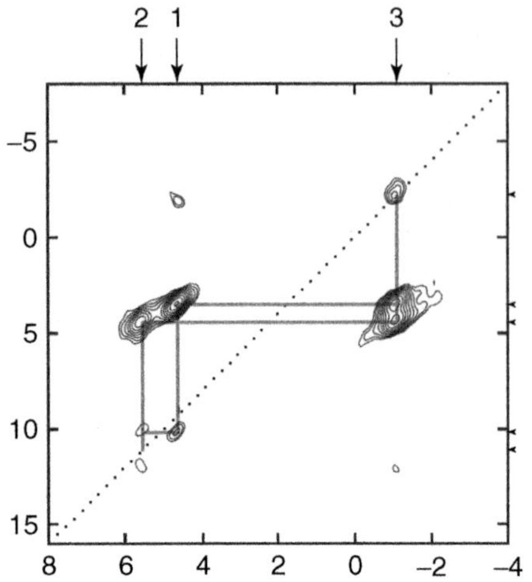

Figure 9.9. 2 DQ DOR ^{23}Na spectrum of Na_2SO_3. As well as the Na1–Na3 and Na2–Na3 cross-peaks, there is a strong DQ peak on the diagonal from Na3–Na3 and a weaker one from Na2–Na2, which are separated by 5.46 Å. (Reproduced from Ref. 14. © American Institute of Physics, 2008.)

Na1–Na3. The three sites are clearly resolved by a DOR experiment with linewidths ~0.5 ppm. For short mixing times no cross-peaks are visible, for 100 ms cross-peaks between the two outer lines appear since there are three times as many Na2–Na3 connections as Na1–Na2. For 500 ms, mixing time cross-peaks between all sites are observed.

A SQ spin-diffusion experiment has the disadvantage that "self" connections, e.g., Na3–Na3, which are separated by 3.76 Å, cannot be observed as they appear on the diagonal. However, Brinkmann *et al.*[14] showed that double-quantum (DQ) coherence can be excited between two dipolar-coupled half-integer quadrupolar nuclei using the $R2^1{}_2\,R2^{-1}{}_2$ recoupling sequence,[15] such that 2D DQ-SQ DOR correlation spectra can be obtained. The 2D DQ DOR spectrum of Na_2SO_3 is shown in Figure 9.9.[14] Clear cross-peaks are observed between Na1–Na3 and Na2–Na3. In addition, there is a strong peak on the diagonal for Na3 and also a weaker one for Na2 where the Na2–Na2 distance is 5.46 Å. While the efficiency of DQ generation was quite high, ~5%, for this sample enabling the full 2D spectrum to be obtained in ~6 h, for ^{27}Al in VPI-5 a rather low efficiency was observed[14] partly because of the fairly large shift difference between the Al^{IV} and Al^{VI} resonances.

9.3.2 Multiple-Quantum DOR

Soon after the introduction of multiple-quantum magic angle spinning (MQMAS) to obtain isotropic spectra from quadrupolar nuclei multiple-quantum double rotation (MQDOR) experiments were performed.[16] The standard experiment in MQMAS, an amplitude-modulated three-pulse z-filtered pulse sequence[17] (with sideband suppression[18]) can be used to record MQDOR spectra and directly provide information that is difficult to derive in any other way. Specifically, as shown by Hung *et al.*,[19] by applying two shear transformations[20] and suitable scaling, it is possible to obtain a 2D spectrum that has only the chemical shift in the direct dimension and only the quadrupolar shift, δ_{QIS} (from which P_Q can be determined using equation 9.3) in the indirect dimension, something unique to DOR. For disordered materials, this enables the distribution in chemical shift and also in P_Q (and thus the EFG) to be directly *independently* determined. An example is shown in Figure 9.10 for vitreous B_2O_3, where the chemical shift distribution is shown as the summed projection above the 2D plot. Figure 9.10 also shows that for the peak centred near 18 ppm (which is the boroxyl peak as shown by a DOR spin-diffusion experiment[13]), the contour lines are nearly circular, i.e., the chemical shift and EFG are only very weakly correlated. However, for the peak near 14 ppm, there is a significant slope in the contour lines indicating that as the chemical shift changes so does the EFG. For crystalline samples, MQDOR enables both the isotropic shift and P_Q to be determined, obviating the need for multifield DOR experiments. For well-ordered crystalline samples, the MQDOR linewidths can be very small; in $RbNO_3$, for instance, the ^{87}Rb CS linewidth is $\leq$0.2 ppm and the width in the quadrupole dimension, $\leq$20 Hz 0.1 ppm, is limited by the intrinsic spin echo relaxation time T_2. Thus, δ_{iso} and P_Q can be determined to high accuracy, ~0.1 ppm in δ_{iso} and 0.01 MHz in P_Q. However, the relatively large size of the outer rotor, and thus the RF coil, limits the maximum C_Q for which MQDOR is currently feasible.

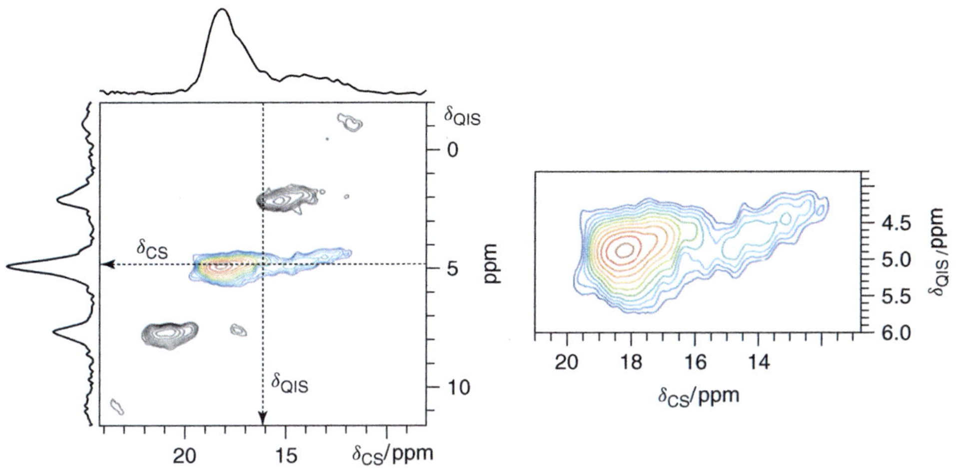

Figure 9.10. Multiple-quantum DOR spectrum of ^{11}B in vitreous B_2O_3 after shearing. The chemical shift (unaffected by the quadrupole interaction), δ_{iso}, is shown on the x-axis and the quadrupolar shift, δ_{QIS}, (from which P_Q can be determined using equation 9.3) on the y-axis.[19] The summed projection of the 2D plot is shown above, from which the chemical shift distribution in B_2O_3 is directly obtained, while the EFG distribution is obtained from the projection onto the y-axis.[13,19]

9.4 CONCLUDING REMARKS

DOR gives high-resolution isotropic spectra of quadrupolar nuclei in the direct dimension with a higher sensitivity, when quadrupolar broadening is dominant, than MAS. From sideband analysis, 1D DOR can give information about the tensorial interactions present such that, in contrast to MAS of quadrupolar nuclei, it could be considered as a routine tool for the extraction of anisotropic NMR interaction parameters. The linewidths under DOR are comparable with those for spin-1/2 nuclei and site-resolved 2D correlation experiments such as spin diffusion can be carried out. In addition, new 2D experiments such as MQDOR can provide unique information about materials. These benefits show that DOR has considerable potential. However, although the present generation of (homebuilt) DOR probes are relatively straightforward to use, the lack of a commercially produced DOR probe has severely limited the use of the technique to date.

RELATED ARTICLES IN THE ENCYCLOPEDIA OF MAGNETIC RESONANCE

Magic Angle Spinning

REFERENCES

1. E. R. Andrew, Magic angle spinning in *Encyclopedia of Magnetic Resonance*, eds R. K. Harris and R. E. Wasylishen, John Wiley & Sons Ltd.: Chichester, 2007, DOI: 10.1002/9780470034590.emrstm0283.
2. A. Llor and J. Virlet, *Chem. Phys. Lett.*, 1988, **152**, 248.
3. A. Samoson, E. Lippmaa, and A. Pines, *Mol. Phys.*, 1988, **65**, 1013.
4. Y. Wu, B. Q. Sun, A. Pines, A. Samoson, and E. Lippmaa, *J. Magn. Reson.*, 1990, **89**, 297.
5. A. Samoson and E. Lippmaa, *J. Magn. Reson.*, 1989, **84**, 410.
6. A. Samoson, *Chem. Phys. Lett.*, 1993, **214**, 456.
7. A. Samoson and J. Tegenfeldt, *J. Magn. Reson. Ser. A*, 1994, **110**, 238.
8. L. M. Bull, B. Bussemer, T. Anupold, A. Reinhold, A. Samoson, J. Sauer, A. K. Cheetham, and R. Dupree, *J. Am. Chem. Soc.*, 2000, **122**, 4948.
9. A. Wong, A. P. Howes, B. Parkinson, T. Anupõld, A. Samoson, D. Holland, and R. Dupree, *Phys. Chem. Chem. Phys.*, 2009, **11**, 7061.
10. I. Hung, A. Wong, A. P. Howes, T. Anupõld, J. Past, A. Samoson, S. P. Brown, G. Wu, X. Mo, M. E. Smith, and R. Dupree, *J. Magn. Reson.*, 2007, **188**, 246.

11. A. P. M. Kentgens, E. R. Hv. Eck, T. G. Ajithkumar, T. Anupold, J. Past, A. Reinhold, and A. Samoson, *J. Magn. Reson.*, 2006, **178**, 212.

12. I. Hung, A. P. Howes, T. Anupold, A. Samoson, D. Massiot, M. E. Smith, S. P. Brown, and R. Dupree, *Chem. Phys. Lett.*, 2006, **432**, 152.

13. I. Hung, A. P. Howes, B. Parkinson, T. Anupold, A. Samoson, S. P. Brown, D. Holland, P. F. Harrison, and R. Dupree, *J. Solid State Chem.*, 2009, **182**, 2402.

14. A. Brinkmann, A. P. M. Kentgens, T. Anupold, and A. Samoson, *J. Chem. Phys.*, 2008, **129**, 174507.

15. M. Edén, D. Zhou, and J. Yu, *Chem. Phys. Lett.*, 2006, **431**, 397.

16. A. Samoson, *J. Magn. Reson. A*, 1996, **121**, 209.

17. J. P. Amoureux, C. Fernandez, and S. Steuernagel, *J. Magn. Reson., Ser. A*, 1996, **123**, 116.

18. A. Samoson and T. Anupold, *Solid State Nucl. Magn. Reson.*, 2000, **15**, 217.

19. I. Hung, A. Wong, A. P. Howes, T. Anupõld, A. Samoson, M. E. Smith, D. Holland, S. P. Brown, and R. Dupree, *J. Magn. Reson.*, 2009, **197**, 229.

20. K. Nagayama, A. Kumar, K. Wuthrich, and R. R. Ernst, *J. Magn. Reson.*, 1980, **40**, 321.

Chapter 10

MQMAS NMR: Experimental Strategies†

Jean-Paul Amoureux[1] and Marek Pruski[2]

[1] *Université de Lille, Villeneuve d'Ascq 59650, France*
[2] *Department of Chemistry, Ames Laboratory, Iowa State University, Ames, IA 50011, USA*

10.1 INTRODUCTION

The ability of NMR to probe the structure of materials at the molecular level hinges upon the availability of high-resolution spectra, which serve as fingerprints of the physicochemical surroundings of the studied nuclei. In powdered solids, the nuclear spins experience various anisotropic interactions, which broaden their spectra and render NMR more difficult for structural determination. For spin-1/2 nuclei these interactions include dipole–dipole coupling between spins, chemical shift anisotropy (CSA), indirect spin–spin interaction (J-coupling), and the interaction with unpaired electrons.[1] The resolution and sensitivity gaps between liquid- and solid-state NMR began to decrease in the 1960s and 1970s with the introduction of radiofrequency (RF) decoupling, magic-angle spinning (MAS), and cross polarization (CP) methods.[2–5] These developments, together with the continuing advances in NMR hardware and software, allowed for increasingly precise measurement of the spin interactions in solid materials.

Until recently, however, these techniques could not overcome the line broadening in NMR spectra of quadrupolar nuclei with spins greater than 1/2. The quadrupolar broadening arises from the coupling of the nonspherical charge distribution of such nuclei with the electric field gradient (EFG) created by the surrounding electrons. Its size can be described by the quadrupolar frequency ν_Q, which is proportional to the quadrupole moment Q of a nucleus and the EFG. Depending upon the size of Q and the local electronic environment, the values of ν_Q can range from zero to several hundred megahertz. In NMR, we generally consider the strong-field case, $\nu_0 > \nu_Q$, where ν_0 is the Larmor frequency in the static field $\boldsymbol{B}_0$. In this approximation, the first- and second-order quadrupolar energies can be easily calculated using perturbation theory. For half-integer spins, the

†This is an abbreviated version of an article "MQMAS NMR: Experimental Strategies and Applications", published in *Encyclopedia of Magnetic Resonance*, eds R. K. Harris and R. Wasylishen, John Wiley: Chichester. DOI: 10.1002/9780470034590.emrstm0211. 2009.

NMR of Quadrupolar Nuclei in Solid Materials
Edited by Roderick E. Wasylishen, Sharon E. Ashbrook and Stephen Wimperis
 ISBN: 978-0-470-97398-1

first-order quadrupolar effect on the satellite transition frequencies ($m-1 \leftrightarrow m, m \neq 1/2$) is on the order of ν_Q. Thus, in a powdered sample, the NMR spectrum is spread over a frequency range that for most nuclei exceeds the range of chemical shifts (CSs) and the bandwidth of the spectrometer under pulse excitation. However, the central transition (CT) ($-1/2 \leftrightarrow 1/2$) is not affected by the first-order quadrupolar terms. In the absence of additional line broadenings from CS and/or dipolar interactions, the width of the powder spectrum for the CT is determined only by the second-order terms. Since this contribution is typically 10^2-10^3 times smaller than ν_Q, the CT yields a narrower and more intense spectrum, which is the subject of most studies on half-integer quadrupolar nuclei. We note that, in the strong-field approximation, the width of the CT spectrum is inversely proportional to the magnitude of the magnetic field $\boldsymbol{B}_0$.

The detection of broad quadrupolar lineshapes can rarely be accomplished with single-pulse excitation because of the loss of magnetization within the dead time of the receiver. This necessitates the use of multipulse excitation, with the two-pulse spin-echo sequences being the most common.[6,7] While static spin-echo experiments continue to be used in the case of strong quadrupolar interactions, most studies of quadrupolar nuclei in the last two decades have utilized the line-narrowing effect of spinning the sample under MAS conditions.[8] MAS eliminates the first-order broadening completely and reduces the quadrupolar contribution to the CT linewidth by a factor of approximately 3. Consequently, the strategies applied in MAS experiments depend on the size of the quadrupolar coupling. For ν_Q values not exceeding few hundred kilohertz, it is often possible to observe the whole spectrum, which appears as a wide manifold of spinning sidebands spaced by the spinning speed ν_R. For large values of ν_Q, it becomes advantageous to use the MAS lineshape corresponding to the CT to determine NMR parameters. Several techniques have been developed that allow for studies of dipolar interactions between the quadrupolar and spin-1/2 nuclei under MAS. These include the CP,[9,10] rotational-echo double-resonance (REDOR),[11,12] transferred echo double-resonance (TEDOR),[12,13] transfer of populations in double resonance (TRAPDOR),[14,15] rotational-echo adiabatic-passage double-resonance (REAPDOR),[16] and symmetry-based resonance-echo saturation-pulse double-resonance (S-RESPDOR).[17]

The second-order quadrupolar interaction is "more anisotropic" than the first-order interactions (including CSA and dipolar coupling) in the sense that it contains higher-order orientational terms of significant magnitude. Thus, as mentioned above, it is not averaged to zero by MAS alone, and more complex motions of the sample, or of the spin magnetization, are needed in order to achieve the highest possible resolution.[18–22] The theoretical foundations and two different experimental realizations[18,19,22] of complete spatial averaging of the second-order broadening of the CT were published in the late 1980s by the groups of Pines and Virlet. They relied on mechanical rotation of the sample around a time-dependent orientation, in 1D double rotation (DOR, see Chapter 9)[18] or 2D dynamic angle spinning (DAS, see Chapter 8) experiments.[19]

In 1995, it has been proposed and demonstrated by Frydman *et al.* that the line narrowing of the central transition can also be obtained with MAS by changing the coherence state of the observed spins. This 2D experiment enables the observation of a purely isotropic echo by correlating the phase evolutions of the multiple-quantum (MQ) and single-quantum (1Q) coherences, and is referred to as *multiple-quantum magic angle spinning* (*MQMAS*) NMR.[20,21] The change of the coherence state of half-integer quadrupolar spins can be effectively accomplished by strong RF irradiation.[23,24] The resolution enhancement that this technique can provide is demonstrated in Figure 10.1, which compares static, MAS, and MQMAS spectra of ^{87}Rb in well-crystallized $LiRbSO_4$.[25]

This chapter describes the fundamental principles of the MQMAS experiment, including schemes for detection of pure-phase spectra, data processing methods (including referencing, quantitative analysis, and interpretation of MQMAS spectra), strategies for maximizing the sensitivity, and MQMAS-based double-resonance experiments. Cross-references to articles describing the theoretical background, other high-resolution methods for quadrupolar nuclei, and applications of MQMAS are given in the section "Related Articles in the Encyclopedia of Magnetic Resonance."

10.2 THEORETICAL BACKGROUND

We consider an ensemble of nuclei with half-integer spin quantum number $S > 1/2$ in a strong magnetic field. There are $2S+1$ Zeeman energy levels

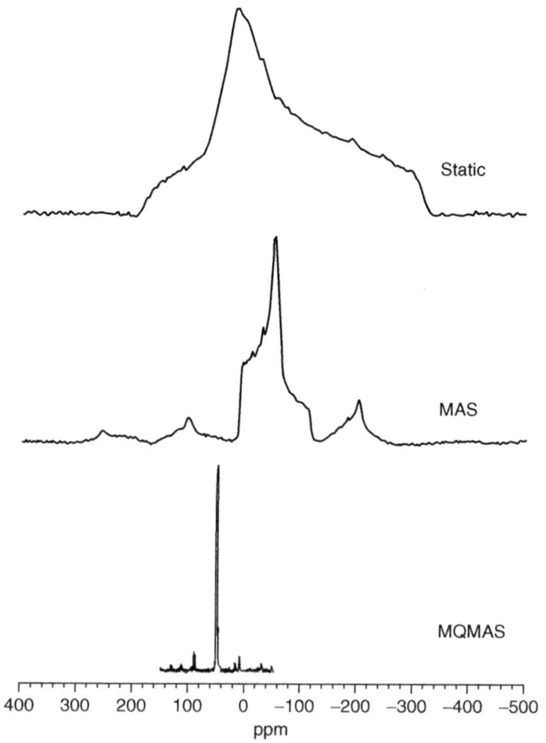

Figure 10.1. Static, MAS, and MQMAS (isotropic projection) spectra of ^{87}Rb in polycrystalline $LiRbSO_4$ taken at 130.88 MHz.

associated with such spins that can be labeled with the magnetic quantum number m. The existence of a nonvanishing (m, m') element in the density matrix implies that there is a p-quantum coherent superposition of the states $|m\rangle$ and $|m'\rangle$ with multiplicity defined by $p = m - m'$. Thus, the diagonal elements of the density matrix (m, m), which represent the relative populations of the Zeeman states $|m\rangle$, correspond to zero-quantum coherences. It can be shown that these are the only coherences present in the density matrix of a system at thermal equilibrium. Under the conditions normally encountered in NMR, a reduced density matrix can be used that describes the relative deviations from a completely uniform distribution. The reduced density matrix at thermal equilibrium is proportional to S_z (the z component of the spin angular momentum operator). In the frame of the Zeeman interaction, i.e., the "rotating frame", the secular second-order Hamiltonian governing the S spin system can be written:

$$H = H_{Q1} + H_{Q2} + [\nu_0 \Delta\delta + m_I J + F(\alpha_s, \beta_s)]S_z \tag{10.1}$$

where

$$H_{Q1} = \bar{\nu}_Q(\alpha_s, \beta_s, \eta_Q)\left[S_z^2 - \frac{S^2}{3}\right] \tag{10.2}$$

$$\begin{aligned} H_{Q2} &= -\frac{\nu_Q^2}{2\nu_0}[V_1V_{-1}(4S^2 - 8S_z^2 - 1) \\ &\quad + V_2V_{-2}(2S^2 - 2S_z^2 - 1)]S_z \\ &= B(\alpha_s, \beta_s, \eta_Q)S_z + D(\alpha_s, \beta_s, \eta_Q)S_z^3 \end{aligned} \tag{10.3}$$

with

$$\begin{aligned} &\bar{\nu}_Q(\alpha_s, \beta_s, \eta_Q) \\ &\quad = 1.5\nu_Q[3\cos^2\beta_s - 1 + \eta_Q \sin^2\beta_s \cos 2\alpha_s] \end{aligned} \tag{10.4}$$

and

$$\nu_Q = \frac{e^2qQ}{4S(2S-1)h} = \frac{C_Q}{4S(2S-1)} \tag{10.5}$$

H_{Q1} and H_{Q2} represent the first and second-order perturbation terms for the quadrupole interaction, $\Delta\delta$ is the resonance offset (in ppm), J is the scalar coupling constant between spin S and another spin I with magnetic number m_I, eq is the largest principal axis value of the EFG tensor, and C_Q is the quadrupole coupling constant. The last term in equation (10.1) accounts for CSA, the anisotropy of J-coupling and heteronuclear dipolar couplings, $\bar{\nu}_Q$ is the quadrupole splitting, ν_Q is the quadrupole splitting parameter, and V_1, V_{-1}, V_2, and V_{-2} are the eigenvalues of the quadrupolar tensor in the laboratory frame. Note that different definitions of the quadrupolar parameters exist in the literature, e.g., ν_Q is often defined as $3C_Q/4S(2S-1)$. The orientation dependencies of F, $\bar{\nu}_Q$, V_1, V_{-1}, V_2, and V_{-2} can be described in terms of two polar angles α_s and β_s, which determine the direction of $\boldsymbol{B}_0$ in the principal axis system of the EFG tensor. Implicit in F is the dependence of the nonquadrupolar anisotropic interactions on their orientation with respect to the quadrupolar tensor, which can be described using additional sets of Euler angles.

The MQ transitions are induced by using RF pulses. Following these excitations, coherences freely precess during the evolution periods. Under fast rotation at an angle χ with respect to $\boldsymbol{B}_0$, the evolution phase $\Phi^{\chi}(p/2, -p/2)$ for the antisymmetric p-quantum coherence $(p/2, -p/2)$ in a single crystallite is a function of the second- and fourth-order

Legendre polynomials:

$$P_2(\cos\chi) = \frac{1}{2}(3\cos^2\chi - 1)$$
$$P_4(\cos\chi) = \frac{1}{8}(35\cos^4\chi - 30\cos^2\chi + 3) \qquad (10.6)$$

Under MAS, i.e., when $\chi = \chi_m = 54.736°$, the evolution phase $\Phi^{MAS}(p/2, -p/2)$ can be written as

$$\Phi^{MAS}\left(\frac{p}{2}, -\frac{p}{2}\right) = \{p[\nu_0\Delta\delta + m_I J] + A_0^Q(\eta_Q)C_0(S, p) + A_4^Q(\alpha_R, \beta_R, \eta_Q)C_4(S, p)P_4(\cos\chi_m)\}t \qquad (10.7)$$

where the polar angles α_R and β_R describe the orientation of the rotor axis with respect to the EFG tensor. Explicit expressions for A_k^Q and C_k can be found in the literature.[26] The remaining anisotropic broadening originates from the last term in equation (10.7), which is scaled with respect to a static case, but not eliminated by MAS alone. The unconventional aspect of the MQMAS approach is that the selective averaging of second-order quadrupolar broadening is completed by forcing the spin system to evolve outside of the "allowed" CT coherence, i.e., by transferring it into MQ coherences $(\mp p/2, \pm p/2)$ with $p > 1$. After letting the spins evolve during time t_1, the spin system is transferred back to the $(-1/2, 1/2)$ CT coherence and allowed to freely evolve again during time t_2. The evolution phases in t_1 and t_2 are then correlated in a 2D experiment. The isotropic echo is observed at

$$t_{2e} = -[C_4(S, p)/C_4(S, -1)]t_1 = R(S, p)t_1 \qquad (10.8)$$

with

$$R(S, p) = \frac{p[36S(S+1) - 17p^2 - 10]}{36S(S+1) - 27} \qquad (10.9)$$

As the sign of $C_4(S, p)$ is reversed when changing that of p, it is always possible to obtain a positive ratio for $R(S, p)$, which leads to a detectable echo at positive time t_{2e}. This echo signal corresponds to the selection of $p < 0$ for $|p| = 2S$ and $p > 0$ for $|p| < 2S$. Therefore, the echo pathway for spin $S = 3/2$ nuclei can be described using the notation $0 \rightarrow p = -3 \rightarrow -1$. Correspondingly, the correlation of $p = 3$ and $p = -1$ coherences ($0 \rightarrow 3 \rightarrow -1$ pathway) represents the so-called "antiecho" pathway, where the anisotropic quadrupolar broadening refocuses at negative values of the acquisition time t_2. The concept of antiecho will be later used in describing experiments that allow for the acquisition of pure absorption spectra (Section 10.3).

A 2D Fourier transformation of the time domain signal provides a correlation spectrum consisting of narrow resonance bands with ridges extended along the direction given by

$$F_1 = R(S, p)F_2 \qquad (10.10)$$

where F_2 and F_1 represent frequencies in the single- and MQ dimensions, respectively. It is convenient to apply a shearing transformation in order to align these ridges with the F_2 direction. The shearing process transforms F_1 into a new isotropic dimension, heretofore denoted F_{ISO} (see Section 10.4.1).

The basic MQMAS experiment consists of four steps[20]:

1. RF excitation of MQ $(\pm p/2, \mp p/2)$ coherences,
2. evolution of these coherences under the internal Hamiltonians H during time t_1,
3. reconversion of the MQ coherences into the observable 1Q CT coherence, and
4. detection of the signal as a function of time t_2.

The pulse sequence used in the original MQMAS experiment contains two strong continuous wave (CW) RF pulses of duration τ_i and phase ϕ_i, used for preparation (MQ excitation) and mixing (MQ $\rightarrow$ -1Q conversion): $H_{RF}(\tau_1, \phi_1) - t_1 - H_{RF}(\tau_2, \phi_2) - t_2$, as shown in Figure 10.2. The role of the excitation pulse (or sequence) is to create the largest possible amount of p-quantum coherences along the antidiagonal of the density matrix ($(\pm p/2, \mp p/2)$, $3 \le |p| \le 2S$), as shown schematically at the top of the figure for $S = 3/2$. The dephasing that this

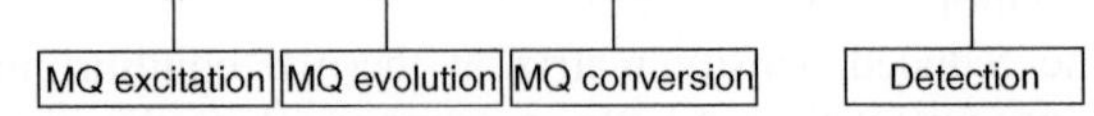

Figure 10.2. Schematic diagram of the original MQMAS experiment.[20] Time evolution of the reduced density matrix of spin-3/2 nuclei is also shown, with darker squares representing the increased amount of (m, m') coherence.

coherence acquires during the evolution time t_1 is then selectively transferred to the observable -1Q CT by the second pulse (or sequence) where it refocuses according to equation (10.8).

Considerable research effort has targeted two main weaknesses of MQMAS, namely, the limited efficiency of excitation and reconversion, and sensitivity to the magnitude of C_Q. The efficiency could be significantly increased by using the highest possible RF magnetic field ν_{RF} (or so-called hard pulses),[27] fast amplitude modulation (FAM),[28] double frequency sweeps (DFS),[29] rotary resonance excitation,[30] rotation-induced adiabatic coherence transfer (RIACT),[31] and soft pulse added mixing (SPAM).[32] FAM, DFS, and RIACT are the least sensitive to C_Q and thus extend the range of quadrupolar couplings that are accessible to MQMAS to larger values. The most commonly used coherence pathways and acquisition schemes include the split-t_1 scheme with shifted (whole) echo,[33–35] the z-filter method,[36] and the SPAM method.[32]

10.3 DETECTION OF PURE-PHASE SPECTRA

The presence of dispersive components in MQMAS spectra leads to lower resolution and obstructs structural analyses. Several strategies for the collection and processing of the data to provide purely absorptive spectra have been proposed. These strategies can be classified into two categories, based on the appearance of the maximum echo signal in the first row (i.e., $t_1 = 0$) of the 2D data set. The first involves a phase-modulated procedure in which the maximum signal (echo) is delayed to $t_{2e} > 0$ by applying one or several $\pm1\text{Q} \rightarrow \mp1\text{Q}$ Hahn echoes.[6] The second category includes amplitude-modulated experiments in which $t_{2e} = 0$ for $t_1 = 0$. We will review these methods briefly in Sections 10.3.1, 10.3.2, and 10.3.5. Sections 10.3.3 and 10.3.4 will describe the methods that allow determination of the sign of frequency in the F_1 dimension for experiments involving amplitude modulation.

10.3.1 Phase Modulation during t_1; Shifted Echo and Antiecho

It is known that the complex Fourier transform of the two halves of a suitably phased symmetrical NMR echo yields a spectrum that is purely absorptive. Thus, the dispersive components can be eliminated from the MQMAS spectrum by introducing a suitably timed refocusing π pulse at the end of the pulse sequence to delay the formation of an echo and/or antiecho in the t_2 domain, such that the entire signal is detected.[35,37] The time domain signals, the pulse sequence, and the corresponding coherence transfer pathways for such experiments, referred to as *whole echo MQMAS*, are schematically shown in Figure 10.3(a). In the time domain, the signal first appears at $t_2 < \tau + t_{2e}$, where t_{2e} is given by equation (10.8) for echo and antiecho. The signal then increases up to a maximum at $t_2 = \tau + t_{2e}$, and decays to zero at $t_2 > \tau + t_{2e}$.

The echo and antiecho signals (Figure 10.3a and b) shift forward and backward, respectively, as the time t_1 increases. Clearly, the delay τ has to be longer in the case of the shifted antiecho, which can result in noticeable loss of signal due to transverse relaxation. A complex 2D Fourier transformation, followed by a first-order phase correction in the F_2 dimension, yields a pure phase spectrum, which is obtained in spite of the fact that the time domain signal is phase modulated. Owing to whole echo acquisition, the signal-to-noise (S/N) ratio is multiplied by $\sqrt{2}\exp(-2\lambda\tau)$, compared to the standard (nonshifted) method, and thus can be increased when the transverse losses, with λ relaxation rate, are minor. An added advantage here is that the phase modulation in itself yields sign discrimination in the F_1 domain. While the above method utilizes sequences for the acquisition of signals via the echo or antiecho pathways in separate experiments, the shifted echo and shifted antiecho signals can be detected simultaneously (Figure 10.3c).

10.3.2 Amplitude Modulation during t_1; z-Filter Method

The phase separation in MQMAS can easily be accomplished by using a combination of the signals from the echo and antiecho coherence pathways. When these two components have equal intensities the resulting time domain signal is amplitude modulated and its hypercomplex 2D Fourier transform yields the spectrum $S(F_1, F_2)$ with a purely absorptive real part. Although the acquisition of echo and antiecho signals can be achieved in two consecutive experiments, e.g., by proper selection of

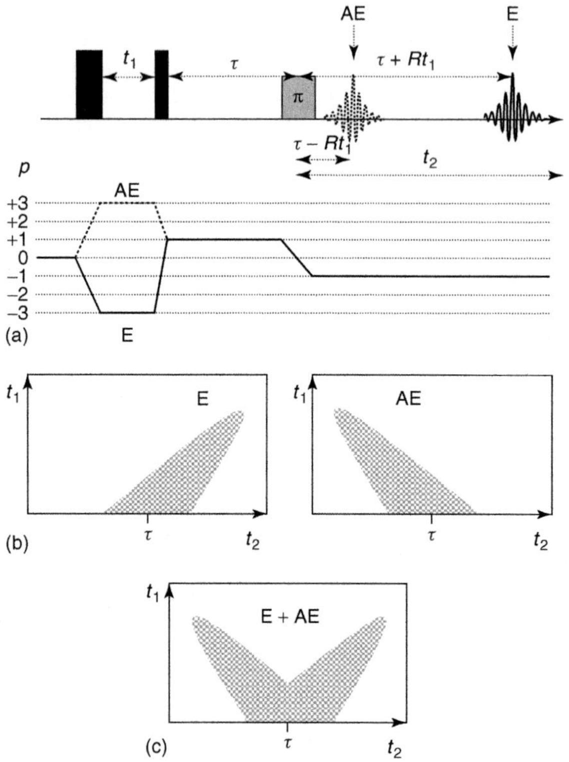

Figure 10.3. (a) Pulse sequence and coherence transfer pathways for the whole echo (antiecho) experiments, with $S = 3/2$ nuclei. Nonselective (hard) and CT-selective (soft) pulses are represented by black and gray rectangles, respectively. (b) Time domain signals for shifted echo (E) and shifted antiecho (AE) phase-modulated experiments. The signal is observed after the third pulse at $t_{2e} = \tau + Rt_1$ for the echo and at $t_{2e} = \tau - Rt_1$ for the antiecho. (c) Similar representation of the whole echo experiment in which shifted echo and shifted antiecho signals are acquired simultaneously.

$0 \to +p \to -1$ and $0 \to -p \to -1$ pathways, it is more suitable for these signals to be acquired simultaneously, which can easily be done with the z-filter.[33,36] The implementation of the z-filter into the 3QMAS experiment utilizing two hard pulses is shown in Figure 10.4 together with the corresponding coherence transfer pathways.[36] The z-filter consists of a short delay τ during which all magnetization is stored along $\boldsymbol{B}_0$, and then transferred to the observable 1Q coherences using a CT-selective "soft" pulse with selective $\pi/2$ flip angle. As a result of the coherence pathway being symmetric $(0 \to \pm p \to 0 \to -1)$, the contributions from the echo and antiecho pathways are equal for all crystallites, leading to cosine amplitude modulation of the time domain signal and a purely absorptive 2D spectrum.

The z-filter is robust and easy to optimize, and thus is widely utilized in MQMAS experiments. The first pulse is adjusted to generate the maximum amount of the desired MQ coherences, and the second pulse maximizes their transfer to the population states associated with the central transition. The delay τ usually lasts a few rotor periods. The choice of a suitable ν_{RF} level for the soft z-filter pulse results from a compromise between the selectivity and completeness of irradiation of the CT; setting ν_{RF} at 10–20 kHz is usually a good choice for the selective pulses.

The disadvantage of amplitude-modulated methods is that the resulting spectra are symmetric with respect to $F_1 = 0$ and that they present two resonances in the p-quantum dimension, at $F_1 = \pm\nu_p$. In principle, this problem can be avoided by locating the carrier frequency on the same side of all spectral lines. However, this requires prior knowledge of the spectrum and increases the offset effects. The so-called time-proportional phase incrementation (TPPI) approach or hypercomplex data acquisition is often used to restore sign discrimination in F_1.

10.3.3 Time-proportional Phase Incrementation

The TPPI method uses linear incrementation of the phase ϕ^{prep} of the RF preparation pulse (or sequence) in t_1 according to equation (10.11)

$$\phi^{\text{prep}} = \frac{\pi t_1}{2p\Delta t_1} \tag{10.11}$$

where Δt_1 is the time increment in t_1.[38] After Fourier transformation, the spectrum is purely absorptive and shifted along F_1 by $+1/4\Delta t_1$ for ν_p, and by $-1/4\Delta t_1$ for its folded resonance at $-\nu_p$. Thus, it becomes possible to place the carrier frequency in the center of the spectral region and discriminate between positive and negative offsets as if the signal were acquired in quadrature mode. It is advisable to decrease Δt_1 by a factor of 2 with respect to experiments performed without TPPI, in order to avoid the frequency fold-over between the two symmetrical lines.

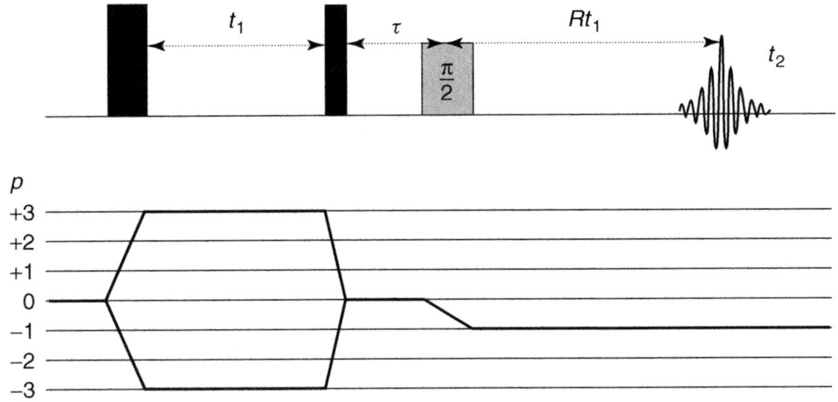

Figure 10.4. Pulse sequence and coherence transfer pathway for *z*-filter 3QMAS.

10.3.4 Hypercomplex Data Acquisition (States Method)

The hypercomplex approach, also known as the *States method*,[39] relies upon acquisition of signals from two complementary experiments, which provide the distinction of real and imaginary parts in t_1 in the same way as the classical quadrature detection does in t_2. Both echo and antiecho MQMAS signals are acquired simultaneously in this method. For each increment in t_1, two time domain signals s_X and s_Y are created and measured with the phases ϕ^{prep} of the preparation sequence equal to 0 and $\pi/2p$, respectively[35]

$$s_X(t_1, t_2) = 2s_p \cos(2\pi\nu_p t_1)\exp(2\pi i\nu_{-1}t_2) \times \exp[-(\lambda_1 t_2 + \lambda_p t_1)] \quad (10.12)$$

$$\begin{aligned} s_Y(t_1, t_2) &= 2s_p \cos(2\pi\nu_p t_1 + p\phi^{\text{prep}})\exp(2\pi i\nu_{-1}t_2) \times \exp[-(\lambda_1 t_2 + \lambda_p t_1)] \\ &= -2s_p \sin(2\pi\nu_p t_1)\exp(2\pi i\nu_{-1}t_2) \times \exp[-(\lambda_1 t_2 + \lambda_p t_1)] \end{aligned} \quad (10.13)$$

We assume here that the echo and antiecho contribute equally to the observed signal. The complex Fourier transformation in t_2 yields

$$s_X(t_1, F_2) = 2s_p \cos(2\pi\nu_p t_1)[a(F_2, \nu_{-1}) + id(F_2, \nu_{-1})]\exp(-\lambda_p t_1) \quad (10.14)$$

$$s_Y(t_1, F_2) = -2s_p \sin(2\pi\nu_p t_1)[a(F_2, \nu_{-1}) + id(F_2, \nu_{-1})]\exp(-\lambda_p t_1) \quad (10.15)$$

where $a(F_i, \nu_j)$ and $d(F_i, \nu_j)$, with $i = 1, 2$ and $j = p, -1$, are the absorptive (real) and dispersive (imaginary) parts of the spectrum along the multiple- and 1Q dimensions, respectively. The real parts of s_X and s_Y are then combined to produce a new complex signal

$$\begin{aligned} s(t_1, F_2) &= s_X^R(t_1, F_2) - is_Y^R(t_1, F_2) \\ &= 2s_p a(F_2, \nu_{-1})\exp(2\pi i\nu_p t_1)\exp(-\lambda_p t_1) \end{aligned} \quad (10.16)$$

A complex Fourier transformation of $s(t_1, F_2)$ with respect to t_1 then produces, for the real part, a pure-phase spectrum

$$S(F_1, F_2) = 2s_p a(F_2, \nu_{-1})[a(F_1, \nu_p) + id(F_1, \nu_p)] \quad (10.17)$$

with a single resonance located at $(F_1, F_2) = (\nu_p, \nu_{-1})$.

The TPPI and hypercomplex methods result in the same S/N ratio per unit of experimental time. However, the hypercomplex method allows for separation of the echo and antiecho signals according to

$$\begin{aligned} s_E(t_1, t_2) &= s_X(t_1, t_2) - is_Y(t_1, t_2) \\ s_A(t_1, t_2) &= s_X(t_1, t_2) + is_Y(t_1, t_2) \end{aligned} \quad (10.18)$$

which can thus be independently processed.

10.3.5 Phase Modulation during t_1; SPAM Echo/Antiecho Acquisition

The *z*-filter and whole echo methods used two hard pulses followed by a soft pulse, with a single coherence level (0Q and +1Q, respectively, see Figures 10.3 and 10.4) being selected between the last two pulses. It has been later proposed to use the 0Q and ±1Q coherence levels simultaneously in the SPAM experiment.[40] To avoid any unwanted dephasing of coherences with $p = \pm 1$, the hard and soft pulses must directly follow each other, forming a composite SPAM conversion sequence.[41] The phase of the hard MQ excitation pulse is cycled to select only the targeted coherence level required for the echo or antiecho pathway (+3Q or −3Q in the case of 3QMAS).[32] Phases of the hard and soft components of the SPAM pulse must be fixed at the same value (e.g., x, x) for $+3Q \rightarrow -1Q$ transfers and at opposite values (e.g., $x, -x$) for $-3Q \rightarrow -1Q$ transfers.

The optimized pulse durations in the echo (or antiecho) SPAM experiments are exactly the same as in the z-filter method, while the resulting intensity is roughly doubled. From the user's standpoint, the difference is only in the phase cycling, as well as the data acquisition and processing. Absorptive SPAM MQMAS spectra can be obtained by using a shifted echo or split-t_1 approach (see below). Alternatively, the dispersive contributions can be eliminated by detecting both $SPAM_E$ and $SPAM_{AE}$ signals. Indeed, the coherence pathways during t_1 for the echo and antiecho are symmetrical, which implies that the resulting phase-modulated $SPAM_E$ and $SPAM_{AE}$ intensities are equal. However, the two signals must be recorded and processed separately, then added to yield a pure-phase $SPAM_{E/AE}$ spectrum with frequency discrimination. Since the antiecho is used solely to eliminate the dispersive contributions from the echo pathway, only its initial part that "overlaps" with the echo (see Figure 10.3c) has to be acquired.[32]

The time performance of the $SPAM_{E/AE}$ experiment is determined by several factors[32]: (i) the roughly twofold increase of echo and antiecho signals due to the SPAM composite pulse, (ii) the loss associated with separate acquisition of the echo and antiecho (or part of the latter), and (iii) the fact that it is a phase-modulated experiment that does not require TPPI or hypercomplex acquisition. It has been demonstrated that the truncated $SPAM_{E/AE}$ method can lead to an overall S/N increase of approximately 3 compared to that of the z-filter method.

10.4 PROCESSING AND INTERPRETATION OF MQMAS SPECTRA

10.4.1 Shearing Transformation; Referencing of MQMAS Spectra

A typical MQMAS spectrum, resulting from a standard 2D Fourier transformation, consists of several extended bands similar to those depicted in Figure 10.5(a). One of the primary benefits of MQMAS is that it allows for separation and measurement of the isotropic chemical and quadrupolar-induced shifts by analyzing a single 2D spectrum.[21] However, proper assignment of the resonance shifts along the F_1 and F_2 dimensions is not immediately obvious. Analysis and interpretation of such an unsheared spectrum is facilitated by introducing the CS, the quadrupole-induced shift (QIS), and the anisotropic (A) axes. When the quadrupole interactions are negligible, the resonances are sharp and located on the CS axis. In the presence of significant quadrupolar interaction, the center of gravity of a given resonance is displaced from the CS axis, along the QIS direction, by the QIS. The A axis is defined by the orientation of anisotropic ridges in MQMAS spectra and has a slope $R(S, p)$ given by equation (10.9).

MQMAS spectra can be more conveniently analyzed by applying a shearing transformation, which places the A axis parallel to the F_2 axis (Figure 10.5b) and creates a new dimension labeled F_{ISO} (or δ_{ISO}). The direct advantage of such an approach is that the orthogonal projection of the MQMAS onto the F_{ISO} axis does not include the contributions of anisotropic quadrupole and CSs. A similar projection onto the F_2 axis yields the distorted pQ-filtered anisotropic MAS powder pattern.

In the sheared spectrum, the frequency F_{ISO} can be suitably expressed as a linear combination of F_1 and F_2,

$$F_{ISO} = F_1 + RF_2 \qquad (10.19)$$

where R is given by equation (10.9). This convention allows the direct determination of the sample rotation rate from the sheared spectrum when it is plotted in hertz.[42] In order to introduce the referencing in ppm, let us first consider a single resonance that is not affected by the quadrupole interaction, located at coordinates $(-p\nu_0\Delta\delta, \nu_0\Delta\delta)$ on the CS axis of an

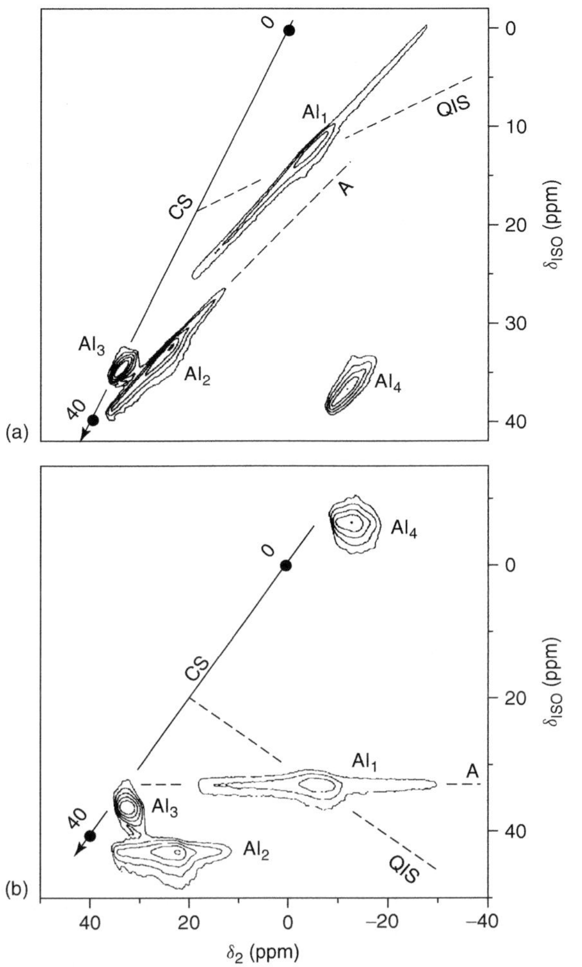

Figure 10.5. (a) Unsheared and (b) sheared 3QMAS spectra of ^{27}Al in $AlPO_4$-14 acquired at $\nu_0 = 104.2$ MHz using the z-filter sequence with rotor synchronization, $\nu_R = 14.5$ kHz and $\nu_{RF} = 200$ kHz. The CS, A, and QIS axes are plotted as described in the text. One fivefold coordinated site (1), two tetrahedral sites (2 and 3), and one sixfold coordinated site (4) are resolved, in agreement with crystallographic data. Note that resonance 4 is folded over in (a) due to the limited spectral width in rotor-synchronized spectra.

unsheared spectrum. After the shearing transformation, its coordinate F_{ISO} in the isotropic dimension becomes

$$F_{ISO} = (R - p)\nu_0 \Delta\delta \qquad (10.20)$$

The ppm scales δ_1, δ_2, and δ_{ISO} use apparent Larmor frequencies, which are equal to ν_0, $-p\nu_0$, and $(R - p)\nu_0$ along F_2, F_1, and F_{ISO}, respectively. In the absence of the quadrupole interaction, this convention results in the shifts being identical and equal to $\Delta\delta$ along the three axes. After the shearing transformation, the QIS in the F_{ISO} dimension becomes always positive and independent of p:

$$\delta_{QIS,ISO} = -\frac{10}{17}\delta_{QIS}(S, -1) = -\frac{10}{17}kC_Q^2(1 + \eta_Q^2/3) \qquad (10.21)$$

with

$$k = -\frac{3}{10}\frac{[4S(S+1) - 3]}{[4S(2S-1)\nu_0]^2}10^6 \qquad (10.22)$$

Thus, the QIS direction in the 2D sheared spectrum expressed in ppm has a constant slope of −10/17, regardless of the values of S and p. Since the CS axis of such a spectrum has a constant slope of 1, the MQMAS sheared spectra displayed using these ppm scales are independent of p for a given spin system. To illustrate this point, the isotropic projections of 3QMAS and 5QMAS spectra of ^{27}Al in aluminophosphate $AlPO_4$-14 are compared in Figure 10.7(a) and (b).[43] It should be noted that the above referencing scheme is directly applicable to the spectra obtained using satellite transitions magic-angle spinning (STMAS) NMR (see Chapter 11).[44] Several other conventions for labeling the F_1 dimension have been proposed in the literature, and subsequently reviewed.[45]

10.4.2 Split-t_1 Method

The split-t_1 method of data acquisition relies upon the splitting of time t_1 into single-quantum and MQ evolution periods in proportion to the ratio $R(S, p)$.[33] Thus, the fourth-order anisotropic term given in equation (10.7) can be always refocused at the end of the t_1 period. After 2D Fourier transformation, the inhomogeneously broadened ridges appear parallel to the F_2 axis, which eliminates the need for the shearing transformation. As is shown below, the split-t_1 method has been combined with the whole echo method to provide a very efficient scheme for MQMAS. This split-t_1 mode of data acquisition can be conveniently used in MQMAS-based heteronuclear correlation spectroscopy experiments and in Carr–Purcell–Meiboom–Gill (CPMG) MQMAS experiments (see Sections 10.5.2 and 10.6).

In the split-t_1 amplitude-modulated methods, for each crystallite, the echo and antiecho coherence pathways must contribute to the processed signal with equal amplitude in order to obtain a pure absorption lineshape. For spin-3/2 nuclei, this can be achieved by using the coherence transfer pathway with the z-filter, $0 \rightarrow \pm 3 \rightarrow \pm 1 \rightarrow 0 \rightarrow -1$, as shown in Figure 10.6(a).[37,46] The efficiency of this four-pulse method can be improved by using FAM,[28] DFS,[29] RIACT,[31,47,48] or SPAM[32,49] methods for the ±3Q to ±1Q conversion. The symmetric split-t_1 pathways $0 \rightarrow \pm 3 \rightarrow \mp 1 \rightarrow 0 \rightarrow -1$ were also proposed for 3QMAS experiments with spins $S > 3/2$ (see Figure 10.6b). However, such experiments are relatively inefficient, as they involve the coherence change of $|\Delta p| = 4$ during $\pm 3 \rightarrow \mp 1$ conversion.

The split-t_1 method can also be used in phase-modulated (shifted echo) experiments, which generate pure phase spectra through the acquisition of the whole echo. The coherence pathway ($0 \rightarrow 2S \rightarrow 1 \rightarrow -1$) displayed in Figure 10.6c for $p = 2S$ requires a short spin-echo interval.[37] For the MQMAS experiments with $|p| < 2S$, the timing must be modified, as shown in Figure 10.6(d). When the loss of magnetization due to T_2' effects is not significant, the MQMAS schemes described in Figure 10.6(a) and (b) have been recommended as the most efficient,[25,37] especially when used with a SPAM, FAM, or DFS conversion. The referencing of split-t_1 spectra can be performed using the same method as described in Section 10.4.1.

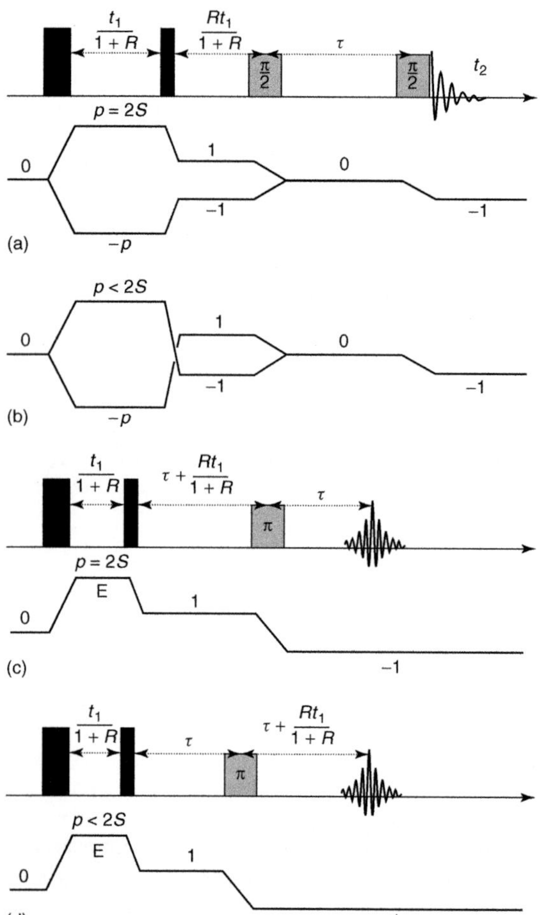

Figure 10.6. (a) Amplitude-modulated split-t_1 pulse sequence and coherence pathway for $p = 2S$. For $p < 2S$, the experiment can be performed using the same pulse sequence and the coherence pathway as shown in (b). (c, d) Phase-modulated split-t_1 pulse sequences and coherence pathways for $p = 2S$ and $p < 2S$.

10.4.3 Resolution of MQMAS Spectra

The isotropic resolution should be consistently achieved by the MQMAS method, at least in the absence of strong dipolar broadening, because the timing of the echo formation, as described by equation (10.8), is independent of the size of quadrupolar and CSA interactions, the orientation of crystallites, and the experimental conditions (ν_{RF} and ν_R). However, several authors have reported an enhancement of resolution upon increasing the MQ order p from 3 to 5 (see Figure 10.7).[43,50,51] It has also been noticed that this increase in the apparent resolution is not uniform among different spin systems. Later investigations have attributed these effects to the interplay between inhomogeneous and homogeneous contributions to the isotropic linewidth.[50,51] In the case of well-crystallized samples, the inhomogeneous contributions may be due to inhomogeneity of the static magnetic field, changes in magnetic susceptibility, incomplete heteronuclear decoupling, and second-order quadrupolar-dipolar couplings.[52,53] Since the Hamiltonians associated with these interactions are linear in S_z, under the p-quantum evolution, the corresponding inhomogeneous broadening is scaled in the same way as the dispersion of the isotropic shifts. Consequently, no apparent increase in resolution should be observed

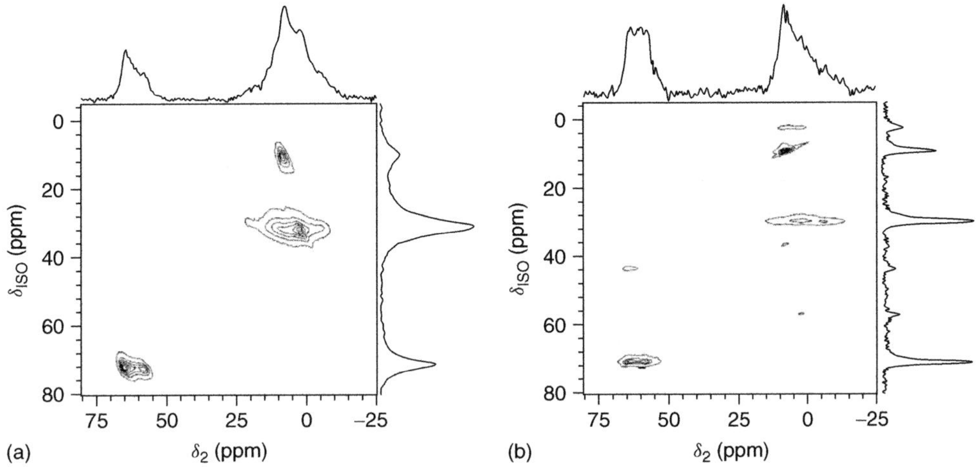

Figure 10.7. Sheared ^{27}Al MQMAS spectra of $SrAl_{12}O_{19}$: (a) 3QMAS and (b) 5QMAS.

in the higher order MQMAS spectra.[42,51] The homogeneous contribution, however, is not expected to increase with p.[51] Therefore, in crystalline samples with a strong homogeneous contribution to the line broadening, considerable enhancement of the isotropic resolution may be observed by moving to higher order MQMAS schemes.[54] This enhancement is less evident in amorphous samples, where the distribution of isotropic shifts dominates the spectra. In such cases, the more sensitive 3QMAS experiments are recommended.

In the presence of strong dipolar interactions, CW decoupling may be insufficient during t_1,[55] and composite decoupling may be necessary to obtain a well-resolved spectrum.[55,56] In the majority of applications, however, the most practical way to achieve efficient homo- and heteronuclear decoupling in MQMAS experiments is to use the highest possible spinning speed. The added advantage of high spinning speed is the increased spectral width along F_1 in the experiments performed with rotor synchronization. The rotor-synchronized spectra have improved S/N ratios, although their spectral width is limited along the F_1 dimension.[57]

10.4.4 Quantification of MQMAS Intensities

Due to the strong dependence of MQ excitation on C_Q, η_Q, and the crystallite orientation, MQMAS spectra are not quantitative in the isotropic dimension and have distorted lineshapes in the MAS dimension. Examples of such effects are shown in Figure 10.7. Strategies have been developed to retrieve quantitative intensities from the MQMAS spectra.

One of these strategies relies upon combining the information given by MQMAS and MAS. In short, MQMAS can be used to obtain highly resolved spectra and to determine the isotropic CSs and the quadrupolar parameters for different sites. This provides a starting set of parameters for simulation of the MAS spectra, recorded in a quantitative way, in order to obtain accurate intensities. We note that this method is suitable for analysis of spectra that are not significantly broadened due to dipolar interactions and/or distribution of CS and quadrupolar parameters.

Quantitative information can be also obtained by calculating the theoretical powder-averaged efficiency curves for MQMAS experiments as a function of $P_Q = C_Q(1 + \eta_Q^2/3)^{1/2}$.[43] Since the parameter P_Q can be easily derived from analysis of the first moment of the MQMAS spectrum, these curves can be used to offset the quantitative distortions of individual resonances. However, in partly disordered or amorphous materials the resonances cannot be described with a unique set of spectral parameters. Instead, the distributions of the isotropic CSs (δ_{CS}) and the quadrupolar parameters (P_Q or C_Q and η_Q) must be considered to properly interpret the MQMAS spectra. Two different strategies, both requiring the use of sophisticated routines,[42,47,58] can be used to provide a complete quantitative determination of such distributions. The so-called direct method relies upon the iterative fit of the distribution to minimize the deviation between experimental and simulated spectra. The second method directly computes the distributions by performing a so-called inversion of the 2D spectrum.[59] The results of these analyses can be further utilized to evaluate other properties, such as internuclear distances and bond angle distributions.

10.5 MEASUREMENTS OF HETERONUCLEAR CORRELATIONS

The development of MQMAS provided new means for analyzing intermediate-range ordering in complex spin systems via the measurements of interactions between the quadrupolar nuclei and their NMR-active neighbors. Most of these measurements utilize indirect excitation via CP for spectral editing of the MQMAS spectra,[60–62] or for MQMAS-based through-space HETCOR experiments.[63–65] Since the spin dynamics of the CP transfer to/from quadrupolar nuclei is very complex,[66–69] the CP-based methods suffer from quantitative uncertainties and are difficult to set up. Thus, alternative methods utilizing dipolar polarization transfer via TEDOR or through-bond transfer via INEPT have been demonstrated, as well.[70,71] Analysis of internuclear interactions between spin-1/2 and quadrupolar nuclei can also be made by reintroducing the heteronuclear dipolar dephasing between these spins using RF pulses that are synchronized with the MAS rotor. One such method, REDOR,[11,12] capable of measuring dipolar interactions and internuclear distances, has been combined with MQMAS.[72,73]

10.5.1 Cross Polarization and MQMAS

The dipolar coupling between two spins can be used to indirectly produce nuclear magnetization of the observed spins via CP transfer from other nuclei. The spin-1/2 → spin-1/2 CP experiments are performed between abundant and rare spins to enhance the nuclear magnetization of the latter.[74] The polarization transfer in such systems is well understood, and obtaining reasonably quantitative spectra is possible. The application of CP transfer to half-integer quadrupolar nuclei is difficult due to the complicated spin dynamics involved in both the spin-locking and the CP processes under MAS. These dynamics are strongly anisotropic with respect to crystallite orientation and depend on the relative size of the quadrupole interaction, the amplitudes of the RF magnetic fields, the spinning speed, and the resonance offsets.[9,10,60,66–69,75–77] CP between spin-1/2 and quadrupolar nuclei does not generally lead to enhanced sensitivity. However, it is of interest as a method for spectral editing. When appropriate line-narrowing methods are applied to both spins, the CP method can be further employed to obtain high-resolution HETCOR spectra.

CP can be coupled with MQMAS in the 1QCP-MQMAS experiment shown in Figure 10.8(a), where the magnetization of spins-1/2 nuclei (I) is transferred into the ±1Q central transition of quadrupolar spins S. The resulting 1Q coherence is

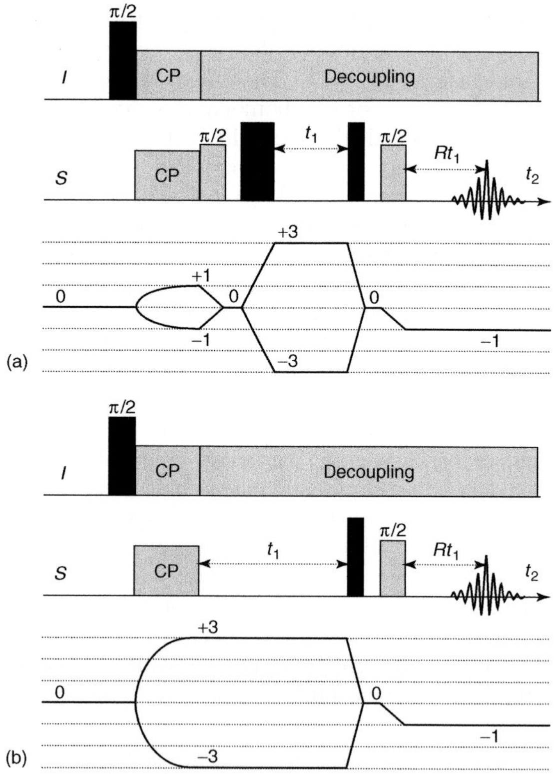

Figure 10.8. Pulse sequences and coherence transfer pathways for (a) 1QCP-3QMAS and (b) 3QCP-3QMAS experiments with z-filter. The phase cycling used in these experiments must include spin temperature inversion, to avoid direct observation of spins S.

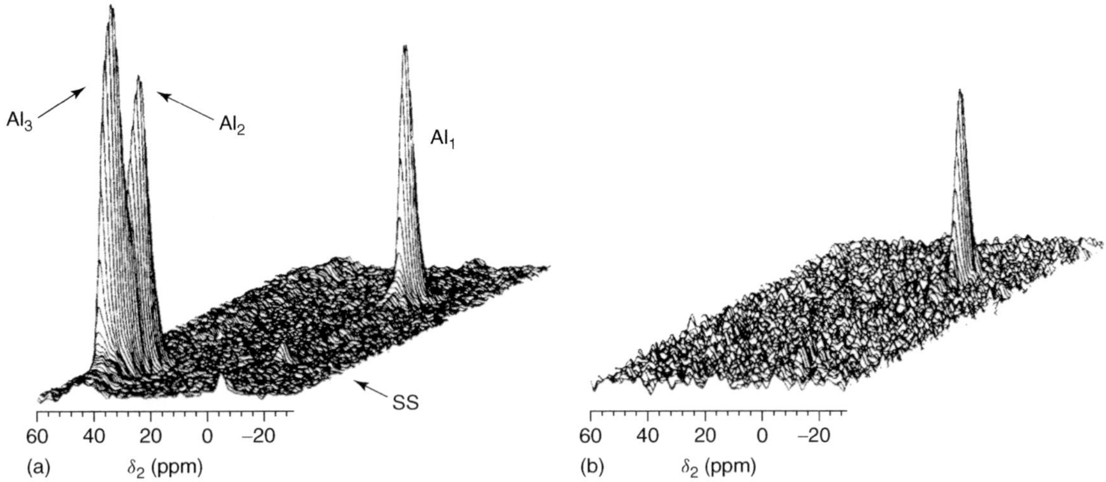

Figure 10.9. 2D NMR spectra of ^{27}Al in $AlPO_4$–CHA obtained using (a) 3QMAS and (b) $^{19}F \rightarrow ^{27}Al$ 1QCP-3QMAS.[60] The experiment was performed using $\nu_0 = 104.2$ MHz, $\nu_R = 20$ kHz, a CP contact time of 0.3 ms, and ν_{RF}^{S} field strengths of approximately 250, 20, and 5 kHz during hard pulses, soft pulses, and CP transfer, respectively. ^{19}F decoupling was applied during evolution and acquisition times using $\nu_{RF}^{I} \approx 80$ kHz. SS denotes the spinning sideband.

converted into a population inversion between the states with $m = \pm 1/2$ by a selective $\pi/2$ pulse. After a short delay, a standard MQMAS is performed, e.g., using a z-filter scheme.[61,78] The 1QCP-3QMAS experiment has been first carried out between ^{19}F and ^{27}Al nuclei in fluorinated triclinic chabazite-like $AlPO_4$ aluminophosphate, which contains one octahedral site Al_1 coordinated to fluorine, and two sites labeled Al_2 and Al_3 representing aluminum in the AlO_4 tetrahedra. The octahedral aluminum site Al_1 that is bonded to fluorine can be easily distinguished by this method (Figure 10.9).[61] Direct polarization transfer to MQ coherences (MQCP-MQMAS), has been successfully implemented by several authors, as well.[66,67,75,76] One of the pulse sequences and coherence transfer schemes proposed for the MQCP-MQMAS experiment is shown in Figure 10.8(b).[76]

In the heteronuclear correlation experiment, it is advantageous to use the polarization transfer from fast relaxing quadrupolar nuclei to obtain a 2D high-resolution through-space *dipolar* MQMAS-*D*-HETCOR spectrum.[64,65] The experiment can be designed for $S = 3/2$ (Figure 10.10a) or $S > 3/2$ (Figure 10.10b), and is performed by spin-locking the isotropic echo that forms at time $Rt_1/(1 + R)$ after the second pulse, followed by CP transfer and acquisition of I-spin signal in the t_2 domain. The 3QMAS-*D*-HETCOR method was first used to correlate the isotropic resonances of ^{23}Na and ^{31}P in $Na_3P_3O_9$ (Figure 10.11).[64] By using a similar scheme with SPAM conversion, the first $^{27}Al \rightarrow ^{29}Si$ 3QMAS-*D*-HETCOR spectrum was obtained from

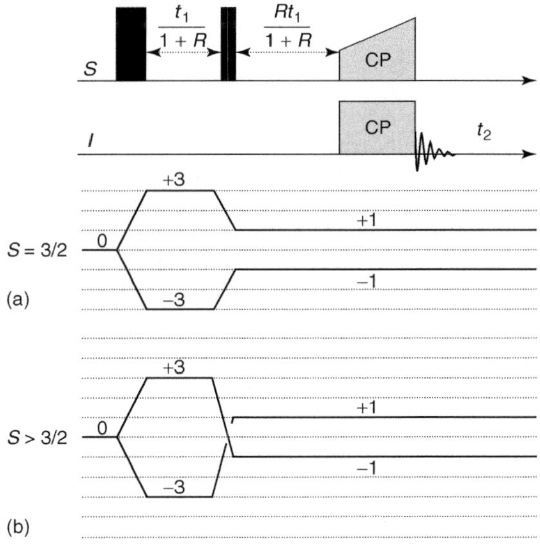

Figure 10.10. Pulse sequence and coherence transfer pathways used in 3QMAS-*D*-HETCOR experiments for (a) $S = 3/2$ and (b) $S > 3/2$.

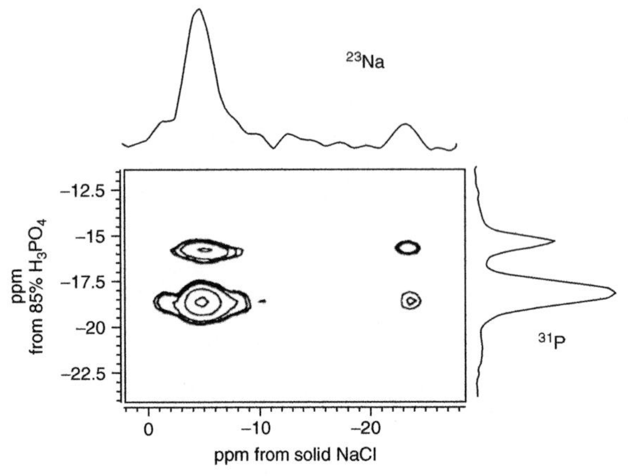

Figure 10.11. $^{23}Na \rightarrow {}^{31}P$ 3QMAS-*D*-HETCOR spectrum of $Na_3P_3O_9$.[63]

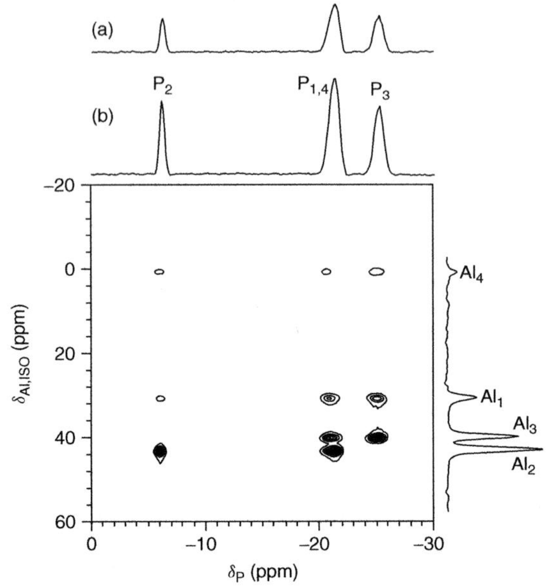

Figure 10.12. ^{27}Al-^{31}P 3QMAS-*J*-HETCOR spectrum of $AlPO_4$-14 obtained using the coherence pathway with SPAM $(0 \rightarrow +3 \rightarrow \{+1, 0, -1\} \rightarrow -1)$. The experiment was performed using $\nu_0 = 156.4$ MHz, $\nu_R = 15$ kHz, $\nu_{RF}^{Al} \approx$ 210 kHz during hard pulses, $\nu_{RF}^{Al} \approx 5$ kHz during the soft pulse and R-INEPT, $\nu_{RF}^{P} \approx 60$ kHz, a delay of 2.5 ms between consecutive pulses during R-INEPT, and a delay of 0.2 s between transients. The resonances are labeled in accordance with Figure 10.5.

a sample of potassium feldspar mineral (microcline silicate) with albite intergrowths.[79]

10.5.2 MQMAS with INEPT, TEDOR, and REDOR

In some cases, the quantitative uncertainties associated with the complex dynamics during spin-locking and CP can be avoided by utilizing experiments based on scalar (J) coupling or by using the through-space TEDOR technique. Although J-couplings are weak compared to dipolar couplings, they have been exploited in several 1D and 2D heteronuclear experiments.[80–82] A 2D spectrum that provides a chemical bond map between quadrupolar (^{27}Al) and spin-1/2 (^{31}P) nuclei in aluminophosphate $AlPO_4$-14 under isotropic resolution is shown in Figure 10.12. This experiment, referred to as MQMAS-*J*-HETCOR, uses a pulse sequence similar to that shown in Figure 10.10, except that the CP transfer is replaced by refocused INEPT (R-INEPT).[71] The spectrum in Figure 10.12, which exhibits all expected Al–O–P connectivities, was acquired using SPAM, which provided an additional sensitivity gain of 1.7.

The TEDOR technique has been successfully used in the studies of HETCOR spectra and dipolar couplings in spin systems involving spin-1/2 and quadrupolar nuclei under MAS.[12,13] In TEDOR, the magnetization is transferred in a way that is similar to that in the INEPT experiment, albeit through space instead of through the chemical bonds. The $I-S$ dipolar interaction, which is refocused every rotor period in the standard MAS experiments, can be reintroduced by using properly spaced rotor-synchronized RF pulses. The TEDOR scheme can be used in concert with MQMAS by replacing the Hartmann–Hahn CP transfer in the sequence of Figure 10.8(a).[70]

Another high-resolution method that avoids the use of CP has been proposed, which combines MQMAS with the direct determination of heteronuclear dipolar couplings via REDOR, leading to the MQ-REDOR experiment (Figure 10.13).[72] This technique employs two strong and two selective RF pulses phase-cycled to select the $0 \rightarrow \pm p \rightarrow 0 \rightarrow +1 \rightarrow -1$ coherence pathway. When spaced by an integer number of rotor periods $(n+1)t_R$, the selective pulses create two windows in which the sequences of π pulses can be applied to the spin-1/2 nuclei. As in the standard REDOR experiment, the role of the π pulses is to prevent MAS from refocusing the dephasing

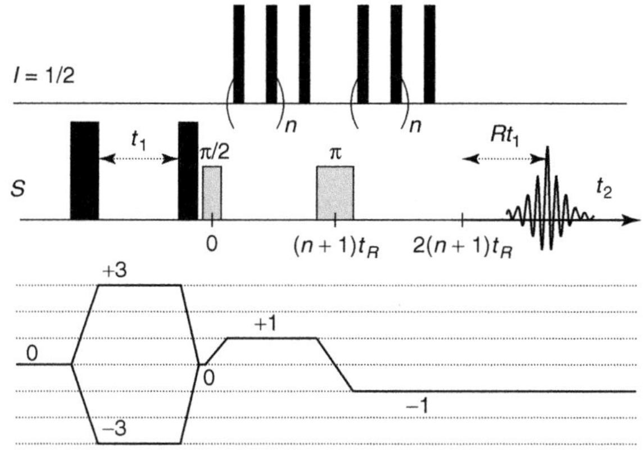

Figure 10.13. Pulse sequence and coherence transfer pathway used in MQ-REDOR experiment.

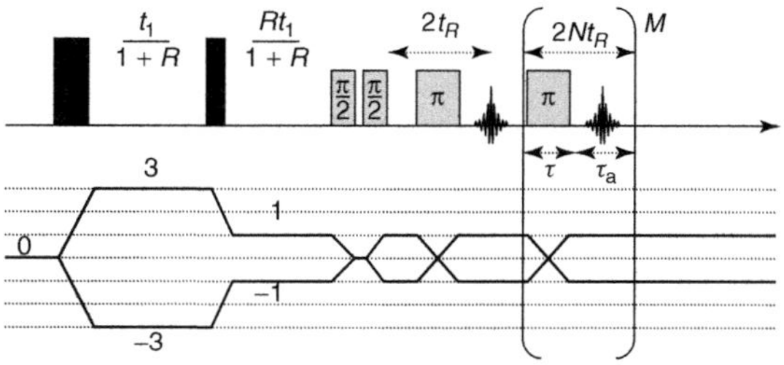

Figure 10.14. CPMG-3QMAS pulse sequence and coherence transfer pathway for spin-3/2 nuclei.[84] The part of the sequence shown in parentheses is rotor-synchronized according to $2Nt_R = \tau + \tau_a$, where τ is the duration of the selective π pulse (including the transient effects) and τ_a is the echo sampling period.

due to the dipolar interaction between spins I and S. The MQ-REDOR sequence has been used to measure distances within pairs of spin-1/2 (^{19}F) and quadrupolar (^{27}Al) nuclei, with both nuclei being observed under high resolution.[72]

10.6 SENSITIVITY ENHANCEMENT VIA CPMG

Advances in manipulating the MQ coherences considerably improved the sensitivity of the MQMAS experiment. An additional gain in the S/N ratio was achieved by applying a CPMG train of rotor-synchronized, selective π pulses during the acquisition of data in t_2.[83–85] The resulting CPMG-MQMAS spectrum is split into periodic combs of spin-echo sidebands that spread along the anisotropic dimension. Since the integrated spectral intensity is not affected by this procedure, the apparent S/N ratio for individual spikes can be significantly enhanced. The gain in the S/N ratio is governed by the periodicity τ_a of the CPMG train and the homogeneous spin–spin relaxation time T_2'. The truncation effects require that τ_a exceeds the net dephasing time of transverse magnetization T_2^*.

An example of the experimental scheme for the CPMG-MQMAS experiment for spin-3/2 nuclei, based on the split-t_1 method, is shown in Figure 10.14.[85] The CPMG sequence uses selective π pulses of the same strength of RF field as the soft pulse in the z-filter method. The sequence in Figure 10.14 utilizes contributions from the echo and antiecho coherence pathways, which leads to cosine-modulated FIDs. Approaches suitable for spin-5/2 nuclei have been also proposed.[86]

The appearance of a CPMG spectrum is determined by its sampling conditions: smaller separation τ_a between the echoes in time domain increases the S/N ratio in frequency domain at the expense of anisotropic lineshape, which is determined by a reduced number of sidebands. Although the spin-echo sidebands are regularly spaced with respect to the carrier frequency, the sideband manifolds corresponding to different sites are usually well separated along the isotropic dimension. This is manifested in Figure 10.15, which shows a series of ^{23}Na 3QMAS and CPMG-3QMAS spectra of an equimolar mixture of Na_2SO_4 and $Na_2C_2O_4$.[85] Although each spectrum was recorded with the same number of transients, spectra b and c correspond to S/N gain factors of 8 and 20, respectively.

10.7 CONCLUSION

MQMAS has become an established method available to the NMR community. Intense research during the first decade following its discovery brought about remarkable advances in MQMAS methodology. Presently, the acquisition of isotropic spectra of most sensitive and quadrupolar nuclei does not pose a serious challenge. While the basic z-filter method is still most often used in standard MQMAS applications, it is to be hoped that the benefits of SPAM will be exploited more routinely. Another interesting alternative to MQMAS is offered by STMAS, which can also be performed on a standard MAS probe.

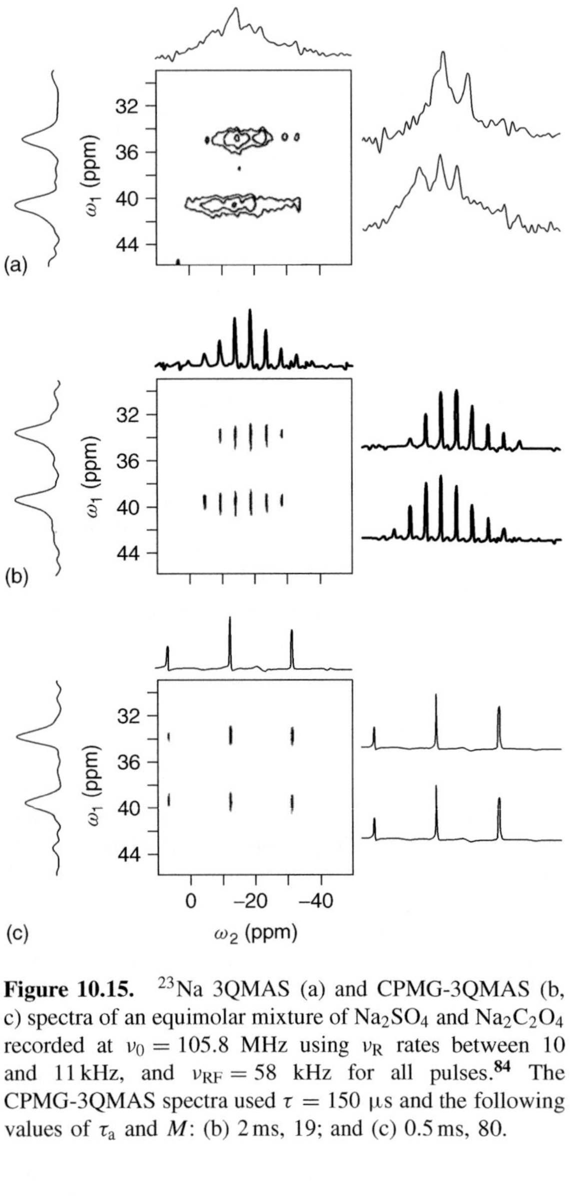

Figure 10.15. [23]Na 3QMAS (a) and CPMG-3QMAS (b, c) spectra of an equimolar mixture of Na_2SO_4 and $Na_2C_2O_4$ recorded at $\nu_0 = 105.8$ MHz using ν_R rates between 10 and 11 kHz, and $\nu_{RF} = 58$ kHz for all pulses.[84] The CPMG-3QMAS spectra used $\tau = 150$ μs and the following values of τ_a and M: (b) 2 ms, 19; and (c) 0.5 ms, 80.

Since the initial excitation in STMAS involves SQ instead of MQ coherences, the experiment is several times more sensitive than MQMAS.[44] The main challenge associated with STMAS is that the timing of pulse sequence must be very accurately synchronized with the pulse sequence and the rotor-axis set perfectly at the magic angle 54.736° (to within a few millidegrees) to avoid reintroducing of the first-order quadrupolar broadening (see Chapter 11). These experimental conditions can be met by controlling the timing of the RF pulses by the MAS tachometer signal and by using a Hall-sensor to adjust the magic angle.[87] In addition, the STMAS method is much more sensitive to microscopic motions,[88] and in that sense is complementary to MQMAS. Most of the high-resolution MQMAS-HETCOR methods can be used with STMAS, as well.[89]

Owing to its simplicity and robustness, the MQMAS experiment is used more often than DOR, DAS, and STMAS. The MQMAS methodology is expected to continue evolving. As high-field NMR instrumentation becomes widely accessible, MQMAS studies of inherently challenging nuclei, such as ^{17}O, ^{25}Mg, and ^{43}Ca, are being reported with increased frequency (see Chapters 17 and 18). Further advances in instrumentation, improvements in sensitivity and data processing, and the development of new types of correlation schemes may offer new perspectives for high-resolution studies of inherently challenging spin systems.

RELATED ARTICLES IN THE ENCYCLOPEDIA OF MAGNETIC RESONANCE

High Speed MAS of Half-Integer Quadrupolar Nuclei in Solids

Line Narrowing Methods in Solids

Multiple-quantum Magic-angle Spinning Experiments on Half-integer Nuclei: Fundamentals

Quadrupolar Metal Nuclides in Bioinorganic Chemistry: Solid-State NMR Studies

Quadrupolar Nuclei in Glasses

Quadrupolar Nuclei in Liquid Samples

Satellite Transition NMR Spectroscopy of Half-Integer Quadrupolar Nuclei under Magic-angle Spinning

REFERENCES

1. A. Abragam, *Principles of Nuclear Magnetism*, Oxford University Press: London, 1961.
2. E. R. Andrew, A. Bradbury, and R. G. Eades, *Nature*, 1958, **182**, 1659.

3. J. S. Waugh, L. M. Huber, and U. Haeberlen, *Phys. Lett.*, 1968, **20**, 180.
4. A. Pines, M. G. Gibby, and J. S. Waugh, *J. Chem. Phys.*, 1972, **56**, 1776.
5. B. C. Gerstein, R. G. Pembelton, R. C. Wilson, and L. M. Ryan, *J. Chem. Phys.*, 1977, **66**, 362.
6. E. L. Hahn, *Phys. Rev.*, 1950, **80**, 580.
7. E. Oldfield, H. K. C. Timken, B. Montez, and R. Ramachandran, *Nature*, 1985, **318**, 163.
8. H. J. Jakobsen, High Speed MAS of Half-Integer Quadrupolar Nuclei in Solids, in *Encyclopedia of Magnetic Resonance*, eds R. K. Harris and R. E. Wasylishen, John Wiley & Sons: Chichester, 2007. DOI: 10.1002/9780470034590.emrstm0212.
9. A. J. Vega, *J. Magn. Reson.*, 1992, **96**, 50.
10. A. J. Vega, *Solid State Nucl. Magn. Reson.*, 1992, **1**, 17.
11. T. Gullion and J. Schaefer, *J. Magn. Reson.*, 1989, **81**, 196.
12. A. W. Hing, S. Vega, and J. Schaefer, *J. Magn. Reson.*, 1992, **96**, 205.
13. C. A. Fyfe, K. T. Mueller, H. Grodney, and K. C. Wong-Moon, *Chem. Phys. Lett.*, 1992, **199**, 198.
14. E. R. H. van Eck, R. Jansen, W. E. J. R. Maas, and W. S. Veeman, *Chem. Phys. Lett.*, 1990, **174**, 428.
15. C. P. Grey and A. J. Vega, *J. Am. Chem. Soc.*, 1995, **117**, 8232.
16. T. Gullion, *J. Magn. Reson.*, 1995, **A117**, 326.
17. (a) L. Chen, Q. Wang, B. Hu, O. Lafon, J. Trébosc, F. Deng, and J.-P. Amoureux, *Phys. Chem. Chem. Phys.*, 2010, **12**, 9395–9405; (b) L. Chen, X. Lu, Q. Wang, O. Lafon, J. Trébosc, F. Deng, and J.-P. Amoureux, *J. Magn. Reson.*, 2010, **206**, 269–273.
18. A. Samoson, E. Lipmaa, and A. Pines, *Mol. Phys.*, 1988, **65**, 1013.
19. K. T. Mueller, B. Q. Sun, G. C. Chingas, J. W. Zwanziger, T. Terao, and A. Pines, *J. Magn. Reson.*, 1990, **86**, 470.
20. L. Frydman and J. S. Harwood, *J. Am. Chem. Soc.*, 1995, **117**, 5367.
21. A. Medek, J. S. Harwood, and L. Frydman, *J. Am. Chem. Soc.*, 1995, **117**, 12779.
22. A. Llor and J. Virlet, *Chem. Phys. Lett.*, 1988, **152**, 248.
23. S. Vega and Y. Naor, *J. Chem. Phys.*, 1981, **75**, 75.
24. N. C. Nielsen, H. Bildsoe, and H. J. Jakobsen, *Chem. Phys. Lett.*, 1992, **191**, 205.
25. M. Pruski, J. W. Wiench, and J.-P. Amoureux, *J. Magn. Reson.*, 2000, **147**, 286.
26. J.-P. Amoureux, *Solid State Nucl. Magn. Reson.*, 1993, **2**, 83.
27. J.-P. Amoureux, C. Fernandez, and L. Frydman, *Chem. Phys. Lett.*, 1996, **259**, 347.
28. P. K. Madhu, A. Goldbourt, L. Frydman, and S. Vega, *Chem. Phys. Lett.*, 1999, **307**, 41.
29. A. P. M. Kentgens and R. Verhagen, *Chem. Phys. Lett.*, 1999, **300**, 435.
30. T. Vosegaard, P. Florian, D. Massiot, and P. J. Grandinetti, *J. Chem. Phys.*, 2001, **114**, 4618.
31. G. Wu, D. Rovnyak, and R. G. Griffin, *J. Am. Chem. Soc.*, 1996, **118**, 9326.
32. J.-P. Amoureux, L. Delevoye, S. Steuernagel, Z. Gan, S. Ganapathy, and L. Montagne, *J. Magn. Reson.*, 2004, **172**, 268.
33. S. P. Brown, S. J. Heyes, and S. Wimperis, *J. Magn. Reson.*, 1996, **A119**, 280.
34. P. J. Grandinetti, J. H. Baltisberger, A. Llor, Y. K. Lee, U. Werner, M. A. Eastman, and A. Pines, *J. Magn. Reson.*, 1993, **A103**, 72.
35. D. Massiot, B. Touzo, D. Trumeau, J. P. Coutures, J. Virlet, P. Florian, and P. J. Grandinetti, *Solid State NMR*, 1996, **6**, 73.
36. J.-P. Amoureux, C. Fernandez, and S. Steuernagel, *J. Magn. Reson.*, 1996, **A123**, 116.
37. S. P. Brown and S. Wimperis, *J. Magn. Reson.*, 1997, **128**, 42.
38. A. Bax, R. Freeman, and S. P. Kempsell, *J. Magn. Reson.*, 1980, **41**, 349.
39. D. J. States, R. A. Haberkorn, and D. J. Ruben, *J. Magn. Reson.*, 1982, **48**, 286.
40. Z. Gan and H. T. Kwak, *J. Magn. Reson.*, 2004, **168**, 346.
41. T. J. Ball and S. Wimperis, *J. Magn. Reson.*, 2007, **187**, 343.
42. J.-P. Amoureux and C. Fernandez, *Solid State Nucl. Magn. Reson.*, 1998, **10**, 211.
43. C. Fernandez, J.-P. Amoureux, J. M. Chezeau, L. Delmotte, and H. Kessler, *Micro. Mater.*, 1996, **6**, 331.
44. Z. Gan, *J. Am. Chem. Soc.*, 2000, **122**, 3242.

45. Y. Millot and P. P. Man, *Solid State Nucl. Magn. Reson.*, 2002, **21**, 21.
46. S. P. Brown and S. Wimperis, *J. Magn. Reson.*, 1997, **124**, 279.
47. T. Charpentier, High-resolution nuclear magnetic resonance of quadrupolar nuclei in solids, PhD thesis, University of Paris XI, France, 1998.
48. K. H. Lim and C. P. Grey, *Solid State Nucl. Magn. Reson.*, 1998, **13**, 101.
49. J.-P. Amoureux, L. Delevoye, G. Fink, F. Taulelle, A. Flambard, and L. Montagne, *J. Magn. Reson.*, 2005, **175**, 285.
50. S. H. Wang, Z. Xu, J. H. Baltisberger, L. M. Bull, J. F. Stebbins, and A. Pines, *Solid State Nucl. Magn. Reson.*, 1997, **8**, 1.
51. K. J. Pike, R. P. Malde, S. E. Ashbrook, J. McManus, and S. Wimperis, *Solid State Nucl. Magn. Reson.*, 2000, **16**, 203.
52. J. McManus, R. Kemp-Harper, and S. Wimperis, *Chem. Phys. Lett.*, 1999, **311**, 292.
53. S. Wi and L. Frydman, *J. Chem. Phys.*, 2000, **112**, 3248.
54. J.-P. Amoureux and J. Trebosc, *J. Magn. Reson.*, 2006, **179**, 311.
55. V. Lacassagne, P. Florian, V. Montouillout, C. Gervais, F. Babonneau, and D. Massiot, *Magn. Reson. Chem.*, 1998, **36**, 956.
56. S. Ganapathy, L. Delevoye, J.-P. Amoureux, and P. K. Madhu, *Magn. Reson. Chem.*, 2008, **46**, 948.
57. D. Massiot, *J. Magn. Reson.*, 1996, **A122**, 240.
58. M. Bak, J. T. Rasmussen, and N. C. Nielsen, *J. Magn. Reson.*, 2000, **147**, 296.
59. J. Zwanziger, *Solid State Nucl. Magn. Reson.*, 1994, **3**, 219.
60. W. Sun, J. T. Stephen, L. D. Porter, and Y. Wu, *J. Magn. Reson.*, 1995, **A116**, 181.
61. M. Pruski, D. P. Lang, C. Fernandez, and J.-P. Amoureux, *Solid State Nucl. Magn. Reson.*, 1997, **7**, 327.
62. S. E. Ashbrook, S. P. Brown, and S. Wimperis, *Chem. Phys. Lett.*, 1998, **288**, 509.
63. T. P. Jarvie, R. M. Wenslow, and K. T. Mueller, *J. Am. Chem. Soc.*, 1995, **117**, 570.
64. S. H. Wang, S. M. De Paul, and L. M. Bull, *J. Magn. Reson.*, 1997, **125**, 364.
65. C. Fernandez, C. Morais, J. Rocha, and M. Pruski, *Solid State Nucl. Magn. Reson.*, 2002, **21**, 61.
66. K. H. Lim and C. P. Grey, *J. Chem. Phys.*, 2000, **112**, 7490.
67. S. E. Ashbrook and S. Wimperis, *Mol. Phys.*, 2000, **98**, 1.
68. S. Vega, *Phys. Rev.*, 1981, **23**, 3152.
69. J.-P. Amoureux and M. Pruski, *Mol. Phys.*, 2002, **100**, 1595.
70. J.-P. Amoureux, J. Trebosc, and G. Tricot, *Magn. Reson. Chem.*, 2007, **45**, S187.
71. J. W. Wiench and M. Pruski, *Solid State Nucl. Magn. Reson.*, 2004, **26**, 51.
72. C. Fernandez, D. P. Lang, J.-P. Amoureux, and M. Pruski, *J. Am. Chem. Soc.*, 1998, **120**, 2672.
73. M. Pruski, A. Bailly, D. P. Lang, J.-P. Amoureux, and C. Fernandez, *Chem. Phys. Lett.*, 1999, **307**, 35.
74. A. Pines, M. G. Gibby, and J. S. Waugh, *J. Chem. Phys.*, 1973, **59**, 569.
75. D. Rovnyak, M. Baldus, and R. G. Griffin, *J. Magn. Reson.*, 2000, **142**, 145.
76. S. E. Ashbrook and S. Wimperis, *J. Magn. Reson.*, 2000, **147**, 238.
77. B. Hu, J.-P. Amoureux, J. Trebosc, and S. Hafner, *J. Magn. Reson.*, 2008, **192**, 8.
78. C. Fernandez, L. Delevoye, J.-P. Amoureux, D. P. Lang, and M. Pruski, *J. Am. Chem. Soc.*, 1997, **119**, 6858.
79. J. W. Wiench, G. Tricot, L. Delevoye, J. Trebosc, J. Frye, L. Montagne, J.-P. Amoureux, and M. Pruski, *Phys. Chem. Chem. Phys.*, 2006, **8**, 144.
80. C. A. Fyfe, K. C. Wong-Moon, Y. Huang, and H. Grodney, *J. Am. Chem. Soc.*, 1995, **117**, 10397.
81. H.-M. Kao and C. P. Grey, *J. Magn. Reson.*, 1998, **133**, 313.
82. D. Massiot, F. Fayon, B. Alonso, J. Trebosc, and J.-P. Amoureux, *J. Magn. Reson.*, 2003, **164**, 160.
83. H. Y. Carr and E. M. Purcell, *Phys. Rev.*, 1954, **94**, 630.
84. S. Meiboom and D. Gill, *Rev. Sci. Instrum.*, 1958, **29**, 688.
85. T. Vosegaard, F. H. Larsen, H. J. Jakobsen, P. D. Ellis, and N. C. Nielsen, *J. Am. Chem. Soc.*, 1997, **119**, 9055.

86. R. Lefort, J. W. Wiench, M. Pruski, and J.-P. Amoureux, *J. Chem. Phys.*, 2002, **116**, 2493.

87. S. Mamone, A. Dorsch, O. G. Johannessen, M. V. Naik, P. K. Madhu, and M. H. Levitt, *J. Magn. Reson.*, 2008, **190**, 135.

88. S. Antonijevic, S. E. Ashbrook, S. Biedasek, R. I. Walton, S. Wimperis, and H. Yang, *Am. Chem. Soc.*, 2006, **128**, 8054.

89. J. Trébosc, O. Lafon, B. Hu, and J.-P. Amoureux, *Chem. Phys. Lett.*, 2010, **496**, 201–207.

Chapter 11
STMAS NMR: Experimental Advances

Sharon E. Ashbrook[1] and Stephen Wimperis[2]

[1]*School of Chemistry and EaStCHEM, University of St Andrews, St Andrews KY16 9ST, UK*
[2]*School of Chemistry and WestCHEM, University of Glasgow, Glasgow G12 8QQ, UK*

11.1 INTRODUCTION

The quest for improved resolution is an ongoing theme in solid-state NMR spectroscopy, with methods such as magic angle spinning (MAS) and decoupling commonly employed to remove shielding anisotropy and dipolar interactions. However, for quadrupolar nuclei (i.e., those with spin quantum number $I > 1/2$) NMR spectra are additionally broadened by the interaction of the nuclear electric quadrupole moment (eQ) with the electric field gradient generated at the nucleus. This quadrupolar interaction is orientation dependent (anisotropic) and may be relatively strong, giving rise to an inhomogeneous broadening often of the order of megahertz. While MAS is able to remove the broadening to a "first-order" approximation, it is unable to remove the "second-order" broadening. Hence, the need for more complex approaches to acquire truly high-resolution or isotropic spectra.

In 2000, Gan introduced a new method for acquiring high-resolution NMR spectra of quadrupolar solids,[1,2] that utilized the satellite transitions (STs) of nuclei with half-integer spin quantum number. This satellite transition magic angle spinning (STMAS)[1–3] experiment is similar in many ways to the commonly used multiple-quantum magic angle spinning (MQMAS) experiment introduced by Frydman and Harwood[4] in 1995, as it employs the correlation of different transitions (the central transition (CT) with either satellite transitions for STMAS or multiple-quantum transitions for MQMAS) in a 2D MAS-based experiment. This enables the complete removal of second-order quadrupolar broadening and yields a high-resolution, isotropic spectrum. However, the single-quantum nature of STMAS promises considerable sensitivity advantages over MQMAS, where filtration through multiple-quantum coherences is an inherently inefficient process.[1–3] In addition,

NMR of Quadrupolar Nuclei in Solid Materials
Edited by Roderick E. Wasylishen, Sharon E. Ashbrook and Stephen Wimperis

unlike older approaches such as dynamic angle spinning (DAS)[5] and double rotation (DOR),[6] STMAS does not require any special, expensive equipment. However, STMAS has been shown to be more technically demanding in its implementation than MQMAS, requiring a very precise setting of the magic angle and stable sample spinning.[1–3]

In this chapter, we discuss the recent advances in STMAS. After a brief review of the theoretical background and basic experimental techniques, we describe a number of improvements and adaptations to the original experiment, including compensated and "ultrafast" versions. We then discuss the effect of molecular-scale dynamics upon STMAS spectra and the effects of higher-order or "cross-term" interactions, before concluding with a review of some of the more useful applications of STMAS to date, many of which exploit its considerable advantage in sensitivity.

11.2 THEORETICAL BACKGROUND

For a full description of the theory behind the quadrupolar interaction and its implications for the STMAS experiment see Chapters 1, 2 and 3, as only a brief summary is presented here.

The electric field gradient (EFG) has the properties of a Cartesian tensor and, in its principal axis system (PAS), is described by three components, with $|V_{zz}| \geq |V_{yy}| \geq |V_{xx}|$. The magnitude of the tensor is given by $eq = V_{zz}$ and its cross-sectional shape by the asymmetry parameter, $\eta_Q = (V_{xx} - V_{yy})/V_{zz}$, where η_Q lies between 0 (for an axially symmetric tensor) and 1. Quadrupolar broadening results from the interaction of the EFG with the nuclear quadrupole moment (eQ), a property that is characteristic of the nucleus. The magnitude of the interaction is commonly parameterized by the quadrupolar coupling constant, $C_Q = e^2qQ/h$ (in hertz). The quadrupolar splitting parameter (in the PAS) can then be defined as $\omega_Q^{PAS} = 3\pi C_Q/(2I(2I-1))$, in radians per second. Although the quadrupolar interaction may be large compared with other internal spin interactions, in most cases it is smaller than the Zeeman interaction and its effect can be described as a perturbation to the Zeeman energy levels. The magnitude of the interaction often necessitates not only a first-order perturbation, but also the consideration of second-order terms. In general, for a sample in a spinning rotor (and assuming an integral number of rotor periods), the averaged frequency of an $m_I = \pm(q-1) \leftrightarrow \pm q$ single-quantum transition (with $q = 1/2, 3/2, 5/2$, etc.) is given by the sum of first- and second-order terms[3]

$$\langle \omega_{\pm(q-1)\leftrightarrow\pm q} \rangle = \langle \omega^{(1)}_{\pm(q-1)\leftrightarrow\pm q} \rangle + \langle \omega^{(2)}_{\pm(q-1)\leftrightarrow\pm q} \rangle \tag{11.1}$$

with

$$\langle \omega^{(1)}_{\pm(q-1)\leftrightarrow\pm q} \rangle = \pm(2q-1)\omega_Q^{PAS} d^2_{0,0}(\chi) d^2_{0,0}(\beta) \tag{11.2a}$$

$$\langle \omega^{(2)}_{\pm(q-1)\leftrightarrow\pm q} \rangle = \left(\frac{\left(\omega_Q^{PAS}\right)^2}{\omega_0} \right) \{A^0(I,q) + A^2(I,q) d^2_{0,0}(\chi) d^2_{0,0}(\beta) + A^4(I,q) d^4_{0,0}(\chi) d^4_{0,0}(\beta)\} \tag{11.2b}$$

where, for simplicity, axial symmetry ($\eta_Q = 0$) has been assumed. The spin-dependent coefficients, $A^l(I, q)$ are given in Table 11.1 and the Wigner reduced rotation matrix elements, $d^l_{m,m'}(\theta)$, can be found in the literature.[7] The angle χ describes the orientation of the rotor axis relative to the external magnetic field (the "spinning angle"), while β describes the orientation of the EFG in the rotor frame.

Two different types of single-quantum transitions emerge from consideration of equations (11.1) and (11.2): a $q = 1/2$ central transition with a frequency unaffected to a first-order approximation (i.e., $\langle \omega^{(1)}_{+1/2\leftrightarrow-1/2} \rangle = 0$) and $q = 3/2, 5/2$, etc. Satellite transition whose frequencies are perturbed. In a static powdered solid the orientation dependence of the first-order quadrupolar interaction produces a large inhomogeneous broadening of the satellite transitions, while the central transition remains relatively sharp and narrow. However, under MAS (i.e., if $\chi = 54.736°$) the first-order interaction can be removed. In contrast, both central transitions and satellite transitions are affected by the second-order quadrupolar interaction. This consists of an isotropic (orientation independent) shift and second- and fourth-rank anisotropic broadening. The second-order interaction is typically an order of magnitude smaller than the first-order interaction, as it is proportional to $(\omega_Q^{PAS})^2/\omega_0$, and not to ω_Q^{PAS}. Although the second-rank term is removed by MAS (as $d^2_{0,0}(\chi = 54.736°) = 0$), the fourth-rank

Table 11.1. Spin(I)- and transition(q)-dependent coefficients, $A^l(I, q)$, for $m_I = \pm(q-1) \leftrightarrow \pm q$ single-quantum transitions and corresponding STMAS ratios, $R(I, q)$

	$A^0(I, q)$	$A^2(I, q)$	$A^4(I, q)$	$R(I, q)$
Spin I = 3/2				
q = 1/2 (CT)	−2/5	−8/7	54/35	
q = 3/2 (ST)	4/5	4/7	−48/35	−8/9
Spin I = 5/2				
q = 1/2 (CT)	−16/15	−64/21	144/35	
q = 3/2 (ST_1)	2/15	−4/3	6/5	7/24
q = 5/2 (ST_2)	56/15	80/21	−264/35	−11/6
Spin I = 7/2				
q = 1/2 (CT)	−2	−40/7	54/7	
q = 3/2 (ST_1)	−4/5	−4	24/5	28/45
q = 5/2 (ST_2)	14/5	8/7	−138/35	−23/45
q = 7/2 (ST_3)	44/5	68/7	−648/35	−12/5
Spin I = 9/2				
q = 1/2 (CT)	−16/5	−64/7	432/35	
q = 3/2 (ST_1)	−2	−52/7	66/7	55/72
q = 5/2 (ST_2)	8/5	−16/7	24/35	1/18
q = 7/2 (ST_3)	38/5	44/7	−486/35	−9/8
q = 9/2 (ST_4)	16	128/7	−240/7	−25/9

broadening is not (i.e., $d^4_{0,0}(\chi = 54.736°) \neq 0$). This results in powder-pattern MAS lineshapes for both central transitions and satellite transitions, dependent upon C_Q and η_Q.

The STMAS experiment[1–3] removes the anisotropic second-order quadrupolar broadening by correlation of satellite transitions- and central transitions in a 2D MAS NMR experiment, with rotor-synchronized acquisition in t_1 to remove the first-order quadrupolar interaction fully from the satellite transitions. The method exploits the fact that the second-order broadenings of the central transitions and satellite transitions are related by a simple numerical factor. Hence, 2D correlation will allow the removal of the anisotropic broadening, while crucially retaining the isotropic shifts and enabling the separation of distinct resonances. In the simplest experiment[1–3] satellite transitions are excited by a single pulse and evolve in t_1, before conversion to central transition coherence, detected in t_2. When $t_2 = (A^4(I, q > 1/2)/A^4(I, 1/2))t_1$ an echo corresponding to the refocusing of the fourth-rank inhomogeneous broadening forms. This ratio, $R(I, q)$, often termed the STMAS ratio, is given in Table 11.1 for nuclei with half-integer spin quantum numbers up to $I = 9/2$.[8,9] Fourier transformation produces a 2D spectrum where refocusing of the anisotropic broadening yields a ridge lineshape lying along an axis inclined at $R(I, q)$.[1–3] A projection onto an axis orthogonal to this generates a spectrum that is free from quadrupolar broadening and yet retains isotropic shift information.

11.3 EXPERIMENTAL IMPLEMENTATION

11.3.1 Basic Pulse Sequences

Although the simple correlation experiment described above refocuses the second-order quadrupolar broadening, it is rarely used in practice as it produces spectra that contain 2D "phase-twist" lineshapes.[10] Pulse sequences for two modifications of STMAS (termed "z-filter"[1] and "shifted-echo"[11] experiments) that produce more desirable pure-absorption lineshapes, are shown in Figure 11.1(a,b), respectively. In the first, an amplitude-modulated data set is recorded, with the selection of both $p = \pm 1$ satellite transition coherences in t_1. These are then combined with equal amplitude through a $p = 0$ population state, or z-filter. The final pulse in the sequence is usually chosen

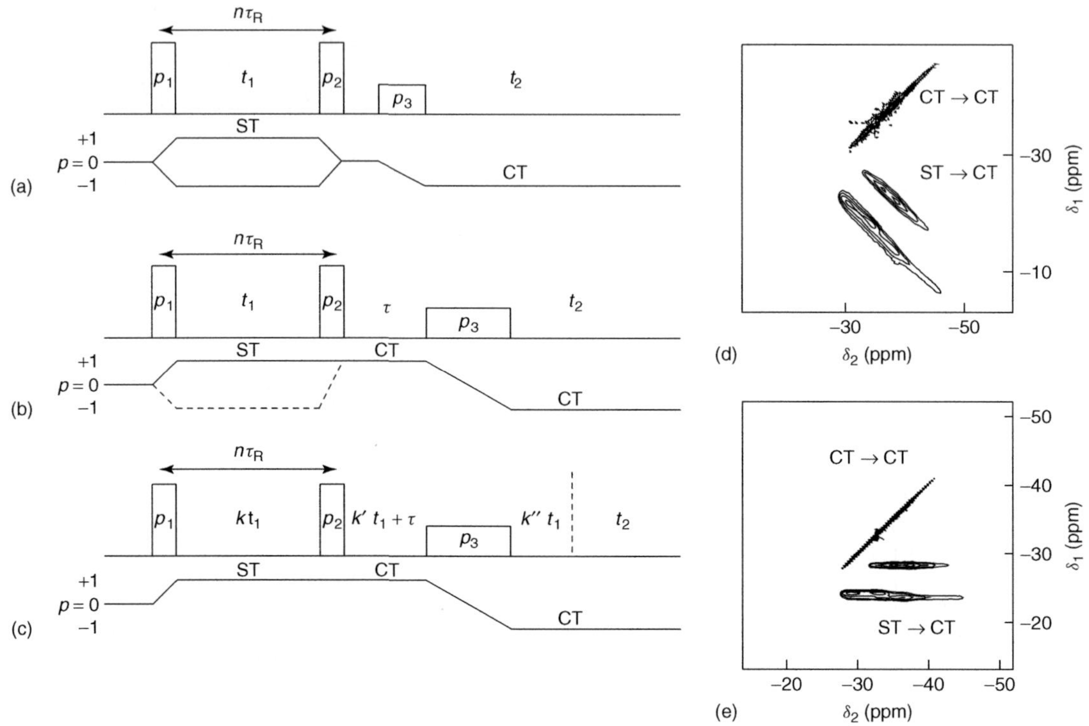

Figure 11.1. Pulse sequences and coherence transfer pathway diagrams for rotor-synchronized (a) amplitude-modulated z-filter,[1] (b) phase-modulated shifted-echo[12] and (c) split-t_1 phase-modulated shifted-echo[11] STMAS experiments. In each case, the third pulse is applied with a low radiofrequency field strength, ensuring it is selective for the central transition. In (b), the dashed coherence transfer pathway is an alternative to the solid one. In (c), the coefficients k, k' and k'' are chosen to refocus the second-order broadening at the end of the t_1 period. (d, e) ^{87}Rb (130.9 MHz) STMAS spectra of $RbNO_3$ acquired with (d) the pulse sequence in (b) and (e) the pulse sequence in (c) with $k = 9/17$, $k' = 8/17$ and $k'' = 0$.[3] Spectra are the result of averaging 32 transients with recycle interval of 250 ms for each of 256 t_1 increments of (d) 50 μs and (e) 94.4 μs. The MAS rate was 20 kHz. All ppm scales are referenced to 1 M $RbNO_3$ (aq). (Reproduced from Ref. 3. © Elsevier, 2004.)

to be selective for the central transition and is applied with a low radiofrequency field strength. The amplitude-modulated data require a hypercomplex 2D Fourier transformation,[10] with sign discrimination in t_1 obtained by additional use of either the States-Haberkorn Ruben or time proportional phase incrementation (TPPI) methods. The second approach in Figure 11.1(b) utilizes the properties of whole echoes to achieve pure-phase lineshapes. The experiment is phase modulated, with only a single pathway selected in t_1 (the choice of $p = +1$ or -1 coherences depending upon the sign of the STMAS ratio, $R(I, q)$).[1–3] After conversion to central transition coherences, a spin-echo is performed, with the τ interval chosen to be of sufficient length that a whole echo signal can be acquired in t_2. Fourier transformation of this symmetrical whole echo yields a real spectral component that is purely absorptive. The final pulse in the sequence is applied with a low radiofrequency field strength pulse to ensure selectivity for the central transition. The phase-modulated signal requires only a complex (i.e., not hypercomplex) Fourier transformation and no additional sign discrimination method is needed.

Both the experiments in Figure 11.1(a,b) produce spectra that contain a series of ridge-like lineshapes lying along an axis $R(I, q)$, as shown in Figure 11.1(d), a ^{87}Rb (130.9 MHz) STMAS NMR spectrum of $RbNO_3$.[3] Here, there are three ridges (two closely spaced together) lying along −8/9 (i.e., $R(I=3/2,\ q=3/2)$) and corresponding to three distinct Rb species. An isotropic spectrum is obtained from a projection orthogonal to this axis. In practice this is more easily obtained by first modifying the 2D spectrum such that the ridges lie parallel to the δ_2 axis. The isotropic spectrum then results simply from a projection onto the δ_1 axis. One method for applying this transformation is shearing, which can be performed in the frequency or mixed time/frequency domains. (See Refs. 3, 13 for further details.) Alternatively, 2D spectra with ridges lying parallel to the δ_2 axis may be obtained directly, without the need for any spectral transformation, through a further modification of the STMAS experiment.[8,11] This "split-t_1" approach,[13] used previously in DAS and MQMAS, involves partitioning t_1 according to the STMAS ratio to include both satellite transition- and central transition evolution periods, ensuring the refocusing of the fourth-rank broadening always occurs at the end of t_1. Split-t_1 versions of both amplitude and phase-modulated experiments are possible. The pulse sequence for a phase-modulated split-t_1 shifted-echo experiment is shown in Figure 11.1(c).[11] The positioning of the central transition t_1 period (either before or after the final pulse) depends upon the STMAS ratio. For correlations with a negative STMAS ratio, the second part of the t_1 evolution is placed before the final pulse (i.e., $k''=0$), allowing correlation of satellite transition- and central transition coherences of the same sign. The relative duration of the two t_1 periods is determined by the magnitude of the STMAS ratio. The satellite transition evolution is set to be $t_1/(1+(R(I,q))$, and that for the central transition to be $1-t_1/(1+(R(I,q))$. Figure 11.1(e) shows a ^{87}Rb (130.9 MHz) STMAS NMR spectrum of $RbNO_3$ acquired with a split-t_1 shifted-echo experiment, with the three ridges now lying parallel to the δ_2 axis.[3]

2D STMAS spectra contain a wealth of information. In addition to an isotropic spectrum yielding information on the number of crystallographically inequivalent species, the quadrupolar information is also retained in one dimension. This information may be extracted by an analysis of the position of the center-of-mass of each lineshape in the δ_1 and δ_2 dimensions, both of which depend upon the isotropic chemical shift, δ_{CS}, and an isotropic second-order quadrupolar shift, δ_Q.[12] The latter is related to the quadrupolar product, $P_Q = C_Q(1+\eta_Q^2/3)^{1/2}$. It should be noted, however, that the exact equations used to define the position of each lineshape depend upon the way in which the chemical shift scale in the isotropic dimension is determined. There are a number of conventions that exist in the literature, some of which calculate this using the actual Larmor frequency, ω_0, others that employ a scaled Larmor frequency. Throughout this chapter, we employ the convention outlined in Ref. 3. In addition, the quadrupolar parameters, C_Q and η_Q, may be extracted individually by fitting the anisotropically broadened lineshapes taken from the 2D spectrum. For a more detailed description of spectral analysis, see Ref. 3.

11.3.2 Experimental Implementation

While similar in concept to MQMAS, the STMAS experiment is technically more demanding to implement, primarily due to the presence of the first-order quadrupolar broadening in the satellite transitions.[1–3] While, in principle, this can be removed by MAS, it is necessary to rotor-synchronize the acquisition in this indirect dimension accurately, setting the t_1 increment equal to the rotor period, τ_R (and thereby restricting the spectral width in this dimension). It has been shown, however, that for the initial t_1 duration it is essential that the finite length of the pulses at either end of t_1 are taken into account (i.e., the initial t_1 duration becomes $\tau_R - (p1/2) - (p2/2)$, rather than simply 0 or τ_R).[3] Figure 11.2(a) shows the result of neglecting this correction for a ^{87}Rb (163.5 MHz) STMAS spectrum of $RbNO_3$, where a large amount of "t_1 noise" is introduced into the spectrum, and both sensitivity and resolution are compromised. Similar problems are observed when the spinning rate is unstable,[1–3] as shown in Figure 11.2(b). In the top spectrum the spinning rate is automatically regulated and controlled by the spectrometer MAS unit (and estimated to be stable within $\pm$1 Hz). In contrast, in the bottom spectrum this regulation has been disabled, and the resulting instability (estimated to be $\pm$10 Hz) introduces considerable t_1 noise into the spectrum.

An essential prerequisite to rotor-synchronized MAS is a spinning angle, χ, accurately adjusted or "set" to the magic angle. It has been shown that to remove the large first-order broadening from the satellite transitions in STMAS an accuracy of $\pm 0.002^\circ$ is required.[1–3,8,9,14] Figure 11.2(c) shows the result of an angle misset of only 0.03° in a ^{87}Rb (130.9 MHz) STMAS NMR spectrum of $RbNO_3$.[3] The resolution

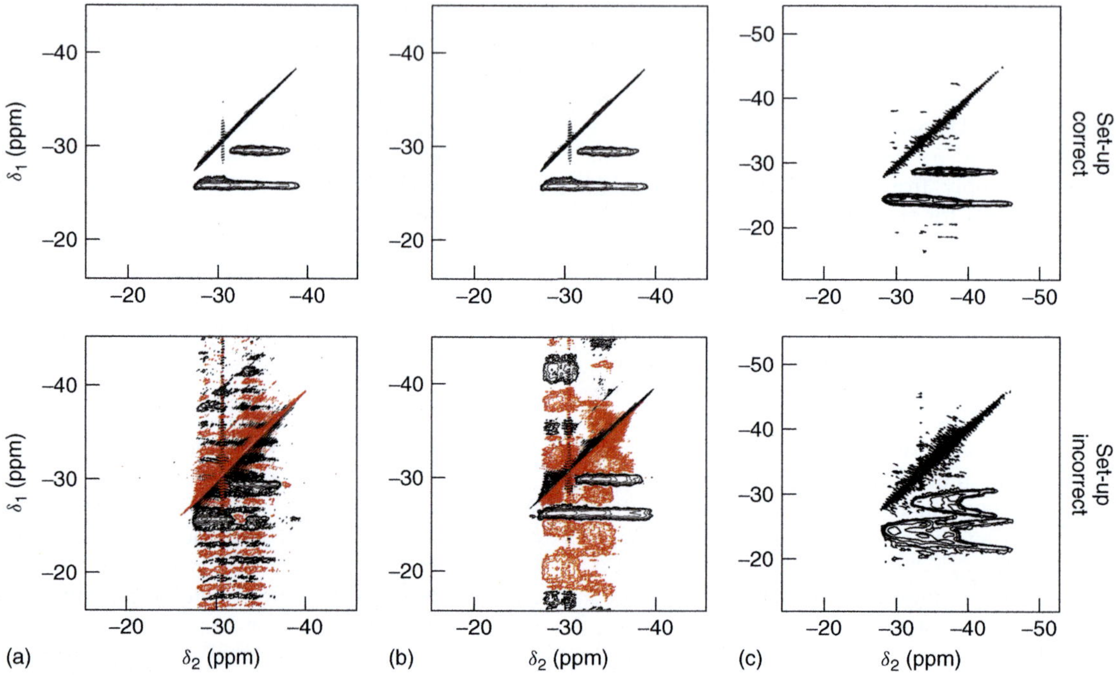

Figure 11.2. ^{87}Rb STMAS spectra of $RbNO_3$, recorded using the pulse sequence shown in Figure 11.1(c) with $k = 9/17$, $k' = 8/17$ and $k'' = 0$, at B_0 field strengths of (a, b) 11.7 T and (c) 9.4 T, with MAS rates of (a, b) 12 kHz and (c) 20 kHz. Spectra are the result of averaging 32 transients with a recycle interval of 250 ms for each of (a, b) 128 and (c) 256 t_1 increments of (a, b) 157.3 and (c) 94.4 µs. In (a), the initial t_1 duration was set to $\tau_R - (p1/2) - (p2/2) = 81.9$ µs (upper) and to $\tau_R = 83.3$ µs (lower). In (b), the automated MAS rate controller was switched on (upper) and off (lower) resulting in a spinning rate deviation estimated to be ±10 Hz. In (c), the spinning angle, χ, was estimated to be 54.736° (upper) and 54.766° (lower).[3] Positive contour levels are shown in black and negative contour levels in red.

in the isotropic spectrum has been severely compromised by the incomplete removal of the first-order quadrupolar interaction, resulting in the introduction of "Pake doublet" lineshapes. One possibility for achieving very high spinning angle accuracy is to set the spinning angle in the STMAS experiment itself by maximizing the signal intensity of the STMAS resonances using a model compound (such as $RbNO_3$)[3] with relatively fast T_1 relaxation. Care should be taken when changing rotors that the accurate setting of the spinning angle is not adversely affected. It should also be noted that the setting of the spinning angle can sometimes have significant effects upon the results of relaxation experiments (and therefore upon linewidths, spin-echo decay and inversion-recovery signals). This has been attributed to a rotational resonance recoupling interaction between the satellite transitions and central transitions in a quadrupolar spin system.[15] The effect depends upon the magnitude of the quadrupolar interaction (and therefore whether any angle misset decreases or increases their overlap). For example, a misset of 0.5° in the angle setting has been shown to lengthen the α-Al_2O_3 ^{27}Al spin-echo decay by 50% and inversion recovery by two orders of magnitude.[15] The possibility of such recoupling obviously has implications for spectral resolution in STMAS, including the ability to employ polarization transfer techniques and the time needed for data acquisition.

11.4 DOUBLE-QUANTUM FILTERED STMAS (DQF-STMAS)

The experimental STMAS spectra shown in Figures 11.1 and 11.2 have all included, in addition

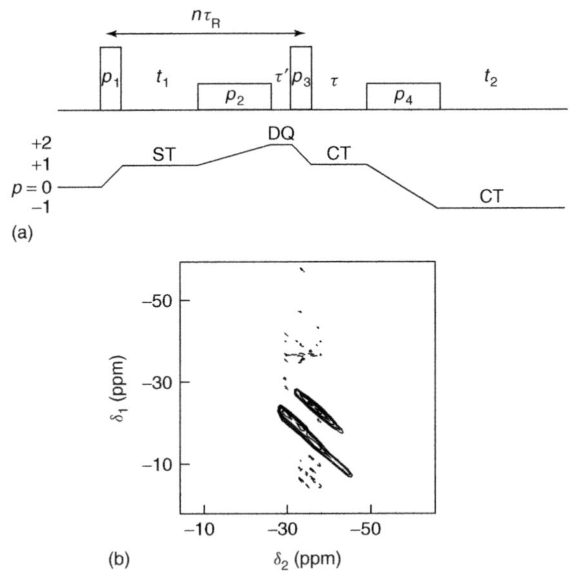

Figure 11.3. (a) Pulse sequence and coherence transfer pathway diagram for a rotor-synchronized double-quantum filtered (DQF) phase-modulated shifted-echo STMAS experiment.[16] The second and fourth pulses are applied with a low radiofrequency field strength, ensuring selectivity for the central transition. (b) ^{87}Rb (130.9 MHz) DQF-STMAS spectrum of $RbNO_3$, recorded using the pulse sequence shown in (a) with 128 transients averaged with a recycle interval of 250 ms for each of 192 t_1 increments of 50 μs. The MAS rate was 20 kHz. (Reproduced from Ref. 3. © Elsevier, 2004.)

to peaks resulting from the correlation of satellite transitions and central transitions, an autocorrelation signal resulting from the evolution of central transition coherences in both t_1 and t_2. This does not offer any additional information, but is impossible to remove by phase cycling, as it also results from single-quantum coherence evolution in t_1. In many cases, the presence of this signal does not affect the interpretation of STMAS spectra (though it can contribute unwanted intensity in the isotropic projection) but its removal would be very beneficial for optimal data acquisition. A number of approaches were employed in early STMAS experiments, including modification of the STMAS pulse durations to minimize the undesired autocorrelation signal[8] and the use of half-rotor synchronization (i.e., setting the t_1 increment to $\tau_R/2$).[11] Although some success can be achieved with these methods, it does come at the cost of sensitivity. The most successful approach employed in early experiments was the application of a long, soft pulse prior to the STMAS experiment, referred to as a "presaturation" of the central transition.[1,2,9] Experimentally, this can be achieved with a good degree of success at relatively little cost (and possibly some benefit) to sensitivity.

In 2003, Kwak and Gan introduced undoubtedly the most promising approach for the removal of unwanted autocorrelation signals, employing a double-quantum filter (DQF).[16] In this DQF-STMAS experiment (shown in Figure 11.3a)[16] an additional CT selective inversion pulse ($p2$) is applied at the end of the t_1 duration. This has the effect of converting the inner satellite transition- ($q = 3/2$ or ST_1) coherences to double-quantum coherences extremely efficiently, at almost no cost to sensitivity. The desired signal may then be selected using phase cycling, with the undesired autocorrelation signal (and any signal resulting from higher-order, $q > 3/2$, satellite transitions) removed from the spectrum. The use of a double-quantum filter, although shown implemented in Figure 11.3(a) in a phase-modulated shifted-echo experiment, is equally applicable to the amplitude-modulated z-filter and phase-modulated split-t_1 shifted-echo experiments shown in Figure 11.1. Figure 11.3(b) shows an ^{87}Rb (130.9 MHz) DQF-STMAS spectrum of $RbNO_3$,[3] recorded using the pulse sequence in Figure 11.3(a). The autocorrelation signal is effectively removed from the spectrum, thereby improving the isotropic spectrum (not shown). An additional advantage to performing the experiment in this manner is the increased ease with which optimization of the STMAS pulse durations can be performed (as the undesired autocorrelation signal no longer contributes to the FID or spectrum). A variation of this approach, where the t_1 evolution is placed after the selective inversion pulse (i.e., a DQ-STMAS experiment), has also been demonstrated.[16]

11.5 STMAS WITH SELF-COMPENSATION FOR ANGLE MISSET (SCAM-STMAS)

Despite the inherent sensitivity advantages of STMAS, the demanding requirement to remove fully the first-order quadrupolar broadening in t_1 has proved the major obstacle to its widespread implementation. In particular, the accuracy with

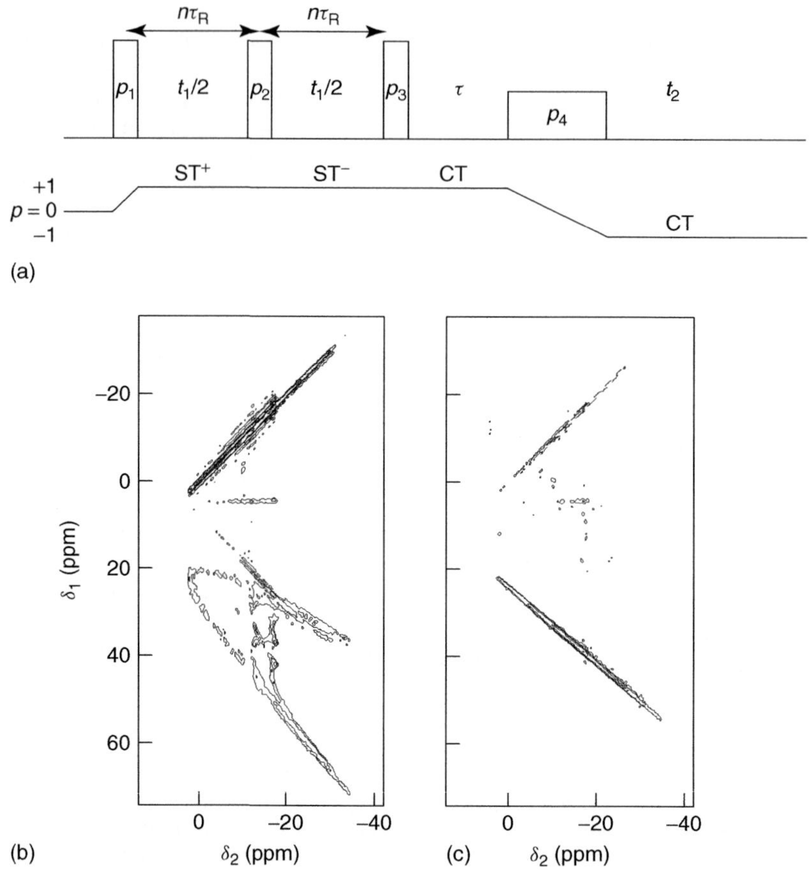

Figure 11.4. (a) Pulse sequence and coherence transfer pathway diagram for a rotor-synchronized phase-modulated shifted-echo SCAM-STMAS experiment.[17] The fourth pulse is applied with a low radiofrequency field strength, ensuring it is selective for the central transition. (b, c) ^{23}Na (105.8 MHz) STMAS and SCAM-STMAS spectra of $Na_2C_2O_4$,[18] recorded using the pulse sequence shown in (a) with (b) 96 and (d) 200 transients averaged with a recycle interval of 250 ms for each of 128 t_1 increments of 66.6 μs. The angle misset was ~0.07°. The MAS rate was 30 kHz. (Reproduced from Ref. 3. © Elsevier, 2004.)

which the spinning angle needs to be set (demonstrated in Figure 11.2c) can be a problem even for probeheads that have been modified to allow finer angle adjustment. It is possible to self-compensate missets of up to about ±1° in the magic angle setting using a novel version of the STMAS experiment, termed SCAM-STMAS.[17,18] Figure 11.4(a) shows a pulse sequence for a phase-modulated shifted-echo SCAM-STMAS experiment. The t_1 period has been split into two halves by the insertion of a pulse that induces coherence transfer between the two $q = 3/2$ satellite transitions satellite transition$^+$ ($m_I = +1/2 \leftrightarrow m_I = +3/2$) and ST$^-$ ($m_I = -1/2 \leftrightarrow m_I = -3/2$), while conserving the sign of the coherence order. As the first-order quadrupolar transitions of these two interactions are of equal magnitude but opposite in sign, the broadening is refocused at the end of the t_1 duration. Figure 11.4(b) shows a ^{23}Na (105.8 MHz) STMAS spectrum recorded using the pulse sequence in Figure 11.1(b), with a spinning angle estimated to be misset by ±0.07° from the magic angle, and displaying considerable splitting of the STMAS resonance that should lie along $R(I = 3/2,\ q = 3/2) = -8/9$.[3] (As expected the autocorrelation peak remains sharp as the central transition is unaffected by the quadrupolar interaction to first-order.) In contrast, Figure 11.4(c) shows a SCAM-STMAS spectrum, recorded using the pulse sequence in Figure 11.4(a), where the quadrupolar broadening has been refocused and a sharp ridge is obtained despite the angle misset. The addition of the extra pulse in SCAM-STMAS can result in a decrease in sensitivity relative to the conventional STMAS experiment by a factor of 2 or 3,[3,17,18] though composite pulses may be used to recover some of this loss.[18] In addition, SCAM-STMAS displays an increased tolerance to instabilities in the MAS rate, thereby reducing the amount of t_1 noise in the spectrum.[18] SCAM-STMAS is a technique best performed when the magic angle is misset, thereby reducing any additional signal from unwanted coherence transfers (i.e., those to higher-order satellite transitions). Alternatively, as with STMAS, both these unwanted signals and the autocorrelation signal arising from the correlation of central transition coherences in t_1 and t_2, can be removed by utilizing double-quantum coherences.[3,18]

11.6 STARTMAS AND OTHER "ULTRAFAST" METHODS

Although the 2D correlations employed in STMAS, MQMAS and DAS allow the complete removal of the quadrupolar broadening, such experiments can be costly in terms of experimental time, as the t_1 signal has to be acquired indirectly. In contrast, in DOR an isotropic spectrum is acquired in real time, i.e., through direct acquisition of a 1D data set. However, the need for specialist hardware (and poor filling factors, low decoupling field strengths, and limited spinning rates) have restricted the application of DOR, although the concept of real-time acquisition remains desirable. In 2006, Wimperis and coworkers introduced a new method for acquiring isotropic $I = 3/2$ NMR spectra in real time, termed STARTMAS (satellite transitions acquired in real-time MAS).[19] This approach exploits the properties of the satellite- and double-quantum transitions and their efficient interconversion through a central transition selective inversion pulse, as demonstrated by Gan in the DQF-STMAS experiment.[16]

The pulse sequence and coherence transfer pathway for STARTMAS is shown in Figure 11.5(a).[19,20] After initial excitation of satellite transition coherences, pairs of selective central transition inversion pulses then refocus the second-order quadrupolar broadening at the end of each τ period, through interconversion of satellite- and double-quantum coherences. The pulses are spaced such that the ratio of the satellite transition and double-quantum evolution is 1:8 and with each rotor-synchronized ($\tau = n\tau_R$) period ending at the midpoint of an acquisition window, where satellite transition coherences are observed. This ensures that first- and second-order quadrupolar broadening are both refocused at the same point. Data can be acquired either during the central portion of the acquisition window, or as single data points at the isotropic echoes.[19,20] The first approach produces a spectrum consisting of isotropic sidebands with intensities spread across the static ST lineshape (and a scaled spectral width reflecting the fraction of the acquisition window sampled), while the latter (the approach commonly used in multiple-pulse CRAMPS experiments) yields an isotropic spectrum. However, as has been used in other methods,[21,22] data acquired using the first method can be reordered into a 2D array, which after Fourier transformation yields a 2D spectrum similar in general appearance to an STMAS or MQMAS spectrum, but acquired in a fraction of the time (sometimes even from just a single transient). This can be seen in Figure 11.5(b) where ^{87}Rb (130.9 MHz) STARTMAS spectra of $RbNO_3$, recorded using the pulse in sequence in Figure 11.5(a), are shown. The spectra have been recorded by averaging either 1024 transients or just a single transient.

As both the double-quantum and satellite transition coherences employed in STARTMAS are affected by the first-order quadrupolar interaction, accurate rotor-synchronization is required throughout, with the technical requirements (spinning speed stability, precise magic angle setting, etc.) similar to those needed for STMAS. For more detail on the practical implementation and analysis of STARTMAS see Ref. 20. STARTMAS is applicable only to $I = 3/2$ nuclei, as nuclei with higher spin quantum numbers would require correlation of satellite transition and double-quantum coherences of opposite sign. In order to achieve good resolution (i.e., a larger number of echoes) the interconversion between single- and double-quantum coherences must be efficient. While this can be achieved for $p = -1$ to $p = -2$ transfer using a central transition inversion pulse, at present there is no equivalently efficient transfer for $p = -1$ to $p = +2$ coherences.

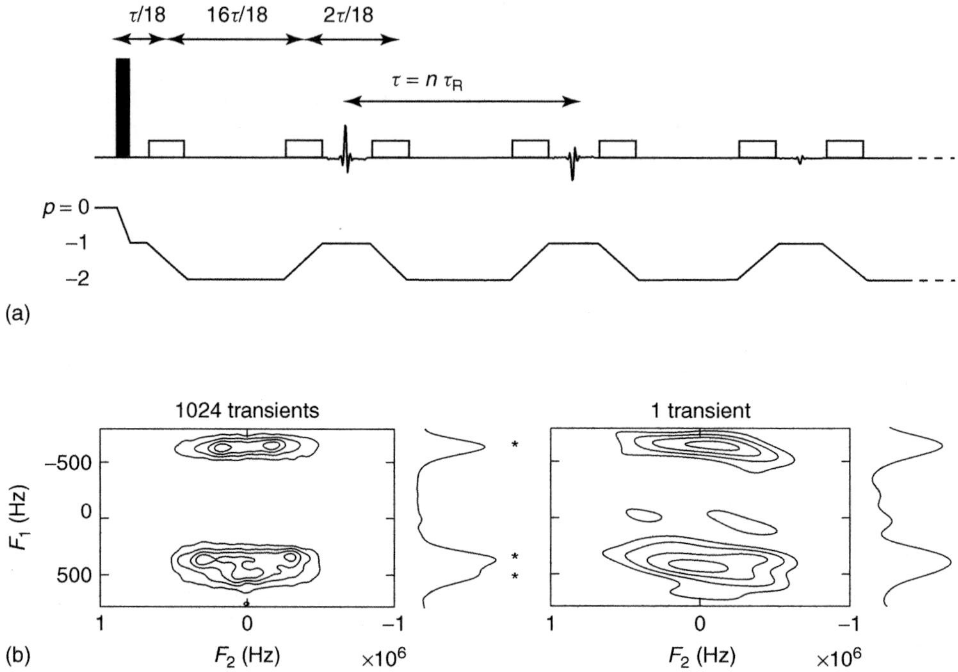

Figure 11.5. (a) Pulse sequence and coherence transfer pathway diagram for a rotor-synchronized STARTMAS experiment for spin $I = 3/2$ nuclei.[19,20] Echoes are acquired during the shorter of the interpulse "windows", as shown. (b) ^{87}Rb (130.9 MHz) STARTMAS spectra of $RbNO_3$, recorded using the pulse sequence shown in (a), with either 1024 transients or a single transient, averaged with a recycle interval of 250 ms. The MAS rate was 14 286 kHz and n (the number of rotor periods within a STARTMAS cycle) was 9. Calculated isotropic frequencies are marked with *. (Reproduced from Ref. 20. © American Institute of Physics, 2008.)

Frydman and coworkers[23] have introduced an alternative approach to the acquisition of "ultrafast" STMAS spectra. The scheme is based on the simple STMAS experiments shown in Figure 11.1 but the initial excitation pulse is replaced by a train of rotor-synchronized selective radiofrequency pulses, applied at different times and frequency offsets. Each pulse produces partial excitation of the sample and so the use of a conventional reconversion pulse results in a train of isotropic echoes from which a 2D spectrum can be reconstructed. While applicable to nuclei with all half-integer spin quantum numbers, this approach suffers from a number of disadvantages, including poor sensitivity (each of the N pulses can, at very best, excite only $1/N$ of the total magnetization) and the difficulty of achieving uniform excitation in multisite spectra where there is a wide range of quadrupolar coupling constants.

11.7 MOTIONAL BROADENING IN STMAS

The isotropic spectra produced by MQMAS and STMAS should, in principle, be very similar. However, it has been shown that STMAS spectra can show significant broadening in this high-resolution dimension, relative to their MQMAS counterparts, as a result of dynamics.[24] An example can be seen in Figure 11.6, that shows ^{27}Al (104.3 MHz) MQMAS and STMAS spectra of *ipa*-AlPO-14 (a microporous aluminophosphate synthesized with an isopropylamine template).[25] Four distinct Al ridges can be seen in the MQMAS spectrum (corresponding to one six-, one five- and two four-coordinated Al sites), and four narrow peaks are resolved in the isotropic spectrum. In contrast, all resonances in the STMAS spectrum display considerable broadening, resulting from the motion

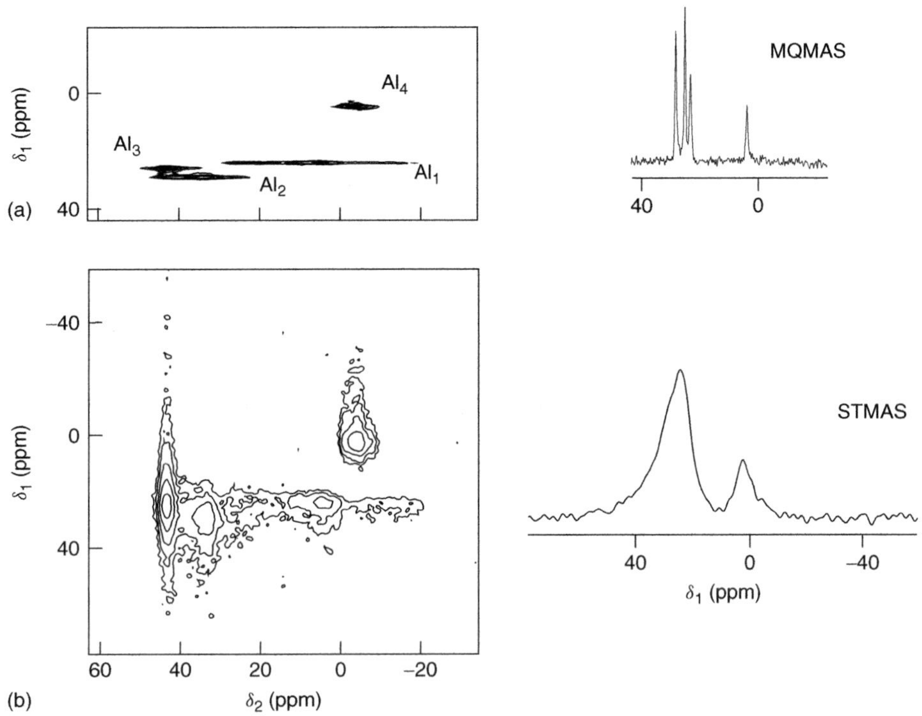

Figure 11.6. 2D ^{27}Al (104.3 MHz) (a) MQMAS and (b) STMAS NMR spectra and corresponding isotropic projections of *ipa*-AlPO-14,[25] recorded using phase-modulated split-t_1 shifted-echo (a) triple-quantum MAS and (b) DQF-STMAS pulse sequences. Spectra are the result of averaging (a) 960 and (b) 1024 transients with a recycle interval of 500 ms for each of (a) 48 and (b) 120 t_1 increments of (a) 64.58 μs and (b) 129.16 μs. The MAS rate was 20 kHz.

of the template and water molecules incorporated into the pores.

In an MQMAS experiment the two transitions correlated have no dependence upon the first-order quadrupolar interaction. Hence, for MAS rates that are fast compared to the smaller second-order broadening, lines will be strongly narrowed and any reorientation of the quadrupole tensor during the rotor period will have little effect upon the spectrum.[24–26] In contrast, in STMAS, the MAS rate is much smaller than the first-order broadening present in the satellite transitions, resulting in a FID that consists of a series of rotary echoes. Consequently, any motion producing a change of the quadrupole tensor will interfere with the formation of the echo and result in a spectral broadening. This broadening will depend (among other things) upon the magnitude of the quadrupolar interaction, the rate constant for the motional process, and the MAS rate.[24–26] Maximum broadening is expected when the rate constant is on the order of the quadrupolar coupling, i.e., typically on the microsecond timescale. It should be noted this effect results not necessarily from direct motion of the nuclei observed by NMR, but from a change in their local environment, i.e., a change in the magnitude or orientation of the interaction tensors, that could be caused by motion elsewhere. For example, the broadening in Figure 11.6 results from the dynamics of the guest species within the pores of the microporous framework (and perhaps compensating changes in the framework).[25] No such broadening is observed in the STMAS spectrum of the calcined material, where both template and water molecules are removed from the structure. Similarly, it has been demonstrated that ^{17}O STMAS NMR spectra of hydrous minerals can be broadened by microsecond timescale dynamics of the nearby ^{1}H nuclei.[24]

While the presence of motional broadening can be viewed as a problem in STMAS with the corresponding implications for spectral sensitivity and resolution, it does, however, (by comparison with MQMAS and utilizing variable-temperature NMR) offer a sensitive, high-resolution probe of motion in solids with great promise for obtaining novel structural information.

11.8 HIGHER-ORDER INTERACTIONS IN STMAS

STMAS enables the complete removal of second-order quadrupolar broadening and therefore (if we assume the smaller dipolar and CSA interactions can be removed by rapid MAS) should result in a spectrum that, in principle, is isotropic. However, while the dipolar and CSA Hamiltonians, H_D and H_{CSA}, yield interactions that are small enough that second-order autocorrelation terms are negligible, they can cross-correlate with the strong quadrupolar hamiltonian, H_Q, giving rise to additional anisotropic broadenings (or "cross-terms") that are second-order in nature and therefore not removed by MAS.[27] While these interactions can be hard to detect (being usually much smaller than the second-order quadrupolar interaction) they can sometimes be observed in MQMAS/STMAS spectra where this interaction is removed.

Anisotropic broadening arising from a correlation between the quadrupolar, $H_Q{}^S$, and dipolar Hamiltonians, $H_D{}^{IS}$, is well known in MAS NMR of $I = 1/2$ nuclei (e.g., the effect on ^{13}C NMR spectra of a dipolar coupling to a nearby ^{14}N ($S = 1$) nucleus[27]), and has also been observed in MQMAS spectra of quadrupolar nuclei that have a significant dipolar coupling to another quadrupolar nucleus.[28,29] In STMAS experiments, with an observed spin I, heteronuclear dipolar coupling to a spin $S = 1/2$ nucleus results in a further quadrupolar-dipolar cross-term, proportional to $H_Q{}^I H_D{}^{IS}$, that yields an anisotropic broadening of the satellite transitions proportional to $(\omega_Q{}^{I,\,PAS}\,\omega_D{}^{IS}/\omega_0{}^I)$.[3] Cross-terms of this type do not affect the central transition or the symmetrical multiple-quantum transitions and hence are not observed in MQMAS. Even, in STMAS, however, they are predicted to be of negligible importance, unless a particularly large dipolar interaction is present, as large values of $\omega_Q{}^{I,\,PAS}$ would also result in difficulties in spectral acquisition. Although predicted theoretically, no such cross-term has yet been observed experimentally.

Cross-terms resulting from the coupling of the quadrupolar interaction with the CSA are also observable in STMAS.[3,30] These are proportional to $H_Q{}^I H_{CSA}{}^I$ and result in an anisotropic broadening proportional to $\omega_Q{}^{I,\,PAS}\,\omega_{CSA}{}^I$, that, interestingly, is independent of B_0. These terms affect only the satellite transitions (i.e., not the central or

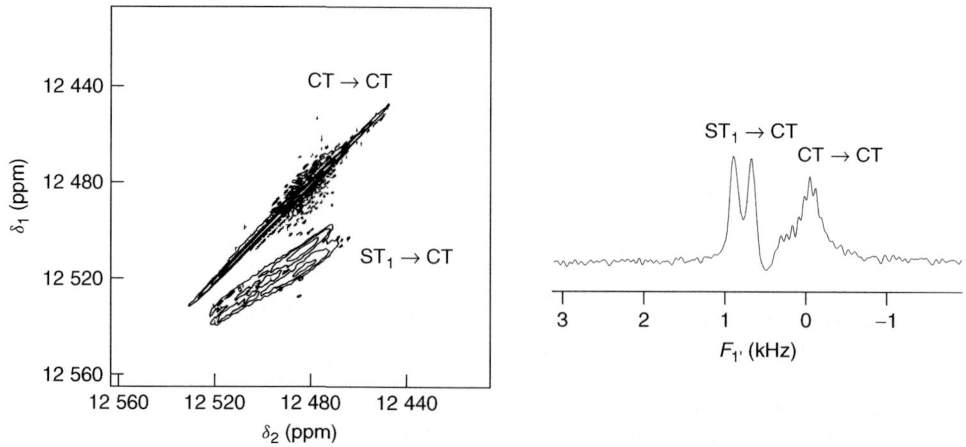

Figure 11.7. 2D ^{59}Co (47.2 MHz) STMAS NMR spectrum and corresponding isotropic projection of Co(acac)$_3$,[3,30] recorded using a phase-modulated shifted-echo pulse sequence (not split-t_1). The spectrum is the result of averaging 960 transients with a recycle interval of 1 s for each of 96 t_1 increments of 100 μs. The MAS rate was 10 kHz. (Reproduced from Ref. 3. © Elsevier, 2004.)

multiple-quantum transitions) and hence are observable only in STMAS spectra (if the CSA is of sufficient magnitude) and not in MAS or MQMAS spectra. Figure 11.7 shows a ^{59}Co ($I = 7/2$) STMAS spectrum of cobalt acetylacetonate (Co(acac)$_3$), where the large CSA (~700 ppm) of the single Co species results in a significant splitting of the STMAS resonance.[30]

STMAS spectra can also exhibit additional anisotropic broadening when the quadrupolar interaction is large, resulting from a third-order perturbation of the Zeeman energy levels.[31] This perturbation is similar in nature to the first-order quadrupolar interaction affecting the satellite transitions but neither the central nor symmetric multiple-quantum transitions. Hence, it can be observed in STMAS but not in MQMAS experiments. The interaction consists of an isotropic component and second-, fourth- and sixth-rank anisotropic broadening and, therefore, is removed neither by MAS nor by 2D STMAS correlation. The broadening is proportional to $(\omega_Q^{PAS})^3/\omega_0^2$ and so is smaller than the first- and second-order interactions, but has been shown to result in a significant splitting of ~1 kHz in the isotropic ^{27}Al STMAS spectrum ($\omega_0/2\pi = 104.3$ MHz) of andalusite (Al_2SiO_5), where the quadrupolar coupling, C_Q, for one of the Al species is 15.3 MHz ($\omega_Q^{PAS}/2\pi = 1.15$ MHz).[3,31] The interaction does show a strong dependence upon the spin quantum number, with the splitting observed for $I = 3/2$ nuclei considerably smaller (e.g., the splitting in the ^{87}Rb ($\omega_0/2\pi = 130.9$ MHz) STMAS spectrum of $RbSO_4$ ($\omega_Q^{PAS}/2\pi = 1.33$ MHz) is predicted to be only 40 Hz).[3]

It should also be noted that both the third-order interaction and the quadrupolar-CSA cross-term for the two $q = 3/2$ satellite transitions (ST^+ and ST^-) have equal magnitude but opposite sign. They are, therefore, refocused by SCAM-STMAS and an isotropic spectrum is obtained once more.[3]

11.9 APPLICATIONS OF STMAS

STMAS offers an alternative approach to MQMAS, DAS and DOR for the acquisition of high-resolution NMR spectra of quadrupolar solids that, although

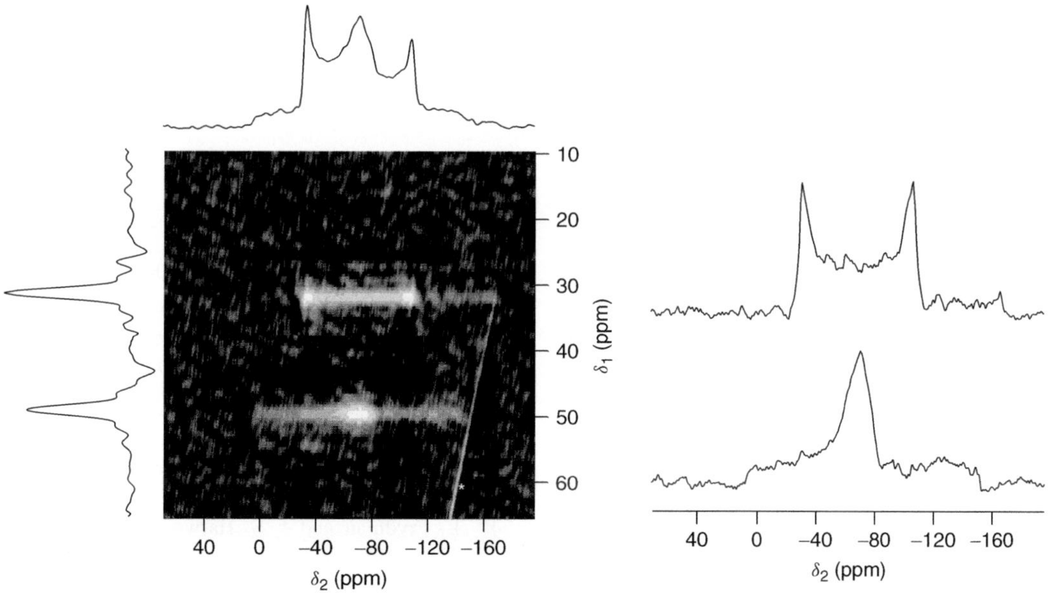

Figure 11.8. 2D ^{39}K (18.7 MHz) STMAS NMR spectrum of K_2SO_4 (recorded using a phase-modulated shifted-echo pulse sequence), with corresponding isotropic projection and δ_2 cross-sections extracted along the ridge lineshapes.[32] The asterisk marks the (aliased) CT → CT ridge. The spectrum is the result of averaging 704 transients with a recycle interval of 2 s for each of 96 t_1 increments of 269.8 μs. The MAS rate was 7 kHz. (Reproduced from Ref. 33. © American Chemical Society, 2003.)

technically more demanding to implement than MQMAS, can be performed using a conventional MAS probehead. Although similar in many ways to MQMAS, it has been shown above that there are significant consequences of utilizing the satellite transitions (including an increased sensitivity to motion on the microsecond timescale and the ability to observe higher-order interactions, such as cross-terms). Of all the possible applications of STMAS, perhaps the greatest area of interest is the exploitation of the intrinsic sensitivity of the technique, that is typically a factor of 2–8 times larger than for MQMAS. In addition, the decrease in sensitivity with increasing MAS rate observed for MQMAS is not apparent in STMAS, where the sensitivity appears to be independent of spinning rate.[8] STMAS has been utilized for the study of "low-γ" nuclei, where the inherently poor sensitivity, large second-order broadenings, and need for fast MAS rates limit the application of MQMAS. In a recent example it has been shown that it is possible to obtain ^{25}Mg (at natural abundance) and ^{39}K STMAS spectra, at a static field strength of only $B_0 = 9.4\,T$.[32] Figure 11.8 shows a 2D ^{39}K (18.7 MHz) STMAS spectrum of K_2SO_4, showing resolution of the two crystallographically distinct K species. In addition, ^{17}O STMAS has also been employed to study high-pressure silicate minerals (phases of Mg_2SiO_4 and $MgSiO_3$)[33,34] where the difficult synthetic route results in only small quantities (5–10 mg) of material, limiting sensitivity.

Despite possessing relatively good inherent sensitivity there has also been recent work aimed at increasing the sensitivity of STMAS, using composite pulses to improve coherence transfer efficiency. A so-called SPAM pulse[35] has been used to improve the transfer of double-quantum to central transition coherences in DQF-STMAS,[36,37] while it has also been shown that the rival FAM pulses,[38] widely used in MQMAS, can also yield a small improvement in the efficiency of satellite transition to central transition coherence transfer in a simple STMAS experiment.[39]

11.10 CONCLUSIONS

Compared with MQMAS, STMAS provides an alternative approach to the acquisition of high-resolution NMR spectra of quadrupolar nuclei and recent improvements (such as DQF-STMAS) have made its implementation much easier. Although, STMAS spectra are generally very similar to MQMAS spectra, it can be shown that they are affected by higher-order interactions (such as cross-terms) that can provide information on interactions usually removed by MAS (e.g., dipolar coupling and CSA). The sensitivity of STMAS to motion on the microsecond timescale also provides (in conjunction with MQMAS or variable-temperature NMR) a sensitive probe of molecular-scale dynamics. The greatest advantage of the STMAS method, however, lies in the improved sensitivity it offers, with the potential to study small sample volumes and nuclei with low γ, low natural abundance or in low concentrations, widening the possible applications of high-resolution methods for quadrupolar nuclei.

RELATED ARTICLES IN THE ENCYCLOPEDIA OF MAGNETIC RESONANCE

Multiple-quantum Magic-angle Spinning Experiments on Half-integer Nuclei: Fundamentals

Satellite Transition NMR Spectroscopy of Half-Integer Quadrupolar Nuclei under Magic-angle Spinning

REFERENCES

1. Z. Gan, *J. Am. Chem. Soc.*, 2000, **122**, 3242.
2. Z. Gan, Satellite transition NMR spectroscopy of half-integer quadrupolar nuclei under magic-angle spinning, *Encyclopedia of Magnetic* , eds D.M. Grant and R.K. Harris, John Wiley & Sons: Chichester, 1996. DOI: 10.1002/9780470034590.emrstm0481.
3. S. E. Ashbrook and S. Wimperis, *Prog. NMR. Spectrosc.*, 2004, **45**, 53.
4. L. Frydman and J. S. Harwood, *J. Am. Chem. Soc.*, 1995, **117**, 5367.
5. A. Llor and J. Virlet, *Chem. Phys. Lett.*, 1988, **152**, 248.
6. A. Samoson, E. Lippmaa, and A. Pines, *Mol. Phys.*, 1988, **65**, 1013.
7. R. N. Zare, *Angular Momentum*, Wiley: Chichester, 1996.

8. S. E. Ashbrook and S. Wimperis, *J. Magn. Reson.*, 2002, **156**, 269.

9. Z. Gan, *J. Chem. Phys.*, 2001, **114**, 10845.

10. R. R. Ernst, G. Bodenhausen, and A. Wokaun, *Principles of Nuclear Magnetic Resonance in One and Two Dimensions*, Clarendon Press: Oxford, 1987.

11. K. J. Pike, S. E. Ashbrook, and S. Wimperis, *Chem. Phys. Lett.*, 2001, **345**, 400.

12. D. Massiot, B. Touzo, D. Trumeau, J. P. Coutures, J. Virlet, P. Florian, and P. J. Grandinetti, *Solid State Nucl. Magn. Reson.*, 1996, **6**, 73.

13. S. P. Brown and S. Wimperis, *J. Magn. Reson.*, 1997, **128**, 42.

14. C. Huguenard, F. Taulelle, B. Knott, and Z. Gan, *J. Magn. Reson.*, 2002, **156**, 131.

15. H. T. Kwak, P. Srinivasan, J. Quine, D. Massiot, and Z. Gan, *Chem. Phys. Lett.*, 2003, **376**, 75.

16. H. T. Kwak and Z. Gan, *J. Magn. Reson.*, 2003, **164**, 369.

17. S. E. Ashbrook and S. Wimperis, *J. Am. Chem. Soc.*, 2002, **124**, 11602.

18. S. E. Ashbrook and S. Wimperis, *J. Magn. Reson.*, 2003, **162**, 402.

19. M. J. Thrippleton, T. J. Ball, S. Steuernagel, S. E. Ashbrook, and S. Wimperis, *Chem. Phys. Lett.*, 2006, **431**, 390.

20. M. J. Thrippleton, T. J. Ball, and S. Wimperis, *J. Chem. Phys.*, 2008, **128**, 034507.

21. A. Bax, A. F. Mehlkopf, and J. Smidt, *J. Magn. Reson.*, 1980, **40**, 213.

22. B. Blümich, P. Blümler, and J. Jansen, *Solid State Nucl. Magn. Reson.*, 1992, **1**, 111.

23. R. Bhattacharya and L. Frydman, *J. Am. Chem. Soc.*, 2006, **128**, 16015.

24. S. E. Ashbrook, S. Antonijevic, A. J. Berry, and S. Wimperis, *Chem. Phys. Lett.*, 2002, **364**, 634.

25. S. Antonijevic, S. E. Ashbrook, S. Biedasek, R. I. Walton, S. Wimperis, and H. Yang, *J. Am. Chem. Soc.*, 2006, **128**, 8054.

26. M. J. Thrippleton, M. Cutajar, and S. Wimperis, *Chem. Phys. Lett.*, 2008, **452**, 233.

27. R. K. Harris, Magic angle spinning: Effects of quadrupolar nuclei on spin-1/2 spectra, in *Encyclopedia of Magnetic Resonance*, eds R. K. Harris and R. E. Wasylishen, John Wiley & Sons: Chichester, 1996. DOI: 10.1002/9780470034590.emrstm0285.

28. G. Wu and R. E. Wasylishen, *Mol. Phys.*, 1999, **313**, 1177.

29. J. McManus, R. Kemp-Harper, and S. Wimperis, *Chem. Phys. Lett.*, 1999, **311**, 292.

30. S. Wi, S. E. Ashbrook, S. Wimperis, and L. Frydman, *J. Chem. Phys.*, 2003, **118**, 3131.

31. Z. Gan, P. Srinivasan, J. R. Quine, S. Steuernagel, and B. Knott, *Chem. Phys. Lett.*, 2003, **367**, 163.

32. N. G. Dowell, S. E. Ashbrook, and S. Wimperis, *J. Chem. Phys. B*, 2004, **108**, 13292.

33. S. E. Ashbrook, A. J. Berry, W. O. Hibberson, S. Steuernagel, and S. Wimperis, *J. Am. Chem. Soc.*, 2003, **125**, 11824.

34. S. E. Ashbrook, A. J. Berry, D. J. Frost, A. Gregorovic, C. J. Pickard, J. E. Readman, and S. Wimperis, *J. Am. Chem. Soc.*, 2007, **129**, 13213.

35. Z. Gan and H. T. Kwak, *J. Magn. Reson.*, 2004, **168**, 346.

36. J. P. Amoureux, A. Flambard, and L. Delevoye, *Chem. Commun.*, 2005, **27**, 3472.

37. J. P. Amoureux, L. Delevoye, G. Fink, F. Taullele, A. Flambard, and L. Montagne, *J. Magn. Reson.*, 2005, **175**, 285.

38. A. Goldbourt, P. K. Madhu, and S. Vega, *Chem. Phys. Lett.*, 2000, **320**, 448.

39. T. J. Ball and S. Wimperis, *J. Magn. Reson.*, 2007, **187**, 343.

[illegible] J. Magn. Reson., 2007, 188, [illegible]

[illegible] J. Phys. Chem. B, 2005, 109, [illegible]

36. J. [illegible] Thrippleton, [illegible] and [illegible]

[illegible]

Chapter 12

Correlation Experiments Involving Half-Integer Quadrupolar Nuclei

Michael Deschamps[1] and Dominique Massiot[2]

[1]*Département de Chimie, Université d'Orléans, BP 6759, 1 Rue de Chartres, 45067 Orléans cedex 2, France*
[2]*CNRS-CEMHTI, Site Hautes Températures, 1D Avenue de la Recherche Scientifique, 45071 Orléans cedex 2, France*

12.1 INTRODUCTION

NMR experiments on quadrupolar nuclei ($S > 1/2$) have to take into account the presence of the quadrupolar interaction which arises from the interaction between their quadrupolar moment with the electric field gradient (EFG) at the position of the nucleus. This interaction, which splits the signal of the quadrupolar nucleus into $2S$ single-quantum transitions, can considerably broaden their NMR lines (up to several megahertz or more) because of second-order field-dependent contributions.

Several methodological developments[1] have opened access to the exploitation of quadrupolar nuclei in NMR experiments.

- The development of high to very high field NMR magnets makes the lines narrower by reducing the second-order quadrupolar broadening, which considerably simplifies the analysis of the corresponding spectra.[2]
- The development of MAS probes with faster spinning speeds has also allowed the spinning sidebands to be separated even further.
- The DOR,[3] DAS,[4] MQ-MAS,[5] or satellite transition-MAS (ST-MAS)[6] experiments[7] allow isotropic resolution to be obtained in 1D or 2D experiments, even at moderate fields, thanks to complex mechanical (DOR, DAS) or spin system (DAS, MQ-MAS, ST-MAS) manipulations, opening the way to enhanced resolution and secure interpretation of the resulting spectra.
- Saturation or inversion of the satellite transitions has led to developments such as rotor assisted population transfer (RAPT),[8] double frequency sweeps (DFS),[9] or inversion of the satellite transition with adiabatic

NMR of Quadrupolar Nuclei in Solid Materials
Edited by Roderick E. Wasylishen, Sharon E. Ashbrook and Stephen Wimperis

ISBN: 978-0-470-97398-1

pulses[10] for the enhancement of the central transition intensity by up to a factor of $S + 1/2$, which can be coupled to quadrupolar Carr-Purcell-Meiboom-Gill (QCPMG)[11] for acquiring spectra of low-sensitivity nuclei.

As most of the spin-bearing nuclei are quadrupolar nuclei, a large number of NMR pulse sequences such as the cross polarization (CP) or heteronuclear multiple-quantum coherence (HMQC) experiments, which aim to correlate quadrupolar nuclei to other nuclei, have been developed over the years. They take advantage of either the J-coupling for through-bond magnetization transfer or the dipolar coupling for the characterization of spatial proximities, and usually can be combined with a high-resolution or enhanced-sensitivity toolbox such as MQ-MAS or DOR (which has been combined with spin diffusion, for example[12]).

12.2 THE SPECIAL CASE OF QUADRUPOLAR NUCLEI

As a single quadrupolar nucleus is intrinsically a multilevel spin system with $2S + 1$ energy levels, the single-quantum spectrum will exhibit $2S$ directly observable transitions and a number of indirectly observed $|-s\rangle\langle+s|$ multiple-quanta transitions (up to $2S$) which can be generated using different excitation schemes. In fact, if we go through the literature, we can see that only a limited number of possible experiments have already been explored and that the complexity of the quadrupolar spin system may well still be a source of new types of experiments in the years to come.

As the quadrupolar interaction can be very significant (up to several megahertz or more) compared to the Larmor frequency, it gives rise not only to a first-order field-independent perturbation (which is the usual case for chemical shift anisotropy, dipolar or indirect J-coupling) but also to a second- or even a third-order field-dependent perturbation that can be seen in MAS, DOR, DAS, MQ-MAS (second order), or ST-MAS (second and third order[13]), and to significant cross terms with other interactions. In very general terms, odd-order terms only affect nonsymmetric transitions (different from $|-m\rangle\langle+m|$), while even-order perturbation terms affect all transitions. As a consequence, a distinction can be made between quadrupolar nuclei with integer spins, where all the single-quantum transitions are affected by the first-order quadrupolar interaction, and half-integer quadrupolar spins, where the central $|-1/2\rangle\langle+1/2|$ single-quantum transition remains unaffected by the first-order quadrupolar interaction, as opposed to the other single-quantum satellite transitions. Since the first-order quadrupolar broadening usually exceeds even the highest available MAS rates, the central transition in half-integer spins is often the preferred transition for observation in an NMR experiment, while the satellite transitions give rise to a spinning sideband manifold, the span of which can often exceed the excitation and detection bandwidths of the NMR probeheads.

Because the quadrupolar interaction can range from far greater to much smaller than the radiofrequency (RF) field strength ω_1 used for the excitation of the spin system, there exist multiple possible interference regimes between the RF field and the quadrupolar interaction itself. The typical case of a half-integer nuclear spin is shown in Figure 12.1,[14] which illustrates the dependence of the nutation regime of the central transition as a function of the strength of the quadrupolar interaction for a given RF field strength. One distinguishes the nonselective excitation regime of the central transition when the RF field dominates a small quadrupolar interaction and a selective excitation regime of the central transition when the quadrupolar interaction dominates the applied RF power, leading to a faster nutation (by a factor $S + 1/2$). In the selective excitation regime, the central transition can be considered as *isolated* and manipulated as a *fictitious* $S = 1/2$ *spin*. This behavior of the central transition can be easily achieved, as soon as the quadrupolar interaction is significant, by lowering the RF power. In between these two extreme regimes lies the intermediate regime, in which the magnetization is also converted into multiple-quantum coherences, allowing, for example, the realization of MQ-MAS experiments but making it difficult or impractical to spin-lock any transition under these conditions.

Finally, and despite a common belief, quadrupolar nuclei can exhibit significantly long to very long T_1 (seconds) and T_2 (hundreds of milliseconds) relaxation times, depending upon the spin system and the experimental conditions, opening the

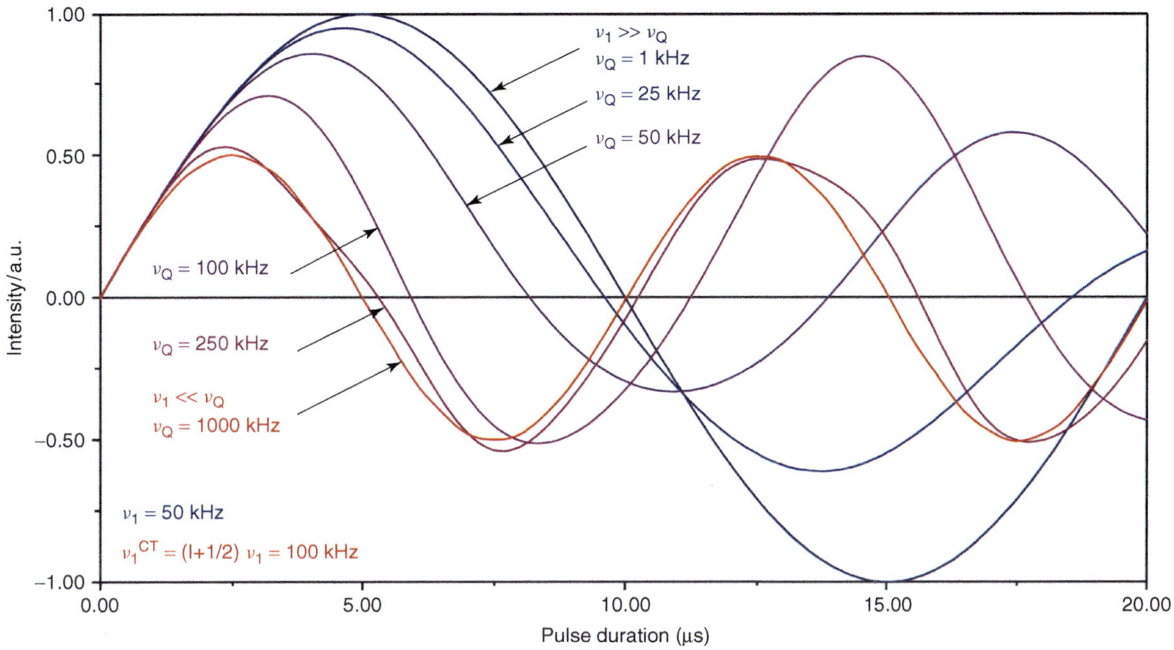

Figure 12.1. Nutation curve for an $S = 3/2$ quadrupolar nucleus, using an RF field strength $\nu_1 = 50$ kHz and various quadrupolar couplings ν_Q with $\eta_Q = 0$.

possibility to perform complex encoding and decoding experiments using small interactions for magnetization transfers.

The ability to select central or satellite single-quantum or multiple-quantum transitions by tuning the RF field or the mechanical manipulation of the sample under MAS or off-MAS provides many possibilities for resolution and sensitivity enhancement when one deals with quadrupolar nuclei. Moreover, long evolution times may be possible to build homonuclear or heteronuclear coherences to probe the chemical connectivity or the spatial proximity in correlation experiments. In this chapter, we will focus on half-integer quadrupolar nuclei, as these represent most of the actual cases in materials science and inorganic chemistry and allow resolution to be obtained from the manipulation of the central transition. It is nevertheless worth mentioning that there exist a number of specific method developments in the case of limited quadrupolar couplings, e.g., for ^{2}H,[15] or of large quadrupolar couplings, e.g., for ^{14}N.[16]

12.3 J-COUPLING AND RESIDUAL SPLITTING

12.3.1 J-Couplings with Quadrupolar Nuclei

J-Couplings occur when two nuclei belong to the same bonding network and are usually detected when the distances between the nuclei do not exceed a few chemical bonds. It must be noted that the J-couplings usually depend upon the bonding network geometry, and that, although J-couplings can be used to characterize chemical bonding differences, the differences in J-coupling values strongly impact the magnetization transfer in correlation experiments (Figure 12.2).[17]

When present, a small but significant J-coupling has very contrasting effects depending on whether one observes the quadrupolar nucleus directly or the nucleus to which it is coupled. In a 1D experiment observing a quadrupolar nucleus, the weak J-coupling usually remains barely detectable in the presence of the strong second-order quadrupolar broadening, while the spectrum of a nucleus coupled to a quadrupolar nucleus will be split into a

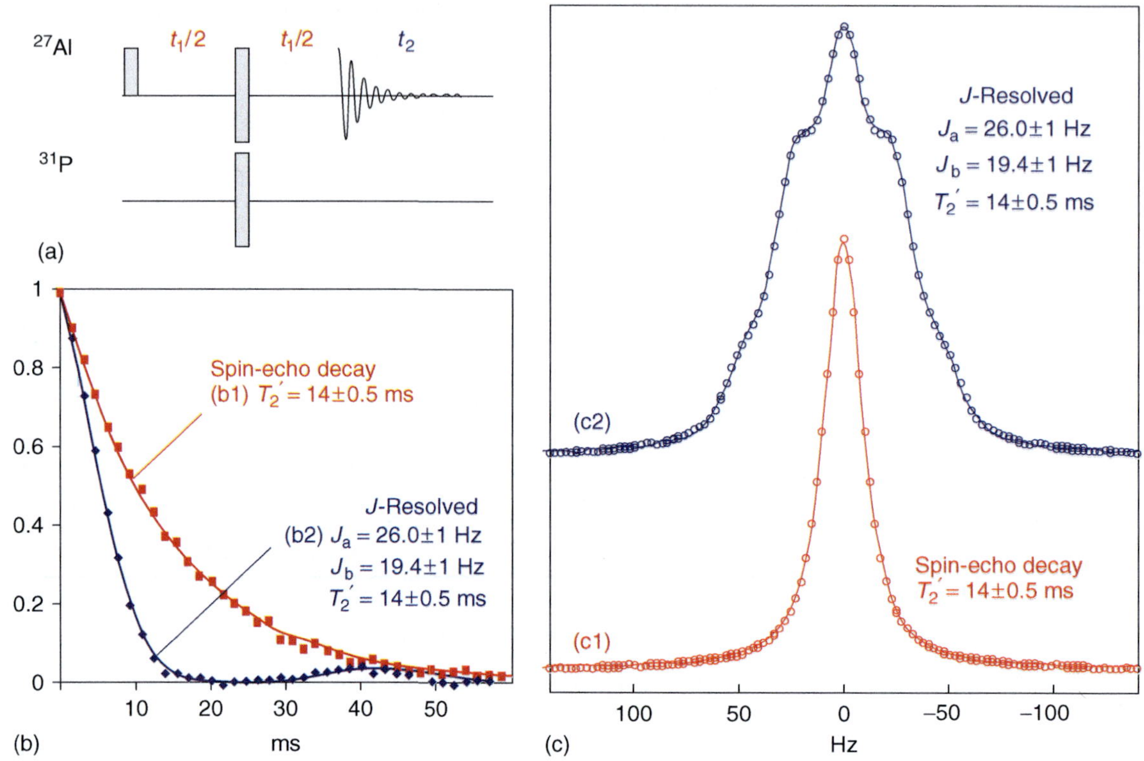

Figure 12.2. $\{^{31}P\}^{27}Al$ J-resolved experiment in $AlPO_4$ berlinite, acquired with an MAS speed $\nu_R = 13$ kHz. (a) Pulse sequence for the J-resolved experiments where the 90° and 180° pulses on the ^{27}Al channel are applied selectively on the central transition (using low RF power). (b) Time domain and (c) frequency domain of the ^{27}Al spin-echo experiment, i.e., with no 180° pulse on the ^{31}P channel (red) and of the J-resolved experiment (blue) and the corresponding fitted curves. (Reproduced from Ref. 17, © Elsevier, 2003.)

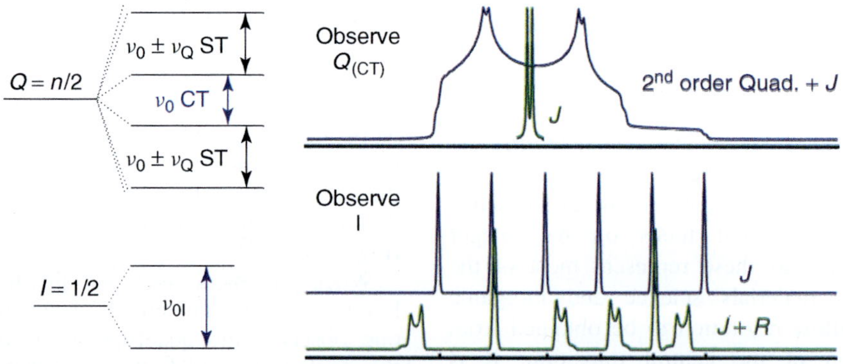

Figure 12.3. Schematic representation of the effect of the J-coupling between a quadrupolar nucleus Q ($S = 5/2$) and a dipolar nucleus I ($I = 1/2$). The central transition of the quadrupolar nucleus Q (top, in blue) is broadened to second order by the quadrupolar interaction and split into two by the J-coupling to I (in green). The signal of the nucleus I is split into a $2S + 1$ (e.g., 6 for $S = 5/2$) multiplet. Additional second-order broadenings and shifts arise when R (the dipolar quadrupolar cross term) becomes significant. (Reproduced from Ref. 19, © Elsevier, 2010.)

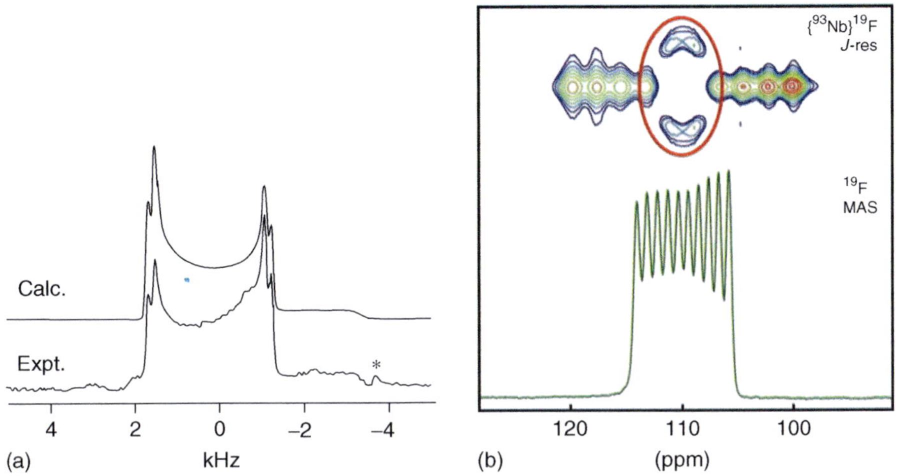

Figure 12.4. (a) Experimental and simulated ^{17}O NMR spectra of triphenylphosphine oxide ^{17}O (37.5%) acquired at $B_0 = 9.4$ T and $\nu_R = 5.2$ kHz and showing the central transition second-order quadrupolar lineshape with a splitting arising from the J-coupling between ^{17}O ($S = 5/2$) and ^{31}P ($I = 1/2$). The asterisk indicates a spinning sideband. (b) 1D ^{19}F MAS and 2D central transition-selective $\{^{93}Nb\}^{19}F$ J-resolved NMR spectra of $CsNbF_6$ showing that only the two centerbands corresponding to the $\langle \pm 1/2 \rangle$ energy levels are manipulated during the 2D J-resolved experiment (on top). The 1D spectrum shows 10 lines stemming from the J-coupling between the ^{19}F spin ($I = 1/2$) and the neighboring ^{93}Nb, which is an $S = 9/2$ spin. The experiment was recorded at $\nu_R = 33$ kHz and $B_0 = 9.4$ T. (Reproduced from Ref. 20, © American Chemical Society, 2003, and from Ref. 19, © Elsevier 2010.)

$2S + 1$ multiplet, eventually shifted and broadened by second-order cross terms (Figure 12.3). The presence of the satellite transitions, which are more difficult to manipulate with RF pulses because they typically span a large frequency range, makes any J-coupling involving a quadrupolar spin more difficult to use in correlation experiments, as the J-couplings to the external spin levels ($m \neq \pm 1/2$) often cannot be refocused by applying RF pulses on the quadrupolar spin. Fortunately, even if these J-couplings are small and may not be detected in the 1D NMR spectrum (as in Figures 12.3 and 12.4),[19,20] they can often be measured in echo-based J-resolved experiments and further used for magnetization transfer.[19]

Measuring the J-couplings can also be difficult in systems with abundant spins such as aluminophosphates where each ^{27}Al ($S = 5/2$) or ^{31}P ($S = 1/2$) atom, both 100% natural-abundance nuclei, is bonded to several (often upto four or six) other spin-bearing atoms via bridging oxygen atoms. In such a case, a selective excitation of the coupled nucleus can allow separate measurements of individual J-couplings of selected pairs of coupled atoms (Figure 12.5).[21]

In nonabundant systems, the HMQC[18] experiment can be used as a spectral editing tool to isolate a pair of coupled spins and observe the doublet stemming from the J-coupling with the selectively excited central transition leading to accurate measurement of the coupling and enhanced resolution (Figures 12.4b and 12.6).[22]

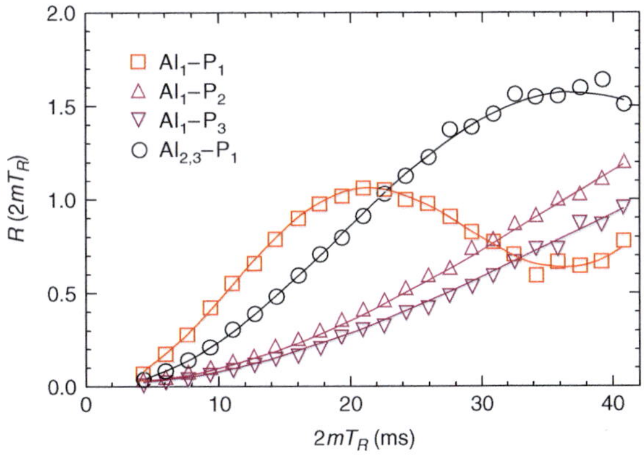

Figure 12.5. Plots and fits of the frequency-selective J-resolved build-up curves measured in $AlPO_4$ VPI-5 with J-values ranging between 15 and 30 Hz. (Reproduced from Ref. 21, © Elsevier, 2005.)

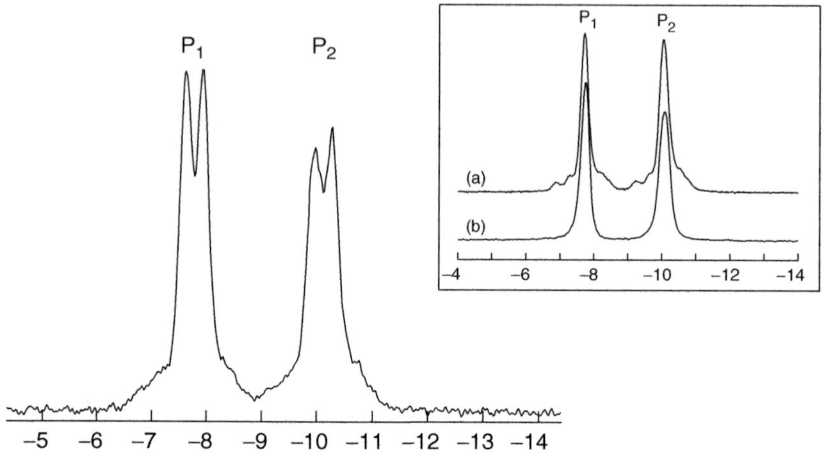

Figure 12.6. $^{31}P\{^{17}O\}$ HMQC spectrum showing the doublet stemming from the J-coupling between the two ^{31}P sites and the $\pm 1/2$ states of ^{17}O ($S = 5/2$) in 30% ^{17}O-enriched α-$Ca_2P_2O_7$ at 17.6 T magnetic field and 14 kHz MAS. Residual signals from ^{31}P spins coupled to two ^{17}O are also detected. (a) The MAS-NMR 1D spectrum. (b) The same NMR spectrum obtained in nonenriched α-$Ca_2P_2O_7$ (no J-coupling). (Reproduced from Ref. 22.)

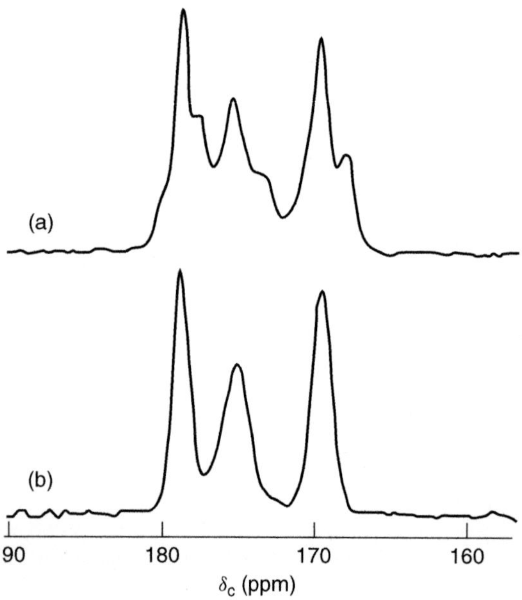

Figure 12.7. ^{13}C spectra of the cyanide carbons of the coordination polymer $[(Me_3Sn)_4Fe(CN)_6]_n$ showing (a) residual dipolar splittings arising from coupling to ^{14}N and (b) their removal by isotopically enriching the molecule in ^{15}N. There are three cyanide environments in the crystal structure. (Reproduced from Ref. 26, © Pergamon Press, 1992.)

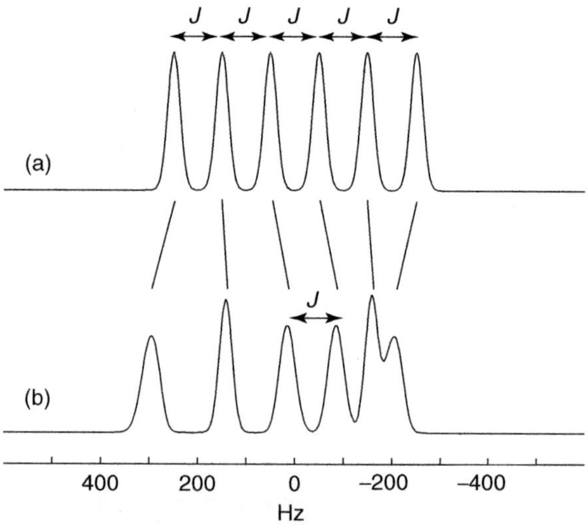

Figure 12.8. Simulated ^{31}P MAS NMR spectra of a ^{31}P–^{17}O pair with a J-coupling $^1J(^{31}P$–$^{17}O) = 100$ Hz with (a) no quadrupolar coupling and (b) a quadrupolar coupling with $C_Q = 4$ MHz and $\eta_Q = 0$ and a ^{31}P–^{17}O dipolar coupling of 1880 Hz [d(P–O) = 1.54 Å] oriented in the quadrupolar principal axes system (PAS) with $\alpha_D = 0°$ and $\beta_D = 20°$ in a 7 T magnetic field. (Reproduced from Ref. 22.)

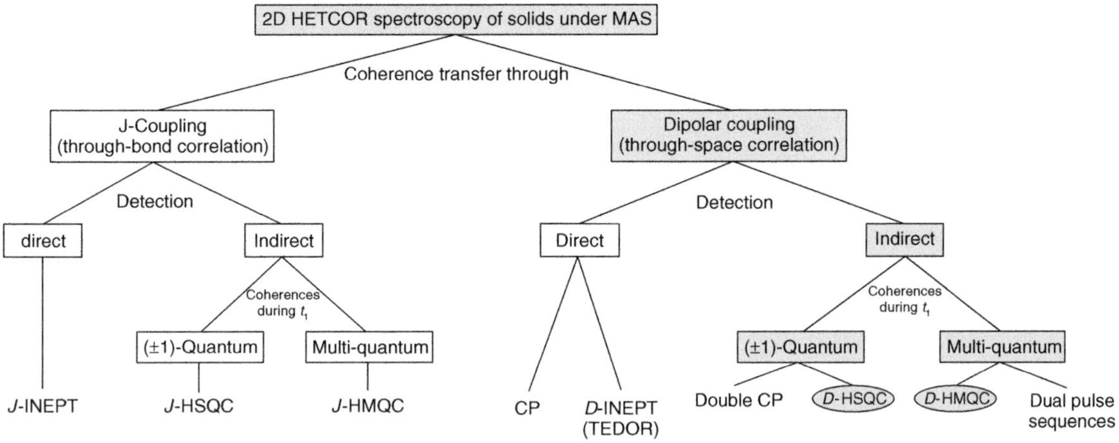

Figure 12.9. Schematic classification of 2D heteronuclear correlations experiments between half-integer quadrupolar spins and spin-1/2 nuclei. The paths corresponding to D-HMQC/D-HSQC using the dipolar interaction are highlighted in red. (Reproduced from Ref. 32. © American Chemical Society, 2009.)

Several possibilities exist when a J-coupling is to be measured between a spin 1/2 and a half-integer quadrupolar spin: either the J-coupling can be directly determined from one of the NMR spectra or one has to use a J-resolved experiment. In such a case, the detected nucleus has to be the quadrupolar spin, keeping the spin-1/2 nucleus in the indirect dimension since 180° pulses on the quadrupolar nucleus are only able to refocus the central transition (Figures 12.4 and 12.5).[23]

12.3.2 Residual Splitting

Unfortunately, the dipolar and quadrupolar Hamiltonians do not commute and therefore residual couplings are sometimes observed.[24] These effects have been notably observed, especially at low principal field, on ^{13}C spectra of amino acids because of the presence of neighboring ^{14}N atoms (as shown in Figure 12.7)[25,26] and also on $^{31}P/^{63/65}Cu$ NMR.[27,28] The effect has also been documented between pairs of quadrupolar nuclei,[29] as well as its effect on spinning sidebands.[30]

Therefore, the multiplet due to the J-coupling may be perturbed by the presence of these effects (Figure 12.8).

Briefly, this effect arises because, in the Hamiltonian, nonsecular parts of the dipolar and quadrupolar Hamiltonian interfere positively, creating cross terms that are not averaged out by MAS, leading the spin-1/2 nucleus to a shift that depends upon the quadrupolar spin state m, the crystallite orientation, and the relative orientation between the EFG and the dipolar interaction.[26] These effects can also be observed in MQ-MAS, ST-MAS, DAS, and DOR to varying degrees.[31]

12.4 EXPERIMENTS USING J-COUPLINGS

12.4.1 Heteronuclear Correlation Experiments

Several possibilities exist for obtaining correlation spectra from the J-coupling or the dipolar interaction in the heteronuclear case. A review of the available strategies is provided in the literature[32] and the possibilities are exemplified in Figure 12.9.

Three types of pulse sequences are primarily used for J-coupling-mediated correlations involving one or two quadrupolar nuclei: insensitive nuclei enhanced by polarization transfer (INEPT) (and distortionless enhancement by polarization transfer (DEPT)), HMQC, and heteronuclear single-quantum coherence (HSQC).[32,33]

The first INEPT experiments with a single quadrupolar nucleus have been realized using the J-coupling between ^{27}Al and ^{31}P

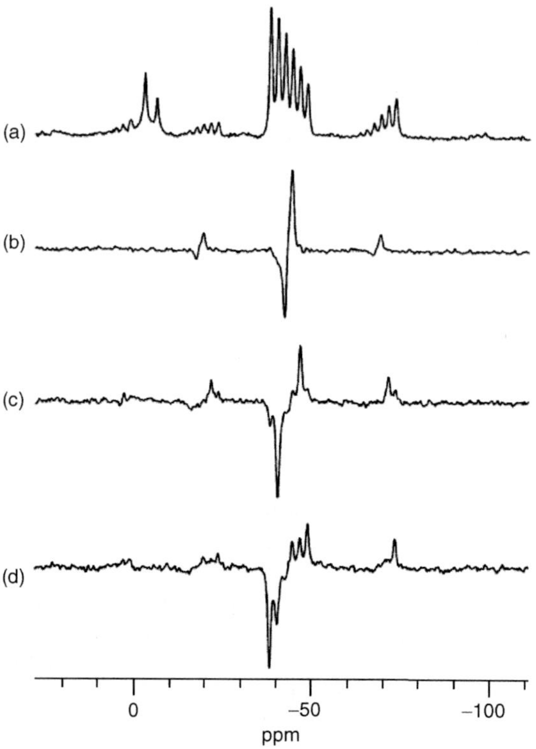

Figure 12.10. Spectra obtained with the INEPT magnetization transfer using variable pulse length on ^{27}Al ($S = 5/2$) resulting in the excitation of higher order terms involving satellite transitions as the pulse becomes longer. (a) ^{31}P one-pulse MAS spectrum, (b)–(d) $^{27}Al \rightarrow {}^{31}P$ INEPT antiphase spectra obtained with values for $\theta = \pi/6$ or 1.5 μs (b), $\pi/3$, or 3.0 μs (c), $\pi/2$ or 4.5 μs (d). The ^{27}Al RF strength is 55 kHz, $\nu_R = 3.6$ kHz, and $\tau = 0.75$ ms. (Reproduced from Ref. 34b. © Elsevier, 1998.)

in aluminophosphates.[34] Usually, the shorter longitudinal relaxation time T_1 and the possibility to use sensitivity enhancement methods (DFS, RAPT, etc.) favor the choice of a magnetization transfer from the central transition of the quadrupolar nucleus to the spin-1/2 nucleus. This is shown in Figure 12.10, where the resulting antiphase signal is shown to vary with the length of the RF pulse applied on the quadrupolar nucleus: only the antiphase doublet connected to the central transition is observed when the θ nutation angle is $\pi/6$ (which is the inherent nutation angle), but corresponds to the selective $(S + 1/2)\theta = \pi/2$ nutation angle on the central transition. For longer RF pulses, more transitions are involved in the magnetization transfer, and the antiphase multiplet becomes more complicated. In a similar manner, DEPT experiments have been performed on the same types of systems.[35]

$$
\begin{array}{ll}
{}^{27}\mathrm{Al} & (\theta)_y - \tau - (2\theta)_y - \tau - (\theta)_{\pm x} \\
{}^{31}\mathrm{P} & \qquad\qquad (\pi)_y - \tau - (\pi/2)_y - \text{Acquire.}
\end{array}
$$

In a similar manner, refocused INEPT experiments, which yield in-phase doublet spectra, have been realized between ^{27}Al and ^{29}Si and between ^{23}Na and ^{29}Si and are shown in Figure 12.11.

HMQC experiments have been performed on similar systems, and also on ^{71}Ga ($I = 3/2$)/^{31}P pairs.[36] Interestingly, the HMQC experiments can also be used to characterize the environments of quadrupolar nuclei S using I spin-counting strategies. The idea is to cycle the phase ϕ of one of the indirect dimension 90° pulses (on I) in an HMQC and to record the corresponding transients (from S) separately. By using FT against the modulated pulse phase ϕ in the second dimension, one can separate the spectra of the S nuclei based on the number of I spins (p) which were involved in the coherence created by the first 90° pulse, as its phase shifts with $exp(\pm ip^{\Phi})$. Therefore, observing a signal in row number p implies that at least pI spins have been involved in the coherence created by the first 90° pulse applied on I in the HMQC experiment.[37] The HMQC experiment can also be performed between pairs of quadrupolar nuclei, as has been done between ^{27}Al ($S = 5/2$) and ^{17}O ($S = 5/2$) in ^{17}O enriched calcium aluminates,[38] and, as in other experiments involving quadrupolar nuclei, a MQ-MAS step can be included.[39]

It must be noted that J-coupling may also contribute notably to the broadenings of the lines of spins neighboring quadrupolar nuclei, and that in some of these cases decoupling may considerably improve the resolution of the spin-1/2 spectrum.[40]

12.4.2 Homonuclear Correlation Experiments

Homonuclear correlation experiments using the J-coupling have seldom been realized for quadrupolar spins. For example, the 115 Hz J-coupling between ^{59}Co spins ($S = 7/2$, 100% natural abundance) has been used to acquire a COSY experiment in a liquid sample of $HFeCo_3(CO)_{11}PPh_2H$.[41] In the solid state, the use of homonuclear J-couplings has not been successful so far, maybe essentially because

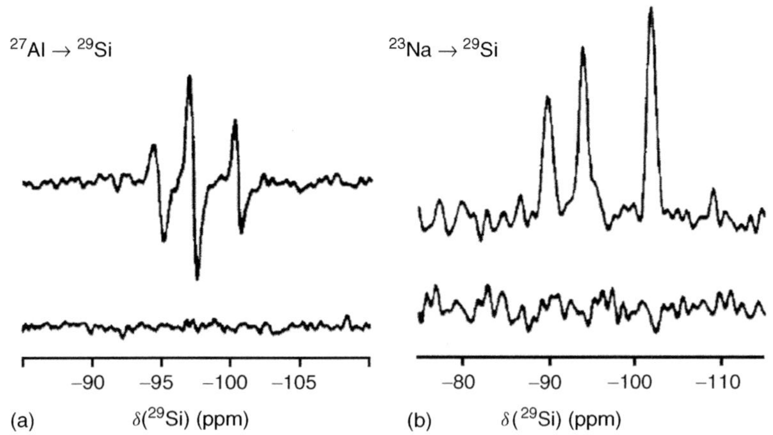

Figure 12.11. (a) $^{27}Al \rightarrow {}^{29}Si$ INEPT spectrum of microcline (720 scans) and (b) $^{23}Na \rightarrow {}^{29}Si$ refocused INEPT spectrum of albite (13 296 scans), with the respective null experiments shown. (Reproduced from Ref. 35, © American Chemical Society, 1995.)

of their small values compared to the available T_2',[42] except for the recently explored possibilities offered by ^{11}B.[42] However, homonuclear correlation experiments have been performed between ^{27}Al atoms using the $^{2}J(^{27}Al{-}O{-}^{31}P)$ in the relayed H-HSQC experiment where the ^{31}P spin ($I = 1/2$) is used as a relay for the magnetization transfer mechanism.[43] An attempt to circumvent the necessity to use a neighboring spin as a relay has been found through a direct covariance processing where a heteronuclear $^{27}Al{-}^{31}P$ correlation spectrum can be processed into a $^{27}Al{-}^{27}Al$ homonuclear correlation spectrum.[44]

12.5 DIPOLAR COUPLINGS

As quadrupolar nuclei are of major interest in a large range of materials, and because the J-coupling values are usually small and the T_2' too short compared to the coherence lifetime of the central transition to allow for efficient magnetization transfer, many schemes to obtain correlation spectra and characterize spatial proximities have been developed using the dipolar coupling.[45–47]

The main obstacle to overcome in experiments involving quadrupolar nuclei arises from the difficulty in manipulating the spin system while avoiding magnetization leakage into unwanted single spin coherences due to the presence of satellite transitions. In the static case, satellite transitions are usually too broad to be easily detected and most reside far off-resonance. However, there is usually not enough resolution in static spectra for most nuclei due to the other broadening mechanisms (dipolar, CSA, etc.) and MAS remains desirable in most cases when one wants to obtain resolution in correlation spectra. Under MAS or close to it, the satellite transition frequencies become modulated and cross the RF window periodically, thereby introducing an additional complexity when dipolar interactions are averaged by MAS and have to be reintroduced. Hence, we will first develop some considerations on the spin-locking of quadrupolar nuclei as a function of the quadrupolar interaction, the MAS rate, and the RF spin-locking power.

12.5.1 Spin-lock and Quadrupolar Nuclei

The theoretical description of spin-locking effects and magnetization leakage during adiabatic passages is available in the literature.[48–50] The interaction occurring between the frequency-modulated satellite transition and the applied RF irradiation is given by the α parameter

$$\alpha = \frac{\omega_1^2}{2\omega_Q^{\mathrm{PAS}}\omega_R}$$

where $\omega_Q^{\mathrm{PAS}} = \frac{3C_Q}{4S(2S-1)}$ C_Q is the quadrupolar coupling constant of the spin S, ω_1 is the RF field strength and ω_R the MAS rate.

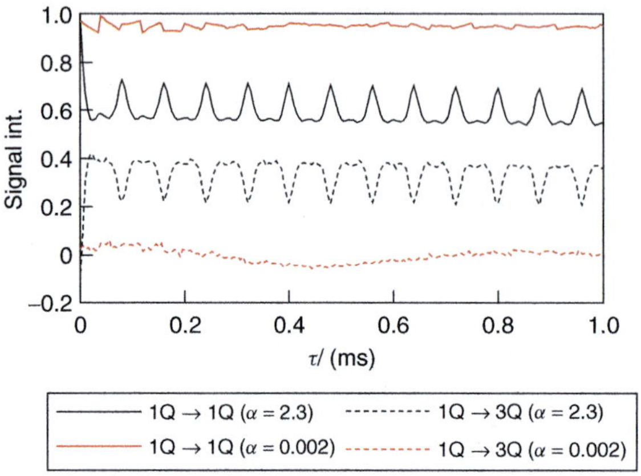

Figure 12.12. Experimental ^{23}Na (132.3 MHz) MAS NMR of powdered $NaNO_2$. The pulse sequence comprises an optimized excitation pulse (for single-quantum coherences (1Q) or triple-quantum coherences (3Q)), followed by a spin-lock and, eventually, a conversion pulse for the 3Q → 1Q reconversion. Central transitions 1Q → 1Q (solid line) and 1Q → 3Q (dotted line) signal intensity as a function of the spin-lock duration are shown. The relative intensity of the 3Q-filtered signal has been scaled to take into account the efficiency of the reconversion pulse. Spin-locking is performed with $\nu_1 = 5$ kHz and $\nu_R = 25$ kHz for $\alpha = 0.002$ (sudden regime, red) and $\nu_1 = 125$ kHz and $\nu_R = 12.5$ kHz, corresponding to $\alpha = 2.3$ (adiabatic passages, black). (Reproduced from Ref. 50. © American Institute of Physics, 2009.)

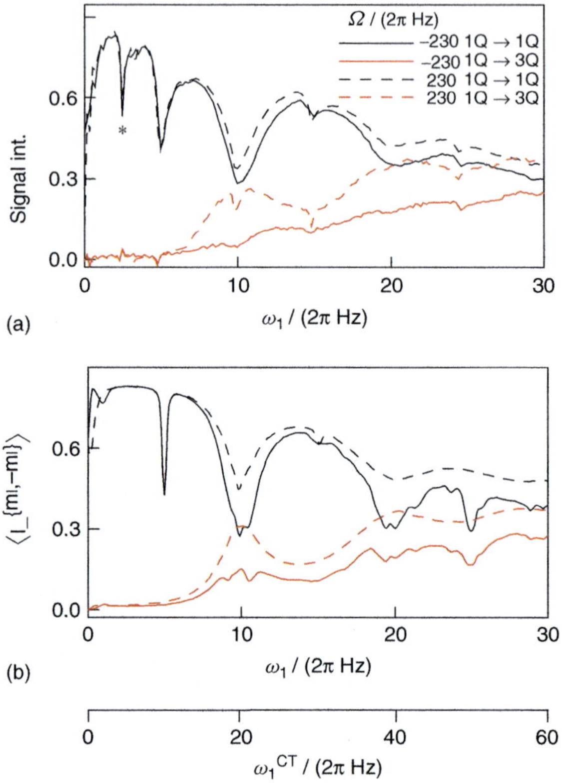

Figure 12.13. (a) Experimental ^{23}Na (132.3 MHz) MAS NMR and (b) simulations (exact density matrix approach) of powdered $NaNO_2$. Central transition 1Q → 1Q (black lines) and 1Q → 3Q (red lines) signal intensities are shown as a function of the radiofrequency field strength ω_1 of the 2 ms spin-locking pulse. Resonance offsets, $\Omega/2\pi$, of −230 Hz from the ^{23}Na chemical shift (on resonance with the 1Q signal, solid lines) and +230 Hz (matched with the 3Q coherences, dashed lines) have been used. The MAS rate is $\omega_R/2\pi = 10.4$ kHz, $\omega_Q/2\pi = 275$ kHz, $\eta_Q = 0$. The dip in spin-locking efficiency observed for the HORROR condition at $\omega_1^{CT} = \omega_R/2$ caused by ^{23}Na–^{23}Na dipolar couplings is highlighted by an asterisk. (Reproduced from Ref. 50. © American Institute of Physics, 2009.)

- If $\alpha \ll 1$, the product $\omega_Q^{PAS}\omega_R$ is large enough (low RF field ω_1, fast MAS, and/or large quadrupolar interaction) and the "sudden" approximation can be made: i.e., the zero-crossings of the frequency-modulated satellite transitions through the RF irradiation window occur fast enough compared to the strength of the spin-locking RF field so that the magnetization remains spin-locked.
- If $\alpha \gg 1$, the MAS-induced zero-crossing of the satellite transitions are adiabatic (see Ref. 48 for more details in spin 3/2).
- if $\alpha = 1$, a zero-crossing will partly convert population differences across spin-locking eigenstates into off-diagonal coherences, spin-locking will not be maintained, and the magnitude of the spin-locked magnetization will decay on a timescale of $1/\omega_R$.

It has been shown that spin-locking transfers the magnetization from single-quantum coherences to higher order coherences and this can be used to promote MQ coherence buildup in MQ-MAS experiments[50] while the best spin-locking conditions are obtained in the sudden regime, i.e., at lower power and faster MAS rates (Figure 12.12).

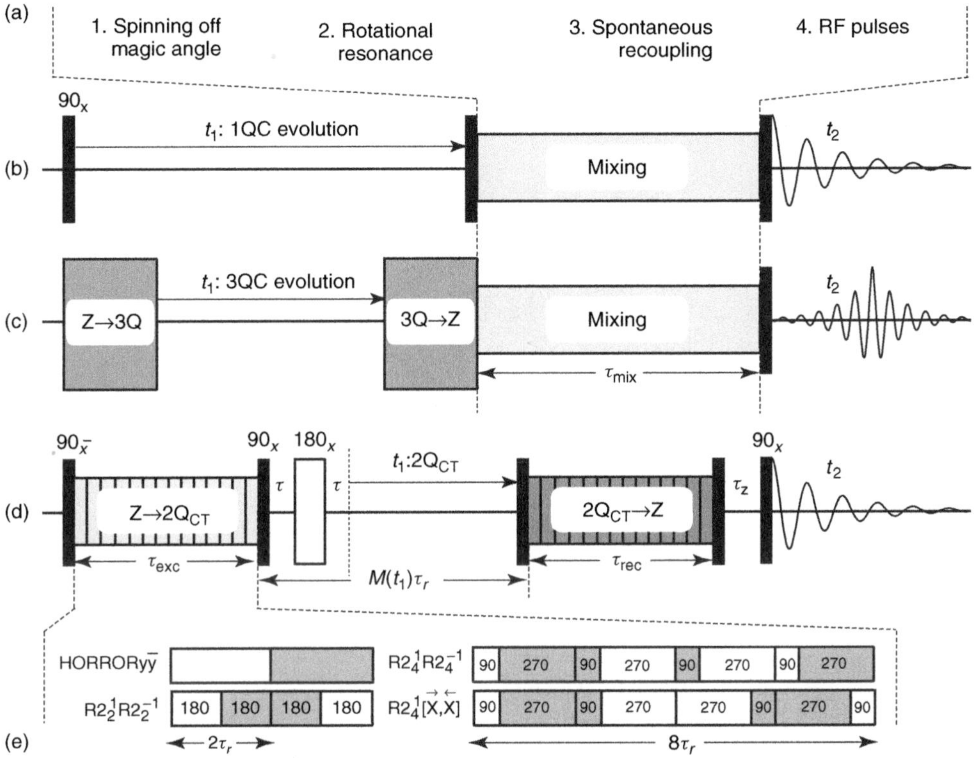

Figure 12.14. (a) Various options for recoupling homonuclear dipolar interactions between half-integer quadrupolar spins in order to obtain interatomic correlation spectra with (b) single-quantum, (c) triple-quantum, and (d) double-quantum coherence evolution during t_1 and single-quantum coherence detection during t_2. CT refers to the coherences involving only the central transitions of the recoupled quadrupolar spins. (e) A selection of double-quantum recoupling sequences. It must be noted that the 90° pulses surrounding the 2Q recoupling symmetry-based, supercycled pulse trains can also be omitted.[60] (Reproduced from Ref. 47. © Elsevier, 2009.)

In the sudden regime, using low spin-locking RF power leads to larger offset effects. In the adiabatic regime, however, the best spin-locking is achieved by using an offset to match the isotropic shift of the unobservable triple-quantum (3Q) coherences and not to match that of the observable single-quantum (1Q) signal (for the particular case of spins $I = 3/2$, this shift in the 3Q frame of reference is of identical magnitude to that for 1Q coherences but opposite in sign).[51]

Under MAS, especially in the sudden regime where MAS is fairly fast, most of the dipolar couplings are averaged out by MAS alone, provided no rotary resonance condition is fulfilled. However, it has been noted by a number of authors that the spin-locking efficiency decreases for particular ratios of ω_1 and ω_R, even if $\alpha \ll 1$.[50–52]

As confirmed by numerical simulations, these rotary resonance conditions are better described using the nutation rate of the central transition in the selective pulse regime, i.e., $\omega_1^{CT} = (S + 1/2)\,\omega_1$ and are found for $\omega_1^{CT} = m'\omega_R$, where $m' = 2, 4, 6, \ldots$ for the increased interchange of 1Q and 3Q coherences seen when only a first-order quadrupolar interaction is to be considered and $m' = 1, 2, 3, \ldots$ for the loss of spin-locking efficiency that is only observed when the second-order quadrupolar interaction cannot be neglected.[50] It must be noted that, when the HOmonucleaR ROtary Resonance (HORROR) condition is fulfilled for the central transition, i.e., $m' = 1/2$, a dip may occur in the spin-locking efficiency due to the recoupling of homonuclear dipolar interactions which opens the way of designing homonuclear dipolar correlation (Figure 12.13).

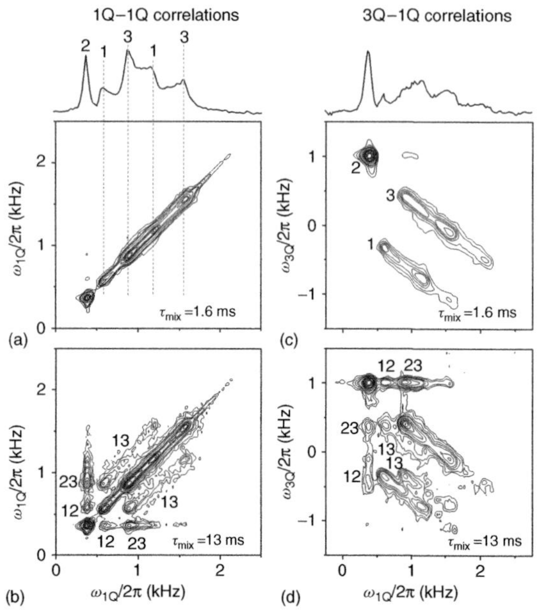

Figure 12.15. Experimental ^{23}Na–^{23}Na correlation spectra from Na_2SO_3 undergoing 10.0 kHz MAS at 4.7 T. The 2D spectra (a and b) and (c and d) were recorded using the pulse schemes of Figure 12.14(b) and (c), respectively, with $R4^1_4$ recoupling pulse trains as shown in Ref. 58b, with short (a and c) and long (b and d) mixing periods. The three inequivalent sites are labeled 1, 2, and 3. The resolution is improved in the (not sheared) 3Q–1Q correlation spectra (c and d). (Reproduced from Ref. 47, © Elsevier, 2009.)

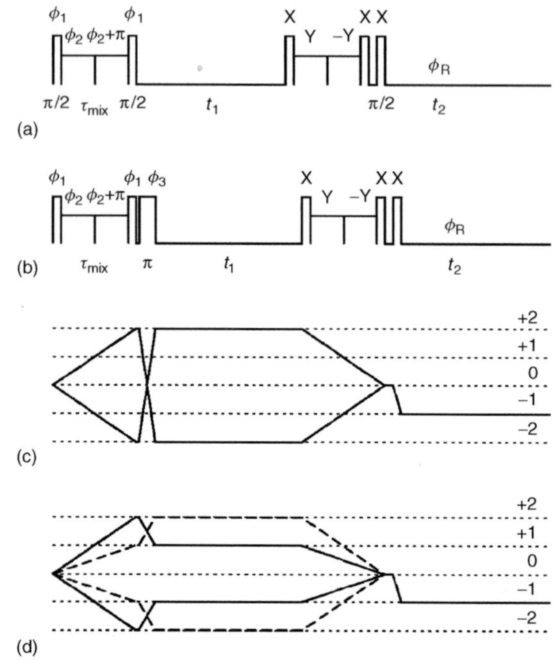

Figure 12.16. 2Q–1Q pulse sequences using the HORROR($y\overline{y}$) recoupling scheme (a) without and (b) with a central transition selective pulse which will induce a ±4Q change in the coherence order if the double-quantum coherence involves (c) two central transitions and only ±1Q change in the coherence order if the double-quantum coherence involves one central transition and a satellite transition which is unaffected by the 180° pulse. The phase Φ_3 of the central transition selective 180° pulse is cycled using 90° steps to ensure the selection of the desired coherence pathway (c) and rejection of the pathway shown in (d). (Reproduced from Ref. 61. © American Institute of Physics, 2004.)

12.5.2 Recoupling and Homonuclear Correlation

Quadrupolar nuclei unfortunately suffer from second-order quadrupolar broadening and from the presence of connected satellite transitions. Most recoupling techniques have been developed with low-power recoupling pulses.[47]

Different recoupling techniques can be used (some are shown in Figure 12.14) and these can be combined with high-resolution blocks (e.g., MQ-MAS), where possible, to obtain better resolution.[53] In addition, one may also use the following strategies:

- Off-MAS reintroduces a residual dipolar coupling, which can promote magnetization transfer between pairs of neighboring spins[54]; however, an NMR probe where the spinning angle can be changed during the experiment is necessary to ensure both fast recoupling during the off-MAS period and adequate resolution during detection.
- Rotational resonance, i.e., when the MAS rate ν_R matches a sub-multiple of the frequency separation between the two resonances to be recoupled, can also be used.[55]
- Spontaneous recoupling, stemming from the noncommutation between the dipolar and the quadrupolar Hamiltonian terms, may provide a mechanism for magnetization transfer; however, it is also strongly dependent upon chemical

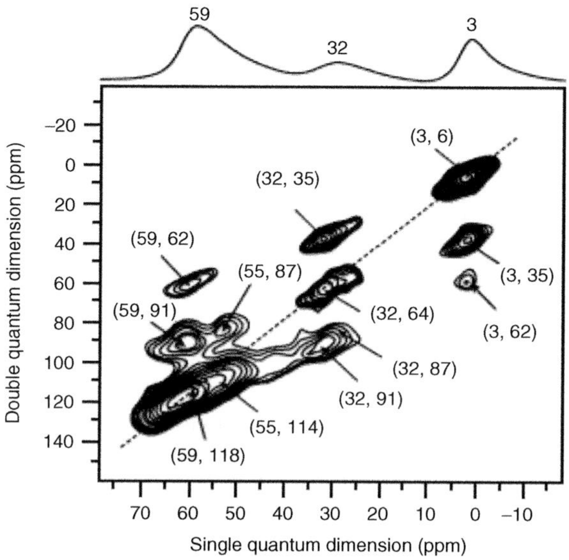

Figure 12.17. ^{27}Al 2Q–1Q correlation spectrum of an HY-700 zeolite (hydrated sample) at 18.8 T with a 3.2 mm probe at a sample rotation rate of 21.5 kHz. About 45 h was required to record this spectrum, using the $BR2^1_2$ experiment described in Ref. 60 with 1.16 ms recoupling delays. (Reproduced from Ref. 63. © Wiley-VCH, 2010.)

shift differences between the recoupled pairs of quadrupolar spins.[56]

However, in order to achieve broadband recoupling, RF pulse trains are usually preferred to the three previously mentioned methods. Most of them are based on the HORROR scheme applied to the central transition of half-integer spins, i.e., when the recoupling RF field $(S + 1/2)\,\omega_1 = \omega_1^{CT} = \omega_R/2$ is used to reintroduce homonuclear dipolar interactions,[57] and many methods now use symmetry-based RF pulse trains to compensate for pulse imperfections and offset effects.[58] Such schemes have also been applied with DOR for optimal resolution.[59]

Two experimental 1Q–1Q and 3Q–1Q ^{23}Na homonuclear correlation spectra are presented in Figure 12.15, highlighting the increased resolution in the 3Q–1Q correlation spectrum.

Double-quantum filtering, however, remains desirable, as it will naturally quench the autocorrelation peaks occurring in 1Q–1Q correlation spectra, which give rise to strong diagonal signals. The main difficulty lies in filtering out the intra-atomic

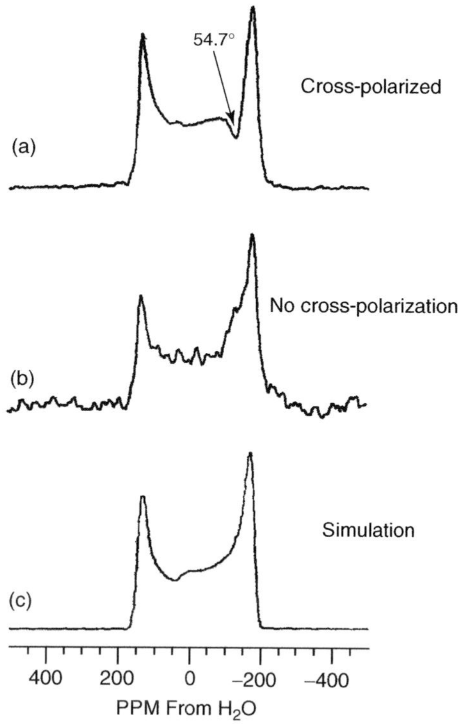

Figure 12.18. Static NMR spectra of the central transition of ^{17}O in polycrystalline 20% ^{17}O-enriched $Mg(OH)_2$ obtained at 67.8 MHz (11.7 T). (a) $^1H \rightarrow {}^{17}O$ cross-polarized spectrum with proton decoupling (200 scans, 1 ms contact time, and 5 s recycle time), showing the dip occurring for ^{17}O–^1H pairs aligned with the magic angle axis. (b) ^1H-decoupled spectrum without CP (200 scans, 5 s recycle time). (c) Simulated second-order quadrupolar powder pattern using $e^2qQ/h = C_Q = 6.8$ MHz, $\eta_Q = 0.0$, and $\delta = 25$ ppm. (Reproduced from Ref. 70. © Elsevier, 1988.)

double-quantum coherences arising between the central transition and the connected satellite transitions while keeping the interatomic double-quantum coherences between central transitions of neighboring spins resulting from the recoupled dipolar interaction. The solution involves the use of a central transition selective 180° pulse, with low RF power, in order to selectively invert the double-quantum coherences involving the central transitions of the recoupled pairs of quadrupolar nuclei without affecting the satellite transitions (Figure 12.16). Phase cycling removes the unwanted coherences.[61] It must be noted that, in these experiments, selectively manipulating the central transition (either for the recoupling pulse trains

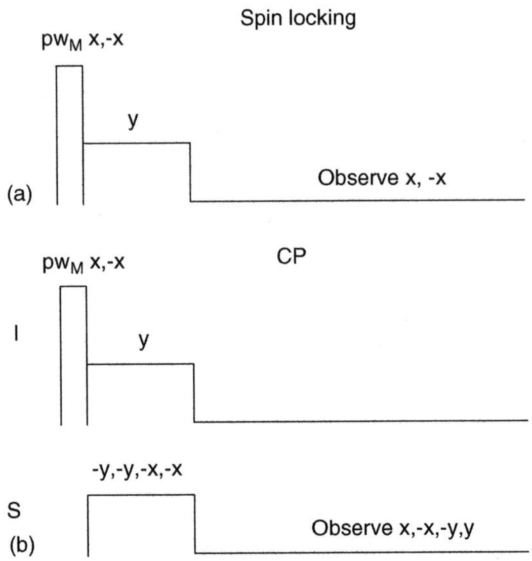

Figure 12.19. (a) Pulse sequence for spin-locking/MAS experiments. (b) Pulse sequence for CP/MAS experiments. pw_M is the smallest pulse width giving the maximum signal. (Reproduced from Ref. 73, © Elsevier, 1999.)

or the 180° pulse) is difficult for nuclei with small quadrupolar couplings.

These correlation experiments give access not only to atomic proximities (Figure 12.17)[62,63] but are also sensitive to the relative orientation of the quadrupolar interaction tensors, which can be derived from the analysis of 2D lineshapes stemming from the second-order quadrupolar broadening. This possibility was first highlighted using a spin-diffusion step in MQ-MAS experiment[64] and further used in 2Q–1Q correlation spectra.[65]

12.5.3 Heteronuclear Correlations Methods

CP was the first method applied to obtain heteronuclear correlation experiments involving a quadrupolar spin.[66] 1D CP is used for spectral editing by selecting a pair of spins, which are close in space and for which a significant dipolar interaction occurs. This method can also be used to increase the signal of a less sensitive spin (due to lower natural abundance—^{29}Si, ^{17}O, etc.—or longer T_1 relaxation times—^{31}P, etc.—for example). If a t_1 evolution delay is placed between the first 90° pulse and the CP transfer HETeronuclear COrrelation (HETCOR), a 2D correlation spectrum can be obtained, as demonstrated on the mesoporous aluminophosphate VPI-5.[67] Other examples can also be found in the literature.[68]

The challenge in performing CP experiments involving quadrupolar nuclei lies in the ability to spin-lock the magnetization of the quadrupolar spin. In the static case, spin-locking of the central transition is not perturbed by the MAS-induced zero-crossings from the satellite transitions. Therefore, the CP transfer is obtained when the Hartmann–Hahn condition is matched, i.e., for

$$\varepsilon_I \omega_{1I} = \varepsilon_S \omega_{1S}$$

where ε_I or $\varepsilon_S = 1$ for spin 1/2 and $I + 1/2$ or $S + 1/2$ if the central transition is manipulated selectively.[69] An example illustrating $^1H \rightarrow ^{17}O$ CP and the potential distortion of the second-order quadrupolar lineshape is provided (Figure 12.18).[70] This strategy has also been applied in aluminas for the study of surface sites.[71]

Under MAS, the best spin-lock conditions are usually obtained for low to very low RF powers, in the "sudden" regime and avoiding any unwanted rotary resonance condition and thus below actual spinning rates (Figure 12.12). To obtain such conditions, one has to first find optimal spin-lock conditions for the quadrupolar spins, and then optimize the CP transfer by fulfilling the Hartmann–Hahn conditions, using the pulse sequences shown in Figure 12.19.[72,73]

The Hartmann–Hahn conditions under MAS matching ω_{1S} (spin-lock RF field on S), ω_{1I} (spin-lock RF field on I), and ω_R (MAS rate) can be found when[73]

$$\varepsilon_I \omega_{1I} - \varepsilon_S \omega_{1S} = \pm n\omega_R \text{ (zero-quantum CP from "flip-flop" terms)}$$

$$\varepsilon_I \omega_{1I} + \varepsilon_S \omega_{1S} = n\omega_R \text{ (double-quantum CP from "flip-flip" and "flop-flop" terms)}$$

where ε_I or ε_S can take on values from 1 (for spins 1/2 and quadrupolar spins when a large spin-lock field ω_1 is used—"adiabatic" regime) to $I + 1/2$ or $S + 1/2$ (when low ω_1 spin-lock fields are used and the central transition is manipulated selectively—"sudden" regime) and n is an integer.

For CP involving a spin-1/2 nucleus, the spin-lock applied to this spin can be chosen arbitrarily, provided no unwanted rotary resonance condition is observed. Hence, the quadrupolar spin can be locked using either low or strong RF fields. When both nuclei are quadrupolar spins and must be spin-locked using low-power RF fields, double-quantum CP is to be

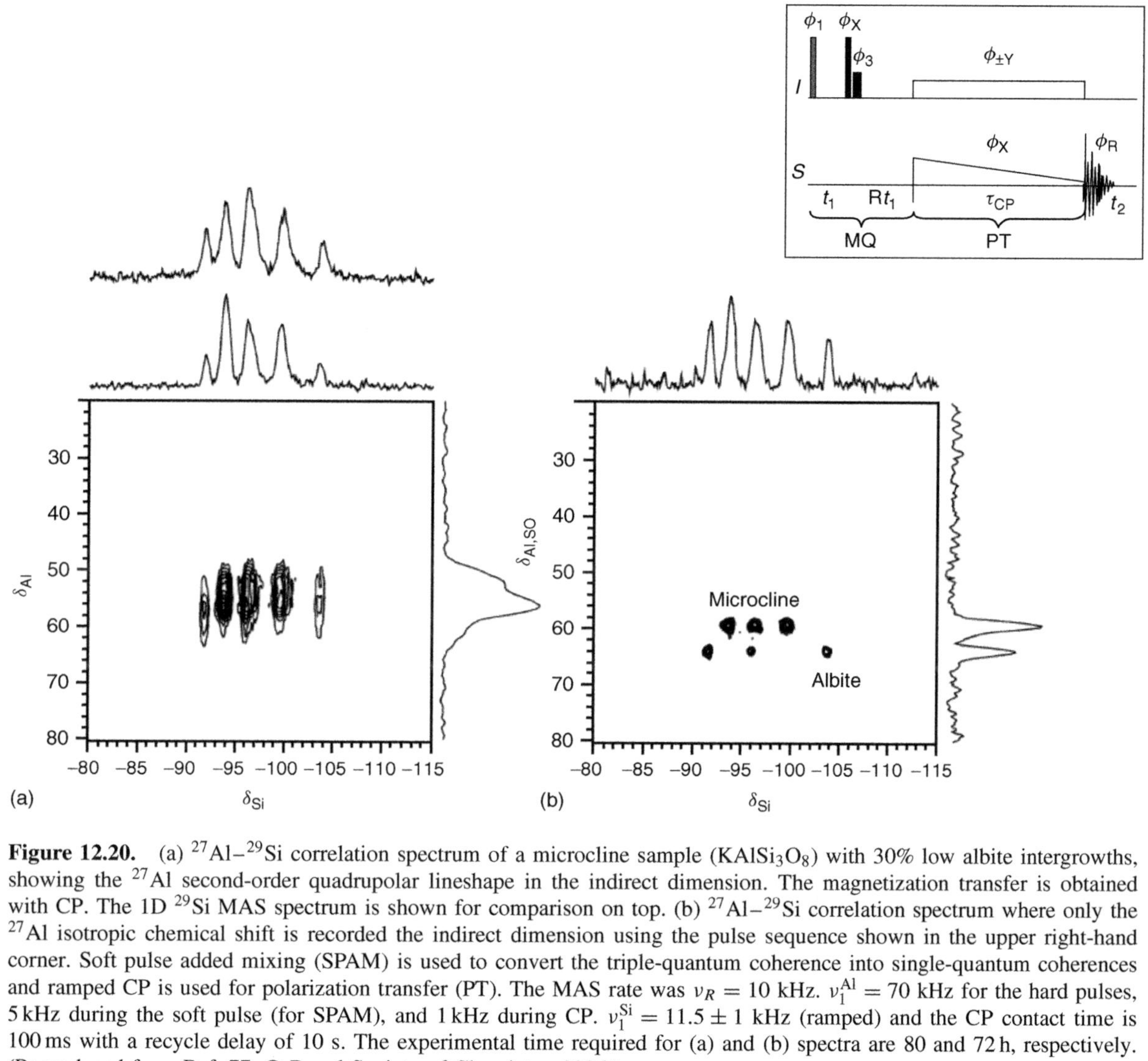

Figure 12.20. (a) ^{27}Al–^{29}Si correlation spectrum of a microcline sample ($KAlSi_3O_8$) with 30% low albite intergrowths, showing the ^{27}Al second-order quadrupolar lineshape in the indirect dimension. The magnetization transfer is obtained with CP. The 1D ^{29}Si MAS spectrum is shown for comparison on top. (b) ^{27}Al–^{29}Si correlation spectrum where only the ^{27}Al isotropic chemical shift is recorded the indirect dimension using the pulse sequence shown in the upper right-hand corner. Soft pulse added mixing (SPAM) is used to convert the triple-quantum coherence into single-quantum coherences and ramped CP is used for polarization transfer (PT). The MAS rate was $\nu_R = 10$ kHz. $\nu_1^{Al} = 70$ kHz for the hard pulses, 5 kHz during the soft pulse (for SPAM), and 1 kHz during CP. $\nu_1^{Si} = 11.5 \pm 1$ kHz (ramped) and the CP contact time is 100 ms with a recycle delay of 10 s. The experimental time required for (a) and (b) spectra are 80 and 72 h, respectively. (Reproduced from Ref. 77. © Royal Society of Chemistry, 2006.)

preferred.[72] In some cases, however, quadrupolar spins, which experience small quadrupolar couplings (like $^{11}BO_4$ units), are best spin-locked using high RF field strengths ("adiabatic" regime), whereas those that experience larger quadrupolar couplings (like BO_3 units) are best spin-locked with low RF field strengths ("sudden" regime).[74] It must be noted that some of the problems associated with using low RF CW spin-locking fields can be alleviated by using trains of spin-locking 180° pulses with the same effective nutation rate.[75]

It is also possible to associate MQ-MAS with CP, either encoding before or after the CP transfer, and eventually to polarize directly the triple-quantum transition.[76] An example where the isotropic chemical shift of ^{27}Al is recorded in the indirect dimension through an initial MQ-MAS step while detecting ^{29}Si in the direct dimension after a CP transfer is provided in Figure 12.20.[77]

A number of other double resonance experiments can be used for either distance measurements or magnetization transfer. A review is provided in the

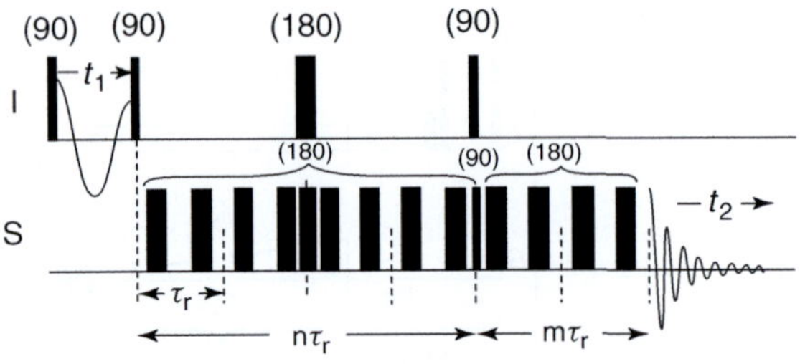

Figure 12.21. Scheme for the 2D Tranfer-Echo DOuble Resonance experiment (TEDOR)with preliminary evolution of the quadrupolar spin I during t_1 and detection of the spin 1/2 during t_2. (Reproduced from Ref. 81. © American Chemical Society, 1995.)

literature.[78] Among them is the Tranfer-Echo DOuble Resonance (TEDOR) [79] experiment (Figure 12.21),[80] which can be used on pairs of spins involving at least one quadrupolar nucleus. Usually, the magnetization is created on the quadrupolar spin (typically shorter T_1) and the recoupling pulses are applied on the spin 1/2 nucleus whenever possible (Figure 12.21).[81]

These experiments have been realized between ^{27}Al and ^{29}Si,[81] ^{27}Al, and ^{31}P.[82] In a similar manner to the experiments described above using CP, one can include an MQ-MAS step in the TEDOR experiment.[83]

Finally, two experiments should also be mentioned: the dipolar-mediated HMQC and HSQC (Figure 12.22).[84]

These two experiments are usually performed with recoupling pulses applied on spin-1/2 nuclei. Most of the recoupling pulse trains use the $\omega_1 = 2\omega_R$ recoupling condition which selectively reintroduces the heteronuclear dipolar couplings, and typically such recoupling pulses lead to magnetization leaks when applied to quadrupolar spins.

Examples of HMQC and HSQC spectra are shown in Figure 12.23.

Several methods have been developed for the recoupling of heteronuclear dipolar couplings and a review is available in the literature.[85] In a similar manner to experiments using J-couplings, spin counting can be achieved with dipolar recoupling in HMQC experiments.[37] One can also include a resolution enhancement step with an ST-MAS scheme within the HMQC experiment.[86]

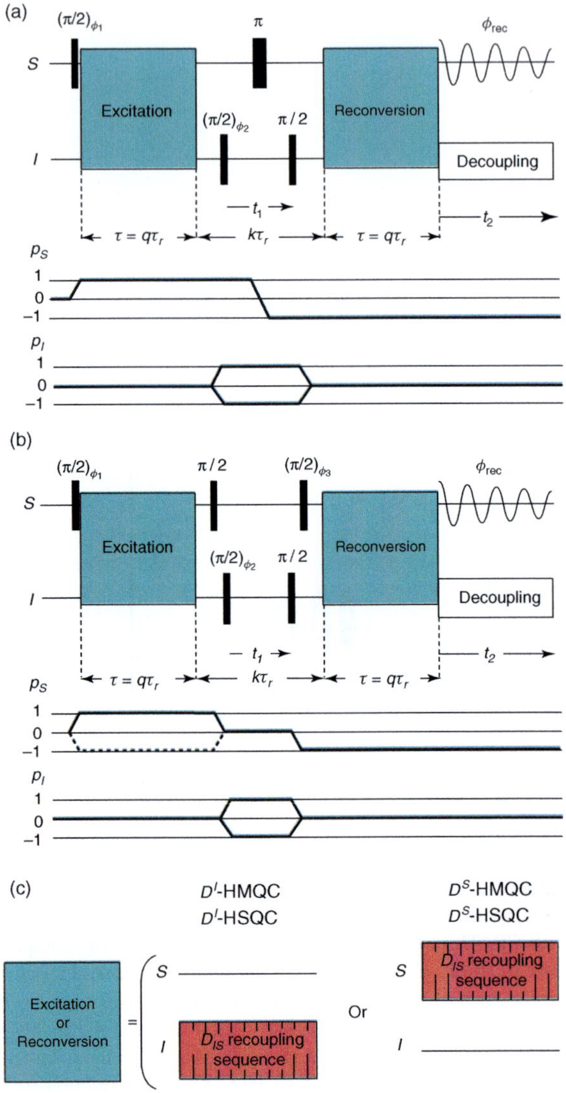

Figure 12.22. Pulse sequences and coherence transfer pathways for (a) dipolar mediated HMQC and (b) HSQC. The recoupling schemes mentioned in (c) are responsible for the reintroduction of dipolar couplings. (Reproduced from Ref. 32. © American Chemical Society, 2009.)

12.6 CONCLUSION

Although quadrupolar spins are usually challenging spin systems, new possibilities have been offered by technical and methodological improvements

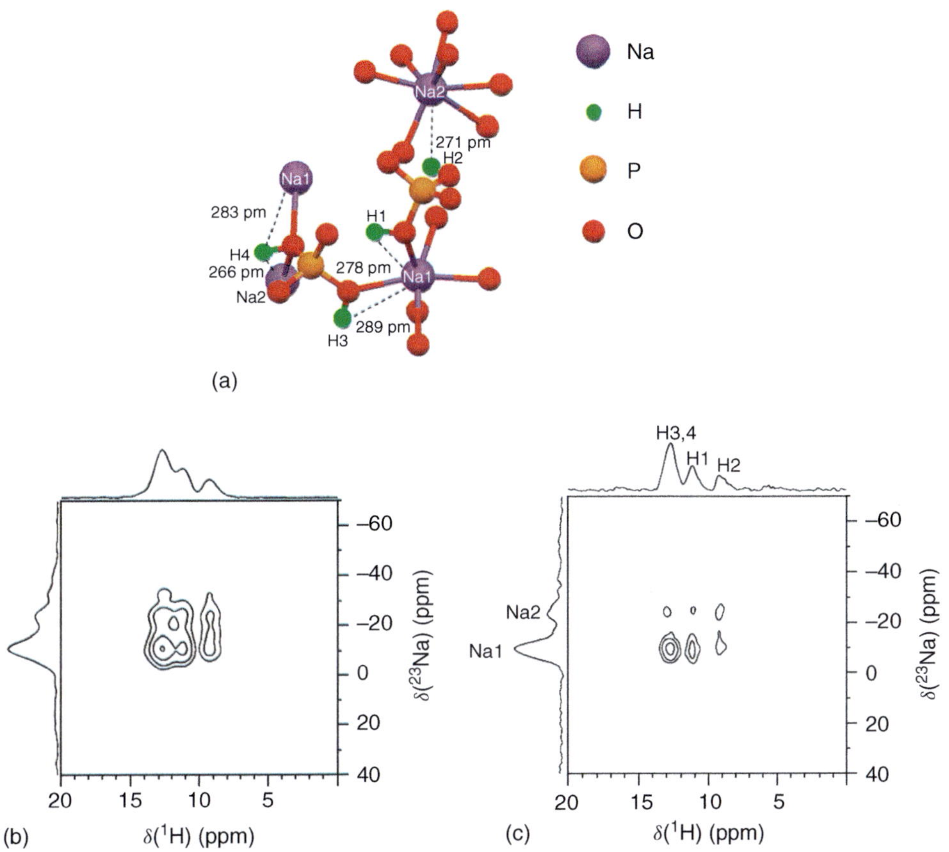

Figure 12.23. (a) Atomic structure of NaH_2PO_4. (b) and (c) Experimental 2D ^{23}Na-^{1}H dipolar-mediated HMQC spectra at $\nu_R = 31.746$ kHz where the ^{23}Na-^{1}H dipolar couplings have been reintroduced by applying $SR4^2_1$ recoupling to the detected ^1H channel. In (c), homonuclear decoupling has been applied on ^1H during t_1 with $C8^4_1$ (half-cos) and during the acquisition t_2 with $wC8^4_1$ (half-cos). (Reproduced from Ref. 32. © American Chemical Society, 2009.)

to retrieve information on their environments using correlation experiments. Furthermore, even though quadrupolar spin systems are more complicated to manipulate, the quadrupolar interaction, which can be measured and compared to computed values arising from structural models, has ceased to be considered only as an annoying broadening phenomenon and is now proven to have advantages of its own.

It might be of interest at this point to analyze the perspectives for future development in terms of methodology. As such the panorama is remarkable because, in the case of half-integer quadrupolar spins, it remains mostly limited to the use of the central transition while recent developments in the use of integer quadrupolar spins make fruitful use of symmetric transitions.[16] This limitation to the central transition is discarding a significant or even the major part of the possible signal and also ignores the magnification factors that could be achieved on dipolar interaction, for example, by the use of multiple-quantum transitions. The single- or multiple-quantum symmetric transitions are also known to have a very different sensitivity to dynamics as compared to satellite transitions[87–89] and this still has to be exploited in the study of real systems. The development of enhanced adiabatic excitation schemes able to manipulate ensembles of spinning sidebands and the increase of MAS spinning rates that would reduce the number of spinning sidebands, increase their individual intensities, and allow their selective excitation are certainly possible avenues of future development.

RELATED ARTICLES IN THE ENCYCLOPEDIA OF MAGNETIC RESONANCE

Double Rotation

High Speed MAS of Half-Integer Quadrupolar Nuclei in Solids

Homonuclear Recoupling Schemes in MAS NMR

Inorganic Solids

Magic Angle Spinning: Effects of Quadrupolar Nuclei on Spin-1/2 Spectra

Multiple-quantum Magic-angle Spinning Experiments on Half-integer Nuclei: Fundamentals

Quadrupolar Nuclei in Glasses

Quadrupolar Nuclei in Liquid Samples

REDOR and TEDOR

Through-Bond Experiments in Solids

REFERENCES

1. S. E. Ashbrook, *Phys. Chem. Chem. Phys.*, 2009, **11**, 6892–6905.
2. Z. Gan, P. Gor'kov, T. A. Cross, A. Samoson, and D. Massiot, *J. Am. Chem. Soc.*, 2002, **124**, 5634–5635.
3. A. Samoson, E. Lippmaa, and A. Pines, *Mol. Phys.*, 1988, **65**, 1013–1018.
4. A. Llor and J. Virlet, *Chem. Phys. Lett.*, 1988, **152**, 248–253.
5. L. Frydman and J. S. Harwood, *J. Am. Chem. Soc.*, 1995, **117**, 5367–5368.
6. Z. Gan, *J. Am. Chem. Soc.*, 2000, **122**, 3242–3243.
7. J. Trebosc, J.-P. Amoureux, and Z. Gan, *Solid State NMR*, 2007, **31**, 1–9.
8. Z. Yao, H. T. Kwak, D. Sakellariou, L. Emsley, and P. J. Grandinetti, *Chem. Phys. Lett.*, 2000, **327**, 85–90.
9. (a) A. P. M. Kentgens and R. Verhagen, *Chem. Phys. Lett.*, 1999, **300**, 435–443; (b) D. Iuga, H. Schäfer, R. Verhagen, and A. P. M. Kentgens, *J. Magn. Reson.*, 2000, **147**, 192–209.
10. (a) R. Siegel, T. T. Nakashima, and R. E. Wasylishen, *Chem. Phys. Lett.*, 2004, **388**, 441–445; (b) R. Siegel, T. T. Nakashima, and R. E. Wasylishen, *J. Magn. Reson.*, 2007, **184**, 85–100; (c) R. Siegel, T. T. Nakashima, and R. E. Wasylishen, *Chem. Phys. Lett.*, 2006, **421**, 529–533; (d) T. T. Nakashima, R. E. Wasylishen, R. Siegel, and K. J. Ooms, *Chem. Phys. Lett.*, 2008, **450**, 417–421.
11. F. H. Larsen, H. J. Jakobsen, P. D. Ellis, and N. C. Nielsen, *J. Magn. Reson.*, 1998, **131**, 144–147.
12. I. Hung, A. P. Howes, T. Anupõld, A. Samoson, M. E. Smith, D. Massiot, S. P. Brown, and R. Dupree, *Chem. Phys. Lett.*, 2006, **432**, 152–156.
13. Z. Gan, P. Srinivasan, J. R. Quine, S. Steuernagel, and B. Knott, *Chem. Phys. Lett.*, 2003, **367**, 163–169.
14. (a) A. Samoson and E. Lippmaa, *Phys. Rev.*, 1983, **B28**, 6567–6570; (b) A. P. M. Kentgens, J. M. M. Lemmens, F. M. M. Geurts, and W. S. Veeman, *J. Magn. Reson.*, 1987, **71**, 62–74.
15. N. Chandrakumar and A. Ramamoorthy, *J. Am. Chem. Soc.*, 1992, **114**, 1123–1124.
16. S. Cavadini, *Prog. Nucl. Magn. Reson. Spectrosc.*, 2010, **56**, 46–77.
17. D. Massiot, F. Fayon, B. Alonso, J. Trebosc, and J.-P. Amoureux, *J. Magn. Reson.*, 2003, **164**, 165–170.
18. A. Bax, R. H. Griffey, and B. L. Hawkins, *J. Magn. Reson.*, 1983, **55**, 301–315.
19. D. Massiot, F. Fayon, M. Deschamps, S. Cadars, P. Florian, V. Montouillout, N. Pellerin, J. Hiet, A. Rakhmatullin, and C. Bessada, *C. R. Chim.*, 2010, **13**, 117–129.
20. D. L. Bryce, K. Eichele, and R. E. Wasylishen, *Inorg. Chem.*, 2003, **42**, 5085–5096.
21. J.-P. Amoureux, J. Trebosc, J. W. Wiench, D. Massiot, and M. Pruski, *Solid State NMR*, 2005, **27**, 228–232.
22. C. Roiland, Etude de l'ordre local dans des phosphates désordonnés modèles par spectroscopies RMN et RAMAN, PhD thesis, Université d'Orléans, 17/10/2007. available from www.cemhti.cnrs-orleans.fr
23. X. Xue, *Solid State NMR*, 2010, **38**, 62–73.
24. D. L. VanderHart, H. S. Gutowsky, and T. C. Farrar, *J. Am. Chem. Soc.*, 1967, **89**, 5056–5057.
25. (a) E. Lippmaa, M. Alla, H. Raude, R. Teeaar, I. Heinmaa, and E. Kundla, In *Proceedings of the XXth Congress Ampere, Tallinn, 1978*, eds E. Kundla, E. Lippmaa, and T. Saluvere, Springer-Verlag: Berlin, Heidelberg, New York, 1979, p. 87; (b) E. Kundla and M. Alla, In *Proceedings of the XXth Congress*

Ampere, Tallinn, 1979, p. 1, 92; (c) M. H. Frey and S. J. Opella, *J. Chem. Soc., Chem. Commun.*, 1980, 474–475; (d) C. J. Groombridge, R. K. Harris, K. J. Packer, B. J. Say, and S. F. Tanner, *J. Chem. Soc., Chem. Commun.*, 1980, 174–175; (e) A. Naito, S. Ganapathy, and C. A. McDowell, *J. Chem. Phys.*, 1981, **74**, 5393–5397.

26. R. K. Harris and A. C. Olivieri, *Prog. Nucl. Magn. Reson. Spectrosc.*, 1992, **24**, 435–456.
27. J. W. Diesveld, E. M. Menger, H. T. Edzes, and W. S. Veeman, *J. Am. Chem. Soc.*, 1980, **102**, 1936–7931.
28. E. M. Menger and W. S. Veeman, *J. Magn. Reson.*, 1982, **46**, 257–268.
29. S. Wi, V. Frydman, and L. Frydman, *J. Chem. Phys.*, 2001, **114**, 8511–8519.
30. A. Davies, R. K. Harris, and A. C. Olivieri, *Mol. Phys.*, 1996, **87**, 669–677.
31. S. E. Ashbrook, J. McManus, M. J. Thrippleton, and S. Wimperis, *Prog. NMR Spectrosc.*, 2009, **55**, 160–181.
32. O. Lafon, Q. Wang, B. Hu, F. Vasconcelos, J. Trébosc, S. Cristol, F. Deng, and J.-P. Amoureux, *J. Phys. Chem. A*, 2009, **113**, 12864–12878.
33. M. Keppert, A. Rakhmatullin, F. Simko, M. Deschamps, G. M. Haarberg, and C. Bessada, *Magn. Reson. Chem.*, 2008, **46**, 803–810.
34. (a) E. R. H. van Eck and W. S. Veeman, *J. Am. Chem. Soc.*, 1993, **115**, 1168–1169; (b) H. M. Kao and C. P. Grey, *J. Magn. Reson.*, 1998, **133**, 313–323; (c) C. M. Morais, V. Montouillout, M. Deschamps, D. Iuga, F. Fayon, F. A. A. Paz, J. Rocha, C. Fernandez, and D. Massiot, *Magn. Reson. Chem.*, 2009, **47**, 942–947.
35. C. A. Fyfe, K. C. Wong-Moon, Y. Huang, and H. Grondey, *J. Am. Chem. Soc.*, 1995, **117**, 10397–10398.
36. V. Montouillout, C. M. Morais, A. Douy, F. Fayon, and D. Massiot, *Magn. Reson. Chem.*, 2006, **8**, 770–775.
37. M. Deschamps, F. Fayon, J. Hiet, G. Ferru, M. Derieppe, N. Pellerin, and D. Massiot, *Phys. Chem. Chem. Phys.*, 2008, **10**, 1298–1303.
38. D. Iuga, C. M. Morais, Z. Gan, D. R. Neuville, L. Cormier, and D. Massiot, *J. Am. Chem. Soc.*, 2005, **127**, 11540–11541.
39. C. M. Morais, V. Montouillout, M. Deschamps, D. Iuga, F. Fayon, F. A. A. Paz, J. Rocha, C. Fernandez, and D. Massiot, *Magn. Reson. Chem.*, 2009, **47**, 942–947.
40. L. Delevoye, J. Trébosc, Z. Gan, L. Montagne, and J.-P. Amoureux, *J. Magn. Reson.*, 2007, **186**, 94–99.
41. P. Kempgens, J. Raya, K. Elbayed, P. Granger, J. Rosé, and P. Braunstein, *J. Magn. Reson.*, 2000, **142**, 64–73.
42. N. S. Barrow, J. R. Yates, S. A. Feller, D. Holland, S. E. Ashbrook, P. Hodgkinson, and S. P. Brown, *Phys. Chem. Chem. Phys.*, 2011, **13**, 5778–5789.
43. (a) M. Deschamps, F. Fayon, V. Montouillout, and D. Massiot, *Chem. Commun.*, 2006, 1924–1925; (b) M. Deschamps and D. Massiot, *J. Magn. Reson.*, 2006, **184**, 13–19.
44. B. Hu, J.-P. Amoureux, J. Trebosc, M. Deschamps, and G. Tricot, *J. Chem. Phys.*, 2008, **128**, 134502.
45. M. E. Smith and E. R. H. van Eck, *Prog. Nucl. Magn. Reson. Spectrosc.*, 1999, **34**, 159–201.
46. J.-P. Amoureux, J. Trebosc, L. Delevoye, O. Lafon, B. Hu, and Q. Wang, *Solid State NMR*, 2009, **35**, 12–18.
47. M. Eden, *Solid State NMR*, 2009, **36**, 1–10.
48. A. J. Vega, *J. Magn. Reson.*, 1992, **96**, 50–68.
49. S. E. Ashbrook and S. Wimperis, *J. Chem. Phys.*, 2004, **120**, 2719–2731.
50. S. E. Ashbrook and S. Wimperis, *J. Chem. Phys.*, 2009, **131**, 194509.
51. Y. Zhang, F. Deng, J. Qiu, and C. Ye, *Solid State Nucl. Magn. Reson.*, 2000, **15**, 209–216.
52. (a) M. Baldus, D. Rovnyak, and R. G. Griffin, *J. Chem. Phys.*, 2000, **112**, 5902–5909; (b) W. Sun, J. T. Stephen, L. D. Potter, and Y. Wu, *J. Magn. Reson. A*, 1995, **116**, 181–188.
53. M. Eden, J. Grinshtein, and L. Frydman, *J. Am. Chem. Soc.*, 2002, **124**, 9708–9709.
54. (a) P. Hartmann, C. Jäger, and J. W. Zwanziger, *Solid State NMR*, 1999, **13**, 245–254; (b) T. G. Ajithkumar and A. P. M. Kentgens, *J. Am. Chem. Soc.*, 2003, **125**, 2398–2399; (c) T. G. Ajithkumar, E. R. H. van Eck, and A. P. M. Kentgens, *Solid State NMR*, 2004, **26**, 180–186.
55. M. Nijman, M. Ernst, A. P. M. Kentgens, and B. H. Meier, *Mol. Phys.*, 2000, **98**, 161–178.
56. M. Eden and L. Frydman, *J. Chem. Phys.*, 2001, **114**, 4116–4123.
57. M. Baldus, D. Rovniak, and R. G. Griffin, *J. Chem. Phys.*, 2000, **112**, 5902–5909.

58. (a) M. Eden, D. Zhou, and J. Yu, *Chem. Phys. Lett.*, 2006, **431**, 397–403; (b) M. Eden, H. Annersten, and A. Zazzi, *Chem. Phys. Lett.*, 2005, **410**, 24–30; (c) A. Y. H. Lo and M. Eden, *Phys. Chem. Chem. Phys.*, 2008, **10**, 6635–6644.

59. A. Brinkmann, A. P. M. Kentgens, T. Anupold, and A. Samoson, *J. Chem. Phys.*, 2008, **129**, 174507.

60. Q. Wang, B. Hu, O. Lafon, J. Trébosc, F. Deng, and J.-P. Amoureux, *J. Magn. Reson.*, 2009, **200**, 251–260.

61. G. Mali, G. Fink, and F. Taulelle, *J. Chem. Phys.*, 2004, **120**, 2835–2845.

62. G. Mali and F. Taulelle, *Chem. Commun.*, 2004, 868–869.

63. Z. W. Yu, A. M. Zheng, Q. A. Wang, L. Chen, J. Xu, J.-P. Amoureux, and F. Deng, *Angew. Chem. Int. Ed.*, 2010, **49**, 8657–8661.

64. N. G. Dowell, S. E. Ashbrook, J. McManus, and S. Wimperis, *J. Am. Chem. Soc.*, 2001, **123**, 8135–8136.

65. M. Eden, *Chem. Phys. Lett.*, 2009, **470**, 318–324.

66. A. J. Vega, *Solid State NMR*, 1992, **1**, 17–32.

67. C. A. Fyfe, H. Grondey, K. T. Mueller, K. C. Wong-Moon, and T. Markus, *J. Am. Chem. Soc.*, 1992, **114**, 5876–5878.

68. S. M. De Paul, M. Ernst, J. S. Shore, J. F. Stebbins, and A. Pines, *J. Phys. Chem. B*, 1997, **101**, 3240–3249.

69. S. Vega, *Phys. Rev. A*, 1981, **23**, 3152–3173.

70. T. H. Walter, G. L. Turner, and E. Oldfield, *J. Magn. Reson.*, 1988, **76**, 106–120.

71. H. Douglas Morris and P. D. Ellis, *J. Am. Chem. Soc.*, 1989, **111**, 6045–6049.

72. (a) J. C. C. Chan, M. Bertmer, and H. Eckert, *Chem. Phys. Lett.*, 1998, **292**, 154–160; (b) J. C. C. Chan, M. Bertmer, and H. Eckert, *J. Am. Chem. Soc.*, 1999, **121**, 5238–5248.

73. M. A. Eastman, *J. Magn. Reson.*, 1999, **139**, 98–108.

74. L. van Wüllen, L. Züchner, W. Müller-Warmuth, and H. Eckert, *Solid State Nucl. Magn. Reson.*, 1996, **6**, 203–212.

75. B. Hu, J.-P. Amoureux, and S. Hafner, *J. Magn. Reson.*, 2008, **192**, 8–16.

76. (a) M. Pruski, D. P. Lang, C. Fernandez, and J.-P. Amoureux, *Solid State Nucl. Magn. Reson.*, 1997, **7**, 327–331; (b) C. Fernandez, L. Delevoye, J.-P. Amoureux, D. P. Lang, and M. Pruski, *J. Am. Chem. Soc.*, 1997, **119**, 6858–6862; (c) J. C. C. Chan, *J. Magn. Reson.*, 1999, **140**, 487–490; (d) S. E. Ashbrook, S. P. Brown, and S. Wimperis, *Chem. Phys. Lett.*, 1998, **288**, 509–517; (e) K. H. Lim and C. P. Grey, *Chem. Phys. Lett.*, 1999, **312**, 45–56; (f) D. Rovnyak, M. Baldus, and R. G. Griffin, *J. Magn. Reson.*, 2000, **142**, 145–152; (g) L. Delevoye, C. Fernandez, C. M. Morais, J.-P. Amoureux, V. Montouillout, and J. Rocha, *Solid State NMR*, 2002, **22**, 501–512.

77. J. W. Wiench, G. Tricot, L. Delevoye, J. Trebosc, J. Frye, L. Montagne, J.-P. Amoureux, and M. Pruski, *Phys. Chem. Chem. Phys.*, 2006, **8**, 144–150.

78. M. E. Smith and E. R. H. van Eck, *Prog. Nucl. Magn. Reson.*, 1999, **34**, 159–201.

79. J. Schaefer, REDOR and TEDOR, in *Encyclopedia of Magnetic Resonance*, eds R. K. Harris and R. E. Wasylishen, John Wiley & Sons: Chichester, 2007. DOI: 10.1002/9780470034590.emrstm0448.

80. A. W. Hing, S. Vega, and J. Schaefer, *J. Magn. Reson.*, 1992, **96**, 205–209.

81. C. A. Fyfe, K. C. Wong-Moon, Y. Huang, H. Grondey, and K. T. Mueller, *J. Phys. Chem.*, 1995, **99**, 8707–8716.

82. C. A. Fyfe, K. T. Mueller, H. Grondey, and K. C. Wong-Moon, *J. Phys. Chem.*, 1993, **97**, 13484–13495.

83. J.-P. Amoureux, J. Trebosc, and G. Tricot, *Magn. Reson. Chem.*, 2008, **45**, S187–S191.

84. (a) J.-P. Amoureux, J. Trebosc, M. Pruski, and J. Wiench, *J. Magn. Reson.*, 2006, **184**, 1–14; (b) J. Trebosc, B. Hu, J.-P. Amoureux, and Z. Gan, *J. Magn. Reson.*, 2007, **186**, 220–227.

85. B. Hu, J. Trebosc, and J.-P. Amoureux, *J. Magn. Reson.*, 2008, **192**, 112–122.

86. J. Trébosc, O. Lafon, B. Hu, and J.-P. Amoureux, *Chem. Phys. Lett.*, 2010, **496**, 201–207.

87. M. Cutajar, S. E. Ashbrook, and S. Wimperis, *Chem. Phys. Lett.*, 2006, **423**, 276–281.

88. S. E. Ashbrook, S. Antonijevic, A. J. Berry, and S. Wimperis, *Chem. Phys. Lett.*, 2002, **364**, 634–642.

89. S. Cavadini, A. Abraham, S. Ulzega, and G. Bodenhausen, *J. Am. Chem. Soc.*, 2008, **130**, 10850–10851.

Chapter 13
Computing Electric Field Gradient Tensors

Josef W. Zwanziger

Department of Chemistry, Dalhousie University, Halifax, NS B3H 4J3, Canada

13.1 INTRODUCTION

Spectra of many types show the effect of the interaction between electric field gradients (EFGs) and electric nuclear quadrupole moments. Examples include nuclear magnetic resonance, nuclear quadrupole resonance, nuclear acoustic resonance, Mössbauer spectroscopy, and gas-phase rotational spectroscopy. The basic interaction is electrostatic in origin, arising from the coupling between the EFG at the nuclear position, which is a second-rank tensor, with the nuclear electric quadrupole moment, also a second-rank tensor. The result includes an isotropic component, which represents an energy shift and is thus observable in the spectrum. Both contributions may be zero, for different reasons. If the nuclear state being probed has zero spin (the state probed is typically the ground state but in Mössbauer spectroscopy may be an excited state), the nucleus must be spherically symmetric and thus will have zero electric quadrupole moment. Moreover, if the nucleus is located at a site of tetrahedral or higher point symmetry in a solid or molecule, then the EFG at the nucleus will be zero. For many common nuclei, however, in typical crystal and molecular environments, the interaction does not vanish and may in fact dominate the spectrum.

We will have little to say about the nuclear part of the coupling. The electric quadrupole moment of a given nuclear state is a complicated and only partially understood function of the protons and neutrons and their couplings, but fortunately is available as an empirical parameter for most nuclear ground states.[1]

The EFG itself can be conveniently expressed starting from the density of charge for the system. Call this distribution $n(\boldsymbol{r})$, and note that it includes the electronic and nuclear charges in the material. This charge distribution determines an electric potential at position $\boldsymbol{R}$ through the Coulomb Law:

$$V(\boldsymbol{R}) = \int d^3\boldsymbol{r}\,\frac{n(\boldsymbol{r})}{|\boldsymbol{r}-\boldsymbol{R}|} \tag{13.1}$$

Hence the EFG is determined through $-\nabla\nabla V$, evaluated at the nuclear site of interest, where the first derivative yields the electric field, and the second the gradient of the field.

The coupling between the EFG and the nuclear electric quadrupole moment is well known. The resulting Hamiltonian is conveniently expressed in terms of nuclear spin degrees of freedom as

$$\mathcal{H}_{\mathrm{Q}} = \frac{eQ}{6I\,(2I-1)}\left[V_{xx}\left(3I_x^2 - I^2\right) + V_{yy}\left(3I_y^2 - I^2\right) + V_{zz}\left(3I_z^2 - I^2\right)\right] \tag{13.2}$$

NMR of Quadrupolar Nuclei in Solid Materials
Edited by Roderick E. Wasylishen, Sharon E. Ashbrook and Stephen Wimperis

Here the nuclear spin quantum number is I, and the z component of the nuclear electric quadrupole moment is given by eQ. The gradient elements are defined by

$$V_{xx} = -\left.\frac{\partial^2 V}{\partial x^2}\right|_{\boldsymbol{R}} \tag{13.3}$$

and similarly for y and z, with V the potential of equation (13.1). The derivatives are evaluated at $\boldsymbol{R}$, the location of the atomic nucleus of interest. Directions x, y, z are chosen such that the 3×3 matrix of second partials is diagonal: this construction defines the so-called principal axes of the interaction. Note that these directions need not coincide with other coordinates of the system, such as those used to describe the unit cell or chemical bonds. Also, due to Laplace's equation the principal components satisfy $\sum V_{ii} = 0$. It is conventional to choose the principal z-axis to lie in the direction of the largest in magnitude principal value of V, and denote this term as $V_{zz} = eq$. The coupling constant between the gradient and the nuclear quadrupole moment is then denoted in frequency units as $C_Q = e^2qQ/h$. The other directions are chosen such that

$$|V_{zz}| \geq |V_{yy}| \geq |V_{xx}| \tag{13.4}$$

The asymmetry of the interaction is defined as

$$\eta = \frac{V_{xx} - V_{yy}}{V_{zz}} \tag{13.5}$$

Owing to the considerations above, the computational effort in determining quadrupole interactions really condenses down to computing the EFG tensor, or more precisely, the charge distribution in equation (13.1). From that quantity, all else follow from simple differentiation. Therefore, the present chapter describes current computational models for finding the charge density, with a special focus on extended systems (solids). As the quadrupole coupling is traceless, due to Laplace's equation, it does not survive the rapid, isotropic averaging of the liquid state and hence is directly observable in NMR only in the solid state. Its effect on relaxation in the liquid state can be profound, but the methods discussed here are also amenable to computation of the EFG in isolated molecules and so could be carried over to that case as well.

13.2 MODELS OF THE CHARGE DISTRIBUTION

13.2.1 Point-charge Models

The simplest model of the charge distribution is a point-charge model, in which the atoms in a solid or molecule are taken to be ions of fixed charge, but no other distribution of electron density is allowed. These models give typically very poor representations of the EFG, exactly because they avoid considering the details of the charge distribution near the nucleus being probed. To see why, consider a nucleus under investigation and a charged particle at location $\mathbf{r}$. Without loss of generality, we may consider the nucleus to be at the coordinate system origin. The Coulomb potential at the origin is then $V(\mathbf{r}) = e/|\mathbf{r}|$, and the EFG components are

$$V_{zz} = -e\left(\frac{1}{r^3} - \frac{3z^2}{r^5}\right) \tag{13.6}$$

$$V_{xx} = -e\left(\frac{1}{r^3} - \frac{3x^2}{r^5}\right) \tag{13.7}$$

$$V_{yy} = -e\left(\frac{1}{r^3} - \frac{3y^2}{r^5}\right) \tag{13.8}$$

Because of the dependence on r^{-3}, the EFG is very sensitive to details of the charge distribution at short range. Therefore, models (such as the point-charge model) that treat the electrons as either localized on the probe nucleus or localized on distant nuclei rarely give accurate results. For example, in the silica polymorph stishovite, the largest component of the EFG at oxygen is 1.08 ± 0.02 in atomic units, experimentally,[1,2] while a point-charge model using +4 on silicon and −2 on oxygen gives 0.262 in atomic units for the EFG at oxygen (1 au $\approx 9.717 \times 10^{21}$ V m^{-2}). These values could be adjusted by including additional ad hoc factors, but that would serve primarily only to reproduce the data, not to explain it or predict new phenomena. An additional issue that arises when point charge models are applied to extended systems is slow convergence. Because the EFG varies as r^{-3}, when it is integrated over space in an extended system (thus the volume increases like r^3) the convergence is very slow. This problem can be accurately treated using an Ewald sum method,[3] but, while this is always done in modern electronic structure codes to treat the ionic part of the EFG, it is almost never done in applications of point-charge models to fragments of solids.

13.2.2 Ab initio Models

Models of the charge density originating from the exact many-body Hamiltonian of the electrons and nuclei in the system may be termed ab initio. This approach is quite commonplace in quantum chemistry, and is implemented in a variety of proprietary and open-source codes, in many styles. While full configuration interaction calculations could reasonably claim to give the exact electron density distribution, free of assumptions and therefore perfectly "ab initio", their extreme computational cost makes them prohibitive for all but the smallest systems (atoms and diatomic molecules). Other ab initio approaches, starting with Hartree–Fock and continuing through more sophisticated treatments of electron correlation and configuration interaction, are best understood as models rather than exact theory. For finite systems such as molecules, this approach is very mature and widely used, and capable of giving excellent results for the charge distribution and hence EFG.

However, only Hartree–Fock, which gives a relatively simplistic treatment of electron correlation, scales with system size in a way that is competitive with density functional theory (DFT), to be discussed below. More accurate treatments of correlation, such as Møller–Plesset perturbation theory and coupled cluster methods, scale much more poorly with system size. For solids with small unit cells, such approaches should nevertheless be practical but still are rarely used.

While all but the simplest ab initio models scale relatively poorly with system size and therefore are not widely used to study extended solids, it is important to bear in mind several areas where they are consistently more accurate than DFT-based methods. One is in systems where dispersion interactions are dominant, because ab initio approaches capture dispersion physics well but only very recently have DFT-based approaches begun to succeed in this area. In solids bound primarily by dispersion forces, a reasonable computational starting point would be to treat a single molecule from the solid with ab initio methods. A second area where DFT methods are difficult to apply is in computation of excited state properties and molecular transition states. As this chapter is focused on the EFG, which is almost always measured for the stable ground state, this relative failing of DFT is not so important. More important is the third point, that it is usually much clearer with ab initio methods what must be done to improve accuracy, due to their variational nature (with respect to the true full configuration interaction energy).

Despite all the above, the fact is that modern DFT methods for solids scale adequately and have adequate accuracy to have become extraordinarily useful tools for interpreting experiments on solids. The remainder of this chapter will be devoted to discussing how DFT is used (and abused) to compute electron densities and EFGs in solids.

13.2.3 Density Functional-based Models

An excellent if somewhat advanced entry into this field is the book by Martin,[4] while the basic solid-state physics is covered, for example, by Ashcroft and Mermin.[5]

13.2.3.1 Overview

DFT is a theory of many-body interactions, which casts the problem in terms of a functional of the particle density. For electrons, the theory is founded on the Hohenberg–Kohn theorems, which state that ground-state electronic properties are unique functionals of the ground-state electronic density distribution and that the ground-state energy is variational as a functional of the density. These beautiful theorems are relatively easy to prove[4] but unfortunately are nonconstructive, meaning that they prove existence but do not show how to carry out the process of finding the density. The Kohn–Sham method provides a partial solution: it maps the many-body problem onto a reference system composed of independent particles, plus a correction. The independent particle problem is broadly like a Hartree–Fock calculation, and this is why DFT scales similarly to Hartree–Fock. The problem, of course, is in the correction. This term is the so-called exchange-correlation functional, and due to the nonconstructive nature of Hohenberg–Kohn theorems, one knows that it exists and is unique but not what it is or how to find it. Many models of this term have been created, some with more or less empirical input—for this reason, DFT methods are often not referred to as ab initio but rather, by way of compromise, as first-principles methods. Be that as it may, for many systems there are models of the exchange-correlation functional that are both highly efficient to use and reasonably accurate, certainly for ground-state properties like the EFG.

13.2.3.2 Basis Sets

Using DFT to find the ground-state density of a system begins from the Kohn–Sham equations, which are the result of mapping the real system onto a fictitious independent particle system. The result is

$$\left(-\frac{\hbar^2}{2m}\nabla^2 + v_{\text{eff}}\right)\phi_i(\boldsymbol{r}) = \epsilon_i\phi_i(\boldsymbol{r}) \qquad (13.9)$$

where ϕ_i is the ith Kohn–Sham orbital, m the electron mass and $\hbar$ is the Planck constant divided by 2π. From the solutions of the Kohn–Sham equations, the density will be constructed from

$$\rho(\boldsymbol{r}) = \sum_{\text{occ}} |\phi(\boldsymbol{r})|^2 \qquad (13.10)$$

The potential v_{eff} consists of a Hartree term (electron–electron Coulomb repulsion), an "external" potential, due both to attraction to the ion cores and any external fields that may be present, and the exchange-correlation potential. It is important to note that the eigenvalues of the Kohn–Sham equations are not the energies of the original system, but the energies of the fictitious independent particle system. The energies of the interacting system are reconstructed from the solutions of the Kohn–Sham equations by using the density generated in equation (13.10).

Solving the Kohn–Sham equations, which are of the Schrödinger equation form, is typically carried out by expanding the unknown Kohn–Sham orbitals in terms of basis functions. One broad category of choices is orbitals localized in real space, for example Gaussians or Slater-type orbitals (STO's) localized at each nucleus. This approach has the advantage that the results are fairly simple to interpret in terms of bonding constructs, and for systems that are also localized in real space, such as molecules, this approach is also conceptually simple. When a solid is treated by a cluster model, that is, by extracting a cluster of atoms from the extended system and terminating the cluster for charge balance such that a molecule is obtained with geometry similar to that of the solid, the method of real space orbitals is almost always used. However, if the solid is treated as truly extended, with periodic boundary conditions, orbitals are not localized in real space but are localized in reciprocal space. It is then in many ways simpler (though not strictly necessary) to use a basis that is also localized in reciprocal space, and by far the simplest choice in this regard are plane waves of the form $e^{i\mathbf{G}\cdot\mathbf{r}}$, where $\mathbf{G}$ is the plane-wave vector. In solids, the translational invariance means as well that the Bloch theorem holds; in other words the crystal momentum $\mathbf{k}$ is a good quantum number, so that the expansion of a wave function takes the form

$$\Psi_{nk}(\boldsymbol{r}) = e^{i\boldsymbol{k}\cdot\boldsymbol{r}}u_{nk}(\boldsymbol{r}) = \sum_{\boldsymbol{G}} e^{i(\boldsymbol{k}+\boldsymbol{G})\cdot\boldsymbol{r}} \qquad (13.11)$$

While it is true that Gaussians and STO's resemble atomic wave functions more closely than plane waves do, and therefore fewer of them are required to represent the Kohn–Sham orbitals as compared to plane wave expansions, the translation of the solid-state problem via the Bloch theorem and plane waves into the expansion equation (13.11) means that the power of the fast Fourier transform can be used to solve the problem and the numerical efficiency becomes extraordinary.

13.2.3.3 Potentials and Pseudopotentials

The Kohn–Sham equations of DFT are solved by expansion in a basis, but a choice must also be made as to how to treat the potential v_{eff}. Recall that the potential includes the Hartree contribution, the external part, and the exchange-correlation term. In treating the potential, one makes effectively two choices: which exchange-correlation model to adopt, and whether and how to approximate the many-body interactions giving rise to the full potential.

Exchange-correlation Potential

Currently there are many, many models for the exchange-correlation potential of DFT. The simplest, the local-density approximation or LDA, takes its name from the fact that it approximates the exchange-correlation energy only from the density $\rho(\mathbf{r})$ itself. The exact form is typically obtained from the exchange energy density and correlation energy density of a homogeneous electron gas. The first of these terms is known exactly, while the second is approximated by fitting to results of quantum Monte Carlo calculations. The resulting potential is very efficient to compute and is available in essentially all codes. It gives reasonable results for many solids but has a few (in)famous failures, for example the prediction that bulk germanium is a metal, rather than a semiconductor.

Models that include the gradient of the density in addition to the density in order to reproduce the exchange-correlation energy can give significant improvements over LDA, at a modest increase in computational cost. These models fall under the heading of generalized gradient approximation (GGA). There are too many to list here—a common choice is the Perdew–Burke–Ernzerhof (PBE) GGA,[6] which is also available in essentially all codes and performs well in many situations.

Common in codes that implement DFT in real space are the so-called hybrid functionals, such as the Becke, Lee, Yang, Parr (BLYP) and PBE0 families. In this approach, the exact exchange energy from the Hartree–Fock approach is combined with exchange-correlation models of the types outlined above, by weighting the contributions. The weights of the two contributions are determined either by fitting to data or from theoretical arguments. The hybrid approach is very widely used in DFT studies of molecules, but less so in extended solids because of numerical difficulties of computing the Hartree–Fock exchange in periodic systems.

Approximating the Potential

Having chosen a model for the exchange-correlation part of the potential, it remains to deal with the full many-body potential of the interacting electrons. Full-potential methods treat all electrons of the problem directly, and in principle this is the most accurate approach. It is however computationally very costly and only relatively small systems (in the sense of small numbers of electrons) can be treated this way. In order to render the problem computationally tractable, typically the system is solved by one method within spheres around each atom and by a second method in the interstitial regions, making sure to impose a proper matching condition at the boundaries. For example, in the linear muffin tin orbital (LMTO) approach, a basis of Bessel or Neumann functions is used within spheres around the atoms while the interstitial region is treated by continuation (or not at all, in the most extreme case). In the augmented plane wave method (APW), the atomic sphere regions are treated by a combination of spherical harmonics and plane waves, while the interstitial region is treated with plane waves only. When this approach is done with the full potential, it is termed FP-LAPW, and is the current state of the art for DFT computations of solids. It is computationally costly however, so alternatives are of great interest.

The most common alternative to the full-potential approach is the use of pseudopotentials. The idea here is to divide the electrons from each atom as arising from core states and valence states, and to treat the core electrons in the full problem indirectly through the use of a "pseudo" potential, which interacts with the valence electrons in the same way as the original core electrons did but which also causes the valence wave functions to have smooth, nodeless radial features near the core, and therefore can be easily expanded in relatively few plane waves. The resulting pseudovalence wave functions therefore are very different near the core from the true valence functions, but nevertheless display the same physics at farther distances. The utility of this approach comes from the fact that the core electrons have little detailed effect on many of the measurable properties of solids, such as band structures, elastic constants, and so forth, so spending considerable computational effort computing their effects is inefficient. On the other hand, the quantity of interest in this chapter, the EFG, does depend sensitively on details of the charge density and hence wave functions near the atomic nuclei, so this approach (in its simplest form) is not useful for EFG calculations.

13.2.3.4 *Projector-augmented Wave Method*

A highly effective compromise between the pseudopotential method, which is fast but neglects details of the charges near the atoms, and the FP-LAPW method, which treats all electrons in a very detailed way but is computationally intensive, is the projector-augmented wave method (PAW).[7–9] Like FP-LAPW, the PAW method separates the electronic problem into spheres around atoms and interstitial regions, but it also makes use of pseudopotentials. Its defining feature is the construction of a linear operator—the PAW transformation—that reconstructs within the spheres the all-electron expectation values of operators by making use of the pseudo-wave functions computed from a pseudopotential calculation. The computational overhead is larger than in a pure pseudopotential calculation but much less than in FP-LAPW; at the same time, for many properties the accuracy is nearly the same.

As the PAW method makes use of pseudopotentials, it relies on separating the states of each atom into core and valence; the valence electrons are treated in the field of the pseudopotentials and each other in the Kohn–Sham equation, and the resulting charge distribution is used to compute the distributions in the spheres. By construction, the sphere corrections vanish outside the spheres, so there (in the interstitial region) the computed density is used directly. Because of the core–valence separation, it is necessary to check that the proper choice was made (that is, that all electrons treated as core really do not contribute beyond their spherical average to the computed observables). This point will be examined in practical cases below. Furthermore, it is necessary to obtain not just pseudopotentials but the full PAW atomic dataset for each atom of interest before starting the calculation. Depending on the code in use, this may or may not be easy. It is also crucial that the sphere radius defined for each atom type do not overlap in the crystal of interest, as the entire method depends on the computation of independent sphere corrections.

The essential PAW transformation operator is typically written as follows:

$$T = 1 + \sum_{i,I} \left[|\phi_i^I\rangle - |\tilde{\phi}_i^I\rangle \right] \langle \tilde{p}_i^I | \qquad (13.12)$$

where I labels the ions in the unit cell, and i the atomic basis functions used to describe the valence electrons within a given sphere. Functions without a tilde, like $|\phi_i^I\rangle$, describe the valence electrons, while those with tilde, for example $|\tilde{\phi}_i^I\rangle$, describe the electrons as interacting with a pseudopotential. The term $\langle \tilde{p}_i^I|$ is the projector from the pseudofunctions to the valence electron functions and is dual to $|\tilde{\phi}_i^I\rangle$, meaning that $\langle \tilde{p}_i^I|\tilde{\phi}_j^{I'}\rangle = \delta_{I'I}\delta_{ij}$. The projectors vanish outside the predefined radius.

The key PAW result is that, *if* the basis of atomic functions for the valence electrons is complete *and* *if* the PAW spheres on different ions do not overlap, then the following identity holds:

$$\begin{aligned}\langle \Psi_{nk}|A|\Psi_{nk}\rangle &= \left\langle \tilde{\Psi}_{nk}|T^{\dagger}AT|\tilde{\Psi}_{nk}\right\rangle = \left\langle \tilde{\Psi}_{nk}|A|\tilde{\Psi}_{nk}\right\rangle \\ &+ \sum_{I,i,j} \left\langle \tilde{\Psi}_{nk}|\tilde{p}_i^I\right\rangle \left[\langle \phi_i^I|A|\phi_j^I\rangle \right. \\ &\left. - \left\langle \tilde{\phi}_i^I|A|\tilde{\phi}_j^I\right\rangle \right] \left\langle \tilde{p}_j^I|\tilde{\Psi}_{nk}\right\rangle \qquad (13.13)\end{aligned}$$

The significance of this result is that the expectation of the observable A may be computed exactly in terms of pseudo-wave functions and atomic quantities that have been precomputed in the construction of the PAW pseudopotentials. Therefore, quality rivaling FP-LAPW should be obtainable at a computational cost similar to that of regular pseudopotential calculations. However, for this approach to be successful, the PAW atomic datasets must be carefully made and checked and, furthermore, the separation between core and valence states must be adequate. In practice, however, once a good PAW atomic dataset is generated for a class of materials (say, oxygen in inorganic oxides), it is usable for many different compounds.

Using the PAW method to compute the EFG is quite straightforward and amounts to a small postprocessing calculation following a typical ground-state PAW determination of electronic structure.[10,11] The charge distribution, which is the result of the PAW ground-state calculation, is written as

$$n(\boldsymbol{r},\boldsymbol{R}) = n_Z(\boldsymbol{R}) + \tilde{n}(\boldsymbol{r}) + n^1(\boldsymbol{r}) - \tilde{n}^1(\boldsymbol{r}) \qquad (13.14)$$

where n_Z are the ionic charges at locations $\boldsymbol{R}$ in the crystal, $\tilde{n}$ is the pseudovalence density, n^1 is the all-electron density within spheres, and $\tilde{n}^1$ is the pseudo density within spheres. As usual in PAW calculations, the pictorial representation is of a pseudo quantity everywhere in the solid, plus the all-electron version in spheres, minus the pseudo version in spheres (see Figure 13.1).

The result is the pseudo value outside spheres and the all-electron version inside spheres, with the spheres large enough such that the two quantities can be smoothly mapped onto one another. The EFG of all terms in equation (13.14) are then computed and summed to give the total. The ion term, n_Z, contains the nuclear and core charges together. Thus, for example, if oxygen is treated with 1s core electrons

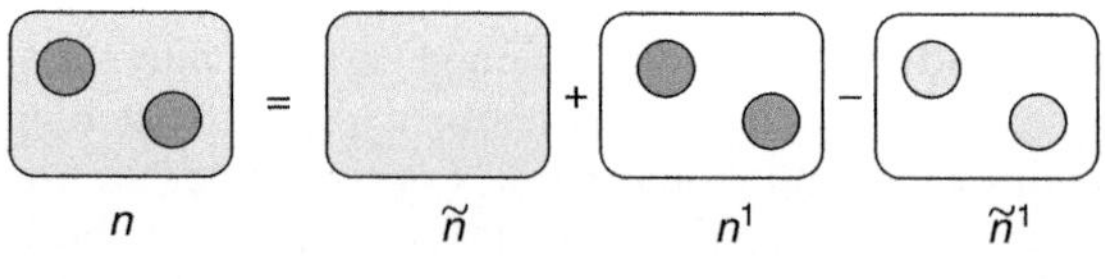

Figure 13.1. Pictorial representation of the PAW treatment of electronic charge density. The total density n is represented as a pseudized density $\tilde{n}$ everywhere, plus an all-electron density within atom-centered spheres n^1 minus the pseudized density within spheres $\tilde{n}^1$. In the schematic, pseudized values are represented by light gray shading, and all-electron values by dark gray shading.

and 2s plus 2p valence, the value of n_Z at each oxygen site would be +6. Due to the slow convergence of the EFG in real space, the EFG due to the ion charges is computed by an Ewald sum method.[3] Next, the EFG due to the pseudovalence charge distribution, $\tilde{n}$, is computed most simply by Fourier-transforming it to reciprocal space and summing the contributions. Finally, the EFGs due to the two on-site densities n^1 and $\tilde{n}^1$ are treated together and computed using the PAW expansion in atom-centered basis functions $|\phi_i\rangle$ and $|\tilde{\phi}_i\rangle$ (see equation (13.12)).

13.2.3.5 Some Comparisons of Methods

As outlined in the sections above, there currently exist a large variety of methods for computing the charge density in a solid or molecule, and hence the EFG. Furthermore, there exist a number of codes, for each method, some of which are open source and some of which are proprietary. Our aim here is not to provide exhaustive comparisons, but just to show broadly how two of the leading methods for solids, PAW and FP-LAPW, compare with each other and with experiment. Note that both are DFT-based. We also note that the technical test of the PAW formalism is reproduction of FP-LAPW, *not* reproduction of experiment, because the PAW formalism should be examined in the same DFT setting with the same exchange-correlation functional approximation but with an all-electron potential. Because the universal and correct exchange-correlation functional is not known, and there are a variety of issues in experimental data that may not be treated in a calculation (such as temperature effects, defects, etc.), agreement with experiment is in general important but not the last word in testing models.

Calculations of EFGs in the PAW formalism are available in at least three codes, including ABINIT,[12] PWSCF,[13] and VASP (cms.mpi.univie.ac.at/vasp/). The first two of these are open source, the third is proprietary. Here we show results from ABINIT, because of our own familiarity with it, using PAW datasets of our own construction using the ATOMPAW code.[14] The FP-LAPW method is likewise available in several packages, including ELK (elk.sourceforge.net) and WIEN2K (www.wien2k.at); here we use ELK, which is open source. All the open-source packages mentioned are available free of charge and can be compiled on many different platforms.

The first set of results we present are for oxygen-17 quadrupole couplings in polymorphs of silica, SiO_2 (Table 13.1).

In these calculations, we used the LDA exchange-correlation functional. The oxygen and silicon valence spaces were chosen as 2s/2p and 3s/3p, respectively, while the core spaces were 1s and 1s/2s/2p. The justification for such a choice comes from the all-electron atomic energy levels, which are computed as part of the PAW atomic dataset construction. These values are shown in Table 13.2.

The gap between the highest core state and the lowest valence state is shown in the table, and we find in practice that gaps greater than about 3 Ha indicate states that may be safely relegated to the core. The core states will be treated as a spherical pseudopotential and will affect the valence electrons indirectly. It is often thought that, for NMR observables, which

Table 13.1. Oxygen quadrupole couplings in several polymorphs of silica. For each site, values obtained from PAW computations using the `Abinit` code, FP-LAPW computations using the `Elk` code, and experimental results are given

Compound	Site	PAW		FP-LAPW		Expt	
		C_Q/MHz	η	C_Q/MHz	η	C_Q/MHz	η
α-Quartz[15]	O1	5.355	0.252	5.398	0.245	5.2 ± 0.2	0.19 ± 0.05
Stishovite[2]	O1	6.530	0.126	6.429	0.126	6.5 ± 0.1	0.13 ± 0.05
Cristobalite[2]	O1	5.372	0.142	5.420	0.139	5.3 ± 0.1	0.13 ± 0.05
Coesite[16]	O1	6.117	0.051			6.05 ± 0.05	0.000 ± 0.005
	O2	5.318	0.195			5.43 ± 0.05	0.166 ± 0.005
	O3	5.429	0.171			5.52 ± 0.05	0.169 ± 0.005
	O4	5.427	0.162			5.45 ± 0.05	0.168 ± 0.005
	O5	5.128	0.293			5.16 ± 0.05	0.292 ± 0.005

Table 13.2. Atomic properties relevant to the PAW atomic datasets used in the present calculations

Atom	Valence	ΔE_{vc}/Ha	r_{PAW}/Bohr
O	$2s^2 2p^4$	35.8	1.45
Si	$3s^2 3p^2$	6.23	1.55
I	$5s^2 5p^5$	2.42	2.40
Cl	$3s^2 3p^5$	12.6	1.95

typically involve factors of r^{-3} and therefore accentuate details of electronic structure near the nuclei, all electrons will participate and cannot be ignored. That view is not correct, however: the interactions probed by NMR are in fact quite weak, so that core states bound by electrostatic interactions of many hartrees are simply not perturbed to any appreciable extent from their atomic spherical shape and so it is both correct and highly efficient to treat them through a pseudopotential construction. Table 13.1 shows this point by the close correspondence between the PAW calculations and the FP-LAPW calculations. We will examine this issue further below.

Another point to consider in the PAW method is the completeness of the PAW states. It was noted above (equation (13.13)) that completeness is required in order for the PAW method to be accurate and well founded. What this means in practice is that the number of states of each angular momentum used to describe the valence must be large enough. Thus for a 2s/2p valence space, there must be at least one s-type and one p-type atomic state within each PAW sphere; practice has shown that one such projector of each type is almost never enough, while two of each type *usually* is. In careful work, this point should be checked by making PAW atomic datasets with 1, 2, and 3 projectors of each angular momentum type in the valence space and examining convergence of properties. This strategy is easy to carry out with open-source codes, usually less so with proprietary codes that provide readymade PAW datasets. Unfortunately, for PAW and other pseudopotential methods, one size does not fit all.

The radius of the PAW spheres must also be chosen properly (Table 13.2). It is essential that the PAW spheres do not overlap; on the other hand, the smaller the spheres, the more the calculations to be handled by the plane wave expansion in the interstitial region, leading to more plane waves and a more expensive calculation. Typically, the bond lengths in the material of interest should be tabulated (at least roughly) and the radii of the available PAW sets compared to these to ensure that there will be no overlap. For instance, in the silica polymorphs treated above, the silicon PAW radius of 1.55 bohr is quite a bit smaller than would be necessary in a calculation of bulk silicon; however, Si–O bonds are short, and so there is no choice here but to use a relatively hard silicon PAW dataset.

The FP-LAPW is free from many of the choices and considerations discussed above for PAW. To be sure, both methods still require a choice of exchange-correlation functional, and both must always be checked on any material for convergence in energy, reciprocal space sampling, and so forth. Nonetheless, the FP-LAPW method relies on parameterizations of all-electron atomic states, which really only need to be done once, and then can be used repeatedly. Given that both methods give essentially the same result, why choose the PAW method, which appears (certainly for first-time users) to be so much more involved to learn and use? The answer is simple: time. The calculations presented in Table 13.1 took about a factor of 30 longer with the FP-LAPW code; thus the simple cells, such as stishovite, required a few minutes with PAW on a desktop PC, and 30–60 min with FP-LAPW; cristobalite took about 30 min with PAW, and 11 h with FP-LAPW, and coesite was considered too large to compute reasonably with FP-LAPW.

As another set of illustrative calculations, we chose the molecule ICl. Iodine is a relatively heavy atom and so is subject to more complex physics than light atoms, such as relativistic effects. The quadrupole coupling at iodine is listed in Table 13.3 for a variety of different approximations within the PAW formalism.

The table summarizes several effects. First, most calculations were made with valence $5s^2 5p^5$

Table 13.3. Computed and experimental quadrupole couplings C_Q for ^{127}I, in MHz, for molecular (gas phase) ICl. In all computations the PAW formalism was used, together with the experimental bond length of 2.3236 Å

Method	C_Q/MHz
5s/5p/rel/LDA	2905
5s/5p/nonrel/PBE	2458
5s/5p/rel/PBE	2949
4 d/5s/5p/rel/PBE	2942
Expt[17]	2928

(see Table 13.2) although the gap between the 4d and 5s states is only 2.42 Ha. The effect of this approximation is seen by adding 4d electrons to the PAW valence, which changed the quadrupole coupling from 2949 to 2942 MHz, other parts of the calculation being unchanged. Much more significant are the choice of exchange-correlation functional and relativistic effects. The LDA exchange-correlation functional gave a result of 2905 MHz, while the PBE GGA yielded 2949, a modest improvement in agreement with experiment although the first is an underestimate while the second is an overestimate. It is tempting but in general poor practice to try to fine-tune results by the choice of exchange-correlation functional; it is much more sound in general to choose a model that is known to reproduce the physics of systems like the one under study and then stick with it.

Relativistic effects are included within PAW implicitly, through the shape of the pseudopotential. The effect in the iodine C_Q is large: 2458 MHz without relativistic corrections, and 2949 MHz with, a change of 20%. Such corrections are included as follows: When the PAW dataset for iodine is constructed,[14] one choice is whether to treat the atomic system with the nonrelativistic Schrödinger equation or the scalar relativistic Koelling–Harmon Equation. For heavy atoms, the second choice is of course preferred and will lead to a different and more correct shape for the pseudopotential. The valence electrons are solved with a nonrelativistic approach, which is correct because they do not move at relativistic speeds, but now they will interact with ionic cores of the correct relativistic shape. This interaction changes their distribution, leading to more accurate observables such as the EFG. For light atoms, even the core electrons are essentially nonrelativistic, so these corrections are negligible, but they become large for heavy atoms.

13.2.4 Beyond Basic DFT

In the Kohn–Sham formalism, DFT is a single-particle theory generated from a map onto the homogeneous electron gas. Perhaps, not surprisingly therefore, there are systems for which basic DFT does not provide a good description. Cases with significant dispersion forces have already been mentioned above, but that is a problem of the current state of the art of the exchange-correlation functional. Here we have in mind systems for which the basic premise of the map onto the electron gas is in some way at fault, or at least too simplistic. A standard set of such materials are the so-called Mott–Hubbard insulators, such as nickel oxide, NiO, and lanthanum titanate, $LaTiO_3$. In these materials, a single-particle picture would suggest that they are metallic, while in fact they are insulators (and typically with antiferromagnetic order).

In Mott–Hubbard insulators, the breakdown of DFT can be detected immediately from the band structure it predicts. Repairing the problem requires some insight into why the system is an insulator, not a metal. Typically, the origin of the physics lies in unusually strong electron–electron correlations on the transition-metal ions, leading to an additional repulsive energy beyond what standard DFT would predict. This energy may be included in DFT using the so-called LDA+U formalism, in which the name indicates that standard DFT is being augmented by the Hubbard U term. The Hubbard Hamiltonian is a simple model of strong electron correlation which involves an on-site energy U and a coupling energy J. Standard DFT can be augmented by these terms, which provides (in an admittedly somewhat ad hoc way) a much improved description, including typically insulating behavior and the proper magnetic ordering.

Such physics matters for the EFG because it has a profound effect on the electron distribution and hence detectable in the quadrupole coupling. For example, in $LaTiO_3$ the coupling at the Ti sites was observed to be quite small compared to what might be expected in a titanate, while the La coupling was normal; we found that by computing the EFG in the PAW formalism together with Hubbard U and J terms, not only was the electronic structure corrected, as expected, but also the EFG values were predicted correctly in accord with experiment.[18] We could show that the reason why the EFG is enhanced in the presence of the Hubbard model physics is that in this system the effect of the electron correlation was to stabilize some of the d orbitals on titanium and to add electron density to these orbitals over and above what simple DFT would predict. Because these orbitals match the symmetry of the EFG tensor itself, adding density to them also has the effect of strongly enhancing the EFG. On the La sites, by contrast, there also is strong electron correlation but it affects the f orbitals; as these do not have the symmetry to match the EFG symmetry, no similar enhancement is seen.

13.3 SUMMARY AND CONCLUSIONS

In this chapter, we have attempted to provide a practical introduction to current techniques for computing the EFG tensor in solids. While the discussion has not been rigorous, hopefully it highlights the areas where these methods are robust and where extra caution is needed. Two methods, both implementations of DFT, were highlighted: the FP-LAPW approach and the PAW formalism. The former is highly accurate (within the limits of current model exchange-correlation functionals) but computationally very demanding. The PAW method, on the other hand, is capable of the same high accuracy in a much shorter time, but requires more effort from the user to construct and test efficient and accurate PAW atomic datasets. Fortunately, once this construction has been accomplished for a class of materials, the datasets can be reused with much less checking than during the initial testing phase.

We showed how relativistic effects are included in the PAW formalism and how they can contribute significantly for heavy atoms. Finally, we discussed typical ways in which DFT can break down, particularly in the presence of strong electron correlations, and how additional terms can be included that account for such effects.

Broadly speaking, outside the realm of systems dominated by dispersion forces, modern DFT is accurate enough to provide a good description of electronic structure and hence the EFG and quadrupole coupling in a very wide range of solids, including insulators (as discussed here) and also semiconductors and metals.

RELATED ARTICLES IN THE ENCYCLOPEDIA OF MAGNETIC RESONANCE

Accurate Determination of Nuclear Quadrupole Coupling Constants and other NMR Parameters in Liquids from the Combination of Molecular Dynamics Simulations and ab initio Calculations

Relaxation Theory for Quadrupolar Nuclei

Shielding: Overview of Theoretical Methods

REFERENCES

1. P. Pyykkö, *Mol. Phys.*, 2008, **106**, 1965.
2. X. Xue, J. F. Stebbins, and M. Kanzaki, *Am. Miner.*, 1994, **79**, 31.
3. A. Honma, *J. Phys. Soc. Jpn.*, 1977, **42**, 1129.
4. R. M. Martin, *Electronic Structure: Basic Theory and Practical Methods*, Cambridge University Press: Cambridge, 2004.
5. N. Ashcroft and N. Mermin, *Solid State Physics*, Saunders College Publishing: Philadelphia, PA, 1976.
6. J. P. Perdew, K. Burke, and M. Ernzerhof, *Phys. Rev. Lett.*, 1996, **77**, 3865.
7. P. E. Blöchl, *Phys. Rev. B*, 1994, **50**, 17953.
8. G. Kresse and D. Joubert, *Phys. Rev. B*, 1999, **59**, 1758.
9. M. Torrent, F. Jollet, F. Bottin, G. Zérah, and X. Gonze, *Comput. Mater. Sci.*, 2008, **42**, 337.
10. M. Profeta, F. Mauri, and C. J. Pickard, *J. Am. Chem. Soc.*, 2003, **125**, 541.
11. J. W. Zwanziger and M. Torrent, *Appl. Magn. Reson.*, 2008, **33**, 4477.
12. X. Gonze, B. Amadon, P.-M. Anglade, J.-M. Beuken, F. Bottin, P. Boulanger, F. Bruneval, D. Caliste, R. Caracas, M. Cote, T. Deutsch, L. Genovese, P. Ghosez, M. Giantomassi, S. Goedecker, D. Hamann, P. Hermet, F. Jollet, G. Jomard, S. Leroux, M. Mancini, S. Mazevet, M. J. T. Oliveira, G. Onida, Y. Pouillon, T. Rangel, G.-M. Rignanese, D. Sangalli, R. Shaltaf, M. Torrent, M. Verstraete, G. Zerah, and J. W. Zwanziger, *Comput. Phys. Commun.*, 2009, **180**, 2582.
13. P. Giannozzi, S. Baroni, N. Bonini, M. Calandra, R. Car, C. Cavazzoni, D. Ceresoli, G. L. Chiarotti, M. Cococcioni, I. Dabo, A. D. Corso, S. de Gironcoli, S. Fabris, G. Fratesi, R. Gebauer, U. Gerstmann, C. Gougoussis, A. Kokalj, M. Lazzeri, L. Martin-Samos, N. Marzari, F. Mauri, R. Mazzarello, S. Paolini, A. Pasquarello, L. Paulatto, C. Sbraccia, S. Scandolo, G. Sclauzero, A. P. Seitsonen, A. Smogunov, P. Umari, and R. M. Wentzcovitch, *J. Phys.: Condens. Matter*, 2009, **21**, 395502.

14. N. A. W. Holzwarth, A. R. Tackett, and G. E. Matthews, *Comput. Phys. Commun.*, 2001, **135**, 329.
15. F. H. Larsen and I. Farnan, *Chem. Phys. Lett.*, 2002, **357**, 403.
16. P. J. Grandinetti, J. H. Baltisberger, I. Farnan, J. F. Stebbins, U. Werner, and A. Pines, *J. Phys. Chem.*, 1995, **99**, 12341.
17. J. B. Davey, A. C. Legon, and E. R. Waclawik, *Chem. Phys. Lett.*, 1999, **306**, 133.
18. J. W. Zwanziger, *Phys. Rev. B*, 2009, **79**, 033112.

PART C
Applications

Chapter 14

Quadrupolar NMR to Investigate Dynamics in Solid Materials

Luke A. O'Dell and Christopher I. Ratcliffe

Steacie Institute for Molecular Sciences, National Research Council of Canada, 100 Sussex Drive, Ottawa, ON K1A 0R6, Canada

14.1 INTRODUCTION

Dynamics, be they vibrational, rotational, or translational, are an important aspect of any solid, as they can play a major role in the overall properties of the material. NMR is perhaps the foremost technique for the study of rotational and translational motions in solid materials, and quadrupolar nuclei have a prominent role in such studies. ^{2}H NMR in particular has been, and will continue to be, a very important tool for the study of molecular dynamics, since the isotope can replace ^{1}H in countless inorganic and organic systems. ^{7}Li has also received a lot of attention, largely for studying motion in ion-conducting materials. With the increased interest in the last decade in quadrupolar nuclei with larger coupling constants (brought about by access to higher magnetic fields and constantly improving experimental techniques and probe technology), there are increasing numbers of reports of dynamics observed using other quadrupolar nuclei, especially ^{17}O, as well as ^{11}B, ^{14}N, ^{23}Na, ^{27}Al, and ^{133}Cs, and also some transition metals involved in tetrahedral oxide ions, such as $^{185,187}Re$ and ^{55}Mn.

Quadrupolar NMR studies of dynamics can yield structural information such as molecular orientation with respect to a crystallographic axis, or local symmetry, and this can sometimes be used to improve or direct structural refinements. It can distinguish between static and dynamic disorder among partially occupied sites and can indicate phase transitions and perhaps give insight into possible mechanisms for these. More directly, the NMR studies can provide rotational rate constants or correlation times, which, if studied as a function of temperature, yield activation energies for the motions.

The underlying behavior which makes the observation of dynamic effects possible in NMR is the modulation of the resonance frequency, brought about by any motion that causes the electric field gradient (EFG) tensor, and hence the quadrupolar interaction tensor, to undergo changes in magnitude and/or orientation.[1–4] These changes can be affected by either motion of the nucleus itself or motion of

NMR of Quadrupolar Nuclei in Solid Materials
Edited by Roderick E. Wasylishen, Sharon E. Ashbrook and Stephen Wimperis
 ISBN: 978-0-470-97398-1

neighboring nuclei. Since most motions are stochastic processes, they are usually described in terms of an autocorrelation function $G(\tau)$:

$$G(\tau) = \overline{f(t)f(t+\tau)} \quad (14.1)$$

where $f(t)$ is a function of the position and orientation of a molecule at time t and the bar represents an ensemble average. In essence, $G(\tau)$ is a measure of the fraction of molecules whose position and orientation are unchanged after a time τ. If there is motion occurring, $G(\tau)$ will decrease over time and, in most cases, this decay can be modeled as an exponential function with a correlation time τ_c that is inversely related to the rate of motion:

$$G(\tau) = \exp\left(-\frac{|\tau|}{\tau_c}\right) \quad (14.2)$$

For thermally activated processes, where the molecule must surmount an energy barrier to move between potential wells, τ_c is assumed to have an Arrhenius dependence on the activation energy (E_a) and on temperature (T):

$$\tau_c = \tau_c^0 \exp\left(\frac{E_a}{RT}\right) \quad (14.3)$$

from which it is seen that as T increases, τ_c decreases (and the rate of a motion increases).

As well as the rate of the motion, the geometry of the dynamic process will also determine how the motion will manifest itself in the NMR response of the quadrupolar nucleus under study, since this affects the change in orientation of the quadrupolar interaction tensor. For example, a 180° rotation about any of the principal axes of the EFG tensor will leave it unchanged and will therefore have no effect, a situation that often arises for twofold molecular reorientations.

Different experimental approaches are generally used to probe motions on different timescales, e.g., 2D exchange experiments for slow rates of motion, lineshape measurements for intermediate rates, and relaxation measurements for fast rates. In this chapter, we provide qualitative descriptions of the underlying mechanisms by which the NMR spectra of various quadrupolar nuclei can be affected by dynamics over an extremely wide range of timescales, as well as discuss the basic principles behind various experimental methods. In Sections 14.2–14.5, we focus on the quadrupolar nuclei ^{2}H, ^{14}N, ^{17}O and $^{6/7}$Li respectively, each of which exhibits different NMR properties and are therefore used in distinct ways to study dynamic processes. In Section 14.6, we consider the effects that dynamics can have on high-resolution, multiple-quantum experiments.

14.2 DEUTERIUM

14.2.1 T_2 Anisotropy and ^{2}H Spin-echo Experiments

Deuterium is perhaps the most commonly exploited quadrupolar nucleus for studying molecular dynamics.[5] It is a nucleus with spin $I = 1$; thus, the first-order quadrupolar interaction is typically the dominant perturbation and smaller interactions may often be ignored. Fortunately, the nucleus has a relatively small quadrupole moment ($Q = 2.86$ mbarn), and the quadrupolar splittings typically only range up to around 250 kHz. Such pattern widths can be uniformly excited and observed relatively easily on modern spectrometers, most commonly by using the quadrupolar-echo pulse sequence ($\pi/2_x - \tau_d - \pi/2_y - \tau_d$–acquire), which allows the signal acquisition to begin at the echo top and thus avoids the spectral distortions due to equipment dead time that are associated with single-pulse experiments. Furthermore, the EFG at the deuterium site and, therefore, the quadrupolar interaction tensor itself are often axially symmetric ($\eta_Q = 0$) with the unique axis aligned along a covalent bond (e.g., a C–H bond in an organic molecule). This can greatly simplify the interpretation of results, since the interaction is then described by only two parameters (C_Q, and the angle θ between the unique axis and B_0). Less fortunately, the ^{2}H isotope has a very low natural abundance (0.012%), making isotopic enrichment necessary. This can, however, be considered an advantage, in some cases, in that it allows the possibility of selective labeling, which, as well as aiding spectral resolution, can be used to probe specific sites in complex systems.

The most common method for extracting dynamic information using ^{2}H NMR is to record the powder pattern from a stationary sample using a spin-echo experiment and then to compare the resultant spectral lineshape to simulations which model a Markovian jump process with a specific geometry and jump rate k.[6,7] Powder lineshapes are generally sensitive to motional rates comparable to their linewidth. At slower rates (or in the absence of motion), the lineshape is characteristic of the static quadrupolar coupling tensor **V**. In the fast motion limit (FML), the lineshape is characteristic of an effective tensor $\mathbf{V}_{\text{eff}}$, which is the population-weighted average of the static tensor components in a common reference frame for all the sites visited during the motion:

$$\mathbf{V}_{\text{eff}} = \frac{\sum\limits_{i=1\to n} \mathbf{V}_i(\alpha, \beta, \gamma) p_i}{\sum\limits_{i=1\to n} p_i} \quad (14.4)$$

where p_i is the population factor of site i and the static tensor components for site i are given by

$$\mathbf{V}(\alpha, \beta, \gamma) = \mathbf{R}\mathbf{V}_{\text{PAS}}\mathbf{R}^{-1} \quad (14.5)$$

where **R** is the Euler angle rotation matrix describing the coordinate transformation and $\mathbf{V}_{\text{PAS}}$ is the static tensor in its principal axis system. Thus, if the static tensor is known, averaged FML lineshapes can be calculated easily for any plausible dynamic model and compared with observation. Intermediate rate dynamics, which cause lineshapes that are distorted away from typical first-order quadrupolar powder patterns, can then be calculated based on feasible model(s) and compared with experimental lineshapes to obtain rate constants or to distinguish models that give identical FML lineshapes. The geometry used for the simulations will usually be based on knowledge of the molecular and crystal structure, e.g., a deuterium located on a methyl group might be expected to undergo C_3 rotation. If the static and FML lineshapes are both available, then a dynamic model (or models, if not unique) can be developed which is compatible with the observed averaging before calculating intermediate rate lineshapes; otherwise, an educated guess has to be made.

The distorted ^{2}H powder patterns that occur in the presence of intermediate motion are caused by anisotropy in the ^{2}H transverse relaxation due to the modulation of the first-order quadrupolar interaction. For such a phenomenon to occur, two conditions must be fulfilled. First, the motion must alter the orientation or magnitude of one or more principal axes of the interaction tensor, in this case the ^{2}H EFG. This alteration can be caused either by motion of the ^{2}H itself or by motion of surrounding atoms. Second, the timescale for the dynamics (k^{-1}) must be comparable to the interpulse delay τ_{d} in the spin-echo pulse sequence used (or comparable to C_{Q}^{-1} in the limit of short τ_{d} values). When these conditions are fulfilled, the resonance frequencies of certain spins will change during the course of the spin-echo sequence and the corresponding signals will therefore fail to be refocused. The extent of this relaxation will be dependent on the jump rate and also the relative orientations (with respect to B_0) or magnitudes of the EFG principal axes at each jump site. The result is that the T_2 relaxation time will differ for each crystallite orientation and will cause the powder pattern to depart significantly from the ideal first-order quadrupolar lineshape.

In typical ^{2}H spin-echo experiments, both C_{Q}^{-1} and τ_{d} are usually on the order of 10^{-5} s; thus, ^{2}H powder patterns are sensitive to motional timescales in the range 10^4–10^7 s^{-1}. Motions occurring on timescales much longer than τ_{d} will be too slow to induce anisotropic relaxation, and standard first-order lineshapes will be observed. In the fast motion regime, where the dynamics occur on timescales much shorter than τ_{d}, the spins will experience the FML EFG tensor, resulting in the observation of an ideal first-order pattern (or a narrow Lorentzian line for motions that average the EFG to zero). In the intermediate timescale, however, the shape of the ^{2}H powder pattern is highly sensitive to both the rate and geometry of the dynamics. For thermally activated dynamics, jump rates can be determined from these intermediate lineshapes over a range of temperatures (with τ_{d} fixed) and then used to determine the activation energy via an Arrhenius plot. Alternatively, τ_{d} can be varied at a constant temperature to observe the effects of the T_2 anisotropy for a fixed jump rate over a range of timescales. Such experiments can potentially provide a broad range of quantitative information on molecular dynamics (number of jump sites, respective site orientations, jump rate constants, and activation energies).

Figure 14.1 shows some simulated ^{2}H powder patterns for a methane molecule undergoing three different types of motion. For the C_2 rotation, the powder pattern shows the axially symmetric ($\eta_{\text{Q}} = 0$) quadrupolar pattern in the slow motion regime, which changes to a $\eta_{\text{Q}} = 1$ pattern in the fast limit. For the C_3 simulations, both the slow and FML lineshapes correspond to axially symmetric EFG tensors. Indeed, rotation around a single axis with C_3 symmetry or higher will always result in an axially symmetric ^{2}H spectrum in the FML. It should be noted here that the signal arising from the deuteron on the axis of C_3 rotation, which is neglected in the simulations, will be unaffected by this motion since this rotation does not alter the EFG tensor for this spin. For the case

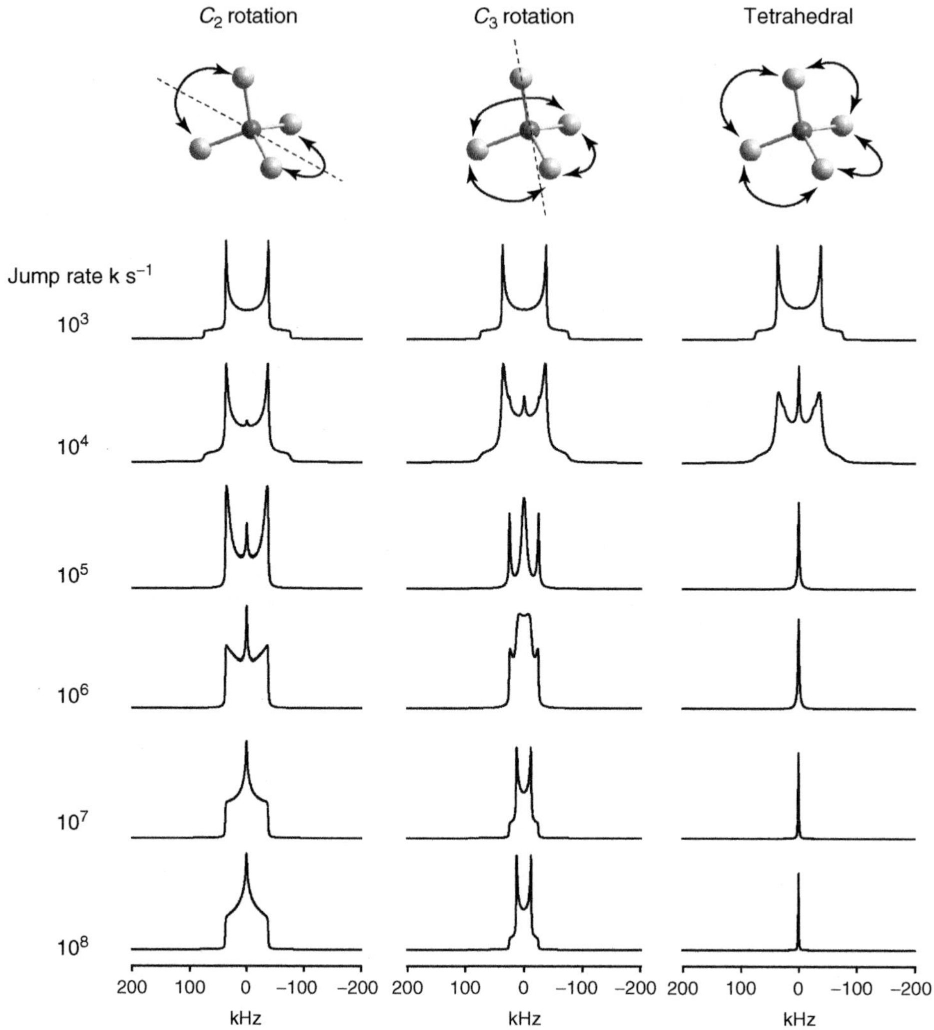

Figure 14.1. Simulated ^{2}H spin-echo powder patterns arising from spins undergoing various types of Markovian jump processes as shown. Simulations were generated with the EXPRESS software,[7] assuming an axially symmetric ^{2}H EFG tensor ($C_Q = 100$ kHz) and a quadrupolar-echo pulse sequence with a 50 μs interpulse delay. For the C_3 rotation, the signal from the axial deuterium site is neglected. Spectral intensities are not to scale.

of tetrahedral exchange, the EFG tensor is averaged completely for jump rates of 10^5 s^{-1} and higher, leaving a narrow Lorentzian lineshape.

One significant problem that can arise in using ^{2}H T_2 anisotropy to probe dynamics in this way is the severe signal loss that occurs in the intermediate regime, where the transverse relaxation is extremely efficient. Figure 14.2 shows the simulated echo intensities for the three dynamic modes in Figure 14.1, and the dramatic signal loss for jump rates of 10^4–10^6 s^{-1} (a decrease in intensity of up to an order of magnitude) is immediately apparent. As mentioned earlier, this intermediate region is where the lineshape shows the highest sensitivity to the dynamics. The resultant spectra can therefore exhibit rather low signal-to-noise ratios. Decreasing the echo delay τ_d can reduce this effect. Alternatively, signal enhancement techniques such as those discussed in

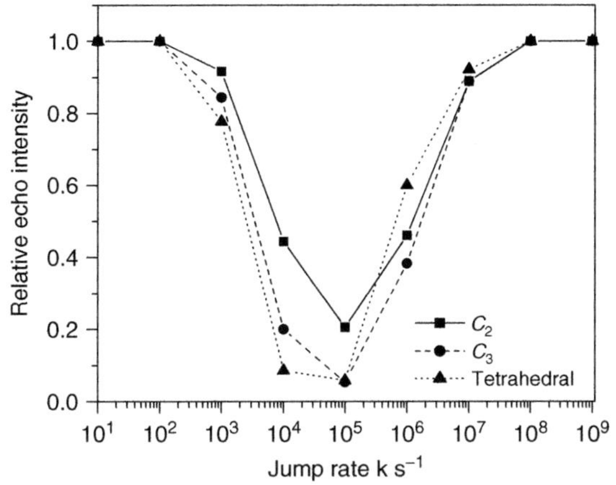

Figure 14.2. Normalized spin-echo intensities vs jump rates for the simulations shown in Figure 14.1.

the following sections can be used to provide a signal boost.

14.2.2 Magic Angle Spinning

In magic angle spinning (MAS) ^{2}H NMR experiments, the first-order quadrupolar broadening is averaged, leaving a manifold of spinning sidebands for each unique deuterium site which can often be resolved on the basis of differences in the second-order quadrupolar and isotropic chemical shifts. Such a method can therefore increase both the sensitivity and resolution of ^{2}H spectra. In studying dynamics, MAS also has the advantage of extending the range of motional timescales that can be probed.[8] In Figure 14.3, the manifold of spinning sidebands in simulated MAS spectra of a system undergoing C_3 rotation (as in Figure 14.1) is seen to be sensitive to a similar range of jump rates as the static powder pattern. On closer inspection, however, the widths of the sidebands themselves are seen to be sensitive to a greater range of jump rates. For example, the width of the most intense spinning sideband is noticeably different for jump rates of 10^2 and 10^3 s^{-1}, while these jump rates are indistinguishable on the basis of the static spin-echo powder patterns. This additional sensitivity of MAS spectra to motional processes arises because the dynamics interfere with the refocusing of the anisotropic interactions by the sample rotation.[9] The disadvantage of this phenomenon is that excessive broadening of the spinning sidebands in the intermediate regime can cause loss of resolution for sites with different isotropic shifts. Spinning sidebands may also provide more precise measurements of dynamics when the sample is spun at a small offset from the magic angle. In this case, a scaled quadrupolar interaction is reintroduced, giving each spinning sideband a characteristic shape that is dependent on the dynamics (particularly those occurring in the kilohertz regime) and may be simulated to extract this information.[10]

14.2.3 QCPMG

Quadrupolar Carr-Purcell-Meiboom-Gill (QCPMG) is now a very commonly used signal enhancement pulse sequence in solid-state NMR. Several variations of this pulse sequence exist, the defining feature being a repeating loop containing a refocusing pulse and an acquisition period, allowing the spin-echo to be re-recorded over the full timescale of the transverse relaxation/dephasing. A Fourier transform of the echo train results in a spectrum consisting of narrow spikelets whose manifold reproduces the approximate shape of the powder pattern. The QCPMG experiment therefore samples a broader range of timescales than a single spin-echo, and so can extend the range of motional timescales that can be studied using ^{2}H based on differences in the linewidths of the spikelets (Figure 14.3).[11] While both MAS and QCPMG offer significant increases in sensitivity, the most significant difference between the techniques is that QCPMG only ever results in a single spikelet manifold for all signals, thus does not offer the shift-based resolution advantage of MAS.

A recent study of methyl group rotation in the hydrophobic core of a protein provides an illustration of the advantages of QCPMG signal enhancement.[12] ^{2}H spectra were recorded from a powder sample of protein at various cryogenic temperatures, with a single methyl group of a hydrophobic core residue selectively labeled with ^{2}H (Figure 14.4). All three motional regimes were observed. In the FML (119 K), the methyl group rotation results in a relatively narrow spikelet manifold corresponding to a motionally averaged EFG tensor. At 84 K, the spikelet manifold narrowed significantly and the amount of signal dropped, indicating motion on the intermediate

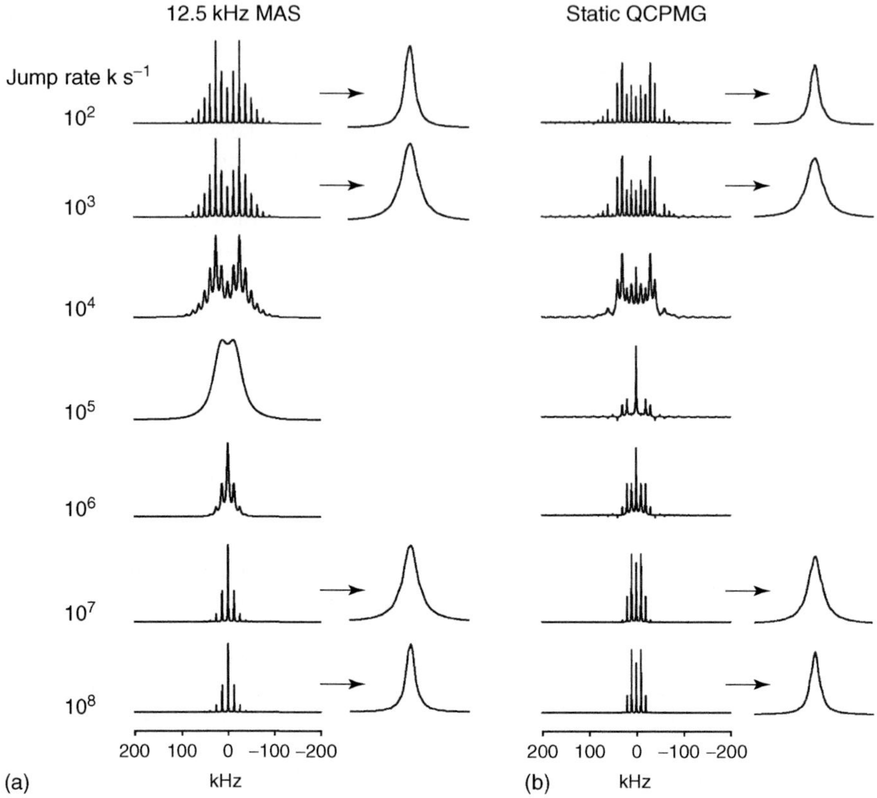

Figure 14.3. (a) Simulated ^{2}H MAS and (b) static QCPMG spectra for the system undergoing C_3 rotations shown in Figure 14.1 (the axial ^{2}H is again neglected). The MAS simulations used a rotor frequency of 12.5 kHz, and the QCPMG simulations have a spikelet separation of 10 kHz. For certain jump rates, horizontal expansions of the most intense sidebands/spikelets are also shown as indicated by the arrows. Simulations were generated using EXPRESS.[7]

timescale. At 31 K, a spikelet manifold corresponding to a relatively broad axial pattern was observed, indicating the freezing-out of the methyl rotation.

14.2.4 T_1 Anisotropy and Inversion Recovery Experiments

Dynamic modes such as those discussed in the previous sections are also capable of causing anisotropies in the (Zeeman) spin-lattice relaxation time T_{1Z}.[13] While transverse relaxation is induced by dynamics whose characteristic frequencies are on the order of the anisotropic interactions present (i.e., the C_Q in the case of ^{2}H, which is on the order of 100 kHz), spin–lattice relaxation is caused by interactions fluctuating at or close to the Larmor frequency (~61 MHz for ^{2}H at a field of 9.4 T). Specifically, T_{1Z} depends on the two spectral densities $J_1(\omega_0)$ and $J_2(2\omega_0)$ (which are themselves dependent on the motional correlation time and are orientation dependent under anisotropic motion), and ^{2}H T_{1Z} measurements can therefore provide access to very fast motional rates (up to ~10^{11} s^{-1}):

$$\frac{1}{T_{1Z}} = \frac{3\pi^2}{2} C_Q^2 (J_1(\omega_0) + 4J_2(2\omega_0)) \qquad (14.6)$$

The simplest way of observing T_{1Z} anisotropy experimentally is using an inversion recovery pulse sequence of the form $\pi - t_1 - \pi/2-$acquire, where the magnetization is stored antiparallel to B_0 for a variable recovery time t_1, and subsequently observed using a $\pi/2$ pulse (or, more typically for ^{2}H, a spin-echo sequence). Figure 14.5 shows a simulated

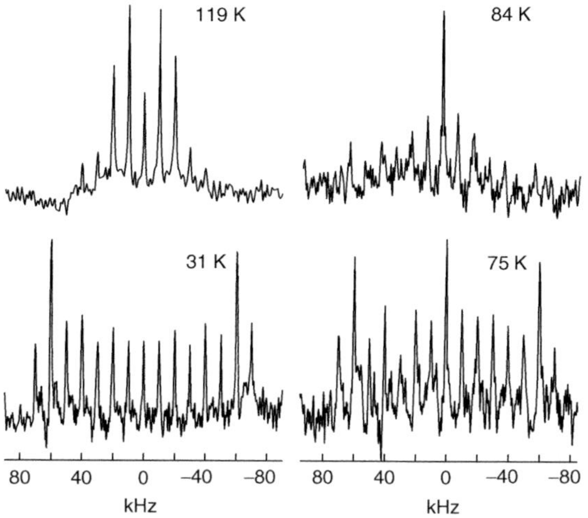

Figure 14.4. ^{2}H QCPMG spectra obtained at 11.7 T and at temperatures shown from a powder sample of chicken villin headpiece subdomain (HP36), selectively ^{2}H-labeled at a single methyl group of residue L69. For the spectra at 119, 84, 75, and 31 K, 1024, 3040, 512, and 128 scans, respectively, were acquired. (Reproduced with permission from Ref. 12. Copyright 2010 American Chemical Society.)

T_{1Z} anisotropy profile for the ^{2}H C_3 rotation discussed above at 9.4 T and with a jump rate of 10^9 s^{-1}. Such a jump rate is too fast to affect the spin-echo lineshape in a standard experiment, resulting in an axially symmetric FML powder pattern. In an inversion recovery experiment, however, the T_{1Z} anisotropy can be measured by recording spectra over a range of t_1 values, and the shape of the T_{1Z} anisotropy will be dependent on both the rate and geometry of the dynamics. We note that these simulations include only the effects of the C_3 dynamics and neglect other possible causes of relaxation such as lattice vibrations, although, in most cases, the relaxation mechanism in ^{2}H will be dominated by the quadrupolar interaction. T_1 behavior can also distinguish between dynamic models that give the same FML lineshape, e.g., Markovian jumps or smooth rotational diffusion about the C_3 axis.

It should also be noted that the anisotropy in the relaxation of quadrupolar order, in which a single transition is inverted, is also sensitive to dynamics. This state can be created using a Jeener–Broekaert pulse sequence.[14] For fast motions, this relaxation can provide complementary information to T_{1Z} measurements since its associated relaxation time T_{1Q} is

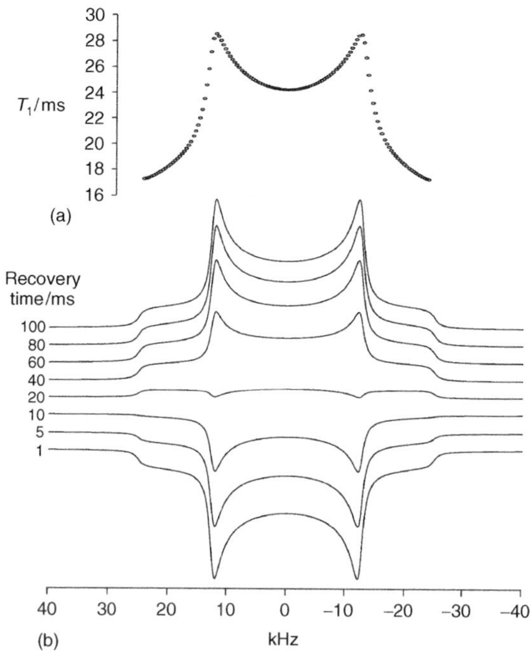

Figure 14.5. (a) Simulated T_1 anisotropy for a ^{2}H system ($C_Q = 100$ kHz, $\eta_Q = 0$, $B_0 = 9.4$ T) undergoing C_3 rotation (as in Figure 14.1) with a jump rate of 10^9 s^{-1} (axial deuteron is again neglected). (b) Corresponding inversion recovery spectra with various recovery times. Simulations were generated using EXPRESS.[7]

dependent on only one spectral density $J_1(\omega_0)$[7]:

$$\frac{1}{T_{1Q}} = \frac{9\pi^2}{2} C_Q^2 (J_1(\omega_0)) \tag{14.7}$$

This allows both spectral densities to be quantified,[15] and also makes T_{1Q} relaxation a useful probe for quantifying very slow reorientations ($\sim$1 s^{-1}), since it is less sensitive to rapid vibrational motions than T_{1Z}.[16] Note also that the QCPMG experiment retains longitudinal relaxation anisotropy while MAS does not[17]; thus, QCPMG can be useful in improving the sensitivity in such experiments.

14.2.5 2D Exchange Experiments

The fitting of simulated spin-echo lineshapes or relaxation anisotropies to experimental ^{2}H data usually necessitates some prior knowledge of the

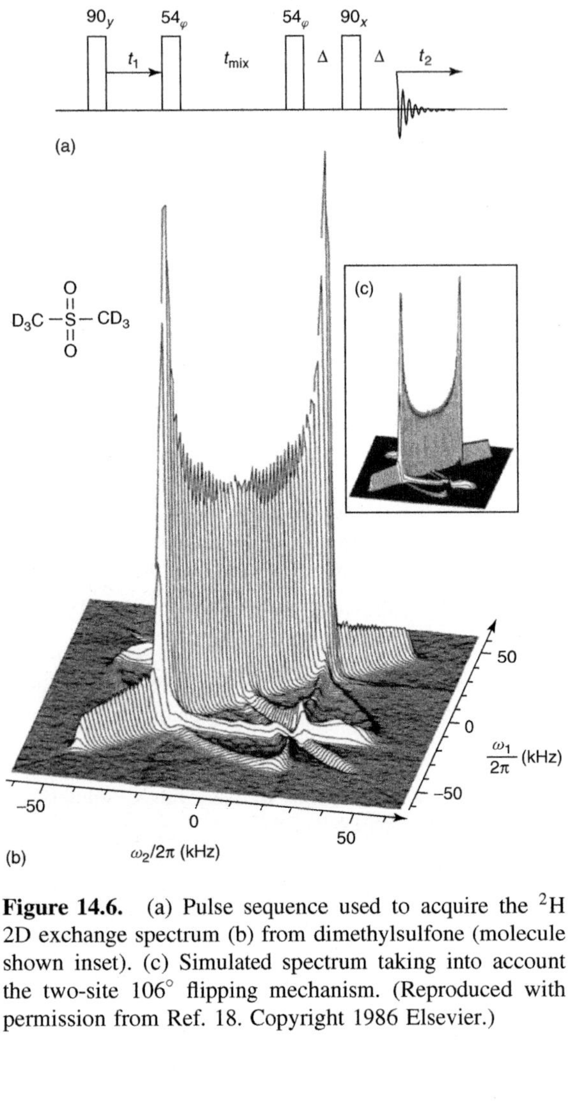

Figure 14.6. (a) Pulse sequence used to acquire the ^2H 2D exchange spectrum (b) from dimethylsulfone (molecule shown inset). (c) Simulated spectrum taking into account the two-site 106° flipping mechanism. (Reproduced with permission from Ref. 18. Copyright 1986 Elsevier.)

molecular structure to guess potential dynamic models. 2D exchange experiments provide an alternative approach where motional rates and geometries can be inferred solely from the experimental data.[18] An example spectrum is shown in Figure 14.6, obtained from dimethylsulfone. In this molecule, the methyl groups undergo rapid rotation such that the ^2H powder pattern corresponds to an axially symmetric EFG tensor aligned along the S–C bond. The molecule also undergoes a slower two-site jump process about its C_2 symmetry axis that alters the orientation of this ^2H EFG by 106°. In the exchange spectrum, the main peaks along the diagonal map out the static, axially symmetric quadrupolar pattern, while the two-site jump process that occurs during the mixing period gives rise to off-diagonal ridges. Crucially, these ridges show characteristic shapes that depend explicitly on the motional geometry, allowing the flip angles and number of jump sites to be deduced directly from the spectrum.[18]

14.3 NITROGEN-14

14.3.1 T_2 Anisotropy and Ultra-wideline Powder Patterns

The ^{14}N isotope is also a spin $I = 1$ nucleus, and so the above techniques are in principle also applicable to this nucleus. Unfortunately, this nucleus has a quadrupole moment that is seven times larger than ^2H (20.44 mbarn); thus, the broadening due to the first-order quadrupolar interaction is usually extremely large (several megahertz) and such broad spectra can be very difficult to observe.[19] Dynamics studies employing this nucleus are therefore rather scarce. Spin-echo experiments using standard excitation pulses have been carried out on systems where the nitrogen sites have a local symmetry that is close to tetrahedral, giving rise to small C_Q values, such as in choline salts.[20] Motionally averaged ^{14}N quadrupolar splittings have also been observed in partially oriented systems such as liquid crystals or lipid bilayers. Such splittings are often directly proportional to the ordering parameter describing the rotational dynamics present in these systems, but again, such cases generally involve nitrogen sites with symmetric environments and small C_Q values. The majority of nitrogen sites feature C_Q values in excess of 1 MHz, and these require more advanced experimental methods.[19]

The jump rate for the molecular flipping mechanism in crystalline urea ($C_Q = 3.86$ MHz and $\eta_Q = 0.28$) has recently been determined using ultra-wideline ^{14}N NMR as $k = 7 \times 10^3\ s^{-1}$ at room temperature.[21] Frequency-swept pulses combined with QCPMG were used to excite a broad region of the powder pattern where the T_2 anisotropy was most pronounced (Figure 14.7). This outermost region of the powder pattern corresponds to crystallite orientations in which the EFG principal axis V_{33} is parallel to B_0; thus,

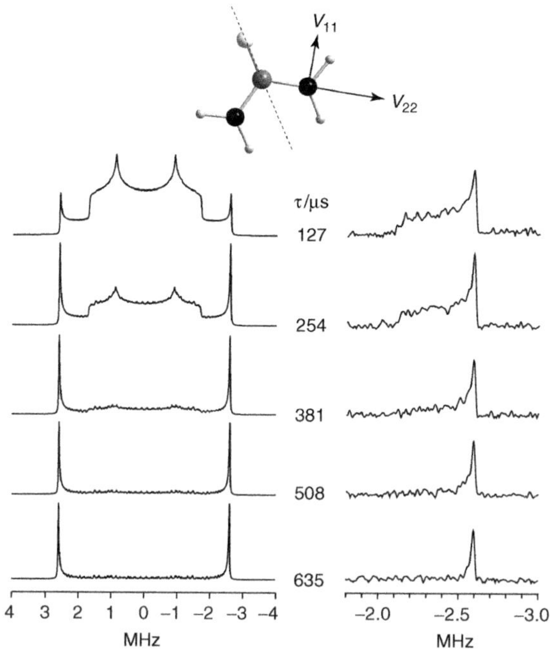

Figure 14.7. (a) Simulated and (b) experimental ^{14}N spectra obtained from urea at 21.1 T and at room temperature. The molecule is shown in the inset, along with the orientation of two of the ^{14}N EFG principal axes (the third, V_{33}, is perpendicular to the molecular plane). Simulations were made using EXPRESS[7] with a jump rate of $7 \times 10^3\ \mathrm{s}^{-1}$ and interpulse delays as shown. Experimental spectra (recorded at the low-frequency edge of the ultra-wideline powder pattern) were made by extracting single echoes from a QCPMG train at corresponding τ values and Fourier transforming these individually. (Reproduced from Ref. 21 by permission of the Royal Society of Chemistry.)

the molecular flipping in these crystallites does not alter the EFG orientation and the motion does not induce transverse relaxation for these spins. The outermost edges of the pattern are therefore selectively enhanced by the QCPMG protocol, as seen in the simulations in Figure 14.7. Fitting of these simulated lineshapes to the experimental spectra provided jump rate constants in good agreement with those measured using other techniques, demonstrating that the simple approach described in Section 2 holds, at least to a good approximation, even for very large quadrupolar interactions.

14.3.2 Hole-burning Methods

Saturation or inversion recovery methods are usually used to study very fast dynamics via their effect on the spin–lattice relaxation rate. However, they may also be used to probe very slow motions occurring at timescales on the order of the T_1 relaxation time itself. The hexamethylenetetramine (HMT) molecule (^{14}N $C_Q = 4.41$ MHz, $\eta_Q = 0$) in its crystalline form undergoes tetrahedral jumps at a rate of $\sim 10\ \mathrm{s}^{-1}$ at room temperature, moving the ^{14}N EFG tensor between four orientations. Such motion is far too slow to be observed as a change in the shape of the ^{14}N powder pattern, but its effects have been observed using saturation–recovery experiments applied to a narrow region of the ultra-wideline powder pattern (referred to as *hole burning*).[22] Specifically, a train of transition-selective pulses can be used to saturate a 4 kHz wide region of the powder pattern, which can then be monitored over time using a conventional spin-echo pulse sequence. After a single train of pulses, a set of ^{14}N spins corresponding to a specific set of EFG orientations are saturated. The EFG orientations of these saturated spins are then altered by the molecular reorientation, which also brings unsaturated magnetization (i.e., ^{14}N spins previously at other EFG orientations) into the hole region. For the four-site jump process in HMT, three quarters of the magnetization is, therefore, quickly recovered due to the dynamics alone, followed by a much slower recovery of the remaining magnetization by ordinary spin–lattice relaxation (Figure 14.8a). Note that the fraction of magnetization recovered in the initial stage provides an immediate quantification of the number of jump sites. Alternatively, all four sets of EFG orientations can be saturated, after which the recovery of magnetization occurs by spin–lattice relaxation only (Figure 14.8b).

14.3.3 T_1 Measurements

T_1 relaxation times can be measured relatively easily from nitrogen sites with high symmetry or fast dynamics where the ^{14}N quadrupolar interaction is absent or extremely small. For example, in ionic solids such as NaCN, the CN^- ions can undergo rapid rotation and therefore exhibit very narrow linewidths despite the relatively high C_Q value of the stationary ion ($\sim$4 MHz). The rapid EFG fluctuations caused by this rotation lead to very efficient relaxation, and

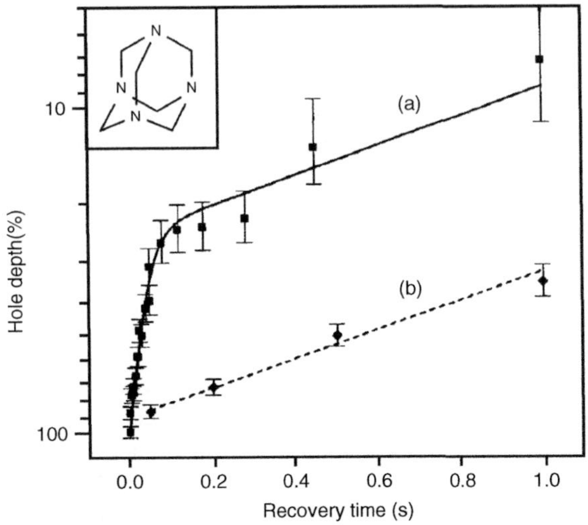

Figure 14.8. Recovery of a narrow region of saturated ^{14}N magnetization in hexamethylenetetramine (molecule shown inset) at 7.05 T as a function of time after (a) a single train of saturation pulses and (b) 16 saturation trains. Solid and dotted lines show fits derived from numerical expressions. (Reproduced with permission from Ref. 22. Copyright 1997, American Institute of Physics.)

the departure from cubic symmetry caused by orientational disorder or defects present in the lattice can leave very small residual quadrupolar broadening which can be observed using the quadrupolar-echo sequence. The spin–lattice relaxation cannot, therefore, be directly inferred from the natural linewidths ($1/T_2^*$), but correlation times for the rotation can still be easily ascertained from inversion recovery T_1 measurements, allowing the very fast ion rotation rate ($\tau \approx 10^{-10}$ s) to be measured.[23]

14.4 OXYGEN-17

14.4.1 Central Transition Lineshapes

Since the majority of quadrupolar nuclei have a half-integer spin number with a corresponding central transition (CT) that is free of first-order quadrupolar broadening, it is of interest to extend the ^2H spin-echo methods outlined above to such nuclei to access dynamics in systems that may not contain hydrogen. In this section, we consider ^{17}O ($I = 5/2$) as an exemplary half-integer quadrupolar nucleus. Oxygen is clearly an element of considerable importance, so it is rather unfortunate that ^{17}O has a very low natural abundance (0.038%) and a relatively low gyromagnetic ratio ($\nu_0 = 54.2$ MHz at 9.4 T). Isotopic enrichment is therefore usually required for studies of this nucleus. On the plus side, the nucleus has only a moderate quadrupole moment (−25.6 mbarn) and a reasonably large chemical shift range (~1000 ppm), meaning that solid-state spectra can often be obtained (under MAS) with a relatively high resolution, particularly at high magnetic field strengths.

^{17}O CT NMR spectra obtained from powder materials under either static or MAS conditions show anisotropic broadening due to the second-order quadrupolar interaction (as well as chemical shift anisotropy in the case of static powders). These anisotropic interactions, like the first-order quadrupolar interaction for ^2H, can be modulated by dynamic processes giving rise to anisotropic relaxation, altering the shape of the powder pattern and thus constituting a mechanism by which motional rate constants and geometries can be measured. Figure 14.9 shows simulated CT lineshapes for ^{17}O spins in an MO_4 unit undergoing various dynamic mechanisms. As with the ^2H experiments discussed earlier, the ^{17}O CT lineshapes are sensitive to motions at intermediate rates ($\sim 10^3 - 10^7$ s^{-1}), which are comparable to the size of the second-order quadrupolar interaction ($C_Q{}^2/\nu_0$). The satellite transitions (STs) are sensitive to motions at much faster rates (on the order of C_Q). Note that the centerband and spinning sidebands in the MAS simulations consist of powder patterns due to incomplete averaging of the second-order quadrupolar interaction and that these powder patterns are also sensitive to the dynamics.

Motions in the slow regime do not affect the CT spectra, leaving standard second-order powder patterns. Interestingly, motions in the fast regime produce powder patterns that, while narrowed, do not correspond to typical second-order quadrupolar lineshapes and cannot be simulated using standard CT powder patterns.[24] This is because the second-order quadrupolar interaction features both second- and fourth-rank terms, which are weighted differently when averaged by fast motion.[25] Such fast regime lineshapes can, therefore, only be reproduced using simulations that take the dynamics into account. A related observation, apparent by

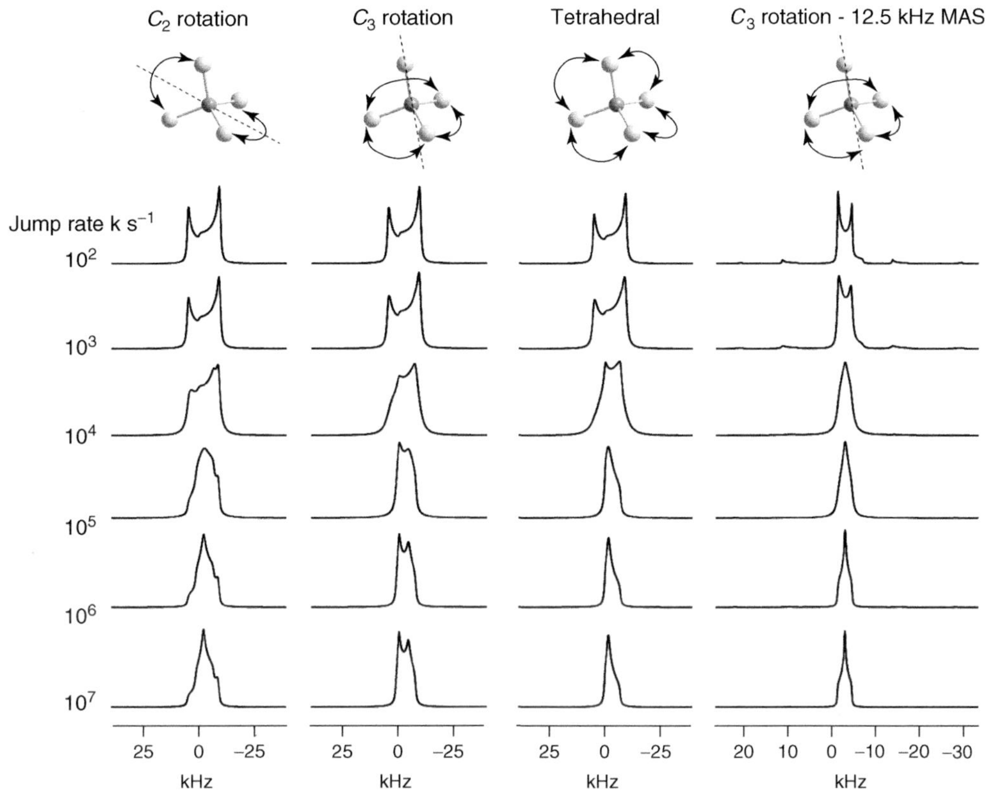

Figure 14.9. Simulated ^{17}O central transition NMR spectra for various dynamic modes of an MO_4 unit shown at the top. Simulations were generated using EXPRESS[7] for a ^{17}O spin at 9.4 T with $C_Q = 5$ MHz and $\eta_Q = 0$. No chemical shielding anisotropy was included. As before, the signal from the axial site in the C_3 rotation model is neglected.

comparing Figures 14.1 and 14.9, is that fast tetrahedral exchange does not completely average away the second-order quadrupolar interaction as it does for the first-order interaction. This is due to the fourth-rank anisotropy terms, which have lower symmetry. For motional rates on the order of ${C_Q}^2/\nu_0$, dynamics with icosahedral or higher symmetry are required for the anisotropic terms to become fully averaged, while the isotropic second-order quadrupolar shift, which has no angular dependence, will remain.[25] However, for much faster rates (on the order of ν_0 or greater), both the anisotropic lineshape and the second-order shift are averaged, so that eventually even averaging over tetrahedral sites will give a narrow lineshape at the true isotropic chemical shift.[26] It should be noted that the chemical shielding anisotropy is neglected from the simulations in Figure 14.9, which, at high magnetic fields, can be similar in size to the second-order quadrupolar broadening and can therefore also play a significant role in determining the shape of the motionally averaged lineshape (except under MAS where this interaction is averaged out).[24] As with ^2H NMR, the use of inversion recovery[27] or QCPMG methods[28] can increase the range of motional timescales accessible by central transition spectra.

^{17}O NMR has been used in this way to characterize the rotation of PO_4^{3-} anions in the low-temperature phase of Na_3PO_4.[29] In this system, the ^{17}O EFG tensor was found to be axially symmetric and oriented along the P–O bond. The static ^{17}O NMR spectrum was observed to undergo dramatic changes in shape on heating (Figure 14.10), leading ultimately to a motionally narrowed Lorentzian lineshape at 585 K which was best acquired using nonselective excitation pulses. The anisotropic lineshapes over the temperature range 425–525 K were shown to fit most accurately to simulations in which the anions undergo

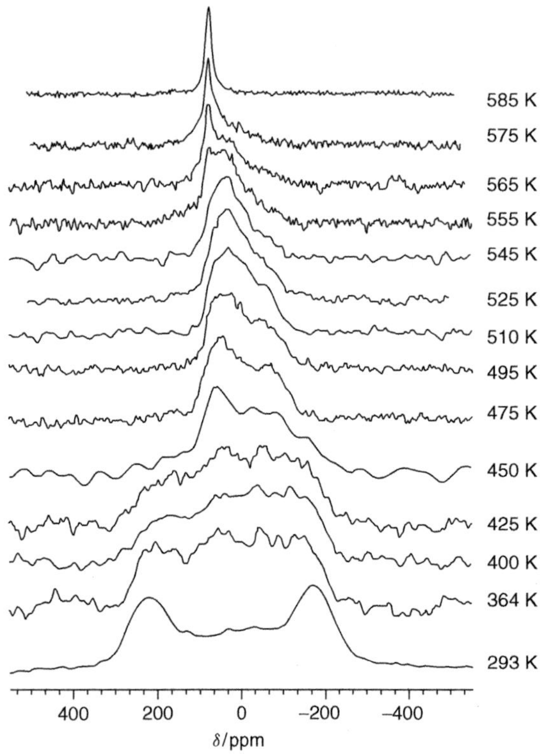

Figure 14.10. Variable-temperature ^{17}O NMR obtained from a static powder sample of Na_3PO_4 at 7 T. (Reproduced with permission from Ref. 29. Copyright 2001 American Chemical Society.)

preferred C_3 rotation around one P–O bond with much slower C_3 rotation around the remaining 3. This allowed the activation energy for this preferred rotation to be obtained, and the phosphate ion motion was then correlated with that of the Na^+ cations (which undergo translational diffusion, studied via 2D nutation spectra).[29]

In another example, the dynamics of water molecules in ice and tetrahydrofuran (THF) clathrate hydrate have been followed using ^{17}O NMR lineshapes.[30] In this system, each O atom has a roughly tetrahedral arrangement of four hydrogen bonds, in which it is bonded directly to two H atoms and accepts H-bonds from two other water molecules. Interestingly, in this particular case, the O atom itself does not move, but the surrounding EFG changes among six octahedral orientations because of the movement of the H atoms. This may be contrasted with ^2H which does move, and the orientation of its EFG tensor changes among four tetrahedral orientations. For ice, ^2H NMR lineshapes as a function of temperature show complete narrowing of the first-order lineshape to a pseudoisotropic lineshape, corresponding to an effective C_Q of 0.[31] ^{17}O NMR, however, shows narrowing of the CT to a new *anisotropic* lineshape which is still present at the melting point, i.e., there is incomplete averaging of the second-order lineshape (Figure 14.11a). In THF clathrate hydrate, the water dynamics are significantly faster, so that lineshapes similar to those seen in ice occur at temperatures about 40–50 K lower, and, in this case, further CT lineshape changes can be followed before melting occurs (Figure 14.11b). It is clear that although the lineshape initially averages toward the expected anisotropic lineshape, it then gradually becomes isotropic in shape, but this "isotropic" line retains a second-order shift as illustrated by the spectrum of the hydrate at 270 K where liquid THF solution and solid THF hydrate coexist (Figure 14.11c). The liquid line is considerably sharper and very close to the shift of water, whereas the broader "isotropic" line of the hydrate is centered at -194 ± 3 ppm, very close to the second-order shift of -193.6 ppm determined from the low-temperature static lineshape. It would seem that the change to an "isotropic" lineshape is associated with the onset of averaging of the first-order interaction (ν_Q is only 964.5 kHz, where $\nu_Q = 3C_Q/2I(2I-1)$ and $2\nu_Q$ gives the total width of the powder lineshape for the $\pm 1/2 \leftrightarrow \pm 3/2$ satellite transitions), and retention of the full second-order shift indicates that the rate is still significantly slower than the resonance frequency ($\nu_0 = 40.68$ MHz).

This same water motion also has an effect on the ^{131}Xe NMR lineshapes of xenon as a guest in clathrate hydrates,[32] where the Xe is an example of a spectator nucleus subjected to changes in EFG due to motions of surrounding molecules. We note that dynamics have also been observed to cause the second-order shift to decrease to zero for ^{23}Na in Na_3PO_4/Na_2SO_4 solid solutions at high temperatures.[33] These examples demonstrate that care must be exercised in judging whether one has the true isotropic chemical shift for apparently isotropic CT lines of noninteger quadrupolar nuclei, and spectra obtained at multiple fields can be used to distinguish second-order shifts from true isotropic shifts.

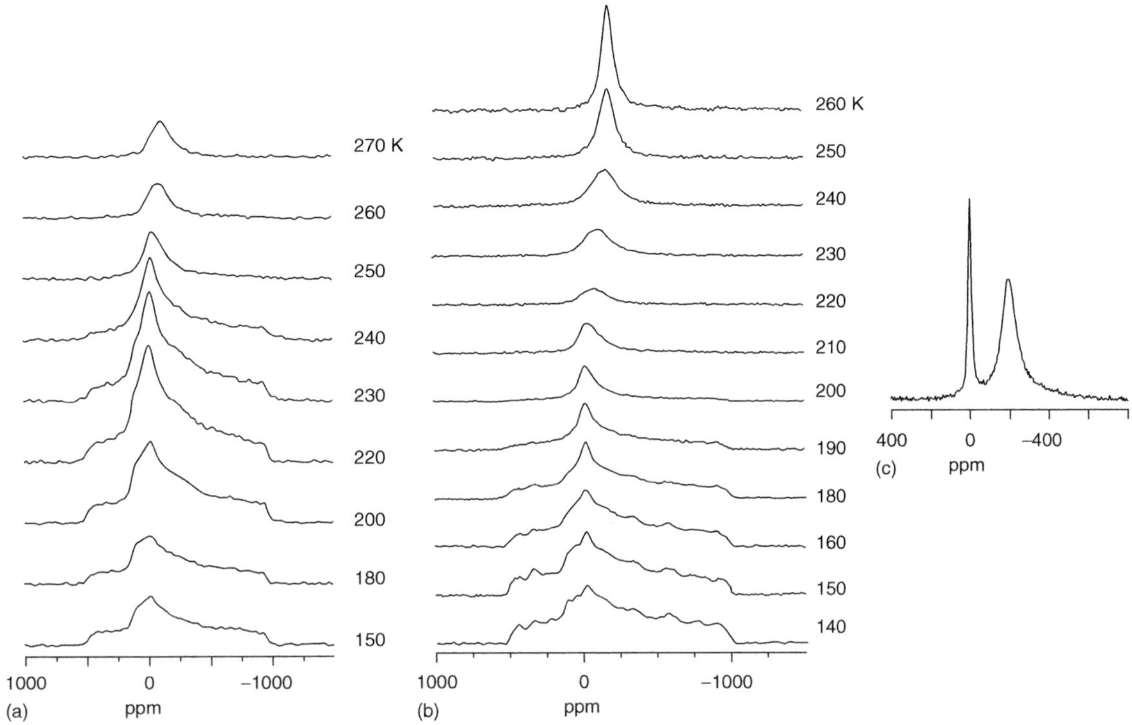

Figure 14.11. Static ^{1}H-decoupled ^{17}O NMR CT transition spectra as a function of temperature at 7T of (a) ice, (b) THF clathrate hydrate, and (c) a spectrum obtained at about 270 K after partial melting of the THF hydrate. The liquid shows an isotropic chemical shift, whereas the apparently "isotropic" hydrate line still also has a second-order quadrupolar shift.[30]

14.4.2 Exchange Experiments and Spin–lattice Relaxation

In systems where second-order quadrupolar broadening is minimal and individual oxygen sites are well resolved by MAS, exchange occurring on a similar timescale to the difference in resonance frequencies can be observed directly in 1D spectra as a broadening (intermediate regime) or coalescence (fast regime) of the two peaks to an averaged position in the spectrum. Such a process has been used to observe the hopping of oxide ions between sites in α-$Bi_4V_2O_{11}$ at rates on the order of $10^4\ s^{-1}$.[34] In the same study, it was also noted that spinning sidebands arising from ^{17}O satellite transitions were also sensitive to dynamics, broadening and disappearing on heating as motions on the order of the C_Q or MAS frequency prevent their formation,[34] a process analogous to the ^{2}H experiments discussed in Section 14.2.2.

Slower exchange can be observed using 2D exchange spectroscopy. An example of this is shown in Figure 14.12, where exchange between all four oxygen sites in an isotopically enriched sample of ZrW_2O_8 is evidenced by the appearance of off-diagonal cross peaks.[35] The mixing time used in this experiment was 50 ms, indicating that the WO_4 groups rotate on a timescale of $\sim 10\ s^{-1}$. Hopping rate constants can in principle be extracted via the relative intensities of the cross peaks, but the quantification of jump rates and activation energies from such magnetization transfer experiments is complicated since both the dynamic exchange process and spin–lattice relaxation effects must be considered. The latter process is multiexponential for half-integer quadrupolar nuclei due to the large number of spin states between which coherences can be exchanged, and analysis is further complicated by the presence of two distinct mechanisms driving the relaxation (modulations

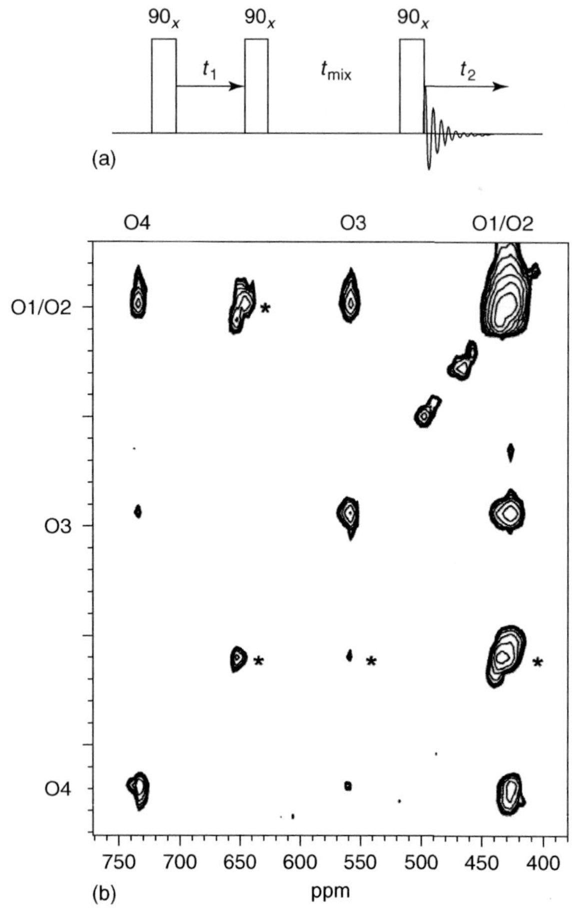

Figure 14.12. (a) The basic 2D exchange (EXSY) pulse sequence, where t_1-encoded chemical exchange occurs between longitudinal components of magnetization during t_{mix}. (b) A ^{17}O EXSY spectrum obtained at 11.7 T from a sample of ^{17}O-enriched ZrW_2O_8 at 57°C and spinning at 14.5 kHz MAS. Off-diagonal peaks indicate exchange between sites O3, O4, and O1/O2, the latter sites being unresolved. Asterisks indicate spinning sidebands. (Reproduced from Ref. 35 by permission of the Royal Society of Chemistry.)

of the quadrupolar interaction as well as the local magnetic field).[36]

For a spin I nucleus, the spin–lattice relaxation can be described by up to $2I$ distinct relaxation times arising from the $2I$ rate equations that define the transition probabilities between the various energy levels.[37] For $I > 1/2$ nuclei experiencing a nonzero quadrupolar interaction and assuming that the first-order quadrupolar interaction is the dominant relaxation mechanism, all $\Delta m = 2$ transitions and $\Delta m = 1$ satellite transitions will each contribute with transition probabilities aW_2 and bW_1, respectively, where a and b are factors specific to the two levels involved in the transition as well as the direction of the transition.[37,38] When the magnetic relaxation mechanism is dominant, relaxation is mediated only by $\Delta m = 1$ transitions, and the central and satellite transitions will each contribute transition probabilities dependent on a single rate process W. In both cases, the relaxation is typically multiexponential. Usually these complex, general solutions can be simplified; for example, in the case of a spin $I = 5/2$ nucleus undergoing quadrupolar relaxation, the description of the spin-lattice relaxation can be reduced to just three relaxation times by assuming that the Boltzmann population differences between all adjacent levels are equal and also noting symmetries in the rate equations, e.g., the $+5/2 \leftrightarrow +3/2$ and $-5/2 \leftrightarrow -3/2$ transitions.[37] Note that W_1 and W_2 are functions of the EFG orientation as well as temperature, which further complicates the description of quadrupolar relaxation in powder samples.

14.5 LITHIUM

With the exception of francium, all the alkali metals have quadrupolar nuclei that can be studied relatively easily using solid-state NMR. The most commonly utilized alkali metal in NMR studies of dynamics is lithium, and this is due both to its favorable NMR properties and the importance of the Li^+ ion in conducting materials. A 2007 review provides a very extensive overview of the various ways in which lithium NMR has been used to provide important information in such systems.[39]

^7Li has a spin $I = 3/2$, a high natural abundance (92.4%), and a moderately sized quadrupole moment (−40.1 mbarn). The second isotope, ^6Li, has a spin $I = 1$, but because of its very small quadrupole moment (−0.81 mbarn), this nucleus is also amenable to study by NMR. Studying dynamics using both nuclei can be advantageous, as it allows dynamic models to be checked against the different properties of each spin. Since isotope effects on the mobility of lithium ions are expected to be negligible, the same dynamic model should apply to both lithium isotopes. For example, lithium ion conduction in Li_3N

has been studied using the second-order quadrupolar shift of ^7Li and the relaxation rates of both ^7Li and ^6Li, with each set of results independently confirming a proposed conduction model.[40]

The second-order quadrupolar shift for ^6Li is negligible due to its very small quadrupole moment; thus, this isotope is best for determining isotropic chemical shifts. Despite a small inherent chemical shift range (~10 ppm for diamagnetic materials), the second-order quadrupolar interaction is small for both isotopes, so that MAS spectra with good resolution can often be obtained from both nuclei and more advanced MAS methods can then be applied (e.g., ^6Li exchange experiments have been used to characterize Li^+ hopping rates in Li_4SiO_4[41]). The resolution can be vastly improved in paramagnetic materials because of the large hyperfine shifts induced by the Fermi contact interaction. Quadrupolar splittings are usually not resolved in ^6Li spectra, but they may be resolved via the satellite transition spinning sidebands in ^7Li MAS spectra. However, many lithium ion conductors exhibit disorder on the lithium sublattice so that, even in the absence of motion, the quadrupolar splittings will be smeared out by distributions in the quadrupolar parameters.

For natural abundance systems, MAS linewidths for ^7Li may be broader than for ^6Li due to homogeneous broadening caused by the homonuclear dipolar interaction, a phenomenon that may be reduced by ^6Li enrichment. The linewidths can also be affected by motional narrowing as the sample temperature is increased. This is caused by the averaging of the dipolar and quadrupolar interactions as the ions become increasingly mobile, and the correlation time for the translational motion can be extracted from the linewidths $\Delta\nu$ using the expression[39,42]

$$\tau_c(T) = \frac{1}{2\pi\alpha.\Delta\nu(T)} \tan\left[\frac{\pi}{2}\left(\frac{\Delta\nu(T)}{\Delta\nu_{RL}}\right)^2\right] \quad (14.8)$$

where $\Delta\nu_{RL}$ is the rigid lattice linewidth (i.e., in the absence of motion) and α is a factor approximately equal to 1.

In general, T_1 relaxation measurements can be used to probe dynamics occurring on the timescale of the Larmor frequency (10^6–$10^8\,s^{-1}$), while $T_{1\rho}$, i.e., the relaxation under a spin-lock field in the rotating frame, can be used to probe slower motions (10^4–$10^6\,s^{-1}$). Extracting information on lithium ion conduction from T_1 relaxation data is complicated by various factors that lead to a departure from simple Bloembergen, Purcell, and Pound theory. Such factors include anisotropy in the diffusion, structural disorder, and cooperative processes due to the Coulomb interaction, and usually theoretical models for the relaxation must be tailored for the system under study. A large number of examples of ^6Li and ^7Li relaxation measurements and analyses have been summarized elsewhere, also including applications of spin-echo alignment experiments using Jeener—Broekaert pulse sequences to study slow ionic motions.[39] The latter experiments, which involve monitoring the amplitudes of stimulated echoes over a range of mixing times, can provide access to jump rates in the range $T_1^{-1} < k < T_2^{-1}$, i.e., slow and ultraslow motions on the order of $1\,s^{-1}$. Figure 14.13 shows a plot of lithium ion jump rates measured from $Li_{0.7}TiS_2$ using three different relaxation mechanisms.[43] These various relaxation processes allowed the measurement of rates over a range of 10 orders of magnitude, a remarkable illustration of the potential sensitivity of NMR relaxation to molecular dynamics.

Lithium ion mobility has also recently been probed in the $Li_3V_2(PO_4)_3$ system using $^6Li\{^{31}P\}$ REDOR

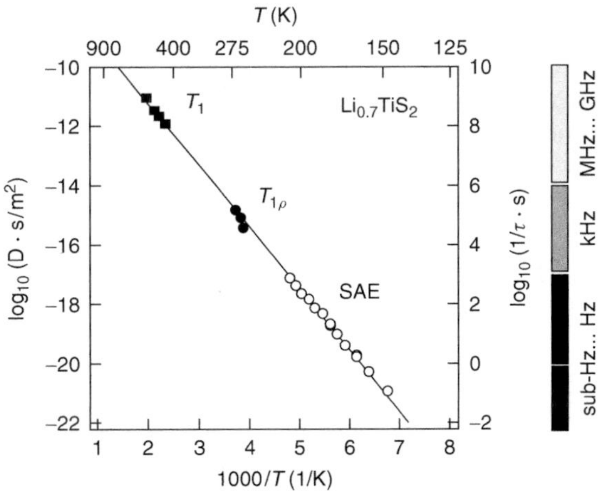

Figure 14.13. Lithium ion jump rates (τ^{-1}) determined from a sample of $Li_{0.7}TiS_2$ at various temperatures using different relaxation mechanisms (open circles, spin alignment echoes; closed circles, relaxation under spin-lock fields; and closed squares, spin–lattice relaxation). (Reproduced with permission from Ref. 43. Copyright 2006 Elsevier.)

(rotational echo double-resonance) experiments under MAS.[44] The pulse sequence involves the observation of a spin echo from the ^{6}Li nuclei while rotor-synchronized π pulses are applied to the ^{31}P nuclei, recoupling the ^{6}Li–^{31}P dipolar interaction which is otherwise averaged by the sample rotation. Typically, the fraction of signal lost by recoupling is monitored as a function of the dipolar evolution time and the dipolar coupling strength is extracted by fitting the resultant REDOR curve. When dynamics are present, the measured dipolar couplings are reduced in magnitude. In the case of $Li_3V_2(PO_4)_3$, the exchange between the lithium sites was frozen out at low temperatures, leaving a "rattling" mode for the ions within voids in the crystal structure, which resulted in a measurable reduction in ^{6}Li–^{31}P dipolar couplings that could be monitored over a range of temperatures as well as correlated with the local structures via multispin simulations.[44]

14.6 MULTIPLE-QUANTUM EXPERIMENTS

Multidimensional pulse sequences that can achieve high-resolution spectra from quadrupolar nuclei, particularly MQ/STMAS and double-quantum experiments, are now being widely applied. As might be expected, phenomena related directly to dynamic processes can be observed in such spectra, and the sensitivity of multiple-quantum experiments to molecular motions is becoming increasingly recognized. In this section, we provide several examples for various quadrupolar nuclei.

14.6.1 Spin $I = 1$ Double-quantum MAS Experiments

Double-quantum (DQ) transitions for $I = 1$ nuclei ($+1 \leftrightarrow -1$) show increased spectral resolution and no first-order quadrupolar broadening. This has been exploited in the case of ^{2}H in a 2D double-quantum MAS experiment to separate the single-quantum (SQ) spinning sideband patterns of chemically resolved deuterons, circumventing the need for selective deuteration. Broadening of the sidebands due to motions of specific deuterons can then be observed for jump rates in the 10^4–$10^6\,s^{-1}$ range.[45] This technique has been further refined to alias the SQ spinning sidebands into a single center line by means of rotor synchronization. The resulting correlation of the SQ spectra with the DQ spectra provides a sensitive means of detecting the presence of dynamics on the microsecond timescale, as illustrated in Figure 14.14 for oxalic acid dihydrate.[46] The acid group ^{2}H atoms are effectively static and appear as sharp lines in both SQ and DQ dimensions. The water deuterons, on the other hand, are dynamic and show a drastic broadening in the SQ dimension but a sharp line in the DQ dimension. Since the SQ static quadrupolar linewidth will be on the order of 250 kHz (much larger than a typical MAS rates), reorientational jumps on the microsecond timescale during the rotor period can change the SQ precession frequency by as much as 10^5 Hz, thereby interfering with the formation of an echo at the end of

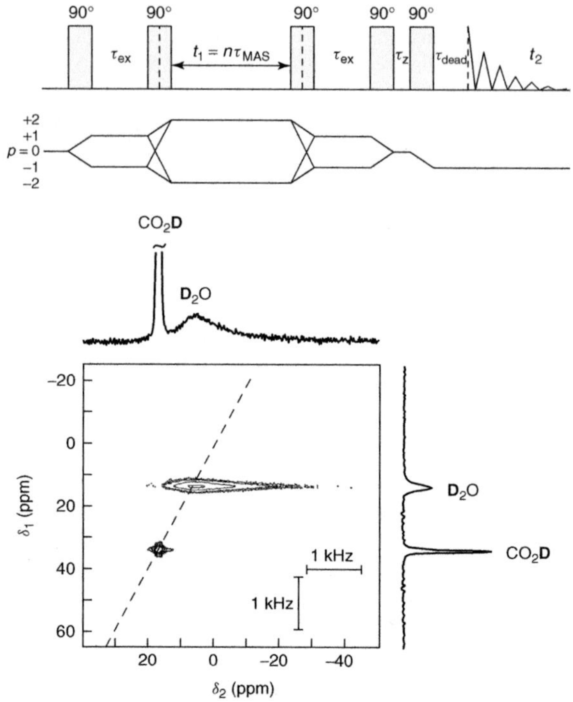

Figure 14.14. 2D rotor-synchronized ^{2}H double-quantum MAS NMR. (a) Pulse sequence and coherence pathways. (b) Spectrum of oxalic acid dihydrate-d_6 at 9.4 T. In the double-quantum dimension, both resonances are sharp, whereas in the single-quantum dimension, the D_2O resonance is broadened by dynamics. (Reproduced from Ref. 46 with permission from Elsevier.)

the rotor period. This reduces the echo intensity and shortens the train of rotary echoes, resulting in broadening. The static DQ linewidth with no quadrupolar broadening is likely to be only a few kilohertz; therefore, any motional jumps will change the frequency by an amount much less than the MAS rate and there is no attenuation of the echo train, hence no broadening.

This particular technique can be more sensitive to certain types of motion than the static linewidth experiments described in Section 14.2.1. For example, the SQ–DQ correlation for hydroxylclinohumite ($4Mg_2SiO_4 \cdot Mg(OD)_2$) at 314 K shows an SQ linewidth of 524 Hz and a DQ linewidth of only 77 Hz, indicative of dynamics on the microsecond timescale.[47] However, static ^2H lineshape studies showed little indication of any motion in this system. This can be rationalized in terms of the motional geometry, which involves a reorientation of the ^2H EFG tensor by 177°. Since this is very close to 180°, its effect on the static lineshape was very small, but the motion was clearly detected via the dynamic broadening of the spinning sidebands in the MAS experiment.

A similar phenomenon has been observed when comparing SQ and DQ linewidths for ^{14}N in indirectly detected ^1H–^{14}N HMQC type experiments.[48] The tripeptide Ala-Ala-Gly contains NH and NH_3^+ groups which show resolved ^{14}N resonances in the DQ $^{14}N/^1H$ correlation spectrum. Significant broadening of the NH_3^+ resonance occurs in the SQ dimension, while it remains sharp in the DQ spectrum, once again indicating motion, this time on the 100 ns timescale due to the larger quadrupolar linewidth of ^{14}N.

14.6.2 MQMAS and STMAS Experiments

The now common MQMAS technique[49] in the case of spin $I = 3/2$ nuclei correlates SQ central transitions ($+1/2 \leftrightarrow -1/2$) with triple-quantum (TQ) transitions ($+3/2 \leftrightarrow -3/2$). In the 2D spectra obtained, the $\delta_1 = 3\delta_2$ diagonal represents the position of the isotropic chemical shift. Resonances with zero quadrupolar coupling will appear on this diagonal, whereas, in the absence of motion or when motions are slower than the Larmor frequency, those with nonzero couplings are shifted off the diagonal by the second-order quadrupolar shift to a lower frequency in δ_2, and by three times this shift to higher frequency in δ_1. As mentioned earlier, if dynamics are occurring with rates comparable to the Larmor frequency, the second-order shift will be affected. At very high dynamic rates, this shift will disappear for isotropic motion and will reach a reduced value corresponding to the effective-averaged coupling for anisotropic motion. Consequently, if the off-diagonal peaks are observed to move toward the diagonal as temperature increases or if the peak is closer to the diagonal than expected based on a known coupling constant, this is clear evidence of very rapid dynamics.

This is illustrated in Figure 14.15 for ^{11}B MQ-MAS of the low-temperature orthorhombic phase of the pseudospherical molecule *ortho*-carborane ($C_2B_{10}H_{12}$).[26] The B and C atoms form an icosahedron, but the presence of the two C atoms (in this case adjacent to each other) reduces the true symmetry, and consequently, there are four chemically distinct types of B, although the peaks for two of these overlap. The MQMAS spectra show that the off-diagonal resonances move toward the diagonal as the temperature is increased (most evidently by comparing the spectra at 223 and 253 K). In fact, by 253 K the peak for B(9,12) lies directly on the diagonal, indicating an effective averaged coupling constant of zero, while the remaining peaks are still off-diagonal. This implies that the molecule is undergoing very rapid anisotropic motion. The room-temperature cubic phase of this material behaves as a plastic crystal, and although MQMAS gave no signals from this phase, a TQ ^{11}B–^1H MAS correlation experiment demonstrated that all the B atoms undergo very rapid isotropic averaging, since the TQ shifts are all three times larger than the SQ ^{11}B shifts observed in a routine ^1H-decoupled MAS spectrum.

Simulations of MQMAS spectra as a function of jump rate also suggest that, in addition to averaging of the second-order lineshape in the CT dimension (typically below $10^6\ s^{-1}$), there will also be a broadening of the TQ linewidths.[50] This is again due to interference of the motion with echo formation but at rates corresponding to the second-order broadening, but, to our knowledge, no clear experimental example corresponding to a single activated process is yet available.

For certain jump rates, STMAS spectra can be subject to significant dynamic broadening, while the MQMAS spectra for the same material do not appear to suffer any line broadening effects.

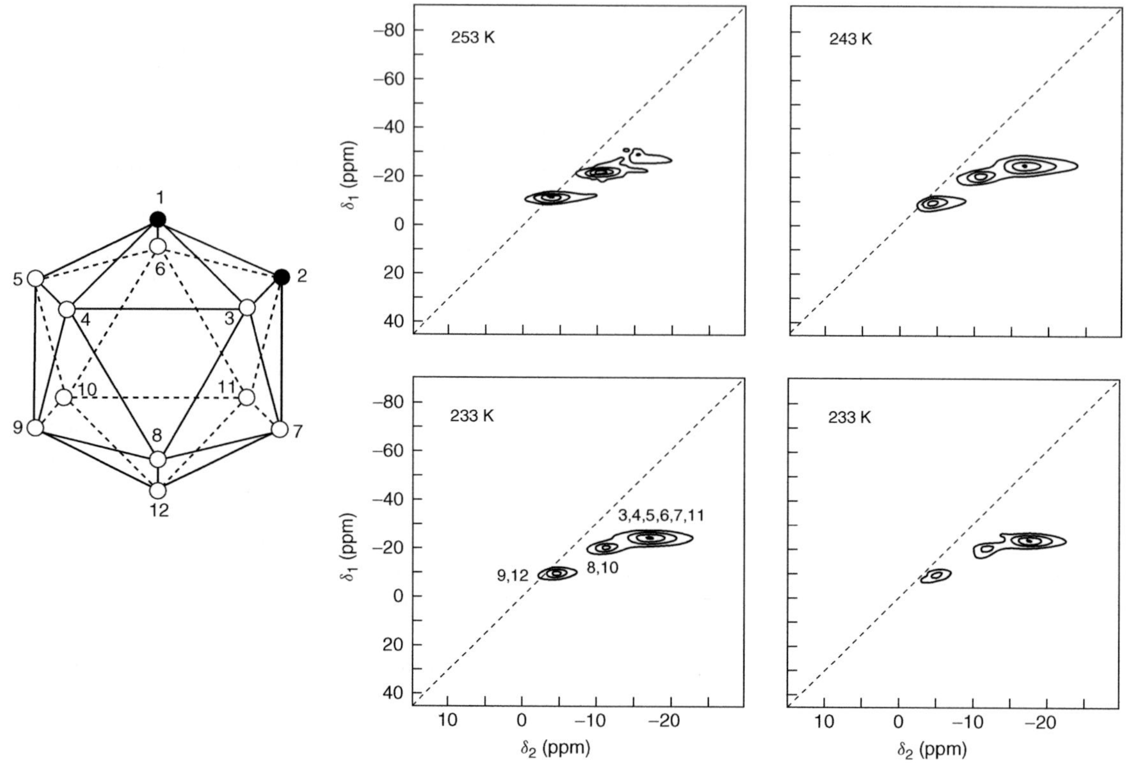

Figure 14.15. Left: Numerical labelling of the atoms in the *ortho*-carborane structure, • = carbon, o = boron. Right: ^{1}H-decoupled ^{11}B MQMAS NMR of *ortho*-carborane in its low temperature orthorhombic phase as a function of temperature. The dashed line indicates the $\delta_1 = 3\delta_2$ diagonal. Assignments of the spectral features to the different B atoms are indicated in the 233 K spectrum. (Reproduced from Ref. 26 with permission from Elsevier.)

Examples can be seen in the ^{17}O spectra of hydroxylchondrodite and hydroxylclinohumite[51,52] where proton-hopping motions affect the ^{17}O spectra, and in the ^{27}Al spectra of AlPO-14 materials[53] where calcined-dehydrated AlPO-14 shows sharp spectra but AlPO-14 as-prepared with isopropylamine shows strong broadening in only the STMAS spectra (Figure 14.16). The explanation for this behavior is similar to that given above for the ^{2}H DQ experiments, where the SQ dimension is subject to the first-order quadrupolar interaction, and fast dynamic jumps interfere with the averaging of this interaction by MAS. In STMAS, the satellite transition is subject to the first-order quadrupolar interaction but the STMAS CT transition and both the TQ and CT transitions in MQMAS are not.

14.7 CONCLUDING REMARKS

Quadrupolar NMR plays a vital role in the study of molecular and ionic dynamics in solids, often providing extensive information that cannot be obtained by other techniques, and which is important for understanding the physical properties of materials. It is also important to be aware of the influences that dynamics can have on the quadrupolar NMR spectra of a solid material, since the presence of motion may not necessarily be expected, and could lead to erroneous interpretation of spectra, for example, in situations where an apparently isotropic line is not at the isotropic chemical shift. This chapter has presented an overview of the current state of this area of research, and with much effort currently focused on the development of solid-state NMR of quadrupolar

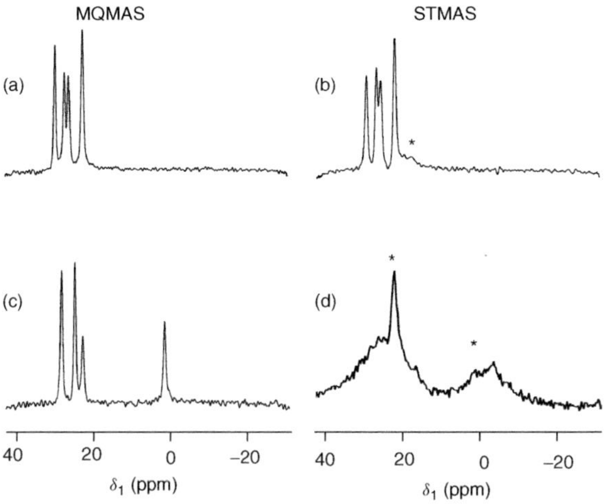

Figure 14.16. Isotropic projections of ^{27}Al MQMAS (TQ) and STMAS spectra of (a, b) calcined-dehydrated AlPO-14 and (c, d) AlPO-14 as-prepared with isopropylamine template, showing dynamic broadening in the STMAS spectrum (d). (Reproduced with permission from Ref. 53. Copyright 2006 American Chemical Society.)

nuclei, there will no doubt be numerous new instances of dynamic effects and new ways of probing them.

RELATED ARTICLES IN THE ENCYCLOPEDIA OF MAGNETIC RESONANCE

Deuterium NMR in Solids

Relaxation of Quadrupolar Nuclei Measured via Multiple-Quantum Filtration

Rotational and Translational Dynamics in Solids

REFERENCES

1. A. Abragam, *The Principles of Nuclear Magnetism*, Clarendon Press: Oxford, 1961, p. 111.
2. M. J. Duer, in *Solid State NMR Spectroscopy Principles and Applications*, ed M. J. Duer, Blackwell Science: Oxford, 2002, Chapter 6.
3. R. Tycko, in *Nuclear Magnetic Resonance Probes of Molecular Dynamics*, ed R. Tycko, Kluwer Academic: Dordrecht, 1994, Chapter. 1.
4. C. I. Ratcliffe, in *NMR Crystallography*, eds R. K. Harris, R. E. Wasylishen, and M. J. Duer, John Wiley & Sons: Chichester, 2009, p. 355, Chapter 24.
5. L. S. Batchelder, Deuterium NMR in solids, in *Encyclopedia of Magnetic Resonance*, eds R. K. Harris and R. E. Wasylishen, John Wiley & Sons: Chichester, 2007. DOI: 10.1002/9780470034590.emrstm0111.
6. E. Meirovitch and J. H. Freed, *Chem. Phys. Lett.*, 1979, **64**, 311.
7. R. L. Vold and G. L. Hoatson, *J. Magn. Reson.*, 2009, **198**, 57.
8. J. H. Kristensen, H. Bildsøe, H. J. Jakobsen, and N. C. Nielsen, *J. Magn. Reson.*, 1992, **100**, 437.
9. M. M. Maricq and J. S. Waugh, *J. Chem. Phys.*, 1979, **70**, 3300.
10. Y. Huang, R. L. Vold, and G. L. Hoatson, *J. Chem. Phys.*, 2006, **124**, 104504.
11. F. H. Larsen, H. J. Jakobsen, P. D. Ellis, and N. C. Nielsen, *Chem. Phys. Lett.*, 1998, **292**, 467.
12. L. Vugmeyster, D. Ostrovsky, J. J. Ford, and A. S. Lipton, *J. Am. Chem. Soc.*, 2010, **132**, 4038.
13. D. A. Torchia and A. Szabo, *J. Magn. Reson.*, 1982, **49**, 107.
14. G. L. Hoatson, *J. Magn. Reson.*, 1991, **94**, 152.
15. G. L. Hoatson, R. L. Vold, and T. Y. Tse, *J. Chem. Phys.*, 1994, **100**, 4756.
16. H. W. Spiess, *J. Chem. Phys.*, 1980, **72**, 6755.
17. R. L. Vold, G. L. Hoatson, L. Vugmeyster, D. Ostrovsky, and P. J. De Castro, *Phys. Chem. Chem. Phys.*, 2009, **11**, 7008.
18. C. Schmidt, S. Wefing, B. Blumich, and H. W. Spiess, *Chem. Phys. Lett.*, 1986, **130**, 84.
19. L. A. O'Dell, *Prog. Nucl. Magn. Reson. Spectrosc.*, 2011, **59**, 295.
20. T. K. Pratum and M. P. Klein, *J. Magn. Reson.*, 1989, **81**, 350.
21. L. A. O'Dell and C. I. Ratcliffe, *Chem. Commun.*, 2010, **46**, 6774.
22. E. A. Hill and J. P. Yesinowski, *J. Chem. Phys.*, 1997, **107**, 346.
23. R. E. Wasylishen, B. A. Pettitt, and K. R. Jeffrey, *J. Chem. Phys.*, 1981, **74**, 6022.
24. J. H. Kristensen and I. Farnan, *J. Chem. Phys.*, 2001, **114**, 9608.
25. R. W. Schurko, S. Wi, and L. Frydman, *J. Phys. Chem. A*, 2002, **106**, 51.

26. T. Kurkiewicz, M. J. Thrippleton, and S. Wimperis, *Chem. Phys. Lett.*, 2009, **467**, 412.
27. J. H. Kristensen and I. Farnan, *J. Magn. Reson.*, 2002, **158**, 99.
28. F. H. Larsen, *J. Magn. Reson.*, 2004, **171**, 293.
29. M. Witschas, H. Eckert, H. Freiheit, A. Putnis, G. Korus, and M. Jansen, *J. Phys. Chem. A*, 2001, **105**, 6808.
30. Y. Ba, J. A. Ripmeester, and C. I. Ratcliffe, *Can. J. Chem.*, 2011, **89**, 1055.
31. R. J. Wittebort, M. G. Usha, D. J. Ruben, D. E. Wemmer, and A. Pines, *J. Am. Chem. Soc.*, 1988, **110**, 5668.
32. I. L. Moudrakovski, C. I. Ratcliffe, and J. A. Ripmeester, *J. Am. Chem. Soc.*, 2001, **123**, 2066.
33. M. Witschas and H. Eckert, *J. Phys. Chem. A*, 1999, **103**, 10764.
34. N. Kim and C. P. Grey, *Science*, 2002, **297**, 1317.
35. M. R. Hampson, P. Hodgkinson, J. S. O. Evans, R. K. Harris, I. J. King, S. Allen, and F. Fayon, *Chem. Commun.*, 2004, 392.
36. P. Hodgkinson and M. R. Hampson, *Solid State Nucl. Magn. Reson.*, 2006, **30**, 98.
37. E. R. Andrew and D. P. Tunstall, *Proc. Phys. Soc.*, 1961, **78**, 1.
38. M. I. Gordon and M. J. R. Hoch, *J. Phys. C: Solid State Phys.*, 1978, **11**, 783.
39. R. Böhmer, K. R. Jeffrey, and M. Vogel, *Prog. Nucl. Magn. Reson. Spectrosc.*, 2007, **50**, 87.
40. D. Brinkmann, M. Mali, J. Roos, R. Messer, and H. Birli, *Phys. Rev. B*, 1982, **26**, 4810.
41. Z. Xu and J. F. Stebbins, *Science*, 1995, **270**, 1332.
42. N. Bloembergen, E. M. Purcell, and R. V. Pound, *Phys. Rev.*, 1948, **73**, 679.
43. M. Wilkening and P. Heitjans, *Solid State Ionics*, 2006, **177**, 3031.
44. L. S. Cahill, C. W. Kirby, and G. R. Goward, *J. Phys. Chem. C*, 2008, **112**, 2215.
45. M. J. Duer and C. Stourton, *J. Magn. Reson.*, 1997, **129**, 44.
46. M. Cutajar, S. E. Ashbrook, and S. Wimperis, *Chem. Phys. Lett.*, 2006, **423**, 276.
47. J. M. Griffin, A. J. Miller, A. J. Berry, S. Wimperis, and S. E. Ashbrook, *Phys. Chem. Chem. Phys.*, 2010, **12**, 2989.
48. S. Cavadini, A. Abraham, S. Ulzega, and G. Bodenhausen, *J. Am. Chem. Soc.*, 2008, **130**, 10850.
49. L. Frydman and J. S. Harwood, *J. Am. Chem. Soc.*, 1995, **117**, 5367.
50. M. Kotecha, S. Chaudhuri, C. P. Grey, and L. Frydman, *J. Am. Chem. Soc.*, 2005, **127**, 16701.
51. S. E. Ashbrook, S. Antonijevic, A. J. Berry, and S. Wimperis, *Chem. Phys. Lett.*, 2002, **364**, 634.
52. J. M. Griffin, S. Wimperis, A. J. Berry, C. J. Pickard, and S. E. Ashbrook, *J. Phys. Chem.*, 2009, **113**, 465.
53. S. Antonijevic, S. E. Ashbrook, S. Biedasek, R. I. Walton, S. Wimperis, and H. Yang, *J. Am. Chem. Soc.*, 2006, **128**, 8054.

Chapter 15

Alkali Metal NMR of Biological Molecules

Gang Wu

Department of Chemistry, Queen's University, Kingston, ON K7L 3N6, Canada

15.1 INTRODUCTION

Alkali metal ions, particularly Na^+ and K^+, are crucial to life. Living cells utilize active and passive ion transport mechanisms to maintain certain Na^+ and K^+ ion concentration gradients across the cell membrane and the resulting transmembrane potential plays a pivotal role in nerve transmission and muscle contraction. Another notable class of biological systems requiring alkali metal ions is nucleic acids (DNA and RNA). Nucleic acids are polyelectrolytes carrying large numbers of negative charges and, as such, require cations (very often Na^+ or K^+) for charge balance to be able to maintain stable structures. In eukaryotic cell nuclei, although charge balance of DNA is partially provided by histone proteins, the presence of K^+ or Na^+ is a key component for DNA structures and functions. There are also many enzymes that require K^+ or Na^+ ions as cofactors to exhibit full activity. Although the concentrations of Li^+, Rb^+, and Cs^+ ions in biological systems are normally negligible, these ions have important pharmacological and clinical applications. For example, Li^+ compounds are used in the treatment of bipolar disorder. Rb^+ and Cs^+ are often used as surrogate probes for studying K^+ in various chemical and biological systems.

Because alkali metal ions are invisible to UV–vis spectroscopy and are also diamagnetic, the major analytical technique for their direct detection in chemical systems in solution is nuclear magnetic resonance (NMR) spectroscopy. However, because the association between alkali metal ions and biological molecules is usually weak, ions often undergo rapid exchange between free and bound states. Consequently, it is generally difficult to use a conventional alkali metal NMR approach to obtain *site-specific* information in biomolecular systems in solution. For crystalline samples of proteins and nucleic acids, crystallography is presently the most reliable experimental technique for localizing alkali metal ions. However, direct detection of light alkali metal ions by diffraction methods sometimes can also be challenging.[1] It is particularly difficult to localize Na^+ ions, because a Na^+ ion has virtually identical X-ray scattering power as a water molecule, making it difficult to distinguish these two species in electron density maps. In many cases, it is also possible that a binding site is shared between alkali metal ions and water molecules. This partial occupancy problem often renders it impossible to identify Na^+ ions even using state-of-the-art diffraction techniques. In

NMR of Quadrupolar Nuclei in Solid Materials
Edited by Roderick E. Wasylishen, Sharon E. Ashbrook and Stephen Wimperis
 ISBN: 978-0-470-97398-1

general, it is also quite difficult to obtain DNA/RNA single crystals that diffract to subatomic resolution.

It is clear that new spectroscopic techniques that can yield direct information about alkali metal ions bound to biomolecules are highly desirable. Our interest in exploring solid-state NMR spectroscopy as a complementary approach to X-ray crystallography for detecting alkali metal ions in biological structures began more than a decade ago.[2] One important advantage of the solid-state NMR approach over crystallography is that solid samples (either polycrystalline powders or lyophilized solids) can be directly used in the NMR experiment. In addition, solid-state NMR parameters are a sensitive measure of the local electronic structure at the ion-binding site. Now solid-state NMR methodologies for studying spin-1/2 nuclei such as ^{13}C and ^{15}N have become a standard method for biomolecular structure determination. In recent years, considerable effort has also been devoted to application of solid-state NMR in the study of alkali metal ions in biological molecules.

In this chapter, we review solid-state NMR results for direct detection of alkali metal ions in biomolecular systems. This is a relatively new area of research. Most studies in the literature to date have focused on detection of Na^+ ions, because ^{23}Na has the highest NMR sensitivity among alkali metal nuclei. This chapter is organized as follows. First we give a very brief description of the nuclear properties of alkali metals and the basic solid-state NMR techniques used to study them. Then we describe recent solid-state alkali metal NMR studies of four classes of biomolecular systems: naturally occurring antibiotics, mononucleotides, double-stranded DNA, and G-quadruplex nucleic acids. Finally we conclude with a prognosis for future directions.

15.2 SOLID-STATE NMR FOR ALKALI METAL IONS

All NMR-active alkali metal isotopes have an atomic nucleus whose nuclear spin quantum number is greater than 1/2 ($I > 1/2$). These nuclei are known as *quadrupolar nuclei*. In this chapter, we will report solid-state NMR results for ^{23}Na ($I = 3/2$, 100% natural abundance (n.a.), $\Xi = 26.451900\%$, $Q = 10.4$ fm^2), ^{39}K ($I = 3/2$, 93.26% n.a., $\Xi = 4.666373\%$, $Q = 5.85$ fm^2), ^{87}Rb ($I = 3/2$, 27.83% n.a., $\Xi = 32.720454\%$, $Q = 13.35$ fm^2), and ^{133}Cs ($I = 7/2$, 100% n.a., $\Xi = 13.116142\%$, $Q = -0.343$ fm^2). The quadrupolar nature of these alkali metal nuclei makes it necessary to use solid-state NMR techniques that are quite different from those developed for more familiar spin-1/2 nuclei such as ^{13}C and ^{15}N. Very often, the resolution in NMR spectra for quadrupolar nuclei is poor, because the quadrupole interactions are usually large (e.g., on the order of several megahertz for ^{23}Na and ^{39}K).

As ^{23}Na, ^{39}K, ^{87}Rb, and ^{133}Cs all are half-integer (or noninteger) quadrupolar nuclei, one can detect only the central transition ($m = +1/2 \leftrightarrow m = -1/2$) in most cases except for ^{133}Cs. Under some favorable conditions such as for quadrupolar nuclei with small quadrupole moments or at very high magnetic fields, reasonable resolution can be obtained in central-transition NMR spectra simply by using a conventional technique known as *magic-angle spinning* (MAS) (see Chapter 4). However, the biggest problem in MAS spectra of half-integer quadrupolar nuclei is that the second-order quadrupole interaction cannot be completely averaged out by MAS. In many cases, the residual line broadening in MAS spectra is still larger than the chemical shift dispersion. Under such circumstances, it is difficult to obtain useful chemical information as different signals are overlapped. Currently, several techniques can be used to overcome this problem. They are double rotation (DOR), dynamic angle spinning (DAS), multiple-quantum MAS (MQMAS), and satellite transition MAS (STMAS) (see Chapters 9–11). These techniques can achieve complete removal of the second-order quadrupole interaction and, as such, produce high-resolution NMR spectra for half-integer quadrupolar nuclei. To date, it appears that only MQMAS has been used in solid-state NMR studies of alkali metal ions in biomolecular systems.

15.3 DETECTION OF ALKALI METAL IONS IN BIOLOGICAL MOLECULES

15.3.1 Naturally Occurring Antibiotic Ionophores

Naturally occurring antibiotic ionophores, which are either neutral macrocycles or monobasic acids, are small or medium-size molecules that are involved in transporting alkali metal ions across biological membranes. The movement of alkali metal ions across

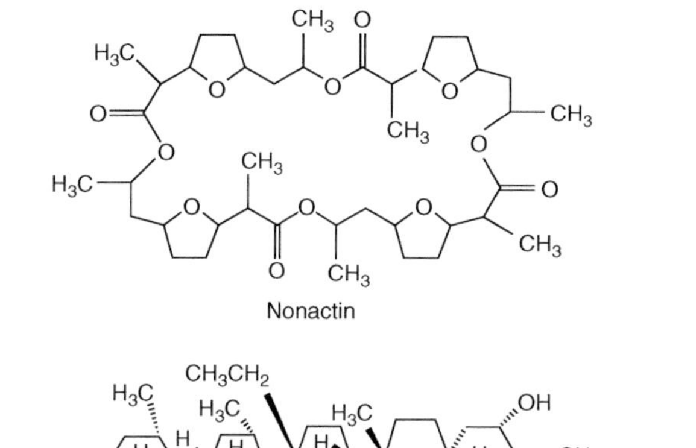

Figure 15.1. Molecular structures of selected naturally occurring antibiotic ionophores.

a membrane can be facilitated by naturally occurring ionophores via two different mechanisms: ion carrier and ion channel. Several common examples of ion carriers are given in Figure 15.1. Among channel-forming natural ionophores, the best known example is perhaps gramicidin A (gA), a linear pentadecapeptide isolated from *Bacillus brevis*. Tabeta *et al.*[3] reported the first solid-state ^{23}Na NMR study of naturally occurring antibiotic ionophores in 1986. They obtained ^{23}Na MAS spectra for Na complexes with valinomycin, nonactin, tetranactin, lasalocid, and monensin. Unfortunately, their analysis of the experimental spectra incorrectly assumed that the second-order ^{23}Na quadrupole interactions are negligible in these systems. As a result, the reported ^{23}Na chemical shifts are inaccurate. Depending on the particular systems, the reported ^{23}Na chemical shifts differ from the correct chemical shifts by values ranging from 2 ppm to as large as 20 ppm. In 2000, Wong and Wu[4] reexamined some of the naturally occurring antibiotic ionophores and reported high-quality ^{23}Na MAS spectra. Several examples from this latter work are shown in Figure 15.2. One can clearly see that, in all cases, typical line shapes from second-order quadrupole interactions are observed. Accurate ^{23}Na NMR parameters (δ_{iso}, C_Q, and η_Q) are then obtained from fitting the experimental ^{23}Na MAS line shapes, as illustrated in Figure 15.2. For example, a spectral analysis yields the following ^{23}Na NMR parameters for Na(nonactin)SCN: $\delta_{iso} = -16 \pm 1$ ppm, $C_Q = 0.58 \pm 0.05$ MHz, and $\eta_Q = 1.00 \pm 0.02$. As naturally occurring antibiotic ionophores exhibit a diverse range of structures, they can potentially be used as models for establishing a relationship between alkali metal ion-binding structure and their NMR parameters. It will be of interest in the future to use alkali metal NMR to examine these ionophores when they are interacting with a biological membrane.

For solid-state NMR studies of alkali metal ions other than Na^+ in the context of natural ionophores, there have been only two reports in the literature. In 1986, Kunwar *et al.*[5] reported a solid-state ^{39}K NMR spectrum for a stationary powder sample of K(monensin). These authors analyzed the static line shape and obtained the following ^{39}K NMR parameters: $\delta_{iso} \approx 0$ ppm, $C_Q = 3.2$ MHz, and $\eta_Q = 0.52$. In 2000, Wong *et al.*[6] reported solid-state ^{133}Cs NMR spectra for Cs(nonactin)SCN and gA/CsCl. They obtained both MAS and static ^{133}Cs NMR spectra for Cs(nonactin)SCN, from which they determined the ^{133}Cs chemical shift tensor components: $\delta_{11} = 27$, $\delta_{22} = 10$, and $\delta_{33} = -26$ ppm. Because ^{133}Cs has a very small nuclear quadrupole moment, the second-order quadrupole effect is usually negligible at moderate magnetic fields. Consequently, to be able to determine the value of C_Q, one has to record satellite transitions. In the same study, Wong *et al.*[6] also obtained solid-state ^{133}Cs NMR spectra of several synthetic macrocyclic compounds, which suggest that C_Q is less than 0.5 MHz for Cs^+ ions bound to most naturally occurring antibiotic ionophores. Wong *et al.*[6] examined a crystalline sample of gA/CsCl complex using both solid-state ^{133}Cs NMR and X-ray crystallography. As shown in Figure 15.3, the ^{133}Cs MAS spectrum of gA/CsCl reveals that the Cs^+ ions residing inside the gA channel give rise to a broad

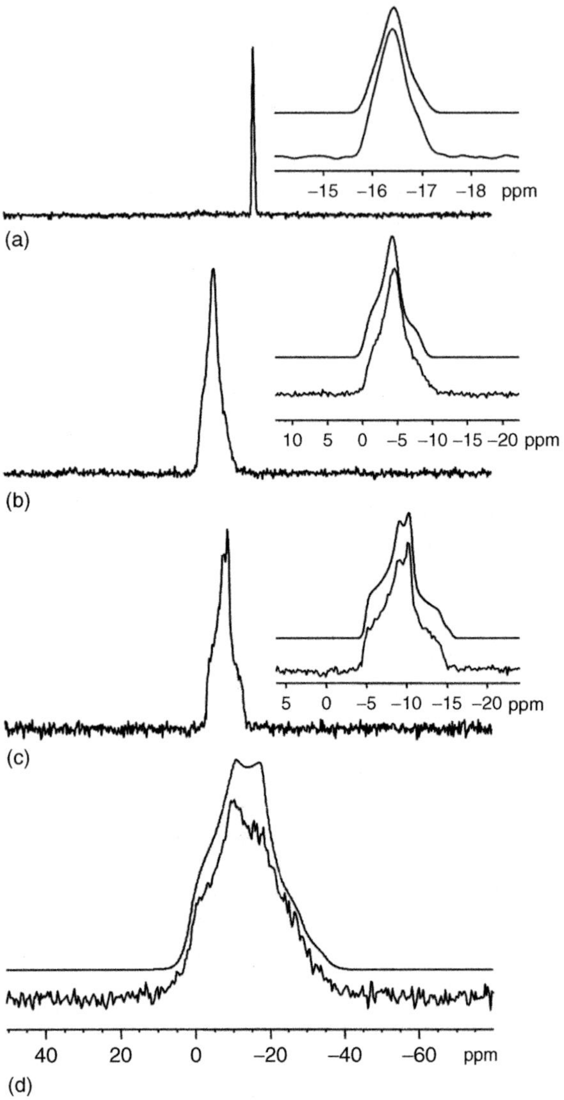

Figure 15.2. Experimental (lower trace) and simulated (upper trace) ^{23}Na MAS spectra of (a) Na(nonactin)SCN, (b) Na(lasalocid A)·MeOH, (c) Na(monesin)Br, and (d) Na(valinomycin)SCN at 11.75 T. (Reproduced from Ref. 4. © American Chemical Society, 2000.)

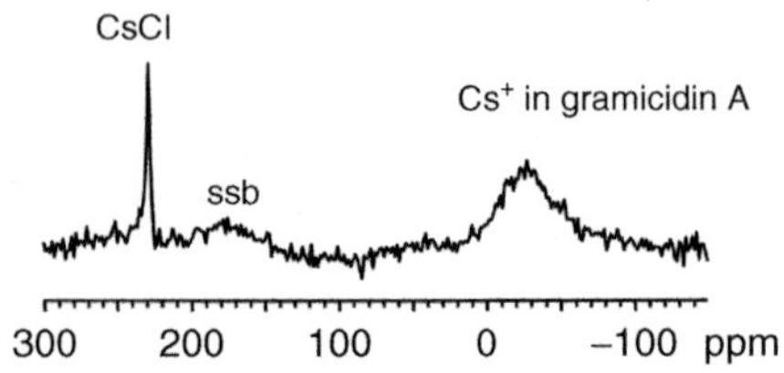

Figure 15.3. Central transition ^{133}Cs MAS spectrum (11.75 T) of a microcrystalline sample of gA/CsCl. (Reproduced from Ref. 6 by permission of NRC Research Press.)

peak. The line width of this broad peak is approximately 3.3 kHz at 11.75 T, which may be due either to the presence of multiple Cs^+ sites within a gA channel or to the presence of multiple independent gA channels in the asymmetric unit. In addition, it is also likely that for each Cs^+ site there exists a distribution of ^{133}Cs chemical shifts due to the presence of disorder. On the basis of the observed relationship between ^{133}Cs chemical shifts and Cs^+ coordination numbers, Wong *et al.*[6] interpreted the observed ^{133}Cs chemical shift in gA/CsCl as due to Cs^+ ions each being surrounded by eight oxygen atoms. This type of binding environment is consistent with those reported for gA crystals in crystallographic studies. For example, a recent X-ray study of gA/CsCl suggests that two gA molecules (a dimer) form a right-handed antiparallel double-stranded β-helix (DSβH$_R$) and that each gA dimer contains three Cs^+ binding sites.[7] On average, each Cs^+ ion has eight closest contacts: six carbonyl oxygens (3.6–4.5 Å) and two water oxygens (2.8–3.1 Å). Therefore, the solid-state ^{133}Cs NMR observation is in agreement with the crystallographic studies with respect to the Cs^+ binding environment in gA, despite the fact that several different overall crystal structures exist. Regardless of whether this particular kind of DSβH$_R$ helical structure is responsible for ion conduction through a gA channel in a lipid environment, this solid-state ^{133}Cs NMR study demonstrates that useful information can be obtained about Cs^+ binding in biological systems. It would be of interest to extend the alkali metal NMR to other gA systems containing Na^+ and K^+ ions.

15.3.2 Mononucleotides

Mononucleotides are the basic units of nucleic acids (RNA and DNA). As shown in Figure 15.4, the potential sites for alkali metal ion binding are phosphate, carbonyl oxygen, and nitrogen from nucleobases, and hydroxyl groups from ribose/deoxyribose. There are a large number of crystallographic studies of mononucleotides in the literature that provide useful information about common modes of alkali metal ion

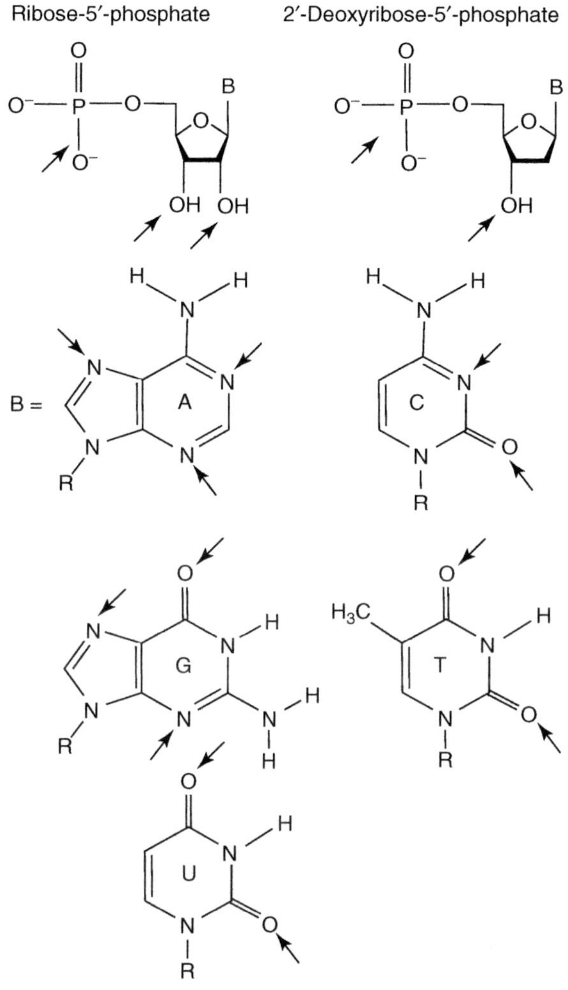

Figure 15.4. Illustration of potential alkali metal ion-binding sites in mononucleotides.

coordination in these systems. Therefore, mononucleotides are natural targets for solid-state NMR studies, because one can obtain NMR parameters for alkali metal ions in a variety of ion-binding environments potentially similar to those found in DNA or RNA oligomers. To date, only a small number of solid-state ^{23}Na and ^{39}K NMR studies have been reported for mononucleotides. Here we first use a hydrated Na salt of guanosine 5′-monophosphate, $Na_2(5'\text{-GMP})\cdot 7H_2O$ (in its orthorhombic form), as an example to illustrate some basic features commonly observed in solid-state ^{23}Na NMR spectra of mononucleotides. Figure 15.5 shows a partial crystal structure of orthorhombic $Na_2(5'\text{-GMP})\cdot 7H_2O$ and solid-state ^{23}Na NMR spectra (MAS and MQMAS) obtained for this compound at 11.75 T. Clearly, the 1D ^{23}Na MAS spectrum does not exhibit any recognizable feature that would allow one to extract information about the number of Na^+ sites and the associated NMR parameters. In contrast, the 2D ^{23}Na MQMAS spectrum displays clearly four spectral regions, suggesting the presence of four distinct Na^+ sites in the crystal lattice. From individual F_1 slice-spectra, one can obtain three ^{23}Na NMR parameters (δ_{iso}, C_Q, and η_Q) for each Na^+ site. Finally, one can compare the simulated total MAS spectrum with the experimental spectrum. For orthorhombic $Na_2(5'\text{-GMP})\cdot 7H_2O$, an analysis of the spectra shown in Figure 15.5 yields the following ^{23}Na NMR parameters: Na_1, $C_Q = 1.30$ MHz, $\eta_Q = 0.7$, $\delta_{iso} = -4.5$ ppm; Na_2, $C_Q = 1.85$ MHz, $\eta_Q = 0.5$, $\delta_{iso} = -2.0$ ppm; Na_3, $C_Q = 1.85$ MHz, $\eta_Q = 0.6$, $\delta_{iso} = -2.0$ ppm; Na_4, $C_Q = 2.30$ MHz, $\eta_Q = 0.7$, $\delta_{iso} = -5.5$ ppm. As the crystal structure of orthorhombic $Na_2(5'\text{-GMP})\cdot 7H_2O$ is known,[8] assignment of these parameters to individual Na^+ sites can be made on the basis of a simple correlation between C_Q and the local ion-binding geometry. Na_1 is assigned to the Na site with four water molecules and two hydroxyl groups from the ribose groups. Na_2 and Na_3 correspond to the two fully hydrated Na sites. Na_4 is the site coordinated to four water molecules and two N7 nitrogen atoms from the pyrimidine moieties. A more reliable method for ^{23}Na spectral assignment is to use ab initio electric field gradient (EFG) and shielding calculations. In 2003, Wong and Wu[9] reported a solid-state ^{23}Na NMR study of four hydrated sodium salts of mononucleotides with an emphasis on penta-coordinate Na ions. Using 1D and 2D ^{23}Na NMR methods, they were able to detect all 12 crystallographically distinct Na sites in these systems. They also showed that spectral assignment can be achieved with the help of quantum chemical EFG calculations at the restricted Hartree–Fock level with a 6-31G(d) basis set (see Chapter 13).

In many cases, if a crystal structure is known for the system under study, one tends to use it automatically to aid solid-state NMR spectral interpretation. However, caution should be exercised in this practice since this analysis is valid only if the sample used in solid-state NMR experiments is identical to that used in the crystallographic study. In the context of mononucleotide or DNA/RNA samples, this may not

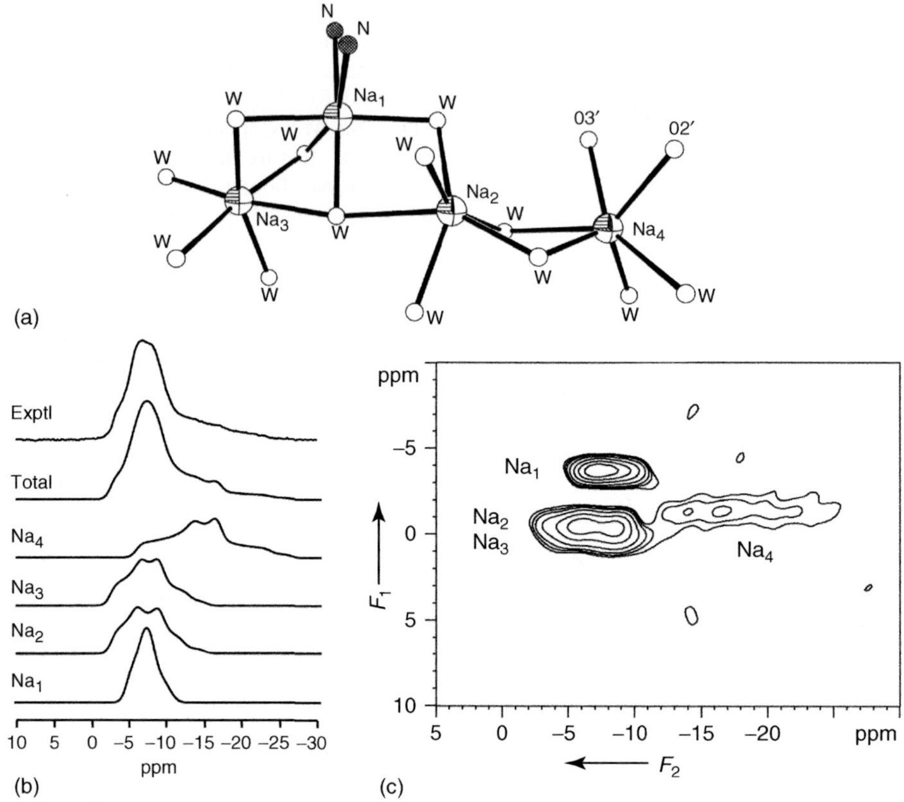

Figure 15.5. (a) Partial crystal structure, (b) ^{23}Na MAS, and (c) MQMAS spectra (11.75 T) of orthorhombic Na_2(5′-GMP)·$7H_2O$.

always be the case. One common problem in dealing with DNA/RNA crystal structures is that structural transitions may occur in the solid state as a function of relative humidity. The best known example is the structural transition of double-stranded DNA between the A and B forms.[10] The B form is a right-handed helix. This form is the most commonly found in aqueous solution and in the solid state at high relative humidity and low salt. At low relative humidity and high salt, DNA adopts the A form right-handed helix. These kinds of structural transitions are also common in simple nucleotides and have contributed to great confusion in the literature regarding solid-state ^{23}Na NMR spectral interpretation. Confusion has largely been caused by the lack of information about the actual nucleotide sample conditions during solid-state NMR experiments. In the discussion that follows, we use solid-state ^{23}Na NMR studies of the Na salt of adenosine 5′-triphosphate (5′-ATP) as an example to highlight this important problem in the study of Na nucleotides.

In 2000, Ding and McDowell[11] reported the first set of ^{23}Na MQMAS spectra for a hydrated Na salt of 5′-ATP. They reported observation of four Na sites with very similar ^{23}Na quadrupole parameters but different chemical shifts; see Table 15.1. Unfortunately, because their spectra were acquired at a relatively low magnetic field of 9.4 T, the spectral quality did not allow the authors to carry out spectral assignment with certainty. A few years later, Wong and Wu[9] re-examined the same compound and also reported observation of four Na sites with quite different ^{23}Na NMR parameters (also shown in Table 15.1). To help with spectral assignment, Wong and Wu used the refined crystal structure of Larson[12] to build molecular clusters and then performed quantum chemical EFG calculations. The calculated EFG results at the HF/6-31G(d) level

Table 15.1. Solid-state ^{23}Na NMR parameters reported for several hydrated $Na_2(ATP)$ samples

System	δ_{iso} (ppm)	C_Q (MHz)	η_Q	References
$Na_2(5'\text{-ATP})\cdot xH_2O$				Ding and McDowell[11]
Na_1	2.8	1.40	0.88	
Na_2	9.0	1.35	0.92	
Na_3	14.6	1.45	0.95	
Na_4	−3.2	1.38	0.85	
$Na_2(5'\text{-ATP})\cdot xH_2O$				Wong and Wu[9]
Na_1	−1.5	1.90	0.60	
Na_2	−1.5	1.80	0.50	
Na_3	−2.5	1.20	0.70	
Na_4	−7.0	2.10	0.50	
$Na_2(5'\text{-ATP})\cdot 3H_2O$				Grant *et al.*[13]
Na_1	0.0	1.1	0.7	
Na_2	0.5	2.1	0.5	
Na_3	−1.9	2.3	0.5	
Na_4	1.6	1.6	0.9	
$Na_2(5'\text{-ATP})\cdot 2H_2O$				Nausner *et al.*[14]
Na_1	−0.20	1.10	0.90	
Na_2	−2.63	2.05	0.43	
Na_3	0.96	2.25	0.72	
Na_4	1.44	1.35	0.80	

for the cluster models were as follows: Na_1, $C_Q = 2.090$ MHz, $\eta_Q = 0.685$; Na_2, $C_Q = 1.437$ MHz, $\eta_Q = 0.500$; Na_3, $C_Q = 0.568$ MHz, $\eta_Q = 0.752$; Na_4, $C_Q = 1.802$ MHz, $\eta_Q = 0.309$. One may argue that the accuracy of these calculated EFG data is not sufficient to allow complete, unambiguous spectral assignment. A more important problem is that, while the crystal structure of Larson[12] is for $Na_2(5'\text{-ATP})\cdot 3H_2O$ (a trihydrate form), the exact hydration state in the solid $Na_2(5'\text{-ATP})$ sample studied by Wong and Wu[9] was not determined. In 2006, Grant *et al.*[13] further examined ^{23}Na MQMAS spectra of hydrated $Na_2(5'\text{-ATP})$ and discovered that different sample conditions resulted in different ^{23}Na MQMAS spectra. Grant *et al.*[13] concluded that different ^{23}Na MQMAS spectra must be associated with different hydration states of $Na_2(5'\text{-ATP})$. By carefully recrystallizing from H_2O/dioxane, these authors obtained a pure $Na_2(5'\text{-ATP})\cdot 3H_2O$. For this pure trihydrate form, they reported the ^{23}Na NMR parameters listed in Table 15.1. Once again, these parameters are clearly different from those reported by Ding and McDowell[11] and by Wong and Wu.[9] Grant *et al.*[13] also attempted to assign these ^{23}Na NMR signals to the four Na sites in $Na_2(5'\text{-ATP})\cdot 3H_2O$. The assignments of Na_3 and N4 were established using the ^{31}P–^{23}Na REDOR experiment, because these two Na ions are directly bound to the phosphate oxygen atoms, placing them in the vicinity of the ^{31}P nuclei. These authors measured the ^{23}Na spin–lattice relaxation times (T_1) for the four Na sites and found that the values of $T_1(^{23}Na)$ for Na_3 and Na_4 are much shorter than those for Na_1 and Na_2. This observation is consistent with the fact that Na_3 and Na_4 are coordinated dynamic water molecules whereas the coordination spheres for Na_1 and Na_2 are more rigid. However, the authors could not make an unambiguous assignment for Na_1 and Na_2.

In 2009, Nausner *et al.*[14] reported a detailed solid-state NMR study of the Na sites in microcrystalline 5′-ATP samples. By carefully controlling the experimental conditions for sample preparation, these authors were able to isolate monohydrate, dihydrate, and trihydrate forms of $Na_2(5'\text{-ATP})$. Figure 15.6 shows the sensitivity of ^{23}Na MQMAS spectra to the degree of hydration of $Na_2(5'\text{-ATP})$. For $Na_2(5'\text{-ATP})\cdot 2H_2O$, they determined the ^{23}Na NMR parameters for all four Na sites; see Table 15.1. The authors made an attempt to assign these four ^{23}Na MQMAS signals by using quantum chemical

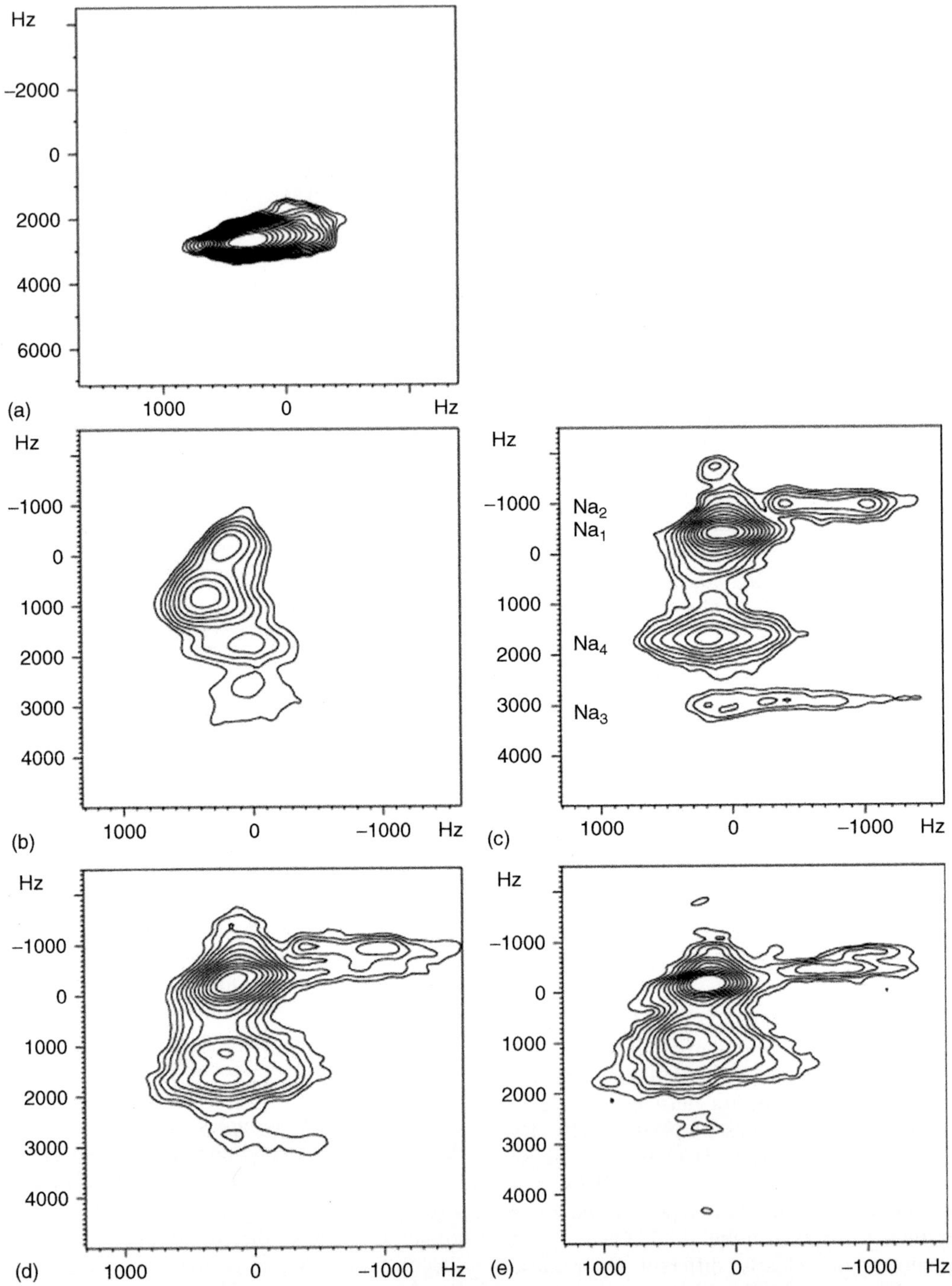

Figure 15.6. Experimental ^{23}Na MQMAS spectra of Na_2(5′-ATP) (a) as purchased and after recrystallization (b) in monohydrate form, (c) dihydrate form, (d) mixed dihydrate/trihydrate form, and (e) trihydrate form. (Reproduced from Ref. 14. © Elsevier B.V. 2009.)

EFG calculations. However, they used the crystal structure of Larson[12] to build cluster models for interpreting their solid-state NMR data obtained for Na_2(5′-ATP)·$2H_2O$. As mentioned earlier, this crystal structure is for the trihydrate form rather than the dihydrate form. Sugawara *et al.*[15] reported the crystal structure of Na_2(5′-ATP)·$2H_2O$, which clearly exhibits some differences in Na coordination as compared with that seen in Na_2(5′-ATP)·$3H_2O$. To this date, a *complete* spectral assignment of the ^{23}Na NMR signals for Na_2(5′-ATP)·$2H_2O$ and Na_2(5′-ATP)·$3H_2O$ still remains a challenging task. It is possible that ^{23}Na chemical shift calculations can be used as an additional tool to help with spectral assignment. Nonetheless, these studies highlight several important aspects of solid-state ^{23}Na NMR for nucleotides especially with regard to sample preparation and spectral assignment.

In addition to ^{23}Na–^{31}P REDOR experiments, T_1(^{23}Na) parameters, and EFG calculations, one may use other NMR spectroscopic properties to aid spectral assignment. For example, Grinshtein *et al.*[16] demonstrated the use of ^{1}H–^{23}Na separate-local-field (SLF) experiments and its combination with MQMAS to help assign the ^{23}Na signals in Na_2(5′-dCMP)·$7H_2O$ and Na_2(3′-UMP)·$4H_2O$. In this case, two well-resolved ^{23}Na signals were observed in the MAS spectrum, in agreement with the presence of two crystallographically distinct Na^+ sites seen in the crystal structure of Na_2(5′-dCMP)·$7H_2O$.[17] According to this crystal structure, one Na^+ site is coordinated to six water oxygen atoms in an octahedral fashion with the Na–O distances ranging from 2.374 to 2.579 Å, while the other Na is bound to five water oxygen atoms with similar Na–O distances. As a result, the hexacoordinate Na^+ should experience stronger ^{1}H–^{23}Na dipolar interactions. The authors concluded that Site II corresponds to the hexacoordinate Na^+ site. This assignment is certainly consistent with the expectation that the hexacoordinate Na site should have a smaller value of C_Q. However, one also notes that the SLF patterns are nonetheless quite similar for the two sites. It is also unclear whether the dynamic nature of water molecules could average out most of the ^{1}H–^{23}Na dipolar interactions. Perhaps, low-temperature experiments would be able to reduce this dynamic effect.

Unlike solid-state ^{23}Na NMR studies of mononucleotides, there has been only one solid-state ^{39}K NMR study in the literature that deals with mononucleotide systems containing K^+ ions. In 2003, Wu *et al.*[18] reported ^{39}K MAS NMR spectra of hydrated K^+ salts of adenosine 2′-monophosphate, K(2′-AMP)·$1.5H_2O$, and adenosine 5′-diphosphate, K(5′-ADP)·$2H_2O$. These spectra were recorded at a high magnetic field of 19.6 T. From these spectra, the authors obtained ^{39}K NMR parameters. The K^+ ion in 2′-AMP is coordinated to two phosphate oxygen atoms, two water molecules, and two hydroxyl groups from the ribose,[19] representing a typical environment for phosphate-bound K^+ ions. In comparison, the K^+ coordination in K(5′-ADP) is quite unusual where the K^+ ion is coordinated to seven ligands: four phosphate groups, one water, one hydroxyl, and a nitrogen atom (N3) from the adenine base.[20] Perhaps this unusual coordination is responsible for the observed low-frequency ^{39}K NMR signal in K(5′-ADP). Even though the chemical shift range for ^{39}K is only ~100 ppm, the spectral resolution observed in the ^{39}K MAS spectra at 19.6 T is encouraging. Further studies are necessary to accumulate more solid-state ^{39}K NMR data for K nucleotides. We anticipate that continuing accumulation of solid-state ^{23}Na and ^{39}K NMR data will lead to a better understanding of the relationship between NMR parameters and ion-binding structure. In the meantime, solid-state NMR of ^{7}Li, ^{87}Rb, and ^{133}Cs should also be explored in mononucleotides where Li^+, Rb^+, and Cs^+ ions are the counterions.

15.3.3 Double-stranded DNA

Alkali metal ion binding is also important in double-stranded nucleic acids. In the literature, there are just a few scattered reports where solid-state ^{23}Na NMR is used to study double-stranded DNA systems. In 1998, He *et al.*[21] attempted for the first time to use solid-state ^{23}Na NMR techniques to probe Na^+ binding to a naturally occurring double-stranded DNA, calf thymus DNA (CT DNA). They reported ^{23}Na MAS NMR spectra obtained with either direct excitation or cross polarization from protons. They also used nutation NMR and measured T_1(^{23}Na) values. They compared the spectra obtained with and without competing species such as Mg^{2+}, ethidium bromide, and $[Ru(phen)_3]^{2+}$. Although there is no concrete conclusion drawn from this study, it appears that they have identified two main classes of Na^+ binding sites in double-stranded

DNA. Some Na^+ ions are located close to DNA, while the others are far from DNA. But they concluded in the end that Na^+ ions are not bound either directly or via a single water molecule to the phosphate group.

Several years later, Madeddu[22] demonstrated that the water content of the CT DNA sample is a key factor for the appearance of its solid-state ^{23}Na NMR spectra. The author showed that, at high relative humidity, each solid-state ^{23}Na NMR spectrum of CT DNA exhibits a single signal. However, when the relative humidity is decreased below 73%, two ^{23}Na NMR signals are observed, with their relative intensities changing with the water content of the CT DNA sample. The author further investigated the T_1 behaviors for both ^{23}Na and ^{31}P nuclei in CT DNA. This work highlights the importance of controlling the water content for DNA samples in order to obtain reproducible solid-state NMR results.

Recently, Wong *et al.*[23] used ^{23}Na MAS and ^{23}Na{^{31}P} REDOR techniques to study Na^+ binding to CT DNA. At high relative humidity, all Na^+ ions are fully hydrated, giving rise to a sharp ^{23}Na NMR signal close to 0 ppm. As the relative humidity is reduced, some Na^+ ions are localized most likely at the phosphate backbone of the DNA and a very small fraction of Na^+ ions form NaCl microcrystallites. At very low relative humidity (dry state), a significant amount of Na^+ ions crystallize as NaCl and about 50% of Na^+ ions are attached to the phosphate groups. For this latter class of Na^+ ions, the ^{23}Na NMR signal exhibits a typical line shape due to second-order quadrupole interactions. The fact that two separate ^{23}Na NMR signals can be observed for CT DNA in the dry state permits a direct measurement of the amount of Na^+ ions that are attached to the DNA backbone. Wong *et al.*[23] further used a ^{23}Na{^{31}P} REDOR method to establish that, at low relative humidity, Na^+ ions are directly bound to the phosphate group of CT DNA and are partially dehydrated. These authors also utilized the spectral separation between bound and free Na^+ ions to determine the relative cation-binding affinity to the DNA phosphate backbone: $Li^+ > Na^+ > NH_4^+ > Cs^+ > Rb^+ > K^+$. This work represents the first determination of cation-binding affinity to dry DNA. The reported experimental ion-binding affinity may be used for testing theoretical treatment of cation–phosphate interactions in DNA.

15.3.4 G-quadruplex Nucleic Acids

The focus of this section is on solid-state NMR studies of alkali metal ion binding to a special type of four-stranded nucleic acid structure known as the *guanine-quadruplex* (i.e., G-quadruplex). The basic structural motif of the G-quadruplex is a guanine tetrad often referred to as a *G-quartet* where four guanine molecules are linked by eight Hoogsteen hydrogen bonds; see Figure 15.7. The G-quartet model was first proposed in 1962 by Gellert *et al.*[24] on the basis of X-ray fiber diffraction data for guanylic acids (3′ and 5′ isomers). Recent discoveries of the existence of the G-quartet motif in many biologically important systems such as telomeres, promoters of many genes, and sequences related to various human diseases have triggered tremendous research interest in this unusual type of nucleic acid structure.[25] Now G-quartet structures are of interest in a diverse range of fields ranging from molecular biology, medicinal chemistry, to supramolecular chemistry and nanotechnology.[26,27]

Alkali metal ions have been found to be critical for the formation, stability, and structural variation of G-quadruplex nucleic acids and related systems. For example, as shown in Figure 15.7, a four-repeat human telomeric DNA sequence, $d(TTAGGG)_4$, can fold into drastically different G-quadruplex structures, depending on the nature of alkali metal ions present in solution or during crystallization.[28–32] The first atomic-resolution crystal structure for a G-quadruplex was reported by Rich and coworkers in 1992 on a single crystal of *Oxytricha nova* telomere DNA repeat $d(G_4T_4G_4)$.[33] They attributed a low electron density found between the two central G-quartets to be due to the presence of a K^+ ion. A more complete crystal structure of $d(G_4T_4G_4)$ was reported by Neidle and coworkers in 2002.[34] In this new crystal structure, five K^+ ions are detected to be bound to the G-quadruplex structure. Three of them reside inside the G-quadruplex structure, each coordinating to eight carbonyl oxygen atoms from two adjacent G-quartets, whereas the other two K^+ ions are located in the diagonal thymine loop regions. The first crystallographic localization of Na^+ ions in a G-quadruplex was reported in 1994 for $d(TG_4T)$.[35] As shown in Figure 15.8, $d(TG_4T)$ forms a parallel-stranded G-quadruplex. A total of seven Na^+ ions are observed to occupy the inner core of the quadruplex, thus giving a striking impression of an ion channel. As will be discussed later, the structural similarity between G-quadruplexes and ion channel

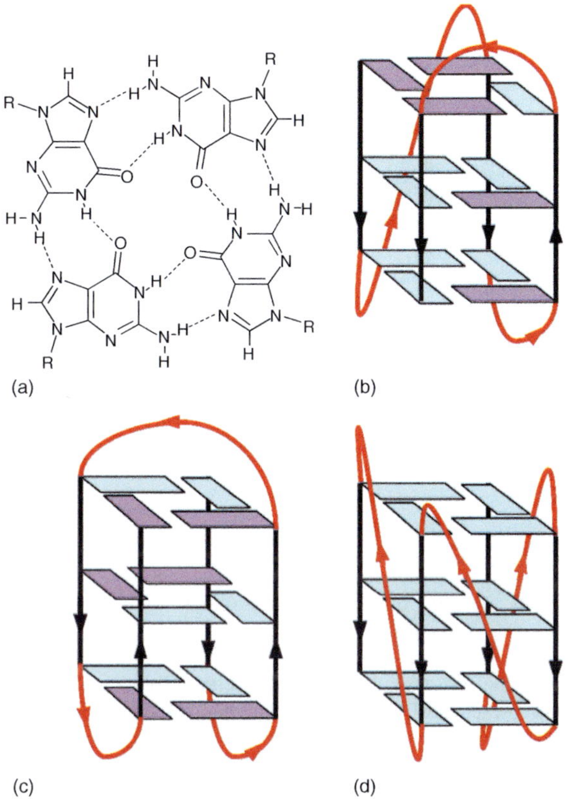

Figure 15.7. (a) G-quartet, (b) K^+ form (NMR structure), (c) Na^+ form (NMR structure), and (d) K^+ form (crystal structure) of a four-repeat human telomeric DNA sequence, d(TTAGGG)$_4$.

proteins also leads to similar ion selectivity between these two classes of systems. Until recently, localization of alkali metal ions in nucleic acids in general has almost exclusively relied on crystallographic techniques. In the past several years, solid-state NMR has emerged as a viable tool for direct detection of alkali metal ions in G-quadruplex structures.

In 2000, Rovnyak *et al.*[36] reported the first solid-state ^{23}Na NMR study of G-quadruplex DNA systems. In particular, they obtained solid-state ^{23}Na NMR spectra for a DNA hexamer, d(TG$_4$T). The G-quadruplex formed by d(TG$_4$T) is an excellent system to study, because both a solution NMR structure[37,38] and a high-resolution crystal structure[35,39] are known. In solution, four d(TG$_4$T) strands form a parallel-stranded G-quadruplex structure containing four stacked G-quartets in the presence of Na^+ ions. The same structure is also

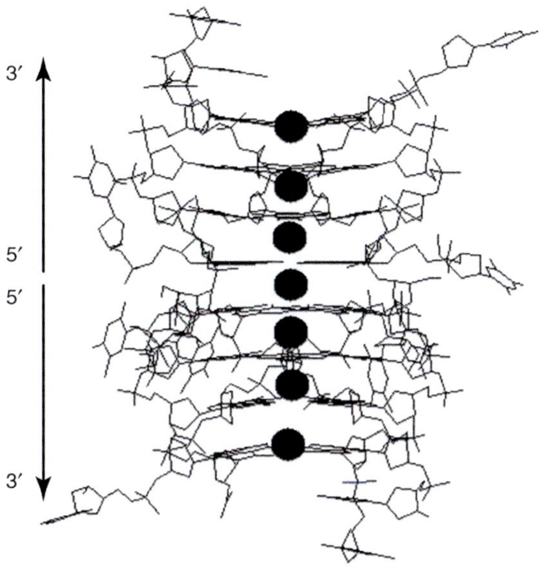

Figure 15.8. Na^+ (filled circles) binding observed in the crystal structure of [d(TG$_4$T)]$_4$ (PDB entry 352D). Note that this particular crystal structure shows a dimeric [d(TG$_4$T)]$_4$ G-quadruplex structure.

formed when Na^+ is replaced by K^+. The crystal structures of the Na^+ form of d(TG$_4$T) showed a four-stranded parallel structure similar to the solution structure. Interestingly, the crystal structure of d(TG$_4$T) also reveals a distinct intermolecular stacking not mentioned in the solution NMR studies. This intermolecular stacking consists of a pair of G-quadruplex structures that are stacked at the 5′ ends. Rovnyak *et al.*[36] observed three signals in the ^{23}Na MAS and MQMAS spectra. Recognizing that there are three classes of Na^+ ions in this G-quadruplex system (Na^+ ions residing inside the G-quadruplex channel, Na^+ ions bound to the phosphate backbone, and free Na^+ ions), the authors attempted to provide an assignment for the three ^{23}Na NMR signals. Although an unfortunate error occurred with respect to spectral assignment, this work represents the first solid-state ^{23}Na NMR study of G-quadruplex DNA. Shortly after, Wu and colleagues[40,41] established the unambiguous ^{23}Na NMR signature for Na^+ ions residing inside a G-quadruplex channel. In particular, Wu and colleagues examined solid-state ^{23}Na NMR of Na$_2$(5′-GMP) self-assembly. As seen from Figure 15.9, three classes of Na^+ ions are clearly observed. The Na^+ ions bound to the

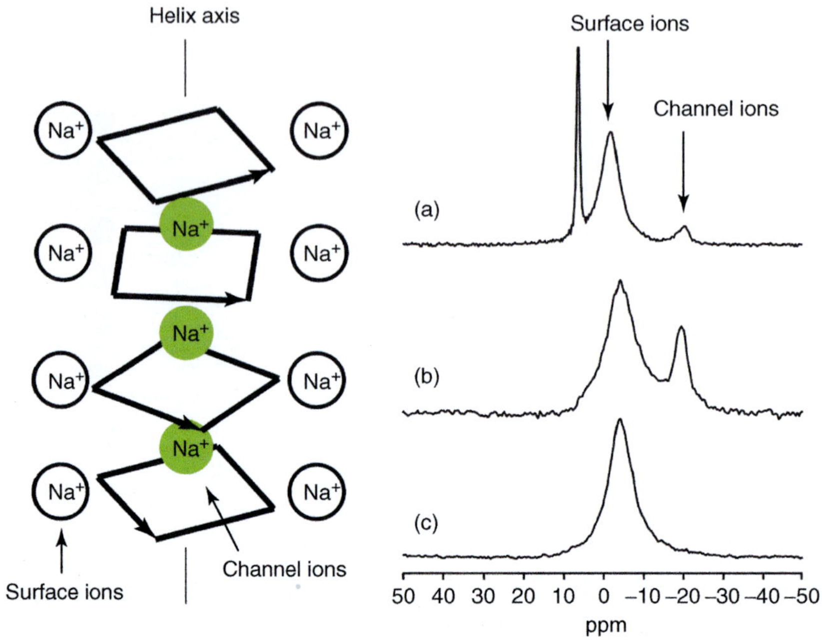

Figure 15.9. (Left) Illustration of two types of Na^+ binding sites to the self-assembled 5′-GMP helix. (Right) ^{23}Na MAS spectra of (a) Na_2(5′-GMP) gel, (b) Na_2(5′-GMP) gel formed in the presence of Cs^+, and (c) Na_2(5′-GMP) gel formed in the presence of K^+.

peripheral phosphate groups (referred to as the surface Na^+ ions) give rise to a ^{23}Na NMR signal at approximately −4 ppm. The Na^+ ions residing inside the G-quadruplex channel (referred to as the channel Na^+ ions) exhibit a signal at −19 ppm. The signal at 7 ppm is due to excessive NaCl salt. This assignment is confirmed by examining changes in solid-state ^{23}Na NMR spectra when CsCl and KCl are added to the 5′-GMP sample. When CsCl is added to the sample, the signal intensity for the surface Na^+ ions is significantly reduced, suggesting that Cs^+ ions do not enter the G-quadruplex cavity but replace partially the surface Na^+ ions. On the other hand, because K^+ ions are preferred over Na^+ ions to occupy the channel sites, the ^{23}Na NMR signal for the channel Na^+ ions disappears completely when KCl is added. These are consistent with the known affinity of the G-quadruplex formed by 5′-GMP for Cs^+ and K^+ ions.[42–44] In another study, Wong *et al.*[41] examined simultaneously solid-state ^{23}Na NMR and the crystal structure of the self-assembly from a liphophilic guansoine nucleoside, 5′-*tert*-butyl-dimethylsilyl-2′,3′-*O*-isopropylidene guanosine (G1). This compound can extract Na^+ and Cs^+ ions from aqueous phase into organic phase to form a G-quadruplex structure consisting of 16 equivalents of G1 as illustrated in Figure 15.10. Because all Na^+ ions in $[G1]_{16}{\cdot}3Na^+{\cdot}Cs^+{\cdot}4pic^-$ reside inside the G-quadruplex channel, the observation of a ^{23}Na NMR signal centered at −19 ppm for G1 establishes unequivocally the spectral characteristics for channel Na^+ ions. The authors further obtained a 2D ^{23}Na MQMAS spectrum for $[G1]_{16}{\cdot}3\,Na^+{\cdot}Cs^+{\cdot}4pic^-$ in which the three crystallographically distinct Na sites are clearly resolved. As the sample of $[G1]_{16}{\cdot}3\,Na^+{\cdot}Cs^+{\cdot}4pic^-$ is in polycrystalline form, each of the three Na^+ ions has a well-defined coordination environment. As a result, the three ^{23}Na MQMAS signals exhibit very high resolution along the F_1 axis. The ^{23}Na MQMAS spectrum of $[G1]_{16}{\cdot}3\,Na^+{\cdot}Cs^+{\cdot}4pic^-$ illustrates the remarkable resolution improvement over 1D MAS spectra. In this case, the spectral resolution in the F_1 axis is less than 0.5 ppm. From analyses of both 1D and 2D solid-state ^{23}Na NMR spectra, the authors obtained the following ^{23}Na NMR parameters for $[G1]_{16}{\cdot}3\,Na^+{\cdot}Cs^+{\cdot}4pic^-$:

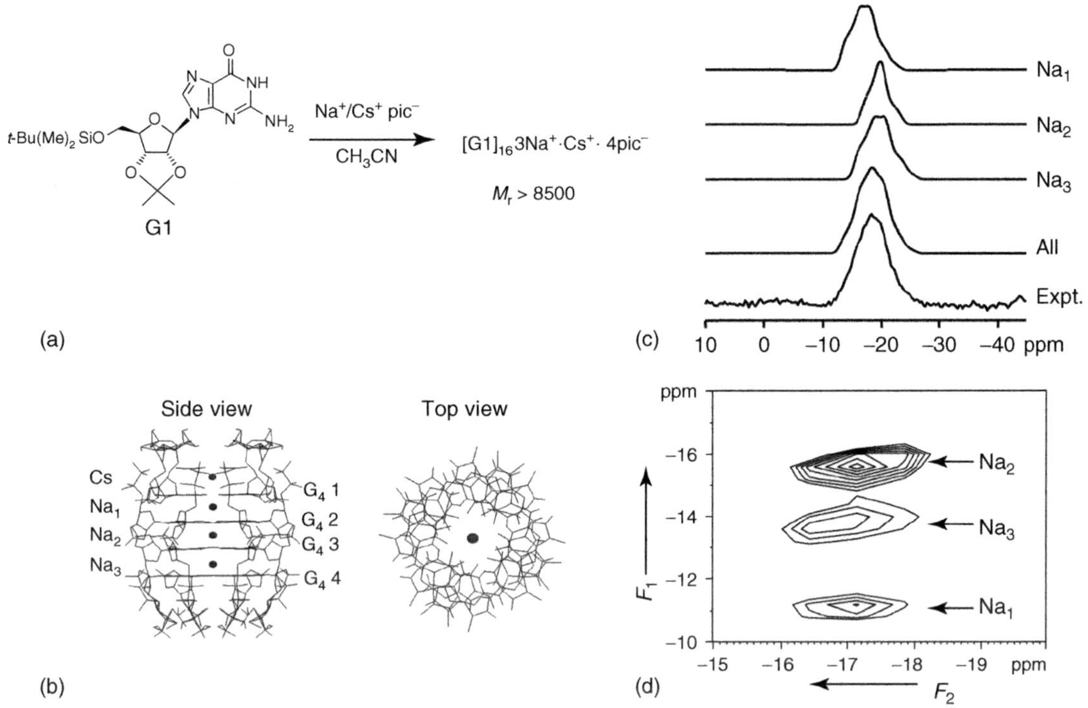

Figure 15.10. (a) Self-assembly, (b) crystal structure, (c) 1D ^{23}Na MAS, and (d) 2D ^{23}Na MQMAS spectra of $[G1]_{16}\cdot[3Na/CsPic_4]$. Picrate molecules and hydrogen atoms are omitted for clarity. Only G-quartets are shown in the top view diagram to illustrate the ion channel structure. (Reproduced from Ref. 41. © American Chemical Society, 2002.)

Na$_1$, $\delta_{iso} = -12.8 \pm 0.2$ ppm, $C_Q = 1.65 \pm 0.05$ MHz, $\eta_Q = 0.60 \pm 0.05$; Na$_2$, $\delta_{iso} = -16.5 \pm 0.2$ ppm, $C_Q = 1.35 \pm 0.05$ MHz, $\eta_Q = 0.80 \pm 0.05$; Na$_3$, $\delta_{iso} = -15.0 \pm 0.2$ ppm, $C_Q = 1.70 \pm 0.05$ MHz, $\eta_Q = 0.60 \pm 0.05$.

On the basis the ^{23}Na NMR spectral signatures for Na^+ ions in G-quadruplexes, Wu and Wong[45] successfully applied this solid-state NMR approach to determine the number and coordination environment of Na^+ ions in $d(G_4T_4G_4)$. In solution, $d(G_4T_4G_4)$ generally adopts an antiparallel, bimolecular quadruplex structure consisting of four stacked G-quartets and two diagonal thymine loops. Although the nature of monovalent cations present in the solution (Na^+, K^+, and NH_4^+) does not appear to affect the overall fold of $d(G_4T_4G_4)$,[46] there is also evidence suggesting that the details of ion binding in $d(G_4T_4G_4)$ may be different for different alkali metal ions. For example, as mentioned earlier, five K^+ ions are found in the K^+ form of $d(G_4T_4G_4)$ in the crystal structure.[34] For the NH_4^+ form of $d(G_4T_4G_4)$, only three channel NH_4^+ ions were observed in a solution NMR study.[47] For the Na^+ form of $d(G_4T_4G_4)$, no crystal structure has been reported to date for the native DNA. However, a crystal structure was reported for a $d(G_4T_4G_4)$/protein complex in the presence of Na^+ ions.[48] In this complex, four Na^+ ions are observed, all of which are located inside or nearly inside the G-quartet plane, i.e., the so-called in-plane mode of binding. In an attempt to provide a definite answer to the question regarding the exact mode of Na^+ binding in native $d(G_4T_4G_4)$, Wu and Wong[45] utilized a solid-state ^{23}Na NMR approach. They observed two distinct NMR signals in the 2D ^{23}Na MQMAS spectrum of $d(G_4T_4G_4)$ with a peak volume ratio of approximately 2:1 at the position associated with channel Na^+ ions; see Figure 15.11. In the dimeric G-quadruplex structure of $d(G_4T_4G_4)$, there are four G-quartets stacking on the top of one another, creating three coordination pockets inside the quadruplex channel. The authors suggested that the signal with twice

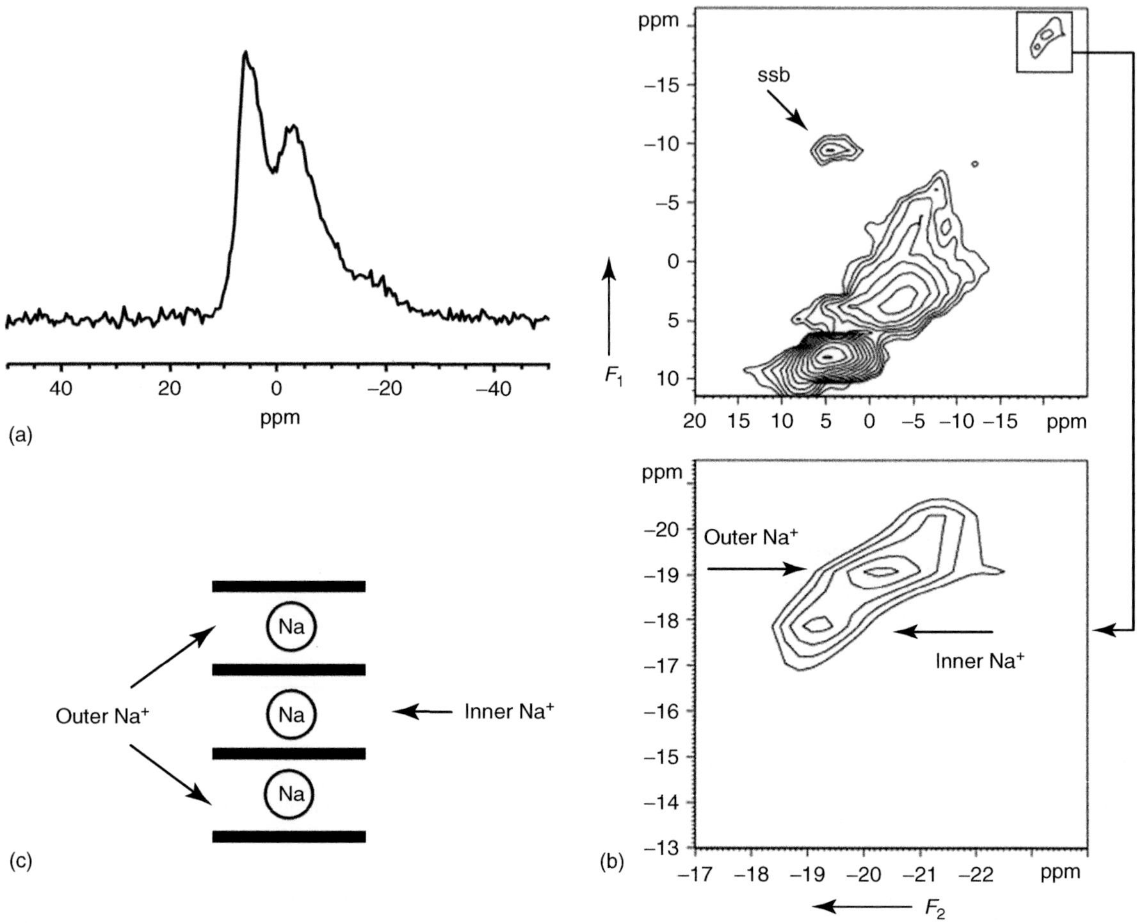

Figure 15.11. (a) Experimental ^{23}Na MAS and (b) 2D MQMAS spectra (11.75 T) of d($G_4T_4G_4$). (c) Illustration of the three Na^+ binding sites inside the G-quadruplex channel.

the intensity is associated with the two Na^+ ions in the symmetry-related outer pockets and the less intense signal is attributed to a Na^+ ion in the central pocket, in analogy with the $NH_4{}^+$ coordination in the $NH_4{}^+$ form of d($G_4T_4G_4$) seen in the solution NMR study.[47] Analyses of the 1D and 2D spectra yielded the following ^{23}Na NMR parameters for the channel Na^+ ions in d($G_4T_4G_4$): outer site, $C_Q = 1.2 \pm 0.2$ MHz, $\eta_Q = 0.4 \pm 0.2$, and $\delta_{iso} = -19.0 \pm 0.2$ ppm; central site, $C_Q = 0.9 \pm 0.2$ MHz, $\eta_Q = 1.0 \pm 0.2$, and $\delta_{iso} = -17.5 \pm 0.2$ ppm. These ^{23}Na NMR parameters are consistent with those observed for the channel Na^+ ions in $[G1]_{16}{\cdot}3Na^+{\cdot}Cs^+{\cdot}4pic^-$ mentioned earlier, suggesting a similar Na^+ coordination between the two systems. In $[G1]_{16}{\cdot}3Na^+{\cdot}Cs^+{\cdot}4pic^-$, each of the channel Na^+ ions is coordinated to eight O6(G) atoms from two adjacent G-quartets, having a square-pyramidal antiprism coordination geometry with an average Na–O distance of 2.81 Å. As also seen from Figure 15.11, each of the two signals for the channel Na^+ ions exhibits a line shape parallel to the F_2 axis, indicating that the coordination environment for the channel Na^+ ions is well defined. In contrast, the signal from the surface Na^+ ions exhibits a diffuse line shape, indicating that the chemical environment for the surface Na^+ ions is not homogeneous.

Now that the mode of Na^+ binding inside the G-quadruplex channel of native d($G_4T_4G_4$) is established, is it possible that Na^+ ions may also reside

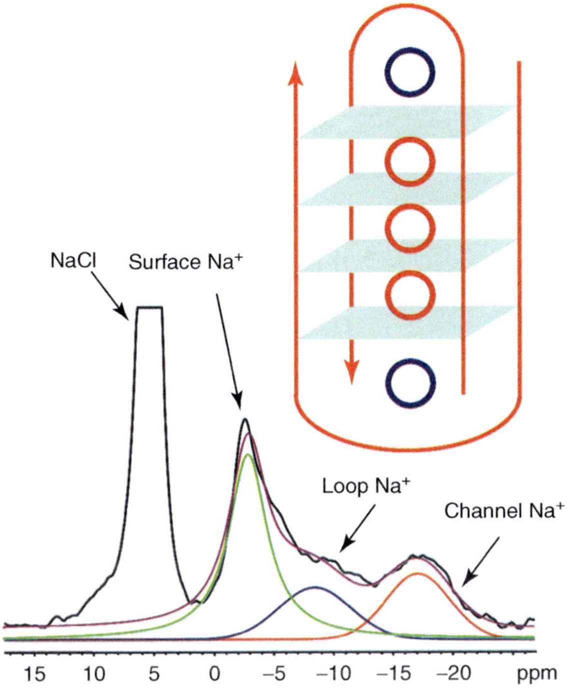

Figure 15.12. Experimental T_2-filtered (upper trace) and fitted (lower trace) ^{23}Na MAS NMR spectra (21.1 T) of a $d(G_4T_4G_4)$ sample prepared under 80% relative humidity. The sample spinning rate was 10 kHz. (Reproduced from Ref. 49. © American Chemical Society, 2008.)

in the T_4 loop region? Unfortunately, this question cannot be answered with the solid-state ^{23}Na spectra reported by Wu and Wong.[45] This is because the ^{23}Na NMR signal arising from phosphate-bound Na^+ ions in that particular DNA sample is so intense and broad that other weak signals between 0 and −15 ppm, if any, would be completely obscured. In a recent study, Ida and Wu[49] managed to circumvent this problem by introducing several modifications to the experimental procedures employed by Wu and Wong.[45] First, Ida and Wu[49] prepared a $d(G_4T_4G_4)$ sample under high relative humidity (80%). Under such a situation, the phosphate-bound ^{23}Na ions are fully hydrated, giving rise to a relatively sharp ^{23}Na signal centered at ~0 ppm. Second, they performed the ^{23}Na MAS experiment at a very high magnetic field of 21.1 T, which resulted in much better spectral separation than that recorded at 11.75 T. Third, they utilized a rotor-synchronized spin-echo experiment (a T_2 filter) to selectively suppress the signal from phosphate-bound Na^+ ions. As seen from Figure 15.12, the new ^{23}Na MAS spectrum of $d(G_4T_4G_4)$ clearly shows another signal centered at −8 ppm, in addition to the channel Na^+ signal at −17 ppm. Ida and Wu[49] further used ^{23}Na chemical shift calculations to establish that the ^{23}Na NMR signal at ca −8 ppm is likely due to Na^+ ions residing inside the T_4 loop region, each coordinating to four carbonyl oxygen atoms from the terminal G-quartet and one carbonyl oxygen from a loop thymine residue and one water molecule. The Na^+ binding site in the T_4 loop of $d(G_4T_4G_4)$ is different from the K^+ binding site in the same region established by X-ray crystallography.[34] In particular, while the Na^+ ion is coordinated to four O6(G) atoms, one water, and one O2(T) atom, the K^+ ion is coordinated to eight ligands: four O6(G) atoms, two O2(T) atoms, and two water molecules. In addition, the Na–O distances are generally shorter than the corresponding K–O distances. As a result, the T_4 loop in the Na form appears more compact than that in the K forms of $d(G_4T_4G_4)$. On the basis of these new ^{23}Na NMR results, Ida and Wu[49] further hypothesized that Na^+ binding to the diagonal T_4 loop is a general phenomenon in G-quadruplex DNA. Several other studies also employed a similar solid-state ^{23}Na NMR approach to obtain information about Na^+ binding in G-qaudruplex related systems.[50–52]

One of the most puzzling problems in alkali metal ion binding to G-quadruplex DNA is related to the mode of Na^+ binding. A statement commonly found in the literature is that Na^+, being a smaller ion than K^+, prefers to reside within the G-quartet plane (i.e., the in-plane mode of binding). Indeed, this mode of Na^+ binding has been observed in $[d(TG_4T)]_4$ and a $d(G_4T_4G_4)$/protein complex by X-ray crystallography. However, solution ^{23}Na NMR studies[49,53] did not find any evidence for in-plane Na^+ binding in both $d(TG_4T)$ and native $d(G_4T_4G_4)$. The solid-state ^{23}Na NMR studies discussed earlier[45] also suggested that the sandwich mode of Na^+ binding is operative in native $d(G_4T_4G_4)$. Is it possible that the in-plane Na^+ binding in these systems observed in crystallographic studies is a crystal packing artefact? Furthermore, what would be the ^{23}Na NMR signature for in-plane Na^+ ions? To search for an answer to this question, Wong *et al.*[54] investigated a calix[4]arene-guanosine conjugate (CG). CG is known to be able to form a dimer, $(CG)_2{\cdot}(H_2O)_n$, in water-saturated $CDCl_3$.[55] This dimer behaves as an ion pair receptor, as it can extract salts such as NaCl from water into organic solvents to give a new complex $(CG)_2 \cdot NaCl \cdot (H_2O)_n$. In this dimer, ^1H NMR spectroscopy showed that a single G-quartet is formed at the center of the structure, providing an ideal preorganized framework for in-plane Na^+ binding to occur. The ^{23}Na MAS spectrum of $(CG)_2 \cdot NaCl \cdot (H_2O)_n$ is shown in Figure 15.13(a). After analyzing these spectra, Wong *et al.*[54] reported the following ^{23}Na NMR parameters: $\delta_{iso} = 1$ ppm, $C_Q = 2.3$ MHz, and $\eta_Q = 1.0$. These ^{23}Na NMR parameters are significantly different from those of the sandwich-type Na^+ ions mentioned earlier. Of course, the observed ^{23}Na NMR parameters alone do not unveil the exact ion coordination in the system. To shed more light onto the nature of ion coordination, Wong *et al.*[54] carried out extensive quantum chemical calculations of ^{23}Na NMR parameters for a variety of molecular models. They concluded that the observed ^{23}Na NMR parameters for $(CG)_2 \cdot NaCl \cdot (H_2O)_n$ can be best reproduced from a model in which the Na^+ ion lies slightly above the G-quartet plane (ca 0.5 Å) and also coordinates to an axial water molecule; see Figure 15.13(b). This pentacoordinate geometry around the Na^+ ion is similar to the two Na^+ sites found at the 3′ end of the $[d(TG_4T)]_4$ quadruplex. For comparison, Wong *et al.*[54] also performed calculations for these two pentacoordinate Na^+ ions using the actual crystal structure (PDB entry 244D). The HF/6-31G(d) results are as follows: Na^+ (G5), $\delta_{iso}(^{23}Na) = 5.1$ ppm, $C_Q = 3.1$ MHz, $\eta_Q = 0.76$; Na^+ (G-5), $\delta_{iso}(^{23}Na) = 1.6$ ppm, $C_Q = 2.2$ MHz, $\eta_Q = 0.11$. These results strongly suggest that the same mode of ion binding is operative in the CG dimer and at the 3′ end of $[d(TG_4T)]_4$ quadruplex. This study represents the first time detection of pentacoordinate Na^+ ions by ^{23}Na NMR in molecular systems containing G-quartets. This study yields benchmark ^{23}Na NMR parameters for pentacoordinate (in-plane) Na^+ ions that can be used in searching for similar Na^+ binding sites in G-quadruplex DNA and in other supramolecular assemblies containing G-quartets.

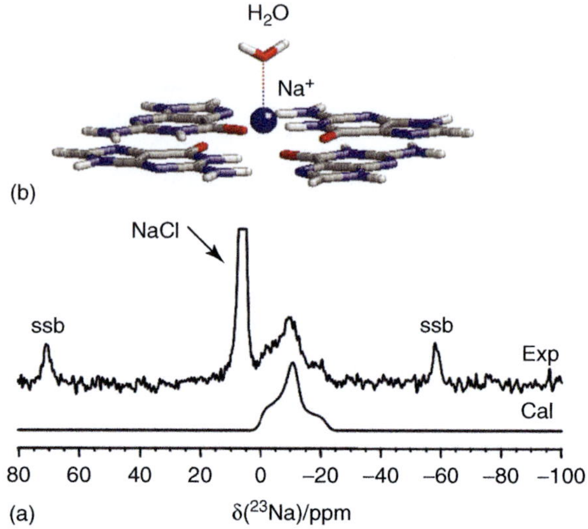

Figure 15.13. (a) Experimental (upper trace) and simulated (lower trace) ^{23}Na MAS NMR spectra of the CG dimer prepared with NaCl. (b) In-plane Na^+ binding mode.

Now we turn our attention to solid-state ^{39}K NMR studies of G-quadruplexes. To obtain the ^{39}K NMR signature for K^+ ions bound to a G-quadruplex structure, Wu *et al.*[18] examined three guanosine derivatives: $[G1]_{16} \cdot 3K^+ \cdot Cs^+ \cdot 4pic^-$, guanosine/KCl gel, and 5′-GMP/K gel. Some of the ^{39}K MAS spectra are shown in Figure 15.14. Similar to the case of solid-state ^{23}Na NMR, because the K^+ binding in $[G1]_{16} \cdot 3K^+ \cdot Cs^+ \cdot 4pic^-$ was determined by crystallography,[56] the ^{39}K MAS spectrum of this K-nucleoside complex can be used as an excellent model for K^+ binding inside a G-quadruplex channel. In particular, the ^{39}K MAS spectrum of $[G1]_{16} \cdot 3K^+ \cdot Cs^+ \cdot 4pic^-$ exhibits a peak centered at 10 ppm with a line width of approximately 1.0 kHz. Again, because all three K^+ ions in this system reside inside the channel,[56] the spectral assignment for channel K^+ ions is unambiguous. Guanosine is also known to form a viscous gel in aqueous solution in the presence of KCl, indicating the formation of a highly ordered molecular assembly. An early X-ray fiber diffraction study confirmed that the guanosine aggregates have a G-quadruplex structure where two adjacent G-quartets are separated by 3.4 Å with a twist of 45° forming a continuous helix.[57] The ^{39}K MAS spectrum of guanosine/KCl gel shows a signal very similar to that observed for $[G1]_{16} \cdot 3K^+ \cdot Cs^+ \cdot 4pic^-$. In addition, a very sharp peak is observed at 46 ppm, which arises from a small excess of KCl. These observations suggest that, in the self-assembly of guanosine, the K^+ ions reside exclusively inside the G-quadruplex channel in a similar fashion as those in $[G1]_{16} \cdot 3Na^+ \cdot Cs^+ \cdot 4pic^-$. In comparison, the ^{39}K MAS NMR spectrum for 5′-GMP/K gel exhibits three groups of signals centered at 46 ppm (free KCl salt), 10 ppm (channel K^+ ions), and −10 ppm (surface K^+ ions). A careful analysis of these ^{39}K MAS NMR spectra yields the

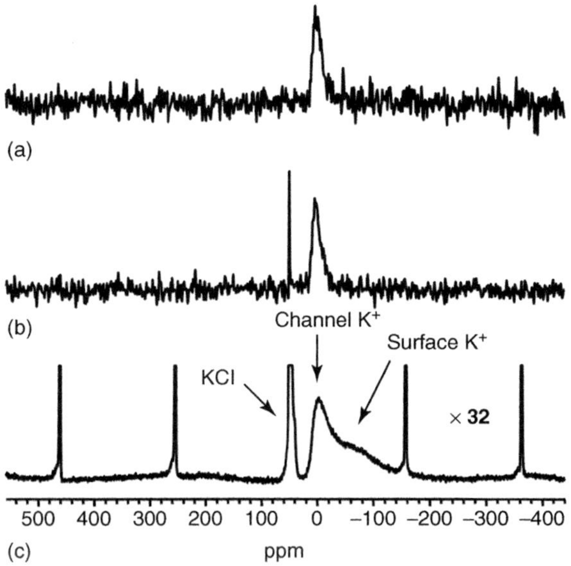

Figure 15.14. 1D ^{39}K MAS spectra (19.6 T) of (a) $[G1]_{16}{\cdot}3K^+{\cdot}Cs^+{\cdot}4pic^-$, (b) guanosine/KCl, and (c) 5′-GMP/K gel. The ^{39}K chemical shift was referenced to the signal of K^+ (aq).

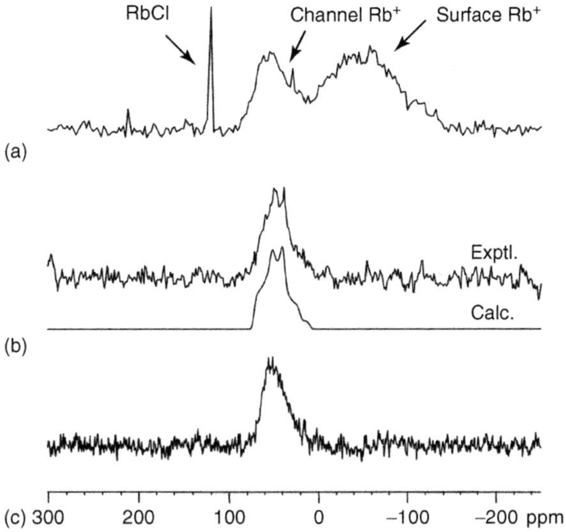

Figure 15.15. Experimental ^{87}Rb MAS NMR spectra (11.75 T) of (a) 5′-GMP/Rb gel, (b) 5′-GMP/Rb gel washed by CsCl(aq), and (c) $[G1]_{16}{\cdot}3Rb/Cs{\cdot}Pic_4$. (Reproduced from Ref. 59. © Royal Society of Chemistry, 2005.)

following ^{39}K NMR parameters for the channel K^+ ions: $[G1]_{16}{\cdot}3K^+{\cdot}Cs^+{\cdot}4pic^-$, $\delta_{iso} = 13$ ppm and $C_Q < 0.7$ MHz; guanosine/KCl gel, $\delta_{iso} = 10$ ppm and $C_Q \approx 0.7\text{–}0.8$ MHz; 5′-GMP/K gel, $\delta_{iso} \approx 15$ ppm and $C_Q < 1.3$ MHz.

It is worth mentioning that the K^+ coordination environment in the G-quadruplex channel is very similar to that of the K^+ ions observed inside the selectivity filter of KcsA K^+ channel protein.[58] It is likely that these two types of K^+ ions have very similar ^{39}K NMR parameters. Although KcsA (~70 kDa) is considerably larger than the largest system examined here, $[G1]_{16}{\cdot}3K^+{\cdot}Cs^+{\cdot}4pic^-$ (8.7 kDa), the current results suggest that the solid-state ^{39}K NMR sensitivity at ultrahigh magnetic fields (19.6 or 21.1 T) is adequate for studying K^+ binding in ion channel proteins.

In addition to Na^+ and K^+, Rb^+ is also known to promote G-quartet formation. Recently, Ida and Wu[59] used solid-state ^{87}Rb NMR to establish the NMR signature for Rb^+ ions inside a G-quadruplex channel. These authors used the same series of guanosine compounds as in the solid-state ^{39}K NMR study. Figure 15.15 shows typical ^{87}Rb MAS spectra of these systems. These authors reported that the ^{87}Rb NMR signatures are as follows: channel Rb^+ ions, $\delta_{iso} = 74$ ppm, $C_Q = 5.1$ MHz, $\eta_Q = 0.6$; phosphate-bound Rb^+ ions, $\delta_{iso} = 5\text{–}10$ ppm, $C_Q = 7.5\text{–}7.7$ MHz. As the authors pointed out, one advantage of solid-state ^{87}Rb NMR is that the sensitivity of detecting ^{87}Rb is much higher than ^{39}K. With ^{87}Rb isotope enrichment to 99%, this sensitivity advantage can represent a factor of as large as 370. Of course, the drawback is that solid-state ^{87}Rb NMR signals are generally four times broader than the corresponding ^{39}K NMR signals. But considering that the ^{87}Rb chemical shift range is also approximately four times larger than that of ^{39}K, the signal dispersion in ^{87}Rb NMR spectra should be as good as that in ^{39}K NMR spectra. As a result, it is possible to use ^{87}Rb as a surrogate NMR probe for studying K binding in nucleic acids and ion channel proteins. This is certainly an area to be further explored. It is quite interesting to note that, on going from Na, K to Rb, the NMR signature for channel ions becomes more positive compared to their respective reference, M^+ (aq). This trend reflects the fact that the paramagnetic shielding contribution becomes increasingly important as the alkali metal ion under observation gets heavier.

Another important aspect of metal ion binding to G-quadruplex is the relative binding affinity of different metal ions. As discussed in the previous sections, the two major types of alkali metal ion binding sites in a G-quadruplex (channel and surface sites) have very different chemical environments. Consequently, these binding sites must also exhibit different affinities for various alkali metal ions. In solution, because alkali metal ions often undergo rapid exchange between all possible binding sites, it is generally difficult to obtain site-specific information about ion affinity by conventional spectroscopic techniques. For this reason, although extensive thermodynamic data on G-quadruplex stability are available in the literature, none of the previous studies has been able to address the question of alkali metal ion affinity for a G-quadruplex structure in a *site-specific* manner.

In 2003, Wong and Wu[60] developed a solid-state NMR method for determining relative ion affinity for different binding sites simultaneously. The method is based on a titration experiment where competitive equilibria between Na^+ and M^+ (M = Li, K, Rb, NH_4, and Cs) ions are monitored by solid-state ^{23}Na NMR. Because Na^+ ions at different binding sites exhibit clearly resolved signals, following individual ^{23}Na NMR peak intensities as a function of the added M^+ salt would yield a thermodynamic equilibrium constant (K_{eq}) for each of the binding competitions between Na^+ and M^+ ions. Figure 15.16 depicts the dependence of $-\Delta G^\circ$ on the reciprocal ionic radius for both binding sites in 5′-GMP. These data suggest that the affinity of monovalent cations for the G-quadruplex channel cavity site follows the order of Eisenman sequence V.[61] The channel cavity site of the G-quadruplex structure prefers K^+ and NH_4^+ over Na^+, whereas Cs^+ is much less favored than Na^+. Wong and Wu[60] obtained the following ion affinity sequence for the channel cavity site: $K^+ > NH_4^+ > Rb^+ > Na^+ > Li^+ > Cs^+$. This sequence is in agreement with the qualitative ranking first reported for 5′-GMP by Pinnavaia *et al.*,[42,43] i.e., $K^+ > Rb^+$, $Na^+ \gg Li^+$, Cs^+. By studying the melting temperatures of 8-bromo-guanosine gels in various salt solutions, Chantot and Guschlbauer[62] also observed a stability sequence $K^+ \gg Rb^+ > NH_4^+ > Na^+ > Cs^+ > Li^+$. Several telomeric DNA sequences show a similar trend in their cation-induced stability, i.e., $K^+ > Na^+ > Cs^+$.[63] In contrast, the sequence of ion affinity for the 5′-GMP surface site is drastically different from that for the channel cavity site.

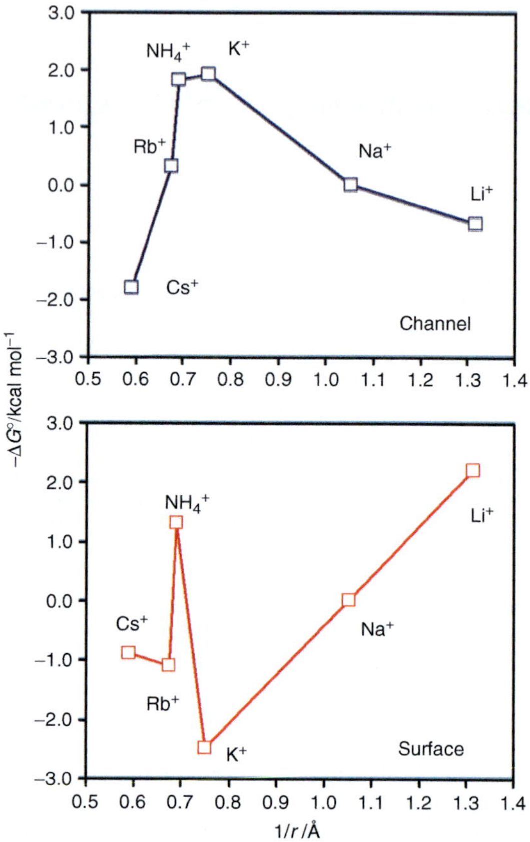

Figure 15.16. Diagrams of free-energy difference versus reciprocal ionic radius for binding of monovalent cations to the G-quadruplex structure formed by 5′-GMP self-assembly. (Reproduced from Ref. 50 by permission of CRC Press.)

According to the observed ΔG° values, the order of cation affinity for the 5′-GMP surface site is $Li^+ > NH_4^+ > Na^+ > Cs^+ > Rb^+ > K^+$. These results clearly indicate that the overall stability of a G-quadruplex structure depends only on the affinity of the monovalent cations present in solution for the quadruplex cavity site. Another striking feature of the data presented in Figure 15.16 is that the affinity sequence observed for the G-quadruplex channel site is remarkably similar to that for K^+ ion channel proteins.[61] The structural basis for the observed similarity in cation selectivity became quite clear after MacKinnon and coworkers published the first high-resolution crystal structure for a K^+ ion channel protein (KcsA) from the bacterium

Streptomyces lividans.[64] Four K^+ binding sites were identified inside the selectivity filter of the KcsA channel, each coordinating to eight carbonyl oxygen atoms from four signature sequences, Thr^{75}-Val^{76}-Gly^{77}-Tyr^{78}. At each binding site, the K^+ ion resides near the center of a square-pyramidal antiprism with a mean K–O distance of 2.85 Å. This type of cation coordination environment is remarkably similar to those found in G-quadruplex structures.

15.4 CONCLUDING REMARKS

As we have shown in this chapter, solid-state NMR has become a viable new method of detecting alkali metal ions bound to biological molecules. Although the current alkali metal NMR studies are largely focused on nucleic acids and related systems, there is no reason to believe that applications should be limited only to these systems. The complementary nature of solid-state NMR to crystallography is also advantageous. The benefit of the solid-state NMR approach ranges from easy sample preparation, freedom of cryogenic artefacts, and the possibility of gaining additional information about the local electronic environment at the binding site. It is also important to address the question as to whether the NMR sensitivity is sufficient for detecting Na^+ or K^+ ions in large biomolecular systems. We have shown that the ^{23}Na NMR signal from a single Na^+ ion out of an 8 kDa G-quadruplex DNA can be detected with confidence at a moderate magnetic field strength of 11.75 T, using MQMAS experiments. The very high spectral resolution (less than 1 ppm) observed in ^{23}Na MQMAS spectra of this system suggests that solid-state ^{23}Na NMR holds great promise to become a practical technique for obtaining detailed information about Na^+ binding sites in even larger biomolecular systems. In the mean time, the ^{39}K NMR sensitivity at magnetic fields greater than 19.6 T is sufficient for collecting 1D ^{39}K MAS spectra. To this date, it remains a challenge to obtain high-quality 2D MQMAS spectra for ^{39}K nucleotides. As very high magnetic fields are becoming available to the NMR community including resistive magnets and hybrid magnets at 40 T or higher, the ^{39}K (or ^{87}Rb) NMR sensitivity will soon reach a level that would allow solid-state NMR studies of large biomolecular systems.

One of the future directions is related to solid-state NMR studies of ion binding/transporting processes in ion channel proteins. An important advantage of solid-state NMR is the feasibility to study ion channel proteins in their membrane-bound state. For many years, structural determination for membrane proteins has been a major driving force for the development of new solid-state NMR methodologies. If information about ion coordination and dynamics can be obtained directly by solid-state alkali metal NMR, it may be possible to link the structural aspect of ion channel proteins to their function, e.g., ion transport. As alkali metal ions are also cofactors of some enzymes, solid-state alkali metal NMR may find useful applications in the characterization of alkali metal ion binding environments in these enzymes.

RELATED ARTICLES IN THE ENCYCLOPEDIA OF MAGNETIC RESONANCE

Biological Systems: Spin-3/2 Nuclei

DNA–Cation Interactions: Quadrupolar Studies

Quadrupolar Nuclei in Liquid Samples

REFERENCES

1. V. Tereshko, C. J. Wilds, G. Minasov, T. P. Prakash, M. A. Maier, A. Howard, Z. Wawrzak, M. Manoharan, and M. Egli, *Nucleic Acids Res.*, 2001, **29**, 1208.
2. G. Wu, *Biochem. Cell Biol.*, 1998, **76**, 429.
3. R. Tabeta, M. Aida, and H. Saito, *Bull. Chem. Soc. Jpn.*, 1986, **59**, 1957.
4. A. Wong and G. Wu, *J. Phys. Chem. A*, 2000, **104**, 11844.
5. A. C. Kunwar, G. L. Turner, and E. Oldfield, *J. Magn. Reson.*, 1986, **69**, 124.
6. A. Wong, S. Sham, S. Wang, and G. Wu, *Can. J. Chem.*, 2000, **78**, 975.
7. B. M. Burkhart, N. Li, D. A. Langs, W. A. Pangborn, and W. L. Duax, *Proc. Natl. Acad. Sci. U.S.A.*, 1998, **95**, 12950.
8. S. K. Katti, T. P. Seshadri, and M. A. Viswamitra, *Acta Crystallogr., Sect. B*, 1981, **37**, 1825.
9. A. Wong and G. Wu, *J. Phys. Chem. A*, 2003, **107**, 579.
10. W. Saenger, *Principles of Nucleic Acid Structure*, Springer: New York, 1984.

11. S. Ding and C. A. McDowell, *Chem. Phys. Lett.*, 2000, **320**, 316.
12. A. C. Larson, *Acta Crystallogr., Sect. B*, 1978, **34**, 3601.
13. C. V. Grant, D. McElheny, V. Frydman, and L. Frydman, *Magn. Reson. Chem.*, 2006, **44**, 366.
14. M. Nausner, J. Brus, M. Hauble, N. Muller, and W. Schoefberger, *Inorg. Chim. Acta*, 2009, **362**, 1071.
15. Y. Sugawara, N. Kamiya, H. Iwasaki, T. Ito, and Y. Satow, *J. Am. Chem. Soc.*, 1991, **113**, 5440.
16. J. Grinshtein, C. V. Grant, and L. Frydman, *J. Am. Chem. Soc.*, 2002, **124**, 13344.
17. J. Pandit, T. P. Seshadri, and M. A. Viswamitra, *Acta Crystallogr., Sect. C*, 1983, **39**, 342.
18. G. Wu, A. Wong, Z. Gan, and J. T. Davis, *J. Am. Chem. Soc.*, 2003, **125**, 7182.
19. G. S. Padiyar and T. P. Seshadri, *J. Biomol. Struct. Dyn.*, 1998, **15**, 803.
20. D. A. Adamiak and W. Saenger, *Acta Crystallogr., Sect. B*, 1980, **36**, 2585.
21. H. He, J. Klinowski, G. Saba, M. Casu, and A. Lai, *Solid State Nucl. Magn. Reson.*, 1998, **10**, 169.
22. M. Madeddu, *Solid State Nucl. Magn. Reson.*, 2002, **22**, 83.
23. A. Wong, Z. Yan, Y. Huang, and G. Wu, *Magn. Reson. Chem.*, 2008, **46**, 308.
24. M. Gellert, M. N. Lipsett, and D. R. Davies, *Proc. Natl. Acad. Sci. U.S.A.*, 1962, **48**, 2013.
25. S. Neidle and S. Balasubramanian, *Quadruplex Nucleic Acids*, The Royal Society of Chemistry: Cambridge, 2006.
26. J. T. Davis, *Angew. Chem. Int. Ed.*, 2004, **43**, 668.
27. J. T. Davis and G. P. Spada, *Chem. Soc. Rev.*, 2007, **36**, 296.
28. Y. Wang and D. J. Patel, *Structure*, 1993, **1**, 263.
29. G. N. Parkinson, M. P. H. Lee, and S. Neidle, *Nature*, 2002, **417**, 876.
30. Y. Xu, Y. Noguchi, and H. Sugiyama, *Bioorg. Med. Chem.*, 2006, **14**, 5584.
31. K. N. Luu, A. T. Phan, V. Kuryavyi, L. Lacroix, and D. J. Patel, *J. Am. Chem. Soc.*, 2006, **128**, 9963.
32. A. Ambrus, D. Chen, J. Dai, T. Bialis, R. A. Jones, and D. Yang, *Nucleic Acids Res.*, 2006, **34**, 2723.
33. C. Kang, X. H. Zhang, R. Ratliff, R. Moyzis, and A. Rich, *Nature*, 1992, **356**, 126.
34. S. Haider, G. N. Parkinson, and S. Neidle, *J. Mol. Biol.*, 2002, **320**, 189.
35. G. Laughlan, A. I. H. Murchie, D. G. Norman, M. H. Moore, P. C. E. Moody, D. M. J. Lilley, and B. Luisi, *Science*, 1994, **265**, 520.
36. D. Rovnyak, M. Baldus, G. Wu, N. V. Hud, J. Feigon, and R. G. Griffin, *J. Am. Chem. Soc.*, 2000, **122**, 11423.
37. F. Aboul-ela, A. I. H. Murchie, and D. M. J. Lilley, *Nature*, 1992, **360**, 280.
38. F. Aboul-ela, A. I. H. Murchie, D. G. Norman, and D. M. J. Lilley, *J. Mol. Biol.*, 1994, **243**, 458.
39. K. Phillips, Z. Dauter, A. I. H. Murchie, D. M. J. Lilley, and B. Luisi, *J. Mol. Biol.*, 1997, **273**, 171.
40. G. Wu and A. Wong, *Chem. Commun.*, 2001, 2658.
41. A. Wong, J. C. Fettinger, S. L. Forman, J. T. Davis, and G. Wu, *J. Am. Chem. Soc.*, 2002, **124**, 742.
42. T. J. Pinnavaia, H. T. Miles, and E. D. Becker, *J. Am. Chem. Soc.*, 1975, **97**, 7198.
43. T. J. Pinnavaia, C. L. Marshall, C. M. Mettler, C. L. Fisk, H. T. Miles, and E. D. Becker, *J. Am. Chem. Soc.*, 1978, **100**, 3625.
44. C. L. Fisk, E. D. Becker, H. T. Miles, and T. J. Pinnavaia, *J. Am. Chem. Soc.*, 1982, **104**, 3307.
45. G. Wu and A. Wong, *Biochem. Biophys. Res. Commun.*, 2004, **323**, 1139.
46. P. Schultze, N. V. Hud, F. W. Smith, and J. Feigon, *Nucleic Acids Res.*, 1999, **27**, 3018.
47. N. V. Hud, P. Schultze, V. Sklenar, and J. Feigon, *J. Mol. Biol.*, 1999, **285**, 233.
48. M. P. Horvath and S. C. Schultz, *J. Mol. Biol.*, 2001, **310**, 367.
49. R. Ida and G. Wu, *J. Am. Chem. Soc.*, 2008, **130**, 3590.
50. G. Wu and A. Wong, Solid-state nuclear magnetic resonance studies of alkali metal ions in nucleic acids and related systems, in *NMR Spectroscopy of Biological Solids*, ed A. Ramamoorthy, CRC Press: Boca Raton (FL), 2006, p. 317.
51. X. Liu, I. C. M. Kwan, S. Wang, and G. Wu, *Org. Lett.*, 2006, **8**, 3685.
52. C. Zhong, J. Wang, N. Wu, G. Wu, P. Y. Zavalij, and X. Shi, *Chem. Commun.*, 2007, 3148.
53. A. Wong, R. Ida, and G. Wu, *Biochem. Biophys. Res. Commun.*, 2005, **337**, 363.

54. A. Wong, F. W. Kotch, I. C. M. Kwan, J. T. Davis, and G. Wu, *Chem. Commun.*, 2009, 2154.
55. F. W. Kotch, V. Sidorov, Y. F. Lam, K. J. Kayser, H. Li, M. S. Kaucher, and J. T. Davis, *J. Am. Chem. Soc.*, 2003, **125**, 15140.
56. S. L. Forman, J. C. Fettinger, S. Pieraccini, G. Gottarelli, and J. T. Davis, *J. Am. Chem. Soc.*, 2000, **122**, 4060.
57. P. Tougard, J. F. Chantot, and W. Guschlba, *Biochim. Biophys. Acta*, 1973, **308**, 9.
58. D. A. Doyle, J. M. Cabral, R. A. Pfuetzner, A. L. Kuo, J. M. Gulbis, S. L. Cohen, B. T. Chait, and R. MacKinnon, *Science*, 1998, **280**, 69.
59. R. Ida and G. Wu, *Chem. Commun.*, 2005, 4294.
60. A. Wong and G. Wu, *J. Am. Chem. Soc.*, 2003, **125**, 13895.
61. G. Eisenman and R. Horn, *J. Membr. Biol.*, 1983, **76**, 197.
62. J. F. Chantot and W. Guschlbauer, *FEBS Lett.*, 1969, **4**, 173.
63. W. Guschlbauer, J. F. Chantot, and D. Thiele, *J. Biomol. Struct. Dyn.*, 1990, **8**, 491.
64. Y. F. Zhou, J. H. Morais-Cabral, A. Kaufman, and R. MacKinnon, *Nature*, 2001, **414**, 43.

Chapter 16

Nitrogen-14 NMR Studies of Biological Systems

Luminita Duma

Ecole Normale Supérieure, Départment de Chimie, Laboratoire des BioMolécules, UMR 7203 CNRS-ENS-UPMC, 24 rue Lhomond, 75005 Paris, France
Université Pierre et Marie Curie Paris 6, 4 Place Jussieu, 75005 Paris, France

16.1 INTRODUCTION

The importance of nitrogen to all life processes cannot be overstated. Nitrogen represents 78.06% of the atmosphere and is embedded in every living tissue. Nitrogen is a vital element for the production of amino acids, proteins, nucleic acids, etc. Nitrogen-14 is the most abundant isotope (natural abundance 99.6%) and has a spin quantum number, $I = 1$. Proctor and Yu made the first observation of the chemical shift for a ^{14}N nucleus in ammonium nitrate in 1950.[1] Despite this important discovery for the development of NMR, further progress in the field of ^{14}N NMR spectroscopy was initially rather modest. In liquids, ^{14}N resonances broaden beyond detection because of fast quadrupolar relaxation except for ^{14}N atoms in rapidly tumbling molecules or in highly symmetric systems. In solids, the large quadrupole coupling constant, which ranges typically from 1 to 5 MHz, and low gyromagnetic ratio ($\gamma_{^{14}\mathrm{N}}/\gamma_{^{1}\mathrm{H}} = 0.07$) made ^{14}N NMR quite unattractive from the perspective of an NMR spectroscopist.

Single-quantum (SQ) central transitions without first-order quadrupolar broadening are typically used for the observation of half-integer quadrupolar nuclei. Since ^{14}N is an integer spin nucleus, such narrow transitions do not occur. Owing to the first-order quadrupole interaction, ^{14}N resonances are generally too broad to allow direct excitation and detection of the whole spectrum simultaneously. For this reason, the ^{14}N spectra of powdered materials have only been obtained indirectly or from compounds with exceptionally small quadrupole coupling constants (e.g., where there is a highly symmetric nitrogen site, such as tetraalkyl ammonium ions). However, nuclear quadrupole resonance (NQR) spectroscopy allows, in the absence of a magnetic field, the quadrupolar coupling constant and asymmetry parameter to be measured directly for many compounds.[2] ^{14}N NMR of single crystals has also been used to characterize the nitrogen environment.[3]

High-resolution ^{14}N magic-angle spinning (MAS) NMR spectroscopy is also relevant for the study of the electrostatic potential at the surface of biological

NMR of Quadrupolar Nuclei in Solid Materials
Edited by Roderick E. Wasylishen, Sharon E. Ashbrook and Stephen Wimperis

 ISBN: 978-0-470-97398-1

model lipid membranes, the electrostatic charge distribution across the membrane interface, and also the structural changes induced by peptide association.[4] Solid-state ^{14}N NMR spectroscopy is limited to membranes containing a phosphatidyl choline lipid. In mixed lipid systems, the ^{31}P and ^{14}N MAS spectra allow generally to directly probe the negatively charged phosphate and positively charged choline groups of the electrostatic $P^{-}-N^{+}$ dipole. Their isotropic chemical shifts can be related to the size of the potential existing at the phosphate and ammonium group of the lipid headgroup while the chemical shielding anisotropy (^{31}P) and anisotropic quadrupole interaction (^{14}N) will report on the changes in headgroup orientation in response to surface potential. The near-symmetric nature of the choline headgroup of a phosphocholine lipid considerably reduces the strength of the quadrupole coupling constant, which does not exceed a few tens of kHz. The quadrupole coupling constant can however vary with the composition of the membrane. Solid-state ^{14}N NMR spectroscopy has also been applied to aligned model membranes to investigate the orientation of the membrane and the oligomerization of membrane-associated peptides.[5]

For spin $I = 1$ nuclei (e.g., ^{14}N or ^{2}H) in solids, double-quantum (DQ) spectra are insensitive to the first-order quadrupole broadening effects and lead to narrow lines.[6] DQ spectra of spin $I = 1$ nuclei in solids can be obtained indirectly via the neighboring spin $S = 1/2$ nuclei (e.g., ^{1}H or ^{13}C).

The extensive studies on the effect induced by the ^{14}N quadrupolar nucleus on the spin $S = 1/2$ spectra and the indirect detection of ^{14}N via "spy" nuclei in amino acids and small peptides represent a prerequired, valuable background in view of an extension of the indirect detection ^{14}N NMR experiments in biomolecules. The great interest in ^{14}N NMR spectroscopy stems from two promising possibilities highly important when studying biological systems. The first is that there will be no need for ^{15}N isotopic enrichment. The second is that ^{14}N NMR could allow for a more sensitive structural and motional characterization of the molecular environments in which ^{14}N is located.

16.2 THEORETICAL BACKGROUND

This section introduces the parameters characterizing the quadrupole interaction and the equations relevant to the experiments described herein. A thorough description of the theory relevant to $I = 1$ solid-state NMR spectroscopy has been recently published and will not be repeated here. The reader is referred to Chapter 3 of the present handbook for details about the theory. Various examples involving ^{14}N and other spin $S = 1/2$ nuclei can be found elsewhere.[7] Quadrupolar nuclei such as ^{14}N have, in addition to their magnetic moment, an electric quadrupole moment, eQ, which interacts with the electric field gradient (EFG) at the nucleus. e denotes the elementary charge and Q is the quadrupole moment. The nonuniform distribution of the electric charge in the nucleus results in an electric quadrupole moment. As for the EFG, the surrounding electric environment creates it. The strength of this interaction is commonly described by the nuclear quadrupole coupling constant, χ:

$$\chi = \frac{e^2 Q q_{zz}}{\hbar} \tag{16.1}$$

where eq_{zz} represents the largest principal component of the EFG tensor with the convention $|eq_{zz}| \geq |eq_{yy}| \geq |eq_{xx}|$. Closely related to the quadrupole coupling constant is "the splitting parameter" ν_Q defined as

$$\nu_Q = \frac{3\chi}{2I(2I-1)} \tag{16.2}$$

The asymmetry of the EFG tensor is defined by

$$\eta = \frac{eq_{xx} - eq_{yy}}{eq_{zz}} \tag{16.3}$$

The quadrupole Hamiltonian of the ^{14}N nucleus in its principal axis system (PAS) is given by the following expression:

$$H_Q = \frac{e^2 q_{zz} Q}{4I(2I-1)\hbar}[(3I_z^2 - I^2) + \eta(I_x^2 - I_y^2)] \tag{16.4}$$

The secular form of the quadrupole Hamiltonian suitable for high-field NMR is independent of the asymmetry parameter. If the orientation of the principal axes of the EFG tensor in the Zeeman frame is known, the expression of the quadrupole Hamiltonian in the Zeeman frame can be easily derived.

For most ^{13}C (S)–^{14}N (I) spin pairs, since the ^{14}N quadrupole energy is small compared to the ^{14}N Zeeman energy, the ^{14}N quadrupole Hamiltonian can be treated as a perturbation and the residual dipolar splitting (RDS) can be extracted using the first-order

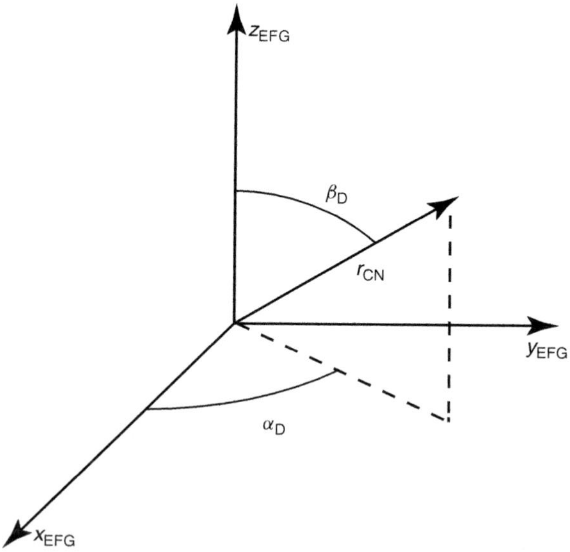

Figure 16.1. Internuclear vector r_{CN} in the coordinate frame where the EFG tensor is diagonal. β_D, α_D represent the polar angles of the dipolar vector.

approximation. The typical asymmetric 1 : 2 splitting observed in the MAS spectrum of spin $S = 1/2$ nuclei directly bonded to nitrogen is as follows[8]:

$$D_{RDS} = \frac{9 D_{CN} \nu_Q}{20 \nu_N} [(3\cos^2\beta_D - 1) - \eta \sin^2\beta_D \cos(2\alpha_D)] \tag{16.5}$$

with $D_{CN} = (\mu_0 \gamma_C \gamma_N \hbar)/(4\pi r_{CN}^3)$ being the dipolar coupling constant and ν_N the ^{14}N Larmor frequency; β_D, α_D stand for the polar angles of the dipolar vector in the EFG tensor principal axis frame (see Figure 16.1).

The ^{14}N isotropic chemical shift obtained via ^{14}N indirect detection methods does not represent the true chemical shift. A second-order quadrupolar shift given by Samoson[9]

$$\delta_Q^{iso} = \frac{3}{40}\frac{3 - I(I+1)}{I^2(2I-1)^2}\left(1 + \frac{\eta^2}{3}\right)\left(\frac{\nu_Q}{\nu_N}\right)^2 = \frac{3+\eta^2}{40}\left(\frac{\nu_Q}{\nu_N}\right)^2 \tag{16.6}$$

shifts downfield the ^{14}N resonances. For nitrogen, this second-order quadrupolar shift can be easily separated from the isotropic chemical shift because the latter can be obtained by a ^{15}N NMR MAS experiment.

16.3 ^{14}N FINGERPRINT IN SPIN $S = 1/2$ SPECTRA

Spin $S = 1/2$ nuclei directly bonded to ^{14}N often exhibit lineshapes that are affected by the incomplete MAS averaging of dipolar coupling to the ^{14}N nucleus. The phenomenon was first observed and in detail studied for the ^{13}C–^{14}N nuclear spin pair.[10] The ^{13}C nuclei close to ^{14}N nuclei show typical asymmetric lineshapes. Theoretical studies have shown that these experimental observations can be explained by the ^{14}N quadrupole effects on the ^{14}N–^{13}C dipolar interactions.[8,11] Owing to the strong ^{14}N quadrupole coupling constant, the quadrupolar interaction has a magnitude comparable to the Zeeman interaction, and hence, the quantization axis of the ^{14}N nucleus is not aligned along the direction of the static magnetic field but is rather oriented along the resultant of the applied field and an axis fixed in the molecular frame, which is determined by the principal axis system of the EFG. Consequently, since the ^{14}N and ^{13}C spins are not quantized along the same axis, the ^{13}C–^{14}N dipolar interaction is not completely averaged out by the MAS. In conclusion, this effect can be visible in the spectra of any spin $S = 1/2$ nuclei close to ^{14}N at low magnetic fields (e.g., where the Zeeman and quadrupole contributions are comparable). The resulting residual dipolar splitting is defined as the difference between the mass center of the more intense component (e.g., $|I = 1, m_I = \pm 1\rangle$ states of ^{14}N) and the mass center of the less intense component (e.g., $|I = 1, m_I = 0\rangle$ states of ^{14}N) of the doublet for a 2 : 1 lineshape. The asymmetry (1 : 2 or 2 : 1) or the lack of asymmetry of the lines can be related to the relative orientations of the principal axis system of the EFG and the molecular frame. A change in the sign of the quadrupolar coupling can inverse the asymmetry of the ^{13}C resonances. Calculated ^{13}C lineshapes are shown in Figure 16.2 for different orientations of the principal axis system of the ^{14}N EFG in glycylglycine.[10] Comparison of the observed ^{13}C lineshapes to the theoretical ones allows the determination of the orientation of the ^{14}N EFG tensor with respect to the molecular frame and the sign of the ^{14}N quadrupolar coupling constant. On the contrary, when the Zeeman term is dominant, the spin $S = 1/2$ lineshapes are unaffected by the dipolar interactions with quadrupolar nuclei. This situation is generally encountered at magnetic fields higher than 10 T. The reader is referred to for details

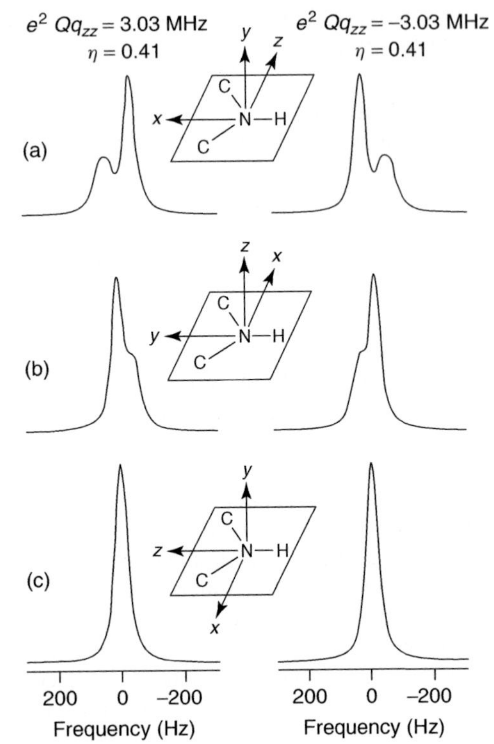

Figure 16.2. Calculated ^{13}C NMR lineshapes for a ^{13}C nucleus directly bonded to a ^{14}N. The parameters of the glycylglycine were used for these calculations. C–N bond length was extracted from an X-ray diffraction study. (a) Large residual dipolar splitting (80–100 Hz) in a well-resolved doublet pattern is observed when the z-axis of the quadrupole coupling tensor of the peptide nitrogen lies in the C–N–C plane and the x-axis is oriented along the N–H bond direction. (b) When the direction of the z-axis is normal to the C–N–C plane and the y-axis lies along the direction of the N–H bond; the calculated lines exhibit an unresolved doublet structure, which appears as a main peak with the distinct shoulder. (c) When the z-axis lies along the direction of the N–H bond, the calculated lineshapes show a narrow singlet pattern. (Reprinted from *Journal of Magnetic Resonance*, **48**, 3, A. Naito, S. Ganapathy, C.A. McDowell, 367–381., Copyright (1982), with permission from Elsevier.)

about the theory and various examples involving ^{14}N and other spin $S = 1/2$ nuclei.

Among the various examples reported, particularly relevant for the NMR of biological molecules are (^{13}C, ^{14}N) and (^{1}H, ^{14}N). The protons bonded to nitrogen nuclei show broad lineshapes, which are brought about by the quadrupole ^{14}N effect. The same characteristic lineshapes as observed for carbons directly bonded to ^{14}N were first detected in ^{1}H CRAMPS (combined rotation and multiple pulse spectroscopy) experiments on amino acids.[12] For highly deuterated samples, the ^{1}H MAS spectra show amine resonances with the same characteristic asymmetric structure due to a partial averaging by the MAS of the ^{1}H–^{14}N dipolar coupling. McDermott *et al.*[13] reported ^{1}H MAS spectra (see Figure 16.3) with asymmetric lines for highly deuterated alanine samples at relatively low and moderate spinning frequencies on a 9.4 T magnet.

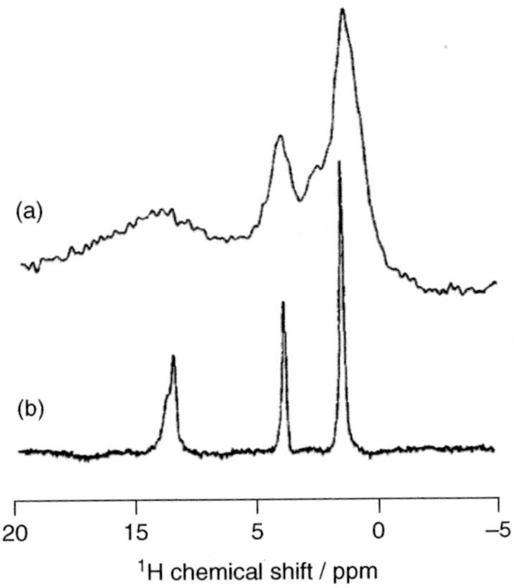

Figure 16.3. ^{1}H NMR spectra of protonated (a) and 99% deuterated (b) L-alanine recorded using a single pulse experiment on a 9.4 T magnet (400 MHz ^{1}H Larmor frequency) and at 11 kHz spinning frequency. The use of multiple pulse homonuclear decoupling schemes such as MREV-8 in combination with MAS (CRAMPS) was necessary to acquire the ^{1}H NMR spectrum of natural abundance L-alanine (a). The lines are assigned to the three ^{1}Hs of the methyl group (1.3 ppm), the ^{1}Hα (4 ppm), and the three protons on the amine (8.5 ppm). The ^{1}H NMR spectrum observed for the 1% protonated sample shows the spectral features of a residual dipolar splitting between the ^{1}H and the bonded ^{14}N, which is only partly averaged by the sample spinning because of the large quadrupole moment of the nitrogen. All chemical shifts are reported in ppm (or ppm divided by the MREV-8 scaling factor) relative to tetramethylsilane (TMS). (Reprinted from *J. Magn. Reson.* **98**, 408., A. E. McDermott, F. J. Creuzet, A. C. Kolbert, R. G., Copyright (1992) with permission from Elsevier.)

The spectra in Figure 16.3 clearly demonstrate that, at least for L-alanine, both the resolution and the sensitivity in extensively deuterated samples are considerably enhanced over those recorded with the CRAMPS methods; this improvement is particularly spectacular for the protons directly bonded to ^{14}N. Decreasing the level of deuteration dramatically affects the spectral resolution at low spinning rates. On the contrary, well-resolved ^1H spectra can be obtained for 1% protonated samples if the spinning frequency exceeds 10 kHz.

16.4 INDIRECT DETECTION OF ^{14}N

16.4.1 ^{13}C TRAPDOR Experiments

Veeman and coworkers[14,15] first used ^{14}N radiofrequency (RF) irradiation under slow MAS rotation in a ^{13}C cross polarization (CP) spin-echo experiment to perturb the ^{14}N spin bath. The ^{14}N quadrupolar interaction becomes time dependent under spinning, and spin transitions between the three $|I, m_I\rangle, m_I = -1, 0, 1$ spin states take place, 2 or 4 times per rotor cycle. The effect of ^{14}N irradiation is monitored indirectly via ^{13}C nuclei, by observing changes in the ^{13}C CP refocused intensity in a spin-echo experiment. A drawing of the pulse sequence dubbed TRAPDOR for transfer of populations in double resonance is given in Figure 16.4(a). The starting ^{13}C transverse magnetization is prepared by standard CP from ^1H to ^{13}C and then allowed to evolve during τ. The π pulse in the middle of the 2τ interval refocuses the evolution of the ^{13}C isotropic chemical shift. ^1H decoupling is active during both the spin-echo ($\tau-\pi-\tau$) and acquisition periods. In practice, two experiments are recorded: one without (reference ^{13}C spectrum

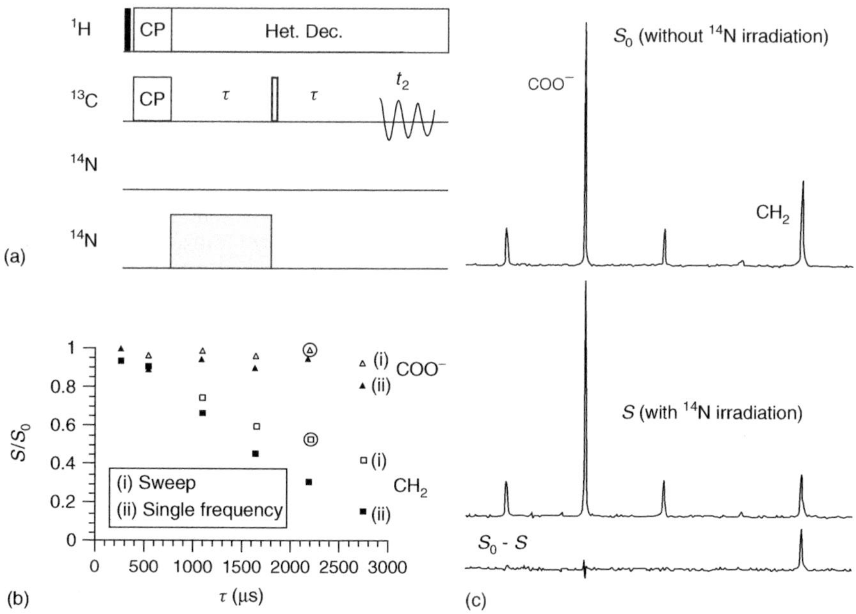

Figure 16.4. (a) $^{13}C-^{14}N$ TRAPDOR pulse sequence scheme, (b) attenuation of the ^{13}C spin-echo intensity as a function of the irradiation interval τ, and (c) ^{13}C spin-echo spectra corresponding to a τ interval of 2.2 ms (highlighted on the decay curve in (b) for a glycine sample enriched to about 20% at C-2 (e.g., $^{13}C^{\alpha}$). (b) The spin-echo decay under ^{14}N irradiation is drawn for two situations: (i) a sweep of 400 kHz was performed in 100 steps through the central carrier position; (ii) a constant irradiation on resonance was applied. Spinning speeds of 3.56 and 3.61 kHz were used for (i) and (ii), respectively. (c) ^{13}C spin-echo intensity without (S_0) and with (S) ^{14}N irradiation when the latter is sweeping through the ^{14}N resonance. These ^{13}C spectra were recorded for an interval $\tau = 2.2$ ms. All the experiments were carried out on a 7 T (300 MHz ^1H Larmor frequency) magnet with a 5 mm probe head. (Reprinted from *Solid State Nucl. Magn. Reson.* **4**, 113., C. P. Grey, A. P. A. M. Eijkelenboom, W. S. Veeman, Copyright (1995) with permission from Elsevier.)

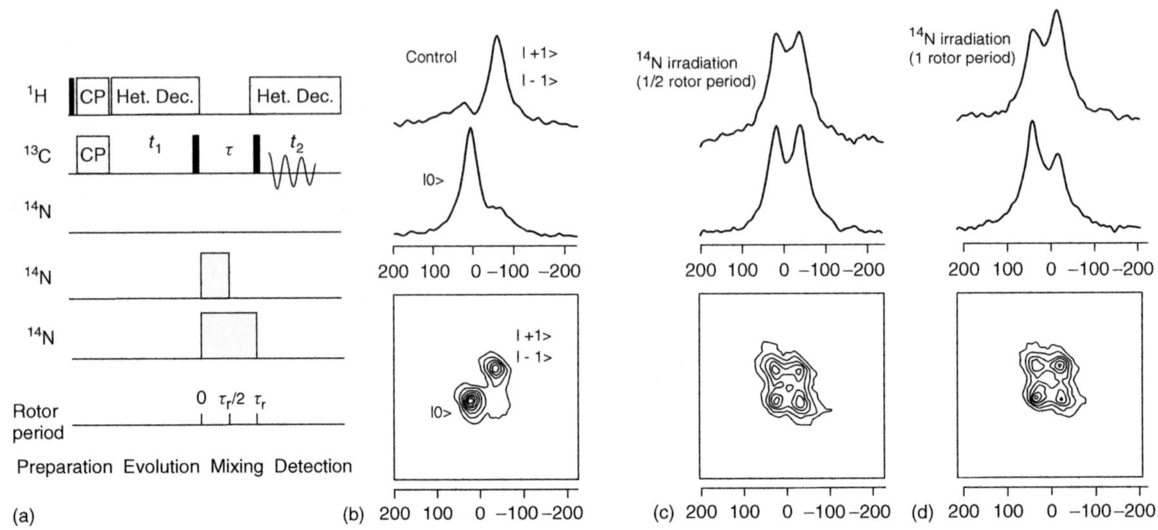

Figure 16.5. 2D $^{13}C-^{13}C$ TRAPDOR pulse sequence (a) and the corresponding spectra (b–d) of *N*-acetyl-D,L-valine showing the CH residual dipolar splitting. The ω_2 cross-sections through the CH doublet are shown above each 2D spectrum. All the experiments were carried out at 1005 Hz on a 7 T magnet (300 MHz ^1H Larmor frequency). (Reprinted from *Chem. Phys. Lett.* **192**, 379, C. P. Grey, W. S. Veeman, Copyright (1992) with permission from Elsevier.)

of signal S_0) and another with (attenuated ^{13}C spectrum of signal S) ^{14}N irradiation. The reduction in ^{13}C CP echo intensity during the spin-echo period when ^{14}N is irradiated depends on the C–N distances: the shorter the internuclear distances, the greater the signal attenuation. The feasibility of the method is illustrated for a glycine sample. The effect is most noticeable for $^{13}C^{\alpha}$, which is directly bonded to the ^{14}N, but is also visible for the $^{13}COO^-$ located at 0.249 nm away from the nitrogen (Figure 16.4b and c). The authors have shown that RF irradiation at constant frequency is more efficient than sweeping through the ^{14}N resonance. Compared to the indirect excitation and detection method of ^{14}N DQ coherence[16] by ^1H Hartmann–Hahn CP via dipolar order, the $^{13}C-^{14}N$ TRAPDOR scheme is the less demanding for the ^{14}N irradiation frequency.

Applied to a 2D $^{13}C-^{13}C$ correlation experiment, the $^{13}C-^{14}N$ TRAPDOR scheme gives off-diagonal correlations for the carbons directly bonded to the nitrogen. Cross peaks are observed between the asymmetric components of the residual $^{13}C-^{14}N$ dipolar splitting which, for *N*-acetyl-D,L-valine, is clearly resolved at about 1 kHz spinning frequency on a 7 T magnet. Such a well-resolved splitting is consistent with the −3.2 MHz quadrupolar coupling constant found for the ^{14}N in *N*-acetyl-L-valine.[3] The pulse sequence of the 2D $^{13}C-^{13}C$ experiment is shown in Figure 16.5a. Here, the ^{13}C transverse magnetization transferred from protons by CP is allowed to evolve during t_1. The magnetization is put along the z-axis by a $\pi/2$ pulse for a duration (e.g., mixing time) equal to either half the rotor period or a full rotor period and then put back into the xy-plane for detection. During t_1 evolution and detection, ^1H decoupling is active. Three different experiments are recorded. In experiment 1, a control experiment, no ^{14}N irradiation is applied during the ^{13}C mixing time (this scheme corresponds to the proton-driven spin diffusion or PDSD experiment). On the contrary, ^{14}N irradiation is applied for half of the rotor period or for the whole rotor period in the experiments 2 and 3, respectively. Only very small cross peaks are observed in the absence of ^{14}N irradiation. The intensity of the cross peaks of the $^{13}C-^{13}C$ correlation experiment increases significantly when the ^{14}N nucleus is irradiated during the mixing period. Their relative intensities reach a maximum when ^{14}N irradiation is active for half of the rotor period and then decrease when ^{14}N irradiation is applied during an entire rotor period. In theory, population transfers induced by passages between the different eigenstates of the Zeeman and quadrupolar Hamiltonian could bring about cross peaks in the $^{13}C-^{13}C$ correlation spectrum. Another phenomenon, which might generate these population transfers, is the ^{14}N longitudinal relaxation (T_1). The low-intensity cross peaks in the control experiment indicate that the contribution from this mechanism is rather small (less than 26%). These cross peaks cannot be the result of DQ transitions because the ^{14}N irradiation was applied with an adequate frequency offset in order to avoid excitation of DQ transitions at any time during the rotor period. The complete mechanism of the ^{14}N-induced polarization transfers under slow spinning conditions is not fully understood.

Such population transfers are limited to the TRAPDOR scheme but can also be applied to rotational-echo adiabatic passage double resonance (REAPDOR)[17] experiments to recover the dipolar interaction between spin $S = 1/2$ nuclei and quadrupolar nuclei under spinning at the magic angle.

16.4.2 HMQC, HSQC Experiments for ^{14}N Indirect Detection

A very promising method is the indirect ^{14}N detection via close $S = 1/2$ nuclei under MAS and at relatively high static magnetic fields. Indirect ^{14}N detection improves not only the resolution but also the sensitivity of the ^{14}N spectra in solids and opens up new horizons for the use of ^{14}N NMR spectroscopy to answer real biological questions. Two groups[18,19] exploited almost simultaneously the second-order quadrupole–dipole cross terms between ^{14}N ($I = 1$) and spy nuclei, such as ^{13}C ($S = 1/2$), to achieve high-resolution $^{13}C-^{14}N$ correlation spectra of amino acids and peptides under MAS. Owing to its inverse relation to the static magnetic field strength, B_0, a 1 : 2 doublet characteristic of a spin $S = 1/2$ bonded to a ^{14}N nucleus is expected to get smaller and smaller at high magnetic fields and even to collapse into a single resonance when the inhomogeneous broadening of the linewidths will exceed the effective residual dipolar splitting. Since the second-order quadrupolar–dipole interaction is orientation dependent, it cannot be averaged out completely by spinning at the magic angle. Typical $^{13}C-^{14}N$ one-bond scalar couplings in amino acids and peptides are of the order of a few hertz: $^1J(^{13}C^{\alpha}, ^{14}N) \approx 3-8$ Hz, $^1J(^{13}C', ^{14}N) \approx 9-11$ Hz.[20] In the most favorable case (e.g., large ν_Q and low B_0), the $^{13}C-^{14}N$ residual dipolar splitting is an order of magnitude larger than the scalar coupling, but it will become comparable with the scalar coupling at high static magnetic fields. It is then expected that both J and residual dipolar couplings will drive heteronuclear correlations. Polarization transfer via these two relatively small interactions is possible because the pulse sequence is built on a spin-echo element on the ^{13}C channel (see Figure 16.6), thus ensuring that the ^{13}C linewidths effective all through the τ transfer intervals correspond to the "refocusable" linewidths. For simple amino acids, these linewidths are as small as a few hertz.

The scheme of the $^{14}N-^{13}C$ pulse sequence, based on the well-known heteronuclear multiple quantum correlation (HMQC) method, is drawn in Figure 16.6(a) together with the coherence transfer pathways for SQ and DQ experiments. Initial SQ ^{13}C coherences are created by CP from ^1H to ^{13}C. During the first τ interval, the buildup of the antiphase ^{13}C coherences generated by the evolution under both J and D_{RDS} splittings will compete with the ^{13}C T_2 relaxation decay. The RF irradiation of the ^{14}N channel that follows leads to partial conversion of the antiphase ^{13}C coherences into ^{14}N SQ and DQ coherences. Prior to reconversion into observable ^{13}C SQ coherences during a second interval τ, the ^{14}N coherences will evolve in t_1. The π pulse in the middle of the t_1 evolution period removes all refocusable interactions. The τ intervals and the increments in the indirect dimension must be *synchronized* with the spinning period. Synchronization in the t_1 dimension ensures complete averaging of the first-order quadrupole interaction when the magic angle is very precisely adjusted. Hence, the predominant broadening of the ^{14}N resonances observed in the indirect dimension is mainly generated by second-order quadrupolar interactions leading to linewidths of only a few kilohertz. The ^{14}N linewidths obtained in the indirect dimension can also present contributions from the third-order quadrupolar interaction.[7,18,21] The Fourier transformation of such a time evolution yields SQ–SQ or DQ–SQ $^{14}N-^{13}C$ correlation spectra depending on the phase of one of the ^{14}N pulses.

Figure 16.6 contains $^{14}N-^{13}C$ SQ–SQ and DQ–SQ HMQC correlation spectra obtained for two simple 2-^{13}C-labeled amino acids[18] at 9.4 T. The standard ^{13}C CPMAS spectra reveal an ill-resolved residual dipolar splitting for L-alanine and no splitting at all for glycine. In the absence of motional

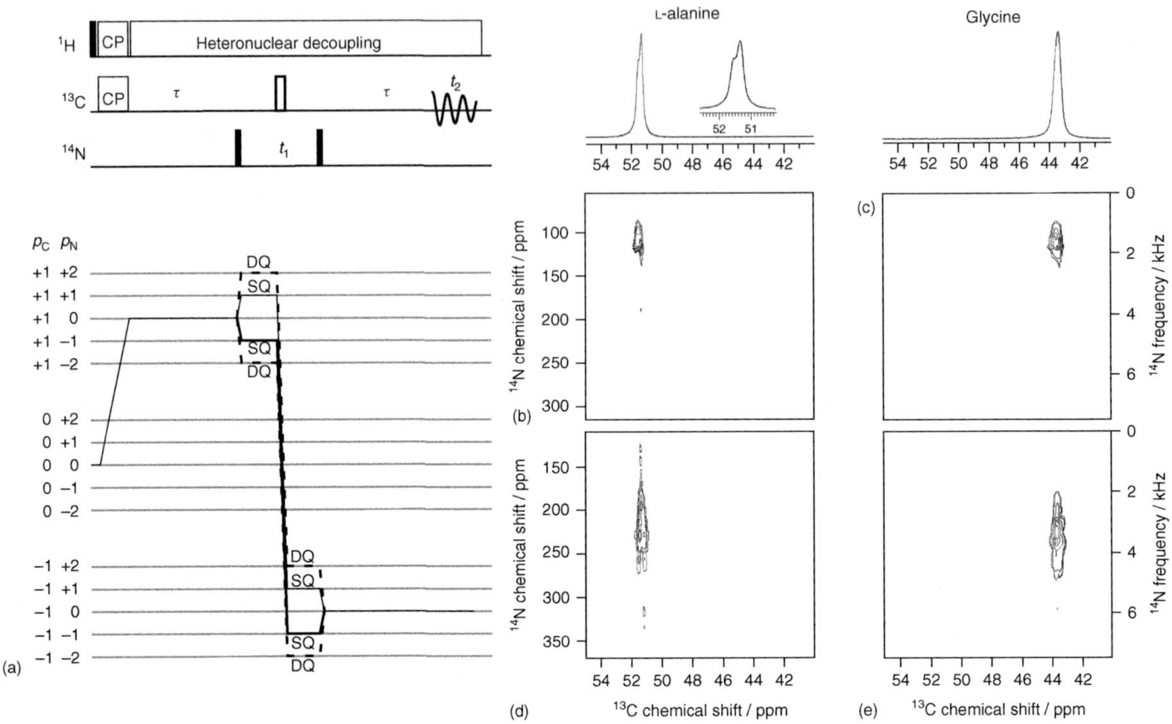

Figure 16.6. $^{14}N-^{13}C$ HMQC pulse sequence together with the coherence transfer pathway diagram (a) and the corresponding 2D SQ–SQ (b, c), and DQ–SQ (d, e) spectra of L-alanine and glycine samples at 30 kHz rotation frequency on a 9.4 T (400 MHz ^{1}H Larmor frequency) magnet. The ^{13}C CPMAS spectra are also shown to illustrate that in the case of L-alanine, the residual dipolar splitting is visible, whereas in the case of glycine, the residual dipolar splitting is masked by the inhomogeneous linebroadening. The following parameters were used: $\tau = 16$ ms, 512 (b, c) and 170 (d, e) t_1 increments of 33.33 μs, 32 (b, c) and 96 (d, e) scans, $\tau_p(^{14}N) = 11$ μs (b, c) and 15 μs (d, e) with $\nu_1(^{14}N) = 50$ kHz, two-pulse phase modulation (TPPM) ^{1}H decoupling with $\nu_1(^1H) = 100$ kHz. (Reprinted with permission from S. Cavadini, A. Lupulescu, S. Antonijevic, G. Bodenhausen *J. Am. Chem. Soc.* 2006, **128**, 7706. Copyright 2006 American Chemical Society.)

broadening and if the magic angle is accurately adjusted, the linewidth of the ^{14}N DQ resonance is twice the linewidth of the SQ peak. Since the increments in the indirect dimension were rotor synchronized, the first-order quadrupole interaction is completely averaged out by the spinning at magic angle. However, the ^{14}N resonances in the indirect dimension should present a broadening and a shift from the isotropic chemical shift position as a result of the second-order quadrupolar effect (see equation 16.6). Typical second-order quadrupolar powder patterns are observed for both the ^{14}N peaks of L-alanine and glycine. For nitrogen, the downfield quadrupolar shift can be obtained experimentally by measuring the difference between the position of the ^{14}N peak in the indirect dimension of the $^{14}N-^{13}C$ correlation experiment and the isotropic chemical shift obtained by recording a ^{15}N MAS spectrum. At 9.4 T, isotropic second-order chemical shifts of about 118 and 135 ppm are expected for L-alanine and glycine, respectively.

The experiment was successfully applied to a model natural abundance polypeptide (AGG)[19] at 14.1 T (see Figure 16.7). At this magnetic field, not only the residual dipolar splitting will get smaller but also the isotropic second-order quadrupolar shift. Moreover, L-alanine and glycine samples would show $^{14}N-^{13}C$ correlations shifted by 52 and 60 ppm, respectively, in the indirect dimension compared to the corresponding isotropic

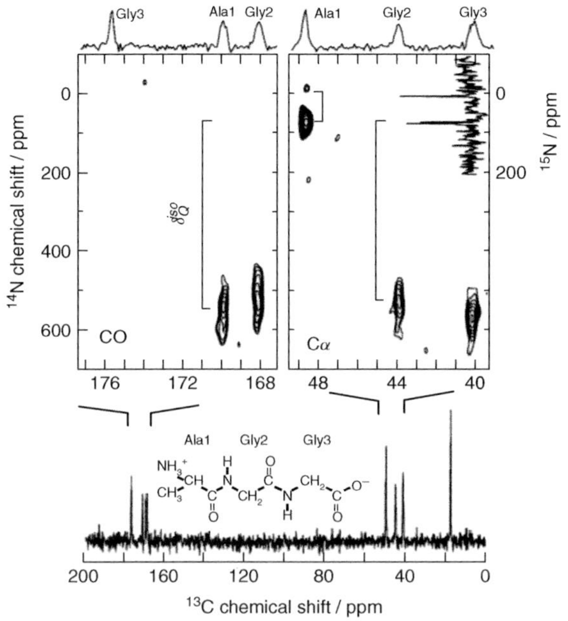

Figure 16.7. ^{14}N–^{13}C SQ–SQ correlation spectrum of natural abundant Ala-Gly-Gly peptide recorded at 25 kHz spinning frequency on a 14.1 T (600 MHz ^{1}H Larmor frequency) magnet. The following parameters were used: $\tau = 15$ ms, 16 t_1 increments, 8192 scans, $\tau_p(^{14}\mathrm{N}) = 2\ \mu\mathrm{s}$ with $\nu_1 = 50$ kHz, small phase increment alternation (SPINAL)-64 ^{1}H decoupling with $\nu_1(^1\mathrm{H}) = 125$ kHz. The vertical brackets indicate the isotropic second-order quadrupolar shift δ_Q^{iso} [see equation (16.6)] from the isotropic chemical shift measured in a separate ^{15}N MAS experiment shown on the top right corner of the 2D spectrum. The nitrogen chemical shift was referenced to solid NH_4Cl. Because of the large second-order quadrupolar shifts, the amide ^{14}N peaks have been dealiased after doubling the spectral window from 25 to 50 kHz. (Reprinted with permission from Z. Gan *J. Am. Chem. Soc.* 2006, **128**, 6040. Copyright 2006 American Chemical Society.)

chemical shift in a ^{15}N MAS spectrum. The tripeptide has three ^{14}N nuclei: one which is very likely an $-NH_3^+$, because the peptide is expected to be more stable as a zwitterion than in the canonical form, and two peptide amides. Five ^{14}N–^{13}C correlation peaks are then expected for this compound: three between the $^{13}C^{\alpha}$ and the ^{14}N of the same amino acid (*intramolecular* correlations) and two between the $^{13}C'$ and the ^{14}N of the next amino acid (*intermolecular* or *sequential* correlations). Five ^{14}N–^{13}C correlations are indeed observed in Figure 16.7. All the peaks correlating carbons with the peptide amide ^{14}N present a significant second-order quadrupolar shift compared to their isotropic chemical shift indicative of large quadrupolar coupling constants. These shifts are between 450 and 500 ppm, suggesting quadrupolar coupling constants of the order of 3 MHz, in agreement with values reported in a previous study of AGG.[22] A fit of the ^{14}N lineshapes is required in order to estimate the asymmetry parameter η from the second-order quadrupolar shift. This would necessitate a fairly good signal-to-noise ratio and a well-resolved spectrum in the indirect dimension. In addition to the second-order quadrupolar shift, the ^{14}N peptide amide peaks look broader compared to the linewidth of the N-terminal $-NH_3^+$ site, which has all the characteristics (e.g., small second-order quadrupolar shift and quite narrow linewidth) of a nucleus with relatively small quadrupolar coupling parameters. This is consistent with the random hopping of the protons around the threefold axis, which partly averages the anisotropic interactions.

The sensitivity of these HMQC-based experiments is determined essentially by the efficiency of the back-and-forth coherence transfer blocks between ^{13}C and ^{14}N.[18,21] In order to assess the efficiency of SQ and DQ coherence transfer elements, the ^{13}C signal amplitude $s(t_1, \omega_2)$ of the heteronuclear 2D experiments when $t_1 = 0$ should be compared to the ^{13}C spectrum obtained after a spin-echo block. For example, an efficiency of about 16% and 8% for SQ and DQ, respectively, is obtained for L-alanine. The following two parameters are crucial for optimizing the overall sensitivity of these ^{14}N–^{13}C HMQC-like experiments: (i) relatively short T_2 decay during the τ intervals; (ii) efficient ^{14}N excitation and flip-back during the two ^{14}N pulses. In practice, the two-spin coherence buildup occurs for long refocusing and defocusing intervals due to the small J and residual dipolar couplings. The signal loss during these long coherence transfer intervals due to T_2 relaxation can be reduced by using ^{1}H decoupling schemes devised for prolonging the T_2'. In the same manner as for direct excitation, efficient ^{14}N excitation in the ^{14}N–^{13}C experiment would require the ^{14}N RF fields at least comparable with the size of the quadrupolar interaction (e.g., a few megahertz). However, the excitation bandwidth can be increased by using ^{14}N pulses with small flip angles ($\beta \approx 35°$) even for the

RF field amplitude of the order of 50 kHz. The acquisition and the processing of the full echo optimize the signal-to-noise ratio (gain of about $\sqrt{2}$).[19]

SQ or DQ coherences of ^{14}N nuclei can also be excited while the spy nuclei are aligned along the static field B_0 during the t_1 evolution period using a slightly modified version of the heteronuclear single-quantum correlation (HSQC) experiment.[23,24] The second-order ^{14}N single- and double-quantum powder patterns thus obtained should not be affected by the inhomogeneous decays of the spy nuclei coherences during the evolution interval. However, during the coherence transfer intervals τ, the evolution under J and D_{RDS} will be affected by the spin $S = 1/2$ T_2 relaxation. The phase cycling required for the HSQC scheme leads to a twofold loss of the signal-to-noise ration compared to the HMQC method for the same experimental time.

These HSQC/HMQC-based heteronuclear experiments are not limited to the spin pair ^{14}N, ^{13}C but can, in principle, be applied to any other spy nucleus, such as ^{1}H, ^{31}P, provided that the T_2 (HSQC) or T_2' (HMQC) of the spin $S = 1/2$ nuclei is long enough compared to $1/D_{RDS}$.

At high B_0 magnetic fields, the residual dipolar coupling decreases, and long coherence transfer intervals are required for the indirect detection of ^{14}N via the combination of scalar and residual dipolar couplings. The signal losses due to spin $S = 1/2$ (e.g., ^{13}C or ^{1}H) T_2 (HSQC) or T_2' (HMQC) relaxation during these long coherence transfer intervals can be reduced by using recoupling methods, which directly exploit the first-order dipolar coupling between the spin $S = 1/2$ and ^{14}N nuclei for establishing heteronuclear correlations under spinning at the magic angle. Recoupling sequences should apply an RF irradiation only to the observed spin $S = 1/2$ nucleus because the ^{14}N quadrupolar interaction is much larger than the available RF amplitudes (a few megahertz compared to a few tens of kilohertz RF amplitude). Various heteronuclear dipole–dipole recoupling schemes were developed for reintroducing the heteronuclear interactions otherwise averaged out by the spinning of the sample at the magic angle. For a detailed study of these recoupling methods, the reader is referred to. Some of these methods are suitable for recoupling the dipolar interaction between spin $S = 1/2$ and ^{14}N nuclei. Rotational-echo double resonance (REDOR),[25] rotary resonance recoupling (R^3) irradiation,[26] continuous rotor-synchronized simultaneous frequency and amplitude modulations ($SFAM_1$ and $SFAM_2$),[27] and symmetry-based sequences[28] have already been tested for ^{13}C or ^{1}H, ^{14}N spin pairs.[29–35] The well-known REDOR sequence consists of applying two π pulses per rotor cycle, whereas the rotary resonance recoupling uses a continuous wave RF field to match the spinning frequency. For SFAM recoupling, the carrier frequency of the RF field is modulated cosinusoidally while the RF amplitude is modulated sinusoidally as a function of the spinning frequency, ω_r. Symmetry-based recoupling schemes are rotor-synchronized and exploit the rotational properties of the nuclear spin interactions. RN_n^{ν} sequence consists of $N/2$ pairs of elements $R_{\phi}R_{-\phi}$ with $\phi = \pi\nu/N$ radians, the overall phase shift. The element R_{ϕ} represents a block of RF pulses which rotates the spins by 180° about the ϕ-axis. Rotor synchronization is ensured by choosing N elements R to span exactly n rotor periods, τ_r. The duration of the element R is given by $\tau_R = n\tau_r/N$. REDOR and R^3 can also be seen as symmetry-based sequences, which have the modern symmetry notation $R4_2^1$ for usual REDOR and $R2_1^0$ ($n = 1$) or $R4_1^0$ ($n = 2$) for R^3.

One potential application of the indirect detection schemes described in this section is the measurement of internuclear distances between ^{14}N and spin $S = 1/2$ nuclei without the need for ^{15}N isotopic labeling. The parameters of the ^{14}N quadrupolar coupling tensor can also be determined by fitting the second-order ^{14}N lineshapes obtained in the indirect dimension. These applications are reported in detail in Section 16.5.

16.4.3 ^{14}N-edited ^{1}H–^{13}C Experiments

The indirect detection techniques of ^{14}N can be combined to obtain a 3D ^{1}H–^{13}C–^{14}N through-space correlation spectrum.[36] The pulse sequence of the 3D experiment is drawn in Figure 16.8(a). A chemical shift labeling of the excited ^{1}H magnetization is performed during the interval t_1. The ^{1}H magnetization is then transferred to ^{13}C by CP. The remaining part of the experiment is exactly the indirect detection of the ^{13}C–^{14}N HMQC element where the recoupling of the ^{13}C–^{14}N dipolar coupling is exploited to transfer the ^{13}C coherence to ^{14}N and then back to the ^{13}C. After the first polarization transfer element, the ^{14}N magnetization is excited and the ^{14}N chemical shift labeling takes place during the t_2 evolution interval. Then, a relatively short irradiation of the ^{14}N should

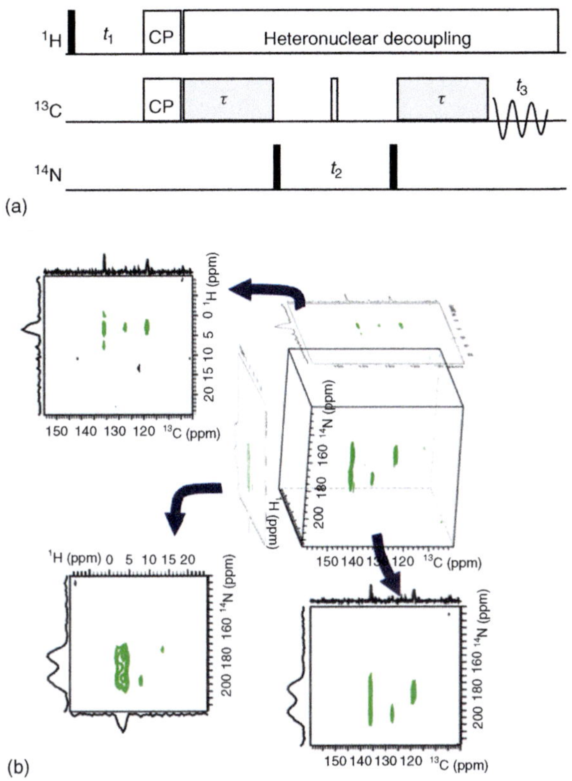

Figure 16.8. $^1H-^{14}N-^{13}C$ pulse sequence (a) and the corresponding correlation spectrum (b) of ^{13}C-enriched L-Histidine HCl H_2O recorded at 12.5 kHz spinning frequency on a 18.8 T (800 MHz ^{1}H Larmor frequency) magnet. The following parameters were used for recording the experiment: $\tau = 240$ µs, 72 t_1 and 32 t_2 increments of 20 and 80 µs, respectively, 32 scans, $\tau_p(^{14}N) = 30$ µs with $\nu_1(^{14}N) = 40$ kHz, TPPM ^{1}H decoupling with $\nu_1(^1H) = 80$ kHz. The recycle delay was 4 s with the total experimental time of 85 h. (Reprinted from *J. Magn. Reson.* **193**, 321., R. Siegel, J. Trébosc, J. P. Amoureux, Z. Gan, Copyright (2008), with permission from Elsevier.)

put it back along the z-axis, and during the second polarization transfer element, the ^{14}N magnetization is transferred to the carbons to be observed during t_3. Not only the polarization transfer elements but also the t_2 evolution period have to be rotor synchronized. Heteronuclear dipolar ^{1}H decoupling was applied during the polarization transfer intervals, t_2 and t_3 evolutions periods.

The feasibility of the experiment was demonstrated on a uniformly ^{13}C-enriched L-histidine hydrochloride monohydrated at 18.8 T (800 MHz ^{1}H Larmor frequency). Histidine has three different nitrogen sites: $-NH_3^+$, which is linked to one carbon (Cα), and two ring nitrogens (δ_1 and ε_2 sites), which are connected to two carbons each (Cγ/Cε_1 and Cδ_2/Cε_1). When protonated, the imidazole ring is positively charged and has two NHs. Two correlation peaks, assigned to the two ring nitrogens, are mainly visible in the 3D experiment. The $-NH_3^+$ group gave a small correlation signal in the 2D $^{14}H-^1N$ HMQC experiment and was not observed in the 2D $^{14}N-^{13}C$ HMQC spectrum. For both experiments, polarization transfer from the spin $S = 1/2$ nucleus to ^{14}N was achieved by R^3. In conclusion, the 3D experiment, when combined with R^3 recoupling scheme to perform the coherence transfer back and forth between the spy nuclei and the ^{14}N, should facilitate the assignment of ^{14}N sites in biomolecules.

Owing to the low efficiency of the $^{13}C-^{14}N-^{13}C$ coherence transfer path (5%), a duration of 85 h was necessary on an 18.8 T magnet to record the 3D experiment described previously. The same information could be gathered in a shorter experimental time if $^{14}N-^1H$ and $^{13}C-^1H$ through-space 2D heteronuclear correlation (HETCOR) spectra were acquired instead of running the 3D experiment. In order to identify the carbons close to the nitrogens, a ^{14}N-edited $^{13}C-^1H$ HETCOR experiment has been recorded. The first selective detection of carbons near ^{14}N has been demonstrated with TRAPDOR and REAPDOR experiments.[17,37] Later, Schmidt-Rohr and Mao have introduced a ^{14}N-selective $^1H-^{13}C$ experiment, termed SPIDER (saturation-pulse induced dipolar exchange with recoupling).[38] This ^{13}C-based approach for investigating the chemical environment of nitrogens is particularly attractive because of the well-defined ^{13}C chemical shift ranges and possibilities of spectral editing. Hence, amide nitrogens bonded to aromatic carbons can be easily differentiated from those bonded to aliphatic groups in a ^{14}N-edited ^{13}C spectrum. The SPIDER experiment is closely related to the technique based on relaxation-induced dipolar exchange with recoupling (RIDER),[39] and as in REDOR and REAPDOR, a series of π pulses applied on ^{13}C channel recouples the heteronuclear dipolar interaction. To saturate the ^{14}N coherences, the SPIDER technique uses a train of short ^{14}N pulses. Amoureux and coworkers[34,35] have introduced another ^{14}N-selective experiment. This MAS ^{14}N-edited method allows for recording high-resolution $^{13}C-^1H$ through-space HETCOR spectra. The pulse sequence of the 2D experiment

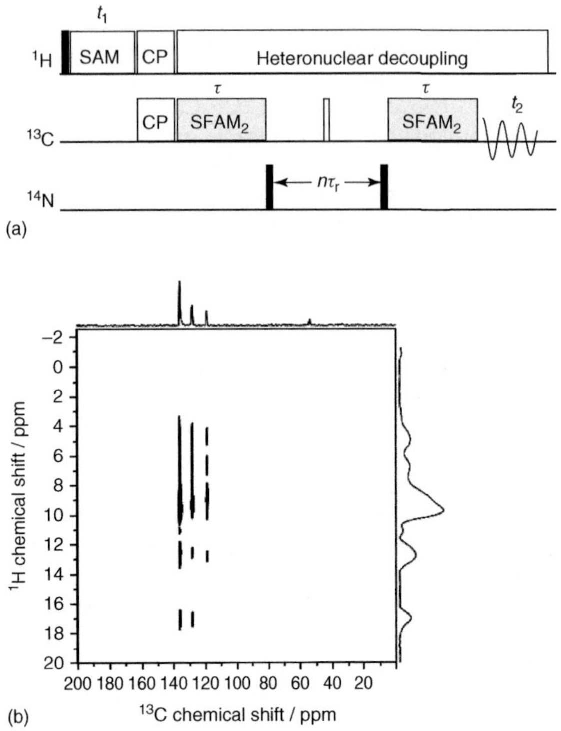

Figure 16.9. Pulse sequence of the $^1H-^{13}C$ CP-HETCOR correlation NMR experiment (a) and corresponding ^{14}N-edited CP spectrum on uniformly ^{13}C-labeled histidine (b) on a 18.8 T (800 MHz ^1H Larmor frequency) magnet and at 17.5 kHz spinning frequency. (a) Smooth amplitude modulation (SAM) homonuclear decoupling was applied during t_1 evolution in order to obtain high-resolution ^1H spectra in the indirect dimension. Simultaneous frequency and amplitude modulated ($SFAM_2$) scheme was applied on ^{13}C nuclei to reintroduce $^{13}C-^{14}N$ dipolar interactions during the excitation and reconversion periods. The t_1 evolution and the two $SFAM_2$ blocks have to be rotor synchronized. Efficient heteronuclear decoupling is required to prevent $^1H-^{13}C$ coherences to be created. (b) The 2D $^1H-^{13}C$ spectra were recorded with the following parameters: 30 t_1 increments, 64 transients per t_1 increment with a relaxation interval of 5 s. The $SFAM_2$ recoupling intervals during excitation and reconversion were $\tau_{rec} = 457$ μs, $\tau_p(^{14}N) = 30$ μs with $\nu_1(^{14}N) = 40$ kHz, and SPINAL ^1H decoupling with $\nu_1(^1H) = 80$ kHz. The total experimental time was 340 min. (Reprinted from *J. Magn. Reson.* **194**, 317., J. P. Amoureux, J. Trébosc, B. Hu, N. Halpern-Manners, S. Antonijevic, Copyright (2008), with permission from Elsevier.)

is drawn in Figure 16.9(a). The ^1H-excited coherences are allowed to evolve during the t_1 period while $^1H-^1H$ dipolar interactions are decoupled in order to obtain high-resolution spectra in the indirect dimension. The ^1H coherences are then transferred by CP to the ^{13}C nuclei. This step is followed by the $^{14}N-^{13}C$ dipolar-driven HMQC filter. The ^{13}C coherences are detected during the t_2 period while heteronuclear ^1H decoupling is applied during both HMQC and acquisition periods. SFAM recoupling method has been used during τ intervals for reintroducing C–N dipolar interactions under spinning at the magic angle. In the SFAM schemes, the RF field amplitude is modulated sinusoidally, whereas the carrier frequency of the RF field is modulated cosinusoidally. When the modulation frequency is equal to the spinning frequency ($SFAM_1$), not only the heteronuclear dipolar interactions are reintroduced but also the homonuclear dipolar interactions. Therefore, $SFAM_1$ is not appropriate for reintroducing $^1H-^{14}N$ dipolar couplings in HMQC-like experiments. The recoupling of the homonuclear dipolar interactions is expected to be less efficient when the modulation frequency is twice the spinning frequency ($SFAM_2$).

The ^{14}N-edited 2D $^1H-^{13}C$ HETCOR spectrum, with ^1H dimension rescaled, of uniformly ^{13}C-labeled histidine is shown in Figure 16.9(b). A 2D $^1H-^{13}C$ HETCOR spectrum containing all carbons can be obtained using the standard CP-HETCOR experiment. Three correlation peaks are obtained instead when the ^{14}N filter is inserted. These peaks correspond to the aromatic ^1H and ^{13}C nuclei in close proximity to nitrogen-14. Again, the $^1H-^{13}C^{\alpha}$ correlation peak is barely visible. The weak efficiency of the HMQC ^{14}N filter has been correlated to the use of ^{14}N hard pulses. The low RF amplitude employed to irradiate the ^{14}N nuclei does not allow to uniformly excite the bandwidths spread over a few megahertz. In order to compensate for this low excitation efficiency, a single ^{14}N pulse has been applied instead of the two ^{14}N hard short pulses, and two 2D spectra have been recorded: one without ^{14}N irradiation (S_0) and another where ^{14}N nuclei are irradiated for one rotor period at the highest available RF field amplitude (S). The ^{14}N-edited $^1H-^{13}C$ HETCOR spectrum is obtained by subtracting the two spectra (e.g., $S_0 - S$) in the manner of TRAPDOR, REDOR experiments. The sensitivity of the ^{14}N-edited $^1H-^{13}C$ HETCOR experiment is 5 times better if a ^{14}N pulse of one rotor pulse is used instead of two hard short pulses. For ^{14}N pulses longer than one rotor period, the spinning will modulate the ^{14}N signal. As for the TRAPDOR or REAPDOR experiments, level anticrossing induced by MAS can occur during the ^{14}N irradiation depending on the adiabaticity factor. Adiabatic passages will efficiently invert the ^{14}N magnetization, whereas fast passages, caused by high spinning frequencies or large quadrupole coupling tensors, will result in less effective inversion of the ^{14}N magnetization.[37]

Indirect detection via spy nuclei also gives access to the first-order ^{14}N quadrupolar lineshapes under MAS if the HMQC-based experiment is performed with a single strong pulse on ^{14}N.[40] Indirect ^{14}N widelines through ^{13}C detection can be obtained if a long ^{14}N pulse replaces the chemical shift encoding pulses, and ^{13}C signal is measured as a function of ^{14}N carrier position. The indirect ^{14}N experiment was obtained as previously by measuring the difference $S_0 - S$, with S_0 and S being the ^{13}C spectra recorded without and with ^{14}N irradiation. The ^{14}N lineshapes thus obtained are very similar to the pattern of a static fist-order quadrupolar spectrum. Low sensitivity and baseline rolls usually observed by direct detection are absent in these ^{14}N indirect detection experiments. Thus, ultrawide ^{14}N spectra of various nitrogen sites can be obtained, thanks to high-resolution ^{13}C spectra under MAS.

16.5 APPLICATIONS

16.5.1 Measurement of ^{14}N Quadrupolar Coupling Constants

The asymmetric line shapes of the spin $S = 1/2$ nuclei directly bonded to ^{14}N offer the possibility of characterizing the bonding geometry and determining the quadrupole parameters for the nitrogen nucleus. These resonances are influenced by a series of parameters (see equation 16.5): the sign, magnitude, and asymmetry parameter of the ^{14}N quadrupole coupling tensor; the internuclear distance; the magnitude of the applied magnetic field; the orientation of the internuclear vector in the principal axis system of the EFG. One or more of these parameters can be determined from the spin $S = 1/2$ spectra if the other parameters are known either by other methods or by molecular symmetry. The physical basis for the influence of ^{14}N on spin $S = 1/2$ resonance line shapes relies on comparing the sign and magnitude of the experimental and calculated asymmetric splitting in the spin $S = 1/2$ spectra. For example, in polycrystalline morphine sulfate, a prototype agonist, 3 out of the 17 carbon resonances show clear evidence of fine structure in the form of the splittings.[41] All the three resonances are directly bonded to ^{14}N. The observed asymmetric doublets were compared with theoretical calculations of the magnitude of the splitting of the ^{13}C resonances as a function of the angle between the ^{13}C–^{14}N internuclear vector and the quantization axis of the ^{14}N EFG. Because the X-ray structure of morphine sulfate was unknown, the authors used for the theoretical calculations the bond lengths from the X-ray structure of morphine hydrochloride trihydrate.

The parameters of the ^{14}N quadrupolar tensor can also be obtained by fitting the ^{14}N line shapes obtained by indirect detection HMQC-type methods.[18,19] The SQ/DQ resonances are affected not only by the isotropic shift but also by the second-order quadrupolar interaction. A contribution from the third-order term is expected to contribute to the SQ signals. For example, simulations with the following parameters fit well the experimental ^{14}N lineshapes of the two magnetically inequivalent ^{13}C sites I and II in powdered L-leucine, which correlate in the ^{14}N–^{13}C HMQC spectrum with two non-degenerate ^{14}N tensors: $\nu_Q = 1.15$ MHz, $\eta = 0.38$ (site I); $\nu_Q = 1.13$ MHz, $\eta = 0.08$ (site II).[18,21] The two magnetically inequivalent sites show different asymmetry parameters, with the ^{14}N EFG tensor of the site II being almost axially symmetric.

16.5.2 Determination of ^{13}C–^{14}N Internuclear Distances

REDOR[25] is one of the most popular methods for measuring dipolar couplings and hence internuclear distances between heteronuclear spin $S = 1/2$ nuclei pairs such as ^{13}C–^{15}N. The regular REDOR experiments are built on a rotor-synchronized sequence of two π pulses per rotor period such as to undo the MAS refocusing of the dipolar dephasing. For $S = 1/2$, $I = 1$ spin pairs, the standard REDOR experiment has to be modified because the π pulses will not be effective in the presence of large quadrupolar interaction. Alternative dipolar recoupling schemes have been developed for $S = 1/2$, $I > 1/2$ spin pairs when the I nucleus is characterized by a large quadrupolar interaction: TRAPDOR,[42] REAPDOR,[17] dipolar exchange-assisted recoupling (DEAR),[43] and RIDER.[39] Dipolar recoupling in TRAPDOR and REAPDOR methods is brought about by a continuous irradiation of the S spins, which induces a so-called adiabatic passage. This adiabatic passage pulse is usually applied for half a rotor period in the TRAPDOR experiments and for a fraction of the rotor cycle in the REAPDOR measurements. The determination of the dipolar coupling from the decay curves requires not only extensive numerical simulations but also the input of several parameters characterizing the spin system, such as the quadrupole coupling constant, its asymmetry parameter, and the relative orientations of the quadrupolar and dipolar tensors. Moreover, a very careful experimental setup and accurate calibration of the pulses are required. REAPDOR has the advantage that its dependence on these parameters is rather weak compared to TRAPDOR.[44] Hence, the dephasing curve obtained for REAPDOR should not differ significantly from the standard dephasing curves, thus allowing the determination of internuclear distances without extensive numerical simulations. All these considerations were demonstrated for half-integer spins with large quadrupolar coupling constants, but they should also apply to ^{14}N. On the contrary, DEAR and RIDER experiments do not require application of specific RF pulses on the *S* spins because they exploit T_1 relaxation of the quadrupolar nuclei for dipolar recoupling. They can therefore be performed on a double-resonance probehead, whereas TRAPDOR and REAPDOR experiments require triple-resonance equipment. Even though DEAR and RIDER make use of the same T_1 relaxation mechanism, the two experiments are quite different because DEAR is a static method, which does not require explicit recoupling of the dipolar coupling as is used in RIDER. The RIDER method uses a π-pulse train to reintroduce the heteronuclear dipolar interaction that is effectively cancelled out by spinning of the sample at the magic angle. Exchange comes about because T_1 relaxation affects the antiphase coherences associated with the heteronucleus. Two conditions are essential for the success of the RIDER experiment: (i) T_1 relaxation time of the passive spin (e.g., ^{14}N) is of the order of 1 s or less, which is generally the case for most quadrupolar nuclei; (ii) slow dynamics or spin diffusion should not occur on the same timescale.

The RIDER method has been demonstrated on a natural abundance L-alanine sample.[39] The exchange due to the ^{14}N T_1 relaxation is very small at short mixing times. Hence, long mixing times are required for observing ^{13}C signals not directly bonded to ^{14}N (e.g., carbonyl and methyl carbons). However, owing to the efficient relaxation behavior of methyl protons in alanine, the condition (i) mentioned previously is not fulfilled, and methyl carbon is completely relaxed at mixing times as long as 1 s. However, internuclear C–N distances can be determined for the C^{α} and C' signals by fitting the normalized dephasing as a function of the recoupling time. The beginning of the dephasing curve fits very well a theoretical model, which generally assumes an isolated spin pair approximation.

The HMQC-based methods devised for indirect detection of ^{14}N can efficiently be exploited for measuring multiple ^{13}C–^{14}N distances in natural abundance solids.[29] The experiment dubbed rotary resonance echo saturation pulse double resonance (R-RESPDOR) is a modified version of REAPDOR where the pulse train of π pulses on carbons was replaced by rotary resonance recoupling and the ^{14}N adiabatic passage pulses by a single saturation pulse. As mentioned previously, an adiabatic passage ^{14}N pulse entails a large adiabaticity parameter for efficient dipolar dephasing and data interpretation. Large ^{14}N quadrupolar couplings under fast MAS would require extremely high RF fields for the ^{14}N to fulfill the adiabatic condition. ^{14}N saturation has the advantage that it can be easily implemented with long pulses, and it ensures efficient dipolar dephasing. Moreover, ^{14}N saturation levels out the orientation and other dependencies hence allowing simple data interpretation with a universal function.

For reintroducing the ^{13}C–^{14}N dipolar interaction, the $n = 1$ or 2 R^3 ($\omega_1 = \omega_R$ or $\omega_1 = 2\,\omega_R$) conditions can be used. Since the ^{13}C–^{13}C dipolar interactions are not dominant, the $n = 1$ R^3 condition is preferred here because it has a more efficient recoupling effect. In addition to the homonuclear dipolar interactions, the chemical shift anisotropy (CSA) of the carbons is also reintroduced during rotary resonance recoupling irradiation at $\omega_1 = \omega_R$. On the contrary, isotropic chemical shift, scalar and residual dipolar splittings are averaged by the RF irradiation. Large CSA makes the rotary resonance condition wider and less prone to RF field inhomogeneity. Even at the highest available spinning speed (e.g., 70 kHz), rotary resonance uses weak ^{13}C RF hence avoiding interference problems with proton decoupling, which is particularly problematic for protonated carbon sites. As in the ^{14}N–^{13}C HMQC experiment, the CSA is refocused by the π pulse in the middle of the pulse sequence thus forming a rotary resonance echo. The saturation pulse changes the ^{14}N spin state hence overcoming the refocusing of the $^{13}C-^{14}N$ dipolar interaction. In the manner of REDOR and REAPDOR methods, the dipolar dephasing can be measured by two experiments: one with (S) and the other without (S_0), the ^{14}N pulse saturation. Then, the normalized rotary resonance echo modulations ($\Delta S/S_0 = (S_0 - S)/S_0$) are compared with the simulation in order to determine the dipolar coupling constant.

The R-RESPDOR method has been applied to the measurement of eight $^{13}C-^{14}N$ distances in natural abundance L-tryptophan with a single rotary resonance echo experiment. One-bond $^{13}C-^{14}N$ dipolar oscillations are observed for the two carbons next to indole (a and h in Figure 16.10). A reasonably good agreement is obtained between the simulations and the $\Delta S/S_0$ results for all the aromatic sites. The dipolar couplings used in the simulations were obtained from a $^{13}C-^{15}N$ REDOR experiment of a ^{15}N-labeled L-tryptophan sample. These dipolar couplings can be used to derive $^{13}C-^{14}N$ internuclear distances for one and two bond pairs. The faster dephasing curves obtained for three bonds and beyond $^{13}C-^{14}N$ pairs suggest that intermolecular effects should be taken into account in the determination of the internuclear distances. The measurement of the corresponding C–N internuclear distances for ^{13}C, $^{13}C^{\alpha}$, and $^{13}C^{\beta}$ would necessitate different experiments because the R^3 scheme is very sensitive to chemical shift offsets.

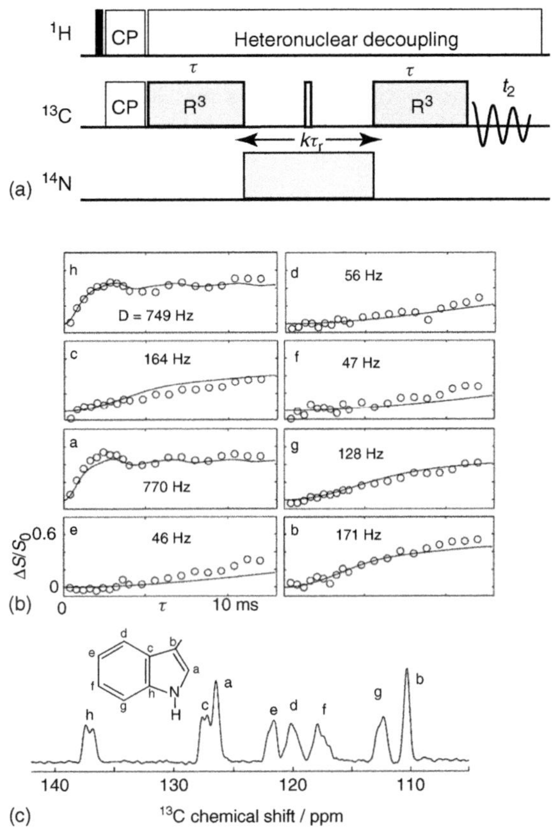

Figure 16.10. Pulse sequence of the $^{13}C-^{14}N$ R-RESPDOR experiment (a), $\Delta S/S_0$ dephasing curves (open circles) with their corresponding simulations (solid lines) using the dipolar coupling frequencies listed (b), and ^{13}C CPMAS spectrum (c) showing the aromatic part and the peak assignment of natural abundant L-tryptophan on a 14.1 T (600 MHz ^1H Larmor frequency) widebore magnet and at 10 kHz spinning frequency. The ^{13}C RF phase was kept constant throughout the pulse sequence and the carrier frequency was set in the middle of the aromatic region (e.g., 124 ppm for L-tryptophan). The structure of the ^{13}C resonances in the ^{13}C CPMAS spectrum indicates the presence of crystallographically nonequivalent molecules in the unit cell. The following parameters were used for recording the experiment: 512 scans, 4 s recycle delay, $\tau_p(^{14}N) = 150$ μs with $\nu_1(^{14}N) = 40$ kHz, $k = 1$, and SPINAL ^1H decoupling with $\nu_1(^1H) = 90$ kHz. (Z. Gan *Chem. Commun.* 2006, 4712. Reproduced by permission of the Royal Society of Chemistry.)

16.5.3 ^{14}N as a Probe of Dynamics in Solids

Molecular dynamic processes can significantly broaden resonances in NMR spectra when the rate of the process is comparable to characteristic frequencies in the NMR experiment. Hence, the interference between motional processes and coherent spin motion induced by decoupling or recoupling multiple pulse sequences can considerably affect the expected efficiency of the applied decoupling or recoupling scheme.[45] Solid-state NMR can probe dynamic processes occurring on the very fast (nanosecond) timescale but also very slow (millisecond to second) timescale. The magnitude of the quadrupole interaction ranging from 0.1 to 4 MHz renders quadrupolar nuclei such as ^{14}N particularly suitable probes to monitor dynamics in solids spun at the magic angle. The indirect detection method of ^{14}N via spy nuclei in HMQC manner has made it possible to extend to ^{14}N, the dynamics method originally demonstrated on ^2H.[46] Hence, the molecular dynamics in a polycrystalline sample of the tripeptide Ala-Ala-Gly (AAG) was probed by studying ^{1}H-detected SQ and DQ ^{14}N MAS spectra.[47] If a dynamic reorientation of the ^{14}N quadrupole tensor occurs, then the SQ ^{14}N spectra will show strong motional broadening, whereas the motional effect is expected to be either absent or much reduced on the corresponding DQ ^{14}N MAS spectra. Three ^{1}H–^{14}N correlations should be

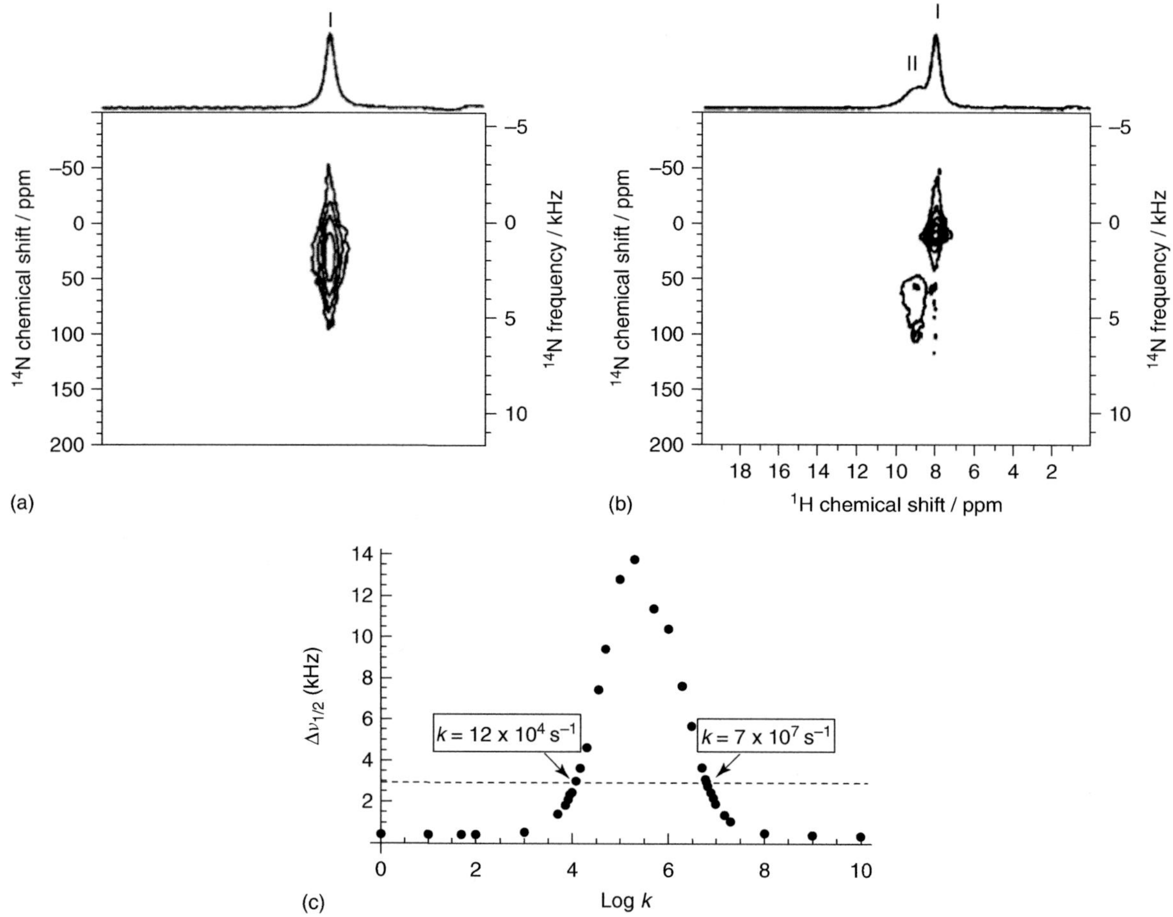

Figure 16.11. Experimental SQ (a) and DQ (b) ^{14}N–^{1}H HMQC correlations of the group (I) and the two rigid amide NH that contribute to the resonance II in the tripeptide AAG on an 18.8 T (800 MHz ^{1}H Larmor frequency) standard bore magnet and at 29.76 kHz spinning frequency. The sample temperature was determined experimentally to be 49 °C using $Sm_2Sn_2O_7$ sample as a chemical shift thermometer. The following parameters were used for recording the two experiments: 256 (SQ) and 512 (DQ) transients for each of 100 Δt_1 increments with $\Delta t_1 = 1/\nu_r = 33.6$ μs, $\tau_p(^{14}N) = 20$ μs (SQ) and 26 (DQ) with $\nu_1(^{14}N) = 80$ kHz. The vertical axis was calibrated using NH_4Cl as a reference with the ^{14}N line set to 0 ppm. (c) Simulations of the full linewidth at half height ($\Delta_{1/2}$) of SQ ^{14}N spectra as a function of the exchange rate k. A two-site jump of ±5° about the y-axis of the quadrupole tensor, $\nu_Q = 1.13$ MHz, $\eta = 0.28$, and 28 656 different crystallite orientations were used for the simulations, and the excitation of the crystallites was assumed to be uniform. A similar curve is obtained for 120° hops around the z-axis of the quadrupole tensor, which is assumed to be parallel to the C–N bond. The dashed horizontal line corresponds to the full linewidth at half height (~3 kHz) of the experimental SQ spectrum after subtracting the contribution coming from the proton–proton dipolar interactions ($\Delta_{1/2}$ ~ 200 Hz) determined from proton spin-echo experiments. (Reprinted with permission from S. Cavadini, A. Abraham, S. Ulzega, G. Bodenhausen *J. Amer. Chem. Soc.* 2008, **130**, 10850. Copyright 2008 American Chemical Society.)

observed for this sample, but only one SQ ^{14}N signal and two DQ peaks appear in the HMQC experiment. The SQ ^{14}N responses of the two peptide NH groups, which have much larger quadrupole couplings than the $-NH_3^+$ group, cannot be observed most likely because of inefficient SQ excitation and/or dynamic effects. On the contrary, since the DQ HMQC experiment is only to a small extent affected by motional broadening and since excitation is less critical,[21] two peaks are visible in the DQ spectrum, one of which should correspond to the overlay of peptide ^{14}N. A narrow linewidth is expected for the $-NH_3^+$ group owing to the threefold rapid hopping of the protons, whereas the resonances of the NH pairs in AAG should not be affected by motional averaging and hence show rather broad resonances. The ^{14}N SQ spectrum of the $-NH_3^+$ group in AAG peptide is broader than the corresponding ^{14}N DQ line suggesting incomplete averaging of the ^{14}N quadrupolar coupling by the threefold jumps of the protons. This can be taken as unambiguous evidence of the presence of dynamics. In single amino acids, narrow SQ ^{14}N resonances were observed[48] for the $-NH_3^+$ group, thus confirming partial averaging of the ^{14}N quadrupole interaction by fast hopping of the protons around the axis defined by the N–C internuclear distance. Other effects in the environment of the ^{14}N might affect the orientation and/or principal values of the quadrupole tensor. For example, the involvement of one of the protons in hydrogen bonding would slow down the threefold hopping process.

In order to extract the reorientational rate constant, theoretical approaches[46,49–51] have to be used to evaluate the effects of dynamics on linewidths in NMR MAS spectra. In the case of ^{14}N, the reorientational motion was modeled by a symmetrical two-site exchange in the manner of McConnell,[52] and the spin dynamics under MAS was described by the time-dependent Liouville equation. A numerical development is feasible if each rotor period is divided into equally spaced intervals, and the propagator is calculated for each interval. The full MAS linewidths at half height can thus be calculated as a function of the exchange rate constant k. Figure 16.11 shows that the single-quantum MAS ^{14}N linewidth fits well the numerical simulations for two possible k values. Variable temperature experiments were carried out to discriminate between these two possible regimes. Since the linewidth decreases with increasing temperature, the lower value can be excluded. Hence, the reorientational motions occur at a rate constant of $k \sim 10^7\ s^{-1}$.

16.6 SUMMARY

Research advances in ^{14}N solid-state NMR methodology are very promising and should lead to further innovative developments and applications of MAS NMR spectroscopy to the study of ^{14}N in biological molecules.

^{14}N nuclei can induce an asymmetric splitting to the closely bonded spin $S = 1/2$ neighbors. Indirect detection in HMQC manner of ^{14}N nuclei via "spy" $S = 1/2$ nuclei such as ^1H or ^{13}C leads to high-resolution and rather sensitive ^{14}N spectra under MAS conditions. Applied to amino acids and small peptides, these methods give access to parameters that provide valuable information about the structure and dynamics of nitrogen-containing solids.

RELATED ARTICLES IN THE ENCYCLOPEDIA OF MAGNETIC RESONANCE

Dipolar Recoupling: Heteronuclear

Magic Angle Spinning Carbon-13 Lineshapes: Effect of Nitrogen-14

Magic Angle Spinning: Effects of Quadrupolar Nuclei on Spin-1/2 Spectra

Proton Chemical Shift Measurements in Biological Solids

Quadrupolar Nuclei in Liquid Samples

REDOR and TEDOR

REFERENCES

1. W. G. Proctor and F. C. Yu, *Phys. Rev.*, 1950, **77**, 717.
2. R. Blinc, M. Mali, R. Osredkar, A. Prelesnik, J. Seliger, I. Zupancic, and L. Ehrenberg, *J. Chem. Phys.*, 1972, **57**, 5087.
3. R. E. Stark, R. A. Haberkorn, and R. G. Griffin, *J. Chem. Phys.*, 1978, **68**, 1996.
4. F. Lindström, P. T. F. Williamson, and G. Gröbner, *J. Am. Chem. Soc.*, 2005, **127**, 6610.

5. A. Ramamoorthy, D. K. Lee, J. S. Santos, and K. A. Henzler-Wildman, *J. Am. Chem. Soc.*, 2008, **130**, 11023.
6. S. Vega, T. W. Shattuck, and A. Pines, *Phys. Rev. Lett.*, 1976, **37**, 43.
7. S. Cavadini, *Prog. NMR Spectrosc.*, 2010, **56**, 46.
8. A. C. Olivieri, L. Frydman, and L. E. Diaz, *J. Magn. Reson.*, 1987, **75**, 50.
9. A. Samoson, *Chem. Phys. Lett.*, 1985, **119**, 29.
10. A. Naito, S. Ganapathy, and C. A. McDowell, *J. Magn. Reson.*, 1982, **48**, 367.
11. R. K. Harris and A. C. Olivieri, *Prog. NMR Spectrosc.*, 1992, **24**, 435.
12. A. Naito, A. Root, and C. A. McDowell, *J. Phys. Chem.*, 1991, **95**, 3578.
13. A. E. McDermott, F. J. Creuzet, A. C. Kolbert, and R. G. Griffin, *J. Magn. Reson.*, 1992, **98**, 408.
14. C. P. Grey and W. S. Veeman, *Chem. Phys. Lett.*, 1992, **192**, 379.
15. C. P. Grey, A. P. A. M. Eijkelenboom, and W. S. Veeman, *Solid State Nucl. Magn. Reson.*, 1995, **4**, 113.
16. M. Reinhold, P. Brunner, and R. R. Ernst, *J. Chem. Phys.*, 1981, **74**, 184.
17. T. Gullion, *Chem. Phys. Lett.*, 1995, **246**, 325.
18. S. Cavadini, A. Lupulescu, S. Antonijevic, and G. Bodenhausen, *J. Am. Chem. Soc.*, 2006, **128**, 7706.
19. Z. Gan, *J. Am. Chem. Soc.*, 2006, **128**, 6040.
20. V. F. Bystrov, *Prog. NMR Spectrosc.*, 1976, **10**, 41.
21. S. Cavadini, S. Antonijevic, A. Lupulescu, and G. Bodenhausen, *Chem. Phys. Chem.*, 2007, **8**, 1363.
22. P. L. Stewart, K. G. Valentine, and S. J. Opella, *J. Magn. Reson.*, 1987, **71**, 45.
23. S. Antonijevic and N. Halpern-Manners, *Solid State Nucl. Magn. Reson.*, 2008, **33**, 82.
24. S. Cavadini, A. Abraham, and G. Bodenhausen, *J. Magn. Reson.*, 2008, **190**, 160.
25. T. Gullion and J. Schaefer, *J. Magn. Reson.*, 1989, **81**, 196.
26. T. G. Oas, R. G. Griffin, and M. H. Levitt, *J. Chem. Phys.*, 1988, **89**, 692.
27. R. Fu, S. A. Smith, and G. Bodenhausen, *Chem. Phys. Lett.*, 1997, **272**, 361.
28. M. Carravetta, M. Edén, X. Zhao, A. Brinkmann, and M. H. Levitt, *Chem. Phys. Lett.*, 2000, **321**, 205.
29. Z. Gan *Chem. Commun.* 2006, 4712.
30. Z. Gan, *J. Magn. Reson.*, 2007, **184**, 39.
31. Z. Gan, J. P. Amoureux, and J. Trébosc, *Chem. Phys. Lett.*, 2007, **435**, 163.
32. S. Cavadini, A. Abraham, and G. Bodenhausen, *Chem. Phys. Lett.*, 2007, **445**, 1.
33. S. Cavadini, V. Vitzthum, S. Ulzega, A. Abraham, and G. Bodenhausen, *J. Magn. Reson.*, 2010, **202**, 57.
34. J. P. Amoureux, J. Trébosc, B. Hu, N. Halpern-Manners, and S. Antonijevic, *J. Magn. Reson.*, 2008, **194**, 317.
35. J. P. Amoureux, Q. Wang, B. Hu, O. Lafon, J. Trébosc, and F. Deng *Chem. Commun.*, 2008, 6525.
36. R. Siegel, J. Trébosc, J. P. Amoureux, and Z. Gan, *J. Magn. Reson.*, 2008, **193**, 321.
37. C. P. Grey, W. S. Veeman, and A. J. Vega, *J. Chem. Phys.*, 1993, **98**, 7711.
38. K. Schmidt-Rohr and J.-D. Mao, *Chem. Phys. Lett.*, 2002, **359**, 403.
39. K. Saalwächter and K. Schmidt-Rohr, *J. Magn. Reson.*, 2000, **145**, 161.
40. Z. Gan, *Chem. Commun.*, 2008, 868.
41. J. G. Hexem, M. H. Frey, and S. Opella, *J. Am. Chem. Soc.*, 1983, **105**, 5717.
42. E. R. H. van Eck, R. Janssen, W. Maas, and W. S. Veeman, *Chem. Phys. Lett.*, 1990, **174**, 428.
43. J. R. Sachleben, V. Frydman, and L. Frydman, *J. Am. Chem. Soc.*, 1996, **118**, 9786.
44. M. Kalwei and H. Koller, *Solid State Nucl. Magn. Reson.*, 2002, **21**, 145.
45. J. R. Long, B. Q. Sun, A. Bowen, and R. G. Griffin, *J. Am. Chem. Soc.*, 1994, **116**, 11950.
46. M. Cutajar, S. E. Ashbrook, and S. Wimperis, *Chem. Phys. Lett.*, 2006, **423**, 276.
47. S. Cavadini, A. Abraham, S. Ulzega, and G. Bodenhausen, *J. Am. Chem. Soc.*, 2008, **130**, 10850.
48. S. Cavadini, S. Antonijevic, A. Lupulescu, and G. Bodenhausen, *J. Magn. Reson.*, 2006, **182**, 168.
49. M. M. Maricq and J. S. Waugh, *J. Chem. Phys.*, 1979, **70**, 3300.
50. S. E. Ashbrook, S. Antonijevic, A. J. Berry, and S. Wimperis, *Chem. Phys. Lett.*, 2002, **364**, 634.
51. M. J. Thrippleton, M. Cutajar, and S. Wimperis, *Chem. Phys. Lett.*, 2008, **452**, 233.
52. H. M. McConnell, *J. Chem. Phys.*, 1958, **28**, 430.

Chapter 17

Oxygen-17 NMR Studies of Organic and Biological Molecules

Gang Wu

Department of Chemistry, Queen's University, Kingston, ON K7L 3N6, Canada

17.1 INTRODUCTION

Oxygen is a key constituent of many important functional groups in organic and biological molecules. Oxygen-17 ($I = 5/2$, $Q = -2.6 \times 10^{-30}$ m^2, natural abundance = 0.037%) is the only naturally occurring, stable oxygen isotope with a nonzero nuclear spin, which makes it suitable for nuclear magnetic resonance (NMR) spectroscopic studies. Despite the enormous importance of oxygen-containing functional groups in chemistry and biology, far fewer ^{17}O NMR studies have been reported in the literature than those of ^{1}H, ^{13}C, and ^{15}N. The primary obstacles in ^{17}O NMR spectroscopy are the low natural abundance of ^{17}O and the large nuclear quadrupolar interactions often associated with oxygen-containing organic functional groups. These make ^{17}O an extremely insensitive NMR nucleus (approximately 10^5 times less sensitive than ^{1}H). In addition, even when the level of ^{17}O can be enriched substantially, the large quadrupolar interactions of ^{17}O often result in rapid nuclear relaxation and thus significant broadening of the ^{17}O NMR signals. For these reasons, it is not surprising that, for many years, ^{17}O NMR applications in organic chemistry were limited almost exclusively to simple molecules.[1]

One remedy for the intrinsic limitation of conventional ^{17}O NMR methods of studying liquid samples is to obtain NMR spectra for solid materials. The general field of solid-state NMR spectroscopy is presently undergoing rapid expansion as a result of the significant advances in both NMR methodology and instrumentation that have accumulated over the past 2 decades. So why is it useful, or even desirable at all, to study ^{17}O NMR spectra of organic and biological solids? What kind of new information can one learn from solid-state ^{17}O NMR studies that cannot be obtained using a more traditional solution NMR approach? The answers to these questions may be summarized as follows. First, greatly restricted molecular motion in solids allows the nuclear Zeeman energy levels to have much longer life times than those in liquid samples. This means that the intrinsic spectral resolution is high in solid-state ^{17}O NMR spectra. Second, solid-state ^{17}O NMR allows direct measurement of the tensor property of

NMR of Quadrupolar Nuclei in Solid Materials
Edited by Roderick E. Wasylishen, Sharon E. Ashbrook and Stephen Wimperis
 ISBN: 978-0-470-97398-1

a particular NMR parameter such as chemical shift (CS) or quadrupole coupling (QC). A complete ^{17}O NMR tensor contains more information about molecular structure and chemical bonding than its averaged value alone. Third, the resolution of solid-state ^{17}O NMR spectra is not limited by the molecular weight of the system under observation. Fourth, some biological systems such as membrane proteins (e.g., ion channels) and insoluble proteins (e.g., amyloid fibrils) cannot be readily studied in their native states by conventional solution NMR methods.

The subject of solid-state ^{17}O NMR of organic molecules was briefly reviewed in 1998 and in 2004.[2,3] A comprehensive review of the subject was published in 2008.[4] A recent review of general ^{17}O NMR spectroscopy also covers some aspects of solid-state ^{17}O NMR.[5] In this chapter, we shall provide an overview of the progress made in recent years in solid-state ^{17}O NMR studies of organic and biological molecules and highlight particularly the latest developments in studying new organic functional groups and large biological molecules.

17.2 SOLID-STATE ^{17}O NMR TECHNIQUES

The major problem associated with solid-state NMR studies of most quadrupolar nuclei, including ^{17}O, stems from the fact that nuclear quadrupolar interactions are often several orders of magnitude larger than other common nuclear spin interactions such as dipolar or chemical shift anisotropy (CSA) interactions. On one hand, because each of these nuclear spin interactions depends on the microscopic orientation of the molecule under observation with respect to the applied magnetic field (i.e., a tensorial property), the resulting NMR spectrum from a polycrystalline sample generally displays broad resonances. On the other hand, these broad resonances in solid-state NMR spectra contain direct information about the NMR tensors involved. Therefore, the dilemma is that, although solid-state NMR spectra in principle contain a wealth of information and have high *intrinsic* resolution, it is a real challenge for experimentalists to invent NMR techniques that can produce high-resolution spectra. In this section, we shall briefly outline the basic NMR techniques that have been used to date in solid-state ^{17}O NMR studies of organic and biological molecules. We chose to describe these techniques in a chronological manner, so that the reader may also benefit from having a historical perspective about the development of the field. For detailed descriptions of these NMR techniques, the reader may consult the related chapters in this book.

While the first ^{17}O NMR signal was detected in 1951 by Alder and Yu for liquid water,[6] the first solid-state ^{17}O NMR study was reported only much later by Rabideau and Waldstein.[7] These latter authors recorded ^{17}O NMR spectra for polycrystalline H_2O ice (enriched with ^{17}O to 22%) using a continuous-wave (CW) NMR method. Reliable ^{17}O CS and QC tensors were determined for ice in subsequent single-crystal NMR studies.[8,9] Single-crystal ^{17}O NMR has also been used by Haeberlen and colleagues[10] and, more recently, by Wittebort and colleagues[11–13] in studying organic molecules.

In 1972, a technique closely related to NMR, known as ^{17}O double-resonance nuclear quadrupole resonance (DR-NQR), was developed by Hahn and colleagues and applied to organic molecules.[14] Quite remarkably, this technique allows detection of ^{17}O NQR signals from organic solids at natural abundance. Throughout the 1970s and 1980s, a large number of organic compounds were examined by ^{17}O NQR spectroscopy. These studies firmly establish the usefulness of ^{17}O quadrupole parameters as a sensitive probe of chemical bonding and molecular structure. A list of ^{17}O NQR studies for organic molecules can be found in the literature.[4] However, because ^{17}O DR-NQR experiments require specialized instrumentation and large amounts of samples, the technique has not been widely used for chemical applications and the interest has gradually faded since the 1980s.

In 1983, Oldfield and colleagues[15,16] demonstrated for the first time that high-resolution ^{17}O NMR spectra can be obtained for solid materials with magic-angle spinning (MAS) and variable-angle spinning (VAS) techniques. This seminal study was focused on inorganic solids. Nearly a decade later, Ando and colleagues applied ^{17}O MAS to organic solids.[17] The main limitation of ^{17}O MAS NMR is that, in many cases, the second-order quadrupole broadening is still substantial. At the end of the 1980s, Fiat and colleagues recorded the first set of solid-state ^{17}O NMR spectra for stationary amino acid samples.[18,19]

Two major breakthroughs occurred in the late 1980s in the general field of solid-state NMR spectroscopy of half-integer quadrupolar nuclei. That was the development of dynamic-angle spinning (DAS)[20,21] and double rotation (DOR)[22–24] techniques. Unlike MAS, these two techniques can completely average the second-order quadrupole interaction, making it possible to obtain truly high-resolution NMR spectra for half-integer quadrupolar nuclei. However, because early DOR probes were not equipped with ^{1}H decoupling capability and had problems with long-term spinning stability, ^{17}O DOR NMR was not used to study organic solids until very recently with the development of a new generation of DOR probes.[25] Similarly, ^{17}O DAS NMR was also not employed to study organic solids, except for a brief report by Gann *et al.*,[26] in which ^{17}O DAS spectra for polycrystalline L-alanine were reported. The authors observed two ^{17}O NMR signals for the carboxyl group of L-alanine, suggesting that the two oxygen atoms are crystallographically distinct. The authors attributed these different ^{17}O NMR parameters to the different intermolecular hydrogen bonding environments in which the two carboxyl oxygen atoms exist in the crystal lattice. These parameters represent the first set of reliable solid-state ^{17}O NMR data that clearly demonstrate the sensitivity of ^{17}O NMR parameters toward hydrogen bonding.

Another breakthrough came about in 1995 when Frydman and colleagues[27,28] introduced the multiple-quantum magic-angle spinning (MQMAS) methodology. Similar to DOR and DAS, this technique can also completely remove the second-order quadrupole interaction. The true advantage of MQMAS over DOR and DAS is that it can be readily implemented with a commercial MAS probe. Applications of MQMAS in ^{17}O NMR studies were first reported for inorganic solids[29,30] and later for organic molecules.[31] In 2000, Gan introduced another technique, termed satellite transition magic-angle spinning (STMAS),[32] which utilizes correlation between satellite and central transitions (CTs) to produce a high-resolution spectrum free of second-order quadrupole broadening (see Chapter 11). However, although ^{17}O STMAS has been applied to inorganic solids,[33] it has not yet been used to date for studying organic solids.

One important application of solid-state NMR is to obtain structural information by directly determining the dipolar coupling constant between two nuclear spins. In 1998, Gullion and colleagues[34] demonstrated the potential of measuring ^{13}C–^{17}O internuclear distances in organic solids. They designed ^{13}C-detected rotational echo double resonance (REDOR) and rotational echo adiabatic passage double resonance (REAPDOR) experiments to reintroduce the ^{13}C–^{17}O dipolar coupling into ^{13}C MAS NMR spectra. This was the first attempt to measure internuclear distances involving ^{17}O. Several years later, they presented a general solid-state NMR approach for internuclear distance measurement between a spin-1/2 and a spin-5/2 using a universal REAPDOR curve.[35] A detailed description of REDOR and REAPDOR techniques can be found in the literature.[36] This approach was recently employed to determine secondary structures of polyamides[37] and to probe glycine incorporation into proteins.[38] Once high spectral resolution is achieved, a variety of correlation spectroscopy experiments can be implemented for establishment of spatial relationships between nuclear spins. For example, Brown and colleagues recently demonstrated heteronuclear correlation ^{17}O NMR spectroscopy for organic solids using both dipole- and J-based methods.[39]

Although REDOR and REAPDOR techniques are useful in distance measurements for spin pairs containing a ^{17}O nucleus and a dilute spin-1/2 nucleus, they cannot be directly used for ^{1}H–^{17}O distance measurement. This is because homonuclear ^{1}H–^{1}H dipolar couplings are usually very strong in organic solids and cannot be effectively removed by MAS alone. Levitt and colleagues[40] showed that a properly designed symmetry-based pulse sequence can simultaneously recouple a ^{1}H–^{17}O heteronuclear dipolar coupling and decouple homonuclear ^{1}H–^{1}H dipolar couplings under the MAS condition. Brinkmann and Kentgens[41,42] introduced two new supercycled symmetry-based pulse sequences to achieve heteronuclear longitudinal two-spin-order $< I_z S_z >$ recoupling. The advantage of their pulse sequences is that the results do not suffer from dipolar truncation and are insensitive to ^{1}H chemical shift anisotropies. Brinkmann and Kentgens[42] further demonstrated a selective version of the ^{1}H–^{17}O distance measurement experiment where a medium ^{1}H–^{17}O distance can be measured in the presence of a short ^{1}H–^{17}O distance. The ability of solid-state

NMR to yield direct information about $^1H-^{17}O$ hydrogen bond length has tremendous potential to be useful in the study of biological systems. One can imagine that if two $^1H-^{17}O$ distances are measured simultaneously in the same molecule, the position of the hydrogen atom can then be determined.

Another critical issue associated with but not limited to solid-state ^{17}O NMR studies of organic and biological molecules is the overall sensitivity of NMR experiments. This is because the very low level of ^{17}O nuclei is often the limiting factor for studying biological macromolecules. The solutions to this general problem currently available are somewhat different for quadrupolar nuclei from those commonly used for spin-1/2 nuclei. For this reason, we shall discuss them in more detail. Here, we divide the experimental conditions into two distinct categories: spinning and stationary. For solid samples undergoing MAS, as the efficiency of cross polarization (CP) from abundant ^1H spins to half-integer quadrupolar nuclei including ^{17}O is relatively low, one usually relies on direct excitation of quadrupolar nuclei. Several methods can be used to partially enhance the signal intensity of the CT through the principle of population transfer. Grandinetti and colleagues[43–45] introduced the rotor-assisted population transfer (RAPT) experiment where population transfer can be achieved by either a train of pulses with alternating phases or frequency-switched Gaussian pulses. Kentgens and colleagues[46,47] developed a double-frequency sweep (DFS) method to enhance the CT signal. Bräuniger and colleagues[48,49] used a similar technique, frequency-swept fast amplitude modulation (SW-FAM) pulses. Wasylishen and colleagues[50–52] demonstrated that population inversion of satellite transitions can be obtained using hyperbolic secant (HS) pulses. If RAPT or DFS is applied many times before waiting for re-equilibration of the spin system, further signal enhancement can be achieved.[41,53] For example, Brinkmann and Kentgens[41] reported an enhancement factor of 4.3 by using 10 DFS transfers to record ^{17}O MAS spectra of $^{17}O^{\eta}$-L-tyrosine·HCl, which is compared with a factor of only 2.8 with a single DFS transfer. Dupree and coworkers also used the RAPT technique to improve the signal-to-noise ratio in ^{17}O DOR spectra.[25] Detailed discussions of sensitivity enhancement techniques for quadrupolar nuclei can be found in the literature.[54]

17.3 CHARACTERIZATION OF ^{17}O NMR TENSORS IN ORGANIC FUNCTIONAL GROUPS

In general, both ^{17}O QC and CS tensors can be determined from solid-state NMR experiments. Historically, for many organic functional groups, ^{17}O QC tensors were first determined from ^{17}O NQR studies, most of which were done in the 1970s and 1980s.[4] In contrast, little was known of the ^{17}O CS tensors in organic functional groups before 2000. Since then, the body of literature on solid-state ^{17}O NMR studies of organic molecules has increased considerably and now both ^{17}O QC and CS tensors are known for many key organic functional groups.[4] These studies have firmly established a correlation between ^{17}O NMR tensors and molecular structure and chemical bonding. Figure 17.1 summarizes common organic functional groups for which solid-state ^{17}O NMR data are available in the literature. As a complete compilation of solid-state ^{17}O NMR data published prior to 2007 for organic functional groups can be found in the literature,[4] we shall focus our discussion in this section only on selected reports that have appeared since 2008.

In 2008, Asakura and colleagues used solid-state ^{17}O NMR to probe the parallel and antiparallel β-sheet structures of an alanine tripeptide, Ala-[^{17}O]Ala-Ala.[55] As seen from Figure 17.2, the parallel and antiparallel structures formed by Ala-Ala-Ala have quite different hydrogen bonding environments at the targeted carbonyl oxygen atom (the central Ala residue). Within each structure, there are also two crystallographically distinct Ala-Ala-Ala molecules, A and B. Figure 17.3 shows the experimental and simulated ^{17}O MAS spectra for Ala-[^{17}O]Ala-Ala. The observed excellent spectral quality can be attributed to the combination of a high magnetic field (21.8 T) used to acquire NMR data and a high level of ^{17}O enrichment (90%) used in sample synthesis. For the antiparallel Ala-[^{17}O]Ala-Ala sample, the authors reported the following ^{17}O NMR parameters: molecule A, $\delta_{iso} = 302$ ppm, $C_Q = e^2q_{zz}Q/h = 8.7$ MHz, $\eta_Q = (q_{xx} - q_{yy})/q_{zz} = 0.40$; molecule B, $\delta_{iso} = 270$ ppm, $C_Q = 8.7$ MHz, $\eta_Q = 0.35$. Here q_{xx}, q_{yy}, and q_{zz} are the principal components of the electric field gradient (EFG) tensor defined according to $|q_{zz}| \geq |q_{yy}| \geq |q_{xx}|$. For the parallel Ala-[$^{17}O$]Ala-Ala sample, the authors reported the same set of ^{17}O NMR parameters for molecules

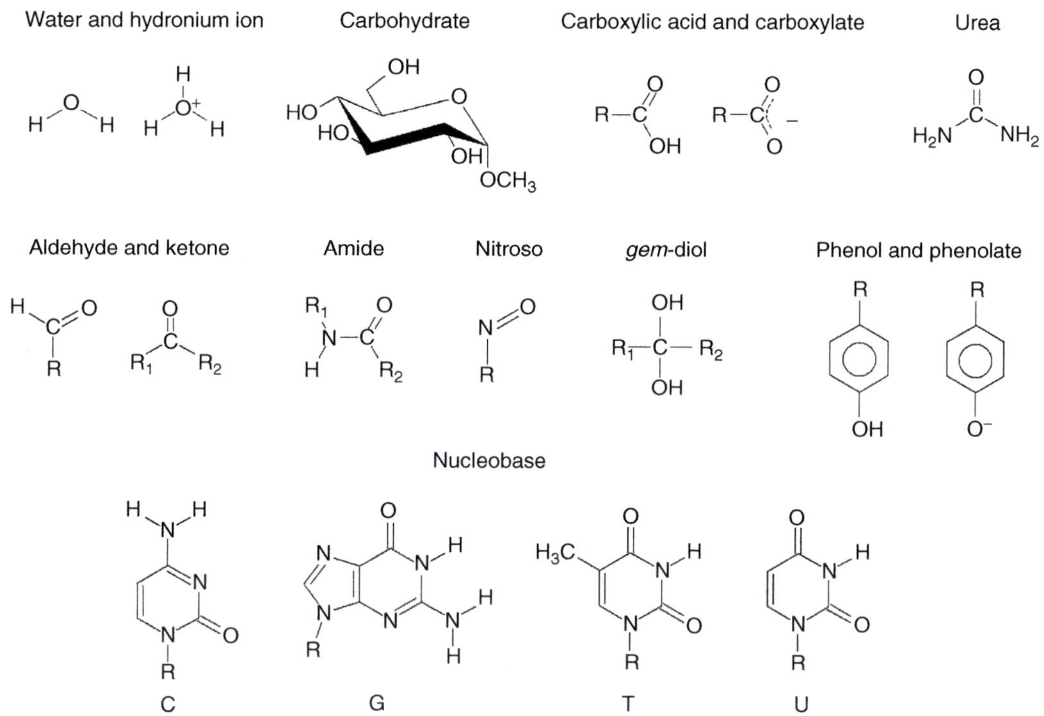

Figure 17.1. Common organic functional groups for which ^{17}O NMR tensors are available in the literature.

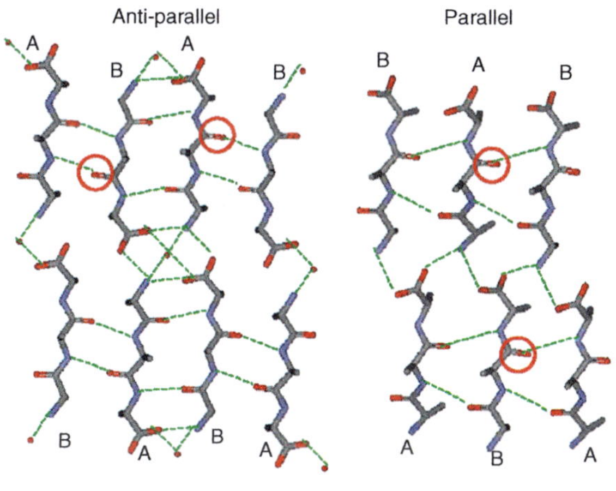

Figure 17.2. Crystal packing in antiparallel β-sheet and parallel β-sheet structures formed by Ala-[^{17}O]Ala-Ala. The circled oxygen atoms are labeled with ^{17}O. (Reproduced from Ref. 55. © Elsevier, 2008.)

A and B: $\delta_{iso} = 293$ ppm, $C_Q = 8.7$ MHz, and $\eta_Q = 0.30$. Using density functional theory (DFT) calculations, the authors attributed the observed differences in ^{17}O CSs between the two structures as well as the difference between the two molecules within each structure to the angle between the C=O and N–H bond vectors. Although one should, in general, be cautious in attempting to link experimental ^{17}O NMR parameters to just one particular aspect of hydrogen bonding, the explanation offered by the authors seems to be reasonable within these well-defined β-sheet structures of Ala-Ala-Ala. It is also interesting to note that the experimental $C_Q(^{17}O)$ values display very little variations among different structures and molecules.

Also in 2008, Wu and colleagues[56] determined the ^{17}O NMR tensors in p-nitro-[1-^{17}O]benzaldehyde. Aldehyde groups (H–C=O) are known to have large values of $C_Q(^{17}O)$ and CSA.[4] Wu and colleagues observed that, at 11.75 and 21.14 T, the ^{17}O NMR spectra of p-nitro-[1-^{17}O]benzaldehyde span a range of over 1200 ppm, suggesting that both ^{17}O QC and CS tensors are highly anisotropic. For p-nitro-[1-^{17}O]benzaldehyde, the authors reported the following ^{17}O NMR tensor parameters: $C_Q = 10.7$ MHz, $\eta_Q = 0.45$; $\delta_{11} = 1050$, $\delta_{22} = 620$,

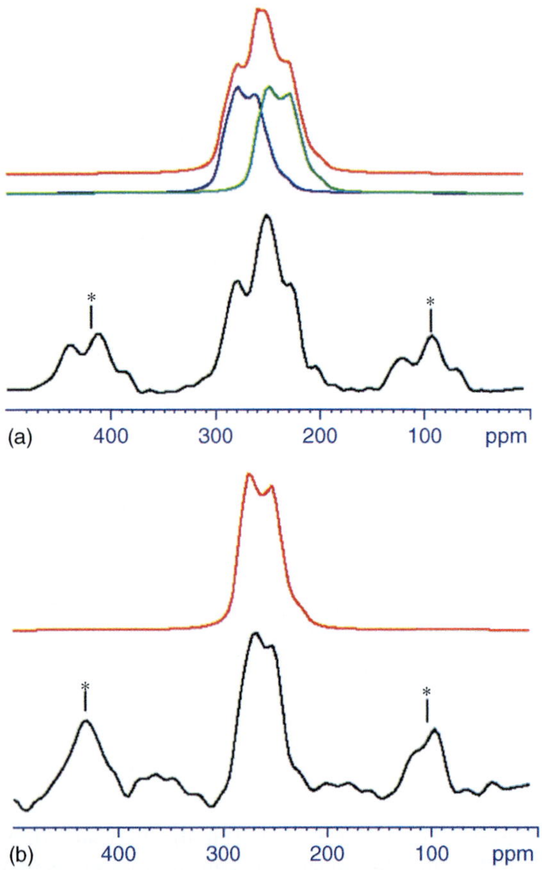

Figure 17.3. Experimental (lower trace) and simulated (upper trace) ^{17}O MAS spectra of Ala-[^{17}O]Ala-Ala samples obtained at 21.8 T. (a) Antiparallel β-sheet structure and (b) parallel β-sheet structure. (Reproduced from Ref. 55. © Elsevier, 2008.)

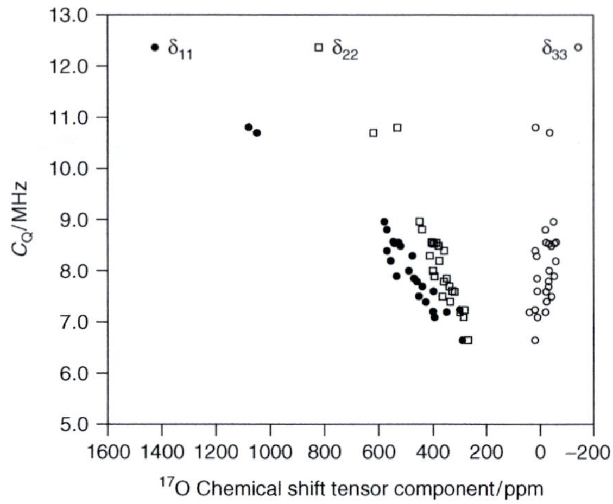

Figure 17.4. Relationship between the ^{17}O quadrupole coupling constant and ^{17}O CS tensor components for carbonyl compounds. (Reproduced from Ref. 56. © American Chemical Society, 2008.)

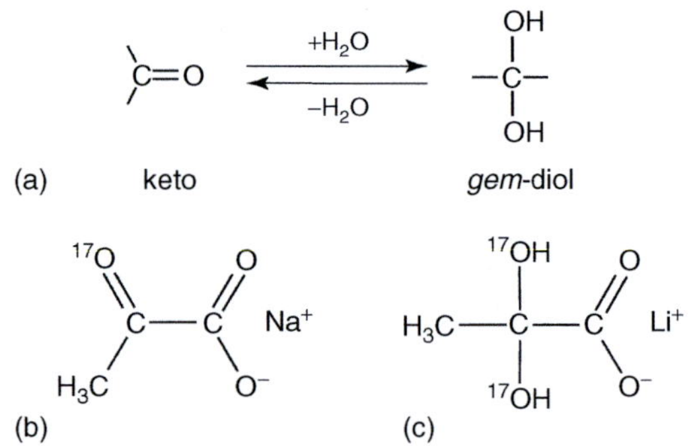

Figure 17.5. Hydration/dehydration equilibrium between keto and *gem*-diol functional group (a) and molecular structures of sodium [2-^{17}O]pyruvate and lithium [2,2′-$^{17}O_2$]pyruvate (b, c). (Reproduced from Ref. 57. © Royal Society of Chemistry, 2009.)

and $\delta_{33} = -35$ ppm. Here δ_{11}, δ_{22}, and δ_{33} are defined according to $\delta_{11} \geq \delta_{22} \geq \delta_{33}$. They also showed that, for carbonyl compounds, individual ^{17}O CS tensor components, especially δ_{11} and δ_{22}, are correlated with C_Q, as depicted in Figure 17.4.

In 2009, Zhu *et al.*[57] showed that solid-state ^{17}O NMR is a sensitive probe for identifying whether the keto group in α-keto acids is in hydrated form (as a *gem*-diol group) or dehydrated form (as a regular keto group), as illustrated in Figure 17.5. In general, α-keto acids including α-keto amino acids and their analogs are of considerable biochemical interest in biosynthesis and degradation of amino acids. Specifically, the keto form of α-keto acids can easily undergo hydration and transform into a *gem*-diol functional group, which can often be found in many enzymatic reaction intermediates. Zhu *et al.*[57] chose to investigate two classic α-keto acid compounds: lithium pyruvate monohydrate and sodium pyruvate. These two compounds are known to have different structures in the solid state. In particular, lithium pyruvate monohydrate is known to take the *gem*-diol form, whereas sodium pyruvate is an anhydrous salt in the normal keto form. Here it is important to note that the "water molecule" in lithium pyruvate monohydrate is covalently bonded to the C2 atom of pyruvate, rather than simply a water of

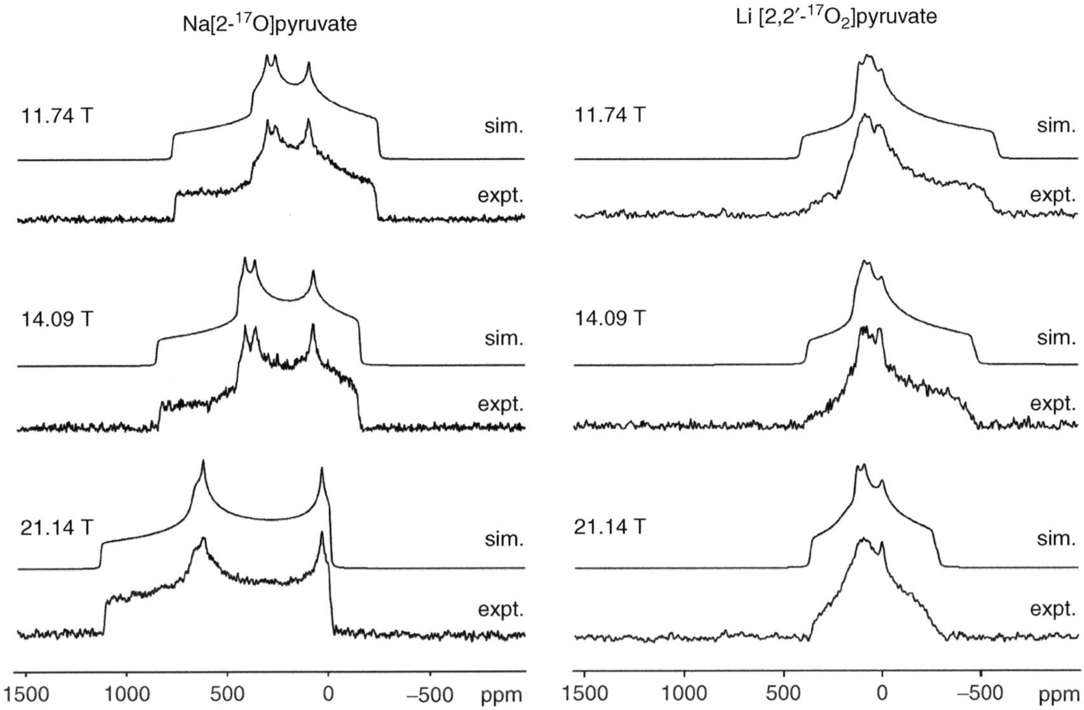

Figure 17.6. Experimental (lower trace) and simulated (upper trace) ^{17}O NMR spectra obtained for stationary powder samples of sodium [2-^{17}O]pyruvate and lithium [2,2′-$^{17}O_2$]pyruvate at different magnetic fields. (Reproduced from Ref. 57. © Royal Society of Chemistry, 2009.)

hydration. Therefore, the two ^{17}O-labeled compounds are denoted as sodium [2-^{17}O]pyruvate and lithium [2,2′-$^{17}O_2$]pyruvate. Figure 17.6 shows the experimental and simulated ^{17}O NMR spectra of sodium [2-^{17}O]pyruvate and lithium [2,2′-$^{17}O_2$]pyruvate obtained under the stationary sample condition at three magnetic fields. The authors also reported ^{17}O MAS NMR spectra for these two compounds. They found that the two functional groups have drastically different ^{17}O QC and CS tensors: for the keto form, $\delta_{iso} = 543$ ppm, $C_Q = 10.8$ MHz, $\eta_Q = 0.48$, $\delta_{11} = 1020$, $\delta_{22} = 640$, $\delta_{33} = -40$ ppm; for the *gem*-diol form, $\delta_{iso} = 62$ ppm, $C_Q = 8.5$ MHz, $\eta_Q = 1.0$, $\delta_{11} = 140$, $\delta_{22} = 45$, and $\delta_{33} = 0$ ppm. It is remarkable that, upon hydration, the ^{17}O CS of the keto group changes more than 480 ppm and the corresponding change in $C_Q(^{17}O)$ is 2.3 MHz. The authors also performed quantum chemical calculations of the ^{17}O QC and CS tensors for sodium [2-^{17}O]pyruvate. Their objective was to examine the importance of intermolecular interactions and crystal packing on ^{17}O NMR tensors. To this end, they utilized two different computational approaches to calculate the ^{17}O QC and CS tensors. One is to employ a molecular cluster approach where several molecular cluster models are built and their NMR properties are computed. The second approach is to consider the entire crystal lattice and to calculate the ^{17}O NMR tensors. Figure 17.7 illustrates a comparison between calculated and experimental ^{17}O CS and QC tensor components for sodium pyruvate. For the ^{17}O CS tensor, both δ_{22} and δ_{33} are reasonably well predicted using the molecular cluster approach. However, δ_{11} is consistently overestimated by approximately 200 ppm. In comparison, the CASTEP results are in much better agreement with the experimental results for both ^{17}O QC and CS tensors. These results suggest that long-range interactions across the crystal lattice are rather important.

In 2009, Brown and colleagues[39] reported a beautiful study demonstrating heteronuclear $^{15}N-^{17}O$ and $^{13}C-^{17}O$ correlation NMR spectroscopy for organic solids. They showed that heteronuclear

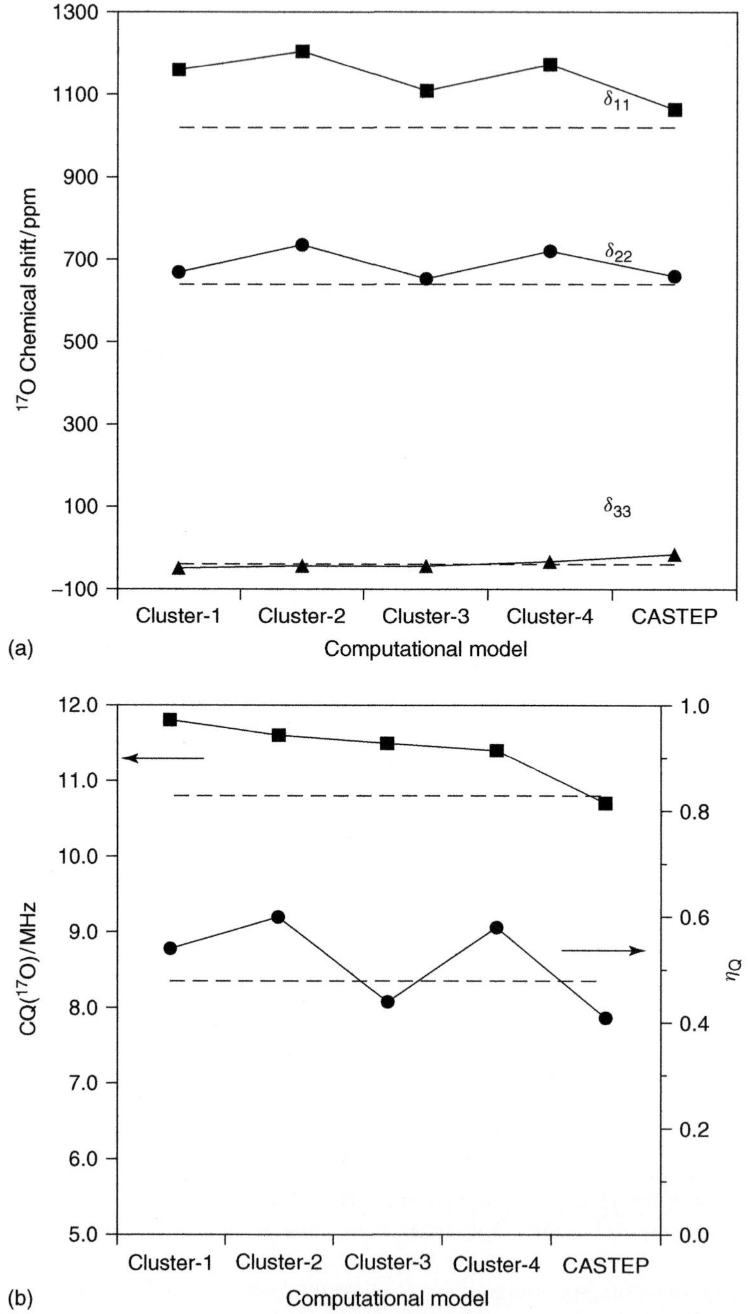

Figure 17.7. Comparison between experimental (dashed lines) and computational (symbols) ^{17}O CS tensor components (a) and quadrupole parameters (b) for sodium [2-^{17}O]pyruvate. (Reproduced from Ref. 57. © Royal Society of Chemistry, 2009.)

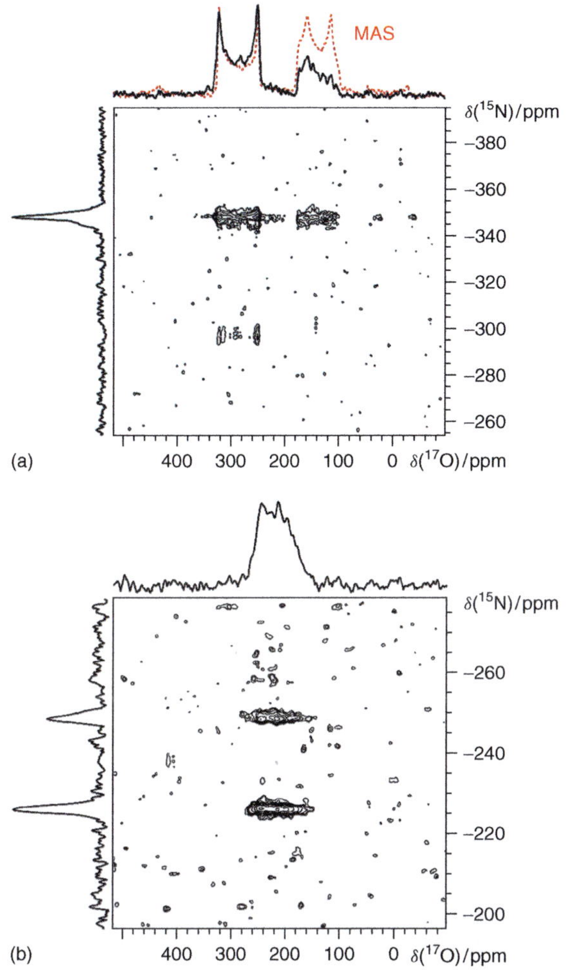

Figure 17.8. 2D $^{17}O/^{15}N$ R^3-HMQC correlation NMR spectra (14.1 T) of (a) [$^2H(NH_3)$,1-^{13}C,^{15}N,$^{17}O_2$]glycine·^{2}HCl and (b) [$^{15}N_2$,$^{17}O_2$]uracil with skyline projections along F_1 and F_2. (Reproduced from Ref. 39. © American Chemical Society, 2009.)

connectivity can be established via either dipolar or J-couplings. For example, Figure 17.8 displays 2D $^{17}O/^{15}N$ R^3-HMQC NMR spectra of [$^2H(NH_3)$, 1-^{13}C,^{15}N,$^{17}O_2$]glycine·^{2}HCl and [$^{15}N_2$,$^{17}O_2$]uracil. Here R^3-HMQC refers to a heteronuclear multiple-quantum correlation (HMQC) experiment performed under the rotary-resonance recoupling (R^3) condition. As seen from Figure 17.8(a), both O1 and O2 resonances show correlation to the ^{15}N resonance, suggesting that they are all in spatial proximity. Indeed, the crystal structure of glycine·HCl indicates that the closest O1–N and O2–N distances are 2.7 and 3.7 Å, respectively. For uracil, two $^{17}O/^{15}N$ correlation peaks are also observed (Figure 17.8(b)). However, because O2 and O4 resonances are not resolved, it is not possible to obtain relative positions of O2 and O4 with respect to each nitrogen atom. In principle, one can combine R^3-HMQC with either MQMAS or STMAS to make a 3D experiment where both ^{17}O and ^{15}N dimensions should have high spectral resolution. In this study, Brown and colleagues also reported the measurement of two very small hydrogen-bond-mediated J-couplings ($^{2h}J_{NO} = 5.8 \pm 0.6, 5.1 \pm 0.6$ Hz in uracil) and three intramolecular J-couplings ($^1J_{CO} = 24.7 \pm 0.2, 25.3 \pm 0.3$ Hz and $^2J_{OO} = 8.8 \pm 0.9$ Hz in glycine·^{2}HCl). These are the first examples of the kind involving ^{17}O in organic solids.

In 2010, Wu and colleagues[58] investigated the ^{17}O NMR tensors in *C*-nitrosoarenes (Ar–N=O). *C*-nitrosoarenes represent an important class of organic compounds widely used in organic chemistry, coordination chemistry, and biochemistry. In the context of biological systems, *C*-nitrosoarenes (as well as H–N=O) can form stable complexes with heme proteins such as cytochrome P450, myoglobin, and hemoglobin. Therefore, *C*-nitrosoarene metal complexes are also of considerable interest. As

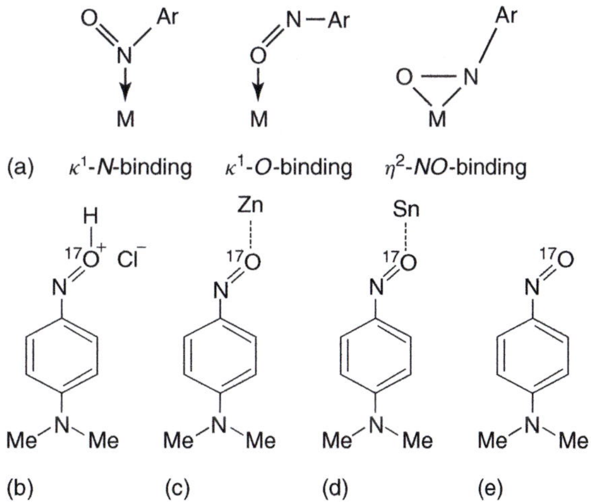

Figure 17.9. (a) Basic binding modes of monometallic *C*-nitrosoarene complexes. Molecular structures of (b) [^{17}O]NODMA·HCl, (c) $ZnCl_2$([^{17}O]NODMA$)_2$, (d) $SnCl_2Me_2$([^{17}O]NODMA$)_2$, and (e) [^{17}O]NODMA. (Reproduced from Ref. 58. © American Chemical Society, 2010.)

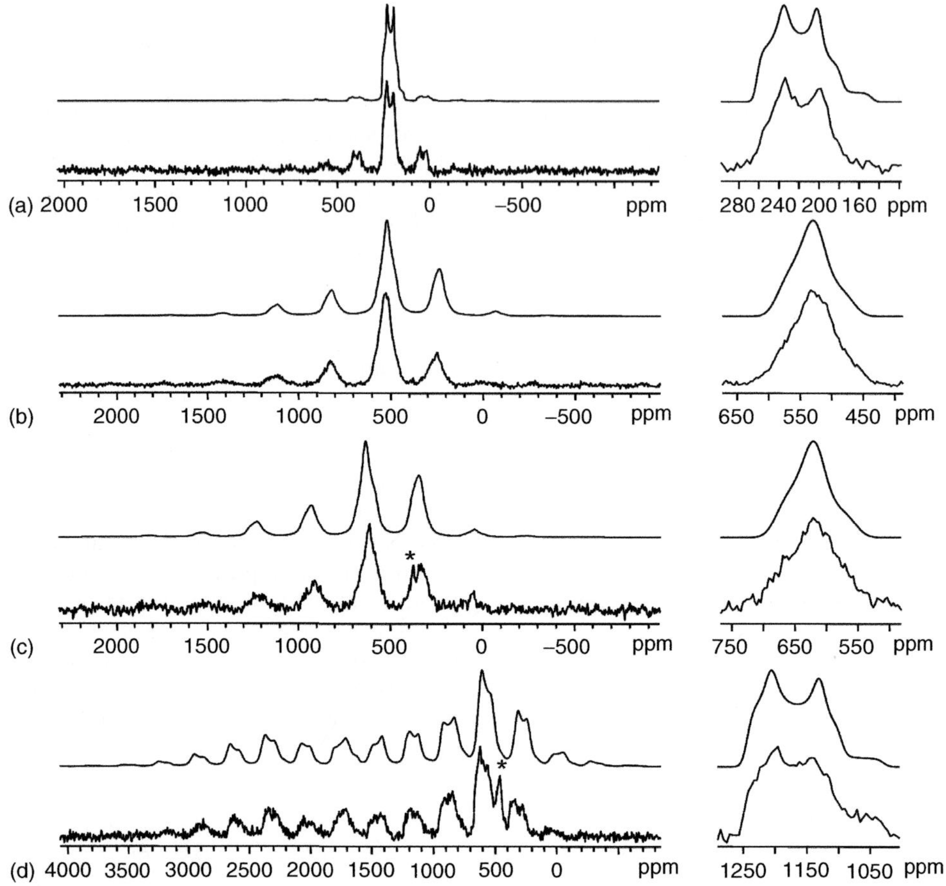

Figure 17.10. Experimental (lower trace) and simulated (upper trace) ^{17}O MAS spectra (left column) and "total" MAS line shapes (right column) of (a) [^{17}O]NODMA·HCl, (b) $ZnCl_2$([^{17}O]NODMA)$_2$, (c) $SnCl_2Me_2$([^{17}O]NODMA)$_2$, and (d) [^{17}O]NODMA. All spectra were recorded at 21.14 T. The sample spinning frequency was (a) 22 kHz, and (b, c, d) 35 kHz. (Reproduced from Ref. 58. © American Chemical Society, 2010.)

illustrated in Figure 17.9, three basic binding modes are known for mononuclear *C*-nitrosoarene metal complexes.[59] In this study, Wu and colleagues[58] studied four ^{17}O-labeled *C*-nitrosoarene compounds (Figure 17.9): *p*-[^{17}O]nitroso-*N*,*N*-dimethylaniline ([^{17}O]NODMA), $SnCl_2$ $(CH_3)_2$([^{17}O]NODMA)$_2$, $ZnCl_2$([^{17}O]NODMA)$_2$, and [^{17}O] NODMA·HCl. Figure 17.10 shows the experimental and simulated ^{17}O MAS spectra for *C*-nitrosoarene compounds at 21.14 T. In most cases, strong spinning sidebands are observed even at a sample spinning frequency of 35 kHz, indicating the presence of significant ^{17}O CSA. For example, the spinning sidebands observed for [^{17}O]NODMA span a range of approximately 3000 ppm (ca 350 kHz at 21.14 T). To analyze these "slow" MAS spectra, the authors first obtained the "total" MAS line shape for each compound by adding all the spinning sidebands onto the central band, as also shown in Figure 17.10. From each "total" MAS line shape, one can immediately see that the observed δ_{iso} values are drastically different among the four *C*-nitrosoarene compounds, reflecting the nature of chemical bonding around the nitroso (N=O) group. For example, protonation of the nitroso group causes a change in the ^{17}O isotropic CS of ca 1000 ppm. The $C_Q(^{17}O)$ values observed for the four *C*-nitrosoarenes are between 9.6 and 15 MHz, which are among the largest so far experimentally measured by solid-state ^{17}O NMR. The authors also recorded ^{17}O NMR spectra under the stationary sample condition and obtained complete information about the ^{17}O NMR tensors in these *C*-nitrosoarenes. Among them, [^{17}O]NODMA has the most anisotropic ^{17}O QC and CS tensors: $C_Q = 15.0$ MHz, $\eta_Q = 0.3$, $\delta_{11} = 2900$, $\delta_{22} = 750$, and $\delta_{33} = 100$ ppm. It is important to point out that the observed value of C_Q(^{17}O) for [^{17}O]NODMA is approaching the theoretical upper limit of all C_Q(^{17}O) values, 20.9 MHz, which is expected for a single electron in the pure 2p atomic orbital of an oxygen atom. This suggests that, at a high magnetic field such as 21.14 T, all oxygen-containing diamagnetic organic molecules should be accessible by solid-state ^{17}O NMR.

Very recently, Wu and colleagues have fully characterized the ^{17}O NMR tensors of the phenolic oxygen in L-tyrosine in different ionization states.[60] This is the first time that these ^{17}O NMR tensors are measured simultaneously for a neutral phenol group and its corresponding phenolate. The phenolic functional group in L-tyrosine is known to have a pK_a of ≈10 in aqueous solution (Figure 17.11). However, a Tyr residue in proteins may exhibit very different values of pK_a. Generally speaking, the ionization equilibrium of Tyr side chains is an important factor in determining the Tyr orientation and dynamics, thus playing a key role in the overall conformation and biological activity of proteins. Wu and colleagues[60] recorded solid-state ^{17}O NMR spectra for L-tyrosine samples prepared under both neutral and very basic conditions. After analyzing ^{17}O MAS and static spectra obtained at three magnetic fields, they were able to obtain complete information about the ^{17}O NMR tensors for the phenolic oxygen in different ionization states. They further examined how the ^{17}O CS tensor orientation changes as a function of phenol ionization and discovered a "cross over" effect between δ_{11} and δ_{22} components for both ^{17}O and ^{13}C CS tensors. As illustrated in Figure 17.12, when the phenol group changes its ionization state, the ^{17}O and ^{13}C CS tensors associated with the phenol group change their orientations in the molecular frame of reference. Consequently, as the authors pointed out, it is more informative to examine changes along directions fixed to the molecular frame rather than to compare apparent changes in individual tensor components, $\Delta\delta_{11}$, $\Delta\delta_{22}$, and $\Delta\delta_{33}$. Such an analysis reveals that, upon phenol ionization, the shielding along the C$^{\zeta}$–O$^{\eta}$ bond direction decreases so much that this direction has the least shielding value (but the highest CS value) and thus corresponds to the CS tensor component of δ_{11}. As the convention for defining the three CS tensor components is $\delta_{11} > \delta_{22} > \delta_{33}$, there is an apparent "cross over" between the δ_{11} and δ_{22} components upon phenol ionization if we link the CS tensor components according to their directions in the molecular frame, as depicted in Figure 17.12. The actual shielding change along the C$^{\zeta}$–O$^{\eta}$ bond direction is approximately 150 ppm, which is much larger than any of the $\Delta\delta_{ii}$ values. In fact, the overall change of ca 60 ppm in δ_{iso}, upon phenol ionization, is primarily caused by the shielding change along the C$^{\zeta}$–O$^{\eta}$ bond direction, because the shielding changes in the other two orthogonal directions are nearly cancelled by one another. The ^{13}C CS tensor at C$^{\zeta}$ has a similar δ_{11}–δ_{22} "cross over" effect, which provides an explanation to the observation made by Herzfeld and colleagues[61] some time ago that only the δ_{22} component of the ^{13}C CS tensor displays weak sensitivity toward phenol ionization, in spite of an apparently drastic change in chemical bonding. It is also interesting to compare the sensitivity of the ^{17}O CS tensor for O$^{\eta}$ toward phenol ionization with that of the ^{13}C CS tensor for C$^{\zeta}$. Upon phenol ionization, the δ_{11} and δ_{22} components of the ^{17}O CS tensor change in the same directions, whereas the corresponding components of the ^{13}C CS tensor change in opposite directions. As a result, the trace of the ^{17}O CS tensor, δ_{iso}(^{17}O), displays a net change nearly six times larger than does δ_{iso}(^{13}C). This increased sensitivity in ^{17}O CS can potentially be utilized as a new NMR probe for monitoring Tyr side chains in proteins.

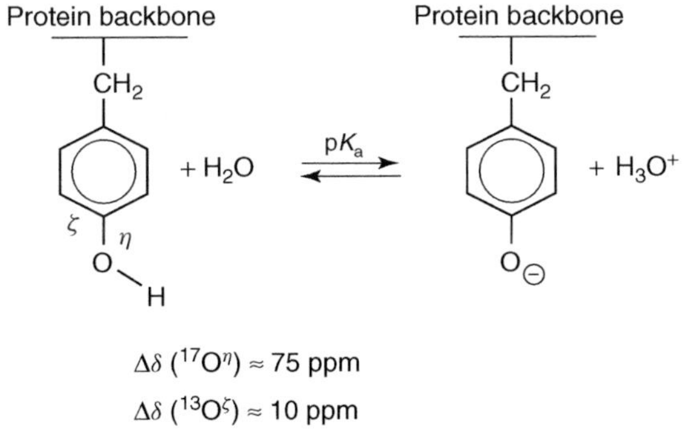

Figure 17.11. Ionization (acid–base) equilibrium of a Tyr side chain and the corresponding ^{17}O and ^{13}C chemical shift changes. (Reproduced from Ref. 60. © American Chemical Society, 2010.)

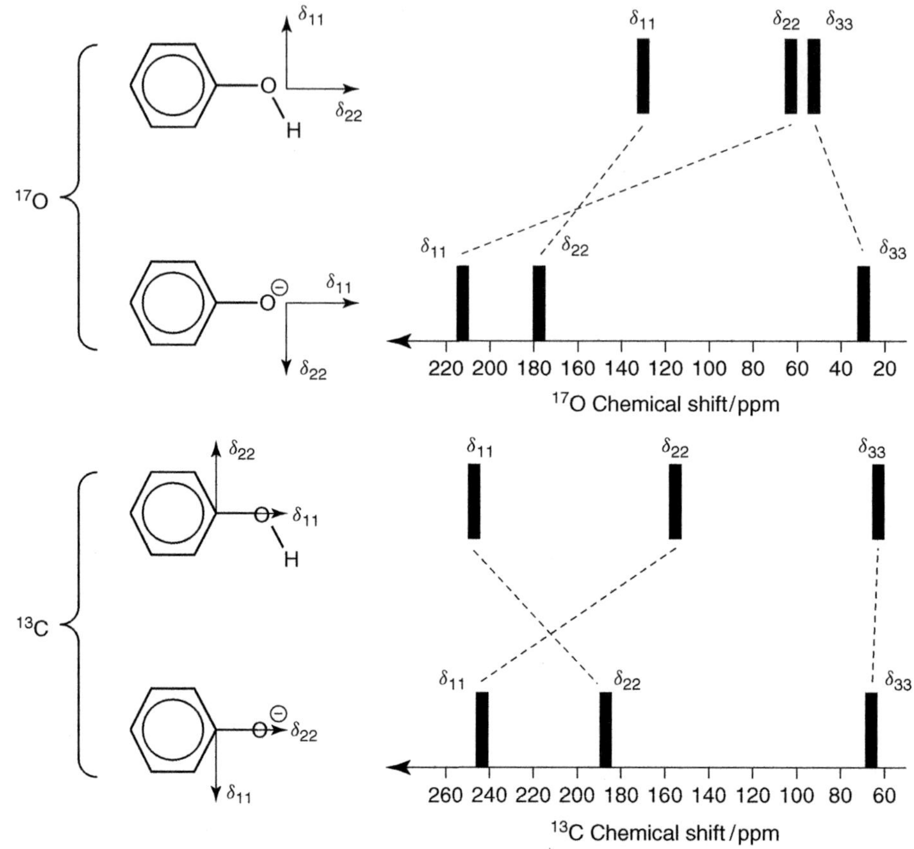

Figure 17.12. Illustration of the ^{17}O and ^{13}C CS tensor orientations in the molecular frame of reference (left) and the δ_{11}–δ_{22} "cross over" effects between protonated and deprotonated phenol groups (right). (Reproduced from Ref. 60. © American Chemical Society, 2010.)

17.4 RECENT ADVANCES IN ^{17}O NMR OF LARGE BIOLOGICAL MOLECULES

As we have shown in the previous section, it has become relatively easy now with modern high-field NMR spectrometers to obtain high-quality solid-state ^{17}O NMR spectra for small or midsize organic molecules (ca molecular weight <1000). It is perhaps safe to claim that all oxygen-containing functional groups in small diamagnetic organic molecules should be accessible now by solid-state ^{17}O NMR. However, obtaining high-quality ^{17}O NMR spectra for large biological systems still represents a considerable challenge. To be able to apply solid-state ^{17}O NMR to large biological molecules, two major problems must be solved. First, we must find a way to introduce ^{17}O labeling (>50%) into biological molecules. Second, the overall sensitivity of the ^{17}O NMR experiment should be sufficiently high. In recent years, significant progress has been made in these fronts and, as such, reports have begun to appear in the literature that deal with large biological molecules. In this section, we shall first give a brief review and then focus on some recent developments.

In 1991, Oldfield and colleagues reported the first set of solid-state ^{17}O NMR spectra for proteins, [$^{17}O_2$]hemoglobin (67 kDa tetramer), and [$^{17}O_2$]myoglobin (17 kDa), under non-spinning (stationary) conditions at two moderate magnetic fields, 8.45 and 11.75 T.[62] This was a pioneering work in solid-state ^{17}O NMR studies of large biological molecules. In 1998, Oldfield and colleagues successfully obtained solid-state ^{17}O NMR spectra

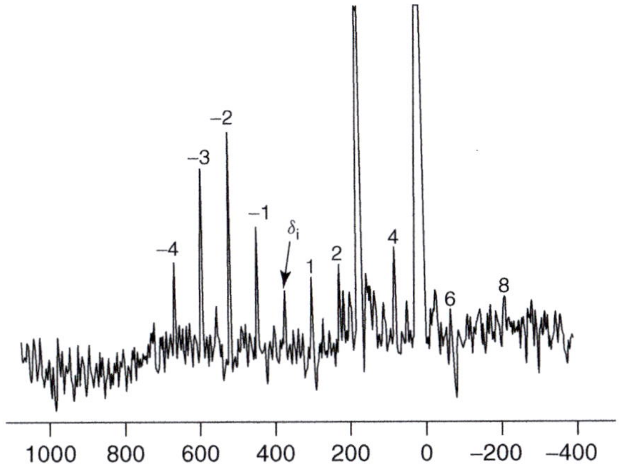

Figure 17.13. Experimental ^{17}O MAS spectrum (11.7 T) of A_1 [$C^{17}O$] horse myoglobin crystals at pH 7 and −24 °C. The spinning sidebands are numbered. (Reproduced from Ref. 63. © American Chemical Society, 1998.)

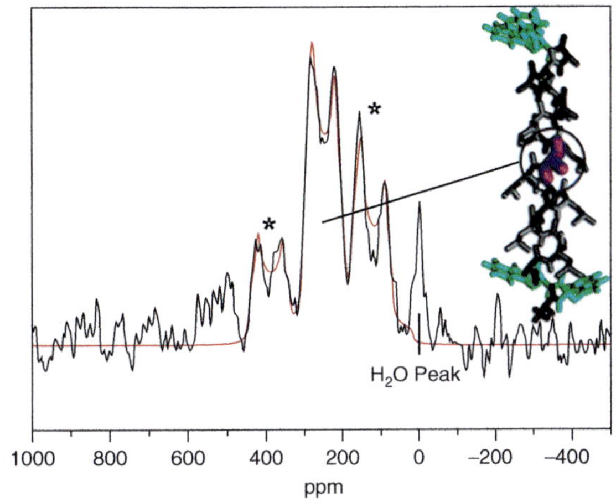

Figure 17.14. Experimental and simulated ^{17}O MAS spectra of [^{17}O]Ala12-WALP23 in hydrated 1,2-distearoyl-sn-glycero-3-phosphatidylcholine (DSPC) vesicles at 14.1 T. The spinning sidebands are marked by asterisks (*). (Reproduced from Ref. 64. © American Chemical Society, 2004.)

for [$C^{17}O$]myoglobin under the MAS condition at 11.7 T.[63] In this case, the experimental ^{17}O MAS spectrum (shown in Figure 17.13) clearly suggests that the value of $C_Q(^{17}O)$ is quite small (<1 MHz) for $C^{17}O$ bound to myoglobin. Under such a circumstance, the ^{17}O CS tensor components can be readily obtained from an analysis of spinning sidebands.

In 2004, Dupree, Watts, and colleagues[64] reported the first solid-state ^{17}O NMR study of a transmembrane 23-residue polypeptide containing only one ^{17}O-labeled amino acid residue, [^{17}O]Ala12-WALP23. Figure 17.14 shows the ^{17}O MAS spectrum of [^{17}O]Ala12-WALP23. The observed ^{17}O NMR parameters for [^{17}O]Ala12-WALP23 are consistent with an α-helical conformation of this polypeptide. These authors also examined the same peptide in a hydrated membrane environment and demonstrated the ^{17}O NMR sensitivity is sufficient even in this dilute environment.

In 2005, Cross and colleagues reported a solid-state ^{17}O NMR study of gramicidin A (gA).[65] In a lipid membrane environment, gA adopts a head-to-head β-helix dimer forming an ion-conducting channel. Inside the 4.5-Å pore of the gA dimer, ion–carbonyl interactions play a key role in controlling the ion-conduction process. Previous studies had been largely focused on using ^{13}C and ^{15}N as probes for ion–carbonyl interactions. As the peptide carbonyl oxygen atom is directly involved in ion binding, ^{17}O is an ideal NMR probe for monitoring ion–carbonyl interactions inside ion channels. Cross and colleagues obtained solid-state ^{17}O NMR parameters by examining a powder sample of ^{17}O-[D-Leu10]gA.[65] More importantly, they also reported ^{17}O NMR spectra of ^{17}O-[D-Leu10]gA uniformly aligned in 1,2-dimyristoyl-sn-glycero-3-phosphatidylcholine (DMPC) bilayers. This oriented sample gave rise to single-crystal-like NMR spectra. As seen in Figure 17.15, when the applied magnetic field is parallel to the normal of the DMPC bilayers, the ^{17}O NMR spectrum exhibits a relatively narrow peak at 470 ppm arising only from the ^{17}O-labeled Leu10 residue. This ^{17}O NMR signal is quite sensitive to the presence of cations such as K^+, suggesting that this residue is directly involved in ion conduction. In a later study, Smirnov *et al.*[66] used solid-state ^{17}O NMR to probe the reversible K^+ ion binding to the gA pore.

In a study published in 2008, Smith and colleagues attempted to use the solid-state ^{17}O NMR spectra obtained for a 28-residue peptide to assess the sensitivity issue.[67] This 28-residue peptide, DYQSLQIGGLVIA[^{17}O-G]ILFILGILVLSRR (MW = 3042), corresponds to the phospholemman transmembrane (PLM) domain and has a single glycine residue labeled by ^{17}O (≈40%), thus

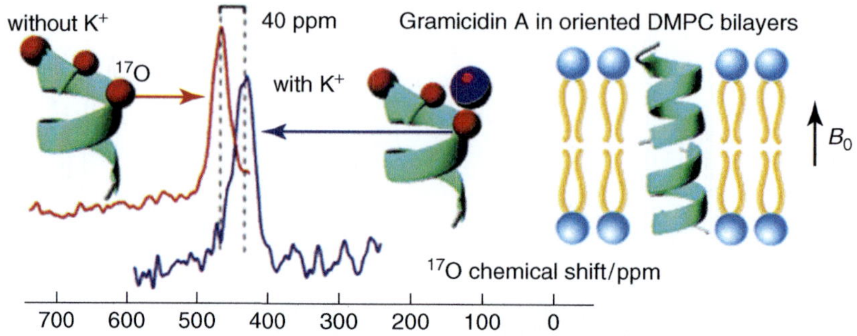

Figure 17.15. Experimental ^{17}O NMR (21 T) spectra of ^{17}O-[D-Leu10]gA uniformly aligned in DMPC bilayers in the presence and absence of K^+ ions. The orientation of the applied magnetic field, B_0, is parallel to the norm of the bilayer. (Reproduced from Ref. 65. © American Chemical Society, 2005.)

denoted as ^{17}O-[gly14]-PLM. The ^{17}O-[gly14]-PLM peptide was reconstituted in a dipalmitoylphosphatidylcholine (DPPC) lipid bilayer membrane with a lipid/peptide molar ratio of 25:1. In the lipid bilayer environment, PLM is known to form an α-helical tetramer unit. The actual peptide/lipid sample has a mass of about 22 mg, among which only 14% ($\approx$3 mg) was ^{17}O-[gly14]-PLM. As the size of PLM tetramer is more than four times larger than WALP23 and gA and the lipid/peptide ratio used in this particular study is also high, the ^{17}O content in the ^{17}O-[gly14]-PLM sample is very low. To circumvent the low sensitivity problem, the authors used frequency-switched Gaussian pulses[54] to enhance central-transition signals. In addition, a large number of transients ($2-3 \times 10^6$) were collected with a short recycle time (80 ms). Under these conditions, it takes about 50 h to obtain an ^{17}O MAS spectrum for ^{17}O-[gly14]-PLM at either 14.1 or 18.8 T.

Very recently, Wu and colleagues presented a comprehensive solid-state ^{17}O NMR study of large biological molecules.[68] In particular, they reported ^{17}O MAS spectra for two large protein–ligand complexes of size ranging from 64 to 80 kDa. One system examined by Wu and colleagues is the avidin–[$^{17}O_2$]biotin complex. Avidin is a glycoprotein isolated from hen egg-white that forms a tetramer with a total molecular weight of about 64 kDa which can bind biotin molecules with extremely high affinity ($K_d = 10^{-15}$ M). Figure 17.16 shows the experimental and simulated ^{17}O MAS spectra of avidin-[$^{17}O_2$]biotin at 21.14 T. Spectral analysis yields the following ^{17}O NMR parameters: $C_Q = 5.8$ MHz, $\eta_Q = 0.4$, and

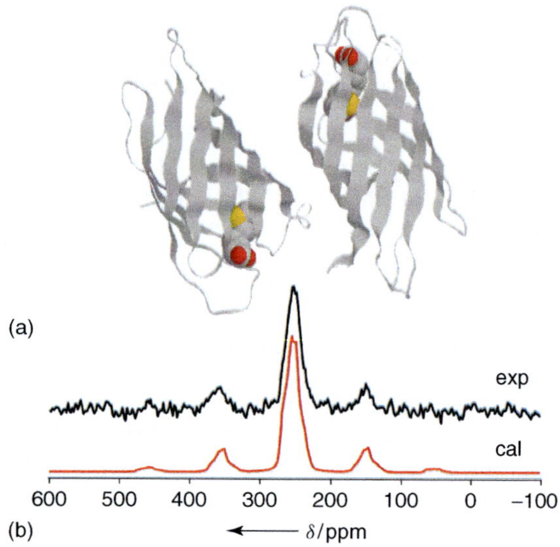

Figure 17.16. (a) Partial crystal structure of avidin/biotin complex (PDB entry 1 avd). Only the asymmetric unit (chains A and B) is shown. (b) Experimental (upper trace) and simulated (lower trace) ^{17}O MAS spectra of avidin/biotin complex obtained at 21.1 T. (Reproduced from Ref. 68. © Wiley-VCH, 2010.)

$\delta_{iso} = 270$ ppm. These are typical of deprotonated carboxylate groups. As each avidin molecule can bind up to four biotin molecules, the dilution level of the complex is approximately 16 kDa per [$^{17}O_2$]biotin. In this case, a relatively large amount of protein–ligand complex (ca 20 mg) was used in the experiment; however, the ^{17}O enrichment level in [$^{17}O_2$]biotin was relatively low, ca 18%. To further test the sensitivity limit at 21.14 T, the authors

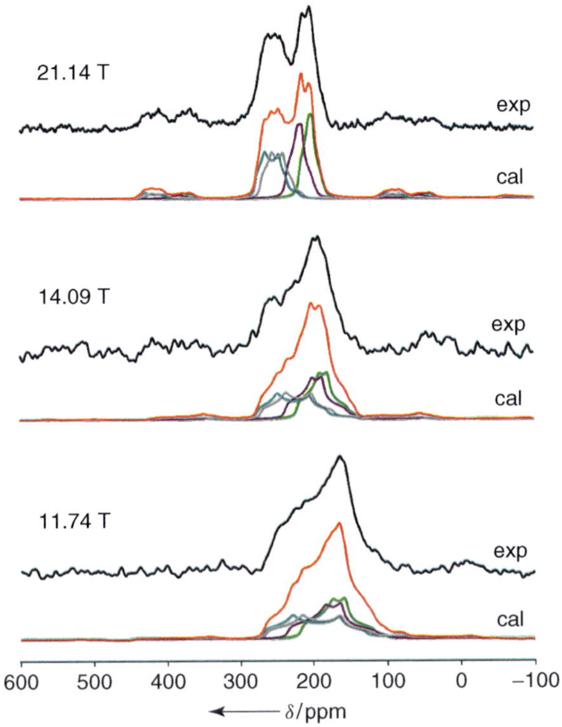

Figure 17.17. Experimental (upper trace) and simulated (lower trace) ^{17}O MAS spectra of OTf-Al-[$^{17}O_4$]oxalate obtained at three magnetic fields. In each case, both the total spectrum and individual components are shown. (Reproduced from Ref. 68. © Wiley-VCH, 2010.)

examined an even larger protein–ligand complex, ovotransferrin–Al(III)–[$^{17}O_4$]oxalate. Ovotransferrin (OTf) is a 80-kDa glycoprotein capable of binding two Al(III) cations and two synergistic anions (e.g., carbonate or oxalate) with high affinity. Thus, the dilution level of OTf-Al-[$^{17}O_4$]oxalate complex is ca 40 kDa per ligand. Figure 17.17 shows the ^{17}O MAS NMR spectra of OTf–Al–[$^{17}O_4$]oxalate obtained at three magnetic fields. It is clearly seen that significant improvement in spectral resolution is achieved at the highest magnetic field, 21.14 T. The authors reported the following ^{17}O NMR parameters: O1, $\delta_{iso} = 219$ ppm, $\xi = \delta_{33} - \delta_{iso} = -160$ ppm, $\eta = (\delta_{22} - \delta_{11})/\xi = 0.6$, $C_Q = 5.75$ MHz, $\eta_Q = 0.70$; O2, $\delta_{iso} = 237$ ppm, $\xi = -160$ ppm, $\eta = 0.6$, $C_Q = 6.30$ MHz, $\eta_Q = 0.70$; O3, $\delta_{iso} = 274$ ppm, $\xi = -240$ ppm, $\eta = 0.2$, $C_Q = 7.70$ MHz, $\eta_Q = 0.45$; O4, $\delta_{iso} = 282$ ppm, $\xi = -240$ ppm, $\eta = 0.2$, $C_Q = 7.90$ MHz, $\eta_Q = 0.35$. The authors also demonstrated that these ^{17}O NMR parameters, together with ^{27}Al and ^{13}C NMR parameters obtained for the same protein–ligand complex, can be used for structural refinement to obtain a more accurate geometry of the ligand molecule. We believe that this aspect of "NMR crystallography" in the context of proteins should be further explored in the future.

The OTf–Al–[$^{17}O_4$]oxalate complex represents the most dilute protein system investigated so far by solid-state ^{17}O NMR. This benchmark case can be used to assess the sensitivity limit of solid-state ^{17}O NMR in studying proteins. The OTf–Al–oxalate sample (ca 25 mg protein) used at 21.14 T contains only approximately 30 µg of [$^{17}O_4$, 50% ^{17}O]oxalate. Three important factors have contributed to the success of obtaining high-quality ^{17}O MAS spectra for such a large biomolecular system. First, the authors utilized a very high magnetic field, 21.14 T. Second, they discovered that solid proteins generally have ^{17}O T_1 values on the order of a few milliseconds (10^{-3} s). The very short ^{17}O T_1 values allowed them to collect ^{17}O NMR data very rapidly (e.g., a recycle delay of 30 ms). Third, they used HS pulses to invert selectively the satellite transitions and obtained an additional gain of ca 2 in sensitivity enhancement (a factor of 4 in time saving) as compared with the single-pulse excitation. The authors also reported that other sensitivity enhancement methods such as RAPT and DFS yielded similar enhancement factors. On the basis of these new solid-state ^{17}O NMR results, Wu and colleagues[68] concluded that one should be able to obtain high-quality ^{17}O MAS NMR spectra at 21.14 T for protein–ligand complexes as large as 300 kDa per ligand with 90% ^{17}O enrichment and 48 h of data acquisition time. This level of sensitivity should make it possible to apply solid-state ^{17}O NMR to many important proteins. It is quite clear from this work that solid-state ^{17}O NMR spectroscopy is a viable technique for probing protein–ligand interactions.

17.5 CONCLUDING REMARKS

In the past several years, we have witnessed a rapid expansion of knowledge with regard to solid-state ^{17}O NMR properties of organic and biological molecules. The selected examples described in this chapter provide a brief survey of what is at the forefront of this research field. These exciting results, although preliminary, strongly indicate that new areas of research are becoming possible. Future applications in this emerging research field may occur along the following directions: (i) characterization of new organic functional groups, (ii) solid-state ^{17}O NMR for large biological molecules, (iii) development of new NMR methods in both dipolar- and in J-based correlation experiments and in possible combination of multiple-pulse experiments with DOR, (iv) studies of paramagnetic proteins, and (v) new synthetic strategies for ^{17}O-labeling of biological molecules. Since oxygen is the key constituent of many biologically important functional groups, solid-state ^{17}O NMR can provide a unique perspective in the study of biological structures and functions. As we have demonstrated in this chapter, significant advances in NMR methodology and continuing development of high-field, high-sensitivity NMR instrumentation have reached a critical point that permits increasingly complex biological systems to be probed by solid-state ^{17}O NMR. We anticipate that more solid-state ^{17}O NMR studies of organic and biological molecules will appear in the very near future.

RELATED ARTICLES IN THE ENCYCLOPEDIA OF MAGNETIC RESONANCE

Magic Angle Spinning

Multiple-quantum Magic-angle Spinning Experiments on Half-integer Nuclei: Fundamentals

Oxygen-17 NMR

Oxygen-17 NMR: Applications in Biochemistry

Tensors in NMR

Tensor Interplay

REFERENCES

1. D. W. Boykin, *^{17}O NMR Spectroscopy in Organic Chemistry*, CRC Press: Boca Raton, FL, 1991, p. 325.
2. G. Wu, *Biochem. Cell Biol.*, 1998, **76**, 429.
3. V. Lemaitre, M. E. Smith, and A. Watts, *Solid State Nucl. Magn. Reson.*, 2004, **26**, 215.
4. G. Wu, *Prog. Nucl. Magn. Reson. Spectrosc.*, 2008, **52**, 118.
5. I. P. Gerothanassis, *Prog. Nucl. Magn. Reson. Spectrosc.*, 2010, **56**, 95; 2010, **57**, 1.
6. F. Alder and F. C. Yu, *Phys. Rev.*, 1951, **81**, 1067.
7. S. W. Rabideau and P. Waldstein, *J. Chem. Phys.*, 1966, **44**, 1304.
8. P. Waldstein and S. W. Rabideau, *J. Chem. Phys.*, 1967, **47**, 5338.
9. H. W. Spiess, B. B. Garrett, R. K. Sheline, and S. W. Rabideau, *J. Chem. Phys.*, 1969, **51**, 1201.
10. W. Scheubel, H. Zimmermann, and U. Haeberlen, *J. Magn. Reson.*, 1985, **63**, 544.
11. Q. W. Zhang, H. M. Zhang, M. G. Usha, and R. J. Wittebort, *Solid State Nucl. Magn. Reson.*, 1996, **7**, 147.
12. Q. W. Zhang, E. Y. Chekmenev, and R. J. Wittebort, *J. Am. Chem. Soc.*, 2003, **125**, 9140.
13. K. W. Waddell, E. Y. Chekmenev, and R. J. Wittebort, *J. Phys. Chem. B*, 2006, **110**, 22935.
14. Y. Hsieh, J. C. Koo, and E. L. Hahn, *Chem. Phys. Lett.*, 1972, **13**, 563.
15. S. Schramm, R. J. Kirkpatrick, and E. Oldfield, *J. Am. Chem. Soc.*, 1983, **105**, 2483.
16. S. Schramm and E. Oldfield, *J. Am. Chem. Soc.*, 1984, **106**, 2502.
17. S. Kuroki, I. Ando, A. Shoji and T. Ozaki, *Chem. Commun.*, 1992, 433.
18. R. Goc, E. Ponnusamy, J. Tritt-Goc, and D. Fiat, *Int. J. Pept. Protein Res.*, 1988, **31**, 130.
19. R. Goc, J. Tritt-Goc, and D. Fiat, *Bull. Magn. Reson.*, 1989, **11**, 238.
20. A. Llor andJ. Virlet, *Chem. Phys. Lett.*, 1988, **152**, 248.
21. K. T. Mueller, B. Q. Sun, G. C. Chingas, J. W. Zwanziger, T. Terao, and A. Pines, *J. Magn. Reson.*, 1990, **86**, 470.
22. A. Samoson, E. Lippmaa, and A. Pines, *Mol. Phys.*, 1988, **65**, 1013.
23. B. F. Chmelka, K. T. Mueller, A. Pines, J. Stebbins, Y. Wu, and J. W. Zwanziger, *Nature*, 1989, **339**, 42.
24. Y. Wu, B. Q. Sun, A. Pines, A. Samoson, and E. Lippmaa, *J. Magn. Reson.*, 1990, **89**, 297.
25. A. P. Howes, T. Anupold, V. Lemaitre, A. Kukol, A. Watts, A. Samoson, M. E. Smith, and R. Dupree, *Chem. Phys. Lett.*, 2006, **421**, 42.

26. S. L. Gann, J. H. Baltisberger, E. W. Wooten, H. Zimmermann, and A. Pines, *Bull. Magn. Reson.*, 1994, **16**, 68.
27. L. Frydman and J. S. Harwood, *J. Am. Chem. Soc.*, 1995, **117**, 5367.
28. A. Medek, J. S. Harwood, and L. Frydman, *J. Am. Chem. Soc.*, 1995, **117**, 12779.
29. G. Wu, D. Rovnyak, B. Q. Sun, and R. G. Griffin, *Chem. Phys. Lett.*, 1996, **249**, 210.
30. G. Wu, D. Rovnyak, P. C. Huang, and R. G. Griffin, *Chem. Phys. Lett.*, 1997, **277**, 79.
31. G. Wu and S. Dong, *J. Am. Chem. Soc.*, 2001, **123**, 9119.
32. Z. Gan, *J. Am. Chem. Soc.*, 2000, **122**, 3242.
33. S. E. Ashbrook and S. Wimperis, *Prog. Nucl. Magn. Reson. Spectrosc.*, 2004, **45**, 53.
34. L. Chopin, S. Vega, and T. Gullion, *J. Am. Chem. Soc.*, 1998, **120**, 4406.
35. A. Goldbourt, S. Vega, T. Gullion, and A. J. Vega, *J. Am. Chem. Soc.*, 2003, **125**, 11194.
36. T. Gullion and A. J. Vega, *Prog. Nucl. Magn. Reson. Spectrosc.*, 2005, **47**, 123.
37. T. Gullion, K. Yamauchi, M. Okonogi, and T. Asakura, *Macromolecules*, 2007, **40**, 1363.
38. T. Gullion, T. Y. Yu, M. Singh, G. J. Patti, G. S. Potter, and J. Schaefer, *J. Am. Chem. Soc.*, 2010, **132**, 10902.
39. I. Hung, A.-C. Uldry, J. Becker-Baldus, A. L. Webber, A. Wong, M. E. Smith, S. A. Joyce, J. R. Yates, C. J. Pickard, R. Dupree, and S. P. Brown, *J. Am. Chem. Soc.*, 2009, **131**, 1820.
40. J. D. van Beek, R. Dupree, and M. H. Levitt, *J. Magn. Reson.*, 2006, **179**, 38.
41. A. Brinkmann and A. P. M. Kentgens, *J. Phys. Chem. B*, 2006, **110**, 16089.
42. A. Brinkmann and A. P. M. Kentgens, *J. Am. Chem. Soc.*, 2006, **128**, 14758.
43. Z. Yao, H. T. Kwak, D. Sakellariou, L. Emsley, and P. J. Grandinetti, *Chem. Phys. Lett.*, 2000, **327**, 85.
44. S. Prasad, H. T. Kwak, T. Clark, and P. J. Grandinetti, *J. Am. Chem. Soc.*, 2002, **124**, 4964.
45. H. T. Kwak, S. Prasad, Z. Yao, P. J. Grandinetti, J. R. Sachleben, and L. Emsley, *J. Magn. Reson.*, 2001, **150**, 71.
46. A. P. Kentgens and R. Verhagen, *Chem. Phys. Lett.*, 1999, **300**, 435.
47. D. Iuga, H. Schafer, R. Verhagen, and A. P. M. Kentgens, *J. Magn. Reson.*, 2000, **147**, 192.
48. T. Brauniger, K. Ramaswamy, and P. K. Madhu, *Chem. Phys. Lett.*, 2004, **383**, 403.
49. T. Brauniger, A. Hempel, and P. K. Madhu, *J. Magn. Reson.*, 2006, **181**, 68.
50. R. Siegel, T. T. Nakashima, and R. E. Wasylishen, *Chem. Phys. Lett.*, 2005, **403**, 353.
51. R. Siegel, T. T. Nakashima, and R. E. Wasylishen, *Chem. Phys. Lett.*, 2006, **421**, 529.
52. R. Siegel, T. T. Nakashima, and R. E. Wasylishen, *J. Magn. Reson.*, 2007, **184**, 85.
53. H.-T. Kwak, S. Prasad, T. Clark, and P. J. Grandinetti, *Solid State Nucl. Magn. Reson.*, 2003, **24**, 71.
54. R. Siegel, T. T. Nakashima, and R. E. Wasylishen, *Concepts Magn. Reson. Part A*, 2005, **26**, 47.
55. K. Yamauchi, M. Okonogi, H. Kurosu, M. Tansho, H. Shimizu, T. Gullion, and T. Asakura, *J. Magn. Reson.*, 2008, **190**, 327.
56. G. Wu, P. Mason, X. Mo, and V. Terskikh, *J. Phys. Chem. A*, 2008, **112**, 1024.
57. J. Zhu, A. J. Geris, and G. Wu, *Phys. Chem. Chem. Phys.*, 2009, **11**, 6972.
58. G. Wu, J. Zhu, X. Mo, R. Y. Wang, and V. Terskikh, *J. Am. Chem. Soc.*, 2010, **132**, 5143.
59. J. Lee, L. Chen, A. H. West, and G. B. Richter-Addo, *Chem. Rev.*, 2002, **102**, 1019.
60. J. Zhu, J. Y. C. Lau, and G. Wu, *J. Phys. Chem. B*, 2010, **114**, 11681.
61. J. Herzfeld, S. K. Das Gupta, M. R. Farrar, G. S. Harbison, A. E. McDermott, S. L. Pelletier, D. P. Raleigh, S. O. Smith, C. Winkel, J. Lugtenburg, and R. G. Griffin, *Biochemistry*, 1990, **29**, 5567.
62. E. Oldfield, H. C. Lee, C. Coretsopoulos, F. Adebodun, K. D. Park, S. T. Yang, J. Chung, and B. Phillips, *J. Am. Chem. Soc.*, 1991, **113**, 8680.
63. M. T. McMahon, A. C. deDios, N. Godbout, R. Salzmann, D. D. Laws, H. B. Le, R. H. Havlin, and E. Oldfield, *J. Am. Chem. Soc.*, 1998, **120**, 4784.
64. V. Lemaitre, M. R. R. De Planque, A. P. Howes, M. E. Smith, R. Dupree, and A. Watts, *J. Am. Chem. Soc.*, 2004, **126**, 15320.

65. J. Hu, E. Y. Chekmenev, Z. Gan, P. L. Gor'kov, S. Saha, W. W. Brey, and T. A. Cross, *J. Am. Chem. Soc.*, 2005, **127**, 11922.

66. E. Y. Chekmenev, P. L. Gor'kov, T. A. Cross, A. M. Alaouie, and A. I. Smirnov, *Biophys. J.*, 2006, **91**, 3076.

67. A. Wong, A. J. Beevers, A. Kukol, R. Dupree, and M. E. Smith, *Solid State Nucl. Magn. Reson.*, 2008, **33**, 72.

68. J. Zhu, E. Ye, V. Terskikh, and G. Wu, *Angew. Chem. Int. Ed.*, 2010, **49**, 8399.

Chapter 18
Oxygen-17 NMR of Inorganic Materials

Sharon E. Ashbrook[1] and Mark E. Smith[2]

[1]*School of Chemistry and EaStCHEM, University of St Andrews, St Andrews KY16 9ST, UK*
[2]*Department of Physics, University of Warwick, Coventry CV4 7AL, UK*

18.1 GENERAL INTRODUCTION

Given the importance of oxide-based materials in nature, technology (e.g., advanced, structural and functional ceramics, catalysts, etc.), solid-state chemistry, and mineralogy, an understanding of the bonding oxygen is involved in is often essential. Direct spectroscopic probes of the local environment around oxygen can provide extremely important information. The three stable isotopes of oxygen have natural abundances of ^{16}O – 99.76%, ^{17}O – 0.037%, and ^{18}O – 0.2%, but only ^{17}O has a nonzero nuclear spin quantum number ($I = 5/2$) and is therefore accessible to NMR. NMR studies of ^{17}O are hampered by two key factors, namely, sensitivity and quadrupolar effects. First, the low sensitivity results from a combination of its modest magnetic moment, such that it resonates at 68% of the frequency of ^{29}Si and produces a comparative receptivity of 0.029, and its low natural abundance. Second, although it possesses only a relatively small quadrupole moment of −25.56 mbarns (a number of papers say ^{17}O has a moderately large quadrupole moment but of the stable noninteger spin quadrupolar nuclei only ^{95}Mo and ^{133}Cs have smaller values), owing to the moderate Larmor frequency and sometimes substantial electric field gradients (EFGs), often significant second-order quadrupolar broadening occurs. In recent years, sensitivity has been aided by enhancement schemes such as rotor-assisted population transfer (RAPT) with application to ^{17}O of silica gel producing an increase in signal of 2.3.[1] A study on forsterite (Mg_2SiO_4) showed a similar factor of 2 increase under RAPT, which combined with the quadrupolar Carr–Purcell–Meiboom–Gill (QCPMG) echo train sequence, producing further enhancement through the summation of signal from the echo train and through the concentration of signal within the "spikelet" spectrum (Figure 18.1).[2] Interpretation of ^{17}O spectra from stationary samples can be complicated by both significant quadrupolar and chemical shift anisotropy (CSA) contributions. The standard technique for improving resolution in solid-state NMR is magic-angle spinning (MAS). Although MAS can often provide sufficient resolution between sites when the quadrupolar interaction is small (see below), for many systems of potentially high interest for study by ^{17}O NMR,

NMR of Quadrupolar Nuclei in Solid Materials
Edited by Roderick E. Wasylishen, Sharon E. Ashbrook and Stephen Wimperis
 ISBN: 978-0-470-97398-1

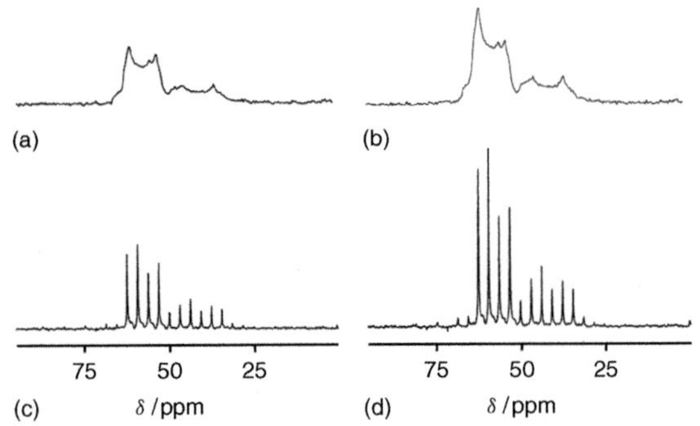

Figure 18.1. (a) Conventional single pulse, (b) RAPT-enhanced, (c) rotor-synchronized CPMG, and (d) RAPT-enhanced rotor-synchronized QCPMG ^{17}O MAS NMR spectra of 35% ^{17}O-enriched forsterite (Mg_2SiO_4). In all cases, 960 transients were averaged with 2 s recycle interval in 32 min, with an MAS rate of 5.8 kHz. In (b, d), 71 sets of +X/−X 1.6 ms pulses were applied prior to a 12.5 ms pulse selective for the central transition. In (c, d), a rotor-synchronized CPMG echo train of 20 × 32 μs 180° pulses was applied, resulting in a spacing of $1/\tau_a = 200$ Hz. All spectra are normalized to (d), with the intensity of those shown in (a) and (b) multiplied by a factor of 2 for clarity. (Reproduced from Ref. 1. © Elsevier, 2003.)

such as aluminosilicates, the much larger quadrupolar interaction means MAS does not really resolve the different sites. However, a range of techniques have been developed that enhance the resolution from such quadrupolar nuclei such as multiple-quantum (MQ) MAS, dynamic angle spinning (DAS), satellite transition (ST) MAS, and double-angle rotation (DOR). These approaches, along with the realization that ^{17}O NMR can provide real insight into inorganic materials, have driven the increase in the number of reports of ^{17}O NMR from such systems.

A small cluster of early reports of solid-state ^{17}O NMR, starting in 1983,[3] focused on high-resolution MAS studies of inorganic materials. It soon became clear from this early work that the quadrupole interaction strongly varies with the nature of the M–O bond,[4] although a range of other factors affect the quadrupolar interaction. In retrospect, a somewhat naïve view in these early papers using simplistic ideas of electronegativity of the bonds produced some correlations which for many cases (but not all) have provided a remarkably accurate guide as to the quadrupolar interaction. Although many of the bonds involving oxygen can be highly ionic, they can still have a covalency associated with them. As a first approximation, the quadrupole coupling constant (χ_Q or alternatively C_Q in some literature) where there are just M–O bonds present can be estimated as

$$\chi_Q\,(\text{in MHz}) = 0.203\, I(\%) + 14.78 \quad (18.1)$$

where I is the ionicity of the M–O bond defined in terms of the Pauling electronegativities EN(M,O) as

$$I(\%) = 100\,[\text{EN(O)–EN(M)}]/\text{EN(O)} \quad (18.2)$$

For asymmetric M–O–M′ bonds, the average ionicity of the two different bonds is employed, or indeed averaging all the direct oxygen contacts for higher coordination numbers. Historically, because of their widespread occurrence and importance, silicates/aluminosilicates were among the first compounds studied by ^{17}O MAS NMR, although they turn out to be one of the toughest problems to tackle. They often contain many crystallographically inequivalent sites, with relatively small chemical shift differences (covering a range of ~70 ppm), but have relatively large quadrupole interactions (~5 MHz), such that the residual second-order quadrupolar broadening in ^{17}O MAS NMR spectra acquired at 11.7 T exhibited broad, overlapping resonances from Si–O–Si and Si–O–Al species, rather than resolving all the crystallographically inequivalent sites. With the very significant advantage of hindsight, this provided an outlook that was unduly pessimistic, since in many cases concerning inorganic materials, the shift range is much larger and the quadrupolar interaction much smaller, thereby making ^{17}O an attractive nucleus for solid-state NMR.

The late 1980s and 1990s brought great advances in the NMR methodology applicable to ^{17}O, e.g., DOR, MQMAS, and STMAS, and through the availability of various hardware and equipment that have improved the quality of spectra. The availability of high magnetic fields e.g., ≥14.1 T, has meant that even oxygen sites with quite large χ_Q can be observed. To be ideally equipped, a range of magnetic fields is extremely desirable. For materials in which chemical shift dispersion is the dominant line broadening mechanism (i.e., M–O bonds with small χ_Q in a disordered material), moderate magnetic fields (7.05–9.4 T) are ideal. For materials such as silicates and aluminosilicates in which there are overlapping second-order quadrupole lineshapes, a relatively limited chemical shift range, and a minor line broadening due to chemical shift dispersion, higher magnetic fields are usually more suitable. In many cases, to be able to extract information more unequivocally, collecting data at several fields can separate sites and interactions because of their differing field dependencies. As an example, Figure 18.2 shows the ^{17}O NMR spectra of $[Ti(acac)O_{1.5}]_2[OSi(C_6H_5)_2]_3{\cdot}2\,C_4H_8O_2$, which has two inequivalent oxygen sites (see structure in Figure 18.2a), at three different magnetic field strengths.[5] The dominance of the second-order quadrupolar broadening in determining the linewidth is clear from the decrease in the linewidth with the increase in the applied magnetic field. The interactions can be readily determined from a fit that is consistent between the three magnetic fields (shown in Figure 18.2b). The magnetic shielding is the key interaction in discriminating the sites in this case, as the χ_Q values are quite similar (3–3.3 MHz), whereas the shift difference between the two is 65 ppm. The site assignment is clear from the relative intensity of the sites following the intensity distribution of $2(O_1){:}1(O_2)$.

Background details of the experimental NMR techniques that have enabled the study of ^{17}O and the recent and developing use of first-principles calculations that allow more structural information to be unambiguously extracted from NMR spectra are examined. This chapter concentrates on applications to inorganic materials, since ^{17}O NMR of organic solids is dealt with in Chapter 17. Note that further discussion of the application of ^{17}O NMR in the earth sciences can also be found in Chapter 22 concerned with the more general application of NMR of quadrupolar nuclei in this area. This chapter can also be read in conjunction with other recent publications that include summaries of ^{17}O applied to inorganic materials.[6–8]

18.2 BACKGROUND AND TECHNIQUE DEVELOPMENTS ENABLING ^{17}O NMR

18.2.1 Enrichment Schemes

Methods for NMR signal enhancement (e.g., ever higher magnetic fields, RAPT, QCPMG, etc.) are resulting in more ^{17}O NMR studies on unenriched materials. However, NMR observation of ^{17}O still usually requires some level of isotopic enrichment. A number of enrichment schemes have been developed, and there is continuing research effort in this area. The two readily available ^{17}O-labeled sources are $^{17}O_2$ and $H_2{}^{17}O$, with typical enrichment levels

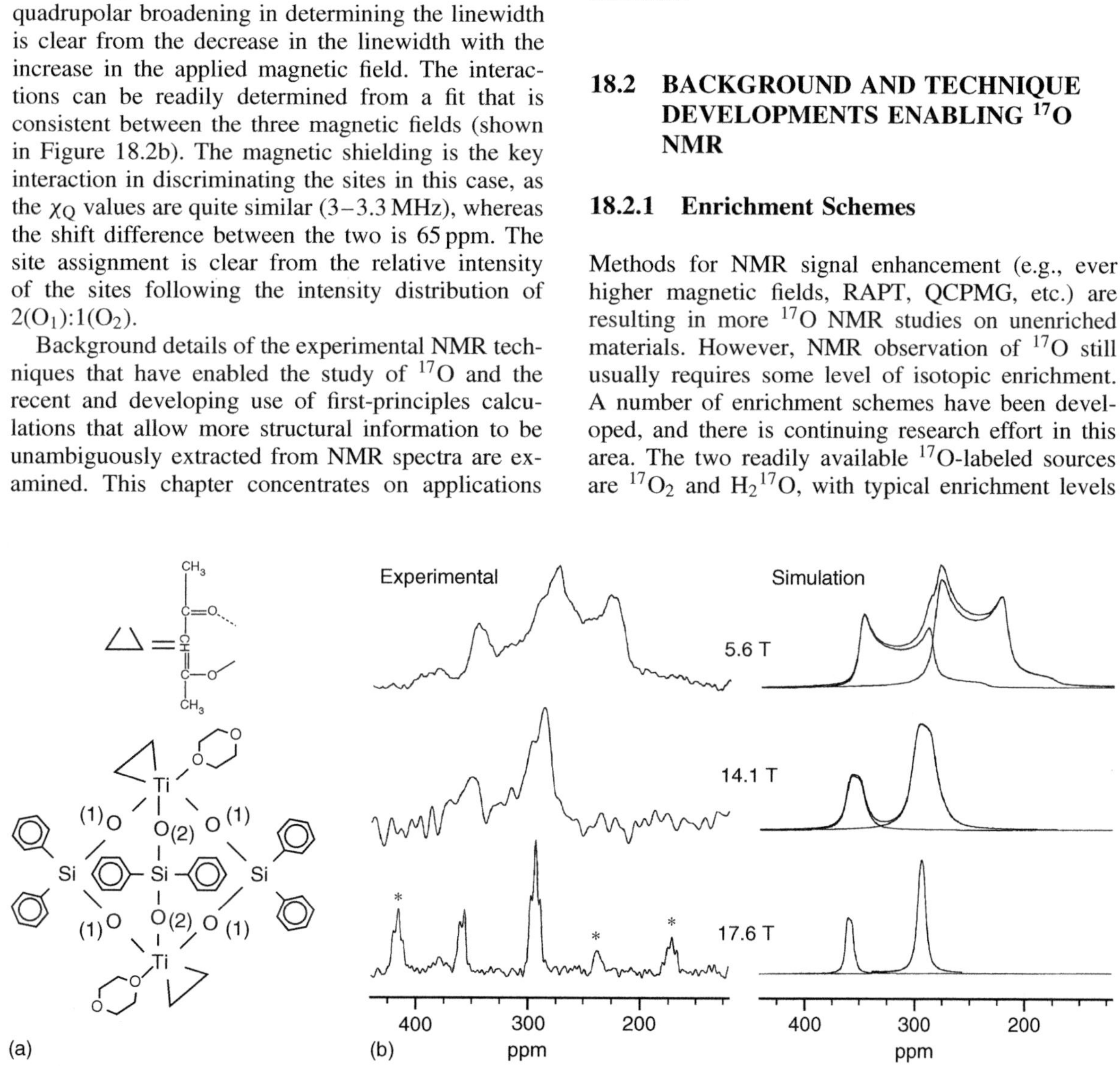

Figure 18.2. Schematic structure of $[Ti(acac)O_{1.5}]_2[OSi(C_6H_5)_2]_3{\cdot}2\,C_4H_8O_2$ showing the inequivalent oxygen sites O1 and O2 (ratio (2:1)) (a) and the ^{17}O MAS NMR at a spinning frequency of 10–14 kHz (with spinning sidebands indicated by *) at three magnetic fields, allowing simulation of the spectrum (b). (Reproduced from Ref. 5. © Elsevier, 2000.)

in these precursors of 10–75 at% (= atomic percent). The enrichment level of choice for any given experiment is a compromise between cost and the required sensitivity. Typically, 20–40 at% enrichment of the precursor is chosen to give sufficient sensitivity for conventional MAS spectroscopy, such that at current prices, 1 g of the labeled oxide product typically costs ~£100 to manufacture. For experiments with less efficient filling factors (e.g., DOR, Section 18.2.3) and especially 2D experiments, higher enrichment levels (≥40 at%) are desirable.

Hydrolysis of a reactive precursor, such as a chloride or an alkoxide, results in efficient incorporation of the oxygen label. After reaction, usually at or below room temperature, samples are heated either to remove organic products or to allow reaction.[6] Recently, some detailed considerations have been given to other synthetic routes so that small amounts of reagents could be employed. As an example, ZrW_2O_8 could be prepared via reaction of the pre-enriched component oxides or via a hydrated intermediate.[9] Thought has to be given to the firing atmosphere for any heat treatment required to form the final product, as exchange can often result in the loss of the isotopic label, and high-purity nitrogen/argon atmospheres are preferred.[10] Other routes to enrichment, including reacting soluble salts to form insoluble hydroxides that are subsequently calcined, exchange of sparingly soluble carbonates via water, or direct reaction of some metals with water, have been reported.[6]

Gas exchange can be used to enrich many oxides, with the sample heated under vacuum to remove the oxygen and the system then back-filled with ^{17}O-labeled O_2 before cooling down to room temperature. This approach has now been widely used for enriching oxides, especially high-temperature ceramic superconductors and zeolites. A detailed study of ^{17}O enrichment by hydrothermal exchange using ^{17}O-enriched H_2O vapor has been carried out for stilbite, a naturally occurring zeolite (with Si/Al = 3). Using high-resolution techniques such as MQMAS, DAS, and DOR (Section 18.2.3), clear separation of Si–O–Si and Si–O–Al resonances was observed, but owing to a distribution of chemical shifts, resolution of individual crystallographically distinct species was not obtained. However, it was shown that when the zeolite was back-reacted, i.e., heated with unenriched water, differences were observed in the relative intensities of the two resonances, indicating that the Si–O–Al oxygen species reacted more quickly. Furthermore, in some stilbites, a third resonance was also observed in MQMAS spectra,[11] which was assigned to Al–O–Al oxygen species. The exchange of ^{17}O-enriched water vapor with a dried gel to give a final microporous solid has also been applied to some aluminophosphates. It was again found that there was differential oxygen exchange with the different sites in $AlPO_4$-11 so that relative intensity information has to be carefully used.[12]

With the ongoing development of ^{17}O-enrichment schemes, several other approaches have been developed. For example, direct hydrothermal reaction of water with aluminum metal using a high-temperature (600 °C) and pressure-sealed platinum capsule was shown to produce single-phase Al_2O_3 without needing to go via a hydroxide or oxyhydroxide intermediate. This means that the enrichment level closely reflects that in the original water such that this is an efficient use of the original label.[13] In the manufacture of ^{17}O-enriched phosphates, such as glassy sodium phosphates, two possible methods are (i) hydrolysis of PCl_5 as a precursor, although this can lead to unwanted residual hydroxyl groups; and (ii) thermal oxidation of red phosphorus into P_2O_5, but scrupulously dry conditions must be used and P_2O_5 requires careful handling because of its strong hygroscopicity. New methods for oxygen exchange in phosphate ions have been proposed using an exchange method by heating the phosphate with water vapor at close to 600 °C. This temperature is chosen as the kinetics of all parts of the process are fast so that homogeneous enrichment is achieved. Effectively, a furnace is used in which ^{17}O-enriched water vapor is introduced. All parts of the furnace are kept hot to ensure that condensation of water does not occur.[14]

18.2.2 Physical Background to Key NMR Interactions that Provide Information for ^{17}O

The ability of NMR to provide information on the local structural environment arises from a range of interactions that produce small changes or perturbations to the basic Zeeman energy levels. For a nucleus with spin quantum number $I = 5/2$, such as ^{17}O, the $(2I + 1)$ possible orientations of the magnetic dipole moment result in a set of six equally spaced Zeeman energy levels, with five degenerate transitions corresponding to the Larmor frequency, ν_0. Interactions with magnetic moments from surrounding nuclei produce dipolar broadening, while modifications of the magnetic field experienced by the nucleus by the surrounding electrons results in a chemical shift. All nuclei experience these interactions, but for nuclei with $I > 1/2$, the nonspherical charge distribution of the nucleus gives rise to an electric quadrupole moment eQ, and the interaction of this with any EFG at the nucleus also results in a perturbation of the energy levels. The EFG can be described by a tensor $\boldsymbol{V}$, which in its principal axis system (PAS) is diagonal, with three components V_{xx}, V_{yy}, and V_{zz}, where $|V_{zz}| \geq |V_{yy}| \geq |V_{xx}|$.[6] The quadrupolar interaction can be described by both its magnitude, given by the quadrupolar coupling constant χ_Q (or alternatively C_Q in some literature)

$$\chi_Q = eQV_{zz}/h \tag{18.3}$$

and its cross-sectional shape described by η_Q, given by

$$\eta_Q = (V_{xx} - V_{yy})/V_{zz} \tag{18.4}$$

such that η_Q lies between 0 (for an axially symmetric tensor) and 1. The effect of this interaction is to lift the degeneracy of the NMR transitions, resulting (to a first-order approximation) in a central transition (for $m_I = 1/2 \leftrightarrow m_I = -1/2$) that remains at the Larmor frequency, and four other transitions, termed *satellite transitions*, with frequencies which now depend on χ_Q.[6,7] The frequency of the STs also depends on the orientation of the quadrupolar PAS with respect to $\mathbf{B}_0$. For the case of $\eta_Q = 0$, this dependence is given by $(3\cos^2\theta - 1)$. In a powdered sample, therefore, where all possible orientations are present, the STs exhibit large anisotropic broadenings, often over many megahertz. Observing these broad lineshapes experimentally may be difficult, if not impossible, and attention is usually focused on the relatively sharp central transition.

As the magnitude of the quadrupolar interaction increases, the use of a first-order approximation for the perturbation of the energy levels may no longer be sufficient to describe the spectrum and the effect of the interaction to second order must be considered. The second-order perturbation is more complicated, consisting of an isotropic (orientation independent) shift, a second-rank broadening (proportional to $(3\cos^2\theta - 1)$), and a fourth-rank term with a more complex orientation dependence. All transitions are affected by the second-order interaction and have frequencies that depend on χ_Q, η_Q, and the orientation of the PAS. The central transition, therefore, is no longer sharp, but exhibits anisotropic second-order broadening in a powdered sample, and is shifted from ν_0 by the isotropic second-order quadrupolar shift. The width of the line depends primarily (though not entirely) on the magnitude of the interaction, proportional to χ_Q^2/ν_0, while the shape is determined by η_Q. The second-order quadrupolar broadening is much smaller than the first-order broadening (proportional to χ_Q), with ^{17}O central transition spectra often only a few tens of kilohertz wide.[6,7]

18.2.3 Comparison of Key Experimental NMR Approaches Relevant to ^{17}O

Solid-state ^{17}O NMR spectra will typically contain lineshapes broadened by the second-order quadrupolar interaction, compromising resolution and leading to difficulties in extracting site-specific information for distinct species. Information can still be obtained in some cases, for example, where only a small number of species are present, where some form of selective enrichment has been used, or where the types of species present are very different. To obtain an undistorted lineshape and to avoid any problems with probe ringdown, spectra are typically recorded using an echo sequence.[6] NMR spectra of stationary samples also contain information on the CSA, and the magnitude of the anisotropy, its asymmetry and the relative orientation of the shielding, and quadrupolar tensor PAS must be taken into account if accurate NMR parameters are to be extracted. Furthermore, any significant dipolar interactions will need to be removed with appropriate decoupling if informative spectra are to be obtained.

It is well known that the use of MAS—rapid spinning of the sample around an axis inclined at 54.736° to $\mathbf{B}_0$—can increase spectral resolution. For $I = 1/2$ nuclei, MAS can produce high-resolution spectra by removing the CSA and dipolar interactions, both proportional to $(3\cos^2\theta - 1)$, providing that the rotation is sufficiently rapid. For quadrupolar nuclei, the presence of quadrupolar broadening poses an additional problem. MAS is able to remove the first-order quadrupolar broadening (present only in the STs and proportional to $(3\cos^2\theta - 1)$), but usually results in an extensive manifold of spinning sidebands, owing to the magnitude of the broadening. However, MAS is not able to remove the second-order quadrupolar broadening (present in both the central and STs), owing to its more complex orientational dependence. Only partial narrowing occurs under MAS and an anisotropic broadening remains in the spectrum. As discussed briefly in a previous section, the sensitivity of NMR spectra of both stationary and spinning samples can be significantly improved through the use of one or more of a wide variety of enhancement techniques, such as from RAPT or double-frequency sweeps (DFS) applied before the initial $\pi/2$ pulse or the use of CPMG echo trains in acquisition.[1,2]

The accurate and site-specific determination of the quadrupolar and chemical shift parameters is often the aim of many NMR experiments, as these exhibit a strong dependence on local environment. In many cases, however, the extraction of this detailed information is hindered by the presence of second-order quadrupolar broadening, particularly if more than one crystallographically distinct species is present. This is clearly shown in Figure 18.3(a), where the ^{17}O MAS NMR spectrum of (75% isotopically enriched) orthoenstatite, $MgSiO_3$, is shown.[15] The overlap of the broadened resonances prevents the determination of the number of distinct species, or the extraction of any detailed information about their NMR parameters and local environment. Given this problem of resolution, there is great interest in the development and improvement of techniques that can produce truly high-resolution NMR spectra, i.e., the complete removal of the anisotropic quadrupolar interaction. These are not discussed in great detail here but are listed below and summarized in Table 18.1. For more information, the reader is referred to other articles in the *Encyclopedia of Magnetic Resonance* and a number of recent reviews.[6,7,16–18]

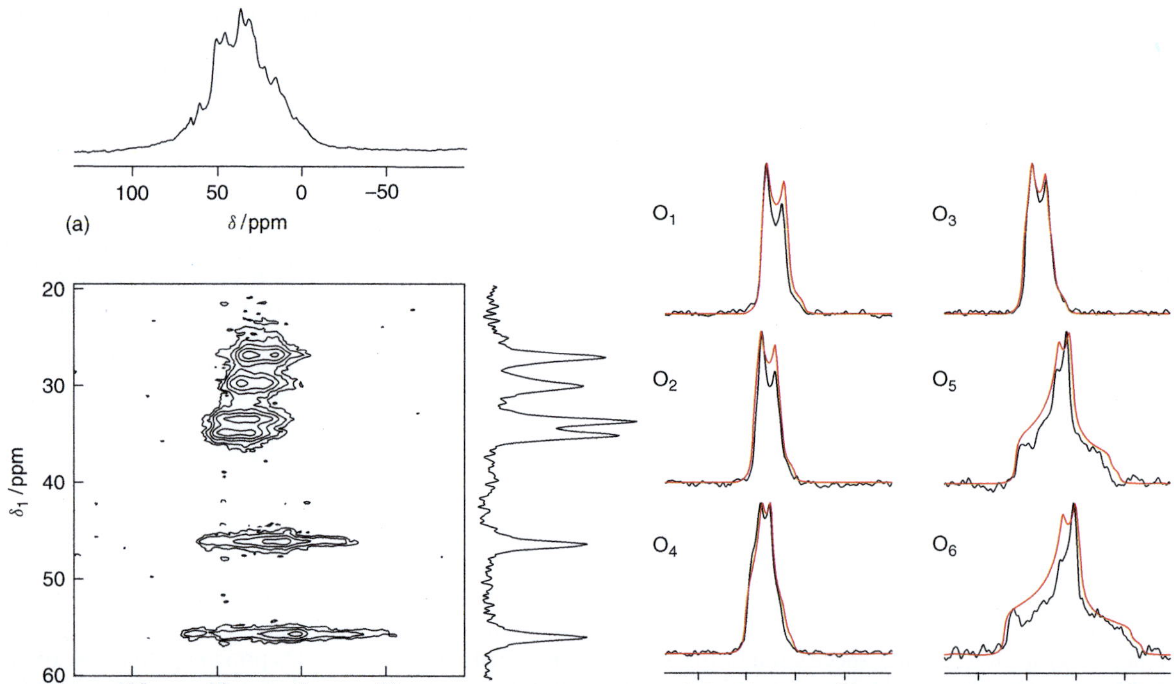

Figure 18.3. ^{17}O (54.3 MHz) (a) MAS and (b) triple-quantum MAS NMR spectrum, recorded using a rotor-synchronized phase-modulated split-t_1 pulse sequence, of orthoenstatite (75% isotopically enriched). The MAS rate was 10 kHz. (c) Cross sections extracted parallel to δ_2, (see (b) from the ^{17}O (54.3 MHz) triple-quantum MAS NMR spectrum of orthoenstatite). Also, shown (in red) are lineshapes simulated using the parameters derived from first-principles calculations. (Adapted from Ref. 15. © American Chemical Society, 2007.)

Table 18.1. Summary of experimental approaches for ^{17}O NMR

	Probe	Resolution	Considerations
Static	Static	Low	May require echo-based experiments but is relatively straightforward to implement
MAS	MAS	Good	Better sensitivity and resolution than spectra on stationary samples
DOR	Specialist	Excellent	Limited MAS rates, a 1D experiment, good sensitivity, but needs high degree of skill to spin
DAS	Specialist	Excellent	A 2D experiment can be limited by T_1 particularly if MAS detected
MQMAS	MAS	Excellent	A 2D experiment relatively straightforward to use but has reduced sensitivity, particularly for high χ_Q
STMAS	MAS	Excellent	A 2D experiment, better sensitivity than MQMAS but does require accurate magic angle setting and very stable MAS rates

1. DOR about two angles (usually 54.736° and 30.56°) simultaneously through the use of a sophisticated double rotor, removing both second- and fourth-rank anisotropic broadening simultaneously.
2. DAS involving sequential rotation around two different angles in a 2D experiment.
3. 2D correlation experiments such as MQ- or STMAS, where removal of the anisotropic quadrupolar broadening is achieved through correlation of either the symmetric ($+m_I \leftrightarrow -m_I$) MQ transitions or the single-quantum STs with the central transition under MAS.

Conceptually, DOR is probably the simplest approach to produce an isotropic spectrum with good sensitivity, although there are relatively few DOR probes available. There have been great improvements in the technology since the original designs when typical outer rotor speeds of only 1 kHz were possible and stability meant that experimental times were short. In more modern designs, the bearing and the drive gas pressures of both rotors are under computer control, and outer rotor speeds up to ~2 kHz are possible with essentially continuous operation. DAS probes, in contrast, are particularly rare. In contrast to DOR and DAS, MQMAS and STMAS can be performed using conventional MAS probes. The utility of MQMAS for ^{17}O is reflected in the number of studies that now employ this approach and several examples can be found in later sections. It is typically a triple-quantum (i.e., $m_I = \pm 3/2$) experiment that is carried out, usually using smaller (2.5 or 3.2 mm) diameter rotors. The limiting factor determining sensitivity for such experiments is often the generation of sufficient RF field (which is a combination of the coil diameter and the voltage applied to the coil), such that the MQ coherences can be generated efficiently. Such limitations in the RF power available also mean that five- or higher quantum experiments are not widely performed. Figure 18.3(b) shows that the six distinct oxygen species in orthoenstatite can be resolved in an ^{17}O MQMAS spectrum, with six sharp, narrow resonances observed in the (indirect) isotropic dimension.[15] Information on the quadrupolar and shielding parameters can be obtained either from the position of the center of gravity of the each ridge lineshape or from cross sections along each, as shown in Figure 18.3(c), where six powder-pattern lineshapes can be observed. STMAS experiments often provide a significant gain in sensitivity, often a factor of 3–8, but are more demanding in their implementation, requiring a very precise adjustment of the magic angle and stable spinning for accurate rotor synchronization. Such approaches are, however, advantageous when sensitivity is limiting.

18.2.4 First-principles Calculations of ^{17}O NMR Parameters

Although NMR provides a sensitive probe of the local structural environment, it can often be difficult to assign, or even simply to interpret, spectra of complex materials. Therefore, much work has focused on understanding the dependence of NMR parameters on a variety of structural parameters, including coordination number, bond distances, and bond angles, with the ultimate aim of providing structural information using NMR spectra of unknown materials. More recently, the development of computational methods has provided an alternative, more flexible, and perhaps ultimately more accurate approach to understanding the dependence of NMR parameters on local structure and for aiding the assignment of complex spectra. The quantum mechanical calculation of NMR parameters is a complex many-body problem that cannot be solved without approximations.[19] One approach, Hartree–Fock (HF) theory, assumes that each electron moves independently in a mean field generated by the other electrons in the system. In an alternative approach, density functional theory (DFT) considers the energy as a functional of the total charge density, reducing the dimensionality of the problem and increasing the efficiency of the calculations. However, further approximations must be made to describe electron correlation and exchange, i.e., the local density or generalized gradient approximations. DFT has been used with great success in a variety of materials, enabling calculations to be performed at a fraction of the computational cost.

The simplest calculations of NMR parameters, often utilizing localized basis sets such as the atomic orbitals, are performed for molecules, while an infinite periodic solid has to be approximated as a cluster, centered on the nucleus of interest, and terminated typically by ^{1}H.[19] As the size of the cluster increases, the calculated parameters approach those of a 3D infinite solid, albeit at increasing computational cost. Tossell and coworkers were among the first to use this approach for the calculation of NMR parameters, with an example demonstrating its application to ^{17}O given in Ref. 20. Although it can be difficult to achieve sufficiently large basis sets (a problem that can restrict the accuracy of this approach), qualitative results can be easily obtained and are often sufficient for spectral assignment. Furthermore, the variation of NMR parameters with local structure can also provide insight, as demonstrated in recent work by Grandinetti[21,22] using a simple cluster-based approach to investigate silicate and aluminosilicate glasses.

In some cases, the level of accuracy achieved by cluster methods is not sufficient for spectral assignment, as demonstrated by Bull *et al.* in their study of the zeolite ferrierite,[23] where many of the ^{17}O species have very similar NMR parameters. For infinite solids, calculations can also be carried out using periodic boundary conditions, exploiting the high translational symmetry found in the solid state.[13] Many such codes use plane wave basis sets, an orthonormal basis set that is computationally simple to implement, although a large number of plane waves may be required for high accuracy. The WIEN code (an augmented plane wave code) allows the accurate determination of EFGs[24]; however, this all-electron approach can be time consuming. More recently, Pickard and Mauri introduced a plane wave pseudopotential method for the calculation of NMR parameters, with the separation of "core-like" and "valence-like" regions.[13] This approach considerably improves the efficiency of the calculation, allowing more complex systems to be investigated on more reasonable timescales. An extension of this theory, the GIPAW (gauge including projector augmented wave) method,[13] "reintroduces" the core electrons, allowing parameters that are dependent on them, such as the chemical shielding, to be computed. Since its introduction, the accuracy of the approach has been demonstrated in a variety of applications, including a full assignment of the 10 oxygen species in the zeolite ferrierite.[13] In another example, the calculated NMR parameters of a variety of silica polymorphs were shown to be in good agreement with experiment, unlike many of the studies using cluster calculations performed previously.[13] Figure 18.3(c) illustrates the good agreement between experiment and calculation attainable for ^{17}O NMR, with lineshapes simulated using the parameters calculated using CASTEP (a GIPAW-based code) in red, compared with those extracted from the ^{17}O MQMAS spectrum of orthoenstatite (shown in Figure 18.3b).[15] The small differences between the distinct oxygen species are accurately reproduced and enable a full assignment of the six species. With more examples of its utility appearing in the literature, this method does appear to offer an accurate and efficient approach to the calculation of ^{17}O NMR parameters.

18.3 SYSTEMS AND MATERIALS

18.3.1 Crystalline Ionic Solids

For ionic oxide materials, high-resolution ^{17}O NMR spectra can be achieved by MAS alone, as the EFGs, which are dominated by the local electron density, tend to be small.[4,6,8] Often, the relatively narrow MAS lines mean that even natural abundance work on such materials can be contemplated.[2,25] Some NaCl-structured oxides (e.g., MgO, CaO, and BaO) produce very narrow resonances that can be used as secondary shift references. In such Group IA and IIA oxides, where a significant quadrupolar interaction is detected, this is presumably attributable to defects in the structure. ^{17}O NMR of sol–gel-produced nanocrystalline MgO produces some very interesting spectra with three peaks at 47, 42–47, and 41–42 ppm, with typical peak widths of 90, 700, and 100 Hz respectively present in the MAS spectrum.[26] Despite the sample clearly containing both residual OCH_3 and OH species, it was shown that the resonances resulting from these species were being filtered out in the MAS experiments via the initial preacquisition delay, as a consequence of their very much greater linewidth. Hence, all the species remaining in the spectrum were associated with Mg–O–Mg. The relative changes of the intensity (and to a lesser extent the peak position) with particle size (Figure 18.4) provided clues as to the identity of each of these peaks. The 47 ppm peak aligns exactly with the position in bulk MgO, with the other narrower peak arising from an MgO-like environment, but with some of the fragments (OCH_3, OH), three positions removed. The broader line was then attributed to Mg–O–Mg environments, but within two atomic layers of the surface—so effectively NMR is in this case discriminating the surface.[26]

In such oxides, oxygen can often adopt a large range of coordination numbers (typically 1–5), with the ^{17}O becoming more shielded with increasing coordination number. For example, from a number of studies, OTi_x coordinations have shifts in the range from 650 to 850 ppm for $x = 2$, from 450 to 650 ppm

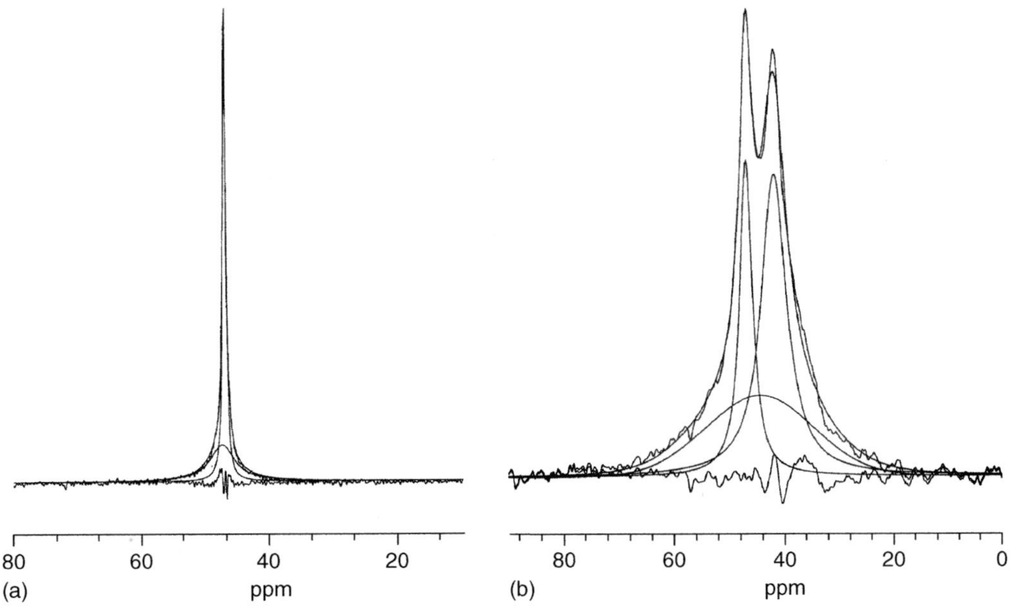

Figure 18.4. ^{17}O MAS NMR of sol–gel-produced nanocrystalline MgO for two different mean particles sizes produced by heat treatment, (a) 13.5 and (b) 3 nm. Three peaks are observed at 47, 42–47, and 41–42 ppm, all corresponding to Mg–O–Mg environments. (Reproduced from Ref. 26. © American Chemical Society, 1998.)

for $x = 3$, from 250 to 450 ppm for $x = 4$, and at <250 ppm for $x = 5$. The total shift range covered by these units is almost 800 ppm, with also large changes in the CSA. The sensitivity of the ^{17}O magnetic shielding to the local bonding can be gauged through the three polymorphs of TiO_2, with all resonances being spread over a shift range of 35 ppm. These materials are all made up of TiO_6 units with the oxygens in OTi_3 coordinations. The polymorphs are structurally distinguished through slight differences in the ways the TiO_6 units are joined (i.e., Ti–O–Ti bond angles). The solid-state conversion of anatase to rutile can be followed by NMR, as the two shifts are separated by ~35 ppm, and the structurally more distorted TiO_6 in rutile shows a small but noticeable χ_Q (Figure 18.5).[27] Hence, both the isotropic and anisotropic contributions to the interactions can be used to discriminate environments. The sensitivity to coordination number is also illustrated in La_2O_3, where there are two inequivalent oxygen sites, an OLa_4 (584 ppm) and an OLa_6 (467 ppm) with a shift difference of ~120 ppm.[25,28]

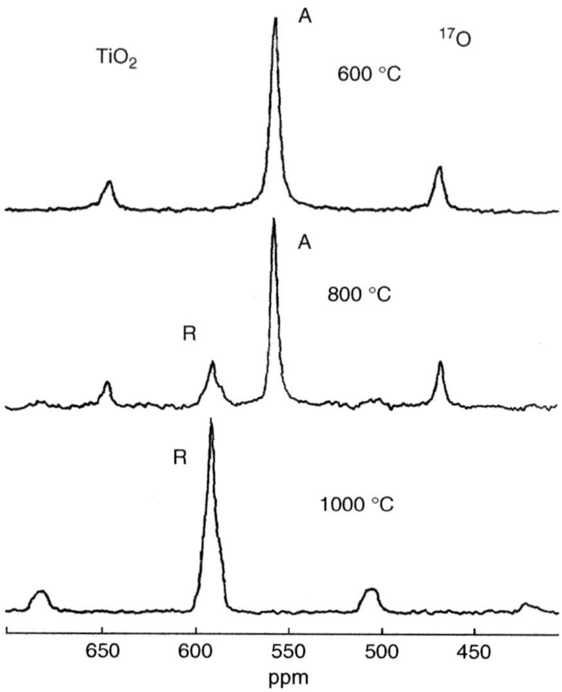

Figure 18.5. ^{17}O MAS NMR at 9.4 T showing the complete resolution of anatase (A) and rutile (R) OTi_3 environments and the conversion between the phases with heat treatment. (Reproduced from Ref. 27. © the Royal Society of Chemistry, 1993.)

^{17}O MAS NMR spectra of ionic ternary oxygen-containing materials show a large variation in linewidth, with some spectra showing extremely high resolution, revealing small shift differences arising from very subtle variations in the bonding around the oxygen site. A selection of typical spectra of titanates and zirconates that illustrate the variation in the number of resonances and the splitting is shown in Figure 18.6. For A_xBO_y compounds, there are now increasingly large datasets of ^{17}O isotropic chemical shifts, with, for example, ^{17}O shift ranges of titanates (from 372 to 564 ppm), zirconates (from 298 to 376 ppm), and hafnates (from 237 to 332 ppm). This clearly indicates that the B ions principally determine these shift ranges for ^{17}O. Comparison of isostructural zirconates and hafnates shows that the ratio of the isotropic shifts for ^{17}O (Hf/Zr) fall in the very narrow range 0.84–0.88, close to the ratio of the polarizing powers of these two elements.[29,30] Cubic perovskites show only a single, sharp oxygen resonance, reflecting the equivalence of all the oxygen sites in the structure. Two resonances typically separated by 5 ppm are observed in materials such as $CaTiO_3$ with the $GdFeO_3$ structure, where oxygen sites become inequivalent. In monoclinic materials of lower symmetry (e.g., Li_2TiO_3), there are only two different ^{17}O NMR signals, but these are now separated by greater shift differences (e.g., ~20 ppm).[29] A number of materials in such systems are important technologically, for instance as nonlinear optical, ferroelectric, and relaxor phases. In $KTiOPO_4$, two distinct classes of oxygen exist. One group links the TiO_6 and PO_4 units (i.e., bridging oxygens (BO) Ti–O–P) and have $\chi_Q \sim 5$ MHz and shifts of 100–150 ppm. The second group is the nonbridging oxygens (NBO) that correspond to short Ti–O bonds, which are very much more ionic so have a smaller χ_Q (~1 MHz) but are markedly shifted at ~750 ppm.[31] The ^{17}O CSA values for such compounds can also be readily characterized, again illustrating that the magnetic shielding interaction dominates the spectrum. The widths of lineshapes resulting from the ^{17}O shielding anisotropy (technically termed the *span*) for titanates tend to spread over larger ranges of shift than for zirconates, reflecting the influence of the B cation on the shielding tensor.

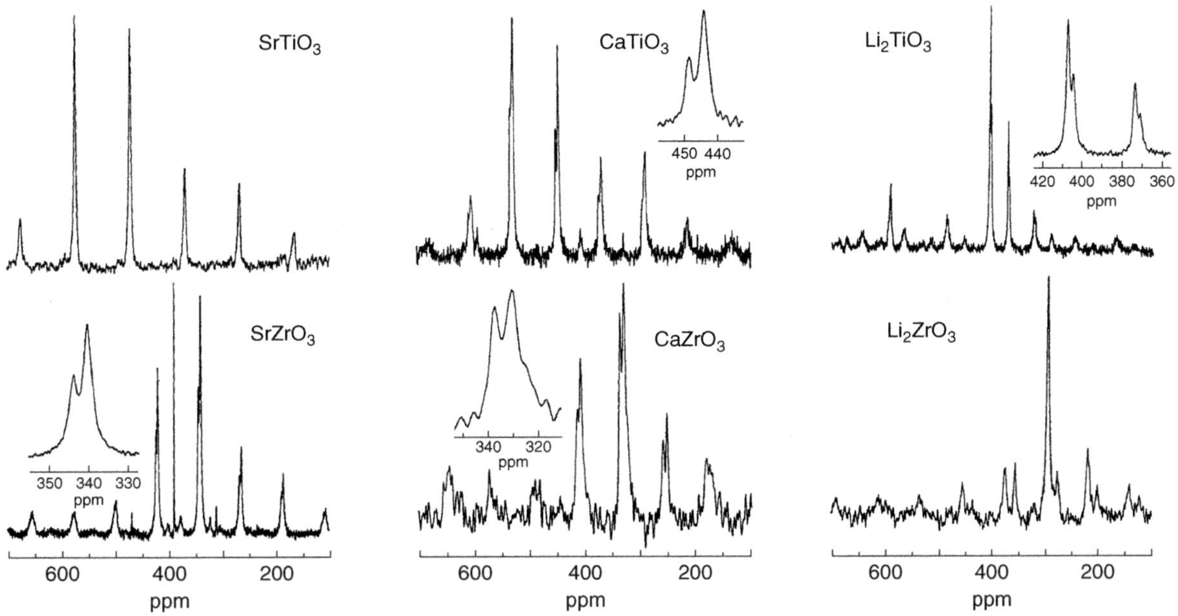

Figure 18.6. Typical ^{17}O MAS NMR spectra from some ionic titanates and zirconates showing the high resolution that is available under MAS alone in crystalline solids when the quadrupole interaction is small. (Reproduced from Ref. 29. © American Chemical Society, 1996.)

18.3.2 Other Simple Ionic Oxides with Larger χ_Q

In other simple oxides, the quadrupolar interaction can be significantly stronger than those examples just discussed. Silicates form such a group and these are dealt with in more detail in Sections 18.3.3 and 18.3.4. A good non-silicate example, however, is GeO_2 that has two crystallographic forms, quartz-like and rutile-like, where the oxygens are OGe_2 and OGe_3, respectively. Both sites have similar, large χ_Q values of 7.3–7.5 MHz, but very distinct shifts that differ by 90 ppm.[32] Another example of such an oxide is that of tellurium with the relatively high covalency of the Te–O bond, which exhibits χ_Q in the range 6.2–7.8 MHz. In comparing the parameters from a variety of tellurites where there is a range of BO and NBO, there was surprisingly little discrimination of the NMR parameters between these species. This was explained in terms of the electronic structure of the TeO_4 units present.[33] In ternary and higher oxides, there is a clear anion involving oxygen (XO_y) that has internally relatively strong covalent bonds and ionic bonds to the metals in the system. For example, in $CaCO_3$, the carbonate anion ($CO_3{}^{2-}$) has strong covalent bonds and the oxygen coordination is C–O–Ca_2. This tricoordinate oxygen has a large χ_Q of almost 7 MHz and an asymmetry parameter (η_Q) of 1.[34]

Aluminates are an intermediate case and ^{17}O NMR of $CaAl_4O_7$ and $CaAl_2O_4$ has provided some useful information with intermediate χ_Q values (of 1.2–1.8 MHz) observed for most Al–O–Al environments. Higher fields ($\geq$11.7 T) do provide resolution of most sites, with eight resonances observed in $CaAl_2O_4$ corresponding to the 12 crystallographically inequivalent oxygens.[35] In $CaAl_4O_7$, there are four distinct oxygen sites with one being an unusual tricluster (OAl_3) site.[36] The aluminums are both AlO_4, but the mixtures of the coordinating oxygens between the OAl_2 and OAl_3 are 3 and 1 for the Al1 site and 2 and 2 for the Al2 site, which is reflected in the increase in χ_Q (of ^{27}Al) from 6.3 to 9.5 MHz. This is unambiguously verified by a ^{17}O–^{27}Al HMQC experiment (Figure 18.7).[37] Further examples of the type of material that have such intermediate values of ^{17}O χ_Q are gallates such as $LaGaO_3$ with a value of 3.3 MHz and $BaIn_2O_5$ where χ_Q is 5.2–5.9 MHz.[38]

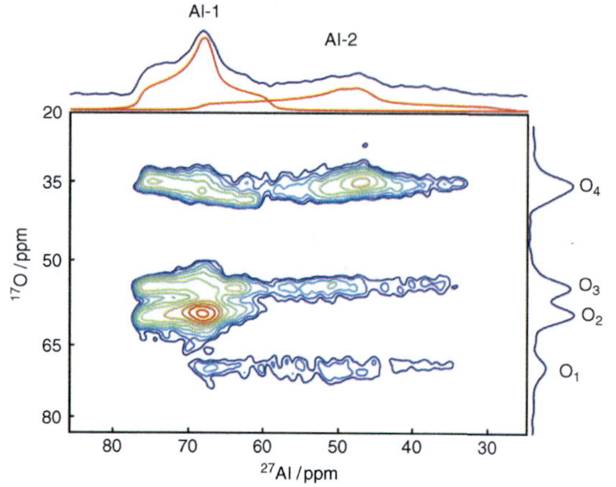

Figure 18.7. $^{17}O-^{27}Al$ HMQC spectrum of $CaAl_2O_7$ with DFS excitation on the ^{27}Al, spinning rate of 15 kHz, and $\tau = 3$ ms, recorded at 17.6 T. (Reproduced from Ref. 37. © American Chemical Society, 2005.)

18.3.3 Synthetic Mineral Phases

Oxygen is the second most abundant element, and the most abundant anion, on Earth, comprising 30% of its mass, and a vital component of the silicate minerals that make up the Earth's crust and mantle. Oxygen NMR can, therefore, offer a sensitive probe of local structure, disorder, and dynamics in the silicate materials that are of importance in determining the physical and chemical properties of the planet. Despite this, relatively little ^{17}O NMR has been reported, in comparison to ^{29}Si (a spin $I = 1/2$ nucleus) or ^{27}Al (a quadrupolar spin with $I = 5/2$, but 100% natural abundance), for example. The Earth's crust is composed principally of aluminosilicate minerals, while the mantle is predominantly magnesium silicate. The structures of these materials are typically constructed from SiO_4 units, which can be isolated (orthosilicates), or may be connected by corner sharing in Si_2O_7 groups (pyro- or sorosilicates) or silicate chains (often termed *inosilicates*). Further polymerization results in silicate sheets or 3D frameworks. When oxygens are linked to two silicons (i.e., a BO), significantly higher χ_Q values (of ~4–6 MHz) are usually observed in comparison to the much smaller values (~1–3 MHz) typically found for NBO. This was demonstrated by early MAS and static ^{17}O NMR studies[39] of alkaline-earth chain silicates, where a dependence of both the magnetic shielding and quadrupolar coupling parameters on the nature of the cations present was observed. However, the lower resolution of these simple experiments hindered the extraction of more detailed information, and often the resolution of all distinct oxygen species was not possible. However, highly crystalline silicate minerals are excellent candidates for the high-resolution NMR techniques described in Section 18.2.3, and many more detailed studies have since followed.

The simplest silicates are perhaps the silica polymorphs themselves. All consist of 3D frameworks formed by corner-sharing silicate units but exhibit a range of structure types, providing an ideal opportunity to determine the effect of structural rather than chemical changes on the NMR parameters. In these frameworks, all oxygen species are bridging (in Si–O–Si linkages) and exhibit large χ_Q, usually between 5 and 6 MHz. Quartz and cristobalite possess only one distinct oxygen and can be studied easily using conventional MAS and/or static experiments.[6] Coesite, however, possesses five distinct oxygens, which were resolved by Grandinetti *et al.* in a 15 day DAS experiment.[40] In contrast to the other polymorphs, stishovite is a high-pressure phase and contains corner-sharing SiO_6 octahedra. This results in an unusual three-coordinated oxygen species that exhibits one of the highest χ_Q values known for an unprotonated oxygen (6.5 MHz).[6] The variation of the isotropic chemical shift χ_Q, as well as η_Q, with structural parameters, particularly with the Si–O–Si bond angle, has been investigated in the silica polymorphs and subsequently used to infer information on the range of local structural environments in silica-based glasses.[6]

Early high-resolution work by Mueller *et al.* exploited ^{17}O DAS and DOR in the study of isotopically enriched magnesium and calcium silicate minerals.[41] In most cases, these approaches enabled the resolution of all inequivalent sites, including eight species in the case of wollastonite ($CaSiO_3$). Site-specific quadrupolar and shielding parameters extracted from the data emphasize the differences between BO and NBO in chain silicates. MQMAS has been employed to study layered aluminosilicate clay minerals, such as pyrophyllite, resolving Si–O–Si, Si–O–$^{[4]}$Al, and Si–O–$^{[6]}$Al oxygens (where the superscript denotes the aluminum coordination).[42] Although similar chemical shifts were observed for many of the oxygens, the BO species, in general, exhibited larger χ_Q values than the NBO species. More recently, further work using MQMAS has probed the local environment in phyllosilicates (among the major components in terrestrial surface environments), distinguishing apical ($^{[6]}$Si–O–2$^{[4]}$Al) and basal ($^{[6]}$Si–O–$^{[6]}$Si and $^{[6]}$Si–O–$^{[6]}$Al) species.[43]

Magnesium silicates have a particular prominence in the earth sciences, as they form the principal chemical constituent of the Earth's mantle. The major component of the upper mantle (to depths of 410 km) is olivine α-(Mg,Fe)$_2$SiO$_4$, an iron-bearing form of forsterite (α-Mg$_2$SiO$_4$). The high-resolution spectrum of synthetic forsterite (35% isotopically enriched in ^{17}O), acquired using MQMAS, displays three resonances with intensities 1:2:1, corresponding to three NBO species.[44] These were assigned in later work using periodic plane wave GIPAW calculations.[45] Deeper in the Earth, as the pressure increases, forsterite undergoes phase transformations to the (iron-bearing) β- and γ-forms of Mg$_2$SiO$_4$, wadsleyite and ringwoodite, respectively, defining a region known as the *transition zone*. For isotopic enrichment, which is necessary to enable some of the NMR experiments, difficult syntheses, requiring high-pressure multianvil apparatus, have to be undertaken. This high-pressure synthesis typically significantly limits the amount of material available, often only to a few milligrams. For such samples, the more sensitive STMAS experiment has been exploited, with a high-resolution spectrum of ~10 mg of 35% enriched wadsleyite exhibiting four distinct resonances.[46] Wadsleyite is a pyrosilicate, i.e., it contains Si$_2$O$_7$ groups, and the spectrum displays resonances corresponding to one BO, two NBOs, and an unusual five-coordinate species, coordinated only by Mg. The latter species exhibits a particularly low χ_Q (of ~1.3 MHz), reflecting the lack of covalent bonds. Interestingly, the χ_Q values observed for the NBO are also slightly higher (3.8–4.4 MHz) than observed for many crustal materials, although these values (and the assignment of the four resonances) were confirmed by subsequent DFT calculations.[45]

It is thought that the mantle contains a vast amount of water, perhaps many more times than that in the Earth's oceans and atmosphere, yet relatively little is known about how or where this water is stored. This results, in particular, in interest in the ability of the nominally anhydrous mantle silicate minerals to incorporate protons (water) into their structure. Owing to the sensitivity of the ^{17}O magnetic shielding and quadrupole parameters to the local environment, ^{17}O NMR offers an ideal probe for the study of both nominally anhydrous and hydrous materials. Hydroxyl oxygens typically possess higher χ_Q values, often between 6 and 8 MHz, and so should be easily distinguished from BO and NBO species in the silicate structures. The humite minerals, nMg$_2$SiO$_4$·Mg(OH)$_2$, have been proposed as models for the storage of water within olivine.[47] High-resolution MQMAS spectra were able to distinguish the distinct NBO species, four for chondrodite ($n = 2$) and eight for clinohumite ($n = 4$). However, the hydroxyl oxygens were not observed in these MAS-based experiments as a result of their increased second-order broadening at 9.4 T and the relatively slow MAS rates used in these experiments. The hydroxyl species in humites can be observed, however, using static experiments, where the ^{17}O spectra revealed a much broader lineshape, similar to that of brucite, Mg(OH)$_2$, in addition to the narrower resonance associated with the NBOs, as shown in Figure 18.8. Only this broad resonance appeared when the spectrum was acquired with cross polarization (CP) (from ^{1}H), confirming its assignment as a hydroxyl oxygen, with χ_Q of ~6.8 MHz for both chondrodite and clinohumite.[47]

Although many investigations have studied synthetic mineral phases, enabling the incorporation of the ^{17}O isotopic label, natural minerals have attained their structural states over geologic time and at geological conditions, and this may result in different

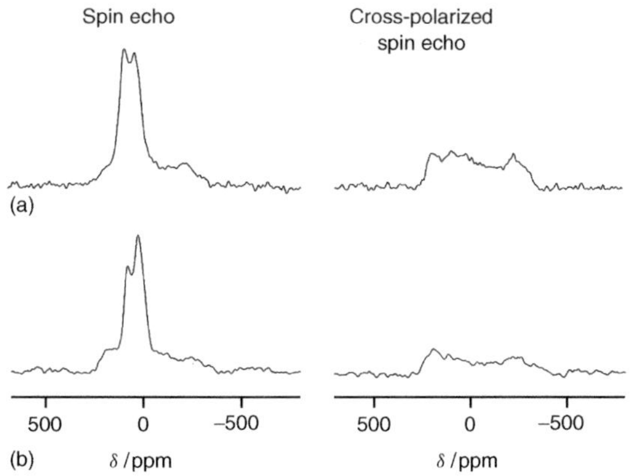

Figure 18.8. Conventional and cross-polarized ^{17}O (54.2 MHz) NMR spectra of 35% isotopically enriched (a) chondrodite and (b) clinohumite, recorded using a spin-echo pulse sequence for stationary samples. (Adapted from Ref. 47. © American Chemical Society, 2001.)

structures at the atomic or macroscopic level. Although natural abundance spectra of simple oxides can be obtained relatively easily (Section 18.3.1), it is much more difficult to acquire spectra for more complex minerals, especially as χ_Q increases. As described in a previous section, it is possible to acquire natural abundance ^{17}O NMR spectra of such systems using combinations of different sensitivity enhancement techniques.[2] For example, CP provides sufficient sensitivity gains to observe the hydroxyl oxygen species in brucite and boehmite (AlO(OH)) at natural abundance levels. For systems that do not contain ^1H, a combination of RAPT and CPMG can also provide a considerable increase in sensitivity, as demonstrated by natural abundance ^{17}O NMR of forsterite.[2] The combination of these techniques was also used in the study of zircon ($ZrSiO_4$), a mineral that has acquired a partially metamict structure over ~570 million years through the radioactive decay of the natural uranium it contains.[2] The natural abundance ^{17}O spectrum of zircon contained, in addition to the nonbridging Si–O–Zr oxygens at ~150 ppm, a considerable amount of bridging (Si–O–Si) oxygen species at ~65 ppm. The presence of these species was attributed to the damage of the structure by the recoil of heavy nuclei that have emitted α particles.[2] The study of natural materials, such as zircon, offers great potential for understanding the long-term storage of radioactive waste within ceramic waste forms.

18.3.4 Zeolites and Mesoporous Oxides

Zeolites are aluminosilicate frameworks based on networks of corner-sharing AlO_4 and SiO_4 tetrahedra, containing cavities and channels of molecular dimensions. This has resulted in a range of applications, such as catalysis, gas storage and as ion-exchange materials and molecular sieves. The extent of the ordering of Al and Si on the tetrahedral sites has a significant effect on chemical reactivity. It is generally assumed, particularly for zeolites with Si/Al $\geq$1, that an aluminum avoidance rule, termed *Lowenstein's rule*, holds, with Al–O–Al linkages forbidden or disfavored. A second, and perhaps more important, aspect of zeolite study is the incorporation of additional or "extra-framework" cations and absorbed molecules; the latter is of particular importance for the understanding of zeolites as heterogeneous catalysts. There has been a great deal of ^{27}Al and ^{29}Si NMR study of zeolitic frameworks,[6] but relatively little ^{17}O NMR, as a consequence of the low natural abundance. However, although ^{27}Al and ^{29}Si NMR may help to determine framework ordering, it is the oxygens that line the pores and cavities and determine the binding of adsorbed species. Therefore, ^{17}O NMR enables detailed study of this binding and, through the sensitivity of the magnetic shielding and quadrupolar parameters to the local environment, provides a probe of the framework structure and order. Many zeolites may be enriched in ^{17}O through simple hydrothermal exchange (Section 18.2.1) although, owing to differences in reactivity of the various oxygen species, care should be taken when interpreting relative peak intensities.

Early ^{17}O NMR of zeolites (using conventional static and MAS experiments)[48] demonstrated that the ^{17}O χ_Q was particularly sensitive to the nature of the local oxygen environment. Clear differences were observed between Si–O–Al (3–4 MHz) and Si–O–Si species (5–6 MHz), reflecting the more covalent nature of the bonding in the latter case. However, the low resolution of these experiments limits the information available, and discrimination between distinct Si–O–Si or Si–O–Al species is often not possible. ^{17}O DAS, DOR, and MQMAS have all been employed to produce high-resolution spectra of a variety of zeolitic materials. Although Si–O–Al and Si–O–Si species are easily distinguished by these approaches, it is not always possible to observe individual distinct species. This results from the large number of such oxygen species in the framework and, in some cases, the presence of a range of local environments. For example, any Si/Al disorder in the framework produces a range of local coordination environments (e.g., different second and fourth nearest neighbors) for a particular oxygen. When differences in T–O–T′ bond angles and different interactions with extra-framework cations are also considered, this can result in a distribution of environments and a broadening of the high-resolution spectrum. In some cases, however, resolution of distinct crystallographic species is possible. This is often the case for zeolites with a high Si/Al ratio, which contain little disorder, and few or no extra-framework cations. Resolution of distinct oxygen species was achieved in siliceous ferrierite[23] with MQMAS (14.1 T), resulting in a spectrum containing five distinct resonances with differing intensities. The combination of this with variable field (11.7–18.8 T) DOR was able to determine the presence of 10 distinct oxygens

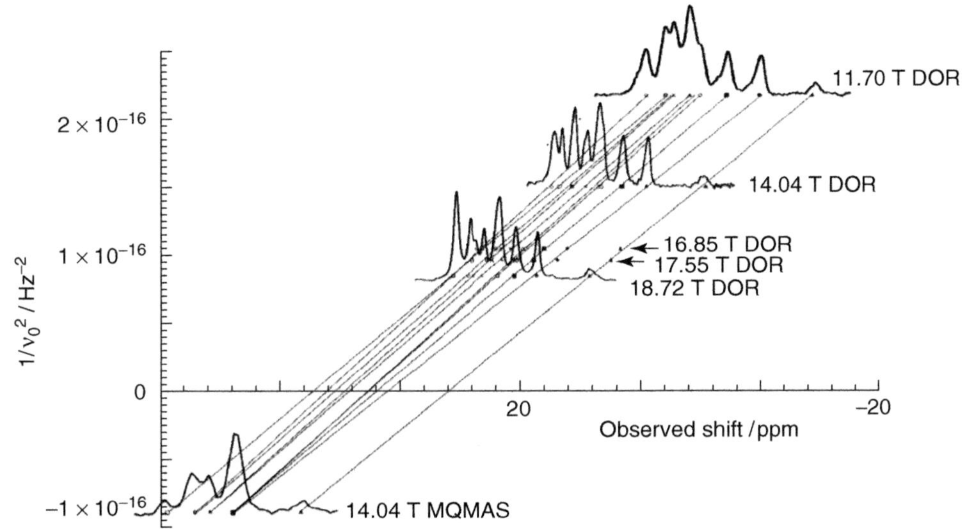

Figure 18.9. Combination of ^{17}O DOR data for ferrierite at five magnetic fields with MQMAS data. The intercept on the x-axis corresponds to the isotropic chemical shift for each site and the gradient is proportional to the quadrupolar product $P_Q[= C_Q(1 + \eta_Q^2/3)^{1/2}]$. (Adapted from Ref. 23. © American Chemical Society, 2000.)

(Figure 18.9), although a full assignment of the spectrum was hindered by the complexity of the system.

Resolution of distinct species may also be possible in aluminosilicate zeolites when Si/Al $\approx$ 1. In this case, "ordering" of the aluminosilicate framework (produced by the apparent preference for aluminum avoidance) reduces the range of possible local environments and, therefore, distributions in NMR parameters. Three crystallographically distinct Si–O–Al environments in hydrated zeolite A were resolved by Pingel *et al.*[49] using MQMAS. The sites exhibited very similar χ_Q (~3.4 MHz) but different isotropic chemical shifts. Similarly, the four oxygen species in zeolite Na-LSX (Si/Al $\approx$ 1) can be resolved using MQMAS. Based on these results, a correlation of ^{17}O isotropic chemical shift with Si–O–Al bond angle was suggested, with the isotropic chemical shift decreasing as the angle increased. However, no such simple correlation was observed for the Si–O–Si species in ferrierite.[23] A more detailed investigation of a range of hydrated and dehydrated ion-exchanged zeolites[50] demonstrated that, although the M–O–M′ angle does affect the chemical shift, no simple general correlation is observed, owing to the effect of the different extra-framework species. Interactions with cations or water can have a dramatic effect on the ^{17}O chemical shift; for example, a shift of ~34 ppm is observed for one oxygen between Na-A and Tl-A.

The catalytic properties of zeolites are determined by the acid sites within the structure. The direct detection of Si–(OH)–Al Brønsted sites was reported only relatively recently,[51] although further studies have since followed. For zeolite HY, the ^{17}O resonance associated with the Brønsted site could not be unambiguously identified by MAS, but its presence was revealed using high-field experiments, and a combination of CP, REDOR, and TRAPDOR experiments. The ^{17}O parameters extracted ($\chi_Q = 6.6$ MHz, $\eta_Q = 0.8$, and $\delta_{iso} = 28$ ppm), revealed a significantly larger quadrupolar interaction than normally associated with Si–O–Al linkages. χ_Q was reduced (to ~5 MHz) on reaction with acetone, which interacts with the zeolite framework via the Brønsted site, and lengthens the O–H bond length from 0.97 to 1.02 Å. Hence, ^{17}O NMR offers much promise as a sensitive probe of Brønsted acidity in zeolites and to the understanding of the structural geometry of these active sites.

Zeolites are not the only type of porous framework materials to be studied by ^{17}O NMR (for example, see also Section 18.3.8). Mesoporous molecular sieves (e.g., M41S and MCM-41) are an important group of related materials. With pores in the 20–100 Å range and large surface areas up to 1400 m^2/g, they can fulfill a number of applications, such as catalysis and small-molecule absorption. An amine template is typically used in the synthesis of transition-metal oxide mesoporous materials. The structure of the wall materials in the initial templating reaction and subsequent changes on removal of the template or aging can be clearly followed by ^{17}O MAS NMR.[52,53] For the cases of four oxides (SiO_2, TiO_2, Ta_2O_5, Nb_2O_5) in the initial templated state, the ^{17}O chemical shifts suggest that all oxygens are coordinated as OM_2. This is highly unusual in these oxides where the oxygen usually exhibits higher coordination, with the exception of SiO_2. The oxides then show very different behaviors as the template is removed, with Ta_2O_5 showing a mixture of OTa_2 and OTa_3 environments, very much as would be seen in the bulk oxide. For TiO_2, the OTi_2 species is lost immediately on removal of the template, and a mixture of OTi_3 and OTi_4 is formed. The OTi_4 becomes dominant with increasing heat treatment, whereas oxygen remains solely as OTi_3 in bulk titanias. In complete contrast, for Nb_2O_5, the oxygen remains in an ONb_2 environment at all times, with a very highly ordered wall structure, evident by the narrow ^{17}O resonances.

18.3.5 Glasses

Structures of glasses at the most local level (i.e., nearest neighbors) closely resemble that observed in related crystalline phases. This was first demonstrated by ^{29}Si MAS NMR, which revolutionized the characterization of silicate-based glasses through the identification of the different local Q^n species.[6] Similarly, ^{17}O NMR of silica glass reveals that the NMR interaction parameters very closely match the values from the crystalline silica polymorphs, quartz and cristobalite, as they are related by the fundamental SiO_4 building block. The key difference between a crystal and a glass is the disorder created by the way these SiO_4 units are joined together to form the extended network structure in the system, i.e., intermediate range order. In SiO_2 glasses, the Si–O–Si linkages show a distribution of bond angles, and this can be related directly to the nature of the disorder. Although ^{29}Si chemical shifts do vary as a result of changes in the bond angle at the oxygen, Si is more remote from the structural changes, and the effect is also averaged over the four corners of the SiO_4 tetrahedra. The ^{17}O NMR parameters provide a much more direct probe of the bond angle distribution, and there has been considerable theoretical effort aimed at relating the observed parameters to the variation in structure, particularly by Grandinetti and coworkers.[22] The bond angle distributions have been extracted in some detail for both SiO_2 and GeO_2 glass.[6]

The possibilities for disorder increase in multicomponent glasses. For example, the addition of network modifiers to a simple silicate glass requires charge balance within the system through the formation of NBOs. ^{17}O NMR spectra of such glasses (using a combination of spectra under stationary, MAS, DAS, and MQMAS conditions) enable discrimination of the BO and NBO and determination of their NMR parameters. Changing the alkali cation affects both the maximum bond angle and the width of the bond angle distribution. The intermediate range order in such metal silicate glasses is determined by both the Si–O–Si bond angle distribution and the ordering of the cations. Cation ordering has profound implications for many properties of a glass, such as the configurational entropy of the system and transport properties. This can be probed by ^{17}O NMR, with the NBO being particularly sensitive to coordination by the metal. A comparison of the isotropic dimension of DAS spectra of $K_2Si_4O_9$ and $KMg_{0.5}Si_4O_9$ glasses showed no significant increase in width in the mixed cation glass, but the peak position was shifted from that in the pure potassium glass.[54] There is still some dispute as to the implication of this observation. It was originally suggested that a high degree of cation ordering was occurring, with the NBO atoms probably coordinated by two potassiums and one magnesium, whereas other more recent studies have interpreted this simply as favoring Mg-rich distributions around the NBO. In other systems, MQMAS and DAS spectra suggest a random distribution of cations.

A further level of complication arises when additional network formers or intermediates are also present in the system, as the ordering of these species can also affect the glass properties. Given the importance of crystalline aluminosilicates in the earth sciences (Section 18.3.3), aluminosilicate glasses, for example, are important as models of melt structure. A key question in this case is whether or not the aluminum avoidance principle, generally found in crystalline materials, applies for glasses and melts. For albite ($NaAlSi_3O_8$) glass, each AlO_4^- is charge balanced by a Na^+ such that the glass network is completely connected (i.e., Q^4). However, there remains the question of the degree of ordering of the

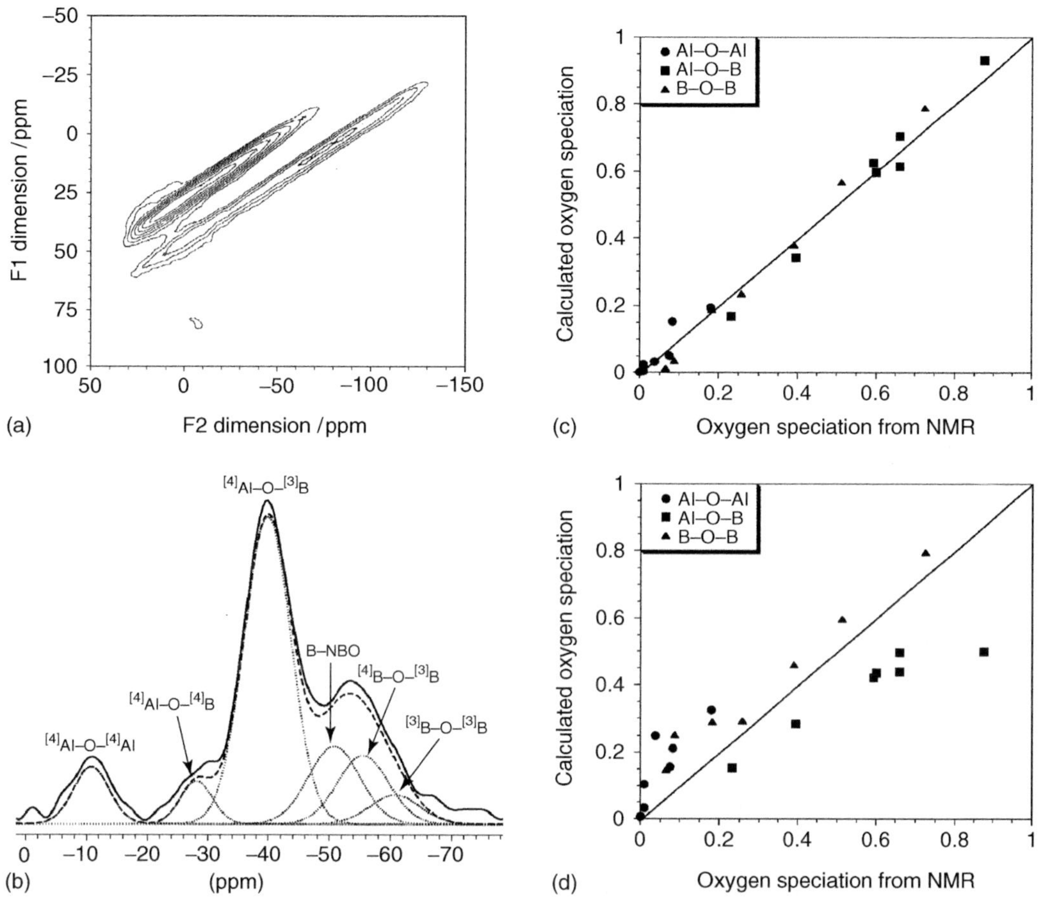

Figure 18.10. Examples of ^{17}O 3QMAS NMR spectra from multicomponent glasses (a) showing the complete resolution of Si–O–Al and Si–O–Si fragments in albite glass and (b) the 3QMAS isotropic projection from an aluminoborate glass with six distinct contributions, with (c, d) intensity distributions predicted for models including (c) tetrahedral avoidance and (d) a random distribution of tetrahedral species. (Reproduced from Refs 55 and 57 © Elsevier Science, 1997 and 2005.)

silicon and aluminum within the network, which, in principle, can have Si–O–Si, Si–O–Al, and Al–O–Al linkages. ^{17}O MQMAS NMR spectra of albite glass revealed only Si–O–Si and Si–O–Al contributions (as shown in Figure 18.10a) and the ratio was, within error, the 1:1 expected for an ordered network with no Al–O–Al groups.[55] The calibration of ^{17}O parameters from Al–O–Al linkages in aluminates allowed the NMR interaction parameters, and hence their probable position in MQMAS spectra, to be deduced. This shows that there will probably be significant overlap of the Al–O–Al and Al–O–Si peaks in MQMAS spectra, meaning that detection of a few percentage of Al–O–Al may be difficult.[56]

Borosilicates are important technological glasses, and the long-standing ^{11}B NMR measurements for characterizing such materials are now being supplemented by ^{17}O to distinguish the B–O–B, Si–O–B, and Si–O–Si groups, along with the NBO present in the system. It appears from initial work that the borate and silicate parts of the network are intimately mixed, with intensities reflecting almost random ordering. ^{17}O MQMAS NMR and its analysis has recently become much more sophisticated for such systems, with separate ^{17}O signals being detected from NBO, Si–O–Si, Si–O–$^{[4]}B$, Si–O–$^{[3]}B$, $^{[4]}B$–O–$^{[3]}B$, and $^{[3]}B$–O–$^{[3]}B$. Hence detailed, quantitative information is available on the fragments that make up the structure. In aluminoborate glasses, the same type of information is available, with the additional complication that the aluminum can have [4], [5], and [6] coordination. There is increasingly extensive data from crystalline model compounds and other simpler glasses, such that the parameters from the different contributions are becoming much better defined. In a recent example investigating sodium aluminoborate glasses, up to six different contributions were fitted (Figure 18.10b).[57] The information obtained for these materials is also supported by ^{11}B and ^{27}Al NMR. There are a number of models that can be used to calculate the oxygen distribution between the different environments, with two of the most straightforward random mixing or maximum $^{[4]}M-O-^{[4]}M$ avoidance within the constraints of the composition. Analysis of the data for sodium aluminoborate glasses indicated that the distribution was far from random (Figure 18.10c and d) and revealed that $^{[4]}Al-O-^{[4]}B$ formation is favored over homonuclear tetrahedral groups. The data also suggested that the oxygen associated with the $^{[5,\,6]}Al-O-^{[4]}Al$ linkages is probably tricoordinate.[57]

The study of the formation, alteration, and the local structure of glasses has benefited considerably in recent years from a combination of experimental NMR and first-principles DFT calculations. NMR parameters can be calculated for a variety of structural environments, either using a range of simple model crystalline compounds or using substitution and supercells to model disordered materials. A recent example by Charpentier and coworkers[58] combined classical molecular dynamics (MD) simulation and DFT calculations to simulate ^{17}O NMR spectra of amorphous $CaSiO_3$. NMR parameters were calculated using the GIPAW approach for two different glass models initially obtained using MD simulations. Simulated spectra were in very good agreement with those obtained experimentally, with clear separation of the Si–O–Ca and Si–O–Si species, and confirmed a clear correlation between ^{17}O NMR parameters with the number of surrounding Ca.

18.3.6 High-temperature Ceramic Superconductors

An area that for a short period saw very significant ^{17}O NMR activity was in the oxide-based ceramic superconductors first discovered in 1986. Oxygen is at the core of these structures and plays a key role in the superconductivity. The large shifts between the oxygen sites within the structure meant that even NMR spectra from stationary samples provided good resolution. The charge correlations that are at the heart of the properties of these materials are reflected in the ^{17}O shifts and relaxation times.[59] ^{17}O NMR spectra have been reported from a range of the ceramic superconductors that include $YBa_2Cu_3O_{7-x}$ and $Tl_2Ba_2CaCu_2O_8$.[60] Typical ^{17}O NMR spectra from stationary samples show a broad asymmetric peak with a shift of 1500–2000 ppm, as well as narrower peaks much closer to the shift range of the related diamagnetic oxides. By comparison with structurally similar compounds, the resonance with the large shift is attributed to the oxygen in the CuO_2 planes.[60] This significant shift was attributed to Knight shift effects due to the charge carriers in these systems. Variable temperature work showed that the large shifts of these oxygen resonances decreased markedly with temperature toward the diamagnetic shift region as the charge-carrier density decreased. The complexity of the charge-carrier dynamics was illustrated by the $n = 3$ bismuth phase, where the shifts of the two CuO_2 planes showed different temperature dependences, indicating that a common spin susceptibility cannot be used to describe the system[61] which was very different behavior to that seen in the yttrium-containing systems.

18.3.7 Sol–Gel-produced Phases

Sol–gel production of materials is becoming an increasingly important approach offering alternative synthetic routes for some materials and, in some cases, allowing novel materials to be made. In this route, precursors are mixed in solution which then, via a series of reactions, cross-link to form a network that increases in extent until it gels. To produce the final solid, the gel is normally heated to stabilize it by the removal of more volatile fragments such as residual organic fragments and hydroxyls. For oxides, a hydrolytic route is often used, which then involves a hydrolysis step and at this point ^{17}O-labeled water can be used. As the condensation process proceeds, various oxo bridges are formed. An important characteristic of the gel is the range of these bridges present and the relative amount of each type. The systems are highly disordered at most points during the processing so that a local probe such as NMR is necessary to get a good understanding of the coordination environments present. ^{17}O MAS NMR studies of a wide range of sol–gel-produced simple oxides have been undertaken (Ref. 6, Section 6.3.2). The large chemical shift range that typifies ^{17}O means that these systems can be straightforwardly studied using MAS at moderate magnetic fields and with modest isotopic enrichment. Resonances resulting from the different coordination environments of oxygen are then usually completely resolved within the spectrum. The oxygen coordinations present in the precursor, and then formed during the structural rearrangements caused by heat treatment, can be readily identified. What has become apparent is that a wide range of structural behavior is seen, with in some cases very subtle differences occurring between oxides, and even for the same oxide depending on the exact nature of the preparation route.

The local environment in the initial gels in many cases closely aligns with that in the final stable oxide. In ionic oxides (e.g., ZrO_2, TiO_2, HfO_2, and Ta_2O_5), the linewidths observed in the gels, while much broader than the crystalline analogs, are still much below those in oxide materials where χ_Q is significant. The linewidths observed here are largely determined by chemical shift dispersion, reflecting the wider range of environments and hence a more disordered structure of the oxide gels at this point. As an example, in the sol–gel formation of ZrO_2 by the hydrolysis of zirconium *n*-propoxide,[62] the NMR spectrum of the dry as-made gel contains two peaks at 405 and 303 ppm that correspond to OZr_3 and OZr_4 environments (Figure 18.11). These resonances are very similar to those observed for the corresponding environments in the crystalline monoclinic form, at 401 and 322 ppm, but with the linewidths a factor of 5 greater, reflecting the structural disorder at this stage. Subsequently, as the heat treatment proceeds, there are small changes in shift and relative intensity, presumably reflecting the changing connectivity of the oxide framework on the removal of residual organic and hydroxyl fragments. At 360 °C, significant structural rearrangement occurs on crystallization, with a third intermediate peak at 374 ppm, which corresponds to tetragonal ZrO_2. This becomes the dominant coordination on heating to 500 °C,

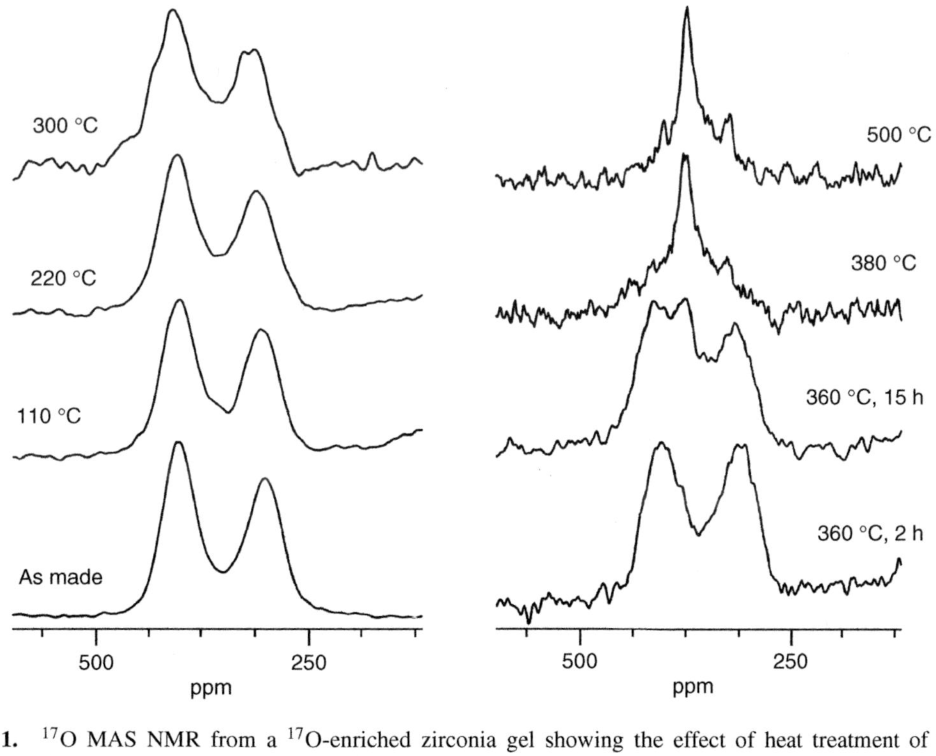

Figure 18.11. ^{17}O MAS NMR from a ^{17}O-enriched zirconia gel showing the effect of heat treatment of the different oxygen coordinations present. (Modified from a figure in Ref. 62. © American Chemical Society, 2001.)

although small monoclinic peaks, corresponding now to crystalline monoclinic ZrO_2, are also seen. ZrO_2 has an interesting phase behavior with the monoclinic form the stable room temperature bulk phase, but with the tetragonal form stabilized by surface effects in small domains. Hence, in the nanosized crystalline domains initially produced, the tetragonal form dominates. As the domains grow, the monoclinic form becomes dominant with peaks at 401 and 322 ppm (not shown),[62] although it is very interesting to note how, while the environments in the initial gel are clearly "monoclinic-like", the shift difference does indicate they are distinct. The behavior of ZrO_2 contrasts nicely with that of TiO_2, where in the initial gel, a mixture of OTi_3 and OTi_4 occurs (bulk TiO_2 only has OTi_3 environments). With heat treatment, there is simply a progressive loss of OTi_4 up to the crystallization temperature, where all the OTi_4 has been removed and a large decrease in the linewidth of the OTi_3 is observed. Studying in detail, the sol–gel formation of TiO_2 has revealed that the structural development is controlled by the hydrolysis ratio and by the use of complexing agents.[63] The nature of the ligand was found to change the local structural units in the titania core.

In more complex chemical systems (e.g., binaries and ternaries), the precursors can have very different reaction rates. In binary gels containing SiO_2 and other metal oxides, this can be a significant problem because of the very different reaction rates. These different reaction rates can impact on the homogeneity of mixing and the local atomic-scale connectivity of the different components. An example that has been much studied because of possible technological applications is TiO_2–SiO_2, with usually TiO_2 added as the minority component. In this system, distinct, well-separated ^{17}O resonances can be seen in the MAS NMR spectra, from Ti–O–Ti, Ti–O–Si, and Si–O–Si. The presence of Ti–O–Si indicates that at least to some extent there must be atomic mixing, with its observation in the absence of Ti–O–Ti indicating that the titanium is dispersed in the silica network. At low concentrations (<15 mol% TiO_2), only Si–O–Si and Ti–O–Si are observed, with the Ti–O–Ti being seen only at concentrations where phase separation is expected.[64] For a whole range of metal oxides added to SiO_2, these different fragments can be observed and hence the phase separation behavior examined. The interesting comparison is that below ~15 mol% added metal oxide, in the region of complete solid solution with SiO_2 (i.e., there is no sign of the formation of M–O–M linkages), there is significant variation in the Q^n speciation revealed by ^{29}Si MAS NMR. This is an indication that although the samples are all effectively homogeneous, the role the added metal is playing within the silica network is very different. The metals fall into two groups, those which form a high fraction of Q^4 (e.g., Ti and Nb) and those which form a spread of Q^n species (e.g., Zr, Ta, and Hf). The difference has been ascribed to the coordination the metal can take. For the former group, the metal is becoming tetrahedrally coordinated as part of the actual framework itself and hence bridging bonds form, whereas the other metals take higher coordinations and create NBOs.

^{17}O NMR work on sol–gel-prepared materials has been extended to systems such as hybrid metal oxide–siloxane materials. Here, the silicon functionality changes such that the four covalent bonds the silicon forms can be to varying numbers of oxygens and carbons. This basis can be used with additions, such as TiO_2. The ^{17}O signals from Si–O–Si fragments showed a ~40 ppm deshielding from the straight TiO_2–SiO_2 materials, reflecting the changed electron density around the central silicon in progressing from the normal SiO_4 unit to SiO_2C_2.[65] In more complex systems where there is a mixture of such functionalities, 3Q MAS has been used to discriminate and quantify the number of oxygens bridging between differently functionalized silicons (i.e., O_3Si–O–SiO_3 will give a different position to CO_2Si–O–SiO_2C); hence, the degree of mixing between the different silicons can be accurately monitored.[66]

18.3.8 Phosphate Crystals and Glasses

In 2006 when we last reviewed ^{17}O solid-state NMR of inorganic materials,[7] there were few reports of ^{17}O spectra of phosphates, which was probably due to a combination of the relative difficulty of designing efficient and straightforward enrichment schemes and the larger χ_Q (typically >4 MHz). Key impetus came from the observation of the ^{17}O NMR signals from P_2O_5.[67] In P_2O_5, there are both bridging and terminal oxygen sites, with the latter exhibiting a smaller χ_Q, of ~4 MHz, compared to the BOs that have $\chi_Q \sim 7.5$ MHz. Such large values only become accessible with the availability of high fields and faster MAS rates. An extensive study that combined experimental approaches from calcium phosphate and calcium silicophosphate phases showed the very wide

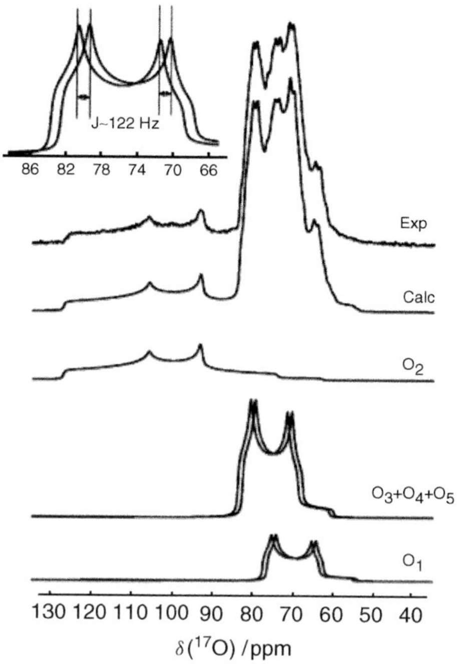

Figure 18.12. ^{17}O MAS NMR spectrum of $Na_5P_3O_{10}$ recorded at 18.8 T. The spectrum shows the five subspectra from the different oxygen sites in the structure, and the high crystallinity of the sample reveals a $^{17}O-^{31}P$ J-coupling of ~122 Hz. (Reproduced from Ref. 69. © American Chemical Society, 2008.)

range of coupling constants present. There was a clear correlation between the PO···Ca bond length and the ^{17}O chemical shift, with a chemical shift change of ~70 ppm observed for a change of bond length on 0.1 Å.[68] The sensitivity of the ^{17}O NMR parameters to the structure as well as the need to combine the experimental data with first-principles calculations is well illustrated with a detailed study of three crystalline sodium phosphates.[69] The 1D ^{17}O MAS NMR spectrum (Figure 18.12) shows clearly the P–O–P (O2) oxygen, distinguished by its much larger χ_Q value, and then a crowded region corresponding to the NBOs, where four lineshapes overlap. Close inspection of the spectrum shows a small splitting that is attributable to $^{17}O-^{31}P$ J-coupling of 120–130 Hz (Figure 18.12). To assign the NBO, the J-coupling can be used in a $^{17}O\{^{31}P\}$ MAS-J-HMQC experiment, where only a single site (O_1) correlates to the Q^2 phosphate unit, whereas the Q^1 phosphate site is correlated to the other three oxygen sites (O_3, O_4, and O_5). This attribution agrees with the intensities obtained from the simulation of the 1D spectrum. In moving away from well-ordered crystalline materials, an interesting comparison was made between ^{31}P and ^{17}O NMR of the ordered compound $BiCd_2PO_6$ and the closely related semiordered crystalline compound $Bi_{1.2}Zn_{1.2}PO_{5.5}$. The ^{31}P NMR data shows much more disorder in the latter system (determined from the linewidth), but even double-quantum NMR experiments do not really pick out any favored correlations in the semiordered sample. Despite the quadrupolar nature of ^{17}O, its much greater chemical shift range means that although linewidths are much greater in both dimensions of the ^{17}O 3Q MAS NMR spectra, several oxygen sites can be seen in the Bi–O–M and P–O–M regions. However, the widths associated with these indicate that the disorder over the associated metal (Bi and Zn) sites is much higher than the ordered cadmium analog.[70]

The natural extension from the semi-disordered crystalline materials is to discriminate BO (P–O–P) and NBO (P–O–M) in phosphate glasses. The overall quantification of BO and NBO has been shown in a number of studies to be consistent between ^{31}P- and ^{17}O-derived NMR data. In the case of sodium borophosphate glasses, a multinuclear approach that included ^{17}O static echo work at 9.4 T and MAS and 3QMAS at 17.6 T showed five distinct oxygen sites (P–O–P, Na···O–P, P–O–B, B–O–B, and Na···O–B). The data showed that not all five were simultaneously present and the P–O–B played a major role in the bonding of the glass.[71] For the interaction of water with phosphate glasses, ^{17}O NMR is also an ideal probe. A detailed study of the weathering mechanism of sodium phosphate glasses first looked at a series of model hydrogen phosphates (see Section 18.3.9). This provided good background as to the interaction parameters for the environments created on the reaction with water. The ^{17}O 3QMAS data (Figure 18.13) shows very clearly that as aging takes place, a resonance that can be attributed to NaH_2PO_4 as a weathering product emerges, as well as much increased overall ordering of the material, as the widths along the isotropic dimension of the 2D 3Q dataset show very significant decreases.[72]

An intriguing extension of ^{17}O MAS NMR to other phosphorus-containing systems is to the binding of self-assembled monolayers of ^{17}O-enriched phosphonic acids deposited on a titania anatase support. A multiple field (9.4 and 17.6 T) approach provided good constraints on the simulations of the

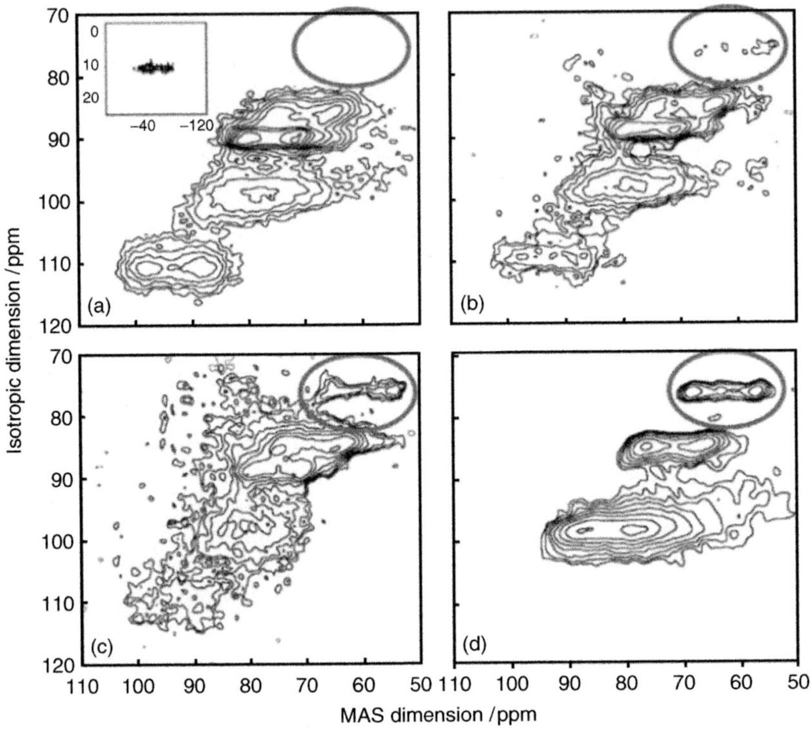

Figure 18.13. ^{17}O 3QMAS NMR spectra of the crystallization process of $NaPO_3$ glass as a result of aging in a water bath at 60 °C for 1 (a), 10 (b), 12 (c), and 2 years (d), with the formation of a NaH_2PO_4 resonance (O_3) highlighted in the circular region. (Taken from Ref. 72. © Royal Society of Chemistry, 2010.)

spectra, with very good resolution provided at 17.6 T. Oxygen species observed included P=O, P–O–H, and P–O–Ti. The strong signal corresponding to P–O–Ti bonds suggest strong surface interaction of TiO_2 with phosphonic acids. Using the data to distinguish the different possible modes of interaction (i.e., bridging vs chelating), the ^{17}O chemical shift of the P–O–Ti sites is consistent with a bridging interaction.[73]

18.3.9 Hydroxy Species

The observation and characterization of hydroxyl (OH) oxygen species is of considerable interest in the study of a wide range of materials, including glasses, sol–gels, and minerals, owing to the importance of quantifying levels of hydration and the investigation of hydrogen bonding. The high degree of covalency of the OH bond results in relatively large χ_Q values, typically in the range 6–8 MHz.[74] The isotropic chemical shift range is again large for hydroxyls, typically −20 to 100 ppm.[75] For the case of Al–O–H linkages, χ_Q values in the region of 5–6 MHz have been observed for simple hydroxides and oxyhydroxides, although larger values (up to 7 MHz) have been observed in aluminosilicate minerals.[74] These large χ_Q values result in broad resonances which may be difficult to observe under MAS (particularly at slow speeds or moderate B_0 field strengths), and in many cases, echo spectra on stationary samples have been employed. Alternatively, the observation of hydroxyl oxygens has been achieved by the use of CP from ^1H, which selectively enhances any oxygen resonances that are spatially close to ^1H nuclei.[75] In CP, the signal intensity depends strongly on the OH distance, and so in some cases (usually where a pair of nuclear spins are fairly isolated), quantitative information can be obtained. This was the case for $Mg(OH)_2$ and $Mg(OH)_x(OCH_3)_{2-x}$. A combination of Lee–Goldberg decoupling and the modeling of the dipolar oscillation in the CP data allowed the $^{1}H-^{17}O$ dipolar coupling to be determined. Extension to a 2D separated local field experiment demonstrated that the ^{17}O quadrupolar and $^{1}H-^{17}O$ dipolar tensors were spatially coincident.[76]

Observation of these hydroxyl species with large χ_Q by MQMAS suffers from a decrease in signal strength as χ_Q increases, owing to the decrease in the excitation efficiency of the MQ coherences. A 2D experiment that combined MQMAS and CP showed relative enhancement of the signal intensity of the hydroxylated species in boehmite.[77] Different types of hydroxyl species were observed by van Eck *et al.*[74] in sol–gel-produced silica. The OH signal was characterized by a T_1 of ~0.1 ms, and an even shorter T_2, restricting the use of echo experiments on stationary samples. It was suggested that this resulted from the high mobility of the hydroxyl hydrogen, creating rapidly fluctuating interactions (e.g., dipolar and quadrupolar) that strongly enhance the oxygen relaxation. This is substantiated by the observation that proton decoupling has little effect and that CP is very inefficient, in contrast to the efficient CP observed for the relatively rigid hydroxyls. Such motion also has implications for the observation of these hydroxyl species by MQMAS, where the excitation of MQ coherences is also inefficient when χ_Q is very small. In these silica gel materials, MQMAS spectra only showed evidence for a small subset of the hydroxyl species, those where motion was presumably slower, with larger χ_Q values of 2.8–3.2 MHz.[74]

A well-resolved Si–O–H (silanol) oxygen species was observed in the spectrum of potassium hydrogen silicate[78] with a chemical shift of 60 ppm and a χ_Q of 3.5 MHz. While smaller than that found in the alkali/alkali-earth silicates, this value is larger than the small χ_Q observed in the sol–gel materials, perhaps implying that rapid ^{1}H motion is not present here. In a range of sodium hydrogen phosphate systems, extensive ^{17}O NMR data was produced for both the POH hydroxyls and those oxygens of the PO_4 groups that have hydrogen bonds to water molecules in the structure. The POH species tend to have larger $\chi_Q \sim 7$ MHz than the hydrogen-bonded species, with χ_Q typically 5 MHz. The observation and accurate quantification of hydroxyl species in glasses is of considerable importance, as the level of glass hydration has significant implications for many of its physical and chemical properties. Water interaction is important to high-temperature, high-pressure properties of magmas, in addition to processes such as surface weathering. It is necessary to determine the possible interactions of water with the glass, i.e., whether any network bonds are broken to form species such as Si–O–H/Al–O–H or whether only exchange of cations (e.g., Na^{+} and K^{+}) for H_2O takes place.

A study of hydrous aluminosilicate glasses revealed significant changes to the ^{17}O NMR spectrum on the addition of water. In particular, a peak corresponding to H_2O was observed, with a fairly large χ_Q, while the chemical shift showed a strong dependence on the nature of the charge-balancing cation (i.e., Na^{+} or K^{+}), suggesting significant cation–H_2O interaction.[79]

18.3.10 Motion and Dynamics via Temperature-dependent NMR

Oxygen ion motion is important in a number of key technological applications of oxides, including solid electrolytes and oxygen sensors, and is often also at the heart of negative thermal expansion behavior. NMR spectroscopy is a highly sensitive probe of dynamics and molecular motion. Different NMR experiments can be used to study motion on a variety of timescales: ranging from very fast motion, probed by changes in relaxation behavior, to very slow motions, typically detected by 2D exchange experiments. Dynamic processes with a timescale intermediate to these extremes can also be probed by solid-state NMR owing to the presence of anisotropic interactions, where a change in molecular orientation is accompanied by changes in the strength of the shielding, dipolar coupling, and, for nuclei with spin $I > 1/2$, such as ^{17}O, quadrupolar coupling. The resulting changes in lineshapes and linewidths can be followed by variable temperature experiments and interpreted to provide information on the nature and rate of the motional process. The timescales that can be probed by such lineshape analysis are determined by the inverse of the magnitude of the coupling perturbed, typically 100 ns–10 μs for lines broadened by the first-order quadrupolar interaction and 100 μs–1 ms for central transition lineshapes subject only to second-order quadrupolar broadening. In addition, if experiments are performed under MAS, the MAS timescale (typically 10–100 μs) is also relevant. ^{17}O NMR spectroscopy, particularly at variable temperature, can be used to investigate the motion of oxygen itself, as in oxide ion conductors for example, or alternatively, a change in the local surroundings (and therefore NMR parameters) resulting from motion of other nuclei.

One of the simplest materials that has been investigated using variable temperature ^{17}O NMR is the SiO_2 polymorph cristobalite. As this material is heated, it transforms from the α- to the β-phase at around 254 °C. At the transition point, there is a significant decrease in T_1 of 1.5 orders of magnitude. Furthermore, the quadrupolar lineshape observed in a ^{17}O spectrum on a static sample at room temperature displays η_Q of ~0.125, but this decreases to <0.04 at the transition temperature. The magnitude of the EFG and CSA remains the same, suggesting the change in η_Q is caused by dynamic rotation of the SiO_4 tetrahedra, which drives the transition.[80]

The oxygen motion in oxygen ion conductors can be readily studied by ^{17}O NMR, with the number of resonances, their position, and their linewidth providing information relating to the nature and rate of any motion. For example, ^{17}O NMR of Nb-doped Bi_2WO_6 shows three distinct O resonances, corresponding to three types of oxygen: (i) oxygen in the bismuth oxide layers and the (ii) equatorial and (iii) axial oxygens in the tungsten oxide perovskite layers of the Aurivillius phase.[81] Variable temperature ^{17}O MAS NMR reveals a coalescence of two of the resonances (due to the equatorial and axial oxygens) at 200 °C. Coalescence is known to occur when the jump frequency between two sites with resonances separated by a frequency ν is higher than $\pi\nu/\sqrt{2}$, revealing jump rates faster than 6.7 kHz (correlation times < 150 μs) at this temperature. There was, however, no indication of exchange between the oxygen atoms in the tungsten and bismuth oxide layers. This process, expected to occur at a significantly lower rate, can be probed using 2D EXSY (exchange spectroscopy) experiments. Cross peaks are observed at 150 °C between the two oxygen resonances in the tungsten layer, while, as the temperature increases, cross peaks start to appear between oxygen species in the two layers, allowing an exchange rate on the millisecond timescale to be determined at 225 °C. Despite such variable temperature studies being relatively few in number, the last 6 years has seen a significant increase in such work, with detailed reports being produced for systems based on $LaGaO_3$, $BiMeVO_x$, $Bi_{26}Mo_{10}O_{69}$, and $La_2Mo_2O_9$.

The cubic AM_2O_8 family of materials, to which ZrW_2O_8 belongs, has attracted considerable attention owing to the unusual phenomenon of negative thermal expansion, i.e., they contract isotropically as they are heated. The room temperature ^{17}O NMR spectra of cubic ZrW_2O_8 exhibits three distinct resonances, corresponding to the four different oxygen sites.[82] As the temperature increases, an order/disorder transition is expected, and the three resonances broaden and coalesce into a single broad resonance, as shown in Figure 18.14(a). This indicates that all sites are in exchange above the phase transition. At lower temperatures, there is clearly some exchange, but the sites that are involved in this process cannot be identified from the MAS spectrum. A 2D EXSY spectrum at 57 °C clearly shows cross peaks between all oxygen species, suggesting that the same motional mechanism is present both above and below the transition. This indicates that the mechanism of the dynamics process is a "ratchet-like" process, ruling out an S_N2-like process, in which case only the O_1 and O_2 oxygens would undergo exchange. By combining NMR data such as the relaxation times and cross-peak intensities, it is possible to quantify rate constants for the exchange; hence, by using variable temperature, activation energies for the different motions can be deduced.

Even under MAS, the extraction of site-specific information can be hindered by the overlap of resonances from distinct species, and high-resolution experiments are often required. Although both the MQMAS and STMAS experiments discussed in Section 18.2.3 produce similar spectra for well-ordered crystalline materials, they exhibit significant differences when dynamics are present. For STMAS, the use of STs within the correlation experiment (i.e., transitions that are affected by the first-order quadrupolar interaction) results in considerable broadening of the isotropic resonances for any motion on the microsecond timescale.[83] However, the MQ transitions used in the corresponding MQMAS spectrum are broadened only by the second-order quadrupolar interaction and remain relatively unaffected by motion on such a timescale. The differential line broadening between the two experiments can therefore be used to detect the presence of dynamics, and the variation of its magnitude with temperature in the STMAS spectrum enables the rate constant and activation energy for the motional process to be estimated. Such line broadening was observed for ^{17}O

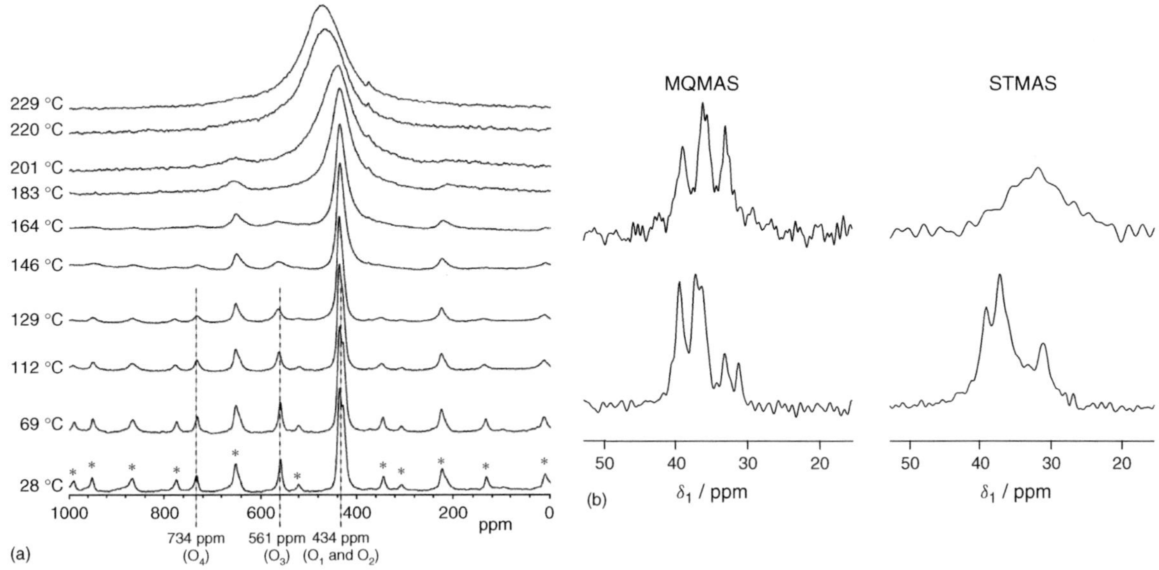

Figure 18.14. (a) Variable temperature ^{17}O MAS NMR spectra of ZrW_2O_8 showing the coalescence of all oxygen resonances, indicating that all are taking part in the exchange process. (Taken from Ref. 82 with permission of the Royal Society of Chemistry). (b) Isotropic ^{17}O 3QMAS and STMAS NMR spectra of chondrodite ($2\,Mg_2SiO_4{\cdot}Mg(OH)_2$, upper) and clinohumite ($4\,Mg_2SiO_4$. $Mg(OH)_2$, lower), demonstrating the sensitivity of the STMAS spectrum to dynamics on the microsecond timescale. (Reproduced from Ref. 83. © Elsevier Science, 2002.)

STMAS NMR spectra of the humite minerals, clinohumite and chondrodite, while the corresponding MQMAS isotropic resonances remained sharp and narrow, as shown in Figure 18.14(b). This was attributed to motion, not of the oxygen species themselves but of the nearby ^1H, which were hopping between two possible sites (H1 and H2) in the crystal structure. This change in the local environment affects the ^{17}O EFG and, correspondingly, the magnitude and orientation of the quadrupolar tensor. Motion in these humite minerals was investigated further in more recent work, with ^{17}O DFT calculations used to interpret the experimental results observed earlier.[84] The results supported the presence of a dynamic, rather than a static, positional disorder of ^1H positions, and were used to explain the appearance of the MQMAS and STMAS spectra. Furthermore, spectra simulated for different jump rates were compared to the experimental spectra, and motion on the microsecond timescale ($k \approx 3.2 \times 10^5\,s^{-1}$) was confirmed. It is anticipated that the combination of calculations and experiment will prove vital in future investigations of solid-state dynamics in complex materials.

18.3.11 Atomic Distributions in Solid Solutions

^{17}O MAS NMR holds much promise for elucidation of the cation distribution in solid solutions as oxygen occurs throughout the structure and the distinct environments can be distinguished by their NMR parameters. The pyrochlore ($A_2B_2O_7$) structure has a rich solid-state chemistry, much of this afforded by the extensive solid solution formation. In such materials, substitution on both the A and B sites can occur. Several important pyrochlores are ionic systems such that the ^{17}O χ_Q is sufficiently small and the shift range sufficiently large that different local coordinations around the oxygen are usually readily distinguished.[6,7,29] For example, in the end member $Y_2Ti_2O_7$, there are two resonances at 385 (O_1 and OTi_2Y_2) and 455 ppm (O_2 and OY_4), with an intensity ratio 6:1. On extension to pyrochlore solid solutions, such as $Y_2B_2O_7$ (B = Ti, Zr, Sn), the local $O(Y,B)_4$ environments could be distinguished, with the range of environments and linewidths measured, indicating that $Y_2(Zr,Ti)_2O_7$ pyrochlores are more disordered than $Y_2(Sn,Ti)_2O_7$.[85] As calcium is doped for yttrium in $Y_2Ti_2O_7$, in addition to the two main resonances, four other resonances are observed with the line at 530 ppm ascribed to an impurity.[86] By following the magnetization exchange at variable temperature, the oxygen motion is shown to involve jumps between O_1 sites and vacancies created by the addition of calcium to the structure.[86]

Fluorite-structured oxides (e.g., ZrO_2 and CeO_2) hold much promise as solid electrolytes. One of the key questions is the vacancy distribution and how this changes with composition. As the Zr^{4+}/Ce^{4+} ion is replaced by lower charged ions, such as Y^{3+}, Sc^{3+}, and Ca^{2+}, charge neutrality is maintained through the introduction of vacancies. In Y_2O_3-doped ZrO_2, both ^{17}O and ^{89}Y MAS NMR have been employed. The distribution of oxygen coordinations can be directly determined from analysis of the intensities in the ^{17}O NMR spectra of both ZrO_2[87,88] and CeO_2,[89] which suggests that the vacancy distribution is probably not random but that there is some pairing of the vacancies. ^{17}O MAS NMR data from doped ceria illustrate very clearly the sensitivity of ^{17}O NMR to the local structural variation that is introduced even in a well-crystalline material due to the atomic disorder of the solid solution.[89] In both the pure CeO_2 and with 5% Y_2O_3 doping, the OCe_4 resonance is dominant (Figure 18.15), but its width increases by almost two orders of magnitude on doping.

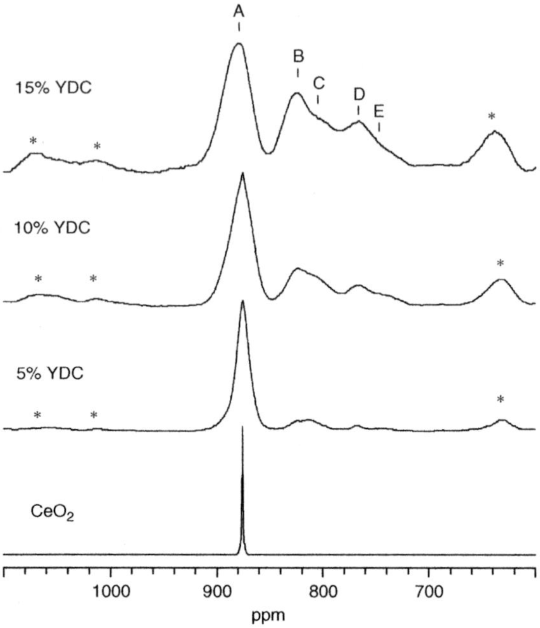

Figure 18.15. ^{17}O MAS NMR of Y_2O_3-doped CeO_2 at 14.1 T with a spinning speed of 20 kHz. Spinning sidebands are marked by asterisks. Peak A is assigned to oxygen with four Ce neighbors, peaks B and C to oxygen with three Ce and one Y neighbor, and peaks D and E to oxygen with two Ce and two Y neighbors. (Taken from Ref. 89. © American Chemical Society, 2007.)

Substitutions into pervoskites (ABO_3) also have many potential applications in, for example, solid oxide fuel cells. The creation of oxygen vacancies through the introduction of M^{3+} ions onto the B^{4+} sites aids oxygen ion motion in the structure. An important material in such applications is $BaZrO_3$, which can have metal ions such as Sc^{3+} introduced to form $BaZr_{1-x}Sc_xO_{3-x/2}$. ^{17}O MAS NMR is a straightforward probe of the distribution of zirconium and scandium in the structure, through the ability to separate contributions from Zr–O–Zr, Zr–O–Sc, and Sc–O–Sc environments. The intensity ratio indicated that the introduction of scandium was close to random. However, the maximum level (i.e., solid solution limit) of scandium introduction was lower than previously supposed and Sc_2O_3 formed above $x = 0.3$.[90] In contrast, samples of lead zirconium titanate, $PbZr_{1-x}Ti_xO_3$ (PZT), across the whole range were observed. In the end members ($x = 0$, 1), all the distinct oxygen sites can be well resolved and assigned on the basis of the relative intensities of the resonances and the correlation of the oxygen chemical shift with Zr/Ti–O bond length. Anisotropy of the behavior of the structure as zirconium is added to $PbTiO_3$ is evidenced by the differing behaviors of the oxygen sites in the Ti–O–Ti chains in different directions. At the morphotropic phase boundary (MPB) composition ($x = 0.48$), a sharp line is observed consistent with a new well-ordered oxygen environment and is linked to the formation of a monoclinic phase.[91]

Another solid solution where substitution has been followed by ^{17}O NMR is $Sc_2(WO_4)_3$–$Sc_2(MoO_4)_3$. The ^{17}O signals show structures close to the two end member Sc–O–W and Sc–O–Mo environments across the whole series, with very little change of peak position or linewidth. This indicates that relatively little change of local structure and hence structural distortion occurs during the solid solution. Variable temperature NMR up to 700 °C suggests that oxygen species might become mobile above 500 °C, with those near molybdenum possibly more mobile than those near tungsten.[92]

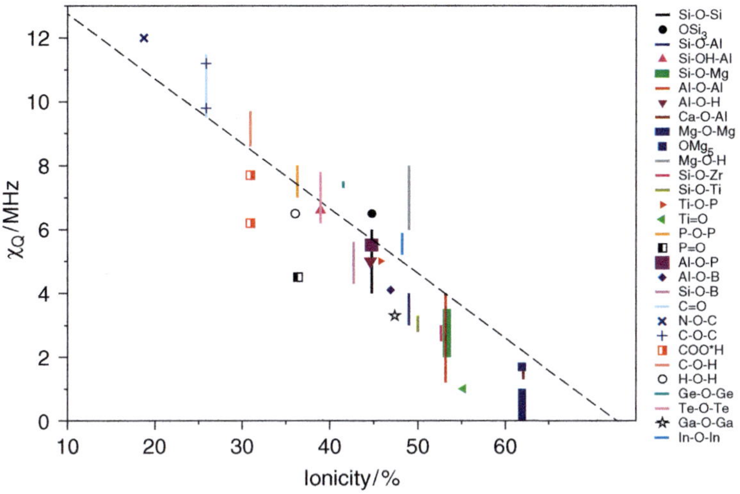

Figure 18.16. Summary of some ^{17}O quadrupole coupling constants (χ_Q) as a function of the bond ionicity for various M–O–M′ fragments. The dashed line shows the trend described by equation (18.1), for comparison. (Adapted from Ref. 7. © Royal Society of Chemistry, 2006.)

18.4 SUMMARY AND PROSPECTS

Oxygen-17 solid-state NMR has certainly come a long way in the last 20 years. It has accelerated in the last 5 as much higher magnetic fields and fast MAS have become much more widely available. This coupled with the more routine use of higher order averaging techniques has meant that real progress in increasing resolution has been made. The progress has meant that, for example, systems with larger χ_Q, such as phosphates and hydroxyl species, have become much more observable. The presence of oxygen in many scientifically and technologically important materials means that it can play a real role as a front-line probe spectroscopy. This article has attempted to show the widespread applicability of ^{17}O to inorganic materials from the range of reports of different systems that have appeared, including probing the spin dynamics of high-temperature ceramic superconductors, the structural changes associated with perovskite solid solutions as solid-state ionic conductors, to the weathering of glasses. Increasingly extensive datasets of ^{17}O NMR parameters are being collected. The large variation in both χ_Q and δ_{iso} shows that ^{17}O NMR is a very sensitive probe of local structure. Furthermore, Figure 18.16 shows that the correlation suggested between χ_Q and ionicity (equation 18.1) more than 20 years ago still reflects well the general trend, although a refinement of the details is probably appropriate. The isotropic chemical shift is also very sensitive to the local oxygen environment, and particularly to the nature of the elements in X–O–Y linkages. However, the chemical shift range for a given fragment (e.g., Ti–O–Ti) can show a large variation, which is sensitive to the actual coordination of the oxygen. It is anticipated that in the years to come there will be an ever-increasing use made of ^{17}O solid-state NMR to probe inorganic materials.

RELATED ARTICLES IN THE ENCYCLOPEDIA OF MAGNETIC RESONANCE

Brønsted Acidity of Solids

Computations of Magnetic Resonance Parameters for Crystalline Systems: Principles

Crystallography and NMR: Applications to Inorganic and Materials Chemistry

Geological Applications

High Speed MAS of Half-Integer Quadrupolar Nuclei in Solids

Inorganic Solids

Materials by NMR of Quadrupoles from Room to High Temperature

Molecular Sieves: Crystalline Systems

Oxygen-17 NMR

Quadrupolar Nuclei in Glasses

REFERENCES

1. P. K. Madhu, K. J. Pike, R. Dupree, M. H. Levitt, and M. E. Smith, *Chem. Phys. Lett.*, 2003, **367**, 150.
2. S. E. Ashbrook and I. Farnan, *Solid State Nucl. Magn. Reson.*, 2004, **26**, 105.
3. S. Schramm, R. J. Kirkpatrick, and E. Oldfield, *J. Am. Chem. Soc.*, 1983, **105**, 2483.
4. S. Schramm and E. Oldfield, *J. Am. Chem. Soc.*, 1984, **106**, 2502.
5. C. Gervais, F. Babonneau, D. Hoebbel, and M. E. Smith, *Solid State Nucl. Magn. Reson.*, 2000, **17**, 2.
6. K. J. D. MacKenzie and M. E. Smith, *Multinuclear Solid State NMR of Inorganic Materials*, Pergamon: Oxford, 2002.
7. S. E. Ashbrook and M. E. Smith, *Chem. Soc. Rev.*, 2006, **35**, 718.
8. I. P. Gerothanassis, *Prog. Nucl. Magn. Reson. Spectrosc.*, 2010, **57**, 1.
9. M. R. Hampson, S. Allen, I. J. King, C. J. Crossland, P. Hodgkinson, R. K. Harris, F. Fayon, and J. S. O. Evans, *Solid State Sci.*, 2005, **7**, 819.
10. P. N. Gunawidjaja, M. A. Holland, G. Mountjoy, D. M. Pickup, R. J. Newport, and M. E. Smith, *Solid State Nucl. Magn. Reson.*, 2003, **23**, 88.
11. J. F. Stebbins, P. D. Zhao, S. K. Lee, and X. Cheng, *Am. Mineral.*, 1999, **84**, 1680.
12. B. Chen and Y. N. Huang, *J. Am. Chem. Soc.*, 2006, **128**, 6437.
13. B. C. Schmidt, F. Galliard, and M. E. Smith, *Solid State Nucl. Magn. Reson.*, 2004, **26**, 197.
14. A. Flambard, L. Montagne and L. Delevoye, *Chem. Commun.*, 2006, 3426.
15. S. E. Ashbrook, A. J. Berry, D. J. Frost, A. Gregorovic, C. J. Pickard, J. E. Readman, and S. Wimperis, *J. Am. Chem. Soc.*, 2007, **129**, 13213.
16. S. E. Ashbrook and M. J. Duer, *Concepts Magn. Reson.*, 2006, **28A**, 183.
17. A. Goldburt and P. K. Madhu, *Annu. Rep. NMR Spectrosc.*, 2005, **54**, 81.
18. A. P. M. Kentgens, E. R. H. van Eck, T. G. Ajithkumar, T. Anupold, J. Past, A. Rienhold, and A. Samoson, *J. Magn. Reson.*, 2006, **178**, 212.
19. C. J. Pickard and F. Mauri, in *Calculation of NMR and EPR Parameters: Theory and Applications*, eds M. Kaupp, Wiley-VCH: Weinheim, 2004, and references therein.
20. J. A. Tossell, *Phys. Chem. Miner.*, 2004, **31**, 41, and References therein.
21. T. M. Clark and P. J. Grandinetti, *Solid State Nucl. Magn. Reson.*, 2000, **16**, 55.
22. T. M. Clark and P. J. Grandinetti, *Solid State Nucl. Magn. Reson.*, 2005, **27**, 233.
23. L. M. Bull, B. Bussemer, T. Anupold, A. Reinhold, A. Samoson, J. Sauer, A. K. Cheetham, and R. Dupree, *J. Am. Chem. Soc.*, 2000, **122**, 4948.
24. P. Blaha, K. Schwarz, P. Sorantin, and S. B. Trickey, *Comput. Phys. Commun.*, 1990, **59**, 399.
25. T. J. Bastow and S. N. Stuart, *Chem. Phys.*, 1990, **143**, 459.
26. A. V. Chadwick, I. J. F. Poplett, D. T. S. Maitland, and M. E. Smith, *Chem. Mater.*, 1998, **10**, 864.
27. T. J. Bastow, A. F. Moodie, M. E. Smith, and H. J. Whitfield, *J. Mater. Chem.*, 1993, **3**, 697.
28. S. Yang, J. Shore, and E. Oldfield, *J. Magn. Reson.*, 1992, **99**, 408.
29. T. J. Bastow, P. J. Dirken, M. E. Smith, and H. J. Whitfield, *J. Phys. Chem.*, 1996, **100**, 18539.
30. T. J. Bastow, M. E. Smith, and H. J. Whitfield, *J. Mater. Chem.*, 1996, **6**, 1951.
31. P. A. Thomas, A. Baldwin, R. Dupree, P. Blaha, K. Schwarz, A. Samoson, and Z. H. Gan, *J. Phys. Chem. B*, 2004, **108**, 4324.
32. R. Hussin, R. Dupree, and D. Holland, *J. Non-Cryst. Solids*, 1999, **246**, 159.
33. R. T. Hart, J. W. Zwanziger, U. Werner-Zwanziger, and J. R. Yates, *J. Phys. Chem. A*, 2005, **109**, 7636.
34. M. E. Smith, S. Steuernagel, and H. J. Whitfield, *Solid State Nucl. Magn. Reson.*, 1995, **4**, 313.
35. J. F. Stebbins, L. S. Du, S. Kroeker, P. Neuhoff, D. Rice, J. Frye, and H. J. Jakobsen, *Solid State Nucl. Magn. Reson.*, 2002, **21**, 105.
36. J. F. Stebbins, J. V. Oglesby, and S. Kroeker, *Am. Mineral.*, 2001, **86**, 1307.

37. D. Iuga, C. Morais, Z. H. Gan, D. R. Neuville, L. Cormier, and D. Massiot, *J. Am. Chem. Soc.*, 2005, **127**, 11540.

38. D. S. Middlemiss, F. Blanc, C. J. Pickard, and C. P. Grey, *J. Magn. Reson.*, 2010, **204**, 1.

39. H. K. C. Timken, S. E. Schramm, R. J. Kirkpatrick, and E. Oldfield, *J. Phys. Chem.*, 1987, **91**, 1054.

40. P. J. Grandinetti, J. H. Baltisberger, I. Farnan, J. F. Stebbins, U. Werner, and A. Pines, *J. Phys. Chem.*, 1995, **99**, 12341.

41. K. T. Mueller, Y. Wu, B. F. Chmelka, J. Stebbins, and A. Pines, *J. Am. Chem. Soc.*, 1991, **113**, 32.

42. S. K. Lee, J. F. Stebbins, C. A. Weiss, and R. J. Kirkpatrick, *Chem. Mater.*, 2003, **15**, 2605.

43. S. K. Lee and C. A. Weiss, *Am. Mineral.*, 2008, **93**, 1066.

44. S. E. Ashbrook, A. J. Berry, and S. Wimperis, *Am. Mineral.*, 1999, **84**, 1191.

45. S. E. Ashbrook, L. Le Polles, C. J. Pickard, A. J. Berry, S. Wimperis, and I. Farnan, *Phys. Chem. Chem. Phys.*, 2007, **9**, 1587.

46. S. E. Ashbrook, A. J. Berry, W. O. Hibberson, S. Steuernagel, and S. Wimperis, *J. Am. Chem. Soc.*, 2003, **125**, 11824.

47. S. E. Ashbrook, A. J. Berry, and S. Wimperis, *J. Am. Chem. Soc.*, 2001, **123**, 6360.

48. H. K. C. Timken, G. L. Turner, J. P. Gilson, L. B. Welsh, and E. Oldfield, *J. Am. Chem. Soc.*, 1986, **108**, 7231.

49. U. T. Pingel, J. P. Amoureux, T. Anupõld, F. Bauer, H. Ernst, C. Fernandez, D. Freude, and A. Samoson, *Chem. Phys. Lett.*, 1998, **294**, 345.

50. D. Freude, T. Loeser, D. Michel, U. Pingel, and D. Prochnow, *Solid State Nucl. Magn. Reson.*, 2001, **20**, 46.

51. L. M. Peng, Y. Lui, N. J. Kim, J. E. Readman, and C. P. Grey, *Nat. Mater.*, 2005, **4**, 216.

52. B. O. Skadtchenko, Y. Rao, T. F. Kemp, P. Bhattacharya, P. A. Thomas, M. Trudeau, M. E. Smith, and D. M. Antonelli, *Angew. Chem. Int. Ed.*, 2007, **46**, 2635.

53. Y. Rao, T. F. Kemp, P. Bhattacharya, P. A. Thomas, M. Trudeau, M. E. Smith, and D. M. Antonelli, *J. Am. Chem. Soc.*, 2008, **130**, 15726.

54. I. Farnan, P. J. Grandinetti, J. H. Baltisberger, J. F. Stebbins, U. Werner, M. A. Eastman, and A. Pines, *Nature*, 1992, **358**, 31.

55. P. J. Dirken, S. C. Kohn, M. E. Smith, and E. R. H. van Eck, *Chem. Phys. Lett.*, 1997, **266**, 568.

56. J. F. Stebbins, J. V. Oglesby, and S. K. Lee, *Chem. Geol.*, 2001, **174**, 63.

57. L. S. Du and J. F. Stebbins, *Solid State Nucl. Magn. Reson.*, 2005, **27**, 37.

58. A. Pedone, T. Charpentier, and M. C. Menziani, *Phys. Chem. Chem. Phys.*, 2010, **12**, 6054.

59. A. Rigamonti, F. Borsa, and P. Carretta, *Rep. Prog. Phys.*, 1998, **61**, 1367.

60. E. Oldfield, C. Coretsopoulos, S. Yang, L. Reven, H. C. Lee, J. Shore, O. C. Han, and E. Ramli, *Phys. Rev. B*, 1989, **40**, 6832.

61. A. P. Howes, R. Dupree, D. M. McKPaul, and S. Male, *Physica C*, 1992, **193**, 189.

62. A. V. Chadwick, G. Mountjoy, V. M. Nield, I. J. F. Poplett, M. E. Smith, J. H. Strange, and M. G. Tucker, *Chem. Mater.*, 2001, **13**, 1219.

63. E. Scolan, C. Magnenet, D. Massiot, and C. Sanchez, *J. Mater. Chem.*, 1999, **9**, 2467.

64. P. J. Dirken, M. E. Smith, and H. J. Whitfield, *J. Phys. Chem.*, 1995, **99**, 395.

65. C. Gervais, F. Babonneau, and M. E. Smith, *J. Phys. Chem. B*, 2001, **105**, 1971.

66. F. Babonneau, C. Bonhomme, C. Gervais, and J. Maquet, *J. Sol.-Gel Sci. Technol.*, 2004, **31**, 9.

67. B. R. Cherry, T. M. Alam, C. Click, R. K. Brow, and Z. H. Gan, *J. Phys. Chem. B*, 2003, **107**, 4894.

68. F. Pourpoint, C. Gervais, L. Bonhomme-Coury, T. Azaïs, C. Coelho, F. Mauri, B. Alonso, F. Babonneau, and C. Bonhomme, *Appl. Magn. Reson.*, 2007, **32**, 435.

69. F. Vasconcelos, S. Cristol, J.-F. Paul, G. Tricot, J.-P. Amoureux, L. Montagne, F. Mauri, and L. Delevoye, *Inorg. Chem.*, 2008, **47**, 7327.

70. M. Colmont, L. Delevoye, and O. Mentre, *New J. Chem.*, 2009, **33**, 19.

71. M. Zeyer-Düsterer, L. Montagne, G. Palavit, and C. Jäger, *Solid State Nucl. Magn. Reson.*, 2005, **27**, 50.

72. N. Forler, F. Vasconcelos, S. Cristol, J. F. Paul, L. Montagne, T. Charpentier, F. Mauri, and L. Delevoye, *Phys. Chem. Chem. Phys.*, 2010, **12**, 9053.

73. F. Brodard-Severac, G. Guerrero, J. Maquet, P. Florian, C. Gervais, and P. H. Mutin, *Chem. Mater.*, 2008, **20**, 5191.

74. E. R. H. van Eck, M. E. Smith, and S. C. Kohn, *Solid State Nucl. Magn. Reson.*, 1999, **15**, 181.

75. T. H. Walter, G. L. Turner, and E. Oldfield, *J. Magn. Reson.*, 1988, **76**, 106.
76. E. R. H. van Eck and M. E. Smith, *J. Chem. Phys.*, 1998, **108**, 5904.
77. S. E. Ashbrook and S. Wimperis, *J Magn. Reson.*, 2000, **147**, 238.
78. J. V. Oglesby, S. Kroeker, and J. F. Stebbins, *Am. Mineral.*, 2001, **86**, 341.
79. J. V. Oglesby, P. Zhao, and J. F. Stebbins, *Geochim. Cosmochim. Acta.*, 2002, **66**, 291.
80. D. R. Spearing, I. Farnan, and J. F. Stebbins, *Phys. Chem. Miner.*, 1992, **19**, 307.
81. N. Kim, R. N. Vannier, and C. P. Grey, *Chem. Mater.*, 2005, **17**, 1952.
82. A. Soleilhavoup, M. R. Hampson, S. J. Clark, J. S. O. Evans, and P. Hodgkinson, *Magn. Reson. Chem.*, 2007, **45**, S144.
83. S.E. Ashbrook, S. Antonijevic, A. J. Berry, and S. Wimperis, *Chem. Phys. Lett.*, 2002, **364**, 634.
84. J. M. Griffin, S. Wimperis, A. J. Berry, C. J. Pickard, and S. E. Ashbrook, *J. Phys. Chem. C*, 2009, **113**, 465.
85. N. Kim and C. P. Grey, *J. Solid State Chem.*, 2003, **175**, 110.
86. N. Kim and C. P. Grey, *Dalton Trans.*, 2004, 3048.
87. T. Viefhaus, T. Bolse, and K. Muller, *Solid State Ionics*, 2006, **177**, 3063.
88. N. Kim, C.-H. Hseih, H. Huang, F. B. Prinz, and J. F. Stebbins, *Solid State Ionics*, 2007, **178**, 1499.
89. N. Kim and J. F. Stebbins, *Chem. Mater.*, 2007, **19**, 5742.
90. L. Buannic, F. Blanc, I. Hung, Z. H. Gan, and C. P. Grey, *J. Mater. Chem.*, 2010, **20**, 6322.
91. A. Baldwin, R. Dupree, and P. A. Thomas, *J. Phys. Condens. Matter*, 2005, **44**, 7159.
92. N. Kim and J. F. Stebbins, *Chem. Mater.*, 2009, **21**, 309.

Chapter 19

Chlorine, Bromine, and Iodine Solid-State NMR

David L. Bryce, Cory M. Widdifield, Rebecca P. Chapman, and Robert J. Attrell

Department of Chemistry and Centre for Catalysis Research and Innovation, University of Ottawa, 10 Marie Curie Private, Ottawa, ON K1N 6N5, Canada

19.1 INTRODUCTION AND NMR PROPERTIES OF THE QUADRUPOLAR HALOGENS

Chlorine, bromine, and iodine all have isotopes that are amenable to study by solid-state nuclear magnetic resonance (SSNMR) spectroscopy.[1–3] In this chapter, we present an overview of the NMR properties of the spin-active isotopes of Cl, Br, and I, details of the experimental aspects of data acquisition and interpretation, and a survey of NMR data available, along with some highlights of important practical applications of chlorine, bromine, and iodine SSNMR. The focus is on powdered samples. Shown in Table 19.1 are the NMR properties of the spin-active isotopes of chlorine ($^{35/37}$Cl), bromine ($^{79/81}$Br), and iodine (^{127}I).[4–6]

In contrast to the more widely studied ^{19}F ($I = 1/2$)[7], all of these nuclides are quadrupolar, with spin quantum numbers of 3/2 ($^{35/37}$Cl, $^{79/81}$Br) or 5/2 (^{127}I). The natural abundances of these isotopes are moderate (24.22% for ^{37}Cl) to excellent (100% for ^{127}I) and the resonance frequencies range from moderately low (^{37}Cl and ^{35}Cl are below ^{15}N) to comparable to ^{13}C (^{79}Br and ^{81}Br). As such, SSNMR studies of these quadrupolar halogen nuclides can be relatively straightforward under favorable conditions. In addition to the magnetic shielding interaction, the NMR spectra of these quadrupolar nuclei are influenced by the quadrupolar interaction (QI) between the nuclear electric quadrupole moment and the electric field gradient (EFG) at the nucleus (see Chapter 2). This interaction is often described using the quadrupolar coupling constant ($C_Q = eV_{33}Q/h$) and asymmetry parameter ($\eta_Q = (V_{11} - V_{22})/V_{33}$). The main limiting factors in recording SSNMR spectra of $^{35/37}$Cl, $^{79/81}$Br, and ^{127}I, as with many quadrupolar nuclei, are the broad line widths, which may be observed for powdered samples when the nucleus of interest sits at a site with a large EFG. Halide anions that are either isolated or located at

NMR of Quadrupolar Nuclei in Solid Materials
Edited by Roderick E. Wasylishen, Sharon E. Ashbrook and Stephen Wimperis

ISBN: 978-0-470-97398-1

high-symmetry crystallographic sites have low (or zero) EFGs, and for this reason, the majority of SSNMR studies of chlorine, bromine, and iodine have been of compounds wherein these elements are present in their ionic halide form (Cl^-, Br^-, and I^-).[1,2] Some data have also been recorded for perhalogenate ions (XO_4^-).[8–13] Nuclear quadrupole resonance (NQR) methods have been applied extensively to measure C_Q (and η_Q in the case of ^{127}I) for halogens in covalent bonding environments.[14] It is also interesting to note that $C_Q(^{35/37}\text{Cl})$ has been measured indirectly by observing the NMR spectrum of spin-1/2 nuclei which are spin coupled (dipolar and J) to chlorine.[15]

The effects of quadrupolar line broadening tend to dominate the SSNMR spectra of Cl, Br, and I, except when the nucleus is in a perfectly symmetric environment. As a result, the spectra (and the corresponding values of C_Q and η_Q) are very sensitive to the local environment as well as to the longer range crystal packing. The diagnostic nature of the spectral parameters has been demonstrated to be useful in crystal structure refinement,[16–18] in the characterization of polymorphs,[19] and in the discrimination of different hydrates of various compounds[17,20,21] (*vide infra*).

In most cases of interest, the QI is sufficiently large to broaden the satellite transitions (STs) to such an extent that they cannot be easily observed, and experimentally only the central transition (CT) spectrum is typically recorded. The CT is not broadened by first-order quadrupolar effects, but is affected by second-order effects. The breadth of the second-order CT lineshape of a stationary powdered sample ($\Delta\nu_{CT}$) may be described with the following equation[5]:

$$\Delta\nu_{CT} = \left[\frac{3C_Q}{2I(2I-1)}\right]^2 \times \frac{(\eta_Q^2 + 22\eta_Q + 25)(I(I+1) - 3/4)}{144\nu_0}$$

The inverse relationship between $\Delta\nu_{CT}$ and ν_0 demonstrates why higher applied magnetic fields are beneficial when acquiring SSNMR spectra of the quadrupolar halogens: increasing B_0 increases the value of ν_0, and therefore results in a decrease in the second-order quadrupolar broadening of the CT. High applied magnetic fields also result in an increase in Boltzmann sensitivity. High fields are also very valuable for the measurement of chlorine, bromine, and iodine chemical shift (CS) tensors.

When interpreting trends in quadrupolar coupling constants for Cl, Br, and I, it is important to consider the Sternheimer antishielding factor, conventionally expressed as $1 - \gamma_\infty$.[5,14,17] This factor accounts for the EFG at the nucleus owing to the polarization of inner-shell electrons.[22,23] That is, the EFG due solely to ionic charges in the lattice, eq_{lattice}, is modified as shown below[24]:

$$eq_{\text{obs}} = (1 - \gamma_\infty)eq_{\text{lattice}}$$

The quantity eq_{obs} must be considered for the accurate discussion of quadrupolar coupling constants and relative line widths. The antishielding factors for Cl, Br, and I are presented in Table 19.1. Consideration of the large value for iodine explains why it can be difficult to observe ^{127}I SSNMR spectra in a series of isostructural halide compounds for which observation of the $^{35/37}$Cl and $^{79/81}$Br SSNMR spectra is feasible (Figure 19.1).

Given that ^{127}I quadrupolar coupling constants in particular can be quite large, thereby leading to particularly broad powder patterns, some discussion is provided below regarding the application of suitable data acquisition methods (e.g., variable radiofrequency (RF) transmitter offsets and signal enhancement). The utility of complementary ^{127}I NQR measurements is noted,[14,17] as well as guidelines pertaining to the validity of second-order perturbation theory in the analysis of spectral lineshapes.[17,25–28]

Table 19.1. NMR properties of the quadrupolar halogens[4–6]

	N.A.^{a} (%)	I	$\gamma/10^7$ rad $T^{-1}s^{-1}$	Q/mb	$1-\gamma_\infty$
^{35}Cl	75.78	3/2	2.624198	−81.65(80)	43.0
^{37}Cl	24.22	3/2	2.184368	−64.35(64)	43.0
^{79}Br	50.69	3/2	6.725616	313(3)	81
^{81}Br	49.31	3/2	7.249776	262(3)	81
^{127}I	100.0	5/2	5.389573	−696(12)	163

^{a}Natural abundance.

19.2 EXPERIMENTAL ASPECTS

19.2.1 Chemical Shift Referencing

19.2.1.1 Chlorine-35/37

The IUPAC CS reference for chlorine is 0.1 mol dm^{-3} NaCl in D_2O.[4] With this standard, the exact conditions of preparation are very important as they

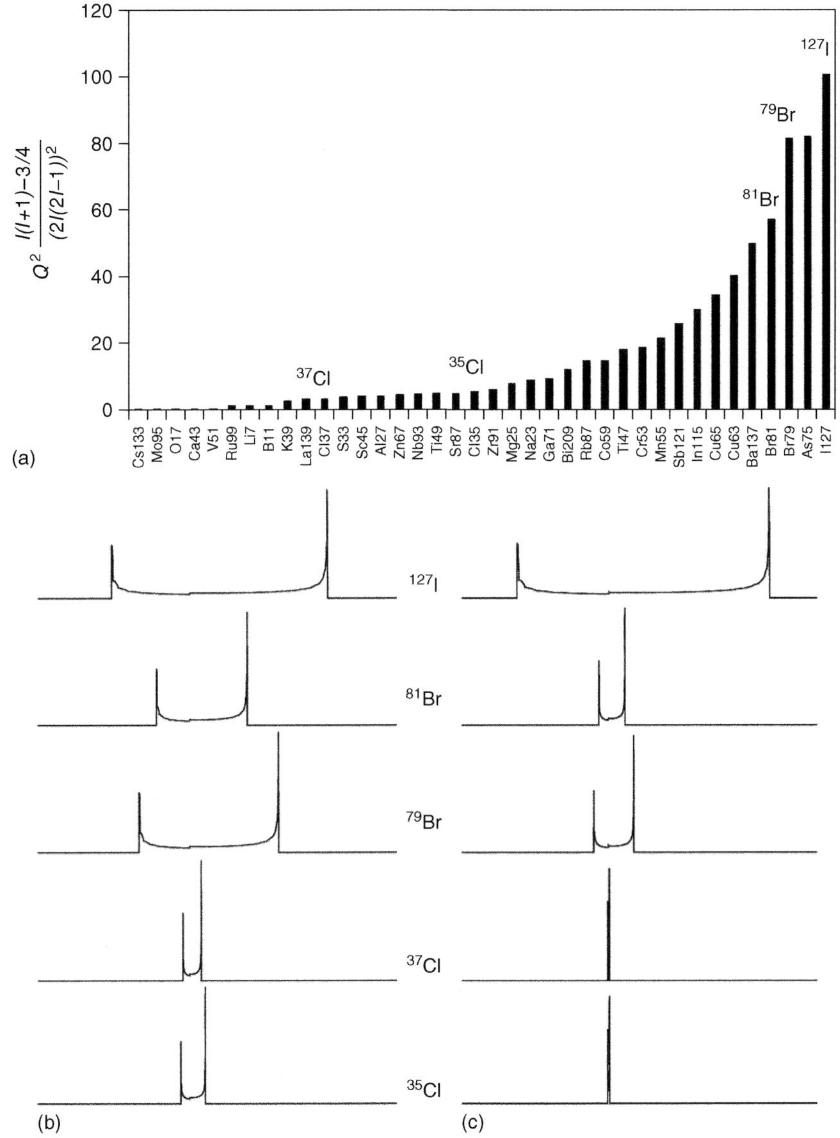

Figure 19.1. Computed central transition line widths for the quadrupolar halogens with axially symmetric EFGs in powder samples. (a) Comparison of the intrinsic CT line width for various quadrupolar nuclei on the basis of their quadrupole moments and nuclear spin quantum numbers only (assuming a constant EFG at the nucleus and a constant Larmor frequency). (b) Comparison of the CT powder patterns for the isotopes of chlorine, bromine, and iodine, under the same conditions as (a) except also including the effect of the relative Larmor frequencies. (c) Same as (b) except also including a scaling of the EFG by the Sternheimer antishielding factor to account for its effect on the effective EFG at the nucleus. Data are plotted in (b) and (c) on ppm scales. Scales are different in (b) and (c), i.e., the breadth of the ^{127}I powder pattern has been arbitrarily held constant in the two sets of spectra for ease of viewing.

may significantly affect the observed isotropic CS. There is a concentration dependence of the chlorine CS as well as strong solvent effects if H_2O is used instead of D_2O.[3,29,30] Solid sodium chloride has been suggested as an alternative reference, as it does not have concentration or solvent issues and has a narrow line width at relatively low magic-angle spinning (MAS) frequencies.[3] An alternative secondary reference is solid potassium chloride, which was found to be ideal as a result of the small dipole–dipole coupling between potassium and chlorine along with the vanishingly small EFG at the chlorine nucleus.[31] The chlorine CSs of all the alkali metal chlorides with respect to the IUPAC reference are listed in Table 19.2.[2] Shown in Figure 19.2 are the known CS ranges for chlorine, bromine, and iodine.

In addition, chlorine is the only quadrupolar halogen for which a reliable and precise absolute shielding scale has been determined. Using a combination of experiment and theory, Gee *et al.* established a value of 974(4) ppm for the chlorine isotropic magnetic shielding constant (σ_{iso}) of an infinitely dilute aqueous NaCl solution.[32]

Table 19.2. Chlorine NMR data for cubic alkali metal chlorides

Compounds	$\delta_{iso}{}^{a}$ (ppm)
LiCl	9.93
NaCl	−41.11
KCl	8.54
RbCl	49.66
CsCl	114.68

Source: From Ref. 2.
[a]From ^{35}Cl SSNMR data; with respect to 0.1 mol dm^{-3} NaCl in D_2O.

19.2.1.2 *Bromine-79/81*

The IUPAC CS reference for bromine is 0.01 mol dm^{-3} NaBr in D_2O.[4] Once again, care must be taken in preparation for the same reasons as noted above

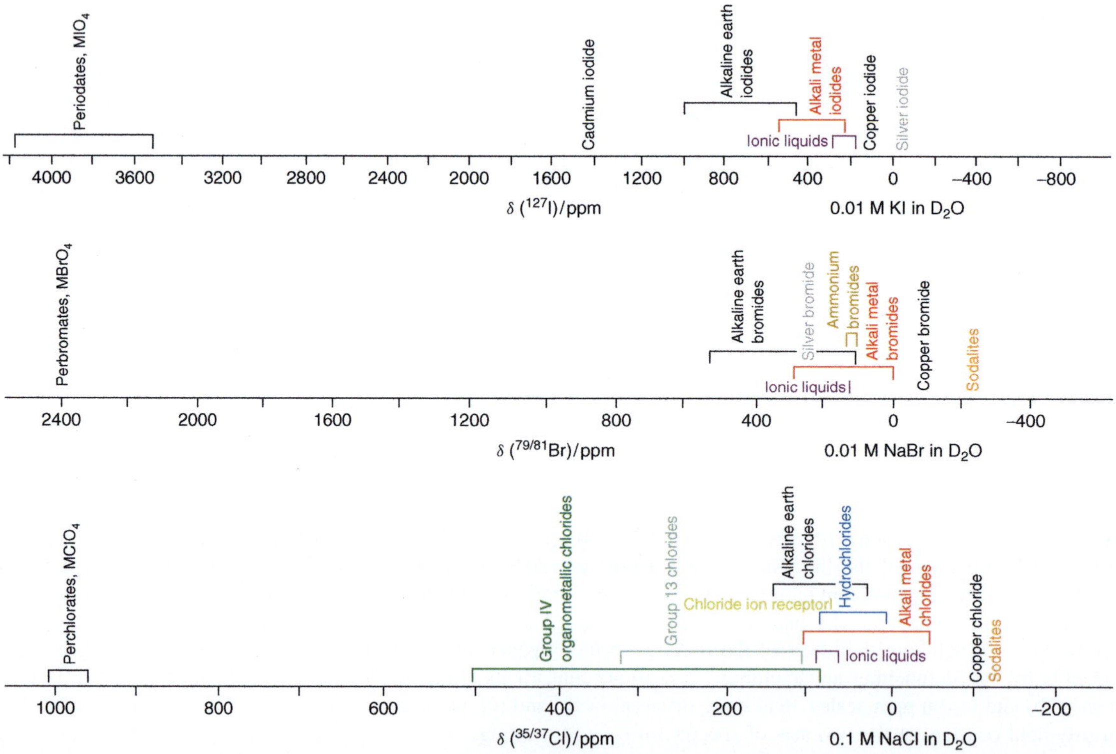

Figure 19.2. Known chlorine, bromine, and iodine chemical shift ranges for solids.

for chlorine and thus a secondary reference of either solid sodium or potassium bromide is convenient. The bromine CSs of all the alkali metal bromides with respect to 0.03 mol dm^{-3} NaBr in D_2O are presented in Table 19.8.[1]

19.2.1.3 Iodine-127

The IUPAC CS reference for iodine is 0.01 mol dm^{-3} KI in D_2O.[4] This suggested standard presents further practical problems in addition to concentration and solvent CS dependences. As a result of the dilute nature of iodine in the IUPAC suggested solution, the time required to obtain an adequate signal makes it undesirable as a setup sample in our experience. Therefore, in addition to the reasons mentioned above for bromine and chlorine, sensitivity issues make it desirable to use solid potassium iodide or sodium iodide as secondary standards. The iodine CSs of all the alkali metal iodides with respect to the IUPAC reference are reported in Table 19.11.[2]

19.2.2 Data Acquisition Techniques

For cases where the lineshape is sufficiently narrow (i.e., maximally on the order of tens of kilohertz), MAS NMR experiments are of great utility for observing the SSNMR signals associated with the quadrupolar halogen nuclei. In fact, the ^{79}Br SSNMR signal of KBr is often used to precisely calibrate the magic angle.[33] When MAS is used, SSNMR lineshape broadening due to the dipole–dipole and chemical shift anisotropy (CSA) mechanisms can be removed. In addition, the residual second-order quadrupolar broadening will be reduced by roughly a factor of 2.43–3.43.[5] Owing to the removal of the additional line broadening mechanisms, lineshape modeling of the MAS SSNMR spectra of the quadrupolar halogen nuclei is greatly simplified, and the EFG tensor information (C_Q, η_Q), as well as the isotropic halogen CS, can often be readily determined. More generally, available MAS rates will not be rapid enough and spectra can be obtained under stationary conditions. Furthermore, in many cases, depending on the magnetic field strength and the chemical environment of the halogen under investigation, spectra of the CT of powdered samples cannot be acquired in a single piece. Potential remedies for data acquisition of very broad lineshapes are discussed in the Section 19.2.2.3.

19.2.2.1 Echoes

Although a variety of advanced pulse sequences now exist to acquire SSNMR data, it is often the case that more traditional echo experiments (especially when the experiments are performed at very high magnetic fields) are acceptable to produce high S/N ratio spectra in a reasonable amount of time. This is especially true for the $^{79/81}$Br and ^{127}I nuclides, as their magnetogyric ratios and natural abundances are rather high. This is also generally true even in the case of exceedingly broad (>1 MHz) lineshapes. For the acquisition of $^{35/37}$Cl SSNMR spectra, however, other pulse sequences, such as quadrupolar Carr–Purcell–Meiboom–Gill (QCPMG), may be of use (see Section 19.2.2.2) because of the lower inherent sensitivity of these nuclides. The 90/90 or "solid echo" experiment[34] is suggested for situations where a somewhat strong signal is being observed, as it can produce more uniform excitation at a given RF nutation frequency relative to the 90/180 or "Hahn echo"[35] experiment. A theoretical and experimental discussion of the optimal conditions for echoes performed on half-integer quadrupolar nuclei has been provided by Bodart *et al.*[36] The relative increase in the uniform signal excitation afforded using the solid echo experiment comes at the expense of reduced signal intensity (by roughly a factor of 2) relative to the Hahn echo experiment; it is important to note that the exact factor will depend on the phase cycling and coherence pathway selection. At a typical magnetic field of 11.75 T, and assuming an RF field of 100 kHz (for $^{79/81}$Br and ^{127}I) and 40 kHz (for $^{35/37}$Cl), solid echo experiments using one transmitter setting are expected to be ideal until C_Q values exceed ~12–14 MHz for bromine, 25 MHz for iodine, and ~5 MHz for chlorine.

19.2.2.2 Signal Enhancement Methods

While echo experiments may be useful for the acquisition of SSNMR signals associated with the quadrupolar halogen nuclei, in many situations sensitivity-enhancing pulse sequences will be required. The QCPMG[37] pulse sequence takes advantage of the potentially significant ratio between the natural transverse (spin–spin) relaxation time constant of the sample (i.e., T_2), and the same relaxation time constant due to B_0 inhomogeneities (i.e., $T_2{}^*$). Through the use of a "train" of refocusing π pulses, the time-domain

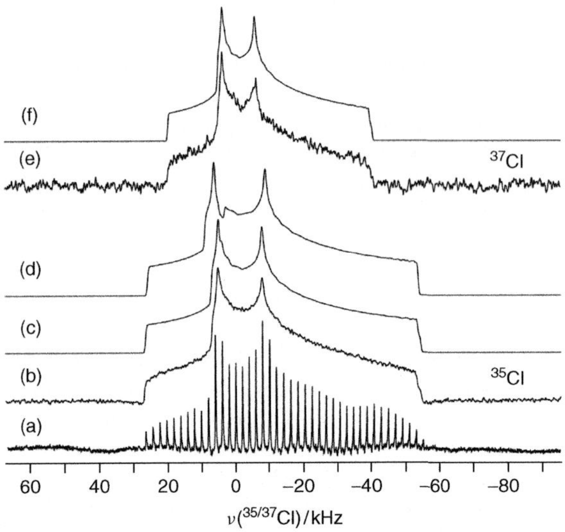

Figure 19.3. (a) ^{35}Cl QCPMG and (b) $\pi/2 - \tau - \pi/2 - \tau$ echo and (e) ^{37}Cl $\pi/2 - \tau - \pi/2 - \tau$ echo NMR spectra of powdered $CaCl_2{\cdot}2H_2O$ acquired at 11.75 T under stationary conditions. Shown in (c) and (f) are the corresponding best-fit simulated spectra. For comparison, shown in (d) is the simulated ^{35}Cl NMR spectrum obtained when the chemical shift tensor span is assumed to be zero. Spectral parameters are given in Table 19.4. (Reproduced from Ref. 20. © Wiley-VCH, 2007.)

system response resembles a series of evenly spaced spikes (spacing = τ). Once subjected to Fourier transformation, the frequency-domain response will also be a series of spikelets, which carry a separation of $1/\tau$ (Figure 19.3). While experimental resolution is reduced using this method, the sensitivity is increased, as multiple echoes are collected per scan.

In addition to the QCPMG experiment, the rotor-assisted population transfer (RAPT),[45] double-frequency sweeps (DFS),[46] and hyperbolic secant (HS)[47] pulse sequences may also prove useful for $^{35/37}$Cl SSNMR experiments. However, as these pulse sequences rely upon the manipulation of the ST populations, one must be able to reasonably excite the full ST manifold. Hence, the RAPT, DFS, and HS pulse sequences are not expected to be of general use for $^{79/81}$Br and ^{127}I unless very high site symmetry is present. The RAPT, DFS, and HS sequences have also been coupled with QCPMG and afford further sensitivity enhancement.[48,49]

19.2.2.3 Wideline Methods

Wideband, Uniform Rate, and Smooth Truncation (WURST) Pulses

The use of wideband, uniform rate, and smooth truncation (WURST)-type pulses in NMR experiments was developed a number of years ago as a solution to the problem of wideband inversion and broadband decoupling in liquid state experiments.[50] Recently, a number of studies have shown that WURST pulses can also be used to uniformly excite broad spectral regions using a single transmitter setting for half-integer quadrupolar nuclei.[51] In addition, they have been coupled with QCPMG.[52] These pulses are so efficient at uniformly exciting broad spectral regions that the probe bandwidths are the limiting factor for uniform signal detection, as was demonstrated during the collection of the ^{127}I SSNMR signal of one of the sites in SrI_2.[17]

Variable-Offset Cumulative Spectrum (VOCS) Data Acquisition

To acquire meaningful SSNMR spectra of the quadrupolar halogen nuclei, data acquisition will often be carried out using multiple RF transmitter settings. Failure to uniformly excite the entire CT powder pattern will result in incorrect spectral analysis and erroneous NMR tensor parameters. Unless high site symmetry (i.e., tetrahedral, octahedral) at the halogen nucleus is known a priori, it is advisable to obtain halogen SSNMR spectra using at least two different transmitter frequencies to confirm that the entire CT spectral pattern has been acquired. A suggested offset for this endeavor (using a high-power amplifier with an echo pulse sequence) would be *ca.* 100–200 kHz. To acquire the CT powder patterns of the quadrupolar halogen nuclei that are clearly broadened beyond what can be uniformly excited using a single transmitter frequency, the variable-offset cumulative spectrum (VOCS) method is useful.[53] This protocol simply involves collecting a set of SSNMR spectra, with each spectrum being acquired at a unique, but uniformly offset, transmitter frequency (Figure 19.4). Each spectrum is to be collected using the same number of scans and should be processed individually. Once this is accomplished, the resulting spectra are coadded in the frequency domain, producing the final VOCS spectrum.

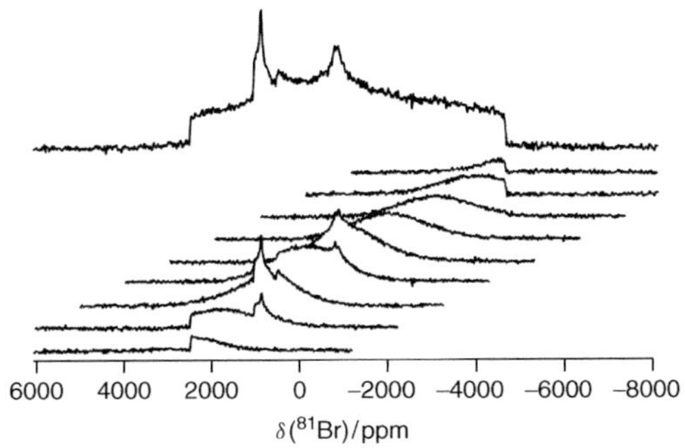

Figure 19.4. Bromine-81 SSNMR spectrum of 3-chloroanilinium bromide acquired at 21.1 T using VOCS data acquisition. 3072 scans were acquired per piece, and high-power proton decoupling was applied during acquisition. Each of the nine slices are shown, as well as their sum (top).

19.2.3 Data Analysis

19.2.3.1 Importance of Acquiring Data at Two Magnetic Fields and Utility of Two Spin-Active Isotopes of Chlorine and Bromine

One of the benefits when extracting information using chlorine and bromine SSNMR spectroscopy is the availability of two NMR-active nuclides for each. In order to accurately extract EFG and CS tensor information using SSNMR experiments on quadrupolar nuclei, data acquisition should be carried out at multiple fields. Owing to the multiple NMR-active isotopes for chlorine and bromine, multiple field data acquisition is not a stringent requirement. This is because each nucleus has a unique magnetogyric ratio and nuclear electric quadrupole moment, Q (as summarized in "Introduction and NMR Properties of the Quadrupolar Halogens").[6] In effect, if one conducts SSNMR experiments on both NMR-active isotopes, then the equivalent of multiple field data acquisition has been performed using the internal nuclear properties rather than by adjusting the external applied field. This assumes that isotope effects on the magnetic shielding tensor are negligible, which is a valid assumption when dealing with SSNMR of powder samples. An ideal method for SSNMR data acquisition using chlorine and bromine would be to acquire spectra of each nuclide within a very high B_0. While the combination of multiple field and multiple nuclide data acquisition is not a requirement to extract the relevant NMR tensor information for chlorine and bromine, it becomes worthwhile when considering samples that possess multiple sites, as demonstrated for $SrBr_2$.[21]

19.2.3.2 Breakdown of Second-Order Perturbation Theory

When fitting the SSNMR spectrum of a half-integer quadrupolar nucleus, it is well known that the QI will at least split the resonance into a multiplet structure, with $2I - 1$ STs flanking the CT. In the case of a very weak QI, first-order perturbation theory has been shown to be adequate. Other than for the cubic halides, however, first-order perturbation theory is rarely sufficient to model the observed SSNMR spectrum associated with a quadrupolar halogen nucleus. In these systems, one generally measures the CT only, which will carry additional second-order broadening that can be used to extract the EFG and CS tensor information. While second-order perturbation theory is often sufficient to model the observed SSNMR lineshapes for quadrupolar nuclei, there exist cases where even second-order perturbation theory does not lead to the correct NMR tensor parameters. Of the quadrupolar halogens, this is most likely to occur when carrying out experiments upon the ^{127}I nuclide. Indeed, an example regarding the interpretation of the ^{127}I SSNMR lineshape of one of the sites in powdered SrI_2 has been reported.[17] It was found that even though the data were acquired at the highest possible applied field (21.1 T), the internal QI was still strong enough (i.e., $C_Q = 214$ MHz) to produce a nonuniform shift in the ^{127}I SSNMR spectrum that could not be accounted for using second-order perturbation theory (Figure 19.5). A recently developed model, which combines the Zeeman and QI effects exactly,[28] was applied to arrive at a significantly different isotropic CS value than was determined using second-order perturbation theory. Since third-order quadrupolar effects are zero for the CT, fourth-order effects on the CT are likely mainly responsible for these observations. Third-order effects are non-zero for the STs, however, and an exact simulation will of course take these and higher order effects into account. This is likely to be important since the ST powder patterns will generally overlap partially with the CT powder pattern for very large values of C_Q.

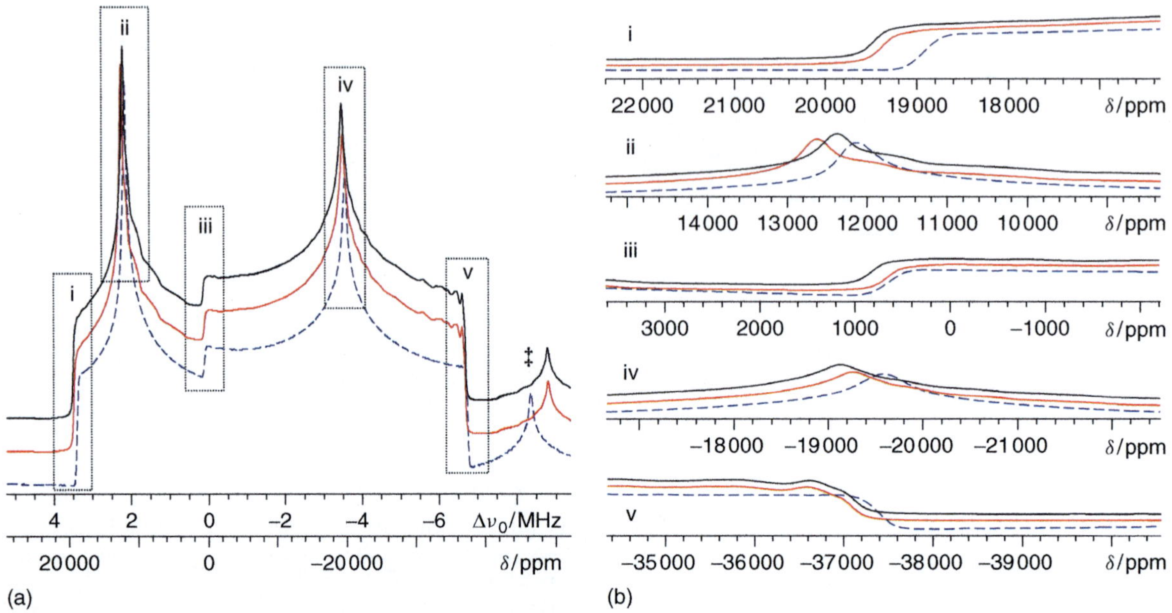

Figure 19.5. Breakdown of second-order perturbation theory. (a) Comparison of ^{127}I SSNMR powder patterns generated using second-order perturbation theory (solid red and black traces) with one calculated using exact theory (dashed blue trace). $C_Q = 214$ MHz; $\eta_Q = 0.316$ (values for one site in SrI_2); $B_0 = 21.1$ T. Iodine CSA ($\Omega = 460$ ppm) is included in the black trace simulation. Relative to the second-order perturbation theory simulations, the exact simulation is nonuniformly shifted to lower frequency. Low-frequency shoulders (‡) are from STs. (b) Horizontal expansions of the regions in (a). These illustrate that the incorporation of a reasonable value for the iodine CSA generally influences the positions of the discontinuities to a lesser extent than do higher order quadrupolar-induced effects in this case. (Reproduced from Ref. 17. © American Chemical Society, 2010.)

19.3 REPRESENTATIVE QUADRUPOLAR AND CHEMICAL SHIFT DATA AND DISCUSSION OF APPLICATIONS

19.3.1 Chlorine

Among the quadrupolar halogens, ^{35}Cl is the nucleus that has been studied most frequently with SSNMR, despite its moderately low resonance frequency.[1–3] This is a result of its smaller quadrupole moment compared to $^{79/81}$Br and ^{127}I which, as mentioned above, results in significantly less line broadening. While both NMR-active nuclei of chlorine have moderately large Q values and low Larmor frequencies, the higher natural abundance of ^{35}Cl makes it the more popular of the two nuclei for study.

A wide variety of materials has been analyzed with chlorine SSNMR, ranging from biologically important model compounds to catalysts to geological samples. The chlorine NMR properties of these different classes of materials have been found to vary significantly, with isotropic chlorine CSs ranging from −100 to over 1000 ppm (with respect to the IUPAC standard) and ^{35}Cl quadrupolar coupling constants for chloride ions varying from 0 to 40.4 MHz. In addition, the increasing availability of high field instruments has allowed for the determination of CS tensor parameters for many materials, with spans ($\Omega = \delta_{11} - \delta_{33}$) up to 800 ppm having been quantified.

Many early studies using chlorine SSNMR focused on cubic salts, as the absence of a QI results in narrow lines. Therefore, the chlorine CSs for the alkali metal chlorides,[2,54,55] along with other cubic chlorides (i.e., AgCl,[56] CuCl[56–58]) are well known and cover a significant range. For example, NaCl appears at −41.11 ppm, whereas CsCl appears at 114.68 ppm (with respect to 0.1 M NaCl in D_2O). For further information, the reader is referred to Refs 1 and 2. See also Table 19.2.

Table 19.3. Selected chlorine NMR data for solid organic hydrochlorides

Compounds	$\|C_Q(^{35}Cl)\|$/MHz	η_Q	$\delta_{iso}{}^{a}$ (ppm)	Ω (ppm)	κ	Euler angles[b]/degrees	References
Cocaine HCl	5.027(0.02)	0.2(0.05)	−41	—	—	—	Yesinowski *et al.*[38]
L-Tyrosine HCl	2.23(0.02)	0.72(0.03)	53.6(0.5)	<150	—	—	Bryce *et al.*[39]
	2.3(0.1)	0.7(0.1)	54(1)	—	—	—	Gervais *et al.*[40]
Glycine HCl	6.42(0.05)	0.61(0.03)	60(5)	100(20)	0.30(0.30)	95(20), 0(4), 0(20)	Bryce *et al.*[41]
L-Valine HCl	5.89(0.05)	0.51(0.05)	49(10)	125(40)	0.35(0.50)	65(20), 0(20), 0(20)	Bryce *et al.*[41]
L-Glutamic acid HCl	3.61(0.01)	0.65(0.02)	61(1)	66(15)	0.0(0.3)	9(20), 77(20), 6(20)	Bryce *et al.*[41]
Quinuclidine HCl	5.25(0.02)	0.05(0.01)	14.0(10.0)	50(20)	—	—	Bryce *et al.*[39]
L-Cysteine ethyl ester HCl	3.78(0.02)	0.03(0.03)	57.5(0.5)	47(4)	−0.8(0.2)	—	Bryce *et al.*[39]
L-Cysteine methyl ester HCl	2.37(0.01)	0.81(0.03)	52.5(0.7)	45(15)	—	—	Bryce *et al.*[39]
Cysteine HCl monohydrate	3.92(0.01)	0.47(0.02)	63.1(0.5)	66(10)	0.12(0.12)	155(20), 0(10), 0(20)	Chapman *et al.*[42]
L-Lysine HCl	2.49(0.01)	0.42(0.02)	64(2)	26(10)	−0.4(0.4)	0(20), 52(20), 0(20)	Bryce *et al.*[41]
L-Serine HCl	3.0(0.3)	0.8(0.2)	79(30)	<150	—	—	Bryce *et al.*[41]
L-Proline HCl	4.50(0.05)	0.63(0.05)	−4(5)	63(5)	−0.54(0.08)	48(20), 69(3), 9(20)	Bryce *et al.*[41]
L-Isoleucine HCl	4.39(0.05)	0.25(0.03)	55(20)	75(30)	>0.85	20(20), 12(20), 0(20)	Bryce *et al.*[41]
L-Phenylalanine HCl	6.08(0.05)	0.52(0.03)	55(5)	129(20)	0.26(0.25)	91(20), 13(20), 10(20)	Bryce *et al.*[41]
L-Tryptophan HCl	5.05(0.04)	0.86(0.03)	63.9(1.0)	72(5)	0.1(0.1)	90(15), 20(15), 2(20)	Bryce *et al.*[18]
D,L-Arginine HCl monohydrate	2.035(0.020)	0.98(0.02)	50.4(1.0)	57.5(3.0)	0.27(0.10)	85(15), 77.5(12.0), 30(30)	Bryce *et al.*[18]

(continued overleaf)

Table 19.3. *Continued*

Compounds	$\|C_Q(^{35}Cl)\|$/MHz	η_Q	$\delta_{iso}{}^a$ (ppm)	Ω (ppm)	κ	Euler angles[b]/degrees	References
L-Alanine HCl	6.4(0.1)	0.75(0.06)	65(5)	60(30)	−0.3(0.5)	90(15), 0(15), 0(15)	Chapman *et al.*[42]
L-Aspartic acid HCl	7.1(0.1)	0.42(0.05)	61(5)	75(30)	−0.9(0.1)	0(20), 30(20), 93(20)	Chapman *et al.*[42]
L-Histidine HCl monohydrate	4.59(0.03)	0.46(0.02)	52(1)	<150	—	—	Chapman *et al.*[42]
L-Methionine HCl	4.41(0.02)	0.35(0.03)	58(1)	100(20)	0.3(0.3)	93(20), 163(15), 7(20)	Chapman *et al.*[42]
L-Threonine HCl	5.4(0.1)	0.94(0.02)	58(10)	95(40)	−0.2(0.5)	95(15), 0(10), 0(15)	Chapman *et al.*[42]
Procaine HCl	4.87(0.07)	0.28(0.04)	55(6)	125(25)	−0.4(0.3)	95(15), 3(2), 32(8)	Hamaed *et al.*[19]
Tetracaine HCl	6.00(0.10)	0.27(0.04)	30(6)	80(15)	0.4(0.3)	60(8), 8(5), 10(10)	Hamaed *et al.*[19]
Lidocaine HCl monohydrate polymorph 1	4.67(0.07)	0.77(0.03)	59(4)	110(15)	−0.85(0.03)	12(3), 40(10), 80(3)	Hamaed *et al.*[19]
Lidocaine HCl monohydrate polymorph 2	Site 1: 2.52(0.12)	0.95(0.05)	44(10)	20(10)	−0.8(0.2)	90(40), 50(50), 60(40)	Hamaed *et al.*[19]
	Site 2: 5.32(0.10)	0.32(0.10)	69(10)	45(10)	0.8(0.2)	5(5), 50(15) 40(40)	
Bupivacaine HCl monohydrate	3.66(0.10)	0.72(0.08)	55(10)	100(25)	0.2(4)	105(20), 90(5), 5(5)	Hamaed *et al.*[19]
1-Butyl-3-methyl imidazolium chloride complex of *meso*-octamethyl calix[4]pyrrole	1.0(0.1)	0.7(0.1)	79(10)	50(5)	0.4(0.4)	15(30), 27(20), 60(15)	Chapman *et al.*[43]

[a]With respect to 0.1 mol dm^{-3} NaCl in D_2O.
[b]Euler angles determined using the Arfken convention. (See Ref. 44)

19.3.1.1 Organic Hydrochlorides

Organic chloride and hydrochloride salts have been extensively studied by chlorine SSNMR, with both EFG and CS tensor parameters being extracted for a variety of materials (see Table 19.3 for selected data).[1,2] The first ^{35}Cl SSNMR study of a hydrochloride salt was carried out by Pines and coworkers on powdered cocaine hydrochloride at 7.0 T under static conditions.[38] At this field, the VOCS method was required to collect the full spectrum, which was modeled to provide a ^{35}Cl quadrupolar coupling constant of 5.027(0.02) MHz. Later, Bryce *et al.* reported a $^{35/37}$Cl NMR study of multiple hydrochloride salts: L-tyrosine hydrochloride, L-cysteine methyl ester hydrochloride, L-cysteine ethyl ester hydrochloride, and quinuclidine hydrochloride.[39] MAS, Hahn echo, and QCPMG techniques were used to collect spectra at 9.4 and 18.8 T, allowing for the determination of EFG tensor parameters and chlorine CSs for all the salts. The authors noted an inverse relationship between the number of hydrogen bonds and the magnitude of the quadrupolar coupling constant.[39]

The study of amino acid hydrochlorides was continued by Gervais *et al.*, who published a multinuclear SSNMR study of four amino acid hydrochlorides.[40] Bryce and coworkers extended the use $^{35/37}$Cl SSNMR to study several other amino acid hydrochlorides, noting that they are biologically relevant models for chloride-binding proteins (Figure 19.6).[18,41,42] A total of 17 amino acid hydrochlorides have been analyzed with $^{35/37}$Cl SSNMR. The spectra were found to vary significantly, depending on the precise nature of the chloride environment, and demonstrated the sensitivity of the technique to small changes in structure. The magnitude of C_Q(^{35}Cl) was found to range from 2.035 to 7.1 MHz, whereas the chlorine CSs varied from −4 to 79 ppm (with respect to the IUPAC standard).[18,41,42] In addition, the collection of spectra at 21.1 T allowed for the reliable determination of the CS tensor parameters and the relative orientation of the CS and EFG tensors for the majority of these amino acid hydrochlorides, with the CS tensor spans ranging from 26 to 129 ppm.[18,41,42] Interestingly, a large quadrupolar coupling constant does not always correlate with a large CS span, demonstrating that the two interactions are governed by different factors and thus offer complementary information. Computational studies using the B3LYP (Becke's three-parameter

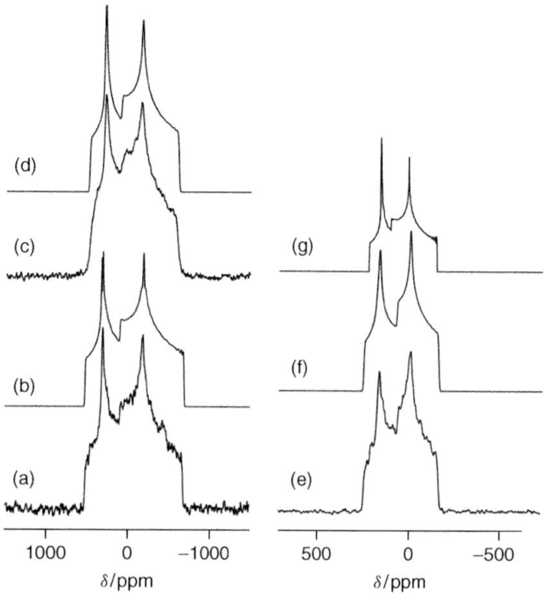

Figure 19.6. Solid-state $^{35/37}$Cl NMR spectroscopy of L-cysteine hydrochloride monohydrate. Experimental spectra of stationary powdered samples: (a) ^{35}Cl at 11.75 T, (c) ^{37}Cl at 11.75 T, (e) ^{35}Cl at 21.1 T. Simulations incorporating EFG and CS tensor parameters listed in Table 19.3 appear in (b), (d), and (f). (g) Depicts a simulation assuming no CSA. (Reproduced from Ref. 42. © Royal Society of Chemistry, 2007.)

Lee–Yang–Parr exchange-correlation functional) hybrid DFT method and the restricted Hartree–Fock (RHF) method also played an important role in these studies, and aided in the interpretation of the SSNMR spectra. The two methods, as well as different basis sets, were tested against the experimental data and the neutron diffraction structures of several amino acid hydrochlorides, and an optimized method for calculating chlorine NMR parameters was determined.[18,41] The optimal method was applied to calculate the expected chlorine NMR parameters of a chloride ion channel and showed that chlorine NMR on complex systems should be feasible, in that prohibitively broad CT lineshapes are not expected.[41] In addition, the importance of optimizing hydrogen atom positions in cases where only an X-ray structure is available and the effectiveness of including point charges for improving the agreement between computational and experimental values were shown.[18,42] The amino acid hydrochloride series of data was also utilized to test the accuracy of the GIPAW–DFT method in calculating chlorine NMR parameters in organic hydrochlorides.[43] The results showed again that the optimization of hydrogen atom positions was essential and that while both the value of $|C_Q(^{35}Cl)|$ and the CS tensor span were overestimated by calculations, the experimental trends in both parameters were reproduced. While both the cluster model calculations using B3LYP/RHF and GIPAW–DFT methods using periodic boundary conditions are useful, it appears that the latter technique is slightly more accurate relative to the presently available experimental chlorine NMR data.[18,42,43]

In 2008, ^{35}Cl SSNMR was shown to be an effective method to distinguish polymorphs in a study of pharmaceuticals by Hamaed *et al.*[19] Four local anesthetic pharmaceuticals were examined at 9.4 and 21.1 T: procaine HCl, tetracaine HCl, monohydrated lidocaine HCl (LH), and monohydrated bupivacaine HCl (BH). The EFG and CS tensors, along with their relative orientations, were reported for all four salts, as well as high temperature polymorphs of LH and BH. The parameters were found to be in a similar range to those observed for the amino acid hydrochlorides, with the CSs ranging from 30 to 77 ppm (with respect to the IUPAC standard), CS tensor spans ranging from 20 to 160 ppm, and $C_Q(^{35}Cl)$ magnitudes ranging from 2.52 to 6.00 MHz. Notably, ^{35}Cl SSNMR was found to be more effective than powder X-ray diffraction and ^{13}C SSNMR at distinguishing polymorphs in these cases.

Owing to the moderately large Q of the two spin-active chlorine nuclei, the QI typically dominates the CT lineshape when the chlorine is in a noncubic environment. A $^{35/37}$Cl SSNMR study[43] carried out at 9.4 and 21.1 T of the 1-butyl-3-methylimidazolium chloride complex of *meso*-octamethylcalix[4]pyrrole, a known anion receptor, provides a counter example. Specifically, at 21.1 T, the $^{35/37}$Cl NMR spectra collected under stationary conditions exhibited lineshapes typical of spin-1/2 nuclei, i.e., the powder pattern is dominated by CSA. The value of $|C_Q(^{35}Cl)|$ was found to be low compared to those of the amino acid hydrochlorides, at only 1.0 MHz, but is within the range observed for perchlorates, ionic liquids (*vide infra*), and a series of several alkylammonium chlorides studied by Honda.[59]

19.3.1.2 Metal Chlorides

Compared to the organic hydrochlorides, there are significantly fewer inorganic noncubic chloride salts that have been analyzed by chlorine SSNMR,[60,61] although the number has grown significantly in recent years (see Table 19.4 for chlorine NMR data for selected metal chlorides). In 2007, a study of several alkaline earth chloride hydrates, in which both EFG and CS tensor information was extracted, was carried out at magnetic fields of 11.7 and 21.1 T (Figure 19.3).[20] The values of $|C_Q(^{35}Cl)|$ observed for the hydrates were in the range 1.41 to 4.26 MHz, whereas the CS tensor spans ranged from 41 to 72 ppm, both on the order of the values observed for the organic hydrochlorides. The chlorine CSs, however, are higher than those for the organic hydrochlorides. In addition, the study demonstrated the power of chlorine SSNMR to distinguish pseudopolymorphs in the case of the anhydrous, dihydrate, and hexahydrate forms of $SrCl_2$, which displayed very different chlorine NMR tensor parameters. It was noted in this study that the CSs decreased as hydration increased. In addition, the study demonstrated that using the GIPAW-DFT method to calculate chlorine NMR tensor parameters yielded good agreement with experimental results for all samples for which a neutron diffraction structure was available.

Calcium chloride and several of its hydrates were further investigated in a subsequent report.[62] The study presents ^{35}Cl NMR spectra of the anhydrous and hexahydrate forms at 11.75 and 21.1 T as well as GIPAW–DFT calculations of the chlorine NMR parameters for other pseudopolymorphs. It was noted that the new data for $CaCl_2$ were consistent with the expected values for $C_Q(^{35}Cl)$ and δ_{iso}, on the basis of its crystal structure and prior correlations between these parameters and local structure.[20] The experimental CS and $C_Q(^{35}Cl)$ values for the anhydrous salt appeared at 105 ppm (with respect to the IUPAC standard) and 8.82(8) MHz, both significantly higher than the values for $CaCl_2{\cdot}6H_2O$.[62]

Rossini *et al.* have studied several organometallic compounds, including Cp_2TiCl_2, $CpTiCl_3$, Cp_2ZrCl_2, Cp_2HfCl_2, $Cp^*_2ZrCl_2$, $CpZrCl_3$, Cp^*ZrCl_3, $Cp_2ZrMeCl$, Cp_2ZrHCl, and $(Cp_2ZrCl)_2(\mu\text{-O})$ (Cp = cyclopentadienyl; Cp* = pentamethylcyclopentadienyl) (Figure 19.7).[61] These compounds exhibit relatively large chlorine QIs and therefore the VOCS–QCPMG method and a very high magnetic field (21.1 T) were employed in many cases to collect the full CT spectrum. The

Table 19.4. Selected chlorine NMR data for solid metal chlorides

Compounds	$\lvert C_Q(^{35}Cl)\rvert$/MHz	η_Q	$\delta_{iso}{}^a$ (ppm)	Ω (ppm)	κ	Euler angles[b]/degrees	References
$CaCl_2{\cdot}2H_2O$	4.26(0.03)	0.75(0.03)	68.9(2.0)	72(15)	0.6(0.2)	90(10), 82(5), 0(20)	Bryce and Bultz[20]
$BaCl_2{\cdot}2H_2O$	Site 1: 2.19(0.08)	0.00	122.3(2.0)	50(25)	−0.8(0.2)	85(20), 32(10), 60(20)	Bryce and Bultz[20]
	Site 2: 3.42(0.08)	0.31(0.10)	115.5(2.0)	50(25)	0.20(0.25)	20(15), 8(10), 0(20)	—
$MgCl_2{\cdot}6H_2O$	3.02(0.05)	0.00	33.9(1.0)	<75	—	—	Bryce and Bultz[20]
$SrCl_2$	~0	n/a	147.1(1.0)	—	—	—	Bryce and Bultz[20]
$SrCl_2{\cdot}2H_2O$	1.41(0.02)	0.80(0.10)	101.0(1.0)	41(10)	0.5(0.2)	86(15), 75(5), 37(10)	Bryce and Bultz[20]
$SrCl_2{\cdot}6H_2O$	3.91(0.05)	0.00	49.3(1.0)	45(20)	−1.0	0(10), 90(10), 0(10)	Bryce and Bultz[20]
$CaCl_2$	8.82(0.08)	0.383(0.015)	105(8)	135(15)	0.0(0.3)	90(20), 90(5), 0(5)	Widdifield and Bryce[62]
$CaCl_2{\cdot}6H_2O$	4.33(0.03)	<0.01	57(3)	40(8)	−1	n/a, 90(7), 0(8)	Widdifield and Bryce[62]
$Cp_2TiCl_2{}^c$	22.1(0.5)	0.61(0.03)	500(500)	—	—	—	Rossini *et al.*[61]
Cp_2ZrCl_2	16.0(0.5)	0.72(0.04)	300(150)	800(500)	0.0(0.5)	2(10), 72(20), −70(20)[d]	Rossini *et al.*[61]
Cp_2HfCl_2	17.1(0.4)	0.65(0.05)	400(500)	—	—	—	Rossini *et al.*[61]
$Cp^*_2ZrCl_2{}^e$	16.7(0.4)	0.73(0.03)	400(400)	—	—	—	Rossini *et al.*[61]
$CpTiCl_3$	15.5(0.4)	0.54(0.05)	500(150)	750(400)	−0.4(0.5)	80(30), 5(15), 5(30)	Rossini *et al.*[61]
$Cp_2ZrMeCl$	13.7(0.4)	0.75(0.10)	400(400)	—	—	—	Rossini *et al.*[61]
$(Cp_2ZrCl)_2\mu$-O	16.3(0.4)	0.43(0.07)	300(400)	—	—	—	Rossini *et al.*[61]
Cp^*ZrCl_3	Site 1: 12.8(0.5)	0.10(0.10)	400(200)	500(400)	0.4(0.8)	10(90), 15(30), 0(90)[d]	Rossini *et al.*[61]
	Site 2: 13.3(0.5)	0.12(0.10)	400(200)	500(400)	0.4(0.8)	10(90), 15(30), 0(90)[d]	—
	Site 3: 14.6(0.5)	0.88(0.10)	200(200)	200(200)	—	—	—
	Site 4: 14.0(0.5)	0.80(0.10)	200(200)	200(200)	—	—	—
$CpZrCl_3$ (multiple sites)	14.8–18.6	0.7–0.8	300	—	—	—	Rossini *et al.*[61]
Cp_2ZrHCl	19.7(0.3)	0.20(0.04)	80(50)	—	—	—	Rossini *et al.*[61]
$AlCl_3$	22.5(1.0)	0.63(0.10)	284(100)	300(200)	−0.5(0.5)	90(20), 90(20), 0(20)	Chapman and Bryce[60]
$InCl_3$	24.5(1.0)	0.52(0.10)	334(100)	500(200)	0.5(0.5)	20(20), 90(20), 30(20)	Chapman and Bryce[60]
$GaCl_2$	31.2(0.7)	0.15(0.15)	159(100)	200(200)	−0.5(0.5)	90(20), 90(20), 0(20)	Chapman and Bryce[60]
	32.0(0.7)	0.2(0.2)	159(100)	200(200)	−0.5(0.5)	90(20), 90(20), 20(20)	—
$GaCl_3$	40.4(2.0)	0.03(0.03)	159(100)	—	—	—	Chapman and Bryce[60]
	38.1(2.0)	0.09(0.05)	109(100)	—	—	—	—
	28.3(2.0)	0.48(0.05)	209(100)	—	—	—	—

[a] With respect to 0.1 mol dm^{-3} NaCl in D_2O.
[b] Euler angles determined using the Arfken convention unless otherwise noted.
[c] Cp = cyclopentadienyl.
[d] Rose convention used for Euler angles.
[e] Cp* = pentamethylcyclopentadienyl.

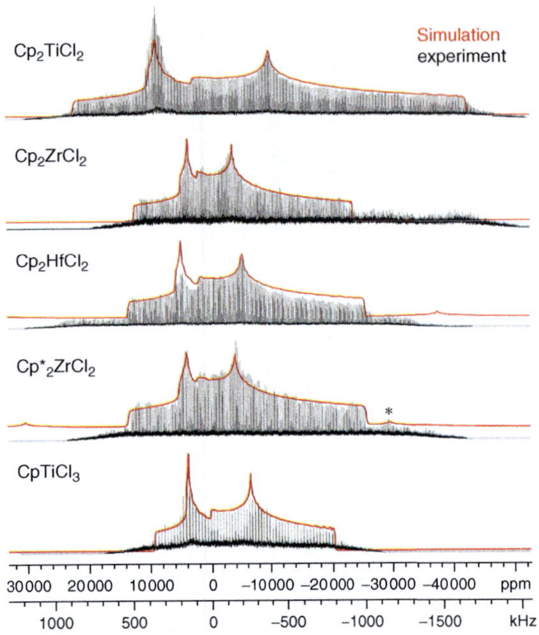

Figure 19.7. ^{35}Cl QCPMG SSNMR spectra and analytical simulations of the spectra (solid traces) for Cp_2TiCl_2, Cp_2ZrCl_2, Cp_2HfCl_2, $Cp^*{}_2ZrCl_2$, and $CpTiCl_3$ ($B_0 = 9.4$ T). See Table 19.4 for parameters. Satellite transitions are visible in the spectra of Cp_2ZrCl_2, $Cp_2{}^*ZrCl_2$, and Cp_2HfCl_2. The asterisk in the spectrum of $Cp^*{}_2ZrCl_2$ denotes a discontinuity of a satellite transition. (Reproduced from Ref. 61. © American Chemical Society, 2009.)

$C_Q(^{35}Cl)$ magnitudes observed are notably higher than those observed for the alkaline earth metal chlorides, ranging from 12.8(5) MHz for one of the four sites in Cp^*ZrCl_3 to 22.1(5) MHz in Cp_2TiCl_2. The authors were also able to extract CS tensor data for three of the complexes. The CS tensor spans observed were very large compared to others in the literature with the largest reported value being 800(500) ppm for Cp_2ZrCl_2.

The range of $C_Q(^{35}Cl)$ magnitudes, and therefore chloride ion environments, which may be practically observed with chlorine SSNMR, was expanded in a study of four group 13 metal chlorides.[60] Again, the VOCS–QCPMG technique and a very high magnetic field (21.1 T) were required to collect the full CT spectra of both ^{35}Cl and ^{37}Cl. The compounds, which included the catalysts $AlCl_3$ and $GaCl_3$, were found to have strong chlorine QIs, with $C_Q(^{35}Cl)$ magnitudes ranging from 22.5 MHz in $AlCl_3$ to 40.4 MHz for one site in $GaCl_3$. The chlorine CSs range from 109 to 334 ppm (with respect to the IUPAC standard), which is consistent with those observed for other chloride-containing organometallics,[61] and significantly higher than the values observed for organic hydrochlorides.[2,42] It was noted that the chlorine CS increased as the M–Cl bond length increased in all cases.[60] CS tensor data were also extracted for three of the chlorides, with CS tensor spans ranging from 200 to 500 ppm.

19.3.1.3 Perchlorates

Perchlorates ($ClO_4{}^-$) have been well studied by chlorine SSNMR due to the nearly tetrahedral symmetry about the chlorine site (Table 19.5).[8,63–71] This arrangement results in a relatively small QI, and consequently small chlorine C_Q values and CT spectral breadths. In addition, the chlorine chemical environment in these materials is interesting as large paramagnetic contributions to the shielding tensor lead to chlorine CSs that are much larger than those for chloride ions.

Jurga *et al.* have published several spectra of multimethylammonium perchlorates.[63] Relatively small ^{35}Cl quadrupolar coupling constants ranging between 0.238 and 1.12 MHz were extracted from spectra collected under MAS and stationary conditions. In addition, one phase of dimethylammonium perchlorate was found to have a chlorine CS highly deshielded at 1012 ppm (with respect to the IUPAC standard). Tarasov *et al.* extended the study of perchlorates through the collection of ^{35}Cl SSNMR spectra of three alkali metal perchlorates, $CsClO_4$, $RbClO_4$, and $KClO_4$, at 7.04 T.[66,67] The ^{35}Cl quadrupolar coupling constants observed were similar to those observed by Jurga, with magnitudes ranging from 0.51 to ~0.63 MHz. In 1999, a thorough study was published by Skibsted and Jakobsen concerning a large series of perchlorates (Figure 19.8).[8] The authors looked at 13 samples of anhydrous and hydrated perchlorates, including those examined by Tarasov. Chlorine-35 MAS, satellite transition spectroscopy (SATRAS), and echo experiments for all of the salts at 14.1 T allowed for accurate determination of the chlorine NMR parameters. In select cases, ^{37}Cl NMR spectra were also collected. Their results were consistent with earlier observations, and chlorine CSs were all greater than 988.5 ppm (with respect to the IUPAC standard), although the relative range of ^{35}Cl quadrupolar coupling constants was more substantial,

Table 19.5. Selected chlorine NMR data for solid perchlorates

Compounds	$\lvert C_Q(^{35}Cl)\rvert$/MHz	η_Q	$\delta_{iso}{}^a$ (ppm)	References
$(CH_3)_3NHClO_4$—phase II	0.318	—	—	Jurga *et al.*[63]
$(CH_3)_3NHClO_4$—phase III	0.32 to 0.35	0.6 to ~1	—	
$(CH_3)_2NH_2ClO_4$—phase I	~0	n/a	1012	Jurga *et al.*[63]
$(CH_3)_2NH_2ClO_4$—phase II	0.238	0	—	
$(CH_3)_2NH_2ClO_4$—phase III	1.12	0	—	
$CH_3NH_3ClO_4$—phase II	0.258	0	—	Jurga *et al.*[63]
$CH_3NH_3ClO_4$—phase III	0.932	0.75	—	
$NaClO_4$	0.887(0.014)	0.92(0.02)	1003.2(0.5)	Skibsted and Jakobsen[8]
$NaClO_4 \cdot H_2O$	0.566(0.009)	0.90(0.02)	998.8(0.3)	Skibsted and Jakobsen[8]
$LiClO_4$	1.282(0.008)	0.34(0.01)	993.1(0.5)	Skibsted and Jakobsen[8]
$LiClO_4 \cdot 3H_2O$	0.695(0.004)	0.00(0.03)	1004.8(0.5)	Skibsted and Jakobsen[8]
$KClO_4$	0.51(0.01) (at 296 K)	0.52(0.10)	—	Tarasov *et al.*[67]
	0.440(0.006)	0.88(0.02)	1008.1(0.3)	Skibsted and Jakobsen[8]
$RbClO_4$	Temperature dependence of C_Q monitored	0.53(0.02)	$-3(5)^b$	Tarasov *et al.*[67]
—	0.537(0.015)	0.87(0.03)	1008.3(0.3)	Skibsted and Jakobsen[8]
$CsClO_4$	Temperature dependence of C_Q and η_Q determined	0.55	—	Tarasov *et al.*[66]
—	0.585(0.008)	0.86(0.02)	1006.6(0.3)	Skibsted and Jakobsen[8]
$Mg(ClO_4)_2$	2.981(0.007)	0.57(0.01)	995.1(0.5)	Skibsted and Jakobsen[8]
$Mg(ClO_4)_2 \cdot 6H_2O$—site 1	0.309(0.006)	0.00(0.08)	1005.5(0.3)	Skibsted and Jakobsen[8]
$Mg(ClO_4)_2 \cdot 6H_2O$—site 2	0.475(0.008)	0.00(0.05)	1004.4(0.3)	
$Ba(ClO_4)_2$	2.256(0.008)	0.58(0.01)	988.5(0.5)	Skibsted and Jakobsen[8]
$Ba(ClO_4)_2 \cdot 3H_2O$	0.383(0.005)	0.00(0.03)	999.5(0.3)	Skibsted and Jakobsen[8]
$Cd(ClO_4)_2 \cdot 6H_2O$	0.328(0.005)	0.00(0.03)	1003.3(0.3)	Skibsted and Jakobsen[8]
$(CH_3)_4NClO_4$	0.307(0.004)	0.00(0.03)	1008.2(0.3)	Skibsted and Jakobsen[8]

[a] With respect to 0.1 mol dm^{-3} NaCl in D_2O, unless otherwise noted.
[b] With respect to 0.1 mol dm^{-3} $RbClO_4$(aq).

at 0.307 to 2.981 MHz. Skibsted and Jakobsen improved the accuracy of the earlier data because of the inclusion of CSA in their analyses and the analysis of the spinning sideband manifolds for the STs.[8]

19.3.1.4 Other Systems

In addition to the classes of compounds discussed above, other materials have been analyzed using chlorine SSNMR (Table 19.6). For example, several chlorine-containing glasses have been studied with ^{35}Cl MAS SSNMR. In a 2004 report, a series of silicate and aluminosilicate glasses were studied and a correlation was noted between the M–Cl bond distance and the chlorine CS.[72] The value of $C_Q(^{35}Cl)$ in these glasses ranged from 2.9 to 4.4 MHz.[72]

A new class of materials was analyzed by Gordon *et al.* in 2008 when they used chlorine SSNMR to examine four ionic liquids, which are solid at room temperature.[73,74] The magnitudes of $C_Q(^{35}Cl)$ extracted were quite small, ranging from 0.805 to 1.500 MHz, and demonstrated the small chlorine QI in these salts. The CSs observed ranged from 60.6 to 91.7 ppm (with respect to the IUPAC standard). In addition, it was shown that the four crystallographic chlorine sites in ethylmethylimidazolium chloride could be resolved in a ^{35}Cl MAS spectrum collected at 21.1 T.

Covalently bound chlorine in organic moieties is characterized by $C_Q(^{35}Cl)$ magnitudes on the order of 60–80 MHz.[14,15] Acquiring the $^{35/37}Cl$ SSNMR spectrum of covalently bound chlorine in a powder sample is technically feasible in a high magnetic

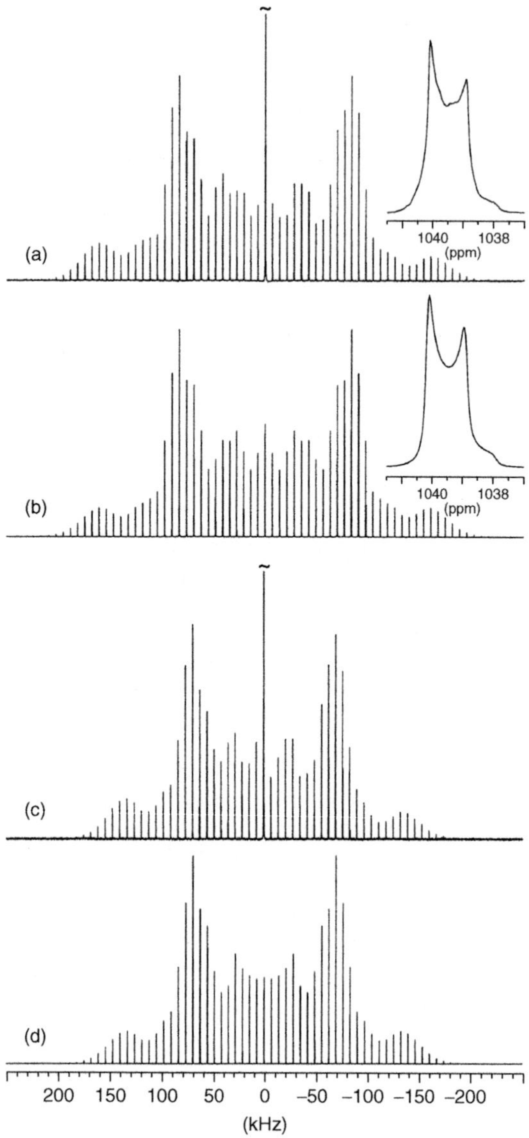

Figure 19.8. ^{35}Cl MAS NMR spectra ($\nu_r = 7.0$ kHz) of the satellite transitions for (a) $Ba(ClO_4)_2{\cdot}3H_2O$ and (c) $Cd(ClO_4)_2{\cdot}6H_2O$ shown with the central transition cutoff at $\sim^1/_{10}$ of its total height. The inset in panel (a) illustrates the lineshape for the central transition for $Ba(ClO_4)_2{\cdot}3H_2O$. Simulated spectra of the spinning sideband manifolds from the satellite transitions are shown in panels (b) and (d) for $Ba(ClO_4)_2{\cdot}3H_2O$ and $Cd(ClO_4)_2{\cdot}6H_2O$, respectively, and employ the ^{35}Cl NMR parameters in Table 19.5. A simulation of the central transition for $Ba(ClO_4)_2{\cdot}3H_2O$ is shown as the inset in panel (b). (Reproduced from Ref. 8. © American Chemical Society, 1999.)

field ($\geq$18.8 T), but is also quite time consuming and typically impractical.[75]

19.3.2 Bromine

Relative to ^{35}Cl and ^{37}Cl SSNMR spectroscopy, significantly fewer ^{79}Br and ^{81}Br SSNMR data are available, although a number of similar systems have been studied.[1,2] Typical ^{81}Br C_Q values for noncubic ionic bromide sites are on the order of 10–20 MHz, which represents a static line width of about 100–500 kHz at 11.75 T. As such, while $^{79/81}$Br SSNMR experiments are generally applicable for the study of bromide-containing materials, the VOCS method will often be required for the acquisition of complete CT $^{79/81}$Br SSNMR signals. Although the experimental observation of bromine CSA is rare, typical values of the CS tensor span for bromide-containing materials are on the order of 100–200 ppm.[1,2,21,74,76]

19.3.2.1 Organic Hydrobromides

A few organic hydrobromide systems have been studied using ^{81}Br SSNMR spectroscopy, primarily using single crystals; however, the effects of bromine CSA were neglected in all cases.[77,78] The EFG tensor data pertaining to these systems are summarized in Table 19.7. Although the data are somewhat sparse, it is observed that $C_Q(^{81}Br)$ values for these systems range from 11.26 to 49.0 MHz, and hence future studies on organic hydrobromide systems should be possible using standard spectrometer systems. Owing to the low symmetry present at the bromide anions in the systems studied to date, η_Q values are seen to strongly deviate from zero (ranging from 0.59 to 0.86). Bromine-81 SSNMR experiments using a single crystal of deuterated glycyl-L-alanine $HBr{\cdot}H_2O$ established that the V_{33} component of the bromine EFG tensor oriented approximately along the shortest H–Br hydrogen bond,[78] which highlights the sensitivity of $^{79/81}$Br SSNMR experiments to weak intermolecular interactions in solids.

19.3.2.2 Inorganic Bromides

The alkali metal bromides, of the general form MBr (M = Li, Na, K, Rb, Cs), were the first compounds to be studied using solid-state $^{79/81}$Br NMR

Table 19.6. Selected chlorine NMR data for other chlorine systems

Compounds	$|C_Q(^{35}Cl)|$/MHz	η_Q	δ_{iso}[a] (ppm)	References
Cl-containing silicate and aluminosilicate glasses	2.9(0.2) to 4.4(0.4)	0.7	−85(11) to 107(22)	Sandland *et al.*[72]
Butyldimethylimidazolium chloride	0.978(0.004)	0.10(0.02)	71.60(0.02)	Gordon *et al.*[73,74]
Butylmethylimidazolium chloride	1.500(0.002)	0.390(0.005)	71.65(0.05)	Gordon *et al.*[73,74]
Ethylmethylimidazolium chloride	Site 1: 0.808(0.004)	0.95(0.01)	91.72(0.03)	Gordon *et al.*[73,74]
	Site 2: 0.805(0.005)	0.20(0.02)	74.98(0.04)	
	Site 3: 0.884(0.004)	0.86(0.01)	71.36(0.04)	
	Site 4: 0.972(0.005)	0.80(0.01)	60.60(0.03)	
Butylmethylpyridinium chloride	Site 1: 0.857(0.008)	0.525(0.005)	83.15(0.07)	Gordon *et al.*[73,74]
	Site 2: 0.889(0.008)	0.08(0.08)	70.56(0.04)	

[a]With respect to 0.1 mol dm^{-3} NaCl in D_2O.

Table 19.7. Selected bromine NMR data for solid organic hydrobromides

Compounds	$|C_Q|$/MHz	η_Q	References
Deuterated glycyl-L-alanine $HBr{\cdot}H_2O$	19.750[a]	0.8328	Kehrer *et al.*[78]
L-Leucine HBr	49.0[b]	0.59	Persons and Harbison[77]
L-Tyrosine HBr	11.26[b]	0.86	Persons and Harbison[77]

[a]From ^{81}Br SSNMR data.
[b]From ^{79}Br SSNMR data.

Table 19.8. Selected bromine NMR data for solid inorganic bromides—cubic systems

Compounds	δ_{iso} (ppm)	References
LiBr	119.33(0.15)[a]	Widdifield *et al.*[1]
NaBr	1.57(0.09)[a]	Widdifield *et al.*[1]
KBr	54.51[a]	Widdifield *et al.*[1]
RbBr	126.13(0.11)[a]	Widdifield *et al.*[1]
CsBr	282.76(0.25)[a]	Widdifield *et al.*[1]
AgBr	223.66(0.07)[b]	Hayashi and Hayamizu[56]
CuBr	−79.83(0.20)[b]	Hayashi and Hayamizu[56]

[a]From ^{81}Br SSNMR data with respect to 0.03 mol dm^{-3} NaBr in D_2O.
[b]From ^{79}Br SSNMR data with respect to 0.03 mol dm^{-3} NaBr in D_2O.

spectroscopy[79] owing to the essentially cubic environment at the bromide anions, which greatly reduces the QI relative to noncubic environments.[2,54,55] The lack of a significant QI makes the alkali metal bromides ideal compounds with which to set up further experiments on either NMR-active bromine nucleus. The CS values of this series relative to 0.03 mol dm^{-3} NaBr in D_2O have been precisely determined under MAS conditions, and are reported in Table 19.8.[1] Additional simple inorganic bromide salts (NH_4Br, AgBr, CuBr, and TlBr) have been studied using $^{79/81}$Br SSNMR, including three crystalline phases of NH_4Br, where the bromine CS and spin-lattice relaxation time were both found to be sensitive to the phase of this material.[80,81]

Many alkaline earth metal bromides and selected hydrates have also been studied using $^{79/81}$Br SSNMR spectroscopy (see Table 19.9 for EFG and CS tensor parameters).[16,21] Owing to the extreme sensitivity of the $^{79/81}$Br nuclei to the EFG, these experiments, when combined with the GIPAW–DFT computational method and ultrasoft core electron pseudopotentials, were used to propose a modified structure for $MgBr_2$.[16] None of the anhydrous MBr_2 series compounds (M = Mg, Ca, Sr, Ba) studied to date possess cubic local symmetry, which leads to the observation of relatively large bromine QIs ($C_Q(^{81}Br)$ ranges from 7.32 to 62.8 MHz). Bromine-79/81 SSNMR experiments were shown to be sensitive to the hydration level of the alkaline earth metal bromides through decreases in the bromine C_Q and δ_{iso} values upon sample hydration.[21] A rare example of $^{79/81}$Br MAS NMR data for a noncubic sample was presented for $BaBr_2{\cdot}2H_2O$, which possessed unusually low $C_Q(^{79/81}Br)$ values. While the QI is expected to dominate the line broadening in the alkaline earth metal bromides, even in high magnetic fields, the effects of bromine CSA in the observed SSNMR spectra were quantified, with Ω values ranging from 50 to 250 ppm (Figure 19.9).[21]

Table 19.9. Selected bromine NMR data for solid alkaline earth metal bromides

Compounds	$\lvert C_Q(^{81}Br)\rvert$/MHz	η_Q	δ_{iso} (ppm)[a]	Ω (ppm)	κ	Euler angles[b]/degrees	References
$CaBr_2$	62.8(0.4)	0.445(0.02)	335(50)	250(150)	0[c]	270[c], 90(20), 180[c]	Widdifield and Bryce[21]
tris-Sarcosine $CaBr_2$	21.9	0.64	—	—	—	—	Erge *et al.*[82]
$MgBr_2$	21.93(0.20)	0.02(0.02)	340(10)	—	—	—	Widdifield and Bryce[16]
$MgBr_2{\cdot}6H_2O$	19.0(0.2)	0.23(0.03)	112(7)	50(20)	0.7(0.3)	170(10), 57(10), 180[c]	Widdifield and Bryce[21]
$SrBr_2$—site 1	10.3(0.3)	0.07(0.04)	477(5)	50(20)	−1[d]	90[c], 90(15), 180(5)	Widdifield and Bryce[21]
$SrBr_2$—site 2	18.1(0.2)	0.03(0.02)	465(8)	85(25)	−1[d]	90[c], 90(10), 180(8)	Widdifield and Bryce[21]
$SrBr_2$—site 3	25.6(0.2)	0.695(0.015)	375(10)	110(30)	0.3(0.4)	42(8), 90(10), 235(20)	Widdifield and Bryce[21]
$SrBr_2$—site 4	53.7(0.6)	0.33(0.02)	355(50)	—	—	—	Widdifield and Bryce[21]
$SrBr_2{\cdot}6H_2O$	27.7(0.3)	<0.01	150(15)	70(30)	−1[d]	210[c], 90(20), 180(10)	Widdifield and Bryce[21]
$BaBr_2$—site 1	23.5(0.3)	0.17(0.02)	335(10)	200(20)	−0.6(0.2)	0[c], 47(7), 180[c]	Widdifield and Bryce[21]
$BaBr_2$—site 2	27.2(0.3)	0.070(0.015)	535(15)	170(30)	0.1(0.2)	180[c], 18(7), 180[c]	Widdifield and Bryce[21]
$BaBr_2{\cdot}2H_2O$	7.32(0.03)	0.76(0.02)	272.7(1.0)	86(5)	−0.20(0.15)	70(5), 95(8), 253(5)	Widdifield and Bryce[21]

[a]From ^{81}Br SSNMR data, with respect to 0.03 mol dm^{-3} NaBr in D_2O.
[b]Given as α, β, and γ according to the "ZYZ" convention. (See Ref. 44.)
[c]Simulated SSNMR lineshape is not sensitive to variation in this parameter.
[d]Assumed on the basis of crystallographic site symmetry.

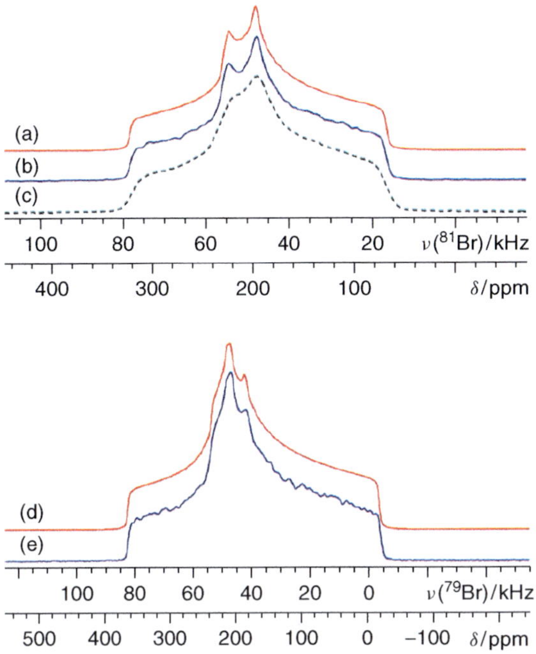

Figure 19.9. Analytical simulations incorporating quadrupolar and CSA effects (a, d, f, h) and experimental static solid echo (b, c, e, g, i) $^{79/81}$Br{^{1}H} SSNMR spectra of powdered $BaBr_2{\cdot}2H_2O$, acquired at $B_0 = 21.1$ T (b, c, e) and 11.75 T (g, i). See Table 19.9 for parameters. In spectrum (c), ^{1}H decoupling is not applied. (Reproduced from Ref. 21. © American Chemical Society, 2010.)

19.3.2.3 Other Systems

Solid-state bromine (primarily ^{81}Br) NMR experiments have also been carried out on a variety of alkali metal perbromates,[9,11,12] *n*-alkyltrimethylammonium bromides,[76] ionic liquids,[74] and mesoporous materials.[83] Selected bromine EFG and CS tensor data for these systems are located in Table 19.10. Owing to the high symmetry present at the bromine sites in the alkali metal perbromates, very small C_Q(^{81}Br) values were measured, ranging from 1.32 MHz in $CsBrO_4$ to 3.35 MHz in $KBrO_4$. Although these experiments were carried out under static conditions, bromine CSA was neglected during the spectral modeling. These systems could thus be of great use as setup samples, if one wished to have an appreciable second-order quadrupolar lineshape (which would not be present in the alkali metal halides).

A series of *n*-alkyltrimethylammonium bromide ($n = 1$, 12, 14, 16, 18) systems have been studied using ^{81}Br SSNMR under both static and MAS conditions at high B_0.[76] As with many of the compounds studied above, the ^{81}Br nuclei were found to be sensitive probes of minute structural variations. For example, the bromine-81 δ_{iso}, C_Q, and η_Q parameters were found to possess weak dependencies upon the alkyl chain length for the $n > 1$ series. The variation in the C_Q(^{81}Br) value was attributed to an increase in the Br–N distance as the alkyl chain length increased. The ^{81}Br SSNMR data were acquired at multiple MAS frequencies, followed by a numerical data fitting procedure, which allowed for the extraction of bromine CS tensor span values around 110 ppm (Figure 19.10). The sensitivity of ^{81}Br SSNMR measurements to molecular dynamics has also been observed using a single crystal of *tris*-sarcosine $CaBr_2$.[82,85]

A series of four bromide-containing ionic liquids were subjected to ^{79}Br SSNMR experiments at both standard and very high magnetic fields.[74] Owing to the rather isolated nature of the bromides in these systems, C_Q(^{79}Br) values were found to lie within a relatively narrow range of 5.12–17.50 MHz, while the CSs varied between 122 and 172 ppm relative to 0.01 mol dm^{-3} NaBr in D_2O. Bromine CSA and

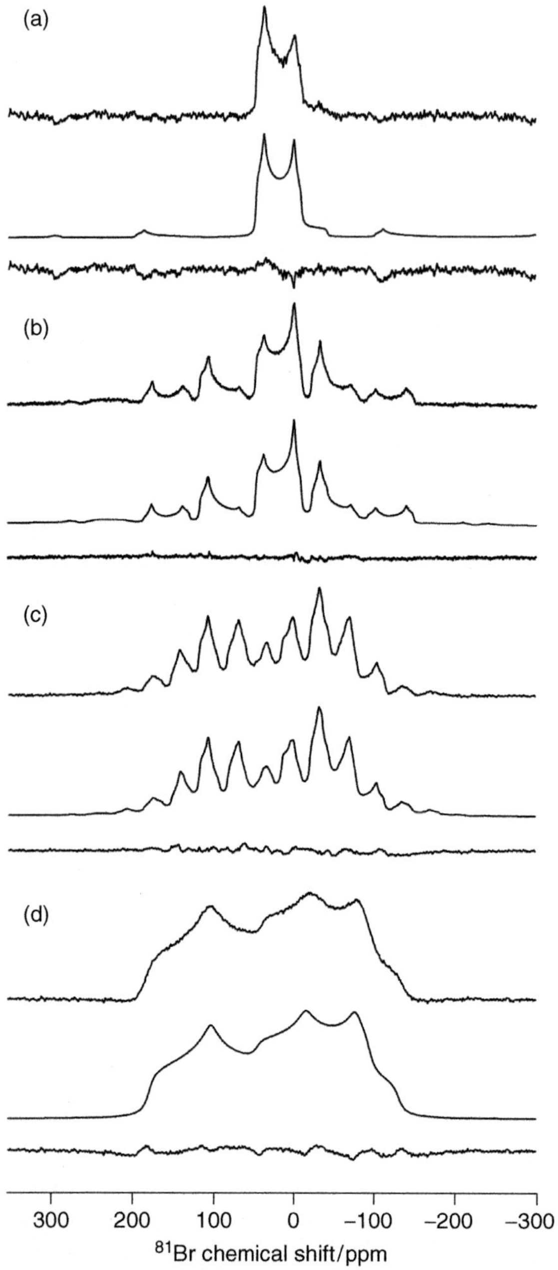

Figure 19.10. Modeling of ^{81}Br spectra of hexadecyltrimethylammonium bromide: (a) $\nu_{MAS} = 30$ kHz, (b) $\nu_{MAS} = 14$ kHz, (c) $\nu_{MAS} = 7$ kHz, and (d) $\nu_{MAS} = 0$ kHz. In each of the four parts, the experimental spectrum is shown at the top, the model in the middle, and the difference spectrum at the bottom. (Reproduced from Ref. 76. © American Chemical Society, 2009.)

CS/EFG tensor interplay were also observed in all samples (Table 19.10).

An account by Alonso *et al.* noted the possibility of carrying out ^{81}Br SSNMR experiments on mesoporous materials;[83] however, it appears as though MAS experiments were not fruitful for the systems that were considered. The authors attribute the difficulty in direct observation to a spatially nonuniform charge distribution (and hence a distribution in the QI parameters) when the probe molecule is within the mesoporous material. At the same time, it was noted that information pertaining to the ^{81}Br environment could potentially be gained via indirect methods, such as ^{1}H{^{81}Br} TRAPDOR experiments. Indeed, it was observed using both ^{81}Br and ^{14}N SSNMR data that the bromide anions were distributed close to the alkyltrimethylammonium head group, the *n*-alkyl chains, and the phenyl groups on the siloxane surface.

Covalently bound bromine in organic systems is typically characterized by $C_Q(^{81}Br)$ magnitudes on the order of 400–600 MHz.[15] Acquiring the ^{81}Br SSNMR spectrum of covalently bound bromine in a powder sample is impractical with currently available magnetic field strengths.

19.3.3 Iodine

Owing to the relatively large quadrupole moment for the iodine-127 nucleus, much of the ^{127}I SSNMR data reported in the current literature involves either ionic iodides or periodates.[1,2,13,17] Although these systems are chosen to minimize the resulting QI, typical $C_Q(^{127}I)$ values for ionic iodides and periodates are still large (up to 214 MHz, which translates into a CT line width of about 10 MHz at 21.1 T). Hence, the VOCS method will often be used for the acquisition of ^{127}I SSNMR signals. The experimental observation of iodine CSA is very rare: there exists only a handful of reliable measurements.[13,17] This is due to the generally small impact of the CSA on the second-order quadrupolar-dominated lineshape (in systems observed to date) as well as the technical difficulties associated with acquiring broad ^{127}I SSNMR spectra.

19.3.3.1 Inorganic Iodides

Owing to the cubic lattice symmetry present for the alkali metal iodides (general form MI, M = Li, Na,

Table 19.10. Selected bromine NMR data for other solid bromine-containing systems

Compounds	$\|C_Q\|$/MHz	η_Q	δ_{iso} (ppm)	Ω (ppm)	κ	Euler angles/degrees	References
$KBrO_4$	3.35(0.03)[a]	0.71(0.05)	—	—	—	—	Tarasov *et al.*[11]
$RbBrO_4$	2.36(0.04)[a]	0.37	—	—	—	—	Tarasov *et al.*[11]
$CsBrO_4$	1.32(0.04)[a]	0	2400(200)[b]	—	—	—	Tarasov *et al.*[9]
NH_4BrO_4	2.27(0.05)[a]	0.99	—	—	—	—	Tarasov *et al.*[12]
$(CH_3)_4NBr$	6.03(0.002)[a]	0.02(0.01)	100.2(0.4)[c]	33(3)	0.6(0.3)	9(9), 5(6), 9(2)[d]	Alonso *et al.*[76]
$(C_{12}H_{25})(CH_3)_3NBr$	7.39(0.10)[a]	0.11(0.02)	113.1(0.9)[c]	117(3)	−0.3(0.3)	39(9), 68(2), 50(28)[d]	Alonso *et al.*[76]
$(C_{14}H_{29})(CH_3)_3NBr$	7.74(0.17)[a]	0.16(0.03)	111.2(0.8)[c]	112(4)	−0.5(0.1)	26(7), 115(2), 47(6)[d]	Alonso *et al.*[76]
$(C_{16}H_{33})(CH_3)_3NBr$	8.03(0.03)[a]	0.18(0.02)	110.4(0.8)[c]	106(6)	−0.5(0.2)	23(20), 116(1), 59(24)[d]	Alonso *et al.*[76]
$(C_{18}H_{37})(CH_3)_3NBr$	8.08(0.07)[a]	0.19(0.01)	108.7(0.5)[c]	105(8)	−0.3(0.1)	29(6), 116(2), 63(21)[d]	Alonso *et al.*[76]
$(C_4H_9)NH_3Br$	17.50(0.02)[e]	0.01(0.01)	137.0(0.5)[f]	75(2)	0.05(0.05)	0, 0, 0,[g]	Gordon *et al.*[74]
1-Ethyl-3-methylimidazolium bromide	12.40(0.01)[e]	0.28(0.01)	122(1)[f]	73(3)	0.95(0.05)	54(3), 81(1), 8(2)[g]	Gordon *et al.*[74]
1-Ethyl-1-methylpyrrolidinium bromide	5.12(0.05)[e]	0.37(0.08)	123(1)[f]	51(1)	0.96(0.04)	97(2), 21(2), 148(2)[g]	Gordon *et al.*[74]

[a] From ^{81}Br SSNMR data.
[b] With respect to 1 mol dm^{-3} KBr.
[c] With respect to 0.03 mol dm^{-3} NaBr in D_2O.
[d] Given as ϕ, χ, and ψ according to Ref. 84.
[e] From ^{79}Br SSNMR data.
[f] With respect to 0.01 mol dm^{-3} NaBr in D_2O.
[g] Given as α, β, and γ according to the "ZYZ" convention. (See Ref. 44.)

Table 19.11. Selected ^{127}I NMR data for solid inorganic iodides—cubic systems

Compounds	δ_{iso} (ppm)^a	References
LiI	408.66	Chapman *et al.*[2]
NaI	226.71	Chapman *et al.*[2]
KI	192.62	Chapman *et al.*[2]
RbI	269.87	Chapman *et al.*[2]
CsI	562.41	Chapman *et al.*[2]
AgI	−37.6(1.0)	Hayashi and Hayamizu[56]
CuI	200.3(0.2)	Hayashi and Hayamizu[56]

^{a}With respect to 0.01 mol dm^{-3} KI in D_2O.

K, Rb, Cs), it should not be surprising that these systems were the first to be characterized using ^{127}I SSNMR, and that these compounds currently serve as useful reference materials.[2,54,55] In addition to the alkali metal iodides, additional cubic iodides, such as CuI and AgI have been studied.[56,58,86] Both CuI and AgI possess iodine CSs that are highly sensitive to temperature (on the order of 0.5 ppm/K). As the resolution of ^{127}I MAS SSNMR experiments for these systems typically results in CSs with errors on the order of 1 ppm, the iodine-127 SSNMR signals of these materials could be used to measure temperatures. The iodine CS values for these cubic systems, relative to 0.01 mol dm^{-3} KI in D_2O, have been precisely determined, and are reported in Table 19.11.

The alkaline earth metal iodides, selected hydrates thereof, and the semiconducting material CdI_2 have also been characterized using multiple-field ^{127}I SSNMR data acquisition and GIPAW–DFT computations (experimental data are in Table 19.12; see also Figure 19.11).[17] The ^{127}I nuclei were found to be useful probes of the hydration state (a decreasing iodine CS results upon sample hydration). Reliable measurements of iodine CSA were able to establish a range of ^{127}I CS tensor span values from 60 to 300 ppm, although owing to the very large line widths (on the order of MHz), very high field data acquisition is a requirement to extract these modest Ω values. Interestingly, high-order (i.e., beyond second-order) quadrupole-induced effects were observed in the CT SSNMR signal when analyzing the spectrum of one of the two iodine sites in SrI_2 (*cf.* Figure 19.5). This was confirmed using both exact lineshape simulation software,[28] and ^{127}I NQR experiments. The high-order effects were such that they lead to an underestimation in both (i) $C_Q(^{127}I)$ and (ii) δ_{iso} when the spectra were modeled using software that includes QI effects to only second order.

19.3.3.2 *Periodates*

In order to perform SSNMR experiments using ^{127}I nuclei that are not present as an iodide anion, very high local site symmetry is required. As mentioned in Section 19.1, this is because of the combination of a large quadrupole moment and a significant Sternheimer antishielding factor. Favorable conditions for observing ^{127}I SSNMR spectra exist in a variety of periodates. Compared to the alkaline

Table 19.12. Selected ^{127}I NMR data for solid inorganic iodides—noncubic systems

Compounds	$\lvert C_Q(^{127}I)\rvert$/MHz	η_Q	δ_{iso} (ppm)^a	Ω (ppm)	κ	Euler angles/degrees
MgI_2	79.8(0.5)	0.02(0.02)	920(50)	120(80)	-1^b	90^c, 90(20), 0^c
CaI_2	43.5(0.3)	0.02(0.02)	755(10)	<50	—	—
SrI_2—site 1	105.2(0.7)	0.467(0.012)	880(70)	—	—	—
SrI_2—site 2	$214.0(0.1)^d$	$0.316(0.002)^d$	$720(150)^e$	—	—	—
$SrI_2{\cdot}6H_2O$	$133.6(0.1)^d$	<0.01	$440(25)^e$	—	—	—
BaI_2—site 1	96.2(0.8)	0.175(0.015)	$650(70)^e$	300(100)	<−0.5	0^c, 45(20), 180^c
BaI_2—site 2	$120.9(0.2)^d$	$0.015(0.015)^d$	$1000(80)^e$	—	—	—
$BaI_2{\cdot}2H_2O$	53.8(0.3)	0.53(0.01)	630(20)	60(15)	>0.5	45(15), 45(15), —
CdI_2 (4H polytype)	~96.6(1.0)	0^b	~1450(100)	—	—	—

Source: From Widdifield and Bryce.[17]
^{a}With respect to 0.01 mol dm^{-3} KI in D_2O.
^{b}Assumed on the basis of crystallographic site symmetry.
^{c}Simulated SSNMR lineshape is not sensitive to variation in this parameter.
^{d}Established using both ^{127}I SSNMR and ^{127}I NQR experimental data.
^{e}Established with the aid of exact simulation software.

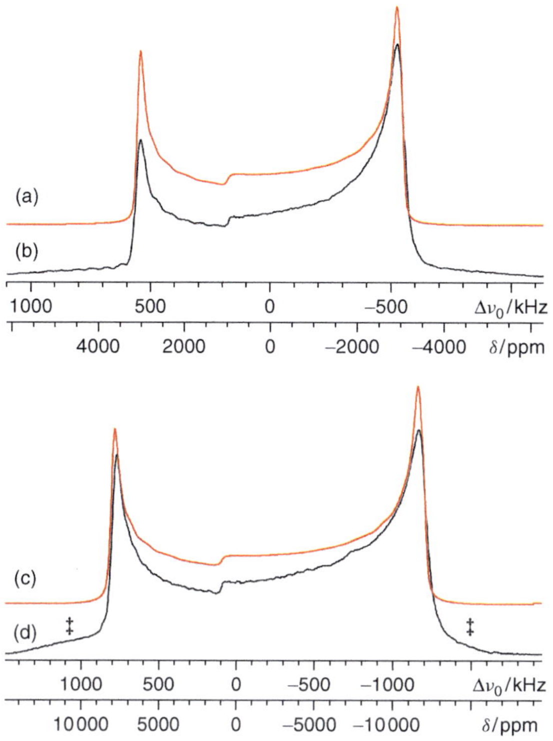

Figure 19.11. Analytical simulations (a and c) and experimental static VOCS Solomon echo (b and d) ^{127}I SSNMR spectra of powdered MgI_2, acquired at (b) $B_0 = 21.1$ T and (d) $B_0 = 11.75$ T. Partially excited STs are denoted with "‡". See Table 19.12 for parameters. (Reproduced from Ref. 17. © American Chemical Society, 2010.)

earth metal iodide systems outlined above, the iodine QIs present in the periodates are much smaller. This is analogous to what has been observed for the chlorine and bromine analogs (vide supra). Observed $C_Q(^{127}I)$ values range from only 1.00 MHz in $CsIO_4$ to 43.00 MHz in HIO_4.[13] A great deal of effort has also been expended to determine the temperature dependence of $C_Q(^{127}I)$ for the periodates, with both positive and negative correlations being noted.[87–93] Although these systems are known to possess very small relative QIs, the corresponding isotropic CS values are the highest known, due to the largely covalent O–I local bonding environment. CSs range from about 3500 to 4200 ppm with respect to 0.01 mol dm^{-3} KI in D_2O, which highlights the high sensitivity of the iodine-127 nucleus to small magnetic (in addition to electric) perturbations. The CS tensor spans for the alkali metal periodates appear to be less than 50 ppm in all cases, with only one value precisely reported ($\Omega = 18(2)$ ppm for $CsIO_4$).[13] Additional instances are reported in the literature[94]; however, further experiments must be carried out to validate these observations. Information pertaining to selected periodate systems near room temperature is summarized in Table 19.13.

19.3.3.3 Other Systems

Few additional systems have been probed using ^{127}I SSNMR, and selected relevant parameters can be found in Table 19.14. Two iodide ionic liquids have been characterized, and they possess rather similar $C_Q(^{127}I)$ and CS values as the iodides noted above.[74] To the best of our knowledge, glycyl-L-alanine hydroiodide monohydrate appears to be the only hydroiodide system studied using ^{127}I SSNMR.[78,95] The relatively small $C_Q(^{127}I)$ in this system (74.04 MHz) suggests that further studies on the amino acid hydroiodide compound class may be worthwhile. A few ^{127}I SSNMR accounts of mixed halogen systems, such as IF_7 and $[IF_6]^+[AsF_6]^-$, have also been reported.[96,97] Variable-temperature ^{127}I SSNMR experiments were found to be very sensitive toward the detection of phase transitions and molecular motion in IF_7.

Covalently bound iodine in organic systems is typically characterized by $C_Q(^{127}I)$ magnitudes on the order of 1700–2000 MHz.[15] Acquiring the ^{127}I SSNMR spectrum of covalently bound iodine in a powder sample is impractical with currently available magnetic field strengths.

19.4 CONCLUSIONS AND FUTURE PROSPECTS

Of the quadrupolar halogens, applications of $^{35/37}$Cl SSNMR spectroscopy and the interpretation of the results in terms of local structure are the most common. Most studies have understandably focused on chloride ions in organic and inorganic environments because of the relatively small QI of the chlorine nucleus. The spectroscopies of $^{79/81}$Br and ^{127}I are less developed owing to the larger quadrupole moments and Sternheimer antishielding factors for these isotopes. Nevertheless, recent work has demonstrated that a combination of very high magnetic fields, modern signal acquisition or enhancement methods,

Table 19.13. Selected ^{127}I NMR data for solid periodates

Compounds	$\vert C_Q(^{127}I)\vert$ /MHz	η_Q	δ_{iso} (ppm)^a	References
HIO_4	43.00(0.01)	0.75	3527(10)	Wu and Dong[13]
NH_4IO_4	10.00(0.01)	0.0	4187(10)	Wu and Dong[13]
$NaIO_4$	42.24(0.01)	0.0	4177(10)	Wu and Dong[13]
KIO_4	20.66(0.01)	0.0	4187(10)	Wu and Dong[13]
$RbIO_4$	15.65(0.01)	0.0	4187(10)	Wu and Dong[13]
$CsIO_4$	1.00(0.01)^b	0.0	4199(2)	Wu and Dong[13]
$(CH_3)_4NIO_4$	15.7	0	—	Klobasa and Burkert[87]
$(C_2H_5)_4NIO_4$	<1.5	—	—	Klobasa and Burkert[88]
$(n\text{-}C_4H_9)_4NIO_4$	3.66	0.67	—	Burkert and Grommelt[89]
$(CH_3)_4PIO_4$	<2.0	—	—	Klobasa and Burkert[88]
$(C_2H_5)_4PIO_4$	5.87(0.03)	0	—	Klobasa *et al.*[90]
$(C_6H_5)_4PIO_4$	4.47(0.05)	0	—	Burkert and Klobasa[93]
$(CH_3)_4AsIO_4$	≤1.8	—	—	Grommelt and Burkert[91]
$(C_2H_5)_4AsIO_4$	5.55(0.03)	0	—	Klobasa *et al.*[90]
$(C_6H_5)_4AsIO_4$	4.53(0.05)	0	—	Burkert and Klobasa[93]
$(C_2H_5)_4SbIO_4$	5.64(0.03)	0.41(0.02)	—	Klobasa and Burkert[92]
$(C_6H_5)_4SbIO_4$	2.63(0.05)	0	—	Burkert and Klobasa[93]

^{a}With respect to 0.01 mol dm^{-3} KI in D_2O.
^{b}For this compound, $\Omega = 18(2)$ ppm and $\kappa = 1$.

Table 19.14. Selected ^{127}I NMR data for other iodine-containing systems

Compounds	$\vert C_Q(^{127}I)\vert$/MHz	η_Q	δ_{iso} (ppm)	References
IF_7	—	—	3040(40)^a	Weulersse *et al.*[96]
$[IF_6]^+[AsF_6]^-$	2.3–2.9	—	—	Hon and Christe[97]
Glycyl-L-alanine HI·H_2O	74.04	0.776	—	Kehrer *et al.*[78,95]
1,2,3-Trimethylimidazolium iodide	36.8(0.9)	0.73(0.01)	143(18)^b	Gordon *et al.*[74]
1,1-Dimethylpyrrolidinium iodide	16.9(0.1)	0.27(0.02)	282.0(0.7)^b	Gordon *et al.*[74]

^{a}With respect to 5 mol dm^{-3} KI (aq).
^{b}With respect to 0.01 mol dm^{-3} KI in D_2O.

and proper data analysis render feasible the study of chlorine, bromine, and iodine anions in noncubic inorganic and bioinorganic compounds. Quadrupolar and CS tensors are now available for halides in a range of chemical environments, and the sensitivity of the tensor parameters to factors such as local site symmetry, hydration state, and polymorphism has been demonstrated. It is likely that the $^{35/37}$Cl isotopes will continue to be the most accessible to study by SSNMR, and it is for chlorine that the most comprehensive understanding of the relationship between local structure and the observed NMR spectrum is currently available. Acquisition of $^{35/37}$Cl SSNMR spectra of covalently bound chlorine is technically feasible in very high magnetic fields, but such experiments are most often impractical and time consuming. Acquisition of $^{79/81}$Br or ^{127}I SSNMR spectra of covalent bromines or iodines is not practical in currently commercially available magnetic fields. Sustained advances and applications in the area of quadrupolar halogen SSNMR are anticipated; however, such studies will likely continue to be focused on chlorine, bromine, and iodine in environments in which they exist predominantly in the ionic form. One interesting future direction is the study of halides that participate in halogen bonds.[98]

RELATED ARTICLES IN THE ENCYCLOPEDIA OF MAGNETIC RESONANCE

Chemical Shift Tensors

Fluorine-19 NMR;

Fluorine-19 NMR of Solids Containing Both Fluorine and Hydrogen

Quadrupolar Nuclei in Liquid Samples

Tensor Interplay

REFERENCES

1. C. M. Widdifield, R. P. Chapman, and D. L. Bryce, *Annu. Rep. Nucl. Magn. Reson. Spectrosc.*, 2009, **66**, 195.
2. R. P. Chapman, C. M. Widdifield, and D. L. Bryce, *Prog. Nucl. Magn. Reson. Spectrosc.*, 2009, **55**, 215.
3. D. L. Bryce and G. D. Sward, *Magn. Reson. Chem.*, 2006, **44**, 409.
4. R. K. Harris, E. D. Becker, S. M. Cabral de Menezes, R. Goodfellow, and P. Granger, *Pure Appl. Chem.*, 2001, **73**, 1795.
5. K. J. D. MacKenzie and M. E. Smith, *Multinuclear Solid-State NMR of Inorganic Materials*, Pergamon: Amsterdam, 2002.
6. P. Pyykkö, *Mol. Phys.*, 2008, **106**, 1965.
7. R. K. Harris and P. Jackson, *Chem. Rev.*, 1991, **91**, 1427.
8. J. Skibsted and H. J. Jakobsen, *Inorg. Chem.*, 1999, **38**, 1806.
9. V. P. Tarasov, S. A. Petrushin, and Y. K. Gusev, *Dokl. Akad. Nauk SSSR*, 1987, **293**, 1423.
10. V. P. Tarasov, V. I. Privalov, Y. A. Buslaev, and U. Eichhoff, *Z. Naturforsch. B*, 1990, **45**, 1005.
11. V. P. Tarasov, V. I. Privalov, K. S. Gavrichev, V. E. Gorbunov, Y. K. Gusev, and Y. A. Buslaev, *Russ. J. Coord. Chem.*, 1990, **16**, 854; [*Translated from: Koordinats. Khim.*, 1990, **16**, 1603.]
12. V. P. Tarasov, S. A. Petrushin, and Y. K. Gusev, *Russ. J. Inorg. Chem.*, 1988, **33**, 452. [*Translated from: Zh. Neorg. Khim.*, 1988, **33**, 804.]
13. G. Wu and S. Dong, *Solid State Nucl. Magn. Reson.*, 2001, **20**, 100.
14. E. A. C. Lucken, *Nuclear Quadrupole Coupling Constants*, Academic Press: London, 1969.
15. K. Eichele, R. E. Wasylishen, J. S. Grossert, and A. C. Olivieri, *J. Phys. Chem.*, 1995, **99**, 10110.
16. C. M. Widdifield and D. L. Bryce, *Phys. Chem. Chem. Phys.*, 2009, **11**, 7120.
17. C. M. Widdifield and D. L. Bryce, *J. Phys. Chem. A*, 2010, **114**, 10810.
18. D. L. Bryce and G. D. Sward, *J. Phys. Chem. B*, 2006, **110**, 26461.
19. H. Hamaed, J. M. Pawlowski, B. F. T. Cooper, R. Fu, S. H. Eichhorn, and R. W. Schurko, *J. Am. Chem. Soc.*, 2008, **130**, 11056.
20. D. L. Bryce and E. B. Bultz, *Chem. Eur. J.*, 2007, **13**, 4786.
21. C. M. Widdifield and D. L. Bryce, *J. Phys. Chem. A*, 2010, **114**, 2102.
22. R. Sternheimer, *Phys. Rev.*, 1950, **80**, 102.
23. R. Sternheimer, *Phys. Rev.*, 1951, **84**, 244.
24. G. Wu and V. Terskikh, *J. Phys. Chem. A*, 2008, **112**, 10359.
25. W. P. Power, R. E. Wasylishen, S. Mooibroek, B. A. Pettitt, and W. Danchura, *J. Phys. Chem.*, 1990, **94**, 591.
26. J. T. Cheng, J. C. Edwards, and P. D. Ellis, *J. Phys. Chem.*, 1990, **94**, 553.
27. P. J. Chu and B. C. Gerstein, *J. Chem. Phys.*, 1989, **91**, 2081.
28. A. D. Bain, *Mol. Phys.*, 2003, **101**, 3163.
29. J. Blaser, O. Lutz, and W. Steinkilberg, *Z. Naturforsch. A*, 1972, **27**, 72.
30. J. Blaser, O. Lutz, and W. Steinkilberg, *Phys. Lett. A*, 1970, **32**, 403; [*Errata: ibid.*, 1972, **38**, 458.]
31. S. Hayashi and K. Hayamizu, *Bull. Chem. Soc. Jpn.*, 1990, **63**, 913.
32. M. Gee, R. E. Wasylishen, and A. Laaksonen, *J. Phys. Chem. A*, 1999, **103**, 10805.
33. J. S. Frye and G. E. Maciel, *J. Magn. Reson.*, 1982, **48**, 125.
34. I. Solomon, *Phys. Rev.*, 1958, **110**, 61.
35. E. L. Hahn, *Phys. Rev.*, 1950, **80**, 580.
36. P. R. Bodart, J. P. Amoureux, Y. Dumazy, and R. LeFort, *Mol. Phys.*, 2000, **98**, 1545.
37. F. H. Larsen, H. J. Jakobsen, P. D. Ellis, and N. C. Nielsen, *J. Phys. Chem. A*, 1997, **101**, 8597.
38. J. P. Yesinowski, M. L. Buess, A. N. Garroway, M. Ziegeweld, and A. Pines, *Anal. Chem.*, 1995, **67**, 2256.

39. D. L. Bryce, M. Gee, and R. E. Wasylishen, *J. Phys. Chem. A*, 2001, **105**, 10413.
40. C. Gervais, R. Dupree, K. J. Pike, C. Bonhomme, M. Profeta, C. J. Pickard, and F. Mauri, *J. Phys. Chem. A*, 2005, **109**, 6960.
41. D. L. Bryce, G. D. Sward, and S. Adiga, *J. Am. Chem. Soc.*, 2006, **128**, 2121.
42. R. P. Chapman and D. L. Bryce, *Phys. Chem. Chem. Phys.*, 2007, **9**, 6219.
43. R. P. Chapman, J. R. Hiscock, P. A. Gale, and D. L. Bryce, *Can. J. Chem.*, 2011, **89**, 822.
44. G. Arfken, *Mathematical Methods for Physicists*, 3rd edn., Academic Press, Orlando, Florida, 1985.
45. Z. Yao, H. T. Kwak, D. Sakellariou, L. Emsley, and P. J. Grandinetti, *Chem. Phys. Lett.*, 2000, **327**, 85.
46. A. P. M. Kentgens and R. Verhagen, *Chem. Phys. Lett.*, 1999, **300**, 435.
47. R. Siegel, T. T. Nakashima, and R. E. Wasylishen, *Chem. Phys. Lett.*, 2004, **388**, 441.
48. R. W. Schurko, I. Hung, and C. M. Widdifield, *Chem. Phys. Lett.*, 2003, **379**, 1.
49. R. Siegel, T. T. Nakashima, and R. E. Wasylishen, *Concepts Magn. Reson. A*, 2005, **26A**, 47.
50. Ē. Kupče and R. Freeman, *J. Magn. Reson. A*, 1995, **115**, 273.
51. R. Bhattacharyya and L. Frydman, *J. Chem. Phys.*, 2007, **127**, 194503.
52. L. A. O'Dell and R. W. Schurko, *Chem. Phys. Lett.*, 2008, **464**, 97.
53. D. Massiot, I. Farnan, N. Gautier, D. Trumeau, A. Trokiner, and J. P. Coutures, *Solid State Nucl. Magn. Reson.*, 1995, **4**, 241.
54. Y. Yamagata, *J. Phys. Soc. Jpn.*, 1964, **19**, 10.
55. T. L. Weeding and W. S. Veeman, *J. Chem. Soc. Chem. Commun.*, 1989, 946.
56. S. Hayashi and K. Hayamizu, *J. Phys. Chem. Solids*, 1992, **53**, 239.
57. B. D. Guenther and R. A. Hultsch, *J. Magn. Reson.*, 1969, **1**, 609.
58. K. D. Becker, *J. Chem. Phys.*, 1978, **68**, 3785.
59. H. Honda, *Z. Naturforsch. A*, 2003, **58**, 623.
60. R. P. Chapman and D. L. Bryce, *Phys. Chem. Chem. Phys.*, 2009, **11**, 6987.
61. A. J. Rossini, R. W. Mills, G. A. Briscoe, E. L. Norton, S. J. Geier, I. Hung, S. Zheng, J. Autschbach, and R. W. Schurko, *J. Am. Chem. Soc.*, 2009, **131**, 3317.
62. C. M. Widdifield and D. L. Bryce, *Can. J. Chem.*, 2011, **89**, 754.
63. S. Jurga, G. S. Harbison, B. Blümich, H. W. Spiess, F. Fujara, and A. Olinger, *Ber. Bunsen-Ges. Phys. Chem.*, 1986, **90**, 1153.
64. S. L. Segel, S. Maxwell, R. D. Heyding, P. Ingman, E. Ylinen, and M. Punkkinen, *Solid State Commun.*, 1988, **66**, 1039.
65. T. J. Bastow and S. N. Stuart, *J. Phys.: Condens. Matter*, 1989, **1**, 4649.
66. V. P. Tarasov, M. A. Meladze, G. A. Kirakosyan, A. E. Shvelashvili, and Y. A. Buslaev, *Phys. Status Solidi B*, 1991, **167**, 271.
67. V. P. Tarasov, M. A. Meladze, and G. A. Kirakosyan, *Russ. J. Coord. Chem.*, 1992, **18**, 707.
68. N. Binesh and S. V. Bhat, *Solid State Ionics*, 1996, **92**, 261.
69. H. Ono, S. Ishimaru, R. Ikeda, and H. Ishida, *Chem. Phys. Lett.*, 1997, **275**, 485.
70. H. Ono, S. Ishimaru, R. Ikeda, and H. Ishida, *Bull. Chem. Soc. Jpn.*, 1999, **72**, 2049.
71. O. Czupiński, M. Wojtaś, J. Zaleski, R. Jakubas, and W. Medycki, *J. Phys.: Condens. Matter*, 2006, **18**, 3307.
72. T. O. Sandland, L. S. Du, J. F. Stebbins, and J. D. Webster, *Geochim. Cosmochim. Acta*, 2004, **68**, 5059.
73. P. G. Gordon, D. H. Brouwer, and J. A. Ripmeester, *J. Phys. Chem. A*, 2008, **112**, 12527.
74. P. G. Gordon, D. H. Brouwer, and J. A. Ripmeester, *ChemPhysChem*, 2010, **11**, 260.
75. V. V. Terskikh, S. J. Lang, P. G. Gordon, G. D. Enright, and J. A. Ripmeester, *Magn. Reson. Chem.*, 2009, **47**, 398.
76. B. Alonso, D. Massiot, P. Florian, H. H. Paradies, P. Gaveau, and T. Mineva, *J. Phys. Chem. B*, 2009, **113**, 11906.
77. J. Persons and G. S. Harbison, *J. Magn. Reson.*, 2007, **186**, 347.
78. A. Kehrer, N. Weiden, and A. Weiss, *Z. Phys. Chem.*, 1992, **178**, 1.
79. R. V. Pound, *Phys. Rev.*, 1950, **79**, 685.
80. J. Itoh and Y. Yamagata, *J. Phys. Soc. Jpn.*, 1962, **17**, 481.

81. J. Itoh and Y. Yamagata, *J. Chem. Phys.*, 1956, **24**, 621.

82. T. Erge, D. Michel, J. Petersson, and F. Engelke, *Phys. Status Solidi A*, 1991, **123**, 325.

83. B. Alonso, T. Mineva, P. Innocenzi, G. Trimmel, K. Stubenrauch, I. Melnyk, Y. Zub, F. Fayon, P. Florian, and D. Massiot, *C. R. Chim.*, 2010, **13**, 431.

84. S. Hayashi, *Magn. Reson. Chem.*, 1996, **34**, 791.

85. D. Michel, U. Häcker, T. Erge, and J. Petersson, *Phys. Status Solidi B*, 1994, **185**, 257.

86. S. Hayashi and K. Hayamizu, *J. Chem. Phys.*, 1990, **92**, 2818.

87. D. G. Klobasa and P. K. Burkert, *Z. Naturforsch. A*, 1984, **39**, 1222.

88. D. G. Klobasa and P. K. Burkert, *Magn. Reson. Chem.*, 1987, **25**, 154.

89. P. K. Burkert and M. Grommelt, *Z. Naturforsch. B*, 1988, **43**, 1381.

90. D. G. Klobasa, P. K. Burkert, and G. Müller, *Z. Naturforsch. A*, 1986, **41**, 330.

91. M. Grommelt and P. K. Burkert, *Z. Naturforsch. B*, 1989, **44**, 1053.

92. D. G. Klobasa and P. K. Burkert, *Z. Naturforsch. A*, 1987, **42**, 275.

93. P. K. Burkert and D. G. Klobasa, *Z. Naturforsch. A*, 1985, **40**, 274.

94. S. L. Segel and H. M. Vyas, *J. Chem. Phys.*, 1980, **72**, 1406.

95. A. Kehrer, S. Q. Dou, and A. Weiss, *Z. Naturforsch. A*, 1989, **44**, 659.

96. J. M. Weulersse, J. Virlet, and P. Rigny, *Mol. Phys.*, 1979, **38**, 923.

97. J. F. Hon and K. O. Christe, *J. Chem. Phys.*, 1970, **52**, 1960.

98. P. Metrangolo and G. Resnati, eds, *Halogen Bonding: Fundamentals and Applications*, Structure and Bonding, Springer, 2008, Vol. 126.

Chapter 20

Quadrupolar NMR of Ionic Conductors, Batteries, and other Energy-Related Materials

Frédéric Blanc[1], Leigh Spencer[2], and Gillian R. Goward[2]

[1]*Department of Chemistry, University of Cambridge, Lensfield Road, Cambridge CB2 1EW, UK*
[2]*Department of Chemistry and Brockhouse Institute for Materials Research, McMaster University, Hamilton, ON L8S 4M1, Canada*

20.1 INTRODUCTION

20.1.1 General Introduction

Energy storage methods have been in the spotlight in recent years[1] owing to the present "energy crisis" and growing environmental concerns. Researchers are continuously searching for the most efficient and cost-effective way to produce and store energy. Ionic conductors are critical to the performance of many energy-storage devices. In particular, the history of battery development includes many types of ionic conductors, including materials that transport ions such as Ag^+, Na^+, O^{2-}, and Li^+.[2]

NMR of Quadrupolar Nuclei in Solid Materials
Edited by Roderick E. Wasylishen, Sharon E. Ashbrook and Stephen Wimperis
 ISBN: 978-0-470-97398-1

Solid-state NMR has played an important role in the study of the local structure and dynamics in many of these novel materials, complementing other methods such as diffraction, electron microscopy, and other strategies. The study of ionic conductors can be considered as two related endeavors: structural evaluation and the study of dynamic processes. NMR is an invaluable technique, capable of answering questions in either of these categories. Moreover, the sensitivity of quadrupolar nuclei to their surrounding environments makes them ideal for structural determination, as outlined extensively in this handbook.

20.1.2 Systems

Among the various energy-related materials that have been studied by NMR spectroscopy, we have chosen to review only the two major types: lithium ion batteries and oxygen conductors.

20.1.2.1 Lithium Ion Batteries

Lithium ion batteries are known to have a high energy density, which makes them a promising energy storage device in electric vehicles.[1,3] First commercialized in 1990 by SONY, lithium ion batteries have made a significant impact on the development of portable electronic devices, from cellular phones to laptops.[3] The most commonly used cathode material is $LiCoO_2$. Unfortunately, it is both toxic and expensive. This has formed part of the motivation to replace this material with a more environmentally friendly and cost-effective alternative. Many candidate materials have been considered, with the target characteristics including high ionic and electronic conductivity, stability under electrochemical cycling, low cost, and environmental benignity. Candidates include the spinel series, for example, $LiMnO_2$, and the rock salt series, for example, $Li[Ni_{1/3}Mn_{1/3}Co_{1/3}]O_2$ and $LiFePO_4$, among others.[1] To address the structural and ion dynamic properties of these paramagnetic materials, NMR investigations have been focused primarily on ^{6}Li and ^{7}Li NMR. In these materials, the primary challenge for solid-state NMR studies is not the quadrupole moment of the nuclei (both ^{6}Li and ^{7}Li have comparatively small quadrupole moments), but rather the paramagnetic interactions with the transition metal centers.[4]

20.1.2.2 Oxygen Conductors

Oxide ion conductors have been increasingly studied because of their application in a range of devices having very important economic and environmental potential, such as solid oxide fuel cells (SOFC), oxygen sensors, catalytic converters, and membrane separators. All these devices rely on anionic conductivity in the solid state, which, despite tremendous progress in the last decade, remains a relatively high-temperature phenomenon. This makes their characterization by NMR quite challenging. Typical conductivities are now observed in the range of sub-siemens per centimeter at temperatures of 800 °C for materials having the fluorite structure (e.g., doped ZrO_2), the pyrochlore structure [e.g., $Y_2B_2O_7$ ($B = Sn$, Ti, Zr)], and the perovskite structure (e.g., doped $LaGaO_3$).

Although one could deduce the dynamics and ionic jump rates from a temperature-dependent change in the NMR line shape of a large range of half-integer nuclei,[5] ^{17}O NMR studies remain the preferred method in the characterization of dynamic processes in anionic conductors.[6] ^{17}O NMR is able to identify both the ions responsible for the conduction process as well as any oxygen atoms in the framework that do not contribute to the conduction processes.

In this chapter, we demonstrate how solid-state NMR spectroscopy of quadrupolar nuclei can be used to study some of the most common phases used as both anionic (O^{2-}) and cationic (Li^+) conductors. We will present examples demonstrating the utility of these NMR methods in characterizing performance–property relationships in a variety of materials for lithium ion batteries. We will highlight how both structural and dynamic information, not readily accessible by diffraction-based techniques and impedance measurements, can be obtained by NMR, and refer the reader to other articles in the *Encyclopedia of Magnetic Resonance* for the description of the quadrupolar interaction and the methods used to study quadrupolar nuclei in solids. Table 20.1 lists the properties of the nuclei included in this chapter.

Table 20.1. Relevant nuclei and their properties [a]

Nucleus	Spin	Natural abundance (%)	Magnetogyric ratio ($\gamma/10^7$ rad T^{-1} s^{-1})	Quadrupole moment (Q/fm^2)
^{6}Li	1	7.59	3.937	−0.0808
^{7}Li	3/2	92.41	10.397	−4.01
^{17}O	5/2	0.038	−3.628	−2.558
^{51}V	7/2	99.75	7.046	−5.2
^{139}La	7/2	99.91	3.808	20.0

[a]Taken from the *Encyclopedia of Magnetic Resonance* article, **Nuclear Spin Properties and Conventions for Chemical Shifts (IUPAC Recommendations 2001)**.

20.2 STRUCTURE DETERMINATION

20.2.1 NMR of Paramagnetic Systems

The study of lithium ion conductors includes the NMR of paramagnetic and diamagnetic materials. A paramagnetic contribution can significantly change the appearance of an NMR spectrum, as well as the relaxation properties of the nuclei. A good illustration of this is seen in the NMR spectra of the layered materials, $La_2Li_{0.5}B_{0.5}O_4$ (B = Cu, Ni), reported by Ganapathy *et al.*[7] The Ni material is paramagnetic due to the unpaired electrons of the low-spin Ni^{3+}. In contrast, the low-spin Cu sample has no unpaired electrons and is therefore diamagnetic. These isostructural samples show very different ^{7}Li static NMR spectra, shown in Figure 20.1.[7]

In Figure 20.1(a), the ^{7}Li static solid-state NMR spectrum of $La_2Li_{0.5}Cu_{0.5}O_4$ shows the typical powder pattern of a spin $I = 3/2$ nucleus in an axially symmetrical electric field gradient. It has a quadrupolar coupling constant of 88.9 kHz, as measured from the satellite transitions apparent in the spectrum.

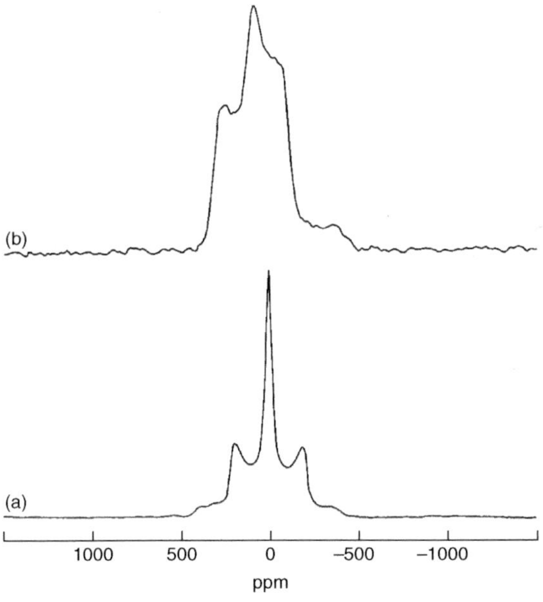

Figure 20.1. ^{7}Li static solid-state NMR spectra of $La_2Li_{0.5}M_{0.5}O_4$ system: (a) M = Cu, (b) B = Ni obtained at 7 T. (Reprinted with permission from S. Ganapathy *et al.*, *J. Phys. Chem.* 2000, **104**, 2007. Copyright 2007 American Chemical Society.)

In contrast, the ^{7}Li static solid-state NMR spectrum of $La_2Li_{0.5}Ni_{0.5}O_4$, shown in Figure 20.1(b), is significantly different. This is due to the influence of the paramagnetic dipolar interaction on the spectrum, which is already broadened because of the quadrupolar interaction. Paramagnetic dipolar interactions, which in some cases are similar in magnitude to the quadrupolar interactions, remove the symmetrical quadrupolar features and complicate the static lineshape. While this material is an illustrative example of the influence of paramagnetism on the Li line shape, it is not a highly relevant battery material. In $^{6,7}Li$ NMR studies of battery materials, the quadrupole parameters of the nuclei themselves are rarely targeted. Rather, the chemical shift resolution is essential to understand local structures. Thus, NMR studies of paramagnetic materials are carried out under optimized experimental conditions, including low magnetic field strength, to minimize the paramagnetic interaction, and fast magic-angle spinning (MAS), to maximize the chemical shift resolution. Using this strategy, details of the local structure may be extracted. These local details are often difficult to detect by diffraction, but they are critical to the performance of the battery material.

The following sections will outline the approaches used to mitigate the paramagnetic influence on materials, and the differences between solid-state NMR of paramagnetic and diamagnetic samples.

20.2.2 Chemical Shift and Structural Determination

The chemical shift and overall appearance of NMR peaks in paramagnetic samples is the result of the combination of one or more of the following phenomena: the Fermi-contact interaction (arising from the hyperfine coupling between electrons and nuclei); through-space dipolar coupling of the nucleus of interest to the unpaired electrons present in the system; and the pseudocontact interaction (the interaction of anisotropic electron g-tensor with nucleus).[8,9]

The use of $^{6,7}Li$ NMR, in the presence of paramagnetic transition metals, to ascertain structural information in materials for lithium ion batteries has been applied to a range of cathode materials. Several illustrative examples are chosen here, where the chemical shift range is enhanced by the paramagnetic interactions, and the assignments are based on the local geometry of the overlap between the paramagnetic

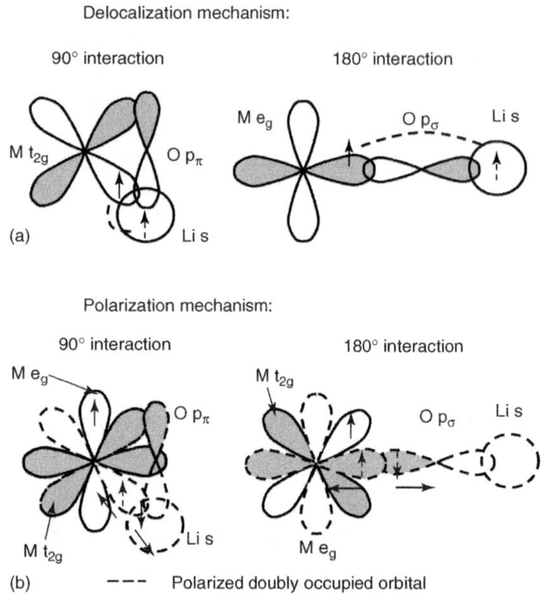

Figure 20.2. The different M–O–Li spin transfer mechanisms illustrated for 90° and 180° interactions. (a) The delocalization mechanism. (b) The polarization mechanism. (Reprinted figure with permission from D. Carlier *et al.*, *Phys. Rev. B* 2003, **67**, 2174103. Copyright 2003 by the American Physical Society http://prb.aps.org/abstract/PRB/v67/i17/e174103).

centers and the lithium 2s orbitals. Details of the concepts behind these chemical shift assignments have been reviewed by others.[4,9,10] Briefly, the concept is based on geometrical arguments and the additive nature of the paramagnetic interactions.[9] There has been a great effort to relate the precise chemical shift of a ^{6,7}Li NMR resonance peak to the orbital interaction of the Li s orbital with a nearby paramagnetic center. This interaction consists of two contributions: *spin delocalization* and *spin polarization*, each of which describes the transfer of unpaired electron spin density from a transition metal center to a Li site through interaction with oxygen p orbitals.[9,10]

To illustrate spin delocalization (or hybridization), Figure 20.2(a) refers to the positive spin transfer that occurs when Li, O, and M (transition metal) orbitals overlap in such as way as to maintain a given spin polarization. This results in a positive chemical shift for peaks in a ^{6,7}Li NMR spectrum.[10] Alternatively, spin polarization, Figure 20.2(b), refers to the negative spin transfer that results from the polarization of the Li s crystalline orbitals by unpaired electrons that are located on the transition metal M. This occurs through the oxygen orbitals, as is shown in Figure 20.2, for both 90° and 180° interactions. This results in a negative chemical shift.

In general, the chemical shift induced by spin delocalization is greater than that produced by spin polarization, and when considering the total spin transfer, both mechanisms are considered.[10] Emphasis is placed on the geometry and orbital orientation of the specific system of interest to determine which mechanism is dominant. The isotropic portion of the interaction can be described by the following Hamiltonian

$$H_c = A_s I_z \langle S_z \rangle \tag{20.1}$$

with A_s the hyperfine coupling constant, which depends on the nature of orbital overlap, I_z the nuclear spin, and $\langle S_z \rangle$ the thermally averaged value of electron spin. $\langle S_z \rangle$ is proportional to the magnetic susceptibility and given by

$$\langle S_z \rangle = -\frac{B_0}{\mu_0 g N_A \mu_B} \chi_M \tag{20.2}$$

where B_0 is the external magnetic field strength, χ_M the magnetic susceptibility, μ_0 the permeability of free space, g the electron g-factor, N_A the Avogadro number, and μ_B the Bohr magneton.

The size and sign of A_s determine the size and direction, respectively, of the chemical shift observed. The total Fermi-contact interaction is additive and the total shifts can be calculated as follows[10]:

$$\delta = \sum_i \frac{\Delta\omega_i}{\omega_0} = -\sum_i \frac{A_i}{\omega_0 \hbar} \langle S_z \rangle \tag{20.3}$$

The dipolar coupling between the electrons and the nucleus of interest is described by the Hamiltonian

$$H_{en} = \langle \mu_e \rangle \cdot \boldsymbol{D}_{en} \cdot \mu_N \tag{20.4}$$

with μ_N the magnetic moment of the nucleus of interest, $\boldsymbol{D}_{en}$ the dipolar coupling tensor and $\langle \mu_e \rangle$ the time-averaged magnetic moment of the electron.[9]

For cases in which the magnetic moment of the electrons is isotropic, the coupling does not result in a chemical shift change, but does create a broad line shape, or spinning side-band manifold under MAS and asymmetric line shapes. Alternatively, coupling to an anisotropic magnetic moment leads to an isotropic pseudocontact shift and also to line broadening in solids.[9] Together, these contributions to the line shape and chemical shift dominate the interpretation of the NMR spectra of the paramagnetic materials discussed here.

20.2.2.1 Local Structure in Paramagnetic Cathode Materials

Early work in this area includes the analysis of the chemical shifts and relative intensities as a function of synthesis temperature, which was used to understand the presence of defects in the $LiMn_2O_4$ spinel structure. The spectra, shown in Figure 20.3, were assigned on the basis of the principles described above, drawing specifically on the oxidation state of the manganese ions as a function of synthesis temperature and the angle and distance of the Fermi-contact interaction between the paramagnetic center and the lithium ions in various crystallographic positions.

Of the lithium ion battery cathode materials, $LiCoO_2$ is currently the most prevalent. In spite of its success, the cost and toxicity of cobalt have motivated a broad search for alternatives with competitive electrochemical performance in the composition range $LiMO_2$, where M is a combination of Ni, Mn, and Co. To this end, ^{6}Li

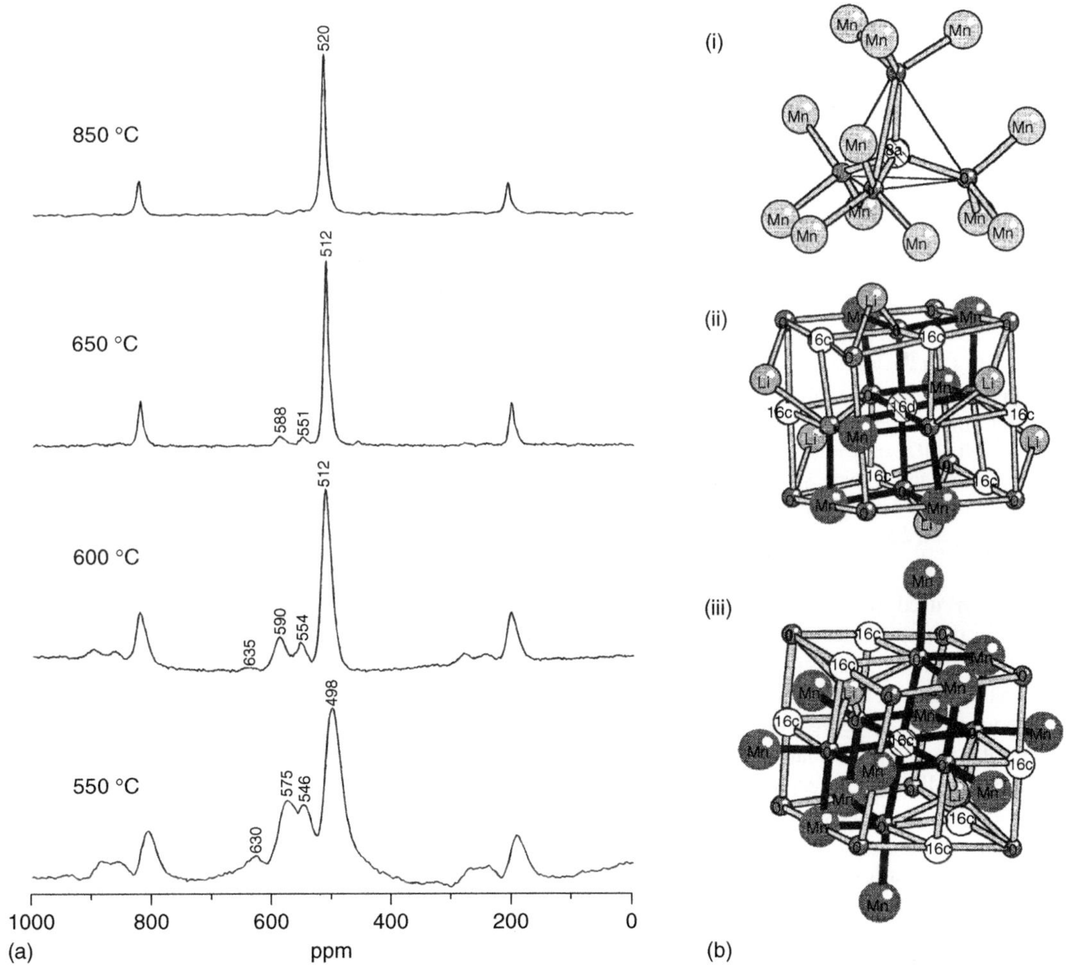

Figure 20.3. (a) ^{6}Li NMR spectra of $LiMn_2O_4$ spinel as a function of the firing temperature obtained at the MAS frequency of 9 kHz. (b) The local coordination environment of the lithium and manganese ions (or vacancies) in the spinel structure. (i) The lithium ion in the 8a site. (ii) The 16 d site typically occupied by manganese. (iii) A 16c octahedral vacancy. The M–O–Mn bonds from the central atom or vacancy (M) 8a (Li), 16 d, or 16c cations are shaded. (Reproduced with permission from Y. J. Lee *et al.*, *J. Am. Chem. Soc.* 1998, **120**, 12601. Copyright 1998 American Chemical Society.)

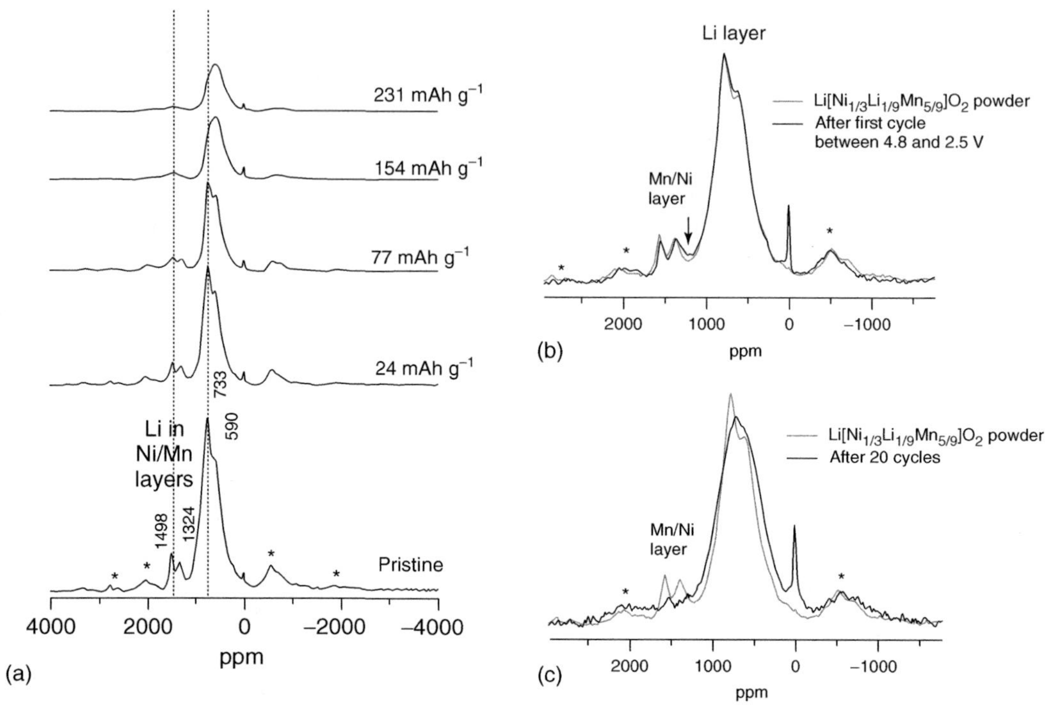

Figure 20.4. ^{6}Li MAS NMR spectra of $Li[Li_{1/9}Mn_{5/9}Ni_{1/3}]O_2$ as a function of state of charge. Spectra obtained (a) during the first charge and (b) following discharge to 2.5 V after (b) 1 and (c) 20 cycles. The capacities given in (a) were calculated from the current passed during charge, and correspond to approximate values of x in the nominal composition $Li_x[Li_{1/9}Mn_{5/9}Ni_{1/3}]O_2$ of $x = 1$ (pristine), 0.9, 0.6, 0.2, and approximately 0 (231 mAh g^{-1}).[12] The peaks marked with asterisks correspond to spinning sidebands and the resonance at approximately 0 ppm in (b) and (c) is due to Li^+ ions in the electrolyte or from the surface–electrolyte interface (SEI). (Reproduced by permission of the Electrochemical Society from C.P. Grey *et al.*, *Elec. and Solid State Lett.* 2004, **7**, A290.)

NMR has been used to determine the location of lithium ions in the lattice. In one example, NMR was able to identify lithium ions occupying spaces within the transition metal layers of the α-$NaFeO_2$ structure and, moreover, to observe the electrochemical extraction of these particular lithium ions and correlate their NMR signatures with electrochemical cyclability. Figure 20.4 shows the ^{6}Li NMR spectra acquired as a function of electrochemical lithium extraction, where the sites assigned to lithium ions in the transition metal layer are clearly removed as the cell is delithiated. Assignment of these resonances is based on the local coordination environments, in particular, the number of Ni^{2+} and Mn^{4+} neighbors for a given lithium nucleus.

Recently, the phosphate-based cathode materials, and in particular the olivine structure $LiFePO_4$, have attracted significant interest on account of the inherent safety of these lower voltage materials. From the NMR perspective, the surprising observation is that a wide range of chemical shifts have been observed for structures that are remarkably similar to one another. This raises the question of how chemical shift assignments can be made, beyond the geometrical arguments that work so well for the oxides. Recent results have used first-principles calculations employing computational methods such as the unrestricted Hartree–Fock method and density functional theory (DFT) to predict the chemical shifts of nuclei in paramagnetic systems.[10,14,15] This growing area is likely to provide significant value to the solid-state NMR of paramagnetic materials in the coming years, in the same manner that DFT and the CASTEP code have been of tremendous value to the broader quadrupolar NMR community in recent years.

20.2.2.2 Diamagnetic Li^+ Electrolyte Materials

Motivated by safety concerns, which are related to the instability of organic electrolytes at high voltage, the development of an all solid-state lithium ion battery has been a goal of many researchers.[1] This has necessitated the development of a solid-state lithium ion electrolyte material. This has taken the form of polymer electrolytes or ceramic electrolytes. These materials are in general insulating materials, with no redox-active centers, and moreover tend to be diamagnetic. Diamagnetic materials offer several challenges for solid-state ^{6,7}Li NMR due to the narrow chemical shift range of lithium (typically ±5 ppm), as well as the long relaxation times. In spite of these challenges, ^{6,7}Li NMR has been used to characterize a variety of solid-state electrolytes, including lithium–lanthanum garnet structures, and numerous polymer electrolytes.

Many of the solid-state ceramic lithium ion electrolytes contain a large amount of lanthanum. The typically diamagnetic nature of these electrolyte systems has made ^{6,7}Li NMR a challenge, and this has prompted the use of other nuclei in the system to study the structure of these materials. Studies of $Li_5La_3Nb_2O_{12}$ revealed a strong annealing temperature dependence of the population of lithium ions occupying tetrahedral versus octahedral sites, as shown in Figure 20.5.[16] This difference, which was not detected in diffraction studies, nevertheless correlated with ion dynamics, and it was established that lithium ions in the tetrahedral sites contributed negligibly to ion transport.

In a related garnet structure, substitution of iron into the framework had a surprising influence on both the local lithium environments, as observed by ^{7}Li NMR, and the two lanthanum sites in the crystal structure. ^{139}La NMR has been performed on the lithium ion conductors $Li_{16}La_{32}Fe_{6.4}O_{67}$ (Figure 20.6). In this material, there was a distribution of NMR parameters for one of the lanthanum sites. This was a result of an atomic-level distribution of elements in the materials that was observed in ^{7}Li MAS NMR, Figure 20.6(a). The sensitivity of the lanthanum nuclei to this distribution was essential in understanding the crystal structure of these materials. In Figure 20.6(c), the lanthanum environments of $Li_{16}La_{32}Fe_{6.4}O_{67}$ are shown.[17] The two lanthanum environments differ not only in the values of the quadrupolar coupling constant C_Q and the symmetry parameter η, but one of the sites also shows a distribution of C_Q and η while the other can be clearly defined by one set of values (Figure 20.6b).[17]

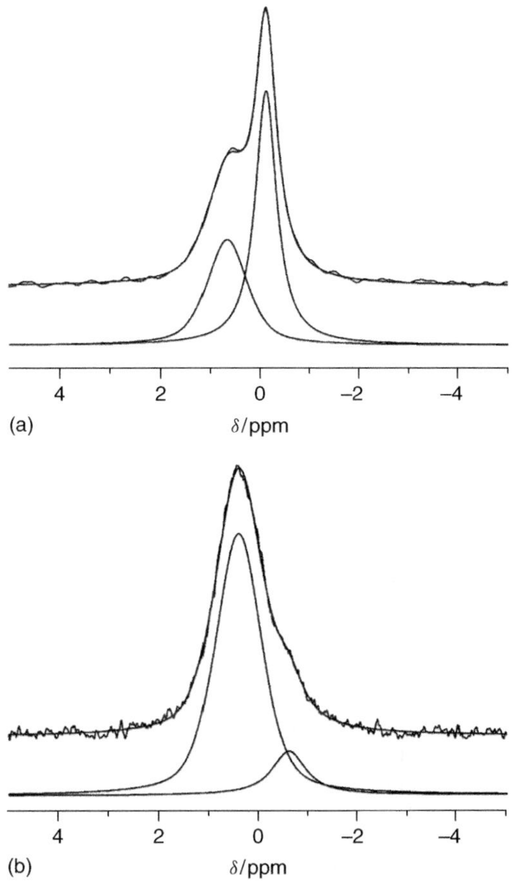

Figure 20.5. ^{6}Li NMR spectra of $Li_5La_3Nb_2O_{12}$ garnet. (a) $Li_5La_3Nb_2O_{12}$ garnet prepared at 850 °C and NMR data collected at room temperature. (b) $Li_5La_3Nb_2O_{12}$ garnet prepared at 900 °C and NMR data collected at −73 °C. (L. van Wüllen *et al.*, *Phys. Chem. Chem. Phys.* 2007, **9**, 3298. Reproduced by permission of the PCCP Owner Societies http://dx.doi.org/10.1039/B703179C).

The wide spectrum obtained by ^{139}La NMR is due to the large quadrupolar coupling constant of one of the La sites, La-2, which was found through simulation of the spectra to be 56 ± 1 MHz. La-2 lies on an axis of fourfold rotational symmetry and is surrounded by oxygen in a nonspherical environment. These factors contribute to the small value of η, 0.05 ± 0.05, and the large quadrupolar coupling constant, respectively. The value of η is responsible for the axial symmetry of the powder pattern of La-2.

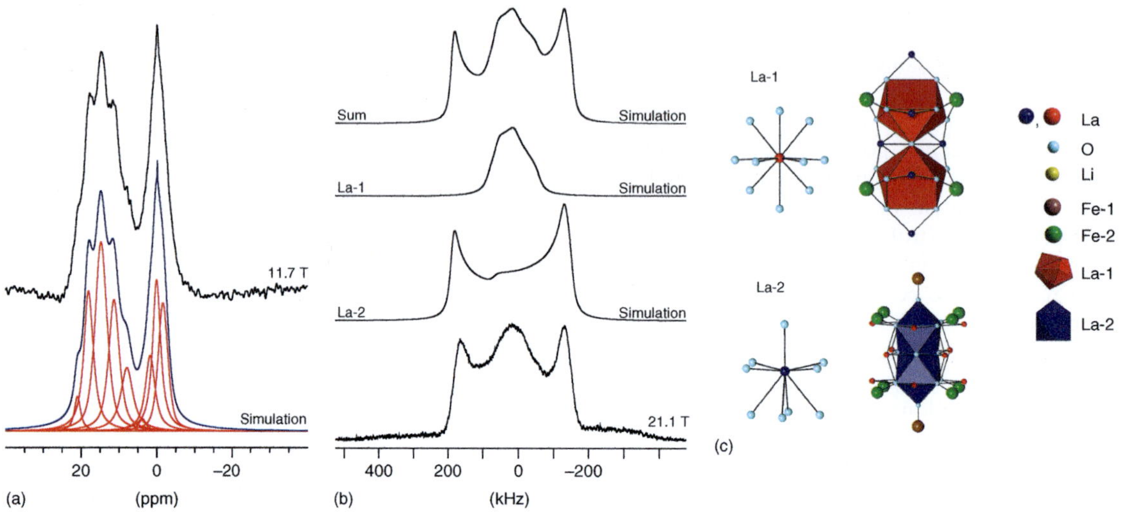

Figure 20.6. NMR spectra of $Li_{16}La_{32}Fe_{6.4}O_{67}$. (a) ^{7}Li MAS spectrum collected at 11.7 T, under MAS of 20 kHz. (b) ^{139}La WURST-echo spectrum of $Li_{16}La_{32}Fe_{6.4}O_{67}$ collected at 21.1 T, shown with simulation. This spectrum can be represented as a superposition of two sites with La-2 having $C_Q = 56.1$ MHz and $\eta = 0.05$; and La-1 having $C_Q = 29.8$ MHz and $\eta = 0.6$. (c) The individual lanthanum sites in $Li_{16}La_{32}Fe_{6.4}O_{67}$ crystal structure showing the coordination to oxygen, as well as the second coordination sphere environment around the La polyhedra. The brown spheres represent Fe-2, which does not participate in the Li/Fe disorder. (Reproduced from L. Spencer *et al.*, *Can. J. Chem.*, 2011, **89**, 770–778.)

The use of high-field instruments was extremely useful for the measurement of the ^{139}La NMR spectrum. La-1 lies on a less symmetrical site, having only twofold rotational symmetry. It is, however, surrounded by oxygen in a more spherical environment. This leads to a lower value of C_Q and a higher value of η when compared to La-2, 29 ± 2 MHz, and 0.6 ± 0.1, respectively. A considerable amount of line broadening was needed to simulate the La-1 powder pattern, whereas negligible line broadening was needed to simulate La-2. This indicates that the C_Q and η values obtained for La-1 are indeed the statistical average of a distribution of values. This is due to the many possible La-1 environments resulting from a variable Li/Fe/Vacant site, which is in close proximity to La-1, but far from La-2.[17,18] Additionally, La-1 is closer to the paramagnetic Fe centers, which may also contribute to the broadness of this line shape.

The assignment of the ^{139}La NMR spectrum is essential in the understanding of the ^{7}Li MAS NMR spectrum for this material. The studies of such broad spectra are assisted considerably by the use of the wideband, uniform rate, and smooth truncation-quadrupolar-Carr–Purcell–Meiboom–Gill (WURST-QCPMG) pulse sequence.[19] This NMR pulse sequence has the ability to excite and collect a broad spectral range in a short period of time, and provides an important first step in measuring these broad NMR spectra, as discussed by R. Schurko in Chapter 5.

20.2.2.3 Diamagnetic Anode Materials

In the case of anode materials for lithium ion batteries, the important candidates are materials that can be differentiated on the basis of their insulating versus conducting natures. In particular, metallic materials give rise to Knight shifts. For example, the Knight shift of metallic lithium is found at 256 ppm. This original anode material is now strongly disfavored, because of safety concerns. Nevertheless, many candidate anode materials exhibit lithium-metal-like NMR shifts, as a result of their metallic nature.

Owing to the extreme sensitivity of the NMR shift to the electronic environment of the nucleus, the degree of the observed Knight shift can thus be used as an indicator of the association of lithium centers into locally conductive domains during the lithium insertion process. For example, the insertion of lithium into tin oxide glasses was followed by ^{7}Li

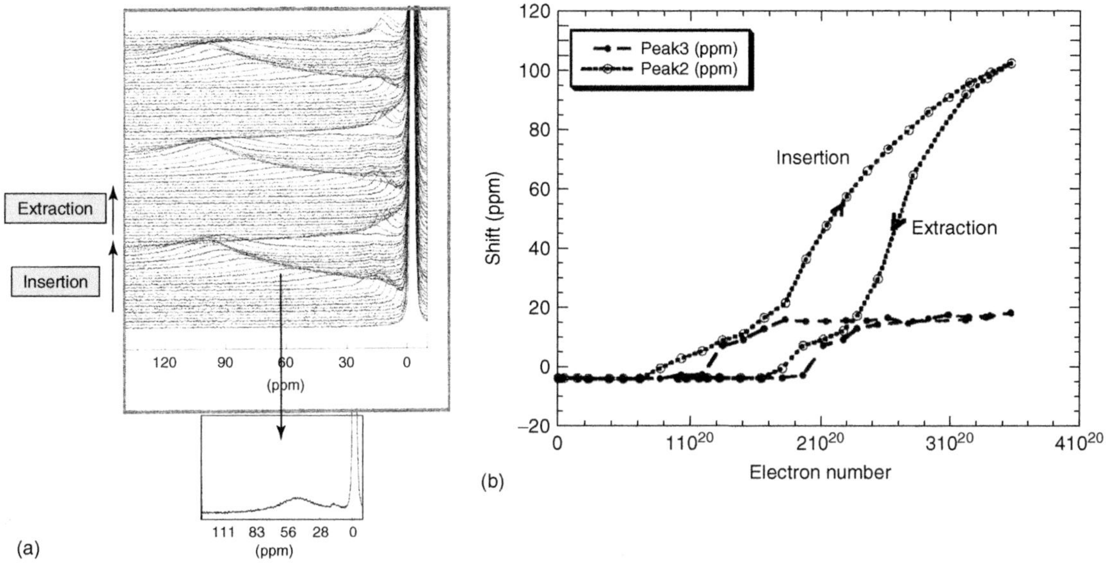

Figure 20.7. (a) In situ ^7Li static NMR spectra recorded during cycling of the plastic Li/C cell. The initial spectrum is at the bottom. The inset represents the detail of one spectrum. (b) Chemical shift as a function of the number of electrons *n* (first electrochemical cycle). (Reprinted with permission from M. Letellier *et al.*, *J. Chem. Phys.* 2003, **118**, 6038. Copyright 2003, American Institute of Physics.)

NMR, which showed that the particle size could be retained at a lower degree of agglomeration in the amorphous glasses than in the crystalline tin oxide itself.[20] In addition, graphitic and other forms of carbon have been excellent alternatives to lithium, offering reasonable capacity and reversibility, as well as an attractive voltage versus many of the cathode candidates.

In situ ^7Li NMR studies of electrochemical systems involve the electrochemical cycling of a lithium ion cell within the NMR coil while the NMR data is acquired. This methodology was originally applied to study lithium insertion into hard carbon, and revealed the smooth and reversible change in lithium chemical shift as a function of the number of electrons and corresponding lithium ions inserted into the anode material (Figure 20.7).[21] Since then, in situ ^7Li NMR has been used to characterize lithium insertion into silicon, where it has been revealed that transient species, which are not detected using ex situ NMR methods, could be detected.[22] This has opened up a new field of NMR research,[21–23] which is likely to have a significant impact on the field of lithium batteries, as the detection of degradation pathways remains a challenging area of active research in the electrochemical community.

20.2.3 Oxygen Phases

The NMR study of anionic conductors often includes ^{17}O as the key nucleus which has both the disadvantages of being present at a very low natural abundance (0.037%), requiring expensive and time consuming isotropic enrichment, as well as having a large quadrupolar moment. Nevertheless, recent methodology developments, as outlined in this handbook, have made the ^{17}O solid-state NMR spectroscopy a very powerful tool.

20.2.3.1 Pyrochlores

A number of oxide materials with the general formula $A_2B_2O_7$ (e.g., $Y_2Sn_2O_7$ and $Y_2Ti_2O_7$) adopt the pyrochlore structure where the A^{3+} cation occupies an eight-coordinate site, and the smaller B^{4+} cation occupy an octahedral site (Figure 20.8a).[24] $A_2B_2O_7$ materials, where the size of the B^{4+} cation approaches that of the A^{3+} cation (e.g., $Y_2Zr_2O_7$), adopt the fluorite structure (Figure 20.8b) with complete mixing of A and B on the cation sites. Such distinction between Y/Zr cation mixing and ordering is challenging for diffraction-based techniques and was solved by ^{17}O solid-state NMR.

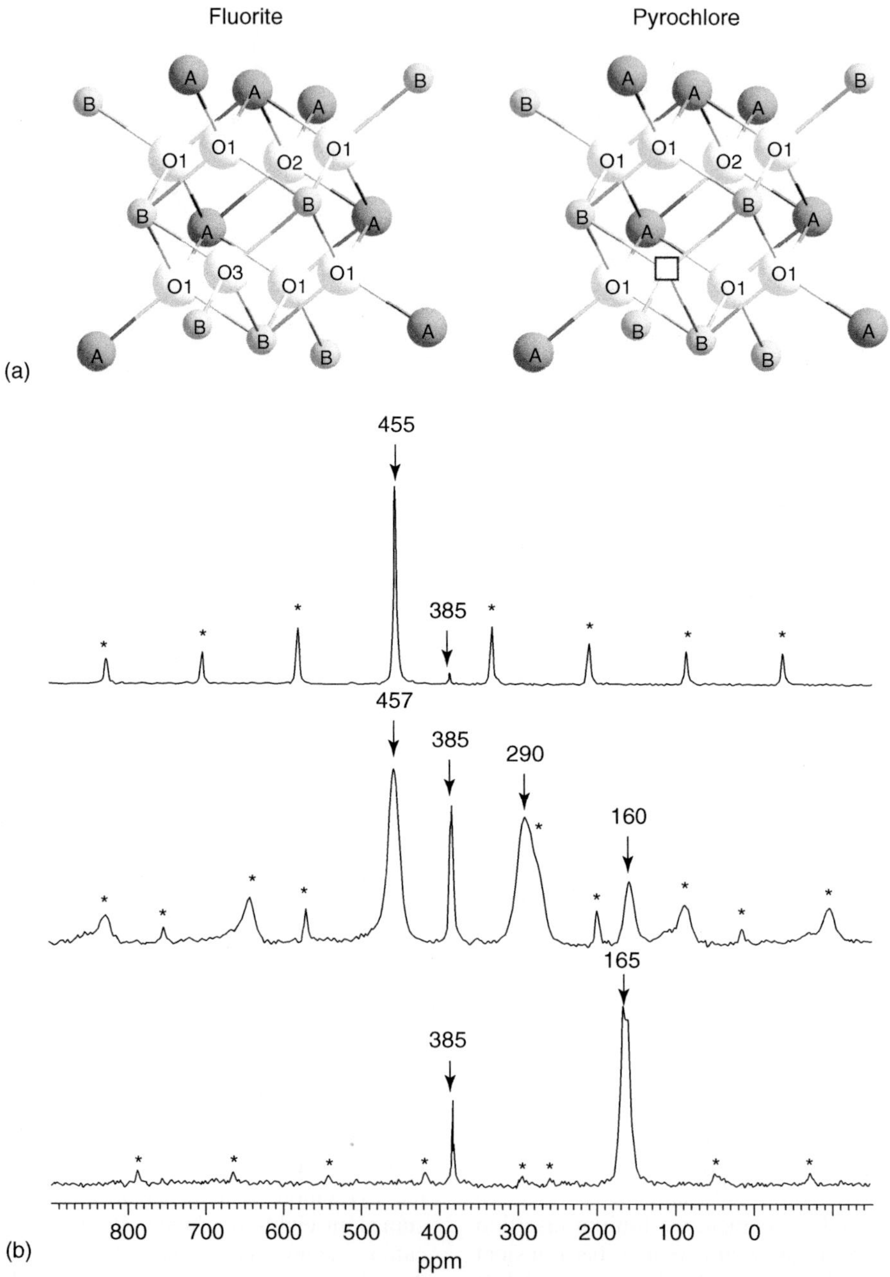

Figure 20.8. (a) Structure of $A_2B_2O_7$ in the fluorite and pyrochlore crystal structures highlighting the various crystallographic different oxygen sites (O1, O2, and O3) and the vacancy ordering (□). (b) ^{17}O MAS NMR spectra of $Y_2Sn_2O_7$ (top), $Y_2(Sn_{0.4}Ti_{0.6})_2O_7$ (middle), $Y_2Ti_2O_7$ (bottom) obtained at 14.1 T with an MAS frequency of 15 kHz. Centerband resonances are marked with arrows, while the asterisks denote spinning sidebands. (Reprinted from N. Kim and C. P. Grey, *J. Solid State Chem.* 2003, **175**, 110. Copyright (2003), with permission from Elsevier.)

Although the pyrochlore structure has been extensively studied by solid-state NMR spectroscopy using spin $I = 1/2$ nuclei such as ^{89}Y and ^{119}Sn, the use of quadrupolar nuclei remains relatively scarce. High-resolution ^{17}O MAS NMR spectra of $Y_2(B_{1-x}B'_x)_2O_7$ (B, B′ = Sn, Ti, Zr, $0 < x > 1$)[25] pyrochlores have shown that one can observe all the crystallographically different oxygen sites and follow the B/B′ site substitution in detail (Figure 20.8b) when the long-range structures of the materials are similar. Using ^{17}O NMR, pyrochlore-like local ordering has been detected in $Y_2Zr_2O_7$, which adopts the overall long-range fluorite structure.

20.2.3.2 *Perovskites*

The perovskite structure of general formula ABO_3 has also been studied extensively and is illustrated in Figure 20.9(a). A very large number of oxide-deficient perovskites obtained via cation substitution of a lower valence cation into the A and B sites have been found to be good anionic conductors. Ishihara *et al.*[26] and Feng and Goodenough[27] discovered that the Sr- and Mg-doped lanthanum gallates, $LaGaO_3$, were pure anionic conductors with a transport number very close to 1 and had the highest oxygen ion conductivity reported. Such characteristics make these perovskites potential electrolytes for SOFCs.[28] While a number of diffraction and computational studies are known and reveal structural information and oxygen migration mechanism, little is yet known about the cation environments.

Using ^{27}Al NMR, Stebbins *et al.*[29] showed that the aluminum substitution in $MgSiO_3$ perovskite occurs on both the A = Mg and B = Si sites as revealed by Al sites in multiple coordination environments. Using a similar approach, Grey *et al.*[30,31] and Tankeyev *et al.*[32] investigated the local structure in other doped perovskites including the anionic conductors Sr- and Mg-doped $LaGaO_3$.

A combination of ^{71}Ga and ^{25}Mg NMR studies has revealed that oxygen vacancies are preferentially located near gallium cations, as shown by the presence of both six- and five-coordinate Ga.[30,32] These results suggest vacancy trapping near the host cation in sharp contrast with NMR studies on other perovskites (for example, Sc-doped $BaZrO_3$) where vacancies are found near the dopant.[31]

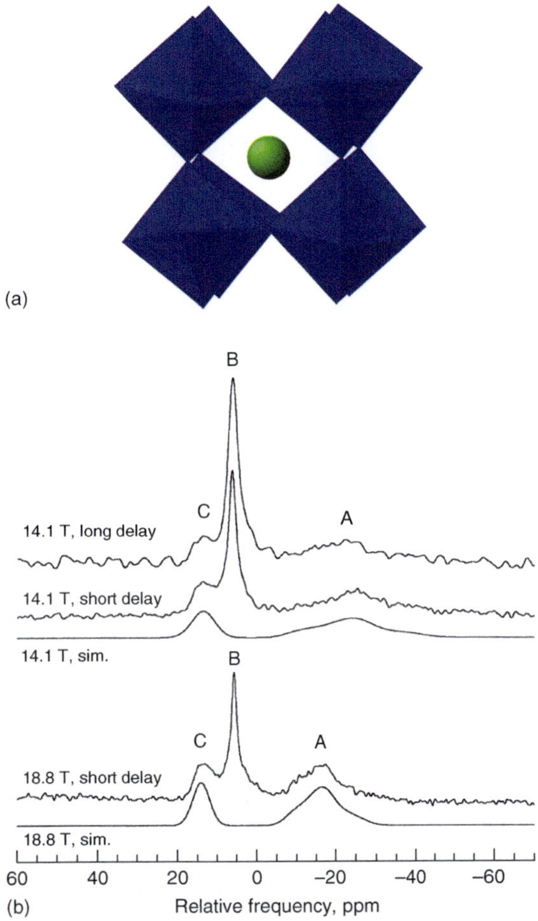

Figure 20.9. (a) ABO_3 perovskite structure in the corner-shared BO_6 octahedral representation (blue) with A centered on 12-coordinate sites (green). (b) ^{27}Al MAS NMR spectra of $MgSi_{0.9}Al_{0.1}O_{2.95}$ obtained at both 14.1 and 18.8 T with an MAS frequency of 18 kHz, together with numerical simulations. Peaks A, B, and C indicate 12-coordinate, 6-coordinate, and 5-coordinate Al sites, respectively (Reproduced with permission from Stebbins, J. F *et al. American Mineralogist* 2003, **88**, 1161.)

20.2.3.3 *The $La_2Mo_2O_9$ Structure and Related β-$SnWO_4$ Phase*

A final example of fast anionic conductors whose structures have been studied by NMR spectroscopy is the LAMOX material based on lanthanum molybdate, $La_2Mo_2O_9$.[33] This material undergoes at 580 °C a monoclinic to cubic first-order phase transition associated with an increase of two orders

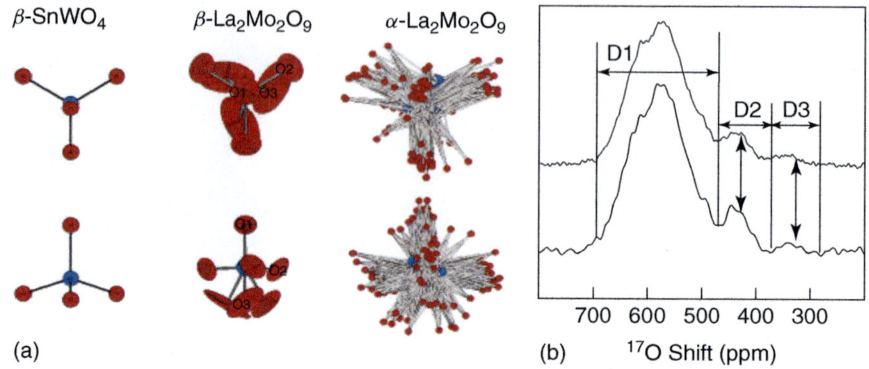

Figure 20.10. (a) Comparison of B site coordination in β-$SnWO_4$, β-$La_2Mo_2O_9$, and α-$La_2Mo_2O_9$ shown in two mutually perpendicular views. For α-$La_2Mo_2O_9$, the picture represents a superposition of all independent Mo atoms and their coordination spheres obtained by transformation of the monoclinic superstructure into the underlying cubic subcell. (Reprinted with permission from I. R. Evans *et al.*, *Chem. Mater.* 2005, **17**, 4074. Copyright 2005 American Chemical Society) (b) ^{17}O MAS synchronized Hahn-echo spectra of the central transition of ^{17}O enriched α-$La_2Mo_2O_9$ obtained at 7.0 T (top) and 9.4 T (bottom). (Reproduced with permission from J. Emery *et al.*, *Magn. Res. Chem.* 2005, **43**, 366. Copyright John Wiley & Sons Ltd.)

of magnitude in the ionic conductivity. The structure of both low-temperature α and high-temperature β-$La_2Mo_2O_9$ phases are known and given in Figure 20.10(a).

The structure of β-$La_2Mo_2O_9$ (space group $P2_13$) is relatively simple and analogous to β-$SnWO_4$, where the metal cation arrangements are essentially the same but the oxygen sublattice is more complex because of the difference in stoichiometry. In contrast, the structure of the monoclinic α phase (space group $P2_1$) of $La_2Mo_2O_9$ is very complex, with 312 atoms in the asymmetric unit cell.[34] The basic cation arrangement of the β phase is preserved in the α phase, and the major differences are found in the oxygen sublattice which yields a mixture of four-, five-, and six-coordinated Mo sites suggesting that variable Mo coordination number is a key factor in providing a low-energy O^{2-} conduction pathway.

Although there are 216 different O crystallographic sites in α-$La_2Mo_2O_9$ (Figure 20.10a), only three oxygen resonances were observed in the ^{17}O NMR spectrum of α-$La_2Mo_2O_9$[35] (Figure 20.10b). These were correlated to the three oxygen sites O1, O2, and O3 of the high-temperature β phase, which was the only phase known at that time. Nevertheless, relaxation studies and 3Q MAS experiments have revealed static disorder associated with a large number of oxygen sites, which were subsequently observed when the crystal structure of α-$La_2Mo_2O_9$ was solved using diffraction.[34] This example illustrates the complementary nature of solid-state NMR and diffraction approaches.

20.3 DYNAMICS

20.3.1 NMR Methods of Studying Dynamics in Ion Conductors

NMR is well known for its ability to probe dynamics over a wide range of timescales.[36] Methods include probes of spin–lattice relaxation as a function of temperature and field, as well as coalescence and line narrowing which can be indicative of changes in the timescale of local dynamics that give rise to spin–spin (T_2) relaxation. Beyond these relaxation phenomena, pulse sequences such as exchange spectroscopy (EXSY)[37] can be used to probe dynamics in a site-specific fashion.

Ionic motion can also be indicated by the behavior of nutation curves for quadrupolar nuclei. For quadrupolar nuclei of spin I, the size of the quadrupole frequency ν_Q affects the frequency at which the spins nutate around an applied radiofrequency (RF) field strength, ω_1, and is given by

$$\nu_Q = \frac{3C_Q}{2I(2I-1)} \tag{20.5}$$

which becomes

$$\nu_Q = \frac{C_Q}{2} \tag{20.6}$$

and

$$\nu_Q = \frac{3}{20} C_Q \tag{20.7}$$

for ^{7}Li ($I = 3/2$) and ^{17}O ($I = 5/2$), respectively. For liquid samples, or solid samples in which rapid motion averages out the quadrupolar interaction, or samples with small quadrupolar couplings, $\nu_Q < \nu_1$, and all the $2I$ allowed transitions can be excited and the nutation frequency is the same as that observed for a spin $I = 1/2$ sample, i.e., ν_1. Alternatively, for rigid sites in which there is no rapid exchange and are associated with a large C_Q, where $\nu_Q \gg \nu_1$, only the central transition is excited and the nutation frequency becomes $I + 1/2$ times ν_1 (with an amplitude $1/(I + 1/2)$ times greater). For example, for ^{17}O ($I = 5/2$), this yield a nutation curve three times faster than ν_1, with a third of the amplitude. A full discussion of the influence of dynamics on the line shapes of quadrupolar nuclei can be found in Chapter 14.

20.3.2 Cation Conductors

20.3.2.1 Studies of Cation Dynamics in Diamagnetic Compounds

An example of studying cation dynamics by variable-temperature ^{7}Li NMR is provided by the work of Jeon *et al.* (Figure 20.11) in which the ionic motion of lithium in a composite polymer electrolyte sample is observed.[38] This is just one example of many such variable-temperature direct-excitation NMR experiments, where the line-narrowing phenomenon of the isotropic resonance is correlated to the activation energy of a dynamic ion transport process.

The temperature dependence of the corresponding correlation time τ_c can be calculated from the ^{7}Li NMR line width data by using the Bloemberg, Purcell and Pound (BPP) equation for polymer electrolytes. In the case of this diamagnetic material, there is no significant change in the ^{7}Li chemical shift as a function of temperature, and the direct correlation between the line width trend and the ion dynamics is based on the assumption that the line width is due to homonuclear dipolar coupling interactions, which are increasingly averaged because of local motion as the temperature increases.

20.3.2.2 Studies of Cation Dynamics in Paramagnetic Compounds

In paramagnetic samples, chemical shift trends as a function of temperature are influenced both by the changes in the magnetic susceptibility with temperature as well as changes in ion dynamics with temperature. Nevertheless, trends in both chemical shift and line width as a function of temperature have been used to evaluate ion dynamics in lithium ion battery materials. One example is thc anode material Li_7MnN_4, where six unique lithium environments are resolved in the ^{6}Li NMR spectrum at ambient temperature, ranging in chemical shift from 1333 to 4 ppm, as shown in Figure 20.12. These resonances undergo a selective coalescence process in the temperature range from ambient to 250 °C, which indicates preferential dynamics pathways. Moreover, the coalescence phenomena can be associated with electrochemical performance, or more specifically, the rate capability of the material.[a] This has been successfully achieved in the comparison of the anode materials Li_7MnN_4, $Li_{7.9}MnO_{3.2}N_{1.6}$ and $Li_4Ti_5O_{12}$.[39]

While the coalescence phenomena and other chemical shift and line width trends can be used to qualitatively assess relative dynamics, it is difficult to determine exact rates of exchange, particularly in paramagnetic systems. This is due to the competing effects of magnetic susceptibility as well as local dynamics, which are responsible for the chemical shift trends. Therefore, many authors have turned to 2D EXSY to quantitatively assess lithium ion mobility. Examples include studies of the hopping semiconductor $LiMn_2O_4$,[40] the lithium transition-metal phosphates,[41,42] and the lithium manganese nitrides described above.[43]

For $Li_3M_2(PO_4)_3$, where M = V or Fe, the EXSY NMR data clearly show that rapid exchange, on the timescale of 2–8 ms, is occurring between all three Li sites (Figure 20.13). The use of variable mixing times

[a] The rate of charge or discharge is defined using the variable C, which is equivalent to a full charge or discharge in 1 h. Rates are defined relative to this, where $C/10$ is equivalent to a charge or discharge process over 10 h

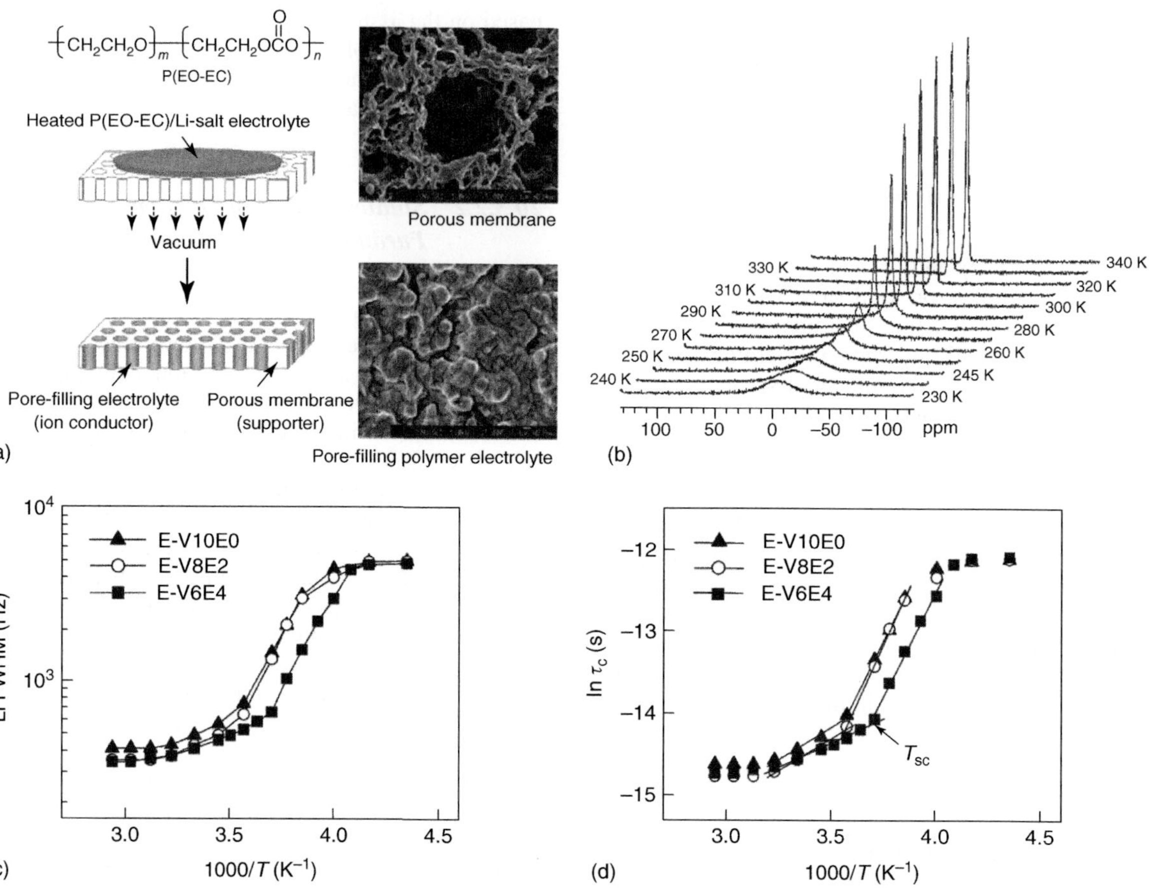

Figure 20.11. (a) Concept of the pore-filling polymer electrolyte based on a porous membrane filled with a viscous polymer/Li-salt electrolyte. The pictures show typical field-emission scanning electron microscopy (FE-SEM) images of a porous membrane (M-V6E4) and a polymer electrolyte (E-V6E4), respectively. (b) Temperature dependence of the ^{7}Li NMR spectra for E-V6E4 with 1.5 mmol-$LiCF_3SO_3$/g-P(EO-EC) at temperatures as shown in the plot. (c, d) Line widths of ^{7}Li NMR spectra for polymer electrolytes as a function of temperature. (Reprinted with permission from J.-D. Jeon *et al.*, *Macromolecules* 2006, **39**, 8027. Copyright 2006 American Chemical Society.)

can determine the rate of exchange between each pair of sites.[41,42] In these materials, the exchange was found to be sufficiently rapid to allow accurate evaluation of the dynamics prior to the spin–lattice relaxation. This was essential to allow the use of the 2D EXSY experiment. Once the timescales are established, the activation energies for the different pairs of exchange partners could also be evaluated, ranging from 0.59 to 0.81 (±0.05) eV, and correlated to the structural constraints of the materials framework.

20.3.3 Oxygen Phases

20.3.3.1 The Fluorite-type Structure: Example of Y-doped CeO_2

Many NMR studies have investigated the oxygen dynamics in the fluorite structure, which we will illustrate with the example of Y-doped CeO_2.[44] Early NMR work on yttria-doped ceria focused on the study of the motion based on ^{17}O spin relaxation measurements of T_1 and T_2. Two motional regimes

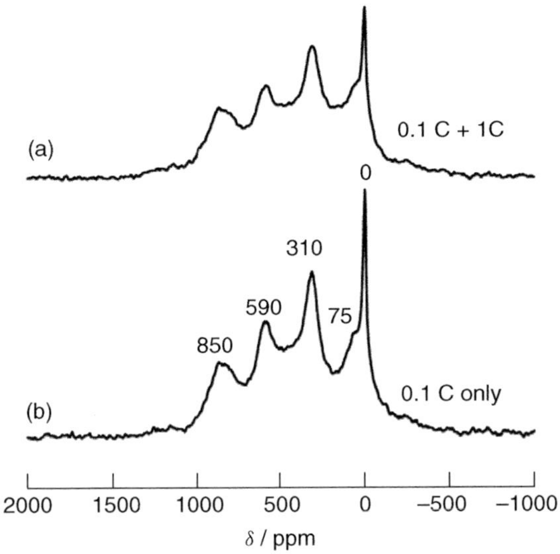

Figure 20.12. ^{6}Li NMR spectra of $Li_{7.9}MnN_{3.2}O_{1.6}$ obtained at 4.7 T and 38 kHz MAS after (a) a charge at *C*/10–discharge at *C* cycle and (b) after one cycle at *C*/10. Representative shift values are provided. A contribution from impurities of the electrolyte salt on the sample is observed on the signals at 0 ppm. (Reprinted from J. Cabana *et al.*, *Eletrochem. Commun.* 2010, **12**, 315. Copyright (2010), with permission from Elsevier.)

were observed and associated with the diffusion of oxygen vacancies (below about 300 °C) and with simultaneous movement of oxygen vacancies and oxygen ions (above about 300 °C).

Using the standard equation for quadrupolar relaxation mechanism, an NMR activation energy of 47 kJ mol^{-1} (Figure 20.14a and b) was found, which was lower than the 76–87 kJ mol^{-1} range determined by AC conductivity measurements (below 300 °C). This is in agreement with the fact that NMR relaxation is sensitive to local single-ion hops while conductivity measurements probe long-range transport where the highest barrier may determine the conductivity.

In the high-temperature regime (above 300 °C), using dynamic Monte Carlo simulation,[45] the asymmetric behavior of the relaxation of yttria-doped ceria (Figure 20.14a) was attributed to multiple quadrupolar relaxation mechanisms for simultaneous movement of oxygen vacancies and oxygen ions.

More recently, Kim and Stebbins[44] reinvestigated both the structure and dynamics of Y-doped CeO_2 using high-resolution ^{17}O and ^{89}Y NMR. Deviation from a random distribution of yttrium atoms in the fluorite CeO_2 structure as well as vacancy trapping near yttrium was observed. High-field and high-temperature ^{17}O NMR experiments (Figure 20.14c) allow the determination of the

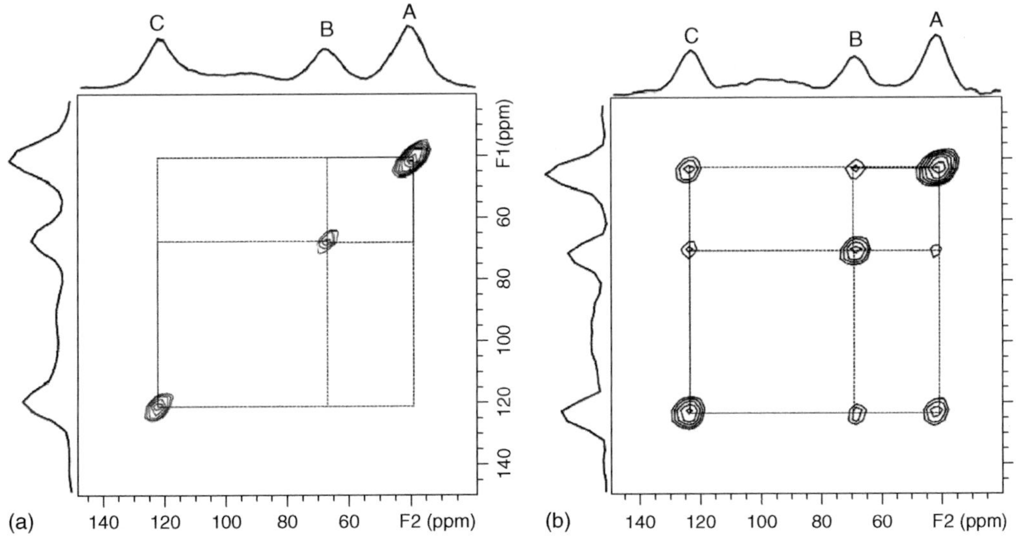

Figure 20.13. ^{6}Li 2D EXSY spectrum of $Li_3Fe_2(PO_4)_3$ obtained under MAS of 25 kHz and at 311 K with an exchange time τ_{mix} of 0.5 ms (a) and 3.0 ms (b). The three Li sites are marked A, B, and C. (Reprinted with permission from L. J. Davis *et al.*, *Chem. Mater.* 2010, **22**, 769. Copyright 2010 American Chemical Society.)

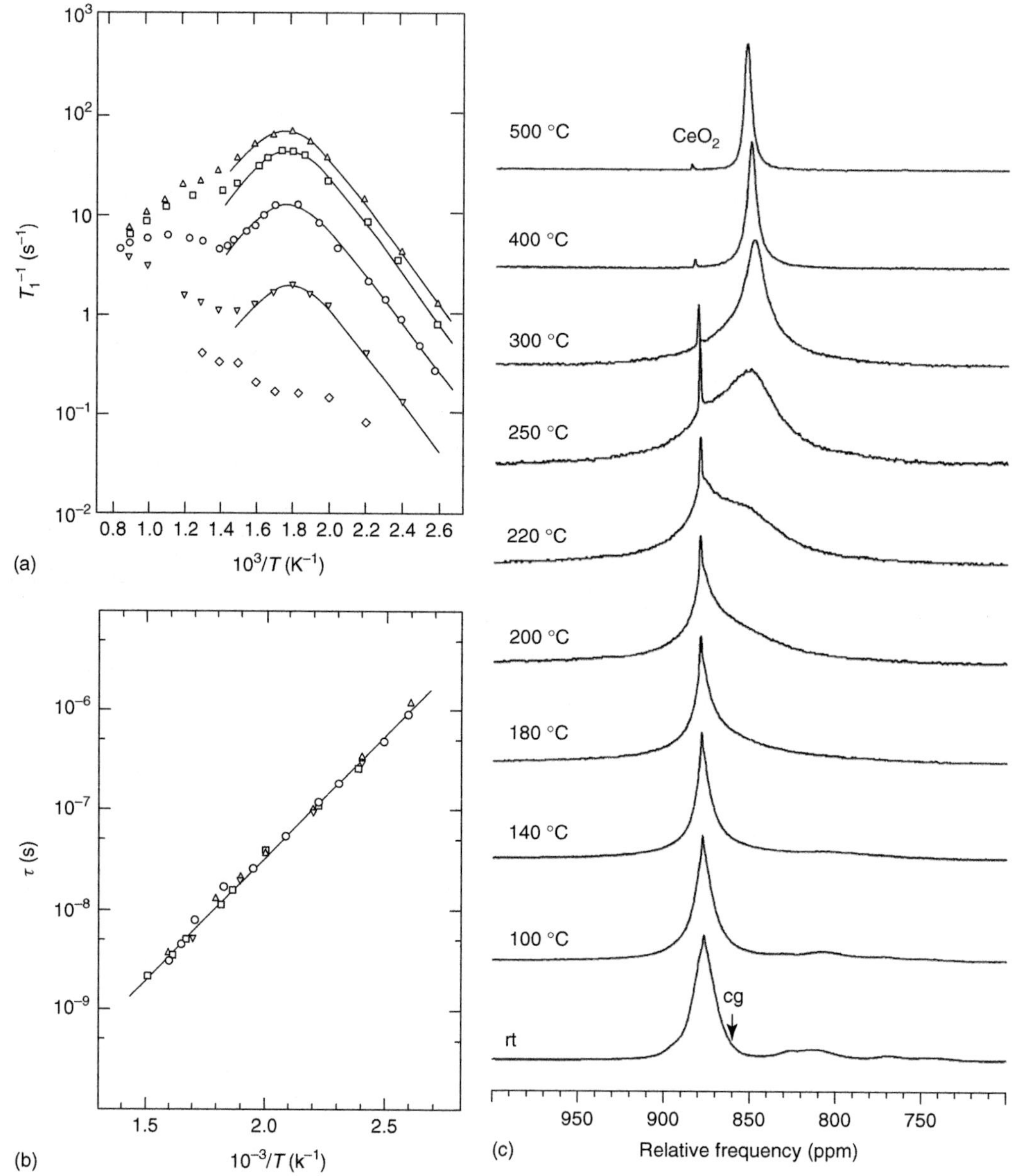

Figure 20.14. (a) ^{17}O Spin–lattice relaxation times T_1 of Y_2O_3-doped CeO_2 as a function of inverse temperature for various amount of Y_2O_3 dopant of 0.02% (∇), 0.06% (○), 0.2% (□), 0.6% (Δ), and undoped CeO_2 (◇). The curves drawn on the data correspond to the theoretical fit assuming electric field gradient fluctuation. (b) Correlation time τ_c of oxygen vacancy motion as a function of inverse temperature. The least-squares fit to an Arrhenius equation is shown. (Reprinted from K. Fuda *et al.*, *J. Phys. Chem. Solids.* 1984, **45**, 1253. Copyright (1984), with permission from Elsevier.)[46] (c) High-temperature ^{17}O MAS NMR spectra of 5% Y_2O_3-doped CeO_2 at 9.4 T. The center of gravity of all the resonances at room temperature (about 860 ppm) is labeled as cg. A very narrow peak for pure CeO_2 impurity is also visible. (Reprinted with permission from N. Kim and J. F. Stebbins, *Chem. Mater.* 2007, **19**, 5742. Copyright 2007 American Chemical Society.)

site-specific oxygen dynamics to be monitored. Exchange between the oxygen resonances containing only Ce neighbors (at about 875 ppm) and the sites containing both Ce and Y neighbors (broad sites between 700 and 850 ppm assigned to oxygen with three Ce and one Y neighbor and with two Ce and two Y neighbors) was identified. In addition, coalescence of all oxygen resonances above 400 °C indicated a rapid exchange of all oxide anions throughout the structure, which are thus responsible for the long-range oxygen conductivity.

20.3.3.2 BIMEVOX Oxides

BIMEVOX (oxidic materials based on the γ-$Bi_4V_2O_{11}$ structure in which V is partially substituted by a metal cation) materials have been widely studied because they have shown moderate to high conductivities at relatively low temperatures (300–600 °C). They are derived from the parent compound $Bi_4V_2O_{11}$ and are related to the m = 1 member of the Aurivillius-type oxides of the general formula $(Bi_2O_5)(A_{m-1}B_mO_{3m+1})$. The structure consist of alternating $Bi_2O_{22}^+$ and perovskite-like $A_{m-1}B_mO_{3m+1}$ layers (Figure 20.15a).

Understanding the conduction mechanism of these phases has been difficult as a result of highly disordered structures. Several reports demonstrate that ^{17}O is an extremely powerful method to understand the anionic conductivity.

Variable-temperature ^{17}O and ^{51}V NMR spectroscopy as well as $^{17}O/^{51}V$ double resonance experiments[47] have been used to study both α-$Bi_4V_2O_{11}$ and γ-$Bi_4V_{1.7}Ti_{0.3}O_{10.85}$ phases (Figure 20.15b and c). Oxygen sites in the room-temperature α phase of $Bi_4V_2O_{11}$ were identified by their ^{17}O chemical shifts and assigned to the bismuth (at 265 ppm), and vanadium layers (between 600 and 1000 ppm).

Oxide ion hops were observed by variable-temperature ^{17}O NMR and occurred via chemical exchange within the V–O layer. These involve both the "equatorial" (V–O–V) and "axial" (V–O–Bi_n) positions in the perovskite layers (Figure 20.15b). In addition, motional broadening of all spinning sidebands have been observed, suggesting that,

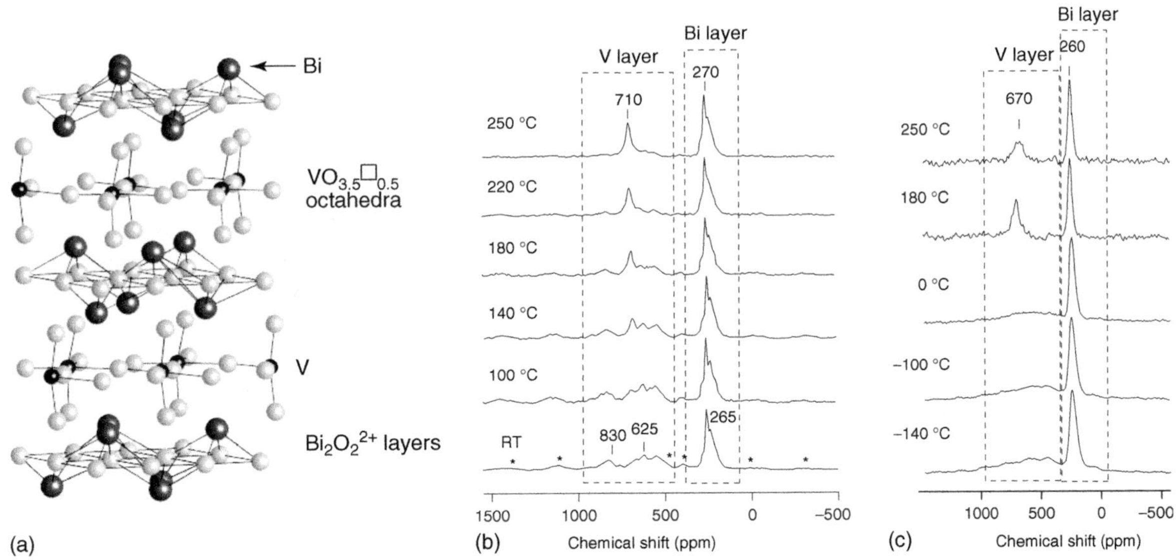

Figure 20.15. (a) Idealized structure of α-$Bi_4V_2O_{11}$. The $VO_{3.5}\square_{0.5}$ groups in the perovskite layers are shown as octahedral, where □ represents oxygen vacancies. (b) ^{17}O MAS NMR spectrum of ^{17}O-enriched α-$Bi_4V_2O_{11}$ obtained with a rotor-synchronized Hahn echo sequence at 8.45 T under an MAS frequency of 15 kHz. (c) ^{17}O MAS NMR spectrum of ^{17}O-enriched γ-$Bi_4V_{1.7}Ti_{0.3}O_{10.85}$ obtained with a rotor-synchronized Hahn echo sequence at 8.45 T under a MAS frequency of 9 kHz (for low-temperature data) and at 18 kHz (for high-temperature data). The isotropic resonances are marked by their isotropic shifts; the other peaks are marked with asterisks and are spinning sidebands. (From N. Kim and C. P. Grey, *Science* 2002, **297**, 1317. Reprinted with permission from AAAS.)

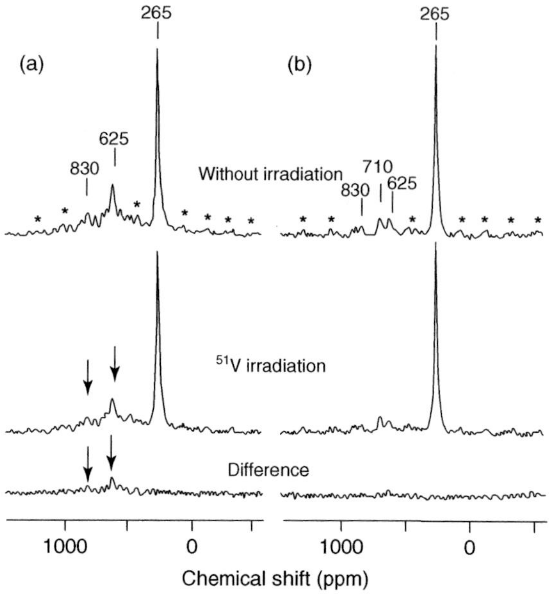

Figure 20.16. $^{17}O/^{51}V$ TRAPDOR experiments at 8.45 T under an MAS frequency of 15 kHz at (a) room temperature and (b) 150 °C. The centerband resonances are marked by their isotropic shifts; the other peaks are marked with asterisks and are spinning sidebands. (From N. Kim and C. P. Grey, *Science* 2002, **297**, 1317. Reprinted with permission from AAAS.)

although the oxygen ions in the $(Bi_2O_2)^{2+}$ layers are not involved in the hopping process, these anions are sensitive to the nearby motion in the V–O layers. A set of variable-temperature $^{17}O/^{51}V$ transfer of population in double resonance (TRAPDOR) experiments[47] observed on both ^{17}O and ^{51}V nuclei were used to probe the V–O connectivity, confirming the assignment. This was also used to investigate the oxygen mobility (Figure 20.16). It was demonstrated that the more mobile oxygen anions O^{2-} are associated with the more distorted vanadium tetrahedra, VO_4.

20.3.3.3 The ZrW_2O_8 Structure

Cubic zirconium tungstate ZrW_2O_8 displays isotropic negative expansion–contraction of volume on heating–and shows dynamic oxygen disorder at unusually low temperatures. This dynamic disorder is responsible for an order–disorder phase transition at about 175 °C from space group symmetry $P2_13$ (α form of ZrW_2O_8) to $Pa\bar{3}$ (β form) arising from oxygen migration.

Oxygen mobility at such low temperature is unusual, and its mechanism has thus been investigated thoroughly by ^{17}O solid-state NMR[49,50] studies. The structure of ZrW_2O_8 (Figure 20.17a and b) contains a network of corner-sharing octahedra and tetrahedra with a W_2O_8 unit as a pair of WO_4 tetrahedra.

The room-temperature ^{17}O MAS NMR spectra of ZrW_2O_8 (Figure 20.17c) showed all three oxygen sites assigned to the four crystallographically unique oxygen sites in α-ZrW_2O_8 (the O1 and O2 oxygen sites are crystallographically inequivalent although chemically similar). Upon heating, the NMR peaks first broaden and then coalesce into one single board peak for β-ZrW_2O_8 at the weighted averaged of the individual lines in α-ZrW_2O_8. This is evidence that all oxygen sites are involved in the exchange.

Further details were obtained by 2D EXSY (Figure 20.17d) performed at 57 °C, where at this temperature oxygen motion already occurs.[50] The 2D EXSY spectrum showed an off-diagonal cross peak, indicating exchange on the 50 ms timescale and full scrambling of all oxygen sites taking place. This is in agreement with oxygen motion happening through a "ratcheting" model involving whole rotations of WO_4 moieties (Figure 20.17b).

A more detailed quantitative analysis of the slow oxygen motion was obtained by combining T_1 relaxation and ^{17}O exchange rate (k_{exch}) measurements.[49] The T_1 measurement is in agreement with a magnetic mechanism (vs a quadrupolar mechanism) and is a strong function of the motional exchange as revealed by the T_1 drop from 32 s at 40 °C to 0.5 s at 226 °C (Figure 20.17e). The ^{17}O exchange rates k_{exch} were then calculated and plotted against inverse temperature (Figure 20.17f). From this, activation energy barrier for the oxygen exchange of both low-temperature α ($E_a = 24.5 \pm 0.5$ kJ mol^{-1}) and high-temperature β ($E_a = 42 \pm 9$ kJ mol^{-1}) phases could be obtained. These were found to be in agreement with dielectric measurements (48 kJ mol^{-1} for the β phase) and powder diffraction experiments ($E_a = 34 \pm 5$ kJ mol^{-1} for the α phase of the related $ZrWMoO_8$ material). A model based on local flips of W_2O_8 groups occurring within an ordered material was proposed in the α phase, while a time-averaged $Pa\bar{3}$ structure model was given in the β phase, which simultaneously explains both NMR and diffraction datasets.

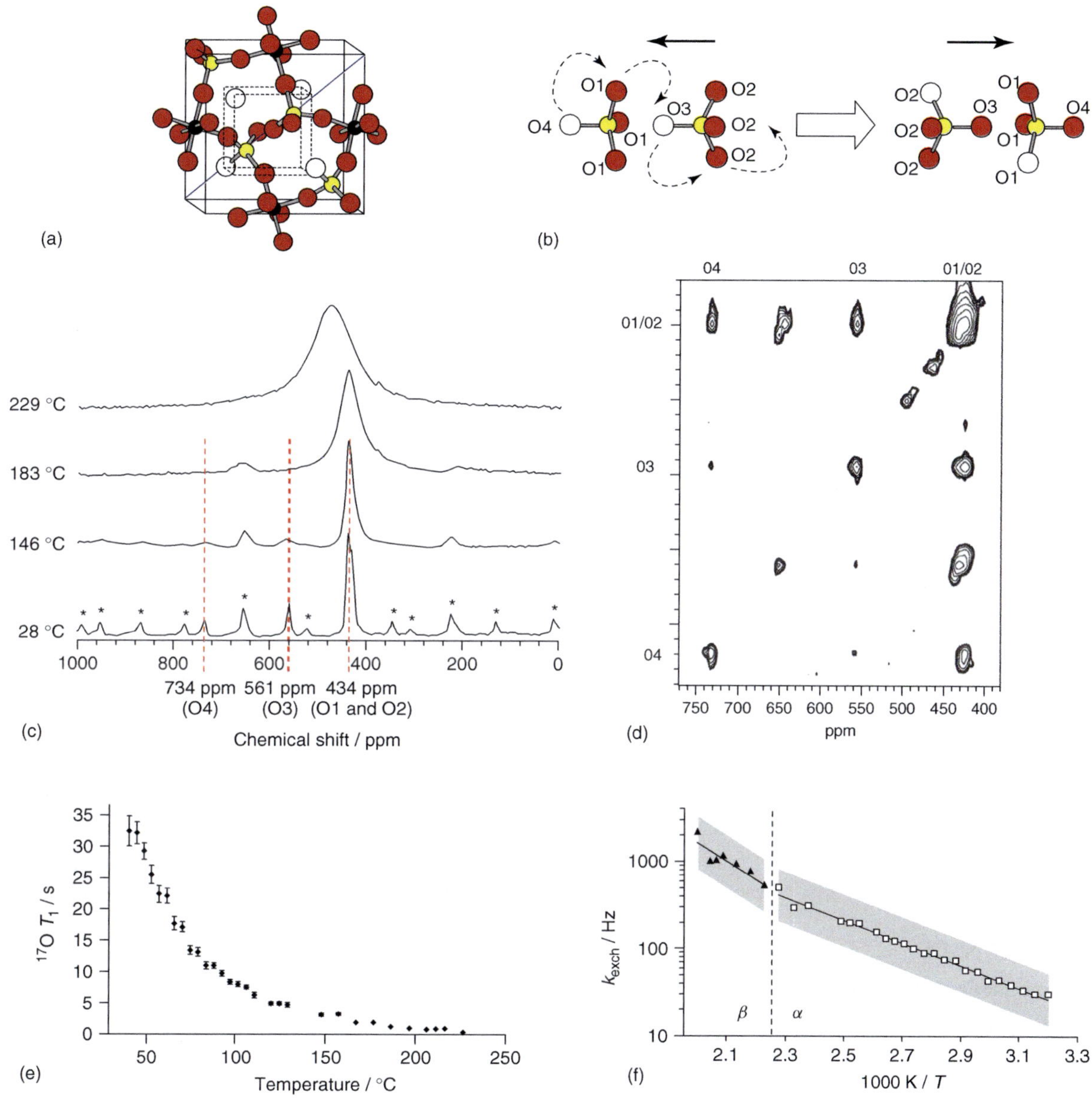

Figure 20.17. (a) Structure of cubic ZrW_2O_8, showing two-coordinate oxygen sites (O1-3) (red), one-coordinate site O4 (white), zirconium atoms (black), and tungsten atoms (yellow). (b) Schematic diagram showing the reversal of the W_2O_8 group direction (black arrow) by a "ratcheting" oxygen exchange motion. (c) Variable-temperature ^{17}O MAS NMR spectra of ZrW_2O_8 along with the spectral assignments.[50] (d) ^{17}O 2D EXSY spectrum of ZrW_2O_8 obtained at 57 °C with a mixing time of 50 ms.[50] Asterisks denote spinning sidebands. (e) Fitted ^{17}O T_1 values as a function of temperature. (f) Arrhenius plot of k_{exch} vs inverse temperature. The gradients of the fitted straight lines give the activation energy for both phases. The uncertainty of the absolute values of k_{exch} lies within the shaded region. (Reprinted with permission from M. R. Hampson *et al.*, *J. Am. Chem. Soc.* 2005, **127**, 15175. Copyright 2005 American Chemical Society and M. R. Hampson *et al.*, Chem. Commun. 2004, **4**, 392. Copyright 2005 Royal Society of Chemistry.)

20.4 SUMMARY

A wide variety of solid-state NMR methods have been applied to the characterization of materials for energy storage and conversion. In particular, the structure and dynamics of lithium ion battery materials and solid-oxide fuel cell materials have been successfully studied using $^{6,7}Li$ NMR and ^{17}O NMR respectively, as well as the NMR of a variety of complementary quadrupolar nuclei found in the relevant structures.

This chapter has focused on presenting various electrochemically important phases and the NMR studies that have elucidated the structural details, as well as the site-specific dynamic profiles of the materials of interest.

In the case of oxygen conductors, the quadrupolar nature of the nuclei plays an important role in the type of NMR experiments conducted. In the case of Li^+ conductors, the paramagnetic nature of many samples, together with the small quadrupole parameters of both naturally occurring nuclear isotopes, precludes the detailed study of the quadrupole properties themselves. Nevertheless, solid-state NMR has been shown to provide unique insight into the materials properties, which can be directly correlated with electrochemical performance.

RELATED ARTICLES IN THE ENCYCLOPEDIA OF MAGNETIC RESONANCE

Knight Shift

Oxygen-17 NMR

Paramagnetic Relaxation in Solution

Quadrupolar Nuclei in Liquid Samples

Two-Dimensional Methods of Monitoring Exchange

REFERENCES

1. J. B. Goodenough and Y. Kim, *Chem. Mater.*, 2010, **22**, 587.
2. P. G. Bruce, *Solid State Electrochemistry*, Cambridge University Press: Cambridge, 1995.
3. J. M. Tarascon and M. Armand, *Nature*, 2001, **414**, 359.
4. C. P. Grey and N. Dupre, *Chem. Rev.*, 2004, **104**, 4493.
5. R. W. Schurko, S. Wi, and L. Frydman, *J. Phys. Chem. A*, 2002, **106**, 51.
6. S. E. Ashbrook and M. E. Smith, Oxygen-17 NMR of Inorganic Materials, in *Encyclopedia of Magnetic Resonance*, eds R. K. Harris and R. E. Wasylishen), John Wiley & Sons: Chichester, 2011. DOI: 10.1002/9780470034590.emrstm1213.
7. S. Ganapathy, P. R. Rajamohanan, P. Ganguly, T. N. Venkatraman, and A. Kumar, *J. Phys. Chem. A*, 2000, **104**, 2007.
8. A. Nayeem and J. P. Yesinowski, *J. Chem. Phys.*, 1988, **89**, 4600.
9. Y. J. Lee and C. P. Grey, *Chem. Mater.*, 2000, **12**, 3871.
10. D. Carlier, M. Menetrier, C. P. Grey, C. Delmas, and G. Ceder, *Phys. Rev. B*, 2003, **67**, 2174103.
11. Y. J. Lee, F. Wang, and C. P. Grey, *J. Am. Chem. Soc.*, 1998, **120**, 12601.
12. W. S. Yoon, N. Kim, X. Q. Yang, J. McBreen, and C. P. Grey, *J. Power Sources*, 2003, **119**, 649.
13. C. P. Grey, W. S. Yoon, J. Reed, and G. Ceder, *Electrochem. Solid-State Lett.*, 2004, **7**, A290.
14. D. Carlier, M. Menetrier, and C. Delmas, *J. Phys. Chem. C*, 2010, **114**, 4749.
15. J. Kim, D. S. Middlemiss, N. A. Chernova, B. Y. X. Zhu, C. Masquelier, and C. P. Grey, *J. Am. Chem. Soc.*, 2010, **132**, 16825.
16. L. van Wullen, T. Echelmeyer, H. W. Meyer, and D. Wilmer, *Phys. Chem. Chem. Phys.*, 2007, **9**, 3298.
17. T.L. Spencer, G.R. Goward, and A.D. Bain, *Can. J. Chem.*, 2011, **89**, 770.
18. D. Mazza, F. Abbatista, M. Vallino, and G. Ivaldi, *J. Less-Common. Met.*, 1985, **106**, 277.
19. L. A. O'Dell and R. W. Schurko, *Chem. Phys. Lett.*, 2008, **464**, 97.
20. G. R. Goward, F. Leroux, W. P. Power, G. Ouvrard, W. Dmowski, T. Egami, and L. F. Nazar, *Electrochem. Solid-State Lett.*, 1999, **2**, 367.
21. M. Letellier, F. Chevallier, C. Clinard, E. Frackowiak, J. N. Rouzaud, F. Beguin, M. Morcrette, and J. M. Tarascon, *J. Chem. Phys.*, 2003, **118**, 6038.
22. B. Key, R. Bhattacharyya, M. Morcrette, V. Seznec, J. M. Tarascon, and C. P. Grey, *J. Am. Chem. Soc.*, 2009, **131**, 9239.

23. R. Bhattacharyya, B. Key, H. L. Chen, A. S. Best, A. F. Hollenkamp, and C. P. Grey, *Nat. Mater.*, 2010, **9**, 504.
24. M. A. Subramanian, G. Aravamudan, and G. V. S. Rao, *Prog. Solid State Chem.*, 1983, **15**, 55.
25. N. Kim and C. P. Grey, *J. Solid State Chem.*, 2003, **175**, 110.
26. T. Ishihara, T. Kudo, H. Matsuda, and Y. Takita, *J. Am. Ceram. Soc.*, 1994, **77**, 1682.
27. M. Feng and J. B. Goodenough, *Eur. J. Solid State Inorg. Chem.*, 1994, **31**, 663.
28. K. Q. Huang, R. Tichy, and J. B. Goodenough, *J. Am. Ceram. Soc.*, 1998, **81**, 2581.
29. J. F. Stebbins, H. Kojitani, M. Akaogi, and A. Navrotsky, *Am. Miner.*, 2003, **88**, 1161.
30. D. S. Middlemiss, F. Blanc, C. J. Pickard, and C. P. Grey, *J. Magn. Reson.*, 2010, **204**, 1.
31. L. Buannic, F. Blanc, I. Hung, Z. Gan, and C. P. Grey, *J. Mater. Chem.*, 2010, **20**, 6322.
32. A. Buzlukov, A. Trokiner, V. Kozhevnikov, S. Verkhovskii, A. Yakubovsky, I. Leonidov, A. Gerashenko, A. Stepanov, I. Baklanova, and A. Tankeyev, *J. Solid State Chem.*, 2011, **184**, 36.
33. P. Lacorre, F. Goutenoire, O. Bohnke, R. Retoux, and Y. Laligant, *Nature*, 2000, **404**, 856.
34. I. R. Evans, J. A. K. Howard, and J. S. O. Evans, *Chem. Mater.*, 2005, **17**, 4074.
35. J. Emery, D. Massiot, P. Lacorre, Y. Laligant, and K. Conder, *Magn. Reson. Chem.*, 2005, **43**, 366.
36. D. Reichert, *Ann. Rep. NMR Spectrosc.*, 2005, **55**, 159.
37. J. Jeener, B. H. Meier, P. Bachmann, and R. R. Ernst, *J. Chem. Phys.*, 1979, **71**, 4546.
38. J. D. Jeon and S. Y. Kwak, *Macromolecules*, 2006, **39**, 8027.
39. J. Cabana, C. M. Ionica-Bousquet, C. P. Grey, and M. R. Palacin, *Electrochem. Commun.*, 2010, **12**, 315.
40. V. W. J. Verhoeven, I. M. de Schepper, G. Nachtegaal, A. P. M. Kentgens, E. M. Kelder, J. Schoonman, and F. M. Mulder, *Phys. Rev. Lett.*, 2001, **86**, 4314.
41. L. S. Cahill, R. P. Chapman, J. F. Britten, and G. R. Goward, *J. Phys. Chem. B*, 2006, **110**, 7171.
42. L. J. M. Davis, I. Heinmaa, and G. R. Goward, *Chem. Mater.*, 2010, **22**, 769.
43. J. Cabana, N. Dupre, G. Rousse, C. P. Grey, and M. R. Palacin, *Solid State Ionics*, 2005, **176**, 2205.
44. N. Kim and J. F. Stebbins, *Chem. Mater.*, 2007, **19**, 5742.
45. S. B. Adler, J. W. Smith, and J. A. Reimer, *J. Chem. Phys.*, 1993, **98**, 7613.
46. K. Fuda, K. Kishio, S. Yamauchi, K. Fueki, and Y. Onoda, *J. Phys. Chem. Solids*, 1984, **45**, 1253.
47. C. P. Grey and A. J. Vega, *J. Am. Chem. Soc.*, 1995, **117**, 8232.
48. N. Kim and C. P. Grey, *Science*, 2002, **297**, 1317.
49. M. R. Hampson, J. S. O. Evans, and P. Hodgkinson, *J. Am. Chem. Soc.*, 2005, **127**, 15175.
50. M. R. Hampson, p. Hodgkinson, J. S. O. Evans, R. K. Harris, I. J. King, S. Allen, and F. Fayon, *F. Chem. Commun.*, 2004, **4**, 392.

Chapter 21
Quadrupolar NMR of Nanoporous Materials

Mohamed Haouas, Charlotte Martineau, and Francis Taulelle

Tectospin, Institut Lavoisier de Versailles, Université de Versailles-St. Quentin en Yvelines, 78035 Versailles, France

21.1 INTRODUCTION

Materials containing holes in their solid network are considered porous when the volume ratio of the pores to the total volume of the material (i.e., the porosity) ranges between 0.2 and 0.95. Depending on the size of the pores, these materials have been classified by the International Union of Pure and Applied Chemistry as microporous (pore size lower than 2 nm in diameter), mesoporous (pore size between 2 and 50 nm), or macroporous (pore size larger than 50 nm). Nanoporous solids combine large porosity (greater than 0.4) with pore diameters in the range 1–100 nm, and thus include the microporous and most of the mesoporous compounds. The presence of holes in the structure of solids confers upon them a wide variety of properties that the corresponding bulk materials do not have, such as, for example, high specific area, fluid permeability, molecular sieving, or shape-selective effects, making them suitable for applications in the fields of adsorption, separation, catalysis, drug encapsulation/delivery, etc. Materials responding to such porosity criteria range from polymers to carbon, glasses, foams, and metals; then to zeolitic minerals,[1] silicates,[2] and phosphates (AlPOs, SAPOs, GaPOs, etc.)[3,4]; and finally to hybrid coordination polymers[5] or metal organic frameworks (MOFs).[6] In the present chapter, we will mostly focus on the latter inorganic and organic–inorganic hybrid compounds.

One major challenge related to these materials is the lack of a clear understanding of structure–property relationships as well as the templating effects of the organic molecules to be able, *in fine*, to tune the chemistry towards the synthesis of solids with the desired topology. As these low-density porous materials are often obtained as powders, may contain isoelectronic atoms distributed randomly, and can have their pores filled with guest molecules (potentially perturbing the long-range periodicity), tackling these challenges with diffraction techniques alone is not possible. Additional and complementary characterization tools that are able not only to provide an assessment of the structural and periodic model (usually proposed by diffraction) but also to complete the description of the structure by analyses of the organization of the extra framework (i.e., the molecules or anions filling the pores) are therefore necessary.

With key methodological developments in pulse sequences,[7] decoupling techniques,[8] and phase

NMR of Quadrupolar Nuclei in Solid Materials
Edited by Roderick E. Wasylishen, Sharon E. Ashbrook and Stephen Wimperis
 ISBN: 978-0-470-97398-1

cycling,[9] as well as the great technological improvements in probes,[10] magnets, and spectrometer hardware,[11] the popularity of solid-state NMR over the past decades has progressively increased and NMR is nowadays considered as an essential tool to address problems related to the characterization of porous solids. Such hybrid organic–inorganic composite materials contain several attractive active NMR nuclei, such as ½ spins ^{1}H, ^{13}C, ^{29}Si, etc., but also quadrupolar nuclei such as ^{23}Na, ^{27}Al, $^{69,71}Ga$, or ^{17}O which can occupy two distinct sets of positions in a crystal lattice: either located at the center of a polyhedron (^{27}Al, ^{23}Na, etc.) or linkers between polyhedra (^{17}O, ^{33}S, etc.). One of the major strengths of NMR lies in the multiplicity and the flexibility of the available experiments: from 1D for various nuclei to two (or more) dimensional experiments, in which one can select the interactions to be separated and/or the nuclei to be correlated with each other. This way, even if a nucleus itself cannot be measured, it can still be indirectly accessed through other, more sensitive neighboring nuclei.

In the field of porous materials, the use of NMR is mostly directed in two main directions: the determination or the validation of a structural model, as was shown for example in zeolitic type aluminosilicates,[12] metallophosphates,[13] or in MOFs;[14–16] and the understanding of structure–property relationships, including the followup of chemical processes occurring inside the pores.[17] Along the first path, the useful information extracted from NMR experiments comprises the number and multiplicity of crystallographically inequivalent positions in the unit cell as well as the determination of distances, through-space proximities, or through-bond connectivities, which can in turn lead to the determination of the structural building unit (SBU). The relative multiplicity for the various atoms determined by NMR represents the crystallochemical formula of the compound, which is a strong constraint on the possible space groups (SGs) since it has to fit into the available Wyckoff positions. 2D correlation NMR spectra can also unveil the assembly of the SBUs in the 3D space. This last building step, i.e., the stacking of SBUs, can be assisted by computational methods such as the automated assembly of SBUs.[18] The purpose of the second path is essentially to deal with (i) the elucidation of the polymerization mechanism and structure of products based on SBUs,[19] (ii) the establishment of structure–property relationships,[20] and (iii) the monitoring of the catalytic activity and chemical processes occurring in materials.[21]

As the nuclear spin I of quadrupolar nuclei is greater than ½, they interact not only with the external magnetic field but also with the electric field gradient (EFG) generated by their surrounding environment. This leads to typically broadened (and often overlapping) NMR resonances,[22] composed, for a half-integer quadrupolar nucleus (or for half-integer quadrupolar nuclei), of a central transition (CT) $+1/2 \leftrightarrow -1/2$ only affected by the second-order quadrupolar interaction, flanked by satellite transitions (STs) also affected by the first-order quadrupolar interaction. The complexity of NMR spectra of quadrupolar nuclei is associated with the fact that the transposition of NMR techniques developed for ½ spins is not necessarily straightforward, which probably explains why solid-state NMR of quadrupolar nuclei has taken longer to become a systematic characterization technique for porous solids. After the development of magic-angle spinning (MAS) in the late 1950s, several routes have been proposed to deal with quadrupolar nuclei with broad CTs, such as double rotation (DOR),[23] quadrupolar magic-angle turning, or 2D experiments such as dynamic angle spinning,[24] quadrupolar phase-adjusted spinning sideband, etc. However, the major breakthrough technique that has propelled NMR of quadrupolar nuclei in the solid state forward is the multiple-quantum magic-angle spinning (MQMAS) experiment. The MQMAS experiment separates the isotropic part of the quadrupolar interaction in a second indirect dimension, in which the individual resonances become resolved through their isotropic quadrupolar shifts.[22] Further improvements have contributed to the expansion of the range of application of this technique, such as satellite transition magic-angle spinning (STMAS),[9] slow central transition magic-angle spinning (CTMAS),[25] the combination of MQ/STMAS with cross-polarization (CP) or related experiments,[26] as well as signal enhancement methods such as double frequency sweeps,[27] wideband uniform rate and smooth truncation,[28] rotor assisted population transfer,[29] hyperbolic secant,[30] or quadrupolar Carr–Purcell–Meiboom–Gill.[31]

In this chapter, a brief introduction to the nuclei of interest for the study of porous materials is made. The methods referred to here are introduced in several articles of the Encyclopedia of Magnetic Resonance and will not be repeated here. We propose to illustrate, through some selected examples of MOFs and zeolites, the variety of structural characterizations that can be carried out with high-resolution solid-state NMR studies.

21.2 MOST USEFUL NUCLEI

The successful implementation of an NMR experiment for a given nucleus depends on the factors affecting the sensitivity of the NMR response, which are mainly the natural abundance, the value of gyromagnetic ratio γ, and the magnitude of the quadrupolar moment. A high value of the quadrupolar moment may lead to more or less broad lines or even prevent any observation at all when it is excessively large. In contrast to the natural abundance and NMR sensitivity of an isotope, which are constant parameters, the effect of quadrupolar interaction can be minimized either by employing a strong external magnetic field, high-speed MAS and DOR techniques, or methodologies based on multiple-quantum transitions. Difficulties arising from quadrupolar properties themselves are becoming more and more surmountable and today the major limitations mostly concern the poor NMR sensitivity of the low-γ quadrupolar nuclides. There remain therefore many nuclei from the periodic table that are involved in various solids but are still unexplored or poorly investigated. Progress in the field of high-resolution solid-state NMR is in demand to further develop this area of scientific research. Most NMR studies of quadrupolar nuclides in porous materials concern nuclei with spin quantum numbers $I = 3/2$ (e.g., ^{11}B, ^{23}Na, ^{87}Rb, ^{71}Ga) or $I = 5/2$ (e.g., ^{17}O, ^{27}Al). A few exceptions include $I = 1$ nuclei such as ^{2}H, ^{6}Li, and ^{10}B, which have small quadrupolar moments and thus mostly behave like $I = 1/2$ spins. In this section, we provide some features, context, and representative results of the most used quadrupolar NMR nuclei for studying crystalline porous solids.

21.2.1 Aluminium-27

The element Al is magnetically active and has been extensively studied. The isotope aluminum-27 is naturally 100% abundant, has a good receptivity ($D = 1170$ relative to ^{13}C), an intermediate-strength quadrupole moment Q (0.14×10^{-28} m^2), and a spin of 5/2. The ^{27}Al chemical shift range is relatively wide, covering nearly 300 ppm from ca. −50 to 250 ppm (relative to the reference, $Al(H_2O)_6{}^{3+}$), but is reduced by half (with a low-field limit down to 100 ppm) in crystalline Al-based porous materials such as aluminosilicate zeolites,[7,21,22] aluminophosphate AlPOs,[8,9,26,32–36] aluminum carboxylate MOFs,[15,16,37] and the respective related compounds[10,18,19,38] for which only oxygen atoms are present within the first coordination shell of the Al atoms. The sensitivity of the ^{27}Al chemical shift to the number of atoms in the first-neighbor coordination change is high enough so that, at 4.70 T for line widths as great as 2 kHz, speciation of four-, five-, and six-coordinated aluminum in AlO_x environments can be achieved from a 1D ^{27}Al NMR spectrum of such materials (Figure 21.1). Three distinct chemical shifts ranges are usually observed, the shifts moving downfield as the coordination number of the Al atoms decreases from six,[15,16,18–22,27,37,39] to five,[19,21,27,39] and four.[12,17,19–22,27,39] An additional shielding effect is often observed when phosphorus atoms are present in the second coordination shell, typically in AlPOs, resulting in systematic upfield shifts of all these ranges by ~30–20 ppm.[8,26,33,34,36,38,40]

Zeolites are tectosilicates (3D silicate structures), i.e., $(Si + Al)/O = 1/2$, with as many equivalents of exchangeable cations as there are atoms of Al in the aluminosilicate frameworks. Owing to their catalytic importance in their protonated form, they are

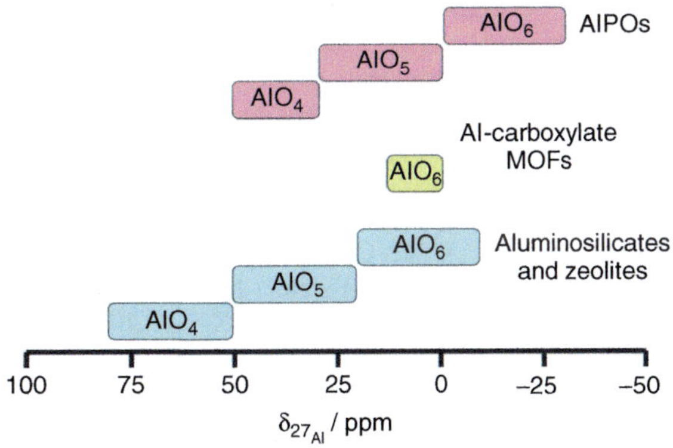

Figure 21.1. A chart showing the gross features of ^{27}Al chemical shift variation in most classes of porous solids.

receiving much attention as evidenced by investigations using solid-state ^{27}Al NMR in conjunction with ^{29}Si NMR. The preparation and dealumination of zeolites can be studied,[17,20,21,27] and aluminum within and without the lattice distinguished[12,17,20,21,39] and its quantitative loss measured.[12,17,20,22,27,39] Since Al in the framework is present exclusively in tetrahedral environments and the Al extra-framework species are often in octahedral coordination, solid-state ^{27}Al MAS has been found to be the technique of choice for their characterization. In order to reduce the EFG, and in turn the quadrupolar coupling constants C_Q, as well as to make the signal detection and quantification easier, samples should be fully hydrated prior to measurements.[39] When surface water is present, the formation of a larger cation (OH_3^+) compensates the framework tetrahedral Al charge and renders a structurally symmetrical environment around the Al–O–Si bridging bonds. The structure of zeolites has also been studied with respect to their catalytic activity[27]; for example, the effect of the flexibility of the lattice aluminum in zeolite beta during the nitration of aromatics has been investigated. The octahedral coordination complex between lattice aluminum and the nitration agent observed during the catalytic process induces the para-selective nitration of toluene through steric hindrance of surface intermediates.

Crystalline aluminophosphate molecular sieves (AlPOs) exhibit not only properties characteristic of zeolites but also novel structural traits that are linked to the ability of aluminum to adopt various coordination states within the lattice: from tetrahedral to octahedral through penta-coordinated aluminum. Thus, their study by solid-state NMR employing ^{27}Al and ^{31}P NMR has been found to be particularly convenient and versatile due to the richness of the chemistry and structural properties occurring in these compounds. As an example, $AlPO_4$-CJ19 has a network of strictly alternating Al units, including AlO_4, AlO_5, AlO_6, and P tetrahedra, including PO_4 and PO_3OH.[36] ^{27}Al MAS NMR, ^{31}P MAS NMR, and $^1H \rightarrow {}^{31}P$ CPMAS NMR experiments have shown that solid-state NMR techniques are an effective complement to X-ray diffraction (XRD) analysis for the structural elucidation of this compound. Numerous pieces of information can be provided by the NMR crystallography of this class of materials, including enumeration of crystallographic sites, determination of their relative population, identification of their coordination state, and establishment of their connectivity schemes. From 1D to 2D and 3D MAS NMR experiments involving ^{31}P and ^{27}Al, it has been shown that the structure of the microporous aluminophosphate $AlPO_4$-40 contains at least four times more sites than expected in various environments including octahedral, penta-coordinated, and two types of tetrahedral Al sites.[35] Solid-state NMR is also capable of detecting the presence of amorphous impurities that diffraction methods cannot detect. This was found, for example, in some aluminum oxyfluorinated samples of $AlPO_4$-CJ2 where signals from amorphous impurities were filtered out using $^1H \rightarrow {}^{27}Al$ CP-MQMAS while signals of the well-crystallized fraction of $AlPO_4$-CJ2 were retained.[26] Similarly, signals of the amorphous "AlO_xF_y" were isolated using $^{19}F \rightarrow {}^{27}Al$ CP-MQMAS.

Porous aluminum-organic frameworks were discovered recently and, since the pioneering work of the groups of Férey and Taulelle in the late 2000s, only a couple of structures have been determined, mostly Materials of the Institute Lavoisier (MIL-*n*) materials.[15,16,37] To the best of our knowledge, they are exclusively aromatic carboxylate-based compounds, reflecting the low reactivity in the coordination chemistry of the trivalent Al^{3+} cation compared to many transition metals. In most cases, the carboxylate functional groups coordinate to Al through a chelating mode, which is quite different from those of Al with phosphates or silicates in AlPOs or zeolites where one phosphate or silicate interacts with a neighboring Al strictly through one oxygen atom. Furthermore, aluminum atoms always are in an octahedral environment as isolated aluminum clusters[37] or infinite corner-[15] or edge-[16] sharing aluminum chains connected to each other through carboxylates. This has a direct consequence on the flexibility of the structure around the aluminum cation, which appears much more rigid in MOF materials in comparison to zeolites and zeotypes constructed with tetrahedral units. Consequently, ^{27}Al NMR spectra often experience line broadening due to the strong quadrupolar interaction in these compounds with quadrupolar coupling constants up to ~15 MHz. Solid-state ^{27}Al NMR in combination with ^{1}H and ^{13}C MAS NMR has been used to assist structure determination by nonconventional methods such as lattice energy minimizations combined with Rietveld refinement or microdiffraction using synchrotron radiation. Most Al-based MOF materials crystallize as polycrystalline powdered solids, preventing any investigation by single-crystal diffraction.

21.2.2 Sodium-23

^{23}Na also has 100% natural abundance with a good receptivity ($D = 525$ relative to ^{13}C) and a relatively small quadrupole moment Q (0.1×10^{-28} m^2) with a spin quantum number I of 3/2. Despite characteristics similar to those of ^{27}Al, ^{23}Na NMR has not received the same interest as ^{27}Al, most probably due to the narrower chemical shift range for the Na^+ cation covering only ca. 30 ppm (from −10 to 20 ppm) with respect to 1 M aqueous NaCl (0 ppm). Nevertheless, valuable information can be obtained from studies of other NMR parameters, such as line shape analysis, relaxation rates, and dynamic behavior. For instance, the dependence of chemical shifts and line widths on temperature can yield information about the motion as well as dynamic and diffusion properties of Na^+ cations.[41] In contrast to aluminum, Na cations are involved in zeolites and related microporous solids as extraframework species, usually mobile and exchangeable, compensating for the lattice charge excess. Their localization by conventional diffraction methods is therefore difficult, but can be achieved by solid-state NMR. There are rare cases where partial substitution by the alkali cations occurs in some metallophosphate frameworks such as the super-sodalite aluminum and gallium phosphates MIL-74 incorporating lithium and sodium in their lattices.[38] In those cases, both single-crystal XRD and solid-state MAS NMR spectroscopies (^{6}Li, ^{7}Li, ^{23}Na, ^{27}Al, ^{31}P) were used to establish the structure.

During the past decades, several MAS-based NMR techniques that utilize the heteronuclear dipolar coupling between quadrupolar and spin-1/2 nuclei were developed to ascertain connectivity and spatial interaction information. For example, solid-state $^{31}P/^{23}Na$ TRAnsfer Population in Double Resonance (TRAPDOR) NMR experiments have been used to investigate the spatial distribution of sodium cations with respect to the phosphate backbone for a series of sodium phosphate glasses.[34] From the $^{31}P/^{23}Na$ TRAPDOR data gathered, information about the medium-range order in these materials was obtained. Alternatively, rotational echo double resonance (REDOR) experiments can also be performed to quantify ^{31}P–^{23}Na dipolar interactions, and, by measuring second moments, interatomic distances have also been calculated.[8] These techniques have extensively been applied on glassy materials and could be transferred to crystalline microporous metallophosphates containing sodium. In the case of MIL-74, to provide further evidence for ^{31}P signal assignments, two additional experiments have been carried out, i.e., $^{23}Na \rightarrow {}^{31}P$ CPMAS and $^{27}Al \rightarrow {}^{31}P$ CPMAS (in both 1D and 2D HETeronuclear CORrelation (HETCOR) experiments), in conjunction with a single-pulse experiment.[38] In the $^{23}Na \rightarrow {}^{31}P$ CPMAS spectrum, only the signal from phosphorus directly connected to sodium through oxygen bonds appears, confirming its assignment. Furthermore, it is only with the $^{23}Na \rightarrow {}^{31}P$ CPMAS spectrum that one has the first evidence that the alkali cations have been really incorporated to the framework.

The study of the layered chiolite structure, $Na_5Al_3F_{14}$, was presented as a model for a general strategy integrating NMR, simulations, and diffraction for solving inorganic structures such as microporous solids.[18] An important feature of the approach lies in the combination of NMR-based structural information together with prior knowledge of the structure of interest (i.e., chemical composition, crystal chemical information) into a computational search of candidate crystal structures. The information that ^{19}F, ^{23}Na, and ^{27}Al MAS NMR measurements can provide concerns the number of inequivalent fluorine, aluminum, and sodium atoms per unit cell. For example, the ^{23}Na MAS spectrum reveals that there are two inequivalent cations with relative multiplicities 1:4. The investigation of the heteroatomic fluorine–aluminum and fluorine–sodium subnetwork can complement the analysis of the simple MAS spectra of each nuclear species. This can be achieved by means of a HETCOR experiment with CP from ^{19}F to ^{23}Na and from ^{19}F to ^{27}Al.

21.2.3 Oxygen-17

With a natural abundance of only 0.037%, a receptivity D that is 0.061 times that of ^{13}C, a moderate quadrupole moment Q (-0.026×10^{-28} m^2), and a spin of 5/2, ^{17}O, the only magnetic isotope of oxygen, seems a priori poorly suited for study by NMR. Nonetheless, the chemical importance of the element oxygen has aroused considerable interest in its NMR spectroscopy, and following the earliest pioneering studies, the advent of modern spectrometers associated with high-field magnets has now established the technique on a routine basis. Oxygen would be a very important molecular probe for investigating structural properties of porous solids; unfortunately, the very low natural abundance of its NMR-active isotope hampers its extensive application, often making a preliminary isotopic enrichment necessary. With a chemical shift range of over 1000 ppm, considerable spectral dispersion is therefore available. For mineral oxides, this range is reduced considerably and resonances occur usually between 30 and 200 ppm with respect to water which is used as standard. From the results on these materials, quadrupole coupling constants C_Q ranging from 1 to 8 MHz were obtained, together with chemical shift anisotropies Δ_{CS} of up to 50 ppm.[9,42–44] Typically, results are those obtained from forsterite, Mg_2SiO_4,[42] in which three crystallographically distinct oxygen sites were perfectly resolved using MQMAS or STMAS techniques providing various data including isotopic chemical shifts (δ_{iso}), quadrupole coupling constant (C_Q), EFG tensor asymmetry parameters (η_Q), and chemical shift anisotropy (Δ_{CS}) and asymmetry (η_{CS}). This represents much more information than available from ^{29}Si NMR since the material has only one type of silicon site.

Thanks to enriched samples and to the development of high-resolution techniques, the amount of information from ^{17}O NMR spectra of silicates is rapidly increasing. Nevertheless, the spectra are still very difficult to interpret, owing in part to the lack of empirical correlations to enable the assignment of chemical shifts to each site. Indeed, recent first-principles prediction of ^{17}O NMR parameters in silica polymorphs by Pedone *et al.*[44] has shown that the convergence with respect to cluster size of the computed ^{17}O isotropic shift and the principal components of the chemical shift tensor, together with ^{17}O quadrupolar coupling constants, is system-dependent. In particular, it is hard to achieve for oxygen atoms that are part of small n-membered rings ($n < 6$). Moreover, in a separate study on sodium silicates, it has also been observed that the isotropic chemical shift could not be simply related to structural parameters of the first and second coordination sphere, as well as of bridging oxygen atoms (BOs).[43] Nevertheless, for BOs, the quadrupolar coupling constant C_Q and asymmetry parameter η_Q are primarily influenced by the <Si–O–Si> bond angle. This dependence can be described by relationships previously derived from ab initio calculations using model clusters. However, no clear influence of the coordinating sodium on the quadrupolar parameters could be established.

21.2.4 Gallium-71

Gallium has two magnetically active isotopes, ^{69}Ga (60.5% natural abundance, receptivity $D = 246$ relative to ^{13}C, quadrupole moment $Q = 0.18 \times 10^{-28}$ m^2) and ^{71}Ga (39.5% natural abundance, receptivity $D = 319$ relative to ^{13}C, quadrupole moment $Q = 0.11 \times 10^{-28}$ m^2), both with a spin quantum number I of 3/2. The least abundant ^{71}Ga is generally preferred for NMR work owing to its higher receptivity and smaller quadrupolar moment Q, which results in signal-to-noise ratio expected from this nucleus about four times greater than for ^{69}Ga. The line widths obtained are much greater than with ^{27}Al and the broader lines mean that the resolution of individual resonances is usually more difficult, though this is aided by the increased chemical shift dispersion of the gallium nucleus. Indeed, chemical shifts cover a wide range from −700 to 700 ppm with respect to the standard 1 M aqueous $Ga(NO_3)_3$, and for compounds of interest here, i.e., fluorides[23] and oxides,[13,14,23,25] this range is contracted to within −60 to 120 ppm. As for ^{27}Al, there is a well-established and confirmed correlation that links gallium chemical shifts for four-, five-, and sixfold coordination in gallo-fluorides, -phosphates, and -carboxylates. Thus, usually the chemical shifts for 4-coordinate Ga falls in the range 90–120 ppm,[13] which, at static magnetic fields higher than 4.7 T for line widths lower than 5 kHz, can be distinguished from 6-coordinate Ga at −60 to 10 ppm,[13,14] and 5-coordinate Ga resonates between these regions at 0–30 ppm.[23] The prevalence of broad lines means also that the problem of non-observable resonances is even more acute for gallium. Nevertheless, with the enormous progress in the field of rapid pulsed Fourier transform instruments and NMR methodologies, these difficulties are becoming nowadays less and less problematic and are going to be resolved progressively, whilst limitations inherent to the nuclear properties could persist.

Recently, a few ^{71}Ga MAS NMR studies were conducted on some MOF-type gallium carboxylates.[14,25] It is interesting to note that the estimated quadrupolar coupling constants vary considerably from one compound to another: in the MIL-96 gallium trimesate, they are much smaller than those observed in MIL-53 and MIL-61 gallium carboxylates, as clearly seen from the comparison of the ^{71}Ga MAS NMR spectrum of MIL-96 recorded at 17.6 T with those of MIL-53(Ga) and MIL-61(Ga) acquired at the same magnetic field.[14] While the width of the former resonance is a few kilohertz, those of the latter are a few hundreds of kilohertz, and thus split into several spinning sidebands. It seems that the gallium environment is less distorted in the case of the MIL-96(Ga) material. In the gallium carboxylates MIL-120 and MIL-124, the two gallium sites have been resolved using the recently developed slow CT-MAS method, which also allowed the determination of the ^{71}Ga quadrupolar parameters.[25] Slow CTMAS is easy to implement, and requires only conventional MAS probes, a low radio frequency field, and low magnetic fields, and pushes even further the limit of detection of quadrupolar nuclei with very large C_Qs. This method should meet larger uses because it is technically less demanding than the usual MQMAS or STMAS experiments, even though some processing efforts are still needed to get it to the level of a routine analytical method.

21.2.5 Other Nuclei

Other attractive NMR nuclei, such as ^{11}B, ^{87}Rb, or ^{2}H, have been found to be of particular interest for studying microporous solids when the corresponding elements are incorporated either in the framework or as extra-framework cations, or are present in occluded species. Though ^{11}B and ^{87}Rb nuclei have relatively high natural abundances (80.4 and 27.9%, respectively), the ^{2}H natural abundance is however much lower (0.015%). Furthermore, NMR of quadrupolar nuclei with non-integer spin such as ^{11}B and ^{87}Rb (i.e., $I = 3/2$) can be more advantageous when compared with ^{2}H ($I = 1$), as under certain MAS conditions the $+1/2 \rightarrow -1/2$ transition can result in a sharp resonance avoiding broadening arising from the second-order quadrupolar interaction. As an example of recent applications, the transformation of porous glass spheres into MFI spheres in a synthesis with low water content (3 moles H_2O per mole TO_2 with T = Si, B, Al) was observed by in situ ^{11}B MAS NMR spectroscopy in two synthesis procedures.[45] The yield of crystalline material could be determined by means of the intensity of a narrow signal at 3.7 ppm in the ^{11}B MAS NMR spectra. In another study, the structure of the intermediate species in the clear sol synthesis of the MFI framework (zeolite ZSM-5) has been investigated using the characteristic ^{13}C, ^{14}N, ^{15}N, and ^{2}H spectra of the tetrapropylammonium (TPA) template ions, as well as the ^{29}Si spectra of the silicate species present in samples which can be isolated by centrifugation.[46] Some of the most informative NMR experiments in terms of detecting the presence of the MFI framework appear to involve ^{2}H NMR, including ^{29}Si, and observe CP REDOR with dephasing by ^{2}H. For these experiments, the TPA was regioselectively deuterated as a labeling approach, allowing a specific interaction to be probed.

21.3 NMR CHARACTERIZATION STRATEGIES FOR NANOPOROUS MATERIALS

There are several major issues in understanding the structure or the specific properties of nanoporous crystals. First, to be nanoporous these crystals must be templated, and subsequently emptied so that their pores become accessible. Second, the nanoporous compounds are usually flexible, with low or very large density variation. Finally, nanoporous crystals can be fully chemically ordered as for some MOFs, or not as in zeolites containing aluminum. All these aspects make the structural resolution from diffraction data only difficult.

Though a wide variety of NMR experiments using a wide range of nuclei have been performed, especially using quadrupolar nuclei directly or in combination with heteronuclear experiments, the main trends, up to now, that make NMR of such materials important can be grouped into a few areas.

The first and largest field of characterization concerns the assessment of the quality of a structural model that has been established by diffraction. It usually combines structure modeling and quantum calculations of NMR observables, which can be compared with experimental measurements. In this case, both diffraction and NMR data can potentially be used as a basis to build the structural model and as cost functions for the structure refinement. A second important area of nanoporous characterization relies on accessing by NMR atoms that are indistinguishable by diffraction such as the isoelectronic pairs Al/Si in zeolites or OH/F atoms in fluoride synthesis of crystals. A third class of problems in structure resolution is the influence of extra-framework entities, which modulate the framework symmetries. Templating structure-directing agents or templating by water molecules plays a role in the differentiation of atoms in the framework leading to a lifting of crystallographic inequivalence in many cases. There is another class of materials that is usually described by diffraction as disordered or lacking long-range order but which is in fact ordered with different degrees of periodicity (one degree for the framework, one degree for the templating agents) which are not necessarily that of the crystal.

A small selection of examples will illustrate the strategies that can be undertaken to access the information in the crystal structure at the periodic and nonperiodic levels. Many others aspects of these characterization strategies are predicted to develop at a high pace, owing to the remarkable efficiency of interweaving different structural description/characterization methods into specifically targeted strategies like "generalized crystallography" or "SMARTER crystallography" (Structure elucidation by coMbining mAgnetic resonance, compuTation modEling and diffRaction, SMARTER).

21.3.1 Assessment of Structural Models

21.3.1.1 NMR Crystallography of Crystals Containing Quadrupolar Nuclei

With the recent advent of quantum periodic computations, and especially those allowing NMR observables to be computed, one of the fast emerging approaches using NMR of quadrupolar nuclei for nanoporous materials is—from a structure model established by XRD, modeling, or NMR—to compare the computed NMR observables to those measured experimentally. This approach is a part of a method termed "NMR crystallography". The quadrupolar interaction is especially useful from this point of view, owing to its intrinsically longer range compared with other interactions (see Chapter 3).[47] Additionally, the NMR line shape for static samples or under MAS can provide access to the full tensor orientation of the point group of any site in the crystallographic frame. From a quantum point of view, calculating the quadrupolar tensor has nowadays reached a high level of confidence with several computational codes.[43] In some recent studies, the structures of different model compounds containing aluminum atoms have been established by NMR crystallography. In the case of chiolite,[18] the structural model could be established by NMR using two complementary strategies: measuring the "Wyckoff spectrum" of the crystallochemical formula, and establishing the proximities and/or connectivities between the central metal atoms of topological polyhedra. In this example, the final cost function chosen was the isotropic chemical shifts of fluorine atoms. In the case of nanoporous compounds, this strategy has not yet been carried to completion, but many studies are considering a fraction of the network to improve the detailed description of the structure.[5,42,48] To achieve a much better structural description of a complex structure, a combination of techniques is required, described in the following part as SMARTER crystallography.

21.3.1.2 SMARTER Crystallography of Nanoporous Compounds

Nanoporous materials are mainly formed with the presence of extra-framework entities and as a consequence exhibit a flexibility of the crystals, leading to more difficult structure elucidation. In such situations, one combines structural modeling, diffraction, and NMR, the so-called SMARTER crystallography protocols. Now, each aspect of structure elucidation plays an approximately equivalent role and alternates in an iterative strategy of structure description. The first noticeable contribution to this combination of methods was given by Wright *et al.* for a methylphosphonate, using powder diffraction data, a simulated annealing method for model structure determination with the Fox software,[49] and further Rietveld refinement. To establish the content of the unit cell, elemental analysis, NMR, thermogravimetric analysis, and density measurements were used.[50] When NMR observables computed from the properly established structure model converge towards the experimental observed values, the structure model is considered as valid. As a current example, an Al fluorinated MOF of MIL-53(Al) topology, $ZnAlF_5.[C_2N_3H_3]$, has been studied by diffraction and NMR, and the model structure provided by diffraction was considered good enough to describe the crystal structure in the first approximation. After optimizing the atomic

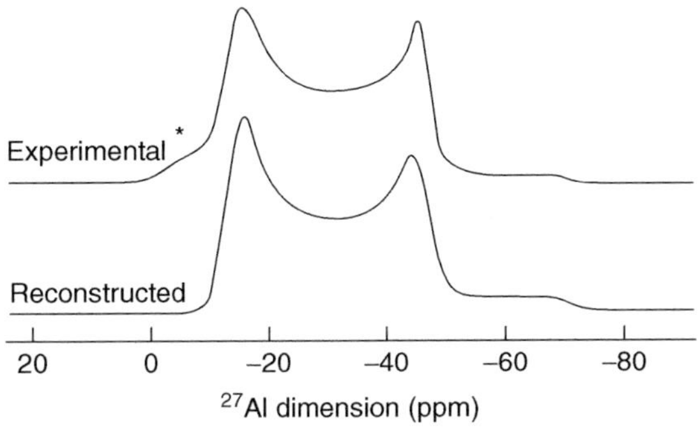

Figure 21.2. Experimental and reconstructed central transition of the ^{27}Al MAS NMR spectrum of $ZnAlF_5 \cdot [C_2N_3H_3]$. The star indicates the presence of an unidentified impurity. (Reproduced from Ref. 51. © The Royal Society of Chemistry, 2011.)

positions, the quadrupolar parameters were recalculated, and as it fitted the experimental NMR measurement (Figure 21.2), the structural model was validated with a much better accuracy for the atomic positions. Computing such NMR observables with full quantum packages is not yet sufficiently fast to be used as a refinement technique; rather it typically leads to an evaluation or assessment of the structural model. The nature of the crystal structure descriptions obtained may contain approximate solutions for the periodic part of the structure with additional descriptions required for the nonperiodic parts.

In the case of ULM-18,[23] the crystal structure can be described by an assembly of units of double 4-rings (D4R), with alternation of Ga and P, in which one of the gallium atoms carries one additional PO_4 group. By ^{71}Ga DOR, it was possible to demonstrate that three Ga atoms were directly connected to the fluorine atom located at the center of the D4R unit. Additional quantum calculations showed that this fluorine is also connected to a hydrogen atom, leading to an activated HF molecule covalently connected to three gallium atoms of the D4R unit. This appears to be a general feature of D4R units in nanoporous materials for Ga as well as for Al. The presence of hydrogen inside the D4R, as well as the three covalent connections to gallium atoms, could not be elucidated from the diffraction experiments. The efficiency of combining diffraction for structure model elaboration with NMR and quantum computations to complete the description of the compound has been demonstrated.

21.3.2 Substitution of Atoms in an Identical Topological Net

21.3.2.1 Zeolites and Si/Al Substitution

Zeolites comprise a class of specific crystals that are often positionally well characterized without knowing exactly which atom occupies which particular crystallographic site. The specific reason for such a situation lies in the fact that silicon and aluminum atoms are isoelectronic and therefore difficult to distinguish by diffraction. However, measuring the ^{29}Si MAS spectra has enabled the indirect measurement of the Si/Al ratio by extracting the intensities of the $Q^{4,n}$, where Q^4 describes a SiO_4 group connected by its four surrounding oxygen atoms to further tetrahedral species and "*n*" indicates the number of aluminum (rather than Si) atoms connected. However, this was perhaps the first field that benefited extensively from the use of the newly established method of MQMAS to measure the nature of the aluminum atomic environments. MQMAS provided much better resolution in the isotropic dimension of aluminum spectra which were not well resolved simply by MAS. This then enables an analysis of the distributions of both chemical shift and quadrupolar coupling.

21.3.2.2 Fluorometallophosphates and F/OH Substitution

Another family of isoelectronic atomic environments is the pair F/OH. This is also an extremely general problem in crystal chemistry, particularly geochemistry. Though it may seem paradoxical to cover this specific problem in the context of quadrupolar nuclei, it is, however, very strongly related. As a matter of fact, for most cases where such substitution can take place, it alters the environment of the metal to which the F atoms or OH groups are bonded, generating a distribution of metal sites. As the quadrupolar nuclei are usually the least resolved, a direct determination is often not the most efficient method. In the case of $AlPO_4$-CJ2,[26] it was possible to demonstrate the level of OH/F substitution on only one bond, the one bridging the two aluminum atoms of the four-membered ring composed of two phosphorus atoms alternating with two aluminum atoms. One of the aluminum species, though six coordinated, has only four connections to the network, the remaining two being the OH/F bridge and a terminal F. The terminal fluorine atom is not substituted, whereas the bridging bond may exhibit, as a function of sample preparation, various degrees of substitution. The $^{19}F/^{27}Al$ HETCOR NMR spectrum provides directly and completely this information. The resolution of the HETCOR experiment can still be improved, in order to get a fully resolved situation, by increasing the magnetic field or by using new pulse sequences, heteronuclear decoupling schemes, or by using MQMAS in the indirect dimension.

21.3.2.3 Fluoroaluminophosphates with Zeolitic Net: The SAPO-34 Case

SAPO-34 is an archetypal zeolitic net, with the CHA topology, found as the natural silicate chabazite. Its silicon framework can be substituted by both aluminum and phosphorus atoms. Diffraction results indicate an invariant CHA net with aluminum and phosphorus substitution as well as fluorine atoms incorporated into the structure. SAPO-34 exhibits therefore a combination of Al–Si substitution, Si–P substitution, and localization of fluorine in the network. This nanoporous compound with exceptional catalytic activity represents a dual facetted material with a well-defined average framework represented by the CHA topology and a nonperiodic arrangement of the Si, Al, P, and F atoms. Several studies have probed its framework organization using advanced quadrupolar-based NMR experiments.[10,51]

In a first in situ study, the species responsible for the crystal formation were elucidated and the sequence of events leading to a proposition of the crystal mechanism formation was determined (Figure 21.3).

In a second study,[51] the full aluminophosphate $AlPO_4$-34, with the chabazite topology, was used to establish to the method of measurement of the heteronuclear dipolar interactions between quadrupolar and spin-1/2 nuclei in solids by MQ-REDOR, providing proximity as well as distance information. The ^{27}Al MQ-REDOR experiment has been demonstrated on a sample of fluorinated triclinic chabazite-like $AlPO_4$-CHA aluminophosphate, which was used in its as-synthesized form with the morpholinium cation $C_4H_{10}NO^+$ as a template. The microcrystalline structure of this compound, which was established using diffraction with a synchrotron radiation source,

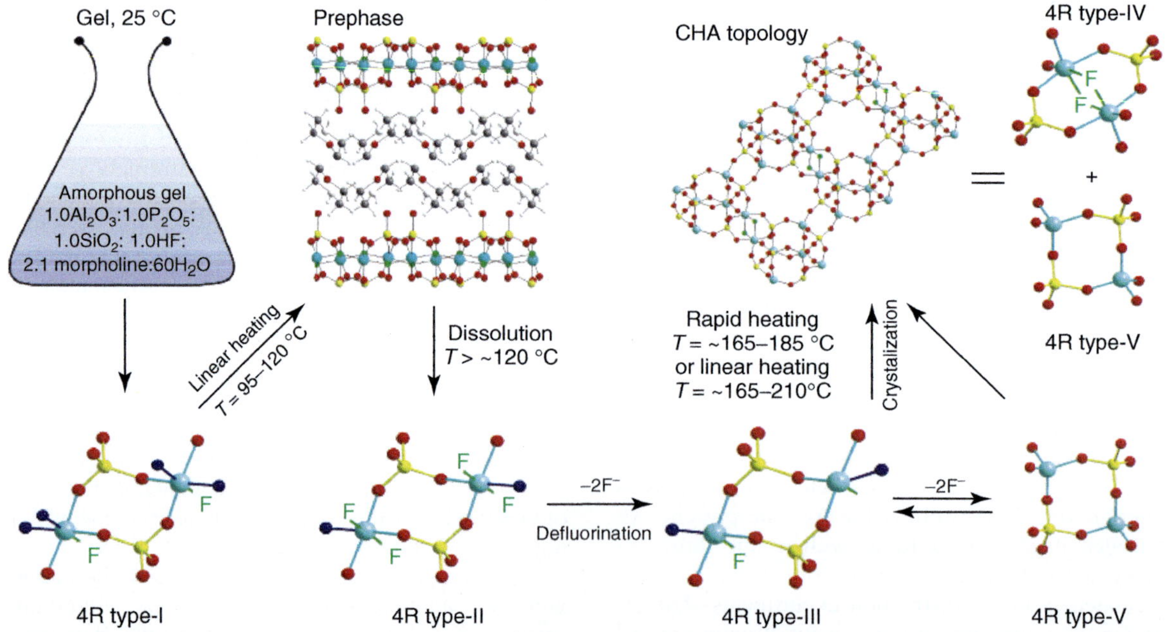

Figure 21.3. Schematic overview of the formation mechanism of SAPO-34 including the most important equilibria and synthesis parameters obtained from in situ NMR and in situ XRD analysis during the hydrothermal synthesis. (Reproduced from Ref. 59. © American Chemical Society, 2003.)

contains three equally populated aluminum sites: an Al1 site octahedrally coordinated to four oxygen and two fluorine atoms located at rAl1–F1 = 1.88 Å and rAl1–F2 = 1.96 Å; and two sites labeled Al2 and Al3 representing tetrahedral aluminum. The nearest neighbor Al–F distances for the tetrahedral sites are rAl2–F = 3.94 Å and rAl3–F = 5.72 Å, respectively. The following values for the isotropic chemical shift δ_{iso} and the second-order quadrupolar effect [SOQE $= C_Q(1 + {\eta_Q}^2/3)^{1/2}$] parameter were measured in earlier MQMAS and CP MQMAS studies of this sample: δ_{iso} − 4.5, 43.9, and 47.6 ppm and SOQE 0.9, 2.0, and 2.6 MHz for sites Al1, Al2, and Al3, respectively.[52]

In a third study,[10] by using several solid-state NMR techniques, the authors were able to characterize the several phases obtained at different stages of SAPO-34 crystallization in a fluoride medium. For triclinic SAPO-34, the results have been interpreted on the basis of the proposed structure and confirmed the assumed structural analogy of the compound to triclinic $AlPO_4$-34. The initial gel, obtained without heating, was also characterized. It seems that a small amount of fluoroaluminophosphate amorphous material forms immediately upon mixing the sources of P, Al, and HF. Si exists largely as unreacted silica.

For the prephase of SAPO-34, the NMR data indicate that there is one tetrahedral Al and two octahedral Al sites. The two octahedral Al sites likely have very similar chemical environments but different local geometry. There is only one P site with a small average P–O–Al angle. The chemical environment of the P is likely to be $P(OAl)_3(OH)$. There is only one terminal fluorine site and its environment is different from that in triclinic SAPO-34. The chemical shift, line width and line shape of the P and octahedral Al at 6 ppm are almost identical to those of P(3) and Al(1) in triclinic SAPO-34, suggesting that the 4Rs involving P(3) and Al(1) may already exist in the prephase and are unaffected by the transformation to SAPO-34.

Combining in situ experiments for identifying the possible units of construction using all nuclei involved in the synthesis, then MQMAS-REDOR to identify the proximity of atoms, and further the analysis of the different solid phases, confirms the evolution of the building units through the different steps of solid formation. This crystal with a chabazite topology exhibits all the aspects of an invariant network with variable atomic substitution.

21.3.3 Extra-framework Parts of Nanoporous Crystals

21.3.3.1 Water Structured Within Pores of Nanoporous Crystals

The MIL-74 structure ($Zn_6Al_{12}P_{24}O_{96}\cdot[N(CH_2CH_2NH_3)_3]_8\cdot(H_2O)_{34}$)[40] is quite a special compound that may be viewed as a super-sodalite network with $Al[Zn_2Al_2P_4O_{24}]$ super-square units (Figure 21.4). This is a beautiful example of the "decoration of the sodalite network" in the sense of O'Keeffe.[53] This structure has the lowest framework density known to date, in which the cationic templates *tren* ($[N(CH_2CH_2NH_3)_3]^{3+}$) are included as part of the framework and cover the hexagonal pores of the sodalite cage. The use of organic molecules as templating agents is widespread in zeolite synthesis.[54] Templating fulfils a threefold role by directing the structure, controlling the shape, and fixing the basicity of the synthesis.[55] Although selectivity between the templating agent and the framework has been established in the case of the ZSM-18 zeolite,[56] it is likely that many other frameworks may be synthesized without such selectivity. Accordingly, even if *tren* cations obstruct the channels, they nevertheless leave the pore cavity accessible to water molecules. This leads to the other specific feature of this structure, that is, the occurrence of a highly symmetrical water cluster comprising 17 water molecules located in the sodalite cage and organized in an amazingly well-defined pattern of pentagons. Determination, by

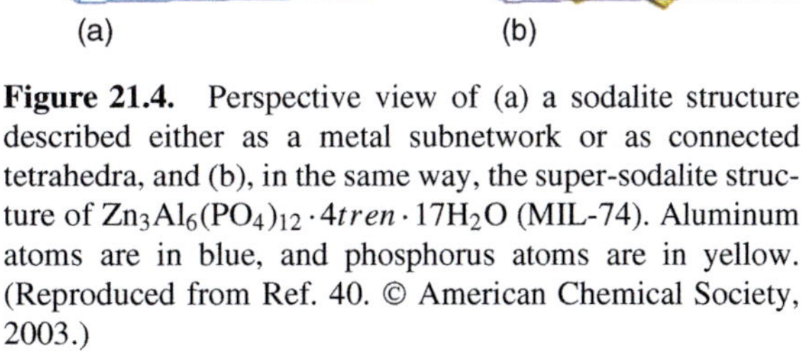

Figure 21.4. Perspective view of (a) a sodalite structure described either as a metal subnetwork or as connected tetrahedra, and (b), in the same way, the super-sodalite structure of $Zn_3Al_6(PO_4)_{12}\cdot 4$*tren*$\cdot 17H_2O$ (MIL-74). Aluminum atoms are in blue, and phosphorus atoms are in yellow. (Reproduced from Ref. 40. © American Chemical Society, 2003.)

diffraction, of the number and the nature of the structure of the water clusters in the pores of nanoporous materials is a challenging issue, owing both to static and motional disorder. Consequently, such water networks have been determined accurately only for very few materials. The structure of bound water in VPI-5 has been successfully determined at low temperature, and shows an exceptional water-based triple helix.[57] The MIL-74 structure is therefore quite instructive with its very well-defined aggregate of 17 water molecules. This observation raises the interesting question of the role played by water molecules during crystal nucleation and growth. In most advanced efforts of template design for better selectivity, water is never considered as a possible template or co-template. It is supposed, at best, to play the role of a void-filling agent.

MIL-50,[3] an open-framework structure, has been obtained by assembling hexameric units of the Ga_3P_3 type and consists of large 18-ring channels encapsulating 1,6-diaminohexane species and water molecules. In addition, common to several already identified structures, rubidium atoms were incorporated in the framework and trapped within the small six-ring channels. This is a configuration that has not been yet observed in comparable open-framework compounds. In the pores of such open frameworks, more space than usual remains, in which some water molecules are localized. The mobile water molecules are inside the 18-ring cavities and move inside these along the *c*-axis. Rubidium, in contrast, is localized in the six-ring channels along the *c*-axis. There is, therefore, no obvious reason why the water should meet rubidium. The latter stays therefore "dry". MIL-50 appears as a fascinating material, made of hexameric units, including "dry" rubidium atoms and hopping water within small ponds attached to the inorganic framework in the vicinity of the phosphate groups P3 and P5. The water sublattice is formed of a periodic pattern of tiny ponds of lower symmetry than that of the framework. The interaction is weak enough to allow water to exchange within the available sites of one small pond. The hopping exchange modulates the DQ ^{31}P signals of P3 specifically, weakening the DQ intensities. Though detected by NMR, this symmetry lowering is not detected in diffraction. MIL-50 is thus a very nice example for demonstrating how different SGs may be invoked for a single structural model depending on the sensitivity of the measurement to the actual experimental parameters, here XRD and NMR. It shows also that hybrid situations where different symmetries can coexist within the same unit cell can very well be encountered, leading to partial symmetry lowering. Solid-state NMR appears well suited for pointing out the existence of such a superposition of SGs within the same unit cell. Actually, this example demonstrates that only a SMARTER crystallography approach is able to handle this situation.

21.3.3.2 Extra-framework Characterization in MOFs

The extra-framework characterization of nanoporous materials is still in its infancy. In fact, from zeolites to MOFs most of the time the structural characterization of the crystals is performed on calcined (for the zeolite) or "activated" (for the MOFs) compounds, meaning that the pores of the nanoporous compounds are emptied to study their structures. In actuality, the templating effect necessary to generate a crystal with a given topology is still today not fully understood because of its multifold aspect. Furthermore, as in many cases an identical topology can be obtained with different structure-directing agents, one expects that the pore-filling may be achieved with loose to tight embedding of the template. This leads in many cases to nonperiodic arrangements of the templating agents. Though it has a specific impact on the quadrupolar nuclei in the framework, a considerable amount of work needs to be performed between modeling, diffraction, and NMR to correlate the nonperiodic order of the templating agents and the NMR parameters of the quadrupolar nuclei of these crystals.

21.3.4 Order–Disorder, Motions, and Flexibility in Crystals

21.3.4.1 Metal Substitution in a Super Sodalite Framework

MIL-74 a zinc-based, aluminum phosphate, $Zn_3Al_6(PO_4)_{12}\cdot 4tren\cdot 17H_2O$, exhibits a super-sodalite topology with the lowest framework density known.[40] This super-sodalite topology has many facets: its ability to accommodate many divalent and monovalent cations in the network, and an intriguing water cluster. In addition, the templating agents are not located inside the pores but belong to the framework. It is, therefore, a true intermediate compound between a classical aluminophosphate nanoporous material and a metal-organic framework, but with cationic organic linkers instead of the classical anionic linkers. The ordering of Zn and Al has been studied in detail by combining XRD and NMR. Both aluminum and phosphorus NMR spectra exhibit distributions of atoms, represented by a fractional occupancy factor of the sites occupied by aluminum or zinc atoms. The ^{31}P NMR spectra were, however, much better resolved than the ^{27}Al NMR spectra. The "disorder" of this structure has been shown to be reproducible and thermodynamically controlled. This type of reproducible homogeneous distribution of atoms may be of utmost importance in the construction of periodic materials with a reproducible deviation from perfect periodicity. The symmetry chosen to describe the crystal structure, here I-43m or P-43n, appears to be a way to take into account the amount of information experimentally collected. The symmetry must be lowered here, from I-43m to P-43n, to accommodate the increased amount of information when adding to XRD the NMR results. To describe the order in this crystal, one must notice that minimizing the enthalpic and entropic part of the crystal energy is not permitted within the same SG, so a compromise solution is found in the crystal between both minima defining unique site occupancy factors. Describing an ordered crystal fulfilling such conditions cannot be achieved within the 230 3D SGs.

21.3.4.2 Motion and Flexibility in MOFs

Motion is an important aspect of nanoporous materials. They exhibit small to large amplitude flexibility, at least as much as in zeolites, but normally considerably more. Different types of motion can take place inside a nanoporous materials: a random motion of the structure with constrained correlated positions to allow for a global bond conservation, comparable to those observed in zeolites, and local

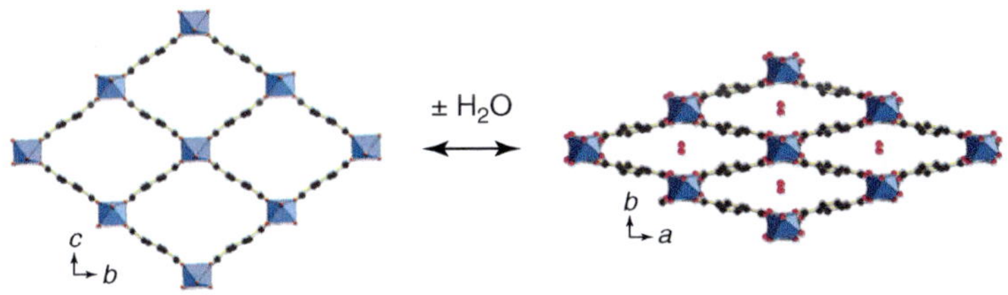

MIL-53*ht* (Al) or Al(OH)[O_2C-C_6H_4-CO_2]

Calcination at 275 °C removes the templates irreversibly and leaves the framework topology unchanged only distances are affected.
Channel dimensions: 8.5 × 8.5 Å^2.
Orthorhombic system, space group *Imma* (n°74), a = 6.608(1) Å, b = 16.675(3) Å, c = 12.813(2) Å.

MIL-53*ht* (Al) or Al(OH)[O_2C-C_6H_4-CO_2].$H_2$0

At room temperature, reversible adsorption of water hydrogen-bonded to the hydrophilic part of the network leads to 2.6 × 13.6 Å^2 channels. Monoclinic system, space group *Cc* (n°9), a = 19.513(2) Å, b = 7.612(1) Å, c = 6.576(1) Å, b = 104.24(1)°.

Figure 21.5. Breathing effect in MIL-53 structure upon water adsorption–desorption.

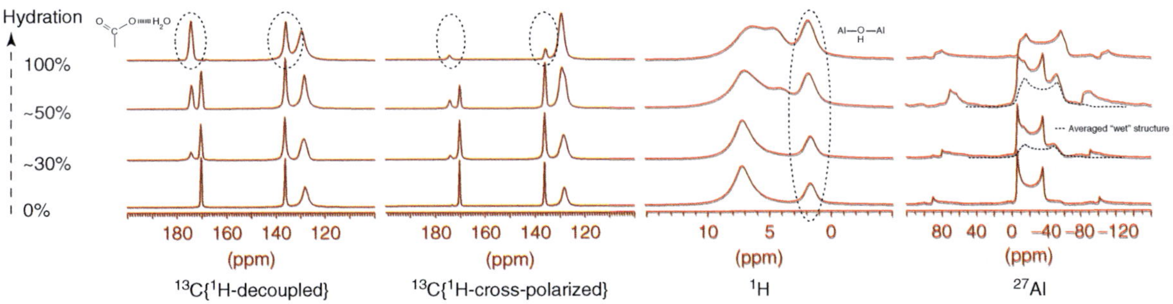

Figure 21.6. Calcination and hydration of the MIL-53 MOF studied by solid-state NMR.

motions such as, for example, vibration of the aromatic part of the linker.[58] The MIL-53 family[6] is built up from the interconnection of infinite trans chains of corner-sharing (via OH groups) $MO_4(OH)_2$ octahedra (M = V, Cr, Al, Ga, Sc) by benzene-di-carboxylic ligands. They can undergo very large amplitude variation of the ⟨ linker–M–linker ⟩ angle. In this latter case, alternatively emptying and filling the pores induces the breathing of the structure (Figure 21.5).

The vibration of the linkers, usually belonging to aromatic carboxylates, can be monitored by quadrupolar ^{2}H (spin $I = 1$) NMR if the benzene ring has been deuterated. It allows therefore the determination of the amplitude and frequency of the angle flip of the benzene ring. The activation by calcination induces the flexibility. The hydration/rehydration process or any adsorption/desorption of small molecules affects the flexibility of the framework. Though only the open and closed forms have been detected by diffraction, use of ^{27}Al MAS during the hydration rehydration process enables the progressive change of a fraction of the structure while it coexists with the dry form to be investigated. Though the control of hydration is kinetically driven, MAS NMR allows an accurate measurement of the proportion of dry and hydrated phases. Beside the dry phase, only one phase is present that exhibits a well-defined second-order quadrupolar pattern, dependant on the hydration level. This unusual behavior indicates a higher activation energy for the insertion of the first amounts of water followed by a succession of intermediate states of partial hydration with almost negligible activation energy for insertion and diffusion (Figure 21.6).

21.4 CONCLUSIONS

Nanoporous materials very often contain metallic atoms with a quadrupolar nucleus. In most cases, NMR characterization involves several techniques as well as all the nuclei contained in the materials. Thanks to the spectacular development of NMR of quadrupolar nuclei, which nowadays yields spectra with very high resolution by direct or indirect detection, it is possible to face the challenges of describing to an unprecedented level the structure of these delicate crystals. The specificity of these compounds lies in the pores that are formed with templates, subsequently emptied, and the crystals are left with an open structure that can enable motion and flexibility. The use of the whole toolbox of NMR experiments to elucidate and describe nanoporous compounds with a high level of accuracy is essential.

RELATED ARTICLES IN THE ENCYCLOPEDIA OF MAGNETIC RESONANCE

Diffusion in Porous Media

Fundamental Principles of NMR Crystallography

High Speed MAS of Half-Integer Quadrupolar Nuclei in Solids

Magic Angle Spinning: Effects of Quadrupolar Nuclei on Spin-1/2 Spectra

Microporous Materials and Xenon-129 NMR

REFERENCES

1. L. P. Cruz, J. M. Savariault, J. Rocha, J. C. Jumas, and J. D. P. de Jesus, *J. Solid State Chem.*, 2001, **156**, 349.
2. C. M. Kowalchuk, F. A. A. Paz, D. Ananias, P. Pattison, L. D. Carlos, and J. Rocha, *Chem.—Eur. J.*, 2008, **14**, 8157.
3. L. Beitone, J. Marrot, T. Loiseau, G. Ferey, M. Henry, C. Huguenard, A. Gansmuller, and F. Taulelle, *J. Am. Chem. Soc.*, 2003, **125**, 1912.
4. V. Munch, F. Taulelle, T. Loiseau, G. Ferey, A. K. Cheetham, S. Weigel, and G. D. Stucky, *Magn. Reson. Chem.*, 1999, **37**, S100.
5. F. A. A. Paz, J. Rocha, J. Klinowski, T. Trindade, F. N. Shi, and L. Mafra, *Prog. Solid State Chem.*, 2005, **33**, 113.
6. T. Loiseau, C. Serre, C. Huguenard, G. Fink, F. Taulelle, M. Henry, T. Bataille, and G. Ferey, *Chem.—Eur. J.*, 2004, **10**, 1373.
7. Y. Millot and P. P. Man, *Solid State Nucl. Magn. Reson.*, 2002, **21**, 21.
8. A. Iuga, C. Ader, C. Groger, and E. Brunner, *Ann. Rep. NMR Spec.*, 2007, **60**, 145.
9. A. Goldbourt and P. K. Madhu, *Ann. Rep. NMR Spec.*, 2005, **54**, 81.
10. Z. M. Yan, B. H. Chen, and Y. Huang, *Solid State Nucl. Magn. Reson.*, 2009, **35**, 49.

11. A. Cadiau, A. Le Bail, A. Hemon-Ribaud, M. Leblanc, M. Body, F. Fayon, E. Durand, J. C. Boulou, and V. Maisonneuve, *Cryst. Growth Des.*, 2010, **10**, 5159.

12. J. P. Tessonnier, B. Louis, S. Walspurger, J. Sommer, M. J. Ledoux, and C. Pham-Huu, *J. Phys. Chem. B*, 2006, **110**, 10390.

13. M. F. Tang, Y. H. Liu, P. C. Chang, Y. C. Liao, H. M. Kao, and K. H. Lii, *Dalton Trans.*, 2007, 4523.

14. C. Volkringer, T. Loiseau, G. Ferey, C. M. Morais, F. Taulelle, V. Montouillout, and D. Massiot, *Microporous Mesoporous Mater.*, 2007, **105**, 111.

15. C. Volkringer, T. Loiseau, N. Guillou, G. Ferey, M. Haouas, F. Taulelle, E. Elkaim, and N. Stock, *Inorg. Chem.*, 2010, **49**, 9852.

16. C. Volkringer, T. Loiseau, M. Haouas, F. Taulelle, D. Popov, M. Burghammer, C. Riekel, C. Zlotea, F. Cuevas, M. Latroche, D. Phanon, C. Knofelv, P. L. Llewellyn, and G. Ferey, *Chem. Mater.*, 2009, **21**, 5783.

17. M. Muller, G. Harvey, and R. Prins, *Microporous Mesoporous Mater.*, 2000, **34**, 135.

18. J. Dutour, N. Guillou, C. Huguenard, F. Taulelle, C. Mellot-Draznieks, and G. Ferey, *Solid State Sci.*, 2004, **6**, 1059.

19. J. Hu, G. S. K. Kannangara, M. A. Wilson, and N. Reddy, *J. Non Cryst. Solids*, 2004, **347**, 224.

20. Y. Fan, X. J. Bao, X. Y. Lin, G. Shi, and H. Y. Liu, *J. Phys. Chem. B*, 2006, **110**, 15411.

21. H. Najar, M. S. Zina, and A. Ghorbel, *React. Kinet. Mech. Catal.*, 2010, **100**, 385.

22. A. Abraham, S. H. Lee, C. H. Shin, S. B. Hong, R. Prins, and J. A. van Bokhoven, *Phys. Chem. Chem. Phys.*, 2004, **6**, 3031.

23. F. Taulelle, A. Samoson, T. Loiseau, and G. Ferey, *J. Phys. Chem. B*, 1998, **102**, 8588.

24. S. Prasad, P. Zhao, J. Huang, J. J. Fitzgerald, and J. S. Shore, *Solid State Nucl. Magn. Reson.*, 2001, **19**, 45.

25. R. Hajjar, C. Volkringer, T. Loiseau, N. Guillou, J. Marrot, G. Ferey, I. Margiolaki, G. Fink, C. Morais, and F. Taulelle, *Chem. Mater.*, 2011, **23**, 39.

26. F. Taulelle, M. Pruski, J. P. Amoureux, D. Lang, A. Bailly, C. Huguenard, M. Haouas, C. Gerardin, T. Loiseau, and G. Ferey, *J. Am. Chem. Soc.*, 1999, **121**, 12148.

27. A. E. W. Beers, J. A. van Bokhoven, K. M. de Lathouder, F. Kapteijn, and J. A. Moulijn, *J. Catal.*, 2003, **218**, 239.

28. E. Kupce and R. Freeman, *J. Magn. Reson., Ser. A*, 1995, **115**, 273.

29. Z. Yao, H. T. Kwak, D. Sakellariou, L. Emsley, and P. J. Grandinetti, *Chem. Phys. Lett.*, 2000, **327**, 85.

30. R. Siegel, T. T. Nakashima, and R. E. Wasylishen, *Chem. Phys. Lett.*, 2004, **388**, 441.

31. F. H. Larsen, H. J. Jakobsen, P. D. Ellis, and N. C. Nielsen, *J. Magn. Reson.*, 1998, **131**, 144.

32. L. Delevoye, C. Fernandez, C. M. Morais, J. P. Amoureux, V. Montouillout, and J. Rocha, *Solid State Nucl. Magn. Reson.*, 2002, **22**, 501.

33. R. D. Gougeon, E. B. Brouwer, P. R. Bodart, L. Delmotte, C. Marichal, J. M. Chezeau, and R. K. Harris, *J. Phys. Chem. B*, 2001, **105**, 12249.

34. D. P. Lang, T. M. Alam, and D. N. Bencoe, *Chem. Mater.*, 2001, **13**, 420.

35. C. M. Morais, V. Montouillout, M. Deschamps, D. Iuga, F. Fayon, F. A. A. Paz, J. Rocha, C. Fernandez, and D. Massiot, *Magn. Reson. Chem.*, 2009, **47**, 942.

36. D. Zhou, J. Xu, J. H. Yu, L. Chen, F. Deng, and R. R. Xu, *J. Phys. Chem. B*, 2006, **110**, 2131.

37. M. Haouas, C. Volkringer, T. Loiseau, G. Ferey, and F. Taulelle, *Chem. Eur. J.*, 2009, **15**, 3139.

38. T. Loiseau, L. Beitone, C. Huguenard, F. Taulelle, M. Henry, and G. Ferey, *J. Phys. Chem. B*, 2004, **108**, 20011.

39. M. Muller, G. Harvey, and R. Prins, *Microporous Mesoporous Mater.*, 2000, **34**, 281.

40. L. Beitone, C. Huguenard, A. Gansmuller, M. Henry, F. Taulelle, T. Loiseau, and G. Ferey, *J. Am. Chem. Soc.*, 2003, **125**, 9102.

41. M. Wiebcke and H. Koller, *Acta Crystallogr., Sect. B Struct. Commun.*, 1992, **48**, 449.

42. S. E. Ashbrook, L. Le Polles, C. J. Pickard, A. J. Berry, S. Wimperis, and I. Farnan, *Phys. Chem. Chem. Phys.*, 2007, **9**, 1587.

43. T. Charpentier, S. Ispas, M. Profeta, F. Mauri, and C. J. Pickard, *J. Phys. Chem. B*, 2004, **108**, 4147.

44. A. Pedone, M. Pavone, M. C. Menziani, and V. Barone, *J. Chem. Theory Comput.*, 2008, **4**, 2130.

45. H. Liu, H. Ernst, D. Freude, F. Scheffler, and W. Schwieger, *Microporous Mesoporous Mater.*, 2002, **54**, 319.

46. C. A. Fyfe, R. J. Darton, C. Schneider, and F. Scheffler, *J. Phys. Chem. C*, 2008, **112**, 80.

47. L. A. O'Dell and C. I. Ratcliffe, *J. Phys. Chem. A*, 2011, **115**, 747.

48. D. Dambournet, A. Demourgues, C. Martineau, E. Durand, J. Majimel, A. Vimont, H. Leclerc, J. C. Lavalley, M. Daturi, C. Legein, J. Y. Buzare, F. Fayon, and A. Tressaud, *J. Mater. Chem.*, 2008, **18**, 2483.

49. P. Grewal, N. Radhamonyamma, J. Gonzalez, P. A. Wright, and P. A. Cox, *Stud. Surf. Sci. Catal.*, 2004, **154**, 1777.

50. M. Edgar, V. J. Carter, D. P. Tunstall, P. Grewal, V. Favre-Nicolin, P. A. Cox, P. Lightfoot, and P. A. Wright, *Chem. Commun.*, 2002, 808.

51. A. Cadiau, C. Martineau, M. Lebianc, V. Maisonneuve, A. Hemon-Ribaud, F. Taulelle, and K. Adil, *J. Mater. Chem.*, 2011, **21**, 3949.

52. C. Fernandez, J. P. Amoureux, L. Delmotte, and H. Kessler, *Microporous Mater.*, 1996, **6**, 125.

53. M. O'Keeffe, M. Eddaoudi, H. L. Li, T. Reineke, and O. M. Yaghi, *J. Solid State Chem.*, 2000, **152**, 3.

54. R. M. Barrer, *Hydrothermal Chemistry of Zeolites*, Academic Press: London, 1982.

55. H. Gies and B. Marler, *Zeolites*, 1992, **12**, 42.

56. S. L. Lawton and W. J. Rohrbaugh, *Science*, 1990, **247**, 1319.

57. G. Cheetham and M. M. Harding, *Zeolites*, 1996, **16**, 245.

58. D. I. Kolokolov, H. Jobic, A. G. Stepanov, V. Guillerm, T. Devic, C. Serre, and G. Ferey, *Angew. Chem. Int. Ed.*, 2010, **49**, 4791.

59. O. B. Vistad, D. E. Akporiaye, F. Taulelle, and K. P. Lillerud, *Chem. Mater.*, 2003, **53**, 1639–1649.

Chapter 22
Quadrupolar NMR in the Earth Sciences

Jonathan F. Stebbins

Department of Geological and Environmental Sciences, Stanford University, Stanford, CA 94305, USA

22.1 INTRODUCTION

NMR studies of quadrupolar nuclides in solid materials of interest to earth sciences are closely related to topics presented in other chapters in this handbook. In this chapter, the emphasis is on applications to the types of problems most commonly encountered in earth sciences, where information on atomic-scale structure and dynamics is sought to help understand large-scale processes. These problems are, of course, by no means unique to earth sciences, and the distinctions among mineralogy, geochemistry, materials science, and solid-state chemistry, have long been blurred.

As emphasized in earlier reviews of NMR in the Earth Sciences,[1–6] most of the minerals that make up the planet are complex, multicomponent solid solutions. Although the long-range, average crystal structures of minerals can generally be determined by X-ray and/or neutron diffraction, details of short-range structure such as cation and anion site occupancies, ordering vs "ideal" random mixing, and local distortion of sites from ideal symmetry can often only be determined by spectroscopic methods such as NMR. Such details are of key importance in accurate models of macroscopic properties, which are in turn essential for quantitative treatments of almost any geological process. For example, variations in the extent of disorder among Al and Si cations on tetrahedral sites in common aluminosilicate minerals can change the configurational entropy (and hence, the free energy) sufficiently to cause shifts of hundreds of degrees or many kilobars in calculated phase equilibria.[7] Even more disordered are silicate glasses, which are widespread products of volcanic eruptions and, more generally, record the structures of magmatic liquids at their glass transition temperatures. Glasses serve as a starting point for relating structure to the properties of molten silicates at high temperatures and pressures in the Earth's interior. Changes in structure with temperature, pressure, and composition have therefore been emphasized in geoscience-based spectroscopy of glasses and melts.[8–11] Given that most of the materials in the Earth exist naturally at high pressures (from a few GPa at the base of the crust to 130 GPa at the base of the mantle), studies of high-pressure mineral structure have also long been pioneered by geoscientists, including the application of NMR.

NMR of Quadrupolar Nuclei in Solid Materials
Edited by Roderick E. Wasylishen, Sharon E. Ashbrook and Stephen Wimperis
 ISBN: 978-0-470-97398-1

Understanding of dynamics and reaction mechanisms, whether these involve diffusion, viscous flow, phase transitions, dissolution in or crystallization from a fluid phase, have special importance in earth sciences. Not only are natural processes generally multiphase and inherently complex but the time scales of geological processes are also often so long that real-time empirical measurements fall short by many orders of magnitude. Models must therefore be based on accurate knowledge of fundamental, atomic-scale processes to have any hope of realistic extrapolation to natural conditions: it is not uncommon, for example, to attempt to integrate diffusion processes in minerals over millions of years in order to estimate rates of heat and mass transport in the crust.

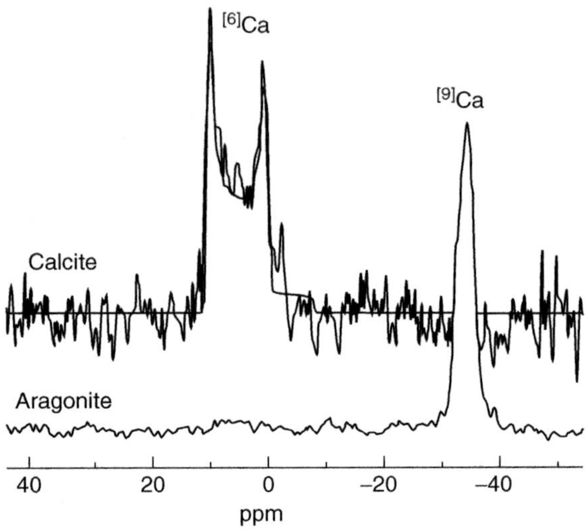

Figure 22.1. Natural abundance (0.15%) ^{43}Ca MAS spectra and lineshape simulation for two polymorphs of $CaCO_3$, collected at 8.5 T with 14 mm MAS rotor. (Reproduced with permission from Ref. 17. © Elsevier, 1997.)

22.2 MINERALS AS MODEL COMPOUNDS

As Nature is often better at growing crystals of silicates and oxides than the laboratory scientist, natural mineral specimens have frequently been used in fundamental NMR studies. In the early days of solid-state NMR, data collected on large single crystals of minerals helped to develop and test theory and to determine basic parameters such as nuclear spin numbers and quadrupolar moments.[12,13] Well before high-field MAS NMR began providing chemical shift data, mineralogists compiled single-crystal ^{27}Al data on quadrupolar coupling constants (C_Q) and asymmetry parameters (η_Q) to develop correlations with polyhedral distortions that could be applied to unknown structures.[14] As reviewed at several stages,[2,4,15,16] more recent, high-resolution NMR on natural minerals and their synthetic equivalents have also been of importance in establishing empirical correlations between chemical shift and local structural parameters such as mean cation–oxygen bond distance. These have often been useful in elucidating the unknown structures of disordered minerals and glasses, for example, for ^{6}Li, ^{11}B, ^{17}O, ^{23}Na, ^{25}Mg, ^{27}Al,^{33}S, ^{35}Cl, ^{43}Ca, and ^{45}Sc. An especially challenging experiment of this sort (Figure 22.1), on natural abundance ^{43}Ca, demonstrates the significant decrease in chemical shift resulting from the increase in the coordination number and Ca–O bond distance from calcite ($CaCO_3$, $^{[6]}Ca$) to aragonite ($CaCO_3$, $^{[9]}Ca$).[17] This type of first-order effect is typical for cations of most NMR nuclides in solid oxides.

Natural and synthetic minerals have also provided ideal test samples for the development of more sophisticated, higher resolution experiments for quadrupolar nuclides, for example, double-rotation (DOR), dynamic-angle spinning (DAS) (see Chapter 8), and multiple-quantum MAS (MQMAS) NMR on various high-temperature and high-pressure silicates.[18–22] Recently, such studies have been combined with high-level theoretical calculations of NMR parameters, again helping to validate these approaches for application to new structures.[23–25] An example of a recent study of this type is ^{17}O MQMAS NMR of majorite garnet ($MgSiO_3$ or $^{[8]}Mg_3[^{[6]}Mg^{[6]}Si]^{[4]}Si_3O_{12}$, an important mineral at the depths in the mantle of a few hundred km) in which signals from five of the six oxygen sites were resolved (Figure 22.2).[23] Natural framework aluminosilicates, and closely related synthetic zeolites, have also played important roles in the development of double resonance methods involving quadrupolar nuclides, such as coherence transfer from ^{27}Al to ^{29}Si in cross-polarization and Transfered Echo DOuble Resonance (TEDOR) experiments.[26,27]

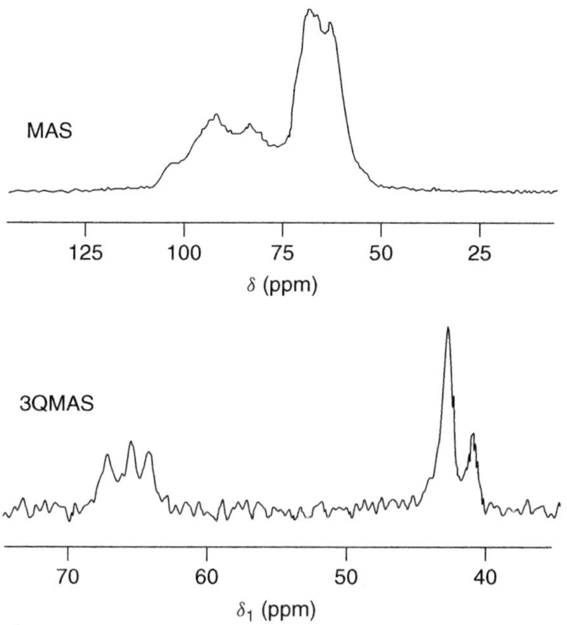

Figure 22.2. ^{17}O MAS and isotropic dimension projection of 3QMAS spectra (14.1 T) for a 16-mg sample of majorite ($MgSiO_3$) garnet synthesized at 19.5 GPa. (Reproduced with permission from Ref. 23. © American Chemical Society, 2007.)

22.3 SITE OCCUPANCIES IN DISORDERED CRYSTALLINE SOLID SOLUTIONS

Framework aluminosilicates, in the form of feldspar and zeolite minerals (see also Chapter 21), are nearly ubiquitous in the Earth's crust. Given the widespread application of ^{29}Si MAS NMR for determining Si/Al site occupancies in both natural zeolites and those synthesized for industry,[15,16] attempting to complement such information with high-resolution ^{27}Al is a natural goal.[28] In ordered framework aluminosilicates, determination of true, isotropic ^{27}Al chemical shifts demonstrated that local structural effects are similar to those well known for ^{29}Si.[29] However, second-order quadrupolar broadening, and the fact that Al tetrahedra in such minerals usually have exactly four Si first neighbors, often reduces the information content. Nonetheless, in some cases, at least partially resolved resonances can be observed in high-field ^{27}Al MAS spectra for Al in different tetrahedral sites; resolution may be enhanced in the spinning sidebands for the inner satellite transitions (Figure 22.3)[29] or by MQMAS.[20,30] When C_Qs are unusually small, as for nepheline solid solutions ([Na,K]$AlSiO_4$), ^{27}Al and ^{29}Si spectra can be similarly well resolved.[31]

In a number of other important groups of minerals, Al can occupy sites with different coordination numbers, most commonly tetrahedra ($^{[4]}Al$) and octahedra ($^{[6]}Al$). Here, ^{27}Al NMR has played a more important role. For example, spinels form a major group of (Mg, Fe, Cr, Ti) oxides that are widely used in calculations of temperatures of crystallization of igneous and metamorphic rocks ("geothermometry"). The distribution of Mg and Al on tetrahedral and octahedral sites in the pure end-member $MgAl_2O_4$ has become an important test of models of energetics in this group.[7] Even in early single-crystal studies, C_Q differences allowed the detection of temperature effects on the extent of Mg/Al disorder (Figure 22.4)[32]; high-resolution ^{27}Al MAS NMR has taken advantage of the large chemical shift difference between $^{[4]}Al$ and $^{[6]}Al$ to refine such measurements (Figure 22.5),[33,34] including high-temperature, single-crystal NMR.[35] Widespread, sheet-structured

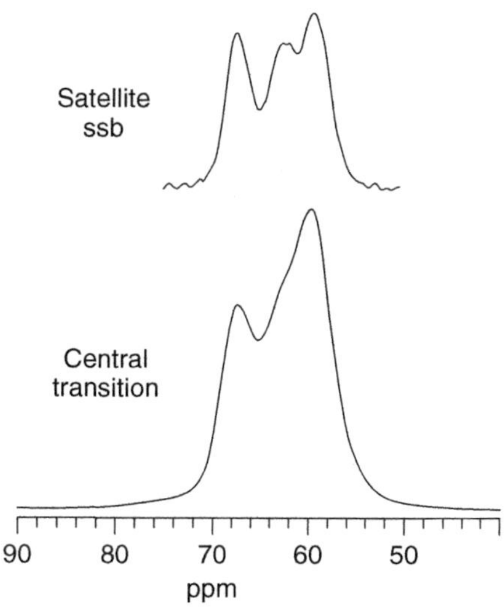

Figure 22.3. ^{27}Al MAS NMR spectra (11.7 T) showing signals from three distinct tetrahedral sites in the framework silicate mineral leucite ($KAlSi_2O_6$). Note improved resolution in 1/2–3/2 satellite spinning sideband (ssb).[29] (Reproduced with permission from Ref. 29. © Mineralogical Society of America, 1994.)

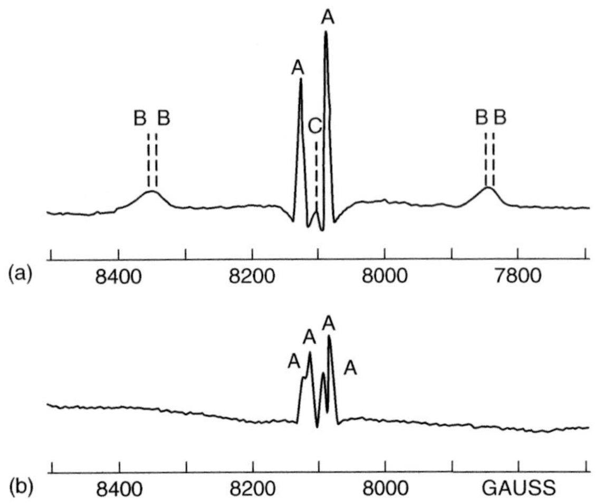

Figure 22.4. Single-crystal ^{27}Al spectra (ca 0.8 T) for a natural spinel ($MgAl_2O_4$). The spectrum in (a) is for H_0 parallel to the fourfold axis and that in (b) is for a general, oblique orientation. A, B: central and satellite transitions for $^{[6]}$Al; C: $^{[4]}$Al. (Reproduced with permission from Ref. 32. © Oldenbourg Wissenschaftsverlag, 1962.)

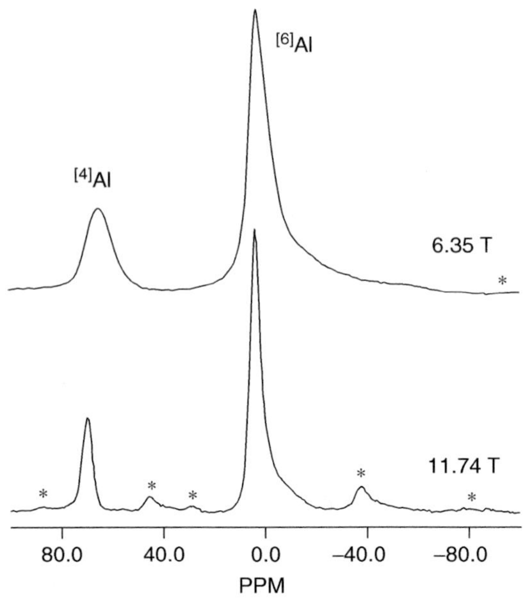

Figure 22.6. ^{27}Al MAS NMR spectra (6.3 vs 11.7 T) of a beidelite clay. Spinning sidebands marked by *. (Reproduced with permission from Ref. 37. © Mineralogical Society of America, 1989.)

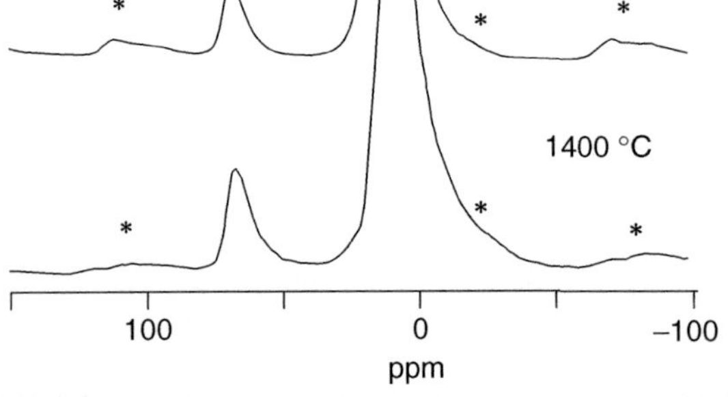

Figure 22.5. ^{27}Al MAS NMR spectra (9.4 T) of synthetic spinel ($MgAl_2O_4$) annealed and quenched from temperatures shown. Spinning sidebands marked by *. (Reproduced with permission from Ref. 33. © Mineralogical Society of America, 1992.)

aluminosilicates such as micas and clays can also contain mixtures of $^{[4]}$Al and $^{[6]}$Al. Particularly for the latter group, where small crystal size and complex stacking sequences can make X-ray structure refinements difficult, ^{27}Al NMR has proven quite helpful in determining site occupancies (Figure 22.6).[16,36,37] Recent studies have involved multiple-quantum and double resonance experiments including HETeronuclear CORrelation (HETCOR) and REDOR.[38]

Although Al is much less abundant in the major minerals of the mantle than in the crust, substitution of minor amounts of this element into the predominant Mg silicates can have important effects on properties needed to constrain geophysical data. ^{27}Al NMR studies of synthetic high-pressure minerals such as $MgSiO_3$ perovskite (the most abundant mineral in the planet) are therefore important in defining mechanisms of solid solution (Figure 22.7).[39] Ever-smaller MAS rotor diameters have greatly facilitated such work, as NMR sample sizes from even a "large volume" high-pressure apparatus can be as small as 1–3 mg. ^{27}Al NMR has also been critical in helping to define new, high-pressure crystal structures

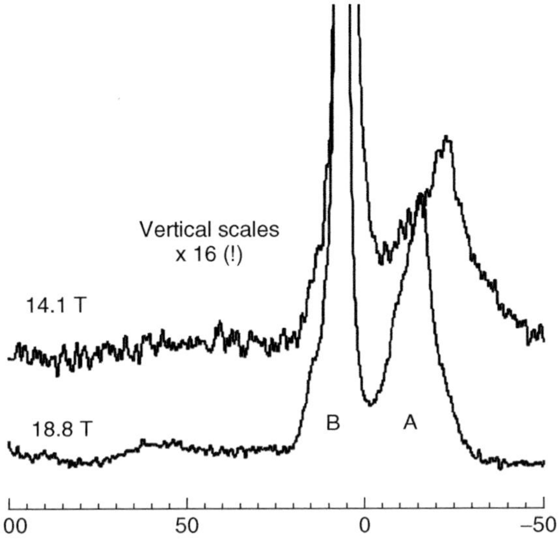

Figure 22.7. ^{27}Al MAS NMR spectra (14.1 vs 18.8 T) of a 4-mg sample of $MgSi_{0.95}Al_{0.05}O_{2.975}$ perovskite synthesized at 27 GPa, showing signal from Al in octahedral "B" site and in larger "A" site. (Reproduced from Ref. 39. © Mineralogical Society of America, 2006.)

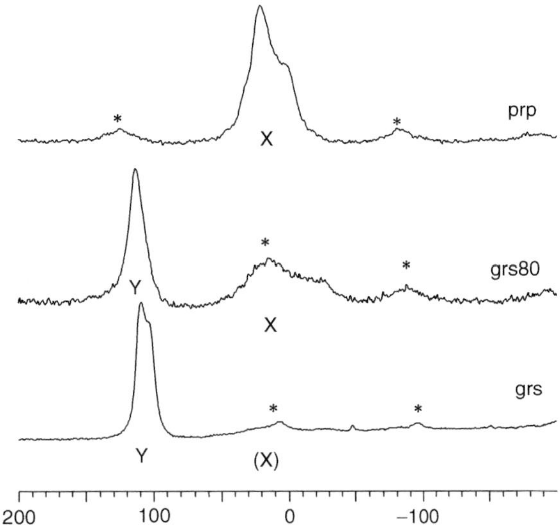

Figure 22.8. ^{45}Sc MAS NMR spectra (18.8 T) of 10- to 20-mg samples of pyrope (prp, $Mg_3Al_2Si_3O_{12}$)—grossular (grs, $Ca_3Al_2Si_3O_{12}$) garnets containing 5 to 8 wt% Sc_2O_3, grown at 3 to 4 GPa. Signals from $^{[8]}$Sc (X site) and from $^{[6]}$Sc (Y site) are visible. "grs80" is the solid solution containing 80% of this component. Spinning sidebands marked by *. (Reproduced from Ref. 41. © Mineralogical Society of America, 2007.)

such as that of a disordered polymorph of topaz-OH ($Al_2SiO_4[OH]_2$).[40]

The higher resolution and sensitivity that have resulted from higher external fields has also begun to allow resolution among cations in different sites for less-commonly studied nuclides, even when they are present in relatively low concentrations, such as ^{45}Sc in high-pressure garnet solid solutions (Figure 22.8).[41] Given that minor and trace element contents serve widely as records of diverse geochemical processes in many groups of minerals, but that few real constraints are known for their substitution mechanisms, such spectroscopic studies of short-range structure should be of growing importance. For "sensitive" nuclides such as ^{11}B, NMR can determine cation coordination numbers even at concentrations below 100 ppm, for example, in biominerals such as calcium carbonates (Figure 22.9).[42,43]

Oxygen is the numerically and volumetrically predominant ion in nearly all silicate and oxide minerals. ^{17}O NMR therefore can provide local structural information relevant to a wide range of problems. Such studies have been done on synthetic equivalents of many natural minerals.[4,15] Empirical and theoretical results for known, ordered structures have yielded correlations among parameters such as C_Q and the isotropic chemical shift (δ_{iso}) and local structural variables such as bond angles and distances, that have proven useful in elucidating the structures of disordered materials, especially silicate glasses.[15] Recently, this type of study has been extended to include 1D, satellite-transition MAS (STMAS) and MQMAS NMR on small samples of synthetic high-pressure minerals such as garnets, pyroxenes, and Mg silicates.[23–25,44]

Relatively few ^{17}O NMR studies have been made on oxide or silicate minerals that are disordered by the effects of temperature or solid solution, and hence where there are major unknowns in the local structure. Both 1D MAS and static spectra can provide useful resolution in such systems, as in disordered spinels[33,45] and zeolites.[46] The enhancements given by MQMAS have in some cases led to unexpected findings, such as the presence of Al–O–Al linkages (violation of "aluminum avoidance") in samples of natural zeolites that had been exchanged with ^{17}O at elevated temperature.[47,48] In other cases, such as pyrope-grossular garnets ($[Mg,Ca]_3Al_2Si_3O_{12}$), ^{17}O

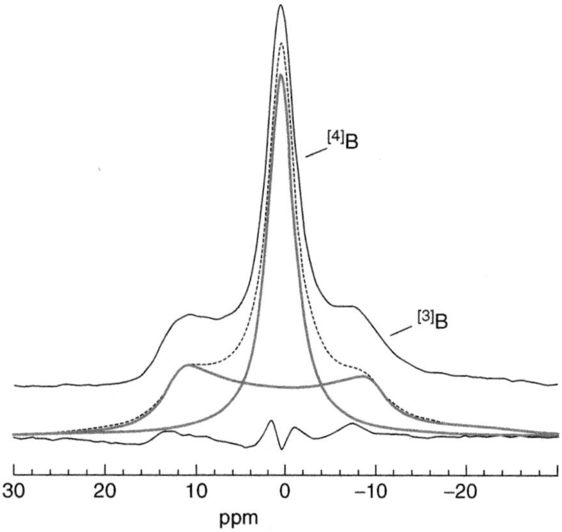

Figure 22.9. ^{11}B MAS NMR spectrum (7.1 T) and lineshape simulation for a trace amount of boron in calcite ($CaCO_3$) shells of foraminifera. (Reproduced with permission from Ref. 43. © Mineralogical Society of America, 2009.)

MQMAS spectra added new constraints to thermodynamic mixing models.[49]

Disorder, often dynamic in character, is a common unknown in hydrous silicates containing either OH groups or molecular H_2O: X-ray diffraction (XRD) generally cannot directly constrain proton sites, and neutron diffraction can be difficult on small, high-pressure mineral samples. For example, ^{17}O NMR on humite minerals (hydrous Mg silicates), analyzed by theoretical calculations, has recently indicated such disorder,[50] which was also characterized by ^{2}H double-quantum NMR.[51]

22.4 SILICATE GLASSES AND MELTS

NMR has had a very important impact on our knowledge of the structure of oxide glasses, as reviewed previously[8,52,53]. Again, the emphasis here will be on questions particularly relevant to the geosciences. NMR studies of spin-1/2 nuclides, particularly ^{29}Si and ^{1}H, have also been quite critical, but are beyond the scope of this chapter, as are the extensive studies of silicate *liquids* by *in situ*, high-temperature NMR.

Silicate glasses, as solid materials, are of major importance at the Earth's surface: a large explosive eruption of a silica-rich magma can blast hundreds or even thousands of cubic kilometers of fine glass particles ("ash") into the atmosphere; the accompanying hot, gas-charged pyroclastic flows can be catastrophically destructive. The properties of the glasses deposited by such processes, for example, their mechanisms of reaction with water during hydrothermal alteration or weathering, depend on the glass structure. As in studies of the corrosion of technological glasses (in particular those being tested for radioactive waste sequestration), these processes have motivated a number of NMR studies (see Section 22.5).

Many studies of glass structure in the geosciences have been motivated by the clues that glasses provide about the structure of the high-temperature molten silicates that not only drive volcanic eruptions but which also dominate many of the heat and mass transfer processes within the Earth's interior. As a rough approximation, a glass formed on cooling of a melt records the structure when the system falls out of equilibrium at the glass transition temperature. This is often a good starting point for evaluating the types of structural species present in the melt, their abundances, the extent of their disorder, and how these change with composition. However, magmatic temperatures are often many hundreds of degrees above glass transition temperatures, and most silicate melts have large "configurational" components to their heat capacities and thermal expansivities, and thereby to their free energies and molar volumes.[11] Major increases in structural disorder must therefore take place with increasing temperature. Quantifying such changes has thus become an important goal in a number of NMR studies. These have involved both *in situ* high-temperature NMR, as well as studies of glasses quenched at different rates, which record the melt structure at different "fictive" temperatures (T_f).[9,54]

An example from a relatively simple aluminosilicate glass composition is shown in Figure 22.10.[55] Here and in previous studies,[56,57] ^{17}O MQMAS NMR allowed the counting of the three distinct types of bridging oxygens that link the tetrahedral network (Si–O–Si, Si–O–Al, Al–O–Al), and thus directly measured the extent of network disorder. In rapidly quenched glasses with T_f increased by about 200 °C, significantly more disorder was measured. ^{17}O NMR has also been pivotal in quantifying the degree of randomness in the mixing of network modifying cations (e.g., Na, K, Ca, Mg), as well as of network-formers

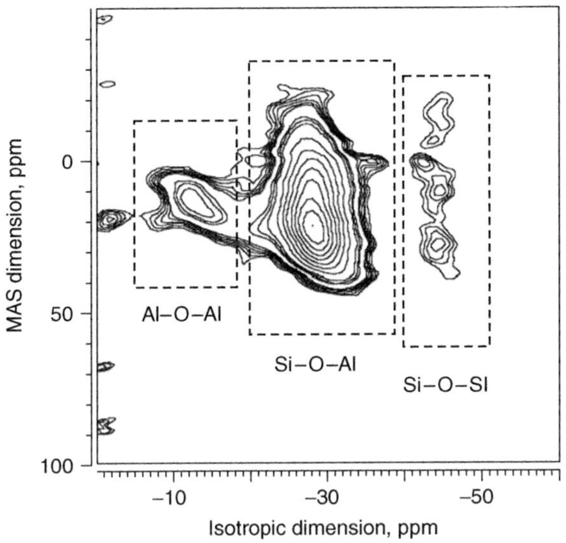

Figure 22.10. ^{17}O 3QMAS spectrum (9.4 T) for a rapidly quenched $NaAlSiO_4$ glass, showing three types of bridging oxygens. (Reproduced from Ref. 55. © Mineralogical Society of America, 2006.)

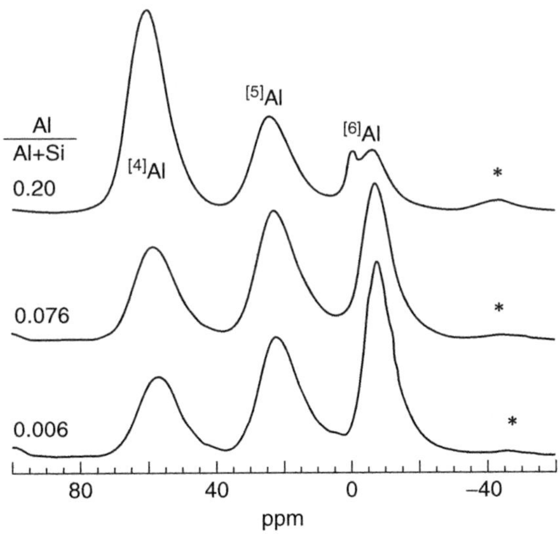

Figure 22.11. ^{27}Al MAS NMR spectra (18.8 T) spectra for Na silicate glasses containing small amounts of Al, melted and quenched at 6 GPa. Spinning sidebands marked by *. (Reproduced from Ref. 68. © Mineralogical Society of America, 2009.)

(B, Al, Si), both of which are critical to thermodynamic models.[58–60]

NMR has been especially useful in discovering and quantifying structural species that are absent in conventional, crystal-structure-based models of glass structure. These include nonbridging oxygens (NBO) in nominally "fully polymerized" compositions such the SiO_2–$CaAl_2O_4$ join,[61] and five-coordinated Al ($^{[5]}Al$) in alkaline earth aluminosilicate glasses.[62–64] MQMAS and very high-field (17.6 to 21 T) MAS NMR have been especially important for the latter. These minor species are again important in models of melt properties, but went undetected through decades of structural studies by other spectroscopic and diffraction methods.

As for minerals, pressure effects on silicate melt structure are of particular interest in the geosciences. NMR studies currently must be done on quenched, decompressed glasses, which may capture only part of the high P/T effects, but which nonetheless provide unique information on compositional effects and mechanisms of structural change. Increases in the abundances of $^{[5]}Al$ and $^{[6]}Al$ with pressure are particularly dramatic (Figure 22.11).[65–68] ^{17}O MQMAS NMR has also revealed dramatic changes in the network linkages as Si and Al coordination numbers increase in densified glasses (Figure 22.12).[67,69,70]

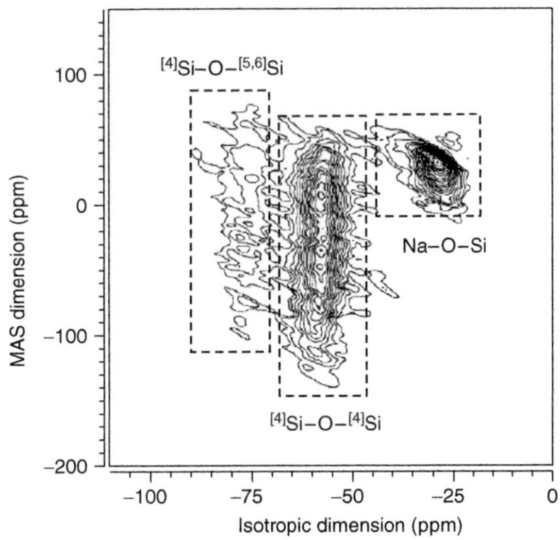

Figure 22.12. ^{17}O 3QMAS spectrum (7.1 T) of $Na_2Si_3O_7$ glass quenched from 8 GPa, showing multiple types of oxygen sites. (Reproduced with permission from Ref. 70. © Elsevier, 2004.)

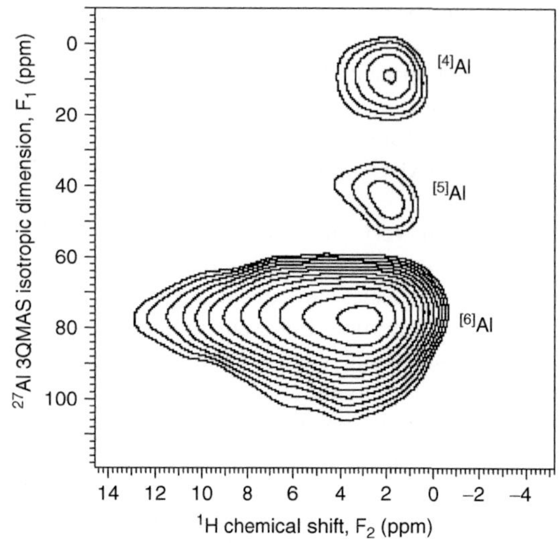

Figure 22.13. ^{27}Al to ^{1}H 3QMAS/HETCOR NMR spectrum (9.4 T) of a Ca, Mg aluminosilicate glass containing 1.9 wt% H_2O. (Reproduced with permission from Ref. 74. © Elsevier, 2008.)

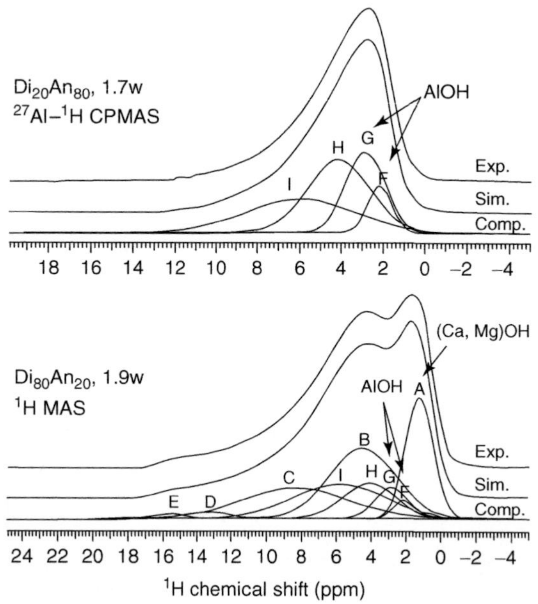

Figure 22.14. ^{27}Al to ^{1}H CPMAS and ^{1}H MAS NMR spectra (9.4 T) of hydrous Ca, Mg aluminosilicate glasses, showing enhancement of signal from AlOH sites by CP. (Reproduced with permission from Ref. 74. © Elsevier, 2008.)

At elevated pressures, silicate melts can dissolve considerable quantities of water, which has huge effects on properties, lowering crystallization temperatures by hundreds of degrees and viscosities by orders of magnitude. Furthermore, it is the degassing of H_2O, which is the major driving force of dangerous explosive volcanism. NMR has thus been widely applied to the surprisingly complex issues of the structural roles of OH and H_2O groups in aluminosilicate glasses. ^{17}O, ^{23}Na, and ^{27}Al (and of course ^{1}H and ^{29}Si) spectra have all provided clues.[71–73] Recently, sophisticated double resonance methods, including cross-polarization from ^{27}Al to ^{1}H and combination of MQMAS and HETCOR methods, have led to key new insights into this long-standing problem (Figures 22.13 and 22.14).[74]

22.5 DYNAMICS AND KINETICS

The dynamics, kinetics, and mechanisms of transitions and reactions in minerals have long been of interest in the Earth sciences.[7,75,76] NMR of quadrupolar nuclides has been applied to a number of such problems, including single-crystal studies of the second-order phase change in anorthite ($CaAl_2Si_2O_8$)[77] and extensive early work on quadrupolar relaxation through phase transitions.[78] More recently, high-resolution NMR work has emphasized the characterization of diffusion processes by the motional averaging of distinct resonances for cations in different structural sites, for example, with high-temperature MAS NMR of ^{23}Na in cryolite (Na_3AlF_6),[79] nepheline ([Na, K]$AlSiO_4$) (Figure 22.15),[80] and sodalites (halide-bearing framework aluminosilicates),[81] and in a single-crystal ^{25}Mg NMR study of forsterite (Mg_2SiO_4) to 1400 °C (Figure 22.16).[82] Changes with increasing temperature of the quadrupolar lineshape can denote either static changes in site symmetry or dynamic exchange among different polyhedral orientations, without the need for high-temperature MAS. Examples include ^{17}O in cristobalite (SiO_2),[83] ^{27}Al in cryolite,[79] ^{23}Na in albite ($NaAlSi_3O_8$),[84] and ^{27}Al and ^{133}Cs in pollucite ($CsAlSi_2O_6$).[85] Detailed analysis of quadrupolar lineshapes in cases of such exchange is a complex problem, where considerable recent progress has been made.[50,86]

Interactions of minerals and glasses with water are part of many geochemical processes, and have been

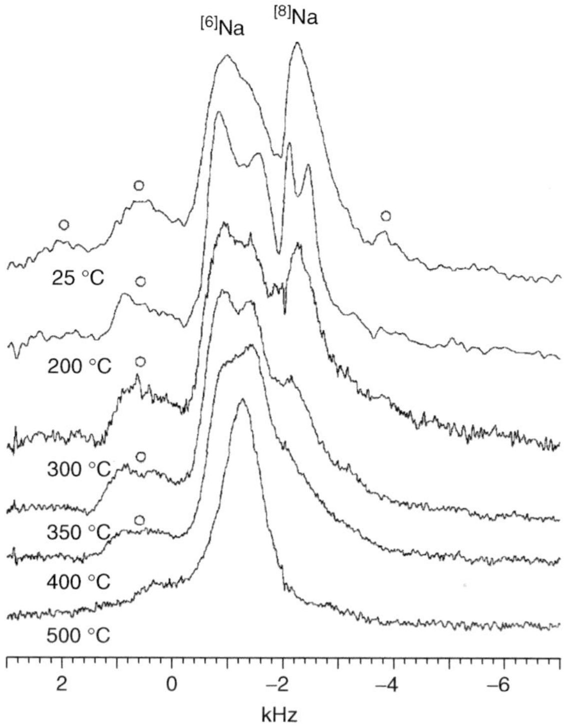

Figure 22.15. High-temperature ^{23}Na MAS NMR spectra (9.4 T) of $Na_{0.75}(Na_{0.20}K_{0.05})AlSiO_4$ nepheline, showing exchange between two Na sites above 300 °C. Circles mark spinning sidebands. (Reproduced with permission from Ref. 80. © Springer, 1989.)

studied by NMR. For example, early high-resolution NMR studies of the corrosion of rhyolite and of $NaAlSi_3O_8$ glass demonstrated the sensitivity of ^{27}Al NMR to the development of secondary phases containing $^{[6]}Al$, and of ^{23}Na NMR to the hydration of the Na^+ cations from the glass (Figure 22.17).[87,88] These and subsequent studies have exploited double and triple resonance CPMAS NMR, and more complex experiments, to select Al and Mg sites adjacent to protons and thus emphasize hydrated surface layers in silicate minerals and glasses.[89,90] For a sample of the zeolite stilbite in which much of the network oxygen had been exchanged with ^{17}O by reaction with $H_2{}^{17}O$ vapor, then reexposed to isotopically normal H_2O for varying lengths of time, ^{17}O MQMAS NMR clearly showed that the three different types of bridging oxygens reacted at different rates, with Si–O–Si being the most, and Al–O–Al being the least stable (Figure 22.18).[46,48] The sorption of metal cations

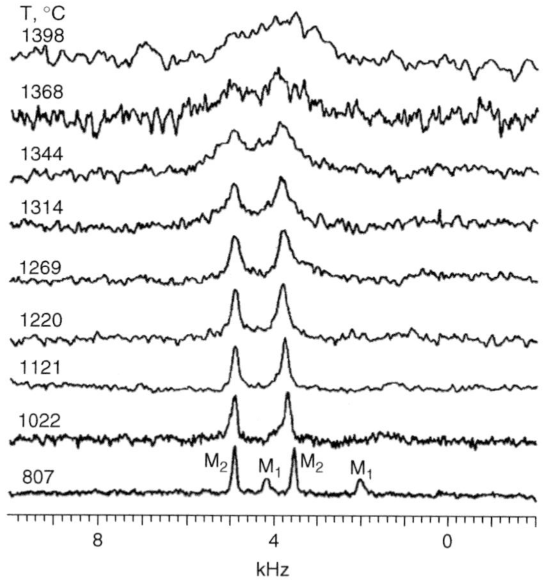

Figure 22.16. High-temperature, single-crystal ^{25}Mg NMR spectra (9.4 T) of forsterite (Mg_2SiO_4) showing exchange among Mg sites. (Reproduced from Ref. 82. © Mineralogical Society of America, 1997.)

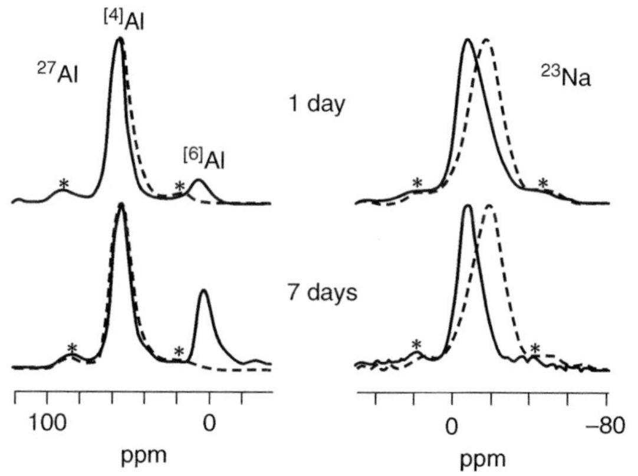

Figure 22.17. ^{27}Al and ^{23}Na MAS NMR spectra (11.7 T) for $NaAlSi_3O_8$ glass reacted with hot acid for times shown. Dashed curves show data for unreacted glass. (Reproduced with permission from Ref. 87. © Mineralogical Society of America, 1989.)

onto mineral surfaces is a major subject of geochemical research; NMR studies of nuclides such as ^{11}B, ^{23}Na, and ^{133}Cs on high surface-area minerals have yielded useful insights into such problems.[91,92]

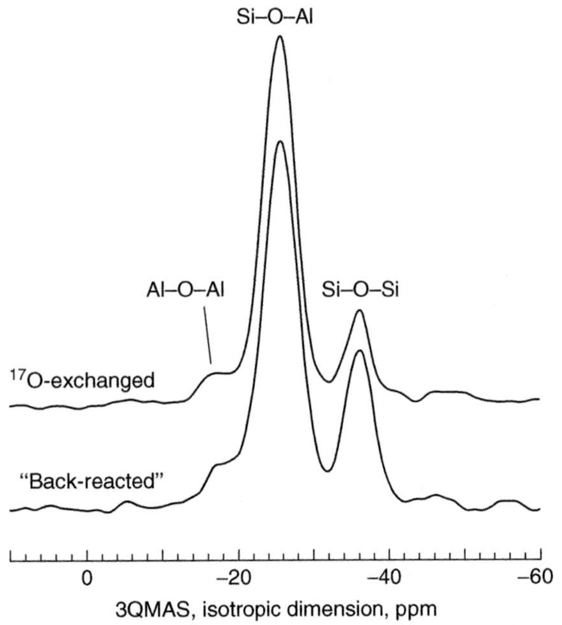

Figure 22.18. Isotropic dimension projections of ^{17}O 3QMAS spectra (14.1 T) of a natural zeolite (stilbite) after initial exchange with $H_2{}^{17}O$ vapor then back-reaction with normal H_2O. (Reproduced from Ref. 48. © Mineralogical Society of America, 1999.)

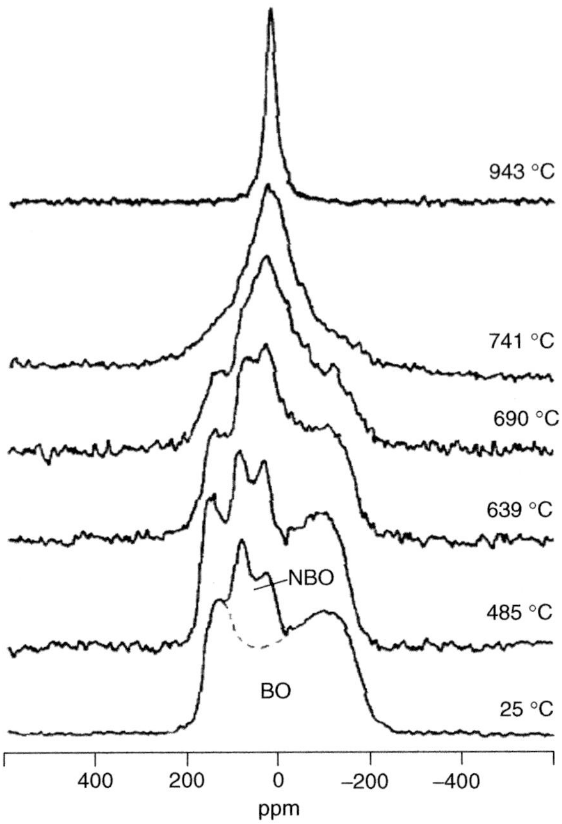

Figure 22.19. High-temperature static ^{17}O NMR spectra (9.4 T) of $K_2Si_4O_9$ glass and liquid, showing exchange of bridging oxygen (BO) and non-bridging oxygen (NBO). (Reproduced with permission from Ref. 93. © Elsevier, 1992.)

Finally, silicate melts just above their glass transitions are indeed thermodynamically liquids, but are nonetheless viscous enough so that exchange among structural species occurs at timescales more commonly associated with solids. An example is *in situ*, high-temperature ^{17}O NMR of $K_2Si_4O_9$ liquid, where the exchange among bridging and NBO leads to complete motional averaging a few 100 °C above the glass transition (Figure 22.19). This process requires the breaking and reforming of Si–O bonds that control much of the diffusion and viscous transport in melts.[3,93] More recent high-temperature MAS NMR on a K-silicate melt with even higher silica content has further refined such dynamics.[94]

22.6 MINERALS CONTAINING UNPAIRED ELECTRON SPINS

Applications of high-resolution solid-state NMR to geological materials have often been hampered by the severe line broadening, and even loss of signal, that is caused by the unpaired electron spins in most transition metal and rare earth element cations. Given that Fe is one of the four most abundant elements in the Earth, this means that NMR studies of many important groups of minerals (e.g., pyroxenes, olivines, garnets, amphiboles, spinels) have generally been limited to unusual, low-Fe end members, or to synthetic, Fe-free compositions. There are a few exceptions in single-crystal ^{27}Al NMR studies of Fe-rich silicates such as pyrope–almandine garnet ($[Mg, Fe]_3Al_2Si_3O_{12}$), in which large frequency shifts caused by electron-nuclear dipolar couplings were noted.[95] Ferric-iron oxyhydroxides are widespread products of both natural weathering and of the mining of Fe-sulfide-containing ore deposits, and can

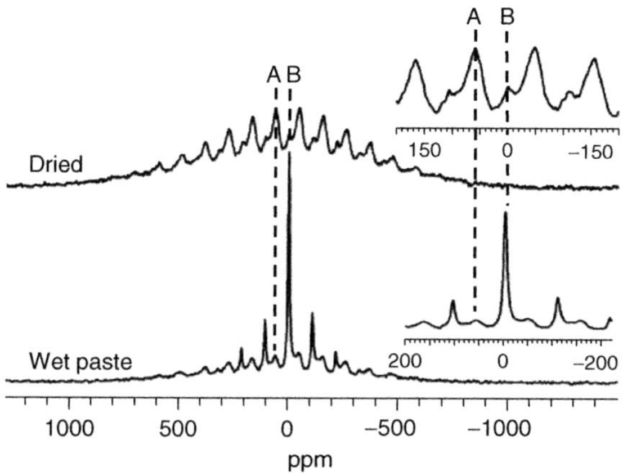

Figure 22.20. ^{7}Li MAS spectra (8.5 T) for Li^{+} sorbed on lepidocrocite (γ-FeOOH). Peak A is the central resonance for an inner-sphere complex (Fe^{3+} in first cation shell of Li^{+}) that has a significant Fermi contact shift; peak B is that of an outer sphere complex (all H_2O first neighbors), which is more abundant in the wet sample. All other peaks are spinning sidebands. (Reproduced with permission from Ref. 96. © American Chemical Society, 2008.)

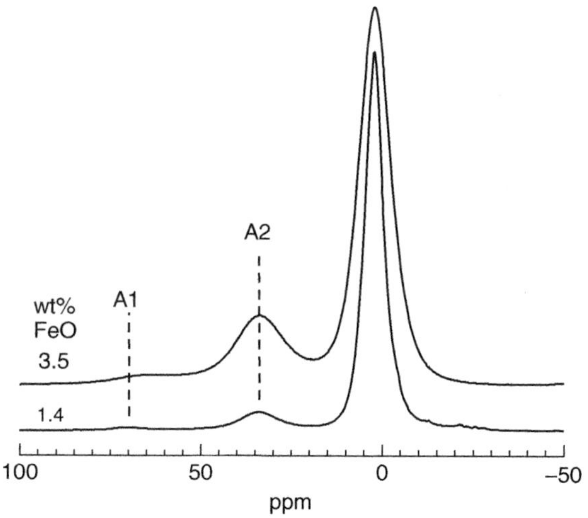

Figure 22.21. ^{27}Al MAS NMR spectra (14.1 T) for natural pyrope ($Mg_3Al_2Si_4O_{12}$) garnets with FeO contents as shown. The A2 and A1 peaks are contact shifted by interactions with the Fe^{2+} cations. (Reproduced from Ref. 99. © Elsevier, 2009.)

play an important role in the sorption and release of toxic metals into the environment. It is therefore encouraging that recent studies of both spin-1/2 and quadrupolar nuclides have demonstrated informative NMR spectra from cations sorbed onto such "impossible" Fe-rich minerals, such as ^{6}Li, ^{7}Li, and ^{2}H on goethite and lepidocrocite (both FeOOH). Here, in addition to large effects on relaxation and peak widths, well-defined contact shifts can sometimes be noted, especially if the sample is above its Néel (or Curie) temperature and thus lacks magnetic ordering (Figure 22.20).[96,97] Finally, in at least a few cases for natural and synthetic silicates such as forsterite and pyrope, ^{29}Si and/or ^{27}Al NMR have revealed small, "anomalous" resonances. These are clearly some type of contact shift (Figure 22.21), as confirmed by recent variable-temperature NMR,[98] although understanding the relative contributions from Fermi (through-bond) vs pseudocontact (through-space) shifts remains a complex problem.[99] A better understanding of such interactions would provide renewed prospects for using NMR to learn about short-range order/disorder in minerals containing significant concentrations of paramagnetic cations.

RELATED ARTICLES IN THE ENCYCLOPEDIA OF MAGNETIC RESONANCE

Amorphous Materials

Crystallography and NMR: Applications to Geochemistry

Double Rotation

Geological Applications

HETCOR in Organic Solids

Multiple-quantum Magic-angle Spinning Experiments on Half-integer Nuclei: Fundamentals

REDOR and TEDOR

Structural Phase Transitions

REFERENCES

1. R. J. Kirkpatrick, in *Spectroscopic Methods in Mineralogy and Geology*, ed. F. C. Hawthorne, Mineralogical Society of America: Washington, DC, 1988.
2. R. J. Kirkpatrick and B. L. Phillips, *Appl. Mag. Reson.*, 1993, **4**, 213.
3. I. Farnan, in *Encyclopedia of Earth System Science*, ed. W. A. Nierenberg, Academic Press: San Diego, CA, 1991.

4. J. F. Stebbins, in *Handbook of Physical Constants*, ed. T. J. Ahrens, American Geophysical Union: Washington, DC, 1995.
5. J. F. Stebbins and I. Farnan, *Science*, 1989, **245**, 257.
6. M. Fechtelkord, in *Spectroscopic Methods in Mineralogy*, eds. A. Beran and E. Libowitsky, Eotvos University Press: Budapest, 2004.
7. A. Putnis, *Introduction to Mineral Sciences*, Cambridge University Press: Cambridge, 1992.
8. J. F. Stebbins, in *Structure, Dynamics, and Properties of Silicate Melts*, eds J. F. Stebbins, P. F. McMillan, and D. B. Dingwell, Mineralogical Society of America: Washington, DC, 1995.
9. J. F. Stebbins, *Chem. Geol.*, 2008, **256**, 80.
10. G. H. Wolf and P. F. McMillan, in *Structure, Dynamics, and Properties of Silicate Melts*, eds J. F. Stebbins, P. F. McMillan, and D. B. Dingwell, Mineralogical Society of America: Washington, DC, 1995.
11. B. O. Mysen and P. Richet, *Silicate Glasses and Melts, Properties and Structure*, Elsevier: Amsterdam, 2005.
12. R. V. Pound, *Phys. Rev.*, 1950, **79**, 685.
13. H. E. Petch, N. G. Cranna, and G. M. Volkoff, *Can. J. Phys.*, 1953, **31**, 837.
14. S. Ghose and T. Tsang, *Am. Mineral.*, 1973, **58**, 748.
15. K. J. D. MacKenzie and M. E. Smith, *Multinuclear Solid-State NMR of Inorganic Materials*, Pergamon: New York, 2002.
16. G. Engelhardt and D. Michel, *High-Resolution Solid-State NMR of Silicates and Zeolites*, John Wiley & Sons: New York, 1987.
17. R. Dupree, A. P. Howes, and S. C. Kohn, *Chem. Phys. Lett.*, 1997, **276**, 399.
18. B. F. Chmelka, K. T. Mueller, A. Pines, J. F. Stebbins, Y. Wu, and J. W. Zwanziger, *Nature*, 1989, **339**, 42.
19. K. T. Mueller, J. H. Baltisberger, E. W. Wooten, and A. Pines, *J. Phys. Chem.*, 1992, **96**, 7001.
20. J. H. Baltisberger, Z. Xu, J. F. Stebbins, S. Wang, and A. Pines, *J. Am. Chem. Soc.*, 1996, **118**, 7209.
21. P. J. Grandinetti, J. H. Baltisberger, I. Farnan, J. F. Stebbins, U. Werner, and A. Pines, *J. Phys. Chem.*, 1995, **99**, 12341.
22. T. M. Clark and P. J. Grandinetti, *Sol. St. NMR*, 2005, **27**, 233.
23. S. E. Ashbrook, A. J. Berry, D. J. Frost, A. Gregorovic, C. J. Pickard, J. E. Readman, and S. Wimperis, *J. Am. Chem. Soc.*, 2007, **129**, 13213.
24. S. E. Ashbrook, A. J. Berry, W. O. Hibberson, S. Steuernagel, and S. Wimperis, *Am. Mineral.*, 2005, **90**, 1861.
25. S. E. Ashbrook, L. Le Polles, C. J. Pickard, A. J. Berry, S. Wimperis, and I. Farnan, *Phys. Chem. Chem. Phys.*, 2007, **9**, 1587.
26. C. A. Fyfe, K. C. Wong-Moon, Y. Huang, H. Grodney, and K. T. Mueller, *J. Phys. Chem.*, 1995, **99**, 8707.
27. S. M. DePaul, M. Ernst, J. S. Shore, J. F. Stebbins, and A. Pines, *J. Chem. Phys.*, 1997, **B 101**, 3240.
28. B. L. Phillips, R. J. Kirkpatrick, and G. L. Hovis, *Phys. Chem. Minerals*, 1988, **16**, 262.
29. B. L. Phillips and R. J. Kirkpatrick, *Am. Mineral.*, 1994, **79**, 1025.
30. F. H. Larsen and I. Farnan, *J. Phys. Chem. B*, 2004, **108**, 9764.
31. G. L. Hovis, D. R. Spearing, J. F. Stebbins, J. Roux, and A. Clare, *Am. Mineral.*, 1992, **77**, 19.
32. E. Brun and S. Hafner, *Zeit. Krist.*, 1962, **117**, 37.
33. R. L. Millard, R. C. Peterson, and B. K. Hunter, *Am. Mineral.*, 1992, **77**, 44.
34. B. J. Wood, R. J. Kirkpatrick, and B. Montez, *Am. Mineral.*, 1986, **71**, 999.
35. H. Maekawa, S. Kato, K. Kawamura, and T. Yokokawa, *Am. Mineral.*, 1997, **82**, 1125.
36. R. A. Kinsey, R. J. Kirkpatrick, J. Hower, K. A. Smith, and E. Oldfield, *Am. Mineral.*, 1985, **70**, 537.
37. D. E. Woessner, *Am. Mineral.*, 1989, **74**, 203.
38. M. Fechtelkord, H. Behrens, F. Holtz, J. L. Bretherton, C. A. Fyfe, L. A. Groat, and M. Raudsepp, *Am. Mineral.*, 2003, **88**, 1046.
39. J. F. Stebbins, L.-S. Du, K. Kelsey, H. Kojitani, M. Akaogi, and S. Ono, *Am. Mineral.*, 2006, **91**, 337.
40. X. Xue, M. Kanzaki, and H. Fukui, *Am. Mineral.*, 2010, **95**, 1276.
41. N. Kim, J. F. Stebbins, S. Quartieri, and R. Oberti, *Am. Mineral.*, 2007, **92**, 1875.
42. S. Sen, J. F. Stebbins, N. G. Hemming, and B. Ghosh, *Am. Mineral.*, 1994, **79**, 819.
43. K. Klochko, G. D. Cody, J. A. Tossell, P. Dera, and A. J. Kaufman, *Am. Mineral.*, 2009, **73**, 1890.
44. K. E. Kelsey, J. F. Stebbins, L.-S. Du, and B. Hankins, *Am. Mineral.*, 2007, **92**, 210.
45. R. L. Millard, R. C. Peterson, and B. K. Hunter, *Am. Mineral.*, 1995, **80**, 885.

46. Z. Xu and J. F. Stebbins, *Geochim. Cosmochim. Acta*, 1998, **62**, 1803.
47. X. Cheng, P. Zhao, and J. F. Stebbins, *Am. Mineral.*, 2000, **85**, 1037.
48. J. F. Stebbins, P. Zhao, S. K. Lee, and X. Cheng, *Am. Mineral.*, 1999, **84**, 1680.
49. K. E. Kelsey, J. F. Stebbins, L.-S. Du, J. L. Mosenfelder, P. D. Asimow, and C. A. Geiger, *Am. Mineral.*, 2008, **93**, 134.
50. J. M. Griffin, S. Wimperis, A. J. Berry, C. J. Pickard, and S. E. Ashbrook, *J. Phys. Chem. C*, 2009, **113**, 465.
51. J. M. Griffin, A. J. Miller, A. J. Berry, S. Wimperis, and A. E. Ashbrook, *Phys. Chem. Chem. Phys.*, 2010, **12**, 2989.
52. H. Eckert, in *Solid-State NMR IV. Methods and Applications of Solid-State NMR*, ed. B. Blümich, Springer-Verlag: Berlin, 1994.
53. J. F. Stebbins, in *Solid-State NMR: Theory and Applications*, ed. M. Duer, Blackwell Scientific: Oxford, 2002.
54. J. F. Stebbins, E. V. Dubinsky, K. Kanehashi, and K. E. Kelsey, *Geochim. Cosmochim. Acta*, 2008, **72**, 910.
55. E. V. Dubinsky and J. F. Stebbins, *Am. Mineral.*, 2006, **91**, 753.
56. P. J. Dirken, S. C. Kohn, M. E. Smith, and E. R. H. van Eck, *Chem. Phys. Let.*, 1997, **266**, 568.
57. S. K. Lee and J. F. Stebbins, *J. Phys. Chem. B*, 2000, **104**, 4091.
58. P. Florian, K. E. Vermillion, P. J. Grandinetti, I. Farnan, and J. F. Stebbins, *J. Am. Chem. Soc.*, 1996, **118**, 3493.
59. J. R. Allwardt and J. F. Stebbins, *Am. Mineral.*, 2004, **89**, 777.
60. S. K. Lee, G. D. Cody, and B. O. Mysen, *Am. Mineral.*, 2005, **90**, 1393.
61. J. F. Stebbins and Z. Xu, *Nature*, 1997, **390**, 60.
62. P. F. McMillan and R. J. Kirkpatrick, *Am. Mineral.*, 1992, **77**, 898.
63. M. J. Toplis, S. C. Kohn, M. E. Smith, and I. J. F. Poplett, *Am. Mineral.*, 2000, **85**, 1556.
64. D. R. Neuville, L. Cormier, and D. Massiot, *Chem. Geol.*, 2006, **229**, 173.
65. J. L. Yarger, K. H. Smith, R. A. Nieman, J. Diefenbacher, G. H. Wolf, B. T. Poe, and P. F. McMillan, *Science*, 1995, **270**, 1964.
66. J. R. Allwardt, J. F. Stebbins, B. C. Schmidt, D. J. Frost, A. C. Withers, and M. M. Hirschmann, *Am. Mineral.*, 2005, **90**, 1218.
67. S. K. Lee, G. D. Cody, Y. Fei, and B. O. Mysen, *Chem. Geol.*, 2006, **229**, 162.
68. K. E. Kelsey, J. F. Stebbins, J. L. Mosenfelder, and P. D. Asimow, *Am. Mineral.*, 2009, **94**, 1205.
69. J. R. Allwardt, B. C. Schmidt, and J. F. Stebbins, *Chem. Geol.*, 2004, **213**, 137.
70. S. K. Lee, G. D. Cody, Y. Fei, and B. O. Mysen, *Geochim. Cosmochim. Acta*, 2004, **68**, 4189.
71. B. C. Schmidt, T. Riemer, S. C. Kohn, H. Behrens, and R. Dupree, *Geochim. Cosmochim. Acta*, 2000, **64**, 513.
72. Z. Xu, H. Maekawa, J. V. Oglesby, and J. F. Stebbins, *J. Am. Chem. Soc.*, 1998, **120**, 9894.
73. Q. Zeng, H. Nekvasil, and C. P. Grey. *J. Phys. Chem. B*, 1999, **103**, 7406.
74. X. Xue and M. Kanzaki, *Geochim. Cosmochim. Acta*, 2008, **72**, 2331.
75. A. Putnis, in *Physical Properties and Thermodynamic Behaviour of Minerals*, ed. E. K. H. Salje, Reidel: Dordrecht, 1988.
76. J. F. Stebbins, *Chem. Rev.*, 1991, **91**, 1353.
77. J. L. Staehli and D. Brinkmann, *Zeit. Krist.*, 1974, **140**, 360.
78. F. Borsa and A. Rigmonti, in *Structural Phase Transitions-II*, eds K. A. Muller and H. Thomas, Springer-Verlag: New York, 1990.
79. D. R. Spearing, J. F. Stebbins, and I. Farnan, *Phys. Chem. Miner.*, 1994, **21**, 373.
80. J. F. Stebbins, I. Farnan, E. H. Williams, and J. Roux, *Phys. Chem. Miner.*, 1989, **16**, 763.
81. M. Fechtelkord, *Sol. St. NMR*, 2000, **18**, 70.
82. J. F. Stebbins, *Am. Mineral.*, 1997, **81**, 1315.
83. D. R. Spearing, I. Farnan, and J. F. Stebbins, *Phys. Chem. Min.*, 1992, **19**, 307.
84. A. M. George and J. F. Stebbins, *Am. Mineral.*, 1995, **80**, 878.
85. S. E. Ashbrook, K. R. Whittle, L. Le Polles, and I. Farnan, *J. Am. Ceram. Soc.*, 2005, **88**, 1575.
86. J. H. Kristensen and I. Farnan, *J. Chem. Phys.*, 2001, **114**, 9608.
87. W.-H. A. Yang and R. J. Kirkpatrick, *Am. Mineral.*, 1989, **53**, 805.

88. W. A. Yang and R. J. Kirkpatrick, *Am. Mineral.*, 1990, **75**, 1009.

89. M. C. Davis, W. J. Brouwer, D. J. Wesolowski, L. M. Anovitz, A. S. Lipton, and K. T. Mueller, *Phys. Chem. Chem. Phys.*, 2009, **11**, 7013.

90. N. Tsomaia, S. L. Brantley, J. P. Hamilton, C. G. Pantano, and K. T. Mueller, *Am. Mineral.*, 2003, **88**, 54.

91. Y. Kim and R. J. Kirkpatrick, *Geochim. Cosmochim. Acta*, 1997, **61**, 5199.

92. Y. Kim and R. J. Kirkpatrick, *Geochim. Cosmochim. Acta*, 2006, **70**, 3231.

93. J. F. Stebbins, I. Farnan, and X. Xue, *Chem. Geol.*, 1992, **96**, 371.

94. S. Sen, *Phys. Rev. B*, 2008, **78**, 100201.

95. S. Ghose, *Sol. St. Commun.*, 1964, **2**, 361.

96. J. Kim, U. G. Nielsen, and C. P. Grey, *J. Am. Chem. Soc.*, 2008, **130**, 1285.

97. U. G. Nielsen, Y. Paik, K. Julmis, M. A. A. Schoonen, R. J. Reeder, and C. P. Grey, *J. Phys. Chem. B*, 2005, **109**, 18310.

98. A. Palke and J. F. Stebbins, *Am Mineral.*, 2011, **96**, 1090–1099.

99. J. F. Stebbins and K. E. Kelsey, *Phys. Chem. Chem. Phys.*, 2009, **11**, 6906.

Chapter 23
Quadrupolar NMR of Superconductors

Nicholas J. Curro

Department of Physics, University of California, Davis, CA 95616, USA

23.1 INTRODUCTION

One of the most active areas of modern condensed matter physics is the study of superconductors. NMR has played a central role in the study of superconductivity for several decades, and continues to be a useful technique for investigating novel materials because it provides microscopic information about the electronic state. Of the 288 naturally occurring primordial isotopes in nature, approximately 41% have nonzero nuclear spins, and approximately 21% have nuclear spins greater than 1/2. Therefore the vast majority of compounds that can be synthesized may contain nuclei that are either NMR active or even quadrupolar active. Quadrupolar NMR is particularly useful in the investigation of superconductivity because it provides greater flexibility in terms of magnetic fields. Superconductivity is a somewhat fragile state of matter that exists only under specialized conditions. The electrons of a normal metal undergo an electronic phase transition below a critical temperature T_c and below a critical magnetic field H_{c2} (Figure 23.1). In order to investigate this state, it is necessary to work with magnetic fields $H < H_{c2}$, which may present a problem for NMR of spin-1/2 nuclei if H_{c2} is small. H_{c2} can vary anywhere from a few tens of milliTeslas to several hundred Teslas, depending on the material. In cases where H_{c2} is small, one can often use a quadrupolar nucleus to perform either zero-field NQR or NMR of one of the quadrupolar satellites in small magnetic fields. Furthermore, quadrupolar NMR provides a unique probe of the local charge density. Other probes such as muon spin rotation (μSR) and electron spin resonance (ESR) provide microscopic information about the local spin environment of superconductors, but no other resonance technique provides simultaneous information about the local charge. This capability offers unique insight into both the normal and superconducting states of the high-temperature superconductors, and enables local spectroscopic probes of impurities in these materials.

Five decades of research has taken place and several excellent review articles have been written on the subject of NMR in superconductors (Refs 2–5). This chapter is not meant to be a comprehensive review, but will provide an overview of some example experiments in which quadrupolar NMR has played

NMR of Quadrupolar Nuclei in Solid Materials
Edited by Roderick E. Wasylishen, Sharon E. Ashbrook and Stephen Wimperis

a central role in observing the physics of superconductors. Sections 23.2 and 23.3 provide overviews of the properties of quadrupolar nuclei in fields and superconductors in general. Section 23.4 reports NQR experiments that probe the spin–lattice relaxation rate in the superconducting state. Section 23.5 discusses the fascinating interplay between antiferromagnetism and superconductivity present in several unconventional superconductors, and Section 23.6 focuses on the unusual changes in the electric field gradient (EFG) brought on by superconductivity.

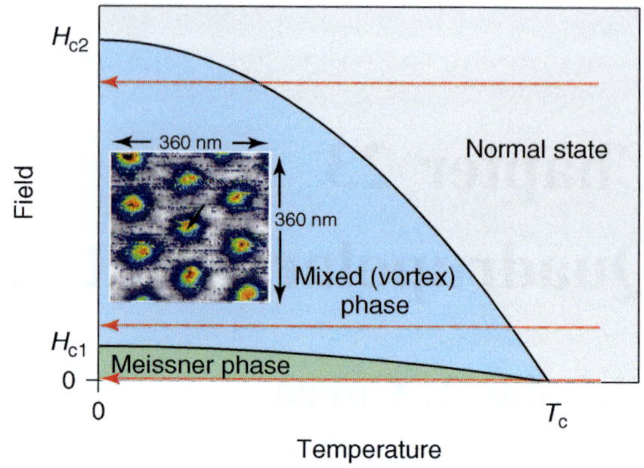

Figure 23.1. Field–temperature phase diagram for a type II superconductor. A typical NMR or NQR experiment tracks the spectra and relaxation rates as a function of temperature along the paths indicated by the red lines. Inset: Image of the vortex phase on the surface of 2H-NbS_2 acquired by scanning tunneling microscopy at a field of 1500 Oe at 0.1 K, over a range of 360 nm × 360 nm. (Inset reproduced from Ref. 1. © American Physical Society, 2008.)

23.2 QUADRUPOLAR SPECTRA

For nuclei with spin $I > 1/2$, the nuclear quadrupolar moment Q couples to the local charge distribution via the electric quadrupole interaction, with the Hamiltonian

$$\mathcal{H}_Q = \frac{eQV_{zz}}{4I(2I-1)}\left[\left(3\hat{I}_z^2 - \hat{I}^2\right) + \eta\left(\hat{I}_x^2 - \hat{I}_y^2\right)\right] \tag{23.1}$$

where e is the electronic charge, $V_{\alpha\beta}$ are the components of the EFG tensor, and $\eta = (V_{xx} - V_{yy})/V_{zz}$ is the asymmetry parameter of the EFG (see Chapter 1). For a site with axial symmetry ($\eta = 0$), the NQR frequency (in zero field) is defined as

$$\nu_Q = \frac{3eQV_{zz}}{2I(2I-1)h} \tag{23.2}$$

and h is the Planck constant. There are $2I - 1$ quadrupolar resonances at frequencies $n\nu_Q$, where $n = 1, \ldots, 2I - 1$. If $\eta > 0$, then the resonances are not equally spaced. The EFG tensor is fully characterized by three parameters: V_{zz}, η, and a unit vector $\hat{q}$ in the direction of the principal axis of the EFG with the largest eigenvalue. The principal axes of the EFG are usually determined by the local symmetry of the charge environment of the nuclei, and may coincide with either the bond axes of the atom of interest or the crystal axes of the unit cell.

For spin I, there are at least $2I + 1$ allowed transitions, each with different frequencies. These resonant frequencies are strong functions of the field orientation. In order to characterize the response of the nuclei, it is convenient to consider the relationship between the applied field and the resonant frequency for each transition. In general, these resonant fields must be determined by numerical diagonalization of the Hamiltonian $\mathcal{H}_Z + \mathcal{H}_Q$, where $\mathcal{H}_Z = \gamma\hbar\hat{\boldsymbol{I}} \cdot \boldsymbol{B}_0$ is the Zeeman term and $\mathcal{H}_Q$ is given by equation (23.1).

As an example, consider the case of ^{115}In ($I = 9/2$) in the heavy fermion superconductor $CeCoIn_5$. Figure 23.2 shows the frequency versus field relationship for three different In sites. This material has tetragonal symmetry, and there are two crystallographically distinct indium sites: In(23.1) has axial symmetry with $\hat{q} \parallel [001]$ along the crystal c-axis; and In(2) with $\eta \sim 0.4$ and $\hat{q}$ in the ab plane perpendicular to the unit cell faces. When the applied field $\boldsymbol{B}_0$ is in the ab plane (i.e., along [100]), then for one-half of the In(2) sites the field lies parallel to $\hat{q}$ (the In(2a) sites indicated in Figure 23.2a). For the other half of the In(2) sites, $\boldsymbol{B}_0 \perp \hat{q}$ (the In(2b) sites indicated in Figure 23.2a). Since the resonance frequency depends on the orientation of $\boldsymbol{B}_0$ with respect to $\hat{q}$, these two sites are distinct in the NMR spectrum. This example demonstrates the versatility offered by quadrupolar NMR because, if the In were spin $I = 1/2$, then these two sites would differ only if their magnetic shifts differed. Since the quadrupolar interaction is so large, the resonant frequencies for these two sites are well separated in frequency, allowing for clean measurements of each site separately. Furthermore, for a spin $I = 1/2$ nucleus there would be only one line through the origin in Figure 23.2 corresponding to a single resonance condition. The quadrupolar nature of the $I = 9/2$ ^{115}In in this case provides several different resonance conditions for any given field or frequency. This property is crucial for investigations of the superconducting state, since it permits one to probe several frequencies without the need to change the magnetic field. It is preferable to do field-cooling for most NMR experiments or, in other words, to cool from the normal to superconducting states in constant applied field (Figure 23.1). If the magnetic field changes in the superconducting state of a type II superconductor, vortices will be swept in or out of the edges of the sample. In the bulk, the vortices will move between pinning sites. Depending on the field, temperature, and the sample properties, the vortices may not be able to form a regular vortex lattice in this case, and may exhibit a nonuniform density throughout the material causing significant line broadening.

Since the EFG is very large in this material, one cannot reliably approximate the solution of the Hamiltonian by simply treating either the Zeeman or quadrupolar terms as perturbations. As shown in Figure 23.2(a), there are several regions of field where the frequency no longer depends linearly on the field. There are also "forbidden" transitions in this region, where $\Delta m \neq \pm 1$. In order to fully characterize such behavior, the intensity of the color in the graph is shown proportional to the square of the transition matrix element $M_{ij} = \langle \phi_i | \hat{I}_+ | \phi_j \rangle$, where $\phi_{i,j}$ is the nuclear spin wavefunction and $\hat{I}_+$ is the raising operator for spin I, because in a pulsed NMR experiment the intensity of a transition will be proportional to M_{ij}^2. Note that we consider selective excitation of only a single transition in this case; in other words, not all of the transitions are excited simultaneously by a single pulse. Nonselective excitation of multiple transitions is possible in materials with much smaller EFGs, but in this case the intensities are modified (see Chapter 2). As seen in Figure 23.2, some transitions tend to fade out because the transition probability vanishes as the field increases. The reason for this behavior is that $\mathcal{H}_Q + \mathcal{H}_Z$ does not commute with $\hat{I}_z$ in general. The intercept on the vertical axis corresponds to NQR, where $H_0 = 0$. As seen in Figure 23.2, the NQR transitions split linearly for small fields when $\boldsymbol{B}_0 \parallel \hat{q}$, and have a parabolic form for $\boldsymbol{B}_0 \perp \hat{q}$.

Clearly, single crystals are critical for such studies. Powder samples would lead to broad powder pattern NMR spectra, spanning more than 100 MHz. By measuring the splitting between the quadrupolar satellites in a single crystal, one can map out the orientation of the field relative to the crystalline axes to achieve alignments to within a degree. Modern

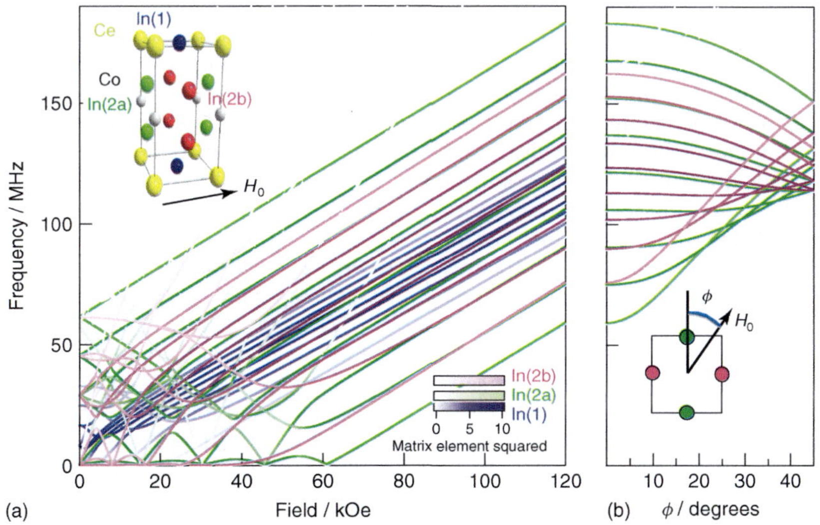

Figure 23.2. (a) Resonance frequency versus applied field along the [100] direction in $CeCoIn_5$. There are three distinct indium sites for this orientation, and each exhibits a different set of transitions. The color intensity of each transition is proportional to the square of the transition matrix element, as described in the text. (b) The resonant frequencies as a function of azimuthal angle ϕ in the *ab* plane for an applied field of 120 kOe.

NMR laboratories equipped with magnets capable of sweeping the magnetic field allow one to probe both the frequency-swept and field-swept spectra. The complete EFG tensor and crystal alignment can be ascertained by matching the positions of resonances in frequency–field space. Probe heads that include single or double axis goniometers are particularly useful for these cases. For many superconductors, the physics of the superconducting condensate depend critically on the field direction in single crystals, in which case the detailed spectra available via quadrupolar NMR are indispensable. Typically, the alignment can be done in situ to within a fraction of a degree in the normal state.

23.3 SPIN–LATTICE RELAXATION IN A SUPERCONDUCTOR

In metals and superconductors, the response of the nuclei typically is dominated by their hyperfine coupling to the electron spins, **S**:

$$\mathcal{H}_{\text{hyp}} = \hat{\boldsymbol{I}} \cdot \mathbb{A} \cdot \mathbf{S} \tag{23.3}$$

where the hyperfine coupling tensor $\mathbb{A}$ can be either a Fermi contact term or a more complicated form involving transferred couplings to nearest neighbor spins.[6,7] This hyperfine coupling leads both to a Knight shift $K = A_\alpha \chi_\alpha$, where χ_α is the static electron spin (Pauli) susceptibility along the α direction, and to the dominant mechanism for nuclear spin–lattice relaxation through time-dependent fluctuations of $\mathbf{S}(t)$. In the superconducting state, the electronic excitations in the vicinity of E_F, the Fermi energy, change dramatically due to the development of a superconducting energy gap and these changes are reflected in both the NMR Knight shift and in the spin-lattice-relaxation rate. This chapter focuses on measurements of the latter, since the most direct measurements of T_1^{-1} are via NQR in zero field, and Knight shift measurements are not unique to quadrupolar NMR. We refer the reader to any of several excellent reviews for a discussion of the Knight shift in the superconducting state, including Refs 2, 3, and 5.

23.3.1 Coherence Peak

In a fully gapped spin-singlet superconductor at $T = 0$ there are no fluctuations of $\mathbf{S}(t)$ to couple to the nuclei through equation (23.3), since the condensate has no net spin and there are no electron states available to scatter with the nuclei. At finite temperatures, quasiparticle excitations couple to the nuclear spins, but since the excitations in the superconducting state are actually *superpositions* of spin-up and spin-down quasiparticles, the hyperfine field experienced by the nuclei is different than in the normal state. This fact has profound consequences for the spin-lattice-relaxation rate in superconductors, and gives rise to the Hebel–Slichter coherence peak. As the temperature grows, there is a greater probability of quasiparticle states to be excited above the superconducting gap, Δ. Eventually, when the temperature is sufficiently large ($T > T_c = 1.76\Delta/k_B$ in the Bardeen–Cooper–Schrieffer (BCS) theory of s-wave weak coupling), the superconducting state collapses and T_1^{-1} becomes equal to that in the normal state of the metal. The temperature dependence of the spin-lattice-relaxation rate in the superconducting state is given by

$$T_1^{-1} \propto A^2 \int C_+(E, E') N_{\text{BCS}}(E) f(E) N_{\text{BCS}}(E')(1 - f(E'))\mathrm{d}E \tag{23.4}$$

where A is the (isotropic) hyperfine coupling, $f(E)$ is the Fermi function, and the density of states of an s-wave ($L = 0$ symmetry of the Cooper pairs) is given by

$$N_{\text{BCS}}(E) = \frac{E}{\sqrt{E^2 - \Delta^2}} N_{\text{N}}(E) \tag{23.5}$$

where $N_{\text{N}}(E)$ is the density of states in the normal state. The coherence factor is given by

$$C_+(E, E') = \frac{1}{2} + \frac{EE'}{2\Delta^2} \tag{23.6}$$

where $E' = E + \hbar\omega_L$, and ω_L is the Larmor frequency. Since $N_{\text{BCS}}(E)$ has a singularity at $E = \Delta$ and $C_+(E, E')$ is nonzero, expression (23.4) leads to a singularity of T_1^{-1} at $T = T_c$. This is the Hebel–Slichter coherence peak, first observed experimentally in superconducting Al as an unexpected peak in T_1^{-1} just below T_c (see dashed line in Figure 23.7).[8] The intensity of the peak is limited in practice by both a smearing of the density of states as well as spin–orbit scattering effects. In practice, the coherence factor can be rather difficult to observe experimentally, particularly in materials with heavier elements where the spin–orbit interaction is larger. The reader is referred to Ref. 2 for further details. It is important to note that this peak is relatively difficult to observe in most materials and that its absence is not necessarily an indication of unconventional superconductivity.

23.3.2 Probing the Density of States

For $T \lesssim T_c/3$ in an s-wave superconductor, equation (23.4) reduces to $T_1^{-1} \sim \exp(-\Delta/k_B T)$, and the spin-lattice-relaxation rate exhibits an activated temperature dependence as expected in the presence of a uniform energy gap. Thus measurements of T_1^{-1} provide a direct measure of the superconducting gap magnitude. This behavior, however, is present only if the gap is isotropic in **k**-space, i.e., for s-wave superconductivity. If $L > 0$ and $\Delta_\mathbf{k}$ develops nodes in **k**-space, then the density of states $N_{BCS}(E)$ is modified. For line nodes in an $L = 2$ (d-wave) superconductor, $N_{BCS}(E) \sim E/\Delta_0$ for $E \ll \Delta_0$, where Δ_0 is the maximum gap value. In this case, equation (23.4) leads to $T_1^{-1} \sim T^3$ for $T \lesssim T_c/3$. For point nodes, as may be the case for $L = 1$ (triplet) pairing, then $T_1^{-1} \sim T^5$.[9] These dramatic differences in temperature dependence arise because there are always quasiparticle excitations available at the gap nodes. Measurements of the temperature dependence of T_1^{-1} can help to identify the presence of such nodes and thus distinguish between these different types of pairing symmetries.

23.4 NQR IN SUPERCONDUCTORS

The expected temperature dependence of the spin-lattice-relaxation rate in the superconducting state outlined above almost always is modified by the presence of a vortex lattice in applied fields. The magnetic vortex lines can be mobile, with a temperature and field dependent hopping between different pinning sites. This hopping and the collective motion of the vortex lattice can give rise to fluctuating fields at the nuclei, contributing to the nuclear spin-lattice relaxation if the power spectrum of fluctuations has sufficient weight at the Larmor frequency.[10] These vortex lattice contributions can obscure the intrinsic behavior of the superconducting condensate. Thus it is advantageous to measure T_1^{-1} in zero field, eliminating all contributions from the vortex lattice. NQR is ideal to probe the low temperature intrinsic properties of superconductors and has played a central role in many of the most important superconductors discovered during the last two decades. Below we outline a selection of experiments which take advantage of the quadrupolar interaction to probe the intrinsic properties of superconductors.

23.4.1 Cu NQR in the High-Temperature Superconductors

23.4.1.1 Spin–lattice Relaxation

The high-temperature superconducting cuprates all contain 2D layers of copper oxide (CuO_2) planes (see Figure 23.4). Both isotopes ^{63}Cu and ^{65}Cu have spin $I = 3/2$ and large quadrupolar moments. The EFG at the planar Cu ($3d^9$ electronic configuration) in the cuprates is largely determined by the charge distribution resulting from the unfilled 3 d shell, and the NQR frequencies are typically in the range of 20–40 MHz. Therefore Cu NQR is an ideal probe of the superconducting state. Figure 23.3 shows the spin-lattice-relaxation rate of the two Cu sites (chain and planar sites) in the prototypical high-T_c superconductor $YBa_2Cu_3O_7$. The temperature dependence is clearly a power law, and shows no coherence peak. These data were among the first clues that the order parameter symmetry in this material was not s-wave (isotropic), which would show activated behavior. The T^3 power law clearly indicated that the density of states was consistent with either a d-wave symmetry of the Cooper pairs, or a strongly anisotropic s-wave. For either an isotropic or anisotropic s-wave gap, the superconducting gap function $\Delta_\mathbf{k}$ retains the same sign for all points in **k**-space. For higher order symmetries ($L > 0$), such as p-wave or d-wave, $\Delta_\mathbf{k}$ changes sign at nodes. In this case, the absolute magnitude is not single-valued, but distributed in some fashion between zero and the maximum gap value, Δ_0. This difference is manifest in the density of states. Measurements of the nuclear spin-lattice-relaxation rate can probe only the magnitude of the gap function, not the change of sign. Later tunneling experiments confirmed the d-wave nature of the superconducting state of this material.

The pairing symmetry of the Cooper pairs is a key piece of information about any superconductor

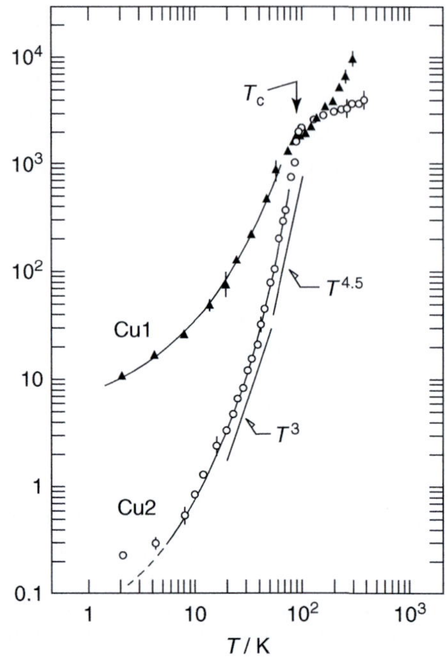

Figure 23.3. T_1^{-1} versus temperature at the chain (Cu1) and planar (Cu2) Cu sites in the high-temperature superconductor $YBa_2Cu_3O_7$ measured by NQR. (Reproduced from Ref. 13 by permission of the Physical Society of Japan.)

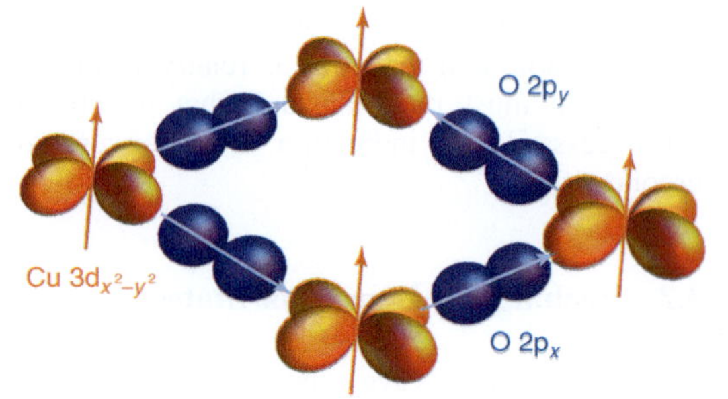

Figure 23.4. Cu $3d_{x^2-y^2}$ and O $2p_x$ and $2p_y$ orbitals in the CuO_2 plane of the cuprates. The arrows indicate the direction of $\vec{q}$, the principal axis of the EFG with the largest eigenvalue. Hole doping removes electrons from the plane, modifying the occupation of the O 2p orbitals, and consequently changing the local EFG at the Cu and O sites.

because it reflects the symmetry of the pairing interaction. In conventional superconductors, the pairing is mediated via phonons, or distortions of the lattice due to the presence of the charged electrons. This phenomenon is well understood, and is captured by the BCS theory of superconductivity. The high-temperature superconducting cuprates cannot be described by a conventional phonon pairing mechanism. Not only does such a theory predict s-wave pairing but it also limits the maximum T_c to around 40 K, a fraction of the actual value in most superconducting cuprates. The abundance of evidence for d-wave pairing, on the other hand, suggest a magnetic pairing mechanism in which the spins of the electrons of the Cooper pairs experience an attractive interaction through the exchange of antiferromagnetic fluctuations.[11] d-wave pairing is also a natural consequence of the so-called resonant valence bond (RVB) state.[12] The RVB state is fundamentally different from a nearly antiferromagnetic Fermi liquid, and reflects a different physical description of these materials. To date, there is no conclusive evidence for either pairing mechanism, although the d-wave nature originally observed by NMR has been widely accepted.

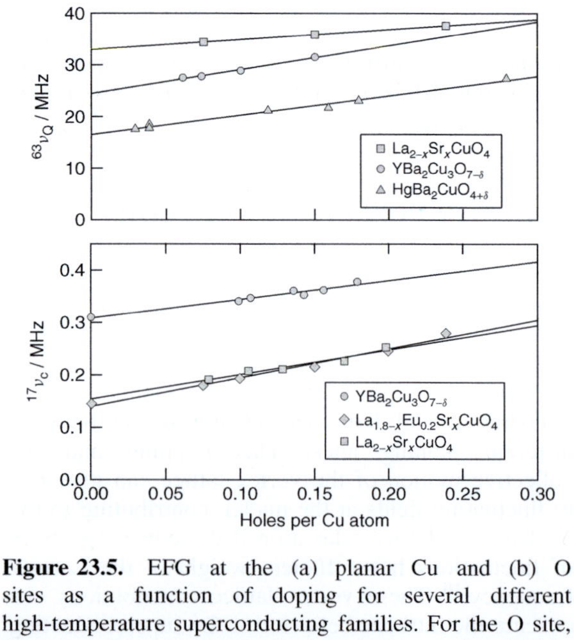

Figure 23.5. EFG at the (a) planar Cu and (b) O sites as a function of doping for several different high-temperature superconducting families. For the O site, the asymmetry parameter η is nonzero, and only the component along the c-axis is shown. (Data compiled from Refs 15, 16.)

23.4.1.2 Charge Ordering

Another distinct advantage of quadrupolar NMR in the high-temperature superconductors is that it provides information about the local *charge* environment. The common structural element of all of the cuprate superconductors is the CuO_2 planes (see Figure 23.4). The number of CuO_2 planes per unit cell and the interplane atoms vary depending on the particular cuprate family. Figure 23.5 shows the dependence of the EFG at the planar ^{17}O ($I = 5/2$) and ^{63}Cu sites as a function of doping in several different cuprate families.[14,15] Substituting Sr for La or adding excess oxygen into the structure removes electrons from (adds holes to) the CuO_2 planes. These holes go primarily into the O 2p orbitals. In the absence of doping, the Cu has nominal electronic configuration $3d^9$ (Cu^{2+}) and the oxygen has configuration $2p^6$ (O^{2-}). The unfilled Cu 3 d shell gives rise to a large EFG even in the absence of hole doping. When holes are doped into the CuO_2 planes, the filled 2p shell of the oxygen begins to empty partially and the EFG at the O nuclei is modified correspondingly. At the Cu site, the $3d^9$ configuration remains largely unaffected, but virtual hopping between the O 2p states and the nominally unoccupied Cu 4s state gives rise to a contribution to the EFG at the Cu site.[16] The EFG at the both the Cu and O sites vary linearly with doping, indicating that the EFG is a sensitive measure of the local hole content. Thus quadrupolar NMR of the O and Cu in the cuprates provides detailed information about the local charge environment in the CuO_2 planes.[17]

Recently, certain classes of the high-temperature superconductors have been observed to exhibit stripe or nematic correlations, in which both the antiferromagnetic order of Cu $3d^9$ ($S = 1/2$) spins and the local oxygen hole content is modulated over a length scale of several lattice constants. This picture contrasts with one in which the doped holes are homogeneously distributed in the CuO_2 plane, and is likely the result of strong electron–electron correlations present in these materials.[18] This inhomogeneous spin and charge configuration manifests itself in the NMR spectra as a dramatic enhancement of the quadrupolar broadening of the NMR resonances.[19] Under certain conditions, the Cu spin fluctuations become very slow at low temperatures, and the modulated antiferromagnetic order becomes quasi-static. In this case, the spin-lattice-relaxation rate at the O sites is enhanced by several orders of magnitude, causing the nuclear spins to relax so quickly that the NMR

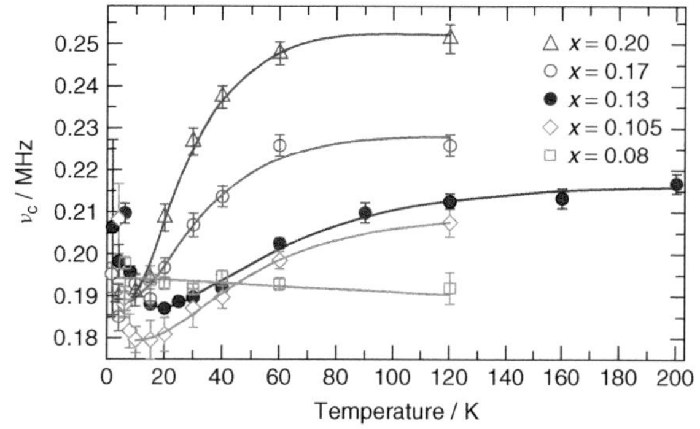

Figure 23.6. EFG at the planar O site in $La_{1.8-x}Eu_{0.2}Sr_xCuO_4$ as a function of temperature and doping. This unusual temperature dependence can be attributed to a spatial correlation between the spin and charge ordering in the CuO_2 plane, and represents a unique piece of information that is available only via quadrupolar NMR in these materials. (Reproduced from Ref. 14. © American Physical Society, 2006.)

signal is wiped out.[14] For oxygen sites located in the antiferromagnetic regions, far from the regions of enhanced hole density at the nodes of the antiferromagnetic order, the spin-lattice-relaxation rate is unaffected and these sites remain observable at low temperatures. For these sites, the EFG is reduced, since they are located in regions with lower hole doping (see Figure 23.6). As a result, the EFG appears to be temperature dependent. In reality, there is a distribution of EFGs, and only the upper portion of the distribution is wiped out by the spin fluctuations at low temperatures, and the splitting between the satellites appears to decrease, as seen in Figure 23.6. These results demonstrate a clear spatial correlation between the local charge and local spin environment in the CuO_2 planes, and exemplifies the unique type of information available with quadrupolar NMR.

23.4.2 Ga NQR in $PuCoGa_5$

NMR/NQR are particularly advantageous as probes of the low-temperature density of states because they are noninvasive techniques and require no leads to touch the sample being investigated. As such, they are ideal probes of radioactive materials, where the sample must be sealed or isolated from the environment. Figure 23.7 shows the temperature dependence of T_1^{-1} measured by ^{69}Ga (I = 3/2) in

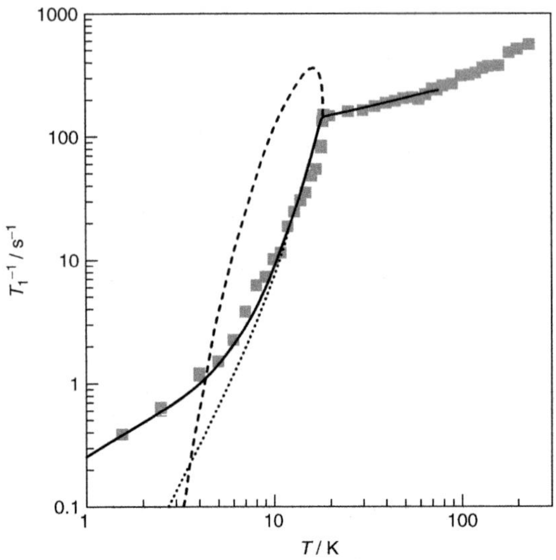

Figure 23.7. T_1^{-1} versus temperature at the Ga(2) site in the heavy-fermion superconductor $PuCoGa_5$ measured by NQR. The dashed line shows the expected behavior for an s-wave superconductor, the dotted line shows that for a pure d-wave superconductor, and the solid line shows the behavior for a dirty d-wave superconductor. (Reproduced from Ref. 20. © Nature Publishing Group.)

the plutonium-based superconductor $PuCoGa_5$.[20] In this case, the emission of α-particles by the ^{239}Pu (half-life 21 400 years) not only renders this material dangerous to handle, but has profound consequences for the superconducting state. This material has a superconducting transition temperature of nearly 19 K, but T_c reduces as the sample ages because the emission of the α-particle gives rise to tremendous lattice damage. Both the α-particle itself and the recoiling heavy ^{235}U nucleus displace thousands of atoms.[21,22] These displaced atoms introduce pair-breaking sites which suppress the superconducting condensate density, and ultimately reduces T_c to zero within a few years of synthesis. The discovery of this unconventional superconductor was a surprise, as the transition temperature was an order of magnitude larger than that of the isostructural heavy-fermion superconductor $CeCoIn_5$. This material is important theoretically because it bridges the gap between two extremes of unconventional superconducting families: the heavy-fermion and high-temperature superconductors.

The solid line in Figure 23.7 shows a theoretical fit to T_1^{-1} assuming both a d-wave superconducting gap, but with pair-breaking impurities that give rise to a finite density of states at energies below the maximum gap value. Impurities in superconductors give rise to low-energy, spatially localized electron states within the gap. For sufficiently large impurity concentrations, the electrons can hop from one site to another, forming an itinerant impurity band.[23] These impurity bands can relax the nuclei through a Korringa process. In this case, the $T_1^{-1} \sim T^3$ power law of the spin-lattice-relaxation rate is modified at low temperature and becomes $T_1^{-1} \sim T$ at a temperature corresponding to the energy scale of the impurity band. In Figure 23.7 this occurs around 8 K. The impurity band in this material is age dependent, as the concentration of impurities increases with time as the Pu undergoes radioactive decay, and gives rise to an impurity scattering rate $\Gamma(t)$, where t is the sample age. Different measures of Γ extracted by spin-lattice-relaxation rate, Knight shift, and T_c versus age are consistent between differently aged samples and indicate that T_c decreases at the rate of 1 K every four months.

23.4.3 As NQR in the Iron Pnictides

After the discovery of superconductivity in the cuprate oxide superconductors by Bednorz and Müller in the late 1980s, many scientists sought similar behavior in other transition-metal compounds. Only copper oxides exhibited high-temperature superconductivity until early 2008, when a new class of materials containing iron and arsenic was discovered with transition temperatures up to 55 K.[24,25] The iron pnictides, or ferropnictides, can be divided into three main classes, with chemical formulas RFeAsO (frequently referred to as "1111") (R = rare-earth element), AFe_2As_2 ("122") (A = Ca, Ba, or Sr), and LiFeAs ("111").[26,27] The main structural element of these materials consists of layers of FeAs planes, similar to the CuO_2 planes of the cuprate high-T_c superconductors. Here ^{75}As ($I = 3/2$) NMR has played a key role in elucidating the properties of these materials, yet the symmetry of the superconducting condensate remains mysterious to date.

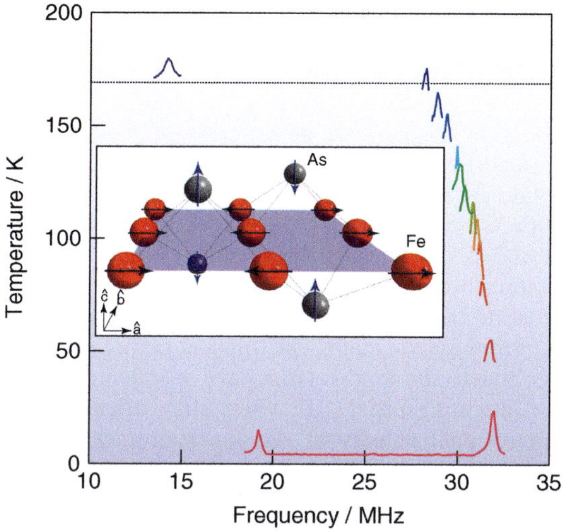

Figure 23.8. ^{75}As spectra as a function of temperature in $CaFe_2As_2$. The dotted line indicates the antiferromagnetic phase transition, which is first order. INSET: The internal field (blue arrows) at the As sites (gray spheres) produced by the ordered moments (black arrows) at the Fe atoms (red spheres) in the FeAs plane. (Reproduced from Ref. 28. © Institute of Physics, 2009.)

23.4.3.1 Antiferromagnetic Order Parameter

Among the different classes of iron pnictide superconductors, the 122 have attracted significant attention because large single crystals are available.[26] The phase diagram of these materials is strikingly similar to that of the cuprates in that the undoped system exhibits antiferromagnetism, and for sufficient doping with either holes (substituting alkali metals at the A site) or electrons (substituting other transition metals at the Fe site) the material becomes superconducting. Superconductivity can also be induced with hydrostatic pressure of several gigapascals in the undoped compounds. Studies of the latter with NMR are particularly useful since there is no doping-induced disorder in the lattice, which gives rise to a broad distribution of local EFGs.

$CaFe_2As_2$ is of particular interest because superconductivity emerges at relatively low pressures. At ambient pressure, without any doping, the material undergoes a first-order structural and magnetic transition below 169 K. Below this temperature, the lattice changes from tetragonal to orthorhombic, and the Fe spins undergo an unusual antiferromagnetic ordering in which the spins are parallel along one direction but antiferromagnetic in the perpendicular directions (see inset of Figure 23.8). In this case, As NMR is particularly useful for determining the magnetic order because the static hyperfine field at the As site does not vanish in the ordered state. Figure 23.8 shows a series of ^{75}As spectra at different temperatures in $CaFe_2As_2$ acquired at zero field.[28] Above the Neel transition, there is a single resonance corresponding to the NQR $\pm1/2 \leftrightarrow \pm3/2$ transition. Below T_N, the static internal hyperfine field lifts the degeneracy of the quadrupolar levels, and there are three transitions (the lower transition at ~9 MHz is not shown). Because of the orthorhombic transition, the asymmetry parameter also becomes nonzero. By measuring the temperature dependence of the splitting between the central and upper transition, one can extract the static hyperfine field, and hence the sublattice magnetization in this material. As seen in Figure 23.8, the internal field changes discontinuously at the first-order transition. This information is important because the antiferromagnetism and superconductivity appear to be closely linked in these materials.[29] Detailed measurements of the temperature and doping dependence of the sublattice magnetization and the nuclear spin-lattice-relaxation

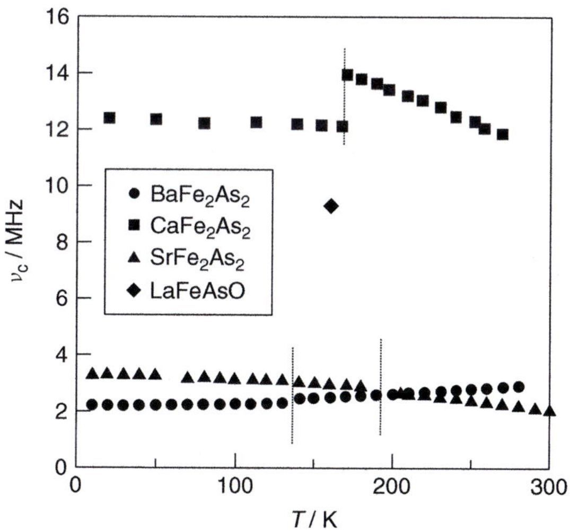

Figure 23.9. *c*-axis component of the EFG at the As site in several different parent antiferromagnetic iron pnictide systems as a function of temperature. Dotted lines indicate the structural/magnetic phase transition. (Data were compiled from Refs 7, 30–32.)

rate provide important clues about the role of antiferromagnetic fluctuations and their relationship with the superconducting state.

The temperature and material dependence of the EFG at the As site in the iron pnictides is highly unusual (see Figure 23.9). Depending on the particular alkaline-earth atom, the EFG varies by more than a factor of six, and the temperature dependence and change in the EFG at T_N differ significantly from one material to another. This material dependence contrasts with that in the high-temperature superconducting copper oxides, where the EFG varies by only a few percent from one material to another (see Figure 23.5). Curiously, Kitagawa has discovered that $\vec{q}$, the principal axis of the EFG with the largest eigenvalue, rotates by 90° below the structural transition.[7] The origin of this strong material and temperature dependence of the EFG in the iron pnictides is poorly understood at present, but certainly reflects dramatic changes in the local electronic environment at the As site.

23.4.3.2 *Superconductivity Under Pressure*

Upon doping, the EFG at the As sites in these materials becomes inhomogeneous, and the quadrupolar

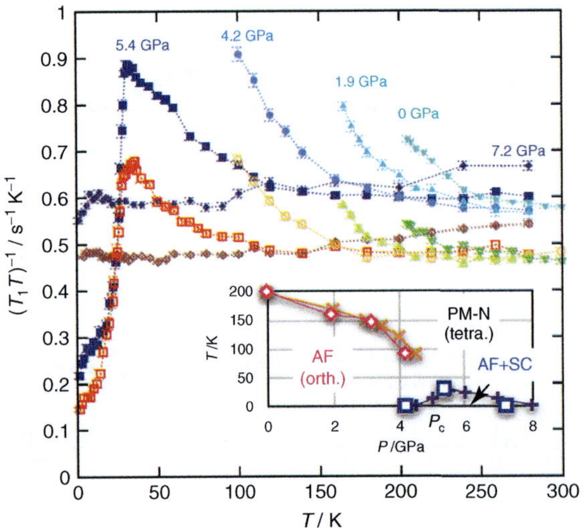

Figure 23.10. $(T_1T)^{-1}$ versus temperature and pressure in $SrFe_2As_2$. The inset shows the phase diagram as a function of pressure. (Reproduced from Ref. 34. © American Physical Society, 2009.)

satellites become too broad to observe.[33] It is therefore advantageous to investigate the phase diagram as a function of pressure in a stoichiometric compound. Figure 23.10 shows the evolution of T_1^{-1} as a function of pressure in $SrFe_2As_2$.[34] In this case, the quadrupolar satellites remain well defined, and provide information about the crystal structure under pressure. For pressures greater than 4 GPa, this material becomes superconducting. For a range of pressures between 4 and 6 GPa, there is a spontaneous formation of a nanoscale hybrid superconducting and antiferromagnetic state. The unusual nature of the superconductivity in this structure is reflected in the T_1^{-1} data in Figure 23.10, which shows that a significant residual density of states remains even as the temperature approaches zero. This residual density of states is manifest as a finite $(T_1T)^{-1}$ at $T = 0$ in Figure 23.10 for 5.4 and 6 GPa. In a conventional superconductor, $(T_1T)^{-1}$ should vanish at $T = 0$ if the entire Fermi surface is gapped by the superconducting state. The novel state that emerges in this system is particularly interesting from a theoretical viewpoint, as both antiferromagnetic and superconducting states are competing for the ground state in this pure stoichiometric compound. In a doped system, one might expect this competition to lead to a spatially inhomogeneous state. Here the electronic system appears to spontaneously phase-separate at the nanometer scale. These experiments depend critically on the capability to observe simultaneously both the presence of an internal static hyperfine field, and the presence of structural orthorhombicity. The latter is determined from the position of the quadrupolar satellites of the As, and demonstrates a clear example in which quadrupolar NMR plays a crucial role in the study of this new state of matter.

23.5 ANTIFERROMAGNETISM AND SUPERCONDUCTIVITY

In nearly all of the unconventional superconductors discovered in the last two decades, there appears to be an intimate connection between antiferromagnetism and superconductivity.[11,35] In such cases, there are regions of phase space in which these two order parameters, anathema to one another in the conventional theory of superconductivity, actually coexist microscopically. The nature of this coexistence can be identified and probed via quadrupolar NMR. The case discussed above of $SrFe_2As_2$ under pressure is one example. Below we discuss the coexistence of these two orders in the heavy fermion system, $CeCoIn_5$. In this material, the Ce 4f electrons hybridize with the conduction electrons, giving rise to a number of unusual low-temperature properties including enhanced specific heat, magnetic susceptibility, and d-wave superconductivity. The 4f electrons exhibit both localized properties with Ce magnetic moments associated with an unfilled f shell typical of rare-earth systems, and itinerant behavior with large Pauli susceptibilities at low temperatures. This dichotomy leads to antiferromagnetism, superconductivity, and a wide range of unusual spin fluctuations associated with the proximity to a quantum phase transition at zero temperature, in which the RKKY (Ruderman–Kittel–Kasuya–Yosida) interaction between localized Ce magnetic moments competes with Kondo screening, where the itinerant conduction electrons form spin singlets with the local Ce magnetic moments.[36] This behavior is poorly understood theoretically, but leads to a rich spectrum of unusual ground state behaviors, including field-induced antiferromagnetism within the superconducting phase.

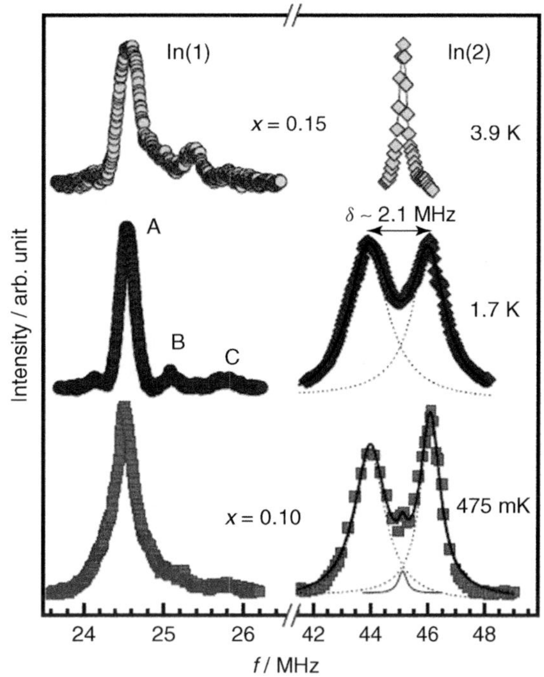

Figure 23.11. Spectra of the In(1) and In(2) in $CeCo(In_{1-x}Cd_x)_5$. The gray points and red data points were obtained on the $x = 0.15$ sample, and the blue points were obtained on the $x = 0.10$ sample. (Reproduced from Ref. 39. © American Physical Society, 2007.)

23.5.1 In NQR in Doped $CeCoIn_5$

$CeCoIn_5$ contains two distinct In sites (see Figure 23.2), with two different symmetries. The In(1) site has axial symmetry with $\nu_Q = 8.173$ MHz; the In(2) site does not have axial symmetry, with $\nu_c = 15.489$ MHz and $\eta = 0.386$.[37] ^{115}In ($I = 9/2$) NQR is ideal to investigate both the superconducting state, as well as the subtle changes induced in this material with either doping or applied magnetic field. When $CeCo(In_{1-x}Cd_x)_5$ is doped with Cd at the In site, the ground state changes from superconducting to antiferromagnetic.[38] For $x = 0.1$, the ground state is both superconducting and antiferromagnetic, and for $x = 0.15$ the ground state is purely antiferromagnetic. Figure 23.11 shows the NQR spectra of both the In(1) and In(2) sites for both $x = 0.1$ and $x = 0.15$ for three different temperatures.[39] In all three cases, the In(1) spectrum remains unaffected by the onset of antiferromagnetic order of the Ce electron spins below T_N, but the In(2) spectrum clearly shows a splitting in the ordered state. The splitting is due to a static transferred hyperfine coupling to In sites from the nearest neighbor Ce spins. Although the In(1) site also experiences a similar transferred coupling to the four nearest neighbor Ce sites, the high symmetry of this crystallographic site leads to a complete cancellation of the static hyperfine field. There is no such cancellation at the low-symmetry In(2) site.

In the $x = 0.1$ sample, antiferromagnetism develops first at higher temperature ($T_N \approx 2.8$ K), and superconductivity emerges at a slightly lower temperature within the antiferromagnetic state ($T_c \approx 1.2$ K). Clearly, the splitting develops below T_N and *remains unchanged* below T_c. The In(2) spectrum reveals a homogeneous splitting, therefore the antiferromagnetic order must remain constant throughout the sample. The antiferromagnetism coexists *microscopically* with the superconductivity, and the two order parameters are neither phase-separated nor inhomogeneous. Measurements of T_1^{-1} also probe both phase transitions and confirm this conclusion. This result is surprising, since it indicates that the same electrons are responsible for both ground states simultaneously, reflecting the duality of the f electrons as both localized and itinerant.

23.5.2 Field-Induced Antiferromagnetism

Pure $CeCoIn_5$ also exhibits field-induced antiferromagnetism within the superconducting phase for applied fields perpendicular to the tetragonal *c*-axis.[40] Figure 23.12 shows a series of spectra of the In(1) $(-5/2 \leftrightarrow -7/2)$ transition and the In(2) $(-1/2 \leftrightarrow -3/2)$ transition acquired in a magnetic field of 11.1 T applied along the (100) direction (see also Figure 23.2).[40] The superconducting phase diagram of this material shows a highly unusual feature at high magnetic fields close to H_{c2}.[41,42] A new thermodynamic phase emerges for fields $H_{AF} < H < H_{c2}$ and below $T < T_{AF} < T_c$, where $H_{AF}(T)$ ($T_{AF}(H)$) describes this new phase boundary. This phase was initially identified as a Fulde–Ferrell–Larkin–Ovchinnikov (FFLO) form of superconductivity.[43,44] FFLO superconductivity was predicted several decades ago, and can emerge in special cases where the effective mass of the quasiparticles is large, as in $CeCoIn_5$. The NMR spectra in Figure 23.12 clearly show that this phase is more complex than a simple FFLO picture. As seen in the figure, the behavior of the In(1) and In(2) resonances in this phase are completely different from each other. This different response cannot be reconciled within any theory of superconductivity alone, since the length scale for changes of the superfluid density is much larger than the size of the lattice. The NMR data clearly show that there are distinct changes *within the unit cell*. The only way to understand the different response of these two sites is the presence of local moment magnetism.

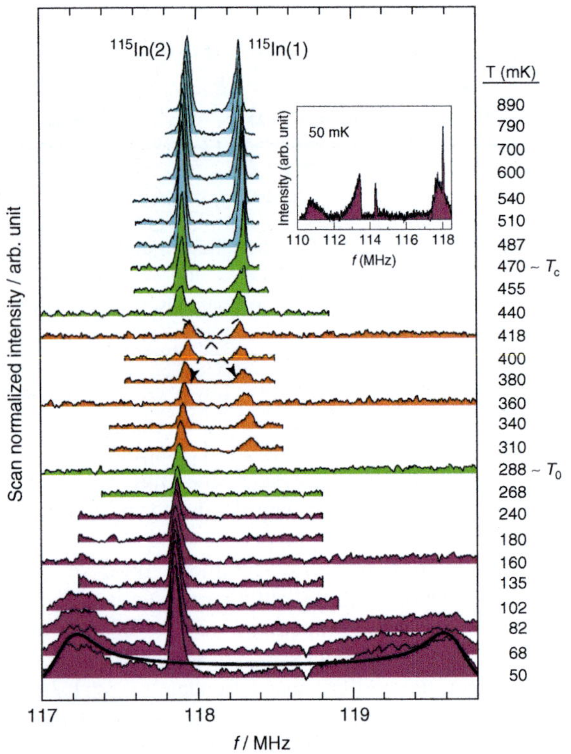

Figure 23.12. NMR spectra of the In(1) and In(2) in $CeCoIn_5$ in a field of 11.1 T along the [100] direction. Cyan spectra are in the normal state, orange spectra are in the superconducting state, and purple spectra are in the field-induced antiferromagnetic state. Green spectra indicate crossover regions from one phase to another. The solid line shows a simulation for an incommensurate antiferromagnetic spin density. (Reproduced from Ref. 40. © American Physical Society, 2007.)

In fact, the In(2) spectra clearly reveal the presence of long-range antiferromagnetic order of Ce moments, similar to that observed in Figure 23.11. The solid line in Figure 23.12 shows the expected spectrum in this case. Once again, as in the case of Cd-doped $CeCoIn_5$ and in $SrFe_2As_2$ under pressure, data from quadrupolar NMR provide evidence for a homogeneous coexistence of antiferromagnetic order and superconductivity. Recent neutron scattering data have confirmed this interpretation in this field-induced phase of $CeCoIn_5$, as well as more complete NMR experiments as a function of field.[45,46] The discovery of field-induced antiferromagnetism in this material has stimulated significant theoretical interest.[47] At present it is unclear whether the superconducting phase itself is purely d-wave, or a more exotic form of FFLO coexisting with the antiferromagnetism. Whichever the case may be, the advantage of quadrupolar NMR is that several distinct nuclear transitions of the same site can be measured *at the same magnetic field*. Not only does this fact eliminate the need to change the magnetic field within the superconducting state, but it also allows confirmation of the effect by checking the response of several different transitions.[48]

23.6 CHARGED VORTICES IN HIGH-TEMPERATURE SUPERCONDUCTORS

We conclude this article with a fascinating discovery in the vortex state of the high-temperature superconductors.[15] Figure 23.13 shows the change of the EFG at the planar Cu site measured in two different high-temperature superconductors as a function of temperature. The NQR frequency of the Cu ($I = 3/2$) was measured first as a function of temperature in zero field. Next, the splitting between the upper ($+3/2 \leftrightarrow +1/2$) and lower ($-3/2 \leftrightarrow -1/2$) satellite transitions was measured in an applied field. The difference between these two frequencies is equal to $2\nu_Q$. The field-induced change $\delta\nu_Q(H,T) = \nu_Q(H,T) - \nu_Q(0,T)$ is shown in Figure 23.13. The EFG apparently is field dependent in the superconducting state. This result is surprising, since the EFG typically remains unchanged by the superconducting order. The superconducting condensate is composed of electron states within $k_B T_c$ of the Fermi energy and does not modify the charge density at the nuclei. The dominant contributions to the EFG include local orbital occupations and lattice charges, which are determined by energy scales much larger than $k_B T_c$.

In fact, the changes observed in Figure 23.13 have been attributed to excess charge present in the vortex cores. A vortex can be thought of as a tube of normal state penetrating a superconducting background along the direction of the applied field. In the vortex core, the superconducting order parameter vanishes within a radius of $\xi \sim 30$ Å for the high-temperature superconductors. The chemical potential within the cores differs from that outside the vortex cores, which gives rise to a net transfer of charge from one subsystem to the other. Since there is a net energy gain for the charge carriers that form the superconducting condensate, it is energetically favorable for the vortices to transfer charge to the superconducting background.[49] In a field of 8–9 T, the volume fraction of vortices is less than 10%, and the dominant contribution to the NMR signal arises from nuclei located outside the vortex cores in the superconducting background. Since the vortex cores become charged, the background charge density also becomes modified in order to maintain charge neutrality over the bulk of the material. This change is reflected in the Cu EFG through changes of the electron occupation of the Cu 4p and 3 d electron orbitals. Since the dominant contribution to the EFG is from the local charge distribution in the unfilled electron shells surrounding the nucleus, it is a sensitive probe of the subtle changes to the charge density of the superconducting background. The vortices present in the materials shown in Figure 23.13 carry excess charges of $|Q| \sim 10^{-2}e$. The magnitude of this value is a factor of 10 larger than theoretical predictions, and may be related to the presence of antiferromagnetic order in the vortex cores, which would enhance the charge imbalance between the superconducting background and the normal vortex cores. The difference between the two materials is likely due to the different doping levels.

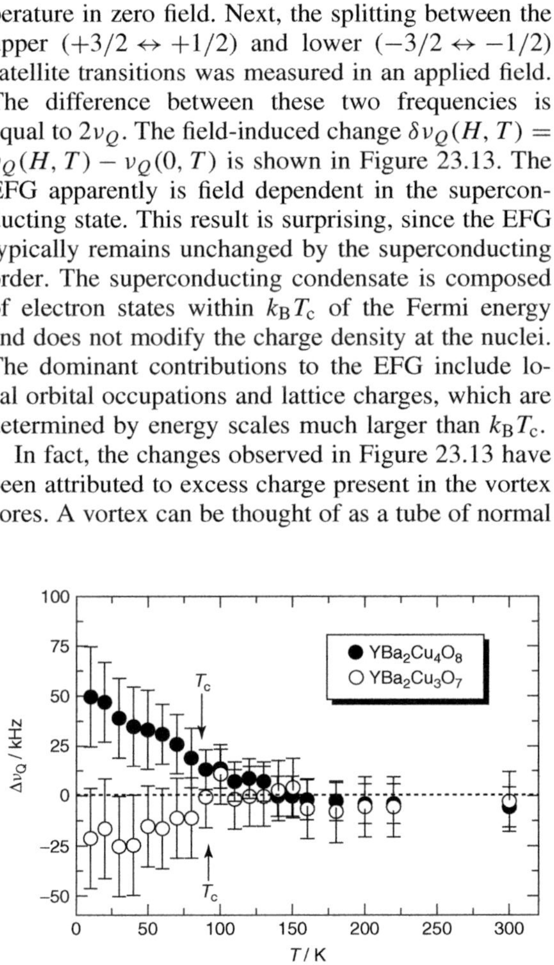

Figure 23.13. Change of the ^{63}Cu NQR frequency $\delta\nu_Q(H,T) = \nu_Q(H,T) - \nu_Q(0,T)$ as a function of temperature for $YBa_2Cu_4O_8$ and $YBa_2Cu_3O_7$. The change in the superconducting state has been ascribed to vortex charging, as discussed in the text. (Reproduced from Ref. 15. © American Physical Society, 2001.)

23.7 SUMMARY AND FUTURE DIRECTIONS

Quadrupolar NMR has contributed enormously to the study of superconductors, both conventional and unconventional. Although this article focuses primarily on recent developments in the newly discovered unconventional superconductors, NQR measurements of the nuclear spin-lattice-relaxation rate played a significant role several decades ago in the conventional BCS superconductor In.[2] In fact, the EFG has been observed to shift below T_c in superconducting In, Ga, and La metals, but the origin of this behavior is not presently understood.[50–52] In the unconventional superconductors, there still remain several open questions that can only be addressed with quadrupolar NMR. In particular, the unusual field-induced magnetism in $CeCoIn_5$ and $CeRhIn_5$ (under pressure), as well as the pressure-induced superconducting phase of the AFe_2As_2 materials, may constitute an exotic new state of matter that has only begun to be investigated. No other technique is capable of probing the microscopic properties of this state under the extreme conditions of high magnetic field, high pressure (in the case of $CeRhIn_5$), and ultralow temperature (down to 10 mK). For further details, we direct the reader to Ref. 5.

In general, dopant atoms in any of the unconventional superconductor families tend to modify the local EFG, which opens the door to investigations of the local physics in the vicinity of impurities. These types of studies have led to important discoveries about the high-temperature superconducting cuprates[4] as well as the heavy fermion superconductors, and will undoubtedly provide new and important information about the behavior of other systems, such as the iron arsenide superconductors. An important open question in the iron arsenides is what determines the large EFG at the As site. Not only does the EFG change by up to factors of six between different families of materials but the temperature dependence of the EFG can vary enormously from one material to the next, even though the crystal structure remains unchanged. A fundamental understanding of the origin of the EFG in these materials, similar to that outlined in Section 23.4.1 for the CuO_2 plane in the cuprates, may provide new information about the electronic structure in this new family of unconventional superconductors. Furthermore, investigations of the local response of the iron pnictides to impurities (which modify the local EFG) may shed new light on the unusual magnetism and superconductivity in these materials.

RELATED ARTICLES IN THE ENCYCLOPEDIA OF MAGNETIC RESONANCE

Cryogenic NMR probes: Applications

High Temperature Superconductors

Knight Shift

Metallic Superconductors

Metals: Pure and Alloyed

Probe Design and Construction

Quadrupolar Nuclei in Liquid Samples

Solid State NMR Probe Design

REFERENCES

1. I. Guillamón, H. Suderow, S. Vieira, L. Cario, P. Diener, and P. Rodière, *Phys. Rev. Lett.*, 2008, **101**, 166407.
2. D. E. MacLaughlin, *Solid State Phys.*, 1976, **31**, 168.
3. Y. Kuramoto and Y. Kitaoka, *Dynamics of Heavy Electrons*, Oxford University Press: Oxford, 2000.
4. H. Alloul, J. Bobroff, M. Gabay, and P. J. Hirschfeld, *Rev. Mod. Phys.*, 2009, **81**, 64.
5. N. J. Curro, *Rep. Prog. Phys.*, 2009, **72**, 026502 (24pp).
6. F. Mila and T. M. Rice, *Phys. Rev. B*, 1989, **40**, 11382.
7. K. Kitagawa, N. Katayama, K. Ohgushi, M. Yoshida, and M. Takigawa, *J. Phys. Soc. Jpn.*, 2008, **77**, 114709.
8. L. C. Hebel and C. P. Slichter, *Phys. Rev.*, 1959, **113**, 1504.
9. M. Sigrist and K. Ueda, *Rev. Mod. Phys.*, 1991, **63**, 239.
10. N. J. Curro, C. Milling, J. Haase, and C. P. Slichter, *Phys. Rev. B*, 2000, **62**, 3473.
11. P. Monthoux, D. Pines, and G. G. Lonzarich, *Nature*, 2007, **450**, 1177.
12. P. Anderson, in *Valence Fluctuations and Heavy Fermions*, eds L. C. Gupta and S. K. Malik, Plenum: New York, 1987, p. 9.
13. T. Imai, T. Shimizu, T. Tsuda, H. Yasuoka, T. Takabatake, Y. Nakazawa, and M. Ishikawa, *J. Phys. Soc. Jpn.*, 1988, **57**, 1771.
14. H. J. Grafe, N. J. Curro, M. Hucker, and B. Buchner, *Phys. Rev. Lett.*, 2006, **96**, 017002.
15. K.-I. Kumagai, K. Nozaki, and Y. Matsuda, *Phys. Rev. B*, 2001, **63**, 144502.
16. J. Haase, O. P. Sushkov, P. Horsch, and G. V. M. Williams, *Phys. Rev. B*, 2004, **69**, 094504.
17. S. Ohsugi, Y. Kitaoka, K. Ishida, and K. Asayama, *J. Phys. Soc. Jpn.*, 1991, **60**, 2351.

18. S. A. Kivelson, E. Fradkin, and V. J. Emery, *Nature*, 1998, **393**, 550.
19. T. Imai, C. P. Slichter, K. Yoshimura, and K. Kosuge, *Phys. Rev. Lett.*, 1993, **70**, 1002.
20. N. J. Curro, T. Caldwell, E. D. Bauer, L. A. Morales, M. J. Graf, Y. Bang, A. V. Balatsky, J. D. Thompson, and J. L. Sarrao, *Nature*, 2005, **434**, 622.
21. J. L. Sarrao, L. A. Morales, J. D. Thompson, B. L. Scott, G. R. Stewart, F. Wastin, J. Rebizant, P. Boulet, E. Colineau, and G. H. Lander, *Nature*, 2002, **420**, 297.
22. B. D. Wirth, A. J. Schwartz, M. J. Fluss, M. J. Caturla, M. A. Wall, and W. G. Wolfer, *MRS Bull.*, 2001, **26**, 679.
23. A. V. Balatsky, I. Vekhter, and J.-X. Zhu, *Rev. Mod. Phys.*, 2006, **78**, 61.
24. H. Takahashi, K. Igawa, K. Arii, Y. Kamihara, M. Hirano, and H. Hosono, *Nature*, 2008, **453**, 376.
25. R. Zhi-An, L. Wei, Y. Jie, Y. Wei, S. Xiao-Li, Zheng-Cai, C. Guang-Can, D. Xiao-Li, S. Li-Ling, Z. Fang, *et al.*, *Chin. Phys. Lett.* 2008, **25**, 2215.
26. P. C. Canfield and S. L. Bud'ko, *Annu. Rev. Condens. Matter Phys.*, 2010, **1**, 27.
27. K. Ishida, Y. Nakai, and H. Hosono, *J. Phys. Soc. Jpn.*, 2009, **78**, 062001.
28. N. J. Curro, A. P. Dioguardi, N. ApRoberts-Warren, A. C. Shockley, and P. Klavins, *New J. Phys.*, 2009a, **11**, 10.
29. I. Mazin and J. Schmalian, *Physica C*, 2009, **469**, 614.
30. S.-H. Baek, N. J. Curro, T. Klimczuk, E. D. Bauer, F. Ronning, and J. D. Thompson, *Phys. Rev. B*, 2009, **79**, 4.
31. G. Lang, H.-J. Grafe, D. Paar, F. Hammerath, K. Manthey, G. Behr, J. Werner, and B. Büchner, *Phys. Rev. Lett.*, 2010, **104**, 097001.
32. K. Kitagawa, N. Katayama, K. Ohgushi, and M. Takigawa, *J. Phys. Soc. Jpn.*, 2009b, **78**, 063706.
33. A. P. Dioguardi, N. apRoberts Warren, A. C. Shockley, S. L. Bud'ko, N. Ni, P. C. Canfield, and N. J. Curro, *Phys. Rev. B*, 2010, **82**, 140411.
34. K. Kitagawa, N. Katayama, H. Gotou, T. Yagi, K. Ohgushi, T. Matsumoto, Y. Uwatoko, and M. Takigawa, *Phys. Rev. Lett.*, 2009a, **103**, 257002.
35. T. Moriya and K. Ueda, *Rep. Prog. Phys.*, 2003, **66**, 1299.
36. P. Coleman and A. J. Schofield, *Nature*, 2005, **433**, 226.
37. N. J. Curro, B. Simovic, P. C. Hammel, P. G. Pagliuso, J. L. Sarrao, J. D. Thompson, and G. B. Martins, *Phys. Rev. B*, 2001, **64**, 180514.
38. L. D. Pham, T. Park, S. Maquilon, J. D. Thompson, and Z. Fisk, *Phys. Rev. Lett.*, 2006, **97**, 4.
39. R. R. Urbano, B.-L. Young, N. J. Curro, J. D. Thompson, L. D. Pham, and Z. Fisk, *Phys. Rev. Lett.*, 2007, **99**, 4.
40. B.-L. Young, R. R. Urbano, N. J. Curro, J. D. Thompson, J. L. Sarrao, A. B. Vorontsov, and M. J. Graf, *Phys. Rev. Lett.*, 2007, **98**, 036402.
41. A. Bianchi, R. Movshovich, C. Capan, P. G. Pagliuso, and J. L. Sarrao, *Phys. Rev. Lett.*, 2003, **91**, 187004.
42. H. A. Radovan, N. A. Fortune, T. P. Murphy, S. T. Hannahs, E. C. Palm, S. W. Tozer, and D. Hall, *Nature*, 2003, **425**, 51.
43. P. Fulde and R. A. Ferrell, *Phys. Rev.*, 1964, **135**, A550.
44. A. I. Larkin and Y. N. Ovchinnikov, *Sov. Phys. JETP*, 1965, **20**, 762.
45. M. Kenzelmann, T. Strassle, C. Niedermayer, M. Sigrist, B. Padmanabhan, M. Zolliker, A. D. Bianchi, R. Movshovich, E. D. Bauer, J. L. Sarrao, *et al.*, *Science*, 2008, **321**, 1652.
46. V. F. Mitrovic, M. Horvatic, C. Berthier, G. Knebel, G. Lapertot, and J. Flouquet, *Phys. Rev. Lett.*, 2006, **97**, 117002.
47. D. F. Agterberg, M. Sigrist, and H. Tsunetsugu, *Phys. Rev. Lett.*, 2009, **102**, 4.
48. N. J. Curro, B.-L. Young, R. Urbano, and M. J. Graf, *J. Low Temp. Phys.*, 2009b, **188**, 89. DOI: 10.1140/epjst/e2010-01299-6.
49. D. I. Khomskii and A. Freimuth, *Phys. Rev. Lett.*, 1995, **75**, 1384.
50. W. W. Simmons and C. P. Slichter, *Phys. Rev.*, 1961, **121**, 1580.
51. R. H. Hammond and W. D. Knight, *Phys. Rev.*, 1960, **120**, 762.
52. W. M. Poteet, R. F. Tipsword, and C. D. Williams, *Phys. Rev. B*, 1970, **1**, 1265.

Chapter 24
Quadrupolar NMR of Semiconductors

James P. Yesinowski

Chemistry Division, Naval Research Laboratory, Washington, DC 20375-5342, USA

24.1 INTRODUCTION

Nuclear magnetic resonance studies of quadrupolar nuclei have played an important role in the study of semiconductors for over half a century. This is an ongoing role, continually refreshed by advances in instrumentation, new NMR techniques, new theoretical approaches, as well as advances in the science of semiconductors itself. The preparation of new types of semiconductor samples by various different methods, especially nanocrystals or quantum dots, has also enlarged the scope of NMR investigations in the past few decades.

The rubric of "quadrupolar NMR of semiconductors" is, however, somewhat artificial, since spin-1/2 nuclei provide the same types of information about chemical and Knight shifts, electron hyperfine interactions, and indirect scalar coupling constants as do the quadrupolar nuclei. This chapter will consider such topics, but with examples drawn from studies of quadrupolar nuclei. It will, at the same time, cover those aspects that are unique to quadrupolar nuclei as a result of their interaction with electric field gradients (EFGs).

NMR of Quadrupolar Nuclei in Solid Materials
Edited by Roderick E. Wasylishen, Sharon E. Ashbrook and Stephen Wimperis

ISBN: 978-0-470-97398-1

24.2 BACKGROUND

24.2.1 Semiconductor Characteristics

Semiconductors are electronic conductors having typical room-temperature resistivity values of 10^{-2}–10^{9} Ω-cm, intermediate between good metallic conductors (10^{-6} Ω-cm) and insulators (10^{14}–10^{22} Ω-cm).[1] This is a consequence of having a bandgap, or energy gap, E_g between the occupied orbitals of a valence band (VB) and an unoccupied (at absolute zero temperature) conduction band (CB), as shown in Figure 24.1(a). An electron can be promoted from the VB into the delocalized states of the CB, either by thermal or optical excitation (the latter being more efficient for direct rather than indirect gap semiconductors). This produces simultaneously an electron carrier as well as a hole (the absence of an electron) carrier in the VB. The quantity known as the *density of states* (DOS) is depicted for the CB.

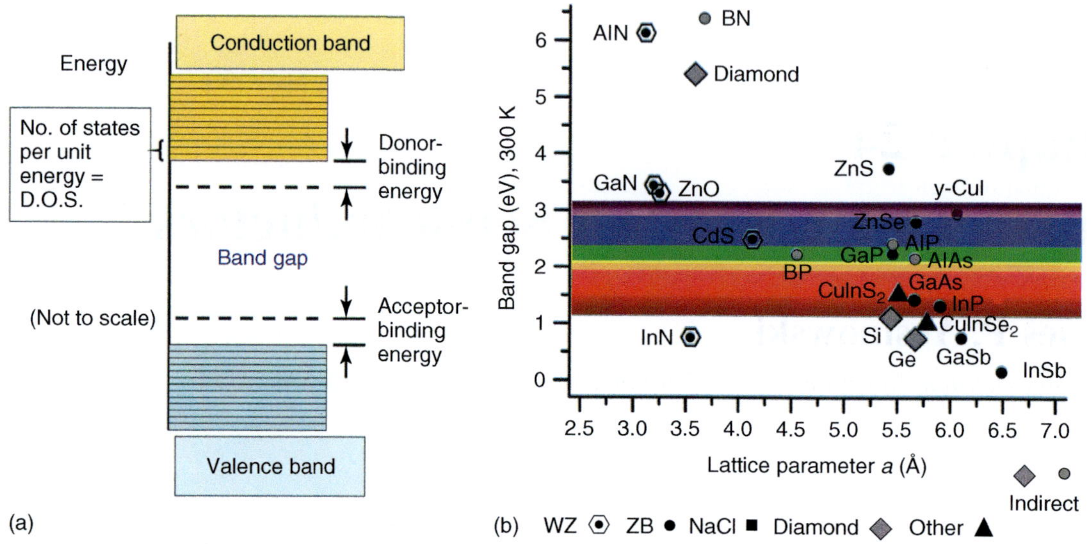

Figure 24.1. (a) Schematic energy levels in a semiconductor. (b) Bandgap energy vs lattice constant for semiconductors, depicting the most common structural type. All are direct gap semiconductors except as indicated. (Data plotted were obtained from Ref. 2.)

Carriers may also be produced by foreign atoms either intentionally or unintentionally introduced (dopants) as well as by structural defects or deviations from ideal stoichiometry (native defects). When any such species have energy levels that are below the CB by no more than a few times the thermal energy (kT), and are capable of donating an electron into the CB, they are known as *shallow donors*. On the other hand, if they have levels within a few kT above the VB and can accept an electron from the VB, they are called *shallow acceptors*. The former produces n-type-doped semiconductors, the latter p-type. When both are present simultaneously, full or partial compensation results. Shallow donor and acceptor energies as well as bandgaps and other physical and electronic properties of semiconductors are given in an extensive handbook.[2]

Figure 24.1(b) depicts the structural information and bandgaps for many of the semiconductors studied by quadrupolar NMR. Wide bandgap semiconductors have energy gaps in the UV, and narrow gap ones in the IR. Not shown are semiconductor alloys, whose bandgaps are generally intermediate between those of the end members.

Above a certain critical concentration of carriers, which differs for each semiconductor, a Mott transition, commonly referred to as the *metal–insulator transition* (MIT), occurs. Above this transition, the electrical conductivity is greater (although still very much less than that of metals such as copper) and has a temperature dependence characteristic of true metals. The nature of this transition and of the intermediate regime has been extensively studied by NMR in silicon and germanium, as summarized in an early review.[3]

24.2.2 Quadrupolar NMR of Semiconductors

Quadrupolar NMR in crystalline inorganic semiconductors has the ability to provide information on both atomic and electronic structures. Atomic structure refers to such aspects as (i) the identification of polytypes (structural forms differing in stacking sequences) or polymorphs; (ii) the presence, concentration, and local structure of defects or dopants in the crystal lattice; (iii) the short-range and longer range structure of semiconductor alloys and the degree of partial ordering. The information is typically conveyed by the effects of electronic charge on the EFGs and resultant nuclear quadrupole coupling constants (NQCCs), which are characterized by the coupling constant parameter C_Q and asymmetry parameter η.

Electronic structure in this context refers to those aspects related to the existence of carriers. Here the effects are primarily those of their *spin* property on the quadrupolar NMR. Delocalized carriers in the CB or VB can produce Knight shifts[4,5] analogous to those seen in metals, but almost always much smaller and often difficult to observe. They may also serve as a source of spin–lattice relaxation for the nuclei via the Korringa mechanism.[4,5]

The NMR of inorganic semiconductors has been the subject of a recent general review.[6] Optical pumping in semiconductors has been covered theoretically and experimentally in a 1984 monograph[7] and in subsequent review articles,[8,9] and so will not be specifically discussed here.

Several factors account for the important role accorded quadrupolar nuclei in studies of semiconductors. With the prominent exception of undoped silicon, many of the most technologically important semiconductors contain quadrupolar nuclei of both the cation and anion in binary compounds. This is true of the important class of III–V semiconductors, typified by GaAs and GaN. Almost all these quadrupolar nuclei have high natural abundances, reasonably large magnetogyric ratios, and modest spin–lattice relaxation times. These three factors result in good NMR detection sensitivity, even at the low magnetic field strengths used in early studies. The receptivity also improves with increasing I to an extent that depends on the detailed assumptions made.

With the exception of ^{14}N ($I = 1$) and ^{10}B ($I = 3$), all these nuclei have half-integral spin $> 1/2$. Furthermore, the most common semiconductor ideal structural types discussed below involve perfect or slightly distorted tetrahedral environments or else octahedral environments. This has the important consequence that the nuclear electric quadrupole interactions are typically modest or even unobservably small. Thus, the observation of a $1/2 \leftrightarrow -1/2$ central transition (CT), as well as in many cases the satellite transitions (ST), is more straightforward than in the case of the larger quadrupolar interactions found in many other classes of materials. For the important case of semiconductor alloys however, this is no longer true, and newer techniques that have been developed since the 1990s to investigate quadrupolar nuclei with significant quadrupole couplings are becoming increasingly relevant.

24.2.3 Common Semiconductor Structures and Quadrupolar NMR Nuclei

In the simple diamond lattice structure adopted by elemental silicon and other Group IV elements, the sole quadrupolar nuclei that have been studied are ^{73}Ge in elemental germanium and $^{10,11}B$ in boron-doped silicon and diamond.

The zincblende (ZB) structure (Figure 24.2) is the stable form for all combinations of the binary III–V semiconductors, except the nitrides, given by the expression ($^{10,11}B$,^{27}Al,$^{69,71}Ga$,$^{113,115}In$)(P,^{75}As,$^{121,123}Sb$), where all the quadrupolar isotopes have been given superscripts. It is also adopted by the II–VI semiconductor combinations (^{67}Zn)(^{17}O,^{33}S,Se,Te), and the I–VII semiconductors ($^{63,65}Cu$)($^{35,37}Cl$,$^{79,81}Br$,^{127}I). Most of the other II–VI semiconductors also adopt the ZB structure

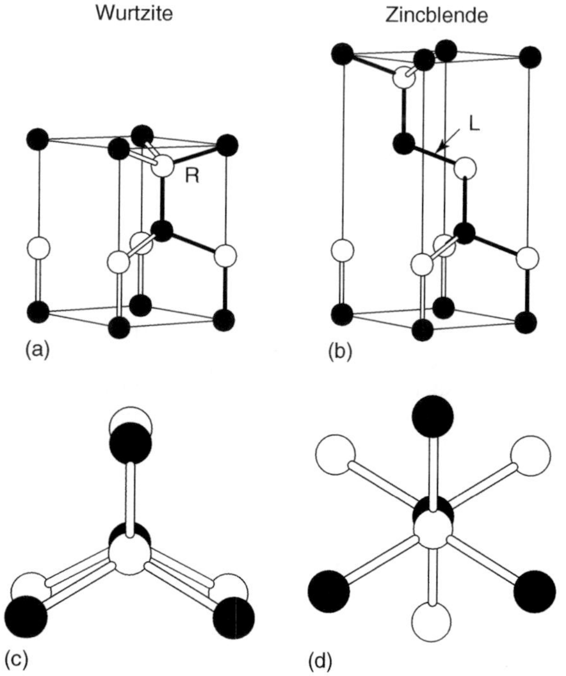

Figure 24.2. Differences between wurtzite (WZ) and zincblende (ZB) polytypes: (a) and (b) show the handedness of the fourth interatomic bond is along the right (R) for WZ and left (L) for ZB (c) and (d) show the eclipsed (WZ) and staggered (ZB) conformation of atoms. (Reprinted figure with permission from Ref. 10. Copyright 1992 by the American Physical Society.)

but involve solely spin-1/2 nuclei. The ZB structure is cubic, consisting of perfect tetrahedra of one element bonded to four atoms of the other element, and stacked in an ABCABC... arrangement.

The wurtzite (WZ) structure (Figure 24.2) is the stable form for the Group III nitrides (^{10,11}B,^{27}Al,^{69,71}Ga,^{113,115}In)(^{14}N), although the slightly less stable ZB form can be made under certain conditions. The WZ structure is hexagonal, leading to an axial distortion of the tetrahedrally bonded atoms along the *c*-axis, and has an ABABAB... stacking arrangement. Thus, the chemical shift anisotropy (CSA) and EFG tensors are axially symmetric, with a principal axis along the *c*-axis in the ideal compounds.

Other common structures include the chalcopyrite structure, related to the ZB structure, adopted by (^{63,65}Cu)(^{113,115}In)(^{33}S,Se)$_2$, and the rocksalt (NaCl) structure adopted by (Cd)(^{17}O). One technologically important wide bandgap semiconductor is titania, (^{47,49}Ti)(^{17}O)$_2$, which commonly exists in rutile or anatase polymorphs.

24.2.4 Experimental Considerations for NMR of Semiconductor Samples

The various growth techniques for preparing semiconductors are discussed in the text by Yu and Cardona.[11] Methods widely used in the semiconductor industry typically produce single-crystal semiconductor films with thicknesses ranging from tens of nanometers to hundreds of micrometers attached to substrates such as sapphire, silicon carbide, or silicon. The levels of doping in film samples can often be well-controlled, and important electrical characterization in terms of carrier concentrations and mobilities can be carried out.

In contrast, many of the more "chemical" approaches to making semiconductors yield polycrystalline powders where the doping levels and native defects are harder to control. Also, electrical characterization of carriers is generally not possible for such samples. However, the amounts of sample available for NMR may be much larger than with the former methods.

Two potential experimental difficulties in NMR experiments on semiconductors should be noted. The noncentrosymmetric ZB and WZ structures lead to piezoelectric behavior, which manifests itself as spurious strong "signals" in the FID over some range of frequencies, and oscillating peaks in the spectrum. The piezoelectric response can be identified by its strong temperature dependence due to lattice dimensional changes or its lack of dependence on the magnetic field. A second phenomenon is the generation of oscillating electrical currents in moderately conductive semiconductors undergoing magic-angle spinning (MAS) that can resistively heat the sample inside a rotor by several hundred degrees.[12] The effect can be essentially eliminated by diluting the sample in an inert matrix such as ground quartz glass.

24.3 NMR SPIN HAMILTONIAN IN SEMICONDUCTORS

The spin Hamiltonian for describing the behavior of an ensemble of possibly different nuclear spins in a semiconductor can be given as

$$\mathcal{H}_{\text{total}} = \mathcal{H}_{\text{Zeeman}} + \mathcal{H}_{\text{magnetic shielding (chemical shift)}} \tag{24.1a}$$

$$+ \mathcal{H}_{\text{magnetic shielding (Knight shift)}}$$
$$+ \mathcal{H}_{\text{electron hyperfine}}$$
$$+ \mathcal{H}_{\text{orbital}} \tag{24.1b}$$
$$+ \mathcal{H}_{\text{exchange}} + \mathcal{H}_{\text{nuclear dipolar}} \tag{24.1c}$$
$$+ \mathcal{H}_{\text{quadrupolar}} \tag{24.1d}$$

All these terms except the main Zeeman interaction can be anisotropic, i.e., they depend on the orientation of the crystal or crystallite axis with respect to the external magnetic field. The exact expressions for these Hamiltonian terms are readily available[4,5,13] and will not be given here. They involve spin operators and angular orientation factors, and, in the case of the dipolar term, internuclear distances as well. They also involve fundamental NMR parameters such as magnetic shielding, indirect spin–spin coupling, and electric quadrupole coupling constants that derive ultimately from the electronic structure of the solid and which can be affected by temperature, pressure, applied electric fields or currents, external stress, and optical irradiation.

The terms in equation 24.1 are grouped to reflect certain commonalities. The terms in equations (24.1a) and (24.1b) are all linear in I_z, i.e., the z-component of nuclear spin; the distinction is that the terms in equation (24.1b) are influenced by the unpaired spin or orbital angular momentum of the electron.

In the case of the Knight shift, the electron (or hole) carrier is delocalized over many nuclei, whereas the electron hyperfine (or "transferred hyperfine") interaction arises from a localized paramagnetic site. The H_{orbital} term describes the coupling of the nuclear spin to the orbital angular momentum of the electron, which is relevant to the interactions with holes in the p-type valence band.[13]

The two terms in equation (24.1c) are bilinear in the nuclear spin and thus describe a coupling between two nuclei that is either direct (through-space) in the case of the dipolar coupling or indirect and electron-mediated in the case of the exchange coupling ("scalar" or J-coupling). The quadrupolar term in equation (24.1d) is also bilinear in nuclear spin, but only one spin is involved.

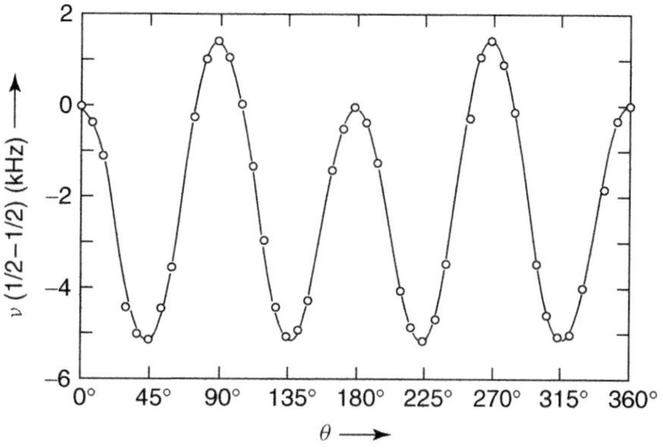

Figure 24.3. ^{67}Zn NMR rotation plot of the CT in ZnO rotated by angle θ about an axis perpendicular to both the c-axis and the magnetic field. The solid line fit includes both CSA and second-order quadrupolar shifts. (Reprinted from Ref. 14. © Wiley-VCH, 1988.)

24.3.1 Chemical Shifts

Chemical shifts of quadrupolar nuclei in many binary semiconductors have been obtained from NMR spectra of stationary polycrystalline samples, since the major source of modest line broadening is typically the dipolar and exchange couplings, and the quadrupole couplings are typically very small (for ZB) or modest (for WZ) and enable at least the CT to be observed. Thus, ^{67}Zn ($I = 5/2$), ^{17}O ($I = 5/2$), and ^{33}S ($I = 3/2$) shifts have been measured for zinc chalcogenides, and CSA parameters have been obtained for the WZ forms of ZnO and ZnS.[14] Accurate measurement of the ^{67}Zn CSA in the former compound was enabled by a single-crystal rotation plot of the CT, as shown in Figure 24.3. Since the CSA and the second-order quadrupolar shift have a different dependence on the angle θ between the crystal c-axis and the magnetic field, fitting the experimental curve yields both C_Q and the chemical shift tensor.

The ZB and WZ polytypes of the same compound can be distinguished on the basis of the isotropic chemical shifts in two compounds where the comparison has been made: ZnS[14] and GaN.[15] The ^{67}Zn chemical shift of ZB ZnS is 15.5 ppm to higher frequency of that in WZ ZnS, similar to the 24.5 ppm shift in the same direction seen for the ^{71}Ga MAS NMR shifts in GaN (right inset of Figure 24.4).[15] The shift differences between polytypes for the "anion" resonances, ^{33}S[14,16] and ^{14}N,[15] are much smaller, <5 ppm. Chemical shifts may be useful in identifying the possible presence of normally unstable polytypes in semiconductor nanocrystals. As an example, a peak has been observed in the ^{71}Ga MAS NMR of nanophase GaAs, which is shifted to low frequency by 27 ppm from the position of the bulk ZB GaAs signal and has been attributed to a WZ form of GaAs in the nanophase material.[17]

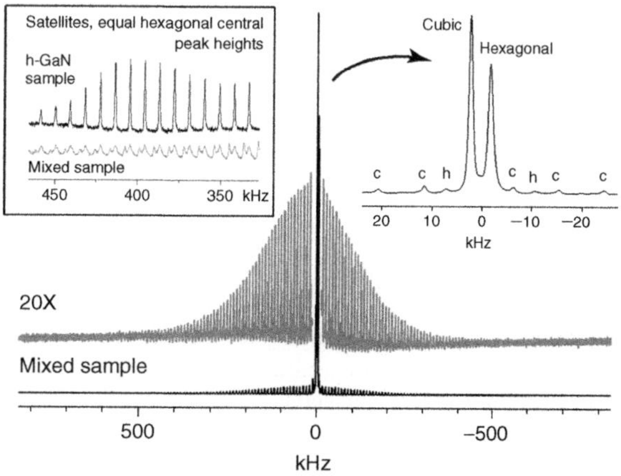

Figure 24.4. ^{71}Ga MAS NMR at 9.0 kHz spinning speed of a mixed sample of GaN containing both ZB (cubic) and WZ (hexagonal) polytypes, shown as resolved CTs in inset at right. For discussion of dynamical effects on ST intensities of h-GaN, see Section 24.5.7. (Reprinted figure from Ref. 15.)

In addition to distinguishing polytypes, chemical shifts obtained from MAS NMR spectra can distinguish crystal polymorphs as well. Titanium dioxide provides a good example, where natural-abundance ^{17}O MAS NMR shows a difference of 32 ppm between the chemical shifts of the rutile and anatase polymorphs.[18]

The ^{33}S chemical shifts of a variety of metal sulfides have also been obtained, with small CSAs measured for the WZ forms of ZnS and CdS.[16] The results were discussed in terms of the changing contribution of the paramagnetic term in the Ramsey perturbation expression for magnetic shielding and changing orbital overlaps in terms of the bond orbital model (BOM) of Harrison. Further discussion of chemical shifts in II–VI and III–V semiconductors as well as their alloys in terms of this model can be found in a review article.[19] Other examples where chemical shifts of quadrupolar nuclei have been obtained and interpreted in similar terms include ^{27}Al shifts in Al–Group V semiconductors,[20] ^{11}B shifts in BN and BP,[21] and polycrystalline III–V semiconductors (^{27}Al,$^{69,71}Ga$,^{115}In)(N,P,As,Sb).[22] In studies of semiconductors with unknown electrical characteristics, unsuspected Knight shifts could be present and affect the accuracy of chemical shift measurements.

24.3.2 Electron Hyperfine Interactions and Knight Shifts

Electron hyperfine interactions between localized electron spins on paramagnetic centers and quadrupolar nuclei can have two types of effects. They can serve as a source of spin–lattice relaxation. They can also produce shifts in the positions of NMR peaks via the isotropic contact interaction ("transferred hyperfine interaction") with the thermally averaged electron spin and possibly orbital magnetic moments or via the anisotropic through-space dipolar interaction with the same moments in the case of single crystals. Semiconductor alloys with reasonably high concentrations of transition-metal ions forming the so-called dilute magnetic semiconductors (DMS) show such effects, but so far they have been reported apparently only for spin-1/2 nuclei.[6] In the case of MAS, the electron hyperfine dipolar interactions would be averaged to nearly zero for electrons with small g-anisotropy but would contribute to strong spinning sideband intensities.[23]

The effects of delocalized electron spins from either electron or hole carriers have been observed for quadrupolar nuclei in semiconductors in a number of cases. As is the case with localized electron spins, there are effects on nuclear spin–lattice relaxation and spectral effects referred to as *Knight shifts*. The basic theory underlying these effects has been described in a number of places,[4,5,24] but there are many complications depending on the band structure, carrier concentration, temperature, and nucleus being studied. These effects have been studied extensively for silicon and for PbTe and PbSe, but unanswered questions remain.[6]

One of the most important factors involved in interpreting NMR results on doped semiconductors is the existence of the MIT mentioned in Section 24.2.1, which occurs over a fairly narrow range of carrier concentrations. In this intermediate region, the shallow donor (acceptor) wave functions begin to overlap to form an impurity band, which at yet higher concentrations moves into the CB (VB). The semiconductor is then said to be in the metallic state, with a degenerate electron gas similar to that in a metal. Although the critical carrier concentration does not seem to be a property that has been determined and tabulated for a wide variety of semiconductors, it can be estimated using the simple Mott criterion as ($0.25/a_H$), where a_H is the Bohr radius of the atomic impurity (shallow donor or acceptor).[3]

24.3.2.1 Examples of Quadrupolar NMR Knight Shifts and Korringa Relaxation

Figure 24.5 shows the ^{71}Ga MAS NMR spectra at room temperature of both a chemically synthesized h-GaN (hexagonal GaN, WZ form) powder containing 0.13% Ge, a known n-type dopant, and an undoped h-GaN sample. The spectrum of h-GaN doped with 0.13% Ge shows profound differences from that of the undoped sample, due to the presence of Knight shifts. One can use the undoped sample as a "zero Knight shift" reference from which to measure the Knight shifts in the doped sample, based on results obtained from a high-quality single-crystal film of undoped GaN.[25] Correctly referencing Knight shifts by subtracting out the true chemical shift can be a significant problem in semiconductors, much more so than in metals. The ^{14}N Knight shift is very similar to the ^{71}Ga Knight shift[6] and provides information about the electronic structure at a different position in the lattice.

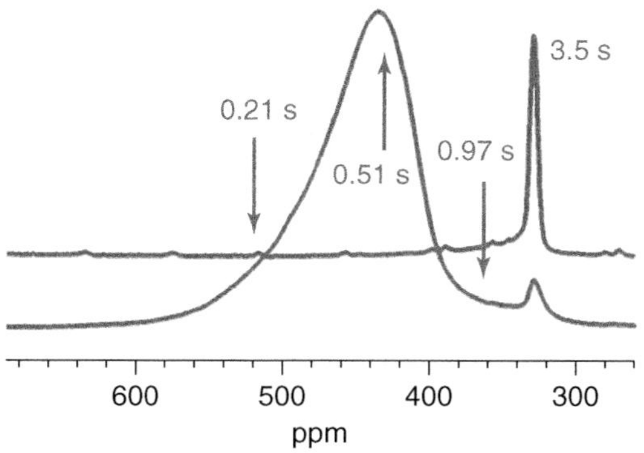

Figure 24.5. ^{71}Ga MAS NMR spectra of the CT of h-GaN (WZ) doped n-type with 0.13% Ge (lower spectrum) and undoped h-GaN (upper spectrum), the latter also exhibiting very weak ST at multiples of the 9.0 kHz spinning speed. The T_1 values measured at various positions indicated by the arrows in the inhomogeneously broadened peak are given as well as the T_1 of the undoped peak. (Reprinted figure from Ref. 26.)

In the metallic regime, the Knight shift K for a mobile system of independent electrons (the assumed carriers in this discussion) coupled to the nuclei by an isotropic exchange interaction is given by the expression

$$K = (8\pi/3)\chi_s\langle|\psi(0)|_k{}^2|\rangle$$

where $\langle|\Psi(0)|_k{}^2\rangle$ is the amplitude of the electron wave function at the nuclear site, averaged over the electrons on the Fermi surface, and χ_s is the electronic spin susceptibility.[3] For a degenerate system of free electrons in a parabolic band, the latter quantity is proportional to $n_c^{1/3}$, where n_c is the electron carrier density. Since this proportionality has been observed to hold in a number of cases, the Knight shift can provide information on carrier concentrations. However, a distribution of Knight shifts is generally seen rather than a single Knight shift, and the interpretation of such distributions is not unambiguous.[6] Although in metals the Knight shift arising from the contact interaction is to a higher frequency, in semiconductors, the sign and magnitude of the Knight shift depend on the effective g-factor, g^*, and can be in either direction.[6]

For the metallic regime described earlier, when the isotropic contact interaction is also responsible for the nuclear relaxation, Korringa showed that the Knight shift K and T_1 are related by

$$K^2 T_1 T[(4\pi k_B/\hbar)(\gamma_n/\gamma_e)^2] = 1$$

where T is the absolute temperature and the expression in brackets contains standard physical constants including the magnetogyric ratios of the nucleus γ_n and of the electron γ_e.[5] The physical basis is that fluctuations in the spin state of the electrons in the CB that give rise to the Knight shift also relax the nuclei. This expression has later been modified to include the effective g-tensor of the electron, g_{eff}, which in the above expression is assumed to have a value of 2.0.[27]

The measurement of T_1 for quadrupolar nuclei with nonzero NQCCs presents some complications, especially under MAS conditions, which are discussed in Section 24.4.1. After these are taken into account, the improved resolution offered by MAS enables T_1 values to be measured across the entire ^{71}Ga Knight shift distribution in doped h-GaN shown in Figure 24.5. The results at the measurement temperature of 304 K, after subtracting a small background contribution to $T_1{}^{-1}$ from quadrupolar relaxation, follow the above Korringa relationship, both for this Ge-doped sample and for an unintentionally doped sample.[26] This indication of a metallic regime is consistent with a reported metal–insulator critical carrier concentration range of 1.6–7 × 10^{18} cm^{-3} and the measured Ge concentration, which is an order of magnitude higher. It also agrees with measurements of a g_{eff} for h-GaN close to 2.0.[28]

Relaxation at higher temperatures is often dominated by the quadrupolar mechanism. By using measurements at lower temperatures down to 4.2 K, Korringa-type relaxation could be observed for ^{115}In in doped InSb and InAs samples, and the magnitude of some Knight shifts was deduced and compared with theoretical predictions.[27] In another example,[29] the ^{71}Ga $T_1{}^{-1}$ relaxation rates versus temperature in doped and semi-insulating GaAs samples, depicted in Figure 24.6(a), show that at temperatures above ~30 K a two-phonon quadrupolar relaxation mechanism dominates and has an approximate T^2 dependence. Below 30 K, the linear relationship between $T_1{}^{-1}$ and T seen in Figure 24.6(b) for the two doped GaAs samples and the use of the modified Korringa relationship with the known $g_{\text{eff}} = -0.44$ enables $|K(^{71}\text{Ga})|$ to be determined. A value of 11 ppm is obtained for the sample with $n_c = 2 \times 10^{18}$. A value for K of −3.3 ppm at room temperature is obtained by comparing the MAS NMR shift of this heavily doped GaAs sample with that

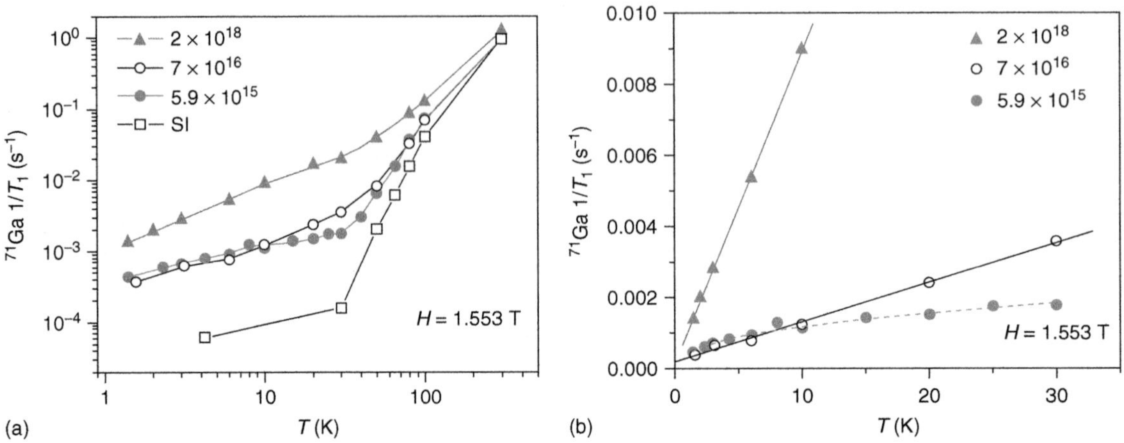

Figure 24.6. ^{71}Ga relaxation rates T_1^{-1} as a function of T for semi-insulating and metallic GaAs samples. (a) Log–log plot showing onset of quadrupolar relaxation mechanism above ~30 K. (b) Linear plot in low-temperature region showing Korringa relaxation with linear fits for the two metallic samples, and a dashed line ~$T^{0.4}$ fit for the most lightly doped sample. (Reprinted figure with permission from Ref. 29. Copyright 2006 by the American Physical Society.)

in a semi-insulating sample. It was concluded that there should be only a small correction to the Knight shift of 7.5% going from 30 to 300 K. Therefore, although the sign of K determined directly from the MAS NMR spectrum agrees with the negative sign for g_{eff}, its magnitude is considerably smaller than estimated from the Korringa relaxation results.

Figure 24.6(b) also reveals that the relaxation rate of the insulating sample, with the lowest n_c, is comparable below 10 K to that in a metallic sample having an order of magnitude higher n_c. This indicates the greater effectiveness of relaxation and rapid spin diffusion from localized paramagnetic shallow donor sites.[29]

Direct measurement of Knight shifts by MAS NMR is likely the most reliable and sensitive method. More Knight shifts of quadrupolar nuclei in semiconductors may be measured in the future by investigating heavily doped samples, employing careful chemical shift referencing to semi-insulating samples, and improving resolution by employing higher magnetic fields and decoupling indirect heteronuclear couplings that broaden the peaks.

24.3.2.2 Impurity Quadrupolar NMR of Electrically Active Dopants

NMR investigation of doping in semiconductors most often involves nuclei of the host semiconductor. However, because of the favorable receptivity of many quadrupolar nuclei, at high enough concentrations NMR signals from the quadrupolar nuclei of the dopant itself may be detected. Such studies have so far been carried out for dopants that are electrically active, i.e., that produce carriers.

The first case involves ^{11}B NMR ($I = 3/2$, NA = 80.42%) studies of B-doped p-type silicon, where low-temperature T_1 measurements showing Korringa behavior yield Knight shifts that surprisingly do not depend on the B concentration.[30] A room-temperature MAS NMR investigation of more heavily B-doped silicon yielded a similar magnitude of Knight shift from the Korringa relaxation.[31] Quadrupolar nutation experiments show that the NQCC is very small, reflecting B substitution in a tetrahedral Si site. A distribution of Knight shifts is also observed, but spin-counting experiments reveal that all the ^{11}B nuclei are being observed, in disagreement with predictions of a model for such distributions.[30] A ^{11}B and ^{29}Si NMR study of this system over a wider range of 100–500 K was analyzed in terms of the impurity band behavior and localization at concentrations near the MIT.[32]

The second case involves several ^{11}B and ^{10}B ($I = 3$, NA = 19.58%) static and MAS NMR 1D and 2D studies of B-doped diamond,[33] which is of interest because samples with high hole concentrations exhibit superconductivity at low temperatures. A 2D ^{11}B NMR on–off decoupling experiment was developed to identify possible ^{1}H–^{11}B dipolar broadening, which showed that a broad peak previously assigned to a boron/hydrogen complex was implausible. The use of very high field strengths (up to 21.8 T) enabled four kinds of boron to be distinguished. One of the peaks having a small NQCC was assigned to tetrahedrally substituted boron, its short T_1 suggesting Korringa relaxation and its relative intensity in different samples correlating with the superconductivity onset temperatures around 4 K. This result is consistent with low-temperature ^{11}B NMR results from thin films. Further $^{11}B-^{11}B$ 2D spin-diffusion measurements and nutation NMR experiments suggested that the excess boron that occurs is trigonally coordinated, possibly in a graphite-like structure.

The impurity NMR of quadrupolar nuclei need not be restricted to cubic lattices such as the diamond or ZB lattice. One striking example of impurity NMR combined with host NMR in a WZ lattice is the "all-quadrupolar" systems of Al-doped and Ga-doped ZnO. Here, the impurity NMR of ^{27}Al and ^{69}Ga and the host NMR of ^{67}Zn were obtained at room temperature and above in powder samples doped to 0.03–3 atom%.[34] Increases in the asymmetry parameter η of ^{67}Zn rather than changes in C_Q were the main effect of increasing Ga substitution. The C_Q of ^{69}Ga could also be measured from the static second-order quadrupolar pattern, and its similarity to the value for ^{67}Zn suggests that Ga occupies the Zn site.

Separating the ^{69}Ga (and ^{27}Al) Knight shifts from the chemical shifts is not trivial, since the observed shift trends with decreasing Ga concentrations do not permit an extrapolation to a chemical shift; rather, development of a narrow impurity band seems to have an enhanced electronic effect on the shifts at low concentrations. Therefore, the $T_1{}^{-1}$ relaxation rates were measured as a function of dopant concentration as well as temperature, from 300–500 K. The rates vary linearly with temperature, typical of Korringa relaxation in a degenerate electron system. In addition, for a fixed temperature, the data were plotted as $(T_1{}^{-1})^{1/2}$ vs observed shift, which for both ^{69}Ga and ^{27}Al give straight lines whose slopes were nearly identical to those predicted by the Korringa relationship. The intercept of these lines, corresponding to a zero relaxation rate from conduction electrons, corresponds to the chemical shifts of both nuclei.

Information about atomic structure was obtained from ^{69}Ga spin echoes on samples with differing degrees of oxygen deficiency or excess (Figure 24.7). A very broad ^{69}Ga peak, seen as the sharp time-domain spin-echo signal whose intensity grows with increasing oxygen content, is attributed to a diamagnetic carrier trap involving gallium in a distorted environment complexed to a defect. The diamagnetic nature of this complex was deduced from the relaxation behavior; its role as a carrier trap was deduced from the decrease in the $T_1{}^{-1}$ rate of the substitutional Ga as the oxygen excess increased and more diamagnetic complex formed. This study shows the feasibility of using quadrupolar NMR to obtain detailed atomic as well as electronic structural information about dopants in semiconductors.

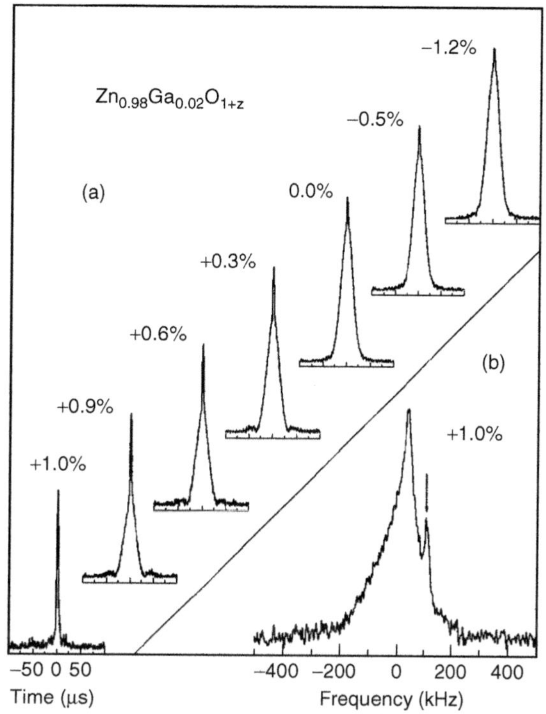

Figure 24.7. (a) ^{69}Ga time-domain spin echoes for $Zn_{0.98}Ga_{0.02}O_{1+z}$ samples of varying oxygen excess or deficiency z. (b) ^{71}Ga frequency domain spectrum of oxygen-rich sample $Zn_{0.98}Ga_{0.02}O_{1.01}$. (Reprinted figure with permission from Ref. 34. Copyright 1998 by the American Physical Society.)

24.3.3 Exchange and Nuclear Dipolar Couplings

The nuclear dipolar coupling has the unique characteristic that it can be exactly calculated from known parameters if the structure is known. In the static NMR of semiconductors, it provides a modest broadening interaction calculable by the Van Vleck method of moments.[4] The broadening effects average to zero under MAS at modest spinning speeds. It also provides a mechanism to redistribute magnetization via spin diffusion and assist relaxation by paramagnetic impurities.[4] Since the detailed arrangement of atoms in semiconductor alloys or glasses is generally not known, the same dipolar-coupling-based NMR methods so useful in the study of glasses[35] may prove useful as well for semiconductors.

The indirect exchange interaction (J-coupling) in semiconductors is a second-rank tensor parameter related to the electronic structure that is difficult to measure accurately and unambiguously. It is also one that is difficult to calculate from the theoretical expressions developed by Bloembergen and Rowland[36] along the lines of Ramsey and Purcell's treatment of indirect coupling in molecules and Ruderman and Kittel's extension to indirect coupling via the conduction electrons in a metal. Therefore, it is not surprising that it has been less useful as a source of information about semiconductors than other interactions, and the isotropic heteronuclear exchange interaction often mainly functions as a source of residual line broadening in MAS NMR spectra.

The earliest observation of exchange effects in semiconductors was for the quadrupolar nuclei in InAs and InSb,[37] and subsequently GaAs and GaSb,[38] where analysis of possible sources of line broadening led to the conclusion that the indirect exchange interaction between unlike nuclei was responsible for the broadened Gaussian line shapes observed. The theoretical discussion predicted a strong falloff with distance between the nuclei, as the inverse fourth power, so that couplings between directly bonded nuclei dominated.

One uncommon magnetic resonance method that has been applied to the study of exchange couplings in single crystals of the four III–V compounds above as well as AlSb is nuclear acoustic resonance (NAR), which relies on the coupling between quadrupolar nuclear spin systems and acoustic phonons via the dynamic nuclear quadrupole coupling.[39] It appears as an attenuation of ultrasound due to elastic strain energy being absorbed by the nuclear spins. Because the selection rules differ from those of NMR, exchange couplings between like nuclei as well as unlike nuclei broaden the NAR peaks. Detailed information about the exchange couplings was obtained and comparisons were made with theory.

More accurate measurements of exchange couplings in a dozen III–V compounds were made possible by using MAS to average out dipolar couplings, anisotropic exchange (pseudodipolar) couplings, as well as first-order ST splittings seen for ZB compounds that arise from defects.[22] The Gaussian line shapes observed in the MAS NMR were analyzed in terms of exchange couplings to the four immediate unlike neighboring spins.

A new technique for studying exchange couplings using MAS was demonstrated in a ^{69,71}Ga and ^{75}As MAS NMR study of GaAs.[40] Detailed analysis shows not only good agreement with previous determinations of the heteronuclear couplings based on analysis of the Gaussian line widths but also how homonuclear exchange couplings can be determined based on their effects on the timing of Hahn spin-echo formation. The comparison between nearest and next-nearest neighbor couplings shows a less rapid falloff with distance than the inverse fourth power from theory.

The anisotropic (pseudodipolar) part of the exchange interaction has been observed in a few cases. The orientationally dependent second moment of the static ^{71}Ga NMR in a single crystal of GaAs is smaller than the calculated value based on dipolar couplings alone due to interference effects with the pseudodipolar term.[41] The pseudodipolar couplings in GaP and GaAs again were measured in a similar manner and the exchange couplings related to the static dielectric constant in terms of Harrison's bond orbital model.[42] Similar effects were also observed in ^{31}P NMR in AlP and InP;[43,44] InP has been the subject of a number of further investigations in this regard. By using the low-abundance ^{113}In isotope to cross-polarize ^{31}P nuclei, two distinct possible values for the pseudodipolar term could be obtained, the ambiguity resulting from the unknown signs involved.[45] In both cases, the anisotropic term is greater (by a factor of ~2 or ~4) than the isotropic (scalar) term, indicating that mechanisms other than the Fermi contact term are involved.

In ^{113,115}InX (X = ^{75}As, ^{123}Sb), using cross polarization from ^{113}In to X and ^{115}In decoupling during MAS enabled the isotropic exchange couplings to be determined.[46] The isotropic indirect couplings for InSb, InAs, and InP (after being corrected for the different magnetogyric ratios) decrease in the same order, consistent with theoretical arguments based on the trend of increasing bandgap.

24.3.4 Quadrupolar Interactions—General

Some manifestations of quadrupolar interactions have already been mentioned in the previous sections on terms in the spin Hamiltonian. The next section will consider the unique features of and problems associated with measuring and interpreting spin–lattice relaxation in multilevel quadrupolar spin systems (see Chapters 2 and 4).

The section after that deals with several fundamental aspects of quadrupolar interactions and provides specific examples of the information obtained. The EFGs in semiconductors are affected by atomic and possibly electronic structure, including internal perturbations such as dopants, defects, and alloyed atoms. They can also be influenced by external perturbations: the effects of strain from applied stress and of electric fields on EFGs have been measured. Strain and electric fields also can exert their influence as internal perturbations when dopants are considered. In addition, temperature (but not pressure) effects on EFGs have been measured in semiconductors, which result not only from changes in lattice parameters but also from thermally activated dynamic processes such as atomic diffusion. Increasingly, accurate theoretical calculations of EFGs by density functional theory (DFT) or other methods are also important in extracting the maximum amount of information from experimental results.

24.4 SPIN–LATTICE RELAXATION OF QUADRUPOLAR NUCLEI

24.4.1 General Mechanisms and T_1 in Quadrupole-perturbed Systems

The mechanisms of spin–lattice, or T_1, relaxation of quadrupolar nuclei in solids can be described by referring to Figure 24.8, which shows the four Zeeman energy levels of a spin $I = 3/2$ nucleus in a magnetic field. We assume that the levels are quadrupole-perturbed, so that STs are resolved from the CT, a situation that arises in imperfect ZB crystals as well as in WZ and other noncubic forms. (When all transitions are near-degenerate, cross relaxation or spin exchange between the overlapping homogeneous line widths of the different levels restores the populations to a common spin temperature, resulting in single-exponential relaxation.[47])

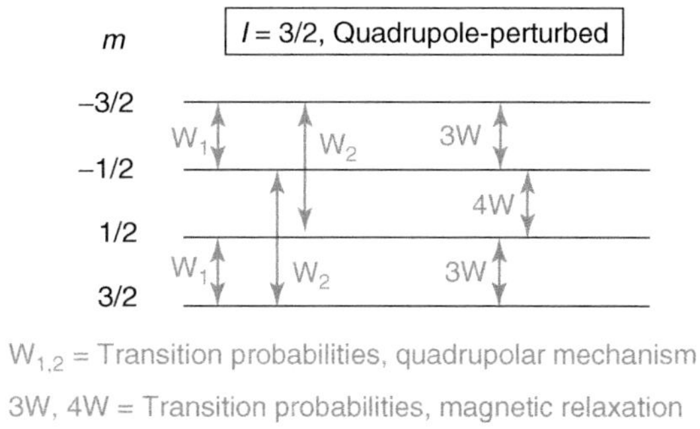

Figure 24.8. Quadrupole-perturbed Zeeman energy levels for $I = 3/2$ nucleus, with transition rates for restoring equilibrium population differences between pairs of levels. Magnetic relaxation mechanisms are single-quantum and involve a single rate constant W. Quadrupolar mechanisms can be single quantum ($W1$) or double quantum ($W2$). (Reprinted from Ref. 48.)

Deviations of the populations of these levels from the Boltzmann distribution corresponding to thermal equilibrium, typically produced by RF pulses or continuous wave (CW) irradiation, will eventually vanish as spin–lattice relaxation returns the spin system to equilibrium. Once the nonequilibrium population distribution has been created, two fundamental types of mechanisms can return the spin system to equilibrium: *magnetic* or *quadrupolar* relaxation mechanisms.

Magnetic relaxation mechanisms are those that can produce T_1 relaxation in spin $I \geq 1/2$ systems via randomly fluctuating magnetic fields affecting the nuclear magnetic dipole. In semiconductors, these include nuclear dipole–dipole interactions that are modulated at the Larmor frequency or twice its value by lattice motions (phonons), or the hyperfine interactions of a nucleus with electron spins whose orientation fluctuates as a result of the T_1 processes of the electron. The electron spin may be localized on a paramagnetic impurity, or it may be delocalized as in the case of an electron carrier in the CB (or a hole carrier in the VB) that leads to the Korringa-type relaxation already discussed in Section 24.3.2.1. Each magnetic mechanism at a given temperature is characterized by a rate constant W, which, with the numerical prefactors shown in Figure 24.8, determines the rate at which the equilibrium population difference between any two levels is restored. The fluctuating magnetic fields serve to drive the transitions, much as applied RF pulses are able to.

The quadrupolar relaxation mechanism, in contrast, occurs through randomly fluctuating EFGs that couple not to the nuclear magnetic dipole moment but rather to its electric quadrupole moment. One important point to note is that such relaxation readily occurs in ZB structures where the EFG is ideally zero. This is so because the lattice motions involving phonons produce deviations of the nuclear positions from their equilibrium values that generate time-dependent EFGs at a given nucleus.

For spin–lattice relaxation in the more general case where there are $2I + 1$ unequally spaced energy levels, there are $2I$ energy spacings. This leads to $2I$ relaxation times characterizing the relaxation behavior that represents solutions to the $2I$ differential rate equations describing population differences between adjacent levels.[47] Calculations for quadrupolar relaxation for $I = 3/2$ and $I = 5/2$ and for magnetic relaxation have been made, and the influence of the $W1$ and $W2$ rate processes in Figure 24.8 has been discussed.[47] The weighting of the various exponentials in the recovery observed for CT or ST peaks depends on the initial conditions and will differ, for example, if all transitions are saturated, or only the CT is saturated.

Although the $W1$ and $W2$ quadrupolar relaxation rates are often assumed to be equal, there is little experimental evidence that this assumption is justified; in fact, they have been shown to be unequal in some of the few cases where they have actually been measured.[48] If they are unequal but comparable, a single exponential may fit a recovery curve, but the rate derived can be considerably greater than the $W1$ or $W2$ rate.

Magnetic and quadrupolar mechanisms can be distinguished by comparison of the relaxation behavior of two isotopes of the same atom having different magnetogyric ratios and nuclear quadrupole moments, since the rates depend on the squares of the magnetogyric ratios or quadrupole moments, respectively. This approach has been used most often with the pair ^{69}Ga and ^{71}Ga. They may also be distinguished by investigating the temperature dependence of the relaxation, as shown for GaAs in Figure 24.6. A method for separating magnetic from quadrupolar mechanisms using only one isotope has been described and applied, and analytical solutions are given for $I = 1$, 3/2, and 5/2.[49] A general strategy for measuring all three rates $W1$, $W2$, and W, for an $I = 3/2$ single crystal has been described.[48] It involves analyzing the recovery when both CT and ST peaks are saturated at different frequency offsets and the recovery when only the CT peak is saturated. The $W1$ and $W2$ rates of ^{71}Ga and ^{69}Ga in the h-GaN sample were unequal in certain crystal orientations, and the conclusion that the magnetic relaxation contribution was much smaller than the quadrupolar could be verified by comparing results from the two isotopes.

For polycrystalline samples, the above strategies cannot be directly applied. Furthermore, measuring T_1 relaxation times under MAS conditions presents the complications that, if a comb of saturating pulses is applied at the CT frequency, the ST may be affected by some of the pulses as the ST transits the center frequency during the rotor cycle.[50] Furthermore, cross-relaxation between the CT and ST may occur when this transit occurs. The only case where a comparison has been made between detailed relaxation results on a static single-crystal and MAS NMR results on a powder is for ^{71}Ga NMR of h-GaN. Applying a rotor-asynchronous train of saturating pulses with specific timing to saturate all transitions led to a nearly single-exponential recovery, which was shown to be in semiquantitative agreement with the detailed single-crystal measurements, although the process of MAS serves to average the orientation-dependent rates and thus provides less detailed information.[48]

24.4.2 Relaxation Behavior of Semiconductors

Quadrupolar relaxation has a temperature dependence given by Van Kranendonk's theory that depends on the Debye temperature θ, which for a number of III–V compounds ranges from 200 to 400 K.[51] For $T > \theta$, the rate ${T_1}^{-1}$ is proportional to T^2, a behavior seen in many different instances. For $T < 0.02\theta$, the rate is predicted to be proportional to T^7, although the long relaxation times at such low temperatures and competition from magnetic relaxation mechanisms make such a limiting temperature dependence difficult to observe. Comparison of the relative ${T_1}^{-1}$ rates in a dozen different III–V compounds was made in connection with the theory and Debye temperatures derived.[51] The theoretical spin dependence of the rate on $(2I + 3)(I^{-2})(2I - 1)^{-1}$ was verified by observing the two isotopes of Sb, which have $I = 5/2$ and $I = 7/2$, and an isotropic character of the relaxation times was derived theoretically and observed experimentally.

Theoretical and experimental study of the relaxation rates in (Ga,In)(As,Sb) from 4 to 300 K used the different nuclear properties of the isotope pairs $^{69}Ga/^{71}Ga$ and $^{123}Sb/^{121}Sb$ to separate magnetic from quadrupolar relaxation mechanisms, as discussed earlier.[52] By comparing the rates of the Group III and Group V nuclei, the quadrupolar relaxation was separated into two parts, namely, relaxation by acoustic phonons at lower temperatures and relaxation by optical phonons at higher temperatures. The proposed mechanism for the latter is generation of electric fields by optical phonons which produce fluctuating electric dipoles at the Groups III and V atoms, which differ depending on their polarizability. These dipoles then produce fluctuating EFGs at the neighboring atoms' nuclei. (Note that the Stark effect on EFGs discussed in Section 24.5.3 involves static displacements of the nuclei as well as the rearrangement of the electronic distribution that is invoked to explain these relaxation results.)

A study of ^{69}Ga and ^{71}Ga temperature-dependent relaxation in two different samples of h-GaN (WZ) provides a clear example of the consequences of either magnetic or quadrupolar mechanisms.[53] In a heavily doped conductive crystal, relaxation is biexponential and the $T_1{}^{-1}$ rate is linear with T from 10 to 50 K as expected for a magnetic mechanism, Korringa relaxation due to CB electrons in this case. Above 150 K, the quadrupolar mechanism begins to become significant. In a nonconducting Mg-doped GaN crystal, the relaxation is single exponential, the ratio of rates for the two isotopes establishes a quadrupolar mechanism, and the temperature dependence can be fitted well with an expression for the rate involving a Debye temperature of 600 K, which agrees well with other measurements.

The magnetic relaxation mechanism arising from localized paramagnetic impurities or dopants has been observed in a number of cases. For example, ^{71}Ga (and ^{31}P) relaxation in GaP doped with Te or S at various levels approaching the MIT has shown effects on the electron spin correlation time arising from donor–donor interactions, as well as signs of diffusion-limited relaxation.[54] The ^{11}B NMR relaxation in hexagonal-BN and c-GaN (ZB) has been measured at different field strengths and temperatures to show that the dominant mechanism is relaxation by paramagnetic centers arising from N-vacancies and subsequent spin diffusion.[55]

24.5 QUADRUPOLAR INTERACTIONS—EFFECTS AND APPLICATIONS

24.5.1 Polytypes and Polymorphs

Differences in EFG values in different polytypes and polymorphs allow these to be identified. The ^{71}Ga MAS NMR of h-GaN (WZ) exhibits ST at multiples of the spinning speed extending over more than 1 MHz that reflect the axially symmetric NQCC with $C_Q = 1.72$ MHz.[15] In contrast, the c-GaN spectrum (ZB) shows a much narrower pattern of ST, with a Gaussian envelope of 297 kHz whose width is attributed to a random distribution of defects in the sample. The ^{14}N ($I = 1$, 99.635% NA) MAS NMR spectrum of the h-GaN polytype has a center peak and spinning sidebands whose envelope resembles the classic $I = 1$ powder pattern for $\eta = 0$ seen often for ^{2}H ($I = 1$) NMR;[15] the very small NQCC and long T_1 are similar to that observed in the ^{14}N MAS NMR of AlN and BN.[56] The spectrum of the c-GaN polytype has a Lorentzian envelope of spinning sidebands that is less extensive than that seen for h-GaN.[15]

The three polymorphs of titania (TiO_2) have very different second-order quadrupole-broadened powder patterns for ^{47}Ti ($I = 5/2$, 7.3% NA) and ^{49}Ti ($I = 7/2$, 5.5% NA) from which C_Q and the small η values could be obtained for the common forms anatase and rutile but not for brookite.[57] One peculiarity of this pair of nuclei is that their resonances overlap because of the nearly identical magnetogyric ratios. The quadrupolar Carr–Purcell–Meiboom–Gill (QCPMG) technique was modified to make the sequence selective for ^{49}Ti based on its different spin compared to ^{47}Ti, and both quadrupolar and CSA parameters were determined at very high field.[58] One example of the utility of static $^{47,49}Ti$ NMR is a study using peak deconvolutions of the conversion of nanophase titania from the anatase to rutile form on heating to higher temperatures.[59]

As found in the $^{47,49}Ti$ QCPMG experiments above, the sensitivity of detection of broad CT second-order patterns can be improved by adding together the individual echoes in a CPMG train and stepping the frequency. This technique was first demonstrated for the CT in ^{93}Nb ($I = 9/2$) NMR,[60] and it was also shown to be useful for transition-selective wideline NMR of the $I = 1$ nucleus ^{14}N.[61] The $^{47,49}Ti$ CT intensities in the static and MAS NMR spectra of anatase can also be enhanced by population transfer from ST peaks,[62] and such techniques are expected to be useful in many other semiconductors.

24.5.2 Electric Field Gradients in WZ or Diamond Structures

24.5.2.1 EFG Calculations for WZ Structures

The WZ structure can be described as two hexagonal close-packed sublattices displaced in the c-direction by uc, and having as structure parameters the lattice constant a, the axis ratio c/a, and the parameter u. Actual semiconductors show deviations of c/a from the ideal value $(8/3)^{1/2}$ and of u from the ideal value of 3/8, meaning that the tetrahedra of next neighbors are slightly distorted and resulting in the finite C_Q values that have been measured for a number of WZ-form semiconductors.[63] Theoretical DFT calculations of EFGs for the WZ forms of AlN, ZnO, BeO, CdS, GaN, CdSe, AgI, and ZnS show a strong influence of the u parameter, whose experimental values have large uncertainties. Theoretical calculation of u by energy minimization using experimental a and c/a values shows a rather flat minimum which agrees with experimental values.[63] The cation and anion EFGs depend linearly on u, with opposite slopes. The experimental values available, whose signs are unknown, were compared with the theoretical calculations.

In general, theoretical calculations of EFGs are not able to reproduce experimental values that have been measured with a high degree of accuracy, but trends may be observed (see Chapter 13). For h-GaN (WZ), DFT calculations of EFGs of both (partially covalent) cation and anion were carried out in an attempt to obtain more accurate structural parameters.[53] The effects of electrons in the CB at high concentrations of $\sim 10^{20}$ to 10^{21} cm^{-3} were also calculated and shown to be small. Experimental measurements of the latter effects are lacking and would be difficult.

24.5.2.2 Temperature Coefficients of EFGs in WZ Structures

The temperature dependence of NQCCs in semiconductors has not been extensively explored, much less the pressure dependence, but the sensitivity of the calculated EFGs in WZ semiconductors to lattice parameters would suggest that nonisotropic changes in lattice dimensions with temperature should play an important role. Although NQCCs in molecular solids typically decrease with increasing temperature due to the averaging effects of increasing librational motion, this picture does not apply to semiconductors. Indeed, the ^{67}Zn NQCC increases in ZnO with increasing temperature,[14] as do the NQCCs of ^{9}Be in BeO,[64] ^{27}Al in AlN,[56] and ^{71}Ga in h-GaN.[25] A simple ionic point-charge model suggested that an increase in u or c/a could account for the positive temperature coefficients of NQCC, but accurate structural parameters at different temperatures are needed.[14] For ^{47,49}Ti NQCCs in the rutile form of TiO_2, C_Q is only weakly temperature-dependent whereas η varies strongly with temperature, an effect attributed to the known nonisotropic thermal expansion of rutile.[65]

24.5.2.3 Isotope Effects on EFGs in Germanium

The presence of different nuclear isotopes having different masses and hence different vibrational averaging characteristics in a solid lattice is generally ignored in discussions of EFGs and NQCCs. However, in the case of ^{73}Ge NMR ($I = 9/2$, 7.7% NA) in a pure single crystal of germanium, which would have EFGs of exactly zero in the perfect diamond lattice of monoisotopic composition, broadening due to random nonzero EFGs as a result of lattice strain from mass fluctuations in the isotopically disordered sample is observed.[66] The broadening could be accounted for by studying dilute ^{73}Ge ($\sim$0.1%) in single crystals having tailored isotopic compositions.

24.5.3 Linear Stark Effect on EFG

The application of an external electric field to a solid lacking an inversion center produces an EFG at the nuclear sites, a nuclear linear Stark effect first observed in NQR. The induced field gradient tensor at the nucleus is given by Gill and Bloembergen[67]

$$eq_{ij}{}^{E} = \Sigma_k R_{ijk} E_k + \Sigma\,\Sigma_{lmk} S_{ijlm} d_{lmk} E_k$$

The indices i, j and l, m, k represent all three Cartesian axes x, y, z for each index. The first term represents the EFG produced by the electric field at constant strain, where the third-rank tensor R has only three nonvanishing components for the T_d symmetry of the ZB or diamond lattice. The second term represents the EFG produced by piezoelectric strain, involving the fourth-rank piezoelectric tensor S, whose components in T_d symmetry are described by four independent values. Since stress also changes EFGs, the effects of electric fields depend on whether they are made at constant stress or constant strain; conversely, the effects of stress depend on whether they are made at constant field or constant displacement.[68]

For a pure GaAs (ZB) crystal with the electric field applied along the [111] body diagonal, the induced EFG was shown to depend primarily on a single component of the R tensor.[67] Splittings of several kilohertz were observed for ^{69,71}Ga and ^{75}As when an electric field of 10^4 V cm^{-1} was applied. The measured R tensor component agreed with a physical model that considered the EFG to arise from a distortion of the valence orbitals by the electric field as well as a relative displacement of the Ga lattice with respect to the As lattice.

The Stark effect has been measured for ^{9}Be in a single crystal of BeO (WZ), and the results show that the unperturbed EFG has a positive sign, a conclusion difficult to reach by other experimental methods.[69] Approaches for measuring the linear Stark effect have been developed that rely on effects of AC electric fields at radio frequencies.[70–72] Finally, although somewhat outside of the scope of this chapter, a promising new general technique called *POWER* (*Perturbations Observed With Enhanced Resolution*) has enabled optically detected NMR (ODNMR) to determine the nanoscale distribution of electric fields (as well as electron density) induced by light irradiation in a single AlGaAs/GaAs heterojunction, relying on the linear Stark effect just discussed.[73]

24.5.4 Stress/Strain Effects on EFGs

Strain in semiconductor materials, i.e., a displacement of atomic positions from their equilibrium values in the absence of stress, is an important phenomenon that can affect the electronic properties. It can arise, for instance, from lattice mismatch in epitaxial growth on substrates or in heterojunctions such as quantum-well structures, from the altered crystalline potentials introduced by defects and dopants substituting for regular lattice ions, and from mechanical stresses such as those generated at interfaces between materials with differing thermal expansion coefficients.

The generation of finite EFGs by elastic axial deformation was first observed by applying weights to a single-crystal InSb sample (ZB, with unstressed NQCC = 0) in the shape of a cylindrical rod and monitoring the broadening of the ^{115}In ($I = 9/2$) resonance.[74] From the measured elastic constants of InSb, it was deduced that strains of less than 10^{-4} produce measurable EFGs at the indium sites. Observing NAR in seven III–V compounds yielded the S-tensor components defined in Section 24.5.3 for various Group III and Group V atom positions. The results were compared with a model that included covalent as well as ionic contributions.[75]

To study strain in GaAs/AlGaAs quantum-well (QW) structures, the sensitive detection method of ODNMR has been used. Calibration of the quadrupole splitting versus strain can be done by bending the structure to a measured radius.[76] Mapping of the strain (and electric fields) in a multiple quantum-well (MQW) structure without applying bending has been carried out using ODNMR of ^{75}As.[77] Figure 24.9 shows the splitting of the ^{75}As resonance in one location. Lateral variations in strain can be detected by varying the laser beam position and depth profiling by varying the wavelength of the detected photoluminescence (PL) to select a given QW in the vertical direction. Strains of the order of 10^{-5} were detected, and a number of possible sources were considered.

24.5.5 Effects of Defects/Dopants and Distributions of EFGs

One of the first detailed studies of quadrupole broadening arising from the presence of defects was carried out by studying the orientation dependence of the quadrupole broadening of ^{69,71}Ga and ^{75}As in a single crystal of undoped n-type GaAs, after taking into account dipolar and pseudodipolar broadening.[41] The crystals were investigated before and after the introduction of controlled and measurable amounts of the defects in the form of As monovacancies by heating to various temperatures and quenching. The As vacancies act as charged defects with an estimated charge of 0.87 e based on the covalent character of the bond. Instead of using a simple point-charge model, the electric fields produced by the charges at

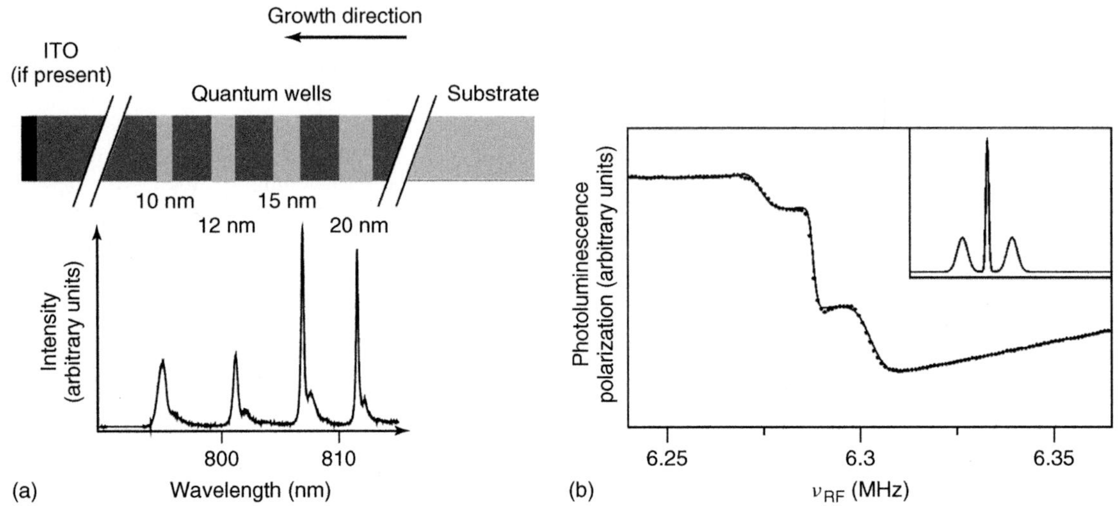

Figure 24.9. (a) Multiple quantum-well (MQW) structure for ODNMR in AlGaAs (dark gray) and GaAs (light gray), with the photoluminescence (PL) spectrum showing peaks placed directly below corresponding QW. (b) ODNMR spectrum obtained by monitoring the PL from the 20 nm QW in (a) while scanning the radiofrequency through the ^{75}As resonances. The three steps correspond to a strain-induced quadrupole splitting of 15 kHz, as depicted in the conventional spectrum simulation shown in the inset. (Reprinted figure with permission from Ref. 77. Copyright 2003 by the American Physical Society.)

the defects were considered to act through a static macroscopic dielectric constant to generate EFGs by means of the Stark effect discussed in Section 24.5.3. The effects of strain due to vacancies as a source of EFGs were argued to be negligible. Summing the effects of defects over random lattice sites yielded angular dependencies that could be compared with experimental measurements of the second moments. The preferred site of the defects (in this case As sites) could be inferred from measurements of the average EFG on the cation and anion sublattices. The defect density was found to be proportional to the second moments from quadrupolar broadening, and the measured values were as small as 2×10^{14} cm^{-3}. This work was extended and compared with earlier theoretical predictions of the effect of defects on the FID, with a consistent prediction that the FID of the undoped material should be scaled by a factor $\exp(-Kt^{3/2})$, where K is proportional to the defect density.[78]

However, the effects of strain alone can be observed in the case of isovalent defects. The nuclei of all three atoms in GaAs heavily doped with In show effects of strain out to about 30 Å from the In atom, and the ^{75}As C_Q and η for the 12 As next-nearest neighbors to an In atom could be measured.[79] The In atoms substitute at the Ga sites, and previous measurements of the relationship between stress/strain and EFGs were used to deduce distortions of bond angles from tetrahedrality, which were greatest not for the nearest neighbor atoms to In but rather for the next-nearest neighbor As atoms.

Some of the discrepancies or shortcomings of these initial studies were clarified in a study of the nuclei in GaAs that used Zn as a dopant to investigate charge effects (Zn^{++} vs Ga^{3+}) and isovalent In as a dopant to study strain effects.[80] Greater accuracy in the observation of quadrupole broadening was also obtained by using spin-echo pulses whose second pulse length was adjusted to give either a maximum echo or an echo for the ST only. The dependence of the strain-induced EFGs on distance from the dopant was obtained, and discrepancies in earlier work were shown to arise from the fact that the strain effects caused the ST to extend over several megahertz and thus to disappear from practical second-moment measurements. Taking this strain effect into account enabled the slightly sublinear relationship between broadening and defect density to be extended to very high defect densities.

The use of MAS offers another approach to separating the ST generated by defects from the CT, since the ST appear as a set of sharp spinning sidebands with consequent improvement in sensitivity compared to a static spectrum. A number of III–V compounds with the ZB structure have been observed to have a Lorentzian envelope of ST in the MAS NMR spectra, in accordance with an early theory assuming randomly distributed point-charge defects and a linear sum of EFGs from each defect.[22] In the case of ^{71}Ga MAS NMR of c-GaN (ZB) however (Figure 24.4), the envelope is Gaussian, although the corresponding envelope of the ^{14}N MAS NMR is Lorentzian.[15] The ^{63,65}Cu and ^{127}I MAS NMR of γ-CuI (ZB) is another example where an extended pattern of ST peaks has been observed.[12] In all these cases, the theoretical linear relationship between the width of the envelope and the defect density could not be established because of the unknown proportionality factor. The effects of defects on EFGs in static NMR of cubic solids have also been extensively discussed.[81]

The possibility of correlating both the *charge* (in the sense of factors that produce changes in the EFG) and the *spin* effects of dopants or defects was revealed by a ^{71}Ga NMR study of h-GaN (WZ) single-crystal films.[82] The CT of one sample shows two peaks, one of which has a Knight shift to high frequency from unintentional dopants or defects and which also has a shorter T_1 due to Korringa relaxation. The ST peaks corresponding to both CT peaks are broadened beyond mere dipolar line width broadening by a distribution of EFGs. However, difference spectra show that the more rapidly relaxing ST, corresponding to the Knight-shifted CT, has a much greater line width than the other ST. This type of correlation between spin effects of a dopant (i.e., the carrier concentrations it generates) and charge effects (how large a distribution of EFGs a given dopant concentration produces) offers a more informative approach to characterizing dopants.

24.5.6 Semiconductor Alloys

The ability to tailor the optical and electronic properties of semiconductors by forming alloys is a keystone of modern semiconductor engineering. Ternary and even quaternary alloys of many binary semiconductors have been grown and studied, and a wide range of compositions is possible when the atoms are not too mismatched in size and bonding characteristics. The goal might be to produce a desired, specific bandgap in a single material, as for instance growing $In_xGa_{1-x}N$ with the right value of x to produce green emission for a laser diode, or growing $Al_xGa_{1-x}N$ with an x value that makes it useful as a UV photodetector. Or, it might involve growing $Al_xGa_{1-x}As$ layers surrounding a layer of GaAs, the alloy's higher bandgap providing the needed potential barrier to form a QW structure.

One of the questions about such alloys relevant to their optoelectronic properties is whether the substitutional atoms enter the structure randomly or otherwise. For certain compositions (e.g., 50/50), several types of structures with long-range ordering have been observed in ZB lattices (Figure 24.10a), whose degree of perfect ordering can be described by an order parameter, as was done in a ^{31}P MAS NMR study of ordering in $In_xGa_{1-x}P$.[83] In addition, local ordering is possible, in which the substitutional atom prefers another substitutional atom in its vicinity ("clustering"), or instead an atom of the other component.

For studying possible ordering in semiconductor alloys that contain only quadrupolar nuclei, obtaining information about EFGs in different structural environments by measuring NQCCs is a useful strategy, particularly since EFGs are sensitive to atom replacements at long distances. For $Al_xGa_{1-x}As$ thin films, the static ^{75}As NMR resolves the symmetric first-shell structures As[Al_4] and As[Ga_4] on the basis of their chemical shift differences.[85] A solid-echo pulse sequence shows a very sharp time-domain echo signal for ^{75}As whose intensity is much greater than the ST of the two symmetric structures above would generate. This indicates that the sharp echo signal arises mainly from the strongly second-order quadrupole-broadened CT of mixed As[Al_nGa_{4-n}] environments with $n = 1, 2, 3$. This conclusion was verified in a study where the broadened patterns were directly observed (Figure 24.10b).[84] From ^{71}Ga nutation spectra and quantitative intensity arguments, the samples are deduced to have no significant ordering.

The ^{75}As static NMR spectrum of $Al_xGa_{1-x}As$ at the top of Figure 24.10(b) was obtained using high RF fields from a small solenoidal coil.[84] Frequency-stepped CPMG spectra of the very broad CT peaks were used to obtain the NQCC parameters for the ^{75}As nuclei without four identical neighbors

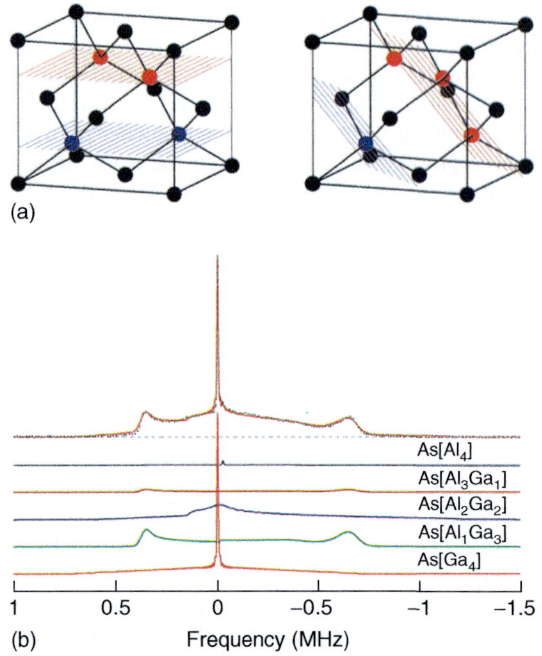

Figure 24.10. (a) Two possible types of ordering in the ZB lattice of AlGaAs. The black atoms are As, the red and blue atoms Al and Ga cations. Left is CuAu ordered with planes alternating along the < 001 > direction and twofold symmetry for As; right is CuPt ordered with planes along < 111 > direction and threefold symmetry for As. (b) Top: ^{75}As Hahn echo spectrum at 18.8 T of $Al_{0.297}Ga_{0.703}As$ (dotted line) and fit (solid line). Bottom: The five spectra show individual contributions from each arsenic site used in the fit. (Reprinted figure from Ref. 84. Reproduced by permission of the PCCP Owner Societies.)

(i.e., for the nonsymmetric environments). These parameters were used to fit the observed quantitative Hahn-echo spectrum using a genetic algorithm to weighted sums of the subspectra for the various nearest shell configurations shown below the experimental spectrum on top. The fit, shown as the solid line in the top spectrum, is very good.

Additional experiments involving ^{69}Ga 3QMAS and nutation experiments exclude the presence of highly symmetric sites and show a distribution of EFGs.[84] The symmetric first-shell As coordinations, As[Al_4] and As[Ga_4], show finite ^{75}As NQCCs due to EFGs produced by disordered Ga or Al occupancy of further removed shells. To describe these NQCCs, the Czjzek distribution can be used,[86,87] which describes a joint probability density function for EFGs in amorphous solids $P(V_{zz}, \eta)$ in terms of two parameters σ and d. For the asymmetric first-shell environments, an "extended Czjzek" model coinciding with an extended Gaussian Isotropic Model was developed, in which the effects of disorder on the EFG from more remote atoms were added to a "base" EFG representing that of the asymmetric first shell. Very extensive DFT calculations involving additivity of EFG tensors from distant charge perturbations reproduce the Czjzek and extended Czjzek distribution found experimentally and show that the ^{75}As NQCC is sensitive to composition up to the seventh shell. Absence of long-range ordering is observed, although for one sample the data could not be fitted to a zero-order parameter, suggesting the possible existence of, perhaps, local order.

Another semiconductor alloy that has been investigated by quadrupolar NMR is partially ordered $Ga_{0.52}In_{0.48}P$, where ^{71}Ga NMR led to an estimate of the order parameter using a point charge model.[88] The point charge was found to be inadequate in DFT calculations,[89] but a tight-binding approach that was computationally more feasible for evaluating EFGs from configurations of 17-atom clusters was used to obtain order parameters in this material.[90]

24.5.7 Observation of Dynamical Effects Using NQCCs

Inorganic semiconductors are generally considered to be rather rigid substances at room temperature, a reflection of their partial covalency and bond directionality. Although atomic diffusion may be an important aspect of semiconductor doping, it generally takes place at the very high growth temperatures used to produce semiconductor films, and techniques exist to measure it. There have been very few NMR studies of semiconductors at high temperatures.

A ^{49}Ti NMR study of defect motion and cation diffusion in single-crystal rutile with the composition TiO_{2-x} based on T_1 and $T_{1\rho}$ measurements over a range 150–1500 K demonstrates the feasibility of very high temperature NMR experiments for investigation of semiconductor dynamics.[91] Another example is ^{63}Cu and ^{115}In NMR in the chalcopyrite compounds $CuInX_2$ (X = S, Se, Te), studied over a range from room temperature to 823 K,[92] and from 123 to 473 K for X = S,Se.[93] Motional narrowing of the ^{63}Cu lines at higher temperatures was observed and yielded activation energies for Cu^+ diffusion, but despite the sizeable NQCC, the narrowing was predominantly due to an averaging of dipolar interactions. Nonstoichiometry in this class of compounds affects the diffusion. As a sidepoint of interest, these chalcopyrites, as well as those with other Group III atoms replacing In, have yielded NQCC values for many different nuclei, and a good correlation of the C_Q of ^{63}Cu with the tetragonal distortion has been observed.[92]

The Cu^+ ion is also fairly mobile in CuI, whose higher temperature phases are copper-ion conductors. The ^{63}Cu MAS NMR in the room-temperature phase γ-CuI (ZB) shows a set of ST peaks at multiples of the spinning speed due to the presence of defects in the cubic lattice, believed to be mostly Cu^+ vacancies in the naturally p-type nonstoichiometric material.[12] These ST peaks broaden as the temperature is raised from 300 to 360 K as a result of Cu^+ hopping from one site to another with a different EFG and hence a different resonance frequency. This process interferes with the formation of rotational echoes responsible for the sharp ST peaks, and was quantitatively analyzed with a simple model to yield an activation barrier that agreed with earlier static ^{63}Cu relaxation time measurements (see Chapter 14).

An unexpected case where a similar type of broadening of ST peaks was observed is the ^{71}Ga MAS NMR of a sample of GaN containing a mixture of hexagonal (WZ) and cubic (ZB) forms.[15] The left inset of Figure 24.4 shows an expanded ST region for two samples: the bottom spectrum is that of the mixed sample, showing weak peaks from the c-GaN as well as weak and broadened peaks from the h-GaN; and the top spectrum shows what the peaks of a pure h-GaN sample should look like after scaling to the same h-GaN CT peak heights in both samples. Clearly, the h-GaN ST peaks in the mixed sample have greatly decreased peak heights due to greater broadening. Some dynamical process is clearly modulating the EFGs at the Ga sites, based on a number of static ^{69}Ga and ^{71}Ga NMR Hahn-echo measurements of T_2 at different temperatures, field strengths, and positions in the powder pattern. Since atomic diffusion in h-GaN is expected to be negligible at room temperature, a possible explanation is that the modulation of the EFGs is produced by a dynamic equilibrium involving partially ionized donor sites which continually donate or receive electrons from the CB and thus alter the EFG at Ga sites within some range of distances.

24.5.8 EFG Measurement by Shallow Donor EPR Overhauser Shifts

As might be surmised from the important role electron spin has had in the NMR studies discussed, magnetic resonance techniques related to EPR have played a very important role in the study of semiconductors, particularly with regard to the identification of point defects. Although EPR per se is outside the intended scope of this chapter, it seems worthwhile to point out how all the EFGs at the quadrupolar sites in ZnO (WZ) and GaN (WZ), ^{67}Zn, ^{17}O, $^{69,71}Ga$ and ^{14}N can be measured by low-temperature EPR of shallow donors in the samples.[94] The nuclei are polarized by dynamic nuclear polarization (DNP) from saturating the electron spins, and their large polarization produces a shift in the EPR signal (Overhauser shift, the inverse phenomenon of the Knight shift in NMR) which is followed as the NMR frequency is swept. The DNP of shallow donors may provide a route to enhance the sensitivity of direct NMR observation, since, for example, very high nuclear polarizations of ^{67}Zn and ^{1}H spins resulting from DNP have been inferred in ZnO nanoparticles with a $Zn(OH)_2$ capping layer.[95]

RELATED ARTICLES IN THE ENCYCLOPEDIA OF MAGNETIC RESONANCE

Computations of Magnetic Resonance Parameters for Crystalline Systems: Principles

Diffusion in Solids

Dipolar and Indirect Coupling: Basics

Dipolar and Indirect Coupling Tensors in Solids

High Speed MAS of Half-Integer Quadrupolar Nuclei in Solids

Indirect Nuclear Spin-Spin Coupling Tensors

Inorganic Nonstoichiometric Crystalline Systems and Atomic Ordering

Knight Shift

Magic Angle Spinning

Magnetic Shielding and Chemical Shifts: Basics

Optically Enhanced Magnetic Resonance

Optically Pumped NMR of Semiconductors and Two-Dimensional Electron Systems

REFERENCES

1. C. Kittel, *Introduction to Solid State Physics*, 8th edn, John Wiley & Sons: Hoboken, NJ, 2005.
2. O. Madelung, *Semiconductors: Data Handbook*, Springer: Berlin, Heidelberg, New York, 2004.
3. M. N. Alexander and D. F. Holcomb, *Rev. Mod. Phys.*, 1968, **40**, 815.
4. A. Abragam, *The Principles of Nuclear Magnetism*, Oxford University Press: London, 1961.
5. C. P. Slichter, *Principles of Magnetic Resonance*, Springer- Verlag: Berlin, 1990.
6. J. P. Yesinowski, in *Topics in Current Chemistry*, Springer: Berlin, 2012, Vol. 306, p. 229.
7. F. Meier and B. P. Zakharchenya, eds, *Optical Orientation*, Modern Problems in Condensed Matter Sciences, Elsevier: Amsterdam, 1984, Vol. 8.
8. R. Tycko and J. A. Reimer, *J. Phys. Chem.*, 1996, **100**, 13240.
9. S. E. Hayes, S. Mui, and K. Ramaswamy, *J. Chem. Phys.*, 2008, **128**, 052203.
10. C. Y. Yeh, Z. W. Lu, S. Froyen, and A. Zunger, *Phys. Rev. B*, 1992, **46**, 10086.
11. P. Y. Yu and M. Cardona, *Fundamentals of Semiconductors: Physics and Materials Properties*, Springer: Heidelberg, New York, 2010.
12. J. P. Yesinowski, H. D. Ladouceur, A. P. Purdy, and J. B. Miller, *J. Chem. Phys.*, 2010, **133**, 234509.
13. W. A. Coish and J. Baugh, *Phys. Status Solidi B*, 2009, **246**, 2203.
14. T. J. Bastow and S. N. Stuart, *Phys. Status Solidi B-Basic Res.*, 1988, **145**, 719.
15. J. P. Yesinowski and A. P. Purdy, *J. Am. Chem. Soc.*, 2004, **126**, 9166.
16. H. Eckert and J. P. Yesinowski, *J. Am. Chem. Soc.*, 1986, **108**, 2140.
17. L. D. Potter, A. A. Guzelian, A. P. Alivisatos, and Y. Wu, *J. Chem. Phys.*, 1995, **103**, 4834.
18. T. J. Bastow and S. N. Stuart, *Chem. Phys.*, 1990, **143**, 459.
19. H. Eckert, K. Moran, D. Franke, and C. Hudalla, In *Nuclear Magnetic Shieldings and Molecular Structure*, ed J. A. Tosell, Kluwer Academic Publishers: Dordrecht, Boston, London, 1993, p. 49.
20. R. E. J. Sears, *Phys. Rev. B*, 1980, **22**, 1135.
21. R. E. J. Sears, *Phys. Rev. B*, 1981, **24**, 4072.
22. O. H. Han, H. K. C. Timken, and E. Oldfield, *J. Chem. Phys.*, 1988, **89**, 6046.
23. A. Nayeem and J. P. Yesinowski, *J. Chem. Phys.*, 1988, **89**, 4600.
24. D. Wolf, *Spin-Temperature and Nuclear-Spin Relaxation in Matter: Basic Principles and Applications*, Clarendon Press: Oxford, 1979.
25. J. P. Yesinowski, *Phys. Status Solidi (C)*, 2005, **2**, 2399.
26. J. P. Yesinowski, A. P. Purdy, H. Q. Wu, M. G. Spencer, J. Hunting, and F. J. DiSalvo, *J. Am. Chem. Soc.*, 2006, **128**, 4952.
27. P. Braun and S. Grande, *Phys. Status Solidi B-Basic Res.*, 1976, **74**, 209.
28. W. E. Carlos, J. A. Freitas, M. A. Khan, D. T. Olson, and J. N. Kuznia, *Phys. Rev. B*, 1993, **48**, 17878.
29. J. Lu, M. Hoch, P. Kuhns, W. Moulton, Z. Gan, and A. Reyes, *Phys. Rev. B*, 2006, **74**, 125208.
30. R. K. Sundfors and D. F. Holcomb, *Phys. Rev. A Gen. Phys.*, 1964, **136**, A810.
31. J. Haase, E. Oldfield, and K. Schmitt, *Chem. Phys. Lett.*, 1992, **193**, 274.
32. S. E. Fuller, E. M. Meintjes, and W. W. Warren, *Phys. Rev. Lett.*, 1996, **76**, 2806.
33. M. Murakami, T. Shimizu, M. Tansho, Y. Takano, S. Ishii, E. A. Ekimov, V. A. Sidorov, and K. Takegoshi, *Diamond Relat. Mater.*, 2009, **18**, 1267, References 1, 2 and 7 therein.
34. N. Roberts, R. P. Wang, A. W. Sleight, and W. W. Warren, *Phys. Rev. B*, 1998, **57**, 5734.
35. H. Eckert, S. Elbers, J. D. Epping, M. Janssen, M. Kalwei, W. Strojek, and U. Voigt, in *Topics in Current Chemistry*, Springer: Heidelberg, Berlin, 2005, Vol. 246, p. 195.
36. N. Bloembergen and T. J. Rowland, *Phys. Rev.*, 1955, **97**, 1679.
37. R. G. Shulman, J. M. Mays, and D. W. McCall, *Phys. Rev.*, 1955, **100**, 692.
38. R. G. Shulman, B. J. Wyluda, and H. J. Hrostowski, *Phys. Rev.*, 1958, **109**, 808.
39. R. K. Sundfors, *Phys. Rev.*, 1969, **185**, 458.
40. L. D. Potter and Y. Wu, *J. Magn. Reson., Ser. A*, 1995, **116**, 107.
41. R. K. Hester, A. Sher, J. F. Soest, and G. Weisz, *Phys. Rev. B*, 1974, **10**, 4262.

42. M. K. Cueman and J. F. Soest, *Phys. Rev. B*, 1976, **14**, 13.
43. Y. S. Hwang and R. E. Norberg, *Phys. Rev. B*, 1976, **14**, 3765.
44. M. Engelsberg and R. E. Norberg, *Phys. Rev. B*, 1972, **5**, 3395.
45. M. Tomaselli, D. deGraw, J. L. Yarger, M. P. Augustine, and A. Pines, *Phys. Rev. B*, 1998, **58**, 8627.
46. T. Iijima, K. Hashi, A. Goto, T. Shimizu, and S. Ohki, *Jpn. J. Appl. Phys.*, 2006, **45**, 651.
47. E. R. Andrew and D. P. Tunstall, *Proc. Phys. Soc. London*, 1961, **78**, 1.
48. J. P. Yesinowski, *J. Magn. Reson.*, 2006, **180**, 147.
49. A. Suter, M. Mali, J. Roos, and D. Brinkmann, *J. Magn. Reson.*, 2000, **143**, 266.
50. D. E. Woessner and H. K. C. Timken, *J. Magn. Reson.*, 1990, **90**, 411.
51. R. L. Mieher, *Phys. Rev.*, 1962, **125**, 1537.
52. J. A. McNeil and W. G. Clark, *Phys. Rev. B*, 1976, **13**, 4705.
53. M. Corti, A. Gabetta, M. Fanciulli, A. Svane, and N. Christensen, *Phys. Rev. B*, 2003, **67**, 064416.
54. D. P. Tunstall and A. Rogerson, *J. Phys. C: Solid State Phys.*, 1979, **12**, 3105.
55. M. Fanciulli and M. Corti, *Phys. Rev. B*, 1995, **52**, 11872.
56. T. J. Bastow, D. Massiot, and J. P. Coutures, *Solid State Nucl. Magn. Reson.*, 1998, **10**, 241.
57. T. J. Bastow, M. A. Gibson, and C. T. Forwood, *Solid State Nucl. Magn. Reson.*, 1998, **12**, 201.
58. F. Larsen, I. Farnan, and A. Lipton, *J. Magn. Reson.*, 2006, **178**, 228.
59. C. Gervais, M. E. Smith, A. Pottier, J. P. Jolivet, and F. Babonneau, *Chem. Mater.*, 2001, **13**, 462.
60. J. H. Ross, Z. Wang, and C. P. Slichter, *Phys. Rev. B*, 1990, **41**, 2722.
61. E. A. Hill and J. P. Yesinowski, *J. Am. Chem. Soc.*, 1996, **118**, 6798.
62. T. Brauniger, P. K. Madhu, A. Pampel, and D. Reichert, *Solid State Nucl. Magn. Reson.*, 2004, **26**, 114.
63. M. Dietrich, J. Kortus, W. Cordts, and S. Unterricker, *Phys. Status Solidi B-Basic Res.*, 1998, **207**, 13.
64. C. A. Sholl and J. A. Walter, *J. Phys. C: Solid State Phys.*, 1969, **2**, 2301.
65. O. Kanert and H. Kolem, *J. Phys. C: Solid State Phys.*, 1988, **21**, 3909.
66. S. Verkhovskii, A. Yakubovsky, B. Malkin, S. Saikin, M. Cardona, A. Trokiner, and V. Ozhogin, *Phys. Rev. B*, 2003, **68**, 104201.
67. D. Gill and N. Bloembergen, *Phys. Rev.*, 1963, **129**, 2398.
68. F. A. Collins and N. Bloembergen, *J. Chem. Phys.*, 1964, **40**, 3479.
69. R. H. Thorland, A. K. Garrison, and R. C. Duvarney, *Phys. Rev. B*, 1972, **5**, 784.
70. E. Brun, R. J. Mahler, H. Mahon, and W. L. Pierce, *Phys. Rev.*, 1963, **129**, 1965.
71. M. R. Tarasek and J. G. Kempf, *J. Phys. Chem. A*, 2010, **114**, 10634.
72. M. R. Tarasek and J. G. Kempf, *J. Phys. Chem. A*, 2010, **114**, 5743.
73. J. G. Kempf, M. A. Miller, and D. P. Weitekamp, *Proc. Natl. Acad. Sci. U.S.A.*, 2008, **105**, 20124.
74. R. G. Shulman, B. J. Wyluda, and P. W. Anderson, *Phys. Rev.*, 1957, **107**, 953.
75. R. K. Sundfors, *Phys. Rev. B*, 1974, **10**, 4244.
76. D. J. Guerrier and R. T. Harley, *Appl. Phys. Lett.*, 1997, **70**, 1739.
77. M. Eickhoff, B. Lenzmann, D. Suter, S. Hayes, and A. Wieck, *Phys. Rev. B*, 2003, **67**, 085308.
78. M. K. Cueman, R. K. Hester, A. Sher, J. F. Soest, and I. J. Lowe, *Phys. Rev. B*, 1975, **12**, 3610.
79. W. E. Carlos, S. G. Bishop, and D. J. Treacy, *Phys. Rev. B*, 1991, **43**, 12512, Reference 11 therein.
80. D. Mao and P. C. Taylor, *Phys. Rev. B*, 1995, **52**, 5665.
81. O. Kanert and M. Mehring, In *NMR Basic Principles and Progress*, eds P. Diehl, E. Fluck, and R. Kosfeld, Springer: Berlin, 1971, p. 1.
82. N. A. Mahadik, S. B. Qadri, M. V. Rao, and J. P. Yesinowski, *Appl. Phys. A*, 2007, **86**, 67.
83. R. Tycko, G. Dabbagh, S. R. Kurtz, and J. P. Goral, *Phys. Rev. B*, 1992, **45**, 13452.
84. P. J. Knijn, P. J. M. van Bentum, E. R. H. van Eck, C. M. Fang, D. Grimminck, R. A. de Groot, R. W. A. Havenith, M. Marsman, W. L. Meerts, G. A. de Wijs, and A. P. M. Kentgens, *PCCP*, 2010, **12**, 11517.
85. C. Degen, M. Tomaselli, B. Meier, M. Voncken, and A. Kentgens, *Phys. Rev. B*, 2004, **69**, 193303.

86. G. Czjzek, J. Fink, F. Gotz, H. Schmidt, J. M. D. Coey, J. P. Rebouillat, and A. Lienard, *Phys. Rev. B*, 1981, **23**, 2513.

87. J. B. D. de Lacaillerie, C. Fretigny, and D. Massiot, *J. Magn. Reson.*, 2008, **192**, 244.

88. D. Mao, P. C. Taylor, S. R. Kurtz, M. C. Wu, and W. A. Harrison, *Phys. Rev. Lett.*, 1996, **76**, 4769.

89. S. H. Wei and A. Zunger, *J. Chem. Phys.*, 1997, **107**, 1931.

90. C. Nelson, P. Taylor, and W. Harrison, *Phys. Rev. B*, 2004, **70**, 224207.

91. H. Kolem and O. Kanert, *Z. Metallkd.*, 1989, **80**, 227.

92. K. D. Becker and S. Wagner, *Phys. Rev. B*, 1983, **27**, 5240.

93. P. E. Stallworth, J. H. Guillemoles, J. Flowers, J. Vedel, and S. G. Greenbaum, *Solid State Commun.*, 2000, **113**, 527.

94. D. Reiser, J. Blomker, G. Denninger, and J. Schneider, *Solid State Commun.*, 1997, **102**, 359.

95. S. Orlinskii, J. Schmidt, P. Baranov, C. Donegá, and A. Meijerink, *Phys. Rev. B*, 2009, **79**, 165316.

Chapter 25

Quadrupolar NMR of Metal Nuclides in Biological Materials

Tatyana Polenova[1], Andrew S. Lipton[2], and Paul D. Ellis[2]

[1]*Department of Chemistry and Biochemistry, 036 Brown Laboratories, University of Delaware, Newark, DE 19716, USA*

[2]*Biological Sciences Division, K8-98, Pacific Northwest National Laboratory, Richland, WA 99352, USA*

25.1 INTRODUCTION

Half-integer-spin quadrupolar metal nuclei are ubiquitous in biological systems. Found in metalloproteins (which constitute approximately one-third to one-half of all proteins[1]), in natural products, and as ionic species either in free forms or bound to various biomolecules in cells, half-integer quadrupoles are essential to multiple biological functions. For example, copper-, zinc-, cobalt-, and vanadium-containing proteins play diverse roles, such as structural, catalytic, binding, storage, and electron transfer.[2] Sodium and potassium ions are charge carriers and are essential in maintaining the osmotic balance[2]; calcium ions participate in signal transduction.[2] Table 25.1 summarizes some of the prevalent biological functions of half-integer-spin quadrupolar metals. Despite decades of intense research aimed at establishing the relationships between the structure of metal sites and biological functions and using various structural and spectroscopic techniques (X-ray crystallography, electron paramagnetic resonance (EPR), optical spectroscopy, extended X-ray absorption fine structure (EXAFS), near-edge X-ray a fine structure (NEXAFS), and others), generalized understanding is still lacking. Investigations of biological metal sites have been hampered by many biologically relevant oxidation states of metals being "spectroscopically silent" and thus not amenable to EPR and UV–vis spectroscopies, which, in turn, precludes understanding of the electronic structures of these states. At the same time, X-ray structures of metal-containing biomolecules are often of insufficient quality with incomplete ligand sets and/or missing hydrogens.[1]

Solid-state NMR spectroscopy of half-integer-spin quadrupolar metal sites in biological systems is an emerging area. Traditionally challenging due to large quadrupolar coupling constants and/or low receptivity and/or low concentration of metal sites in

NMR of Quadrupolar Nuclei in Solid Materials
Edited by Roderick E. Wasylishen, Sharon E. Ashbrook and Stephen Wimperis
 ISBN: 978-0-470-97398-1

Table 25.1. Magnetic properties and biological functions of half-integer-spin quadrupolar metals [a]

Metal	Magnetically active isotopes	Nuclear spin I	Quadrupole moment $Q(\times 10^{24}\ cm^2)$	Gyromagnetic ratio γ $(\times 10^{-7}\ rad\ T^{-1}\ s^{-1})$	Natural abundance (%)	Receptivity vs ^1H	Some biological functions
Na	^{23}Na	3/2	0.12	7.0704	100	9.25×10^{-2}	Charge carrier, osmotic balance
Mg	^{25}Mg	5/2	0.22	−1.6389	10.13	2.72×10^{-4}	Hydrolase; isomerase; structural
K	^{39}K	3/2	0.055	1.2499	93.1	4.73×10^{-4}	Charge carrier; osmotic balance
	^{41}K	3/2	0.06	−1.75544	6.88	5.78×10^{-6}	
Ca	^{43}Ca	7/2	0.2 ± 0.1	−1.8028	0.145	9.27×10^{-6}	Charge carrier; signal transduction; structural
V	^{50}V	6	0.21	2.6721	0.24	1.76×10^{-4}	Nitrogen fixation; haloperoxidases
	^{51}V	7/2	−0.043	7.0492	99.76	0.381	
Cr	^{53}Cr	3/2	−0.15	−1.5077	9.55	8.62×10^{-5}	Essential trace element; possible involvement in glucose tolerance
Mn	^{55}Mn	5/2	0.4	6.6453	100	0.175	Photosynthesis; oxidases; structural
Co	^{59}Co	7/2	0.42	6.3015	100	0.277	Alkyl group transfer; oxidases
Ni	^{61}Ni	3/2	0.16	−2.3943	1.19	4.00×10^{-5}	Glyoxalase; hydrogenase; hydrolase; superoxide dismutase
Cu	^{63}Cu	3/2	−0.22	7.1088	69.09	6.42×10^{-2}	Dioxygen transport; electron transfer; oxidases
	^{65}Cu	3/2	−0.195	7.6104	30.91	3.54×10^{-2}	
Zn	^{67}Zn	5/2	0.15	1.6778	4.11	1.17×10^{-4}	Carbonic anhydrase; hydrolase; neurotransmitter; structural
Mo	^{95}Mo	5/2	0.16	−1.7514	15.72	5.13×10^{-4}	Nitrogen fixation; oxidases; oxo transfer
	^{97}Mo	5/2	0.17	−1.7884	9.46	3.28×10^{-4}	

[a] Reproduced from "**Quadrupolar Metal Nuclides in Bioinorganic Chemistry: Solid-State NMR Studies**".

biomacromolecules, direct detection of quadrupolar metal nuclides has become possible in the recent years with the advent of high magnetic fields (17.6 T or higher), as well as rapid advances in hardware (very fast MAS technology and broadbanded probes) and pulse sequences for sensitivity or resolution enhancement and/or broadband excitation of quadrupolar powder patterns under static or spinning conditions. As discussed in this chapter, metal nuclei that were for many years deemed impossible to study, such as ^{43}Ca or $^{63/65}$Cu, are now accessible to the experimentalist, and the range of biological systems containing quadrupolar nuclides that are amenable to solid-state NMR detection continues to grow.

In this chapter, we discuss the various solid-state NMR methods for the characterization of half-integer-spin quadrupolar metal sites in biological and bioinorganic systems. The experimental NMR observables, such as quadrupolar and shielding tensor elements, report with great sensitivity on the local geometric and electronic structure of the metal site, including coordination and hydrogen-bonding environment, ionization and protonation states of coordinated ligands, as well as internal dynamics of the metal. Detailed knowledge of the geometry and electronic structure of the metal sties is necessary for understanding the biological function of the corresponding molecule, such as catalytic mechanisms of metalloenzymes or redox properties of bioinorganic complexes. To relate the experimental NMR parameters to the molecular structure, quantum chemical calculations are commonly employed where quadrupolar and shielding tensor elements are calculated for various possible metal site environments and compared against the experimental data. Following the discussion of the contemporary solid-state NMR experimental and quantum chemical computational approaches, we review the current examples of studies of specific metal nuclei in biological systems.

As discussed in other chapters of this handbook and other articles of the Encyclopedia of Magnetic Resonance (see section Related Articles in the Encyclopedia of Magnetic Resonance), the main observables in the solid-state NMR spectra of half-integer-spin quadrupolar nuclei are components of the quadrupolar tensor (quadrupole coupling constant C_Q and quadrupole asymmetry parameter η_Q) and, in the presence of substantial chemical shift interaction, components of the shielding tensor (isotropic chemical shift δ_{iso}, reduced anisotropy δ_σ, anisotropy $\Delta\delta$, and asymmetry parameter of the chemical shift anisotropy (CSA) tensor η_σ defined according to the Haeberlen convention,[3] see below), as well as the Euler angles describing the relative orientations of the quadrupolar and CSA tensors. These NMR parameters are defined as follows.

The quadrupole coupling constant is directly proportional to the electric field gradient (EFG) at the nucleus and is expressed as

$$C_Q = \frac{e^2qQ}{h} = \frac{eV_{zz}^{PAS}Q}{h} \tag{25.1}$$

Q is the quadrupole moment of the nucleus and $eq = V_{zz}^{PAS}$ is defined as the largest absolute value of the field gradient tensor in the principal axis system (PAS) described by the diagonalized field gradient tensor $\boldsymbol{V}$. The traceless field gradient tensor[4] in its PAS frame can be described in terms of V_{zz} and its asymmetry parameter, η_Q,

$$|V_{zz}| \geq |V_{yy}| \geq |V_{xx}| \qquad \eta_Q \equiv \frac{V_{xx} - V_{yy}}{V_{zz}} \tag{25.2}$$

The SI units for V_{zz} and Q are V m^{-2} and m^2, respectively. For a compilation of nuclear quadrupole moments, see Pyykko.[5]

The various derived components of the shift tensor can be defined according to the Haeberlen convention[3]:

$$\Delta\delta = \delta_{zz} - \frac{1}{2}(\delta_{xx} + \delta_{yy})\ \delta_\sigma = \delta_{zz} - \delta_{iso};$$
$$\eta_\sigma = \frac{\delta_{yy} - \delta_{xx}}{\delta_{zz} - \delta_{iso}}$$
$$\delta_{iso} = \frac{1}{3}(\delta_{xx} + \delta_{yy} + \delta_{zz})|\delta_{zz} - \delta_{iso}| \geq |\delta_{xx} - \delta_{iso}| \geq |\delta_{yy} - \delta_{iso}| \tag{25.3}$$

For a detailed discussion of the basic spin properties including energy levels and NMR frequencies and lineshapes in half-integer-spin quadrupolar nuclei, see other chapters of this handbook and also the section Related Articles in the Encyclopedia of Magnetic Resonance. In the current review, we restrict the discussion to the points essential for understanding the salient features of the solid-state NMR spectra of half-integer-spin quadrupolar nuclei. For the reader to appreciate the challenges in detecting solid-state NMR signals of half-integer-spin quadrupolar nuclei in biological systems, we briefly turn our attention to the most important factors dominating the spectra: the breadth of the spectral lineshape and the sensitivity of the NMR experiment for a particular nucleus.

The quadrupolar frequency ν_Q, which determines the size of the quadrupolar splittings in the spectra, is expressed in frequency units as

$$\nu_Q = \frac{3C_Q}{2I(2I-1)} \tag{25.4}$$

For solid powders, the first-order quadrupolar frequency is orientation dependent and can be expressed in the laboratory frame as a function of polar and azimuthal angles θ and ϕ, which define the Zeeman field orientation with respect to the PAS of the EFG tensor:

$$\Delta\nu_Q(\theta,\phi) = \frac{1}{2}\nu_Q(3\cos^2\theta - 1 - \eta_Q \sin^2\theta \cos 2\phi) \tag{25.5}$$

And the corresponding first-order frequency shifts of the $m \leftrightarrow m+1$ transitions are

$$\Delta\nu^{(1)}_{m\leftrightarrow m+1}(\theta,\phi) = \Delta\nu_Q(\theta,\phi)\left(m+\frac{1}{2}\right) \tag{25.6}$$

The equation (25.6) describes the observable first-order quadrupolar lineshapes in the laboratory frame. Note that in solid powders the individual $m \leftrightarrow m+1$ transitions overlap and are broadened because of the orientational dependence of the quadrupolar frequency, except for the central transition. The central transition (CT), $1/2 \leftrightarrow -1/2$, is independent of the quadrupolar interaction to first order, which results in a relatively narrow central transition lineshape.

The typical quadrupolar coupling constants in half-integer-spin quadrupolar metals are in the megahertz range, giving rise to the lineshapes encompassing central and satellite transitions that are often as broad as several megahertz. These broad lineshapes make acquisition of the full spectral envelope challenging, and thus necessitating frequency-stepped approaches for excitation and detection of the lineshape, as discussed later in this chapter.

An important factor is the dependence of the quadrupolar frequency $\Delta\nu_Q(\theta,\phi)$ on the spin quantum number I. According to equation (25.4), the breadth of the spectral lineshape encompassing the central and satellite transitions decreases significantly with the increase in I. For example, assuming $C_Q =$ 200 kHz and disregarding the second-order quadrupolar interaction, the breadth of the static powder envelope for $I = 3/2$, 5/2, and 7/2 is 200, 120, and 85.6 kHz, respectively, according to equations (25.4–25.6). Similarly, the breadth of the central transition is proportional to

$$\Delta\nu_{CT} \propto \frac{C_Q^2}{\nu_0[2I(2I-1)]^2} \tag{25.7}$$

and assuming the same values of the quadrupolar coupling constant and Larmor frequency ν_0, the breadth of the central transition for $I = 3/2$, 5/2, and 7/2 is proportional to a factor of 1/9, 1/100, and 1/441, respectively.

For biological systems, one has to also consider that quadrupolar nuclei typically constitute a very small fraction of the sample, e.g., metalloproteins typically contain one or at most several metal atoms per protein molecule. For example, for carbonic anhydrase, the weight percentage of zinc is ~0.2%, and in vanadium chloroperoxidase the weight percentage of vanadium is ~0.075%, making detection of these metals challenging in the respective proteins. For comparison, in the inorganic model compounds, the weight percentage of the metal is 10- to 1000-fold higher than that in proteins. Furthermore, many of the biologically important half-integer-spin quadrupolar metals possess low natural abundance and/or low gyromagnetic ratios, which in conjunction with inherently broad lines and high dilution necessitates application of various sensitivity-enhancement techniques, which are introduced in this and other chapters of this handbook as well as in the section Related Articles in the Encyclopedia of Magnetic Resonance.

25.2 EXPERIMENTAL APPROACHES FOR QUADRUPOLAR METAL NMR IN BIOLOGICAL SYSTEMS

25.2.1 Magic-angle Spinning

Generally, the magnitude of the dominant quadrupolar interaction of metals in biological systems exceeds the range of the currently available MAS frequencies, and the complete averaging of the first-order quadrupolar term is not feasible. Additionally, MAS methods cannot average the second-order quadrupole interaction to an isotropically narrow resonance; MAS can only reduce the width of the lineshape. As a result, MAS is useful in recording spectra of half-integer-spin quadrupolar nuclei only when (i) the quadrupolar interaction is sufficiently small for acquisition of spectra of central and satellite transitions

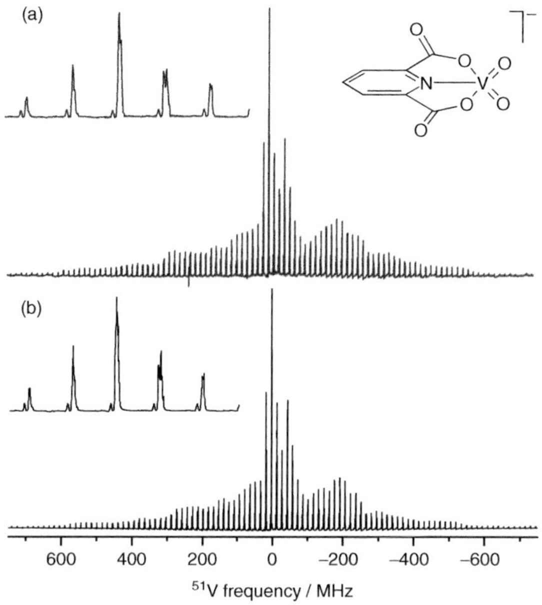

Figure 25.1. 9.4 T ^{51}V MAS spectra of a vanadium(V)O_2-dipicolinic acid derivative $NH_4[VO_2(dipic)]$. (a) Experimental. (b) Simulated in SIMPSON. The expansions illustrate the second-order lineshapes of the individual spinning sidebands, which are reproduced correctly in the simulated spectra. The MAS frequency is 15 kHz. Numerical simulations in SIMPSON of this spectrum together with spectra acquired at 11, 13, and 17 kHz yield the following best fit parameters: $C_Q = 5.8 \pm 0.1$ MHz, $\eta_Q = 0.4 \pm 0.05$, $\delta_{iso} = -537 \pm 5$ ppm, $\delta_\sigma = -550 \pm 30$ ppm, $\eta_\sigma = 0.0 \pm 0.05$. (Adapted from Ref. 6 © 1997 American Institute of Physics.)

(see below) and (ii) the quadrupolar interaction is prohibitively large for the detection of the complete manifold of spinning sidebands but the central transition is sufficiently narrow and can be broken into a series of spinning sidebands under MAS.[6]

Figure 25.1 shows an experimental and simulated ^{51}V MAS spectrum encompassing central and satellite transitions for vanadium (V) O_2-dipicolinic acid derivative $NH_4[VO_2(dipic)]$.[7] The spinning sideband manifold represents the first-order quadrupolar interaction ($C_Q = 5.8$ MHz, $\eta_Q = 0.4$), while the individual spinning sidebands are broadened by the second-order quadrupolar interaction, resulting in a characteristic asymmetric lineshape. The shape of the spinning sideband manifold corresponding to the central transition is dominated by the CSA. It is worth noting that the magnitudes of the quadrupolar and CSA interactions dominating the lineshape are typically very different. Therefore, numerical analysis of the spinning sideband intensities together with the analysis of the second-order quadrupolar lineshapes of the individual sidebands in the spectra acquired at two or more MAS frequencies permits the extraction of the NMR observables describing the EFG and CSA tensors and of their relative orientations[8,9]: C_Q, η_Q, δ_σ, δ_{iso}, η_σ, α, β, γ.

When the second-order quadrupolar interaction dominates the width of the central transition, or when multiple sites are present, complete removal of the second-order quadrupolar interaction is necessary to attain narrow spectral lines and resolve the resonances corresponding to the individual sites. Several approaches have been developed that permit averaging of the spatial terms to the fourth order by either simultaneous or sequential spinning of the sample about two different axes in the double rotation (DOR) and dynamic angle spinning (DAS) (see Chapter 8) experiments. Alternatively, a combined spatial/spin averaging method has been implemented in a 2D multiple-quantum MAS (MQMAS) experiment and its latter variants. Despite their ubiquitous applications in the studies of inorganic materials, these experiments have found only limited use for biological solids because of the inherently low sensitivity of the latter and because often it is impossible to spin the biological sample rapidly enough to average the second-order quadrupolar interaction using DOR, DAS, and MQMAS.

25.2.2 Sensitivity-enhancement Protocols

In biological systems, sensitivity is the main challenge in the detection of half-integer-spin quadrupolar metal nuclei. The typically low concentration of the metal site in a protein coupled with the inherently low receptivity for most of the half-integer-spin quadrupolar metals makes it imperative (with few rare exceptions) to employ various sensitivity-enhancement strategies for studies of half-integer-spin quadrupolar metals in biology.

Cross polarization (CP) is most commonly used to transfer magnetization from abundant spin-1/2 nuclei to dilute spin-1/2 nuclei, but it can also be used to transfer polarization from spin-1/2 nuclei to half-integer-spin quadrupolar nuclei. In this case, the

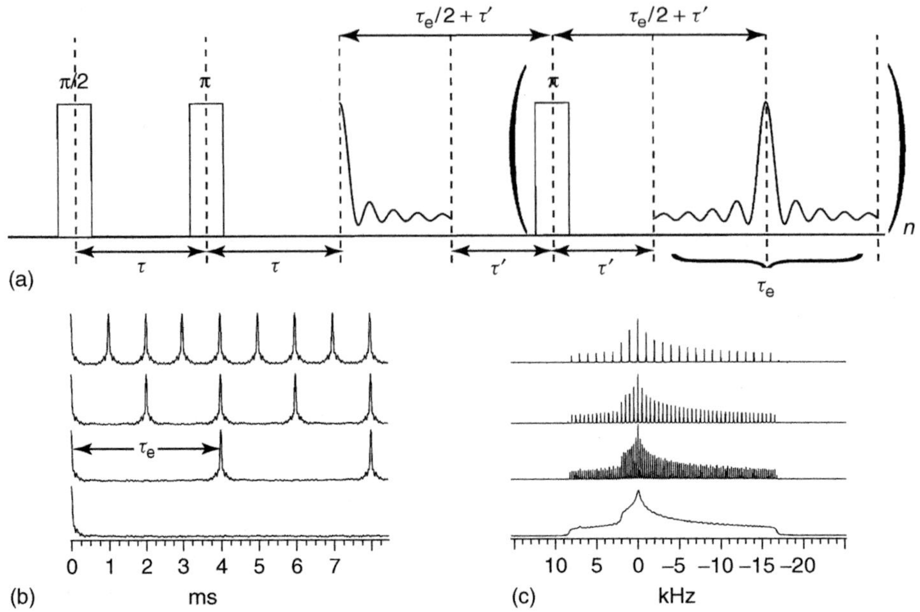

Figure 25.2. The QCPMG pulse sequence is shown as (a) the timing diagram with the consequences of τ_e in (b) the time domain and (c) the frequency domain. The data in (c) is the Fourier transform of the simulated signals of (b) showing the correlation between echo spacing and spike separation.

Hartmann–Hahn matching condition for stationary samples becomes

$$\gamma_H B_{1,H} = [I(I+1) - m(m-1)]^{1/2} \gamma_X B_{1,X} \quad (25.8)$$

where I the spin of the quadrupolar nucleus X with magnetorgyric ratio of γ_X and m the magnetic quantum number and B_1 is the radio frequency field applied on H or X nucleus. CP is usually used for the signal enhancement of the central transition, $m = 1/2$, where maximum sensitivity gain is expected.

It is worth noting that in the case of cross polarization and magic-angle spinning (CPMAS) between a spin-1/2 and a quadrupolar nucleus, the polarization transfer dynamics are complex and, in the general case, there may be no sensitivity enhancement. Furthermore, in the CPMAS experiments involving polarization transfer to half-integer quadrupolar nuclei, distorted lineshapes may be obtained.[10] Even though CPMAS-based experiments are performed for spectral editing in inorganic solids,[11] in biological systems they have not yet found utility because of the sensitivity limitations.

Another approach that works very well for sensitivity enhancement in biological half-integer-spin quadrupolar metal-containing solids is quadrupole Carr–Purcell–Meiboom–Gill (QCPMG) or "spikelet" spectroscopy. A QCPMG detection sequence consists of a train of π pulses timed so that the full echoes are formed and acquired (Figure 25.2). A Fourier transform of the resulting signal gives rise to a manifold of sidebands that retain the anisotropic quadrupolar and shielding interactions. In the QCPMG spectra, the linewidth of the individual sidebands is encoded by the overall duration of the echo train, while the anisotropic tensor information is encoded by the individual echoes. From QCPMG spectra, the magnitudes and relative orientations of the EFG and shielding tensors can be extracted by numerical simulations, as was originally demonstrated by Larsen *et al.*[12] QCPMG can be combined with MAS in a QCPMG-MAS experiment,[13] yielding further gains in sensitivity.

A promising strategy that has been increasingly applied to half-integer-spin quadrupolar nuclei for excitation of broad lines and yields sensitivity enhancement is frequency and amplitude modulation of excitation pulses. The early DFS[14] and hyperbolic secant[15] approaches are advantageous for studies of inorganic solids; more recently, optimal control theory (OCT) has been applied for constructing complex waveform excitation profiles that yield enhancements of the central transition signals of a factor of 2 or more, as demonstrated on $RbClO_4$.[16] These sensitivity-enhancement protocols await implementation in biological systems.

Finally, significant increase in sensitivity can be realized by performing experiments at cryogenic temperatures, as low as 10 K. Since the equilibrium spin polarization is determined by the Boltzmann distribution, and is proportional to the inverse of the temperature, a signal gain with temperature is realized by going from higher to lower temperatures (e.g., a factor of 30 gain in signal is expected at 10 K compare to 300 K). Additional improvements are observed by eliminating thermal noise by cooling the tuning components of the probe and the sample coil. If thermal noise is the dominant noise source, then the signal-to-noise ratio is proportional to $T^{-3/2}$,[17–19] and an additional factor of 164 is expected from cooling the electronics. Cryogenic probe technology for solid-state NMR applications is rapidly developing and is expected to become widely accessible and indispensable for biological applications.

It is important to note that, while experiments at cryogenic temperatures are advantageous from the nominal sensitivity standpoint, in many systems the spin–lattice relaxation rates become too small for such conditions to be useful in practice. In some cases, paramagnetic doping[20] or CP from fast-relaxing methyl protons[21] (if applicable) can alleviate the long T_1 problem.

25.3 INTERPRETATION OF NMR PARAMETERS IN TERMS OF MOLECULAR STRUCTURE: QUANTUM CHEMICAL CALCULATIONS

To interpret the experimentally determined NMR parameters corresponding to the EFG and shielding tensors in terms of geometric and electronic structure of the metal site, either empirical or quantum chemical approaches are commonly used. When suitable synthetic bioinorganic models are available with a variety of metal coordination environments to mimic the possible geometry of the biomolecule, it is possible to construct a database of experimental NMR parameters for these models and infer the possible metal coordination environment of the protein by comparison against the database. However, only approximate information can be gained following this procedure. The most rigorous approach is to perform first-principles calculations of the NMR observables using quantum chemistry. The first quantum mechanical calculations of shielding tensors and chemical shifts date back to the early and mid-1970s,[22–25] but it was not until the early 1990s that reliable theoretical protocols became available opening the door to chemical shift calculations in biological molecules.[26–29] These calculations can be performed by (i) treating the entire metal-containing biological molecule quantum mechanically; (ii) treating the metal and its immediate coordination environment quantum mechanically with the rest of the molecule being represented classically (the so-called QM/MM approach); or (iii) building a tractable model of the active site that contains the metal and the part of the biomolecule necessary to recapitulate the NMR parameters correctly, and performing quantum chemical calculations on this truncated model. All calculations require a starting model of the protein, which is generally obtained from an X-ray structure. From this starting model, subsequent models are built, where protonation states, and/or coordination geometry, and/or ligand sets are varied to sample the possible environments in the biomolecule. For each of these models, geometry optimization followed by the calculation of NMR parameters is performed, and the most likely coordination environment is assumed to be the one that yields the closest agreement between experiment and theory.

The first protocol, while the most rigorous, is limited to relatively small molecules where the computations are not prohibitively time consuming. The second approach, QM/MM, is rapidly gaining momentum in metalloproteins because the electronic structure of the metal site can be obtained at the most rigorous level while retaining the coordinates of the entire molecule including the waters of hydration and any cofactors or bound small molecules. In QM/MM calculations, the protein (or biomolecule in general) is divided into two regions: the quantum and the classical. The quantum region is treated quantum chemically (using wavefuntion- or density functional theory (DFT) based methods), while the classical region is described using molecular mechanics. The quantum region recapitulates all the critical interactions, such as hydrogen bonding, and contains the complete ligand sets that define the electronic and geometric properties of the metal (ultimately dictating the NMR observables). The third approach, while being minimalistic, works surprisingly well in some instances, indicating that it may be sufficient to focus only on small regions of the biomolecule encompassing the active site and its vicinity and ignore the rest of the structure. The main advantage of this approach is its computational cost: a substantial number of models can be analyzed relatively quickly to gain at the very least qualitative insight on the possible geometries and narrow the subset of the models for the subsequent expensive QM/MM calculations.

The validity of this combined experimental NMR–theoretical quantum chemical approach to probe metal coordination environments has been tested over the past decade on multiple bioinorganic models for which high-quality X-ray structures are available.[20,30–34] While solid-state NMR studies of metal sites in metalloproteins are still scarce, these reported investigations also demonstrate that an integrated NMR/theory protocol provides an accurate description of the geometry and electronic structure in large biomolecular systems.

For additional information and reviews of the field of molecular theory and quantum chemical calculations, and also two reviews[35,36] that also contain references to the existing software packages. The most popular packages for NMR parameter calculations are NWChem[37] and the commercial packages Gaussian,[38] ChemShell, Amsterdam Density Functional (ADF), Jaguar, and GAMESS. Some of these programs are developed for cluster calculations, while others allow periodic structure calculations that take into account the crystal lattice.

25.4 EXAMPLES OF APPLICATIONS IN BIOLOGICAL SYSTEMS

As discussed in the background sections above and illustrated below in examples of applications to specific biological systems, the choice of the experimental approach for detection of a particular half-integer-spin quadrupolar metal nucleus is dictated by its receptivity and the magnitude of the quadrupolar coupling constant. A large number of studies of metalloproteins to date required the use of cryogenic temperatures and sensitivity enhancement through CP from protons in QCPMG experiments. In several cases, because of favorable NMR properties, the detection of metal sites could be performed at ambient temperatures and with MAS spectroscopy alone. It is very important to note that in all the studies to date the interpretation of the NMR parameters in terms of the molecular structure relied on an integrated experimental NMR and theoretical quantum chemical approach, and critical insights were also gained from the studies of small synthetic bioinorganic models.

25.4.1 Characterization of Metal Sites under Cryogenic Conditions

25.4.1.1 ^{67}Zn Solid-state NMR of Zinc-binding Proteins

Zinc-containing proteins are ubiquitous in biology with zinc sites playing catalytic and structural roles (Table 25.1). Traditional characterization methods of Zn(II) sites in proteins relied on X-ray/EXAFS or on the use of surrogate probes, with both approaches yielding limited information on zinc coordination environments. Thus, the development of sensitive site-specific spectroscopic probes of zinc was needed for an understanding of the mechanism of the zinc-containing proteins.

Despite the challenges of NMR-based detection of zinc, ^{67}Zn solid-state NMR study of a DNA-binding domain of human nucleotide excision repair protein XPA emerged as the observation of a half-integer-spin quadrupolar metal site in a metalloprotein.[39] A single Zn center in this 14.7 kDa protein was detected at 9.4 T and a temperature of 25 K with CP/QCPMG spectroscopy. The C_Q of 4.9 MHz was in good agreement with the C_Q of 3.15 MHz for the tetrakis(thiourea) zinc nitrate complex, indicating similar coordination environments in the protein and the bioinorganic model. This first study validated the use of ^{67}Zn solid-state NMR spectroscopy as a probe of zinc coordination environments and paved the way for subsequent characterization of zinc proteins and bioinorganic zinc-containing complexes.

Among several examples of applications of ^{67}Zn solid-state NMR to probe zinc proteins, the most insightful from the mechanistic standpoint is perhaps the ^{67}Zn solid-state NMR investigation of human carbonic anhydrase II (hCAII), illustrated in Figure 25.3.[40] In this work, the 18.8 T ^{67}Zn CP-QCPMG spectra acquired at two pH values (5.0 and 8.5) and temperatures of 10 and 20 K, respectively, revealed the independence of the zinc

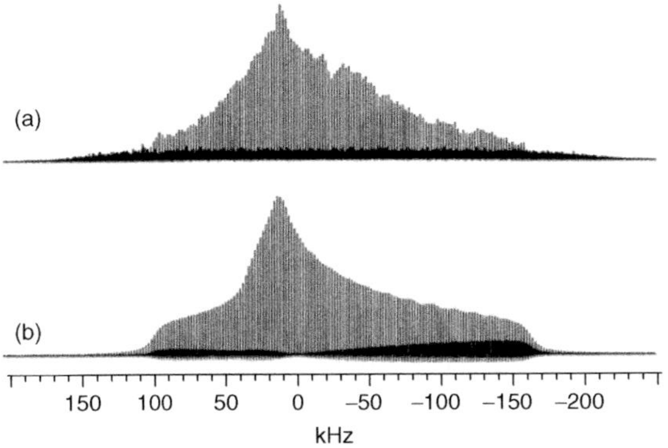

Figure 25.3. (a) The sky projection of the ^{67}Zn NMR spectra of CAII at pH 7 acquired at 9.4 T and 10 K. (b) A simulated QCPMG experiment using ideal pulses and a C_Q of 10.5 MHz, η_Q of 1.0, a matched apodization function of 10 kHz, and conventional line broadening of 10 Hz. (Adapted from Ref. 40 © 2004 American Chemical Society.)

environment on pH. On the basis of the comparison of the experimentally derived quadrupolar coupling constant (9.8 ± 0.2 MHz) with the theoretically predicted values for various active site models (methyl imidazole for histidine), it was concluded that the solvent molecule bound to the zinc remains in the hydroxide form under all observed conditions. This in turn led to a proposed refinement of the catalytic mechanism of the enzyme. As a side note, subsequent QM/MM calculations show that the active site of hCAII ameliorates the developing charge at the metal during the ionization between water and hydroxide, perhaps accounting for the incredible turnover rate of this enzyme's reaction mechanism.

Similarly, a ^{67}Zn solid-state NMR study of a zinc-dependent deacetylase LpxC corroborated the essential features of the enzymatic mechanism, through the analysis of the experimental NMR parameters in light of the coordination environment of the metal.[41] At pH 6, two environments with C_Q of 10 and 12.9 MHz were found, while at pH 9 a third species with C_Q of 14.3 MHz was dominant. These results suggested that the two ionizations of LpxC observed kinetically correspond to deprotonation of Glu78 at low pH followed by deprotonation of His 265 at higher pH while a water ligand is coordinated to zinc throughout the entire pH range. It should be noted that although pK_a's were derived experimentally from these data, it is difficult to interpret pH in buffered solutions. Most buffers have pK_a, and therefore pH, values that vary with temperature. Even allowing for a change in pH with varying temperature, one still needs to determine the change in pK_a of the residue in question within the protein with a given temperature change.

Finally, it was demonstrated that ^{67}Zn can also be employed as a surrogate probe to derive detailed coordination environments of nonreactive structural metal sites.[42] Zn^{2+}-substituted rubredoxin was chosen as a model protein for the development of reliable QM/MM protocols for interpreting the NMR parameters in terms of molecular structure. This work highlighted the fact that accurate QM/MM predictions of quadrupolar and CSA tensors require that the quantum region of the active site contains residues participating in the critical hydrogen-bonding and hydrophobic interactions with the metal-coordinated ligands.

25.4.1.2 ^{25}Mg Investigations of DNA Repair Protein Apurinic/Apyrimidic Endonuclease 1

Magnesium is another biologically important metal whose magnetically active isotope (^{25}Mg, $I = 5/2$) has been traditionally challenging to probe by NMR spectroscopy because of its low natural abundance and large quadrupole moment (Table 25.1). With the current technology, direct detection of ^{25}Mg in metalloproteins requires isotope enrichment, cryogenic temperatures, and sensitivity enhancement. The only reported example of ^{25}Mg solid-state NMR spectroscopy in a metalloprotein is the study of magnesium-binding environments of a 35.5 kDa DNA repair protein apurinic/apyrimidic endonuclease 1.[43] Before this work, any metal site studies of the protein were performed using surrogate probes (Pt, Pb, and Sm), and conflicting accounts of the binding stoichiometry and the catalytic mechanism appeared in the literature. The most unusual "moving metal" mechanism was proposed on the basis of molecular dynamics and surrogate Pb^{2+} probe studies involving shuttling of a single Mg^{2+} metal between two binding sites during catalysis. The ^{25}Mg CP-QCPMG spectra recorded under a variety of conditions revealed that the protein contains a single Mg^{2+} binding site where the metal is disordered. The disorder is not due to the shuttling of the metal between two different binding sites but is rather of a static nature because of the presence of cis and trans isomers of protein ligands about the Mg^{2+}. This study has resolved the controversy regarding the binding stoichiometry and catalytic mechanism of apurinic/apyrimidic endonuclease 1 and once again highlighted the importance and the sensitivity of NMR parameters of the half-integer-spin quadrupolar metal as structural and functional probes.

25.4.1.3 ^{65}Cu NMR Spectroscopy of Azurin

The largest quadrupolar coupling constant detected to date in a metal of a protein belongs to the Cu(I) state in azurin, $C_Q = \pm 71.2\,\text{MHz}$.[44] Blue copper azurin is one of the most thoroughly investigated metalloproteins, yet its Cu^{+} state is spectroscopically silent and thus not amenable to characterization by EPR spectroscopy or related techniques. The ^{65}Cu CP-QCPMG spectrum acquired at 18.8 T, illustrated in Figure 25.4, was the first to reveal the electronic environment of the Cu(I) state. Despite the large quadrupolar coupling constant, the spectrum was relatively easy to detect because of the surprisingly high sensitivity and short T_1 relaxation time even at cryogenic temperatures. Acquisition of the full spectral envelope corresponding to the central transition required collection of several frequency-shifted datasets. High sensitivity of the predicted NMR parameters using MP2 level of theory to the details of the model (inclusion of axial ligands and hydrogen bonding) indicates that experimental NMR observables should be a sensitive probe of the ligand geometry and electronic structure of the metal site in copper proteins.

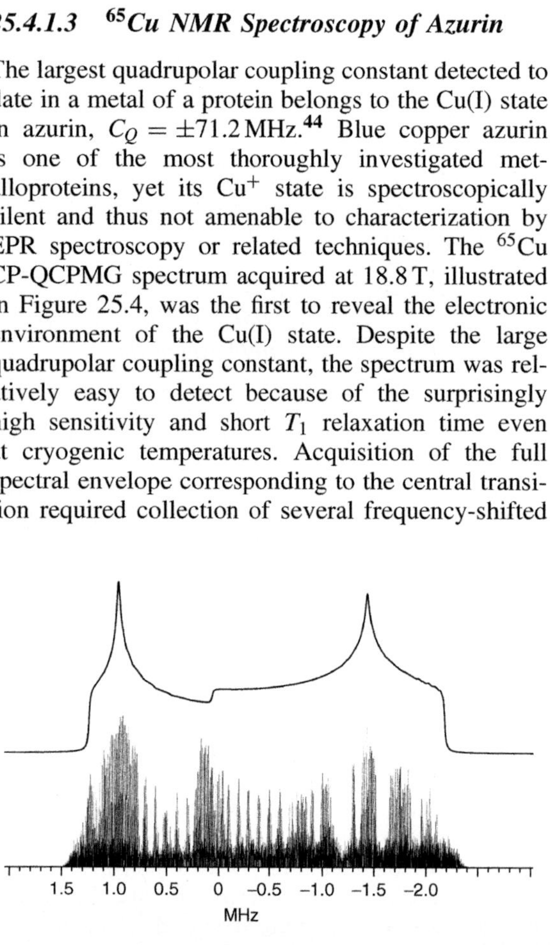

Figure 25.4. Skyline projection of the various frequency offsets of the ^{65}Cu NMR spectra of azurin at 18.8 T and 10 K, with a simulated lineshape calculated from the following parameters: $C_Q = \pm 71.2 \pm 1$ MHz, $\eta_Q = 0.2 \pm 0.1$. Each individual experiment required 4.2 h with a total time of approximately 3.5 weeks. (Adapted from Ref. 44 © 2009 American Chemical Society.)

25.4.1.4 ^{55}Mn Studies of a Model of Oxygen-evolving Complex in Photosystem II

A recent and potentially exciting direction is application of ^{55}Mn solid-state NMR spectroscopy for probing antiferromagnetically coupled states in manganese-containing biological systems. Mn-containing redox-active clusters are at the core of oxygen-evolving complex in photosystem II (PSII). Despite decades of spectroscopic, structural, biophysical and biochemical studies, and multiple synthetic bioinorganic manganese cluster models developed, accurate modeling of the oxygen-evolving complex is still out of reach. ^{55}Mn solid-state NMR coupled with DFT calculations has the potential to shed light on the electronic structure of the metal in PSII. The recent report of the temperature-dependent ^{55}Mn CP/QCPMG spectra of a bis(μ-oxo)dimanganese(IV) complex $[Mn_2O_2(salpn)_2]$ represents the first example of direct detection of an antiferromagnetically coupled half-integer-spin quadrupolar metal state in a bioinorganic system,[45] paving the way for studies of PSII and other manganese-containing biological systems. Broken-symmetry DFT calculations employing relativistic zeroth-order regular approximation (ZORA) yielded structural and NMR parameters in very good agreement with the experimental values ($C_Q{}^{expt} = 24.7$ MHz vs $C_Q{}^{calc} = 23.4$ MHz). The experimental and simulated spectra are illustrated in Figure 25.5.

25.4.2 Studies of Metal Sites at Ambient Temperatures

25.4.2.1 ^{59}Co NMR Investigations of Biological Systems

The seminal reports demonstrating the feasibility of ambient-temperature solid-state MAS spectroscopy in half-integer-spin quadrupolar metals

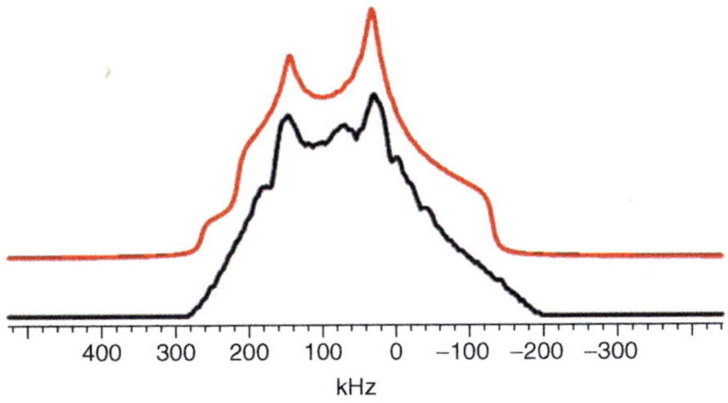

Figure 25.5. The bottom trace (black) is the experimental ^{55}Mn NMR spectrum of bis(μ-oxo)dimanganese(IV) complex [Mn_2O_2(salpn)$_2$] at 8.5 K. The top trace (red) is the simulation to the experimental data achieved using parameters discussed in the text. Zero frequency occurs at −1162.5 ppm from MnO_4^{1-}. (Adapted from Ref. 45 © 2010 American Chemical Society.)

in biological molecules are from Frydman and colleagues. As early as 1997, they demonstrated the experimental protocols for ^{59}Co solid-state NMR spectroscopy in diamagnetic porphyrin complexes[6] and cobalamins.[46] Subsequently, the same group of researchers conducted ^{59}Co solid-state studies of cobaltophthalocyanines,[47] polymorphs of vitamin B-12,[48] and polynucleotides.[49] Owing to the generally large quadrupolar coupling constants in biological cobalt sites, usually only central-transition ^{59}Co spectra are recorded. These early ^{59}Co investigations of the various classes of biological complexes revealed the exquisite sensitivity of the NMR parameters to the local geometric and electronic structure of the metal sites as well as to the presence of structural disorder and polymorphism.

25.4.2.2 ^{23}Na and ^{87}Rb Investigations of Metal Ions Bound to Proteins and Nucleic Acids

Several of the biologically important half-integer-spin quadrupolar metals, such as ^{23}Na, ^{39}K, and ^{43}Ca, are found predominantly as ionic species in the living organisms. These ions are typically bound to proteins, peptides, DNA, proteoglycans, or other biological molecules. Ionic species are typically expected to have relatively small quadrupolar interaction and in principle be directly detectable. Of the three metals, ^{23}Na ($I = 3/2$) possesses the most favorable NMR properties and has been studied quite extensively in the various biological environments, bound to DNA,[50–55] nucleotides, and nucleosides.[56,57] It has been found in the majority of the studies that there are multiple, distinct sodium environments. These species exhibit different quadrupolar coupling constants and chemical shifts, and to resolve the individual sites, MQMAS and its variants were employed.[52] In a related study, ^{87}Rb ($I = 3/2$) was used as a surrogate probe, and different cation environments in DNA duplexes were detected, corresponding to free as well as surface- and channel-bound ions.[58]

25.4.2.3 ^{51}V Spectroscopy of Vanadium Haloperoxidases

^{51}V is among the half-integer-spin quadrupolar metal nuclei with favorable NMR properties (Table 25.1). Possessing close to 100% natural abundance, high spin ($I = 7/2$), and a relatively small nuclear quadrupole moment, ^{51}V is surprisingly well amenable to direct detection in large vanadium-containing proteins. Owing to short T_1 and high receptivity, no sensitivity enhancement is needed, and ^{51}V MAS NMR spectra encompassing the central and satellite transitions of vanadium-containing proteins can be readily detected, even at moderate magnetic fields (14.1 T) with 10–15 mg of protein sample, in an overnight to a 5-day long experiment. To date, ^{51}V spectroscopy has been reported in vanadium-containing 67.5 kDa monomeric chloro- and 120.4 kDa dimeric bromoperoxidases.[59,60] Understanding the protonation states of the oxygen atoms coordinated to the vanadium is important for elucidating the chlorinating activity in haloperoxidases, and extensive work on bioinorganic complexes suggested that ^{51}V solid-state NMR parameters would be highly sensitive to the coordination geometry and protonation states of the metal. For vanadium chloroperoxidase, C_Q of 10.5 MHz and δ_σ of 520 ppm were obtained from the ^{51}V MAS NMR spectra shown in Figure 25.6. DFT calculation of NMR parameters for an extensive series of active site models revealed that the most likely vanadium coordination environment involves an anionic vanadate cofactor with the axial hydroxo ligand, while one hydroxo and two oxo groups are present in the equatorial plane. The detailed protonation environment was unavailable from the X-ray structure or any spectroscopic measurements.

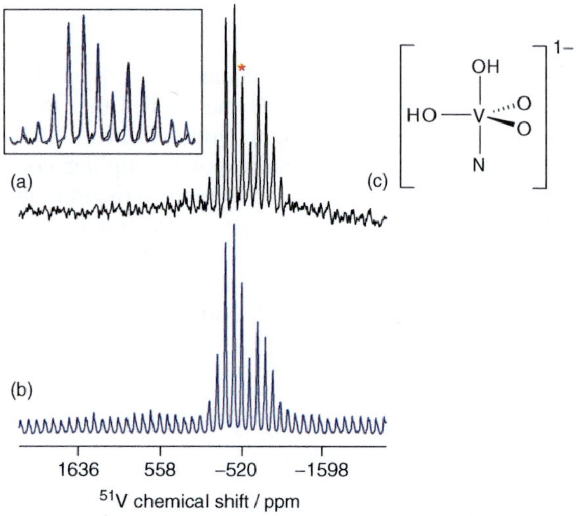

Figure 25.6. (a) Experimental. (b) Simulated in SIMPSON using the best fit parameters: ^{51}V MAS (17 kHz) NMR spectra of vanadium chloroperoxidase. $C_Q = 10.5$ MHz, $\eta_Q = 0.55$, $\delta_\sigma = -520$ ppm, $\eta_\sigma = 0.4$. The isotropic shift is -507.5 ppm. In the simulated spectrum (b), the magic-angle misset of 0.2° was taken into account. The inset is an overlay of the experimental and simulated central transition spinning sidebands to demonstrate the quality of the fit. (c) The protonation environment of the vanadate cofactor derived on the basis of ^{51}V solid-state NMR experiments and DFT calculations of VCPO active site models. (Adapted from Ref. 59 © 2006 American Chemical Society.)

The detailed mechanism of substrate specificity in haloperoxidases is still an open question, and studies of the resting state and intermediates in the two proteins and their mutants are ongoing.

25.5 CONCLUSIONS

Solid-state NMR spectroscopy of half-integer-spin quadrupolar metal nuclei in biological systems remains a challenge, but rapid progress has been made in the past decade in the development of high-field magnets, cryogenic MAS probe technologies, and sensitivity-enhancement protocols. These developments have paved the way for the detection of spectroscopically silent metal sites in metalloproteins including metal sites with huge quadrupolar coupling constants and antiferromagnetically coupled states and generally low receptivity nuclei in dilute environments. In parallel, molecular theory has been rapidly developing to enable high-level quantum chemical predictions of NMR parameters in large biological molecules. These current results are highly promising, and we envision that further technological breakthroughs (such as construction of superconducting magnets with fields of 28.2 T and higher and hyperpolarization methods) will permit further studies of metal environments in a wide variety of biological systems.

RELATED ARTICLES IN THE ENCYCLOPEDIA OF MAGNETIC RESONANCE

Double Rotation

Internal Spin Interactions and Rotations in Solids

Magic Angle Spinning

Magic-angle Spinning Extensions

Multiple-quantum Magic-angle Spinning Experiments on Half-integer Nuclei: Fundamentals

Oxygen-17 NMR

Oxygen-17 NMR: Applications in Biochemistry

Quadrupolar Metal Nuclides in Bioinorganic Chemistry: Solid-State NMR Studies

Quadrupolar Nuclei in Liquid Samples

Relativistic Computation of NMR Shieldings and Spin-Spin Coupling Constants

Relaxation Theory for Quadrupolar Nuclei

Satellite Transition NMR Spectroscopy of Half-Integer Quadrupolar Nuclei under Magic-angle Spinning

Shielding Calculations

Shielding Calculations: GIAO Methods

Shielding Calculations: IGLO Method

Shielding Calculations: LORG and SOLO Approaches

Shielding Calculations: Perturbation Methods

Shielding: Overview of Theoretical Methods

Shielding Tensor Calculations

REFERENCES

1. R. H. Holm, P. Kennepohl, and E. I. Solomon, *Chem. Rev.*, 1996, **96**, 2239.

2. S. J. Lippard and J. M. Berg, *Principles of Bioinorganic Chemistry*, University Scienc Books: Mill Valley, 1994.
3. U. Haeberlen, in *Advances in Magnetic Resonance*, ed J. S. Waugh, Academic Press: Orlando, 1976, Suppl 1.
4. M. H. Cohen and F. Reif, *Solid State Phys. Adv. Res. Appl.*, 1957, **5**, 321.
5. P. Pyykko, *Mol. Phys.*, 2001, **99**, 1617.
6. A. Medek, V. Frydman, and L. Frydman, *J. Phys. Chem. B*, 1997, **101**, 8959.
7. S. E. Bolte, K. J. Ooms, T. Polenova, B. Baruah, D. C. Crans, and J. J. Smee, *J. Chem. Phys.*, 2008, **128**(4), 0523175, 1.
8. H. J. Jakobsen, J. Skibsted, H. Bildsøe, and N. C. Nielsen, *J. Magn. Reson.*, 1989, **85**, 173.
9. J. Skibsted, N. C. Nielsen, H. Bildsoe, and H. J. Jakobsen, *Chem. Phys. Lett.*, 1992, **188**, 405.
10. P. J. Barrie, *Chem. Phys. Lett.*, 1993, **208**, 486.
11. J. P. Amoureux, J. Trebosc, L. Delevoye, O. Lafon, B. Hu, and Q. Wang, *Solid State Nucl. Magn. Reson.*, 2009, **35**, 12.
12. F. H. Larsen, H. J. Jakobsen, P. D. Ellis, and N. C. Nielsen, *J. Phys. Chem. A*, 1997, **101**, 8597.
13. F. H. Larsen, H. J. Jakobsen, P. D. Ellis, and N. C. Nielsen, *J. Magn. Reson.*, 1998, **131**, 144.
14. D. Iuga, H. Schafer, R. Verhagen, and A. P. M. Kentgens, *J. Magn. Reson.*, 2000, **147**, 192.
15. R. Siegel, T. T. Nakashima, and R. E. Wasylishen, *Chem. Phys. Lett.*, 2004, **388**, 441.
16. L. A. O'Dell, K. J. Harris, and R. W. Schurko, *J. Magn. Reson.*, 2010, **203**, 156.
17. Y. W. Kim, W. L. Earl, and R. E. Norberg, *J. Magn. Reson. A*, 1995, **116**, 139.
18. P. Styles, N. F. Soffe, and C. A. Scott, *J. Magn. Reson.*, 1989, **84**, 376.
19. P. Styles, N. F. Soffe, C. A. Scott, D. A. Cragg, F. Row, D. J. White, and P. C. J. White, *J. Magn. Reson.*, 1984, **60**, 397.
20. A. S. Lipton, T. A. Wright, M. K. Bowman, D. L. Reger, and P. D. Ellis, *J. Am. Chem. Soc.*, 2002, **124**, 5850.
21. R. A. Wind, S. Emid, D. J. Ligthelm, and J. F. J. M. Pourquie, *Bull. Magn. Reson.*, 1984, **6**, 71.
22. F. Birnstoc, *Mol. Phys.*, 1973, **26**, 343.
23. R. Ditchfield and P. D. Ellis, in *Topics in Carbon-13 NMR Spectroscopy*, ed G. C. Levy, Wiley: New York, 1974, p. 18.
24. M. Hladnik, D. Pumpernik, M. Zaucer, and A. Azman, *Chem. Phys. Lett.*, 1976, **42**, 361.
25. M. Zaucer and A. Azman, *Croat. Chem. Acta*, 1975, **47**, 17.
26. J. D. Augspurger, C. E. Dykstra, and E. Oldfield, *J. Am. Chem. Soc.*, 1991, **113**, 2447.
27. D. B. Chesnut and C. G. Phung, *Chem. Phys. Lett.*, 1991, **183**, 505.
28. J. E. Combariza, M. Barfield, and J. H. Enemark, *J. Phys. Chem.*, 1991, **95**, 5463.
29. A. M. Orendt, R. R. Biekofsky, A. B. Pomilio, R. H. Contreras, and J. C. Facelli, *J. Phys. Chem.*, 1991, **95**, 6179.
30. D. L. Bryce and G. D. Sward, *J. Phys. Chem. B*, 2006, **110**, 26461.
31. R. P. Chapman and D. L. Bryce, *Phys. Chem. Chem. Phys.*, 2007, **9**, 6219.
32. D. Rehder, T. Polenova, and M. Buhl, *Annu. Rep. NMR Spectrosc.*, 2007, **62**, 49.
33. Y. Zhang, S. Mukherjee, and E. Oldfield, *J. Am. Chem. Soc.*, 2005, **127**, 2370.
34. A. S. Lipton and P. D. Ellis, *J. Am. Chem. Soc.*, 2007, **129**, 9192.
35. M. Head-Gordon, *J. Phys. Chem.*, 1996, **100**, 13213.
36. J. Cuny, S. Messaoudi, V. Alonzo, E. Furet, J. F. Halet, E. Le Fur, S. E. Ashbrook, C. J. Pickard, R. Gautier, and L. Le Polles, *J. Comput. Chem.*, 2008, **29**, 2279.
37. M. Valiev, E. J. Bylaska, N. Govind, K. Kowalski, T. P. Straatsma, H. J. J. Van Dam, D. Wang, J. Nieplocha, E. Apra, T. L. Windus, and W. d e Jong, *Comput. Phys. Commun.*, 2010, **181**, 1477.
38. M. J. T. Frisch, G. W. H. B. Schlegel, G. E. Scuseria, M. A. Robb, J. R. Cheeseman, J. A. Montgomery Jr, T. Vreven, K. N. Kudin, J. C. Burant, J. M. Millam, S. S. Iyengar, J. Tomasi, V. Barone, B. Mennucci, M. Cossi, G. Scalmani, N. Rega, G. A. Petersson, H. Nakatsuji, M. Hada, M. Ehara, K. Toyota, R. Fukuda, J. Hasegawa, M. Ishida, T. Nakajima, Y. Honda, O. Kitao, H. Nakai, M. Klene, X. Li, J. E. Knox, H. P. Hratchian, J. B. Cross, V. Bakken, C. Adamo, J. Jaramillo, R. Gomperts, R. E. Stratmann, O. Yazyev, A. J. Austin, R. Cammi, C. Pomelli, J. W. Ochterski, P. Y. Ayala, K. Morokuma, G. A. Voth, P. Salvador, J. J. Dannenberg, V. G. Zakrzewski, S. Dapprich, A.

D. Daniels, M. C. Strain, O. Farkas, D. K. Malick, A. D. Rabuck, K. Raghavachari, J. B. Foresman, J. V. Ortiz, Q. Cui, A. G. Baboul, S. Clifford, J. Cioslowski, B. B. Stefanov, G. Liu, A. Liashenko, P. Piskorz, I. Komaromi, R. L. Martin, D. J. Fox, T. Keith, M. A. Al-Laham, C. Y. Peng, A. Nanayakkara, M. Challacombe, P. M. W. Gill, B. Johnson, W. Chen, M. W. Wong, C. Gonzalez, and J. A. Pople, Gaussian, Inc: Wallingford (CT), 2004.

39. A. S. Lipton, G. W. Buchko, J. A. Sears, M. A. Kennedy, and P. D. Ellis, *J. Am. Chem. Soc.*, 2001, **123**, 992.

40. A. S. Lipton, R. W. Heck, and P. D. Ellis, *J. Am. Chem. Soc.*, 2004, **126**, 4735.

41. A. S. Lipton, R. W. Heck, M. Hernick, C. A. Fierke, and P. D. Ellis, *J. Am. Chem. Soc.*, 2008, **130**, 12671.

42. A. S. Lipton, R. W. Heck, G. R. Staeheli, M. Valiev, W. A. De Jong, and P. D. Ellis, *J. Am. Chem. Soc.*, 2008, **130**, 6224.

43. A. S. Lipton, R. W. Heck, S. Primak, D. R. McNeill, D. M. Wilson, and P. D. Ellis, *J. Am. Chem. Soc.*, 2008, **130**, 9332.

44. A. S. Lipton, R. W. Heck, W. A. d e Jong, A. R. Gao, X. J. Wu, A. Roehrich, G. S. Harbison, and P. D. Ellis, *J. Am. Chem. Soc.*, 2009, **131**, 13992.

45. P. D. Ellis, J. A. Sears, P. Yang, M. Dupuis, T. T. Boron, V. L. Pecoraro, T. A. Stich, R. D. Britt, and A. S. Lipton, *J. Am. Chem. Soc.*, 2010, **132**, 16727.

46. A. Medek, V. Frydman, and L. Frydman, *Proc. Natl. Acad. Sci. U.S.A.*, 1997, **94**, 14237.

47. A. Medek, V. Frydman, and L. Frydman, *J. Phys. Chem. A*, 1999, **103**, 4830.

48. A. Medek and L. Frydman, *J. Am. Chem. Soc.*, 2000, **122**, 684.

49. C. V. Grant, V. Frydman, J. S. Harwood, and L. Frydman, *J. Am. Chem. Soc.*, 2002, **124**, 4458.

50. R. Ida, I. C. M. Kwan, and G. Wu, *Chem. Commun.*, 2007, **8**, 795.

51. R. Ida and G. Wu, *J. Am. Chem. Soc.*, 2008, **130**, 3590.

52. D. Rovnyak, M. Baldus, G. Wu, N. V. Hud, J. Feigon, and R. G. Griffin, *J. Am. Chem. Soc.*, 2000, **122**, 11423.

53. A. Wong, J. C. Fettinger, S. L. Forman, J. T. Davis, and G. Wu, *J. Am. Chem. Soc.*, 2002, **124**, 742.

54. A. Wong, Z. M. Yan, Y. N. Huang, and G. Wu, *Magn. Reson. Chem.*, 2008, **46**, 308.

55. G. Wu and A. Wong, *Biochem. Biophys. Res. Commun.*, 2004, **323**, 1139.

56. A. Wong and G. Wu, *J. Phys. Chem. A*, 2003, **107**, 579.

57. G. Wu and A. Wong, *Chem. Commun.*, 2001, **24**, 2658.

58. R. Ida and G. Wu, *Chem. Commun.*, 2005, **34**, 4294.

59. N. Pooransingh-Margolis, R. Renirie, Z. Hasan, R. Wever, A. J. Vega, and T. Polenova, *J. Am. Chem. Soc.*, 2006, **128**, 5190.

60. T. Polenova, N. Pooransingh-Margolis, R. Renirie, R. Wever, and D. Rehder, in *Vanadium, The Versatile Metal*, eds K. Kustin, D. C. Crans, and J. Costa Pessoa, American Chemical Society, Washington, DC, 2007, p. 178.

Chapter 26

Nuclear Waste Glasses: Insights from Solid-State NMR

Scott Kroeker

Department of Chemistry, University of Manitoba, Winnipeg, MB R3T 2N2, Canada

26.1 INTRODUCTION

Increasing population, decreasing fossil fuel resources, and fears of pollution-driven climate change drive an urgent search for new global energy solutions. Nuclear fission is a traditional energy source that has been undergoing revitalization, at least as a temporary measure until renewable sources become more established. Despite the tremendous amount of energy that can be obtained from nuclear fission, it consumes uranium, a limited natural resource. Hence, it is prudent to reclaim the fissile material remaining in the partially spent fuel and reprocess it into new fuel rods. This requires an extraction procedure to separate the uranium and plutonium from a multitude of fission products formed upon neutron bombardment. While this process is chemically straightforward, it results in a waste liquor containing many types of highly radioactive ions. Disposing of this high-level waste (HLW) to ensure that the long-lived radioactive components can decay to safe levels without being dispersed into the biosphere has been the subject of much research.

The worldwide consensus about the most effective way to dispose of HLW appears to be vitrification. Glasses are ideal for long-term immobilization of radioactive ions because their high-temperature melts are excellent solvents, dissolving large concentrations of many different types of ions, and can be cooled to a chemically durable solid that incorporates the waste ions into a homogeneous glassy matrix. Vitrification is a mature technology used in many countries, but waste streams are highly variable, placing stringent demands on glass to accommodate different levels and types of fission products without compromising its durability. Furthermore, vitrification requires high temperatures and is a correspondingly energy-intensive process, motivating efforts to maximize waste loading. Owing to the enormous long-term health risks associated with failure, nuclear waste glasses are overengineered. For these reasons, research on all aspects of their composition, production, properties, and performance continues to be essential.

Many properties must be simultaneously optimized for effective waste immobilization. To facilitate waste

NMR of Quadrupolar Nuclei in Solid Materials
Edited by Roderick E. Wasylishen, Sharon E. Ashbrook and Stephen Wimperis

ISBN: 978-0-470-97398-1

mixing and processing, the melt viscosity must be relatively low at accessible working temperatures (e.g., <1100 °C) to reduce energy costs and prevent elemental volatilization. The melt must not be corrosive to the melter, but be able to dissolve a wide variety of waste ions without becoming heterogeneous upon solidification. The glass transition temperature (T_g) must be high enough to prevent ionic mobility and phase transformation due to radiogenic heating during storage. Chemical durability must be high to ensure that radioisotopes are retained in the glass even in the event of contact with groundwater. Mechanical properties and radiation resistance must also be considered to accommodate structural, morphological, and compositional changes occurring from radioactive decay of the waste constituents.[1]

The most durable and versatile HLW glass would be SiO_2, but the high melting temperatures required for pure silica glass are impractical. Most commercial HLW glasses are borosilicates, where the addition of B_2O_3 reduces the melting temperature and improves the thermal properties without compromising other essential characteristics such as mechanical integrity and chemical durability. A typical base glass composition is given in Table 26.1. Once loaded with radioactive waste, however, the final composition may contain over 30 oxide components. Although facilities exist for experiments on radioactive materials, the vast majority of structural studies, property characterization, and development research are performed on inactive analogs in which non-radioactive isotopes are substituted for the fission products (Table 26.2). Such experiments are generally valid but cannot account for radiation damage, compositional alterations due to transmutation, and macroscopic morphological changes arising from He bubble formation. Nevertheless, essential fundamental information about approved nuclear waste glasses and novel materials and processes under development can be practically obtained only on inactive materials.[2]

Solid-state nuclear magnetic resonance (NMR) spectroscopy has played an important role in characterizing nuclear waste materials. As an isotope-selective structural probe, NMR can provide valuable information about glassy and microcrystalline environments not readily available by diffraction-based methods. Dynamics can also be studied using NMR methodology, leading to species-specific insight about ionic mobility and local motions (see Chapter 14). The isotope selectivity of NMR is also advantageous for studying complex, multicomponent materials, where a systematic multinuclear approach can lead to a complete picture of the structure. All these general features make NMR a powerful and unique tool for the study of current and developing nuclear waste glasses and technologies.

Table 26.1. Composition of mixed-alkali borosilicate base glass frit "MW" from the United Kingdom's National Nuclear Laboratory[3]

Oxide	mol%
SiO_2	60.61
B_2O_3	18.57
Na_2O	10.53
Li_2O	10.29

Table 26.2. Composition of French reference glass "SON68"

Oxide	mol%	Oxide	mol%
SiO_2	52.57	BaO	0.27
B_2O_3	13.99	Cr_2O_3	0.23
Na_2O	11.05	Pr_2O_3	0.23
CaO	5.00	SrO	0.22
Li_2O	4.60	Ce_2O_3	0.20
Al_2O_3	3.34	La_2O_3	0.19
ZnO	2.13	P_2O_5	0.14
ZrO_2	1.49	CoO	0.11
Fe_2O_3	1.27	TeO_2	0.10
MoO_3	0.82	Y_2O_3	0.06
NiO	0.69	CdO	0.02
MnO_2	0.58	Ag_2O	0.01
Cs_2O	0.35	SnO_2	0.01
Nd_2O_3	0.33	Sb_2O_3	<0.01

26.2 NUCLIDES OF INTEREST IN NUCLEAR WASTE GLASSES

Nuclear waste glasses are attractive from an NMR perspective because the network-forming cations are receptive and lend themselves readily to analysis by multinuclear magnetic resonance spectroscopy. The following is an overview of the nuclei that have proved useful in studies of nuclear waste glasses and their salient characteristics.

26.2.1 Boron-11

Boron is a key component in nuclear waste glasses, and is responsible for many of the properties that make these materials effective. ^{11}B ($S = 3/2$, natural abundance (NA) = 80.1%) NMR is easy to observe spectroscopically, and its behavior in borate-based glasses is well documented.[4,5] The most common use of ^{11}B NMR in oxides is the measurement of the ratio of three- and four-coordinate boron polyhedra. Four-coordinate boron ($^{[4]}B$) exists in near-tetrahedral geometry, resulting in a small electric field gradient (EFG) at boron. This is generally quantified by the product of the quadrupole moment (eQ) and the largest component of the EFG tensor (V_{zz}), $C_Q = (eQ)(V_{zz})/h$, where h is the Planck constant. This quadrupolar coupling constant is relatively small for the highly symmetrical $^{[4]}B$, $C_Q < 600$ kHz, whereas three-coordinate boron ($^{[3]}B$) is essentially trigonal planar and possesses a much larger C_Q of about 2.6 MHz. This difference produces ^{11}B NMR spectra in which the narrow $^{[4]}B$ peaks are readily distinguishable from the broader $^{[3]}B$ peaks, and accurate peak intensities can be measured to provide reliable fractions of these two species. This simple structural parameter is significant because it is intimately related to several properties that affect the performance of nuclear waste glasses. While straightforward spectral deconvolution can be used for nonspinning spectra and magic-angle spinning (MAS) spectra at low fields, ^{11}B MAS NMR at field strengths of 11.7 T and higher produces full resolution of these two signals. Accurate quantification of $^{[3]}B$ and $^{[4]}B$ concentrations requires that small tip angles (<30°) be used to ensure that their different nutation rates do not inadvertently skew the observed intensities. Furthermore, direct comparison of integrated center band intensities should also take into account that some additional intensity from satellite transitions may be present in the $^{[4]}B$ signal, and that some of the central-transition intensity of the $^{[3]}B$ may be lost to first-order spinning sidebands.

^{11}B NMR can also provide evidence of network depolymerization through the detection of three-coordinate boron bonded to one or more nonbridging oxygens (NBOs). Such species reduce the network connectivity and degrade the chemical durability of the glass. While there is ample precedent for the identification and quantification of NBOs, reliable measurements are not necessarily routine, as severe peak overlap between all types of three-coordinate boron is generally observed. Spectral deconvolution requires prior knowledge of likely parameters for NBO-bearing $^{[3]}B$ and is subject to large errors and possible misinterpretations. Low-field nonspinning ^{11}B NMR studies and ultrahigh field ^{11}B MAS NMR studies (i.e., >21.1 T) are most likely to yield reliable values for these species. Multiple-quantum (MQ) MAS NMR, a 2D method that yields enhanced resolution for half-integer quadrupoles, has also been used to resolve NBO-bearing $^{[3]}B$ units,[6] but quantitative accuracy is compromised by C_Q-dependent excitation and conversion efficiencies (see Chapter 10).

The immense amount of ^{11}B NMR that has been carried out on borosilicate glasses and crystals provides ample data to permit slight variations in peak position to be interpreted in terms of next-nearest neighbors, and hence network-former connectivity. This is most easily done for $^{[4]}B$, where the peak position can be reasonably approximated as the isotropic chemical shift. Numerous studies have correlated $^{[4]}B$ peak positions with bonding through oxygen to some number of silicate polyhedra, with the general guideline that a greater number of bonded silicons drives the peak to lower chemical shift. At some magnetic fields, multiple $^{[4]}B$ peaks may be detected (though rarely fully resolved), and provide a valuable indication of the degree of network mixing. Such data can be related to phase separation, known to occur in some borosilicate glasses of similar composition, yet very difficult to quantify from a molecular perspective. Chemical shift correlations amongst three-coordinate boron are complicated by spectral overlap, but relationships between chemical shift and ring formation, depolymerization, charge-balancing cations, and bonded network formers have been documented.[7]

26.2.2 Silicon-29

As a relatively accessible nonquadrupolar nucleus, ^{29}Si ($S = 1/2$, $NA = 4.7\%$) NMR is often used to obtain information about network connectivity in borosilicate nuclear waste glasses. The most common use to which ^{29}Si NMR is applied is the assessment of the relative degree of polymerization of the silicate network in terms of the concentrations of silicate polyhedra bonded to different numbers of neighboring network-forming polyhedra. Chemical shift ranges for Q^4, Q^3, and Q^2 species (i.e., silicate tetrahedra bonded through oxygen to four, three, or two network formers, respectively) are distinct, and can, in principle, provide a direct measure of their ratios. In practice, however, ^{29}Si MAS NMR of multicomponent glasses yields severely overlapping peaks that span more than one "Q" region, demanding deconvolution based on assumptions about their peak positions and widths, with quantitatively unambiguous results rarely obtained. Comparisons between ^{29}Si NMR spectra with different compositions or processing history can be qualitatively informative, especially in conjunction with complementary NMR measurements of other nuclei, to provide insight into the relative degree of polymerization. The reason for the broad peak widths is partly simply due to the geometric disorder intrinsic to the glassy state, but may also be attributed to chemical shift effects from next-nearest neighbors. For example, network-forming cations, such as Si, B, and Al, will subtly influence the ^{29}Si peak positions of a given Q^n species, as will network-modifying cations such as Na^+, Cs^+ and Ca^{2+}. While these effects have the potential to offer structural insight, it is more common for them to obscure even the accurate measurement of Q-speciation.

26.2.3 Aluminum-27

Aluminum is a minor component in many nuclear waste glasses, and its role in the glass structure is easily probed using ^{27}Al ($S = 5/2$, $NA = 100\%$) MAS NMR. While most aluminum exists in nuclear waste glasses as anionic, four-coordinate species, the presence of five- and six-coordinate Al can be readily detected in some materials as distinct peaks in characteristic chemical shift regions. (Caution must be exercised, however, since many "zirconia" rotors contain small amounts of high-coordinate Al, which can be mistakenly attributed to the sample.) Experiments at low magnetic fields may yield unresolved peaks due to a combination of structural disorder and second-order quadrupolar broadening, but moderate to high magnetic fields can generally be relied upon to produce resolved peaks from which dependable relative intensities can be measured. MQMAS can be profitably employed to achieve higher resolution, however, bearing in mind that quantitative accuracy may be compromised by the different and unknown quadrupolar parameters for different coordination numbers. Except perhaps at the highest magnetic fields, ^{27}Al peak widths in glasses are typically too broad and chemical shifts too insensitive to support assignments based on network connectivity or charge compensation.

26.2.4 Oxygen-17

As a ubiquitous network-forming anion, ^{17}O ($S = 5/2$, $NA = 0.04\%$) NMR provides a direct probe of connectivity between network-forming cations and depolymerization. To the extent that the NMR parameters differ for oxygens linking different network formers and for NBOs charge-compensated by different modifier cations, these important structural features can be monitored and related to key properties such as depolymerization and phase homogeneity. An extensive body of work has been developed on this premise using MQMAS and ultrahigh field MAS NMR in a wide variety of borosilicate glasses closely related to nuclear waste glasses.[8] Limiting the general applicability of ^{17}O NMR to nuclear waste glasses is the need to isotopically enrich samples relative to its very low *NA*. While sensitivity enhancement methods may eventually render this requirement obsolete, it currently restricts sample sizes owing to the high cost of ^{17}O-enriched reagents, which may in turn influence the microscopic and macroscopic structural characteristics of the glass due to quench rate effects and/or elemental volatility. Furthermore, the requirement of 2D MQMAS experiments to achieve adequate spectral resolution makes further stringent sensitivity demands.

26.2.5 Cesium-133

Fission products ^{137}Cs ($t_{1/2} = 30$ years) and ^{135}Cs ($t_{1/2} = 2.3$ M years) are present in waste-loaded glasses, the latter being responsible for a significant proportion of the long-term residual radioactivity. ^{133}Cs ($S = 7/2$, $NA = 100\%$) is a nonradioactive NMR-active nucleus which is intrinsically receptive and serves as a valuable probe of fission product incorporation. Despite being a spin-7/2 nucleus, its quadrupole moment is relatively small, and

C_Q values typically do not introduce appreciable second-order quadrupolar effects in oxide solids at routine NMR fields. On the other hand, the ^{133}Cs chemical shift is highly sensitive to its local environment. Together, these effects result in broad ^{133}Cs MAS NMR peaks for glasses and sharp peaks for crystalline phases, which can be useful to detect devitrification. Fast spinning (>30 kHz) is generally required to resolve glassy spinning sidebands from the center band and obtain reliable peak positions and relative intensities. While the resolution of multiple amorphous environments in a single material has not been observed, systematic changes in line widths and peak positions have been documented as a function of composition and interpreted in terms of local structural ordering and cesium coordination number, respectively.[9]

Drawbacks to ^{133}Cs NMR spectroscopy of nuclear waste glasses include the low levels of Cs loading in typical waste glasses (<1 mol%) and long spin–lattice relaxation times, making quantitatively reliable experiments time consuming (e.g., days). These obstacles can be overcome by increasing the Cs content in model waste glasses and by adding paramagnetic ions to the material to shorten T_1s. While the former is probably justifiable owing to the high alkali content in waste-loaded glasses, the latter introduces ambiguities due to unknown partitioning of the paramagnetic ions and the possibility of preferential relaxation, skewing relative intensities and peak widths.

26.2.6 Sodium-23

The high sodium content of most nuclear waste glasses and ease of spectral acquisition make ^{23}Na ($S = 3/2$, $NA = 100\%$) NMR popular. However, moderate quadrupolar interactions and a small chemical shift range render ^{23}Na NMR peaks in glasses broad and insensitive to structural changes. A dependence of isotropic chemical shift on coordination number has been discerned in crystalline compounds,[10] but applicability to amorphous materials is limited by the broad, featureless peaks arising from highly disordered cationic environments. Ultrahigh applied fields partially remedy this shortcoming. As with ^{133}Cs NMR, systematic comparisons in a series of closely related compositions reveal subtle peak shifts, which can be interpreted generally as changes in the local coordination number and/or nearby anionic units.

26.2.7 Molybdenum-95

Molybdenum is one of several fission products present at low levels in many waste streams, which are poorly soluble in borosilicate glasses and can cause devitrification. Despite the attractiveness of ^{95}Mo ($S = 5/2$, $NA = 15.9\%$) NMR for directly studying molybdenum-rich amorphous and crystalline separated phases, its nuclear properties conspire against easy acquisition. Low NA, low resonance frequency (39 MHz at 14.1 T), a moderate quadrupole moment, and slow spin–lattice relaxation contribute to long NMR acquisition times, exacerbated by the low levels of Mo in typical nuclear waste glasses (<3 mol%). ^{97}Mo ($S = 5/2$, 9.55%) is also spin-active, but its NMR properties are even less favorable. Since fast MAS is rarely required, large rotors can be used to obtain spectra from which crystalline phases can be identified and quantified, provided the said phases have been previously characterized by ^{95}Mo NMR. Although very little work has been done on ^{95}Mo in glasses, distinct ranges of different coordination numbers have been established,[11] which provide evidence that Mo^{6+} exists as four-coordinate molybdate units in both amorphous and crystalline phases in model nuclear waste glasses.[12] Isotopic enrichment is an option for overcoming the low sensitivity in model glasses, but has not yet been attempted for these applications.

26.2.8 Lithium-6/7

Lithium is sometimes added to nuclear glass frits to facilitate glass formation. ^{7}Li ($S = 3/2$, $NA = 92.5\%$) is a receptive quadrupolar nucleus, which has played a key role in understanding structure and dynamics in many applications, but has not yet played a part in nuclear waste research. While its chemical shift range is small, fast MAS results in narrow central-transition peaks, which may be resolved for different species under some circumstances. The satellite transitions are easily measured by the spinning sidebands and can be related with local environments. ^{6}Li ($S = 1$, $NA = 7.5\%$) is less receptive, but can yield narrower peaks owing to its lower NA and magnetogyric ratio. Both nuclei are well known and may prove to be valuable for studies of nuclear waste glasses.

26.2.9 Other Nuclides

Several important constituents of industrial nuclear waste glasses could potentially be studied by NMR. For example, calcium is an additive used to precipitate insoluble molybdate phases, zirconium is a poorly soluble component of the fuel rod cladding, strontium is a long-lived fission product, and lanthanum is a common surrogate for minor actinides in model systems. While NMR studies of ^{43}Ca ($S = 7/2$, $NA = 0.135\%$), ^{91}Zr ($S = 5/2$, $NA = 11.2\%$), ^{87}Sr ($S = 9/2$, $NA = 7\%$), and ^{139}La ($S = 7/2$, $NA = 99.9\%$) in well-behaved systems have been published, their unfavorable NMR characteristics include low *NA*s and magnetogyric ratios, and moderate to large quadrupole moments, mitigating against success in complex nuclear waste materials. However, with increases in magnetic field strength as well as improvements in sensitivity enhancement methodologies and isotopic enrichment, NMR studies of such nuclei could one day prove useful in applications to nuclear waste glasses.

26.3 CHARACTERIZATION OF NUCLEAR WASTE GLASSES

Solid-state NMR has been applied to correlate aspects of glass structure with essential properties of nuclear waste glasses as well as to specific problems related to nuclear waste vitrification and the long-term behavior of these materials.

26.3.1 Glass Structure

NMR studies of typical base glass formulations and waste-loaded glasses have provided valuable information about the influence of waste ions on glass structure. Owing to the compositional and structural complexity of "real" nuclear waste glasses, most studies of these materials involve model systems simplified to reproduce the essential features of the materials of interest. Ternary alkali borosilicate glasses are the simplest model system bearing relevance to nuclear waste glasses and have been extensively studied in the compositional regimes of interest. Increasing alkali content–representing waste loading–causes the transformation of neutral three-coordinate boron species to anionic four-coordinate borons. This can be conveniently monitored by ^{11}B MAS NMR at sufficiently high applied fields, or by ^{11}B NMR of nonspinning samples at lower fields, and readily yields the ratio of "trigonal" and "tetrahedral" boron, often reported as the fraction of four-coordinate boron over all boron species, N_4 (Figure 26.1). This relationship has been extensively documented in many borate glass systems and is approximately described in borosilicates by the "Dell and Bray model".[13] Several recent studies have observed deviations from this model, particularly in the region of high alkali loading, where N_4 is overestimated, but the general features appear reliable as a reference.[13,14] A high $^{[4]}$B concentration is associated with a high degree of network connectivity and correlates with higher T_g values,[15] a key parameter in the production of nuclear waste glasses. There is, however, some evidence that ionic mobility may also be facilitated by high concentrations of $^{[4]}$B in some compositions, which is a contributing factor to increased leach rates.[16]

A competing charge-balance mechanism in borosilicates is the formation of silicate tetrahedra with one or more NBOs. This occurs with alkali loading and represents a reduction in the degree of network connectivity (i.e., depolymerization). ^{29}Si MAS NMR has been used to estimate the fraction of Q^3 in particular (i.e., four-coordinate silicon with one NBO), which appears to be positively correlated with elemental losses from the melt during glass formation due to volatility.[17] Volatilization introduces uncertainty in the final glass composition—and hence, properties—leading to poorly characterized products. It is also a major concern for industrial plants, as the escape of radioactive species from the melt results in a buildup of radioactive deposits in the hot cells, thereby creating secondary decommissioning problems.

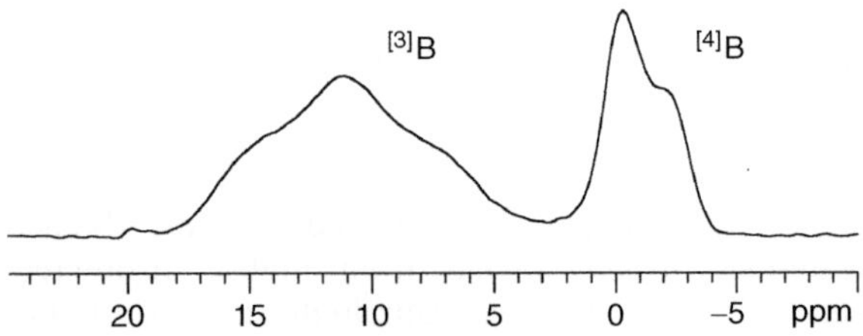

Figure 26.1. ^{11}B MAS NMR ($B_0 = 14.1$ T) of a cesium borosilicate glass representing a highly simplified nuclear waste glass (~58 mol% SiO_2; 29 mol% B_2O_3; 13 mol% Cs_2O).

High Q^3 concentrations have also been shown to correlate with degraded chemical durability.[3] Long-term geologic storage of canisters containing nuclear waste glasses is the eventual fate of all HLW. In the event that the stainless steel canister is breached, the waste-loaded glass is expected to retain its composition and morphological integrity, even upon interaction with groundwater. As such, understanding the structural factors associated with high chemical durability is essential for optimizing the long-term performance of nuclear waste glasses. A reduction in network connectivity, as measured by elevated Q^3 levels, should be expected to result in lower chemical durability, as measured by increased leach rates in solution. Indeed, semiquantitative correlation between Si Q^3 and leach rates has been observed, confirming intuition.[3] On the basis of the foregoing, it should be clear that performance optimization of nuclear waste glasses involves maximizing connectivity through $^{[4]}B$ while minimizing depolymerization occurring via silicon Q^3 species. As four-coordinate boron competes effectively for alkali/alkaline-earth cations, increasing the B_2O_3 content results in a higher degree of silicate polymerization.

In addition to network transformations, NMR can provide valuable information about the distinct roles of cations as charge compensators and network modifiers in more complex glasses. Most nuclear waste glasses have high concentrations of alkali and alkaline-earth cations. For example, rubidium, cesium, strontium, and barium make up about 17 wt% of all fission products in French nuclear waste streams.[18] Moreover, sodium and lithium are often added to facilitate glass formation, and calcium is a common additive for enhanced durability. ^{11}B and ^{29}Si NMR have shown that the substitution of calcium for sodium increases the fraction of $^{[3]}B$ and Q^3 at the expense of $^{[4]}B$ and Q^4, respectively, with Na^+ preferentially charge-compensating $^{[4]}B$ and Ca^{2+} modifying the silicate network to charge-balance NBOs.[18] This can be rationalized by their respective cation *field strengths*, an electrostatic parameter involving the ratio of the cation charge to the square of the average metal–oxygen bond distance: the low-field-strength Na^+ prefers to form longer bonds to the less polarized oxygens of $^{[4]}B$, whereas the higher field strength Ca^{2+} tends to participate in stronger bonds with the negatively charged NBOs. Within each group, it appears that the larger alkaline-earth cations stabilize the formation of $^{[4]}B$, whereas the reverse is true for the alkali cations (except for Li^+, which produces a lower N_4 than Na^+). When aluminum is present in the glass, ^{27}Al MAS NMR experiments reveal changes in the peak positions and breadths with different mixed-cation ratios, which imply that sodium preferentially charge compensates $^{[4]}Al$. These results show that partitioning of alkali and alkaline-earth cations in multicomponent glasses is dominated by size and charge factors, and can be used to compositionally tune glass properties.

Even more complex glasses can be analyzed by NMR to determine systematics specific to particular additives. Given the complexity of real nuclear waste glasses, it is necessary to define the behavior of one component while controlling other compositional effects, and thereby find a comprehensive understanding of interdependencies. For example, whereas alkaline-earth cations increase the fraction of four-coordinate boron, Pb^{2+} reduces N_4 and appears to participate in network formation, possibly as three-coordinate species that compete with silicate and borate units for charge-balancing alkali ions.[19] This type of "intermediate" behavior is roughly analogous to that of Al_2O_3 and exerts a profound influence on the structural characteristics of the glasses by altering the connectivity through a redistribution of NBOs amongst network formers and charge-balancing cations amongst anionic species.

Two important classes of elements present in nuclear waste glasses are lanthanides and actinides. Fission products (La, Ce, Pr, Nd), minor actinides (Np, Am, Cm), as well as Pu, U, and Th are all possible components in waste-loaded glasses from various fuels and present numerous experimental challenges. Not only does their radioactivity hamper laboratory studies, but the accessibility of multiple oxidation states influences their solubility in typical oxide melts. Furthermore, the applicability of NMR to disordered paramagnetic materials is limited. With rare exception, the influence of actinides on glass structure is assayed by incorporating diamagnetic, nonradioactive surrogate lanthanide oxides. Progressive La_2O_3 incorporation both depolymerizes the silicate network (thus serving as a modifier oxide) and suppresses four-coordinate boron (thereby acting as an intermediate oxide, like Al_2O_3 and PbO).[3,20] In rare-earth-rich glasses, RE^{3+} are thought to exist in highly depolymerized regions of the glass, sequestering charge modifiers from the borate regions. However, this conclusion is based on X-ray absorption spectroscopic studies of Nd^{3+}, and its applicability to La^{3+} and other f-block constituents has yet to be verified.

26.3.2 Devitrification

A major challenge in the design of high-performance nuclear waste glasses is the ubiquity of highly charged, poorly soluble ions. Typified by fission product Mo^{6+}, even small quantities of these waste ions (~2.5 mol%) can cause phase separation in oxide melts, leading to crystallization of molybdate phases during cooling. Such phases are highly alkaline and at high temperatures can accelerate melter corrosion. The resulting crystalline phases incorporate (possibly radioactive) alkali and alkaline-earth ions and may be water soluble, raising the risk of radioisotope release. In simulated complex waste glasses, comprehensive structural and compositional characterization of these phases by diffraction-based methods has been hampered by the sheer number of ionic substituents that may be present. The use of NMR is particularly attractive because of its isotope specificity and sensitivity to short-range structure; however, NMR methods are inherently limited by the paramagnetic character imparted on the sample by many transition metals and f-block ions. Nevertheless, NMR has proven to be effective for detecting and identifying devitrification products in simplified molybdenum-bearing glasses. ^{95}Mo MAS NMR readily resolves sharp peaks corresponding to various mixed-alkali crystalline phases and broad signals arising from four-coordinate Mo in amorphous environments in devitrified composites (Figure 26.2).[21] Integration of the respective signals leads to quantification of the molybdenum phase partitioning, although the low Mo levels, inherently poor ^{95}Mo NMR sensitivity, and the slow relaxation rates of ^{95}Mo require long experimental times to ensure precise measurements.

The phase partitioning of fission product Cs^+ can be conveniently probed by ^{133}Cs MAS NMR. The relatively small quadrupolar moment and high spin of this nuclide translate into peaks broadened only by chemical shift distributions at moderate applied fields. Hence, Cs^+ located in glassy environments appears as a broad Gaussian peak, whereas Cs^+ in crystalline phases is clearly evident as sharp peaks

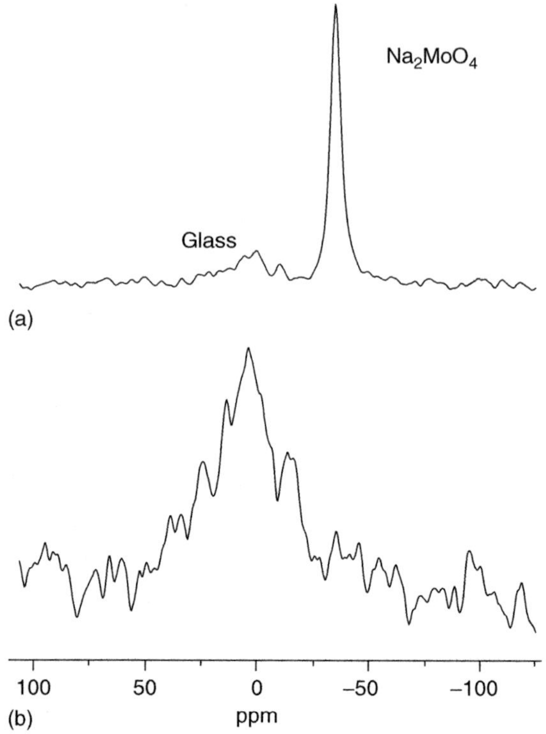

Figure 26.2. ^{95}Mo MAS NMR ($B_0 = 11.7$ T) of a molybdenum-bearing sodium borosilicate model waste glass (62 mol% SiO_2; 16.5 mol% B_2O_3; 19 mol% Na_2O; 2.5 mol% MoO_3). (a) Slow-cooled sample containing Na_2MoO_4 fraction. (b) Fast-cooled sample with no detectable crystalline components.

(Figure 26.3). The high polarizability of Cs confers on this nucleus a strong sensitivity of chemical shift to local environments, producing well-resolved peaks for different chemical species. In a sodium–cesium borosilicate glass with Mo, the devitrified product showed only the formation of sodium molybdate by X-ray diffraction, but ^{133}Cs MAS NMR revealed the presence of two previously uncharacterized crystalline mixed-alkali molybdate phases.[21] Once again, low levels of Cs combined with long spin–lattice relaxation times make these measurements time consuming, especially since small rotors are required to obtain sufficiently fast spinning rates to resolve the spinning sidebands from the broad glassy peak.

Even more powerful for the identification and quantification of devitrification phases too dilute or too complex for X-ray diffraction is a full

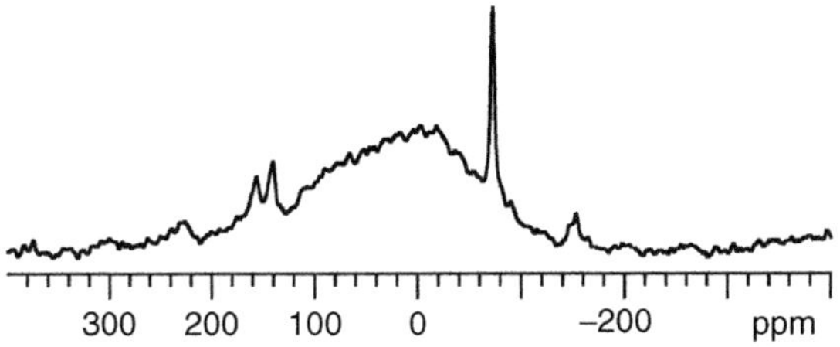

Figure 26.3. ^{133}Cs MAS NMR ($B_0 = 14.1\,T$) of a slow-cooled molybdenum-bearing mixed-alkali borosilicate model waste glass (61.9 mol% SiO_2; 16.5 mol% B_2O_3; 16.1 mol% Na_2O; 3.0 mol% Cs_2O; 2.5 mol% MoO_3). Broad peak corresponds to Cs in disordered environments and sharp peaks have been assigned to crystalline $Cs_3Na(MoO_4)_2$ and $CsNaMoO_4$ phases.

multinuclear magnetic resonance approach. For example ^{133}Cs, ^{23}Na, ^{95}Mo, ^{17}O, and ^{29}Si NMR spectroscopy are all sensitive to crystallinity in predominantly amorphous composites and can be used to fingerprint the precipitates.

Owing to the complexity of nuclear waste glasses and their propensity to phase-separate, the common practice of adding paramagnetic ions to glasses to accelerate relaxation rates is risky. For example, the addition of first-row transition metals such as manganese, chromium, iron, or cobalt raises the possibility of substitution into metal-rich regions similar to molybdates or zirconates, which would not only sequester alkali cations from the glass network but may be invisible to NMR observation due to rapid relaxation. The addition of rare-earth oxides (e.g., Nd_2O_3 or Gd_2O_3) as relaxation agents is similarly problematic because of the unique role of such ions in alkali-rich depolymerized regions, which would preferentially relax parts of the structure while leaving others unaffected. In either case, such nonhomogeneous distribution effects are likely to go undetected.

26.3.3 Glass-in-glass Phase Separation

A full understanding of the phase separation process involves insight about the structural role of poorly soluble ions at levels below the devitrification threshold. This is particularly challenging because coexisting amorphous phases seldom have distinct enough spectroscopic signatures to provide resolved features. An exception is ^{17}O NMR, where B–O–B, Si–O–B, and Si–O–Si bonds are often resolvable by either high-field or MQMAS NMR, depending on the other cations in the system (Figure 26.4). The relative proportions of these oxygen species can be predicted for a homogeneous borosilicate glass according to statistical probabilities and by comparing with the measured quantities. Elevated levels of B–O–B and Si–O–Si relative to Si–O–B linkages imply some degree of macroscopic separation of borate- and silicate-rich glassy phases.[22] It should be noted that this interpretation does not account for the possibility of preferential polyhedral connectivity defining medium-range order, but assumes a

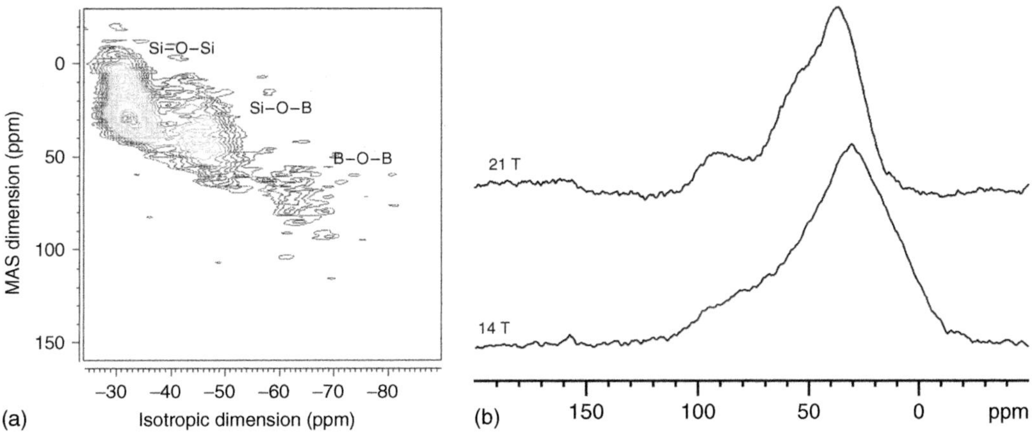

Figure 26.4. ^{17}O NMR spectra of a phase-separated cesium borosilicate glass (~58 mol% SiO_2; 29 mol% B_2O_3; 13 mol% Cs_2O). (a) MQMAS (B_0 = 14.1 T). (b) MAS at B_0 = 14.1 and 21.1 T. Peaks representing B–O–B (≈90 ppm), Si–O–B (≈50 ppm), and Si–O–Si (≈30 ppm) can be discerned.

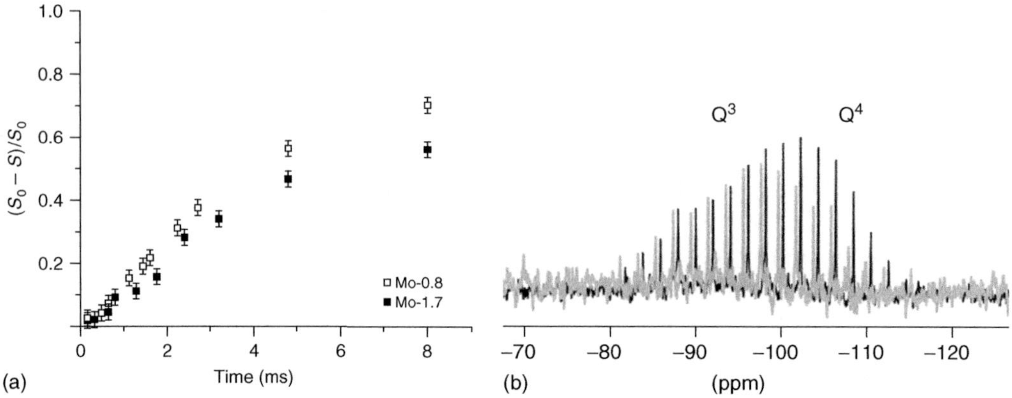

Figure 26.5. ^{29}Si MAS NMR data ($B_0 = 14.1$ T) for molybdenum-bearing mixed-alkali borosilicate model waste glasses ("Mo-0.8": 63.6 mol% SiO_2; 17.6 mol% B_2O_3; 16.4 mol% Na_2O; 1.6 mol% Cs_2O; 0.8 mol% MoO_3. "Mo-1.7": 61.6 mol% SiO_2; 16.7 mol% B_2O_3; 18.1 mol% Na_2O; 2.0 mol% Cs_2O; 1.7 mol% MoO_3). (a) ^{29}Si{^{11}B} REDOR-CPMG dephasing curves. (b) ^{29}Si CPMG MAS (black) and ^{29}Si{^{11}B} cross-polarization CPMG (gray) of "Mo-1.7"—offset horizontally for clarity—showing reduced intensity in Q^4 region, consistent with phase separation. (Figure modified from Ref. 23).

continuous random network model of glass formation. This approach has proven successful for the study of borosilicate glasses in general but has not yet been applied to phase separation due to poorly soluble ions in model nuclear waste glasses.

Glass-in-glass phase separation in a solid sample can also be probed by double-resonance NMR methods using the additional distance dimension afforded by dipolar interactions.[23] Rotational-echo double-resonance (REDOR) NMR between ^{11}B and ^{29}Si revealed reduced dipolar fields in a phase-separated glass containing 1.7 mol% Mo relative to a non-phase-separated glass with 0.8 mol% Mo. ^{29}Si{^{11}B} cross-polarization confirmed this result, additionally showing attenuation of borate mixing with silicon Q^4 species, in particular (Figure 26.5). Analogous REDOR experiments between the alkali cations (^{23}Na, ^{133}Cs) and network formers (^{11}B, ^{29}Si) provided insight into the partitioning of low-field-strength cations amongst different amorphous phases. Taken together, these results present a consistent picture of the phase-separated Mo-bearing glass as comprising amorphous sodium-rich molybdate clusters–presumably precursors to crystallization–along with regions enriched in silicon Q^4 and $^{[3]}$B, and regions possessing augmented $^{[4]}$B concentrations. Hence, the addition of MoO_3 sequesters alkali cations from the borosilicate network, effectively nudging the bulk composition toward the liquid–liquid immiscibility region, where further phase separation occurs. This may account for the common observation of increased Si Q^4 with Mo content.

The addition of Ca^{2+} to Mo-bearing borosilicate melts inhibits the formation of low-durability sodium molybdate phases in favor of the less soluble calcium molybdate, namely, powellite. ^{95}Mo NMR exhibits a distinct peak shift in quenched glasses upon replacing sodium by calcium, confirming that Ca^{2+} preferentially charge-compensates the $MoO_4{}^{2-}$ units.[24] Increasing the boron content in these systems appears to further stabilize against sodium molybdate formation, as four-coordinate boron is preferentially charge-compensated by sodium, and powellite levels are enhanced. ^{23}Na NMR is insensitive to Mo content in Ca-bearing systems, but spectral changes with boron additions may be interpreted as corresponding to a shift in its role from network modifier (i.e., silicate depolymerization) to charge compensator (i.e., balancing four-coordinate boron).[25] This relationship is more complicated in the presence of rare-earth elements where silicate phases cocrystallize, necessitating further research in progressively more complex systems using NMR and X-ray diffraction as complementary tools.

The sensitivity of phase separation to both composition and processing conditions means that sample preparation is an important parameter and must be taken into account when comparing results from different laboratories. For example, the small samples sizes usually prepared in academic laboratories or for isotopically enriched materials may have different macroscopic properties than those prepared on industrial scales. Since the quench rate required to produce a homogeneous glass depends on the thermal conductivity of the material and the sample size (i.e., surface area/volume), compositionally equivalent specimens prepared on a subgram scale may differ from those prepared on 100 g or kg scales. Commercial waste glasses poured into canisters cool at about 1° min^{-1}, whereas many laboratory glasses are cooled much more quickly. Size effects may also be significant for elemental boiloff.[26]

26.3.4 Elemental Leaching

Structural alterations due to glass leaching can be directly probed using NMR spectroscopy. Early studies used ^{11}B NMR to assess borosilicate glass network changes resulting from dissolution under a variety of conditions, showing that both alkali borate-rich regions and depolymerized sodium silicate-rich regions are especially susceptible to aqueous attack.[27] By contrast, sodium cations associated with four-coordinate boron are unlikely to exchange. This conclusion was supported by a comparison of the leaching behavior and ^{23}Na NMR parameters in the French nuclear glass "R7T7" and a typical basaltic glass, which found the more distorted sodium sites in R7T7 at greater risk of leaching.[28] These structural changes were monitored even more directly by examination of the oxygen linkages in a sodium aluminoborosilicate glass before and after alteration by water. Peaks associated with boron (B–O–B and Si–O–B) in the unaltered glass were not observed in an (unenriched) glass sample altered by ^{17}O-enriched water, implying both that boron loss can be used to measure glass dissolution and that the resulting aluminosilicate alteration gel retains a structure similar to that of the original bulk glass.[29]

26.4 FUTURE OUTLOOK AND CHALLENGES

While the foregoing has focused entirely on ex situ studies, where samples are cooled to ambient temperature before NMR analysis, additional information about the phase separation processes may be available from high-temperature in situ studies. High-temperature NMR instrumentation capable of probing melts in industrially relevant temperature regimes is limited. Home-built probes for nonspinning samples can access temperatures in excess of 1000 °C, but lack the requisite spectral resolution afforded by MAS to obtain site-specific information. Commercial high-temperature MAS probes are available, but spinning rates are generally low (<5 kHz) and the temperature range is limited to about 700 °C. Nevertheless, many phase separation and crystallization events are thought to occur well below the melting temperature, providing opportunities for further research and a stimulus for hardware development. Structural and dynamic NMR studies of key transformations around T_g may provide more direct insight into the particular species involved in macroscopic changes. For example, it has been shown that temperature induces a shift in the equilibrium from four-coordinate to three-coordinate boron, which may be relevant to nuclear waste vitrification.[30] A more comprehensive understanding of vitrification will require an expansion of focus from products to process.

The ubiquity of paramagnetic constituents in real nuclear waste glasses presents a fundamental challenge to NMR spectroscopy. While paramagnetic crystalline phases can often be studied by fast MAS NMR, disordered paramagnetic systems suffer from a distribution of electron–nuclear interactions, which cannot be averaged by MAS. While electron paramagnetic resonance can be used to monitor paramagnetic species, simplification of nuclear waste materials for NMR spectroscopic analysis remains a standard practice, partly to isolate certain compositional effects and partly to render samples amenable to NMR observation. Substitution of paramagnetic ions by chemically similar diamagnetic ions is necessary, and its validity will continue to be tested as multiple analytical techniques are brought to bear upon these complex materials.

Nearly all NMR studies of nuclear waste immobilization materials are done on nonradioactive analogs of the active materials. While this is chemically valid, it cannot account for the effects of radiation damage from decay processes. For example, local structural modifications due to the atomic recoil associated with high-energy α-emissions can be significant.[2] In ceramic materials, the resulting amorphization can severely degrade durability, but even in glasses, such events can alter the network structure. Furthermore, elemental transmutation modifies the chemical composition of waste-loaded materials over time, giving rise to unpredictable chemistry. A few NMR facilities exist that can accommodate radioactive samples (e.g., Pacific Northwest National Lab, Karlsruhe Institute for Transuranium Elements). For solids requiring MAS to achieve resolution, safety concerns demand multiple containment to prevent sample dispersion in case of rotor damage.[31] Progress in this area would offer important insight into the long-term behavior of glasses subject to radioactive decay.

RELATED ARTICLES IN THE ENCYCLOPEDIA OF MAGNETIC RESONANCE

Amorphous Materials

Quadrupolar Nuclei in Glasses

REFERENCES

1. I. W. Donald, B. L. Metcalfe, and R. N. J. Taylor, *J. Mater. Sci.*, 1997, **32**, 5851.
2. W. J. Weber, R. C. Ewing, C. A. Angell, G. W. Arnold, A. N. Cormack, J. M. Delaye, D. L. Griscom, L. W. Hobbs, A. Navrotsky, D. L. Price, A. M. Stoneham, and M. C. Weinberg, *J. Mater. Res.*, 1997, **12**, 1946.
3. D. Holland, B. G. Parkinson, M. M. Islam, A. Duddridge, J. M. Roderick, A. P. Howes, and C. R. Scales, *Appl. Magn. Reson.*, 2007, **32**, 483.
4. P. J. Bray, *Inorg. Chim. Acta*, 1999, **289**, 158.
5. P. J. Bray, Quadrupolar Nuclei in Glasses, in *Encyclopedia of Magnetic Resonance*, eds R. K. Harris and R. E. Wasylishen, John Wiley & Sons: Chichester, 2007. DOI: 10.1002/9780470034590.emrstm0430.
6. P. M. Aguiar and S. Kroeker, *J. Non-Cryst. Solids*, 2007, **353**, 1834.
7. S. Kroeker and J. F. Stebbins, *Inorg. Chem.*, 2001, **40**, 6239.
8. L. S. Du and J. F. Stebbins, *J. Phys. Chem. B*, 2003, **107**, 10063.
9. V. K. Michaelis, P. M. Aguiar, and S. Kroeker, *J. Non-Cryst. Solids*, 2007, **353**, 2582.
10. X. Y. Xue and J. F. Stebbins, *Phys. Chem. Miner.*, 1993, **20**, 297.
11. S. H. Santagneli, C. C. de Araujo, W. Strojek, H. Eckert, G. Poirier, S. J. L. Ribeiro, and Y. Messaddeq, *J. Phys. Chem. B*, 2007, **111**, 10109.
12. S. Kroeker, I. Farnan, S. Schuller, and T. Advocat, *Scientific Basis for Nuclear Waste Management XXXII*, eds R. B. Rebak, N. C. Hyatt, and D. A. Pickett, Warrendale, Materials Research Society, 2009, Vol. 1124, p. 153.
13. W. J. Dell, P. J. Bray, and S. Z. Xiao, *J. Non-Cryst. Solids*, 1983, **58**, 1.
14. L. S. Du and J. F. Stebbins, *J. Non-Cryst. Solids*, 2005, **351**, 3508.
15. J. M. Roderick, D. Holland, and C. R. Scales, *Phys. Chem. Glasses*, 2000, **41**, 392.
16. M. M. Islam, D. Holland, and C. R. Scales, *Phys. Chem. Glasses-Eur. J. Glass Sci. Technol. Part B*, 2008, **49**, 229.
17. B. G. Parkinson, D. Holland, M. E. Smith, A. P. Howes, and C. R. Scales, *J. Non-Cryst. Solids*, 2007, **353**, 4076.
18. D. Caurant, A. Quintas, O. Majerus, T. Charpentier, and I. Bardez, *Glass – The Challenge for the 21st Century*, 2008, **39–40**, 19.
19. M. M. Islam, D. Holland, A. P. Howes, and C. R. Scales, *Phys. Chem. Glasses-Eur. J. Glass Sci. Technol. Part B*, 2010, **51**, 137.
20. X. Deschanels, S. Peuget, J. N. Cachia, and T. Charpentier, *Prog. Nucl. Energy*, 2007, **49**, 623.
21. S. Kroeker, C. S. Higman, V. K. Michaelis, N. B. Svenda, and S. Schuller, *Scientific Basis for Nuclear Waste Management XXXIV*, eds K. L. Smith, S. Kroeker, B. Uberuaga, and K. R. Whittle, Warrendale, Materials Research Society, 2010, Vol. 1265, p. 95.
22. L. S. Du and J. F. Stebbins, *J. Non-Cryst. Solids*, 2003, **315**, 239.
23. C. Martineau, V. K. Michaelis, S. Schuller, and S. Kroeker, *Chem. Mater.*, 2010, **22**, 4896.
24. M. Magnin, S. Schuller, D. Caurant, O. Majérus, D. de Ligny, and C. Mercier, *Environmental Issues and Waste Management Technologies in the Materials and Nuclear Industries XII*, John Wiley & Sons, 2009, p. 59.
25. D. Caurant, O. Majerus, E. Fadel, A. Quintas, C. Gervais, T. Charpentier, and D. Neuville, *J. Nucl. Mater.*, 2010, **396**, 94.
26. V. K. Michaelis, and S. Kroeker, *Phys. Chem. Glasses*, 2009, **50**, 249.

27. B. C. Bunker, G. W. Arnold, D. E. Day, and P. J. Bray, *J. Non-Cryst. Solids*, 1986, **87**, 226.

28. F. Angeli, P. Faucon, T. Charpentier, J. C. Petit, and J. Virlet, *Scientific Basis for Nuclear Waste Management XXI*, eds I. G. McKinley, and C. McCombie, Materials Research Society: Warrendale, 1998, Vol. 506, p. 71.

29. F. Angeli, T. Charpentier, S. Gin, and J. C. Petit, *Chem. Phys. Lett.*, 2001, **341**, 23.

30. J. F. Stebbins and S. E. Ellsworth, *J. Am. Ceram. Soc.*, 1996, **79**, 2247.

31. I. Farnan, H. Cho, W. J. Weber, R. D. Scheele, N. R. Johnson, and A. E. Kozelisky, *Rev. Sci. Instrum.*, 2004, **75**, 5232.

Chapter 27

Quadrupolar Metal NMR of Oxide Materials Including Catalysts

Olga B. Lapina[1] and Victor V. Terskikh[2]

[1]*Boreskov Institute of Catalysis, Russian Academy of Sciences, Prospect Lavrent'eva 5, Novosibirsk 630090, Russia*
[2]*Steacie Institute for Molecular Sciences, National Research Council Canada, Ottawa, ON K1A 0R6, Canada*

27.1 INTRODUCTION

The ubiquitous nature of metal oxides is only amplified by their presence in all aspects of our everyday lives. In the context of this chapter, metal oxides are perhaps the most important players in materials science and not the least in heterogeneous catalysis.

NMR of Quadrupolar Nuclei in Solid Materials
Edited by Roderick E. Wasylishen, Sharon E. Ashbrook and Stephen Wimperis

The need for a deeper understanding of the molecular level structure of metal oxides does not require a special justification here, as even slightest improvements in the materials or catalytic performance bring immediate and substantial social benefits. The local structure is known to be a key factor determining the properties of many materials including activity and selectivity of heterogeneous catalysts. Among the many advanced tools in the arsenal of modern chemists, nuclear magnetic resonance (NMR) spectroscopy is one of the most important, owing to its sensitivity to the local molecular structure and interactions between molecules. Often combined with other materials characterization techniques, including diffraction and optical spectroscopies, the use of NMR is beneficial in studying disordered systems and dynamic processes. NMR applications are still challenging for the study of solid heterogeneous catalysts and surface chemistry since this research involves major methodological difficulties due to significantly broader NMR linewidth as compared with solutions. Strong line broadening sharply reduces the sensitivity of the method and, in most cases, results in the loss of information about important characteristics of the NMR spectra such as chemical shifts and spin–spin interactions. Reduction in sensitivity usually renders observation of many magnetic isotopes with low natural abundance or low resonance frequency nearly impossible if present in catalysts in small amounts, for example, as promoters and modifiers, or when dealing with adsorbed molecules on the catalyst surface.

The greater availability of modern high-field NMR spectrometers and the continuing development of line-narrowing and sensitivity-improving techniques have in recent years dramatically expanded the range of solid-state NMR applications, including those in heterogeneous catalysis. This allowed researchers to move to a more systematic investigation of the nuclei of different elements that make up the catalysts as well as to examine in more detail the surface centers of catalysts and adsorbed molecules. High-resolution NMR spectroscopy of solids greatly benefits from the development of magic-angle spinning (MAS) at very high spinning speeds, up to 70 kHz, and various line-narrowing techniques, which open up fundamentally new possibilities in the study of heterogeneous catalysis.

Considering the extremely wide adaptation of solid-state NMR in modern catalytic research, it is next to impossible to review all the aspects of the field in one short chapter. The purpose of this work is rather to educate the reader about general approaches in quadrupolar metal NMR, as it pertains to the subject, and to outline the main challenges and problems that still exist. For this, we have chosen three important metal nuclei, ^{51}V, ^{93}Nb, and ^{95}Mo, fully illustrating the research methodology, which can be further expanded to other quadrupolar nuclei of interest. The choice of these three nuclei is due to their nuclear properties, typical among other quadrupolar metals, demonstrating a complex interplay between quadrupolar and chemical shielding interactions, which strongly depend on the coordination environments of the nucleus. Vanadium, niobium, and molybdenum are also some of the most important transition elements in catalysis, which are frequently found at the very heart of the catalytically active sites. Vanadium oxide and molybdenum oxide catalysts are used in redox catalytic processes, while niobium oxide-based catalysts are employed for acidic catalytic reactions. Mixed oxide catalysts, V–Nb, V–Mo, and VNbMoTe, are some of the most important in catalytic transformation of hydrocarbons.

Of the three nuclei discussed in this chapter, ^{51}V NMR is perhaps the most explored field in solid-state quadrupolar NMR of heterogeneous catalysis, not only because of the favorable NMR properties of ^{51}V but also because of the significance of vanadium-based catalytic systems in the chemical industry, from fine organic synthesis to large-scale installations for sulfuric acid production, and environment protection for sustainable society. Owing to the larger nuclear quadrupole moment of ^{93}Nb, the niobium solid-state NMR is more challenging but steadily gaining in popularity as more high-field NMR spectrometers become available for routine research. The last example, ^{95}Mo NMR, has recently appeared as a potential research tool in heterogeneous catalysis. Molybdenum-95 NMR applications in the solid state are hampered by a very low resonance frequency even at ultrahigh magnetic fields. However, several recent examples of ^{95}Mo NMR in solids show its great potential and will help to further the progress in this field.

27.2 METAL COORDINATION ENVIRONMENT AND NMR PARAMETERS IN OXIDES

27.2.1 Vanadium-51

Solid-state ^{51}V NMR spectroscopy has a long tradition, in many respects, owing to the favorable NMR properties of the ^{51}V nucleus. High natural abundance (99.8%) coupled with a large magnetogyric ratio (7.046×10^7 rad s^{-1} T^{-1}) and relatively small quadrupole moment ($-5.2\,\text{fm}^2$) make solid-state ^{51}V NMR feasible even at low and moderate magnetic fields. There are several recent reviews covering the subject.[1,2] Applications of ^{51}V NMR in heterogeneous catalysis and oxide systems were previously summarized by us in Ref. 3 and more recently in Ref. 4. A separate article in the Encyclopedia of Magnetic Resonance is dedicated to ^{51}V NMR applications in biostructural chemistry.

Because the ^{51}V spin number is high, $I = 7/2$, and the quadrupole moment is small, the central transition ^{51}V resonances are often narrow and are dominated by magnetic shielding interactions. For the same reasons, the satellite transitions, three pairs for ^{51}V, are also quite compact and overlap heavily. For example, the complete ^{51}V spectrum having $C_Q(^{51}V) = 4.5$ MHz will extend just over 1.5 MHz, including the central and all six satellite transitions. As a result, it

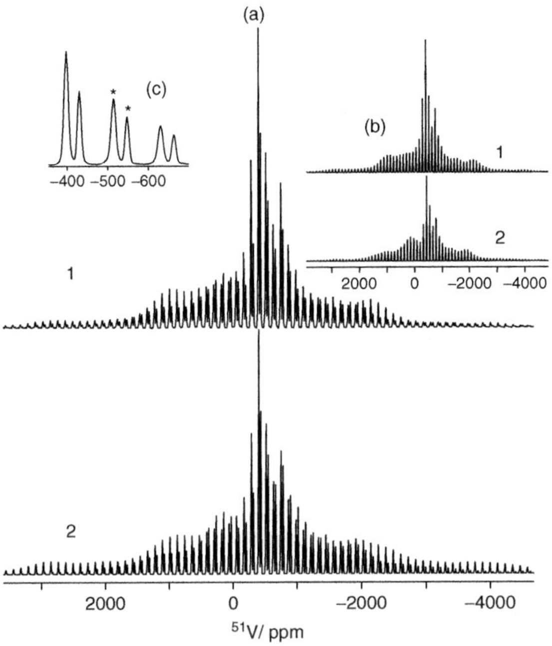

Figure 27.1. Vanadium-51 MAS NMR spectra of a powder rubidium hexavanadate sample, $Rb_2V_6O_{16}$, recorded at 9.4 T and 14 kHz MAS. (a1) Experimental spectrum. (a2) Calculated spectrum composed of two subspectra corresponding to two vanadium sites in the structure. (b1) First subspectrum calculated with $C_Q = 2.66$ MHz, $\eta_Q = 0.99$, $\zeta = 448$ ppm, $\eta_\sigma = 0.16$, $\delta_{iso} = -514$ ppm. (b2) Second subspectrum calculated with $C_Q = 2.34$ MHz, $\eta_Q = 0.30$, $\zeta = 400$ ppm, $\eta_\sigma = 0.05$, $\delta_{iso} = -547$ ppm. (c) Isotropic region of the MAS spectrum with asterisks indicating isotropic lines for the two vanadium sites. (Reprinted from Ref. 4 © 2008 with permission from Elsevier.)

is usual while recording the solid-state ^{51}V spectra, either static or MAS, to observe not only the central transition but also some or all of the satellite transitions, which may complicate interpretation of spectra with multiple sites. A typical example of a ^{51}V MAS NMR spectrum is shown in Figure 27.1 for rubidium hexavanadate, $Rb_2V_6O_{16}$, which has two nonequivalent vanadium sites in its crystal structure.

It is common for ^{51}V in oxide systems to show quadrupolar coupling constants in the range from 2 to 6 MHz, and magnetic shielding anisotropies from 100 to 500 ppm. Interplay between the two interactions contributes to the complexity of the spectra. Moderate magnetic fields, 9–11 T, are the most suitable for recording ^{51}V NMR spectra in oxide systems. For a comprehensive analysis of ^{51}V NMR parameters, it is often necessary not only to acquire MAS spectra at high spinning speeds of 35 kHz or higher and to perform the satellite transition spectra (SATRAS) analysis of the MAS spectra, obtained at low spinning speeds, but also to obtain correct static spectra without having significant distortions. Simultaneous analysis of static and MAS spectra, MASSA,[4] helps in determining reliable parameters for the chemical shift anisotropy (CSA) and electric field gradient (EFG) tensors including their relative orientation.

In oxide systems, vanadium is found in a variety of different VO_x environments, with the vanadium coordination number ranging from 4 to 6. Depending on the system, the same vanadium coordination number may represent several different coordination environments; for example, 4-coordinated VO_4 species are found either as tetrahedral sites or as trigonal pyramids. To discriminate associated and isolated species, it is common to describe them as Q^n, where n indicates the number of shared oxygen atoms. For example, tetrahedral VO_4 sites can be present in the crystal structure as isolated species (Q^0), dimers (Q^1), or extended 1D chains (Q^2). An even higher level of association is possible for VO_5 and VO_6 polyhedra. For example, VO_6 species can exist as isolated (Q^0) and dimeric (Q^1) units, but can also create chains ($n = 2, 3, 4$) and layers ($n = 4$), and form 3D polyoxoanions at $n = 6$. In general, there are eight distinct coordination environments for vanadium in oxide systems (Table 27.1). A large body of ^{51}V NMR data is now available for many individual vanadium compounds with well-known crystal structures supplied by X-ray or neutron diffraction. A thorough analysis of this data has allowed for different types of vanadium environments to be identified either by their CSA or EFG parameters or by a combination of both. This finding is directly transferable to studying vanadium catalysts where the molecular level structures often are unknown from diffraction experiments due to high levels of disorder.

There are several empirical correlations between ^{51}V NMR parameters and vanadium coordination environments in oxide systems. Since ^{51}V NMR spectra of solid samples are often dominated by the magnetic shielding interactions, the most straightforward was to correlate the observed chemical shielding anisotropy (ζ) and the CSA asymmetry parameter (η_σ) as shown in Figure 27.2. As these values had been compiled for over 100 individual

Table 27.1. Eight types of VO_x sites that can be recognized by solid-state ^{51}V NMR

Type of VO_x sites	Structure of VO_x sites	ζ/ppm	η_σ	$-\delta_\perp$/ppm	C_Q/MHz	Key NMR parameters
Tetrahedral Q^0 sites		<100	—	—	0–6	ζ, η_σ
Tetrahedral Q^1 sites	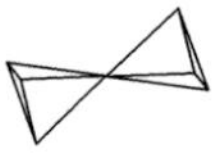	100–200	0.1–0.9	—	2.5–10	ζ, η_σ
Tetrahedral Q^2 sites	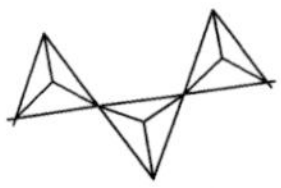	200–500	0.6–0.8	—	2–7	ζ, η_σ
Associated nonaxial VO_5 and VO_6 sites	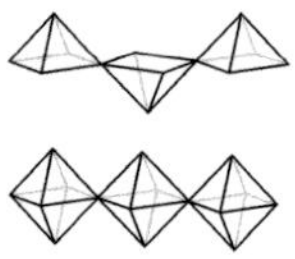	200–400	0.3–0.6	—	3.5–6	ζ, η_σ
Isolated octahedral VO_6 and tetragonal VO_5 pyramids	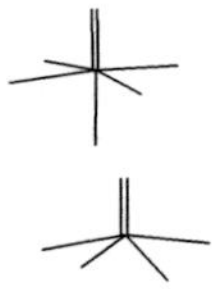	>700	0–0.1	200–350	0–2	ζ, η_σ, $\delta_\perp$, C_Q
Associated pyramids VO_5 and VO_6	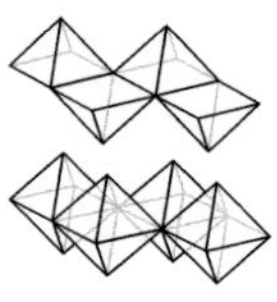	200–400	0–0.2	200–400	2–3	ζ, η_σ, $\delta_\perp$, C_Q
Strongly associated octahedra VO_6	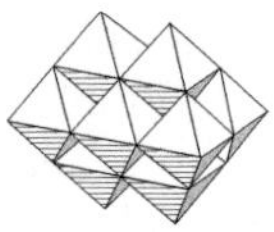	300–400	0–0.2	200–350	3–4	ζ, η_σ, $\delta_\perp$, C_Q
Trigonal pyramids VO_4		400–500	0–0.2	400–500	1–4	ζ, η_σ, $\delta_\perp$, C_Q

EFG tensor conventions: $|V_{33}| \geq |V_{22}| \geq |V_{11}|$, $C_Q = e^2qQ/h = eV_{33}Q/h$, $\eta_Q = (V_{11} - V_{22})/V_{33}(1 \geq \eta_Q \geq 0)$. CSA tensor conventions: $\delta_{ij} = (\sigma_{iso}(\text{ref}) - \sigma_{ij})/(1 - \sigma_{iso}(\text{ref}))$, $\delta_\perp = (\delta_{11} + \delta_{22})/2$, $|\sigma_{zz} - \sigma_{iso}| \geq |\sigma_{xx} - \sigma_{iso}| \geq |\sigma_{yy} - \sigma_{iso}|$, $\sigma_{iso} = (\sigma_{xx} + \sigma_{yy} + \sigma_{zz})/3$, $\zeta = \sigma_{zz} - \sigma_{iso}$, $\eta_\sigma = (\sigma_{yy} - \sigma_{xx})/\zeta$.

compounds, it is clear that similar types of VO_x species usually have similar CSA parameters. Tetrahedral Q^0, Q^1, and Q^2 sites are most obvious to fall into separate groups. This was one of the earliest correlations discovered while first studying ^{51}V NMR spectra of individual metal vanadates.[1,5] In addition, nonaxial VO_5 and VO_6 species and axial VO_n ($n = 4$, 5, 6) species are also well separated by their CSA parameters. The situation is not as obvious for isolated and associated VO_6 octahedra and VO_5 tetragonal pyramids, as well as for trigonal VO_4 pyramids, which are often difficult to distinguish using only CSA parameters. This can be done, however, by also considering EFG parameters namely, the magnitude of the quadrupolar coupling constant, C_Q. As shown in Figure 27.2, in cases with a large axial CSA, $\delta_\sigma < 200$ ppm, and $\eta_\sigma < 0.2$, the EFG tensor becomes very sensitive to

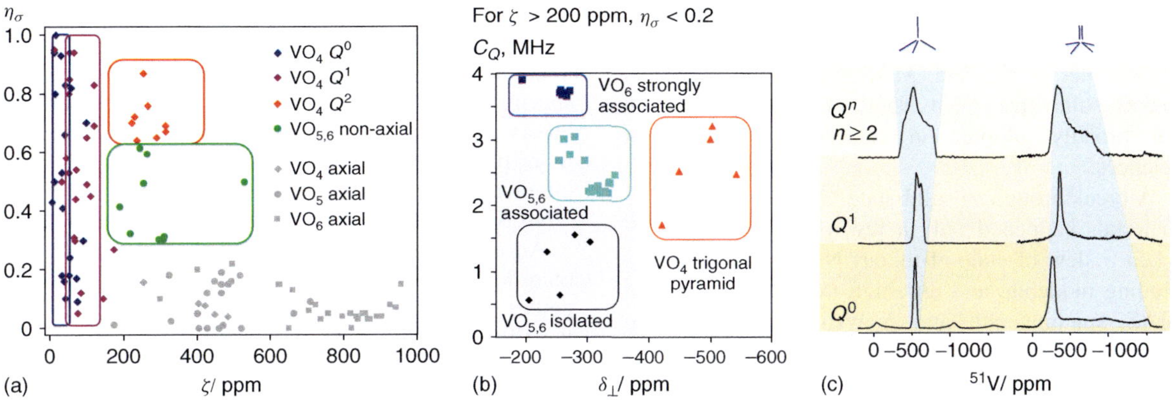

Figure 27.2. (a) Correlation between parameters of the ^{51}V chemical shielding tensor, asymmetry, η_σ, and anisotropy, ζ, obtained for various vanadium compounds. (b) Correlation between the value of perpendicular component of the ^{51}V NMR chemical shielding tensor, $\delta_\perp$, and the quadrupolar coupling constant, C_Q, for vanadium polyhedra with one short V=O bond characterized by $\zeta > 200$ ppm and $\eta_\sigma < 0.2$. (c) Correlation between the local structure of the vanadium sites and association of VO_x polyhedra with the type and value of CSA. The value of $\delta_{||}$ is determined by the V=O distance; variation of this distance from 1.56 to 1.68 Å leads to $\delta_{||}$ variation from −1600 to −800 ppm. The value of $\delta_\perp$ corresponds to the different types of VO_x pyramids: for VO_5 and VO_6 species, it ranges from −200 to −350 ppm and for VO_4 species, it is from −450 to −600 ppm. (Reprinted from Ref. 4 © 2008 with permission from Elsevier.)

the coordination type in axial vanadium coordination environments.[4]

There are some other interesting empirical tendencies that are useful in correlating experimental ^{51}V NMR parameters with the molecular level structure of vanadium species in oxides. For example, in metal vanadates, the absolute value of the isotropic shift in Q^0 sites strongly depends on the metal type and increases with the metal atomic number in the series of alkali metavanadates and alkali earth vanadates. In alkali vanadates, the quadrupolar coupling constant is usually found within 1.0–3.5 MHz range, while for M^{2+} vanadates, the C_Q falls in a rather narrow range of 0.5–1.0 MHz. The C_Q values are largest in M^{3+} vanadates and can reach up to 3.5–5.6 MHz.

The degree of association of vanadium species in the solid state is also an important factor in defining the corresponding ^{51}V NMR parameters. Isolated VO_4 and VO_5 pyramids usually exhibit significant magnetic shielding anisotropy, while associated pyramids tend to produce smaller anisotropies; see Figure 27.2. The opposite trend is observed for tetrahedral VO_4 species, where the magnetic shielding anisotropy is smallest for isolated tetrahedra and increases in dimeric species and chains.

Accumulating a sufficiently large library of data for individual compounds with known crystal structures and well-defined coordination environments is usually the first step in studying such complex and multicomponent systems as heterogeneous oxide catalysts. In the simplest cases, the spectra of individual compounds can be used for fingerprinting of individual catalysts' constituents, and to follow transformations of these constituents in the process of catalysts preparation and further in the catalytic reactions and processes.

27.2.2 Niobium-93

Niobium-93 is one of the most NMR-receptive nuclei because of its 100% natural abundance and large magnetogyric ratio (6.567×10^7 rad s^{-1} T^{-1}), a winning combination for NMR leading to a theoretical receptivity of 0.482 relative to ^1H. At the same time, ^{93}Nb is a quadrupolar nucleus with spin $I = 9/2$ and a moderately large quadrupole moment, −32 fm^2, resulting in solid-state ^{93}Nb NMR spectra often dominated by significant second-order quadrupolar broadening. An additional complication arises from a large ^{93}Nb NMR chemical shift range, in excess of 4000 ppm, which, in disordered systems, can result in significant inhomogeneous line broadening due to chemical shift dispersion, both under static and MAS conditions. All this makes solid-state ^{93}Nb NMR challenging and not yet broadly adopted in catalysis and materials sciences.

A breakthrough in solid-state ^{93}Nb NMR in oxide materials happened only a few years ago, in 2005, when a slew of state-of-the-art NMR techniques including moderate and ultrahigh magnetic fields, fast MAS, spinning sideband analysis of selected transitions in MAS spectra (SSTMAS), and simultaneous computer modeling of MAS and static spectra recorded at different fields was employed to analyze a series of inorganic niobates with known crystal structures.[6] For the first time, a comprehensive set of accurate ^{93}Nb quadrupolar (C_Q, η_Q) and chemical shift (δ_{iso}, ζ, η_σ) parameters was published, which allowed spectroscopists to suggest an empirical ^{93}Nb chemical shift scale depending on the NbO_x coordination environment.[6] This first work encouraged further systematic research in this field, and since then, the number of niobium compounds characterized by ^{93}Nb NMR has been growing steadily.[4] Utilization of multiple magnetic fields to acquire static spectra supplemented by ab initio density functional theory (DFT) calculations has recently allowed to more precisely discriminate between CSA and EFG interactions in several niobates[7] and to confirm the ^{93}Nb NMR chemical shift scale suggested earlier.

An example of a ^{93}Nb SSTMAS NMR spectrum recorded for Li_3NbO_4 is shown in Figure 27.3 together with its simulation. The interplay between the EFG and CSA tensors results in a complex pattern of overlapping spinning sidebands originating from the central and four pairs of satellite transitions (see Figure 27.3). For ^{93}Nb MAS spectra (spin 9/2), the narrowest spinning sidebands, due to the $+3/2 \leftrightarrow +5/2$ and $-5/2 \leftrightarrow -3/2$ satellite transitions, are important in the SSTMAS analysis. The relative orientation of the EFG and CSA tensors is determined from modeling static NMR spectra recorded at different magnetic fields (Figure 27.4).

The niobium coordination number in niobium-containing oxide compounds varies from 4 to 8.

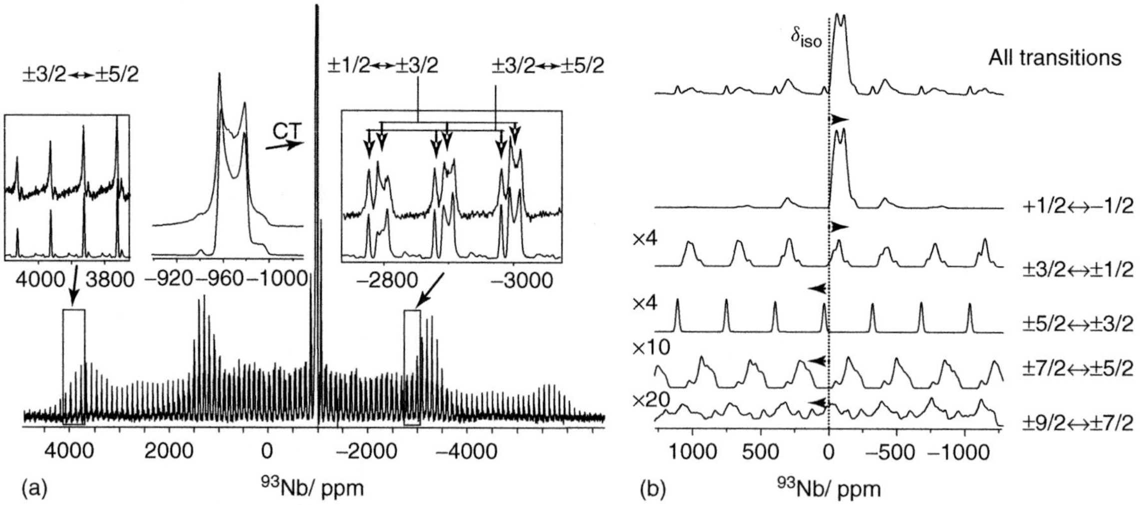

Figure 27.3. (a) Niobium-93 MAS NMR spectrum of a powder lithium orthoniobate sample, Li_3NbO_4, acquired at 9.4 T and 10 kHz MAS. Details of the central transition $+1/2 \leftrightarrow -1/2$ and two satellite transitions, $\pm 3/2 \leftrightarrow \pm 1/2$ and $\pm 5/2 \leftrightarrow \pm 3/2$, are shown as insets. Simulated spectra shown below experimental spectra were calculated using the following parameters: $C_Q = 11.5$ MHz, $\eta_Q = 0.1$, $\delta_{iso} = -950$ ppm, $\eta_\sigma = 0.3$, $\zeta = 135$ ppm. (b) Calculated MAS NMR spectra for a spin-9/2 system illustrating fine details of the spinning sidebands in the region of the isotropic chemical shift. The full spectrum due to superposition of the satellite transitions and the central transition is shown above. Spectra were calculated at a resonance frequency of 122.4 MHz (^{93}Nb at 9.4 T) with the simulation parameters as follows: $C_Q = 5$ MHz, $\eta_Q = 0.4$, $\zeta = 200$ ppm, $\eta_\sigma = 0.7$, $\alpha = 20°$, $\beta = 20°$, $\gamma = 0°$, $\gamma_r = 10$ kHz. (Reprinted from Ref. 4 @ 2008 and Ref. 6 @ 2005 with the permission from Elsevier.)

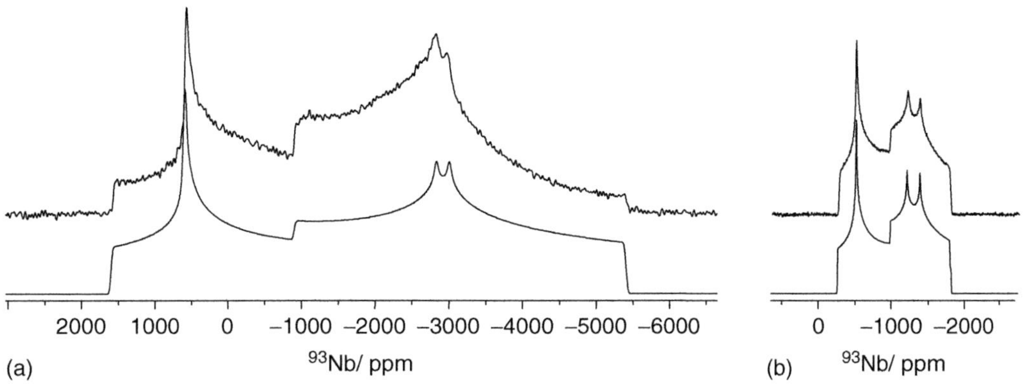

Figure 27.4. Niobium-93 NMR spectra of a stationary powder sample of $YNbO_4$ obtained with a quadrupolar solid-echo pulse sequence at 9.4 T in (a) and at 21.1 T in (b). The experimental spectra are shown at the top. Two model spectra at the bottom were calculated using the same single set of parameters: $C_Q = 82$ MHz, $\eta_Q = 0.38$, $\delta_{iso} = -845$ ppm, $\eta_\sigma = 0.48$, $\zeta = 200$ ppm, $\alpha = 165°$, $\beta = 16°$, $\gamma = 100°$. (Reprinted from Ref. 4 © 2008 with permission from Elsevier.)

The most typical are compounds with 6-coordinated niobium, among them many alkali metal and alkali earth metal niobates, several lead niobates (cubic, tetragonal, and rhombic symmetry), and mixed cation niobates (Table 27.2). Compounds with the tetrahedral Nb coordination are not very common; most of them are rare earth niobates of the $MNbO_4$ type, with M = Y, La, Pr, Nd. In these compounds, the NbO_4 tetrahedra are strongly distorted. Compounds having niobium in coordination state 5 are also rare, one example being Na_5NbO_5. Compounds having 7- and 8-coordinated Nb sites would always have several nonequivalent Nb sites in their crystal structure; for example, $LaNb_5O_{14}$ has three sites, two of them NbO_6 and one NbO_7. In $NaNb_3O_8$, there are two sites, NbO_7 and NbO_8. Only in Cs_3NbO_8, there is a single 8-coordinated site. ^{93}Nb NMR parameters for various NbO_x environments, $x = 4$, 5, 6, 7, and 8, are summarized in Table 27.2.[4,6–10]

The fact that ^{93}Nb NMR parameters appear to correlate with the niobium coordination environment is significant and is of immediate practical use in studying other niobium oxide compounds and catalysts. By analyzing the data summarized in Table 27.2, it becomes clear that the "strong" correlation exists between the isotropic ^{93}Nb NMR chemical shifts and the niobium coordination environment (Figure 27.5). For example, for the 4-coordinated Nb sites, the isotropic chemical shifts occur within a rather narrow range from −700 to −900 ppm with significantly large quadrupolar coupling constants, $C_Q > 70$ MHz. For the 6-coordinated NbO_6 sites, δ_{iso} varies within a wider range from −900 to −1600 ppm, C_Q from 0 to 100 MHz, and δ_σ from 0 to 700 ppm.

It is useful to identify distinct NbO_6 species having different ^{93}Nb NMR parameters. A range of chemical shifts from −900 to −1000 ppm is typical for species with cubic symmetry. For 6-coordinated Nb sites with noncubic symmetry, the ^{93}Nb isotropic chemical shifts are influenced by the nature of the cation. In M(+1, +2, +3) niobates, the niobium sublattice has a more anionic character, and the isotropic shifts fall in the range from −1000 to −1100 ppm. When M(+4) or M(+5) cations are present, the niobium sublattice has a more cationic character, and the isotropic chemical shifts are more negative, and range from −1200 to −1600 ppm. The values of the quadrupolar coupling constants found for 6-coordinated Nb sites vary significantly and would mostly depend on the site symmetry. The CSA for these species is significant, up to 700–800 ppm. For 7-coordinated niobium sites, the isotropic chemical shifts range from −1200 to −1600 ppm, while those for 8-coordinated NbO_8 niobium sites occur at δ values less than −1500 ppm (i.e., at lower frequencies).

27.2.3 Molybdenum-95

Molybdenum has two magnetic isotopes, ^{95}Mo and ^{97}Mo, both are spin-5/2 nuclei with very low magnetogyric ratios (-1.751×10^7 rad s^{-1} T^{-1} and -1.788×10^7 rad s^{-1} T^{-1}), which leads to significantly decreased sensitivity and low resonance

Table 27.2. Niobium-93 NMR parameters for individual niobium compounds

Compound	References	C_Q/MHz	η_Q	δ_{iso}/ppm	η_σ	ζ/ppm	κ	Ω/ppm	α,β,γ
	4-coordinated								
$YNbO_4$	4	82	0.38	−845	0.48	200	0.45	348	165, 16, 100
$LaNbO_4$	4	89	0.15	−800	0.3	200	0.64	330	145, 9, 103
$PrNbO_4$	6	87	0.25	−400	0	0	—	—	—
	6-coordinated								
Na_3NbO_4	6	11.1	0.01	−903	0.6	+120	−0.33	216	0, 31, 90
$Pb(Mg_{1/3}Nb_{2/3})_3$	8	<0.8	—	−900	—	—	—	—	—
$Mg_4Nb_2O_9$	7	46	0	−921	0	−259	−1	389	0, 0, 0
$CsBiNb_2O_7$	7	24.1	0.473	−936	0.15	480	0.81	756	0, 0, 0
Li_3NbO_4	6	12.0	0.1	−954	0.3	−135	−0.64	223	14, 24, 117
α-$BiNbO_4$	4	20.7	0.51	−977	0.56	151	0.37	269	24, 22, 78
La_3NbO_7	4	49	0.28	−968	0.69	−113	−0.25	208	50, 27, 72
$CaNb_2O_6$	6	50	0.8	−990	0.5	300	0.43	525	—
	7	50	0.79	−975	0.62	291	0.31	526	43, 143, −10
$LiNbO_3$	7	22.3	0	−988	0	−115	−1	172	0, 0, 0
$Cd_2Nb_2O_7$	7	24.8	0	−988	0	6	1	9	0, 0, 0
$SnNb_2O_6$	7	40.1	0.48	−1014	0.75	−155	−0.21	291	90, 157, 90
$MgNb_2O_6$	7	53.8	0.25	−1020	0.39	345	0.54	584	26, 63, −15
$NaNbO_3$	7	19.96	0.79	−1064	0.38	69	0.54	116	11, 99, 95
$NaBa_2Nb_5O_{15}$	9	21	—	−940	—	—	—	—	—
2 sites		24	—	−1079	—	—	—	—	—
$KNbO_3$	7	22.3	0.82	−1019	0.3	71	0.64	117	23, 94, 81
$Te_3Nb_2O_{11}$	6	22	0.6	−1176	0.1	250	0.87	388	—
PNb_9O_{25}	9	12	—	−1177	—	—	—	—	—
3 sites		25	—	−1202	—	—	—	—	—
VNb_9O_{25}	4	20	—	−1167	—	—	—	—	—
3 sites		29	—	−1204	—	—	—	—	—
		68	—	−1250	—	—	—	—	—
$NbOPO_4$	9	21	—	−1316	—	—	—	—	—
$NbVO_5$	10	16.5	0.9	−1338	—	—	—	—	—
$Na_4Nb_8P_4O_{32}$	9	29	—	−1285	—	—	—	—	—
4 sites		30	—	−1338	—	—	—	—	—
$Nb_3(NbO)_2(PO_4)_7$	9	30	—	−1583	—	—	—	—	—
3 sites									
	7-coordinated								
$LaNb_5O_{14}$	6	NbO_6	—	−1200	—	—	—	—	—
3 sites		NbO_6	—	−1230	—	—	—	—	—
		NbO_7	—	−1267	—	—	—	—	—
$NaNb_3O_8$	6	NbO_7	—	−1250	—	—	—	—	—
2 sites		NbO_8	—	−1500	—	—	—	—	—
	8-coordinated								
Cs_3NbO_8	4	NbO_8	—	−1600	—	—	—	—	—

EFG tensor conventions: $|V_{33}| \geq |V_{22}| \geq |V_{11}|$, $C_Q = e^2qQ/h = eV_{33}Q/h$, $\eta_Q = (V_{11} - V_{22})/V_{33} (1 \geq \eta_Q \geq 0)$. Two CSA tensor conventions: Herzfeld-Berger, $\delta_{11} \geq \delta_{22} \geq \delta_{33}$, $\delta_{iso} = (\delta_{11} + \delta_{22} + \delta_{33})/3$, $\Omega = (\delta_{11} - \delta_{33})$, $\kappa = 3(\delta_{22} - \delta_{iso})/(\delta_{11} - \delta_{33}) (1 \geq \kappa \geq -1)$. Haeberlen–Mehring–Spiess, $\delta_{ij} = (\sigma_{iso}(\text{ref}) - \sigma_{ij})/(1 - \sigma_{iso}(\text{ref}))$, $|\sigma_{zz} - \sigma_{iso}| \geq |\sigma_{xx} - \sigma_{iso}| \geq |\sigma_{yy} - \sigma_{iso}|$, $\sigma_{iso} = (\sigma_{xx} + \sigma_{yy} + \sigma_{zz})/3$, $\zeta = \sigma_{zz} - \sigma_{iso}$, $\eta_\sigma = (\sigma_{yy} - \sigma_{xx})/\zeta$. α, β, γ-Euler angles.

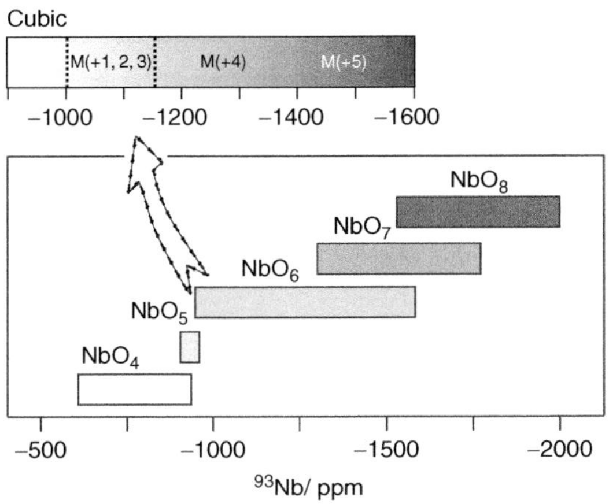

Figure 27.5. Niobium-93 NMR chemical shift scale for NbO_x polyhedra. (Reprinted from Ref. 4 © 2008 with permission from Elsevier.)

frequency. Even at a moderately high magnetic field of 11.7 T, the Larmour frequency for ^{95}Mo is only 32.6 MHz, and for ^{97}Mo it is 33.3 MHz. The low resonance frequency is known to cause additional complications during spectra acquisition due to acoustic ringing. Of the two magnetic Mo isotopes, ^{95}Mo is most often chosen because of its lower quadrupole moment, which leads to substantially smaller second-order quadrupolar effects. At the same time, ^{97}Mo has an advantage of faster relaxation in cases where the quadrupolar relaxation mechanism is dominant. The latter becomes important since the T_1 relaxation in Mo NMR is often in the range of several seconds to minutes. The natural abundance of ^{95}Mo (15.9%) is comparable to that for other popular NMR nuclei such as ^{29}Si. The quadrupole moment of ^{95}Mo ($-2.2\,fm^2$) is close to that for ^{17}O ($-2.6\,fm^2$), but it is smaller than that for ^{27}Al ($14.7\,fm^2$). All this makes solid-state ^{95}Mo NMR spectroscopy practical even at moderate magnetic fields, i.e., by using isotope-enriched samples and by taking necessary steps to minimize the acoustic ringing via spin-echo experiments. Further improvement in sensitivity can be achieved using signal enhancement techniques and Carr–Parcell–Meiboom–Gill “QCPMG”-based pulse sequences.

The range of ^{95}Mo NMR chemical shifts for samples in solution exceeds 8000 ppm, with diamagnetic Mo(II) species observed in the range from 4030 to 3700 ppm, Mo(IV) species from 1160 to 920 ppm, Mo(V) from 570 to 520 ppm, and Mo(VI) in the range from 200 to −20 ppm. Solid-state ^{95}Mo NMR research has been mostly done with Mo^{6+} compounds and metallic systems, with Mo^0 metal having a Knight shift of 5848 ppm. The coordination environment of Mo in solid molybdates is surprisingly diverse and often results in significantly distorted tetrahedral $MoO_4{}^{2-}$ and octahedral $MoO_6{}^{6-}$ species.

Considering the growing importance of molybdenum chemistry, it is not surprising that solid-state ^{95}Mo NMR research has been ongoing in inorganic chemistry,[11] materials chemistry,[12–14] and in catalysis.[15–18] Earlier ^{95}Mo NMR research had been conducted at low magnetic fields, often below 12 T, and thus suffered from the strong acoustic ringing and insufficient signal intensity. Results of such studies were often biased and controversial and prevented quantitative measurements. As high-field NMR instruments are becoming increasingly available, the quality of Mo solid-state NMR spectra has improved considerably, making even challenging systems amendable for a detailed NMR analysis often using chemicals with natural isotope abundance.[19–21] We describe below some of the recent ^{95}Mo solid-state NMR results pertaining to oxide systems and catalysis.[20–24]

27.2.3.1 Tetrahedral Mo(VI) Compounds

In A_nMoO_4 molybdates, the oxygen packing and cation coordination vary significantly from one compound to another. This leads to a range of possible crystal structures, most of which can be described as being constructed of $MoO_4{}^{2-}$ tetrahedra and cation polyhedra.

Monomolybdates of alkali metals of the type $A(I)_2MoO_4$ represent an interesting series of compounds to assess the sensitivity of ^{95}Mo NMR to its local environment. In all these compounds, molybdenum is tetrahedrally coordinated. The coordination about molybdenum is much more symmetric in the case of cesium molybdate, Cs_2MoO_4, forming an almost perfect MoO_4 tetrahedron with a deviation angle of less than 1° and a variation in the Mo–O distance of no more than 0.01 Å. In the series of alkali cations, Li^+, Na^+, K^+, Rb^+, and Cs^+, the cation coordination is different. Lithium cations occupy the tetrahedral vacancies in the oxygen close-packed structure alternating with molybdenum cations. In sodium molybdate, Na_2MoO_4, Na^+ cations are 6-coordinated. As the size of the cation increases from K to Rb and to Cs, the size of the cation coordination polyhedron also increases resulting in a gradual deviation from

Table 27.3. Molybdenum-95 NMR parameters for individual molybdenum compounds with tetrahedral MoO_4 sites

Compound	References	C_Q/MHz	η_Q	δ_{iso}/ppm	ζ/ppm	η_σ	Ω/ppm	κ	α,β,γ
A(I)									
Li_2MoO_4	20	1.27	0.65	−26.0	79.4	0.73	148	0.22	5, 33, 40
Na_2MoO_4	21	0	—	−33.5	—	—	—	—	—
α-K_2MoO_4	20	1.27	0.65	−17.8	50	0.28	82	0.66	90, 5, 0
Rb_2MoO_4	20	1.12	0.69	−21.5	42	0.43	72	0.5	90, 7, 0
Cs_2MoO_4	20	0.34	0.45	−22.0	−15.2	0.43	26	−0.5	—
$(NH_4)_2MoO_4$ mS60	22	1.2	0.44	−23.6	−48	—	—	—	—
$(NH_4)_2MoO_4$ mP60	22	1.45	0.82	−26.3	−80	—	—	—	—
A(II)									
$CaMoO_4$	20	2.89	0	46	−58.7	0	88	−1.0	0, 90, 0
$SrMoO_4$	20	2.26	0	−1.0	−42.0	0	63	−1.0	0, 90, 0
$BaMoO_4$	20	1.68	0.08	−35.4	−36.7	0	55	−1.0	0, 90, 9
$CdMoO_4$	20	3.05	0.0	131	−63.3	0	95	−1.0	0, 90, 0
$PbMoO_4$	20	2.03	0.0	118.5	−38.7	0	58	−1.0	0, 90, 0
α-$ZnMoO_4$	21	2.47	0.6	−40	—	—	—	—	—
		1.15	0.65	−165	—	—	—	—	—
		2.04	0.6	−81	—	—	—	—	—
A(III)									
$Al_2(MoO_4)_3$	21	—	—	−254, −261, −270, −279, −314, −325	—	—	—	—	—

EFG tensor conventions: $|V_{33}| \geq |V_{22}| \geq |V_{11}|$, $C_Q = e^2qQ/h = eV_{33}Q/h$, $\eta_Q = (V_{11} - V_{22})/V_{33}(1 \geq \eta_Q \geq 0)$. Two CSA tensor conventions: Herzfeld-Berger, $\delta_{11} \geq \delta_{22} \geq \delta_{33}$, $\delta_{iso} = (\delta_{11} + \delta_{22} + \delta_{33})/3$, $\Omega = (\delta_{11} - \delta_{33})$, $\kappa = 3(\delta_{22} - \delta_{iso})/(\delta_{11} - \delta_{33})(1 \geq \kappa \geq -1)$. Haeberlen-Mehring-Spiess, $\delta_{ij} = (\sigma_{iso}(\text{ref}) - \sigma_{ij})/(1 - \sigma_{iso}(\text{ref}))$, $|\sigma_{zz} - \sigma_{iso}| \geq |\sigma_{xx} - \sigma_{iso}| \geq |\sigma_{yy} - \sigma_{iso}|$, $\sigma_{iso} = (\sigma_{xx} + \sigma_{yy} + \sigma_{zz})/3$, $\zeta = \sigma_{zz} - \sigma_{iso}$, $\eta_\sigma = (\sigma_{yy} - \sigma_{xx})/\zeta$. α, β, γ-Euler angles.

the densest packing motif. And yet, despite these significant structural differences, the ^{95}Mo NMR spectra of alkali metal molybdates are all characterized by resonances in a very narrow range of chemical shifts (δ_{iso}) between −17 and −34 ppm (Table 27.3).[20–22]

An almost perfect Lorentzian lineshape was observed for the ^{95}Mo resonance of α-Na_2MoO_4. In contrast, the central transition in lithium and potassium monomolybdates shows the characteristic splitting arising from second-order quadrupolar effects, with very similar quadrupolar parameters (C_Q, η_Q), reflecting a similar distortion of the Mo coordination tetrahedra in these two compounds. The observed $C_Q(^{95}Mo)$ values for the Li, K, and Rb molybdates do not appear to depend exclusively on local distortions from T_d symmetry of the MoO_4^{2-} anion, but rather suggest that long-range effects may play a more significant role in determining the EFG parameters at these Mo sites. Another possibility is that the MoO_4^{2-} anion is undergoing either translational motion or it may be experiencing rapid C_2 or C_3 jumps, or both; either of these would result in motional-averaged $C_Q(^{95}Mo)$ values.[20] The Mo chemical shielding tensor in alkali metal monomolybdates deviates from axial symmetry and has a considerable anisotropy of about 150 ppm.[20] In a similar fashion, there are two polymorphs of the ammonium monomolybdates $(NH_4)_2MoO_4$ (Table 27.3). In each polymorph the unit cell contains 60 atoms, and therefore, the two polymorphs have been denoted *m*S60 and *m*P60. The two structures differ by their lattice constants and in the arrangements of the ammonium ions relative to the MoO_4^{2-} tetrahedra, i.e., they possess different hydrogen bonding patterns.[22]

Molybdates of divalent metals A(II)MoO_4 usually crystallize in the same scheelite space group, where each Mo atom is coordinated by four equivalent oxygen atoms in an almost ideal tetrahedral symmetry. In turn, each oxygen is bonded to two A(II) cations and one Mo cation. The compounds $CaMoO_4$, $SrMoO_4$, $BaMoO_4$, $CdMoO_4$, and $PbMoO_4$ all crystallize in the scheelite-type structure, and their crystal structures were refined using neutron diffraction. However, the relative simplicity of the scheelite-type structure is

Table 27.4. Typical ^{95}Mo NMR parameters for molybdenum compounds with tetrahedral MoO_4 sites

Compound	δ_{iso}/ppm	Ω/ppm	C_Q/MHz	η_Q
$A(I)_2MoO_4$	−17 to −35	0 to 150	0 to 1.3	0.4 to 0.7
$A(II)MoO_4$	130 to −170	50 to 100	1.2 to 3.0	0
$A(III)_2(MoO_4)_3$	−250 to −350	0	0	—

EFG tensor conventions: $|V_{33}| \geq |V_{22}| \geq |V_{11}|$, $C_Q = e^2qQ/h = eV_{33}Q/h$, $\eta_Q = (V_{11} - V_{22})/V_{33}(1 \geq \eta_Q \geq 0)$. CSA tensor conventions: $\delta_{11} \geq \delta_{22} \geq \delta_{33}$, $\delta_{iso} = (\delta_{11} + \delta_{22} + \delta_{33})/3$, $\Omega = (\delta_{11} - \delta_{33})$.

achieved via significant distortion of the coordination $MoO_4{}^{2-}$ tetrahedron. This leads to a greater range of chemical shift values than found for alkali metal molybdates, exceeding 160 ppm. Moreover, in the MAS spectra, there is considerable broadening due to the second-order quadrupolar effects. In addition, since all the oxygen atoms are nearly equivalent, the asymmetry is relatively small compared with those for $A(I)_2MoO_4$ compounds. The crystal structure of α-$ZnMoO_4$ is much more complicated than that of scheelite. In α-$ZnMoO_4$, there are three different 4-coordinated Mo sites with significantly different quadrupolar parameters, which is manifested in a complex spectrum with three overlapping sites.

Only one molybdate of the type $A(III)MoO_4$ has been studied thus far via ^{95}Mo NMR, α-$Al_2(MoO_4)_3$. This compound is of considerable interest for heterogeneous catalysis, as it is formed in catalysts containing molybdenum deposited on alumina. The low-temperature form of α-$Al_2(MoO_4)_3$ has alternating AlO_6 octahedra and MoO_4 tetrahedra sharing common edges and forming a 3D lattice. Small structural distortions lead to the appearance of four nonequivalent Al sites and six nonequivalent Mo sites. A single-pulse ^{95}Mo MAS NMR spectrum of α-$Al_2(MoO_4)_3$ recorded at 19.6 T is composed of six heavily overlapped resonances.[21]

Analysis of the data in Table 27.4 allows us to discern several tendencies in the isotropic chemical shift values and in the range of chemical shifts depending on the cation type. As the charge of cation increases, the isotropic shift decreases (i.e., the shielding increases) while the range of chemical shifts increases. The quadrupolar coupling constants reported for all studied molybdates are small, with the largest values observed for bivalent cations. Interestingly enough, the CSA for one- and two-valent cations is in the range of 50–100 ppm, with the maximum value observed for Li_2MoO_4 and the minimum for Cs_2MoO_4.

27.2.3.2 Octahedral Mo Compounds

The highest level of molybdenum coordination in oxides is six with octahedral MoO_6 species tending to associate and form compact polyanionic structures. It is common to describe the first coordination shell of Mo in these polyanions in terms of terminal (O_t), internal (O_i), and bridging (from μ_2 to μ_6) oxygen atoms (Table 27.5). Depending on the degree of association, MoO_6 octahedra can have quite large deviations from the ideal structure, which results in ^{95}Mo NMR spectra significantly different from tetrahedral monomolybdates discussed above. First of all, the characteristic range of chemical shifts is now significantly broader, from 200 to −100 ppm, which overlaps with the chemical shift range observed for Mo in tetrahedral coordination. Thus, these two types of molybdenum species, MoO_4 and MoO_6, cannot be distinguished by using only their ^{95}Mo chemical shift values. In addition, owing to a significant distribution of angles and bond lengths in the coordination sphere, the spectra are often characterized by much less defined and often featureless lineshapes. ^{95}Mo NMR parameters for octahedral molybdenum compounds are summarized in Table 27.5.[21–24]

Ba_2CaMoO_6 is unusual in the sense that its structure is close to that for perovskite (cubic space group Fm3m) with Mo located at the center of a highly symmetric isolated octahedron with zero EFG at the Mo site and, accordingly, the ^{95}Mo MAS NMR spectrum does not show any indication of the second-order quadrupolar lineshape. Unlike other octahedral Mo compounds, the ^{95}Mo NMR resonance is shifted to low field comparing with the tetrahedral compounds.[21]

MoO_3 is an important precursor for many Mo-based catalysts. The lamellar structure of α-MoO_3 is described by the octahedral Mo coordination; however, the Mo–O distances are such that the structure can also be described by tetrahedral coordination, i.e., this structure is intermediate

Table 27.5. Typical 6-coordinated molybdenum compounds

Sample crystal structure	MoO_6 site symmetry	Oxygen sites	δ_{iso}/ppm	C_Q/MHz	η_Q	η_σ	ζ/ppm	References
Ba_2CaMoO_6 Perovskite	(×1) O_h	$6O_t$	168.5					21
$(NtBu_4)_2Mo_6O_{19}$ Lindquist	(×3) C_{4v}	$1O_t$, $4\mu_2$, $1\mu_6$	112	2.7			400	21
$H_4SiMo_{12}O_{40}$ Keggin		$1O_t$, 4μ, $1\mu_4$	−5					23
$[(NH_4)_6Mo_7O_{24}]4H_2O$ Evans	(×3) C_{2v}	$2O_t$, $2\mu_2$, $1\mu_3$, $1\mu_4$	29	0.9				21
		$2O_t$, $3\mu_2$, $1\mu_4$	44	2.9				
		$2\mu_2$, $2\mu_3$, $2\mu_4$	205	3.0				
$(NH_4)_2Mo_2O_7$		$3O_t$, 1μ	30.2	3.12	0.88			22
			−19.8	2.61	0.32			
$AlMo_6(OH)_6O_{18}(NH_4)_3$ Anderson		$2O_t$, $4\mu_2$, $2\mu_3$	−30					23
α-MoO_3 Lamellar	(×1) C_S	$1O_t$, $2\mu_2$, $2\mu_3$, $1\mu_4$	−73	2.76	0.3	0.55	310	24

(×N) is the number of nonequivalent Mo sites.
EFG tensor conventions: $|V_{33}| \geq |V_{22}| \geq |V_{11}|$, $C_Q = e^2qQ/h = eV_{33}Q/h$, $\eta_Q = (V_{11} - V_{22})/V_{33}(1 \geq \eta_Q \geq 0)$. CSA tensor conventions: $\delta_{ij} = (\sigma_{iso}(\mathrm{ref}) - \sigma_{ij})/(1 - \sigma_{iso}(\mathrm{ref}))$, $|\sigma_{zz} - \sigma_{iso}| \geq |\sigma_{xx} - \sigma_{iso}| \geq |\sigma_{yy} - \sigma_{iso}|$, $\sigma_{iso} = (\sigma_{xx} + \sigma_{yy} + \sigma_{zz})/3$, $\zeta = \sigma_{zz} - \sigma_{iso}$, $\eta_\sigma = (\sigma_{yy} - \sigma_{xx})/\zeta$. α, β, γ-Euler angles.

between octahedral and tetrahedral symmetry. The ^{95}Mo NMR lineshape is typical of that for tetrahedral compounds influenced by the quadrupolar interaction although there is a sizable contribution from the CSA observed in the MAS sideband pattern.[21]

Another molybdenum compound important in catalysis, a Lindqvist-type polyoxomolybdate $(NtBu_4)_2Mo_6O_{19}$, is an example of an octahedral structure where the presence of terminal oxygen atoms results in significant distortion of the cubic O_h symmetry around Mo. The structure of the $Mo_6O_{19}{}^{2-}$ anion consists of six MoO_6 octahedra with a common vertex forming a super octahedron. One central oxygen atom, μ_6, is linked to all Mo. All MoO_6 octahedra with one terminal oxygen atom (O_t) and one μ_6 atom in the trans-position are nearly equivalent. The remaining four μ_2 oxygen atoms are bridging between two Mo. Mo–O_t bonds are shorter (1.68 Å) than its opposite Mo–O(μ_6) (2.32 Å). The ^{95}Mo NMR central transition in this compound has no definite shape although there is a systematic shift of 0.9 ± 0.1 kHz downfield between the center of rotation of the satellites and the position of the central transition, which allows an estimate of C_Q, the isotropic chemical shift, and the CSA. Another common polyoxomolybdate compound, $(NH_4)_6Mo_7O_{24}$ $4H_2O$, is stable in acidic environments and is a precursor in many molybdenum catalysts prepared via impregnation. This compound is described by the Evans structure as having three different Mo environments. Chemical shifts for the three ^{95}Mo resonances were determined via 3QMAS.[21]

Although potentially useful in characterizing molybdenum catalysts, at the moment there is no obvious empirical correlation between ^{95}Mo NMR parameters and the local coordination environment of molybdenum. No specific range of NMR resonances could be identified to discriminate molybdenum oxide compounds based only on the number of terminal oxygen atoms.[21,23]

Molybdenum-95 NMR spectroscopy benefits from high magnetic fields, which can significantly increase the resolution of spectra compared with those obtained from low-field NMR experiments. Increasing the operating frequency to 58.7 MHz (21 T) and beyond almost completely eliminates problems associated with acoustic ringing and simplifies analysis of spectra dominated by quadrupolar interactions. In many cases, the nonequivalent sites can be resolved by employing simple one-pulse sequences, or even MQMAS (see Chapter 10) and other advanced techniques. At high magnetic fields, it becomes possible to determine the quadrupolar parameters, the chemical shift tensor, and the relative orientation of the two tensors. With the wider availability of high-field NMR spectrometers, ^{95}Mo NMR spectroscopy is becoming more common in studying such complex and multiphase systems as heterogeneous catalysts.

27.3 AB INITIO CALCULATIONS

While the empirical correlations between coordination environments of metals in oxides and their NMR parameters, similar to the ones described above, have proved useful in characterizing catalysts and disordered materials, and also in identifying the structure of many catalytically active sites, complementary quantum chemical computations are often employed to provide these experimental findings a necessary level of reliability.

While DFT methods are more and more frequently used in combination with solid-state NMR spectroscopy, challenges remain in applying these methods to transition metals and other heavy elements. For example, until very recently, there were practically no reliable computational results on chemical shielding parameters in transition metals since earlier approaches were based on localized atomic orbitals and a finite number of atoms. Only with the introduction of the gauge including projected augmented wave (GIPAW) pseudopotential approach[25] is it now possible to compute both EFG and CSA parameters in transition metals with sufficiently high reliability.

One such example for ^{51}V was reported in Ref. 26 where the authors employed periodic calculations to predict EFG and CSA tensors for $AlVO_4$ as well as their relative orientation. In Ref. 27, GIPAW computations were performed for a series of vanadates, and the correlation between theoretical and experimental values of ^{51}V isotropic chemical shifts was reported. The immediate use of this correlation was to positively assign the ^{51}V resonances for $AlVO_4$, $Mg_2V_2O_7$ and $Ca_2V_2O_7$ compounds, the assignment of which was previously somewhat ambiguous. For ortho- and pyrovanadates agreement was found between experimental and theoretical chemical shielding anisotropy parameters.

The combination of first-principles calculations and ^{51}V NMR experiments has been successful in identifying the location and hydrogen bonding between surface hydroxyl groups in cesium decavanadate $Cs_4[H_2V_{10}O_{28}]{\cdot}4H_2O$.[28] Five possible protonated structures have been suggested for the decavanadate cation from diffraction data, and only by using ^1H and ^{51}V NMR and DFT calculations was the correct cation structure found. First-principles calculations of ^{51}V NMR parameters have been shown to be highly sensitive to local intermolecular hydrogen bonding interactions in the decavanadate crystal structure and accurately reproduced the experimental ^{51}V quadrupolar parameters. Theoretical modeling indicated a relationship between the level of octahedral distortion of the VO_6 species and the ^{51}V NMR parameters (C_Q and ζ), which will be of potential interest when studying other oxide vanadium systems containing octahedral VO_6 sites.

Periodic all-electron DFT calculations of ^{93}Nb NMR parameters were used to clarify the structure of $RbSr_2Nb_3O_{10}$.[29] ^{93}Nb NMR parameters were calculated for a series of 15 niobates and a good agreement was found with experimental NMR data.[7] It is interesting that, at the moment, computational GIPAW results for ^{93}Nb are often in better agreement with experimental results than those for ^{51}V, especially when dealing with the chemical shielding anisotropy, which is most likely due to more accurate pseudopotentials available for ^{93}Nb. We also note that calculated NMR parameters for transition metals may be sensitive to the inclusion of relativistic effects.

Developing structural correlations for molybdenum will also most likely require detailed theoretical quantum chemical calculations of ^{95}Mo EFG and CSA parameters based on available crystallographic data. This approach seems to be working for other transition 4 d elements, and it has also recently been applied to a series of molybdates.[30] In this work, the GIPAW-DFT formalism accurately reproduced isotropic ^{95}Mo chemical shifts and anisotropic CS parameters. The accuracy of computed ^{95}Mo EFG and CSA parameters was found to be strongly dependent on the quality of the available crystal structures.

27.4 PARAMAGNETIC EFFECTS

Vanadium has rich redox chemistry, and the V^{5+} oxidation state is only one of several other possible oxidations states, with V^{3+} and V^{4+} being as common and relatively stable in the solid state. Since V^{5+} is diamagnetic, in majority of cases, it is the only oxidation state of vanadium that can be directly probed by ^{51}V NMR spectroscopy. This somewhat limits the applications of ^{51}V NMR in materials science and particularly in catalysis research, where paramagnetic vanadium species in low oxidation states are often present and are thought to be involved in many catalytic transformations.

It is still possible, however, to study paramagnetic vanadium centers by recording NMR spectra of nuclei that are in close proximity to paramagnetic centers. For example, a ^{31}P spin-echo mapping technique was used to study paramagnetic vanadium centers in V–P–O oxidation catalysts.[31,32] While in compounds containing only diamagnetic V^{5+} centers, the ^{31}P NMR chemical shifts are usually in the range of 20–40 ppm; for phosphorus atoms neighboring paramagnetic V^{4+} centers, the ^{31}P NMR chemical shift is significantly larger, often in the region of 1600–2600 ppm. This effect is even more dramatic for V^{3+} species, where ^{31}P NMR signals are found at about 4650 ppm. This paramagnetic ^{31}P NMR line shift is directly proportional to the density of unpaired electrons at the position of the nucleus, which in cases with low and intermediate concentration of paramagnetic centers is mostly due to pseudocontact contributions.[33–35]

Studying paramagnetic shifts is particularly useful in discovering medium and long-range structural information. The magnitude of a paramagnetic shift can provide valuable information about the number of paramagnetic vanadium species in the first and second coordination sphere of the atom, as well as on the oxidation state of such species. In dilute paramagnetic systems, pseudocontact paramagnetic shifts can be experimentally measured even for nuclei located as far as 40 Å from the paramagnetic center.

In paramagnetic systems containing diamagnetic vanadium species, the ^{51}V NMR spin-echo mapping technique can effectively be used to probe paramagnetic sites, even though very fast MAS and very large spectral windows are often necessary. In rare earth vanadates and in vanadates of some 3 d elements, where diamagnetic vanadium species are strongly affected by paramagnetic cations, ^{51}V paramagnetic shifts in excess of 15000 ppm have been reported.[36] Similar effects are observed in vanadium bronzes where some of the vanadium species are in paramagnetic V^{4+} state, and also in cubane-like vanadium compounds and in VO_2 oxides, where closely spaced paramagnetic centers form diamagnetic pairs in persistent magnetic fields. Some typical examples of paramagnetic ^{51}V NMR lineshapes and characteristic paramagnetic chemical shifts are shown in Figure 27.6. Clearly, the very large breadth of the paramagnetic ^{51}V NMR chemical shift scale is already an indication of a significant potential of this technique in studying heterogeneous catalysts containing paramagnetic centers, either located on the surface or incorporated in the bulk.

The formal valence state of vanadium in VO_2 oxide is V^{4+}; however, there is a possibility that the vanadium sites in this oxide exist as mixed V^{5+}–V^{3+} pairs. This problem was solved

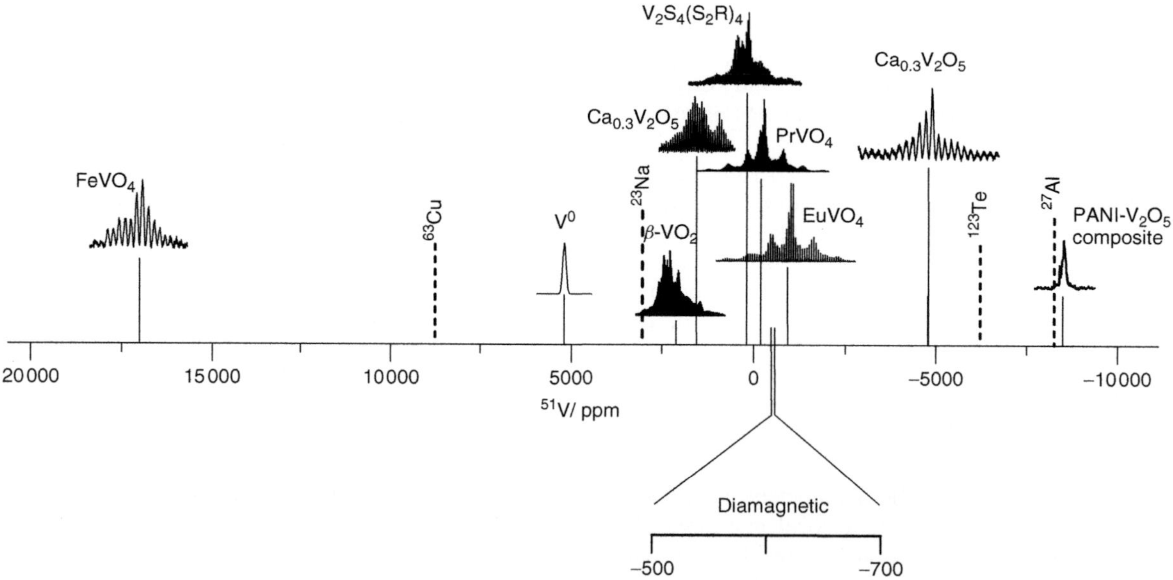

Figure 27.6. Vanadium-51 NMR chemical shift scale for vanadium systems with paramagnetic centers. Solid vertical lines correspond to the isotropic ^{51}V NMR chemical shifts in the compounds which spectra are shown; the dotted lines correspond to the resonance frequencies of the neighboring nuclei: ^{63}Cu, ^{123}Te, ^{23}Na, and ^{27}Al. (Reprinted from Ref. 36 © 2009 with permission from Elsevier.)

by applying high-speed ^{51}V MAS NMR at two different magnetic fields, 9.4 and 14.1 T, where the low-temperature nonmetallic β-form of VO_2 was found to contain only a single V^{4+} site.[37] The variable temperature ^{51}V MAS NMR was also used to study the phase transition in this oxide from nonmetallic (β) to metallic (α), which was accompanied by a dramatic change in the ^{51}V Knight shift value, from 2115 to −4788 ppm.

Michalaka and Mohammad were able to detect ^{51}V NMR spectra from several valence states of vanadium present simultaneously in the low-temperature phase of α'-NaV_2O_5 bronze.[38] Three separate resonances found in ^{51}V NMR spectra for this bronze were attributed to three different vanadium valences, which agreed with the structural models suggested for this compound.

There is only limited experimental data available on paramagnetic effects in ^{93}Nb NMR spectra of oxides. Isotropic ^{93}Nb NMR chemical shifts found in paramagnetic rare earth niobates, i.e., in $PrNbO_4$, have been described as a sum of two contributions, $\delta_{iso} = \delta_0 + \delta_p$, where δ_0 is the chemical shift originating from the electronic structure of filled orbitals, and δ_p is the shift from interactions of niobium with unpaired electrons of M^{3+} rare earth ions. When $\delta_{iso} = \delta_0$, as found in diamagnetic niobates, i.e., $LaNbO_4$ or $YNbO_4$, the isotropic chemical shifts are usually in the range typical for diamagnetic niobates, and δ_p for $PrNbO_4$ is equal to 400 ppm.

The mixed niobium–vanadium bronze $H_{1.4}NbVO_5$ was studied by ^{51}V and ^{93}Nb NMR and compared with similar spectra recorded for $NbVO_5$.[39] The authors reported almost identical spectra for the two systems, even though the bronze was expected to contain a significant number of paramagnetic V^{4+} and V^{3+} species due to charge transfer from hydrogen. The reported NMR results were explained as a result of electron-pairing effect, frequently observed in mixed-valence vanadium compounds.

27.5 EXAMPLES OF QUADRUPOLAR METAL NMR IN MATERIALS SCIENCE

27.5.1 Layered Materials and Intercalation

Layered materials based on transition metal oxides are under consideration owing to their often unique ionic and electronic properties with potential applications in battery technologies, as electrochromic devices, and in heterogeneous catalysis. For example, many vanadium oxide compounds demonstrate a well-defined layered crystal structure, which permits the reversible intercalation using a variety of guest molecules.

The layered structure of vanadium oxide foams prepared via mixing V_2O_5 with hydrogen peroxide and surfactant solutions was studied.[40] Using solid-state ^{51}V MAS NMR, the authors were able to identify several different polyoxovanadate species in the structure of these foams and to suggest the likely mechanism for these foams being formed. The layered structure of foams was governed by a self-assembly of amines while vanadium polyanions intercalated into interlayer spaces. Similarly, the structure of macroporous vanadium oxide foams was investigated by ^{51}V NMR.[41] The local structure of the vanadium sites in these foams is similar to that for V_2O_5 xerogels.[42]

Layered vanadates containing lithium are of significant interest for designing new generation rechargeable lithium batteries with improved qualities. ^{51}V NMR provides a direct means to monitor phase transformations in these materials during synthesis and also to follow changes during charge/discharge cycles, with a possibility of in situ monitoring. For example, Grey *et al.*[43,44] used variable temperature ^{51}V MAS NMR to study the synthesis of $Li_xV_3O_8$ materials via a xerogel precursor route. They found that these materials consist of V_3O_8 layers containing octahedral and penta-coordinated vanadium atoms with intercalation occurring between the layers. During synthesis, the xerogel underwent a series of phase transitions involving layered phases with decreasing water content.

A process of gradual dehydration of a layered α-niobium phosphate, α-$NbOPO_4$, was studied by solid-state wide-line ^{93}Nb NMR combined with X-ray diffraction, thermogravimetric analysis, and ^{31}P MAS NMR.[45] NMR parameters determined from ^{93}Nb NMR spectra recorded at different magnetic fields were found to be highly sensitive to the dehydration state of niobium phosphate. Several individual hydrated phases were identified depending on the dehydration temperature.

The same group has investigated the intercalation of layered α-phase vanadyl phosphate, α-$VOPO_4$, with alkylamines by ^{51}V solid-state NMR.[46] They reported that the short chain amines (C_3–C_6) are tilted with respect to the inorganic basal plane, while the long chain amines (C_{12}–C_{16}) tend to arrange in a manner perpendicular to the layer, and in amines with an intermediate chain length (C_8), both orientations coexist. Both ^{51}V MAS and static spectra of intercalated vanadyl phosphate were found to be sensitive to intercalation.

Layered calcium niobium perovskites, $Ca_2Nb_3O_{10}$, intercalated with partially hydrolyzed aluminum cations were studied using ^{93}Nb and ^{27}Al NMR.[47] In these materials, the study suggests structurally unaltered perovskite layers with interlayer regions occupied by polymeric aluminum hydroxylcations. In another niobium perovskite study, a triple-layered perovskite, $RbSr_2Nb_3O_{10}$, and its acid-exchanged version, $HSr_2Nb_3O_{10}$, were investigated by using multinuclear solid-state magnetic resonance and DFT computational methods.[29] Two types of niobium sites, exterior and interior, were identified, and ^{93}Nb EFG and CSA tensor values were determined for both sites. The high sensitivity of ^{93}Nb NMR parameters to protonation implied significant changes in the niobium coordination environment, which was confirmed by DFT calculations. Upon protonation, the exterior NbO_6 octahedra undergo changes in both bond length and bond angles, while the interior sites only undergo a minor elongation of the octahedra.

27.5.2 Ferroelectric Materials

Owing to its high sensitivity to the NbO_x coordination environment, solid-state ^{93}Nb NMR spectroscopy is becoming increasingly popular to study disorder effects in a variety of materials, including ferroelectric materials, usually in combination with other characterization techniques, neutron and X-ray diffraction experiments. Of particular interest is the state and ordering of the B site in ABO_3 perovskite-type ferroelectrics. For example, ^{93}Nb MAS and nutation NMR experiments at 9.4 and 14.1 T were performed to study polycrystalline solid solution relaxor ferroelectrics $Pb(Mg_{1/3}Nb_{2/3})O_3$ (PMN), with the B sites occupied by Mg or Nb.[8] In these and similar systems, ^{93}Nb NMR spectra are very sensitive not only to the local configuration of the Nb sites but also to the long range ordering effects. Thus, the most symmetric cubic $Nb(OMg)_6$ configuration, with all six neighboring B-sites occupied by Mg in Mg-rich ordered domain regions of the sample, shows very sharp ^{93}Nb NMR signals little affected by quadrupolar interactions. Other possible configurations of Nb sites will depend on the variable B-site occupancies, which generally can be described as $Nb(ONb)_{6-x}(OMg)_x$, where x may vary from 1 to 5. All these configurations will correspond to lower symmetry NbO_6 octahedra. As a result for such configurations, two distinct types of ^{93}Nb NMR signals are usually found having intermediate and high values of quadrupolar coupling constants.[8] Some of these configurations will have axial local symmetry, and others will belong to a lower symmetry type.

B sites can also be substituted with other elements, for example, Zr, Sc, Ba, and La. Such substitution results not only in a different degree of B-site ordering specific for each element but also in a different motional behavior of the B-site cations. All these effects can be directly detected via ^{93}Nb NMR.[48] For example, while ^{93}Nb MAS NMR spectra indicate significant B-site disorder in the Mg-substituted systems as described above, in the Sc-substituted materials, B-site ordering was detected with Nb occupying one single type of axially distorted B site with an intermediate value of C_Q. Using a combination of ^{207}Pb, ^{45}Sc, and ^{93}Nb NMR and diffraction experiments performed between 77 and 420 K, a first-order phase transition was found in partially ordered PbScNb materials at 360 K.[49]

In a related study illustrating high sensitivity of ^{93}Nb NMR to its coordination environment, ^{93}Nb NMR spectra were recorded for a mixed cation $KTa_{(1-x)}Nb_xO_3$ single crystal.[50] Niobium-93 satellite transitions observed in the spectra ruled out niobium occupying the high-symmetry perovskite sites. Instead, it was suggested, and later confirmed, that niobium cations were actually displaced along the [111] body diagonals and were occupying off-center positions of rhombohedral symmetry. In another single-crystal ^{93}Nb NMR study, the ^{93}Nb EFG parameters were determined in Nb-doped titanium oxide.[51]

27.5.3 Silicates

The use of ^{51}V and ^{93}Nb NMR spectroscopy is of general interest while studying silicates containing these metals and other related systems, e.g., zeolites and microporous silicates. Quadrupolar metal NMR in silicates is, in most cases, straightforward and is combined with ^{29}Si MAS NMR and diffraction experiments. As an example, the structure of mixed cation nenadkevichite silicate containing niobium was characterized by using multinuclear ^{23}Na, ^{29}Si, and ^{93}Nb magnetic resonance.[52] The latter experiments confirmed distorted octahedral coordination of niobium sites in agreement with the known crystal structure.

What is unusual about quadrupolar NMR in silicates, however, and which is worth mentioning in this separate section, is that for the niobium silicate $Rb_4(NbO)_2(Si_8O_{21})$, heteronuclear two-bond J-coupling between a quadrupolar nucleus, in this case ^{93}Nb, and a spin-1/2 nucleus (^{29}Si) was first observed. In this silicate, Kao and coauthors reported $^{2}J(^{93}Nb–^{29}Si) = 64$ Hz.[53] This work has shown new research opportunities in the possible relationships between J-coupling parameters for quadrupolar nuclei and structure in the solid state. Later, Bonhomme *et al.*. determined that $^{2}J(^{51}V–^{29}Si) = 28$ Hz in a $[AsW_9O_{33}(^tBuSiO)_3(VO)](^nBu_4N)_3$ complex in the solid state via triple resonance $^{51}V/^{29}Si/^{1}H$ experiments.[54]

27.6 APPLICATIONS IN GLASSES

27.6.1 Vanadium Oxide Glasses

Sulfuric acid production is traditionally carried out using alkali metal pyrosulfate–vanadium oxide catalysts, $M_2S_2O_7–V_2O_5$ (M = Na, K, Cs), supported on silica and other inert supports. Oxidation of SO_2 to SO_3 is performed at 400–600°C with the active component of the catalysts present in the molten state. Similar catalytic systems are also used in the environmentally important processes of treatment of flue gases. Since the active component in these catalysts is a high-temperature hot melt, studying their molecular structure represents a certain challenge. One of the approaches is to examine melt-quenched glasses obtained by fast cooling of the melts. Melt-quenched $M_2S_2O_7–V_2O_5$ glasses are very dark or black solids, which are difficult to study with conventional optical spectroscopy. Owing to their highly disordered structure, such glasses are only of limited access to X-ray diffraction techniques. However, multinuclear solid-state magnetic resonance spectroscopy has proved very successful in directly probing interactions of alkali metal pyrosulfate with vanadium pentoxide and in elucidating the possible structure of the catalytically active species for a wide range of compositions.[55,56] Variable temperature multinuclear magnetic resonance allows the study of different types of motional processes in $M_2S_2O_7–V_2O_5$ glasses, including atomic diffusion of alkali cations, reorientation of molecular fragments of the glass framework, the dynamic glass behavior below and above the glass transition temperature, and ultimately building the phase diagrams in these multicomponent systems.

For example, ^{133}Cs and ^{51}V NMR spectra were recorded for the catalytically important $Cs_2S_2O_7–V_2O_5$ glasses in the temperature range 20–550°C, covering both the solid and liquid states.[55,56] At ambient temperature, the static ^{51}V NMR spectra of melt-quenched glass samples were found to be similar for a wide range of glass compositions. The spectra are broadened and dominated by the chemical shielding anisotropy typical for vanadium atoms in a distorted octahedral oxygen environment with one short V=O bond. The broadening is caused mostly by the Gaussian-like distribution of the quadrupolar coupling parameters. As the vanadium concentration in the glass increases, the distribution of the chemical shielding tensor parameters also increases owing to the local disorder of the vanadium sites. Heating of the glass samples results in gradual changes in the anion framework formed by vanadium oxide and pyrosulfate toward crystallization, which can be detected in situ by ^{51}V NMR.

In the studied systems, two different pathways of crystallization were identified depending on the glass composition and its structure at the molecular level. If the glass structure is similar to the final crystal state, the crystallization occurs at the glass transition temperature, T_g, as a solid-solid transformation since it requires only moderate changes in the anion framework. However, if the glass composition is significantly different from the crystal structure, the glass framework must undergo greater rearrangements, resulting in the glass-to-crystal transition via a metastable liquid state. For example, in the $2Cs_2S_2O_7–V_2O_5$ glass, the crystallization occurs via the first-order solid–solid phase transition resulting in the crystalline $Cs_4(VO)_2O(SO_4)_4$ phase at around 220°C. During this transformation, the axial CSA pattern in the ^{51}V NMR spectra does not change significantly. However, for the $4Cs_2S_2O_7–V_2O_5$ glass, the crystallization proceeds through a metastable liquid phase characterized by narrow symmetric lines usually occurring in liquids. This isotropic averaging of ^{51}V chemical shielding anisotropy is a clear indication of high mobility of vanadium-containing species in the metastable liquid. At 550°C in the molten state, the ^{51}V NMR spectra indicate that the coordination of vanadium changes from distorted octahedral in the solid to tetrahedral in the liquid state.

Detailed analysis of the variable temperature ^{51}V NMR spectra and complementary NMR spectra of alkali metal cations has resulted in a convincing molecular model for the pyrosulfate–vanadia and the related hydrogensulfate–vanadia systems describing the structure of their melt-quenched glasses as replicas of the molten state.[55,56] For both systems, the glass structure is represented by a 3D network of vanadium-oxo-sulfate and/or pyrosulfate complexes, with cations randomly distributed within this network. At medium V_2O_5 content, the vanadium species in the glass are present as $[(VO)_2O(SO_4)_4]^{4-}$ dimers, also found in the crystalline phases. At lower V_2O_5 concentrations, vanadium-oxo-sulfate species are further separated, resulting in the different dynamics observed in these systems upon crystallization.

In a similar study, the molecular level structure of V_2O_5–TeO_2 glasses was investigated by ^{125}Te and ^{51}V static NMR spectroscopy to characterize the local coordination environments of the Te and V species.[57] The structure of these and similar glasses would strongly depend on the glass composition. Thus, with the increasing V_2O_5 content, the fraction of TeO_3 trigonal pyramids was found to increase while the content of TeO_4 trigonal bipyramids decreases. Similar trends were detected in ^{51}V NMR spectra for molar fractions of VO_4 tetrahedra and VO_5 trigonal bipyramids with increasing V_2O_5 content, indicating that both types of species, consisting of tellurite and vanadate, contribute to the formation of the glass network in V_2O_5–TeO_2 glasses.

27.6.2 Niobium Oxide Glasses

Niobium-93 NMR spectroscopy in niobium-containing oxide glasses is more challenging because of the significantly larger ^{93}Nb nuclear quadrupole moment. Even moderate distortions in the niobium coordination may result in large quadrupolar interactions rendering the most distorted and/or disordered niobium sites virtually undetectable by NMR. Nevertheless, examples of the successful application of ^{93}Nb NMR spectroscopy for glasses have been reported. For example, Nb_2O_5–$NaPO_3$ glasses were characterized by ^{31}P, ^{93}Nb, and ^{17}O NMR.[58] As in many other phosphate glasses, in this case, ^{31}P MAS NMR was the most informative, indicating that Nb_2O_5 depolymerises the sodium metaphosphate network with the formation of Q^1 and Q^0 species. ^{93}Nb NMR spectra indicated two niobium sites assigned to corner-shared NbO_6 octahedra resonating at -1500 ppm and edge-shared NbO_6 octahedra at -1000 ppm. The amount of corner-shared octahedra decreases as the content of Nb_2O_5 in the glass increases. The interaction of niobium species with phosphate was more pronounced at the lower niobium concentrations. Heteronuclear P–O–Nb species were found in ^{17}O MQMAS NMR spectra recorded for the ^{17}O-enriched $6Nb_2O_5$–$94NaPO_3$ glass at 18.8 T; however, no Nb–O–Nb species could be detected, most likely owing to their much smaller quantity.

27.6.3 Molybdenum Oxide Glasses

In a related study, binary MoO_3–$NaPO_3$ glasses were investigated using ^{95}Mo NMR, among other techniques.[59] ^{95}Mo NMR spectroscopy of glasses is greatly facilitated by the fact that the glass density is usually higher than in other oxide materials, which allows for significantly improved sensitivity. Molybdenum-95 MAS NMR data obtained for MoO_3–$NaPO_3$ glasses indicated that the majority of molybdenum species are 4-coordinated with the corresponding ^{95}Mo NMR resonance detected at around 0 ppm. Although a minor constituent, a 6-coordinated molybdenum with a ^{95}Mo NMR chemical shift of -500 ppm is also present. This finding indicates a more complex structure for these glasses than originally thought, and it most likely involves sharing and partial transfer of sodium as a network modifier between two network formers, phosphorus and molybdenum, depending on the glass composition. This idea was supported by rotational echo double resonance (REDOR) experiments showing smaller ^{23}Na–^{31}P dipole–dipole interactions with the higher Mo_2O_3 content. Glasses rich in MoO_3 show decreased thermal and crystallization stability and are characterized by isolated phosphate species and MoO_x clusters with the extended Mo–O–Mo connectivity.

The local environments of molybdenum in ternary AgI–Ag_2O–MoO_3 glasses was investigated by ^{95}Mo MAS NMR.[14] In the first series of glass samples having the composition ratio $Ag_2O/MoO_3 = 1$, only isolated orthomolybdate MoO_4^{2-} anions were found. In the second series of glass with a higher molybdenum content, $Ag_2O/MoO_3 < 1$, the glass structure was much more complex, containing tetrahedral MoO_4^{2-} ions and polynuclear MoO_x species. The latter resembled the structure of crystalline disodium dimolybdate, $Na_2Mo_2O_7$, in which MoO_6 octahedra and MoO_4 tetrahedra are linked together.

27.7 APPLICATIONS IN HETEROGENEOUS CATALYSIS

The investigation of catalytic phenomena at the molecular level is a challenging task and involves the study of the chemical nature of the active component of catalysts and their active sites, changes in the chemical bonding involved in the interactions of reagents with the catalyst and their subsequent transformations. Solving these problems involves a number of experimental difficulties, especially in the case of heterogeneous catalysis. This is because, heterogeneous catalysts, in most cases, are highly dispersed and multicomponent systems, either polycrystalline or amorphous. Similar to oxide glasses, the advantages of solid-state NMR spectroscopy over diffraction techniques are recognized in catalysis research, although the two methods are frequently used together, and are often complemented by optical spectroscopy and computer modeling. We can recommend several recent reviews on the subject.[60,61]

27.7.1 Vanadium Catalysts

27.7.1.1 Surface Vanadium Oxide Complexes

Highly dispersed isolated and associated surface vanadium oxide complexes are formed on inert support surfaces at loadings not exceeding a theoretical monolayer. The molecular structure of these surface species depends on the support type, on the number of chemical bonds connecting vanadium to the support, and on the treatment conditions. To better understand the complexity of the topic, let us first discuss a model system prepared via chemical grafting of a vanadium-active component in the form of vanadium oxytrichloride, $VOCl_3$, onto the surface of titanium oxide TiO_2. The latter acts not only as a support for the active component, but often plays an important role in forming the catalytically active species. The grafting technique involves direct chemical reaction between surface hydroxyl groups of titanium oxide and $VOCl_3$ molecules as

$$VOCl_3 + n(HO{-}Ti{-}) \rightarrow VOCl_{3-n}(OTi{-})_n \quad (27.1)$$

Depending on the reaction conditions, three types of isolated surface vanadium species can form with $n = 1, 2, 3$. All three types can be readily identified in corresponding ^{51}V MAS NMR spectra by their isotropic chemical shifts and parameters of the chemical shielding anisotropy.[62] For example, under mild conditions, i.e., gas-phase grafting at room temperature, tetrahedral $VOCl_2(OTi{-})$ species form, with each vanadium having only a single V–O–Ti bond with the surface and the two chlorine atoms remaining in the first coordination sphere. These species are characterized by a slightly off-axial CSA tensor ($\eta_\sigma = 0.2$, $\zeta = -230$ ppm) and the isotropic chemical shift of $\delta_{iso} = -350$ ppm. When vanadium oxytrichloride is grafted onto titania at 110°C, the stronger surface reaction results in a $VOCl(OTi{-})_2$ species with two V–O–Ti bonds and one chlorine atom in the first coordination sphere. Stronger bonding produces a larger negative isotropic chemical shift, $\delta_{iso} = -433$ ppm, and the nonaxial chemical shielding tensor ($\eta_\sigma = 0.4$, $\zeta = -220$ ppm). At 250°C, all three chlorine atoms from $VOCl_3$ are substituted by V–O–Ti bonds. The resulting $VO(OTi{-})_3$ species resonate at even more negative shifts, $\delta_{iso} = -520$ ppm. The chemical shielding tensor reverses the sign while remaining nonaxial ($\eta_\sigma = 0.3$, $\zeta = 310$ ppm). At the same time, the quadrupolar interactions become large enough to be detected and characterized, $C_Q = 2.2$ MHz and $\eta_Q = 0.3$.

This first and somewhat simplified example already fully illustrates the advantages of ^{51}V NMR spectroscopy not only in an unambiguous identification of the surface vanadia species but also in its ease in following their chemical transformations on the surface, something not many other techniques can achieve in such a straightforward manner.

The chemical grafting model described above is only the first step in preparing real VO_x/TiO_2 catalysts. The grafting is usually followed by a cycle of hydration and dehydration at 350°C leading to the formation of the active component. Besides grafting, several other preparation techniques can be used to prepare VO_x/TiO_2 catalysts, including spray drying, impregnation with nonaqueous solutions containing

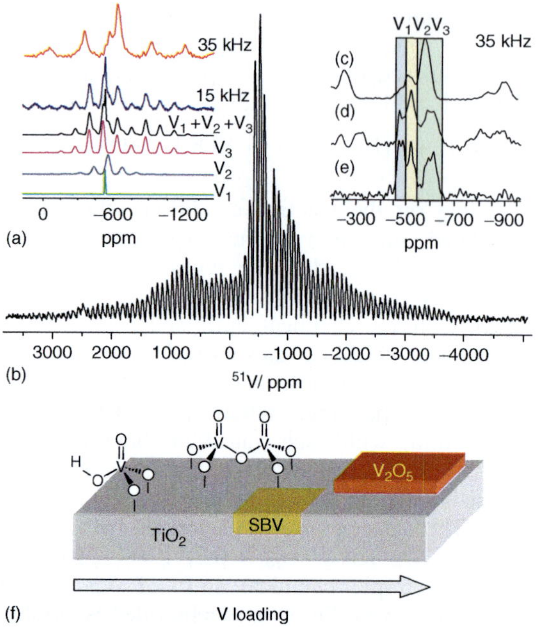

Figure 27.7. Vanadium-51 MAS NMR spectra of VO_x/TiO_2 catalysts obtained by (a) spray drying, (b) Topsøe industrial-type catalyst, (c) gas-phase deposition of $VOCl_3$, (d) impregnation by nonaqueous VO(O*i*−Pr)$_3$, and (e) impregnation by aqueous solution of NH_4VO_3. NMR experimental details and the vanadium site assignment are as follows: (a) MAS spectra (9.4 T) recorded at 35 and 15 kHz. The latter is shown together with deconvolution into three components, V_1—isolated species loosely bound with the support, V_2—low-associated and weakly bound species, V_3—strongly-bound distorted octahedral. (b) The wide-sweep 13 kHz MAS spectrum (14.1 T) showing all satellite transitions. Only V_3 species were detected. (c, d, e) The isotropic region of 35 kHz MAS spectra (9.4 T) recorded for three catalysts prepared by different techniques. All three types of surface vanadium species V_1, V_2, and V_3 are observed. (f) Schematic representation of various vanadium species on TiO_2 surface depending on vanadium loading (see text for discussion). (Reprinted from Ref. 4 © 2008 with permission from Elsevier and from Ref. 66 © ACS, 2004.)

VO(O*i*−Pr)$_3$, or impregnation with aqueous solutions of inorganic salts containing vanadium, for example, NH_4VO_3. Regardless of the preparation technique, the final catalyst will almost always contain three distinct types of similar surface vanadia species, V_1, V_2, and V_3, as can be seen in the corresponding ^{51}V MAS NMR spectra (Figure 27.7). While recording such spectra, special care should be taken to ensure that the correct spectra are recorded due to spectral crowding. Usually, this can be achieved by performing experiments at moderate magnetic fields, e.g., at 9.4 T, but necessarily using very high MAS spinning frequencies, 35 kHz or higher, to eliminate overlap between spinning sidebands and the central transition resonances, which would result in significant loss of information as can be seen in Figure 27.7(b).

The assignment of V_1, V_2, and V_3 groups of signals is made easier owing to the extensive ^{51}V NMR information available for individual vanadium compounds with a variety of vanadium coordination environments described above. For example, from their ^{51}V NMR parameters it is straightforward to assign V_1 species having the less negative isotropic chemical shift of about −530 ppm to trigonal VO_4 pyramids loosely attached to the titania surface via one or two V−O−Ti bonds. There are several individual resonances that can be resolved in the V_1 group, most likely owing to different crystal TiO_2 planes exposed at the surface, not only the densest plane (001) but also several cleavage planes. The V_2 surface sites at about −570 ppm appear to be from associated tetrahedral Q^2 sites, while the V_3 sites having the largest negative isotropic chemical shift of −650 ppm and the most significant quadrupolar coupling constant of $C_Q = 7$ MHz are from monomeric distorted VO_6 octahedral species with one short V=O bond, which are interacting strongly with the support via three or more V−O−Ti bonds.[63] This site assignment is also supported by complementary Raman data for the same series of samples.

Three types of vanadium species on the TiO_2 surface behave differently under the catalytic reaction conditions. For example, in the DeNO$_x$ denitrification process, the most weakly bound V_1 sites change the most to become more symmetric.[2] Their relative content also increases. At the same time, the V_2 sites are practically unaffected under these conditions, and only minor changes can be seen for octahedral V_3 sites. These changes indicate that the V_1 sites are most likely to participate in the catalytic reaction.

The molecular structure of isolated vanadia surface species depends on the support type. For example, chemical grafting of $VOCl_3$ onto a silica surface followed by hydration/dehydration at 350°C almost exclusively results in a VO(OSi−)$_3$ species having one short V=O bond and that is tightly bound to silica via three V–O–Si bonds. In ^{51}V NMR spectra, these surface species are characterized by an axial CSA tensor, with parameters typical for trigonal pyramids,

$\delta_{iso} = -710$ ppm, $\delta_{\perp} = -460$ ppm, $\eta_{\sigma} = 0$, $\zeta = 480$ ppm, and $C_Q = 2.5$ MHz. Surface vanadium species in hydrated VO_x/SiO_2 catalysts are similar to those found in $V_2O_5{\cdot}nH_2O$ gels.[64]

27.7.1.2 Strongly Bound Vanadium

A special type of vanadium oxide species called *strongly bound vanadium* (SBV) has recently been identified in vanadium catalysts prepared by many different techniques, including spray-drying, grafting, impregnation and mechanochemical activation, and using a variety of supports, TiO_2, monoclinic ZrO_2, and Nb_2O_5. This type of vanadium is thought to play role in many catalytic reactions.

The SBV was first identified in V_2O_5/TiO_2 catalysts prepared via spray-drying using a mixture of TiO_2 with vanadyl oxalate followed by thermal treatment. After washing excess V_2O_5 and other soluble vanadium complexes, the samples retained significant amounts of vanadium, thus the "strongly bound vanadium" term. In ^{51}V NMR spectra, these vanadium sites are characterized by unusually large ^{51}V quadrupolar coupling constants of about 10 MHz, while the chemical shielding parameters are very similar to those for a pure V_2O_5 phase.[65] It is understood that during V_2O_5/TiO_2 catalyst preparation, a coherent interfacial boundary is formed between two crystalline phases, TiO_2 and V_2O_5, consisting of closely spaced vanadium and titanium atoms in a 1:1 mixed cation layer. The vanadium sites inside this interfacial boundary experience significant EFGs resulting in large C_Q's. At the same time, the vanadium coordination by oxygen in this layer is similar to that for V_2O_5. The structural arrangement of this interface does not depend on vanadia loading and remains unchanged during catalytic reactions and other treatments. Vanadium-51 NMR parameters for SBV, mostly the quadrupolar coupling constant values, may vary to some extent depending on the preparation procedure.[66] The most symmetric SBV species were found in VO_x/TiO_2 catalysts prepared via chemical grafting.

27.7.1.3 Large Surface Vanadium Oxide Species, Aggregates, and Binary Phases

When vanadium loading exceeds the monolayer coverage, or when the dispersion level of the vanadia is not sufficiently high because of specific surface properties or due to the low density of accessible hydroxyl OH groups, progressively large aggregates of vanadium oxide complexes form on the surface. The most commonly observed are vanadium polyanions, such as decavanadates, $V_{10}O_{28}{}^{6-}$. Vanadium polyanions can often be detected in ^{51}V MAS NMR spectra of VO_x/Al_2O_3 catalysts with vanadium content exceeding the monolayer coverage, but only when such spectra are recorded under sufficiently high MAS spinning frequencies (Figure 27.8). Two other vanadium-containing phases that are readily identifiable in these spectra are V_2O_5 and $AlVO_4$ compounds. The latter is formed because of a chemical reaction between the vanadium precursor and the alumina support; such a reaction is known to readily occur. The relative amounts of the different large surface vanadium aggregates depends not only on the vanadium loading but also on the preparation conditions.[67] When the vanadium oxide phase is formed on the support surface, its crystal structure is almost always distorted due to disorder and surface defects. In ^{51}V NMR spectra recorded for stationary samples, this disorder manifests itself as a broadening of the satellite transitions due to long-range EFG effects; however, the chemical shielding parameters remain mostly unaffected.[68]

The chemical interaction between vanadium and the Al_2O_3 support to form $AlVO_4$ is not unique. Similar reactions have been observed in other systems where the chemistry supports such processes. For example, $NbVO_5$ has been detected in V_2O_5/Nb_2O_5 catalysts, and ZrV_2O_7 vanadate can form in V_2O_5/ZrO_2 systems.[69,70] More inert supports, including titania, silica, and porous silicates do not react with vanadium as easily.

Other chemical interactions occurring during catalyst synthesis are known to affect the structure and catalytic properties of vanadia catalysts. Often this is done intentionally to improve or alter the catalytic properties, i.e., by adding a modifier. Thus in VO_x/TiO_2 catalysis modified with sodium, several new sodium-containing phases have been found (Figure 27.9).[71] When phosphorus is present during synthesis of V_2O_5/TiO_2 catalysts via spray drying, the structure of the SBV phase is affected owing to phosphorus incorporation. In the resulting V–P–Ti species, vanadium exists in a tetrahedral coordination found for isolated or weakly associated tetrahedra (Figure 27.9).[72]

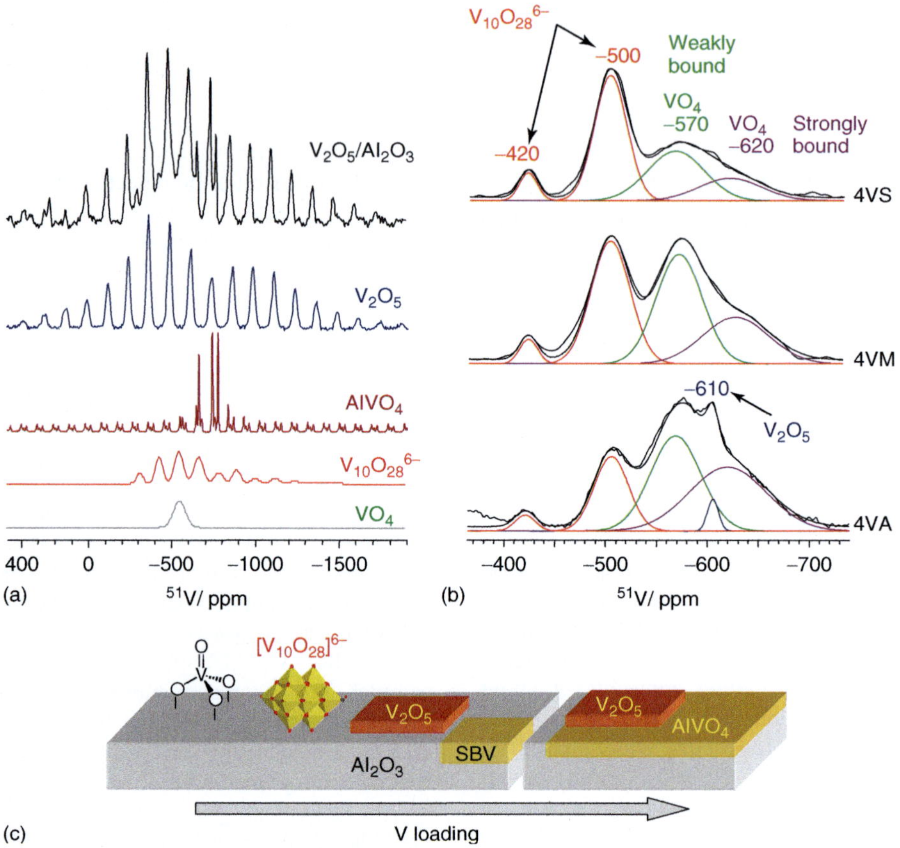

Figure 27.8. (a) Experimental ^{51}V MAS NMR spectrum of VO_x/Al_2O_3 catalyst containing 10.8 wt% of V_2O_5 recorded at 9.4 T and 13 kHz MAS is shown at the top. Below are the spectra of individual vanadium compounds illustrating the relative content of different vanadium phases in this catalyst sample (see the scheme below). (b) ^{51}V MAS NMR spectra of VO_x/Al_2O_3 samples prepared from vanadyl sulfate (VS), ammonium metavanadate (VM), and vanadyl acetylacetonate (VA) recorded at 9.4 T and 33 kHz MAS. (c) Schematic representation of various vanadium species on Al_2O_3 surface depending on vanadium loading (see text for discussion). (Reprinted from Ref. 4 © 2008 with permission from Elsevier and Reprinted with permission from Ref. 69 © 2009 American Chemical Society.)

Vanadium-51 NMR spectroscopy has been successfully applied not only to characterize vanadium species in various solid catalysts but also to probe their catalytic activity. In one recent example, the V–P–O catalysts supported on mesoporous SBA-15 supports were studied by multinuclear solid-state magnetic resonance spectroscopy with the aim of correlating their structure with the catalytic activity in the selective oxidation of *n*-butane to maleic anhydride.[73,74] MAS NMR results indicated that *n*-butane is primarily adsorbed at the surface sites of vanadium phosphate phases similar to δ-$VOPO_4$. The relative concentration of various species containing vanadium and phosphorus found in these catalysts by NMR spectroscopy was used to estimate the turnover frequency of their catalytically active surface sites.

27.7.2 Niobium Catalysts

Solid catalysts containing niobium are not very common in heterogeneous catalysis. Some examples include hydrated niobium pentoxide (niobic acid, $Nb_2O_5 \times nH_2O$) and niobium phosphate in acid-catalyzed reactions; layered niobium compounds containing metals are being tested in

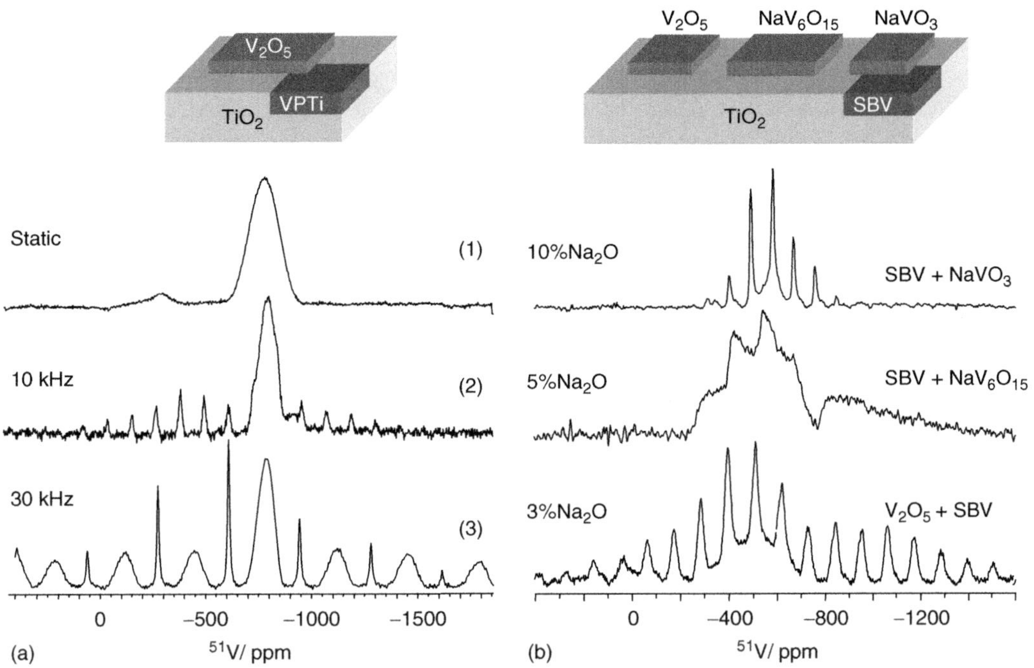

Figure 27.9. (a) Vanadium-51 NMR spectra obtained at 9.4 T for a $5\%P_2O_5/10\%V_2O_5/TiO_2$ catalyst prepared by spray drying. (a1) Stationary sample. (a2) 10 kHz MAS. (a3) 30 kHz MAS. (b) ^{51}V MAS NMR spectra obtained at 9.4 T for $Na_2O/20\%V_2O_5/TiO_2$ catalysts prepared by spray drying. The Na_2O loading (wt%) is shown with each spectrum. (Reproduced from Ref. 4 © Elsevier, 2008 and Ref. 71 by permission of the PCCP Owner Societies. http://dx.doi.org/10.1039/B001282N.)

photocatalytic reactions.[75] Niobium is frequently mixed with other oxides in multicomponent oxide systems as supports for other catalysts, for example, $Nb_2O_5-SiO_2$, $Nb_2O_5-Al_2O_3$, $Nb_2O_5-TiO_2$. Small amounts of niobium added to other catalytic oxide systems are known to improve their performance. In such systems, niobium additives act as promoters of catalytic reactions. For example, a mixed metal oxide catalyst (MoVNbTeO) has been found to have high activity and selectivity in the conversion of light alkanes. These and similar multicomponent oxides are usually prepared and screened in catalytic reactions via a combinatorial approach, with little attention paid to their molecular level structure or the exact nature of their catalytic activity. Multinuclear solid-state magnetic resonance spectroscopy can certainly aid in our understanding of structure–property relationships in these complex systems.

Niobium-93 NMR spectroscopy is far less common in heterogeneous catalytic research than ^{51}V NMR owing to larger nuclear quadrupole moment of ^{93}Nb. Solid oxide catalysts are generally disordered systems, which is known to induce significant quadrupolar broadening in ^{93}Nb NMR spectra. As an illustration, some representative static ^{93}Nb NMR spectra recorded at 21.1 T for a series of NbO_x/Al_2O_3 samples are shown in Figure 27.10.[4] Even at this high magnetic field, the spectra are very broad, in excess of 150 kHz. MAS has had little effect in narrowing these broad lines. By performing experiments at different magnetic fields, two contributions to the linewidth were recognized, from the second-order EFG effects and from the chemical shift distribution. Regardless of the large linewidth, these spectra are still informative in providing direct information about the niobium coordination on the alumina surface. At low niobium loading, the coordination of niobium in NbO_x/Al_2O_3 is most likely octahedral in distorted NbO_6 species possibly incorporating aluminum atoms of the support. As the niobium content increases, a new resonance can be clearly seen in the ^{93}Nb NMR spectra recorded

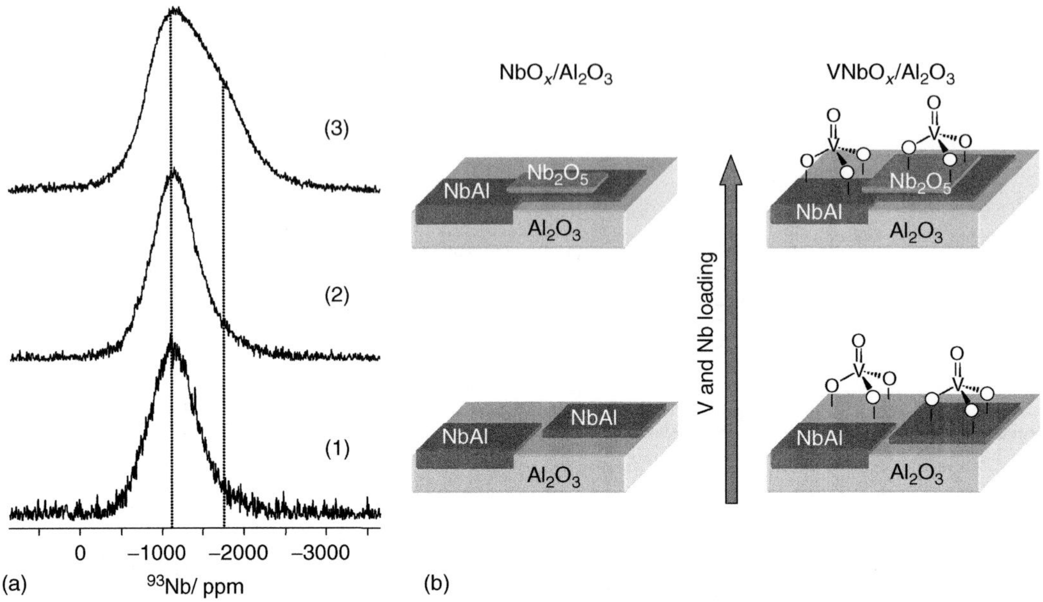

Figure 27.10. (a) Niobium-93 NMR spectra of stationary NbO_x/Al_2O_3 samples recorded at 21.1 T. (1) 4% Nb_2O_5/Al_2O_3. (2) 8% Nb_2O_5/Al_2O_3. (3) 16% Nb_2O_5/Al_2O_3. (b) Schematic representation of various vanadium and niobium species on Al_2O_3 surface depending on loading (see text for discussion). (Reprinted from Ref. 4 © 2008 with permission from Elsevier.)

at 21.1 T (Figure 27.10), which by its chemical shift value can be assigned to 6- or 8-coordinated Nb species. The latter are typical for bulk Nb_2O_5 oxide. This new resonance cannot be resolved at lower magnetic fields where the spectra are much broader.

Because ^{93}Nb NMR is of limited use for NbO_x/Al_2O_3 catalysts, complementary ^{27}Al and ^{1}H MAS NMR spectra are often recorded to study interactions between the supported niobium and the alumina support. Spectra recorded for both nuclei indicate stronger interactions of niobia with alumina than observed in VO_x/Al_2O_3 catalysts. In ^{27}Al MAS NMR spectra of NbO_x/Al_2O_3, characteristic resonances were detected from 5-coordinated aluminum species.[67] The number of such species gradually increases with the niobium loading, indicating an interaction between the alumina and niobia. At the same time, in ^{1}H MAS, NMR spectra even at high niobium loadings indicated a significant amount of residual surface OH groups, contrary to the case for VO_x/Al_2O_3 systems. These observations led the investigators to suggest that the active component is more dispersed in VO_x/Al_2O_3 samples, while in NbO_x/Al_2O_3 it tends to aggregate forming large polynuclear species and islands of Nb_2O_5 (Figure 27.10), in agreement with ^{93}Nb NMR data.

The picture becomes even more complicated when vanadia and niobia are simultaneously deposited onto alumina to prepare mixed $(Nb-V)O_x/Al_2O_3$ catalysts. These catalysts are of interest as they show a synergetic catalytic behavior superior to binary VO_x/Al_2O_3 and NbO_x/Al_2O_3 systems. This synergetic effect can be studied by employing ^{51}V NMR spectroscopy, as it is a very sensitive tool for the elucidation of the vanadium coordination environment, as discussed above. For example, in $(Nb-V)O_x/Al_2O_3$ samples, a new ^{51}V NMR resonance with an isotropic chemical shift of −650 ppm was observed. The relative intensity of this resonance increases with niobium loading, which makes it reasonable to attribute this resonance to the mixed cation $(Nb-V)O_x$ oxide species but not to the $NbVO_5$ phase since the latter has an isotropic ^{51}V NMR chemical shift of −790 ppm. Almost no changes have been observed in the corresponding static ^{93}Nb NMR spectra. Aluminum-27 MAS NMR spectra recorded for $(Nb-V)O_x/Al_2O_3$ samples are similar to NbO_x/Al_2O_3 samples and show no interaction between the supported vanadia and the alumina support even at high vanadium loading. ^1H MAS NMR spectra show a gradual decrease in the concentration of surface OH groups with loading of the active components. All these observations support the island model of the active component on the surface suggested for these systems (Figure 27.10). In this model, the niobium precursor interacts strongly with the alumina forming islands of mixed (Nb–Al)O_x oxides, while the vanadium precursor disperses more randomly on the surface, interacting equally with both oxides.

27.7.3 Molybdenum Catalysts

27.7.3.1 Supported Mo/Al_2O_3 Catalysts

Molybdenum catalysts supported on alumina find many useful applications in the chemical industry. These catalysts are often prepared via impregnation using aqueous solutions of molybdates and polyoxomolybdates. While solution-state chemistry of polyoxomolybdates has been intensively studied using liquid-state ^{95}Mo NMR spectroscopy,[76] studies of supported solid catalysts are still rare. One recent study has outlined general research strategies necessary for the successful application of solid-state ^{95}Mo NMR spectroscopy to study such complex catalytic systems.[23] Primary attention in preparing supported molybdenum catalysts via impregnation is paid to the acidity (pH) of the impregnating solution. Thus, the acidification of the Na_2MoO_4 solution (pH = 6) results in a variety of polynuclear oxomolybdate ions.[76]

$$[Mo_7O_{24}]^{6-} \rightarrow \alpha\text{-}[Mo_8O_{26}]^{4-} \rightarrow \beta\text{-}[Mo_8O_{26}]^{4-} \rightarrow [Mo_{36}O_{112}]^{8-} \quad (27.2)$$

This equilibrium determines the predominant molybdenum species on the catalyst surface after the impregnation. To establish the nature of interaction of the active ingredient with the support, it is often necessary to employ a multinuclear magnetic resonance approach. For example, in the case of Mo/Al_2O_3 catalysts, ^{27}Al NMR spectroscopy helped in identifying surface $[Al(OH)_6Mo_6O_{18}]^{3-}$ species formed as a result of the interaction of paramolybdate with alumina. Comparing these data with ^{95}Mo NMR data allowed the authors to establish a most likely interaction mechanism between the active ingredient and the support as shown in Figure 27.11.[23]

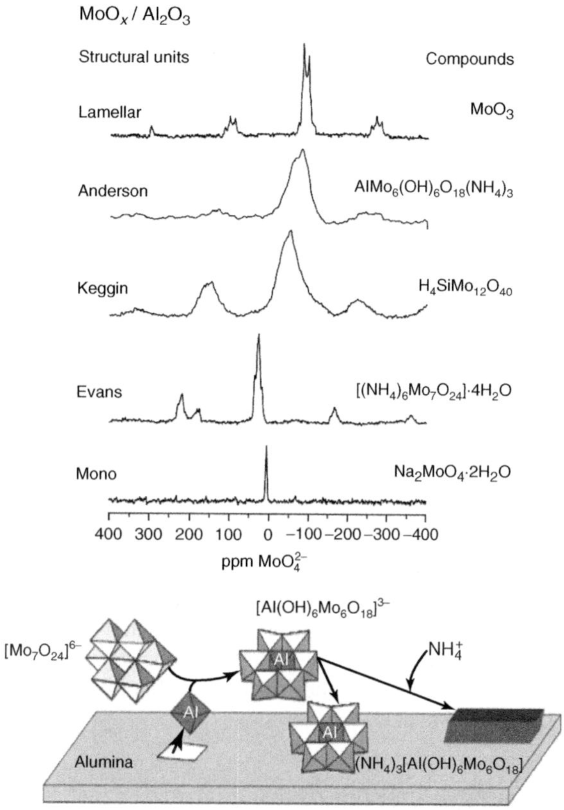

Figure 27.11. Molybdenum-95 single-pulse MAS NMR spectra of individual molybdenum compounds thought to be present on the surface during catalyst preparation, including the two molybdenum salts, sodium molybdate and ammonium heptamolybdate, commonly used in impregnating solutions, as well as the two molybdenum heteropolyanions known to form on the surface of silica or alumina during impregnation. The reaction scheme derived from the multinuclear NMR data is shown below. (Reproduced from Ref. 23 © Springer, 2007.)

27.7.3.2 Catalytically Active Mo Sites in HZSM-5 Zeolite

On the basis of earlier ^{27}Al MQMAS NMR studies of the Mo/HZSM-5 catalyst for methane dehydroaromatization, it has been suggested that the exchanged molybdenum species may be bonded with Brönsted acidic aluminum sites in zeolite via two oxygen bridges. This idea has recently been proven by the direct observation of the active Mo centers in zeolite using ultrahigh field ^{95}Mo NMR spectroscopy at 21 T and ^{95}Mo-enriched MoO_3 reagents.[77]

The research significantly benefited from the fact that the spin–lattice relaxation time, T_1, corresponding to exchanged molybdenum species in Mo/HZSM-5 catalysts was relatively short, about 100 ms at a magnetic field strength of 21.1 T. This is about two orders of magnitude shorter than that measured for small MoO_3 crystallites (about 30 s).[24] Such a large difference has allowed investigators to differentiate exchanged Mo species in Mo/HZSM-5 catalysts from MoO_3 agglomerates on the surface, and to readily estimate their relative fractions present in catalysts with varying Mo loading. A reasonable correlation between the amount of zeolite exchanged species and the catalyst's activity in the catalytic methane dehydroaromatization was obtained. Analysis of the lineshapes obtained from both MAS and static ^{95}Mo NMR spectra indicates that the exchanged sites are heterogeneous, leading to significantly broadened MAS spectra, with essentially a featureless but nearly symmetric static lineshape for the exchanged Mo species. A heterogeneous line broadening mechanism has been suggested to be primarily responsible for the observed MAS and static lineshapes for the exchanged species.

RELATED ARTICLES IN THE ENCYCLOPEDIA OF MAGNETIC RESONANCE

Chemical Shifts and Solid-State Molecular-Level Structure

Chemical Shift Computations for Crystalline Molecular Systems: Practice

Chemical Shift Tensor Measurement in Solids

Chemical Shift Tensors

Computations of Magnetic Resonance Parameters for Crystalline Systems: Principles

Dipolar and Indirect Coupling: Basics

Echoes in Solids

Fast Magic-Angle Spinning: Implications

Ferroelectrics and Proton Glasses

High Speed MAS of Half-Integer Quadrupolar Nuclei in Solids

Inorganic Chemistry Applications

Inorganic Nuclei: Low Sensitivity Transition Metals

Knight Shift

Line Narrowing Methods in Solids

Low-γ Nuclei in Solid Samples

Magic Angle Spinning

Magnetic Shielding and Chemical Shifts: Basics

Measurements of Nuclear Spin Interaction Tensors from Wideline Spectra and by Spinning Sideband Analysis

Multiple-quantum Magic-angle Spinning Experiments on Half-integer Nuclei: Fundamentals

Multiple Quantum NMR in Solids

Quadrupolar Metal Nuclides in Bioinorganic Chemistry: Solid-State NMR Studies

Quadrupolar Nuclei in Glasses

Satellite Transition NMR Spectroscopy of Half-Integer Quadrupolar Nuclei under Magic-angle Spinning

Semiempirical Chemical Shift Calculations

Shielding Calculations

Shielding Tensor Calculations

Supported Metal Catalysts

Tensors in NMR

Tensor Interplay

Vanadium Catalysts: Solid State NMR

REFERENCES

1. O. B. Lapina, V. M. Mastikhin, A. A. Shubin, V. N. Krasilnikov, and K. I. Zamaraev, *Prog. Nucl. Magn. Reson. Spectrosc.*, 1992, **24**, 457.
2. O. B. Lapina, A. A. Shubin, D. F. Khabibulin, V. V. Terskikh, P. R. Bodart, and J. P. Amoureux, *Catal. Today*, 2003, **78**, 91.
3. V. M. Mastikhin and O. B. Lapina, Vanadium Catalysts: Solid State NMR, in *Encyclopedia of Magnetic Resonance*, eds R. K. Harris and R. E. Wasylishen, John Wiley & Sons: Chichester, 2007. DOI: 10.1002/9780470034590.emrstm0585.
4. O. B. Lapina, D. F. Khabibulin, A. A. Shubin, and V. V. Terskikh, *Prog. Nucl. Magn. Reson. Spectrosc.*, 2008, **53**, 128.

5. V. M. Mastikhin, O. B. Lapina, V. N. Krasilnikov, and A. A. Ivakin, *React. Kinet. Catal. Lett.*, 1984, **24**, 119.
6. O. B. Lapina, D. F. Khabibulin, K. V. Romanenko, Z. H. Gan, M. G. Zuev, V. N. Krasil'nikov, and V. E. Fedorov, *Solid State Nucl. Magn. Reson.*, 2005, **28**, 204.
7. J. V. Hanna, K. J. Pike, T. Charpentier, T. F. Kemp, M. E. Smith, B. E. G. Lucier, R. W. Schurko, and L. S. Cahill, *Chem.—Eur. J.*, 2010, **16**, 3222.
8. J. J. Fitzgerald, S. Prasad, J. Huang, and J. S. Shore, *J. Am. Chem. Soc.*, 2000, **122**, 2556.
9. A. Flambard, L. Montagne, L. Delevoye, and S. Steuernagel, *Solid State Nucl. Magn. Reson.*, 2007, **32**, 34.
10. J. Davis, D. Tinet, J. J. Fripiat, J. M. Amarillo, B. Casal, and E. Ruiz-Hitzky, *J. Mater. Res.*, 1991, **6**, 393.
11. A. Proust, R. Thouvenot, S. G. Roh, J. K. Yoo, and P. Gouzerh, *Inorg. Chem.*, 1995, **34**, 4106.
12. T. J. Bastow, *Solid State Nucl. Magn. Reson.*, 1998, **12**, 191.
13. T. J. Bastow, *Solid State Nucl. Magn. Reson.*, 2003, **23**, 116.
14. N. Machida and H. Eckert, *Solid State Ionics*, 1998, **107**, 255.
15. V. M. Mastikhin, O. B. Lapina, and R. I. Maksimovskaya, *Chem. Phys. Lett.*, 1988, **148**, 413.
16. J. C. Edwards, R. D. Adams, and P. D. Ellis, *J. Am. Chem. Soc.*, 1990, **112**, 8349.
17. J. C. Edwards and P. D. Ellis, *Langmuir*, 1991, **7**, 2117.
18. O. H. Han, C. Y. Lin, N. Sustache, M. Mcmillan, J. D. Carruthers, K. W. Zilm, and G. L. Haller, *Appl. Catal. A Gen.*, 1993, **98**, 195.
19. M. E. Smith, *Annu. Rep. NMR Spectrosc.*, 2001, **43**, 121.
20. M. A. M. Forgeron and R. E. Wasylishen, *Phys. Chem. Chem. Phys.*, 2008, **10**, 574.
21. J.-B. d'Espinose de Lacaillerie, F. Barberon, K. V. Romanenko, O. B. Lapina, L. Le Pollès, R. Gautier, and Z. H. Gan, *J. Phys. Chem. B*, 2005, **109**, 14033.
22. A. R. Hove, H. Bildsøe, J. Skibsted, M. Brorson, and H. J. Jakobsen, *Inorg. Chem.*, 2006, **45**, 10873.
23. J. B. d'Espinose de Lacaillerie and Z. Gan, *Appl. Magn. Reson.*, 2007, **32**, 499.
24. J. Z. Hu, J. H. Kwak, Y. Wang, C. H. F. Peden, H. Zheng, D. Ma, and X. Bao, *J. Phys. Chem. C*, 2009, **113**, 2936.
25. C. J. Pickard and F. Mauri, *Phys. Rev. B: Condens. Matter*, 2001, **63**, 245101.
26. L. Truflandier, M. Paris, C. Payen, and F. Boucher, *J. Phys. Chem. B*, 2006, **110**, 21403.
27. L. Truflandier, M. Paris, and F. Boucher, *Phys. Rev. B: Condens. Matter*, 2007, **76**, 035102.
28. L. A. Truflandier, F. Boucher, C. Payen, R. Hajja, Y. Millot, C. Bonhomme, and N. Steunou, *J. Am. Chem. Soc.*, 2010, **132**, 4653.
29. X. F. Wang, J. Adhikari, and L. J. Smith, *J. Phys. Chem. C*, 2009, **113**, 17548.
30. J. Cuny, E. Furet, R. Gautier, L. Le Pollès, C. J. Pickard, and J. B. d'Espinose de Lacaillerie, *ChemPhysChem*, 2009, **10**, 3320.
31. J. Li, M. E. Lashier, G. L. Schrader, and B. C. Gerstein, *Appl. Catal.*, 1991, **73**, 83.
32. A. Tuel, L. Canesson, and J. C. Volta, *Colloids Surf., A*, 1999, **158**, 97.
33. J. P. Jesson, in *NMR of Paramagnetic Molecules: Principles and Applications*, eds G. N. La Mar, W. D. Horrocks Jr, and R. H. Holm, Academic Press: New York and London, 1973, p. 1.
34. I. Bertini and C. Luchinat, *Coord. Chem. Rev.*, 1996, **150**, 1.
35. I. Bertini, C. Luchinat, and G. Parigi, *Prog. Nucl. Magn. Reson. Spectrosc.*, 2002, **40**, 249.
36. A. A. Shubin, D. F. Khabibulin, and O. B. Lapina, *Catal. Today*, 2009, **142**, 220.
37. U. G. Nielsen, J. Skibsted, and H. J. Jakobsen, *Chem. Phys. Lett.*, 2002, **356**, 73.
38. R. Michalaka and H. A. H. Mohammad, *Solid State Nucl. Magn. Reson.*, 2004, **26**, 187.
39. A. W. Glynn, A. L. Ponce, X. Lin, and J. J. Fripiat, *Solid State Ionics*, 1996, **84**, 213.
40. O. Durupthy, M. Jaber, N. Steunou, J. Maquet, G. T. Chandrappa, and J. Livage, *Chem. Mater.*, 2005, **17**, 6395.
41. F. Carn, N. Steunou, J. Livage, A. Colin, and R. Backov, *Chem. Mater.*, 2005, **17**, 644.
42. C. J. Fontenot, J. W. Wiench, M. Pruski, and G. L. Schrader, *J. Phys. Chem. B*, 2001, **105**, 10496.
43. N. Dupre, J. Gaubicher, D. Guyomard, and C. P. Grey, *Chem. Mater.*, 2004, **16**, 2725.

44. M. Dubarry, J. Gaubicher, D. Guyomard, N. Steunou, J. Livage, N. Dupre, and C. P. Grey, *Chem. Mater.*, 2006, **18**, 629.

45. J. F. Zhu and Y. N. Huang, *Inorg. Chem.*, 2009, **48**, 10186.

46. J. Zhu and Y. Huang, *Langmuir*, 2010, **26**, 10115.

47. S. Hardin, D. Hay, M. Millikan, J. V. Sanders, and T. W. Turney, *Chem. Mater.*, 1991, **3**, 977.

48. J. J. Fitzgerald, J. Huang, and H. Lock, in *Piezoelectric Materials: Advances in Science, Technology and Applications*, eds C. Galassi, M. Dinescu, K. Uchino, and M. Sayer, NATO Sci. Ser Kluwer Academic Publishers: Norwell, 2000, p. 203.

49. V. V. Laguta, M. D. Glinchuk, I. P. Bykov, R. Blinc, and B. Zalar, *Phys. Rev. B: Condens. Matter*, 2004, **69**, 54103/1.

50. S. Rankel, B. Zalar, V. V. Laguta, R. Blinc, and J. Toulouse, *Ferroelectrics*, 2005, **314**, 165.

51. K. Sato, S. Takeda, S. Fukuda, T. Minamisono, M. Tanigaki, T. Miyake, Y. Maruyama, K. Matsuta, M. Fukuda, and Y. Nojiri, *Z. Naturforsch., A: Phys. Sci.*, 1998, **53**, 549.

52. J. Rocha, P. Brandao, Z. Lin, A. P. Esculcas, A. Ferreira, and M. W. Anderson, *J. Phys. Chem.*, 1996, **100**, 14978.

53. H. M. Kao and K. H. Lii, *Inorg. Chem.*, 2002, **41**, 5644.

54. C. Bonhomme, C. Coelho, T. Azais, L. Bonhomme-Coury, F. Babonneau, J. Maquet, and R. Thouvenot, *C. R. Chim.*, 2006, **9**, 466.

55. O. B. Lapina, V. V. Terskikh, A. A. Shubin, V. M. Mastikhin, K. M. Eriksen, and R. Fehrmann, *J. Phys. Chem.*, 1997, **101**, 9188.

56. O. B. Lapina, V. V. Terskikh, B. S. Bal'zhinimaev, K. M. Eriksen, and R. Fehrmann, in *Green Industrial Applications of Ionic Liquids*, NATO Science Series II, eds R. D. Rogers, K. R. Seddon, and S. Volkov, Kluwer: Dordrecht, 2002, Vol. 92, p. 85.

57. S. Sakida, S. Hayakawa, and T. Yoko, *J. Phys.: Condens. Matter*, 2000, **12**, 2579.

58. A. Flambard, L. Montagne, L. Delevoye, G. Palavit, J. P. Amoureux, and J. J. Videau, *J. Non-Cryst. Solids*, 2004, **345&346**, 75.

59. S. H. Santagneli, C. C. de Araujo, W. Strojek, and H. Eckert, *J. Phys. Chem. B*, 2007, **111**, 10109.

60. J. McGregor, in *Metal Oxide Catalysis*, eds S. D. Jackson and J. S. J. Hargreaves, Wiley-VCH, Verlag GmbH & Co. KGaA: Weinheim, Germany, 2009, Vol. 1, p. 195, Chapter 5. DOI: 10.1002/9783527626113.ch5.

61. I. E. Wachs, *Appl. Catal., A Gen.*, 2011, **391**, 36.

62. L. G. Pinaeva, O. B. Lapina, V. M. Mastikhin, A. V. Nosov, and B. S. Balzhinimaev, *J. Mol. Catal.*, 1994, **88**, 311.

63. D. Courcot, B. Grzybowska, Y. Barbaux, M. Rigole, A. Ponchel, and M. Guelton, *J. Chem. Soc., Faraday Trans.*, 1996, **92**, 1609.

64. G. Deo and I. E. Wachs, *J. Phys. Chem.*, 1991, **95**, 5889.

65. A. A. Shubin, O. B. Lapina, and V. M. Bondareva, *Chem. Phys. Lett.*, 1999, **302**, 341.

66. U. G. Nielsen, N. U. Topsøe, M. Brorson, J. Skibsted, and H. J. Jakobsen, *J. Am. Chem. Soc.*, 2004, **126**, 4926.

67. A. E. Lewandowska, M. A. Bañares, M. Ziolek, D. F. Khabibulin, and O. B. Lapina, *J. Catal.*, 2008, **255**, 94.

68. A. A. Shubin, O. B. Lapina, and D. Courcot, *Catal. Today*, 2000, **56**, 379.

69. A. E. Lewandowska, M. A. Banares, D. F. Khabibulin, and O. B. Lapina, *J. Phys. Chem. C*, 2009, **113**, 29648.

70. C. Martin, D. Klissurski, J. Rochac, and V. Rives, *Phys. Chem. Chem. Phys.*, 2000, **2**, 1543.

71. V. V. Terskikh, O. B. Lapina, and V. M. Bondareva, *Phys. Chem. Chem. Phys.*, 2000, **2**, 2441.

72. O. B. Lapina, D. F. Khabibulin, A. A. Shubin, and V. M. Bondareva, *J. Mol. Catal. A Chem.*, 2000, **162**, 381.

73. J. Frey, Y. S. Ooi, B. Thomas, V. R. Marthala, A. Bressel, T. Scholkopf, T. Schleid, and M. Hunger, *Solid State Nucl. Magn. Reson.*, 2009, **35**, 130.

74. J. Frey, C. Lieder, T. Scholkopf, T. Schleid, U. Nieken, E. Klemm, and M. Hunger, *J. Catal.*, 2010, **272**, 131.

75. K. Tanabe, *Catal. Today*, 2003, **78**, 65.

76. R. I. Maksimovskaya and G. M. Maksimov, *Inorg. Chem.*, 2007, **46**, 3688.

77. H. Zheng, D. Ma, X. H. Bao, J. Z. Hu, J. H. Kwak, Y. Wang, and C. H. F. Peden, *J. Am. Chem. Soc.*, 2008, **130**, 3722.

Chapter 28

Quadrupolar NMR of Intermetallic Compounds

Frank Haarmann

Institute of Inorganic Chemistry, RWTH Aachen University, Aachen D-52074, Germany

28.1 INTRODUCTION

Since various definitions of intermetallic compounds exist among scientists and in the literature, this class of compounds is introduced first. Phases formed by different metal atoms as well as those formed by metal and metalloid atoms possessing metallic conductivity will be considered here as intermetallic compounds. Using this definition, semiconducting materials are neglected but disorder of the atoms and the formation of homogeneity ranges are included. A comprehensive definition of intermetallic compounds is given by Westbrook and Fleischer.[1]

The growing interest in intermetallic compounds is due to their technological applications such as superconductivity,[2–4] spintronics,[5] thermoelectric,[6,7] and magnetic[8,9] and refractory materials,[8] as well as from the point of view of fundamental research.[10,11] Finally, a target-oriented synthesis of materials with the required properties is highly desirable for commercial and environmental reasons. The key to this will be a better understanding of the relationship of structure, properties, and chemical bonding in intermetallic compounds. Unfortunately, the current knowledge about chemical bonding in intermetallic compounds is rather limited because of the complex interactions[10,11] that cannot be described by the simple pictures of ionic, covalent, or metallic bonding. Thus, strategies and methods for the systematic investigation of intermetallic compounds are needed.

Two features of NMR spectroscopy are most important for the research of intermetallic compounds: NMR is sensitive to the bonding of atoms. Furthermore, NMR is a local probe that allows the investigation of the disorder phenomena frequently observed in intermetallic compounds. For these reasons, NMR spectroscopy is an ideal technique to compensate for the lack of short-range information available from diffraction experiments.

The performance of NMR for the investigation of local arrangements in disordered materials has clearly been demonstrated in nonconducting materials.[12] The experiments successfully applied to these compounds have to be carefully evaluated with respect to materials possessing metallic conductivity due to the NMR relevant couplings arising from interaction of nuclear spins and conduction electrons. A review of ^{45}Sc NMR on intermetallic compounds has recently been given by Eckert and Pöttgen.[13]

NMR of Quadrupolar Nuclei in Solid Materials
Edited by Roderick E. Wasylishen, Sharon E. Ashbrook and Stephen Wimperis
 ISBN: 978-0-470-97398-1

The use of NMR spectroscopy for the investigation of nonmagnetic intermetallic compounds will be reviewed, paying special attention to the benefit arising from quadrupole coupling. First, the background concepts are briefly introduced. The next section deals with various applications. Finally, a summary and outlook is given.

28.2 BACKGROUND CONCEPTS

28.2.1 Line Shape of NMR Signals

28.2.1.1 Regular Powder Samples

The line shape of an NMR signal contains valuable information about the local environments of the atoms. In order to analyze the experimental results quantitatively, line shape analysis is performed by least-squares fitting of the signals considering the relevant interactions of the nucleus with its surroundings. The line shape of the NMR signal for intermetallic compounds, as for nonconducting materials, is determined by the orientation dependence of the magnetic and electric couplings.[14,15] The latter is relevant only for nuclear spins $I > 1/2$. Both couplings are described by a second-rank tensor in its principal axis system (PAS).

The magnetic interactions are usually dominated by the coupling of nuclear and electron spins, e.g., Knight shift (K) for nonmagnetic intermetallic compounds. Additional contributions result from magnetic shielding (σ), which is frequently designated as orbital coupling. Both Knight shift and magnetic shielding are simultaneously present[16] and cannot be resolved experimentally, because the nucleus is influenced by the resulting magnetic field. Thus an additional set of parameters indicated by Δ have to be introduced for the line shape analysis to describe the complete shielding interaction. The average magnetic field experienced by a nucleus is modified by the complete isotropic shielding ($\Delta_{\text{iso}} = \sigma_{\text{iso}} + K_{\text{iso}}$). The orientation dependence of the complete shielding interaction is characterized by the anisotropy parameter (Δ_{aniso}) and the asymmetry parameter (η_Δ).

$$\Delta_{\text{iso}} = 1/3(\Delta_{XX} + \Delta_{YY} + \Delta_{ZZ}) \quad (28.1)$$

$$\Delta_{\text{aniso}} = \Delta_{ZZ} - \Delta_{\text{iso}} \quad (28.2)$$

$$\eta_\Delta = \frac{\Delta_{YY} - \Delta_{XX}}{\Delta_{ZZ} - \Delta_{\text{iso}}} \quad (28.3)$$

The sequence of the principal axis values of the complete shielding tensor Δ_{II} follows the definition of Mehring,[17] $|(\Delta_{ZZ} - \Delta_{\text{iso}})| \geq |(\Delta_{XX} - \Delta_{\text{iso}})| \geq |(\Delta_{YY} - \Delta_{\text{iso}})|$. With these parameters, the signal shift can be calculated for a single site using the polar angles θ and ϕ of the magnetic field with respect to the PAS of the coupling tensor

$$\nu(\theta,\phi) = \Delta_{\text{iso}} + \frac{1}{2}\Delta_{\text{aniso}}\left[3\cos^2\theta - 1 + \eta\sin^2\theta\cos 2\phi\right] \quad (28.4)$$

The above-mentioned situation is found in a single crystal with only one atom in the unit cell. The signal from a regular powder composed of randomly oriented crystallites can be simulated by varying the polar angles according to one of the powder averaging schemes described in the literature.[18] Spinning of the sample about the magic angle (MAS) averages the coupling and results in significant changes of the line shape. Examples for the characteristic line shapes of the NMR signals with and without sample rotation can be found in the book by Duer.[19]

The electric, e.g., quadrupole, coupling for $I > 1/2$ introduces additional changes of the line shape of the NMR signals. For noncubic environments of the nuclei, $2I$ single-quantum transitions can be observed. The signal is principally different for $I = n/2$ ($n \geq 2$) with n even and n odd. Considering half-integer nuclei (odd n), the signal is composed of the central transition $m_I = 1/2 \leftrightarrow m_I = -1/2$ and the satellite transitions $m_I \leftrightarrow m_{I-1}$ with $m_I \neq 1/2$. Various definitions of quadrupole coupling parameters are given in the literature.[20] We use the quadrupole coupling constant C_Q or the quadrupole frequency ν_Q.

$$C_Q = \frac{V_{ZZ}\cdot e\cdot Q}{h} \quad \text{using } V_{ZZ} = eq \quad (28.5)$$

$$\nu_Q = \frac{3\cdot C_Q}{2I\cdot(2I-1)} \quad (28.6)$$

The asymmetry parameter is defined as

$$\eta_Q = \frac{V_{XX} - V_{YY}}{V_{ZZ}} \quad (28.7)$$

with $|V_{ZZ}| \geq |V_{YY}| \geq |V_{XX}|$. V_{XX}, V_{YY}, and V_{ZZ} are the eigenvalues of the traceless symmetric quadrupole coupling tensor.

Only the satellite transitions are influenced by the orientation dependence of the signals given by first-order perturbation theory for $\nu_Q \ll \nu_0$ with ν_0

corresponding to the Larmor frequency.[15]

$$\nu^{(1)}_{\text{quad}}(\theta,\phi) = -\frac{\nu_Q}{2}\cdot\left(m_I - \frac{1}{2}\right) \cdot\left(3\cos^2\theta - 1 + \eta_Q\cdot\sin^2\theta\cdot\cos 2\phi\right) \quad (28.8)$$

For large quadrupole coupling ($\nu_Q < \nu_0$), second-order perturbation theory has to be applied. The second-order contribution to the central transition are discussed in Ref. 15.

$$\nu^{(2)}_{\text{quad}}(\theta,\phi) = \frac{1}{96}\left(I(I+1) - \frac{3}{4}\right)\cdot\frac{\nu_Q^2}{\nu_0} \cdot\left\{\begin{array}{l} 6\sin^2\theta\left(1 - 9\cos^2\theta\right) - 4\cdot\eta_Q\cdot\cos 2\phi \\ \cdot\sin^2\theta\cdot\left(\cos^2\theta + 1\right) + \eta_Q^2\cdot\left(-\frac{16}{3}\right. \\ \left. + 8\cos^2\theta + 6\cos^2 2\phi\cdot\sin^4\theta\right) \end{array}\right\} \quad (28.9)$$

In order to simulate the NMR signal of a regular powder, appropriate powder averaging[18] has to be performed. More details about the quadrupolar interacion and its influence on NMR line shapes in stationary and spinning samples can be found in Chapters 1 to 4 of this book.

Both magnetic and electric interactions are simultaneously present for $I > 1/2$ nuclei. This has to be considered by adding the individual contributions for the simulation of the line shape of the NMR signal. Principally, the PAS of both couplings are different and may have different orientations with respect to the external magnetic field. The relationship of the orientation of the PAS is given by the Euler angles (α, β, γ).[17,19]

28.2.1.2 Aligned Powder Samples

Significant changes of the NMR line shape are observed for partial alignment of the crystallites. The preferred orientation of the crystallites with respect to the external magnetic field may be due to different reasons such as the shape of the crystallites, anisotropic electrical conductivity,[21–23] or the magnetism of the samples.[24] Two limiting cases of the influence of anisotropic conductivity are briefly discussed with respect to the increased information available by analysis of the line shape of the NMR signals.

Anisotropic conductivity is due to different Fermi velocities of the electrons for varying directions in the crystallites that align in the magnetic field by a minimization of the Lorentz force ($\vec{F} = q\,\vec{v}\times\vec{B}$). This is achieved for an orientation of the crystallites with highest Fermi velocities parallel to the external magnetic field. The orientation of the crystallites can be fixed for further experiments using a suspension of fine powdered sample and glue for the alignment in an external magnetic field (B_{or}).

The alignment of the crystallites results in fixed orientations of the coupling tensors with respect to the magnetic field. Two limiting cases are discussed with respect to quadrupole coupling assuming $\eta_Q = 0$: (I) $V_{ZZ}||B_{or}$ and (II) $V_{ZZ}\perp B_{or}$ (Figure 28.1). An extension of powder-averaging has to be performed for $\eta_Q \neq 0$. A single orientation of V_{ZZ} with respect to the external magnetic field used for the NMR experiments is observed for each angle χ between the B_0 and B_{or} in case I. This results in narrow NMR signals for the central transition and the satellite transitions. The frequency of the signals is given by equations 28.8 and 28.9 with $\theta = \chi$ as the angle between V_{ZZ} and B_0. More complex orientations of V_{ZZ} are observed in case II where a disk-like distribution of V_{ZZ} is found in a plane perpendicular to the magnetic field used for the alignment of the crystallites. This results in a narrow NMR signal

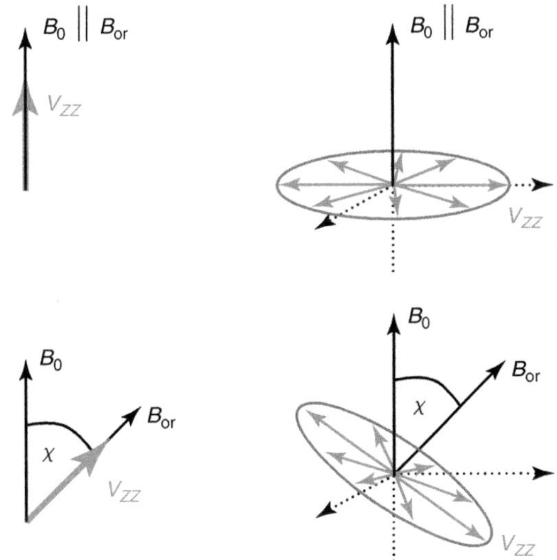

Figure 28.1. Orientation of V_{ZZ} with respect to the external magnetic fields used for alignment of the crystallites B_{or} and the NMR experiments B_0. χ defines the angle between B_0 and B_{or}. $V_{ZZ}||B_{or}$ and $V_{ZZ}\perp B_{or}$ are shown on the right and left, respectively. Parallel orientation of B_0 and B_{or} are depicted on top and $\chi = 45°$ at bottom.

for $\chi = 0^\circ$ corresponding to $\theta = 90^\circ$. A distribution of angles between V_{ZZ} and B_0 is obtained for $\chi \neq 0$, which causes broad frequency distributions of the NMR signals (Figure 28.2). Usually, the central transition signal is analyzed because the width of the satellite transitions signal is too large for experimental determination. For simulation of the signals, equation 28.9 can be used by substitution of the angle θ by $\arccos(\cos(\alpha) \times \cos(\varphi))$, with φ being uniformly distributed between 0° and 180° and $\alpha = 90^\circ - \chi$.

The advantage of partial alignment of the crystallites for case I is a higher resolution of the NMR experiment still allowing the determination of the anisotropy parameter by orientation-dependent measurements. More characteristic features of the line shape are observed in case II. Thus, by performing orientation-dependent experiments, an increased data base is obtained for parameter determination by least-squares fitting of the NMR signals.

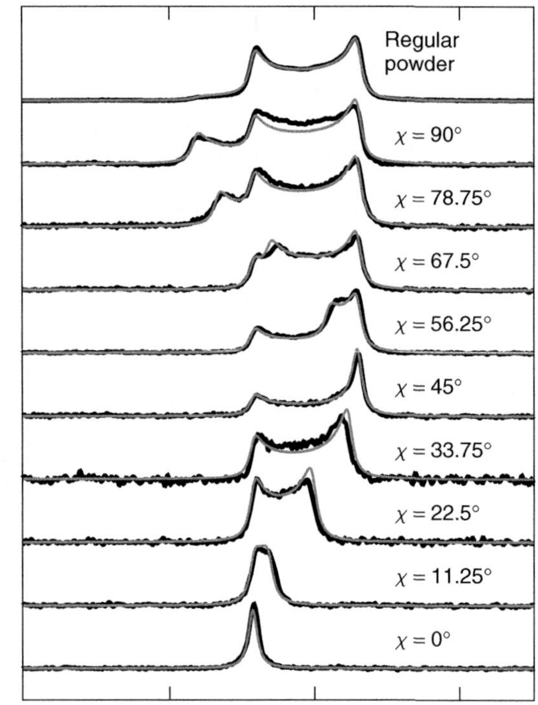

Figure 28.2. Orientation dependence of the ^{71}Ga NMR central transition signal of $BaGa_2$ powder aligned in magnetic field and a regular powder. Full lines represent the observed pattern. The simulated signals are shown as gray lines. χ is defined as the angle between B_0 and B_{or}. (Modified from Ref. 25. © Wiley-VCH Verlag GmbH & Co, 2009.)

28.2.1.3 *Symmetry Considerations*

The orientation dependence of the various interactions in single crystals, regular powders, or aligned powders gives valuable information about the environment of the observed nucleus. This is due to symmetry constraints arising from the fact that the coupling tensor has to obey the symmetry operations of the site under consideration. Thus, for cubic site symmetry such as tetrahedral, octahedral, cubic, or icosahedral coordination, no anisotropic interactions can be observed. Nuclei located on n-fold rotation axis with $n \geq 3$ have axially symmetric coupling tensors with $\eta = 0$. This results in a characteristic orientation dependence of the NMR signals. No constraints on the signals are given for sites of lower symmetry. But the orientation of the various PAS may still be influenced by mirror planes forcing one of the principal axes to be perpendicular to the mirror plane, or by two-fold rotation axis forcing parallel alignment of one of the principal axes. If several interactions are simultaneously present for the nucleus under observation, all the various PAS have to obey the symmetry restrictions, thus also Euler angles constraints may exist. These are summarized in Table 26.1 of *NMR Crystallography*, pp. 396.[26] A comprehensive introduction to symmetry considerations can also be found in Ref. 27.

28.2.1.4 *Experimental Considerations*

The electrical conductivity of intermetallic compounds influences the NMR experiments and have to be taken into account for the experimental setup. The quality factor of the resonant circuit of a probe is reduced by the conductance of the sample. This can be avoided by dilution of the sample with an inert material such as SiO_2 or GeO_2. The gain of intensity by the increased quality factor often prevails over the loss of intensity by the smaller amount of the sample. The problems of eddy currents under MAS, such as heating of the samples or unstable rotation, can be suppressed in the same way.

The skin effect reduces the penetration depth of the RF field in metallic materials.[28] For that reason, measurements of powder samples instead of single crystals are preferable. To increase the available sample volume, powders of small particle size have to be used. The samples have to be carefully annealed to reduce the number of defects introduced by the grinding process.[21] In cases of special interest on single crystal experiments, these can be done by accepting increased experimental duration due to the reduced penetration depth.[29–37]

The spectral width of the NMR signals also has to be considered when planning the experiments, especially if quadrupole coupling is involved and the spectra frequently cover ranges of several megahertz. Above 200–300 kHz width of the signals, distortions from finite pulse length and the quality factor of the probe arise. Therefore special techniques such as frequency or field sweeps are recommended to detect undistorted signals. Since variable fields are rarely available in modern NMR laboratories, the frequency sweep technique is briefly explained. The entire signal is detected in this case by variation of the frequency and performing a series of experiments using long pulses of low power for selective excitation of narrow frequency ranges. The NMR signal is obtained by plotting the intensity of the single experiments versus their carrier frequency.[38]

28.2.2 Calculation of NMR Coupling Parameters

Ab initio calculations of NMR coupling parameters are often performed at the density functional theory (DFT) level, providing reliable results with reasonable computational efforts.[39,40] The aim of such calculations is to provide supporting information for interpretation of the NMR experiments. These can be used for the assignment of the signals due to atomic sites of equal multiplicity where no assignment based on signal intensity is possible. On the other hand, a validation of the calculations and the derived interpretations can be obtained. Furthermore, the calculations can be used to check the existing structure models.

28.2.2.1 Magnetic Coupling

The NMR signal shift due to magnetic shielding can be calculated with high precision using quantum mechanical programs for electrically nonconducting materials.[41–46] Magnetic shielding results from the orbital motion of the electrons and a mixing of ground and excited states, causing diamagnetic and paramagnetic contributions, respectively.[14,15,19]

The situation differs for materials possessing semiconducting or metallic conductivity. Here, the signal shift is influenced in addition to magnetic shielding by the Knight shift. The width of the bandgap and the temperature of the NMR experiment determine the contribution of the Knight shift in semiconducting materials.[47] No contribution of the Knight shift results for large bandgaps and/or low temperatures, enabling calculations of the magnetic shielding to model the system. Quantum mechanical calculations of the signal shift considering both Knight shift and magnetic shielding have not been established in standard codes. This results in ambiguities with the assignment of the NMR signals to certain atomic positions of equal multiplicity based on their signal shift.

28.2.2.2 Quadrupole Coupling

The situation for quantum mechanical calculations of quadrupole coupling differs considerably from that of the calculation of the signal shift for electrically conducting materials.[48,49] In order to calculate the quadrupole coupling parameter, the electric field gradient (EFG) has to be determined. It is defined as the second partial derivative with respect to the spatial coordinates of the electrostatic potential $\vec{v}(r)$ at nuclear site

$$V_{ij} = \left.\frac{\partial^2 \vec{v}(r)}{\partial x_i \partial x_j}\right|_{r=0} \tag{28.10}$$

The advantage of the calculations is that the entire EFG including its orientation is obtained for each site of the structure. The largest principal component of the EFG, V_{ZZ}, and the asymmetry parameter, η_Q, are generally reported in the literature.

The program WIEN2k[48] is often used for calculations of the EFG. Full-potential augmented plane wave plus local orbitals (APW + lo) are used in the code to describe the electronic structure. The use of local orbitals allows an interpretation of the results with respect to chemical bonding.[22,25,36,37,50–52]

The EFG can be split into its contributions. According to Ref. 51, the EFG is composed of the so-called valence and lattice contributions. The former is calculated for the atomic sphere with the local orbitals of the atoms under consideration and those of the neighboring atoms overlapping with this atomic sphere. The lattice contribution results from the joint fraction of the interstitial region described by plane waves and the other atoms without the part already considered in the valence contribution.

The information concerning chemical bonding can be obtained by decomposition of the valence part into its orbital contributions. Evaluation of two orbitals is required to calculate V_{ij}. Thus according to the selection rules, pp, dd, sd, pf, and ff combinations have to be considered.

The entire EFG corresponding to the results of the NMR experiment is given by the sum of its contributions

$$V_{ZZ}^{\text{tot}} = V_{ZZ}^{\text{val}} + V_{ZZ}^{\text{lat}} \quad (28.11)$$

$$V_{ZZ}^{\text{val}} = V_{ZZ}^{\text{val}}(\text{pp}) + V_{ZZ}^{\text{val}}(\text{dd}) + V_{ZZ}^{\text{val}}(\text{sd}) + \dots \quad (28.12)$$

The dominating parts of the EFG can be determined using this procedure. The main contribution results from the valence contribution in many cases. This indicates that the EFG is a very local source of information. An analysis of the orbital characters of the electrons dominating the EFG provides valuable information about chemical bonding.

28.3 APPLICATIONS

The constitution of the investigated material—regular powder, aligned powder, and single crystal—is used in the following to introduce the applications of NMR for the investigation of intermetallic compounds. Finally, the combined application of NMR spectroscopy and quantum mechanical calculations for the investigation of chemical bonding in intermetallic compounds is presented.

28.3.1 Regular Powder Samples

Experiments on regular powders are performed in most studies because the samples are available as powders only. Furthermore, the experiments on single crystals are technically more difficult and result in a loss of intensity (Section 28.2.1.4). Various types of experiments such as wide-line,[22,25,53–55] nutation,[56,57] MAS, Multi-Quantum MAS (MQMAS),[58–60] and Rotational Echo DOuble Resonance (REDOR)[57] experiments have been successfully applied to intermetallic compounds. Crystallographic aspects related to local ordering of the atoms e.g., disorder, are usually the focus of these types of NMR investigations. The lack of short-range information or the missing contrast of the scattering length of diffraction methods is often compensated by NMR spectroscopy. But also studies involving dynamical disorder have been performed.[33,61–63]

Experimentally, the simplest way to measure NMR signals are wide-line experiments without MAS. At the same time, this is the only way to determine the signals influenced by large quadrupole coupling in moderate magnetic fields when MAS experiments fail because of limited rotation frequencies. Unfortunately, low resolution and low intensity hamper these wide-line experiments. Thus, MAS measurements are in general preferred because of the gain in signal-to-noise ratio and resolution. Furthermore, they still permit line shape analysis for determination of the coupling constants, providing valuable information about the bonding situation of the atoms under consideration.

The experimental resolution can be further improved by application of MQMAS; thus a differentiation of overlapping MAS signals becomes possible (see Chapter 10). Furthermore, local ordering of the atoms is indicated by the formation of shift distributions or distributions of quadrupole coupling constants.

The compounds Sc_3TSi_3 with T = Ru, Rh, Ir, Pt are isotypic.[58] This favorable situation allows a systematic investigation of the influence of the transition metal T on the bonding situation of the Sc atoms by comparison of the different compounds. Hence valuable information about the chemistry of the compounds can be obtained. The EFG was studied by ^{45}Sc NMR and quantum mechanical calculations for this purpose.

Three Sc sites of equal multiplicity are located in the crystal structure. For that reason, three ^{45}Sc NMR signals are expected. MQMAS measurements were employed because MAS experiments were not sufficient to resolve the three signals (Figure 28.3). An analysis of the signal contributions of the MQMAS measurements results in three significantly different Sc environments indicated by the variation of the EFG. An assignment of the signals to the respective sites of the crystal structure can only be performed by quantum mechanical calculations

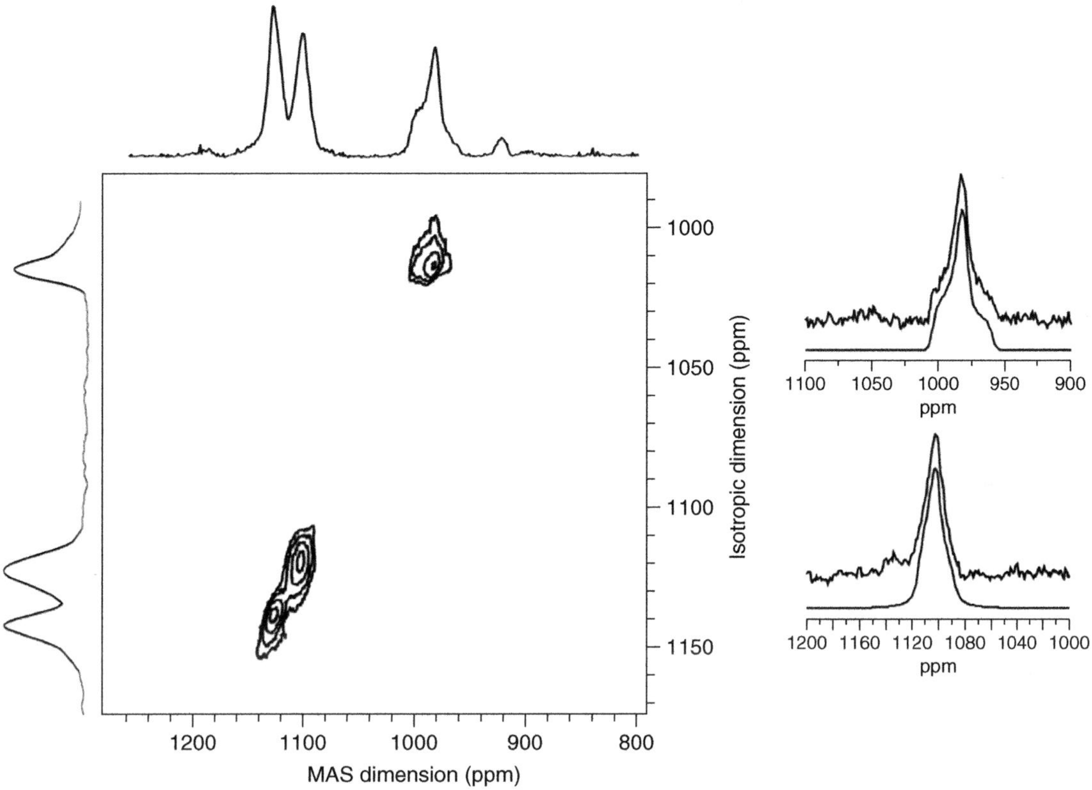

Figure 28.3. ^{45}Sc MQMAS NMR spectrum of Sc_3RuSi_3 and projections of the separate sites on the F2 dimension together with line shape simulations. (Reproduced from Ref. 58. © Wiley-VCH Verlag GmbH & Co, 2010.)

of the EFG because the samples possess metallic conductivity.[58]

Interestingly, the three sites are rather different with respect to the bonding situation of the Sc atoms but the influence of the substitution of the transition metal T is rather small.

$Ba_{8.0}B_{0.1679}Al_{14.3}Si_{31.4}$ realizes a clathrate-I type of structure[13] with cations embedded into voids of the framework formed by the group 13 elements. Three crystallographic positions belong to the framework. The ^{27}Al NMR signals are a proof of intrinsic disorder being expected by a first inspection of the chemical formula. Two overlapping signals and a shoulder were observed by wide-line experiments of the central transition signal using different magnetic fields (Figure 28.4a, top). The width of the signals does not significantly change with the magnetic field. This indicates that quadrupole coupling does not dominate the line shape of the signals. MAS measurements result only in a small reduction of the line width and provide a few more details of the signal. This indicates the influence of the local ordering of the atoms, which is in agreement with the nonobservable satellite transitions. MQMAS experiments result in a narrow signal aligned along the diagonal in the 2D plot after sheared Fourier transformation (Figure 28.4b). The full-width at half-maximum (FWHM) of the signal indicates only a small contribution of quadrupole coupling which is rather unexpected by inspection of the crystal structure. Three regions of the ^{27}Al MQMAS signals can be clearly identified by an analysis of the FWHM. These correspond to the number of nonequivalent Al sites in the crystal structure possessing slightly different quadrupole couplings. The slope of the diagonal of the 2D plot is in agreement with the distribution of the signal shifts. This renders the local ordering of the atoms and results from a distribution of chemically slightly different positions of the Al

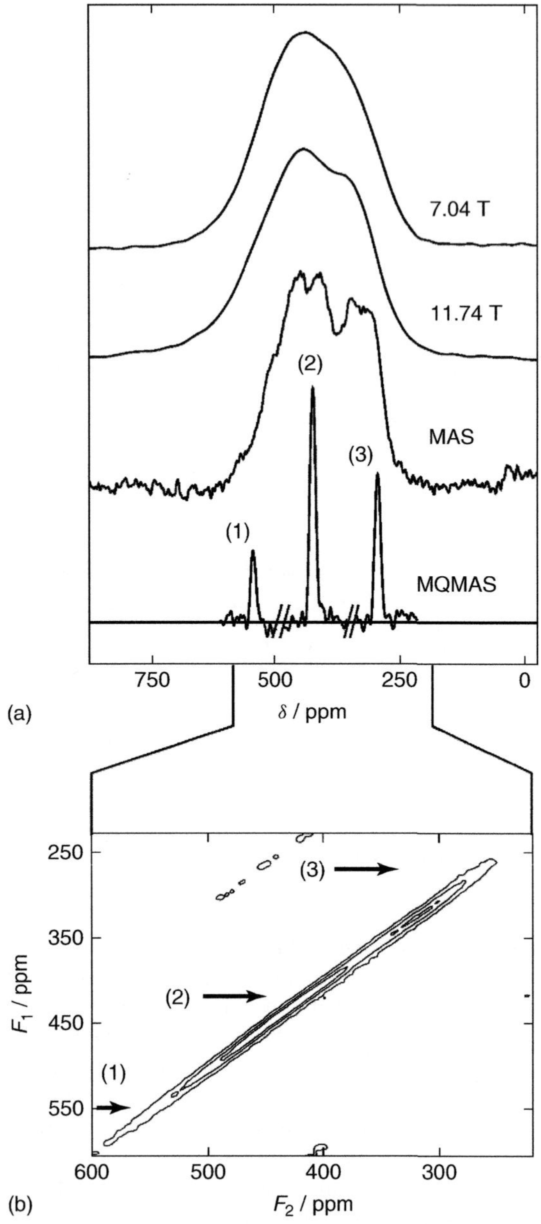

Figure 28.4. ^{27}Al NMR signals of $Ba_{8.0}B_{0.1679}Al_{14.3}Si_{31.4}$. (a) From the top: wide-line experiments measured magnetic fields $B_0 = 7.04\,T$ and $B_0 = 11.74\,T$; MAS spectrum for $\nu_{rot} = 35\,kHz$ rotation frequency and $B_0 = 11.74\,T$; slices extracted from the MQMAS experiment. (b) MQMAS spectrum using $\nu_{rot} = 35\,kHz$. The F1 frequencies corresponding to the extracted slices in (a) are marked by arrows. (Reproduced from Ref. 59. © American Chemical Society, 2008.)

atoms. Similar results were obtained for other compounds having the clathrate structure, demonstrating the flexibility of the framework atoms with respect to substitution.[64–67]

28.3.2 Aligned Powder

Alignment of nonmagnetic crystallites in the magnetic field was used to study local atomic arrangements in $CuAl_2$. The material is of technological interest since it is discussed with respect to the hardening of Al. The chemical formula $CuAl_2$ suggests a fully ordered stoichiometric composition of 1 : 2, which is in conflict with the phase diagram of Cu–Al. Considering the phase diagram, the compound can be understood as $Cu_{1-x}Al_2$. The idealized crystal structure of $CuAl_2$ contains only one crystallographic position for Cu and Al, respectively. Several conflicting mechanisms of local ordering of the atoms are reported in the literature[68,69] (model I: occupation of Cu sites by Al;[68] model II: Cu vacancies;[68] model III: occupation of Al sites by Cu[69]). Thus, NMR spectroscopy was applied to determine the relevant mechanism of the disorder.[21]

The NMR experiments were performed on powder samples obtained by careful grinding of a large single crystal[70] to ensure single-phase material for the NMR investigation. Two narrow ^{63}Cu NMR signals were observed for powder aligned in the magnetic field (Figure 28.5b). The alignment of the crystallites results from the anisotropic conductivity of the sample. The two NMR signals correspond to two Cu sites in the crystal structure Cu(A) and Cu(B) (Figure 28.5a). The orientation dependence of the signals proves the alignment of the crystallites according to $V_{ZZ}||B_{or}$. The c-axis has to be parallel to V_{ZZ} of Cu in the tetragonal crystal structure of $Cu_{1-x}Al_2$ for symmetry reasons. Furthermore, axial symmetry of both Cu sites is indicated by the orientation dependence of the signals. The quadrupole coupling constants $C_Q[Cu(B)] \sim 1/3C_Q[Cu(A)]$ were determined by rotation of the sample around the axis perpendicular to the axis of alignment (Figure 28.6). The large difference of the quadrupole coupling constants proves the significantly different environments of Cu(A) and Cu(B).

The higher resolution obtained by the alignment of the crystallites allows selective excitation of the signals corresponding to the different Cu sites in the crystal structure. Orientation-dependent Spin

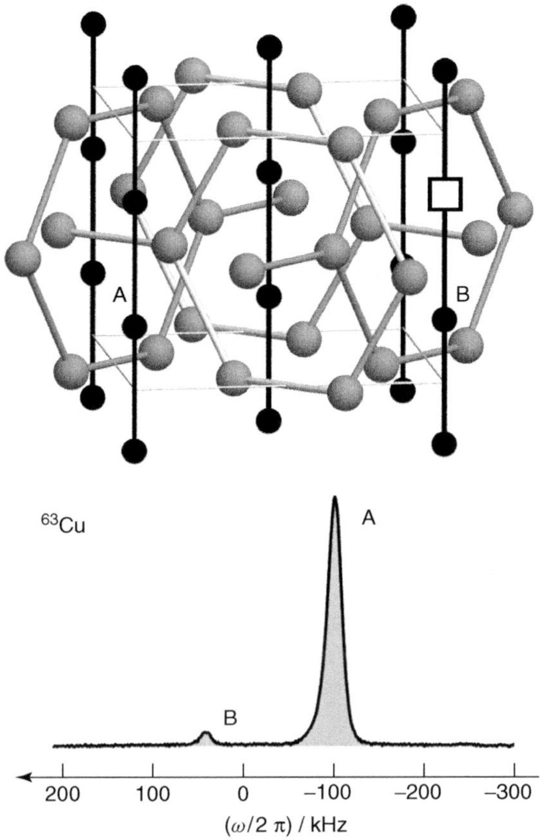

Figure 28.5. (a) Crystal structure of $Cu_{1-x}Al_2$. A Cu vacancy is marked by a square. Cu sites with different environments are marked by A and B. Covalently bonded interpenetrating Al nets are shown with gray Al–Al bonds. (b) ^{63}Cu NMR signal of $Cu_{1-x}Al_2$ aligned in a magnetic field. Signal contributions are marked by A and B corresponding to the sites marked in the crystal structure. (Modified from Ref. 21. © American Chemical Society, 2007.)

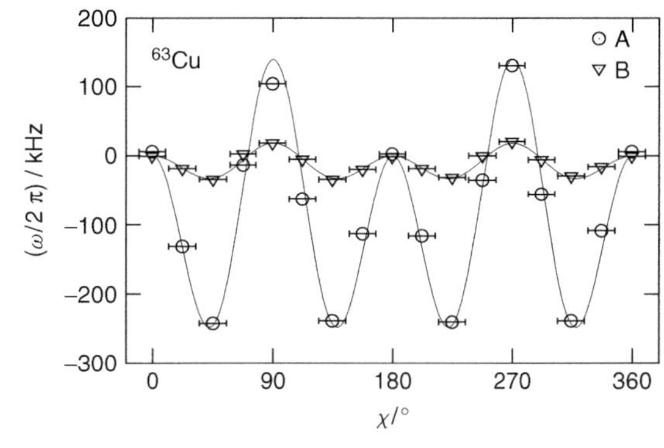

Figure 28.6. Frequency shift of the ^{63}Cu NMR signals of $Cu_{1-x}Al_2$ aligned in a magnetic field as a function of the angle χ between the magnetic fields used for the alignment of the crystallites and for the measurements. Circles represent the main signal A and triangles the minor signal B. The lines show the calculated frequency shifts for $C_Q = 15.4\,\text{MHz}$ and $C_Q = 5.6\,\text{MHz}$ for a magnetic field of $B_0 = 7.04\,\text{T}$. (Modified from Ref. 21. © American Chemical Society, 2007.)

Echo Double Resonance (SEDOR) experiments were performed for a constant interpulse delay of $\tau = 200\,\mu s$ to identify the mechanism of the local ordering of the atoms by investigating the ^{65}Cu↔^{63}Cu and ^{65}Cu↔^{27}Al internuclear dipole—dipole couplings (Figure 28.7). Calculations of the heteronuclear second moments M_2 for the different models of local ordering of the atoms finally reveal the formation of Cu vacancies (model II) as the relevant mechanism of the disorder in $Cu_{1-x}Al_2$ (Figure 28.5a). However, the agreement of the SEDOR experiments and model II can nicely be shown in a descriptive manner. The Cu atoms are aligned in a row along the *c*-axis, resulting in a minimum of the dipole–dipole coupling for an angle $\chi = 54.74°$, the magic angle. The ^{65}Cu↔^{63}Cu experiments show, as expected for this particular orientation of the Cu–Cu internuclear vectors, only a small dipole–dipole coupling for an orientation of $\chi = 45°$. The signal decay S/S_0 with S_0 corresponding to the full intensity and S corresponding to the reduced intensity of Cu(A) is larger than that of Cu(B) for $\chi = 22°$. The Cu(A) atoms with two neighboring Cu atoms are subject to a larger dipole–dipole coupling than the Cu(B) atoms with only one neighboring Cu atom. The ^{65}Cu↔^{27}Al experiments demonstrate the equivalence of Cu(A) and Cu(B) with respect to their Al environment.

28.3.3 Single Crystal Samples

The highest degree of information such as coupling constants and orientation of coupling tensors can be obtained by investigation of single crystals although a loss of intensity has to be tolerated for intermetallic compounds possessing metallic conductivity (Section 28.2.1.4). NMR line widths in single crystal NMR experiments are generally much smaller than those observed for regular powder samples because one

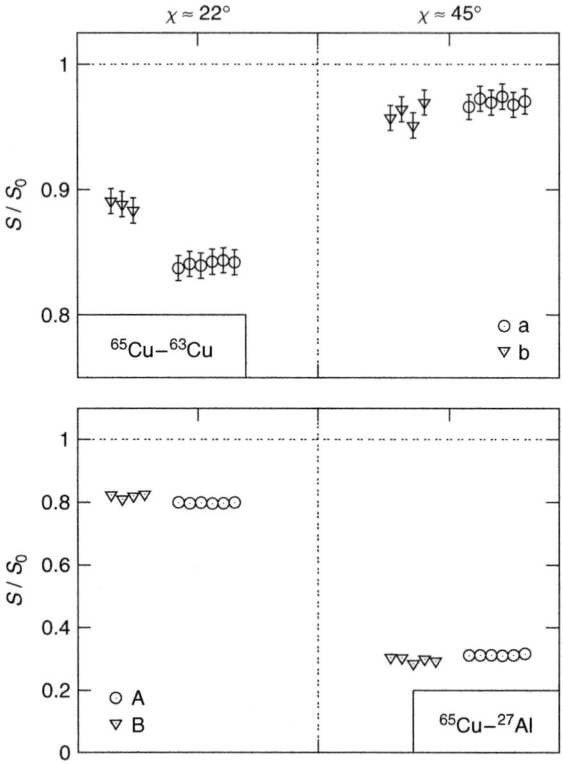

Figure 28.7. Ratio S/S_0 of the reduced signal intensities S with respect to the full signal intensity S_0 of the ^{65}Cu central transition measured by Spin Echo DOuble Resonance (SEDOR). A constant interpulse delay of $\tau = 200$ μs was used for two orientations of $Cu_{1-x}Al_2$ aligned in a magnetic field (left: $\chi \approx 22°$ and right: $\chi \approx 45°$). Top: ^{65}Cu-^{63}Cu and bottom: ^{65}Cu-^{27}Al heteronuclear experiments. Measurements of signal *A* are represented by circles and signal *B* by triangles. (Modified from Ref. 21. © American Chemical Society, 2007.)

is observing the spectrum of the crystal at only one particular orientation in the applied magnetic field as opposed to a powder which contains crystallites at all orientations in B_0. Therefore, these experiments are suited to study crystal structures with many atoms in the unit cell, such as complex metallic alloys[30,31] or quasicrystalline material.[32–34]

Several NMR investigations have been performed on regular powders[52,71,72], and single crystals[35–37] of transition-metal borides. These studies involve quantum mechanical calculations of the EFG for the interpretation of the results and an analysis of the bonding situation of the B atoms.

The crystal structure of YB_{12} has two sites: one for Y and one for B. It is composed of a face-centered cubic arrangement of the Y atoms, and the B atoms forming B_{12} cuboctahedra which are located on all edges of the unit cell. It has to be stressed that the B–B distances within the B_{12} cuboctahedra are longer then the shortest distances between B atoms of the neighboring B_{12} units. This indicates stronger bonding interactions between the cuboctahedra than within the cuboctahedra.

Boron-11 orientation-dependent NMR measurements on single crystals were performed to determine the EFG and its orientation with respect to the crystal axis system (CAS).[35] Furthermore, the EFG including orientation was investigated by theoretical methods.[52]

The orientation-dependent NMR experiments feature an interesting detail. The central transition signal is centered by up to three pairs of satellites depending on the orientation of the crystal with respect to the magnetic field (Figure 28.8). This is due to the fact that crystallographically equivalent sites need not be magnetically equivalent. Thus different orientations of the EFG tensors result in resolvable NMR signals. The orientations of the tensors and the absolute values of the principal axes were determined by these experiments. The principal axes of the coupling tensor are aligned along [1, 0, 0], [0, 1/2, 1/2], and [0, −1/2, 1/2]. This is in agreement with symmetry considerations for the point symmetry of the B atoms (*mm*2) (Section 28.2.1.3).

The tensor elements, including the sign, were determined by quantum mechanical calculations. They reveal that the principal axis of the most negative value is aligned along the B–B contact of neighboring B_{12} cuboctahedra. This is a proof of bonding interactions between the B_{12} units, which are indicated by inspection of the interatomic distances in the crystal structure. Furthermore, the model of the crystal structure is confirmed by the agreement between experimental and theoretical results.

28.3.4 Chemical Bonding

A systematic investigation of the digallides MGa_2 with M = Ca, Sr, Ba and the tetragallides MGa_4 with M = Na, Ca, Sr, Ba was performed to determine reliable NMR parameters for the investigation of intermetallic compounds.[22,25] It turned out in these investigations that the EFG is the desired

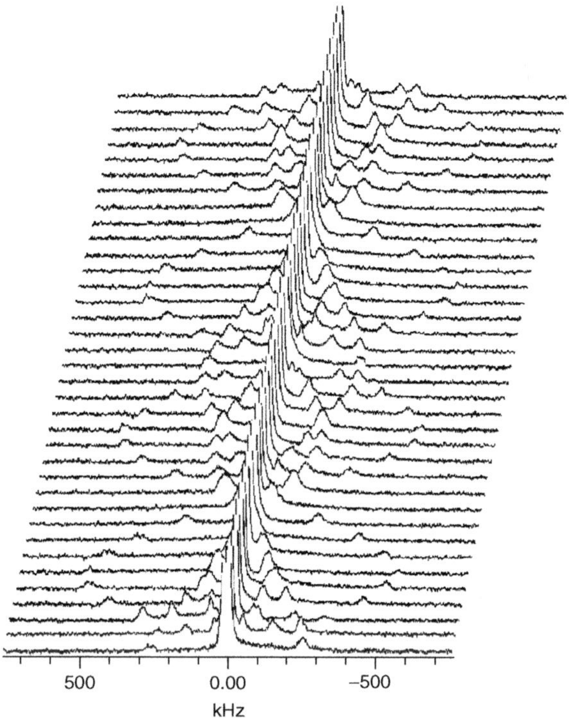

Figure 28.8. ^{11}B NMR spectra obtained for different rotation angles for a single crystal of YB_{12}. (Reproduced from Ref. 35. © American Physical Society, 2007.)

quantity because it is accessible by line shape analysis of the NMR signals and quantum mechanical calculations with high precision for compounds possessing metallic conductivity. Careful evaluation of the results of these quantum mechanical calculations supported by the experimental data provides valuable informations about the chemical bonding. The complete isotropic and anisotropic shielding has been determined in these studies. Both quantities are less informative compared to the EFG, which seems to be mainly related to the lack of appropriate computational methods for theoretical analysis of the experimental results.

The crystal structures of the model compounds are rather simple with one or two crystallographic positions of the Ga atoms. $CaGa_2$ crystallizes in the $CaIn_2$ type with puckered Ga layers resulting in 3+1 Ga–Ga contacts. For MGa_2 with M = Sr, Ba, a modified AlB_2 type with slightly puckered Ga layers and three Ga–Ga contacts are observed.[25] Two crystallographically independent Ga sites are realized in MGa_4 with M = Na, Ca, Sr, Ba. The $CaGa_4$ type is observed for MGa_4 with M = Ca and the $BaAl_4$ type for M = Na, Sr, Ba. The two crystallographically independent Ga atoms are coordinated by four and five Ga atoms in the tetragallides. Bonding interactions are indicated by interatomic distances being below the average distance in α-Ga. The Ga atoms of MGa_2 and MGa_4 are therefore referred as Ga(3b), Ga(3b + 1b), Ga(4b), and Ga(5b).

The main components of the EFG V_{ZZ} determined by line shape analysis of the ^{69,71}Ga NMR and quantum mechanical calculations are compared in Figure 28.9. Good agreement of theoretical and experimental values is obtained, allowing an unambiguous assignment of the NMR signals to the two crystallographically independent Ga sites in MGa_4.

Positive and negative signs of V_{ZZ} are obtained by the quantum mechanical calculations of the EFG for di- and tetragallides, respectively. The decomposition of the EFG proves the dominant part of the valence contribution (Figure 28.10). The lattice contributions are negligible for all investigated compounds. This shows the sensitivity of quadrupole coupling on the

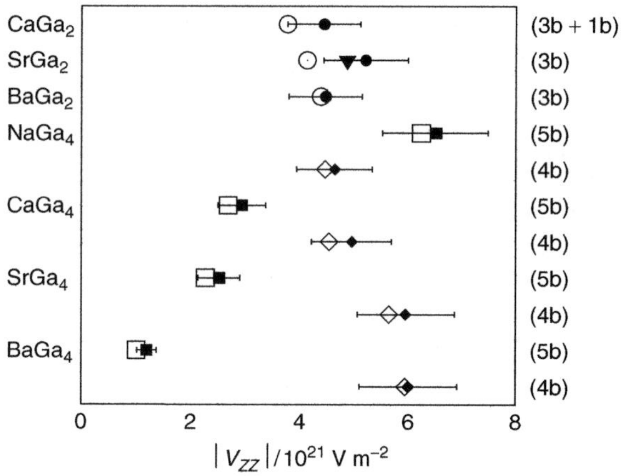

Figure 28.9. Absolute values of V_{ZZ} obtained by quantum mechanical calculations (open symbols) and NMR experiments (error bars) for the Ga atoms of MGa_2 with M = Ca, Sr, Ba, and MGa_4 with M = Na, Ca, Sr, Ba. Circles, diamonds, and squares correspond to Ga(3b), Ga(4b), and Ga(5b), respectively. The full triangle corresponds to $SrGa_2$ considering puckered Ga layers.[25] (Reproduced from Ref. 22. © Wiley-VCH Verlag GmbH & Co, 2011.)

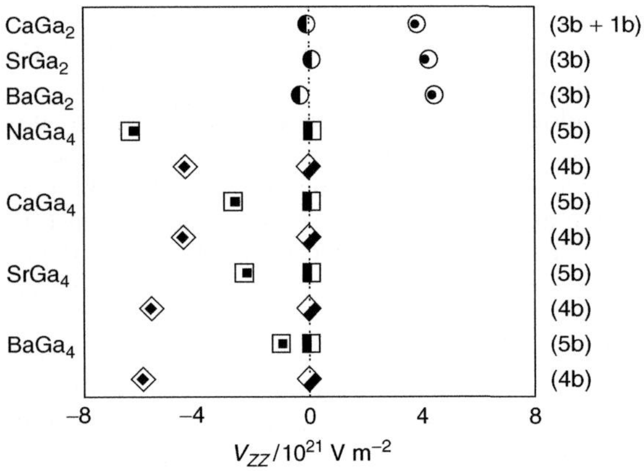

Figure 28.10. Contributions to V_{ZZ} of the Ga species in $M\text{Ga}_2$ with M = Ca, Sr, Ba, and MGa_4 with τM = Na, Ca, Sr, Ba obtained by quantum mechanical calculations. Circles, diamonds, and squares correspond to Ga(3b), Ga(4b), and Ga(5b), respectively. Total, lattice, and valence contributions are indicated by filled, half-filled, and empty symbols, respectively. (Reproduced from Ref. 22. © Wiley-VCH Verlag GmbH & Co, 2011.)

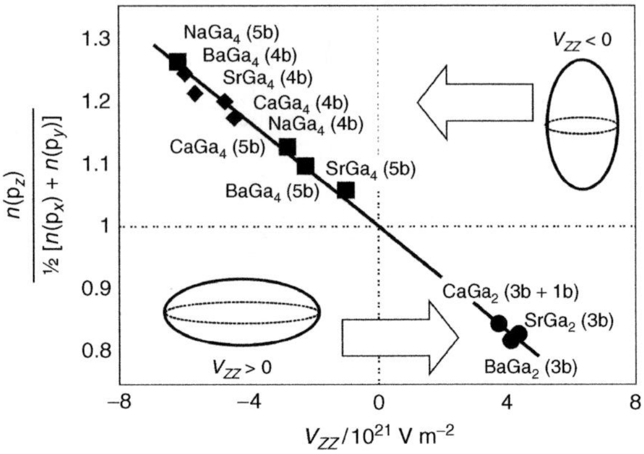

Figure 28.11. Anisotropy ratio $n(\text{p}_z)/(1/2\left[n(\text{p}_x)+n(\text{p}_y)\right])$ vs V_{ZZ} for $M\text{Ga}_2$ with M = Ca, Sr, Ba, and $M\text{Ga}_4$ with M = Na, Ca, Sr, Ba. Circles, diamonds, and squares correspond to Ga(3b), Ga(4b), and Ga(5b), respectively. The straight line serves as a guide to the eye. The prolate-like and oblate-like charge distributions of the Ga atoms are sketched on left and right, respectively. (Reproduced from Ref. 22. © Wiley-VCH Verlag GmbH & Co, 2011.)

local bonding situation of the atoms in intermetallic compounds. An analysis of the orbital contributions to the valence part of V_{ZZ} accounts for the dominant participation of the p-like electrons on the formation of the EFG.

The origin of the anisotropy of the charge distribution causing the EFG can be shown by an analysis of the population difference of the various p-states by an integration of the electronic density of states (DOS) of the p-like electrons. A linear correlation of V_{ZZ} and the anisotropy ratio $n(\text{p}_z)/(1/2\left[n(\text{p}_x)+n(\text{p}_y)\right])$ with $n(\text{p}_i)$ corresponding to the population of states by p_i-like electrons is observed (Figure 28.11). An oblate-like charge distribution of the Ga atoms is determined for the digallides $M\text{Ga}_2$ with M = Ca, Sr, Ba possessing a positive sign of V_{ZZ}. The charge distribution of Ga is, in contrast, prolate-like for the tetragallides $M\text{Ga}_4$ with M = Na, Ca, Sr, Ba having a negative sign of V_{ZZ}. Furthermore, a larger variation of the charge distributions is observed in $M\text{Ga}_4$ than in $M\text{Ga}_2$.

In summary, the variation of the shape of the charge distribution of the Ga atoms was determined by the combined application of NMR spectroscopy and quantum mechanical calculations of the NMR coupling parameter.

28.3.5 Conclusion and Outlook

Basically similar NMR experiments as in insulators can be applied to intermetallic compounds. Difficulties arising from the metallic conductivity of the sample (e.g., sample rotation, resonant circuit tuning) can be overcome by dilution of the samples with nonconducting materials.

Quantum mechanical calculations of EFG tensors are of fundamental importance for interpreting experimental solid-state NMR data from quadrupolar nuclei in intermetallic compounds. They assist by assignment of the NMR signals and provide details about the chemical bonding by analysis of the contributions of the EFG. Especially, the last point is of great interest since there is currently a lack of understanding of the chemical bonding in intermetallic compounds.

Furthermore, doors for structural research of disordered materials are opened by the quantum mechanical calculations. This is schematically sketched in Figure 28.12. Basic structural information can be obtained by diffraction methods. These are limited by the missing short-range information, which can be compensated by a local probe such as NMR spectroscopy. The crystallographic data provide the initial structure model. Ideas concerning deviations

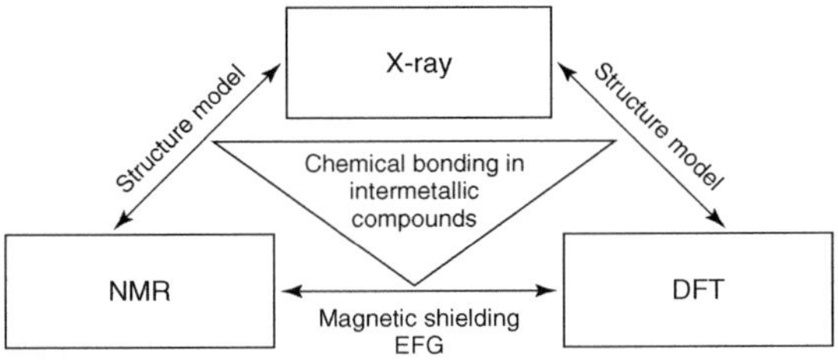

Figure 28.12. Triangle of methods and their interplay.

of the real structure from the initial model can be obtained by careful analysis of the diffraction data. Thus, variations of the structure model can by derived by super lattice structures. On the basis of this, the expected NMR signals can be compared with the experimental spectrum. Furthermore, an improved model of the structure can be used as input for quantum mechanical calculations at the DFT level. The agreement of the experimental and theoretical results can be used to support the improved structure model. Finally, the results of the calculations can be used to analyze the contributions to the EFG with respect to the chemical bonding of intermetallic compounds.

Currently, the EFG for electrically conducting and nonconducting compounds and magnetic shielding for nonconducting compounds and semiconductors possessing a large band gap can be used as link between theory and experiment. Theoretical effort concerning the calculation of signal shifts for metallic compounds will hopefully complete the toolbox of quantum mechanical methods for the investigation of intermetallic compounds soon.

RELATED ARTICLES IN THE ENCYCLOPEDIA OF MAGNETIC RESONANCE

Chemical Shift Computations for Crystalline Molecular Systems: Practice

Computations of Magnetic Resonance Parameters for Crystalline Systems: Principles

Knight Shift

Magnetic Shielding and Chemical Shifts: Basics

Quadrupolar Nuclei in Glasses

Quadrupolar Nuclei in Liquid Samples

REFERENCES

1. J. H. Westbrook and R. L. Fleischer, *Intermetallic Compounds: Principles and Applications*, Wiley-VCH Verlag GmbH & Co. KGaA: 1995. John Wiley & Sons, Ltd: The Atrium/Southern Gate, Terminus Road, Chichester, West Sussex PO19 8SQ.
2. A. Simon, *Angew. Chem. Int. Ed.*, 1997, **36**, 1788–1806.
3. Z.-A. Ren and Z.-X. Zhao, *Adv. Mater.*, 2009, **21**, 4584–4592.
4. R. Pöttgen and D. Johrendt, *Z. Naturforsch.*, 2008, **63b**, 1135–1148.
5. C. Felser, G. Fecher, and B. Balke, *Angew. Chem. Int. Ed.*, 2007, **46**, 668–699.
6. J. Sootsman, D. Chung, and M. Kanatzidis, *Angew. Chem. Int. Ed.*, 2009, **48**, 8616–8639.
7. G. S. Nolas, *Thermoelectrics: Basic Principles and New Materials Developments*, Springer: Berlin, 2001.
8. G. Sauthoff, *Intermetallics*, Wiley-VCH Verlag GmbH & Co. KGaA: New York, 1995.
9. J. H. Westbrook and R. L. Fleischer, *Intermetallic Compounds, Volume 4 — Magnetic, Electrical and Optical Properties and Applications of Intermetallic Compounds*, Wiley-VCH Verlag GmbH & Co. KGaA: New York, 2000.
10. Yu. Grin, U. Schwarz, and W. Steurer, *Alloy Physics: A Comprehensive Reference–Crystal Structure and Chemical Bonding*, Wiley–VCH Verlag GmbH & Co. KGaA: Weinheim, 2007, pp. 62–68.
11. R. Nesper, *Angew. Chem.*, 1991, **103**, 805–834.
12. H. Eckert, *Bunsen-Magazin*, 2008, **10**, 159–179.
13. H. Eckert and R. Pöttgen, *Z. Anorg. Allg. Chem.*, 2010, **636**, 2232–2243.
14. C. P. Slichter, *Principles of Magnetic Resonance*, Springer-Verlag: Berlin, Heidelberg, New York, Third Enlarged and Updated Edition, 1990.
15. A. Abragam, *Principles of Nuclear Magnetism*, Oxford University Press: Oxford, New York, 1961.
16. M. d'Avezac, N. Marzari, and F. Mauri, *Phys. Rev. B*, 2007, **76**, 165122.
17. M. Mehring, *Principles of High Resolution NMR in Solids*, Springer-Verlag: Berlin, Heidelberg, New York, 1983.
18. M. Bak and N. C. Nielsen, *J. Magn. Reson.*, 1997, **125**, 132–139.

19. M. J. Duer, *Introduction to Solid-State NMR Spectroscopy*, Blackwell Publishing: London, 2004.

20. D. Freude and J. Haase, in *NMR, Basic Principles and Progress: Special Applications*, eds P. Diehl, E. Fluck, and R. Kosfeld, Springer-Verlag: Berlin, Heidelberg, New York, 1993, pp. 3–90, Chapter Quadrupole Effects in Solid-State Nuclear Magnetic Resonance.

21. F. Haarmann, M. Armbrüster, and Yu. Grin, *Chem. Mater.*, 2007, **19**, 1147–1153.

22. F. Haarmann, K. Koch, P. Jeglič, O. Pecher, H. Rosner, and Yu. Grin, *Chem. Eur. J.*, 2011, **17**, 7560–7568.

23. O. Pecher and F. Haarmann, *Z. Anorg. Allg. Chem.*, 2010, **636**, 2089.

24. E. Kerscher, M. T. Kelemen, K. H. Diefenbach, O. Stockert, H. Löhneysen, and E. Dormann, *Eur. Phys. J. B*, 2001, **19**, 241–247.

25. F. Haarmann, K. Koch, D. Grüner, W. Schnelle, O. Pecher, R. Cardoso-Gil, H. Borrmann, H. Rosner, and Yu. Grin, *Chem. Eur. J.*, 2009, **15**, 1673–1684.

26. K. Jeffrey and G. Penner, in *NMR Crystallography*, eds R. K. Harris, R. E. Wasylishen, and M. J. Duer, John Wiley & Sons, Ltd: Chichester, West Sussex, 2009.

27. S. Ashbrook and S. Wimperis, in *NMR Crystallography*, eds R. K. Harris, R. E. Wasylishen, and M. J. Duer, John Wiley & Sons, Ltd: Chichester, West Sussex, 2009.

28. S. Reich and A. J. Vega, *J. Appl. Phys.*, 1985, **58**, 2759–2760.

29. P. Jeglič, M. Komelj, M. Klanjsek, U. Tkalec, S. Vrtnik, M. Feuerbacher, and J. Dolinšek, *Phys. Rev. B*, 2007, **75**, 014202.

30. P. Jeglič, M. Heggen, M. Feuerbacher, B. Bauer, P. Gille, and F. Haarmann, *J. Alloys Compd.*, 2009, **480**, 141–143.

31. P. Jeglič, S. Vrtnik, M. Bobnar, M. Klanjšek, B. Bauer, P. Gille, Yu. Grin, F. Haarmann, and J. Dolinšek, *Phys. Rev. B*, 2010, **82**, 104201.

32. P. Jeglič and J. Dolinšek, *Phys. Rev. B*, 2005, **71**, 014204.

33. J. Dolinšek, T. Apih, P. Jeglič, M. Feuerbacher, M. Calvo-Dahlborg, U. Dahlborg, and J. M. Dubois, *Phys. Rev. B*, 2002, **65**, 212203.

34. P. Jeglič and M. K. J. Dolinšek, *Philos. Mag.*, 2007, **87**, 2687–2692.

35. Z. Fojud, P. Herzig, O. J. Żogał, A. Pietraszko, A. Dukhnenko, S. Jurga, and N. Shitsevalova, *Phys. Rev. B*, 2007, **75**, 184102.

36. O. J. Żogał, Z. Fojud, P. Herzig, A. Pietraszko, A. B. Lyashchenko, S. Jurga, and V. N. Paderno, *J. Appl. Phys.*, 2009, **106**, 033514.

37. P. Herzig, Z. Fojud, O. J. Żogał, A. Pietraszko, A. Dukhnenko, S. Jurga, and N. Shitsevalova, *J. Appl. Phys.*, 2008, **103**, 083534.

38. M. E. Smith and E. R. H. van Eck, *Prog. Nucl. Magn. Reson. Spectrosc.*, 1999, **34**, 159–201.

39. T. Charpentier, *Solid State Nucl. Magn. Reson.*, 2011, **40**, 1–20.

40. J. Cuny, S. Messaoudi, V. Alonzo, E. Furet, J.-F. Halet, E. Le Fur, S. E. Ashbrook, C. J. Pickard, R. Gautier, and L. Le Polles, *J. Comput. Chem.*, 2008, **29**, 2279–2287.

41. D. Sebastiani and M. Parrinello, *J. Phys. Chem.*, 2001, **105**, 1951–1958.

42. D. Sebastiani, *Mod. Phys. Lett. B*, 2003, **17**, 1301–1319.

43. C. J. Pickard and F. Mauri, *Phys. Rev. B*, 2001, **63**, 245101.

44. J. R. Yates, C. Pickard, and M. Payne, *J. Chem. Phys.*, 2003, **118**, 5746–5753.

45. J. R. Yates, C. J. Pickard, and F. Mauri, *Phys. Rev. B*, 2007, **76**, 02441.

46. M. Profeta, F. Mauri, and C. J. Pickard, *J. Am. Chem. Soc.*, 2003, **125**, 541–548.

47. H. Selbach, O. Kanert, and D. Wolf, *Phys. Rev. B*, 1979, **19**, 4435–4443.

48. P. Blaha, K. Schwarz, G. Madsen, D. Kvasnicka, and J. Luitz, WIEN2k: An Augmented Plane Wave Local Orbitals Program for Calculating Crystal Properties, Technische Universität Wien, 2001.

49. K. Koepernik and H. Eschrig, *Phys. Rev. B*, 1999, **59**, 1743–1757.

50. P. Blaha, K. Schwarz, and P. Herzig, *Phys. Rev. Lett.*, 1985, **54**, 1192–1195.

51. P. Blaha, K. Schwarz, and P. H. Dederichs, *Phys. Rev. B*, 1988, **37**, 2792–2796.

52. B. Jäger, S. Paluch, O. J. Żogał, W. Wolf, P. Herzig, V. B. Filippov, N. Shitsevalova, and Y. Paderno, *J. Phys.: Condens. Matter*, 2006, **18**, 2525–2535.

53. T. J. Bastow and S. Celotto, *Solid Stat Nucl. Magn. Reson.*, 2009, **35**, 217–222.

54. U. Burkhardt, V. Gurin, F. Haarmann, H. Borrmann, W. Schnelle, A. Yaresko, and Yu. Grin, *J. Solid State Chem.*, 2004, **177**, 389–394.

55. G. Glatz, T. Schmalz, T. Kraus, F. Haarmann, G. Motz, and R. Kempe, *Chem. Eur. J*, 2010, **16**, 4231–4238.

56. T. Harmening, H. Eckert, and R. Pöttgen, *Solid State Sci.*, 2009, **11**, 900–906.

57. T. Harmening, L. van Wüllen, H. Eckert, U. C. Rodewald, and R. Pöttgen, *Z. Anorg. Allg. Chem.*, 2010, **636**, 972–976.

58. T. Harmening, D. Mohr, H. Eckert, A. A. Alam, S. F. Matar, and R. Pöttgen, *Z. Anorg. Allg. Chem.*, 2010, **639**, 1839–1850.

59. C. L. Condron, S. M. Kauzlarich, T. Ikeda, G. J. Snyder, F. Haarmann, and P. Jeglič, *Inorg. Chem.*, 2008, **47**, 8204–8212.

60. C. P. Sebastian, L. Zhang, C. Fehse, R.-D. Hoffmann, H. Eckert, and R. Pöttgen, *Inorg. Chem.*, 2007, **46**, 771–779.

61. T. Dinges, U. C. Rodewald, S. F. Matar, H. Eckert, and R. Pöttgen, *Z. Anorg. Allg. Chem.*, 2009, **635**, 1894–1903.

62. Z. Wu, R.-D. Hoffmann, D. Johrendt, B. D. Mosel, H. Eckert, and R. Pöttgen, *J. Mater. Chem.*, 2003, **13**, 2561–2565.

63. Z. Wu, B. D. Mosel, H. Eckert, R.-D. Hoffmann, and R. Pöttgen, *Chem. Eur. J.*, 2004, **10**, 1558–1564.

64. K. A. Kovnir, M. M. Shatruk, L. N. Reshetova, I. A. Presniakov, E. V. Dikarev, M. Baitinger, F. Haarmann, W. Schnelle, M. Baenitz, Yu. Grin, and A. V. Shevelkov, *Solid State Sci.*, 2005, **7**, 957–968.

65. W. Jung, J. Lörincz, R. Ramlau, H. Borrmann, Y. Prots, F. Haarmann, W. Schnelle, U. Burkhardt, M. Baitinger, and Yu. Grin, *Angew. Chem. Int. Ed.*, 2007, **46**, 6725–6728.

66. J. V. Zaikina, K. A. Kovnir, F. Haarmann, W. Schnelle, U. Burkhardt, H. Borrmann, U. Schwarz, Yu. Grin, and A. V. Shevelkov, *Chem. Eur. J.*, 2008, **14**, 5414–5422.

67. J. V. Zaikina, K. A. Kovnir, U. Burkhardt, W. Schnelle, F. Haarmann, U. Schwarz, Yu. Grin, and A. V. Shevelkov, *Inorg. Chem.*, 2009, **48**, 3720–3730.

68. H. Zogg, *J. Appl. Cryst.*, 1979, **12**, 91–94.

69. T. J. Bastow and S. Celotto, *Acta Mater.*, 2003, **51**, 4621–4630.

70. Yu. Grin, F. R. Wagner, M. Armbrüster, M. Kohout, A. Leithe-Jasper, U. Schwarz, U. Wedig, and H. G. von Schnering, *J. Solid State Chem.*, 2006, **179**, 1707–1719.

71. B. Jäger, S. Paluch, W. Wolf, P. Herzig, O. J. Żogał, N. Shitsevalova, and Y. Paderno, *J. Alloys Compd.*, 2004, **383**, 232–238.

72. B. J. Suh, X. Zong, Y. Singh, A. Niazi, and D. C. Johnston, *Phys. Rev. B*, 2007, **76**, 144511.

Index

Abbreviations and Acronyms

1Q	Single-Quantum
2D	Two Dimensional
2QF-COSY	Double-Quantum-Filtered Correlation Spectroscopy
AAG	Ala-Ala-Gly
ABMS	Anisotropy of the Bulk Magnetic Susceptibility
ADF	Amsterdam Density Functional
ADRF	Adiabatic Demagnetization in the Rotating Frame
APW	Augmented Plane Wave Method
ARP	Adiabatic Rapid Passage
BCS	Bardeen–Cooper–Schrieffer
BLEW	Burum, Linder & Ernst (Windowless pulse sequence)
BLYP	Becke, Lee, Yang, Parr
BO	Bridging Oxygen
BOM	Bond Orbital Model
BPP	Bloembergen–Purcell–Pound
BR-24	Burum & Rhim (pulse sequence)
CAS	Crystal Axis System
CB	Conduction Band
CG	Conjugate
CODEX	Centerband-only Detection of Exchange Experiment
COSY	Correlation Spectroscopy
CP	Cross Polarization
CPMAS	Cross Polarization and Magic Angle Spinning
CPMG	Carr–Purcell–Meiboom–Gill
CRAMPS	Combined Rotation and Multiple-Pulse Spectroscopy
CS	Chemical Shift
CSA	Chemical Shift Anisotropy
CST	Chemical Shift Tensor
CT	Central Transition
CT	Contact Time
CTMAS	Central Transition Magic Angle Spinning
CW	Continuous Wave
CYCLOPS	Cyclically Ordered Phase Sequence
D	Dipolar
DAH	Dynamic Angle Hopping
DANTE	Delays Alternating with Nutations for Tailored Excitation
DAS	Dynamic Angle Spinning
DD	Dipole-Dipole
DEAR	Dipolar Exchange-Assisted Recoupling
DEISM	Direct Enhancement of Integer-Spin Magnetization
DEPT	Distortionless Enhancement by Polarization Transfer
DFS	Double Frequency Sweeps
DFT	Density Functional Theory
DMS	Dilute Magnetic Semiconductors
DNP	Dynamic Nuclear Polarization
DOR	Double Rotation
DOS	Density of States
DPPC	Dipalmitoylphosphatidylcholine
DQ	Double-Quantum
DQC	Double-Quantum Coherence
DQF	Double-Quantum Filter
DR-NQR	Double-Resonance Nuclear Quadrupole Resonance
DRSE	Dipolar-Rotational Spin Echoes
DSPC	Distearoyl-*sn*-glycero-3-phosphatidylcholine
EFG	Electric Field Gradient
ENDOR	Electron-Nucleus Double Resonance
EPR	Electron Paramagnetic Resonance
ESR	Electron Spin Resonance
EXAFS	Extended X-Ray Absorption Fine Structure
FAM	Fast Amplitude Modulation
FC	Fermi-Contact
FFLO	Fulde–Ferrell–Larkin–Ovchinnikov
FFT	Fast-Fourier Transform
FID	Free Induction Decay
FML	Fast Motion Limit
FOQI	First-Order Quadrupolar Interaction

FSLG	Frequency-Switched Lee Goldburg
FT	Fourier Transform
FWHM	Full-Width at Half-Maximum
GGA	Generalized Gradient Approximation
GIPAW	Gauge Including Projector Augmented Waves
HDOR	Heteronuclear Dipolar-Order Rotor-Encoding
HETCOR	Heteronuclear Correlation
HF	Hartree–Fock
HLW	High-Level Waste
HMQC	Heteronuclear Multiple-Quantum Coherence
HMQC	Heteronuclear Multiple Quantum Correlation
HMT	Hexamethylenetetramine
HOHAHA	Homonuclear Hartman-Hahn
HORROR	Homonuclear Rotary Resonance
HS	Hyperbolic Secant
HSQC	Heteronuclear Single-Quantum Coherence
IBMS	Isotropic Bulk Magnetic Susceptibility
INADEQUATE	Incredible Natural Abundance Double Quantum Transfer Experiment
INEPT	Insensitive Nuclei Enhanced by Polarization Transfer
INEPT-HSQC	Insensitive Nuclei Enhanced by Polarization Transfer-Heteronuclear Single-Quantum Correlation
KSAs	Knight Shifts and Associated Anisotropies
LG-CP	Lee–Goldberg CP
LMTO	Linear Muffin Tin Orbital
MAH	Magic Angle Hopping
MAS	Magic Angle Spinning
MAS-J-HMQC	Magic Angle Spinning-J-Heteronuclear Multiple Quantum Coherence
MAS-J-HSQC	Magic Angle Spinning-J-Single Quantum Coherence
MD	Molecular Dynamics
MIL	Materials of the Institute Lavoisier
MIT	Metal–Insulator Transition
MLEV-4	A broadband decoupling sequence
MPB	Morphotropic Phase Boundary
MQ	Multiple-Quantum
MQDOR	Multiple-Quantum Double Rotation
MQMAS	Multiple-Quantum Magic Angle Spinning
MQW	Multiple Quantum Well
MREV	Mansfield, Rhim, Elleman, & Vaughan (pulse sequence)
MRFM	Magnetic Resonance Force Microscopy
NA	Natural Abundance
NAR	Nuclear Acoustic Resonance
NBO	Non-Bridging Oxygen
NEXAFS	Near-Edge X-Ray a Fine Structure
NMR	Nuclear Magnetic Resonance
NOE	Nuclear Overhauser Effect
NOESY	NOE Spectroscopy
NQCC	Nuclear Quadrupole Coupling Constant
NQR	Nuclear Quadrupolar Resonance
OCT	Optimal Control Theory
ODESSA	One-Dimensional Exchange Spectroscopy by Sideband Alternation
ODNMR	Optically Detected NMR
OTf	Ovotransferrin
PAC	Perturbed Angular Correlation
PAS	Principal Axis System
PAW	Projector Augmented Wave
PBE	Perdew–Burke–Ernzerhof
PDSD	Proton-Driven Spin Diffusion
PFG	Pulsed Field Gradient
PL	Photoluminescence
PLM	Phospholemman Transmembrane
PMLG	Phase-Modulated Lee Goldburg

Continued on the back end papers

PSD	Proton Spin-Diffusion
PSII	Photosystem II
Q	Quadrupolar
QC	Quadrupole Coupling
QCPMG	Quadrupolar Carr–Purcell–Meiboom–Gill
QF	Quadrupole moment/Field gradient (interaction or relaxation mechanism)
QI	Quadrupolar Interaction
QIS	Quadrupole-Induced Shift
QW	Quantum Well
RAPT	Rotor Assisted Population Transfer
RDC	Residual Dipolar Coupling
RDS	Residual Dipolar Splitting
REAPDOR	Rotational Echo Adiabatic Passage Double Resonance
REDOR	Rotational Echo Double Resonance
RELM	Rotor-Encoding of Longitudinal Magnetization
REPT-HMQC	Recoupled Polarization-Transfer Heteronuclear Multiple-Quantum Coherence
REREDOR	Rotor-Encoded Rotational Echo
RF	Radiofrequency
RFDR	Radio Frequency Driven Recoupling
RHF	Restricted Hartree–Fock
RIACT	Rotation-Induced Adiabatic Coherence Transfer
RIDER	Relaxation-Induced Dipolar Exchange with Recoupling
R-INEPT	Refocused INEPT
RVB	Resonant Valence Bond
S/N	Signal-to-Noise
SA	Shielding Anisotropy
SAM	Smooth Amplitude Modulation
SATRAS	Satellite Transition Spectroscopy
SBU	Structural Building Unit
SBV	Strongly Bound Vanadium
SD	Spin-Dipolar
SEDOR	Spin Echo Double Resonance
SEFT	Spin Echo Fourier Transform
SFAM	Simultaneous Frequency and Amplitude Modulations
SG	Space Group
SLF	Separate-Local-Field
SOFC	Solid Oxide Fuel Cells
SOQI	Second-Order Quadrupolar Interaction
SPAM	Soft Pulse Added Mixing
SPI	Selective Population Inversion
SPIDER	Saturation-Pulse Induced Dipolar Exchange with Recoupling
SPINAL	Small Phase Increment Alternation
SPT	Selective Population Transfer
SQ	Single-Quantum
SR	Spin-Rotation (interactive or relaxation mechanism)
S-RESPDOR	Symmetry-Based Resonance-Echo Saturation-Pulse Double-Resonance
ssb	Spinning Sidebands
SSNMR	Solid-State Nuclear Magnetic Resonance
SSTMAS	Selected Transitions in MAS Spectra
ST	Satellite Transition
STARTMAS	Satellite-Transition Acquired In Real Time Magic Angle Spinning
STEAMER	Slow-Turning Echo Amplitude Modulation and Echo Reduction
STMAS	Satellite-Transition Magic Angle Spinning
STO	Slater-Type Orbital
TEDOR	Transferred Echo Double-Resonance
THF	Tetrahydrofuran
TMS	Tetramethylsilane
TOSS	Total Suppression of Sidebands
TPA	Tetrapropylammonium
TPPI	Time Proportional Phase Incrementation
TPPM	Two-Pulse Phase Modulation
TQ	Triple-Quantum
TRAPDOR	Transfer of Populations in Double Resonance

UE	Unpaired Electron relaxation mechanism
UHF	Ultra-High Field
UW	Ultra-Wideline
VAS	Variable Angle Spinning
VB	Valence Band
VOCS	Variable-Offset Cumulative Spectrum
WAHUHA	Waugh, Huber, & Haeberlen (pulse sequence)
WALTZ-16	A broadband decoupling sequence
WISE	Wide-Line Separation
WURST	Wideband, Uniform Rate, and Smooth Truncation
WZ	Wurtzite
XRD	X-Ray Diffraction
ZB	Zincblende
ZORA	Zeroth-Order Regular Approximation
ZQ(C)	Zero Quantum (Coherence)

Printed and bound by CPI Group (UK) Ltd, Croydon, CR0 4YY
16/09/2026
14963480-0001